PHYSICAL MODELLING IN GEOTECHNICS

PROCEEDINGS OF THE 9TH INTERNATIONAL CONFERENCE ON PHYSICAL MODELLING IN GEOTECHNICS 2018 (ICPMG 2018), LONDON, UK, 17–20 JULY 2018

Physical Modelling in Geotechnics

Editors

Andrew McNamara, Sam Divall, Richard Goodey, Neil Taylor,
Sarah Stallebrass & Jignasha Panchal
City, University of London, UK

VOLUME 2

CRC Press
Taylor & Francis Group
Boca Raton London New York Leiden

CRC Press is an imprint of the
Taylor & Francis Group, an **informa** business

A BALKEMA BOOK

CRC Press/Balkema is an imprint of the Taylor & Francis Group, an informa business

© 2018 Taylor & Francis Group, London, UK

Typeset by MPS Limited, Chennai, India

All rights reserved. No part of this publication or the information contained herein may be reproduced, stored in a retrieval system, or transmitted in any form or by any means, electronic, mechanical, by photocopying, recording or otherwise, without written prior permission from the publishers.

Although all care is taken to ensure integrity and the quality of this publication and the information herein, no responsibility is assumed by the publishers nor the author for any damage to the property or persons as a result of operation or use of this publication and/or the information contained herein.

Published by: CRC Press/Balkema
Schipholweg 107C, 2316 XC Leiden, The Netherlands
e-mail: Pub.NL@taylorandfrancis.com
www.crcpress.com – www.taylorandfrancis.com

ISBN: 978-1-138-55975-2 (Hbk set + USB)
ISBN volume 1: 978-1-138-34419-8 (Hbk)
ISBN volume 2: 978-1-138-34422-8 (Hbk)

ISBN: 978-0-203-71227-6 (eBook set)
ISBN volume 1: 978-0-429-43866-0 (eBook)
ISBN volume 2: 978-0-429-43864-6 (eBook)

Table of contents

Preface XVII
International advisory board XIX
Local organising committee XXI
Manuscript reviewers XXIII
Sponsors XXV

VOLUME 1

Keynote and Themed lectures

Modelling tunnel behaviour under seismic actions: An integrated approach 3
E. Bilotta

An example of effective mentoring for research centres 21
C.E. Bronner, D.W. Wilson, K. Ziotopoulou, K.M. Darby, A. Sturm, A.J. Raymond, R.W. Boulanger, J.T. DeJong, D.M. Moug & J.D. Bronner

Geotechnical modelling for offshore renewables 33
C. Gaudin, C.D. O'Loughlin & B. Bienen

Physical modelling applied to infrastructure development 43
R.J. Goodey

The role of centrifuge modelling in capturing whole-life responses of geotechnical infrastructure to optimise design 51
S. Gourvenec

Development of geotechnical centrifuges and facilities in China 77
Y.J. Hou

Physical modelling of structural and biological soil reinforcement 87
J.A. Knappett

Current and emerging physical modelling technologies 101
W.A. Take

1. Sample preparation and characterisation

Investigation into 3D printing of granular media 113
O. Adamidis, S. Alber & I. Anastasopoulos

Undrained shear strength profile of normally and overconsolidated kaolin clay 119
A. Arnold, W. Zhang & A. Askarinejad

LEAP GWU 2017: Investigating different methods for verifying the relative density of a centrifuge model 125
R. Beber, S.S.C. Madabhushi, A. Dobrisan, S.K. Haigh & S.P.G. Madabhushi

Centrifuge modelling of Continuous Compaction Control (CCC) 131
B. Caicedo & J. Escobar

Shear wave velocity: Comparison between centrifuge and triaxial based measurements 137
G. Cui, C.M. Heron & A.M. Marshall

Development of layered models for geotechnical centrifuge tests 143
S. Divall, S.E. Stallebrass, R.J. Goodey & E.P. Ritchie

The influence of temperature on shear strength at a soil-structure interface — 149
J. Parchment & P. Shepley

Development of a 3D clay printer for the preparation of heterogeneous models — 155
L.M. Pua, B. Caicedo, D. Castillo & S. Caro

2. Engineered platforms

Centrifuge modelling utility pipe behaviour subject to vehicular loading — 163
S.M. Bayton, T. Elmrom & J.A. Black

Experimental model study on traffic loading induced earth pressure reduction using EPS geofoam — 169
T.N. Dave & S.M. Dasaka

Physical modelling of roads in expansive clay subjected to wetting-drying cycles — 175
S. Laporte, G.A. Siemens & R.A. Beddoe

Scaled physical modelling of ultra-thin continuously reinforced concrete pavement — 179
M.S. Smit, E.P. Kearsley & S.W. Jacobsz

The effect of relative stiffness on soil-structure interaction under vehicle loads — 185
M.S. Smit, E.P. Kearsley & S.W. Jacobsz

Plate bearing tests for working platforms — 191
G. Tanghetti, R.J. Goodey, A.M. McNamara & H. Halai

Geotechnical model tests on bearing capacity of working platforms for mobile construction machines and cranes — 197
R. Worbes & C. Moormann

1g physical modelling of the stoneblowing technique for the improvement of railway track maintenance — 203
A.A. Zaytsev, A.A. Abrashitov & A.A. Sydrakov

3. Physical/Numerical interface and comparisons

Millisecond interfacing of physical models with ABAQUS — 209
S. Idinyang, A. Franza, C.M. Heron & A.M. Marshall

Verification and validation of two-phase material point method simulation of pore water pressure rise and dissipation in earthquakes — 215
T. Kiriyama, K. Fukutake & Y. Higo

Centrifuge and numerical investigations of rotated box structures — 221
T.A. Newson, O.S. Abuhajar & K.J.L. Stone

Multibillion particle DEM to simulate centrifuge model tests of geomaterials — 227
D. Nishiura, H. Sakaguchi & S. Yamamoto

Trapdoor model test and DEM simulation associated with arching — 233
M. Otsubo, R. Kuwano, U. Ali & H. Ebizuka

4. Scaling

Variability of small scale model reinforced concrete and implications for geotechnical centrifuge testing — 241
J.A. Knappett, M.J. Brown, L. Shields, A.H. Al-Defae & M. Loli

Modelling experiments to investigate soil-water retention in geotechnical centrifuge — 247
M. Mirshekari, M. Ghayoomi & A. Borghei

Studies on the use of hydraulic gradient similitude method for determining permeability of soils — 253
K.T. Mohan Gowda & B.V.S. Viswanadham

A new insight into the behaviour of seepage flow in centrifuge modelling — 259
W. Ovalle-Villamil & I. Sasanakul

Applicability of the generalised scaling law to pile-inclined ground system *K. Sawada, K. Ueda & S. Iai*	265
Permeability of sand with a methylcellulose solution *T. Tobita*	271

5. Sensors

Investigation of an OFDR fibre Bragg system for use in geotechnical scale modelling *R.D. Beemer, M.J. Cassidy & C. Gaudin*	279
Free fall cone tests in kaolin clay *A. Bezuijen, D.A. den Hamer, L. Vincke & K. Geirnaert*	285
A new shared miniature cone penetrometer for centrifuge testing *T. Carey, A. Gavras, B. Kutter, S.K. Haigh, S.P.G. Madabhushi, M. Okamura, D.S. Kim, K. Ueda, W.Y. Hung, Y.G. Zhou, K. Liu, Y.M. Chen, M. Zeghal, T. Abdoun, S. Escoffier & M. Manzari*	293
Shear wave velocity measurement in a large geotechnical laminar box using bender elements *J. Colletti, A. Tessari, K. Sett, W. Hoffman & J. Coleman*	299
Low cost tensiometers for geotechnical applications *S.W. Jacobsz*	305
A field model investigating pipeline leak detection using discrete fibre optic sensors *S.I. Jahnke, S.W. Jacobsz & E.P. Kearsley*	311
Development of an instrumented model pile *A.B. Lundberg, W. Broere & J. Dijkstra*	317
New method for full field measurement of pore water pressures *M. Ottolini, W. Broere & J. Dijkstra*	323
Ambient pressure calibration for cone penetrometer test: Necessary? *Y. Wang, Y. Hu & M.S. Hossain*	329

6. Modelling techniques

Development of a rainfall simulator in centrifuge using Modified Mariotte's principle *D. Bhattacherjee & B.V.S. Viswanadham*	337
Development of model structural dampers for dynamic centrifuge testing *J. Boksmati, S.P.G. Madabhushi & N.I. Thusyanthan*	343
Experimental evaluation of two-stage scaling in physical modelling of soil-foundation-structure systems *A. Borghei & M. Ghayoomi*	349
Development of a window laminar strong box *S.C. Chian, C. Qin & Z. Zhang*	355
Ground-borne vibrations from piles: Testing within a geotechnical centrifuge *G. Cui, C.M. Heron & A.M. Marshall*	359
A new Stockwell mean square frequency methodology for analysing centrifuge data *J. Dafni & J. Wartman*	365
Novel experimental device to simulate tsunami loading in a geotechnical centrifuge *M.C. Exton, S. Harry, H.B. Mason, H. Yeh & B.L. Kutter*	371
A new apparatus to examine the role of seepage flow on internal instability of model soil *F. Gaber & E.T. Bowman*	377
Centrifuge model test on the instability of an excavator descending a slope *T. Hori & S. Tamate*	383
Transparent soils turn 25: Past, present, and future *M. Iskander*	389

Application of 3D printing technology in geotechnical-physical modelling: Tentative experiment practice *Q. Jiang, L.F. Li, M. Zhang & L.B. Song*	395
Scaling of plant roots for geotechnical centrifuge tests using juvenile live roots or 3D printed analogues *T. Liang, J.A. Knappett, G.J. Meijer, D. Muir Wood, A.G. Bengough, K.W. Loades & P.D. Hallett*	401
Revisit of the empirical prediction methods for liquefaction-induced lateral spread by using the LEAP centrifuge model tests *K. Liu, Y.G. Zhou, Y. She, P. Xia, Y.M. Chen, D.S. Ling & B. Huang*	407
Physical modelling of atmospheric conditions during drying *C. Lozada, B. Caicedo & L. Thorel*	413
Centrifuge model tests on excavation in Shanghai clay using in-flight excavation tools *X.F. Ma & J.W. Xu*	419
Effect of root spacing on interpretation of blade penetration tests—full-scale physical modelling *G.J. Meijer, J.A. Knappett, A.G. Bengough, K.W. Loades & B.C. Nicoll*	425
Development of a centrifuge testing method for stability analyses of breakwater foundation under combined actions of earthquake and tsunami *J. Miyamoto, K. Tsurugasaki, R. Hem, T. Matsuda & K. Maeda*	431
Modelling of rocking structures in a centrifuge *I. Pelekis, G.S.P. Madabhushi & M.J. DeJong*	437
A new test setup for studying sand behaviour inside an immersed tunnel joint gap *R. Rahadian, S. van der Woude, D. Wilschut, C.B.M. Blom & W. Broere*	443
3D printing of masonry structures for centrifuge modelling *S. Ritter, M.J. DeJong, G. Giardina & R.J. Mair*	449
A mechanical displacement control model tunnel for simulating eccentric ground loss in the centrifuge *G. Song, A.M. Marshall & C.M. Heron*	455
Preliminary results of laboratory analysis of sand fluidisation *F.S. Tehrani, A. Askarinejad & F. Schenkeveld*	461
Rolling test in geotechnical centrifuge for ore liquefaction analysis *L. Thorel, P. Audrain, A. Néel, A. Bretschneider, M. Blanc & F. Saboya*	465
Design and performance of an electro-mechanical pile driving hammer for geo-centrifuge *J.C.B. van Zeben, C. Azúa-González, M. Alvarez Grima, C. van 't Hof & A. Askarinejad*	469
A new heating-cooling system for centrifuge testing of thermo-active geo-structures *D. Vitali, A.K. Leung, R. Zhao & J.A. Knappett*	475
Physical modelling of soil-structure interaction of tree root systems under lateral loads *X. Zhang, J.A. Knappett, A.K. Leung & T. Liang*	481

7. Facilities

A new environmental chamber for the HKUST centrifuge facility *A. Archer & C.W.W. Ng*	489
Upgrades to the NHRI – 400 g-tonne geotechnical centrifuge *S.S. Chen, X.W. Gu, G.F. Ren, W.M. Zhang, N.X. Wang, G.M. Xu, W. Liu, J.Z. Hong & Y.B. Cheng*	495
A new 240 g-tonne geotechnical centrifuge at the University of Western Australia *C. Gaudin, C.D. O'Loughlin & J. Breen*	501
Development of a rainfall simulator for climate modelling *I.U. Khan, M. Al-Fergani & J.A. Black*	507
The development of a small centrifuge for testing unsaturated soils *K.A. Kwa & D.W. Airey*	513

Full scale laminar box for 1-g physical modelling of liquefaction 519
S. Thevanayagam, Q. Huang, M.C. Constantinou, T. Abdoun & R. Dobry

8. Education

Using small-scale seepage physical models to generate didactic material for soil mechanics classes 527
L.B. Becker, R.M. Linhares, F.S. Oliveira & F.L. Marques

Centrifuge modelling in the undergraduate curriculum—a 5 year reflection 533
J.A. Black, S.M. Bayton, A. Cargill & A. Tatari

Geotechnical centrifuge facility for teaching at City, University of London 539
S. Divall, S.E. Stallebrass, R.J. Goodey, R.N. Taylor & A.M. McNamara

Development of a teaching centrifuge learning environment using mechanically stabilized earth walls 545
A.F. Tessari & J.A. Black

9. Offshore

Development of a series of 2D backfill ploughing physical models for pipelines and cables 553
T. Bizzotto, M.J. Brown, A.J. Brennan, T. Powell & H. Chandler

Capacity of vertical and horizontal plate anchors in sand under normal and shear loading 559
S.H. Chow, J. Le, M. Forsyth & C.D. O'Loughlin

A novel experimental-numerical approach to model buried pipes subjected to reverse faulting 565
R.Y. Khaksar, M. Moradi & A. Ghalandarzadeh

Wave-induced liquefaction and floatation of pipeline buried in sand beds 571
J. Miyamoto, K. Tsurugasaki & S. Sassa

Surface pipeline buckling on clay: Demonstration 577
R. Phillips, J. Barrett & G. Piercey

Centrifuge modelling for lateral pile-soil pressure on passive part of pile group with platform 583
G.F. Ren, G.M. Xu, X.W. Gu, Z.Y. Cai, B.X. Shi & A.Z. Chen

Centrifuge model tests and circular slip analyses to evaluate reinforced composite-type breakwater stability against tsunami 589
H. Takahashi, S. Sassa, Y. Morikawa & K. Maruyama

10. Offshore – shallow foundations

Centrifuge tests on the influence of vacuum on wave impact on a caisson 597
D.A. de Lange, A. Bezuijen & T. Tobita

Physical modelling of active suction for offshore renewables 603
N. Fiumana, C. Gaudin, Y. Tian & C.D. O'Loughlin

Cyclic behaviour of unit bucket for tripod foundation system under various loading characteristics via centrifuge 609
Y.H. Jeong, H.J. Park, D.S. Kim & J.H. Kim

Physical modelling of reinstallation of a novel spudcan nearby existing footprint 615
M.J. Jun, Y.H. Kim, M.S. Hossain, M.J. Cassidy, Y. Hu & S.G. Park

Reduction in soil penetration resistance for suction-assisted installation of bucket foundation in sand 623
A.K. Koteras & L.B. Ibsen

Evaluation of seismic coefficient for gravity quay wall via centrifuge modelling 629
M.G. Lee, J.G. Ha, H.J. Park, D.S. Kim & S.B. Jo

Sleeve effect on the post-consolidation extraction resistance of spudcan foundation in overconsolidated clay 635
Y.P. Li & J.Y. Shi

Measuring the behaviour of dual row retaining walls in dry sands using centrifuge tests *S.S.C. Madabhushi & S.K. Haigh*	639
Verification of improvement plan for seismic retrofits of existing quay wall in small scale fishing port *K. Mikasa & K. Okabayashi*	645
Visualisation of mechanisms governing suction bucket installation in dense sand *R. Ragni, B. Bienen, S.A. Stanier, M.J. Cassidy & C.D. O'Loughlin*	651
Recent advances in tsunami-seabed-structure interaction from geotechnical and hydrodynamic perspectives: Role of overflow/seepage coupling *S. Sassa*	657
Evaluation of seismic behaviour of reinforced earth wall based on design practices and centrifuge model tests *Y. Sawamura, T. Shibata & M. Kimura*	663
Centrifuge tests investigating the effect of suction caisson installation in dense sand on the state of the soil plug *M. Stapelfeldt, B. Bienen & J. Grabe*	669
Centrifuge model tests on stabilisation countermeasures of a composite breakwater under tsunami actions *K. Tsurugasaki, J. Miyamoto, R. Hem, T. Iwamoto & H. Nakase*	675
Interaction between jack-up spudcan and adjacent piles with non-perfect pile cap *Y. Xie, C.F. Leung & Y.K. Chow*	681

11. Offshore – deep foundations

Centrifuge modelling of long term cyclic lateral loading on monopiles *S.M. Bayton, J.A. Black & R.T. Klinkvort*	689
Centrifuge modelling of screw piles for offshore wind energy foundations *C. Davidson, T. Al-Baghdadi, M.J. Brown, A. Brennan, J.A. Knappett, C. Augarde, W. Coombs, L. Wang, D.J. Richards, A. Blake & J. Ball*	695
General study on the axial capacity of piles of offshore wind turbines jacked in sand *I. El Haffar, M. Blanc & L. Thorel*	701
Dynamic load tests on large diameter open-ended piles in sand performed in the centrifuge *E. Heins, B. Bienen, M.F. Randolph & J. Grabe*	707
Centrifuge model tests on holding capacity of suction anchors in sandy deposits *K. Kita, T. Utsunomiya & K. Sekita*	713
A review of modelling effects in centrifuge monopile testing in sand *R.T. Klinkvort, J.A. Black, S.M. Bayton, S.K. Haigh, G.S.P. Madabhushi, M. Blanc, L. Thorel, V. Zania, B. Bienen & C. Gaudin*	719
Experimental modelling of the effects of scour on offshore wind turbine monopile foundations *R.O. Mayall, R.A. McAdam, B.W. Byrne, H.J. Burd, B.B. Sheil, P. Cassie & R.J.S. Whitehouse*	725
Centrifuge tests on the response of piles under cyclic lateral 1-way and 2-way loading *C. Niemann, O. Reul, Y. Tian, C.D. O'Loughlin & M.J. Cassidy*	731
Physical modelling of monopile foundations under variable cyclic lateral loading *I.A. Richards, B.W. Byrne & G.T. Houlsby*	737
Centrifuge model testing of fin piles in sand *S. Sayles, K.J.L. Stone, M. Diakoumi & D.J. Richards*	743
Dynamic behaviour evaluation of offshore wind turbine using geotechnical centrifuge tests *J.T. Seong, J.H. Kim & D.S. Kim*	749

An investigation on the performance of a self-installing monopiled GBS structure under lateral loading *K.J.L. Stone, A. Tillman & M. Vaziri*	755
Model tests on the lateral cyclic responses of a caisson-piles foundation under scour *C.R. Zhang, H.W. Tang & M.S. Huang*	761
Comparison of centrifuge model tests of tetrapod piled jacket foundation in saturated sand and clay *B. Zhu, K. Wen, L.J. Wang & Y.M. Chen*	767
Author index	773

VOLUME 2

12. Tunnel, shafts and pipelines

Study of the effects of explosion on a buried tunnel through centrifuge model tests *A. De & T.F. Zimmie*	779
Uplift resistance of a buried pipeline in silty soil on slopes *G.N. Eichhorn & S.K. Haigh*	785
Modelling the excavation of elliptical shafts in the geotechnical centrifuge *N.E. Faustin, M.Z.E.B. Elshafie & R.J. Mair*	791
Shaking table test to evaluate the effects of earthquake on internal force of Tabriz subway tunnel (Line 2) *M. Hajialilue-Bonab, M. Farrin & M. Movasat*	797
Effect of pipe defect size and maximum particle size of bedding material on associated internal erosion *S. Indiketiya, P. Jegatheesan, R. Pathmanathan & R. Kuwano*	803
Modelling cave mining in the geotechnical centrifuge *S.W. Jacobsz, E.P. Kearsley, D. Cumming-Potvin & J. Wesseloo*	809
Experimental modelling of infiltration of bentonite slurry in front of shield tunnel in saturated sand *T. Xu & A. Bezuijen*	815

13. Imaging

Flow visualisation in a geotechnical centrifuge under controlled seepage conditions *C.T.S. Beckett & A.B. Fourie*	823
A new procedure for tracking displacements of submerged sloping ground in centrifuge testing *T. Carey, N. Stone, B. Kutter & M. Hajialilue-Bonab*	829
Identification of soil stress-strain response from full field displacement measurements in plane strain model tests *J.A. Charles, C.C. Smith & J.A. Black*	835
Imaging of sand-pile interface submitted to a high number of loading cycles *J. Doreau-Malioche, G. Combe, J.B. Toni, G. Viggiani & M. Silva*	841
Image capture and motion tracking applications in geotechnical centrifuge modelling *P. Kokkali, T. Abdoun & A. Tessari*	847
A study on performance of three-dimensional imaging system for physical models *B.T. Le, S. Nadimi, R.J. Goodey & R.N. Taylor*	853
Visualisation of inter-granular pore fluid flow *L. Li, M. Iskander & M. Omidvar*	859

A two-dimensional laser-scanner system for geotechnical processes monitoring 865
M.D. Valencia-Galindo, L.N. Beltrán-Rodriguez, J.A. Sánchez-Peralta, J.S. Tituaña-Puente,
M.G. Trujillo-Vela, J.M. Larrahondo, L.F. Prada-Sarmiento & A.M. Ramos-Cañón

14. Seismic – dynamic

Dynamic behaviour of model pile in saturated sloping ground during shaking table tests 873
C.H. Chen, T.S. Ueng & C.H. Chen

Investigation on the aseismic performance of pile foundations in volcanic ash ground 879
T. Egawa, T. Yamanashi & K. Isobe

Effective parameters on the interaction between reverse fault rupture and shallow foundations: Centrifuge modelling 885
A. Ghalandarzadeh & M. Ashtiani

Seismic amplification of clay ground and long-term consolidation after earthquake 891
Y. Hatanaka & K. Isobe

Evaluation of period-lengthening ratio (PLR) of single-degree-of-freedom structure via dynamic centrifuge tests 897
K.W. Ko, J.G. Ha, H.J. Park & D.S. Kim

Centrifuge modelling of active seismic fault interaction with oil well casings 903
J. Le Cossec, K.J.L. Stone, C. Ryan & K. Dimitriadis

Centrifuge shaking table tests on composite caisson-piles foundation 909
F. Liang, Y. Jia, H. Zhang, H. Chen & M. Huang

Dynamic behaviour of three-hinge-type precast arch culverts with various patterns of overburden in culvert longitudinal direction 915
Y. Miyazaki, Y. Sawamura, K. Kishida & M. Kimura

Dynamic centrifuge model tests on sliding base isolation systems leveraging buoyancy 921
N. Nigorikawa, Y. Asaka & M. Hasebe

Centrifugal model tests on static and seismic stability of landfills with high water level 929
B. Zhu, J.C. Li, L.J. Wang & Y.M. Chen

15. Seismic – liquefaction

Partial drainage during earthquake-induced liquefaction 937
O. Adamidis & G.S.P. Madabhushi

Centrifuge modelling of site response and liquefaction using a 2D laminar box and biaxial dynamic base excitation 943
O. El Shafee, J. Lawler & T. Abdoun

Experimental simulation of the effect of preshaking on liquefaction of sandy soils 949
W. El-Sekelly, T. Abdoun, R. Dobry & S. Thevanayagam

Dynamic centrifuge testing to assess liquefaction potential 955
G. Fasano, E. Bilotta, A. Flora, V. Fioravante, D. Giretti, C.G. Lai & A.G. Özcebe

Experimental and computational study on effects of permeability on liquefaction 961
H. Funahara & N. Tomita

The importance of vertical accelerations in liquefied soils 967
F.E. Hughes & S.P.G. Madabhushi

The effects of waveform of input motions on soil liquefaction by centrifuge modelling 975
W.Y. Hung, T.W. Liao, L.M. Hu & J.X. Huang

Horizontal subgrade reaction of piles in liquefiable ground 981
S. Imamura

Experimental investigation of pore pressure and acceleration development in static liquefaction induced failures in submerged slopes 987
A. Maghsoudloo, A. Askarinejad, R.R. de Jager, F. Molenkamp & M.A. Hicks

Centrifuge modelling of earthquake-induced liquefaction on footings built on improved ground A.S.P.S. Marques, P.A.L.F. Coelho, S.K. Haigh & G.S.P. Madabhushi	993
Investigating the effect of layering on the formation of sand boils in 1 g shaking table tests S. Miles, J. Still & M. Stringer	999
Centrifuge modelling of mitigation-soil-structure-interaction on layered liquefiable soil deposits with a silt cap B. Paramasivam, S. Dashti, A.B. Liel & J.C. Olarte	1005
Centrifuge modelling of the effects of soil liquefiability on the seismic response of low-rise structures S. Qi & J.A. Knappett	1011
Liquefaction behaviour focusing on pore water inflow into unsaturated surface layer Y. Takada, K. Ueda, S. Iai & T. Mikami	1017

16. Dams and embankments

Performance of single piles in riverbank clay slopes subject to repetitive tidal cycles U. Ahmed, D.E.L. Ong & C.F. Leung	1025
Load transfer mechanism of reinforced piled embankments M.S.S. Almeida, D.F. Fagundes, M.C.F. Almeida, D.A. Hartmann, R. Girout, L. Thorel & M. Blanc	1031
Experiments for a coarse sand barrier as a measure against backwards erosion piping A. Bezuijen, E. Rosenbrand, V.M. van Beek & K. Vandenboer	1037
Load transfer mechanism of piled embankments: Centrifuge tests versus analytical models M. Blanc, L. Thorel, R. Girout, M.S.S. Almeida & D.F. Fagundes	1043
Physical model testing to evaluate erosion quantity and pattern M. Kamalzare & T.F. Zimmie	1049
Physical modelling of large dams for seismic performance evaluation N.R. Kim & S.B. Jo	1055
Centrifuge model tests on levees subjected to flooding R.K. Saran & B.V.S. Viswanadham	1061
Centrifuge model test of vacuum consolidation on soft clay combined with embankment loading S. Shiraga, G. Hasegawa, Y. Sawamura & M. Kimura	1067

17. Geohazards

Effects of viscosity in granular flows simulated in a centrifugal acceleration field M. Cabrera, P. Kailey, E.T. Bowman & W. Wu	1075
Using pipe deflection to detect sinkhole development E.P. Kearsley, S.W. Jacobsz & H. Louw	1081
Model tests to simulate formation and expansion of subsurface cavities R. Kuwano, R. Sera & Y. Ohara	1087
Centrifuge modelling of a pipeline subjected to soil mass movements J.R.M.S. Oliveira, K.I. Rammah, P.C. Trejo, M.S.S. Almeida & M.C.F. Almeida	1093
Effects of earthquake motion on subsurface cavities R. Sera, M. Ota & R. Kuwano	1099
Preliminary study of debris flow impact force on a circular pillar A.L. Yifru, R.N. Pradhan, S. Nordal & V. Thakur	1105

18. Slopes

Centrifuge modelling of earth slopes subjected to change in water content P. Aggarwal, R. Singla & A. Juneja	1113

Centrifuge and numerical modelling of static liquefaction of fine sandy slopes   1119
A. Askarinejad, W. Zhang, M. de Boorder & J. van der Zon

Modelling of MSW landfill slope failure   1125
Y.J. Hou, X.D. Zhang, J.H. Liang, C.H. Jia, R. Peng & C. Wang

Effects of plant removal on slope hydrology and stability   1131
V. Kamchoom & A.K. Leung

Centrifuge model test on deformation and failure of slopes under wetting-drying cycles   1137
F. Luo & G. Zhang

Centrifuge model studies of the soil slope under freezing and thawing processes   1143
C. Zhang, Z.Y. Cai, Y.H. Huang & G.M. Xu

An experimental and numerical study of pipe behaviour in triggered sandy slope failures   1149
W. Zhang, Z. Gng & A. Askarinejad

19. Ground improvement

Investigation of nailed slope behaviour during excavation by Ng centrifuge physical model tests   1157
A. Akoochakian, M. Moradi & A. Kavand

Relative contribution of drainage capacity of stone columns as a countermeasure against liquefaction   1163
E. Apostolou, A.J. Brennan & J. Wehr

Observed deformations in geosynthetic-reinforced granular soils subjected to voids   1169
T.S. da Silva & M.Z.E.B. Elshafie

Analytical design approach for the self-regulating interactive membrane foundation based on centrifuge-model tests and numerical simulations   1175
O. Detert, D. König & T. Schanz

Earthquake-induced liquefaction mitigation under existing buildings using drains   1181
S. García-Torres & G.S.P. Madabhushi

Deformation behaviour research of an artificial island by centrifuge modelling test   1187
X.W. Gu, Z.Y. Cai, G.M. Xu & G.F. Ren

Effect of lateral confining condition of behaviour of confined-reinforced earth   1193
H.M. Hung & J. Kuwano

An experimental study on the effects of enhanced drainage for liquefaction mitigation in dense urban environments   1199
P.B. Kirkwood & S. Dashti

Influence of tamper shape on dynamic compaction of granular soil   1205
S. Kundu & B.V.S. Viswanadham

Behaviour of geogrid reinforced soil walls with marginal backfills with and without chimney drain in a geotechnical centrifuge   1211
J. Mamaghanian, H.R. Razeghi, B.V.S. Viswanadham & C.H.S.G. Manikumar

Centrifuge model tests on effect of inclined foundation on stability of column type deep mixing improved ground   1217
S. Matsuda, M. Momoi & M. Kitazume

Large-scale physical model GRS walls: Evaluation of the combined effects of facing stiffness and toe resistance on performance   1223
S.H. Mirmoradi & M. Ehrlich

Deep vibration compaction of sand using mini vibrator   1229
S. Nagula, P. Mayanja & J. Grabe

Dynamic centrifuge tests on nailed slope with facing plates   1235
S. Nakamoto, N. Iwasa & J. Takemura

Influence of slope inclination on the performance of slopes with and without soil-nails
subjected to seepage: A centrifuge study 1241
V.M. Rotte & B.V.S. Viswanadham

Performance of soil-nailed wall with three-dimensional geometry: Centrifuge study 1247
M. Sabermahani, M. Moradi & A. Pooresmaeili

Behaviour of geogrid-reinforced aggregate layer overlaying poorly graded sand
under cyclic loading 1253
A.A. Soe, J. Kuwano, I. Akram, T. Kogure & H. Kanai

Physical modelling of compaction grouting injection using a transparent soil 1259
D. Takano, Y. Morikawa, Y. Miyata, H. Nonoyama & R.J. Bathurst

Centrifuge modelling of remediation of liquefaction-induced pipeline uplift using
model root systems 1265
K. Wang, A.J. Brennan, J.A. Knappett, S. Robinson & A.G. Bengough

Comparative study of consolidation behaviour of differently-treated mature fine
tailings specimens through centrifuge modelling 1271
G. Zambrano-Narvaez, Y. Wang & R.J. Chalaturnyk

Physical modelling and monitoring of the subgrade on weak foundation and its
reinforcing with geosynthetics 1277
A.A. Zaytsev, Y.K. Frolovsky, A.V. Gorlov, A.V. Petryaev & V.V. Ganchits

20. Shallow foundations

Effect of spatial variability on the behaviour of shallow foundations: Centrifuge study 1285
L.X. Garzón, B. Caicedo, M. Sánchez-Silva & K.K. Phoon

1g model tests of surface and embedded footings on unsaturated compacted sand 1291
A.J. Lutenegger & M.T. Adams

Experimental study on the coupled effect of the vertical load and the horizontal load
on the performance of piled beam-slab foundation 1297
L. Mu, M. Huang, X. Kang & Y. Zhang

Determining shallow foundation stiffness in sand from centrifuge modelling 1303
A. Pearson & P. Shepley

The effect of soil stiffness on the undrained bearing capacity of a footing on a layered
clay deposit 1309
A. Salehi, Y. Hu, B.M. Lehane, V. Zania & S.L. Sovso

Centrifuge investigation of the cyclic loading effect on the post-cyclic monotonic
performance of a single-helix anchor in sand 1315
J.A. Schiavon, C.H.C. Tsuha & L. Thorel

Bearing capacity of surface and embedded foundations on a slope: Centrifuge modelling 1321
D. Taeseri, L. Sakellariadis, R. Schindler & I. Anastasopoulos

21. Deep foundations

Performance of piled raft with unequal pile lengths 1329
R.S. Bisht, A. Juneja, A. Tyagi & F.H. Lee

Pile response during liquefaction-induced lateral spreading: 1-g shake table tests with
different ground inclination 1335
A. Ebeido, A. Elgamal & M. Zayed

Effect of the installation methods of piles in cohesionless soil on their axial capacity 1341
I. El Haffar, M. Blanc & L. Thorel

Model testing of rotary jacked open ended tubular piles in saturated non-cohesive soil 1347
D. Frick, K.A. Schmoor, P. Gütz & M. Achmus

Model tests on soil displacement effects for differently shaped piles *A.A. Ganiyu, A.S.A. Rashid, M.H. Osman & W.O. Ajagbe*	1353
Comparison of seismic behaviour of pile foundations in two different soft clay profiles *T.K. Garala & G.S.P. Madabhushi*	1359
Issues with centrifuge modelling of energy piles in soft clays *I. Ghaaowd, J. McCartney, X. Huang, F. Saboya & S. Tibana*	1365
Centrifuge modelling of non-displacement piles on a thin bearing layer overlying a clay layer *Y. Horii & T. Nagao*	1371
Rigid pile improvement under rigid slab or footing under cyclic loading *O. Jenck, F. Emeriault, C. Dos Santos Mendes, O. Yaba, J.B. Toni, G. Vian & M. Houda*	1377
Pull-out testing of steel reinforced earth systems: Modelling in view of soil dilation and boundary effects *M. Loli, I. Georgiou, A. Tsatsis, R. Kourkoulis & F. Gelagoti*	1383
Pile jetting in plane strain: Small-scale modelling of monopiles *S. Norris & P. Shepley*	1389
Influence of geometry on the bearing capacity of sheet piled foundations *J.P. Panchal, A.M. McNamara & R.J. Goodey*	1395
Kinematic interaction of piles under seismic loading *J. Pérez-Herreros, F. Cuira, S. Escoffier & P. Kotronis*	1401
Behaviour of piled raft foundation systems in soft soil with consolidation process *E. Rodríguez, R.P. Cunha & B. Caicedo*	1407
Displacement measurements of ground and piles in sand subjected to reverse faulting *C.F. Yao, S. Seki & J. Takemura*	1413

22. Walls and excavations

Centrifuge simulation of heave behaviour of deep basement slabs in overconsolidated clay *D.Y.K. Chan & S.P.G. Madabhushi*	1421
Soil movement mobilised with retaining wall rotation in loose sand *C. Deng & S.K. Haigh*	1427
Lateral pressure of granular mass during translative motion of wall *P. Koudelka*	1433
Deflection and failure of self-standing high stiffness steel pipe sheet pile walls embedded in soft rocks *V. Kunasegarm, S. Seki & J. Takemura*	1439
A new approach to modelling excavations in soft soils *J.P. Panchal, A.M. McNamara & S.E. Stallebrass*	1445
1g-modelling of limit load increase due to shear band enhancement *K.-F. Seitz & J. Grabe*	1451
Concave segmental retaining walls *D. Stathas, L. Xu, J.P. Wang, H.I. Ling & L. Li*	1457
A combined study of centrifuge and full scale models on detection of threat of failure in trench excavations *S. Tamate & T. Hori*	1463
Dynamic behaviour on pile foundation combined with soil-cement mixing walls using permanent pile *K. Watanabe, M. Arakawa & M. Mizumoto*	1469
Centrifuge modelling of 200,000 tonnage sheet-pile bulkheads with relief platform *G.M. Xu, G.F. Ren, X.W. Gu & Z.Y. Cai*	1475
Author index	1481

Preface

The International Conference on Physical Modelling in Geotechnics is held under the auspices of Technical Committee 104 (*TC104: Physical Modelling in Geotechnics*) of the International Society for Soil Mechanics and Geotechnical Engineering (ISSMGE). Early workshops on physical modelling were held in Manchester, California and Tokyo in 1984 and, as the physical modelling community grew, the first international conference was held only 30 years ago in Paris in 1988. The possibilities offered by physical modelling became apparent around the world and the conference has developed into a quadrennial event that regularly attracts researchers from over 30 countries. The last meeting of the global community was in Perth, Western Australia; a veritable feast to sate the appetite of the hungry faithful, under the very capable leadership of Professor Christophe Gaudin. Regional conferences have also become established following the first Eurofuge held at City, University of London in 2008 followed by European regional conferences at TU Delft and IFSTTAR, Nantes and Asian regional conferences at IIT Bombay and Tongji University, Shanghai. These conferences bring together a community of great innovators; the most practical and capable engineers, in an exciting and specialist field.

TC104 selected London as the destination for the 9th International Conference (ICPMG 2018) which was held at City, University of London, in July 2018. The United Kingdom is a hotspot for physical modelling activity; centrifuges are established at Cambridge University, City, University of London, University of Dundee, University of Nottingham and University of Sheffield.

The conference coincided with the 4th Andrew Schofield Lecture, established by TC104 and named after Professor Andrew Schofield, the great pioneer of geotechnical centrifuge modelling. As the highest honour that can be bestowed upon a member of our community it is fitting that the lecture was delivered by Professor Neil Taylor of City, University of London and Secretary General of ISSMGE; a former doctoral student of Professor Schofield.

The conference programme was a physical modelling extravaganza divided into plenary and parallel sessions running over four days, 17th–20th July. Four keynote lectures were given in the areas of seismic behaviour, design optimisation, new facilities and environmental engineering representing significant areas of interest of the assembled audience. Themed lectures in the areas of education, new technology, urban infrastructure and offshore engineering addressed a key aim of TC104 in showcasing research opportunities to industry who attended a specific half day event. A total of 138 oral presentations were made from 230 papers submitted, originating from over 30 countries, and included in the conference proceedings in 22 chapters. All papers that were not presented orally were presented as posters. The conference gave delegates an opportunity to experience exciting and historic aspects of London that are normally inaccessible to those visiting the city. A welcome reception was held at the historic Skinners' Hall, home to one of the Great Twelve City livery companies and delegates enjoyed a sumptuous gala dinner at the spectacular Middle Temple Hall dating from 1573; one of the four Inns of Court exclusively entitled to call their members to the English Bar as barristers. A pleasant afternoon and evening was spent on a visit to Greenwich on the River Thames, home to the Meridian Line, the famous Cutty Sark, the Royal Observatory, the National Maritime Museum and the Old Royal Naval College.

Physical modelling has come of age and advances in all areas of technology, from digital imaging to computing, electronics and materials offer exciting opportunities to push boundaries well beyond the early experimental work. Visionary and adventurous physical modellers developed the basic techniques and important scaling laws that are the backbone of our work today. Such research made possible important contributions to the understanding of complex soil/structure interaction problems long before numerical modelling was capable of even attempting to establish such insight. Present day physical modellers are just as ambitious and adventurous as their forefathers and are anxious to build ever larger facilities and undertake increasingly complex experimental work. To this end, plans are underway for a 1000 g/tonne 'megafuge' capable of modelling the very largest of geotechnical structures. Physical modelling enjoys increasing popularity as a powerful means of exploring geotechnical problems. However, it rarely finds favour over numerical modelling in the eyes of industry where results of experimental studies are required soon after commissioning the work; regardless of accuracy and at minimal cost. Current work that focuses on exploring the interface between physical modelling and numerical modelling is a particularly exciting development and has the potential to yield important new knowledge applicable to both fields.

The organisation of a major international conference is a massive undertaking. My thanks go to the Local Organising Committee and the International Advisory Board and to everyone who participated in the very thorough review process. Particular thank are due to my colleagues, Sam Divall, Richard Goodey, Jignasha

Panchal, Sarah Stallebrass and Neil Taylor at City, University of London who rolled up their sleeves to help with all aspects of the conference; but notably in managing and editing the huge volume of poorly formatted papers. For anyone reading this far, please do not alter the template when writing your conference papers.

Andrew McNamara
Chair, Technical Committee 104 on Physical Modelling in Geotechnics, 2014 – 2018
International Society for Soil Mechanics and Geotechnical Engineering

International advisory board

Adam Bezuijen
Emilio Bilotta
Jonathan Black
Miguel Cabrera
Bernado Caicedo
Jonny Cheuk
Michael Davies
Jelke Dijkstra
Mohammed Elshafie
Vincenzo Fioravante
Christophe Gaudin
Susan Gourvenec
Stuart Haigh
SW Jacobsz
Ashish Juneja
Diethard König
Dong Soo Kim

Jonathan Knappett
Bruce Kutter
Jan Laue
Colin Leung
Xianfeng Ma
Alec Marshall
Tim Newson
Dominic Ek Leong Ong
Ryan Phillips
Kevin Stone
Andy Take
Jiro Takemura
Luc Thorel
David White
Daniel Wilson
Varvara Zania

Local organising committee

Andrew McNamara
Sam Divall
Neil Taylor
Richard Goodey
Sarah Stallebrass
Joana Fonseca
David White
Susan Gourvenec
Stuart Haigh
Mohammed Elshafie

Manuscript reviewers

O. Abuhajar
A. Ahmed
M. Alheib
I. Anastasopoulos
J. Barrett
A. Bezuijen
E. Bilotta
J.A. Black
M. Blanc
M. Bolton
A.J. Brennan
J. Breyl
L. Briancon
A. Broekman
M.J. Brown
M. Cabrera
Q. Cai
B. Caicedo-Hormaza
T. Carey
D. Chang
J. Cheuk
D. Chian
S.C. Chian
U. Cilingir
P.A.L.F. Coelho
G. Cui
T. da Silva
C. Dano
C. Davidson
M.C.R. Davies
A. Deeks
D. deLange
L. Deng
O. Detert
R. di Laora
T. Dias
J. Dijkstra
S. Divall
H. El Naggar
I. El-Haffar
G. Elia
S. Escoffier
V. Fioravante
J. Fonseca
T. Fujikawa
T. Gaspar
C. Gaudin
A. Gavras
L. Geldenhuys

R.J. Goodey
S.M. Gourvenec
G.J. Ha
J.G. Ha
S. Haigh
H. Halai
A. Hashemi
F. Heidenreich
H. Hong
K. Horikoshi
S.W. Jacobsz
A.J. Jebeli
A. Juneja
G. Kampas
T. Karoui
E. Kearsley
E.Y. Kencana
M.H. Khosravi
D.S. Kim
J.H. Kim
J.A. Knappett
D. Koenig
B. Kutter
R. Kuwano
L.Z. Lang
G. Lanzano
J. Laue
A. Lavasan
B.T. Le
S.W. Lee
F.H. Lee
C.F. Leung
L. Li
L.M. Li
T. Liang
H.I. Ling
M. Loli
C. Lozada
A. Lutenegger
F. Ma
A. Marshall
M. Masoudian
R. McAffee
A.M. McNamara
G.J. Meijer
M. Millen
H. Mitrani
A. Mochizuki
S. Nadimi

T. Newson
M. Okamura
D.E.L. Ong
J.P. Panchal
H.J. Park
J. Perez-Herreros
R. Phillips
G. Piercey
C. Purchase
M. Qarmout
S. Qi
M. Rasulo
S. Ravjee
A. Rawat
S. S.Chian
F. Saboya Junior
A. Sadrekarimi
M. Silva Illanes
G. Smit
S.E. Stallebrass
S.A. Stanier
K. Stone
A. Takahashi
A. Take
J. Takemura
G. Tanghetti
R.N. Taylor
L. Thorel
I. Thusyanthan
T. Tobita
K. Ueno
R. Uzuoka
R. Vandoorne
K. Wang
D.J. White
D. Wilson
K.S. Wong
H. Wu
Y. Xie
J. Yang
J. Yu
V. Zania
C. Zhang
G. Zhang
L. Zhang
Z. Zhang
B.L. Zheng
Y.G. Zhou
B. Zhu

Sponsors

PLATINUM SPONSOR – Actidyn http://www.actidyn.com/

ANDREW SCHOFIELD LECTURE RECEPTION SPONSOR – http://www.broadbent.co.uk/

CONFERENCE SPONSOR – Tekscan, Inc. – http://www.tekscan.com/

12. Tunnel, shafts and pipelines

Study of the effects of explosion on a buried tunnel through centrifuge model tests

A. De
Manhattan College, Bronx, New York, USA

T.F. Zimmie
Rensselaer Polytechnic Institute, Troy, New York, USA

ABSTRACT: The effects of a surface explosion on underground structures were studied in a series of centrifuge model tests. All tests were conducted at $70\,g$, on 1:70 scale models and the effects observed were those due to a prototype explosion caused by 0.9 tons of TNT equivalent at $1\,g$. Strains measured at different locations of the model structure showed dependence on the thickness and the nature of the cover material separating the structure from the explosion. The use of a compressible protective barrier, composed of polyurethane geofoam, was found to mitigate the effects of explosion to some extent. Tests with underwater explosion showed that strains induced on the structure, as well as excess pore water pressure generated in the soil due to the explosion, both increased with increasing height of free water.

1 INTRODUCTION

1.1 Background

An explosion on the ground surface has the potential to cause significant damage to an underground tunnel. It is important to assess quantitatively the level of damage caused by an explosion, in order to design better protection measures.

When studying the effects of explosions, centrifuge model tests have a distinct advantage over other methods, such as full-scale tests (due to logistical challenges when using large masses of explosives) or numerical modeling (which need calibration against physical test results).

The effects of an explosion scale according to Hopkinson's or the cube-root scaling law. Shock waves generated by two explosive charges (with same geometry and type of explosives, but different mass), located at two different scaled distances, are in proportion to the cube-root of the weight of explosives (Baker et al., 1973; Taylor, 1995).

Thus, when a 1:N scale model is tested at an acceleration of Ng in a centrifuge test, the effects of the explosion are the same as those generated by an explosive with a mass equal to N^3 times that of explosives tested under normal $1\,g$ gravity. In the specific case of experiments reported here, models were prepared to a scale of 1:70 scale and tested at $70\,g$ on the centrifuge. The effects of explosion created in the centrifuge experiment were those of a prototype explosion that is 70^3 or 343,000 times larger.

The scaling relations applicable to explosion in centrifuge tests have been validated through extensive testing, reported by Charlie et al. (2005), Goodings et al. (1988), Kutter et al. (1988), Schmidt and Holsapple (1980), and Schmidt and Housen (1995). Davies (1991), De and Zimmie (2007), De et al. (2017), and Liu and Nezili (2016) have presented results of centrifuge tests where the effects of explosion on underground structures were studied.

1.2 Scope

The primary focus of the research reported here was to study the effects of a surface explosion on an underground structure, such as a tunnel or a pipeline. The effects of an explosion on an underground structure are controlled both by the nature of the explosion (type, mass, and orientation of the explosive), as well as on the nature and dimensions of the surrounding medium and the intervening medium (which separates the structure from the explosion location).

A total of 17 centrifuge tests were conducted as part of this research. The type, mass, and orientation of the explosive, described in a later section, were kept constant in all the tests. A summary of test conditions used in this series is presented on Table 1. The same soil, placed at the same relative density, was used in all the tests. In 13 tests, the sand was placed dry, while saturated condition was modeled in the remaining four tests. As can be seen in Table 1, the main variables in the different tests were the following:

- Thickness of soil cover between ground surface and underground structure
- Presence and thickness of protective cover

Table 1. Test configurations.

Condition	Material	Total cover (m)	Test #
Dry	Soil only	1.8	Test 2
Dry	Soil only	2.7	Test 9
Dry	Soil only	3.6	Test 1
Dry	Soil + geofoam (fully covered)	1.8 (0.9 soil + 0.9 foam)	Test 5
Dry	Soil + geofoam (fully covered)	3.6 (1.8 soil + 1.8 foam)	Test 6
Dry	Soil + geofoam (fully covered)	3.6 (2.7 soil + 0.9 foam)	Test 4
Dry	Soil + geofoam (only on top)	2.25 (0.9 soil + 1.35 foam)	Test 12
Dry	Soil + geofoam (only on top)	3.15 (1.8 soil + 1.35 foam)	Test 11
Dry	Soil + geofoam (only on top)	3.6 (0.9 soil + 2.7 foam)	Test 7
Dry	Soil + geofoam (only on top)	2.0 (0.9 soil + 1.1 concrete)	Test 13
Dry	Soil + geofoam (only on top)	2.5 (0.9 soil + 1.6 concrete)	Test 10
Dry	Soil + geofoam (only on top)	2.7 (1.8 soil + 0.9 concrete)	Test 8
Saturated	Soil only	2.7 (water at gs)	Test 15
Saturated	Soil only	2.7 (water 2.7 above gs)	Test 14*
Saturated	Soil only	2.7 (water 2.7 above gs)	Test 17*
Saturated	Soil only	2.7 (water 5.5 above gs)	Test 16

gs: ground surface.
* Tests 14 and 17 had identical conditions and demonstrated repeatability of results.
In all tests listed above, explosion was located directly above centerline.
Test 3, not shown on the table, used explosion above the springline of the tunnel and is not used for comparison.

- Nature of the protective cover (compressible vs. rigid)
- Effect of soil saturation (dry vs. saturated)
- Height of free water above ground surface

2 EXPERIMENTAL SETUP

2.1 Model configuration

A 760 mm long section of a copper pipe was used to model a circular tunnel. At 1:70 scale, it represented a 53 m long tunnel, with a diameter of 5.3 m. The model tunnel had a flexural stiffness (EI) equal to 13×10^6 kN-m^2. Bending and wall buckling were expected to the primary modes of deformation under transient load from an explosion located above the centerline. No specific prototype structure was modeled in this study; rather, the dimensions were selected such that they represent those of a typical single-track transit or roadway tunnel, or a large diameter pipeline.

The model tunnel was placed in a clean, uniform sand, designated as Nevada #120 with a dry unit weight of 15.7 kN/m^3, which represented a relative density of approximately 60%. The soil was compacted uniformly, ensuring consistency between the different tests. Any settlement during spinup and at target

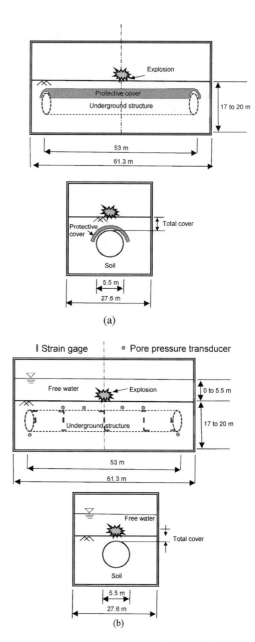

Figure 1. Experimental setups, showing model configuration for tests on (a) dry soil, with protective cover on top and (b) saturated soil, with free water above ground surface. Instrument locations are only shown in (b) for clarity. Strain gage locations also apply to (a). (All dimensions are in prototype scale; based on 1:70 scale model).

g-level was assumed to be roughly the same for all the tests and was ignored. The static and dynamic properties of this sand has been extensively characterized by Arulmoli et al. (1992). Schematic diagrams of the model setup are shown in Figure 1. Figure 1 (a) shows the setup with a protective cover on the top of the structure and Figure 1 (b) shows that with free water above the ground surface.

2.2 Instrumentation and data acquisition

Up to 19 strain gages (uniaxial and biaxial) were used to measure axial and circumferential strains at different locations on the model structure. In addition, up to six pore water pressure transducers were used in the tests where saturated conditions were modeled. Instrument locations are shown in Figure 1(b). The strain gage locations also apply to Figure 1(a).

Data was acquired at a rate of 15,000 readings per second (15 kHz) for a duration of 15 s, starting prior to explosion and continuing after explosion, until instrument readings reached a steady state. Additional details about the instrumentation are provided in De et al. (2017).

Figure 2 shows a photograph of the model structure, with instrumentation installed and located in sand, prior to introduction of water.

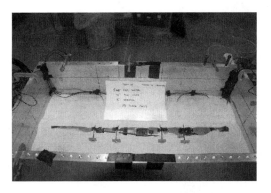

Figure 2. Photograph of test setup from Test 16, showing model structure placed in sand inside model container. Strain gages are covered by black protective patches and four pore water pressure transducers can be seen in the foreground.

2.3 Explosives

The same mass of explosive, installed in the same configuration, was used in each test. Two exploding bridgewire (EBW) charges, with a total TNT-equivalent of 2.6 g, were used on a model scale. Following the cube-root scaling relation discussed earlier, each test, conducted at 70 g on a 1:70 scale model, utilized the equivalent of 2.6 g times 70^3 g or 888 kg, i.e., approximately 0.9 tons of TNT equivalent on a prototype scale.

The charges were connected through the centrifuge slip rings to the EBW control unit and firing module. When the model reached the target g-level, the data acquisition was first started and then a safety key was inserted to complete the circuit to trigger the explosion.

Adequate safety measures were taken to protect the centrifuge and its associated equipment from shrapnel and debris caused by the explosion. It was important to select a protective system which would prevent flying debris from reaching the equipment, while, at the same time, not interfere with the effects of the explosion. A total confinement of the explosion zone would prevent damage, but also distort the shock waves and air flow, which are important characteristics resulting from the explosion.

After careful testing, a flexible, multilayer configuration was adopted, which consisted of layers of wire mesh and non-woven geotextile. The layers were given progressively more slack, going from inside to outside. This ensured that shrapnel were contained, while allowing air to vent and pressure to dissipate. The protective covers were replaced at the end of each test.

3 RESULTS

3.1 Effects of thickness of soil cover

In all the tests, a soil cover separated the crown of the structure from the ground surface. The effect of the explosion on the model structure (measured as strains) would depend on the thickness of the soil cover. This was investigated by comparing results from Tests 1, 2, and 9. Plots of axial strains measured at crown at quarterspan are shown in Figure 3. As shown in the figure, the strain reaches a peak value immediately following the explosion and then reduces. Comparing the magnitudes of peak strain, one can see that the highest strain was recorded for a 1.8 m soil cover (in prototype scale), while the smallest was recorded for 3.6 m. Thus, the peak strain appears to bear an inverse relation to the thickness of soil cover, as would be expected. Also, the smallest soil cover (1.8 m) resulted in a residual strain, whereas there was no significant residual strains under steady state, following explosion, for the other two cases.

The results of this series of tests indicates that (a) an explosion on the ground surface may induce significant strains on an underground structure and (b) the magnitude of strain generally decreases as the thickness of soil cover increases. Whether the peak or the residual strain can result in critical damage to the tunnel would be a consideration for the design engineers, based on the material properties and serviceability limits on induced strain. The authors have not yet found a quantitative relation between the decrease in strain and increase in soil cover thickness.

3.2 Effects of geofoam cover

A protective cover may be used over the underground structure to help mitigate the effects of a surface explosion. A compressible barrier, made of polyurethane geofoam was used in these tests. The geofoam barrier was installed around the model structure by applying a spray-on foam sealant. Detailed description of the procedure and material tested are provided in De and Zimmie (2007).

The effect of including the compressible barrier between the ground surface and the crown of the structure can be studied by comparing the strains measured in Tests 1 and 4. In Test 1, a uniform soil cover of thickness 3.6 m (prototype scale) was used. In Test 4, the total cover thickness was also 3.6 m, but it consisted of

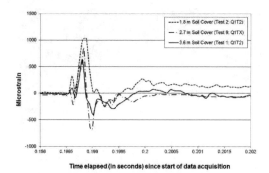

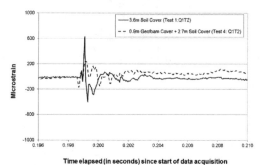

Figure 3. Axial strain time histories measured at crown at quarterspan for three different soil covers (values of time elapsed since data acquisition were adjusted for display purposes).

Figure 4. Axial strain time histories measured at crown at quarterspan for soil cover, vs. soil and geofoam cover, with same total thickness (values of time elapsed since data acquisition were adjusted for display purposes).

2.7 m of soil directly below the ground surface, with a 0.9 m thick geofoam barrier placed in intimate contact with the outer surface of the structure.

Figure 4 shows plots of axial strain at the crown at quarterspan from Tests 1 and 4. There is a reduction in axial strain of more than 60% in Test 4, where the geofoam compressible barrier was utilized, as compared with Test 1, with a homogenous soil cover of the same total thickness. The compressible barrier underwent deformation under load from the explosion, thereby reducing the stress transferred to the structure, resulting in a reduction in induced strain. Results of numerical model analyses to further investigate the effects of geofoam barriers were reported by De et al. (2016).

3.3 Underwater explosion: Effects of water level

The effects of explosion on an underground tunnel, which is itself located below a body of free water, were investigated in this portion of the research study (Tests 14–17). The explosives were located directly on the ground surface, below the water body. A constant height of 3.6 m of soil cover was used in all centrifuge tests in this series.

Four tests were conducted, with three different heights of water. Strains were measured at different locations of the model structure. Pore water pressures in the soil were measured at different distances from the explosion.

Figure 5 shows the plots of axial strain with elapsed time measured at invert at centerspan from all four tests.

The highest strain is measured for a height of free water equal to 5.5 m (top graph) and the lowest strain for water at the ground surface, which is comparable to strain for dry soil (bottom graph). The strain measured in two tests for water at 2.7 m (middle graph) are in between the two extreme cases and quite close to each other, indicating repeatability of results.

When an explosion takes place under water, the rapidly expanding mass of gaseous products is confined by the presence of water, whereas it would have

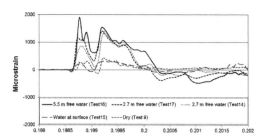

Figure 5. Axial strain time histories measured at invert at centerspan for four different heights of free water (values of time elapsed since data acquisition were adjusted for display purposes).

dissipated into air if no water was present. Because of this confinement, a portion of the energy is directed towards the ground surface and, subsequently, on the buried underground structure.

Larger strains are induced in the structure when the height of free water is increased, since better confinement results in greater energy being directed into the subsurface. De et al. (2017) presented comparisons of results from centrifuge model tests and numerical analyses to study the effects of an underwater explosion. According to De et al. (2017), there is actually a critical height of water beyond which additional increases in water height do not result in any consequent increase in induced strains.

Figure 6 shows the plots of pore water pressure with elapsed time, measured near the base of the test container, immediately below the endspan of the structure. In prototype scale, the pore water pressure measurement point was located at a vertical depth of 14.7 m and a horizontal distance of 26.5 m from the explosion in each test. The model underground structure was located between the explosion and the measurement location.

Looking at each pore pressure time history, it can be seen that the initial readings are fairly constant, corresponding to the calculated value of hydrostatic

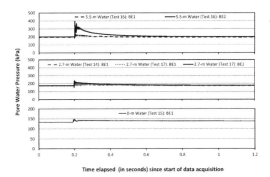

Figure 6. Pore water pressure time histories measured below the invert at endspan for four different heights of free water (values of time elapsed since data acquisition were adjusted for display purposes) Note: the vertical scale of bottom graph is different from the other two.

pore water pressure at the location. The arrival of the first shock wave from the explosion is marked by an instantaneous rise in pore water pressure, followed by a gradual reduction that follows a classic pore pressure dissipation curve.

The highest value of peak pore water pressure was recorded for the case with 5.5 m of free water above the ground surface (Test 16). The total peak pore water pressure was 394 kPa, with an excess pore pressure of 194 kPa. The lowest value of peak pore water pressure at this same location was recorded when the water level was at the ground surface (Test 15). The peak pore water pressure was 150 kPa, with an excess pore pressure equal to 16 kPa.

The peak pore water pressures corresponding to a 2.7 m height of free water were between 211 and 236 kPa, measured at a total of three locations in Tests 14 and 17. The results were comparable, indicating repeatability of results. The corresponding excess pore pressures were between 39 and 61 kPa.

The observed trend in pore water pressure is consistent with that of axial strain, discussed in the preceding section. A larger height of free water provides greater confinement and results in a higher portion of the energy being directed into the subsurface, resulting in a higher peak pore pressure. When the height of free water is small or water is at the surface, the resultant peak pore pressure is smaller.

4 CONCLUSIONS

The results of a series of physical model tests presented here illustrate that centrifuge modelling represents a viable method to study the effects induced by an explosion on an underground structure, as well as on the subsurface soil medium surrounding the structure. Taking advantage of the cube root scaling law, approximately 2.6 g of TNT equivalent was utilized at 70 g to produce the effects of 0.9 tons of explosives on a prototype scale.

The peak strains measured on the model underground structure decreased when the thickness of soil cover, which separates the explosion from the crown of the underground structure, was increased. The strains were also found to decrease when a compressible cover (made of polyurethane geofoam) was used, in addition to the soil cover. This represents a possible method to mitigate damage due to surface explosions.

The effects of underwater explosion were also studied through centrifuge model tests. The peak strains measured in the underground model structure increased when the height of free water above the ground surface (and above the explosion) was increased. The value of peak excess pore water pressure, measured in the saturated soil, was also found to increase when the height of free water above the ground surface was increased.

The increases in peak strain and peak excess pore water pressure values with height of water are believed to be caused by the confining effects of the free water. The presence of free water tends to prevent the expanding gases produced by an explosion from readily dissipating into the atmosphere, and instead, directs the energy into the subsurface. A higher free water surface results in a better confinement, thus resulting in larger strains and pore pressures.

In a separate series of numerical analyses (De et al., 2017), it has been demonstrated that there is, in fact, a maximum critical height of free water, above which any further increase in water height does not result in additional strains.

Centrifuge model tests, such as those reported here, can be used to study the effects of explosions on new and existing tunnels and pipelines. The effectiveness of suitable mitigation measures to reduce damage (such as protective cover) can be investigated. The results of centrifuge model tests can also be used to calibrate and validate numerical models, which can then be used to study different configurations and design variations.

REFERENCES

Arulmoli, K., Muraleetharan, K.K., Hossain, M.M., Fruth, L.S. 1992. *VELACS: Verification of Liquefaction Analyses by Centrifuge Studies, Laboratory Testing Program, Soil Data Report*. Earth Technology Corp. Project No. 90-0562. Irvine, California.

Baker, W.E., Westine, P.S., Dodge, F.T., 1973. *Similarity Methods in Engineering Dynamics*. Spartan Books, Rochelle Park, New Jersey.

Charlie, W. A., Dowden, N. A., Villano, E. J., Veyera, G. E., and Doehring, D. O., 2005. "Blast-Induced Stress Wave Propagation and Attenuation: Centrifuge Model Versus Prototype Tests", *Geotechnical Testing Journal*, ASTM, Vol. 28, No. 2, pp. 1–10.

Davies, M. C. R. 1991. "Buried Structures Subjected to Dynamic Loading", Chapter 8 in *Structures Subjected to Dynamic Loading*, R. Naryanan and T. M. Roberts (Editors), Elsevier, pp. 271–302.

De, A. and Zimmie, T. F. 2007. "Centrifuge Modeling of Surface Blast Effects on Underground Structures", *Geotechnical Testing Journal*, ASTM, 30(5), 427–431.

De, A., Morgante, A. N., and Zimmie, T. F. 2016. "Numerical and physical modeling of geofoam barriers as protection against effects of surface blast on underground tunnels", *Geotextiles and Geomembranes*, 44, 1–12.

De, A., Niemiec, A., and Zimmie, T. F. 2017. "Physical and Numerical Modeling to Study Effects of an Underwater Explosion on a Buried Tunnel", *Journal of Geotechnical and Geoenvironmental Engineering*, ASCE, doi: 10.1061/(ASCE)GT.1943-5606.0001638.

Goodings, D. J., Fourney, W. L., and Dick, R. D. 1988. "*Geotechnical Centrifuge Modeling of Explosion-induced Craters – A check for Scaling Effects*", U.S. Air Force Office of Scientific Research, Washington D.C., Report No. AFOSR-86-0095.

Kutter, B.L., O'Leary, L.M., Thompson, P.Y., and Lather, R. 1988. "Gravity-scaled Tests on Blast-induced Soil-structure Interaction", *J. Geotech. Engrg.*, 10.1061/(ASCE)0733-9410(1988), 114:4(431), 431–447.

Liu, H. and Nezili, S. 2015. "Centrifuge modeling of underground tunnel in saturated soil subjected to internal blast loading". J. Perform. Constr. Facil., 10.1061/(ASCE)CF.1943-5509.0000760.

Schmidt, R. and Housen, K. 1995. "Problem Solving through Dimensional Analysis", *Industrial Physicist*, 1(1), 21–24.

Schmidt, R. M. and Holsapple, K. A. 1980. "Theory and Experiments on Centrifuge Cratering", *Journal of Geophysical Research*, 85(1), 235–252.

Taylor, R.N. 1995. *Geotechnical Centrifuge Technology*, Blackie Academic & Professional, Chapman and Hall, Glasgow.

Uplift resistance of a buried pipeline in silty soil on slopes

G.N. Eichhorn & S.K. Haigh
Department of Engineering, University of Cambridge, UK

ABSTRACT: Pipeline routing is required for successful regulatory approval and construction of a transmission or distribution pipeline. As part of the routing process, engineers and geoscientists must consider the geohazards present along proposed pipeline route. Limited guidelines exist for operators to build high pressure transmission lines across varying geophysical environments. Current practice involves discritising a continuum model into soil-spring pairs to model the soil-structure interaction. The model was developed in the 1980's assuming a clean sand on flat ground, with no consideration to the effect of natural soil properties such as fines content. A series of uplift resistance tests were conducted on a drum centrifuge to measure the non-dimensional load-displacement response for varies configurations of a buried pipe on a sloping ground. The results do not match the proposed guideline for soil-structure interaction used by regulators and industry.

1 INTRODUCTION

Recent pipeline rupture events involving landslides (Barlow & Richmond, 2016) have identified a need to improve the understanding of landslide interaction with buried linear infrastructure. The continuum model to represent the soil structure interface is modelled by four non-linear Winkler type springs: uplift resistance, downwards vertical bearing, lateral, and axial. While this idealization may be practical, the underlying assumptions of the relative soil-spring factors are designed for horizontal ground. The distribution of loading in space and time for large ground deformations such as landslides is largely unknown (PRCI, 2008).

The current state of practice for building buried natural gas and liquids pipelines across hillsides varies significantly from one pipeline operator to another. In North America, the American Lifelines Alliance produced a best practice approach to managing risk from landslides (ALA, 2001), which was reviewed and expanded upon by other research and regulatory agencies (PRCI, 2008). However, they are limited in that they do not account for site specific geohazards and merely recommend that a competent geotechnical engineer be consulted. Geotechnical modelers currently rely upon a soil-spring representation of the pipeline-soil interface to determine the expected forces acting on a pipeline from the soil, and the forces on the soil from the pipeline.

This study has been undertaken to examine the soil-structure interaction between landslides and buried onshore pipelines using centrifuge modelling.

2 PREVIOUS WORK

A non-linear Winkler structural-type analysis is typically conducted within an FE model to impose soil loads on pipelines. The soil loads are transferred to the pipeline by springs.

The individual soil springs rely on non-dimensional force-displacement coefficients to quantify the reaction force from the soil. The non-dimensional peak resistance force has been based on empirical studies and takes the form of a hyperbolic load-displacement curve for vertical uplift and lateral movement. The general form of this reaction force is:

$$\frac{F}{F_{max}} = \frac{x}{Ax_u + Bx} \quad (1)$$

where F = soil reaction force; F_{max} = peak soil resistance; x = displacement of the pipe; x_u = displacement of pipe at peak resistance; A & B are empirical coefficients.

Vertical uplift resistance of buried conduits was first carried out by Martson (1930) who proposed that the uplift resistance of a buried conduit could be estimated by two vertically projecting slip surfaces extending to the ground surface. Cavity expansion theory was applied to the upward movement of buried objects and was one of the first theoretical approaches applied specifically to buried pipes (Vesic, 1972).

Experimental work in the 1980's focused on the non-dimensional load-displacement response of buried pipelines (Trautmann et al., 1985). Challenges identified in those experiments included the increase and rapid decrease in resistance as the soil sheared past the pipeline into the void beneath the pipe.

A large amount of the attention given to breakout resistance of pipelines was advanced due to interest in offshore pipelines. More recently, studies have been done to include breakout resistance and upheaval buckling (Yimsiri et al., 2004), and have been investigated for a wider range of cohesive materials compared to the traditional clean sand approach (Baumgard, 2000). The current guideline (C-CORE et al., 2009) for calculating the peak force component of the vertical soil-spring is:

$$Q_u = N_{cv}cD + N_{qv}\bar{\gamma}HD \qquad (2)$$

where N_{cv} and N_{qv} are uplift factors for clay and sand respectively; c = cohesion; D = pipe outer diameter; $\bar{\gamma}$ = effective unit weight of soil; and H = depth from ground surface to the springline of the pipe. The uplift factors for clay and sand in equation 2, are closed form empirical solutions based largely on experimental results by Trautmann (1985). The peak uplift resistance, Q_u, occurs at displacement, Δq_u, and has been reported to range from 0.01 H to 0.02H, for dense to loose sands, and 0.1 H to 0.2 H for stiff and soft clays (PRCI, 2008; ALA, 2001).

All the experimental studies for the vertical resistance soil-spring have been isolated to pipelines buried under flat ground, with little consideration to any interdependency of the other soil springs (lateral and axial), and with no consideration of shear of soil past the pipe as displacement of the soil body progresses, such as in a landslide on a hill. The complexities of a landslide environment may make it possible for uplift to occur but no void created beneath the translating pipe, due to the redistribution of the larger soil body. This complexity cannot be accommodated for by the current peak force calculation described above. A simplified study of uplift resistance on a hill side is one component of examining the soil-springs for pipelines in landslide, and is what this research has focused on to date.

3 PIPE DESIGN

To conduct centrifuge experiments on a pipeline, an appropriate model representing the pipe must be selected for use in the centrifuge. To reflect the prevailing size found in high pressure transmission pipelines, nominal pipe size (NPS) 18, or 456 mm outer diameter (OD) was selected for this study for the work to date. In future work, it will be necessary to investigate a variety of pipe sizes. In an experiment that is accelerated at 100 times Earth's gravity, N = 100 g, the scaling law for dimensions of length or width would indicate a model pipe outside diameter of approximately D = 4.56 mm.

The corresponding pipe wall thickness of an NPS 18 pipeline is approximately 10 mm. This dimension at model scale would be only 0.1 mm. This wall thickness of model pipe is not readily available, and so for the ease of obtaining a model pipe and for ease of use of this material, another material must be selected to represent the steel pipeline being studied. Another important reason for choosing a gravity field of near N = 100, is that the maximum height of the model container can be used to exploit the largest possible hillside.

The model pipe was designed using stress-strain equations for thin-walled pressure vessels, or cylinders (Young & Young, 1989). A pipeline typically has a diameter to wall thickness ratio (D/T) of 20 to 100 (Liu, 2003). Thin-shelled stress-distribution assumptions for cylinders are applicable for D/T ratios above 20 (Young & Young, 1989). Work was carried out to determine the optimal model pipe to represent a prototype steel pipe of OD = 456 mm. As the model pipe material is different from the prototype, certain structural properties become critical to replicate in the model pipe. A review of the axial stiffness (EA) and bending stiffness (EI) was critical as the pipeline is most often in axial tension or bending.

The OD of the model was set at 4.56 mm to match the prototype pipe to avoid adjustments to the scaling. The only other properties of the pipe that could vary therefore was the modulus of elasticity (E) and the wall thickness (WT). Several materials were considered including: aluminum, copper, brass, and nylon. A wall thickness of 0.5 mm was used for preliminary design. A series of analyses were done to assess different materials to determine the correct wall thickness given a value of E. Copper and brass which have values of E of 117 GPa and 107 GPa respectively, produced scaled results which closely matched the material response of a steel pipe at full scale. The two most common modelling pipe diameters available these materials were 0.45 mm and 0.225 mm.

The analysis in Figure 1 shows that brass with wall thickness of 0.225 mm, at an acceleration of N = 90 g, produces an EI and EA variance between the real world and prototype of −2.2% and +2.3% respectively. Brass tubing was selected as the material of choice for the model pipe due to the ease of obtaining the pipe in the correct wall thickness. Copper may offer advantages in that it more closely matches the ultimate strength of some steels, as compared to brass, however obtaining copper tubing of this dimension is not economical. For these reasons, brass was obtained and used for this research.

The axial, radial, and bending response of the prototype compared to the real-world pipe analysis were individually considered, and a theoretical load-displacement response was calculated for a range of applied loads, over the elastic region of the material. The variance of EI and EA was constant over the range of applied loads, and therefore the model was assumed to be valid for the elastic response of the brass material. The brass pipe at model scale was then used in a series of vertical uplift resistance tests, to simulate a buried steel pipeline at prototype, with closely

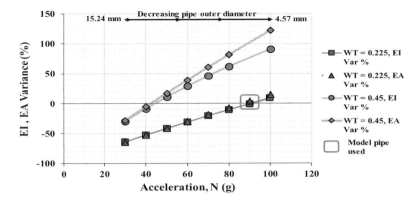

Figure 1. Brass pipe parametric analysis of EI and EA variance.

Table 1. Comparison of pipe properties.

Property (Units)	Full Scale	Model	Prototype
Material (–)	Steel	Brass	Brass
NPS (inch)	18	–	18
OD (mm)	457.2	5.08	457.2
WT (mm)	10	0.225	20.25
E (GPa)	20	10	107
Poisson Ratio, v (–)	0.27	0.39	0.39
Yield Strength (MPa)	30	45	45
Tensile Strength (MPa)	46	60	60
D/T	46	23	23
EI	0.07	–	0.071
EA	2.90	–	2.974

Table 2. Soil properties of KW15.

Property	Value	Units
Passing No. 200 Sieve	16.6	%
Average Particle Size, D_{50}	0.5	mm
Maximum Void Ratio, e_{max}	1.02	–
Minimum Void Ratio[1], e_{min}	0.35	–
Specific Gravity, G_s	2.68	–
Moisture Content[2], w_{opt}	9.8	%
Dry Unit Weight[2], $\gamma_{d,max}$	19	kN/m^3
Friction Angle[3], φ	39	degrees
Cohesion, c	7.4	kPa
Soil Classification, USCS	SM	–
Soil Classification, AASHTO	A-3	–

1 – Modified e_{min} ratio test described in full by Bowman (2017).
2 – Based on a modified proctor test.
3 – Taken at modified proctor maximum density.

matching material response properties. The property comparison between real-world, model, and prototype are summarized in Table 1.

A steel pipe at full scale is typically made of high strength steel, X52 line pipe, specified by the American Petroleum Institute (API) as a standard line pipe made of mid-range high quality steel, corresponding to a yield strength of 52 ksi. The brass pipe used was alloy 260, corresponding to a copper content of 70%.

4 MATERIAL PROPERTIES

Many geotechnical modelers use manufactured soil analogues in physical modelling. Given the gap in research described previously, a natural silty sand, known as KW Sand, was selected for the tests described herein. This sand originates in Kazakhstan and was being studied by others for a transmission pipeline project, and so offered a good opportunity to experiment on a soil that had some relevance to a transmission pipeline. For the tests described here, the 'KW15' sand was selected. The soil properties were measured by others (Bowman, 2017) and are summarized below in Table 2.

5 EXPERIMENTAL PROCEDURE

To investigate the problem of buried pipelines in slopes, a model container occupying a portion of the whole mini-drum ring channel was used. The container bottom is curved to match the radius of the drum inner diameter, r = 0.5 m. The typical arrangement is a container divided down the long axis of the box by a glass window. The larger side of the container is reserved for the soil body and instrumentation, and the narrower side is used for placing cameras to photograph the experiment. The interior dimensions of the soil side of the container are approximately 150 mm × 220 mm × 300 mm (H × W × L). A photograph of the model container is shown in Figure 2.

Slopes were constructed in the model container by first mixing the KW15 soil to optimum water content. Samples of the moist soil were collected before compaction and post-testing to verify moisture content. Soil was added to the container in approximately

Figure 2. Split-box mini-drum model container.

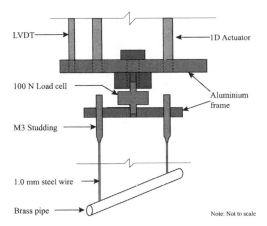

Figure 3. Schematic of 1D actuator to pipe connection.

20 mm lifts and compacted by hand, using a tamping rod with circular footing.

The soil surface of each 20 mm lift was scarified to ensure proper bonding of each lift to the next. This was completed sequentially in the slope side of the model container, until a slope profile was achieved. The model hillside was approximately 100 mm in height and a slope length of 150 mm. The slope profile for these series of experiments was targeted at 30 degrees from horizontal.

Soil density was identified in the previous work as being a key component to producing good results for the soil-spring formulations (Trautmann et al., 1985). The compacted density was verified by using a rotary-coring tube, specially designed to collect a core sample from the small model container, while minimizing sample disturbance. This method was able to produce a density measurement within 3% of a control sample.

The mini-drum centrifuge is equipped with a 1-D actuator. This is driven by a lead screw, connected to a gear box and a DC electric motor. The 1D-actuator is seated at the centre of the rotating axis of the drum, and actuates perpendicular to the axis of rotation, and normal to the top of the model container. The 1-D actuator provides a mechanism to move the pipe within the soil body. To measure the resistance created between the soil and the pipe, a 100 N load cell was attached to the 1-D actuator, between the load frame of the actuator, and the vertical steel wires that connect to the pipe. A schematic of the setup between the actuator, load cell, and pipe is shown in Figure 3. The displacement of the 1D actuator was measured using a linear variable displacement transducer (LVDT), which is affixed to the actuator. The stroke length of the LVDT matches the full range of the 1D actuator, or approximately 200 mm. The steel wires were pre-tensioned such that the strain in the wire was less than 0.01%.

The model pipe was installed in a trench dug into the model slope at various depths. The model pipeline was placed parallel to the fall line of the hill. The centrifuge was accelerated such that the model hill and pipe experienced as stress-state of N = 90 g at mid-slope. The 1-D actuator and pipe were made to displace vertically at a rate of 0.5 mm/min at model scale, corresponding to 4 m per day at prototype scale, which is extremely rapid in the scale of landslide movement rates (Cruden & Varnes, 1996). The pipe was displaced vertically to approximately 3 to 5 pipe diameters, which is well beyond previous studies. The pipe and soil were placed against a vertical glass window such that the pipe and soil displacement could be captured with a camera in cross-section. While not described here, the soil texture was tracked using direct image correlation (DIC), utilizing the open-source image processing package Geo-PIV-RG (Stanier et al. 2016).

6 CENTRIFUGE TEST RESULTS

Centrifuge experiments were carried out in the mini-drum, using various configurations of pipe alignment, burial depth, and material compaction.

The prototype scale load-displacement results for tests MD-02, 04, 05, and 06 are shown in Figure 4, as non-dimensional force and displacement. The non-dimensional results were calculated by taking the dimensionless force, N_v, from equation 3, and the dimensionless displacement Z/D, where Z is the displacement. A comparison of the peak uplift resistance measured to the predicted value for equation 2, shows that the predicted resistance for this length of pipe should range from 280 kN to 960 kN for the different burial depths and compaction. The results are within range of the prediction.

$$N_v = \frac{F}{\gamma HDL} \quad (3)$$

where F_{max} is the peak uplift resistance, γ = soil unit weight, L = length of pipe section being tested, and H and D are previously described. The results are shown with the H/D ratio for each pipe, the arrow indicating the peak resistance. The results show that,

Table 3. Centrifuge test parameters at model scale.

Test Name	Pipe OD (mm)	WT (mm)	Location of pipe	H (mm)	Scale Factor, N	Pipe Alignment* (°)	Slope Angle (°)	W (%)	ρ (kg/m^3)	γ (kN/m^3)
MD-02	5.98	0.48	window	20	90	0	30	11	1500	14.7
MD-04	4.98	0.24	window	25	90	0	30	11.3	1911	18.7
MD-05	4.98	0.24	window	35	90	0	30	11.4	1911	18.7
MD-06	4.98	0.24	centre	22	90	0	30	10.2	1963	19.3

*-0° = Parallel to fall line of hill, 90° = transverse across hill.

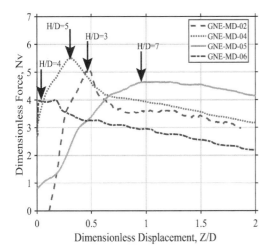

Figure 4. Non-dimensional load-displacement results.

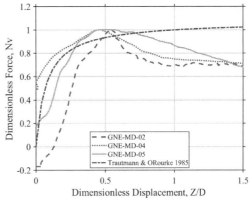

Figure 5. Prototype load-displacement results.

except for MD-02, the peak force increases with greater H/D ratios. This is to be expected since a greater burial depth would require more mobilized soil above the pipe to allow it to displace. Tests MD-04, 05, and 06 had similar bulk densities, compared to MD-02, which had a relatively low bulk density compared to the other tests. This may explain the difference in the peak force, and the observed softening of the soil in MD-02.

The results are typically normalised to peak force and displacement at peak force (Trautmann et al. 1985). Equations 4 and 5 below are used to determine the normalized force, F", and normalized displacement, Z" respectively. This is shown in Figure 5.

$$F'' = \frac{\left(\frac{F}{\gamma HDL}\right)}{N_v} \quad (4)$$

$$Z'' = \frac{\left(\frac{Z}{D}\right)}{\left(\frac{Z_f}{D}\right)} \quad (5)$$

The normalised results are plotted with the hyperbolic curve suggested by Trautmann and O'Rourke (1985) who proposed coefficients of 0.07 and 0.93 for A and B in equation 1. The peak normalised forces match well with this predicted curve, however the load-displacement response of the KW15 soil shows that the rate of increase of force with displacement is less than the curve proposed by Trautmann and O'Rourke. Additionally, the residual force decreases in this series of tests, counter to the predicted curve. The previous research typically only displaces pipes to one pipe diameter and does not account for behaviour at critical state. This series of experiments extended that range.

Another way of examining the results is to compare the dimensionless peak uplift resistance values measured to the curves proposed by Trautmann and O'Rourke (1985), shown in Figure 6.

They propose curves for loose, medium, and dense sand. The results from this study follow an opposite trend to Trautmann's results, in that the loose soils from this work occur at higher values of Nv and dense soils at low values of Nv. These results may indicate that the bulk density measurement technique for this work needs further improvement. One difference between these curves and the results shown are that Trautmann and O'Rourke used a clean sand, where these experiments had a fines content of 16%. This may be contributing to the difference in results. Test MD-06 was the only test conducted in the centre of the slope, away from the sidewall and glass window. This test was done to determine if a difference exists in the load-displacement curve compared to a test at the window. The test is most comparable to test MD-04, which

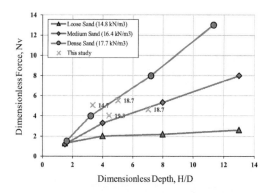

Figure 6. Normalised load-displacement results (Modified Trautmann and O'Rourke, 1985).

had very similar bulk density and burial depth. The peak uplift resistance for test MD-04 and MD-06 were 310 kN and 50 kN respectively, a difference of six times. While this may indicate boundary effects at the glass, there is not enough data yet to confirm the effect of sidewall friction causing higher resistance at the glass window, and will be addressed in future work.

7 CONCLUSION

The results from the current work have shown that the mini-drum centrifuge is a proven experimental apparatus capable of rapid parametric testing. The soil springs relied upon for soil-structure interaction problems involving buried pipe need to be reassessed to understand the interdependency of springs, and in natural soil.

The model pipe design used for this work matches the axial and bending stiffness of a prototype steel pipe typically used in large diameter pipeline construction. Future work will address instrumenting the model pipe to measure strain response from uplift resistance.

The uplift displacement results from the first four tests have shown that the current peak uplift resistance equations used by industry match the experimental results for peak uplift force, but that residual force on the pipe is not captured. The loading path to reach peak force does not match the bilinear model proposed by Trautmann, which is applicable for a clean sand. The loading path assumed by the current models may not capture the strain redistribution in a full landslide, for the locations of a landslide that would produce uplift resistance on a buried pipe. Future work will need to address these issues.

REFERENCES

American Lifelines Alliance, ALA, 2001. Guidelines for the design of buried steel pipe. American Socoiety of Civil Engineers (ASCE).

Barlow, J.P. & Richmond, J.A., 2016. The Cheecham Landslide Event. In 11th International Pipeline Conference. Calgary, pp. 1–10.

Baumgard, A. J., 2000. Monotonic and cyclic soil responses to upheaval buckling in offshore buried pipelines. PhD dissertation, Cambridge University, UK.

Bowman, A., 2017. Performance of Silty Soils and Their Use in Flexible Airfield Pavement Design. PhD dissertation, Cambridge University, UK.

Ferris, G., Newton, S. & Porter, M., 2016. Vulnerability of Buried Pipelines to Landslides. In 11th International Pipeline Conference. Calgary, pp. 1–8.

Martson, A., 1930. The Theory of External Loads on Closed Conduits in the Light of the Latest Experiments, Bulletin 96, Iowa Engineering Department Station, Ames, Iowa.

Pipeline Research Council International, PRCI, 2008. Pipeline Integrity for Ground Movement Hazards PR-07-082-459. Catalogue No. L52291. (PRCI).

Stanier, S.A. et al., 2016. Improved image-based deformation measurement for geotechnical applications. Canadian Geotechnical Journal, 53: 727–739.

Trautmann, C.H., O'Rourke, T.D. & Kulhawy, F.H., 1985. Uplift Force-Displacement Response of Buried Pipe. Journal of Geotechnical Engineering, 111(9), pp.1061–1076.

Vesic, A.S., 1972. Expansion of Cavities in Infinite Soil Mass, Journal of the Soil Mechanics and Foundations Division, ASCE, 98(SM3): 265–290.

Yimsiri, S. et al., 2004. Lateral and upward soil-pipeline interactions in sand for deep embedment conditions. Journal of Geotechnical and Geoenvironmental Engineering, 130: 830–842.

Young, W.C & Young, W. C., (6th ed.) 1989. Roark's formulas for stress and strain, New York: McGraw-Hill.

Modelling the excavation of elliptical shafts in the geotechnical centrifuge

N.E. Faustin, M.Z.E.B. Elshafie & R.J. Mair
University of Cambridge, UK

ABSTRACT: Vertical shafts are an essential component of tunnelling projects in urban environments and elliptical plan geometries are sometimes constructed. However, little is known about the performance of elliptical shafts and the associated adjacent ground movement during excavation; centrifuge testing of elliptical shaft excavation can provide valuable insight. Traditionally, excavation simulations in centrifuge testing mainly rely on the use of heavy fluid which is drained in-flight. This method assumes that the horizontal stress changes within the excavation are equal to the vertical stress changes which may not be the case in reality. More realistic stress changes could be achieved by carefully extracting sand from the centre of an elliptical shaft during excavation. This paper describes such centrifuge tests conducted at Cambridge University. The centrifuge model and the mechanism developed to excavate the shaft in-flight are described. Measurements are presented for strains in the elliptical shaft lining and associated ground displacements.

NOMENCLATURE

ϕ_{cs} Angle of friction at critical state
σ_θ Circumferential or hoop stress
σ_b In-plane bending stress
ϕ_{cs} Angle of friction at critical state
G_s Specific gravity
H Shaft excavation depth
R_d Relative density
S_v Settlement
SG   Strain gauge
x Distance from shaft

1 INTRODUCTION

Underground tunnel schemes offer considerable economic and environmental benefits for transportation and infrastructure development in built-up cities. Large vertical shafts are a major component of such tunnelling schemes. They enable access of equipment, personnel and material to the tunnel horizon during tunnel construction and may also provide ventilation and/or emergency access to the completed tunnel.

Typically, circular plan geometries are preferred because they are inherently more stable than other plan geometries. Although less common than their circular counterparts, elliptic shaft geometries may be constructed if there is limited space available. Elliptical shafts have been constructed in the UK for the Green Park London underground station and in Portugal for the Porto Light Rail Metro, as shown in Figure 1. Further details of elliptical shafts constructed for the Porto Light Rail Metro are reported by Topa Gomes et al. (2008) and Pedro (2013).

In urban environments, shafts can be constructed very close to existing buildings and buried pipelines for utilities. Therefore, two important design considerations are how strong to make the shaft lining and how to mitigate the potential effect of ground movement on nearby infrastructure. Consequently, design criteria for elliptical shafts include detailed assessments of (i) the stresses in the shaft linings and (ii) the adjacent ground movement due to shaft construction. European design standards for geotechnical structures offer limited guidance recommending that "a cautious estimate of the distortion and displacement of retaining walls, and the effects on supported structures and services, shall always be made on the basis of comparable experience" (BS:EN1997-1 (2004)). However, few published case studies exist for elliptical shafts. The field observations are also vital to validate any numerical analysis of elliptical shaft excavation.

The uncertainty in the design procedure for elliptical shafts and any conservative predictions of ground movements can have a direct effect on the cost of shaft construction. Thick shaft linings that may not necessarily be needed are adopted and protective measures that also may not be needed are sometimes implemented for buildings and services located near the shaft.

This paper describes the development of small-scale centrifuge model tests, conducted at Cambridge University, to investigate the behaviour of elliptical shaft linings and the adjacent ground during excavation. A major feature of the centrifuge tests is the new technique developed to extract sand from the centre of the model shaft in-flight. The model shaft and the surrounding sand were sufficiently instrumented to capture deformation of the shaft lining and displacement of the adjacent ground during excavation. Full

(a) Marquês Station, Lisbon, Portugal (Pedro, 2013)

(b) Green Park Station, London, UK (Image courtesy Transport for London)

Figure 1. Examples of recent elliptical shaft construction.

details of the centrifuge tests and the results are given in Faustin (2017).

2 CENTRIFUGE MODELLING OF ELLIPTICAL SHAFT EXCAVATION IN SAND

Centrifuge testing of an elliptical shaft excavation in dense Leighton Buzzard Fraction E sand was carried out using the 10 m beam centrifuge at Cambridge University. The tests were carried out in a 700 mm diameter and 620 mm deep aluminium model container. On average, the total height of the centrifuge model was 1570 mm and it weighed approximately 725 kg. A typical cross-section of the centrifuge model is shown in Figure 2 and a photograph of the centrifuge model before testing is shown in Figure 3.

The main apparatus of the centrifuge model were a circular container, an aluminium model shaft, dry Fraction E sand and an in-flight shaft excavation system. The latter comprised a series of stacked steel trays connected to a lifting device (1D actuator) via a lift rod and a rigid bridge assembly.

The new apparatus and centrifuge testing procedure are briefly described below. Further details are provided by Faustin (2017).

2.1 Model shaft

The elliptical model shaft was machined from a solid block of aluminium alloy 6082-T6, at the Cambridge University workshop. Model and corresponding prototype dimensions of the elliptical shaft excavation at a scale of 1:80 are given in Table 1.

2.2 Fraction E sand properties

Leighton Buzzard Fraction E sand with an average relative density of 80% was used for the centrifuge

Table 1. Dimensions of elliptical model shaft.

Parameter	Shaft dimensions	
	model scale (mm)	1:80 (m)
Minor axis length	120	9.6
Major axis length	180	14.4
Wall length	260	20.8
Wall thickness	2	0.16
Max. excavation depth	193	15.4

Table 2. Properties of Fraction E silica sand (Tan, 1990).

Property	Particle Size (mm)			Void Ratio			
	D_{10}	D_{50}	D_{60}	e_{min}	e_{max}	G_s	ϕ_{cs}
Values	0.095	0.14	0.15	0.613	1.014	2.65	32°

tests. Its properties, as reported by Tan (1990), are summarised in Table 2.

2.3 In-flight shaft excavation system

A key requirement for the centrifuge tests was to excavate the shaft in-flight. Faustin et al. (2017) describe different methods to simulate excavations in centrifuge testing. The most common method is to replace the soil at the centre of the shaft with a heavy fluid that is then drained in-flight. This method assumes that the horizontal and vertical stress changes on the shaft lining are equal during excavation.

More realistic changes in soil stresses during excavation were achieved by designing a system to remove sand from the centre of the shaft during the centrifuge

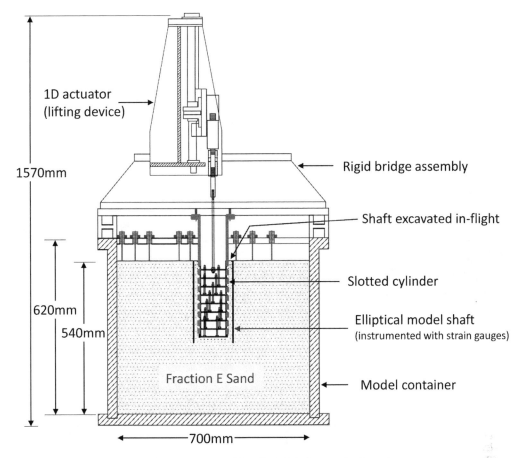

Figure 2. Cross-section Y-Y through centrifuge model for elliptical shaft excavation in sand.

test. The device comprised four components which are shown in Figure 2 and described below:

1. a rigid bridge assembly which spanned the model container to provide support to the main excavation system;
2. a 104 mm diameter slotted cylinder positioned at the centre of the shaft, and supported entirely from above by the bridge assembly;
3. nine steel trays stacked above each other, positioned inside the slotted cylinder at the centre of the shaft;
4. a lifting device or 1D actuator with movement in the vertical direction which was positioned on top of the bridge assembly and connected to the stacked trays via stainless steel rod and a load cell. The lifting device is described by Silva (2005) and its performance specification is given in Table 3.

2.4 Centrifuge model preparation

A uniform dense sand layer with an average relative density of 80%, was pluviated into the model container using a robotic sand pouring machine (Zhao et al. (2006)).

After 280 mm of sand had been poured, the sand pouring operation was paused and the instrumented model shaft and associated excavation system were placed at the centre of the model, as shown in Figure 4. Sand was then pluviated around the model shaft in the annulus between the model shaft and the slotted cylinder. Excavation was subsequently carried out during the centrifuge test by removing the sand in the annulus. The elliptical shaft therefore represented a Support Before Excavation construction, observed with diaphragm walled shafts where the shaft is excavated after the diaphragm walls panels have been installed.

Table 3. Performance specification for the 1D actuator.

Setting	Stroke mm	Maximum velocity mm/s	Load capacity kN
vertical direction	300	10	±10

2.5 Monitoring instruments

Two sets of monitoring instruments were placed on the shaft lining and on the sand surface to monitor

Figure 3. Centrifuge model being lowered via a crane onto the rotor arm of the beam centrifuge, located 3m below ground level.

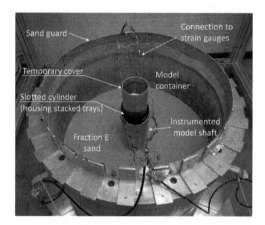

Figure 4. Model preparation.

the behaviour of the shaft and the surrounding ground during the centrifuge tests.

Four arrays of strain gauges (Arrays A, B, C and D) configured in Wheatstone full bridge and half bridge circuits were secured to the internal and external surface of the elliptical shaft to measure longitudinal bending and hoop strains respectively. The strain gauge measurements were correlated to hoop stresses and curvature of the shaft lining.

Miniature linear variable displacement transducers (LVDTs) were positioned on the sand surface, at varying distances from the shaft, to measure ground surface movement, as shown in Figure 2.

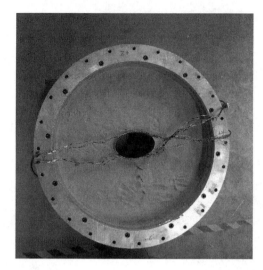

Figure 5. Elliptical shaft excavation (post test).

2.6 Centrifuge testing procedure

The centrifuge model was swung up to 80g followed by a staged excavation, which was activated by lifting each tray sequentially by 20 mm.

At the start of the excavation, the stacked trays covered the openings in the slotted cylinder. When each tray was lifted, during excavation, the opening was exposed thereby allowing sand in the annulus to flow through the opening and into the underlying tray. The lifted trays were supported from above by the rigid bridge assembly (see Figure 2) and remained suspended for the remainder of the excavation. A photograph of the elliptical shaft at the end of a centrifuge test is shown in Figure 5.

3 TYPICAL CENTRIFUGE TEST RESULTS

3.1 Hoop stresses in the elliptical shaft lining

The hoop strain gauges responded to a compressive hoop stress in the elliptical shaft lining (σ_θ) and an in-plane bending stress (σ_b), as illustrated in Figure 6.

Results of the combined hoop and bending stress i.e. cross-sectional stress, $(\sigma_\theta + \sigma_b)_{external}$, at a depth of 80 mm are shown in Figure 7. A compressive cross-sectional stress of -4.9 MPa was derived on the minor axis (Array C) and a tensile cross-sectional stress of 0.8 MPa on the major axis (Array D). This distribution of hoop strain indicates that points of minimal curvature on the original elliptical shaft cross-section deform towards the centre of the ellipse during excavation.

The longitudinal bending strains were very small in comparison to the hoop strains. Derived curvature values were in the order of 10^{-9} per metre.

3.2 Ground surface movements

The location of the sand surface displacement transducers and the ground movement recorded at different

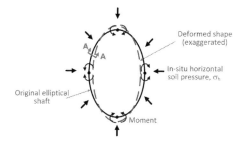

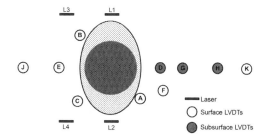

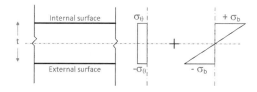

Figure 6. Stresses in an elliptical shaft lining.

Figure 8. Layout of displacement transducers.

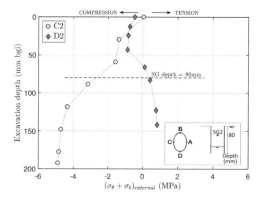

Figure 7. Derived cross-sectional stresses in an elliptical shaft lining.

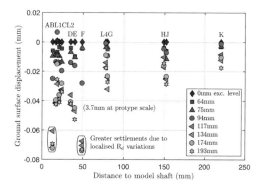

Figure 9. Ground movement observed during excavation of an elliptical shaft in sand at 80g.

shaft excavation depths are shown in Figures 8 and 9 respectively.

Relatively greater settlement of approximately 70 μm recorded by transducers A and F in Figure 9 are due to localised variations in sand density which occurred when the automatic sand pourer overheated during preparation of the model. Neglecting the two measurements at A and F, a maximum settlement of 46 μm was measured close to the shaft wall. This is equivalent to 3.7 mm of settlement at prototype scale. Settlements reduced to less than 20 μm (1.6 mm prototype scale) at a distance of 240 mm from the shaft.

The observed ground movement indicates that the elliptical shaft elongates during excavation, which is consistent with the deformation of the elliptical shaft indicated by the strain gauge measurements. Relatively smaller ground movements were recorded by transducers L1 and L2 on the major axis, where the shaft lining tries to push outwards during excavation. Greater ground movements were recorded at transducer locations B, C, D, E and G where the shaft lining moved towards the centre of the elliptic cross-section during excavation.

4 COMPARISON WITH CIRCULAR EXCAVATIONS

The strain measurements and ground movements from the elliptical excavation were compared with a similar greenfield centrifuge test for a 140 mm diameter circular excavation under similar test conditions. Both greenfield tests (one elliptical and one circular) were excavated using the new system involving stacked trays, described above. Full details of these tests and the results are provided in Faustin (2017).

Negligible longitudinal bending strains were observed for both the circular and elliptical greenfield excavations. However, greater magnitudes of hoop strains were observed for the elliptical excavation compared with a greenfield circular excavation due to the cross-sectional bending component.

Figure 10 compares ground movements observed during excavation of the elliptical shaft and circular shaft in terms of settlement normalised by shaft excavation depth (S/H) versus distance from shaft also normalised by shaft excavation depth (x/H).

The maximum settlement observed during excavation of the elliptical shaft was slightly greater than the circular shaft (0.028%H compared with 0.02%H). The zone of influence extends to a distance of 1.0H and 1.5H from the circular and elliptical shaft linings respectively. The greater settlement and zone

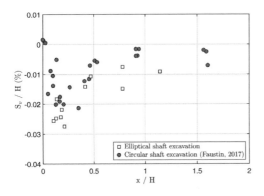

Figure 10. Normalised settlement during elliptical and circular shaft excavation in sand.

of extent due to excavation of the elliptical shaft can be attributed to either the larger shaft geometry or the dominance of the in-plane bending stresses which develop in elliptical shafts. The latter causes the elliptical shaft to elongate during excavation generating relatively greater movements on the minor axis compared with the major axis.

5 CONCLUSIONS

Centrifuge testing is beneficial to investigate the behaviour of elliptical shafts and the adjacent ground, particularly as there are few shaft case histories available. The novel apparatus and centrifuge testing procedure reported in this paper allows more realistic stress changes on the shaft lining to be modelled, compared with traditional centrifuge excavation techniques. This was achieved by extracting sand from the centre of the shafts at high centrifugal accelerations.

The test results showed that both hoop stresses and in-plane bending stresses developed in the elliptical shaft lining to resist in-situ horizontal soil stresses. The cross-sectional stress in the shaft lining, $(\sigma_\theta + \sigma_b)_{external}$, was negative on the minor axis and positive on the major axis indicating that the elliptic cross section elongates during excavation. Similar to the circular shaft excavations, negligible longitudinal bending was observed during excavation of the elliptical shaft. Concrete is well-known to be strong in compression and weak in tension so the practical implication of this behaviour is that elliptical shafts must be appropriately reinforced to account for the in-plane bending.

The distribution of ground movements was consistent with the deformation of the elliptical shaft indicated by the strain gauges. The maximum settlement measured by transducers was approximately 0.028% times the shaft excavation depth (0.028%H).

ACKNOWLEDGEMENTS

The authors are grateful to the Engineering and Physical Sciences Research Council (Award Reference 1220514) and Geotechnical Consulting Group for financial support. Special thanks are given to Neil Houghton for his contribution to the detailed design of the apparatus. The invaluable contribution from the technicians at the Schofield Centre (Cambridge University) is gratefully acknowledged.

REFERENCES

BS:EN1997-1 (2004). *Eurocode 7: Geotechnical Design - Part 1: General rules.* Technical report.

Faustin, N. E. (2017). *Performance of circular shafts and ground behaviour during construction (PhD thesis).* University of Cambridge.

Faustin, N. E., M. Elshafie, & R. J. Mair (2017). Centrifuge modelling of shaft excavations in clay. In *Proc. of the 9th Int. Symposium on Geotechnical Aspects of Underground Construction in Soft Ground, TC204 ISSMGE - IS-SAO PAULO*, Sao Paulo.

Pedro, A. M. (2013). *Geotechnical investigation of IVENS shaft in Lisbon (PhD thesis).* Phd thesis, Imperial College, London.

Silva, M. F. (2005). *Numerical and physical models of rate effects in soil penetration (PhD thesis).* University of Cambridge.

Tan, F. S. C. (1990). *Centrifuge and Theoretical Modelling of Conical Footings on Sand (PhD thesis).* University of Cambridge.

Topa Gomes, A., A. Silva Cardoso, J. Almeida e Sousa, J. Andrade, & C. Campanhã (2008). Design and behaviour of Salgueiros station for Porto Metro. In *6th International Conference on Case Histories in Geotechnical Engineering*, Vancouver, USA.

Zhao, Y., K. Gafar, M. Z. E. B. Elshafie, a. D. Deeks, J. a. Knappett, & S. P. G. Madabhushi (2006). Calibration and use of a new automatic sand pourer. *6th International Conference of Physical Modelling in Geotechnics* 65(Figure 3), 265–270.

Shaking table test to evaluate the effects of earthquake on internal force of Tabriz subway tunnel (Line 2)

M. Hajialilue-Bonab, M. Farrin & M. Movasat
University of Tabriz, Tabriz, Iran

ABSTRACT: A series of 1g shaking table tests were performed to investigate the response of Tabriz subway tunnel, a circle-type tunnels embedded in dry sand, under sinusoidal excitation. Effects of various parameters, including peak ground acceleration and frequency content of input motion on the behavior of tunnel was investigated. Tests were performed in two peak ground accelerations, 0.35g and 0.50g, and different frequencies. The experimental data are presented in terms of dynamic lining forces. Results show that the tunnel moves from a static equilibrium to a dynamic equilibrium state as soon as the earthquake starts. In addition, measurement of shaking table experiments shows that by increasing the frequency of loading, induced maximum strain almost remains constant or decreases little for $A = 0.35g$ but decreases sharply for $A = 0.50g$.

1 INTRODUCTION

Historically, underground facilities have experienced a lower rate of damage than surface structures. Nevertheless, some underground structures have experienced significant damage in recent large earthquakes, including the 1995 Kobe, Japan earthquake, the 1999 Chi-Chi, earthquake and the 1999 Kocaeli, Turkey earthquake (Hashash et al. 2001).

Underground structures show a different seismic response compared to the above ground structures, as the kinematic loading imposed on the structure by the surrounding ground overcomes over inertial loads, resulting from the oscillation of the structure itself (Hashash et al. 2001).

Physical modeling may help to obtain important information and quantitative data on the seismic behavior of underground tunnels. Some physical tests of tunnel models under seismic actions are described in the researches (Ounoe et al. 1998; Yamada et al. 2002; Cilingir and Madabhushi 2011a and 2011b); however, only few of them included measurements of internal forces (Yang et al. 2004). Cao and Huang (2010) have shown the strain time histories of a model tunnel under shaking and Lanzano et al. (2010) investigated the seismic behavior of tunnels by a series of plane-strain centrifuge tests with dynamic loading on a model tunnel. They studied the response of four samples of dry uniform fine sand were prepared at two different densities, in which an aluminum-alloy tube was installed at two different depths.

Up to now, because of the numerous parameters affecting the seismic behavior of underground structures, there is not a comprehensive conclusion in this area of practice in the literature. In this regard, current study aims to perform a series of shaking table tests on a model tunnel to develop some dimensions of the problem. Effects of various parameters, including input motion frequency content and peak ground acceleration on the seismic behavior of subway tunnel were investigated in this study. The tests were undertaken in the framework of a Tabriz university research project funded by Tabriz Urban Railway Organization.

2 TEST APPARATUS AND MODEL PREPARATION

2.1 Shaking table device

The 3×2 m shaking table device in Soil Mechanics Laboratory of Tabriz University was used to induce the desired excitations to models. The Table has 1 degree of freedom with the maximum displacement of 200 mm.
Also, it can sustain a model up to 6 tons weight.

2.2 Shear box

A laminar shear box designed in Tabriz University includes 25 aluminum layers; each having the dimensions of $1460 \times 1000 \times 40$ mm. In order to reduce the friction between the layers and to simulate the displacement of soil layers, ball bearings were used between the two adjacent layers.

2.3 Soil characteristics

Uniform sand provided from Qomtapeh was used in this study. The main index properties of this uniform sand are summarized in Table 1 and the gradation curve of this soil are shown in Figure 1.

Table 1. Main index properties of Qomtapeh sand.

USCS Classification	SP
Specific gravity	2.64
Angle of internal friction	33°
Cohesion (kN/m^2)	0
Coefficient of uniformity	2.17
Coefficient of curvature	1.04
Minimum unit dry weight (kN/m^3)	14.60
Maximum unit dry weight (kN/m^3)	17.38

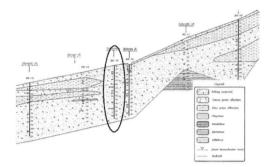

Figure 2. Geological profile of Tabriz metro line 2 in the study area.

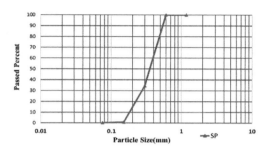

Figure 1. Gradation curve for Qomtapeh sand.

Table 2. Characteristics of the tunnel in prototype and model.

Characteristic	Prototype	Model
Inner Diameter (mm)	8500	197
Outer Diameter (mm)	9200	194
Thickness (mm)	350	1.5
Module of Elasticity (GPa)	30.2	70
Poisson ratio	0.2	0.3

2.4 Tunnel characteristics

Tabriz Urban Railway Line 2 (TURL2) with 22 km in length will connect eastern part of the city to the western part. This study is focused on the central part of line 2, between E2 and H2 stations. The line 2 tunnel has been constructed using one earth pressure balance EPB-TBM with a cutting-wheel diameter of 9.49 m and a shield with external diameter of 9.46 m in front of shield. The precast concrete segments of tunnel lining are installed just behind the shield (Katebi et al. 2015). In the intended part of the path, based on geotechnical studies in the corridor of TURL2, the soil type is mainly ML and SM. A geological cross section along the tunnel path is shown in Figure 2.

The simulation laws for 1g shaking table tests proposed by Iai (1989) were utilized in the current study. Based on the size of the laminar shear box and the tunnel, prototype to model scale factor was considered to be 45. Therefore, 45 m of prototype soil layer was modeled by 1 m in the model tests. By taking all above mentioned considerations, the tunnel model, manufactured from aluminum alloy, is 195.5 mm diameter and 820 mm length, having a thickness of 1.5 mm. According to the scale factor (N = 45), the model corresponds to a circular tunnel having the external diameter of segment is 9.20 m and its thickness and ring length are 0.35 and 1.5 m, respectively (Table 2).

It is obvious that a segmental lining has a flexural stiffness which is not like a continuous tube nor a tube with a fixed number of hinges in particular positions. Several methods are proposed to considering the influence of segmental joints on the flexural stiffness of tunnel lining. In modified usual calculation method for considering decrease of rigidity at segment joints, a transfer ratio of bending moment (ζ) is introduced. This aspect is transferred to analyses with correction of the elastic modulus of the ring. In this method, after correction of the elastic modulus of the ring, according to a factor ζ, the acting loads on lining are calculated. Then the value of the bending moment is modified by increasing and decreasing the value for the segment and joint, respectively, by the same ζ factor. The value of the parameter ζ varies between 0.3 and 0.5 as a function of the number of segments and the stiffness of the surrounding ground (Guglielmetti et al. 2007). The Japanese Society of Civil Engineering descriptively recommends reducing the rigidity of the continuous liner structure by 20–40%. Taking into account the modification factor $\zeta = 0.3$ for Young's modulus of concrete, the real modulus must be considered according to Eq. 1 (Koyama 2003):

$$E_C = (1 - \zeta) \times E_{CLS} = (1 - 0.3) \times E_{CLS} = 0.7 E_{CLS} \quad (1)$$

where E_C is the virtual modulus of the ring and E_{CLS} is the concrete modulus.

2.5 Model preparation

The sand model was made using an automatic hopper system. During the construction, the tunnel and all the embedded transducers were positioned in the model. Several phases of the model construction are presented in Figure 3. To avoid any interaction of the tunnel with the laminar shear box, the tunnel was shorter than the

3 EXPERIMENTAL PROCEDURE

Using the above mentioned method for model preparation, seven models were prepared. The physical model was shaken with a sinusoidal base acceleration having a frequency of 1.0, 3.0, 5.0 and 8.0 Hz and maximum acceleration amplitude of 0.35 g and 0.50 g (because of the maximum displacement limitation of the shaking table, the case of f = 1Hz, A = 0.5g was not applied). Duration of the base excitation was 5.0 sec. It should be noted that these input motions simulate the in-plane shear wave excitation in the tests with respect to the tunnel section.

4 TEST RESULTS AND DISCUSSION

As mentioned, seven models were prepared and subjected to dynamic excitations. The data obtained from data acquisition system were converted to physical parameters, strains. Figures 5 and 6 demonstrate some of the typical recorded results related to strain gauges No. 1 and 4 on the tunnel lining for A = 0.35g, f = 3Hz and A = 0.50g, f = 5Hz.

Although the time histories recorded by the horizontal accelerometer located at the rigid base of the models have been nominally programmed as single frequency sine waves, the recorded signals are actually pseudo-harmonic and some peculiar features can be observed from time domain representations.

The experimental values of internal forces in one section of the lining were derived from the strain gauge records during each seismic event. The typical time histories of dynamic strains are shown in Figures 5 and 6 at the model scale during model shaking for A = 0.35g, f = 3Hz and A = 0.50g, f = 5Hz. The dynamic strain is defined as the additional strain applied to the tunnel lining after the start of the earthquake.

The measured values generally show that the top part of the tunnel experiences larger fluctuation of dynamic strains than the bottom part of the tunnel during the tests, as it may be expected because of lager displacement in the top part of the tunnel. During the shaking, the lining forces generally increased from the static amounts measured before the shaking. Such increase, accumulating after the seismic event, is consistent with the sand densification and the accumulation of irreversible strains of the soil layer around the tunnel. Moreover, the lining was subjected to cycles of increase–decrease of the internal forces because of the reversible component of the soil shear strain during shaking. This means that both the reversible and the residual components of such increments of the internal forces are related to a mechanism of ovalisation of the transverse section of the tunnel, caused by the deformation of the ground layer. Therefore, the distribution of the dynamic strain depends on the pattern of deformation the tunnel experiences after the start of the earthquake and its magnitude seems to be a function of depth. In addition, the values of strain during

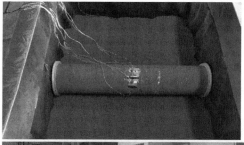

Figure 3. Model preparation.

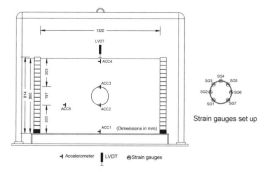

Figure 4. Model configuration – instrumentation scheme.

box width. Two PVC circular plates were placed at both the tunnel ends to avoid the sand entrance into the tunnel-model. To simulate the effects of friction on the soil–tunnel interaction, the outside surface of tunnel was covered by sand particles using epoxy coatings around the tunnel.

During construction, the bulk unit weight was controlled to be constant. To achieve the same target relative density ($D_r = 65\%$) for all layers, the height of sand draining to the box remained fixed equal to 950 mm. Completed models are illustrated in Figure 3.

2.6 Model set up and instrumentation

Seven strain gauges were used in the experiments to better monitor the behavior of the tunnel and soil which installed in the model. Figure 4 presents the final model layout and the instrumentation scheme.

Strain gauges were glued on the outer face of the tunnel to measure the model strains at several locations. A dynamic data logger recorded and transferred all the measured data to a personal computer.

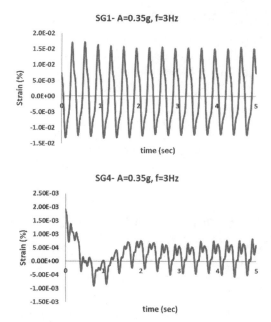

Figure 5. Recorded strain time histories in the locations of SG1 and SG4 of model for A = 0.35g, f = 3Hz.

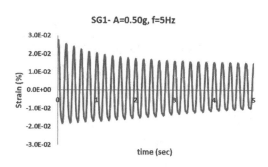

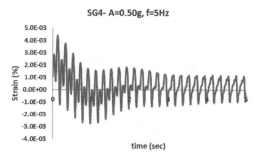

Figure 6. Recorded strain time histories in the locations of SG1 and SG4 of model for A = 0.50g, f = 5Hz.

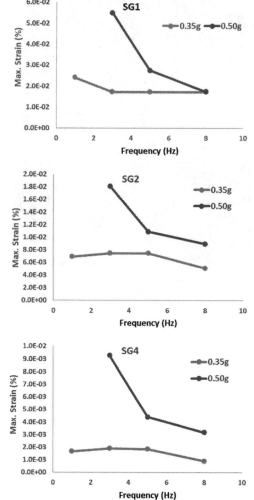

Figure 7. Absolute of induced maximum strain in the lining versus input frequency for SG1, SG2 and SG4.

shaking are higher in the bottom part of the tunnel lining than in the upper.

According to Figures 5 and 6, the time histories of dynamic strain can be divided into two main sections: transient stage and steady-state cycles. The tunnel deforms towards an equilibrium stage during the first few cycles. The cyclic stage begins, which continues until the end of the earthquake.

The maximum dynamic strain on the lining was found by calculating the absolute value of the strain and selecting the largest one. This way, both the negative and positive strain were evaluated. The maximum strain was measured near the crown and the shoulders of the model tunnels.

Dynamic strain in the prototype can be calculated according to Eq. 2 (Iai 1989):

$$\lambda_\varepsilon = \lambda/\left[(V_s)_p/(V_s)_m\right]^2 \qquad (2)$$

which $(V_s)_p$ and $(V_s)_m$ are the shear wave velocities of the prototype ground and model ground, λ and λ_ε are geometrical and strain scaling factor, respectively. According to the field and laboratory measurements $(V_s)_p = 590$ m/sec and $(V_s)_m = 90$ m/sec, therefore, λ_ε

is approximately one. It means that induced strains in the model and prototype are almost equal.

By comparing the results of seven tests, the maximum strain is caused in the location of SG5 that is related to loading of A = 0.50g and f = 3Hz.

In Figure 9, the created strain in the locations of strain gauge No. 1, 2 and 4 is plotted versus different frequencies. As shown in diagrams, for A = 0.35g, maximum strain is constant or reduces little by increasing frequency but for A = 0.50g, maximum strain reduces sharply by increasing frequency for all of the strain gauges and it may be expected because of the distortions of the tunnel cross section. Therefore, high-amplitude acceleration with low frequency cases high-induced strain and deformation in the underground tunnel.

5 CONCLUSION

Seismic response of circular-type tunnels in dry sand was investigated by means of dynamic shaking table testing. Crucial parameters that affect the seismic response of tunnels such as the peak ground acceleration and predominate frequency were investigated within the experimental program. The recorded data highlighted significant aspects of the dynamic response for the above type of underground structures. The main conclusions of the study may be summarized as follows.

Results of shaking table experiments show that the dynamic behavior of circular tunnels can be split into two stages: transient stage and steady-state cycles. During the transient stage, which lasts for the first few cycles, the tunnel structure reaches a dynamic equilibrium configuration. The transient stage is followed by the steady-state cycles, during which the forces in the tunnel lining oscillate around a mean value.

For all of the tunnel tests, strain and lining deformations increase with increase in maximum base acceleration, but the location of the highest and the lowest amounts stays the same.

In addition, measurement of shaking table experiments shows that the tunnel shifts from a static equilibrium to a dynamic equilibrium state immediately after the earthquake starts.

The measured values generally show that the top part of the tunnel experiences larger fluctuation of dynamic strain than the bottom part of the tunnel during the tests.

Largest values for the dynamic strain appear approximately midway between shoulder and the crown of a circular tunnel.

For A = 0.35g, maximum strain is constant or reduces a little by increasing frequency but for A = 0.50g, maximum strain reduces sharply by increasing frequency.

ACKNOWLEDGMENTS

The research program was carried out in the framework of Tabriz University, as a part of a Research Project funded by Tabriz Urban Railway Organization. Their continuous support and helps are acknowledged.

REFERENCES

ACI. 2015. Building Code Requirements for Structural Concrete and Commentary, ACI 318M-14, American Concrete Institute (ACI) Committee 318.

Cao, J. & Huang, M. S. 2010. Centrifuge Tests on the Seismic Behavior of a Tunnel, Proceedings of the 7th International Conference on Physical Modelling in Geotechnics, 7th ICPMG, Zurich, Switzerland, 1: 537–542.

Cilingir, U. & Madabhushi, S. P. G. 2011a. Effect of Depth on the Seismic Response of Circular Tunnels, Canadian Geotechnical Journal, 48(1): 117–127.

Cilingir, U. & Madabhushi, S. P. G. 2011b. A Model Study on the Effects of Input Motion on the Seismic Behavior of Tunnels, Journal of Soil Dynamics and Earthquake Engineering, 31: 452–462.

Guglielmetti V., Grasso, P., Mahtab A. & Xu, S. 2007. Mechanized Tunnelling in Urban Areas: Design methodology and construction control, Taylor & Francis Group, London, UK.

Hashash Y. M. A. & Hook J. J. 2001. Schmidt B., Yao, J.I-C., Seismic design and analysis of underground structures, Journal of Tunnelling and Underground Space Technology, 16: 247–293.

Iai S. 1989. Similitude for shaking table tests on soil-structure-fluid modelin1g gravitational field, Journal of Soils Foundations, 29(1): 105–118.

Katebi H., Rezaei A. H., Hajialilue-Bonab M. & Tarifard A. 2015. Assessment the influence of ground stratification, tunnel and surface buildings specifications on shield tunnel lining loads (by FEM), Journal of Tunnelling and Underground Space Technology, 49: 67–78.

Koyama Y. 2003. Present status and technology of shield tunneling method in Japan, Journal of Tunnelling and Underground Space Technology, 18: 145–159.

Lanzano, G., Bilotta E., Russo G., Silvestri F. & Madabhushi G. 2012. Centrifuge Modeling of Seismic Loading on Tunnels in Sand, Geotechnical Testing Journal, Vol. 35, No. 6.

Onoue, A., Kazama, H., Hotta, H., Kimura, T. & Takemura, J. 1994. Behaviour of Stacked-Drift-Type Tunnels, Proceedings of the International Conference Centrifuge 94, Defense Technical Information Center, Singapore, Malaysia, 1: 687–692.

Yamada, T., Nagatani, H., Igarashi, H. & Takahashi, A. 2002. Centrifuge Model Tests on Circular and Rectangular Tunnels Subjected to Large Earthquake-Induced Deformation, Proceedings of the 3rd Symposium on Geotechnical Aspects of Underground Construction in Soft Ground, Toulouse, France, 1: 673–678.

Yang, D., Naesgaard, E., Byrne, P. M., Adalier, K. & Abdoun, T. 2004. Numerical Model Verification and Calibration of George Massey Tunnel Using Centrifuge Models, Canadian Geotechnical Journal, 41: 921–942.

Effect of pipe defect size and maximum particle size of bedding material on associated internal erosion

S. Indiketiya, P. Jegatheesan & R. Pathmanathan
Swinburne University of Technology, Hawthorn, Victoria, Australia

R. Kuwano
Institute of Industrial Science, The University of Tokyo, Meguro-ku, Tokyo, Japan

ABSTRACT: Sinkhole formation due to internal erosion around defective sewer pipes is identified as a serious problem. One of the effective methods of minimising this type of sinkhole formation is by enhancing the performance of pipe embedment materials against erosion. In this paper, the influence of the crack width with respect to the maximum particle size of the pipe embedment material was examined for a widely used pipe bedding material using a laboratory scale physical model apparatus. The erosion due to pipe defect was evaluated by means of eroded soil mass, ground deformation, scale of the cavity size and damage induced for three different crack widths. Results for maximum particle size of 4.75 mm indicates that once the crack width exceeds the maximum particle size of the bedding material, cavity progression and soil erosion rate accelerated rapidly. Furthermore, it was identified that particles less than 0.3 mm were highly susceptible to erosion.

1 INTRODUCTION

Internal erosion around defective sewer pipes has been identified as the main cause for the formation of underground voids, which ultimately lead to sinkholes (Guo et al., 2013, Davies et al., 2001). The scale of soil erosion through pipe defects depend on many local factors such as the size of the pipe defect, gradation and plasticity characteristics of backfill, the degree of compaction of pipe bedding and the backfill, fluctuation of the groundwater table, ground water infiltration, and exfiltration through pipe defects (Davies et al. 2001; Rogers 1986; Kuwano et al. 2006; Indiketiya et al. 2017).

In that case, the relationship between the crack width of the pipe defect and the grading of the embedment fill is an important factor, which decides the soil loss through the defect. Rogers (1986) proposed a relationship between the soil loss and the ratio of B/D_{85}, where, D_{85} is the size of sieve through which 85% by weight of a soil sample will pass and "B" is the crack width. In his study onset of critical soil flow through pipe defect under monotonic flow was experienced when the crack width was 2.5D_{85} to 4.5D_{85} for cohesionless soils (e.g. silts, sands, and gravels). The same aspect was observed again for a similar type of soil by Mukunoki et al. (2012) and the critical crack width (B) for the cyclic water flow behaviour was presented as 5.9D_{max}, where D_{max} is the maximum particle size.

Ground arching with associated deformation is a key parameter, which needs to be examined in this erosion process and it has been ignored in these studies. Particle Image Velocimetry PIV has been recently identified as a successful, non-destructive method of evaluating ground deformation (White et al. 2001) using image correlation. Therefore, in this paper, the effect of pipe defect size compares to the maximum particle size of the pipe embedment was evaluated for one of the Australian sewer embedment material using three laboratory model tests while associated ground deformation was evaluated by incorporating PIV.

2 METHODOLOGY

2.1 Material properties

A pipe embedment material compatible with Type-360 and 361 of the specified materials in Water Services Association Australia (2002) was chosen for the study. It is a poorly graded sand (ASTM, 2011), known as "Dromana sand" and D_{max} is 4.75 mm. The particle-size distribution of the Dromana sand with approved material types 360 and 361 are shown in Figure 1 to visualise the compatibility. Specific gravity (G_s), e_{max} and e_{min} values for Dromana sand are 2.52, 0.96, 0.59 respectively and the maximum proctor density is 1931 kg/m^3 at the optimum moisture content of 11%.

2.2 Experimental setup

A schematic diagram of the test apparatus is shown in Figure 2. It is a rectangular box with overall dimensions of 920 mm × 450 mm × 100 mm. The exact size

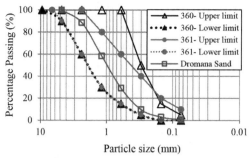

Figure 1. Particle size distribution of Dromana sand.

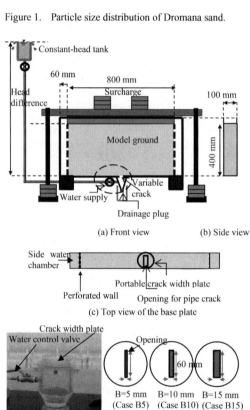

Figure 2. Schematic diagram of the experimental setup.

of the model ground is 800 mm × 400 mm × 100 mm. The front view, side view, and the view from the top are shown in Figure 2 (a), (b) and (c), respectively. A variable pipe defect was incorporated at the base by a detachable circular plate with a diameter of 100 mm (Figure 2c, 2d) and different pipe defects were simulated by switching the plate. As can be seen in Figure 2e, three crack width (B) configurations were chosen as B = 5 mm (Case B5), 10 mm (Case B10) and 15 mm (Case B15). The testing apparatus and the procedure have been comprehensively reported elsewhere (Indiketiya et al. 2017).

2.3 Testing procedure

Once the desired crack width plate was precisely positioned, conical shape soil collection unit (Fig. 2 (d)) was fixed to the tank. Water inflow into the tank and the soil outflow from the tank through the defect is facilitated by the same component by connecting water input valve and a drainage plug. To begin with, the soil with the optimum moisture content of 11% was compacted to 25 mm thick layers to achieve 80% of uniform relative dry density. Shortly after the compaction, 7.5 kPa of overburden pressure was applied to the ground surface to take account the actual vertical and lateral earth pressures in the field.

Prior to starting the test, two digital single-lens reflex cameras (DSLR) were set from either side of the tank with a distance of 1500 mm from the tank to acquire images to evaluate the ground displacement by PIV. Then, the water was supplied into the model using the valve shown in Figure 2 (d). At the start, water flow rate of 11 ml/s was achieved by maintaining 1 m head difference with the constant-head water tank. As this test is to investigate the influence of cyclic water flow, water inflow was stopped after 30 s and then waited two minutes for the water level to stabilise in the tank. Just then, drainage valve was gently opened so that the soil and water flow out through the opening can be collected. Next, the model ground was left for 10 minutes to let the excess water flow to stop prior to inserting water again. In this study, water supply, stabilisation period, and drainage period are defined as one cycle. In order to observe the effect of exfiltration volume, the period of water inflow was increased by 30 s for every other cycle until the ground is subjected to significant erosion, which was unique for each test. Meanwhile, eroded soil and drained water in each cycle were collected into separate containers, and post-sieve analysis for each cycle was performed separately after oven drying.

3 RESULTS AND DISCUSSION

3.1 Void formation and ground failure

Table 1 illustrates the final shape of the model ground in each cycle, indicating the maximum water level corresponds to that particular cycle and the following cycle. Erosion voids were observed for all three test cases and cavity initiation was witnessed at 9th, 12th, and 7th cycle for case B5, case B10 and case B15 respectively (Table 1). Cumulative eroded mass variation with the number of cycles for three test cases is also presented in Figure 3. The cumulative eroded masses at the time of the void initiation for case B5, B10 and B15 were 0.03%, 0.05%, and 0.08% of the total soil mass of the original model ground and these eroded masses are corresponding to 8th, 11th, and 6th cycles respectively. This depicts that the cumulative soil loss at the time of void formation is smaller for smaller pipe defects and on the other hand larger pipe defects will lose more soil from the ground immediately before the cavity formation.

Table 1. Initiation and evolution of erosion voids with repetitive cyclic water flow.

Cycle No (N)	Crack width B = 5 mm (Case B5)	B = 10 mm (Case B10)	B = 15 mm (Case B15)
7			$C7_{max} = 20.9$ $C8_{max} = 22.0$
8	$C8_{max} = 23.0$ $C9_{max} = 24.3$		$C8_{max} = 22.0$ $C9_{max} = 14.8$
9	$C9_{max} = 24.3$ $C10_{max} = 24.8$		$C9_{max} = 14.8$ $C10_{max} = 12.0$
10	$C10_{max} = 24.3$ $C11_{max} = 26.5$		$C10_{max} = 12.0$ $C11_{max} = 12.8$
11	$C11_{max} = 26.5$ $C12_{max} = 23.0$	$C11_{max} = 25.1$ $C12_{max} = 25.2$	$C11_{max} = 12.8$ $C12_{max} = 12.5$
12	$C12_{max} = 23.0$ $C13_{max} = 23.1$	$C12_{max} = 25.2$ $C13_{max} = 24.6$	$C12_{max} = 12.5$ $C13_{max} = 13.4$
13	$C13_{max} = 23.1$ $C14_{max} = 23.8$	$C13_{max} = 24.6$ $C14_{max} = 17.0$	$C13_{max} = 13.4$ $C14_{max} = 13.0$
14	$C14_{max} = 23.8$ $C15_{max} = 25.9$	$C14_{max} = 17.0$ $C15_{max} = 14.3$	$C14_{max} = 13.0$ $C15_{max} = 15.7$
15	$C15_{max} = 25.9$ $C16_{max} = 26.0$	$C15_{max} = 14.3$	$C15_{max} = 15.7$ $C16_{max} = 14.0$
16	$C16_{max} = 26.0$ $C17_{max} = 24.9$		$C16_{max} = 14.0$

―――― Maximum Water level for the current cycle (mm),
‐ ‐ ‐ ‐ Maximum Water level for the next cycle (mm),
$C(i)_{max}$ is the maximum water level correspond to i^{th} cycle, where "i" is the cycle number

Comparing case B10 and B15, cumulative soil loss towards the last cycle is almost equal and is about 20% of the total mass of the model. However, the rates of the cavity expansion are different. In case B10, cavity first appears in the 12th cycle and it expands dramatically reaching almost the final shape within two consecutive cycles whereas, in case B15, cavity growth is progressing. It was also noticed that in case

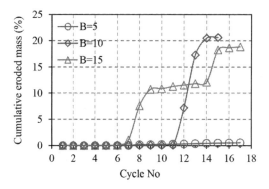

Figure 3. Variation of cumulative eroded soil mass for different crack widths.

Table 2. Ratio of B/Dmax and B/D85 for test cases.

Property	Crack width (mm)			
	5	10	15	20
B/D_{max}	1.05	2.1	3.15	4.2
B/D_{85}	2.38	4.76	7.14	9.5
B/D_{90}	2.12	4.2	6.35	8.47
Cycle number corresponds to cavity initiation	9	12	7	4

B10, the original cavity was formed during stabilisation period (undrained) while the water input was stopped and left to stabilise the water table. During this undrained period, when the water input valve is closed, water in the central region of the soil tank flows downwards and towards side tanks until hydrostatic pressure comes to equilibrium. Roof of the void easily become unstable and fail due to this downward seepage and saturation around the sides and the roof of void which further triggered a rapid expansion of the cavity. To compare these results with previous findings, B/D_{max} and B/D_{85} ratios for each case is presented in Table 2. According to Rogers (1986) critical soil flow was experienced when the crack width was $2.5D_{85}$ to $4.5D_{85}$ for cohesion-less soils and this study shows critical soil flow in case B10 where B/D_{85} is 4.7. However, further studies are required to establish a relationship between the soil flow and the crack width. Despite the crack width, the geometry of the void and the ground water table fluctuation through the void were the key factors, which controlled the stability of the cavity.

The shape of the void and its expansion corresponds to case B5 is different from other two. 5 mm opening is not wide enough to drain the soil and water smoothly without causing a blockage. This material blockage supports to cease the soil flow by forming a self-filtering zone which prevents complete void expansion towards the base of the tank and the same fact has been revealed by Rogers (Rogers 1986). Therefore, void initiation occurred at a higher elevation from the pipe. On the other hand, in other two cases, soil loss was critical and the opening size was wide enough to allow total soil and water to drain through which stops self-control mechanism. Observation of Table 1 reveals that after few consecutive water cycles, the soil is steadily and symmetrically eroded along a smooth sheared surface if the water table does not rise over the spine line of the void in case B10 and B15.

3.2 Erosion susceptibility of different particle sizes

Eroded soil corresponds to each cycle was separately collected, dried and particle size distribution of each was individually evaluated. Measured soil masses, corresponds to different particle size range were plotted against the expected masses for the same range based on the original particle size distribution. Expected mass was evaluated by simply multiplying the total eroded soil mass for a particular cycle by corresponding percentage from original particle-size distribution. Those plots are given in Figure 4. When comparing the particle size distribution of eroded mass with the original distribution, it can be seen that, despite the crack width size, particles less than 0.3 mm are highly susceptible to erosion witnessing that the suffusion mechanism has taken place during the seepage.

3.3 Erosion-induced ground settlement

Acquired images were correlated to obtain the ground settlement for each drainage cycle using PIVlab (Thielicke & Stamhuis 2014), a Graphical User Interface (GUI) based open-source tool in MATLAB (The Mathworks Inc 2014). The detailed procedure of evaluation of the ground deformation is explained in elsewhere (Indiketiya et al. 2017).

For easy representation of the settlement, the total model height of 400 mm was represented by eight layers where the thickness of each layer is 50 mm.

Layers were named from 1 to 8 subsequently starting from bottom to top. Vertical settlement of each layer at 50 mm horizontal distances was calculated for each drainage cycle by using PIV. Even though it is possible to measure each layer deformation for every cycle at any location, the settlement at the center of the top three layers are compared in this paper because the rest of the bottom layers have been eroded at the end of the test. Individual and the cumulative settlement corresponds to each cycle is plotted in Figure 5 along with the corresponding soil loss for three tests.

Results indicate that case B5 has resulted in a cumulative settlement of 0.8 mm at 6th layer (L6) and 0.5 mm at 8th layer (L8) even after 17 cycles. Specially, the soil loss is not directly proportional with the corresponding settlement for case B5, though in case B10 and B15, layer deformation and soil flow are correlated. The reason is that considering the void size, location and the soil loss at each cycle in case B5 is comparatively smaller than the other two cases and the sensitivity of soil loss on top most layers is smaller. Yet again the cumulative settlement induced in case B10 is almost twice of the settlement in B15 for L7. Ground displacement induced by soil flow through pipe defects is not observed in detail in the past and hence, there

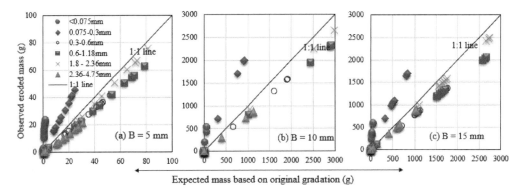

Figure 4. Post-erosion particle size distribution (a) B = 5 mm, (b) B = 10 mm, (c) B = 15 mm.

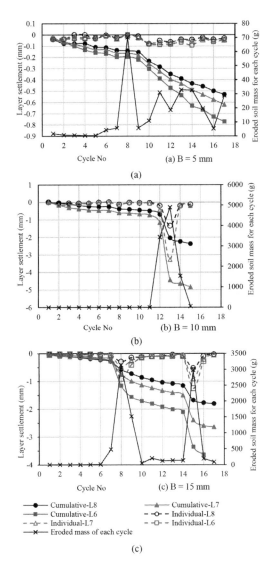

Figure 5. Ground settlement and soil loss (a) B = 5 mm, (b) B = 10 mm, (c) B = 15 mm.

is no reference to compare these values. However, it can be indicated that, once crack width exceeds the D_{max}, settlement or soil loss induced are not a function of crack width, but controlled by a combination of several other local factors such as geometry of the existing void, ground water table with respect to the void, and soil properties which probably need further studies.

4 CONCLUSIONS

- For poorly graded sand with D_{max} of 4.75 mm, erosion voids were formed with significant soil loss, and particles less than 0.3 mm were highly susceptible to erosion through pipe defects with B = 5, 10 and 15 mm.
- Soil flow through the pipe defect occurred if there was water flow into the defective pipe. If there was no water flow towards the pipe, erosion void seems stable providing that the ground water table has not risen over the roof of the existing void.
- Critical soil flow was observed for B = 10 mm case, where B/D_{85} is 4.7 and B/D_{max} is 2.1.
- Cavity initiation and progression mechanisms for $B = D_{max}$ and $B > 2D_{max}$ were different due to clogging and self-healing effect in $B = D_{max}$.
- A clear relationship between the soil loss or ground settlement with pipe defect size was not observed. It seems that the stability of the ground with an existing erosion void is decided by a combination of key factors. Those are the geometry of the existing void, the size of the pipe defect, particle size distribution of soil, ground water table with respect to the existing cavity, and the seepage into the defective pipe.

REFERENCES

ASTM. 2011. Standard Practice for Classification of Soils for Engineering Purposes (Unified Soil Classification System). ASTM standard D2487. ASTM International, West Conshohocken, PA.

Davies, J., Clarke, B., Whiter, J. & Cunningham, R. 2001. Factors influencing the structural deterioration and collapse of rigid sewer pipes. *Urban Water*, 3: 73–89.

Guo, S., Shao, Y., Zhang, T.Q., Zhu, D.Z. & Zhang, Y.P. 2013. Physical Modeling on Sand Erosion around Defective Sewer Pipes under the Influence of Groundwater. *Journal of Hydraulic Engineering*, 139: 1247–1257.

Indiketiya, S., Jegatheesan, P. & Rajeev, P. 2017. Evaluation of defective sewer pipe–induced internal erosion and associated ground deformation using laboratory model test. *Canadian Geotechnical Journal*, 54: 1184–1195.

Kuwano, R., Hiorii, T., Kohashi, H. & Yamauchi, K. 2006. Defects of Sewer Pipes Causing Cave-ins' in the Road. *5th International Symposium on new technologies for urban safety of mega cities in Asia (USMCA)*. Phuket, Thailand.

Mukunoki, T., Kumano, N. & Otani, J. 2012. Image analysis of soil failure on defective underground pipe due to cyclic water supply and drainage using X-ray CT. *Frontiers of Structural and Civil Engineering*, 6: 85–100.

Rogers, C.J. 1986. Sewer deterioration studies the background to the structural assessment procedure in the sewerage rehabilitation manual. *WRc Report* 2ed., Water Research Centre.

The Mathworks Inc (2014) MATLAB Ver. 8.3 Release 2014b, Natick, Massachusetts, USA.

Thielicke, W. & Stamhuis, E. 2014. PIVlab – Towards User-friendly, Affordable and Accurate Digital Partical Image Velocimetry in MATLAB. Journal of Open Research Software 2(1): 30.

Water Services Association Australia. 2002. Sewerage code of Australia, Melbourne retail water agencies edition, Version 1.0, (WSA 02-2002-2.3). Melbourne, Australia.

White, D., Take, W. & Bolton, M. 2001. Measuring soil deformation in geotechnical models using digital images and PIV analysis. *10th International Conference on Computer Methods and Advances in Geomechanics, Tucson, Arizona*.

Modelling cave mining in the geotechnical centrifuge

S.W. Jacobsz & E.P. Kearsley
University of Pretoria, Pretoria, South Africa

D. Cumming-Potvin & J. Wesseloo
University of Western Australia, Perth, Australia

ABSTRACT: A model study was undertaken using centrifuge modelling to investigate the mechanism of cave propagation occurring in cave mines. Cave mining involves controlled undercutting of ore bodies in deep mines, allowing the rock mass to fracture under in situ stress. The paper describes a weak artificial rock mass that was developed and presents results from tests in which caving was observed under conditions of high and low horizontal stress. Comments on appropriate scaling relations is presented. It was found that rock fracture occurred by means of extensional fracture banding, which differs from the Duplancic conceptual model, commonly accepted to describe processes associated with cave advancement.

1 INTRODUCTION

Cave mining involves controlled undercutting of ore bodies in deep mines by drilling and blasting, allowing the rock mass to fracture under its own weight and horizontal in situ stresses. Once a sufficiently large area has been undercut, caving initiates and propagates through the ore body (Bartlett & Nesbitt, 2000). Bulking of the rock mass occurs during the fracturing process and continuous removal of the fractured material from below the undercut is necessary to create the conditions necessary for the self-sustaining caving of the ore-body. The caving zone propagates upwards through the rock mass. Block and panel cave mining are efficient mass mining methods as minimal waste rock is mined and once caving is initiated, the need for blasting is also minimal.

The shape and size of the cave that forms is dependent on the size of the void formed under the ore body, the composition of the ore body and the strength of the abutment zone formed around the caving area. It is generally assumed that the damage ahead of the cave decreases with increasing distance from the cave back and caving takes place primarily as a result of slip along pre-existing discontinuities (Duplancic, 2001). Due to the lack of access to the caving zone, direct observation of the cave propagation process is not possible. As a result, this generally accepted model has not been rigorously scrutinized and verified (Cumming-Potvin et al. 2016a). Duplancic (2001) proposed a model describing the zone of influence affected by cave mining divided into five sub-zones:

1. Caved zone – A zone of caved material that has fallen from the cave back, providing some support to the walls.
2. An air gap between the cave back and the caved zone.
3. A zone of loosening – This comprises loose rock not supporting the overlying rock mass and represents a zone where disintegration occurs.
4. Seismogenic zone – This is a stressed front ahead of the cave back where seismic fracture of the rock mass occurs by means of slip via pre-existing discontinuities.
5. Pseudo-continuous domain – This represents the rock mass ahead of the seismogenic zone where only elastic deformations occur.

A small number of researchers have attempted physical model studies to understand draw control, i.e. how the removal of fractured material from the base of the caved zone affects movement within the fractured rock mass (Kvapil, 1965; Castro et al., 2007; Paredas and Pineda, 2014). The authors are aware of only one study examining the process of cave propagation itself, i.e. McNearny & Abel (1993), who used a two-dimensional model of layers of bricks to examine the effect of drawpoint spacing on cave propagation. There were, however, certain shortcomings associated with this model such as the absence of a realistic horizontal stress field and a joint pattern dictated by the brick geometry, which was not necessarily representative of an actual rock mass.

There are very few centrifuge model studies described in the literature investigating mining problems. Prof Evert Hoek, when Head of the Rock Mechanics Division of the National Mechanical Engineering Research Institute at the South African Council for Scientific and Industrial Research in South Africa during the early 1960s designed and supervised the construction of a 9 ft (2.74 m) diameter centrifuge

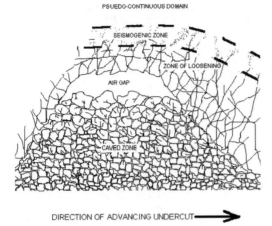

Figure 1. The Duplancic model in concept (Duplancic, 2001).

capable of accelerating models weighing up to 100 lb (45.4 kg) to 1000 g (centrifuge capacity therefore 45 g-ton). The centrifuge was used to simulate gravitational force fields in mine models. Hoek's models generally involved photo-elastic materials that were 'stress frozen' at high acceleration by means of heating elements so that the models could be studied after the tests (Hoek, 1965). Hoek demonstrated the need to simulate gravitational body forces when investigating stress and fracture of a rock mass around mine excavations. Although the equipment Hoek used has been abandoned, the scaling laws he published provide a starting point for the development of scaling law refinement to study cave propagation in the centrifuge.

Based on the principles of similitude to relate the behaviour of a model and a prototype, incorporating conditions of elasticity theory, Hoek (1965) derived the following relationship:

$$\frac{L_p}{L_m} = \alpha \times \frac{E_p}{E_m} \times \frac{\rho_m}{\rho_p} \times \frac{g_m}{g_p} \quad (1)$$

where p and m refer respectively to the prototype and the model, L to length dimensions, E to the modulus of elasticity, ρ to material density and g to acceleration. The parameter α refers to the ratio between the stress in the prototype and that at an equivalent point in the model, i.e. $\alpha = \sigma_p / \sigma_m$. Hoek (1965) derived this scaling relationship (Equation 1) based on the assumption that deformation prior to fracture will be purely elastic, a reasonable assumption for rock. He also assumed that the fracture strength of a rock can be described by a Mohr-Coulomb-type criterion containing a frictional component μ and a strength parameter. He assumed the uniaxial compressive strength (UCS) of the material to adequately represent this strength parameter. In his analysis he considered rock fracturing to result in the loss of the strength parameter, with only the frictional component remaining. Graphically this amounts to shifting the failure envelope to pass through the origin. The implication is that for equation (1) the value of α can be taken to be the ratio between the UCS of the prototype and model materials, i.e. $\alpha = \sigma_p / \sigma_m$. When the model is made from the same material as the prototype the value of α is 1, as are the values of the quotients containing ρ and E, resulting in the traditional scaling relationship:

$$\frac{g_m}{g_p} = \frac{L_p}{L_m} \quad (2)$$

This scaling relationship is not feasible for modelling typical deep mining problems because of the limitations of most centrifuges. It is therefore necessary to use a weaker material than the prototype material and use equation (1) as scaling relationship. In addition, when considering the experimental observation that material strength seems to reduce with the square root of the sample size, Hoek (1965) modified equation (1) further to:

$$\left[\frac{L_p}{L_m}\right]^{3/2} = \alpha \times \frac{E_p}{E_m} \times \frac{\rho_m}{\rho_p} \times \frac{g_m}{g_p} \quad (3)$$

This implies that in order to fracture a small sample, it needs to be accelerated to a higher acceleration compared to when scale effects are ignored.

During cave mining cave propagation is generally monitored by means of micro-seismic monitoring, downhole dipping and borehole cameras and is analysed using numerical methods. However, due to the mining process being hidden below the ground, opportunities allowing for models to be calibrated are rare.

Due to complexities modelling fracture processes theoretically, physical model studies are an attractive means to investigate the rock mass fracturing and cave propagation in this type of mining. This paper presents a study using centrifuge modelling to study some aspects of cave propagation during cave mining.

2 CENTRIFUGE MODEL

2.1 Test frame

The model study was undertaken at the University of Pretoria Geotechnical Centrifuge Facility (see Jacobsz et al. 2014). To allow visual observation of the caving process, a purpose-built two-dimensional model frame, assembled from aluminium channel sections, was constructed. The test frame is illustrated in Figure 2. The cave mining process was modelled using a bank of five 50 mm wide trapdoors next to each other supporting a model rock mass (caving medium). The trapdoors were mounded on pistons (Festo DZF-50-100-P-A) that could be individually retracted to lower the trapdoors and remove support from the caving medium by releasing fluid via solenoid valves. The pistons stroke was 100 mm. The rate of settlement was limited by using a flow restrictor set to the desired flow

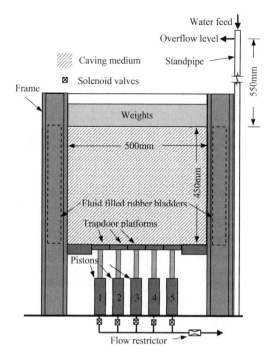

Figure 2. Centrifuge model.

Table 1. Composition of caving medium panels.

Material	Content kg/m^3	Specific gravity	Volume %	Weight %
Water	350	1.00	35.0	18.7
Cement	25	3.14	0.8	1.3
Fly ash	998	2.20	45.3	53.3
Sand	500	2.65	18.9	26.7

Table 2. Mechanical properties of caving material panels.

Property	Caving medium	Quartzite rock*
	Model property	Prototype property
Density (kg/m^3)	1875	2650
Compressive strength (MPa)	0.5	188
Tensile strength** (MPa)	0.03	10
E-value (GPa)	0.5	50
Critical extensional strain ($\mu\varepsilon$)	1000	100

*Source: Bieniawski (1974)
**Split cylinder strength.

rate. The settlements of the trapdoors were individually monitored using LVDTs. The model is illustrated in Figure 2 and also included:

- A glass window to allow the model to be observed to enable image analysis;
- A sample space 500 mm wide, 450 mm high and 50 mm deep to allow plane-strain modelling of the mining process;
- Fluid-filled rubber bladders on the sides of the model rock mass to apply horizontal stress;
- A standpipe in which a constant water level was maintained to exert the horizontal pressure via the fluid-filled bladders;
- Weights, two steel bars of 250 mm length, placed end-to-end and sized to exert 108 kPa at 80 g, resting on polystyrene padding on top of the caving medium to impose overburden stress.

2.2 *Caving medium*

Hoek (1965) mentioned that accelerations as high as practically possible should be used to study mining problems in a centrifuge in order to replicate the high stresses present in deep mines. In South Africa the deepest mines have now advanced to depths in excess of 4 km (Suorineni, 2017), implying rock stresses well above 100MPa. Even with the use powerful centrifuges these stresses are not practical in model studies. It was therefore necessary to develop a material that could be used to model caving at accelerations that could be achieved in the centrifuge. Due to constraints imposed by the monitoring equipment, predominantly the cameras and solenoid valves controlling the lowering of the trapdoors, it was decided to carry out the model study at 80 g.

A cementitious material comprising of sand, flyash, cement and water was developed to provide a material that would cave in a brittle fashion under its own weight at the test acceleration. The mix composition indicated in Table 1 was chosen as the most suitable composition and was determined by means of experimentation through a process of trial and error. Panels of the caving medium measuring 450 × 500 × 50 mm were cast. The panels were removed from the moulds after 24 hours and cured at 60° for another 24 hours before being tested. A strict testing regime was adopted in terms of the time of testing after preparation of the caving medium to ensure consistent material properties in all tests. This was necessary because the material properties were found to change relatively quickly over time.

To confirm material properties, the unconfined compressive strength of the caving material was measured at the time of testing. Cylindrical samples measuring 50 mm in diameter by 100 mm in height, carved from a duplicate panel prepared when casting material for every test, were used. The samples for testing were cut to have the same orientation as the panels that were tested to avoid problems resulting from possible anisotropy. Sample compression and radial strain were measured using local strain measurement. Typical values of the material properties of the caving medium are presented in Table 2.

The caving process was initially found to occur in an unpredictable fashion when casting homogeneous slabs, giving poor repeatability between tests. A series of quasi-randomly orientated joints were therefore cut into the caving material using a purpose-made template just after the material had set. Partial self-healing of the material resulted in a random set of slightly weaker joint planes. The panels were allowed to cure for approximately 3.5 hours after casting before the discontinuities were cut.

2.3 Test sequence

Prior to each test, a caving medium panel was inserted into the frame and the glass window bolted into position. A Teflon sheet at the base of the caving panel was used to minimize horizontal friction. The flexibility of the glass window did allow some out-of-plane strains to occur during testing, but conditions are thought to have reasonably approximated plane-strain conditions. The rubber bladders were filled with water and the standpipe was filled to the overflow level, which was set to exert the desired horizontal stress. The standpipe water level was maintained at the overflow level over the entire test duration by continuously adding water and allowing the excess to spill. Once the test acceleration of 80 g was reached, the pistons supporting the trapdoors were retracted in approximately 1 mm increments in the following sequences: 1, 1-2, 1-2-3, 1-2-3-4, 1-2-3-4-5. The numbers refer to piston numbers in Figure 2.

Some of the retraction sequences listed were repeated before proceeding to the next, based on visual observation of how fracturing of the cave panel proceeded. The undercutting sequence allowed support to be progressively removed from the base of the caving panel, advancing from the left towards the right to mimic the way in which cave mining would advance in practice as shown by the direction of the advancing undercut in Figure 1. The entire undercutting sequence was repeated a number of times, always from left to right, until maximum reach of the pistons or catastrophic failure of the cave panel occurred. The process was monitored visually by taking photographs every 4 seconds using a Canon D100 DSLR camera fitted with a 40 mm fixed focal length lens and web cam video camera.

3 EXPERIMENTAL RESULTS

3.1 Model of cave propagation

As the trapdoors were lowered, fracturing of the rock mass initiated. The lowering of the first trapdoor usually did not result in any fracturing, but fracturing normally followed soon once the second and third trapdoors were lowered. A sequence of images from a typical test is presented in Figure 3. Fracturing occurred as a series of approximately parallel fracture lines forming suddenly and rather unpredictably in a step-like fashion as the trapdoors were lowered. An air gap formed as described by the Duplancic model. However, the gradual fracturing, loosening

Figure 3. A sequence of images illustrating an example of cave propagation in a centrifuge test.

and break-up of the rockmass above the airgap as described by Duplancic was not observed. Instead, the sub-parallel fracturing of the caving medium resulted in slabs being released from the cave back which then broke up subsequently as further undermining occurred due to uneven removal of support and due to the load of newly formed slabs released from above. The fracturing appeared to resemble extensional failure occurring due to the release of vertical stress while elevated horizontal stresses were maintained in the model. More detail is presented by Cumming-Potvin et al. (2016a). The fractures did not appear to occur along the pre-existing weakened randomly orientated joint sets created by cutting the caving medium shortly after casting.

3.2 Influence of pre-existing fractures

The slabs of caving medium prepared measured 450 × 500 mm and weighed approximately 21 kg. They were rather difficult to handle and were very fragile. Despite care being taken not to damage the slabs during installation in the test frame, damage did occasionally occur, causing cracking of the slabs before testing. This, however, allowed the effect of pre-existing cracks on cave propagation, considered to be representative of geological faults, to be observed. Figure 4 illustrates two images showing cave propagation in a test with two approximately parallel curved sub-vertical cracks (shown by the dotted lines). The figure illustrates how the presence of the faults directed

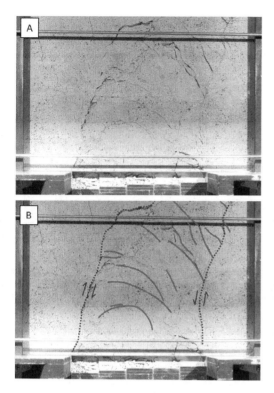

Figure 4. Caving directed by presence of pre-existing fracture formed prior to testing.

the caving process. Large shear movements occurred along the faults, leaving the rock mass outside the zone confined between the cracks largely intact despite the continued application of increased horizontal stress from the water-filled bladders. Extensional fracturing between the faults became sub-horizontal and remained approximately perpendicular to the direction of the faults.

3.3 Influence of applied horizontal stress

The use of water-filled rubber bladders installed in recesses in the sides of the model frame adjacent to the caving medium allowed horizontal stress to be applied to the caving medium. The pressure in the bladders were controlled at the desired level by adjusting the height of the standpipe linked to them. The use of a heavy fluid (e.g. sodium poly-tungstate with a density of 2850 kg/m^3) allowed horizontal stresses well in excess of the vertical stress to be applied to model a situation where the coefficient of lateral rock stress exceeds unity, a common occurrence in mines (Suorineni, 2017). This aspect, however, falls outside the scope of the current paper.

Tests were carried out in which the bladders were filled with water to give a horizontal stress at sample mid-depth 20% higher than the vertical (see Cumming-Potvin et al., 2016b) and some tests were also carried out where the bladders were omitted and

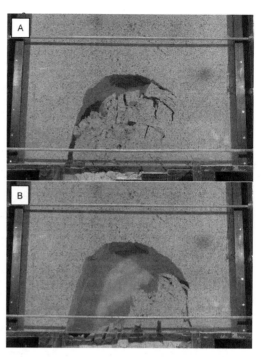

Figure 5. An example of caving stalling in instances where low horizontal stresses were applied.

the space filled with fine sand to exert a smaller horizontal stress.

Caving by means of extensional fracture banding developed readily in instances where large horizontal pressure was applied. When the applied horizontal stress was small, caving seemed to stall as stress redistribution occurred in the caving medium to allow a stable arch to form as illustrated in Figure 5.

4 CONSIDERATION OF THE SCALE FACTOR

Similitude between the prototype and the model materials requires that the Poisson ratio of the two materials must be the same to ensure that horizontal stress response due to changes in vertical stress is comparable in situations where lateral deformation is restricted (Hoek, 1965). Assuming elastic conditions, this, in turn, controls shear stress magnitude, Poisson ratios for rock commonly range between 0.1 and 0.2 (Hoek, 1965) which agree well with values measured for the artificial caving material.

It is of interest to back-calculate the scale factor associated with the model tests conducted. Using equation 1 and typical assumed properties of quartzite rock common in deep South African mines (see Table 2) and the properties measured for the caving medium used for the model tests (see Table 2), a scale factor of 3 400 000 is calculated. This is obviously unrealistically high. Taking into account bulk material strength scale effects and using equation 3, lowers the factor substantially to 22 600. Caving behaviour in the model tests described below suggested that fracture of the caving medium occurred in extension. Following Stacey (1981) and applying a critical extensional strain criterion instead of compressive strength ratio at failure for the calculation of α in equation 3, results in equation 3 being modified to:

$$\left[\frac{L_p}{L_m}\right]^{3/2} = \frac{e_{3p}}{e_{3m}} \times \left(\frac{E_p}{E_m}\right)^2 \times \frac{\rho_m}{\rho_p} \times \frac{g_m}{g_p} \qquad (4)$$

where e_{3p} and e_{3m} are the critical extensional strains for the prototype and model materials respectively. Based on material properties for the caving medium after curing (see Table 2), this reduces the scale factor to 1400 which appears much more realistic. Further improvement of the scaling factors can probably be derived by more rigorously applying similitude principles, considering the theory of fracture. More work is required in this regard.

5 CONCLUSIONS

Despite of the use of a centrifuge it is not practical to recreate in a model study the enormous stresses at play in deep mines. However, this study demonstrates that by using a weak artificial rock mass, centrifuge models can give insight into rock behaviour in caving mines. Similar potential exists to investigate other mining problems. More work is required to develop scaling laws to allow model observation to be related to a prototype of specific dimensions.

The centrifuge models demonstrated a mode of fracturing deviating somewhat from that proposed by the commonly accepted Duplancic model, with fracturing resembling an extensional type failure mechanism above the air gap and not the gradual fracturing and loosening proposed by Duplancic. The majority of the breakup process in the models occurred below the airgap. As the Duplancic model has not been verified independently, these observations may suggest that this is necessary.

The extensional failure mode observed further differs from the Duplancic hypothesis advocating shear deformation along pre-existing joint sets as the major cause of break up. Preformed randomly orientated weakened joints did not seem to have a significant influence on the way in which the rock mass failed ahead of the airgap, but significant pre-existing fractures or faults did control the direction of cave propagation. Caving by means of extensional fracture banding developed more readily in instances where large horizontal stresses were applied in the model.

REFERENCES

Bartlett, P.J. & Nesbitt, K. 2000. Stress induced damage in tunnels in a cave mining environment in kimberlite. *Journal of the South African Institute of Mining and Metallurgy*, 100(10): 341–345.

Bieniawski, Z.T. 1974. Estimating the strength of rock materials. *Journal of the Southern African Institute of Mining and Metallurgy*, March 1974: 312–320.

Castro, R., Trueman, R. & Halim, A. 2007. A study of isolated draw zones in block caving mines by means of a large 3D physical model, *International Journal of Rock Mechanics and Mining Sciences*, 44(6): 860–870.

Cumming-Potvin, D. Wesseloo, J. Jacobsz, S.W. & Kearsley, E.P. 2016a. Fracture banding in caving mines. *Journal of the Southern African Institute of Mining and Metallurgy*, 116(8): 753–761.

Cumming-Potvin, D. Wesseloo, J. Jacobsz, S.W. & Kearsley, E.P. 2016b. Results from physical models of block caving. *Proceedings of the 7th International Conference and Exhibition on Mass Mining (MassMin2016)*, Sydney, Australia, 9–11 May 2016. Australasian Institute of Mining and Metallurgy, Melbourne: 329–340.

Duplancic, P. 2001. *Characterisation of caving mechanisms through analysis of stress and seismicity*. PhD thesis, University of Western Australia.

Duplancic, P. & Brady, B.H. 1999. Characterisation of caving mechanisms by analysis of seismicity and rock stress. *Proceedings of the 9th ISRM Congress*, Paris, August 1999. International Society for Rock Mechanics: 1049–1053.

Hoek, E. 1965. The design of a centrifuge for the simulation of gravitational force fields in mine models, *Journal of the South African Institute of Mining and Metallurgy*, 65(9): 455–487.

Jacobsz, S.W., Kearsley, E.P. & Kock, J.H.L. 2014. The geotechnical centrifuge facility at the University of Pretoria. *Proc 8th International Conference on Physical Modelling in Geotechnics*, Perth. CRC Press: 169–174.

Kvapil, R. 1965. Gravity flow of granular material in hoppers and bins Part 1. *International Journal of Rock Mechanics and Mining Sciences* (2): 35–41.

McNearny, R.L. & Abel, J.F. 1993. Large-scale two-dimensional block caving model tests. *International Journal of Rock Mechanics and Mining Sciences* 30(2): 93–109.

Panek, L.A. 1981. Ground movements near a caving stope. *Design and Operation of Caving and Sublevel Stoping Mines* (ed: D R Stewart), (Society of Mining Engineers, American Institute of Mining, Metallurgical and Petroleum Engineers: Littleton): 329–354.

Paredas, P.S. & Pineda, M.F. 2014. An analysis of the lateral dilution entry mechanisms in panel caving. *Proceedings of Third International Symposium on Block and Sublevel Caving* (Universidad de Chile: Santiago): 118–127.

Stacey, T.R. 1981. A simple extension strain criterion for fracture of brittle rock. *International Journal of Rock Mechanics, Mineral Sciences and Geomechanics* 18: 469–474.

Suorineni, F.T. 2017. Myths of deep and high stress mining – reality checks with case histories. *Deep Mining 2017: 8th International Conference on Deep and High Stress Mining* – J Wesseloo (ed.) Australian Centre for Geomechanics, Perth: 843–860.

Experimental modelling of infiltration of bentonite slurry in front of shield tunnel in saturated sand

T. Xu & A. Bezuijen
Laboratory of Geotechnics, Ghent University, Ghent, Belgium
Deltares, Delft, The Netherlands

ABSTRACT: Bentonite slurry pressured with air pressure has been widely used to support the face of shield tunnels, particularly in saturated sandy ground. Due to the pressure difference between the bentonite slurry and the pores in the soil, the bentonite slurry will infiltrate into the pores in the soil. Partial support pressure applied through bentonite slurry will thus be present as an excess pore water pressure rather than be carried directly onto the soil grains. In this case, the support pressure at the tunnel face will be reduced. The infiltration, therefore, is of importance for the stability of tunnel face. In this paper, a series of tests of bentonite slurries, pressured against saturated sand were presented. Mud spurt and consolidation characteristics of the bentonite slurry were examined. It appears that the permeability of the sand for the bentonite slurry is considerable lower than that for water. The filter cake will impede the fluid discharged from the bentonite slurry, and thus will exponentially reduce the permeability of the sand for the bentonite slurry. It also indicates adding bentonite decreases the permeability of the sand for the bentonite slurry.

1 INTRODUCTION

During shield tunnelling in saturated sandy ground, bentonite slurry is extensively used to support the tunnel face. Because of the pressure difference between bentonite slurry and pore water in soil bentonite slurry will flow into the surrounding soil. This process is known as infiltration leading to the loss of support at the tunnel face. One example is the Second Heinenoord Tunnel in Netherlands, in which the loss of support pressure accounts for 10~20% of the slurry pressure (COB 2000). The similar conditions have also been found at Green Hart Tunnel and North/South Metro Line in Netherlands (Bezuijen et al. 2006; Kaalberg et al. 2014). The mechanism of the loss of support pressure is that the partial support pressure is transferred into pore water rather than onto the soil skeleton. This is the case of the drilling of the tunnelling boring machine (TBM). While the standstill of the TBM, because the channels for fluid flow will be blocked by the bentonite particles, the filter cake will be formed. Consequently, the support pressure is directly transferred onto the soil skeleton.

To understand the mechanism of the infiltration of bentonite slurry at the tunnel face during tunnelling with a slurry shield, various types of infiltration tests on bentonite slurry have been carried out (e.g. Talmon et al. 2013; Steeneken 2016). Unlike filtration of bentonite grout, for infiltration of pure bentonite slurry the bentonite particles included in the fluid will be pushed into the sand matrix, which was described as leak-off by Bezuijen et al. (2009). The infiltration of pure bentonite slurry thus is more complicated. Except the permeability of the sand for the slurry, there is also the permeability of the slurry for the water, when the slurry has formed a filter cake. This should be taken into account as an extra flow which is not in series as the ones above, but which is parallel to the slurry flow. The theory of infiltration of grout (e. g. McKinley & Bolton, 1999), therefore, cannot be employed to determine the permeability of pure bentonite slurry in sand. To explain the characteristics, the infiltration theory of bentonite slurry has to be introduced.

This paper describes the theory of infiltration, the set-up for experiment, the behaviours of infiltration of bentonite slurry, and end with a discussion of the results and conclusions.

2 PRINCIPLE OF PRESSURE INFILTRATION OF BENTONITE SLURRY

Assume the situation sketched in Figure 1; the difference in piezometric head over the sand is $\Delta\phi$ (m). Using Darcy's law, the relation between the difference in piezometric head and the discharge can be written as:

$$\Delta\phi = \frac{\Delta L_s Q}{k_s \pi (D_1/2)^2} \quad (1)$$

where ΔL_s (m) = the length of sand column, k_s (m/s) the permeability of sand, Q (m^3/s) = the discharge, D_1 (m) = the inner diameter of Perspex cylinder, and $\Delta\phi$ (m) = the difference in piezometric head over the sample.

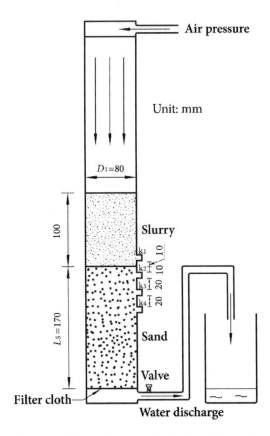

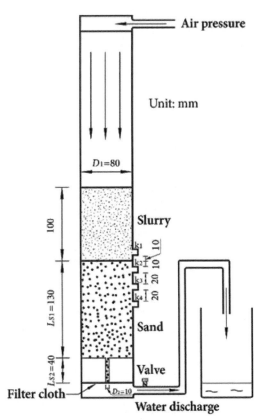

Figure 1. Principle of pressure infiltration of bentonite slurry.

Figure 2. Revised set-up in test.

Replace the item of k_s to k_b, Equation (1) can be also employed to calculate the permeability (k_b) of sand for bentonite slurry:

$$k_b = \frac{\Delta L_s Q}{\Delta \phi \pi (D_1/2)^2} \quad (2)$$

where ΔL_s' (m) = the length of sand between two adjacent pore water pressure transducers (k1 to k4 in Figure 1).

It is should be noted that the calculated result will be unreliable when the discharge becomes too low.

3 EXPERIMENT

3.1 Set-up

For the set-up in Figure 1, if a constant air pressure of 50 kPa is applied on the bentonite slurry surface the hydraulic gradient over the sand column will be $i = \Delta\phi/L_s = 5/0.17 \approx 30$. According to Bezuijen (2006), however, for a tunnel with radius of e.g. 5 m, the hydraulic gradient at the tunnel face $i = \Delta\phi/L_s = 5/5 = 1$. As a result, the infiltration velocity of bentonite slurry in laboratory set-up will be much faster than that at a real tunnel face. To eliminate the effect of the extra hydraulic gradient the set-up was revised. As shown in Figure 2, a Perspex cylinder with smaller inner diameter was placed in the bottom of the large cylinder. The relation between difference in piezometric head and discharge can be approximated with:

$$\Delta\phi = \frac{Q}{\pi k_s}\left(\frac{4L_{s1}}{D_1^2} + \frac{4L_{s2}}{D_2^2} + \frac{1}{D_2} - \frac{1}{D_1}\right) \quad (3)$$

where L_{s1} (m) = the length of sand column in large cylinder, D_1 (m) = the inner diameter of large cylinder, L_{s2} (m) = the length of sand column in small cylinder, D_2 (m) = the inner diameter of small cylinder. In this case the L_s changes to:

$$L_s = L_{s1} + L_{s2}\left(\frac{D_1}{D_2}\right)^2 + \frac{D_1^2}{4D_2} - \frac{D_1}{4} \quad (4)$$

In Equation (4), the first 2 terms on the right hand side follow directly from Darcy's law and are identical to the Equation (3) for the large and small diameter section. The last two terms $1/D_2$ and $1/D_1$ describe the resistance that follows from the contraction from the flow lines when the flow has to change from the large diameter cylinder to the small diameter cylinder. It is

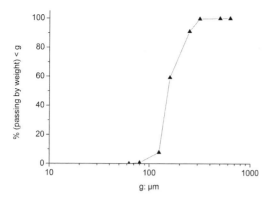

Figure 3. Grain size distribution of Sibelco M32 sand.

Table 1. Properties of the sand for the different tests.

Test	Bentonite amount (g/l)	Sand thickness (cm)	n	k_s (m/s)
1	40	17.1	0.37	4.1×10^{-4}
2	50	17.3	0.37	4.0×10^{-4}
3	60	17.0	0.37	4.1×10^{-4}

Table 2. Rheological properties of the bentonite slurry in each test.

Bentonite amount (g/l)	Yield Point (Pa)	Apparent Viscosity (cP)	Plastic viscosity (cP)	Yield Strength (Pa)
40	1.0	4.0	3	0.5
50	1.5	5.5	4	1
60	3.6	7.5	4	2

assumed that the same sand is used in all cylinders and the same piezometric head is applied in both tests and therefore the ratio at the end of calculation of permeability (k_s) should be 1. The discharge velocity is expected to be about 1/10 of the set-up in Figure 1. It is also assumed that the sand has a lower permeability (k_b) for the bentonite slurry than the permeability (k_s) for the water and the slurry only infiltrates in the large diameter cylinder.

A valve on the top cap was for a constant air pressure of 50 kPa applied on the bentonite slurry surface. Four pore water pressure transducers (PPTs) were installed in the tips on the cylinder in order to monitor the pore water pressure in the bentonite slurry and the sand. k1 was positioned 1 cm above the slurry-sand interface to measure the pore water pressure in the bentonite slurry. k2, k3 and k4 were located at 1, 3 and 5 cm below the slurry-sand interface, respectively. A filter cloth was placed on top of a perforated plate a few centimeters above the bottom of the cylinder. Using this set-up, the slurry infiltration into the soil and the filter cake formation on the soil surface can be observed. Meanwhile the water discharge was measured continuously with an electronic balance. The pore water pressures are measured at a comparable frequency as the discharge.

3.2 Test procedure

A dense layer (40 mm thick in the small diameter cylinder and 130 mm thick in the large diameter cylinder) consisted of Sibelco M32 sand, a uniform medium fine sand with a $d_{15} = 130\,\mu m$. The grain size distribution of the sand is shown in Figure 3. The sand was compacted in the Perspex cylinders by tamping the saturated sand under water on a wire mesh covered by a filter cloth. A relative density of approximately 90% is reached. And the porosity of sand was $n = 0.37$. The filter cloth allowed water flow but blocked the sand grains. To check the consistency of the sand, the porosity and the permeability of every sand column were measured, and the results are shown in Table 1.

Then a bentonite slurry layer of 10 cm thickness was placed on surface of the saturated sand column. The Colclay D90 bentonite and the water are mixed in a mixer for 20 minutes and stiffened for 24 hours according to API (2003). It was mixed for 5 minutes again before use. The properties of bentonite slurries were determined prior to the start of each experiment, the results are shown in Table 2.

At the start of the test the bentonite slurry was pressurised with an air pressure of 50 kPa. The infiltration of bentonite slurry started immediately the valve at the bottom was opened. Meanwhile, the pore water pressure at different depths was measured by PPTs. The pore water pressures are measured at a comparable frequency as the discharge. Two hours later, the valve was closed and the air pressure was released.

4 RESULTS AND DISCUSSIONS

Three bentonite slurries with various amount of bentonite 40, 50 and 60 g/l were tested. Figure 4 gives the plot of discharge volume against the square root of time. It shows a good comparison with the plots provided by Talmon et al. (2013). The infiltration is clearly separated into two sub-processes: mud spurt and consolidation of the bentonite slurry. It can be seen that the discharge volume shows a linear and slight increase in square root of time during the consolidation of bentonite slurry. Due to the formation of the filter cake the slurry pressure will be directly applied on the sand skeleton, thus there will be no flow caused by pressure gradient to push the bentonite particles into and through the filter cake. As a result, the volume rate of water flow becomes slight. When only water filtrates from the bentonite slurry, therefore, the filter cake determines the consolidation rate of the bentonite slurry.

Combining the discharge (Q) and the corresponding difference in piezometric head ($\Delta\phi$), the permeability of sand for bentonite slurry was determined with Equation (3). It should be noted that, herein the Ls used for calculation is the distance between two adjacent PPTs. The value at different time varied.

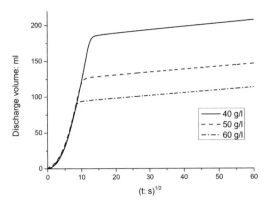

Figure 4. Discharge volume against square root of time.

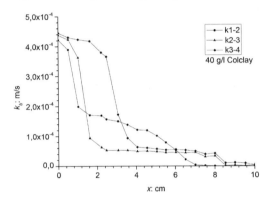

Figure 5. Permeability of sand for 40 g/l Colclay bentonite slurry.

It can be seen from Figure 5 that for Test 1 the permeability of the sand is around 4.4×10^{-4} m/s (see the permeability calculated between the 2 lowest gauges k3 and k4 at the very beginning). The permeability of the slurry in the sand is more or less 6.0×10^{-5} m/s (the k2-3 after 3 cm of infiltration and the k3-4 after 5 cm of infiltration). The k1-2 is the permeability in the upper part of the sand and 1 cm above the sand. After 10 cm of infiltration of the bentonite slurry in the sand the values become lower, because now the cake formation starts at the sand surface. It is very clear that this cake formation is at the sand surface because the permeability between k2 and k4 is not affected by this cake formation. After 10 cm of infiltration the discharge becomes low and the results will be less reliable.

For Test 2, the permeability of the sand for water is 4.4×10^{-4} m/s and for bentonite slurry approximately 3.0×10^{-5}. For Test 3, the permeability of the sand for water for is 4.2×10^{-4} m/s, and the permeability for the bentonite slurry is about 2.0×10^{-5} m/s.

As Figure 6 shown, the measured value of k_b reduced with higher bentonite content. The bentonite content 30 g/l was not used because water would separate from the slurry after stiffening. The k_b of sand for slurry with high amount of bentonite (≥ 70 g/l) was expected to be zero.

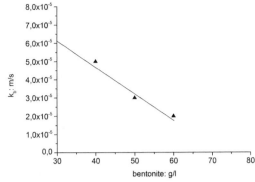

Figure 6. Effect of amount of bentonite on permeability of sand for bentonite slurry.

5 CONSEQUENCES FOR PRACTICE

The experiments described in this paper show that there are two sub-processes during slurry infiltration: mud spurt and formation of filter cake. This is in agreement with previous research (Talmon et al. 2013; Xu et al. 2017). Bezuijen et al. (2017) took into account only the mud spurt to calculate the excess pore water pressures in front of the slurry shield and found reasonable agreement with the measurements. This seems to indicate that the formation of cake formation is not relevant for real tunnelling conditions. In Xu et al. (2017), the pore water pressure distribution in depth was presented (see Figure 7). It was shown that the pore water pressure rapidly decreased with the formation of filter cake and then became steady. This means that there will be no or very limited excess pore water pressures in front of the TBM when filter cake formation starts. Furthermore, the filter cake will be removed immediately when drilling starts. This means that to describe the excess pore water pressures in front of a tunnel face only the mud spurt phase is of importance.

6 CONCLUSIONS

An experimental approach to the plastering of the bentonite slurry at the tunnel face during slurry shield tunnelling in the saturated sand has been presented. The experimental data is analysed and discussed with the related theory. The following conclusions are allowed to be drawn:

- The permeability (k_b) of sand for bentonite slurry is considerable lower than the permeability (k_s) of sand for water. For 40, 50 and 60 g/l Colclay D90 bentonite slurries, the ratio of k_b to k_s is between about 7 and 22. It is also shown that adding bentonite reduces the distance of infiltration, and the permeability (k_b) of sand for bentonite slurry.
- A filter cake was formed in each test. The permeability (k_b) of sand for bentonite slurry is little affected immediately the formation of filter cake starts. Meanwhile, the discharge volume shows a

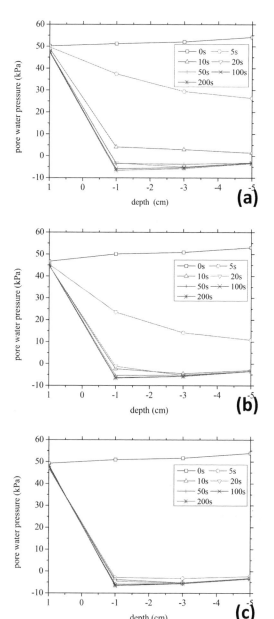

Figure 7. Pore water pressure distribution in Test 1 (a), Test 2 (b), and Test 3 (c) (after Xu et al., 2017).

linear increase in square root of time, but the rate becomes very low. This may be described by the theory of consolidation. It will be discussed in future study.
− There will be no or very limited excess pore water pressures in front of the TBM when the filter cake starts to be formed. And the formation of filter cake is not relevant for real tunnelling conditions.

ACKNOWLEDGEMENT

The first author would like to acknowledge the scholarship funded by China Scholarship Council.

REFERENCES

API. 2003. Recommended Practice Standard Procedure for Field Testing Water-Based Drilling Fluids, 13B-1, 3rd ed., American Petroleum Institute, Washington, DC.

Bezuijen, A., Pruiksma, J. P. & van Meerten, H.H. 2006. Pore pressures in front of tunnel, measurements, calculations and consequences for stability of tunnel face, *Tunnelling. A Decade of Progress. GeoDelft 1995-2005*, 35–41.

Bezuijen, A., Sanders, M. P. M. & den Hamer, D. 2009. Parameters that influence the pressure filtration characteristics of bentonite grouts, *Geotechnique* 59(8): 717–721.

Bezuijen, A., Xu, T. & Dias, T. G. S. 2017. Pore pressures in front of a slurry shield: development and decline. *EURO:TUN 2017*, Innsbruck, Austria.

COB, 2000. Second Heinenoord Tunnel Evaluation Report, COB report K100-06. Gouda, the Netherlands.

Kaalberg, F.J., Ruigrok, J.A.T. & Nijs, R.D. 2014. TBM face stability & excess pore presssures in close proximity of piled bridge foundations controlled with 3D FEM, *Proceeding of 8th International Symposium on Geotechnical Aspects of Underground Construction in Soft Ground*, Seoul, South Korea.

McKinley, J.D. & Bolton, M.D. 1999. A geotechnical description of fresh cement grout: filtration and consolidation behaviour, *Magazine of Concrete Research*, 51(5): 295–307.

Steeneken, S.P. 2016. Excess pore pressures near a slurry tunnel boring machine Modelling and measurements. MSc Thesis, Delft University of Technology, Netherlands.

Talmon, A.M., Mastbergen, D.R. & Huisman, M. 2013. Invasion of pressurized clay suspensions into granular soil, *Journal of Porous Media*, 16: 351–365.

Xu, T., Bezuijen, A. & Dias, T.G.S. 2017. Slurry Infiltration Ahead of Slurry TBM's in Saturated Sand: Laboratory Tests and Consequences for Practice. *9th International Symposium on Geotechnical Aspects of Underground Construction in Soft Ground*, Sao Paulo, Brasil.

13. Imaging

Flow visualisation in a geotechnical centrifuge under controlled seepage conditions

C.T.S. Beckett
School of Engineering, Institute for Infrastructure and Environment, The University of Edinburgh, Edinburgh, Scotland, UK

A.B. Fourie
School of Civil, Environmental and Mining Engineering, The University of Western Australia, Perth, WA

ABSTRACT: Image analysis is a powerful tool to obtain high-resolution displacement data from centrifuge models non-destructively. However, 'invisible' features, for example the phreatic surface, cannot be captured. Rather, analysis must rely on traditional measurement techniques, e.g. pressure transducers. Depending on the geotechnical complexity of the model, such discrete technologies might be insufficient. In this paper, we describe the processes used to inject a tracking fluid to visually identify flow patterns through a model slope. The merits of three tracing fluids were assessed: acrylic-resin ink ("artist's ink"); food-grade dye; and a fluorescent, low-viscosity dye (sodium fluorescein). Results showed that the developed injection technique was able to deliver the fluid without otherwise affecting seepage conditions. Depending on the fluid selected, the technique was equally able to examine the migration of a dense contaminant, assess model homogeneity or identify hidden flaws.

1 INTRODUCTION

Digital image analysis techniques are now commonplace when examining the behaviour of geotechnical centrifuge models (Stanier et al. 2015). The key advantage of such techniques is that high resolution data pertaining to the observed phenomenon can be obtained non-destructively. However, image analysis is clearly limited to visible phenomena: the position of the phreatic surface, for example, must be measured using more traditional, discrete measurement devices e.g. pressure transducers. For hydrostatic problems, such a system might be sufficient. However, such limited measurements might represent a considerable drawback when the flow conditions are unknown.

This work formed part of a larger study studying seepage characteristics of tailings storage facilities (TSFs): man-made geotechnical structures, often hundreds of metres high, formed by the progressive deposition of tailings slurry behind a retaining dam (Beckett et al. 2016, Beckett and Fourie 2016). Accurate control of the water balance in such systems is critical; of 22 TSF failures shortlisted by Blight and Fourie (2005), 7 could be said to be due to seepage-related phenomena, accounting for 268 deaths. The need to improve seepage characterisation in these structures is therefore obvious. In this paper, we discuss the development of a technique to visually track flow through a centrifuge model, under controlled seepage conditions, by injecting a tracing fluid; capturing the entire phreatic surface via image analysis would significantly improve confidence in groundwater level calculations and so those for TSF stability.

2 EXPERIMENTAL PROGRAMME

2.1 Benchtop testing

The fluid injection system discussed in this paper was designed to interface with centrifuge apparatus shown in Figure 1, described in Beckett et al. (2016). In that work, a syringe pump was used in a closed loop to control and monitor downstream seepage from a model embankment subjected to a hydraulic gradient. Here, we built on that apparatus to examine the ability of an additional syringe pump to inject a tracing fluid upstream of the model via the centrifuge's water supply. Critically, the developed system had to deliver the tracing fluid without affecting seepage conditions within the model: fluctuations in the delivered flow rate would result in lengthy re-equilibration times and, if significant, excessive centrifuge imbalance and potential safety concerns.

Benchtop testing was used to validate the proposed system under normal gravity (i.e. 1g), shown in Figure 2. Water was fed into a measuring tank directly from the centrifuge water supply to ensure a realistic delivery pressure. Flow into the tank was controlled by a needle valve and measured via a pressure transducer

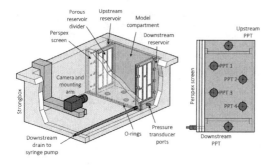

Figure 1. Centrifuge outer strongbox, model container and camera mounting (left) and pore pressure transducer (PPT) numbering (right). Note: downstream syringe pump not shown.

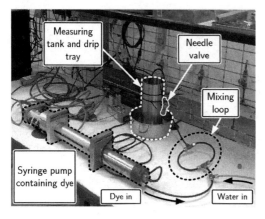

Figure 2. Injection system benchtop testing, showing key components.

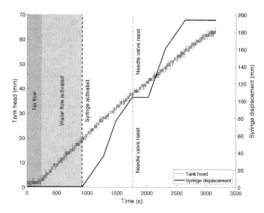

Figure 3. Change in tank head level with time during injection.

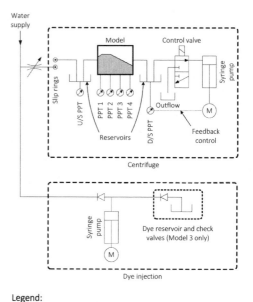

Figure 4. Centrifuge water supply hydraulic diagram incorporating dye injection system.

mounted in the tank's base. Tracer fluid (here, acrylic-resin ink) was injected upstream of the needle valve (via a mixing loop) into the water supply at various rates; head levels in the tank during injection are shown in Figure 3. Head levels increased nominally linearly with time, indicating that injection did not affect flow rate into the tank for all of the examined dye delivery rates: the system was capable of delivering dye to the model at different concentrations without affecting flow conditions. The setup was therefore incorporated into the centrifuge apparatus according to the hydraulic diagram shown in Figure 4.

2.2 Centrifuge modelling

Injection testing was carried out in the 1.8 m radius beam centrifuge based at the National Geotechnical Centrifuge Facility at the University of Western Australia (UWA). Centrifuge models were housed within an outer 650×390×325 mm strongbox in a custom-designed container comprising two flanking reservoirs, a central model compartment and base-mounted pressure transducers (Figure 1). The container was equipped with a 25 mm Perspex screen, fitted with photogrammetry tracking markers at 50 mm vertical and horizontal intervals (photogrammetry was not used as part of the injection study but was used elsewhere). A small-format, five-megapixel resolution machine vision camera (Allied Vision Technologies Prosilica GC2450C) coupled with a Goya C-Mount 8 mm focal length lens captured images of the models in-flight. Again, the centrifuge apparatus is discussed in detail in Beckett et al. (2016).

The full study comprised 18 steady-state and drawdown seepage tests, completed over three rounds on models of increasing geotechnical complexity: 1) (nominally) uniform material; 2) two (nominally) uniform strata of differing permeabilities; 3) heterogeneous material. Flow tracing methods were examined

Table 1. Model layer properties. e_0 void ratio under no effective stress; C_v coefficient of consolidation.

Model	Layer	Fine sand (%)	Silt (%)	Kaolin (%)	k_{sat} (μm/s)	e_0	C_v (m²/year)
1	Embankment	0	100	0	3.71–4.55	1.59	2.88×10^3–1.08×10^5
2	Embankment	0	100	0	3.71–4.55	1.59	2.88×10^3–1.08×10^5
	Base	10	90	0	0.31[a]	0.60	11.85
3	Embankment	31	43	26	3.14×10^{-3}–4.25×10^{-2}	1.06	12.53–118.04

[a] constant over stress range of interest;

in the concluding test of each round, to attempt to visualise the phreatic surface. Soil properties for each of the models are given in Tables 1 and 2. Materials were selected to be increasingly representative of real tailings; provided that the model Reynolds number is sufficiently low, using the same material in the model as in the prototype ensures similar seepage conditions between the two (Hensley and Schofield 1991). Models were manufactured from a slurry (45% water content) of the given materials (representative of field deposition), poured into the model compartment within the strongbox. Material was then centrifugally consolidated and dewatered at an acceleration of 100g, removed from the centrifuge and cut to the desired profile, shown in Figure 5 (at prototype scale); methods used to convert lengths and head levels to prototype scale are described in Beckett et al. (2016). Sand filters of average hydraulic conductivity (k_{sat}) 3.49 mm/s and void ratio (e) 0.62 abutted the models (within the model compartment) to protect the upstream and downstream reservoir membranes. Reservoirs were filled with gravel to provide support to the membranes without hindering flow.

It is noted that traditional tailings deposition methods impart a highly stratified profile whose hydraulic properties can vary by orders of magnitude (Bussière 2007). Layering could not be recreated in the model, however, as scaled layers of representative thickness could not be constructed (can be <1 mm at model scale). The models therefore represented bulk TSF seepage behaviour.

Each test assessed the suitability of one tracing fluid: 1) concentrated acrylic-resin ink (AR); 2) concentrated food-grade dye (FD); and 3) sodium fluorescein (FL). All fluids were non-toxic as fluid had to be discharged to the centrifuge chamber during testing. AR and FD are regularly used as flow tracers in flume testing. Fluorescent tracers are commonly used in medical applications as an alternative to coloured dyes; FL is one such tracer, used as it is non-toxic and water soluble. FL angiography can image structures of the order of 1 μm, which is typical of tailings pore sizes (Lee et al. 2014). Fluorescein concentrations of up to 100 ppm in water (0.1 g/L) cause negligible changes to density, viscosity or surface tension (Timmons et al. 1971, Palladini et al. 2005): 100 ppm FL was used in this study. Dry fluorescein is an orange powder but forms a yellow-green liquid when mixed with water. It fluoresces green under UV light, which provides a good distinction between it and a nominally-red

Table 2. Individual material component properties. d_{10}, d_{50} & d_{60}: 10, 50 and 60% mass passing particle diameters; PL: Plastic limit; LL: Liquid limit.

Material	d_{10} (μm)	d_{50} (μm)	d_{60} (μm)
Filter sand	300	497	529
Fine sand	124	195	211
Silt	3.2	19.7	27.1

Material	PL (%)	LL (%)
Kaolin	27[a]	61[a]

[a] (Cocjin et al. 2014);

background, e.g. geotechnical materials. Fluorescein detection can be enhanced by filtering the absorbed (495 nm) and emitted (525 nm) wavelengths (Keith 1968). Filters were obtained from the UWA Lions Eye Institute and mounted to a UV light source (installed above the camera) and camera respectively.

Tracer fluid injection commenced from a condition of steady-state seepage at 100g; processes used to establish this seepage regime are described in detail in Beckett et al. (2016). Upstream and downstream water levels at the point of injection are shown in Figure 5 (at prototype scale). Fluid was injected into the centrifuge water supply, which fed into the top of the upstream reservoir. In Models 1 and 2, a single syringe barrel was injected to observe pulse migration. In Model 3, multiple (consecutive) barrels, filled via an additional reservoir (shown in Figure 4), were used to flush the model with FL solution to observe the entire phreatic surface. Fluids were injected at a pressure equal to that of the supply to limit dilution on entering the stream; under such conditions, the water supply was effectively cut off and replaced with the dye stream for the duration of injection.

3 RESULTS AND DISCUSSION

Changes in head level during injection are shown in Figures 6a, c and e for Models 1, 2 and 3 respectively, alongside example images showing the fluid's progress. Note that test times are at model scale; timing began at the point of injection into the water supply, i.e. some time elapsed before fluids appeared in the model.

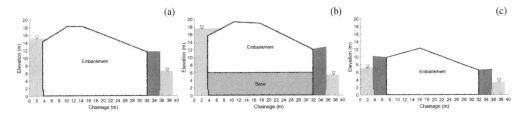

Figure 5. a–c: Model 1–3 profiles at prototype scale. Chainage originates at the upstream reservoir. Light grey zones show upstream (leftmost) and downstream (rightmost) water levels. Dark grey zones show sand filters.

3.1 *Model 1: Acrylic-resin ink*

AR was visible in the upstream and downstream reservoirs roughly 1000s and 3000s after injection respectively. The downstream reservoir was clear after roughly 5000s. Head levels were nominally constant at all points throughout the test: AR was successfully injected without affecting steady-state seepage conditions. However, regular head 'beats' were detected on all PPTs (Figure 6a). These occurred as flow was collected in the downstream reservoir whilst purging the downstream pump, the effect of which was communicated back up the hydraulic gradient. Consequently, these beats diminished with distance from the downstream reservoir. Figure 6a shows that head levels on all PPTs rapidly returned to steady-state conditions after purging was complete.

AR was visible (faintly) in the downstream sand filter, indicated in Figure 6b, indicating that it might be suitable for examining flow phenomena in granular materials. Unfortunately, it was not detected in the main body of the model, either visually or by analysing changes in pixel colour intensities (using a similar method to that described in Beckett et al. (2016)). AR was therefore not suitable for tracing the phreatic surface or the migration of a single pulse for this model.

3.2 *Model 2: Food-grade dye*

PPT responses during FD injection are shown in Figure 6c; the injected pulse partway through testing is shown in Figure 6d. Given the higher viscosity, FD pulse migration took significantly longer than AR: roughly 7 hours at model scale. Such times might be prohibitive given the high cost of centrifuge testing.

Unlike AR, FD was visible in the model material. Colouration was the most intense around the lower, less permeable base stratum, due to lower seepage velocities. Figure 6d also shows dispersed dye close to the embankment crest due to capillary uptake and flow above the phreatic surface (flow in the saturated and unsaturated regions was discussed in (Beckett and Fourie 2016)).

Notably, introducing FD into the water feed reduced the delivered flow velocity, causing a drop in head throughout the model (most strongly at the upstream reservoir). Upstream head level dropped dramatically on the pulse entering the reservoir, at 1500s, the effect of which propagated down the hydraulic gradient. Upstream head levels recovered to steady-state values as the pulse passed into the model. The pulse's transition is seen as subsequent 'rebounding' of head levels at roughly 8000, 1200, 15000 and 18000s in Figure 6c for PPTs 1, 2, 3 and 4. It should be noted, however, that changes in head level during rebound events was less than 2% that occuring due to pump purging in Model 1 for AR. Head level changes were due to the FD's high (with respect to water) viscosity, which reduced local flowrate, and so head, when upstream of a given transducer. Times between rebound events and rebound amplitudes reduced as the pulse progressed: dispersion (reducing the pulse's local viscosity) and the model's tapering shape increased local flowrate (approaching that of water) towards the downstream reservoir. Clearly, diffusion and viscosity effects therefore preclude FD from tracing steady-state groundwater phreatic surfaces. However, capturing unequal seepage velocities may be useful when examining migration of a dense contaminant, for example DNAPLs.

3.3 *Model 3: Fluorescein*

Head level changes during FL injection are shown in Figure 6e. Multiple consecutive syringe barrels were injected (facilitated by an additional dye reservoir, Figure 4) to capture the entire phreatic surface. Excepting in the downstream reservoir, PPT responses showed little fluctuation during injection, indicating steady-state conditions throughout the test. Small downstream head fluctuations (roughly ±0.1 mm at model scale) were due to a feedback error on the syringe control loop which generated excessive signal noise.

Available camera exposure times (≤ 5 s) were insufficient to capture fluorescence when the adsorption filter was fitted; the filter was therefore removed, increasing model illumination but creating an additional glow above the model due to the visible light component of the UV light source. The glow notwithstanding, FL provided an excellent contrast between the wet and dry portions of the sand filters, seen to the left and right hand sides of Figure 6f, using a 4s exposure time. FL also highlighted a large transverse crack within the embankment; the crack formed during steady-state testing but was thought closed.

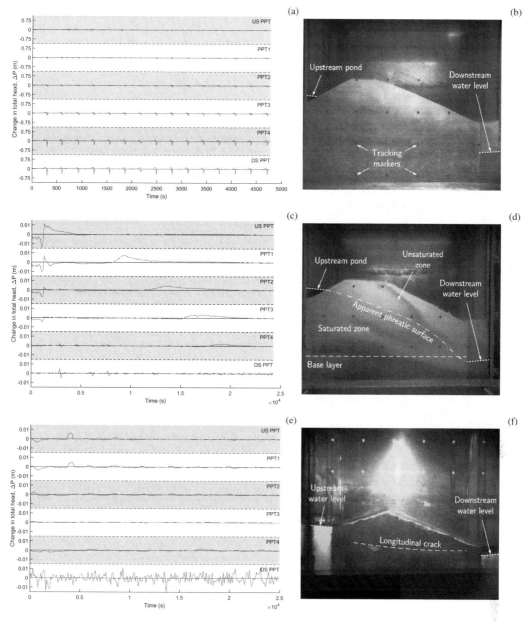

Figure 6. Change in total head (prototype scale), recorded at base-mounted pore pressure transducers (PPTs) during injection (left) and example image of injected fluid (right). Times are at model scale. Top: Model 1 (AR); middle: Model 2 (FD); bottom: Model 3 (FL). "US": upstream; "DS", downstream

Rather, FL demonstrated that it was filled with coarser debris, forming a preferential seepage path (compromising the model). The phreatic surface in the main body, however, could not be identified: what fluorescence there was was masked by other sources. Re-incorporating the light filter may alleviate this issue, however exposure times greater than those permitted using the current software (limited to 5 s) must be accommodated. FL's low viscosity and ready detection may, however, provide an excellent tool to visually verify steady-state phreatic surfaces and model integrity.

4 CONCLUSIONS

This paper presented a method to visually trace seepage flows through a geotechnical centrifuge model. Three models of increasing geotechnical complexity were tested under stress levels representative of full-scale tailings dams. Hydraulic gradients were imposed by controlling water levels upstream and downstream of the models and internal head levels monitored via base-mounted pressure transducers. Tracing dyes were injected on reaching steady-state conditions. Three fluids were investigated: acrylic-resin ink; food-grade dye; and fluorescein. The first two fluids were readily available and all three are non-toxic. Fluorescein is regularly used in medical applications to examine the structures of micron-sized features.

Benchtop testing at 1g demonstrated the developed apparatus' ability to deliver a tracing fluid without affecting flow rates into the model. It was therefore incorporated into the main centrifuge apparatus.

Acrylic-resin ink was visible in the highly-permeable sand filters but was unable to create sufficient colour contrast between saturated and unsaturated regions in the main model body. Contrariwise, food-grade dye provided a good contrast throughout the model but its higher viscosity detrimentally affected local flow velocities and head levels: it was not suitable to trace the phreatic surface but might prove a useful tool if examining dense contaminant migration, for example DNAPLs.

Fluorescein provided excellent distinction between saturated and unsaturated regions of high permeability without affecting steady-state head levels. A previously undetected longitudinal crack within the model was also highlighted. Any fluorescence from the main body was, however, masked by brighter regions; incorporating camera and light filters and increasing image exposure times would alleviate this issue. This study therefore demonstrated the potential for fluorescein injection to identify seepage surfaces or pathways in centrifuge models.

ACKNOWLEDGEMENTS

The authors gratefully acknowledge funding awarded from the Minerals Research Institute of Western Australia (MRIWA) and from the P1087, Integrated Tailings Management Project, funded through AMIRA International by Anglo American, FreeportMcMoRan, Gold Fields, Total E&P Canada, Newmont, Shell Canada Energy, BASF, Nalco and Outotec. They also thank the Lions Eye Institute at the University of Western Australia for use of the fluorescein filters.

REFERENCES

Beckett, C. T. S. & A. B. Fourie (2016, 15–17 March). Centrifuge modelling of drawdown seepage in tailings storage facilities. In A. B. Fourie and M. Tibbett (Eds.), *Mine Closure 2016*, International Conference on Mine Closure, Perth, WA., pp. 271–284. Australian Centre for Geomechanics.

Beckett, C. T. S., A. B. Fourie, & C. D. O'Loughlin (2016). Centrifuge modelling of seepage through tailings embankments. *International Journal of Physical Modelling in Geotechnics 16*(1), 18–30.

Blight, G. E. & A. B. Fourie (2005). Catastrophe revisited — disastrous flow failures of mine and municipal solid waste. *Geotechnical and Geological Engineering 23*, 219–248.

Bussière, B. (2007). Colloquium 2004: Hydrogeotechnical properties of hard rock tailings from metal mines and emerging geoenvironmental disposal approaches. *Canadian Geotechnical Journal 44*(9), 1019–1052.

Cocjin, M., S. Gourvenec, D. White, & M. Randolph (2014). Tolerably mobile subsea foundations — observations of performance. *Géotechnique 64*(11), 895–909.

Hensley, P. J. & A. N. Schofield (1991). Accelerated physical modelling of hazardous-waste transport. *41(3)*(3), 447–465.

Keith, C. G. (1968). Fluorescence ophthalmoscopy. *British Journal of Ophthalmology 52*, 862–863.

Lee, J. K., J. Q. Shang, & S. Jeong (2014). Thermo-mechanical properties and microfabric of fly ash-stabilized gold tailings. *Journal of Hazardous Materials 276*, 323–331.

Palladini, L. A., C. G. Raetano, & E. D. Velini (2005). Choice of tracers for the evaluation of spray deposits. *Sci. Agric. (Piracicaba, Braz.) 62*(5), 440–445.

Stanier, S., J. Blaber, W. Take, & D. White (2015). Improved image-based deformation measurement for geotechnical applications. *53*, 727–739.

Timmons, D. R., C. K. Mutchler, & E. M. Sherstad (1971). Use of fluorescein to measure the composition of water-drop splash. *Water Resources Research 7*(4).

A new procedure for tracking displacements of submerged sloping ground in centrifuge testing

T. Carey, N. Stone & B. Kutter
University of California Davis, Davis, California

M. Hajialilue-Bonab
University of Tabriz, Tabriz, Iran

ABSTRACT: Measuring the displacement of a sloping ground using contact sensors such as potentiometers or LVDTs is problematic because the direction of movement is not known, and the contacting forces can either reinforce the soil or affect the measurements. This is especially true if there is a possibility that sensor attachments might move in liquefied soil. Others have used image-based methods to determine displacements but capturing clear photos of an underwater soil surface has proved challenging. This paper describes the development of a new wave suppressing window and camera setup for recording displacements of a submerged slope during earthquake-induced liquefaction. The bottom of the wave suppressing window was located beneath the water surface and acted like a glass bottom boat. Five GoPro cameras recorded movement of surface markers located on the slope. The videos were converted to displacement time histories using GEOPIV and the process described herein. The displacement time histories from the cameras is consistent with relative displacements calculated by double integration of accelerometer data and with residual displacements from before-and-after hand measurements of the surface markers. Results from this analysis have shown this method for tracking displacements is extremely accurate and can be used to better understand how liquefied slopes displace during strong shaking. The camera data in turn, lend credence to a proposed method to estimate relative displacement time histories from a hybrid of accelerometer measurements, Integrated Positive Relative Velocity (IPRV) and independently measured permanent displacements.

1 INTRODUCTION

Typically, displacement transducers (LVDT), have been used to record displacements of a liquefied soil. Although, as Fiegel & Kutter (1994) observed, these sensors can reinforce soil and resist displacements during liquefaction.

GEOPIV (e.g., White et al. 2003) has been applied extensively to determine deformation patterns from images of geotechnical models. Their typical approach is to set up a camera that views a cross section of a plane-strain model through a window. Thus, friction on the window might affect the movement of soil. Kutter et al. (2017) showed several experiments where the measured displacement on the central plane was different from that on the window boundaries.

For the present project, our goal was to measure surface displacements of curved and sloping submerged ground surfaces. This raised new issues that the target surface was not entirely in the same plane nor would remain in its initial plane. Refraction of light at the air-water interface with water waves generated by shaking would distort images. Kokkali et al. (2017), successfully tracked displacements of a submerged slope and markers using a high resolution, high speed camera, noted the water waves would form at container boundaries. During especially strong shaking, the interference from waves became so severe the markers were difficult to track.

The LEAP project (Liquefaction Experiments and Analysis Projects) is an international collaboration amongst researchers to verify and validate numerical models that predict liquefaction with centrifuge experiments, and blind prediction exercises (Manzari et al. 2015). The current phase of the project, LEAP-UCD-2017 consists of 24 experiments performed at 9 different research facilities.

This paper describes a wave suppressing window developed for the LEAP-UCD-2017 experiments performed on the 1 m centrifuge at the University of California, Davis. Five GoPro cameras were mounted on a camera and lens holder above the wave suppressing window and recorded liquefaction induced ground deformations. The recorded videos were converted to images, and using GEOPIV analysis, displacement time histories were generated. The time histories determined from the GEOPIV analysis are compared to relative displacements computed from accelerometer

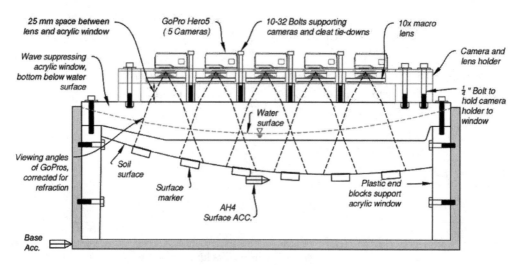

Figure 1. Elevation view of model geometry, and camera setup for UCD's LEAP-UCD-2017 centrifuge tests.

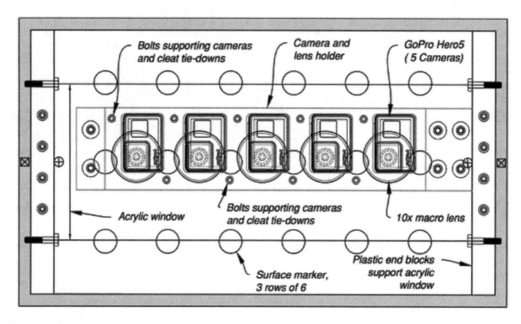

Figure 2. Plan view of model geometry, and camera setup for UCD's LEAP-UCD-2017 centrifuge tests.

data and a hybrid method that combines relative displacement, Integrated Positive Relative Velocity (IPRV) and hand measurements of markers.

2 WAVE SUPPRESSING WINDOW AND CAMERA SETUP

The wave suppressing window, and the camera setup mounted above a submerged slope are sketched in Figures 1 and 2 and photographed in Figure 3 for the LEAP-UCD-2017 test geometry. Five GoPro Hero5 cameras, each recording at 240 frames per second (fps) at 480 × 848p resolution, are mounted on an acrylic plate that also serves as a macro lens holder. 10× macro lenses were used to record sharply focused images at distances as small as 12 cm from the camera. The lenses also permit the cameras to be positioned close to the model container allowing the cameras and lenses to be rigidly secured. By affixing the cameras close to the container, the mm/pixel resolution is improved. The proposed setup consists of three main components that were designed and manufactured from clear acrylic: 1) the wave suppressing window, 2) the lens holder, and 3) the camera holder. These components are described in detail herein.

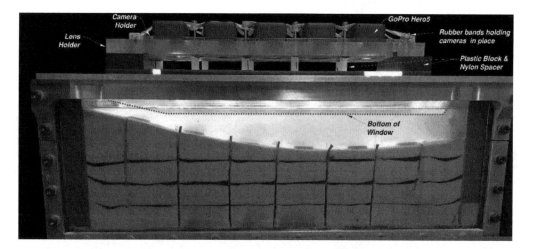

Figure 3. Photo of test geometry and camera set up on the centrifuge arm.

2.1 Wave suppressing window

The bottom of the wave suppressing window is below the water surface, which eliminates the air-water interface and image distortion caused by surface water waves, a similar concept to a glass bottom boat. The window is manufactured from 50 mm thick acrylic. 50 mm thickness was chosen for two reasons: first to control deflection of the window (due to the g-field and strong shaking), and to prevent an air gap from developing when the water surface curves in the radial g-field. A 12-degree taper was machined on the bottom left-hand side of the window (see Figure 1) to avoid interference with the sloping ground surface in this specific experiment. The tapered surface was polished for optical clarity. To alleviate dynamic water pressures, caused by sloshing of water during shaking, the wave suppressing window does not cover the entire model. The window and longitudinal walls of the container are separated by a 57 mm gap. On each end of the container are 25 mm thick plastic end blocks that support the window.

2.2 Camera setup

Figure 4 shows a detailed view of a single GoPro camera and lens. The lens holder was machined to provide a snug fit for the macro lens and prevent movement of the lens during shaking.

The GoPro cameras rest atop the lens holder plate and are retained laterally by a 3 mm thick acrylic camera holder plate. Gasket tape was applied to the holder to prevent the cameras from moving in the holder. Rubber bands (Figure 3) are laced over the cameras to hold them snuggly in the camera holder against the top of the lens holder plate.

The camera and lens holder are located 25 mm above the window, a distance determined to provide the desired field of view, accounting for refraction of the lens, and at the air/plastic/water interfaces. The experiment required 18 surface makers to be placed

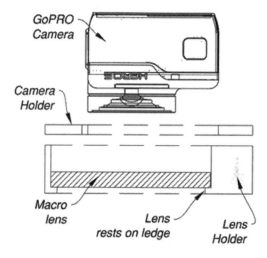

Figure 4. Detailed view of camera and lens holder.

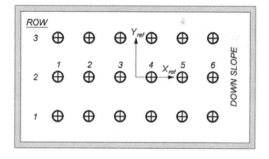

Figure 5. Surface marker locations for the LEAP-UCD-2017.

in 3 rows of 6 markers. A schematic of the locations of the 25 mm diameter markers is shown in Figure 5. The setup was designed to allow markers 2, 3, 4, and 5 in all three rows to be viewed by two cameras located up and down slope of the marker. For example, the middle camera will record the movement of markers 3 and 4, and the camera to the right will record markers 4 and 5. So, these two cameras both provide independent images of marker 4. Markers 1 and 6 are only recorded by the far left and far right cameras, respectively.

GoPro cameras were selected for their ruggedness, size, and low cost compared with other high-speed cameras. The GoPro cameras record at 240 fps, roughly corresponding to 5.5 frames per cycle of shaking of the model earthquake, which, for this experiment was a tapered sine wave with predominant frequency of 43.75 Hz, model scale. 5.5 frames per cycle of shaking was considered sufficient to quantify the cyclic component of dynamic deformation.

An instrumented calibration test was performed to confirm that the camera/lens/window mounting was rigid enough. Accelerometers were placed on the container, wave suppressing window, and camera holder measuring both vertical and horizontal components. The movements of the cameras and waver suppressing window, relative to the model container, were determined to be negligible.

3 PIV ANALYSIS

GEOPIV-RG is an image-based tool for measuring displacements and has been extensively used for geotechnical and centrifuge applications (Stanier et al. 2015). The current version of GEOPIV-RG is more precise by a factor of ten compared with its predecessors (Stanier et al. 2015), providing subpixel resolution.

Figure 6 is a typical image from one of the cameras. The gray areas around the edges of the photos are exclusion zones for the present analysis. The region that is not grayed out is the region of interest, which is divided into 150×150 pixel zones, spaced at 100 pixels. Overlapping of the zones ensures full coverage of the surface. A seed point, indicated in Figure 6, serves a fixed reference location for the zones. The seed point was selected as an LED reflection off the water-window interface. The LEDs and camera were firmly attached to the top of the suppressing window, so the LED reflection point is not expected to move relative to the camera. Analysis was performed using the Eulerian analysis mode.

Using Eulerian analysis method, GEOPIV provides the displacement of each zone in units of pixels. To scale pixels to mm, a conversion factor for each marker in each camera was found by determining the length, to the nearest whole pixel, of the surface marker from the image similar to Figure 6 prior to shaking. For example, if the 25 mm diameter surface marker in Figure 6 is 100 pixels across, the scale factor would be 0.25 pixel/mm. The typical scale factors ranged from about 0.2 to 0.25 pixel/mm for all the cameras depending on the distance between the markers and the camera. The GEOPIV analysis was performed over the entire region shown in Figure 6, but only patches corresponding to the locations of surface markers were processed. This

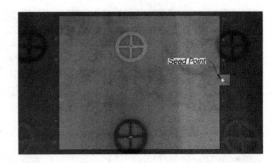

Figure 6. A typical view of a camera looking through the window. The gray area is the exclusion zone, the potions of the frame the GEOPIV software does not track displacements.

allows the use of discrete conversion factors rather than a continuous function across the image, which have been used by others (White et al. 2003) for conversion. This method of scaling introduces small errors from pixel counting.

4 RESULTS

As part of the LEAP-UCD-2017 exercise, the research team at UC Davis performed 3 centrifuge experiments, with multiple shaking events per experiment. The data presented herein is from one of those shaking events. The model consisted of a uniform Ottawa F-65 sand profile inclined at 5 degree slope. The depth of the model at the midpoint of the container is 4 m in prototype scale. The scale factor ($L^* = L_{model}/L_{prototype}$) for this experiment is 1/43.75.

Figure 7 compares the relative displacement obtained from processing accelerometer data to that from the GEOPIV analyses. The relative displacement from the accelerometers is the difference of the double integrated acceleration time histories of the base and AH4 accelerometers (Figure 1) and is reliable for use as validation of the GEOPIV analysis. Following integration, calculated displacement drift was removed by filtering the displacement time histories with a high pass filter with a corner frequency of 8.75 Hz, a process known as "baseline correction". Baseline correction of displacements obtained from accelerometers is necessary because accelerometers are insensitive to long duration, slow displacements. The displacement data from the GEOPIV analysis was also filtered using a high pass filter with a corner frequency of 8.75 Hz. The predominant frequency of the model earthquake was 43.75 Hz, so the 8.75 Hz filtering of the GEOPIV analysis and relative displacement from the accelerometers would not affect the cyclic displacements and the analysis methods could be compared in Figure 7.

Good agreement can be observed for the two methods for determining relative displacements. Surface markers 3 and 4 are the best comparison of the two methods since the surface accelerometer AH4 is located between the two markers. Markers 1 and 5 are

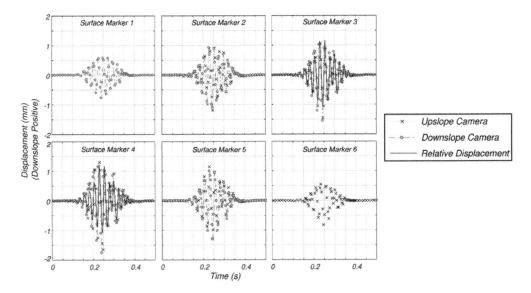

Figure 7. Relative displacement calculated from accelerometers and GEOPIV for markers 3 and 4.

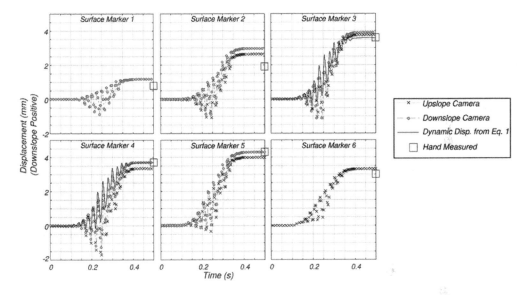

Figure 8. GEOPIV displacement time histories and displacement from IPRV and accelerometer data.

expected to move less than Markers 3 and 4, because they are closer and more restricted by the end wall boundaries of the rigid model container.

Shown in Figure 8 are the unfiltered GEOPIV displacement time series with the residual displacements. Independent hand measurements of marker displacements using rulers and calipers after stopping the centrifuge are indicated by the large square on the right side of each figure. In most cases, the residual displacement from the GEOPIV analysis agrees with the hand measurements of surface markers and are within 0.5 mm, consistent with the accuracy of the hand measurements. Also, in Figure 8, the curve labeled "From IPRV & Accelerometers" uses a process described by Kutter et al. (2017). The IPRV (Integrated Positive Relative Velocity) is defined as,

$$IPRV = \int_0^\infty \chi[v_{rel}(t)]dt \quad (1)$$

$$\text{where } \chi = \begin{cases} 0 & \text{if } v_{rel}(t) < 0 \\ 1 & \text{if } v_{rel}(t) > 0 \end{cases}$$

where v_{rel} is the relative velocity. This function produces a reasonable shape of the accumulation ramp of the permanent displacements. The IPRV ramp function is then scaled to make it agree with the hand measurements of the surface marker displacement. The "From IPRV & Accelerometers" curves were obtained by adding the cyclic displacement from the accelerometers (data shown in Figure 7) to the scaled IPRV ramp.

Overall, good agreement can be observed with GEOPIV and curves determined using accelerometer data with the IPRV ramp. It appears that the IPRV ramp predicts that displacements accumulate earlier in the time series then the PIV analysis. In other words, the first few significant cycles of displacement from the GEOPIV analysis show negligible accumulation of permanent displacement, but the IPRV ramp does accumulate residual displacements in every cycle of relative displacement. The IPRV function does not account for the effect of liquefaction on the shape of the residual displacement ramp, but nevertheless, the IPRV function does allow a reasonable visualisation of the displacement time series data that could be useful in the absence of image analysis data.

5 CONCLUSION

A new device and procedure was developed to track dynamic and residual displacements of a submerged sloping liquefied ground using the software GEOPIV. The device consisted of a wave suppressing window, functioning like a glass bottom boat, that allowed clear images to be obtained by looking vertically down at a submerged sloping surface. Attached to the window was a camera holder, with five GoPro Hero5 cameras recording at 240 fps, and a lens holder with 10x macro lenses, allowing the cameras to be located a short distance from the soil surface. Video recorded during shaking was converted to pixel displacements using GEOPIV. Pixel displacements of surface markers were converted to units of mm using a variable scale factor determined by the length of the surface markers (25 mm) and the number of pixels occupied by the surface markers. This accounted for distortion of the image and variable distance of the markers to the camera.

The cyclic components of displacements from the GEOPIV analysis were filtered to remove permanent deformations and compared well with cyclic displacements calculated from the accelerometer data. The permanent displacements from unfiltered GEOPIV displacements also compared well (usually within about 0.5 mm) to the hand measurements of displacements of surface markers. A method to generate displacement time histories from acceleration data, using the Integrated Positive Relative Velocity (IPRV) to determine the shape of the ramp defining the accumulation or permanent displacement provided a reasonable approximation to the data from GEOPIV.

It has been difficult in centrifuge model testing to measure permanent displacements of submerged liquefying slopes. The use of several cameras viewing through a wave suppressing window along with techniques described herein allows high resolution dynamic displacement data to be obtained at a multitude of points across a submerged sloping ground surface.

ACKNOWLEDGMENTS

Funding for this work was provided by the National Science Foundation under CMMI grant 1635307. The authors would also like to thank Dr. Dan Wilson and the CGM stuff for their review, advice and assistance with this work.

REFERENCES

Fiegel, G.L. and Kutter, B.L., 1994. Liquefaction mechanism for layered soils. *Journal of geotechnical engineering*, 120(4), pp. 737–755.

Kokkali, P., Abdoun, T., Zeghal, M., 2017. Physical modeling of soil liquefaction: Overview of LEAP production test 1 at Rensselaer Polytechnic Institute. *Soil Dynamics and Earthquake Engineering*.

Kutter, B.L., Carey, T.J., Hashimoto, T., Zeghal, M., Abdoun, Madabhushi, S.P.G., Haigh, S.K., Burali d'Arezzo, F., Madabhushi, S.S.C., Hung, W.-Y., Lee, C.-J., Cheng, H.-C., Iai, S., Tobita, T., Ashino, T., Ren, J., Zhou, Y.-G., Chen, Y., Sun, Z.-B., & Manzari, M.T. 2016. LEAP-GWU-2015 Experiment Specifications, Results and Comparisons. *Soil Dynamics and Earthquake Engineering*.

Manzari, M., B. Kutter, M. Zeghal, S. Iai, T. Tobita, S. Madabhushi, S.K. Haigh, L. Mejia, D. Gutierrez, R. Armstrong, M. Sharp, Y. Chen, & Y. Zhou. 2014. Leap projects: concept and challenges. In *Geotechnics for catastrophic flooding events*, pp. 109–116. CRC Press.

Stanier, S.A., Blaber, J., Take, W.A. and White, D.J., 2015. Improved image-based deformation measurement for geotechnical applications. *Canadian Geotechnical Journal*, 53(5), pp.727–739.

White, D.J., Take, W.A. and Bolton, M.D., 2003. Soil deformation measurement using particle image velocimetry (PIV) and photogrammetry. *Geotechnique*, 53(7), pp. 619–631

Identification of soil stress-strain response from full field displacement measurements in plane strain model tests

J.A. Charles, C.C. Smith & J.A. Black
Department of Civil and Structural Engineering, University of Sheffield, UK

ABSTRACT: The significant amount of image data that can be generated during physical model tests can provide a useful alternative and direct route to determining the stress-strain response characteristics of the soil used in the model without recourse to e.g. sampling and triaxial testing. This paper builds on previous work that has shown how using external loading data and Particle Image Velocimetry (PIV) derived full-field displacement data allows the use of optimisation to reconstruct the stress-strain curve of the soil body in a piecewise manner. In this paper the technique is applied to both artificial and physical plane strain test scenarios to illustrate the potential of the method and discusses some of its advantages and challenges, particularly with respect to handling imperfect or noisy physical test data.

1 INTRODUCTION

Determination of the parameters describing the constitutive behaviour of soils is typically done through a number of physical tests (e.g. 1-D compression tests, triaxial tests etc.) in which the stress and/or strain field is assumed to be simple and known. These methods assume a uniform response throughout the soil and can provide useful data to build up a constitutive model of the soil.

The development of Particle Imaging Velocimetry (PIV), first adapted for geotechnical usage by White et al. (2001), allows access to full-field displacement data of a sample undergoing testing. This process uses high resolution digital cameras along with Digital Image Correlation to track the movement of specified patches of soil based on their texture, either natural or artificially applied using floc. PIV is a non-intrusive method that provides no further disturbance to the soil sample undergoing testing.

An Identification Method (such methods are described extensively by Avril et al. 2008), is a method developed for testing solids in order to recover the constitutive parameters of the material based on recorded loading data and image derived displacement data. Although numerous such methods are described in the aforementioned paper, with varying algorithms and data requirements, the principle is that constitutive parameters are to be found such that a modelled material response, using said parameters, is to be optimally close to the physically measured material response.

Most work in the field of identification methods has been applied to metallic samples in the field of material science and focusses primarily on purely elastic material models. Although some work has been done on plastic deformation in this field, (Grédiac & Pierron 2006), the difficulties of dealing with deformations within a soil, the practicalities of measurement, and the constitutive models used to describe the process, will require novel approaches. This paper reports work in progress towards this goal.

The current work is based upon the Virtual Fields Method (VFM), described by Grédiac & Pierron (1998), which utilises full field displacement measurements, along with the principles of the conservation of energy and of virtual work to recover the parameters of a chosen constitutive model.

Work by Gueguin et al. (2015) demonstrated an Identification Method based on a simplification of VFM developed with the goal of recovering and reconstructing the stress-strain response of a soil sample undergoing loading, in addition to the constitutive parameters.

This contribution builds on the work by Gueguin et al. by investigating the issues relating to the application of the method to physical test data. Initial work indicates a number of issues that must be solved for the method to be of practical use. This contribution analyses both artificial generated data as well as physical test data such that these issues can be identified and solutions proposed.

2 AN ENERGY BASED IDENTIFICATION METHOD FOR UNDRAINED SOILS

2.1 *Principles of the identification method*

The principle of this Energy based Identification Method is that for any valid physical system the stress field σ and the strain field ϵ must satisfy the following equation:

$$W_{int}(\sigma,\varepsilon) - W_{ext} = 0 \qquad (1)$$

where W_{int} and W_{ext} represent the internal work (i.e. energy expended by deformation etc.) and external work (i.e. energy imparted by loading etc.)

Internal work W_{int} can be written as an integration of local energy across the whole field with the following equation:

$$W_{int} = \int_{area} \left[s\varepsilon_v + 2t\varepsilon_s \left(1 - 2\sin^2(\theta_\sigma - \theta_\varepsilon)\right) \right] dA \qquad (2)$$

in which θ_σ and θ_ε are the orientations of the principle stress and principle strain respectively, and in which the other terms, calculated from eigenvalues of the stress field (σ_1, σ_3) and the strain field $(\varepsilon_1, \varepsilon_3)$, are: mean stress s, deviatoric stress t, volumetric strain ϵ_v, and shear strain ϵ_s.

For the purposes of illustrating the viability of this method, application will currently be limited to monotonically loaded, plane strain tests on undrained soils, allowing a number of simplifications to be made. Volumetric strain will be taken as nil and associative flow will be assumed, allowing equation 2 to be formulated as:

$$W_{int}(\sigma,\varepsilon) = \int_{area} 2t\varepsilon_s \, dA \qquad (3)$$

2.2 Numerical implementation of the identification method

Due to the limitations of PIV, such as the discretisation of the displacement field, noise, and resolution issues, along with the inherent complexity of plastically deforming soils, it should be assumed that no stress-strain response can be found such that internal work is exactly equal to external work. As such, following discretisation into n PIV images or time-steps, equation 1 should be reformulated as:

$$W_{int}^{(j)} + localgap^{(j)} = W_{ext}^{(j)} \qquad (4)$$

and as such, the "best" stress-strain response can be found with the following conic minimisation:

$$\text{Minimize} \sqrt{\sum_{j=1}^{n_{images}} \left(localgap^{(j)}\right)^2} \qquad (5)$$

The stress-strain curve to be reconstructed will be split into a number of intervals where each interval may chosen arbitrarily by the user. Intervals distributed logarithmically between smallest and largest observed strain will be the approach adopted in this work. The

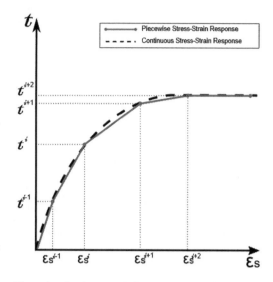

Figure 1. Continuous and piecewise stress-strain response.

number of these specified strain intervals is unrelated to the number of PIV images available. Figure 1 demonstrates the relationship between the piecewise stress-strain curve that will be obtained, and the continuous curve.

The shape of the stress-strain curve will be presupposed in order to provide constraints for the optimisation. For example, strain hardening at a progressively reducing rate can be enforced by requiring the stiffness for each interval to be less than or equal to the previous interval. Further modification of the constraints would allow the modelling of strain softening if required.

A number of minor adjustments have been made since the contribution by Gueguin et al., most notably the ability to constrain the curve to a pre-determined shape, and changing the calculation of the local gap, formerly:

$$localgap^{(j)} = \frac{W_{int}^{(j)}}{W_{ext}^{(j)}} - 1 \qquad (6)$$

to the current form shown in equation 4. The previous use of a relative energy gap rather than an absolute energy gap resulted in a weighting towards the earlier stages of the stress-strain curve.

The method as described was implemented using the Mosek (2017) mathematical optimiser in the Matlab 2016b environment.

3 VALIDATION OF IDENTIFICATION METHOD USING SIMULATED TEST DATA

By generating an arbitrary set of artificial full field displacement data, and an arbitrary stress-strain curve and corresponding force settlement curve, it is possible to test and validate the current identification method for simple test cases. Displacement data was derived

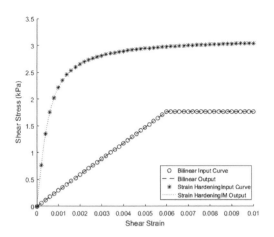

Figure 2. Comparison of input stress-strain curves with the output produced by the Identification Method for both bilinear and strain hardening cases.

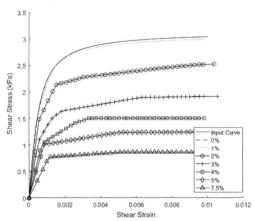

Figure 3. Effect of applying random noise to the displacement field.

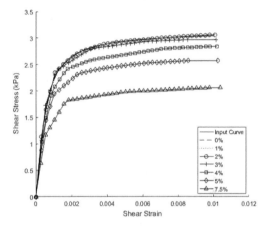

Figure 4. Effect of applying random noise to the shear-strain field. Note that the Input Curve, 0%, 1% and 2% curves effectively overlie each-other.

from a uniform shear strain field. The field was split into multiple elements to facilitate the addition of noise and to better simulate PIV data. A grid of 64 displacement nodes resulting in 98 triangular elements and 51 time steps was used, and specified force-settlement data was taken as both a linear elastic-perfectly plastic case and a case including strain hardening, modelling arbitrarily chosen peak shear strengths of 1.75 kPa and 3 kPa respectively.

For these simple inputs, in which there is no noise and very simple mechanisms, good results were obtained, as shown in Figure 2. As such, the proposed Identification Method is capable of almost exactly reproducing the expected stress-strain curves of idealised data.

Real data is highly unlikely to be perfectly uniform and noise free, as such an investigation into the effects of noise on the output curves was carried out. Figures 4 and 3 show the input and output stress-strain curves with random noise applied to the displacement field, and the strain field respectively. Random noise was applied by individually adjusting each point or element at each time step (image) by up to a specific percentage of the global average displacement or strain values.

These plots demonstrate the effect noise can have on the effectiveness of the proposed Identification Method. Adding noise to the displacement field in particular can have significant negative effects, with noise more than approximately 1% significantly changing the output stress-strain curve. In this example, the generated displacement data represents a uniform simple shear case, in which all movement is in the x-direction, with no movement in the y-direction. As such is is clear that noise, especially as it was applied to both x and y components of displacement, would radically distort such a displacement field.

Where the 'perfect' displacement field is used and instead noise is artificially applied to the shear strain field, the effect is similar but less pronounced. Each element is still shearing in the same direction, only by varying amounts.

In both cases, the result is a shallower stress-strain curve, with both a less stiff response, and lower peak stress. As stated by equation 3, internal work is a function of both the strain field and the stress field. A noisy strain field is likely to have larger changes in strain between time steps, and as such require lower values of shear stress to produce the required internal work. Work is ongoing to develop methods that can robustly deal with these issues.

4 PHYSICAL EXPERIMENTAL DATA AND DISCUSSION

4.1 Experimental background

To demonstrate the viability of the Identification Method, a preliminary physical modelling test was carried out. This test consisted of a simple footing test

Figure 5. Experimental test rig for footing test.

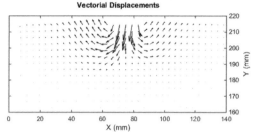

Figure 6. Example quiver plot of observed displacements as difference between first and last frame.

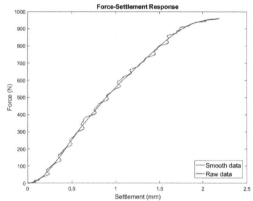

Figure 7. Load-settlement response of the footing, showing both raw and processed data.

on clay under undrained conditions, in which images of the soil were recorded for use with PIV. The test was carried out in an aluminium rig of internal dimensions 200 × 350 × 200 mm, with a perspex viewing window marked with calibration targets. The soil used was Kaolin clay mixed from powder and consolidated within the testing rig. PIV images were recorded using an off-the-shelf digital camera at a rate of 1 frame per second. The experimental setup can be seen in Figure 5.

The images were analysed using GeoPIV-RG (Stanier et al. 2015). The data was converted between image space and object space using the GeoPIV-RG in-built calibration function and the targets marked on the viewing window. As full-field displacement data is necessary for the Identification Method, a bespoke Matlab routine was used to interpolate data missing due to obstruction by calibration targets or removal of anomalies.

PIV displacement data was generated for the upper area of the model around the footing using a 34 by 16 grid of nodes for a total of 544. As such, a total of 990 triangular strain elements were generated for use with the Identification Method. As stated previously, Images were taken at a rate of 1fps, and the first 60 frames were used in the analysis described in this section. Figure 6 shows an example of the observed soil response.

A strain based load actuator was used to apply force, recorded with a load cell, to an aluminium footing of width 20 mm. The actuator moved with a velocity of 0.05 mm/s. The actuation rate was chosen to ensure undrained behaviour such that the assumption of nil volumetric strain is valid. The recorded force settlement response of the footing is provided in Figure 7.

Smoothing, using a moving average function, was applied to the footing settlement data. Significant noise was recorded in the settlement data, likely due to the resolution of the displacement gauge. This was removed such that the processed settlement data increased roughly linearly with time, as would be expected due to use of the strain based actuator.

Additionally, unconsolidated-undrained triaxial tests were carried out on the sample such that the parameters obtained via the identification method can be compared to parameters obtained using conventional methods.

4.2 Experimental results and discussion

The current Identification Method was applied to the data resulting from the described methodology. 200 shear-strain intervals were utilised based on the distribution of element shear strain values throughout the test. The interval distribution is shown in Figure 8.

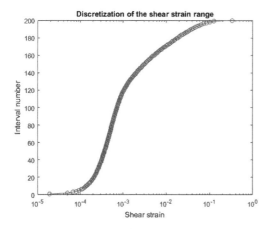

Figure 8. Distribution of strain intervals. Each interval will correspond to a point on the output stress-strain curve (Figure 1).

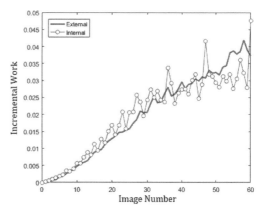

Figure 10. A plot showing how incremental internal and external work vary throughout the test. Work is measured in joules.

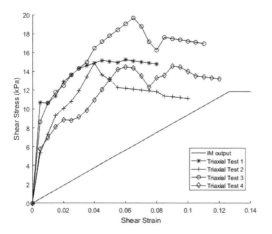

Figure 9. Comparison between the Identification Method derived stress-strain curve, and the curves obtained via a number of triaxial tests.

The output curve generated by the Identification Method was bilinear with a much shallower slope, and hence smaller shear stiffness than was expected based on the triaxial data. The comparison is shown in Figure 9.

As real PIV data was used to produce this output there is of course noise that is likely to affect the output in the same manner described in the section discussing artificial data. Of particular note is the fact that significantly higher shear strains were observed in the PIV data than were measured in the triaxial data, Figure 9 is trimmed for legibility, but the horizontal section continues on to a peak shear strain of approximately 0.34, as seen in Figure 8. The discrepancy is likely to be due to the triaxial tests being terminated either upon the development of a shear-band or simply reaching a pre chosen peak strain value, whereas the footing data continues to show higher plastic strains.

A comparison between internal and external energy throughout the test is provided in Figure 10. The values plotted are increments of work done, i.e. the additional work carried out between images. As such at no point during the test is there negative external work done to the soil, i.e. there is no unloading stage of the test.

Although unlikely due to this problem type, noise or other anomalies may cause individual elements to show negative incremental work due to a decreasing shear strain value. In the current data set this was found to be negligible, and as such, the few negative elemental incremental work values were taken as zero, though in general it is important to ensure positive work is done in all cases.

The lines for both incremental internal and external work are not smooth with excess internal work in the first half, and not enough internal work in the second half. Note that it is absolute energy differences that are minimised during the optimisation and as such the Identification Method does not try to cancel out early differences with opposite later differences.

The issues with the method could additionally stem from the decision to discount volumetric strain. Although for undrained tests, volumetric strain is generally assumed to be negligible, the PIV data indicated that this is not the case, with a peak element volumetric strain of 0.144 (compared to a peak shear strain of 0.34), and an average element volumetric strain of -0.0012 (compared to an average element shear strain of 0.0079). This will certainly affect the energy calculations, either because it is omitted in the formulation or because it has arisen due to inaccuracies in the PIV data.

The PIV mesh resolution is an additional factor that may affect the results and would impact both the volumetric and shear strain interpretation. A full analysis of mesh sensitivity has not been carried out at the time of writing but is work in progress.

Work is ongoing to determine the contribution of this effect in more detail, and in implementing procedures to robustly filter out noise without compromising accuracy. It is intended that the latest findings will be presented at the conference.

5 CONCLUSIONS

A method by which the constitutive parameters and stress-strain response of a soil sample undergoing testing can be obtained from the load-settlement data and PIV obtained full-field displacement data has been described and demonstrated using artificial data.

Initial physical data analysis has demonstrated that imperfect 'real' data can cause loss of accuracy. In general, an increase in robustness is required in order to make the leap from functioning with idealised artificial data to genuine physical modelling data. In particular noise in PIV data causing an underestimate of shear strain values, or artificial volumetric strain, can affect the energy balance calculations.

Work is ongoing to implement procedures to robustly filter out noise without compromising accuracy.

ACKNOWLEDGEMENTS

The primary author acknowledges the PhD Studentship provided by the UK Engineering and Physical Sciences Research Council (EPSRC).

REFERENCES

Avril, S., M. Bonnet, A. S. Bretelle, M. Grédiac, F. Hild, P. Ienny, F. Latourte, D. Lemosse, S. Pagano, E. Pagnacco, & F. Pierron (2008). Overview of identification methods of mechanical parameters based on full-field measurements. *Experimental Mechanics 48*(4), 381–402.

Grédiac, M. & F. Pierron (1998). A T-shaped specimen for the direct characterization of orthotropic materials. *International Journal for Numerical Methods in Engineering 41*(September 1996), 293–309.

Grédiac, M. & F. Pierron (2006). Applying the Virtual Fields Method to the identification of elasto-plastic constitutive parameters. *International Journal of Plasticity 22*(4), 602–627.

Gueguin, M., C. Smith, & M. Gilbert (2015). *Use of digital image correlation to directly derive soil stress-strain response from physical model test data*, pp. 3881–3886.

Mosek (2017). Mosek. https://www.mosek.com/.

Stanier, S. A., J. Blaber, W. A. Take, & D. J. White (2015, may). Improved image-based deformation measurement for geotechnical applications. *Canadian Geotechnical Journal 13*(October 2015), 1–35.

White, D. J., W. A. Take, & M. Bolton (2001). Measuring soil deformation in geotechnical models using digital images and PIV analysis. *10th International Conference on Computer Methods and Advances in Geomechanics*, 997–1002.

Imaging of sand-pile interface submitted to a high number of loading cycles

J. Doreau-Malioche, G. Combe, J.B. Toni & G. Viggiani
CNRS, Université Grenoble Alpes, 3SR, Grenoble INP (Institute of Engineering Université Grenoble Alpes), Grenoble, France

M. Silva
Departamento de Obras Civiles, Universidad Técnica Federico Santa María, Valparaíso, Chile

ABSTRACT: The mechanisms occurring on a micro-scale at sand-pile interface during axial displacement-controlled cyclic loading were analysed quantitatively using x-ray tomography and a grain-based approach of three dimensional digital image correlation. The tests were performed in a mini-calibration chamber using a range of values of cyclic amplitudes and number of cycles. The results were found to be consistent with those obtained in a previous study carried out in a large calibration chamber, at Laboratoire 3SR, in France. The macroscopic response of sand-pile interface showed a two-regime evolution during cycles, with a non-negligible increase of shaft resistance in the latter regime. The test conditions are not representative of real engineering applications, where piles supporting bridges, tidal or wind turbines have to safely sustain severe load-controlled cycles. However, advanced image analysis sheds light on the mechanisms controlling the macroscopic behaviour of the sand-pile interface in each regime.

1 INTRODUCTION

The mechanisms controlling the macroscopic behaviour of sand-pile interface during pile installation and cyclic loading are complex and are difficult to fully understand from field observations. Jardine & Standing (2000, 2012) reported results of multiple axial cyclic loading tests conducted on steel open-ended pipe piles driven in sand, at Dunkerque, northern France. The authors identified three kinds of responses (stable, unstable and meta-stable), depending on the mean shaft load, the shaft cyclic amplitude and the number of cycles. They also noted that a large number of low-level stable cycles can have beneficial effects on shaft capacity, whereas high-level cycles can lead to shaft failure and halving of the axial capacity within a few tens of cycles.

Numerous laboratory investigations have been reported on related topics of sand kinematics, grain crushing, local porosity changes and macroscopic interface behaviour (White & Bolton 2002, Yang et al. 2010, Silva et al. 2013, Arshad et al. 2014). The mechanisms controlling the macroscopic response of the interface were observed either *post-mortem* or in plane strain and mainly during pile installation.

The present work focuses on the sand grains behaviour in the vicinity of the pile during axial cyclic loading. The tests were performed in a mini-calibration chamber installed in an x-ray scanner. The pile was installed by jacking and submitted to 1000 displacement-controlled cycles under constant radial stress. The tests do not represent accurately real engineering applications, where piles supporting bridges, tidal or wind turbines are subjected to load-controlled cycles due to their environment. However, the macroscopic response of the sand-pile interface was compared and found to be consistent with the one obtained in a previous study using a pressurised calibration chamber (1.20 m internal diameter) conducted in the framework of a combined research effort by Laboratoire 3SR and Imperial College London (Tsuha et al. 2012, Silva 2014) in Grenoble, France. The consistency of the results shows their reproducibility for similar testing conditions at different scales of physical modelling. In addition, the combined use of three-dimensional (3D) tomographic imaging and advanced tools such as 3D-digital image correlation (3D-DIC), allows us to obtain quantitative information at the grain scale.

2 PHYSICAL MODEL

2.1 *Mini-calibration chamber and model pile*

The tests were carried out inside the x-ray scanner of Laboratoire 3SR. The mini-calibration chamber consists of a cylindrical cell transparent to x-rays in order to image phenomena while running a test (Fig. 1). The soil used for this study is Glageon sand, whose index properties are summarised in Table 1. Glageon sand is a calcareous sand derived from limestone rock crushed in Bocahut quarry, France, with $D_{50} = 1.125$ mm and a relatively uniform grading (between 1 mm and 1.25 mm). The angular and elongated shape of the grains of Glageon sand makes them highly crushable, which is the main reason they were

3 VISUALIZATION OF THE INTERFACE AND 3D-DIC

3D images were acquired throughout the cyclic loading after a certain number of cycles: 1, 50, 100, 500 and 1000 when the applied amplitude was ±0.5 mm. Restricted by the sample size, scans were taken in 'local' tomography, i.e. with a field of view focused on the tip and the shaft, for a voxel size of 40 μm (which means that there are about 25 voxels across a grain diameter). The x-ray beam was set to a tension of 150 kV and a current of 200 μA. Figure 2a shows a 3D rendering of the sand-pile interface after pile embedment.

3D fields of displacement and strain were obtained using the 3D-DIC code, TomoWarp2 (Tudisco et al. 2017). 3D-DIC is essentially a powerful tool for assessing the spatial transformation between two digital images, here tomographies. For fine sands, grains are too small with respect to voxel size and they cannot be tracked individually. However, a group of grains within a subdomain (containing about 8 grains in the present case), constitutes a speckle pattern and can be followed from one configuration to another. 3D-DIC was successfully used to study the installation of a pile in sand using the mini-calibration chamber (Silva & Combe 2014, Silva et al. 2015). The results, fully three-dimensional, showed distinct regions where the rearrangement of the grains concentrates. A 'recirculation' of sand grains was also observed close to the tip during the penetration of the pile.

In this work, a discrete version of TomoWarp2, developed at Laboratoire 3SR, was used, which allows for measuring the kinematics of each individual sand grain in the sample.

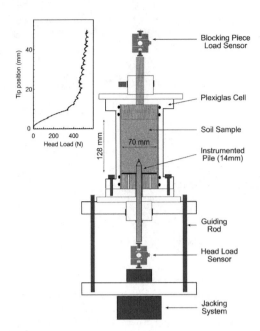

Figure 1. Schematic arrangement of the mini-calibration chamber and a typical head load profile during pile installation (top-left).

selected for this study. The samples were prepared dry with an initial relative density ranging from 70% to 90%. At this relative density, triaxial compression tests gave φ' values of 48°.

An aluminium tubular model pile was used, with an external diameter of 14 mm. The pile tip was conical (60°) and instrumented with strain gauges to record pile tip resistance. A load cell was mounted on the pile head to measure the total load (or head load) applied on the pile. Shaft resistance was estimated by subtracting the tip load from the total load. The model pile had a smooth surface with a roughness of about 0.7 μm. The roughness of sand-pile interface is known to be one of the most important factors affecting the unit shaft resistance (Fioravante 2002, Hebeler et al. 2015, Tehrani et al. 2016). Therefore, in the present study, the friction mobilised by the shaft is lower than the one measured in the field. Direct shear tests on sand-aluminium interface gave $\delta' = 15°$ whereas field piles have a roughness that leads to a typical δ' of about 30°.

2.2 Pile installation and cyclic loading

The pile was installed at a constant displacement rate of 25 μm/s until an initial embedment depth (either 50 or 60 mm) under a confining pressure of 100 kPa. For practical reasons, the pile was installed from the bottom of cell, i.e. it moves upwards. Following the installation, the pile was submitted to a thousand axial displacement-controlled cyclic loadings. The cycles were performed at the same displacement rate as for the installation with an amplitude of ±0.5 mm or ±1.0 mm, alternating between compression and tension phases (two-way cycles).

4 RESULTS AND DISCUSSION

4.1 Evolution of shaft resistance

Figure 3 shows the shaft resistance during the cycles. Two different regimes can be identified. For the first 50 to 100 cycles, shaft resistance slightly decreases (of about 15 N), whereas it increases continuously and significantly for the subsequent load cycles. When the tip is at ±0.5 mm the curves show a peak, which becomes increasingly marked with increasing number of cycles. As part of the national project SOLlicitations CYcliques sur Pieux de fondation (SOLCYP), Tali (2011) and Bekki et al. (2013) also reported an initial phase of 'cyclic softening' followed by a phase of 'cyclic hardening' for displacement-controlled cyclic loading tests carried out on a pile-probe jacked into large size samples of sand.

In the case of an amplitude of cycles of ±1.0 mm the same trends as the one obtained for ±0.5 mm can be observed but the transition between the two regimes occurs for a smaller number of cycles: between 10 to 20 cycles. These results suggest that reducing the imposed amplitude of displacement increases the number of cycles required to reach the increase in shaft

Table 1. Glageon sand properties.

Grain shape	Volumetric weight kN/m^3	D_{50} mm	Coefficient of uniformity (C_u)	e_{max}	e_{min}
Angular	26.54	1.125	1.25	1.070	0.839

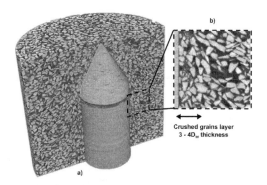

Figure 2. a) 3D rendering after pile installation inside the mini-calibration chamber. b) Zoom in the interface and detection of different phases (low grey level = pores, high grey level = grains, intermediate grey level = fines).

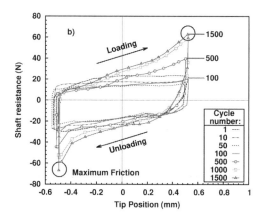

Figure 3. Mini-calibration chamber – Evolution of shaft resistance during cycles with an amplitude of ±0.5 mm for GLAG-C1 (black arrows show the loading path).

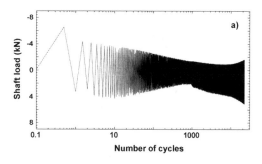

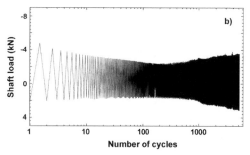

Figure 4. Large calibration chamber – Shaft load evolution during cyclic loading according to the applied amplitude of cycle (tests performed under constant confining pressure, Silva 2014): a) ±0.5 mm, b) ±1.0 mm.

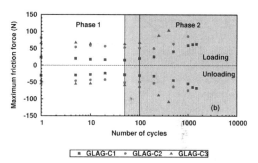

Figure 5. Mini-calibration chamber – Identification of two regimes in shaft resistance evolution (transition between 50 to 100 cycles noted by the shaded region).

resistance. From displacement-controlled calibration chamber model pile tests, Silva (2014) obtained similar results, as shown in Figure 4. The difference in the required number of cycles to pass from one regime to another is likely due to size effects regarding the ratio of the diameter of the chamber to the diameter of the pile as well as the ratio of the pile diameter to the mean particle size.

Figure 5 presents the force induced by friction measured at the peak for various tests performed on Glageon sand, in the mini-chamber, with identical testing parameters and similar relative density. The shaft resistance at the end of the cycles is twice as big as the value measured during the first cycle. The transition between the two regimes occurs at 50 to 100 cycles (shaded region in the figure).

4.2 Grain kinematics

In the following, only the results from test GLAG-C2 are shown, as the results from the other three tests are essentially the same. Typical fields of displacement from 3D-DIC are presented in Figure 6. Two pairs of

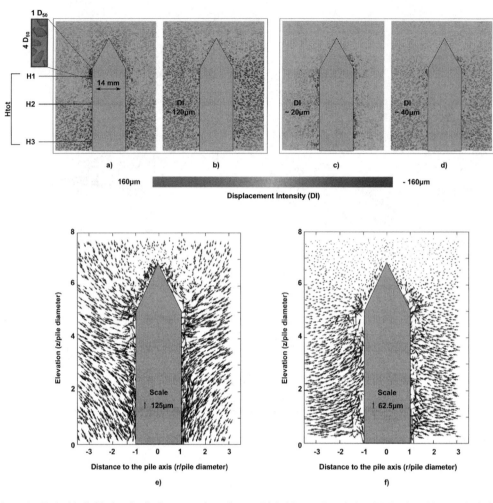

Figure 6. Typical individual grain displacements from discrete Digital Image Correlation (DIC), plotted in a vertical plane passing through the pile axis. Evolution of displacement intensity during cycles: a) vertical displacements, b) horizontal displacements between Cycle 10 and Cycle 50; c) vertical displacements, d) horizontal displacements between Cycle 500 and Cycle 1000. Individual displacement vectors e) between Cycle 10 and Cycle 50 and f) between Cycle 500 and Cycle 1000 (note that the scale is not the same for e) and f)).

3D images were analysed, one from 10 to 50 cycles and one from 500 to 1000 cycles. It should be noted that these two increments respectively fall in the first and second regime of behaviour defined in Subsection 'Evolution of shaft resistance'. Cyclic loading induces significant displacements, mainly in the horizontal direction. Grains move globally towards the shaft for both loading steps leading to a significant radial contraction (Figs 6b, d). The sand mass undergoes very small vertical displacements – less than 20 μm (Figs 6a, c). In this study, sand grains tend to slide alongside the shaft (mainly due to the low roughness of pile shaft); however, relative displacements between grains, that is, rearrangement of the grains occurs and is measured around the pile. By comparing vertical and horizontal displacement fields after 50 and 1000 cycles, two different behaviours can be observed. The first 50 cycles cause higher grain movements within a larger area around the shaft.

Individual displacement vectors for both loading increments are shown in Figures 5e, f. Ahead of the pile tip, the displacement vectors are nearly vertical and relatively minute, while around the shaft the displacement vectors have a much larger radial component. The displacement fields also reveal, in both increments, a thin layer around the shaft where the grains are highly disturbed and difficult to track from one image to another. In the first of the two increments shown, (Fig. 5e), grains are moving downwards. This is likely due to a reduction of the hoop stresses created during pile installation. This effect is erased in the later increment. More details about the grain kinematics analysis during cyclic loading can be found in Doreau-Malioche et al. 2018.

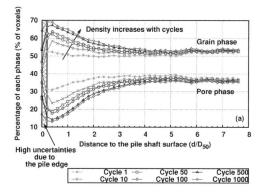

Figure 7. Local densification: radial evolution of intact grain (without fines) and pore phases during cycles for GLAG-C2 (scale: $d = 7D_{50} = 0.5 \times$ pile radius). The two first data points should not be taken into account due to the effect of the edge of the pile on the images.

4.3 *Porosity analysis*

X-ray images were analysed by studying the grey levels in order to quantitatively assess the effect of cycles on soil density (Desrues et al. 1996). The study was carried out by using the open source image processing software Fiji (Schindelin et al. 2012). Three phases were identified in the images as shown in Figure 2b: pores (low grey level), intact grains (high grey level) and fines produced by grain breakage. These fines are smaller than the pixel size, therefore, they cannot be resolved at the spatial resolution of the images. However, they can be associated with an intermediate mean grey level. Choosing an appropriate threshold, the percentage of voxels belonging to each phase was estimated within 5-voxels-thickness hollow cylinders centred on the pile axis. The radial evolution of the density of each phase during cyclic loading is plotted in Figure 7.

For one cycle, each point represents one measurement at a given distance from the shaft. The results show a significant increase in the percentage of voxels associated with intact grains (without taking fines into account) with increasing number of cycles. Close to the interface, the quantity of grains is 20% higher after 1000 cycles, which indicates a local densification at the interface. It can also be observed that the region affected is about $2D_{50}$ for Cycle 1 and $4D_{50}$ for Cycle 1000. In the rest of the sand mass, the proportion of grains and pores remains constant. This result indicates that the thickness of the band adjacent to the shaft is related to the displacements that the grains undergo and to the local shearing loading history. Yang et al. (2010) also report a shear zone $2 \cdot 4D_{50}$ 8 $\cdot 6D_{50}$ wide, after cyclic jacking installation, in their large calibration chamber tests on NE34 Fontainebleau sand ($D_{50} = 0.21$ mm). Similar to Silva et al. (2013), in their post-mortem analysis, Yang et al. (2010) suggest that the shear zone thickness grows with the vertical distance from the pile tip, especially when the installation is not monotonic, and is augmented by later static or cyclic loading.

5 CONCLUSIONS

X-ray tomography and 3D image analysis were used to study cyclic load effects on the sand-pile interface at the grain scale, within a mini-calibration chamber. Two distinct regimes were identified in the evolution of shaft resistance according to the number of applied cycles, with a non negligible increase of shaft resistance in the latter regime. For these two regimes, the measurement of grain kinematics revealed two different responses of the sand mass associated with a significant densification at the interface.

In this experimental study, the test conditions are admittedly not representative of true field conditions. In fact, some of the obtained results cannot (and should not) be directly extrapolated to field cases. For example, the positive effect of cyclic loading on shaft resistance is a clear consequence of the very specific way loading cycles are applied (full failure displacement-controlled) which is rarely the case in the field. However, the macroscopic response of the interface in the mini-calibration chamber is consistent with the one obtained in a previous study carried out within the large calibration chamber of Laboratoire 3SR. These results indicate that the size effects do not affect the main mechanisms occurring at the grain scale and controlling the macroscopic behaviour of the sand-pile interface. The small scale study offers new possibilities in terms of quantitative analysis of the behaviour of the sand grains at the interface.

A second experimental campaign, conducted under constant normal stiffness, would provide further results that could be compared to the ones obtained for field tests.

ACKNOWLEDGEMENTS

The authors gratefully acknowledge Pascal Charrier, Christophe Dano and Edward Andò, all from Laboratoire 3SR, for their contributions. Laboratoire 3SR is part of the LabEx Tec 21 (Investissements d Avenir grant agreement number ANR-11-LABX-0030).

REFERENCES

Arshad, M. F., Tehrani, S. Prezzi, M. & Salgado, R. (2014). Experimental study of cone penetration in silica sand using digital image correlation. *Géotechnique 64 (7)*, 551–569.

Bekki, H., Canou, J., Tali, B., Dupla, J. C. & Bouafia, A. (2013). Evolution of local friction along a model pile shaft in a calibration chamber for a large number of loading cycles. *Comptes Rendus Mcanique 341 (6)*, 499–507.

Desrues, J., Chambon, R., Mokni, M. & Mazerolle, F. (1996). Void ratio evolution inside shear bands in triaxial sand specimens studied by computed tomography. *Géotechnique 46 (3)*, 529–546.

Doreau-Malioche, J., Combe, G., Viggiani, G. & Toni, J. B. (2018). Shaft friction changes for cyclically loaded displacement piles: an x-ray investigation. *Géotechnique Letters*, DOI: 10.1680/jgele.17.00141.

Fioravante, V. (2002). On the shaft friction modelling of non-displacement piles in sand. *Soils Found. 42 (2)*, 23–33.

Hebeler, G. L., Martinez, A. & Frost, J. D. (2015). Shear zone evolution of granular soils in contact with conventional and textured CPT friction sleeves. *KSCE J. Civil Engng. 20 (4)*, 1267–1282.

Jardine, R. J. & Standing, J. R. (2000). Pile load testing performed for HSE cyclic loading study at Dunkirk, France, Vols 1 and 2. *SOffshore Technology Report OTO 2000 007* London, UK: Health and Safety Executive, two volumes 60p, 200p.

Jardine, R. J. & Standing, J. R. (2012). Field axial cyclic loading experiments on piles driven in sand. *Soils Found. 52 (4)*, 723–737.

Schindelin, J., Arganda-Carreras, I. & Frise, E. (2012). Fiji: an open-source platform for biological-image analysis. *Nature methods 9 (7)*, 676–682, PMID 22743772, doi:10.1038/nmeth.2019.

Silva, M. (2014). Experimental study of ageing and axial cyclic loading effect on shaft friction along driven piles in sand. *PhD Thesis*, Université de Grenoble, France.

Silva, M. & Combe, G. (2014). Sand displacement field analysis during pile installation using x-ray tomography and digital image correlation. *International Symposium on Geomechanics from Micro to Macro, Cambridge, UK*, CRC Press/Balkema, vol. 1, pp. 1599–1603.

Silva, M., Combe, G., Foray, P., Flin, F. & Lesaffre, B. (2013). Postmortem Analysis of Sand Grain Crushing From Pile Interface Using X-ray Tomography. In *AIP Conf. Proc. Powders and Grains, Sydney, Australia*. UNSW, vol. 1542, pp. 297-300.

Silva, M., Doreau-Malioche, J. & Combe, G. (2015). Champs cinématiques dans un sable lors de l'enfoncement d'un pieu par tomographie RX: comparaison des corrélations numériques continue et discrète. *22 me Congrès Français de Mécanique*, Lyon, France.

Tali, B. (2011). Comportement de l'interface sols-structure sous sollicitations cycliques: application au calcul des fondations profondes. *PhD Thesis*, Université de Paris Est, France.

Tehrani, F. S., Han, F., Salgado, R., Prezzi, M., Tovar, R. D. & Castro, A. G. (2016). Effect of surface roughness on the shaft resistance of non-displacement piles embedded in sand. *Géotechnique 66 (5)*, 386–400.

Tsuha, C. H. C., Foray, P. Y., Jardine, R. J., Yang, Z. X., Silva, M. & Rimoy, S. (2012). Behaviour of displacement piles in sand under cyclic axial loading. *Soils Found. 52 (3)*, 393–410.

Tudisco, E., Andò, E., Cailletaud, R. & Hall, S. A. A local Digital Volume Correlation code. *SoftwareXC 6*, 267–270.

White, D., & Bolton, M. (2002). Observing friction fatigue on a jacked pile. In Springman S. M. (ed.) *Constitutive and Centrifuge Modeling: Two Extremes*, Rotterdam/Balkema, pp. 347–354.

Yang, L., Jardine, R., Zhu, B., Foray, P. & Tsuha, C. (2010). Sand grain crushing and interface shearing during displacement pile installation in sand. *Géotechnique 60 (6)*, 469–482.

Image capture and motion tracking applications in geotechnical centrifuge modelling

P. Kokkali
WSP USA, New York, NY, USA

T. Abdoun
Rensselaer Polytechnic Institute, Troy, NY, USA

A. Tessari
University at Buffalo, Buffalo, NY, USA

ABSTRACT: Over the past decades centrifuge use in geotechnical physical modeling has grown significantly. The physical modeling community has coupled advanced imaging technologies and sophisticated analysis tools to illuminate soil behavior within complex experiments. Visualization of soil settlements, soil-structure system interaction and underground failure surfaces has provided novel insight to researchers and practitioners. The geotechnical centrifuge facility at Rensselaer Polytechnic Institute (RPI) is equipped with several miniature high definition cameras and one overhead high-speed camera. The on-board cameras were previously used to monitor the progress of an experiment and ensure the safe operation of in-flight equipment. Motion tracking software has transformed the cameras role from monitoring to data acquisition. During highly dynamic tests, the high-speed camera can be employed to acquire high frame rate videos of moving soil or objects. This paper describes some of the applications of image capture and analysis techniques in RPI experiments.

1 INTRODUCTION

Over the past decades the advances in geotechnical physical modeling have been remarkable and the use of physical modeling for validation and verification of numerical methods has grown significantly. Advanced sensor technology, improved event simulation, and sophisticated analysis tools have been introduced to the physical modeling community and researchers have been able to conduct complicated experiments that in turn have provided valuable insights to practitioners. The development of new image capture and processing technologies has significantly contributed to this field. High spatial or high temporal resolution cameras are often used to visualize soil settlements, soil-structure system interaction and underground failure surfaces and illuminate soil behavior. Experimental techniques for image analysis of geomaterials such as transparent soils have considerably advanced and methods such as infrared thermography and X-ray computed tomography are widely used for identification of material properties (Liang et al. 1997, Welker et al. 1999, Zhang et al. 2006, Zhang et al. 2009, Hall et al. 2010, Iskander & Liu 2010, Liu & Iskander 2010, Zandomeneghi et al. 2010, Stanier et al. 2012).

The geotechnical centrifuge facility at RPI has often employed image analysis methods to acquire soil and structure movements in experimental configurations where traditional sensors cannot provide the desired information. The facility is equipped with several miniature high definition cameras and one overhead high-speed camera. The on-board cameras were previously used to monitor the progress of an experiment and ensure the safe operation of in-flight equipment. Sophisticated motion tracking software has transformed the cameras role from monitoring to data acquisition. During highly dynamic tests, such as earthquake simulation or explosive testing, high-speed cameras can be employed to acquire high frame rate videos of moving soil or objects. This paper describes some of the applications of image capture and analysis techniques during static and dynamic event simulations.

2 IMAGE DATA ACQUISITION AT RPI

In addition to traditional instrumentation commonly used in centrifuge testing, the RPI Centrifuge Facility is equipped with several types of advanced sensors. Among them are miniature high definition cameras, high-speed cameras, and motion tracking software. These have intrinsic properties that prevent them from interfacing with traditional data acquisition (DAQ) systems.

The on-board high definition cameras are small and can be easily mounted to various locations on the centrifuge platform or the centrifuge model. The on-board cameras currently used at RPI record 1280 by 720 pixels at 30 frames/sec, however, the system is flexible and can be adapted to incorporate next generation equipment as it becomes available. The cameras on the centrifuge contain lenses that have moderate barrel distortion. Therefore, the recorded videos need to undergo a correction procedure. In order to correct for lens distortion and field of view perpendicularity, the camera is focused on a grid placed in front of the area of interest and a short video is recorded. Using external software, correction parameters are manually modified until the grid is corrected back to square. Corrected videos can then be exported for motion tracking.

The high-speed camera is often used during highly dynamic experiments such as earthquake simulation and explosive testing. Since typical earthquake simulations contain frequencies ranging from 10 to 350 Hz in centrifuge model scale a sampling capacity of 100 to 3000 frames per second is necessary in order to prevent aliasing of the data. This provides a smooth mapping of targets within every cycle. The high-speed camera at RPI is a Phantom v5.1 HI-G manufactured by Vision Research. It can record videos at 1200 frames per second at a full resolution of 1024 by 1024 pixels. The frame rate may be increased up to 95,000 frames per second by lowering the resolution of the video stream. It contains 8 GB of non-volatile internal memory and can capture 6.67 seconds at maximum resolution and frame rate.

The camera is permanently fixed using a mounting system that can be installed on the centrifuge boom cross-member and provides an overhead view of the model that is being tested. The hinges connecting the basket to the boom isolate the camera from vibrations. For high intensity events, such as explosive testing, the mounting system can be fitted with a high-strength transparent shield. The high-speed camera constantly records an 8 second rolling buffer. In many instances, external triggering of the camera is preferred. In that case, the earthquake simulator produces a motion, which is being recorded by the data acquisition system (DAQ). Once a programmable threshold is exceeded by a selected sensor monitored by the DAQ (for example an acceleration level), a trigger signal is sent to the high-speed camera, which then records the buffered video for a specified duration before and after the trigger.

Structured lighting is a critical component of video analysis. The illumination should be constructed with the intent to maximize contrast between the targets and the surrounding soil. Camera sensors consist of a matrix of light measuring pixels, separated into the 3 primary colors with color filters. Video tracking software is analyzing a matrix of intensities of these primary colors and tracking the displacement of the intensities. The ideal target provides high contrast with the surrounding colors and should be black or one of the primary colors. The purpose of lighting is to maximize light reflected from the desired targets, and to minimize the relative level of light from reflections and unintentional light sources. A high intensity light source provides a high signal to noise ratio for video tracking software, a critical component for both tracking speed and accuracy.

The videos recorded by the on-board cameras and the high-speed camera are analyzed with the motion tracking software TEMA Automotive Lite 3.5-016 by Image Systems. The video from the on-board cameras must first undergo a corrective procedure before being loaded into the tracking software. The tracking software operates by comparing the boundaries of objects between multiple frames. The contrast information present in the camera video data is used to develop the boundaries. Therefore, the color of the tracking targets in the soil or on the structures is selected so that the contrast is maximized. The targets tracked by the software produce data in terms of pixels. The fixed distance between two points is used to calculate a scale factor that correlates pixels to model distance. The software can produce displacement, velocity, and acceleration time histories for the points in both the vertical and horizontal axes.

3 IMAGE ANALYSIS IN SIMULATION OF NATURAL HAZARD EFFECTS ON FLOODWALLS IN CLAY

The following application of image analysis is for use in long-term consolidation and settlement analysis and tracking of the undrained behavior of clay soils during extreme loading conditions. The test described herein was conducted in collaboration with the United States Army Corps of Engineers (USACE) and research partners at the Virginia Polytechnic and State University (VT), as part of a program exploring the effects of flood loading on pile-founded concrete floodwalls through centrifuge experiments and numerical simulations.

The experimental set up shown in this paper consisted of a floodwall without pile foundation elements. The floodwall rested on a clay layer of Kaolinite ASP-600 overlaying a small drainage layer. The dimensions of the rigid container used in testing represent a 60L × 16.5W × 9H prototype-meter deposit of soil with a floodwall spanning the width of the container. In-flight consolidation of the clay layer was simulated for several prototype years in order for the clay layer to obtain the desired strength properties before the application of flood loading. The hydraulic loading (undrained loading phase) was performed using a membrane containment system on one side of the model (flood side). The water level on the flood side of the system was gradually increased over the course of the undrained loading phase.

One long side of the container was composed of acrylic and permitted visual observation of the soil strata. Tracking of a clay material using conventional PIV techniques would not be appropriate due to its continuous appearance at the macroscopic level. Therefore, targets were installed in the clay during construction on the clear face of the container to allow for image tracking and analysis during consolidation and hydraulic loading.

The principal challenges in obtaining accurate data from these experiments are due to the intersection between the physical boundaries of the model, lighting, and wide-angle lens distortion of the recorded images. The aforementioned techniques to address proper lighting and camera distortion are likewise complicated by the presence of thick acrylic, which easily reflects point light sources and causes exposure issues in the recorded images. The testing program for consolidation and long-term settlement extended up to 16 model-hours of continuous measurement. The required lighting source needed to provide a wide spectrum of visible light at an appropriate intensity while minimizing glare. It also needed to satisfy heat dissipation requirements over the long-term use. Although LEDs rapidly transition between the on-off states as previously discussed, this inherent operational behavior is only an issue when the exposure time is limited as in high-speed applications. Diffused LEDs arrays were used in these experiments and were positioned such that reflections were outside the frame of the recording camera.

Figure 1 shows a cross-sectional view of the model and the displacement targets before and after the experiment. Data was recorded for both the consolidation phase and the hydraulic loading phase. The initial tracking target positions are indicated by red squares and are only present where the start and finish positions of the targets are different. An overview of the motion tracing data is shown in Figure 2 in terms of lateral versus depth displacement coordinates over the entire course of the hydraulic loading phase. The failure zone extended to a depth of approximately 6 prototype meters, as interpolated between rows of the displacement targets. The displacement at each individual stage was also used to identify the transition between low to high strain and the corresponding characteristics.

The data sets from these experiments were used to calibrate a numerical model using the modeling-of-a-model technique. Indicatively, the response of the calibrated numerical model is depicted in Figure 3 in terms of lateral displacement at a constant depth under the floodwall. The numerical model was calibrated for different ratios of E/S_u (Modulus of Elasticity to Undrained Shear Strength Ratio) to match the centrifuge tracking data. Data from this and future experiments, which included a wide variety of loading conditions and foundation elements, used this methodology as the base case for identifying and quantifying the effects of those additional variables. The numerical modeling is further discussed in Varuso (2011) and the physical modeling results are described in Tessari (2012).

4 IMAGE ANALYSIS IN EARTHQUAKE SIMULATION AND LIQUEFACTION ANALYSIS

The following section describes the application of image analysis in the simulation of an earthquake

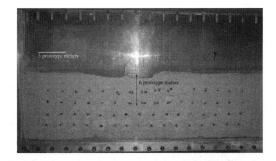

Figure 1. Image analysis in simulation of natural hazard effects on floodwalls in clay: Before (red markers) and after comparison of the position of the soil due to hydraulic loading without foundation elements.

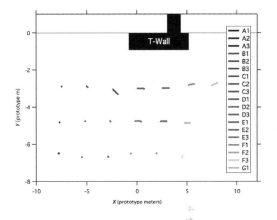

Figure 2. Lateral versus vertical displacement data of tracking targets obtained through software processing (A1 target on the top left corner of the model, F3 target on the bottom right corner).

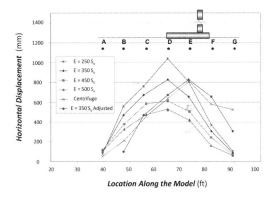

Figure 3. Comparison of the physical modeling results with potential numerical modeling analogs and the resulting calibration.

event. The centrifuge test that is presented was conducted in the framework of the Liquefaction Experiments and Analysis Project (LEAP). LEAP aims at providing numerical modelers with high quality experimental data that can be used in a series of prediction exercises. In early 2015 several centrifuge tests were conducted at different facilities all over the world and were followed by numerical simulations. The centrifuge tests performed at the Centrifuge Facility at RPI employed advanced modeling techniques such as the high-speed video acquisition (Kokkali 2017).

The centrifuge test aimed at exploring the seismic response and liquefaction potential of a sloping deposit of 5 degrees. The test was performed in a one-dimensional laminar container simulating a deposit of 20L×8W×4.875H prototype meters. The instrumentation plan included traditional sensors such as pore pressure transducers and accelerometers. Surface markers (modified zip tie heads) were distributed along the soil surface and served as tracking targets. The model was subjected to 5 shaking events. All motions were ramped sine waves of prototype frequency of 1 Hz and various amplitudes. The shaking was parallel to the centrifuge axis and in the direction of the 5-degree slope. Data from the motion with maximum acceleration amplitude equal to 0.15 g are shown in this section.

The high-speed camera was used to record videos at a sampling rate of 1000 frames per second. A lighting system consisting of halogen bulbs was safely mounted on the centrifuge basket to enhance the quality of the video recording. A top view of the model container on the centrifuge at 1g is provided in Figure 4. The high-speed camera and the mounting system are shown in the top right corner. The blue targets labeled E0, E1, etc. were located across the middle of the slope, while targets A, B, C, D, E0, F, G, H, I were located along the centerline of the slope. Positive displacement signifies motion towards the top of the slope, while negative displacement shows motion towards the bottom of the slope.

Figure 5 provides a comparison of acceleration time histories obtained through traditional accelerometers and image analysis. In specific, Figure 5a compares the input motion as recorded by an accelerometer at the bottom of the container to the acceleration recorded at the top of the container (tracking target at the top of the rigid container 8.3 m above the base of the model). The compared curves exhibit very good agreement in amplitude and frequency. Figure 5b shows a similar comparison of the tracked accelerations at the soil surface (marker E0) and an accelerometer measurement close to soil surface at 0.6 m beneath the marker E0. The acceleration at the soil surface follows the trends established by the surface accelerometer with slight attenuation of the larger strain spikes.

Figure 6 provides an overview of the surface acceleration, velocity, and lateral displacement at the center of the model container obtained through tracking of the displacement of the surface target (E0). The acceleration and velocity time histories are derived from the displacement time history directly through the tracking software. The acceleration time history follows the response close to the soil surface, the velocity time history is in agreement with the input motion, and the displacement time history indicates permanent soil deformation towards the bottom of the slope.

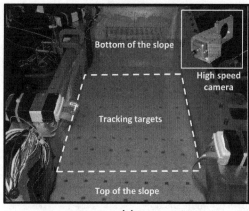

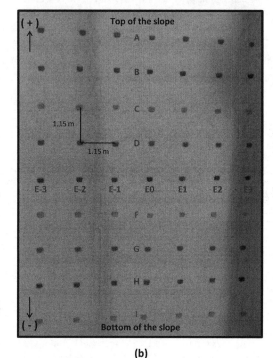

Figure 4. Image analysis in earthquake simulation and liquefaction analysis: (a) Top view of LEAP model on the centrifuge before testing, (b) high speed camera field of view and tracking targets.

The readers should refer to Kokkali et al. (2017) for additional data and details on the movement of the soil surface.

5 SUMMARY AND CONCLUSIONS

This paper presented two applications of image capture and analysis techniques in RPI centrifuge experiments as a means to obtain information about the motion of a soil mass.

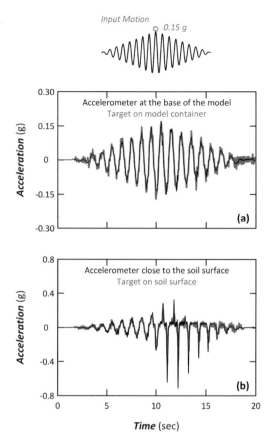

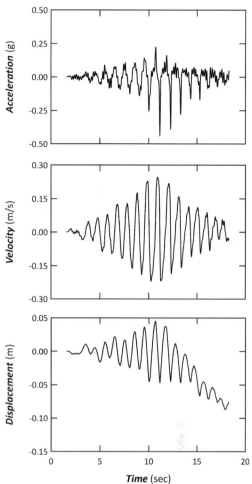

Figure 5. Validation of the image analysis method using the high-speed camera: (a) comparison of acceleration time histories recorded at the base of the rigid container and obtained through image tracking of a surface marker on the top of the container, (b) comparison of acceleration time histories recorded in the center of the container close to the soil surface (top accelerometer in the middle of the container) and obtained through image tracking of the surface marker E0.

Figure 6. Acceleration, velocity and lateral displacement at the center of the soil surface (target E0) obtained through image tracking.

The use of in-flight cameras to track long-term consolidation and settlement in addition to short-term undrained flood loading provided valuable stress-strain relationships that enhanced the understanding of the physical modeling data while enabling meaningful numerical modeling of the results. The application of image analysis in the simulation of an earthquake event provided a plethora of displacement, velocity and acceleration information at the soil surface that would be challenging to obtain through traditional sensing techniques.

The results and efficiency of the image capture and analysis method encourage the physical modeling community to apply image analysis in more complex simulations.

ACKNOWLEDGMENTS

The authors would like to thank the USACE and the LEAP Participants. The support of the Centrifuge Facility Staff and undergraduate researchers at RPI is gratefully acknowledged.

REFERENCES

Hall, S. A., Bornert, M., Desrues, J., Pannier, Y., Lenoir, N., Viggiani, G., & Bésuelle, P. 2010. Discrete and continuum analysis of localized deformation in sand using X-ray μCT and volumetric digital image correlation. *Geotechnique*, 60(5), 315–322.

Iskander, M., & Liu, J. 2010. Spatial deformation measurement using transparent soil. *Geotechnical Testing Journal*, 33(4), 1.

Liang, L., Saada, A., Figueroa, J. L., & Cope, C. T. 1997. The use of digital image processing in monitoring shear band development. *ASTM geotechnical testing journal*, 20(3), 324-339.

Liu, J., & Iskander, M. G. 2010. Modelling capacity of transparent soil. *Canadian Geotechnical Journal*, 47(4), 451-460.

Kokkali, P., Abdoun, T. & Zeghal, M. 2017. Physical modeling of soil liquefaction: Overview of LEAP production test 1 at Rensselaer Polytechnic Institute. Soil Dynamics & Earthquake Eng., doi.org/j.soildyn.2017.01.036.

Stanier, S. A., Black, J. A., & Hird, C. C. 2012. Enhancing accuracy and precision of transparent synthetic soil modelling. *International Journal of Physical Modelling in Geotechnics,* 12(4), 162-175.

TEMA Automotive Lite 3.5-016 [Computer software]. Sweden, Image Systems.

Tessari, A. 2012. Centrifuge Modeling of the Effects of Natural Hazards on Pile-Founded Concrete Floodwalls. Doctoral Dissertation, Rensselaer Polytechnic Institute, Troy, NY.

Varuso, R.J. 2010. Influence of Unstable Soil Movement on Pile-Founded Concrete Floodwalls and a Resulting Design Methodology. Doctoral Dissertation, Louisiana State University, Baton Rouge, LA.

Welker, A. L., Bowders, J. J., & Gilbert, R. B. 1999. Applied research using a transparent material with hydraulic properties similar to soil. *ASTM geotechnical testing journal,* 22(3), 266–270.

Zandomeneghi, D., Voltolini, M., Mancini, L., Brun, F., Dreossi, D., & Polacci, M. 2010. Quantitative analysis of X-ray microtomography images of geomaterials: Application to volcanic rocks. *Geosphere,* 6(6), 793–804.

Zhang, G., Liang, D., & Zhang, J. M. 2006. Image analysis measurement of soil particle movement during a soil–structure interface test. *Computers and Geotechnics,* 33(4), 248–259.

Zhang, G., Hu, Y., & Zhang, J. M. 2009. New image analysis-based displacement-measurement system for geotechnical centrifuge modeling tests. *Measurement,* 42(1), 87–96.

A study on performance of three-dimensional imaging system for physical models

B.T. Le
Ho Chi Minh City University of Transport, Ho Chi Minh City, Vietnam (formerly City, University of London, London, UK)

S. Nadimi
University of Leeds, Leeds, UK, (formerly City, University of London, London, UK)

R.J. Goodey & R.N. Taylor
City, University of London, London, UK

ABSTRACT: A study by Le et al. (2017) reported the application of computer vision techniques structure from motion (SfM) and multi-view stereo (MVS) to measure three-dimensional soil displacements at the surface of physical models. However, little information exists on the significance of the camera resolution and the number of images to the measurement performance. This study assesses the measurement performance of the SfM-MVS, provided by an open source software Micmac, with input images taken by two different types of camera including DSLR (18Mega-pixel) and mobile phone cameras (12Mega-pixel). Rigorous quantifications were carried out to examine the precision of the image analysis, in measuring vertical and horizontal displacements, over a region of interest of 420×200 mm. The measurement precision, achieved by different numbers of images, ranged from 0.06 mm to 0.03 mm. The results from this paper can be useful for researchers to select appropriate camera that satisfies their measurement requirements.

1 INTRODUCTION

The three-dimensional imaging system in this research combines the computer vision technique "structure from motion" and "multi-view stereo" (SfM-MVS) delivered by the open source software Micmac (Galland et al. 2016) with 2D PIV (Stanier et al. 2015) to measure displacements in three dimensions (3D).

SfM is a technique from computer vision and photogrammetry field that analyses input images to produce a high quality 3D point cloud (Ullman 1979). It has wide range of applications on both large and small scales. For large scale, Smith et al. (2015) reported SfM-MVS were used in 3D topographic surveys, monitoring glacier movements, observing and tracking lava movements and landslide displacements. For small scale such as experiments in laboratory, Galland et al. (2016) and Le et al. (2016) used SfM-MVS to measure three-dimensional displacements.

Simply put, SfM-MVS processes a minimum of two images to reproduce the 3D point cloud of the observed scene. The SfM-MVS algorithm is described in detail by Robertson & Cipolla (2009) and Le et al. (2016). The fundamental principles are illustrated in Figure 1 and described briefly below.

1.1 Structure from Motion

Firstly, the identical features known as keypoints in each image are detected and assigned with a unique

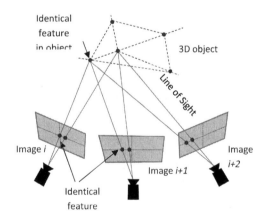

Figure 1. Structure from Motion (SfM) principle (Le et al. 2016).

identifier. An identical feature is a set of pixels that are invariant to changes in scale and orientation and can be detected in other images. The feature detection used in this research is the Scale Invariant Feature Transform (SIFT) algorithm (Lowe 2004).

The locations of the features in multiple images determined in the first step are used in a process named bundle adjustment to estimate the parameters of the scene including individual positions of the cameras, orientation of the cameras, intrinsic camera parameters and relative locations of the features in object space. Images taken from different positions add more data to the bundle adjustment process which improve the precision of the parameters estimated (Triggs et al. 1999) which is confirmed later in this paper.

1.2 Multi-View Stereo image matching

The 3D point cloud obtained from SfM has a coarse density as only identical features were included. Additional matching algorithm named Multi-View Stereo (MVS) is normally carried out after SfM which can increase the density of the 3D point cloud by at least two orders of magnitude (Furukawa & Ponce 2010, James & Robson 2012, Smith et al. 2015). During MVS analysis, most of noise data points (outliers) will be removed.

1.3 Georeference

After the MVS step, the obtained 3D point cloud is in image space (i.e. unit: pixel) and needs to be transformed to object space (e.g. unit: mm) by the Georeferencing process. Basically, georeferencing process uses the provided positions of the Ground Control Points, in image and physical space, to transform the 3D point cloud to physical space. The minimum number of Ground Control Points is three but more points provide better precision in the transformation process.

2 REVIEW ON SFM-MVS TECHNIQUE AND MICMAC SOFTWARE

There are several software that feature SfM-MVS technique such as commercial software Agisoft Photoscan and free software Bundler (Snavely et al. 2006), VisualSfM and Micmac. Smith et al. (2015) reported that MicMac, with sophisticated self-calibration camera models, outperformed Agisoft Photoscan and VisualSfM.

Galland et al. (2016) used Micmac to analyse four images taken by 24 Mega-Pixel cameras and the achieved measurement precision was 50μm. Similarly, Le et al. (2016) reported a precision of 50μm was achieved by using Micmac to analyse three images, taken by three 2Mega-pixel cameras.

Despite the technique SfM&MVS and the software Micmac were known to be able to produce high quality 3D point clouds, there was no guidance on the effects of the camera resolution and the number of images to the measurement performance. This paper aims to provide a clearer insight into these two factors.

3 EXPERIMENT AND RESULTS

Two cameras used in the following experiments are Canon EOS 700D DSLR Camera (18Mega-pixel sensor, 18-55mm lens) and Iphone 6s (12Mega-pixel sensor, 4mm focal length lens).

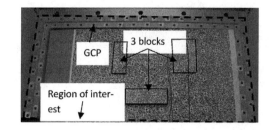

Figure 2. Experiment set up to determine precision of vertical measurement.

3.1 Experiment setup to determine precision of vertical measurement

Figure 2 illustrates the set up that includes a reference plate with 59 Ground Control Points (GCPs) and three blocks with known heights obtained from micrometer. A paper sheet with speckle texture was fixed to the observed flat objects to aid the SfM-MVS process. This is because SfM-MVS requires textures to detect identical features as plain surfaces can not be distinguished.

Five photos were sequentially taken by each camera for image processing purpose. For each camera, there were three different analyses using three, four and five photos to investigate the effect of number of images, in addition to the effect of the camera resolutions, to the performance of the measurements.

The precision of the measurement is determined by comparing the heights of the blocks in the 3D point clouds with the known heights as described by the value mean absolute error (MAE) (Equation 1);

$$MAE = \frac{1}{n}\sum_{i=1}^{n}\left(\left|H_i^{3D} - H_i^M\right|\right) \quad (1)$$

where

H_i^{3D} is the height of block i in the 3D point cloud,

H_i^M is the height of block i measured by a micrometer,

n is the number of the objects, for the experiment for vertical measurements n = 3 blocks.

Tables 1 and 2 shows the performance of the measurement, in terms of number of data points and precision, using photos from the two different cameras. Over a ROI of 420×200 mm, the number of data points obtained from the DSLR camera with a 18 Mega-pixel sensor was approximately 1.5 times more than that for the phone camera with a 12 Mega-pixel sensor (Table 1). This is thought to be analogous with the ratio of the number of pixels in the sensors of the two cameras. Interestingly, the number of the images does not have considerable impact to the number of data points.

The best precision was approximately 30μm achieved in analyses utilised five photos for both cameras (Table 2). For analysis that utilised only three or four photos, the precision was decreased to approximately 50-60μm for both cameras. All the experiments, apart from the abnormally high MAE in iPhone 6s with four photos, shows that more images

Table 1. Number of data points.

Camera	Number of photos		
	3	4	5
DSLR Canon (18Mega-pixel)	2,645,196	2,117,084	2,114,078
IPhone 6s (12Mega-pixel)	1,345,463	1,351,630	1,348,967

Table 2. Mean absolute error in vertical direction (Unit: μm).

Camera	Number of photos		
	3	4	5
DSLR Canon (18Mega-pixel)	57	55	30
IPhone 6s (12Mega-pixel)	44	61	32

yielded higher measurement precision. This is because more images are beneficial for the camera calibration process to determine more precise camera parameters.

The effect of camera resolution seems to be not significant to the measurement precision as IPhone 6s (12Mega-pixel) and DSLR Canon (18Mega-pixel) yielded similar precisions which are in line with the results reported by Galland et al. (2016) and Le et al. (2016).

It can be seen that camera resolution is not the only factor that governs the quality of images and hence the quality of 3D point clouds. Even though high resolution images enable more identical keypoints to be detected and corresponded in SfM&MVS process, the quality of lens is also important. If a low quality lens is used with a high resolution camera, then the obtained image may decrease the quality of the 3D point clouds (Furukawa & Hernández 2015). The lens also controls the depth of field and the sharpness of the image across the whole field of view.

Other factors that also needs to be taken into account when considering the quality of the camera parameters and SfM-MVS process are the type of the camera sensor that affects the noise in the obtained image.

CCD sensors are known to be able to capture high quality images with low noise but they are more expensive than traditional CMOS sensors which are susceptible to noise. However, recent developments in imaging technology allow new CMOS sensors to capture images with comparable quality to those obtained by CCD sensors but at a more affordable price.

The EMVA (European Machine Vision Association) data of the sensors are normally available and is useful for comparison on the characteristics of the sensors. In addition to sensor types, pixel size and sensor size are also important factors. Larger sensor size and pixel size allow larger amount of light into the sensor hence better quality images.

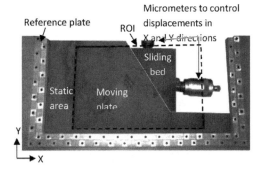

Figure 3. Experiment set up to quantify horizontal displacement measurement precision (after Le et al. 2016).

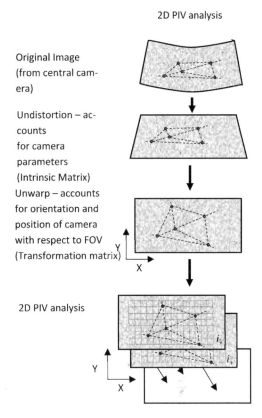

Figure 4. PIV analysis procedure to determine horizontal displacements (after Le at al. 2016).

3.2 *Experiment setup to determine precision of horizontal measurement*

Figure 3 presents the experiment set up that comprises Ground Control Point and a ROI which can be displaced in a precise manner using a sliding bed controlled by two micrometers.

Four experiments have been conducted using the controlled movement of a sliding bed to estimate the measurement precision in the horizontal directions. In each experiment, the sliding bed was moved by 1 mm

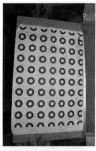 
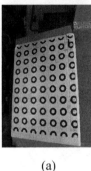 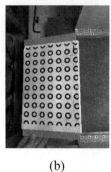

(a) (b)

Figure 5. Typical calibration images of the ring pattern of a) Canon camera and b) iPhone camera.

Table 3. Mean absolute error in the horizontal direction (Unit: μm).

Experiment	DSLR Canon (18Mega-pixel)	IPhone 6s (12Mega-pixel)
1	22	30
2	12	47
3	5	13
4	14	23

in either X or Y directions. At each displacement, one image (test image) was captured for later analysis.

The test images were undistorted and unwarped before being analysed by conventional 2D PIV (Stanier et al. 2016). The detailed procedure is described by Le et al. (2016) and illustrated in Figure 4.

Apart from the test images, a separated set of calibration images containing a series of ring patterns were taken by each camera. These calibration images were used for camera calibration process to determine the camera parameters. A minimum of three calibration images are normally required. In this study, five images were used for calibration purpose (Figure 5). The ring patterns were used as their centres can be determined precisely in comparison with square and circular patterns. The centre of each ring is determined based on outer circle and is refined based on inner circle of the ring. This procedure is similar to the camera calibration procedure using the checkboard pattern described by Zhang (2000).

The determined camera parameters were then used in an in-house Matlab code to remove the distortion in the test images.

The upwarp step performing on the undistorted test images requires the positions of the Ground Control Points in physical space and image space in order to determine the position and orientation of the camera, to correct the images. Finally, the PIV analysis is performed on the rectified images to determine horizontal displacements.

The error is calculated by comparing the displacements determined from the PIV analysis and the movements caused by the micrometers. Table 3 presents the MAE in horizontal displacement measurements in from the four experiments. As can be seen from these experiments, higher camera resolution offers slight improvement on the measurement precision (lower MAE).

The error in this method accumulates from camera calibration, undistortion, unwarp and PIV analysis. Therefore, minimising the error in each step will reduce the error hence improve the measurement precision.

4 CONCLUSIONS

The paper presented simple setups that determined the measurement precision of a 3D imaging system, featuring SfM&MVS and 2D PIV analysis. The IPhone and DSLR cameras were chosen in this study because similar cameras are relatively widely available that allows researchers to quickly carry out simple experiments to ensure the technique is suitable to the intended experiments before purchasing expensive industrial cameras and lenses.

The results show that larger number of images yields higher measurement precision in vertical direction in SfM-MVS analysis. The camera resolution is an important factor that governs the number of data points in the obtained 3D point cloud and hence to the measurement resolution. The number of data points appears to be linear with the number of pixel in the camera sensor. The sensor resolution has a more significant impact to the precision of the proposed measurement system in both horizontal direction than that for vertical directions. Other factors that need to be considered to improve the precision of the measurements are the quality of the camera lens and type of sensor and the depth of field.

REFERENCES

Furukawa, Y. and Hernández, C. 2015. Multi-view stereo: A tutorial. Foundations and Trends®in Computer Graphics and Vision, 9(1-2), pp.1–148.

Furukawa, Y. & Ponce, J. 2010. Accurate, dense, and robust multiview stereopsis. IEEE Trans. Pattern Anal. Mach. Intell. 32, No. 8, 1362–1376.

Galland, O. Bertelsen, H. S. Guldstrand, F. Girod, L. Johannessen, R. F. Bjugger, F. Burchardt, S. & Mair, K. 2016. Application of open-source photogrammetric software

MicMac for monitoring surface deformation in laboratory models. J. Geophys. Res. Solid Earth 121, 2852–2872.

James, M. R. & Robson, S. 2012. Straightforward reconstruction of 3D surfaces and topography with a camera: Accuracy and geoscience application. Journal of geophysical research 117, F03017.

Le, B. Nadimi, S. Goodey, R. Taylor, R.N. 2016. System to measure three-dimensional movements in physical models. Géotechnique Letters 6, 256–262

Lowe, D.G. 2004. Distinctive image features from scale-invariant keypoints. International Journal of Computer Vision 60, 91–110.

Robertson, D.P. & Cipolla, R. 2009. Structure from motion. In: Varga, M. (Ed.), Practical Image Processing and Computer Vision. John Wiley and Sons Ltd. New York.

Smith, M.W. Carrivick, J. & Quincey, D. 2016. Structure from motion photogrammetry in physical geography. Progress in Physical Geography 40: 247–275.

Snavely, N, Seitz. S, Szeliski, R. Photo Tourism: Exploring image collections in 3D. ACM Transactions on Graphics (Proceedings of SIGGRAPH 2006), 2006.

Stanier, S. Dijkstra, J. Leśniewska, D. Hambleton, J. White, D. and Wood, D.M. 2016. Vermiculate artefacts in image analysis of granular materials. Computers and Geotechnics, 72, pp.100–113.

Ullman, S. 1979. The interpretation of structure from motion. Proc. R. Soc. Lond. B, Biol. Sci. 203, No. 1153, 405–426

Zhang, Z. "A Flexible New Technique for Camera Calibration." IEEE Transactions on Pattern Analysis and Machine Intelligence. Vol. 22, No. 11, 2000, pp. 1330–1334.

Visualisation of inter-granular pore fluid flow

L. Li & M. Iskander
New York University, Brooklyn, New York, USA

M. Omidvar
Manhattan College, Riverdale, New York, USA

ABSTRACT: Coupled flow through porous media is important in the study of many geotechnical problems, including filter and drain design, study of internal erosion and piping, and liquefaction, among others. In this study, a novel method is presented to visualise key components of flow of fluid through pores of a saturated granular media. Combined with particle detection and tracking algorithms, the method allows for study of coupled flow problems in geomechanics. In addition, the high fidelity measurements of flow field resulting from this study allow for development and validation of Navier-Stokes type solutions for coupled flow in discrete element model simulations.

1 INTRODUCTION

Flow through porous media is an important problem of interest to many fields of science and engineering. In civil engineering applications, internal erosion and design of filters, solute and contaminant transport, energy geotechnics, and problems where rapid generation and dissipation of pore water pressures are important, such as liquefaction and cavitation, all benefit from advances in the study of flow through porous media.

Flow through the pores of a granular media can be studied at the macro scale or at the micro/pore scale. Empirical studies often rely on macro scale experiments with measurements made at test boundaries to solve problems of engineering significance. Filter design to prevent erosion or clogging has conventionally been studied using these empirical approaches.

The advent of high fidelity imaging systems has made it possible to observe kinematics of granular media at unprecedented micro scales (e.g. Dijksman et al. 2012, Iskander et al. 2015). A plethora of knowledge has been produced by investigating behaviour of soils at the underlying meso-scale and micro scales in recent years (Viggiani & Hall 2008). However, the majority of these studies have investigated granular kinematics; flow through pores has not been the focus of micro scale studies on granular media.

In studies of internal erosion in soils, the macro scale hydraulic gradient that corresponds to a micro scale state where the weight of the soil particles is balanced by the flow-induced forces is an important parameter long studied in the literature (Terzaghi 1939, Skempton & Brogan 1994). There is an information gap between observed macro scale phenomena and micro scale soil-fluid interactions. Nguyen et al. (2017) recently developed a method in an effort to bridge this gap at the meso-scale. The current study presents methods, along with experimental results, that can be used to inform numerical and analytical studies at the micro scale.

In this study, experiments were performed to visualise steady state flow through porous media. The main goal of the study is to produce high fidelity information regarding flow of fluid through the pores of saturated granular media, which can then be used to inform numerical models of flow through porous me-dia. It is anticipated that the results of this study can be used in both calibration and verification of Navier-Stokes type numerical solutions of coupled flow through porous media. In the next sections, details of the experimental setup and the image analysis techniques used are described. Sample results from steady state flow tests through granular media are presented in order to demonstrate the applicability of the method to the study of flow through soils.

2 EXPERIMENTAL SETUP

2.1 Experimental methods

The experimental setup was designed to visualise steady state flow of fluid through the pores of a granular material. A schematic diagram of the setup used is shown in Figure 1. It consisted of (1) an acrylic chamber filled with water saturated hydrogel as granular media, (2) a green laser source and optical elements to illuminate seeded particles in the sample, (3) a pump and water reservoir to circulate water through the sample and produce steady state flow conditions,

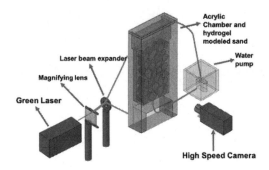

Figure 1. Schematic diagram of the experimental setup used to visualise fluid flow through the pore of a granular material.

and (4) a high speed camera to capture images of the flow field. Each of the aforementioned components are described in the next section.

2.2 Preparation of hydrogel sample to model sand

In this study, flow through porous media was imaged by means of laser illumination of a refractive index (RI) matched granular media. The granular media was fabricated in-house from a hydrogel (Bio-Rad laboratories), following procedures similar to those de-scribed by Byron et al. (2013). The hydrogel was chosen among other hydrogel materials for its stiffness, and because it readily allowed for optical RI matching with water. The Polyacrylamide hydrogel particles were fabricated with morphologies consistent with those of two natural sands, i.e., Ottawa sand 20/30 and irregular sand 30–60. This was achieved by first capturing two-dimensional projections of typical sand particles using a dynamic image analysis-based particle size identifier (Sympatec QicPic). Representative 2D projections of typical natural sand particles are shown in Figure 2. Further details of operation, accuracy, and analysis procedures for capturing representative 2D projections of natural sand particles can be found in Altuhafi et al. (2012).

Representative images captured by the dynamic particle analyser were used to fabricate 3D printed moulds for casting Polyacrylamide hydrogel. A set of 3D printed moulds from 20/30 Ottawa sand and irregular sand up scaled by a factor of 20, 40, and 60 respectively are shown in Figure 3, for demonstration.

Hydrogel was made by mixing 26.7 ml of 30% acrylamide solution and 72.7 ml of De-ionized water to produce 100 ml 8% Polyacrylamide (PAC). 0.5 ml 10% aqueous ammonium persulfate (APS) was then added to the solution as well as 0.1 ml of catalyst tetramethylethylenediamine (TEMED). The solution was then mixed and was poured into the 3D printed moulds (Fig. 3). Prior to pouring the solution, the inside walls of the moulds were coated with mineral oil. Particles cured and became stiff after 2–3 hours in the moulds and then removed and stored in de-ionized water. Particles were kept in a refrigerator to prevented

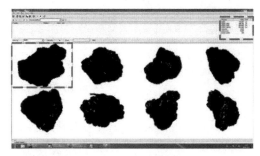

Figure 2. Representative particle gallery of irregular 30–60 sand captured by the QicPic dynamic image analyser.

microbial growth. After 24 hours of refrigeration, particles expanded by approximately 10% in volume. No further increase in volume was observed for longer periods of refrigeration, as previously reported by Byron et al. (2013). The resulting particles resembled extruded two dimensional soil particles which were stacked randomly in a planar chamber to permit study of 2D flow.

The RI of the hydrogel particles fabricated following methods described above was 1.349. As a result, water (RI = 1.33) can be readily used as a matching pore fluid in producing transparent soil samples. Water is preferred over commonly used refractive-index matching fluids in conducting seepage tests, as complications resulting from viscosity scaling are largely eliminated.

2.3 Testing apparatus

The design of the testing apparatus was originally based on ASTM constant head permeability test (D2434 2006), which is a downward seepage test system. The testing apparatus consisted of a cuboid, planar, permeameter and a water reservoir. The permeameter chamber was made of transparent acrylic with internal dimension of 4 in. by 1 in. by 6 in (length by width by height). The plane strain chamber was chosen to accommodate imaging of steady state flow through the sample. The dimensions of the setup were selected so as to mimic the proportions of the high speed camera sensor used to capture images of the setup.

Figure 3. Up scaled 3D printed Ottawa sand 20/30 and irregular sand 30–60 moulds used in this study.

A metal wire mesh (with an opening size of 2 mm), a perforated plastic plate, and gravel-sized particles were placed at the ends of the chamber to enable uniform downward seepage and to evenly distribute inflow of fluid.

A water pump with the maximum capacity of 4.7 l/min was used to circulate water through the sample. The water was seeded with tracer particles to accommodate imaging of flow. The tracer particles comprised silver coated microspheres having a mean diameter of 40 μm. A magnetic mixer was used to stir the water in the reservoir to ensure recirculation of seeded particles during testing. The elevation of the water reservoir was adjusted to maintain a constant head. The hydraulic gradient was maintained at a constant level during testing. The filters at the ends of the sample were selected based on filter design criteria and were adjusted by means of trial and error to ensure that flow of tracer particles through the system was not impeded during testing. Testing apparatus with polyacrylamide hydrogel particles can be seen in Figure 4.

2.4 Optical setup

RI-matched methods require that both the fluid and soil particles have a similar RI. When laser light shines through the chamber there is no light distortion or shadows resulting from interaction of the light with the saturated fluid/particle system. Only the silver coated tracer particles are visible, and they can be clearly recorded by camera and tracked using Particle Image Velocimetry (PIV).

The optical setup used is schematically shown in Figure 1. A diode-pumped solid-state green laser with a maximum output of 2 W was used for illumination of the sample during testing. A beam expander was also

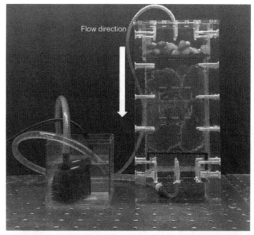

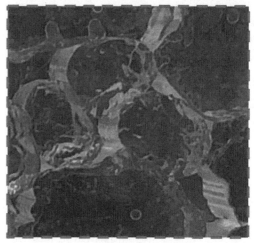

Figure 4. Image of testing apparatus showing Polyacrylamide hydrogel particles saturated with water (top). Close up of area in the middle of the test set-up (bottom).

used to produce a planar laser sheet. A magnifying lens was placed in series with the green laser and the beam expander to enhance the laser light intensity. A NAC HX-5 high-speed camera, which can operate at a maximum resolution of 5 MP, was employed to capture images of the sample during testing with a Tamron 90-mm macro lens.

The interaction of laser light and the tracer particles within the laser sheet plane sparkled and was recorded (Fig. 5). It represents two-dimensional water flow around the boundary of the synthetic sand particles.

2.5 Image analysis

The experimental workflow is summarised in Figure 6. Images captured from the tests were analysed to produce flow fields within the pores of the granular media. PIV was performed on consecutive image pairs. PIV is a pattern matching algorithm used extensively in the

0.004 s using the high-speed camera. A typical test consists of approximately 2s, and over 500 images acquired. Consecutive image pairs were analysed using the PIV techniques described in previous sections. An image acquired from a portion of the test setup is shown in Figure 5. It simulates a mixture of Ottawa sand 20/30 and irregular sand 30–60. Hydraulic gradient was maintained at 2 during photographing. The irregular particle boundaries are clearly discernible due to the slight mismatch between the pore fluid and the Polyacrylamide particles.

Results of PIV analyses on images acquired at two times during a steady state flow test are shown in Figures 7–8. The analyses shown correspond to test times of $t = 0.308$ s and $t = 1.508$ s, respectively. It can be seen that the general pattern of flow within the porous media subject to steady-state flow at the macro scale is maintained at different times during the test. The formation of vortices within the pores points to highly turbulent flow at the pore scale despite steady state laminar flow taking place at the macro scale. Furthermore, it can be seen that despite the establishment of downward flow through the sample, the general flow pattern through the particle pores is often not in the same direction as that of the general flow, and can at times even be in opposing direction to the direction of general flow. These results may have been influenced by the flexibility of hydrogel grains, whose smooth and glossy contact surface along with gravity loading and the hydraulic gradient of water may have caused the hydrogel grains to compress against each other more than natural soils; thus closing the path of water in the vertical direction more than the horizontal direction. This may have contributed to formation of vortices in pores as well as the observed flow patterns in a direction opposite to that of the global direction of fluid flow.

A video demonstrating complex particle-fluid interactions at the pore scale, which are often oversimplified in simple analytical models is available at Li et al. (2017). It is anticipated that the experimental results produced as part of this study can be used to calibrate particle scale discrete element models of flow through porous media.

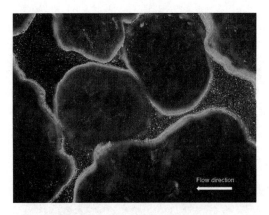

Figure 5. Flow visualisation in porous media by high speed camera at the start of the test.

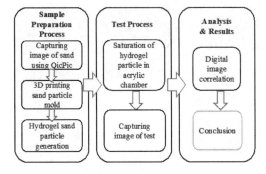

Figure 6. Experiment workflow.

literature. The core components of the algorithm consist of computing the displacement of the square region by constructing a vector from the centre of the square to the peak of the correlation function. The process is then repeated for all squares in the grid to produce the displacement field for the image pair. The process can be repeated on multiple image pairs to capture the evolution of the displacement field. Additional features of the PIV algorithm used consisted of implementation of a multi pass and multi grid interrogation algorithm, as well as sub-pixel estimation. Further details of these features can be found in Iskander (2010) and Omidvar et al. (2014).

An initial window size of 96×96 pixels was selected for the analysis, with a final window size of 32×32 pixels, with a 50% overlap of interrogation windows. As a result, displacement fields were produced on a 16-pixel square grid. The image-to-physical unit conversion was 39.8 μm per pixel, resulting in displacement with field values on a 0.63 mm square grid.

3 MICROSCALE OBSERVATIONS OF FLOW IN GRANULAR MEDIA

Images of the field of view were acquired at a speed of 250 frames/second and regular time intervals of

4 CONCLUSIONS

A successful analysis of water flow by PIV in hydrogel representing scaled natural sand were presented. Hydrogel modelled sand included Ottawa sand 20/30 and irregular sand 30–60, which were manufactured by extruding two-dimensional sand particle images. The test was performed using a downward seepage test conducted in an acrylic chamber. An amplified green laser sheet was used to illuminate two-dimension flow around particle boundary and high-resolution images with clear tracer particles were captured using a high-speed camera.

The study produced highly resolved information regarding flow of fluid through granular media by

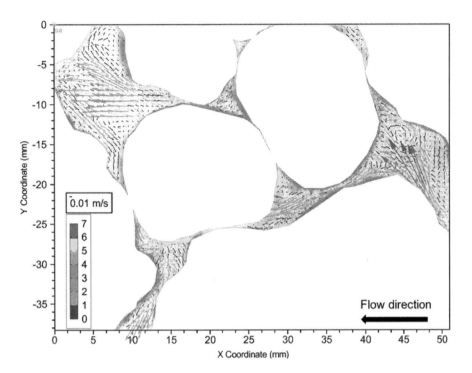

Figure 7. PIV analysis result at t = 0.308 s.

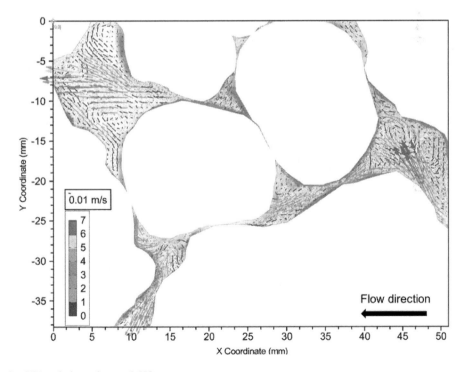

Figure 8. PIV analysis result at t = 1.508 s.

RI matching and PIV. PIV results illustrated variation in the flow velocity and direction under steady state flow conditions. The observed flow velocity and direction were surprisingly turbulent despite the globally established steady state conditions, and the small head employed.

It is envisioned that the high-fidelity measurements of flow field resulting from this study will allow for development and validation of Navier-Stokes type solutions for coupled flow in discrete element model simulations.

REFERENCES

Altuhafi, F., O'Sullivan, C. & Cavarretta, I. 2012. Analysis of an image-based method to quantify the size and shape of sand particles. *Journal of Geotechnical and Geoenvironmental Engineering* 139(8): 1290–1307.

ASTM D2434. 2006. Standard Test Method for Permeability of Granular Soils (Constant Head), *Annual Book of ASTM Standards*, ASTM International, West Conshohocken, PA.

Byron, M.L. & Variano, E.A. 2013. Exp Fluids *54: 1456. https://doi.org/10.1007/s00348-013-1456-z.*

Dijksman, J.A., Rietz, F., Lorincz, K.A., van Hecke, M. & Losert, W. 2012. Refractive Index Matched Scanning of Dense Granular Materials, *Rev. Sci. Instrum.,* 83(1), 011301.

Iskander, M. 2010. Modeling with Transparent Soils, Visualizing Soil Structure Interaction and Multi Phase Flow, Non-Intrusively, *Springer,* New York.

Iskander, M., Bathurst, R.J., & Omidvar, M. 2015. Past, Present, and Future of Transparent Soils, *Geotechnical Testing Journal,* 38(5); 1–17.

Li, L., Omidvar M. & Iskander, M. 2017. Flow visualization using PIV, www.youtube.com/watch?v=WbW6r8kIOsg

Liu, J. & Iskander, M. 2004. Adaptive cross correlation for imaging displacements in soils. *Journal of computing in civil engineering* 18(1): 46–57.

Nguyen, C.D., Benahmed, N., Philippe, P. & Gonzalez, E.V.D. 2017. Experimental study of erosion by suffusion at the micro-macro scale In: *Proceedings of: Powders & Grains 2017, Montpollier,* France, June 3–7, 2017, EPJ Web of Conferences 140, 09024.

Omidvar, M., Chen, Z., & Iskander, M. 2014. Image Based Lagrangian Analysis of Granular Kinematics, *Journal of Computers in Civil Engineering,* 04014101.

Omidvar, M., Iskander, M. & Bless, S. 2016. Soil–projectile interactions during low velocity penetration. *International Journal of Impact Engineering* 93: 211–221.

Skempton, AW, & Brogan, JM. 1994. Experiments on piping in sandy gravels. *Geotechnique,* 44(3), 449-460.

Terzaghi, K. 1939. 45th James Forrest Lecture, 1939. Soil Mechanics-A New Chapter In Engineering Science. *Journal of the ICE,* 12(7); 106–142.

Viggiani, G. & Hall, S.A. 2008. Full-Field Measurements in Experimental Geomechanics: Proceedings, Historical Perspective, Current Trends and Recent Results, *Deformation Characteristics of Geomaterials,* Burns et al. eds., Taylor & Francis, London; 3–68.

A two-dimensional laser-scanner system for geotechnical processes monitoring

M.D. Valencia-Galindo
Pontificia Universidad Católica de Chile, Santiago, Chile

L.N. Beltrán-Rodriguez, J.A. Sánchez-Peralta, J.S. Tituaña-Puente, M.G. Trujillo-Vela,
J.M. Larrahondo, L.F. Prada-Sarmiento & A.M. Ramos-Cañón
Pontificia Universidad Javeriana, Bogotá, Colombia

ABSTRACT: Complex geotechnical processes such as debris or tailings flows, rain-induced landslides, foundation instability, and soil-structure interaction require advanced laboratory techniques, including remote-sensing technology. The purpose of this paper is to describe the implementation of a laboratory-scale LIDAR-type system installed on an environmental flume for geotechnical applications under 1-g conditions. A two-dimensional laser scanner device, which is commonly used in mining operations to measure conveyor-belt volumes, was adapted to move axially along the flume. The scanner measures cross-sectional elevation data on any physical model constructed in the flume to produce a digital elevation model. The flume is an instrumented chamber comprising sprinklers to simulate precipitation (controlled intensity and flow rate), a hydraulic system to induce rise in groundwater level, and a jack to vary model inclination, if required. A calibration protocol and post-processing code were developed to efficiently handle the scanner-recorded elevation data. To test and validate the experimental setup, tailings flow experiments were performed. Results show that laboratory-scale laser sensor technology is a useful and robust tool for physical modelling.

1 INTRODUCTION AND BACKGROUND

Laser-assisted, remote-sensing technology is currently a widely used tool for engineering process monitoring at different scales, e.g., landslide or fault mapping and building information modelling, BIM (Abellán, Vilaplana, & Martinez 2006, Hunter et al. 2011, Kim et al. 2016, Xiong et al. 2013). In addition, two-dimensional laser scanner devices are used in the mining industry in several countries to determine the volume of material transported on conveyor belts (SICK, 2015). In research, laser-scanner systems have been employed to evaluate structural engineering processes like beam deflection and metal frame connections (Cabaleiro et al. 2014, Gordon & Lichti 2007). Furthermore, two-dimensional laser scanners have been recently applied in hydraulic and coastal engineering studies to monitor the evolution of free-surface profiles of regular and irregular waves (Blenkinsopp et al. 2012, Harry et al. 2012, Harry et al. 2011, Streicher et al. 2013).

This article describes the laboratory implementation of a two-dimensional laser scanner system for geotechnical process monitoring. The described system delivers laboratory-scale digital elevation models based upon elevation data obtained from physical models constructed in an instrumented environmental flume.

A number of previous studies describe the use of instrumented chamber flumes to physically simulate landslide processes, particularly those that are rain-induced (Barazzetti et al. 2013, Chen et al. 2010, Fang et al. 2012, Jia et al. 2009, Ling & Ling 2012, Moriwaki et al. 2004). Precipitation of artificially-induced rain is normally generated on the soil physical models via sprinklers uniformly distributed across the flume. To monitor geotechnical processes, these studies use various instruments, including inclinometers, vibrating-wire cell pressures, high-resolution/high-velocity cameras, infrared cameras, and dynamic sensors.

For example, Chen et al. (2010) set a chamber between two sprinklers that spray upward, which allow them to study soil failure and debris flow. On the other hand, Fang et al. (2012) simulate natural-slope failures by varying the slope dip; in their case, the failure is in fact reached once the precipitation event is reproduced. Moriwaki et al. (2004) also present a variable-dip, full-scale slope model, that yields debris flow; their sprinkler system is set in the laboratory ceiling to distribute precipitation uniformly.

In addition to describing the laser scanner system, this article presents the main features of the environmental flume where the scanner is mounted.

2 METHODS

2.1 Laser scanner

The two-dimensional laser scanner device used was a Bulkscan® LMS511 laser volume flowmeter (SICK

Figure 1. Bulkscan® LMS511 laser scanner.

AG, Waldkirch, Germany; see Fig. 1). This scanner is frequently employed in the mining industry to measure bulk materials on conveyor belts or in piles via the time-of-flight principle and multi-echo technology (SICK, 2015). If required, the laser-scanned data captured by this instrument also allows calculation of level, centre of gravity, and loading position, as well as belt monitoring. The scanner uses infrared, eye-safe laser on a two-dimensional plane with aperture angle of 190°, acquiring elevation data on polar coordinates with 0.5° resolution, equivalent to 12.72-mm resolution at the flume's bottom. The scanner works with an operating range from 0.5 to 20 m, at scanning frequencies varying from 35 to 75 Hz. According to the manufacturer, for conveyor-belt applications, the scanner can effectively measure volumes at up to 30 m/s of conveyor-belt speed.

The laser scanner was installed at the top of the environmental flume located at Pontificia Universidad Javeriana's Geotechnical Laboratory. Instead of the typical conveyor-belt operating method, in the laboratory the target physical model remains stationary in the flume while the scanner moves along the main axis of a flume. For this purpose, the laser scanner was mounted on an aluminium frame (Colsein-Item, Colombia) specifically designed for the moving operation of the scanner (see Fig. 2). The frame withstands acceleration/deceleration and cruise vibrations produced by the scanner during travel. The scanner was controlled via a servo-motor, an electronic interface, and a data acquisition system.

2.2 Environmental flume

The environmental flume was constructed with support from the Geophysical Institute and the Department of Civil Engineering of Pontificia Universidad Javeriana. The flume comprises both permanent instruments (including the laser scanner) and project-specific instruments.

The flume features a hydraulic system composed by three sprinklers at the top and a drainage system at

Figure 2. Environmental flume and laser-scanner mounting frame.

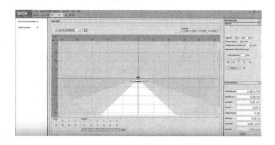

Figure 3. Screenshot of SOPAS Engineering Tool V3 during operation.

base level, and the latter can also act as inlet to yield bottom-up pressure, if required. The hydraulic system is equipped with manometers for pressure control during both rain simulations and bottom-up pore pressure generation. In addition, it includes a hydraulic pump to regulate the flow rate in the system, as well as several valves to control pressures in the sprinklers and at base level.

The flume also includes a jack system on one end, which allows control of the inclination of the models constructed in the flume. This feature is particularly important for slope stability studies under 1-g conditions.

Project-specific instruments that may be coupled to the flume include pore-pressure transducers, pressure cells, and strain sensors for real-time process monitoring.

2.3 Scanned data acquisition and post-processing

The scanner's data acquisition system is the freeware program SOPAS Engineering Tool V3 (SICK AG, Waldkirch, Germany) which is the main computer interface of the system (see Fig. 3).

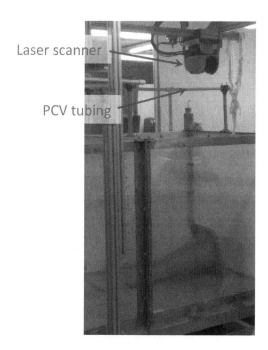

Figure 4a. Traffic cone to be scanned.

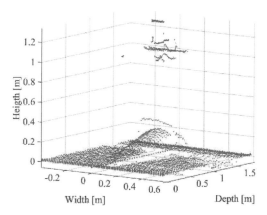

Figure 4b. Traffic-cone elevation data points acquired using the laser scanner system. Data points above 1.0 of height are PVC tubing.

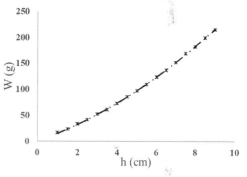

Figure 5. Sprinklers calibration. Top: plastic-cup grid. Bottom: Calibration curve (measured relationship between weight and height of water in the plastic cups).

Because the laser scanner records elevation data in polar coordinates, it was decided to translate the data into Cartesian coordinates. Also, because the amount of output data is rather large, post-processing may be tedious. Hence, a MATLAB code was developed for converting the coordinates of every data point while handling the data efficiently.

The code written also required to define the servomotor position vector for determining the along-the-axis Cartesian coordinate. Thus, the code has the capability to determine height, width, and depth (length) of the scanned physical model with a length resolution depending on the chosen time-step, and a resolution of 0.5° at polar coordinates.

During code writing and validation, several laser-scanning tests were performed using simple objects of known geometry (traffic cones, boxes, etc.; see Fig. 4).

2.4 Sprinklers calibration

Calibration of the sprinklers was undertaken via a facile procedure using plastic cups set on a grid of 1×1 m, separated every 0.1 m from each other. The plastic cups were placed over a stowage whose height with respect to the flume floor was variable, from 0.1 m to 0.5 m, with steps of 0.1 m (see Fig. 5). The purpose of testing the sprinklers was to measure the influence area of induced precipitation depending on the height of fall of the water, by determining the amount of collected water in each plastic cup.

The artificial-rain calibration test was performed by opening the valve of the central sprinkler for 40 seconds, while trying to keep the pressure system on 80 psi (about 550 kPa). Once the test was finished, the height of the water in each cup was measured with a caliper. The weight on every cup was indirectly determined using a previously-prepared calibration curve for one typical plastic cup. Such curve correlated the height of water with measured water weight as shown in Figure 5. For convenience, the calibration curve was fit to a second-degree polynomial regression.

2.5 Tailings flow tests

A series of tailings flow experiments were designed and prepared as part of an ongoing investigation aimed at studying the fundamental flow behaviour of such

■ 0-6 ■ 6-12 ■ 12-18 ■ 18-24 ■ 24-30
■ 30-36 ■ 36-42 ■ 42-48 ■ 48-54 ■ 54-60

Figure 6. Areal distribution of induced-rain precipitation below the central sprinkler (units: cm³/s).

non-newtonian fluid. Filtered tailings produced at a recently-opened gold mine in the Antioquia region of Colombia were used to simulate uncontrolled flow in the laboratory.

The tailings were 73.4% silt-, 23.7% sand-, and 2.9% clay-sized particles. The measured average properties of the tailings were $G_s = 2.77$, $LL = 31.7\%$, $LP = 17.9\%$, and the solid concentration of the tailings water-based paste tested was 70%.

An aluminium-and-acrylic box, with dimensions $0.3 \times 0.3 \times 0.5$ m, provided with a 0.1 m-wide 0.3 m-high gate, was constructed and installed inside the flume. This box contained the paste until the gate was opened to yield flow on a flat aluminium surface.

3 RESULTS

3.1 Sprinklers

As the stowage elevation changes and the sprinklers tests are performed, the areal distribution of precipitation also changes. One contour plot of induced-rain distribution for water falling from a height of 1 m is shown in Figure 6. It can be seen that the distribution of water in the surface is not as uniform as it might be expected. In fact, the water concentrates in the central part of the grid, as compared with the sprinklers' manufacturer predictions (see Fig. 7).

3.2 Tailings flow tests

Figure 8 shows the results from one tailings flow experiment, as well as the digital elevation model reconstructed from laser-scanned data. A total of 187,676 data points were collected to create the digital surface.

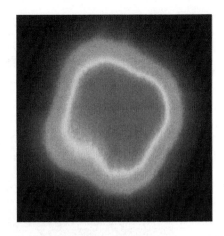

Figure 7. Sprinkler manufacturer's catalogue "shadow". ('Spraying Systems Co. Catalog 75', n.d.).

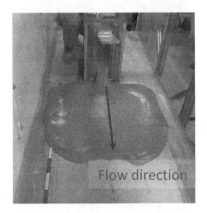

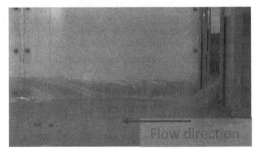

Figure 8a. Tailings flow experiment: photo and video screenshot of the experiment.

These data are currently being used to calibrate and validate computational fluid mechanics simulations of these tailings flows.

4 CONCLUSIONS

Results confirm that laboratory-scale laser sensor technology is a useful and robust tool for physical modelling. Particularly, two-dimensional laser-scanner systems are powerful tools for monitoring geotechnical engineering processes.

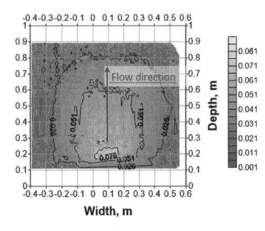
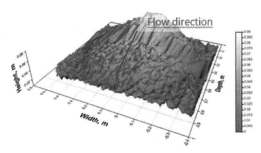
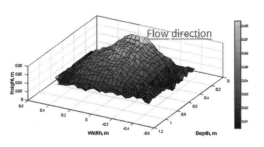

Figure 8a. Tailings flow experiment: laser-scanner-derived digital elevation model (187,676 data points). Top and centre: elevation contours and DEM plot using SURFER®; Bottom: DEM plot using MATLAB®.

By coupling the scanner with an environmental flume and instrumentation, it is possible to simulate geotechnical processes while controlling rain-precipitation parameters and eventually slope dip and pore pressures, all factors affecting geotechnical stability.

The scanner/flume monitoring system will allow future work focusing on experimental modelling of geotechnical processes, as well as validation or calibration of computational models.

ACKNOWLEDGEMENTS

The authors are grateful to the Geophysical Institute, the Department of Civil Engineering, and the Master's in Civil Engineering Program of Pontificia Universidad Javeriana for supporting this research. The authors also appreciate the help of Dr. Diego Cobos of Dynami Geoconsulting, Colombia, who provided the mine tailings and valuable comments.

REFERENCES

Abellán, A., Vilaplana, J. M., & Martínez, J. 2006. Application of a long-range Terrestrial Laser Scanner to a detailed rockfall study at Vall de Núria (Eastern Pyrenees, Spain). *Engineering Geology* 88(3):136–148.

Barazzetti, L., Scaioni, M., Feng, T., Qiao, G., Lu, P., Tong, X., & Li, R. 2013. Photogrammetry in Experiments for Hydrogeological Risk Assessment. *ISPRS-International Archives of the Photogrammetry, Remote Sensing and Spatial Information Sciences* (3):145–152.

Blenkinsopp, C. E., Turner, I. L., Allis, M. J., Peirson, W. L., & Garden, L. E. 2012. Application of LiDAR technology for measurement of time-varying free-surface profiles in a laboratory wave flume. *Coastal Engineering* 68:1–5.

Cabaleiro, M., Riveiro, B., Arias, P., Caamaño, J. C., & Vilán, J. A. 2014. Automatic 3D modelling of metal frame connections from LiDAR data for structural engineering purposes. *ISPRS Journal of Photogrammetry and Remote Sensing* 96: 47–56.

Chen, N. S., Zhou, W., Yang, C. L., Hu, G. S., Gao, Y. C., & Han, D. 2010. The processes and mechanism of failure and debris flow initiation for gravel soil with different clay content. *Geomorphology* 121(3): 222–230.

Fang, H., Cui, P., Pei, L. Z., & Zhou, X. J. 2012. Model testing on rainfall-induced landslide of loose soil in Wenchuan earthquake region. *Natural Hazards and Earth System Sciences* 12(3): 527–533.

Gordon, S. J., & Lichti, D. D. 2007. Modeling terrestrial laser scanner data for precise structural deformation measurement. *Journal of Surveying Engineering* 133(2): 72–80.

Harry, M., Zhang, H., & Colleter, G. 2012. Remotely sensed data for wave profile analysis. *Coastal Engineering Proceedings* 1(33): 45.

Harry, M., Zhang, H., Lemckert, C., Colleter, G., & Blenkinsopp, C. 2011. Remote sensing of water waves: wave flume experiments on regular and irregular waves. *Proceedings of Coasts and Ports, Perth, Australia*.

Hunter, L. E., Howle, J. F., Rose, R. S., & Bawden, G. W. 2011. LiDAR-assisted identification of an active fault near Truckee, California. *Bulletin of the Seismological Society of America* 101(3): 1162–1181.

Jia, G. W., Zhan, T. L., Chen, Y. M., & Fredlund, D. G. 2009. Performance of a large-scale slope model subjected to rising and lowering water levels. *Engineering Geology* 10 6(1): 92–103.

Kim, M.-K., Wang, Q., Park, J.-W., Cheng, J. C., Sohn, H., & Chang, C.-C. 2016. Automated dimensional quality assurance of full-scale precast concrete elements using laser scanning and BIM. *Automation in Construction* 72: 102–114.

Ling, H., & Ling, H. I. 2012. Centrifuge model simulations of rainfall-induced slope instability. *Journal of Geotechnical and Geoenvironmental Engineering* 138(9): 1151–1157.

Moriwaki, H., Inokuchi, T., Hattanji, T., Sassa, K., Ochiai, H., & Wang, G. 2004. Failure processes in a full-scale landslide experiment using a rainfall simulator. *Landslides* 1(4): 277–288.

SICK, S. intelligence. 2015. Sick AG, Bulkscan LMS511, Non-Contact and Maintenance-Free Sensor for Measuring Volume Flow. Sensor Intelligence.

Streicher, M., Hofland, B., & Lindenbergh, R. C. 2013. Laser ranging for monitoring water waves in the new Deltares Delta Flume. In *ISPRS Annals of the photogrammetry, remote sensing and spatial information sciences, Volume II-5/W2. ISPRS Workshop Laser Scanning 2013, 11–13 November 2013, Antalya, Turkey.* Editor (s): M. Scaioni, RC Lindenbergh, S. Oude Elberink, D. Schneider, and F. Pirotti. ISPRS.

Xiong, X., Adan, A., Akinci, B., & Huber, D. 2013. Automatic creation of semantically rich 3D building models from laser scanner data. *Automation in Construction* 31: 325–337.

14. Seismic – dynamic

Dynamic behaviour of model pile in saturated sloping ground during shaking table tests

C.H. Chen
National Center for Research on Earthquake Engineering, Taipei, Taiwan

T.S. Ueng & C.H. Chen
Department of Civil Engineering, National Taiwan University, Taipei, Taiwan

ABSTRACT: Shaking table tests on model pile in saturated inclined ground were conducted at the NCREE, Taiwan to study the soil-structure interaction, especially lateral spreading, in a liquefiable ground during earthquake. The model pile made of aluminium alloy was placed in an inclined biaxial laminated shear box filled with saturated clean sand. The pile tip was fixed at the bottom of the shear box. Input shakings including sinusoidal and recorded earthquake accelerations were mainly imposed perpendicularly to the slope direction. Densely instrumented pile and the transducer arrays inside soil specimen measured during the shaking. Lateral spreading displacements of the soil and pile behaviour were observed while soil liquefaction. The kinematic loading on the model pile due to lateral spreading during shaking can be separated in a suitable way by using the biaxial shear box. In addition, the stiffness of the soil would recover with the dissipation of pore water pressure.

1 INTRODUCTION

Extensive foundation failures have been found in all major disaster earthquakes, such as the 1964 Niigata Earthquake, 1989 Loma Prieta Earthquake, 1995 Kobe Earthquake, 1999 Chi-Chi Earthquake, 2011 Christchurch Earthquake, and the 2011 Great East Japan Earthquake. Therefore, many researches have been studied to understand the mechanism of the dynamic loading on the piles (soil-pile interaction) and their responses under earthquake loading.

Lateral loading tests in the field or in the laboratory and shaking table tests on model piles within soil specimens, under either 1 g or centrifugal conditions, have been used to investigate the pile behaviors and soil-pile interaction in liquefiable soils (e.g. Dobry et al. 2003, Tokimatsu et al. 2005, Brandenberg et al. 2005, Ashford et al. 2006, Cubrinovski et al. 2006, Madabhushi et al. 2010, Ueng and Chen 2010a). The results of these studies, including failure mechanisms, bending moments of pile in laterally spreading ground, pore water pressure variation around the piles, p-y relations for soil-pile interaction, and pile cap effect, can provide information on performance criteria for aseismic design of structures with pile foundations.

However, there are still some uncertainties concerning the soil-pile interaction issues in laterally spreading ground, including: (1) the kinematic loading on pile foundation due to lateral spreading; (2) the relation between the lateral pressure on pile foundation and the ground displacement and (3) the transient responses of the surrounding soil and pile during soil liquefaction. This research used the large biaxial laminar shear box developed at the National Center for Research on Earthquake Engineering (NCREE) as the soil container and an instrumented aluminium model pile was installed inside the shear box filled with saturated sand. The biaxial shear box with the model pile in a saturated sloping ground was placed on 1 g shaking table and the sinusoidal and recorded earthquake accelerations were applied perpendicularly or parallel to the slope direction. The soil and pile responses and their interaction, including the inertial and kinematic actions on the model pile, under these types of shakings were studied.

2 MODEL PILE AND SAND SPECIMEN

The model pile was made of an aluminium alloy, with a length of 1600 mm, an outer diameter of 101.6 mm, a wall thickness of 3 mm and its flexural rigidity ($EI = 75$ kN·m^2) was obtained by flexural test. The shear box was inclined 2° to the horizontal, simulating a mild infinite slope and the sloping direction of this test was defined as X direction. The pile was fixed vertically at the bottom of the shear box. Hence, this physical model can be used to simulate the condition of a vertical pile embedded in sloping rock or within a sloping firm soil stratum. In addition, 6 steel disks (226.14 kg) were attached to the top of the model pile to simulate the superstructure. The model pile was instrumented and installed inside the shear box before preparation of the sand specimen, as shown in Figure 1.

Figure 1. The instrumented aluminum pipe inside the shear box.

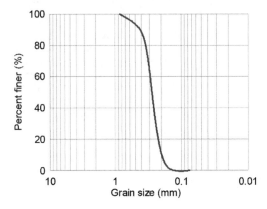

Figure 2. Grain size distribution of Vietnam sand.

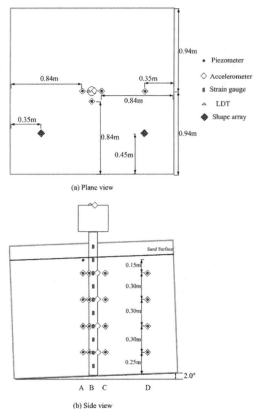

Figure 3. Instrumentation on the model pile and within the sand specimen.

A clean fine silica sand ($G_s = 2.65$, $D_{50} = 0.31$ mm) from Vietnam was used in this study for the sand specimen inside the laminar shear box. This sand has been used in the shaking table tests for liquefaction studies at NCREE (Ueng et al. 2006). The representative grain size distribution of the sand in this test is shown in Figure 2. The maximum and minimum void ratios are 0.918 and 0.631 respectively, according to ASTM D4253 Method 1B (wet method) and ASTM D4254 Method A. The sand specimen was prepared using the wet sedimentation method after placement of the model pile and inside instruments in the shear box. The sand was rained down into the shear box filled with water to a pre-calculated depth. The size of the sand specimen was 1.880 m × 1.880 m in plane and about 1.31 m in height before shaking tests. The relative density of the sand was about 10 %. Details of biaxial laminar shear box and the sand specimen preparation were described in Ueng et al. (2006).

3 INSTRUMENTATION

Two magnetostriction type linear displacement transducers (LDTs) were set up to measure X- and Y-displacements of the pile top. They were mounted to the reference frames outside the shaking table. The waterproof resistance-type strain gauges were placed on the pile surface to measure the bending strains of the model pile. There are 10 different depths with 15 cm spacing along the pile axis as shown in Figure 3. At each depth, two pairs of strain gauges were orthogonally attached on opposite sides of the pile in X- and Y-directions. Vertical acceleration arrays along the pile were also installed on the model pile in X- and Y-directions for acceleration measurements. In addition, in order to observe the build-up and dissipation of the pore water pressures and accelerations in the sand specimen (the region near the pile and that far from the pile), mini-piezometers and mini-accelerometers were installed inside the box at different locations and depths.

4 SHAKING TABLE TESTS

Shaking table tests were first conducted on the model pile without the sand specimen in order to evaluate the dynamic characteristics of the model pile itself. Sinusoidal and white noise accelerations with amplitudes from 0.03 to 0.05 g were applied in X and/or Y directions. The model pile in saturated sloping ground was then tested under one dimensional sinusoidal (1–8 Hz) and accelerations at the Chi-Chi Earthquake and the

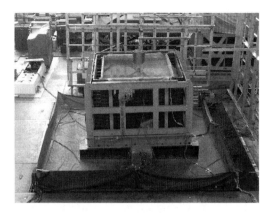

Figure 4. A model pile with 6 steel disks in a saturated sloping ground on the shaking table.

Kobe Earthquake were with amplitudes ranging from 0.03 to 0.15 g. Input motions were mainly imposed perpendicularly to the slope direction for the study of kinematic effect on the pile foundation only, and also tested in another direction parallel to the slope to investigate the resultant force on the pile foundation including the inertial and kinematic effect. White noise accelerations with amplitude of 0.03 g were also applied in both X- and Y-directions to evaluate the dynamic characteristics of the model pile within soil and the sand specimen. Figure 4 shows a picture of shaking table test on the model pile with 6 steel disks on its top. The height of the sand surface after each shaking test was measured to compute the settlement and density of the sand specimen. Soil samples were taken using short thin-walled cylinders at different depths and locations, after completion of the shaking tests, to obtain the densities of the sand specimen.

5 TEST RESULT ANALYSES

5.1 Dynamic characteristics of model pile

Shaking table tests on the model pile without sand specimen were conducted to evaluate the dynamic characteristics of the model pile itself. The amplification curve was obtained from the Fourier spectral ratio of the measured acceleration of the pile top to that of the input motion. The predominant frequency of the model pile with 6 steel disks of mass on its top was identified at about 2.08. Table 1 lists the predominant frequencies of the model pile according to the test data. The average damping ratio of the model pile is about 3% according to test data of forced vibration by the half power method.

5.2 Dynamic characteristics of soil and soil-pile system under small amplitude of shakings

The dynamic characteristics of soil stratum and soil-pile system were evaluated by a series of shaking table tests on the model pile within the saturated sand

Table 1. Predominant frequencies of the model pile.

Mass on pile top	Aluminium pile
	Freq., Hz
No mass	22.9
6 steel disks	2.08

Table 2. Predominant frequencies of soil and the aluminium pile in the soil specimen of different relative densities.

Density of soil	Predominant frequency, Hz	
Dr, %	Pile in soil	Soil
11.9	4.61	10.92
26.0	4.64	11.7
42.4	4.65	12.7
70.1	4.67	13.8

specimen with small amplitude. Table 2 lists the predominant frequencies of the soil and soil-pile system for the model pile (with 6 steel disks on its top) in soil of various relative densities. It can be seen that the predominant frequency of soil increases with the relative density, but that of pile increases only slightly with the relative density. In addition, the predominant frequency of soil-pile system is significantly lower than that of the soil specimen. Comparing the predominant frequencies of the model pile without and within soil specimen (Table 1 and Table 2, respectively), one can find that, except for the case without mass on the pile top, the predominant frequencies of the model pile in the soil specimen were higher than those without soil due to the confinement of the soil on the pile.

5.3 Responses of model pile in liquefiable soil

A shaking table test under one-dimensional sinusoidal acceleration with a frequency of 8 Hz and an amplitude of 0.068 g in Y direction (i.e. the input motion was imposed perpendicularly to the slope direction) was conducted to study the kinematic effect on the pile foundation in a saturated sloping ground with a relative density of 13.6%. The depth of liquefaction was determined based on the measured pore water pressures in the sand specimen and accelerometers on the frames (Ueng et al. 2010b). In this test, the liquefied depth of the sand specimen reached about 112.6 cm. Figure 5 shows a distinct lateral spreading displacement after the shaking (Compared with Fig. 4).

The four piezometer arrays were placed in the sand specimen, near the pile and in the free field. Figure 6 shows the time histories of excess pore water pressure ratios (ru) near the model pile and in the free-field location. It can be observed the measured excess pore water pressures, that the sand at a shallower depth liquefied firstly at about 2.8 seconds and then at greater depth also liquefied at around 3.6 seconds. It indicated that liquefaction progressed from the shallower

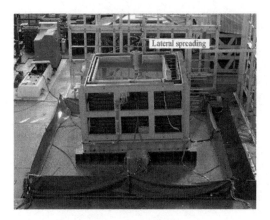

Figure 5. Liquefaction-induced lateral spreading displacement in X direction.

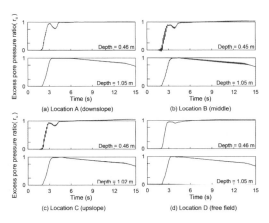

Figure 6. Excess pore water pressure time histories at various locations.

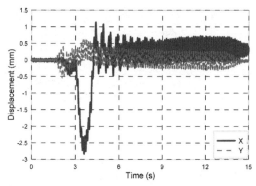

Figure 7. The time histories of relative displacement of the pile top.

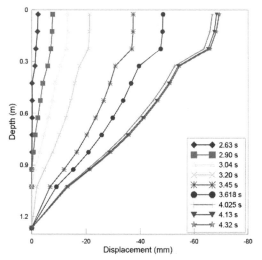

Figure 8. Profiles of free-field ground of soil specimen at various times in X direction.

depth to the greater depth of the specimen in a short period during the shaking. In addition, the pore water pressures at a shallower depth exhibit an impermanent reduction phenomenon near the model pile, especially in the downslope side. At the greater depth, the pore water pressures did not present such kind of responses.

The time histories of relative displacement of the pile top were shown in Figure 7. The displacements in X-direction of the model pile were mainly caused by lateral spreading of the soil, and the displacements in the other direction were induced by the shaking. Hence, the force exerted on the model pile due to lateral spreading in X-direction could be extracted from this kind of test. It is also observed that the pile response in X direction can be divided into three stages during the shaking: In the first stage (i), there were only small movements and rebounded during 2.2 to 3 seconds. This is due to the small amount of movement and softening in the shallow depth of soil. In the second stage (ii), the model pile suffered the majority of liquefaction-induced lateral ground displacement (Fig. 8), and it had a maximum displacement at about 3.618 seconds. After this time, the pile rebounded back because of the reduction of lateral force on the pile when the specimen was liquefied. Pile response at this stage (iii) demonstrated a free vibration motion during 4.2 to 8 seconds. The predominant frequency of the acceleration on the pile top in X direction was about 2 Hz. Comparing this result with the predominant frequency of the model pile without soil specimen (Table 1), one can find that the predominant frequency of the model pile within liquefied soil was almost the same as that of model pile without soil specimen. This infers that the stiffness of the soil nearly vanished when soil liquefaction occurred.

In order to investigate the soil-pile interaction in a lateral spreading condition during the shaking table test, the pile response can be back-calculated from the measured bending moments based on the beam theory. The profile of pile deflection can be computed by integrating twice the fitted function of the measured bending moments, and soil reaction profile is obtained as the second derivative of the measured bending moments. A polynomial function of the fourth order was fitted for the measured bending moments

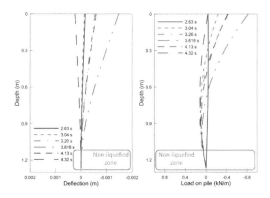

Figure 9. The Profiles of pile response at various times in X direction.

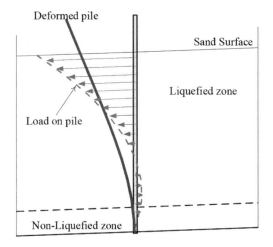

Figure 10. The deflection of pile and kinematic forces on the pile due to lateral spreading in X direction.

along the pile. The shear at sand surface are given according to those generated by the mass at the pile top. The soil pressure at the pile tip was set to zero because no displacement was allowed at the pile tip. With these constraints, the undetermined coefficients of the polynomial can be computed based on the optimization technique of least squares with Lagrange multipliers. Figure 9 illustrates the profiles of deflection and load on the pile at various times during the shaking.

Comparing the profile of pile deflection with that of free-field ground (Fig. 8) at the same instants, it can be seen that the profiles of free-field lateral displacement moved monotonically toward the downstream direction while the pile deflection rebounded after it reached its maximum at about 3.618 seconds. Figure 10 shows the deflection of model pile and the forces on the model pile at maximum pile displacement at about 3.618 seconds. It can be seen that the kinematic forces acting on the model pile due to lateral spreading exhibited an inverted triangular distribution and the direction of pile deflection and that of the lateral spreading are the same.

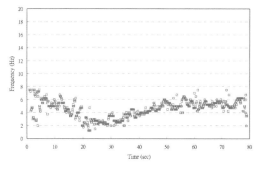

Figure 11. The time-dependent predominant frequency of model pile within saturated sand during earthquake shaking.

5.4 *Changes of dynamic characteristics of model pile in saturated sand during liquefaction*

Shaking table test on model pile in a sloping ground under one-dimensional recorded acceleration in Chi-Chi earthquake, with an amplitude of 0.10 g in Y direction, was conducted to study the changes of dynamic characteristics of model pile in saturated sand during liquefaction. The relative density of sand specimen was about 27% and the liquefied depth of the sand specimen reached about 45 cm.

In order to quantify the time-dependent predominant frequency of soil-pile system during liquefaction, a method of system identification technique, so-called short-time transfer function (STTF) (Chen et al. 2016), was used to identify the predominant frequency of soil-pile system. Figure 11 shows the time-dependent predominant frequency of model pile within saturated sand during earthquake shaking. It can be seen that the predominant frequency of the soil-pile system gradually decreases before liquefaction (from 10–20 sec), and minimum occurs at the onset of initial liquefaction (about 20 sec). After a while, the predominant frequency increase with time at the period of 30–50 sec. After 50 sec, the predominant frequency remains constant.

Based on the average effective stress ratio approach (Chen et al. 2016), the predominant frequency time history of soil-pile system and the average effective stress ratio time history of sand specimen are shown in Figure 12. It can be seen that the trend of the predominant frequency is similar to that of the average effective stress ratio. The result also indicated that the predominant frequency of soil-pile system would increase with the dissipation of pore water pressure after liquefaction. In other word, the stiffness of the soil would increase with the dissipation of pore water pressure and the recoverable increment of soil stiffness is related to the decrease of excess pore water pressure.

Based on the previous data, the relation between the predominant frequency of soil-pile system and the average pore pressure ratio can be obtained as shown in Figure 13. The trend of the predominant frequency of soil-pile system decrease with the increase of the average pore pressure ratio due to the loss of the effective stress of the soil stratum.

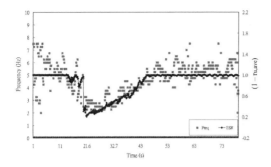

Figure 12. The predominant frequency vs. average effective stress ratio time history.

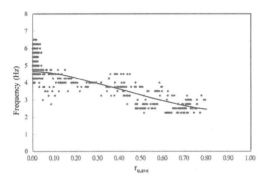

Figure 13. The relation between the predominant frequency of soil-pile system and the average pore pressure ratio.

6 CONCLUSIONS

Shaking table tests were conducted on an aluminium model pile in saturated sloping ground by using the biaxial laminar shear box. Lateral spreading displacements of the soil and the pile behaviour were observed while soil liquefaction was triggered under shakings. It was found that the kinematic loading on the model pile due to lateral spreading during shaking can be separated in a suitable way by using the biaxial shear box. Analyses of the dynamic behaviour of the model pile and the soil stratum were conducted during the shaking tests according to test results.

The test results show that liquefaction was progressive from the shallower depth to the greater depth of the specimen in a short period during the shaking. Based on the back-calculation of the measured data, the kinematic force on the model pile due to lateral spreading exhibited an inverted triangular distribution. Furthermore, the stiffness of the soil nearly vanished during the period of liquefaction. It was also found the stiffness of the soil would increase with the dissipation of pore water pressure, and the recoverable increment of soil stiffness is directly related to the decrease of excess pore water pressure. Further analyses of the test data will be performed to obtain more information on the relation between stiffness of the soil and excess pore water pressure and a model will be set up to assess the seismic behaviour of liquefied soil for more reasonable seismic design.

ACKNOWLEDGEMENTS

This study is supported by NCREE, Taiwan. The technical supports and operational assistances in the shaking table testing, including large specimen preparation by the graduate students from National Taiwan University and National Cheng Kung University and the engineers at NCREE, are gratefully acknowledged.

REFERENCES

Ashford, S.A., Juirnarongrit, T., Sugano, T. & Hamada, M. 2006. Soil–pile Response to Blast-induced Lateral Spreading I: Field Test. *Geotechnical and Geoenvironmental Engineering* 132(2): 152–162.

Brandenberg, S. J., Boulanger, R. W., Kutter, B. L. & Chang, D. D. 2005. Behavior of pile foundations in laterally spreading ground during centrifuge tests. *Journal of Geotechnical and Geoenvironmental Engineering* 131 (11): 1378–1391.

Chen, C.H., Ko, Y.Y., Chen, C.H. & Ueng, T.S. 2016. Time-dependent dynamic characteristics of model pile in saturated sand during soil liquefaction. *Geotechnical Engineering Journal of the SEAGS & AGSSEA* 47(2): 89–94.

Cubrinovski, M., Kokusho, T. & Ishihara, K. 2006. Interpretation from large-scale shake table tests on piles undergoing lateral spreading in liquefied soils. *Soil Dynamics and Earthquake Engineering* 26: (2–4): 275–286.

Dobry, R., Abdoun, T., O'Rourke, T.D. & Goh, S.H. 2003. Single piles in lateral spreads: field bending moment evaluation. *Journal of Geotechnical and Geoenvironmental Engineering* 129 (10): 879–889.

Madabhushi, G., Knappett, J. & Haigh S. 2010. *Design of pile foundations in liquefiable soils*. Imperial College Press: U.K.

Tokimatsu, K., Suzuki, H. & Sato, M. 2005. Effect of Inertial and Kinematic Interaction on Seismic Behavior of Pile with Embedded Foundations. *Soil Dynamics and Earthquake Engineering*. 25(7-10): 753–762.

Ueng, T.S., Wang, M.H., Chen, M.H., Chen, C.H. & Peng, L.H. 2006. A Large Biaxial Shear Box for Shaking Table Tests on Saturated Sand. *Geotechnical Testing Journal* 29(1): 1–8.

Ueng, T. S. & Chen, C. H. 2010. Multidirectional shaking table tests on model piles in saturated sand. *Seventh International Conference on Physical Modelling in Geotechnics* Vol II: 1445–1450.

Ueng, T.S., Wu, C.W., Cheng, H.W. & Chen, C.H. 2010. Settlements of Saturated Clean Sand Deposits in Shaking Table Tests. *Soil Dynamics and Earthquake Engineering* 30(1–2): 50–60.

Investigation on the aseismic performance of pile foundations in volcanic ash ground

T. Egawa & T. Yamanashi
Civil Engineering Research Institute for Cold Region, Sapporo, Japan

K. Isobe
Hokkaido University, Sapporo, Japan

ABSTRACT: Various volcanic ash ground layers are widely and complicatedly deposited in Hokkaido depending on the location. To effectively conduct aseismic reinforcement of pile foundation in volcanic ash ground, it is necessary to establish an evaluation method for classifying the area priority. In order to obtain the basic information for the above purpose, a series of centrifuge model tests on aseismic performance of pile foundations was conducted varying volcanic ash soil sedimentation condition. The test results have been comprehensively analyzed and discussed.

1 INTRODUCTION

Tremendous amounts of money and time are required for inspecting and repairing pile foundations of highway bridges that have suffered damage from earthquakes. Highways affected by inspections and repair work of bridges cannot be fully utilized as traffic routes for a considerable length of time. Aseismic retrofitting of piers and superstructure of highway bridges has been promoted in Japan. However, earthquake safety has not been improved in the pile foundations of highway bridges. Development of technologies for evaluating aseismic capacity of highway bridge pile foundations as well as for aseismic retrofitting is an urgent issue for Japan.

As Japan is located along volcanic belts, volcanic sediments are present widely across the nation. In Hokkaido, the northernmost of the four main islands of Japan, coarse-grained volcanic ash soil has a wide, intricate distribution. Volcanic ash soil is different from common sandy soil in engineering property (Miura et al. 2003). It is known that volcanic ash ground is different from sand ground in terms of the rate of decrease in the coefficient of horizontal subgrade reaction of piles in the ground being affected by an earthquake and liquefaction (Egawa et al. 2016).

In view of this, technologies for aseismic reinforcement of pile foundations need to be developed by taking into account the liquefaction characteristics and the mechanical behavior of volcanic ash ground during an earthquake. To efficiently retrofit a pile foundation on volcanic ash ground, which expands in a wide area and forms a complex structure, a method is needed for evaluating the aseismic performance of the pile foundation, in order to identify the sites and priorities for implementation of aseismic reinforcement.

In the study reported herein, a centrifugal model test was conducted for understanding how differences in the depositional structure of volcanic ash soil, which is subject to liquefaction, affect the behavior of a pile foundation during an earthquake and liquefaction. The aim of the test was to accumulate basic data necessary for proposing a method for evaluating the aseismic performance of pile foundations, in order to identify the sites and priorities for implementing aseismic reinforcement.

2 OVERVIEW OF EXPERIMENT

Table 1 shows experiment cases of the centrifugal model test. In the test, volcanic ash soil layers having a relative density of $D_r = 85\%$ were used as soil layers which were to be subject to liquefaction. Sandy soil layers having a relative density of $D_r = 95\%$ were also used in the test as soil layers which were not assumed to be liquefied. The liquefied volcanic ash soil layers in Cases 1, 2 and 3 were different in thickness. A liquefied volcanic ash soil layer was between the two non-liquefied sand soil layers in Case 4, and a non-liquefied sand soil layer was between the two liquefied volcanic ash soil layers in Case 5. In this test, a liquefaction strength ratio (R_{L20}) is a repeated stress amplitude ratio $\sigma_d/2\sigma'_0$ (σ_d: cyclic deviator stress, σ'_0: effective confining pressure) when the double amplitude axial strain DA is 5% and the number of cycles Nc is 20 in a cyclic undrained triaxial test (JGS0541-2009) of soil. Figure 1 shows the Experiment Cases of the centrifugal model test. Five Cases are shown in section for the sake of comparison. Figure 2 shows an outline of the model used for Experiment Case 2.

In the centrifuge model test, a 1:50 scale model underwent a 50g centrifugal acceleration. A static

Table 1. Experiment cases and formation of model grounds.

Case	Model ground			Thickness	Input seismic motion
1	Volcanic ash soil	$D_r=85\%$	$R_{L20}=0.242$	15.0 m	20 sine waves Frequency 1.5 Hz Max. Acc. 200 cm/s^2
2	Volcanic ash soil	$D_r=85\%$	$R_{L20}=0.242$	10.0 m	
	Toyoura sand	$D_r=95\%$	–	5.0 m	
3	Volcanic ash soil	$D_r=85\%$	$R_{L20}=0.242$	5.0 m	
	Toyoura sand	$D_r=95\%$	–	10.0 m	
4	Toyoura sand	$D_r=95\%$	–	5.0 m	
	Volcanic ash soil	$D_r=85\%$	$R_{L20}=0.242$	5.0 m	
	Toyoura sand	$D_r=95\%$	–	5.0 m	
5	Volcanic ash soil	$D_r=85\%$	$R_{L20}=0.242$	5.0 m	
	Toyoura sand	$D_r=95\%$	–	5.0 m	
	Volcanic ash soil	$D_r=85\%$	$R_{L20}=0.242$	5.0 m	

* in Prototype scale

Table 2. Physical properties of materials for each model ground.

	Volcanic ash soil	Toyoura sand
Sand fraction (%)	67.9	99.8
Silt fraction (%)	26.3	0.2
Clay fraction (%)	5.8	0.0
Fine fraction content FC (%)	32.1	0.2
Max. grain size D_{max} (mm)	0.850	0.425
50% grain size D_{50} (mm)	0.136	0.169
10% grain size D_{10} (mm)	0.013	0.127
Coefficient of uniformity U_c	14.80	1.42
Coefficient of curvature U'_c	1.92	0.97
Soil particle density ρ_s (g/cm^3)	2.366	2.646

horizontal load test was performed at pile heads before conducting a dynamic vibration experiment. Input seismic motion had 20 sine waves, 1.5 Hz frequency in prototype scale and up to 200cm/s^2 acceleration.

A model pile was made of steel (SS400) with an outside diameter of D = 10.0 mm, a thickness of t = 0.2 mm and a length of L = 400 mm (D = 500 mm, t = 10 mm and L = 20 m in prototype scale.) As Figure 2 illustrates, two piles were arranged in two lines with a pile center distance of 3D. A plate-shaped weight was applied to the pile heads for connecting and fixing them. Strain gauges were attached to one out of four combined piles, two at every eleven different depths.

Volcanic ash soil used in the test was Shikotsu pumice-flow deposit (Spfl), volcanic coarse-grained soil typical of Hokkaido, which was sifted through a 0.85mm sieve. To make sandy soil, Toyoura sand was employed, which is commonly used as standard test sand in Japan. Regarding the geomaterials used in each Experiment case, Table 2 shows the characteristics of soil grains and the soil particle density. Both volcanic ash soil and Toyoura sand are classified as a sandy soil ($FC \leq 35\%$, $D_{50} \leq 10$ mm, and $D_{10} \leq 1$ mm) that should to be analyzed for determining the likelihood of liquefaction according to Japanese standards.

Accelerometers and piezometers were placed in the model ground as shown in Figure 2. For pore fluid in each model ground, silicon oil of dynamic viscosity 50 times greater than water was used and fully saturated in a deaerating tank.

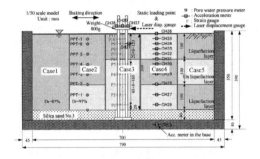

Figure 1. Experiment cases of the centrifugal model test. (Five Cases are shown in section for the sake of comparison).

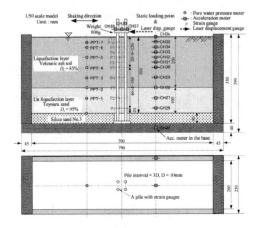

Figure 2. Outline of the model used for Case 2.

3 TEST RESULTS AND DISCUSSION

Data obtained in the centrifuge model test conducted under the aforementioned conditions are organized for analysis below. The numerical data hereafter are values converted in prototype scale.

The time history of $\Delta u/\sigma'_v$ shown in Figure 3 is the time history of excess pore water pressure (Δu) which was measured at each depth in a dynamic vibration experiment and was divided by the initial effective overburden pressure (σ'_v) at each depth. The input vibration in the base was started at 2.0 seconds and

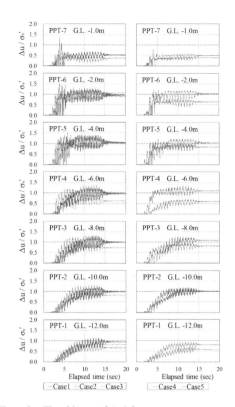

Figure 3. Time history of $\Delta u/\sigma'_v$.

stopped at 15.3 seconds, and thus the input vibration duration was 13.3 seconds. In the layers that were subject to liquefaction in all cases, the value of $\Delta u/\sigma'_v$ reached roughly 1.0, indicating that liquefaction took place in these layers. In Case 4, the value of $\Delta u/\sigma'_v$ was large at G.L.-4.0 m in the non-liquefied layer. It is probably considered that the excess pore water pressure (Δu) increased rapidly at G.L.-6.0 m in the intermediate liquefied layer and propagated to the non-liquefied layer. At G.L.-1.0 m, the value of Δu didn't increase significantly in all cases.

Figure 4 shows the response acceleration measured at G.L.-2.0 m, G.L.-8.0 m, and G.L.-12.0 m in all cases in the dynamic vibration experiment, the bending moments calculated from the measured bending strains of the piles, and the time history of input acceleration in the base. At G.L.-12.0 m, the response acceleration amplitude was equivalent to the input acceleration amplitude, and the bending moment became negative with the increase of the excess pore water pressure (Δu). At G.L.-8.0 m, the response acceleration was amplified in the middle of the vibration duration, and was attenuated when the value of Δu reached a peak. These changes in the response acceleration and the excess pore water pressure were particularly noticeable in Case 3 and Case 5 in which a non-liquefied layer was at G.L.-8.0 m. It should be noted that the response acceleration was also amplified in Case 1, which consists of a liquefied layer alone. The value of the bending moment was large at G.L.-8.0 m in all cases. In Cases 1, 3 and 5, large vibration amplitudes observed during the initial stage of vibration was maintained until these began to attenuate when the excess pore water pressure (Δu) reached a peak. In Case 2 and Case 4, the vibration amplitude attenuated with the increase of Δu. At G.L.-2.0 m, no response acceleration was observed in Case 1. In other cases, response acceleration was observed in the initial stage of vibration. In Case 2, the maximum response acceleration was nearly 600 cm/s^2, and it attenuated with the increase of Δu. In Case 4 where a non-liquefied layer was at G.L.-2.0 m, similar attenuation was observed. This attenuation is likely to have been influenced by the excess pore water pressure (Δu) that rapidly increased in the intermediate layer. Bending moments in Case 1 and Case 2 were slightly amplified during the initial stage of vibration, and attenuated with the increase of Δu. In Cases 3, 4 and 5, large amplitudes of the bending moment were maintained after Δu reached a peak, and the bending moment became positive.

4 EVALUATION OF DECREASE RATE OF THE COEFFICIENT OF HORIZONTAL SUBGRADE REACTION OF PILES

Measurements taken in the dynamic vibration test helped to understand that the response amplitude of the ground acceleration and of the bending moment of piles attenuated with the increase of the excess pore water pressure (Δu), or the progress of liquefaction, although the response varied depending on the depth in the ground in all cases. It is probably considered that the ground was increasingly softened with the progress of liquefaction. In analyzing interactions between the ground and pile foundation during an earthquake, changes caused by liquefaction in the coefficient of horizontal subgrade reaction of piles should be identified. Thus, results of a dynamic vibration test and of a static horizontal load test performed prior to the dynamic vibration test are used for evaluating the decrease rate of the coefficient of horizontal subgrade reaction of piles in association with liquefaction as follows.

Figure 5 shows the procedure for calculating the coefficient of static horizontal subgrade reaction of piles on the basis of measurements taken in a static horizontal load test before conducting a dynamic vibration test. Figure 6 shows the procedure for calculating the coefficient of dynamic horizontal subgrade reaction of piles in the liquefaction process on the basis of measurements taken in the dynamic vibration test. These calculation procedures are based on the method used in the past studies (Tokimatsu et al. 2004) Specifically, various experiments are carried out to obtain bending moments from bending strains of piles measured at depths. Then, resultant bending moments are treated with second-order differential or integral calculus in the direction of depth to figure out horizontal subgrade reactions and horizontal displacements of piles.

In this study, the authors adopted a method for evaluating the horizontal subgrade reaction distribution

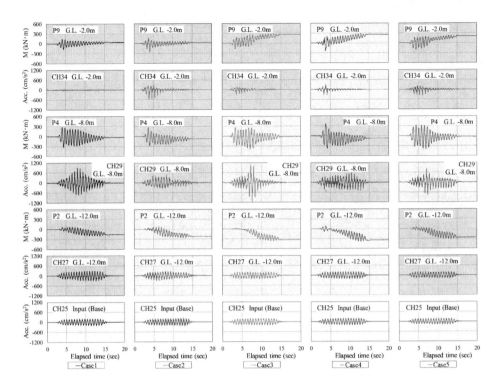

Figure 4. Time history of response acceleration in the ground at G.L.-2.0 m, G.L.-8.0 m and G.L.-12.0 m, and of bending moments of piles.

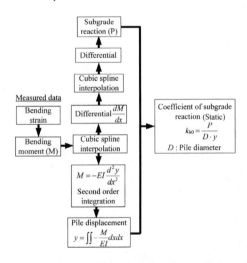

Figure 5. Procedure for calculating the coefficient of static horizontal subgrade reaction of piles on the basis of measurements.

of piles more accurately. In this method, the bending moment distribution of piles interpolated in the cubic spline interpolation method and then differentiated was re-interpolated and re-differentiated once again. Although the accuracy increases as the number of points measured by the strain gauges increases, the number of strain gauges are arranged in the realistic way. Horizontal displacements of piles in the dynamic vibration test can also be obtained by integrating

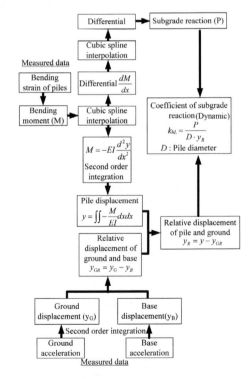

Figure 6. Procedure for calculating the coefficient of dynamic horizontal subgrade reaction of piles on the basis of measurements.

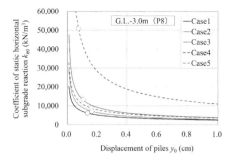

Figure 7. Coefficients of static horizontal subgrade reaction of piles (k_{h0}) correlated with horizontal displacements of piles (y_0) (G.L.-3.0m (P8)).

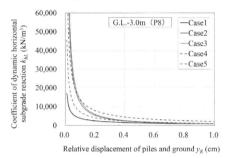

Figure 8. Coefficients of dynamic horizontal subgrade reaction of piles (k_{hL}) correlated with relative displacements of piles and ground (y_R) (G.L.-3.0 m (P8)).

the acceleration of piles at depths with respect to time. Prior to the dynamic vibration test, however, the authors learned that because piles would sway due to the vibration for causing liquefaction, response acceleration would be disturbed and thus horizontal acceleration could not be measured accurately. It was also known to the authors that bending strains could be measured accurately with swaying piles. Therefore, bending strains of piles were used for calculating horizontal displacements of piles.

Figure 7 shows relations between the coefficients of static horizontal subgrade reaction of piles (k_{h0}), calculated by using the measurements taken at G.L.-3.0m before the dynamic vibration test, and the horizontal displacements of piles (y_0). Figure 8 shows relations between the coefficients of dynamic horizontal subgrade reaction of piles (k_{hL}), calculated by using the measurements taken at G.L.-3.0m in the process of liquefaction, and the relative displacements of piles and ground (y_R).

In this study, their relations are expressed with the following relational equations.

The coefficient of static horizontal subgrade reaction

$$k_{h0} = k_0 \cdot B_0 \cdot \left(\frac{y_0}{y_a}\right)^{A_0} \tag{1}$$

where k_0 is unit coefficient of subgrade reaction ($=1 \text{ kN/m}^3$), y_a is unit displacement ($=1 \text{ m}$), A_0 and B_0 are the coefficients in a static state.

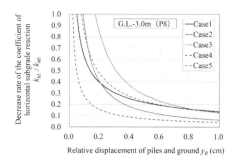

Figure 9. Decrease rates of the coefficient of horizontal subgrade reaction of piles in the liquefaction process of volcanic ash ground (k_{hL}/k_{h0}) (G.L.-3.0 m (P8)).

The coefficient of dynamic horizontal subgrade reaction

$$k_{hL} = k_0 \cdot B_L \cdot \left(\frac{y_R}{y_a}\right)^{A_L} \tag{2}$$

Where, A_L and B_L are the coefficients during liquefaction.

Based on Equations 1 and 2, the decrease in the horizontal subgrade reaction coefficient of piles following ground liquefaction shown in Figure 9 is expressed with Equation 3.

$$\frac{k_{hL}}{k_{h0}} = B \cdot \left(\frac{y_R}{y_a}\right)^{A} \tag{3}$$

where $B = \dfrac{k_0 \cdot B_L}{k_{h0}}, A = A_L$

In the Equation 3, k_{h0} is a calculated value at each depth for the time when the displacement of piles on the ground surface reached 0.5cm, or 1% of the pile's diameter. The values of k_{h0} are shown by ○ in Figure 7. In the same equation, k_{hL} is a value calculated for each depth according to the amount of displacement during the input seismic motion having 20 sine waves.

In the upper layer between G.L.-1.0m and G.L.-4.0 m where strain gauges P10-P7 were attached, horizontal subgrade reactions and horizontal displacements of piles were confirmed in the static horizontal load test. Regarding this layer, the Equations 1, 2 and 3 are used for evaluating the coefficients A and B which are associated with the decrease in the coefficient of horizontal subgrade reaction due to liquefaction of volcanic ash ground.

Figure 10 shows depth distributions of the coefficient A in all cases. It is illustrated that even within the same soil layer, the value of A varies depending on the depth in all cases. In designing pile foundations, soil layers having similar physical properties are considered to be uniform. In view of this, the mean of values at four different depths in each case is also shown in Figure 10. The mean value of the coefficient A, indicating a decrease in the coefficient of horizontal subgrade reaction of piles according to the amount of horizontal displacement, is between -0.8 and -0.9 in Cases 1, 4

883

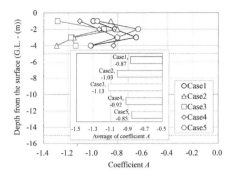

Figure 10. Depth distributions of the coefficient A.

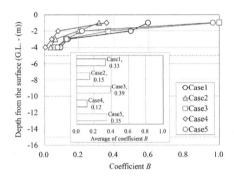

Figure 11. Depth distributions of the coefficient B.

and 5. In Case 2 and Case 3, the mean value is between -1.0 and -1.1, suggesting that the coefficient of horizontal subgrade reactions decreased more steeply than in other three cases.

Figure 11 shows depth distributions of the coefficient B in all cases. The coefficient B, indicating the rate of decrease in the coefficient of horizontal subgrade reaction of piles due to liquefaction, became smaller in the direction of depth. These distributions indicate that the coefficient of horizontal subgrade reaction of piles decreased more significantly at depths than on the ground surface. As in Figure 10, the mean of values at four different depths in each case is also shown in Figure 11. In Cases 1, 3 and 5, the mean value is between 0.3 and 0.4. In Case 2 and Case 4, the value is between 0.1 and 0.15, indicating larger decreases in the coefficient than in other three cases. In Case 4, the upper layer is a non-liquefied layer, and the coefficient of static horizontal subgrade reaction of piles (k_{h0}) before conducting the dynamic vibration test is larger than in other cases (Fig. 7). Between G.L.-2.0 m and G.L.-4.0 m in the upper layer, the values of $\Delta u/\sigma'_v$ (i.e., the excess pore water pressure divided by the effective overburden pressure) are relatively large although these values are less than 1.0 (Fig. 3). These results suggest that the initial rigidity of the upper layer decreased significantly in Case 4. In Case 2, the coefficient of horizontal subgrade reaction of piles decreased more significantly than in Cases 1, 3 and 5 in which the upper layer was liquefied as in Case 2. The coefficient largely decreased in Case 2 probably because acceleration response characteristics and shear strains in the ground are different from those in other cases due to the differences in the natural frequency associated with the soil layer thickness and the layer structure. Additional studies and evaluation need to be conducted in the future by analyzing effective stress for understanding seismic responses of the ground in more detail.

5 CONCLUSIONS

To obtain basic data required for establishing a method for evaluating the aseismic performance of pile foundations in volcanic ash ground, the authors conducted centrifuge model experiments of volcanic ash soil layers in which the deposition of ash soil was varied and liquefaction occurred. The findings are summarized below:

– Although the ground acceleration and the bending moment of piles varied depending on the experiment case, responses were relatively large in the intermediate layer. These responses attenuated with the progress of soil liquefaction, or with the increase in the excess pore water pressure (Δu).
– In Case 4 where the intermediate layer alone was subject to liquefaction, the coefficient of horizontal subgrade reaction in the upper non-liquefied layer significantly decreased. This phenomenon took place probably because the excess pore water pressure (Δu) increased rapidly in the intermediate layer and propagated to the upper layer, resulting in a significant decrease in the initial rigidity of the upper layer soil.
– In Case 2 where the upper layer and the intermediate layer were subject to liquefaction, the coefficient of horizontal subgrade reaction of piles decreased more significantly than in other cases where the upper layer was subject to liquefaction. The coefficient largely decreased in Case 2 probably because acceleration response characteristics and shear strains in the ground are different from those in other cases due to the differences in the natural frequency associated with the soil layer thickness and the layer structure.

The authors will verify the test results in detail and will conduct quantitative evaluation by simulating test results with effective stress analysis.

REFERENCES

Egawa, T., Hayashi, T. and Tomisawa, K. 2016. Centrifuge model test on coefficient of horizontal subgrade reaction of piles in liquefied volcanic ash ground. *2nd Asian Conference on Physical Modelling in Geotechnics*, 244–249.

Miura, S., Yagi, K. and Asonuma, T. 2003. Deformation-strength evaluation of crushable volcanic soils by laboratory and in-situ testing. *Soils and Foundations*, 43(4): 47–57.

Tokimatsu, K. and Suzuki, H. 2004. Pore water pressure response around pile and its effects on p-y behavior during soil liquefaction. *Soils and Foundations*, 44(6): 101–110.

Effective parameters on the interaction between reverse fault rupture and shallow foundations: Centrifuge modelling

A. Ghalandarzadeh
School of Civil Engineering, College of Engineering, University of Tehran, Tehran, Iran

M. Ashtiani
School of Civil Engineering, Babol Noshirvani University of Technology, Babol, Iran

ABSTRACT: Although the performance of surface foundations has been investigated against a large dislocation from a dip-slip fault, the parameters effecting on the fault rupture – foundation interaction should be examined. This paper presents a series of centrifuge model tests to investigate the parameters affecting the interaction between reverse fault rupture and shallow foundations embedded at a depth of D such as the foundation embedment depth, surcharge load of the foundation and dip angle of the fault. The interaction mechanisms and the effect of parameters on the behaviour of the foundation are described.

1 INTRODUCTION

Surface fault ruptures resulting from the 1999 earthquakes in Turkey and Taiwan and the 2008 earthquake in China have provided numerous examples of their devastating effects on buildings and infrastructure. Minor to severe damage was observed in these earthquakes on buildings subjected to fault-induced ground movement. The Alquist-Priolo Earthquake Fault Zoning Act of the State of California prescribes a fault avoidance zone (about 15 m from the fault trace) to avoid or mitigate surface fault hazard risks; however, the Chi-Chi earthquake demonstrated that this avoidance zone may be too broad (Kelson et al. 2001). It seems that damage caused by permanent displacement of earthquakes has shown that fault zoning and avoidance are unavoidable, particularly in urban areas (Konagai 2005).

Extensive researches have been performed to evaluate the response of buildings subjected to dip-slip faulting and the effect of parameters affecting the foundation-fault interaction mechanism. Several centrifuge and numerical studies have been conducted on surface foundations and reverse fault rupture interactions (Ahmed and Bransby 2009; Anastasopoulos et al. 2010). The present study examined the effect of different parameters such as the foundation embedment depth, surcharge load of the foundation and dip angle of the fault on the interaction mechanism of shallow embedded foundations and reverse faulting. Accordingly, several centrifuge tests were performed at 50g acceleration using a geotechnical centrifuge.

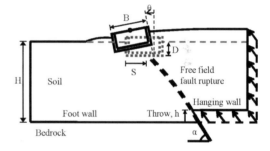

Figure 1. Schematic configuration of a typical centrifuge model.

2 PROBLEM DEFINITION

The models used in this study are schematically illustrated in Figure 1. A uniform soil deposit with the thickness of H is overlain on a reverse fault with a dip angle of α. A rigid shallow foundation with the breadth of B = 170 mm and surcharge load of q is placed near the surface at the depth of D. In each test, the bedrock moves upward in the faulting direction with a vertical component of h (throw) until a rupture surface is developed through the soil layer and interacts with the foundation. When the fault rupture reaches the foundation, rigid-body movement occurs, including rotation, θ.

Free-field fault rupture can be used as a reference for centrifuge tests. This rupture is used to define the different positions of the foundation with respect to its emergence. The position of the foundation is defined

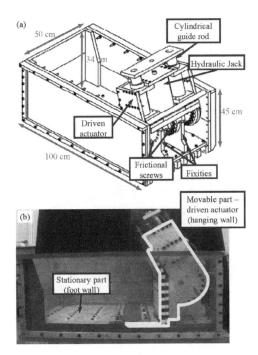

Figure 2. Centrifuge split box: (a) perspective; (b) front view.

Table 1. The lists of centrifuge model tests.

Test No.	Dip angle of fault (°)	Footing embedment, D/B	Footing pressure, q (kPa)	Footing position, s/B
MA-04	60	–	–	–
MA-12	75	–	–	–
MA-05	60	0	81	0.75
MA-06	60	0.3	81	0.75
MA-08	60	0	81	0.4
MA-09	60	0.3	81	0.4
MA-10	60	0.3	40	0.4
MA-25	75	0	81	0.75
MA-13	75	0.3	81	0.75

by the parameter s (Figure 1). This parameter indicates the distance between the left corner of the foundation and the point where the free-field rupture crosses the base.

3 CENTRIFUGE MODELING AND TEST APPARATUS

The experiments were carried at the centrifuge facility of the University of Tehran (Moradi & Ghalandarzadeh 2010). This device comprises a beam centrifuge with a 3.5 m radius that can accelerate a model package of up to 1500 kg at 100 g of centrifugal acceleration. The faulting tests were designed at 50 g acceleration and performed by a centrifuge apparatus (i.e. split-box). The fault simulator (split-box- see Figure 2) was constructed with outer dimensions of $100 \times 50 \times 45$ cm (length × width × height). The split box container was 63 cm in length, 50 cm in width, and 34 cm in height. The maximum allowable offset of the simulator was 50 mm. This box was composed of a stationary footwall, a movable hanging wall, an 18-ton hydraulic jack and a cylindrical guide rod. This split box can be adjusted to dip angles of 45°, 60°, 75° and 90°. Further information about the split-box can be found in Ashtiani et al. (2015).

4 CENTRIFUGE MODEL TESTS

The tests were conducted with the same soil conditions to allow measuring the effect of the parameters on the shallow embedded foundations and reverse faulting interaction. The scaled models were designed using the general descriptions of centrifuge modelling scaling laws provided by the International Technical Committee TC2 (2005).

The basic characteristics of the centrifuge model tests are summarised in Table 1. These tests were divided into two categories: free-field tests (without a foundation) and interaction tests. Tests MA-04 and MA-12 were conducted to provide the free-field fault rupture position for a dip angle of 60° and 75°, respectively. Test MA-13 was carried out to compare the performance of the proposed mitigation measures with the unmitigated scenario. Seven other centrifuge tests were run on the different parameters to investigate the effect of the parameters on the shallow embedded foundations and reverse faulting interaction.

All dimensions are presented at prototype scale unless stated otherwise.

5 RESULTS

5.1 Fault ruptures in the free field

The deformed physical models resulting from fault dislocation are shown in Figures. 3a and 3b for fault dip angles of 60° and 75°, respectively. The dip angle of rupture plane decreases progressively as it propagates towards the soil surface. This behaviour is similar to the results of experimental and numerical studies (Bransby et al. 2008; Anastasopoulos et al. 2007). In addition, by increasing the dip angle of fault, the distance between the fault rupture trace and the projection of fault discontinuity on the ground surface is reduced. The distance is shown with the dimensionless quantity W/H which H is the thickness of the soil layer. The W/H for dip angles 60° and 75° are 0.92 and 0.49, respectively. This is consistent with the findings from Lin et al. (2006). They concluded that an increase in dip angle of fault would lead to a smaller W/H and narrower fault zone.

5.2 Effect of embedment depth, D

Figure 4 shows the interaction tests of reverse faulting for two different embedment depths (D/B = 0

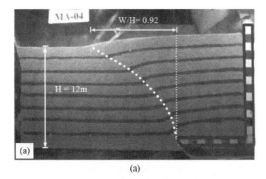

(a)

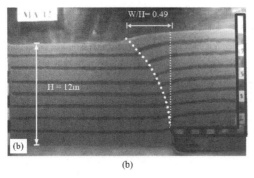

(b)

Figure 3. Typical free field rupture induced by reverse faulting with different dip angles: (a) $\alpha = 60°$; (b) $\alpha = 75°$.

and 0.3) and two different positions (s/B = 0.75 and 0.4) of foundations. For the surficial foundations (i.e. D/B = 0), the position of foundation changes from s/B = 0.75 (in Test MA-05) to s/B = 0.4 (in Test MA-08), thus changing the failure mechanism from footwall to gapping. This observation qualitatively concurs with Ahmed and Bransby (2009). For the foundation positions of s/B = 0.75 and 0.4, increasing the embedment depth of the foundation leads to changing the interaction mechanism, significant foundation rotation and unfavourable performance.

As shown in Figure 5, the embedded foundation (D/B = 0.3) experiences more rotations compared to the surficial foundation (D = 0). The embedded foundation is forced to experience more rotation because their vertical walls and the surrounding soil act as kinematic constraints, forcing the foundation to follow the rotation and translation induced by faulting.

5.3 *Effect of surcharge load, q*

Comparing Figures 6a and 6b shows that decreasing the surcharge load from 81 kPa to 40 kPa (in tests MA-09 and MA-10) leads to increased gap between the soil and the foundation. The reduced-weight foundation appears unable to divert the fault rupture emergence away from the foundation. Therefore, the increased gap length causes a large bending moment and substantial distress.

The effects of surcharge load were studied on the rotations of two embedded foundations positioned at

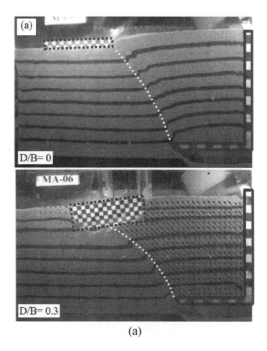

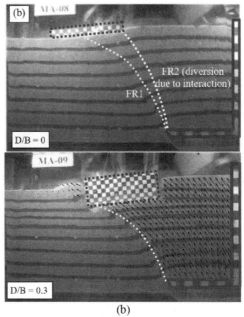

Figure 4. Effect of embedment depth on the fault-foundation interaction; (a) s/B = 0.75; (b) s/B = 0.4.

s/B = 0.4 with q = 81 kPa (as a heavy foundation) and q = 40 kPa (as a light foundation). The results show that the rotation of the heavier foundation was greater than that of light foundation (see Figure 7). Although, the light foundation may experience substantial distress because its gap length is more than the heavy one.

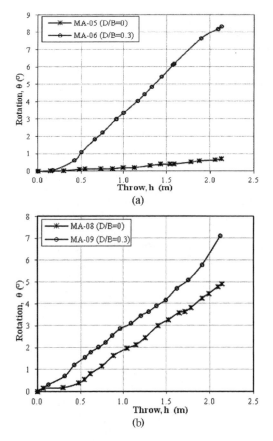

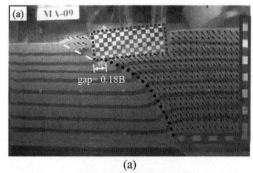

(a)

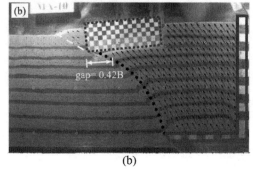

(b)

Figure 6. Effect of surcharge load on the fault-foundation interaction; (a) q = 81 kPa; (b) q = 40 kPa.

Figure 5. Effect of embedment depth on the foundation rotation; (a) s/B = 0.75; (b) s/B = 0.4.

According to Anastasopoulos et al. (2010) and the results presented here, the effect of surcharge load on the interaction between reverse fault ruptures and embedded foundations may depend on the position of the foundation relative to the free-field fault rupture emergence.

5.4 Effect of fault dip angle, α

Figure 8 shows the comparison between the shallow foundation-reverse fault rupture interaction for different fault dip angles. In these tests, the foundations with different embedment depths (D/B = 0 and 0.3) and q = 81 kPa (equivalent to an 8-storey building) are placed at the position of s/B = 0.75. For two different dip angles of fault, increasing the embedment depth of the foundation leads to changing the interaction mechanism from footwall to gapping one. Also, the interaction behaviour of foundations during the fault rupturing approximately is the same for two different dip angles of fault.

To identify the effect of fault dip angle on the interaction of reverse fault and shallow foundations, the foundation rotations were studied for two different fault dip angle (i.e. α = 60° and 75°) and two different embedment depths (D/B = 0 and 0.3) of foundation

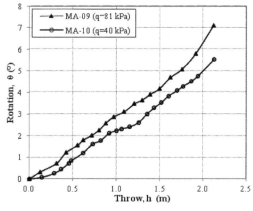

Figure 7. Effect of surcharge load on the foundation rotation; (a) q = 81 kPa; (b) q = 40 kPa.

(Figure 9). The results show that the embedded foundation (D/B = 0.3) experiences more rotations compared to the surficial foundation (D = 0) for two different dip angles of fault.

6 CONCLUSION

In this paper, a series of centrifuge model tests were conducted to study the effects of different parameter such as the embedment depth of foundation,

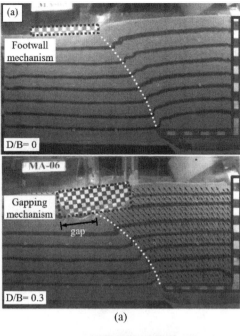

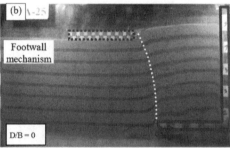

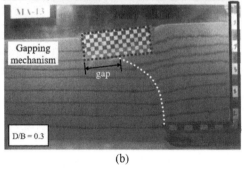

Figure 8. Effect of fault dip angle on the mechanism of foundation-fault interaction; (a) $\alpha = 60°$; (b) $\alpha = 75°$.

surcharge load of foundation and fault dip angle on the interaction of shallow embedded foundations and reverse fault ruptures. The major results obtained are summarised as follows:

- The embedment depth of the foundation (D) leads the foundation to experience significant rotations and substantial distress and unfavourable performance. In this regard, aside from all other factors

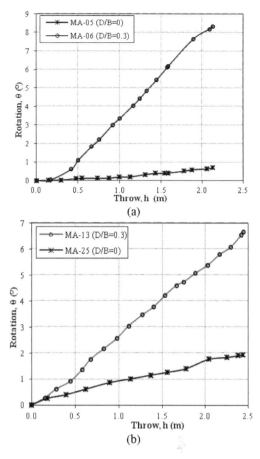

Figure 9. Effect of fault dip angle on the foundation rotation; (a) $\alpha = 60°$; (b) $\alpha = 75°$.

(e.g., rigidity of foundation and foundation position relative to the fault rupture), the embedment depth is an effective parameter on the shallow foundation-reverse fault rupture interaction.
- The effect of surcharge load (q) on the rotation of embedded foundations (θ) depends on the foundation position relative to the fault rupture outcropping. The increase of q causes the fault rupture to deviate from beneath the foundation.
- The different dip angles of fault cause to the same behaviour for the shallow foundations during interaction with reverse fault rupture. Also, the interaction mechanisms are relatively the same at different dip angles.
- The results refer to the cases that were studied in this paper. They suggest that complementary tests are required to explore the effects of parameters on the foundation-fault rupture interaction.

REFERENCES

Ahmed, W. & Bransby, M.F. 2009. Interaction of shallow foundations with reverse faults. *Journal of Geotechnical and Geoenvironmental Engineering*, 135(7): 914–924.

Bransby, M.F., Davies, M.C.R., El Nahas, A. & Nagaoka, S. 2008b. Centrifuge modeling of reverse fault-foundation interaction. *Bulletin of Earthquake Engineering*, 6(4): 607–628.

Anastasopoulos, I., Gazetas, G., Bransby, M.F., Davies, M.C.R. & El Nahas, A. 2007. Fault Rupture Propagation through Sand: Finite-Element Analysis and Validation through Centrifuge Experiments. *Journal of Geotechnical and Geoenvironmental Engineering*, 133(8): 943–958.

Anastasopoulos, I., Antonakos, G. & Gazetas, G. 2010. Slab foundation subjected to thrust faulting in dry sand: Parametric analysis and simplified design method. *Soil Dynamics and Earthquake Engineering*, 30(10): 912–924.

Ashtiani, M., Ghalandarzadeh, A. & Towhata, I. 2015. Centrifuge modeling of shallow embedded foundations subjected to reverse fault rupture. *Canadian Geotechnical Journal*, 53(3): 505–519.

International Technical Committee TC2. 2005. Catalogue of scaling laws and similitude questions in centrifuge modeling [online]. Available from http://geo.citg.tudelft.nl/allersma/tc2/TC2%20Scaling.pdf.

Kelson, K.I., Kang, K.H., Page, W.D., Lee, C.T. & Cluff, L.S. 2001. Representative styles of deformation along the Chelungpu fault from the 1999 Chi-Chi (Taiwan) earthquake: geomorphic characteristics and responses of man-made structures. Bull*etin of the Seismological Society of America*, 91(5): 930–952.

Konagai, K. 2005. Data archives of seismic fault-induced damage. *Soil Dynamics and Earthquake Engineering*, 25(7-10): 559–570.

Lin, M.L., Chung, C.F. & Jeng, F.S. 2006. Deformation of overburden soil induced by thrust fault slip. Engineering Geology, 88(1-2): 70–89.

Moradi, M. & Ghalandarzadeh, A. 2010. A new geotechnical centrifuge at the University of Tehran, I.R. Iran. *In Conference on Physical Modeling in Geotechnics*, ETH Zurich, 28 June–1 July 2010, Switzerland: 251–254.

Seismic amplification of clay ground and long-term consolidation after earthquake

Y. Hatanaka & K. Isobe
Hokkaido University, Hokkaido, Japan

ABSTRACT: In the Great East Japan earthquake of 2011, long-term settlement in the alluvial clay ground was observed in Miyagi Prefecture. And, a similar settlement was also observed in Niigata Prefecture Chuetsu-Oki Earthquake of 2007. The ground investigation revealed that this alluvial clay ground is composed of structured clay. It is thought that long-term settlement is caused by deterioration of this structure. In addition, ground motion may be amplified in the structured clay ground. Therefore, centrifuge model tests were conducted to reveal the mechanism of long-term settlement due to earthquakes and seismic amplification of clay ground.

1 INTRODUCTION

At the time of an earthquake, it is well known that liquefaction is caused in the sandy ground. So far, various research for liquefaction countermeasure in sandy ground has been conducted. On the other hands, in clayey ground, it is thought that the large-scale damage such as liquefaction seldom happen. But, recently, some damages involved with clay ground were reported (Yasuhara et al. 2011; Isobe & Ohtsuka 2013). Specifically, it is guessed that long-term settlement after the earthquakes was caused by deterioration of structure in the structured clay ground. And also, there is possibility that structured clay with large stiffness affect the earthquake wave response. However, the influence on earthquake wave response properties from the viewpoint of degree of structure is not still elucidated. Therefore, centrifugal model tests are conducted using the model ground of the 1/50 scale to clarify mechanism of long-term settlement after earthquakes and evaluate earthquake wave response properties of clay ground in this study.

2 EXPERIMENT OUTLINE

2.1 Model grounds and their properties

Kaolin clay ($\rho_s = 2.63$ (g/cm³), $W_L = 43.50$ (%), $W_P = 28.12$ (%), $I_P = 15.38$ (%)) was used, adding an ordinary Portland cement with dry mass ratio of 0.5% to the clay to simulate structured clay ground. Higher degree of structured clay ground was made curing for 3 days, and less degree of structured ground was made by remixing and reconsolidating after cured for 28 days more. As shown in Figure 1, void ratio of kaolin clay cured for 3 days is higher than that of reconsolidated after cured for 28 days more at the same consolidation pressure. Thus, structured clay ground

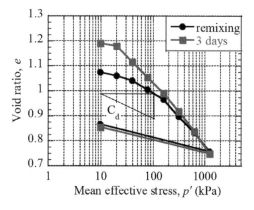

Figure 1. Consolidation curve of Kaolin clay.

can be reproduced. Undrained cyclic compression triaxial tests were conducted for above two types of specimens. As shown in Figure 2, liquefaction strength of the kaolin clay without cement is 0.15. The number of cycles at double amplitude of axial strain reaches 5 % is 35 times (remixing) and 32 times (3 days) in Figure 3. Figure 4 shows the relationship between deviator stress and axial strain for both specimens. A dynamic compression index, C_d, which is defined by Equations (1) and (2), was 0.512 (remixing) and 0.555 (3 days), respectively. It indicates that the structured clay (3 days) has higher compressibility than the less structured clay (remixing) against cyclic shear loading.

$$C_d = \frac{\Delta u}{\log \text{SRR}} \tag{1}$$

$$\text{SRR} = \frac{p'}{p' - \Delta u} = \frac{1}{1 - \Delta u / p'} \tag{2}$$

where, Δu: excess pore water pressure, p': effective stress.

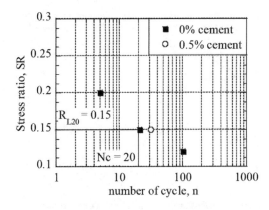

Figure 2. Liquefaction intensity curve of Kaolin clay.

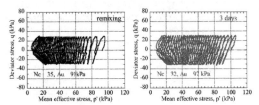

Figure 3. Mean effective stress path for the specimens.

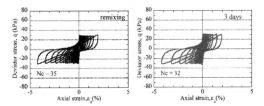

Figure 4. Axial strain and deviator stress for the specimens.

In the centrifugal model experiments, ground motions were given for model grounds with the scale of 1/50 from the ground bottom under centrifugal acceleration of 50 G. The model grounds are made in laminar boxes (40 mm × 9 stages) attaching membrane with a thickness of 0.3 mm on the inside walls, and the depth is 10 m in total with 2.5 m per one layer in prototype scale as shown in Figure 5. Each laminar box can move in the horizontal direction with less friction by attaching bearings between laminar boxes. The model grounds were prepared by self-weight consolidation under centrifuge acceleration of 50 G layer by layer for about an hour. Sand layers (silica sand No. 7) were used in bottom and top ends of the model ground by air pluviation method. Excess pore water pressure, acceleration and settlement of each layer during and after shakes were measured in the series of tests. The outline of model ground and layout of instruments are shown in Figure 5.

2.2 Shaking conditions and test cases

After having been stable in the centrifuge field with an acceleration of 50 G, the ground motions were

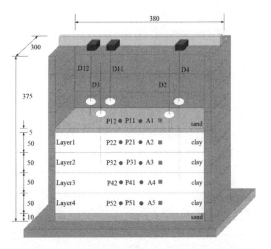

Figure 5. Model grounds in model type scale (unit: mm), P : Hydraulic pressure gauge, A : Accelerometer, D : Displacement gauge.

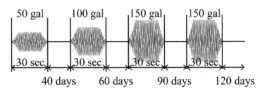

Figure 6. Scenario of the input wave (prototype scale).

generated, changing the magnitude of the input acceleration progressively after the excess pore water pressure was dissipated. The input waves used in the experiment were all sine waves with tapers, and the number of waves per an earthquake is 30. As shown in Figure 6, the target input accelerations at each step were 50 gal, 100 gal, 150 gal and 150 gal, respectively. During and after the shake, the surface of the ground was under drained condition, the side and bottom of the ground were under undrained condition.

Two cases were conducted in this series of tests. For the model ground in Case 1, kaolin clay with less structure was used from Layer 1 to Layer 4. In Case 2, less structured kaolin clay is used for Layer 1, 2 and 4 in the same manner as Case 1, but high degree of structured clay cured for 3 days is used for Layer 3 to simulate the real ground in which was really damaged by Great East Japan earthquake of 2011 and Niigata-ken Chuetsu-oki Earthquake of 2007 (Isobe & Ohtsuka 2013).

3 EXPERIMENT RESULTS

3.1 Acceleration response

The time histories of response acceleration in the ground for both cases are shown in Figures 7 and 8. But, for the convenience of the space, that of 100 gal and second 150 gal are omitted. As an overall trend,

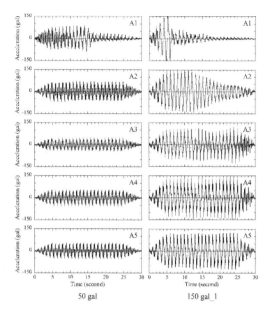

Figure 7. Time history of the response acceleration of the ground (Case 1).

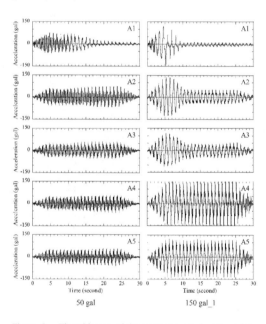

Figure 8. Time history of the response acceleration of the ground (Case 2).

the response acceleration tended to be amplified larger for the shallower ground against the small ground motions. Therefore, the seismic waves are amplified against small ground motions in the clay ground. However, the response acceleration in the surface layer (A1) sharply decreased as the input motion and response acceleration increased. In addition, the sudden reduction of the response acceleration in other upper part of the ground were seen against the larger ground

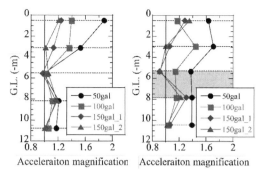

Figure 9. Maximum values of the response acceleration normalised by input acceleration (left: Case 1, right: Case 2).

motions, and the period of the response acceleration were prolonged generating sudden reduction, resulting from decrease of the stiffness due to generating excess pore water pressure and larger shear strain which is discussed later part.

The maximum values of response acceleration at each measurement point normalised by the input acceleration are shown in Figure 9 As mentioned above, in both cases (Case 1 and Case 2), the maximum value of the response acceleration tended to increase at the upper part of the ground when the amplitude of the input acceleration was less than 100 gal. However, the maximum value of the response acceleration decreased when the amplitude of the input acceleration was more than 100 gal for the following presumptive reasons; degradation of the stiffness of the ground due to excess pore water pressure and shear strain.

In addition, the maximum value of the response acceleration of Layer3 in Case 2 was larger than that in Case 1 for the first (50 gal) and second steps (100 gal). It means that the seismic wave is amplified more in the high degree of structured clay ground with larger stiffness.

3.2 *Excess pore water pressure ratio*

The maximum values of excess pore water pressure ratio in depth during and after shaking is shown in Figure 10. The maximum values of excess pore water pressure ratio in Case 2 were larger than that in Case1 for the first and second steps (50 gal and 100 gal).

It is thought to be caused by the result that seismic wave is amplified more in the highly-structured clay ground. However, the excess pore water pressure ratio of both cases was almost identical for the third and fourth steps (150 gal). It can be thought that the ground for Case 2 became similar condition as Case 1 due to the deterioration of the structure against larger ground motions.

3.3 *Settlement*

The settlement of each layer and total settlement are shown in Figure 11. The settlement of each layer in Case 1 came to be small sequentially from the surface

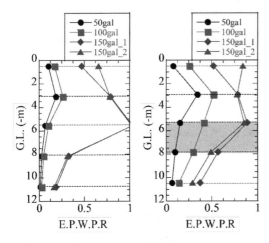

Figure 10. Maximum values of the excess pore water pressure ratio (left: Case 1, right: Case 2).

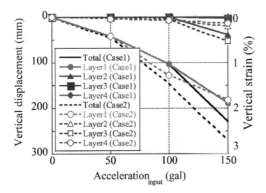

Figure 11. Vertical displacement after each step.

layer to the depth direction. However, the settlement of Layer 3 with highly-structured clay in Case 2 was relatively larger than the other layers. It is the reason why the seismic waves were amplified in highly-structured clay ground with larger shear stiffness, and then fragile behaviour was observed by deterioration of structure with large shear deformation and excess pore water pressure. As a result, total settlement of Case 2 was larger than that of Case 1. It is consistent with the result that the dynamic compression index, C_d of the structured clay is larger.

3.4 Shear stress – shear strain relationship

The shear stress normalized by the initial effective stress and the shear strain in each layer are shown in Figure 12. The shear stress at the midpoint of nth layer was calculated by the eq. (3) (Kazama et al., 1996). Also, the shear strain of nth layer was calculated by the eq. (4).

$$\tau_n = \sum_{i=1}^{n-1}(\bar{a}_i H_i \rho_i) + \frac{\rho_n H_n}{8}(3a_n + a_{n+1}) \quad (3)$$

$$\gamma_n = (d_n - d_{n+1})/H_n \quad (4)$$

where $\bar{a}_i = (a_i + a_{i+1})/2$, a_i: response acceleration; H_n: thickness; ρ_i: density and d_n: displacement obtained by twice integrating the measured acceleration. Shear stiffness is estimated at the unloading point ($\dot{\gamma}_n = 0$) to eliminate the influence of damping force.

Based on this method, the shear stiffness at Layer 1 in all shakes was estimated as negative value because the response acceleration observed at A1 decayed. Therefore, the data after decaying was not shown in Figure 12. In Case 1, the shear strain is generated more in Layer 1, and the shear stiffness normalised by the initial vertical effective stress remarkably decreased during shakes. On the other hands, in Case 2, such a trend was seen in not only Layer 1 but also Layer 3 with high soil structure. This result supports that structured clay ground is vulnerable to large-scale earthquakes.

3.5 Cumulative shear strain

The relationship between cumulative shear strain and excess pore water pressure ratio at each shake is shown in Figure 13. Cumulative shear strain is calculated by summation of the double amplitude of shear strain generated by each wave. The fourth shake (150 gal_2) was almost the same as the third shake (150 gal_1), so it was excluded in this figure. There was no significant difference in the rate of the rise of excess pore water pressure ratio against the increase of cumulative shear strain in each layer for Case1. However, in Case 2, a rapid increase in cumulative shear strain due to deterioration of soil structure was observed during the third step in Layer 3. In the upper layers such as Layer 1 and Layer 2, amplification of the response acceleration due to the interposition of the clay ground with the developed soil structure and the increase of the shear deformation resulted in the great decrease of the stiffness. Thus, the cumulative shear strain did not increase as in Case 1 even if strong earthquake motion was applied. Based on the above, although the highly-structured clay ground has high initial stiffness, it amplifies the seismic waves against small-scale ground motions. If the soil structure deteriorates due to large-scale ground motions, the stiffness is reduced and it is considered to behave fragile.

3.6 Response acceleration magnification

The relationship between the response acceleration magnification and the shear stiffness ratio of each layer for both cases is shown in Figure 14. There is a tendency that the shear stiffness ratio of Case 2 is larger than that of Case 1 across the board. In the range where the shear stiffness ratio is small, it is considered that excess pore water pressure is generated by large shear deformation, resulting in the degradation of the shear stiffness of the ground, and the response acceleration magnification rapidly decreased, and it finally became difficult for the acceleration to propagate. On the other hands, in the range where the shear stiffness ratio is

894

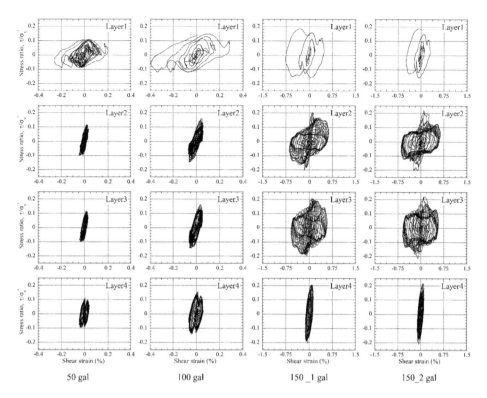

Figure 12(a). Relationship between shear stress normalized by initial effective stress – shear strain at each layer (Case 1).

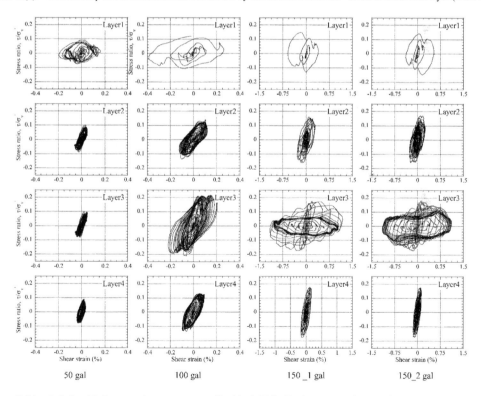

Figure 12(b). Relationship between shear stress normalized by initial effective stress – shear strain at each layer (Case 2).

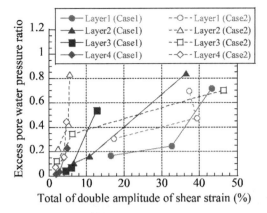

Figure 13. Relationship between cumulative shear strain and excess pore water pressure ratio.

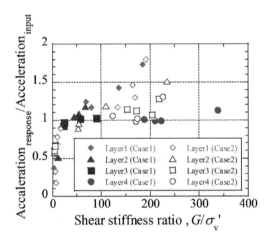

Figure 14. Relationship between amplified ratio of response acceleration and shear stiffness normalised by initial vertical effective stress.

large, the amplification of acceleration occurs and the amplification effect is recognised, but it is presumed that the response acceleration magnification finally converges to 1 as the shear stiffness ratio increases further. In the range where the shear stiffness ratio is intermediate, response acceleration magnitude is more than 1, and it is suggested that there is a shear stiffness ratio at which the response acceleration magnification reaches the peak value. In Case 2, the shear stiffness ratio in Layer 3 with the structured soil and the upper layer which is strongly influenced by Layer 3 is in near the shear stiffness ratio which indicates the peak. It is guessed to be a factor showing a large acceleration response in Case 2.

4 CONCLUSION

I. It was confirmed that the response acceleration of the ground with the developed soil structure (Case 2) increased largely at the early stage, comparing to the ground with the less structure (Case 1). It induced large strain in the layer with high soil structure of Case 2.

II. The excess pore water pressure increased with increase in cyclic loading, the response acceleration becomes smaller and the shear strain increased more, resulting in that the shear rigidity ratio decreases. This trend was notably observed in the ground with high soil structure (Case 2).

III. In the ground with less soil structure (Case 1), a large settlement is observed in the upper layer after the earthquakes. On the other hands, in the clay ground with the developed structure (Case 2), fragile behaviours due to deterioration of soil structure generated, resulting in a large final settlement amount after the earthquakes. This result is consistent with the cyclic triaxial test results which shows the dynamic compression index, C_d of the structured clay is larger than that of the less structured clay.

REFERENCES

Isobe, K. & Ohtsuka, S. 2013. Study on long-term subsidence of soft clay due to 2007 Niigata Prefecture Chuetsu-Oki Earthquake, Proceedings of the 18th International conference on soil mechanics and geotechnical engineering, Paris, France, 1: 1499–1502.

Kazama, M., Toyota, H., Towhata, I. & Yanagisawa, E. 1996. Stress strain relationship of sandy soils obtained from centrifuge shaking table tests. Journal of JSCE. 535(III-34): 73–82. (in Japanese)

Yasuhara, K., Murakami, S., Masuda, K. Sonobe, T. & Saito, O. 2011. Geotechnical damage features in Mito during the 2011 off the Pacific coast of Tohoku Earthquake. Disaster investigation information of the Tohoku-Pacific Ocean Earthquake. (In Japanese)

Evaluation of period-lengthening ratio (PLR) of single-degree-of-freedom structure via dynamic centrifuge tests

K.W. Ko, J.G. Ha, H.J. Park & D.S. Kim
Department of Civil and Environmental Engineering, KAIST, Dajeon, Korea

ABSTRACT: Understanding the soil-foundation-structure interaction (SFSI) is crucial for the seismic design. Two important parameters of SFSI, the period-lengthening ratio (PLR) and the foundation-damping ratio (FDR), have been studied by both analytical and experimental methods. However, the present formulas of PLR considering SFSI could not reflect the non-linear characteristics of soil. In this study, to reflect the non-linear characteristics of soil in PLR, shear modulus of soil was back-calculated by using measured PLR from centrifuge test results. The three structure models which have different natural periods and two shallow foundation models which have different mass were used to perform the dynamic centrifuge tests. From the test results, the PLR of the single-degree-of-freedom (SDOF) structure was evaluated by considering the seismic intensity and frequency contents of seismic waves. The reduction coefficient of shear modulus, which can predict the PLR by using present formula, was suggested as a function of peak ground acceleration at surface.

1 INTRODUCTION

Determination of seismic load on the structure is important for seismic design. Currently, by assuming the SDOF structure with fixed-base condition, maximum response of SDOF structure corresponding to the natural period of structure is evaluated from the spectral acceleration. However, in practice, most of structure would be constructed on the soil, and the structure condition no longer conforms the fixed-base condition. Since the soil increases the degree of freedom for whole system with rocking and sliding motion, two significant phenomena, period lengthening and damping increase, would be generated, and these effects are called SFSI effect. Due to these effects, the seismic load on structure would be different from the response spectrum. Therefore, it is significant to evaluate the period lengthening and damping increasing accurately for a reliable seismic design.

The PLR is ratio of structural natural period considering SFSI to natural period on fixed-base condition (Wolf 1985). The PLR depends on the structural effective stiffness, rocking and horizontal stiffness of foundation. Also, previous studies have suggested the formulas of foundation rocking and sliding stiffness (Gazetas 1991, Gazetas et al. 2013). Suggested formulas of horizontal and rocking stiffness for shallow foundation is governed by shear modulus of soil. On the other hand, shear modulus has non-linear characteristics with strain level. During the earthquake, the non-linear characteristics of shear modulus of soil would affect the rocking and horizontal stiffness of foundation, and the variation of each foundation stiffness value would change the PLR. Consequently, the formulas of PLR should reflect the non-linear characteristics of soil during the earthquake, but present formulas could not reflect the non-linear characteristics of soil. In case of the previous studies, FEMA 440 (2005) provided PLR and FDR formulas. Stewart et al. (1999a,b) pointed out general SFSI formulas didn't reflect the strong motions transmitted to structures and the structural response to these motions and analysed these problems by using analytical and empirical methods. However, these approaches did not be compared with experimental results.

The objective of this study is to evaluate the period-lengthening ratio (PLR) depending on the seismic intensity and frequency of earthquake loading. Three structure models with different natural periods and two foundation models with different mass were used to perform the centrifuge tests. Five different seismic waves (e.g. Ofunato, Hachionhe, Northridge, Sin 4 Hz and Sin 2 Hz) were inputted for dynamic centrifuge tests. The accelerometers on structure, foundation and soil measured the dynamic response of each part, respectively. The PLR from the test results was compared with the PLR from analytical approach. Finally, to reflect the non-linear characteristics of soil easily, the reduction coefficient of shear modulus, which can predict the PLR, was suggested as a function of peak ground acceleration at surface. Of course, the system nonlinearity could contribute to the non-linear characteristics of PLR, but, in this study, the non-linear characteristics of PLR would be only focused on the shear modulus of soil.

Table 1. Natural frequency and period of the structure models.

Structure models	Natural frequency Hz	Natural period S
SDOF1	3.7	0.27
SDOF2	2.7	0.37
SDOF3	2.3	0.44

Table 2. Properties of foundation models.

Foundation models	Mass Tonne	Material
FND1	4.36	Aluminium
FND2	18.72	Steel

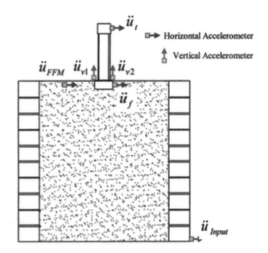

Figure 1. Schematic diagram of dynamic centrifuge test (Kim et al. 2015).

2 CENTRIFUGE TEST PROGRAM

In this study, centrifuge test was conducted for simulating the soil stress condition. A centrifuge apparatus in KAIST KOCED centre was used to do experiment and all tests were performed in 20 g centrifugal acceleration. The maximum payload capacity of the KAIST beam centrifuge with 5 m radius is between 2,400 kg for 100 g and 1300 kg for 130 g centrifugal acceleration for static test condition (Kim et al. 2013a). The loading frequency range of the earthquake simulator at KAIST is 40 to 300 Hz and 40 to 200 Hz for random vibration and sinusoidal wave, respectively, and the other specifications of the earthquake simulator is described in Kim et al. 2013b. All results and model description are presented in prototype units by applying scaling law.

2.1 SDOF structure and shallow foundation model

Table 1 presents the natural period and frequency of the SDOF structure models. The structure and foundation behaviour during earthquake depends on the slenderness ratio (h/L), structure height (h) over foundation length (L) (Gajan & Kutter 2009). To have rocking behaviour of system motion as a dominant motion during the earthquake, all of the height of SDOF structure were made with same height value, and the slenderness ratio of all of testing models are equal to 1.75. Although the effective stiffness (k_s) of structure models are same as 60.5 kN/m, the natural period of each SDOF structures are different since the different mass were located on the top of the structure.

Table 2 presents properties of two foundation models. One of the shallow foundation model was made with an aluminium box shape. The other model was made with a steel box shape. External dimension of these foundations are the same, however aluminium foundation is a hollow foundation. Because of that, mass of foundation is largely different. The dimensions of rectangular shallow foundation model in prototype scale are $2 \times 2 \times 0.6$ m (length × width × height).

2.2 Ground model

The silica sand layers with 60% of relative density ($\gamma_d = 1.48$ t/m^3) were made using a sand raining system. The silica sand thickness was 580 mm in model scale and 11.6 m in prototype scale. The internal friction angle of silica sand is 41 degrees, and Poisson's ratio is 0.26. The site period of soil is about 0.22 sec. In this study, the equivalent shear beam (ESB) box, which reduces the boundary effect, was used to perform the test (Lee et al. 2013).

2.3 Dynamic centrifuge test

Figure 4 shows the test setup and the sensor arrangement of the dynamic tests in 20 g. The thickness of soil layer was about 20 times of foundation height. Shallow foundation was fully embedded same as the height of the foundation model. The accelerometer was used to measure the acceleration of each part during dynamic tests.

Each testing models were subjected to the sweep signal, real earthquake motions (Ofunato, Hachinohe & Northridge), and sinusoidal waves (Sin 4 Hz & Sin 2 Hz). To investigate the natural frequency of the soil, foundation, and structure system, small intensity sweep signal which has uniform energy in various frequency range was applied as a base motion. Ofunato earthquake motion which represents the short period dominated earthquake signal, and Hachinohe earthquake motion which represents the long period dominated earthquake signal, were used as input base motion. Also, the frequency of sinusoidal wave was similar to the natural frequency of structure. Initially, the sweep signal was applied to the testing model, then strong earthquake signals of real earthquake (around 0.35 g) were applied. After that, intensity of input motions increased in stages from small to large acceleration. The number of input earthquakes and

Table 3. The number of input earthquake and peak acceleration of each testing models.

Testing models	The number of input earthquake	Acceleration range (min-max) g
SDOF1	51	0.05–0.37
SDOF2	49	0.06–0.58
SDOF3	45	0.06–0.38

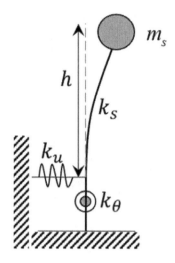

Figure 2. Simplified model of SDOF structure with foundation rocking and sliding spring.

peak acceleration range of each case are presented in Table 3.

3 PERIOD LENGTHENING RATIO: ANALYTICAL APPROACH

Figure 2 presents SDOF structure on the flexible base considering foundation rocking and sliding. Due to the foundation motion, SDOF structure has not only the structural bending motion but also foundation horizontal, and rotational motion during earthquake. From the structural dynamics, the undamped natural period (T), of the structure on the fixed condition are given by following equation.

$$T = 2\pi \sqrt{\frac{m_s}{k_s}} \quad (1)$$

where, m_s = effective mass of SDOF structure.

However, foundation-structure system has rocking stiffness (k_θ) and sliding stiffness (k_u) and it causes period lengthening phenomenon for the first mode of structural natural period. Also, lengthened natural period ($\tilde{T}$) of structure is determined by Equation (2).

$$\tilde{T} = 2\pi \sqrt{m(\frac{1}{k} + \frac{1}{k_u} + \frac{h^2}{k_\theta})} \quad (2)$$

By combining Equation 1 and 2, period-lengthening ratio (PLR) could be derived as follows,

$$PLR = \frac{\tilde{T}}{T} = \sqrt{(1 + \frac{k}{k_u} + \frac{kh^2}{k_\theta})} \quad (3)$$

From the Equation 3, if the height and effective stiffness of SDOF structure are identical, the foundation sliding and rocking stiffness would govern the PLR. The foundation rocking and sliding stiffness are determined by soil and foundation parameters such as shear modulus (G), poisson's ratio (ν), and foundation effective radius for sliding (r_u) and rocking (r_θ), which are expressed as follows,

$$r_u = \sqrt{A_f / \pi} \quad (4)$$

$$r_\theta = \sqrt[4]{4I_f / \pi} \quad (5)$$

where, A_f = area of foundation; and I_f = moment of inertia of foundation. The foundation sliding and rocking stiffness are provided in Gazetas (1991).

$$k_u = \frac{8}{2 - \nu} G r_u \quad (6)$$

$$k_\theta = \frac{8}{3(1 - \nu)} G r_\theta^3 \quad (7)$$

Through the Equation 1 to 7, Table 4 summarized the PLR and sliding, rocking stiffness of structure and foundation of testing models. Although three different structure models had different natural period, PLR of those models are same for same foundation model since the height and effective stiffness of structure models are same. It means present formulas for PLR could not reflect the effect of frequency dependency, so that this study would discuss the PLR depending on the natural period of structure. On the other hand, the identical structure model with different foundation had different PLR value and FND1, made by aluminum, generates larger period lengthening phenomenon than FND2. Light foundation (FND1) has much lower value of moment of inertia, then it causes lower value of foundation radius (r_θ). Finally, due to difference of the foundation stiffness, PLR of FND1 is larger than those of FND2.

The notable results from analytical estimation are the contribution of rocking motion to PLR, which is larger than those of sliding motion. When the effective radius of FND1 for sliding motion is changed into infinite value, which indicates that the sliding motion of foundation is restricted, PLR of FND1 would be changed into 1.052. In contrast to that, if the effective radius of FND1 for rocking motion is changed into infinite value, which indicates the rocking motion of foundation is restricted, then PLR of FND1 is changed into only 1.004. Therefore, the rocking motion of foundation contributes to the PLR more than 90%, and it

Table 4. Analytical estimation of sliding, rocking stiffness and PLR of structure and foundation testing models.

Testing models	FND1	FND2
r_u (m)	0.92	1.13
r_θ (m)	0.93	1.14
k_u (kNm)	1.30×10^5	1.52×10^5
k_θ (kNm)	1.40×10^5	2.33×10^5
Percent of contribution to PLR (%)		
Sliding motion	7.83	11
Rocking motion	92.17	89
PLR		
SDOF1	1.06	1.04
SDOF2	1.06	1.04
SDOF3	1.06	1.04

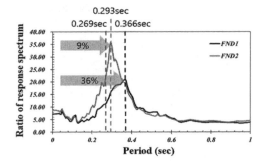

Figure 3. Ratio of response spectrum of SDOF1 on FND1 and FND2 for Northridge earthquake (peak ground acceleration at bedrock: 0.18 g).

signifies that rocking motion of foundation is much dominant for determining the PLR. However, analytical estimation did not reflect the shear modulus degradation with strain level by seismic loading intensity. So the following section discussed the PLR with seismic intensity through the centrifuge test results.

4 CENTRIFUGE TEST RESUTLS

4.1 Period lengthening with foundation model

From the centrifuge test results, natural period of SDOF structure model was evaluated with ratio of response spectrum (RRS). Ratio of response spectrum was determined by ratio the response spectrum of top of structure motion to the response spectrum of surface motion measured by accelerometers.

Figure 3 shows the RRS of SDOF1 for Northridge earthquake (peak ground acceleration of bedrock: 0.18 g). As shown in Figure 3, FND1 has larger PLR than those of FND2, since the rocking stiffness of FND2 is much larger than FND1. This tendency matches analytical results in Section 3. Unfortunately, experimental PLR value of foundation models is different from analytical solution which provides PLR of FND1 and FND2 as 1.06 and 1.04, respectively. PLR from the centrifuge test results presented 1.36 and 1.09 for FND1 and FND2, respectively. As previously discussed, the PLR from analytical approach could not reflect the non-linear characteristics of shear modulus during the earthquake. In addition, the shear modulus of soil varies with the seismic loading intensity. As a result, to determine the accurate PLR, the seismic intensity has to be considered.

4.2 PLR considering non-linear characteristics

Figure 4 shows PLR with peak ground acceleration (PGA) at surface for SDOF1. PLR of FND1 is much larger than those of FND2 regardless of PGA. However, PLR for each foundation models presents a tendency with seismic intensity (PGA at surface). Strong earthquake generated significant nonlinearity

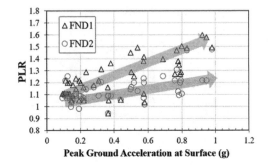

Figure 4. PLR of SDOF1 on FND1 and FND2 with PGA at surface.

in soil and it causes larger shear modulus degradation. Therefore, since the rocking and sliding stiffness of foundation decreased with PGA at surface, the stronger earthquake motion causes the larger period lengthening phenomenon.

To improve the analytical approach for determining the PLR, non-linear characteristics of soil has to be reflected. Through the Equation 6 and 7, shear modulus degradation coefficient (I_0), which matches the PLR between analytical and experimental approaches, was back-calculated from the measured PLR of centrifuge test results. As a result, the Equation 6 and 7 were modified as Equation 8 and 9 considering shear modulus degradation.

$$k_u^* = \frac{8}{2-v}(I_0 G) r_u \qquad (8)$$

$$k_\theta^* = \frac{8}{3(1-v)}(I_0 G) r_\theta^3 \qquad (9)$$

where, k_u^* = sliding stiffness of foundation considering non-linear characteristics of soil; and k_θ^* = rocking stiffness of foundation considering non-linear characteristics of soil.

Figure 5 shows the back-calculated shear modulus degradation coefficient (I_0) with peak ground acceleration at surface for all of the earthquake events on SDOF1, and SDOF2. Lower than 0.4 g and 0.5 g with PGA at surface, shear modulus degradation coefficient

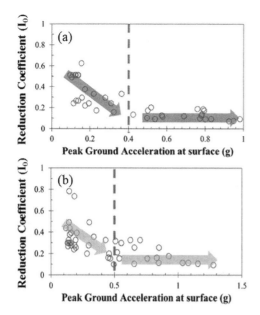

Figure 5. Shear modulus degradation coefficient (I_0) for all of earthquake events with peak ground acceleration at surface: (a) SDOF1; (b) SDOF2.

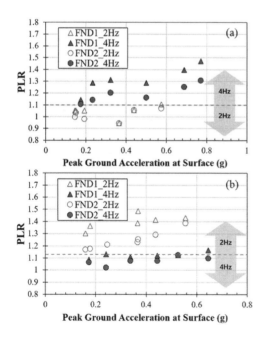

Figure 6. The PLR with PGA at surface during sinusoidal waves: (a) SDOF1; and (b) SDOF3.

(I_0) decreases with PGA at surface. However, when the PGA at surface is larger than 0.4 g and 0.5 g for SDOF1 and SDOF2, respectively, the coefficient (I_0) was converged to 0.1. Since the degradation of coefficient (I_0) reflect the shear modulus degradation, the tendency of Figure 5 is similar to shear modulus degradation with strain level. As a result, by using suggested coefficient, the PLR considering the non-linear characteristics of soil could be determined.

4.3 PLR considering frequency effect

As previously discussed, the present PLR formulas could not reflect the frequency effect. In this section, to observe the frequency effect on PLR, the PLR of SDOF1 and SDOF3, which have different natural period of structure, was evaluated with sinusoidal wave, close to the natural period of structure model.

Figure 6 shows the PLR of SDOF1 and SDOF3 structure models with PGA at surface for sin 2 Hz and sin 4 Hz. In case of SDOF1, since the natural frequency of SDOF1 is much close to the 4 Hz, sin 4 Hz wave generated much larger PLR than sin 2 Hz. In contrast to that, since the natural frequency of SDOF3 is much close to the 2 Hz, sin 2 Hz wave generated much larger PLR than sin 4 Hz. During the earthquake, which has similar dominant frequency with natural frequency of structure, the structure would behave much significantly due to resonance, then it would cause the larger degradation of shear modulus. Therefore, the PLR of structure was larger during the seismic wave which had similar dominant frequency with natural frequency of structure.

5 CONCLUSIONS

To evaluate the period-lengthening ratio (PLR) for SDOF structure by considering SFSI effects, analytical methods and dynamic centrifuge tests were conducted. Three SDOF structure models which had different natural periods and two foundation models which were made into two different materials were used to perform centrifuge tests. The important findings are summarized as follows:

1. Through the analytical methods, PLR of aluminum foundation had larger value than those of steel foundation. These effects were governed by rocking stiffness difference. Also, PLR is always same value regardless of SDOF structure models. However, analytical methods was not able to reflect the shear modulus degradation and frequency effects.
2. From the centrifuge test results, PLR was determined depending on the seismic intensity (PGA at surface). Accordingly, shear modulus degradation coefficient (I_0) was suggested for matching the analytical PLR with PLR from the test results. The shear modulus degradation coefficient decreased with PGA at surface. However, the shear modulus degradation coefficient was converged to 0.1 in large PGA at surface.
3. PLR was also affected by seismic wave frequency contents. If seismic wave frequency is similar to natural frequency of structure, it caused much bigger period lengthening phenomenon regardless of peak ground acceleration at surface.

ACKNOWLEDGEMENT

This work was supported by the National Research Foundation of Korea (NRF) grant funded by the Korea government (MSIT) (No. 2017R1A5A1014883).

REFERENCES

Federal Emergency Management Agency (FEMA) 2005. *Improvement of nonlinear static seismic analysis procedures*. Washington, D.C.: Dept. of Homeland Security.

Gajan, S. & Kutter, B. L. 2009. Effects of moment-to-shear ratio on combined cyclic load-displacement behavior of shallow foundations from centrifuge experiments. *Journal of Geotechnical and Geoenvironmental Engineering* 135(8): 1044–1055.

Gazetas, G. 1991. Formulas and charts for impedances of surface and embedded foundations. *Journal of Geotechnical and Geoenvironmental Engineering* 117(9): 1363–1381.

Gazetas, G., Anastasopoulos, I., Adamidis, O., & Kontoroupi, T. 2013. Nonlinear rocking stiffness of foundations. *Soil Dynamics and Earthquake Engineering* 47: 83–91.

Kim, D.S., Kim, N.R., Choo, Y.W. & Cho, G.C. 2013a. A newly developed state-of-the-art geotechnical centrifuge in Korea. *KSCE Journal of Civil Engineering* 17(1): 77–84.

Kim, D.S., Lee, S.H., Choo, Y.W. & Rames, D. 2013b. Self-balanced earthquake simulator on centrifuge and dynamic performance verification. *KSCE Journal of Civil Engineering* 17(4): 651–661.

Kim, D.K., Lee, S.H., Kim, D.S., Choo, Y.W., & Park, H.G. 2015. Rocking effect of a mat foundation on the earthquake response of structures. *Journal of Geotechnical and Geoenvironmental Engineering* 141(1): 1–13.

Lee, S.H., Choo, Y.W. & Kim, D.S. 2013. Performance of an equivalent shear beam (ESB) model container for dynamic geotechnical centrifuge tests. *Soil Dynamics and Earthquake Engineering* 44: 102–114.

Stewart, J.P., Fenves, G.L. & Seed, R.B. 1999a. Seismic soil-structure interaction in building. I: Analytical methods. *Journal of Geotechnical and Geoenvironmental Engineering* 125(1): 23–37.

Stewart, J.P., Seed, R.B. & Fenves, G.L. 1999b. Seismic soil-structure interaction in building. I: Analytical methods. *Journal of Geotechnical and Geoenvironmental Engineering* 125(1): 23–37.

Wolf, J.P. 1985. *Dynamic soil-structure interaction*. Englewood Cliffs: Prentice-Hall.

Centrifuge modelling of active seismic fault interaction with oil well casings

J. Le Cossec* & K.J.L. Stone
School of Environment and Technologies, University of Brighton, Brighton, UK
*currently: Centre for Energy and Infrastructure Ground Research, University of Sheffield, Sheffield, UK

C. Ryan & K. Dimitriadis**
Tullow Oil plc, London, UK
**currently: Sasol, London, UK

ABSTRACT: Interaction of active seismic faults with oil well casings has been investigated to evaluate the interaction process and determine approaches to mitigate damage and deformation of the well casing. The methodology used was a 2-way approach consisting of (1) undertaking a series of unscaled physical model tests in a geotechnical centrifuge using on simplified geological analogues and (2) validation of 3D numerical models by verification and calibration to the experimental test results. Two type of geological scenarios are presented: an unconsolidated cohesionless deposit and a bedded rock mass deposit with 60° dip bedding. A basement discontinuities of displacement (normal faulting) and of slope and displacement (sloped base deformation) were introduced. The interaction with the propagating faults through the model was studied with post-experiment dissection and PIV analysis. These information were used to calibrate a numerical models and draw appropriate conclusions with regard to mitigation methods.

1 INTRODUCTION

1.1 Preamble

The Macondo disaster in the Gulf of Mexico in 2010 illustrated the economic and environmental consequences of a critical failure in a production oil well. In this case the issues were magnified by the inaccessibly of the wellhead located on a deep ocean bed. The research reported in this document is an example of approach to understanding and managing risks associated with the resource development industry. Specifically the research is focused on the subsurface interaction of oil exploration or development wells with discontinuous ground movement associated with either seismic faulting and/or formation compaction. In both cases discontinuities of displacement develop and propagate through the ground.

1.2 Literature review

Casing shear in a reservoir is generally the result of a localized ground (soil or rock) displacement. In normal circumstances it is the production process that triggers such displacements. The production process induces volume changes inside the reservoir rock which modifies the stress field (Dusseault et al. 2001). By extracting oil and/or gas from reservoir rocks, the fluid pressure is reduced (in the absence of pressure support) and so the interstitial pore pressure is also reduced, from the principal of effective stress, the load on the rock matrix is therefore increased. Depending on the rock strength, the formation will first compact to adapt to the change of stress field, and eventually fail when the stress exceeds the rock strength. In both cases, rock displacements occur which can damage wells.

Casing damage has been observed in various settings: inside the overburden with shear damage localized on horizontal planes; shearing at the top of production and injection intervals; well compression and buckling damage around perforation within the production interval (Peng et al. 2007, He et al. 2005, Bruno 2001, Dusseault et al. 2001, Bruno 1992, Bruno 1990, McCauley 1974).

Displacements can also take place along newly created, steeply inclined, fault planes or along reactivated pre-existing faults (Chan & Zoback 2007) as a result of seismic activity.

The mitigation techniques are currently focused on two points: reducing the amount of shear slip on the casing, or avoiding highly stressed areas (Dusseault et al. 2001). Increasing the casing-grade/strength on specific horizons will retard shearing, and an accurate 2D-3D geomechanical model of the reservoir can identify areas of high shearing potential (Bruno 2001). With this approach it may then be possible to avoid areas prone to high shear by a careful selection of well locations or by the use of inclined drilling.

Table 1. Oil well/fault interaction scaling factors.

Rock	
Formation	Sandstone, Chalk, limestone
Thickness (m)	$n \times 100$

Oil well	
Diameter (cm)	18–34
Wall thickness (mm)	4–16
Grade	L80 steel

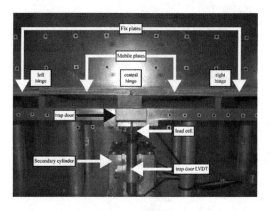

Figure 1. University of Cambridge centrifuge trap door experimental box. Configuration for the slope and displacement discontinuity.

2 EXPERIMENTAL APPROACH

The oil well/fault interaction is typically a problem to be addressed over several hundred meters of the subsurface (Tab. 1). To ensure similarity with the prototype, the model has to be scaled down using the same geometric factor on all dimensions. These constraints lead to an unsolvable problem. To accommodate the rock thickness, high G factor are required resulting in milimetric oil well analogue.

With a reduced scale model respecting similarity criterion out of reach, the experimental approach is the following: (1) producing, not at scale, simplified physical models without any attempt at maintaining the casing strength/rock stress ratio, nor the wellbore length, (2) reproducing the experimental model, using 3D numerical simulation, validating the software and (3) turning to a full-scale simulation and integrating field data into the numerical model.

This paper will principally cover the physical modelling aspect of this research project with minor highlights on the numerical simulation.

Two criterions on the casing/fault interaction are investigated: the fault network density and the type of geological formation.

Centrifuge modelling technique is used, providing more flexibility with the casing analogue material selection.

2.1 Fault generation mechanism

Introducing a discontinuity of displacement in the model will allow creation of faults.

Two type of displacement discontinuity have been deployed, a perfectly punctual discontinuity producing a dense fan of normal faults and a basement discontinuity of slope and displacement allowing for a sloped based deformation (Fig. 1).

The displacement discontinuity is produced by having a mobile boundary next to a fix one. A trap door configuration has been used were the mobile boundary is guided and controlled to only move vertically. The interface between the mobile and fix part will then anchor the faults.

The mobile boundary displacement speed is controlled with a primary/secondary hydraulic cylinder closed loop. The primary plunger position being finely controlled by being attached to an electric linear actuator.

2.2 Centrifuge equipment

The centrifuge experiment was performed at Schoffield Centre, University of Cambridge, using the 150 gT centrifuge. The experimental box allow for maximum soil dimensions of 300 mm thick × 790 mm long × 220 mm wide.

The configuration for the perfect displacement discontinuity is a 100 mm long × 220 mm wide mobile plate with a 25 mm vertical travel capacity. The mobile plate is positioned at the center of the box allowing for the development of two fault network; one at each edge of the mobile plate.

For the slope and displacement configuration (Fig. 1), the trap door equipment is enhanced with a three hinged basement. Two fixed plates (240 mm long × 200 mm wide) are set on the left and right end of the container. The mobile boundary is made of two 150 mm long × 200 mm wide plates. The central hinge rest above the secondary cylinder, once retracted the basement geometry assume a V shape.

The secondary cylinder capacity restricted the achievable acceleration to low G values (<50 G).

2.3 Analog material

2.3.1 Rock deposit
Two geological scenarios have been investigated: (1) an unconsolidated cohesionless deposit and (2) a jointed rock mass deposit with 60° inclined bedding.

Fraction E dry sand is the analogue material for the cohesionless unconsolidated deposit.

9.5 mm thick gypsum plasterboard is used to model the fractured rock mass.

2.3.2 Oil well
The model casings are made of a solid copper wire coated with gypsum plaster to simulate a grout

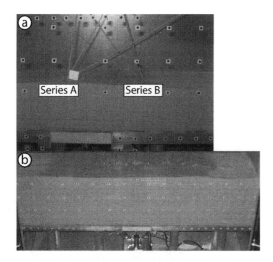

Figure 2. a) Model casings arrangement during pluviation. b) Model with surface instrumentation.

annulus. Plaster coating was done manually with emphasis on achieving a uniform coating thickness along the wire and to have the same thickness on all the wires.

3 UNCONSOLIDATED COHESIONLESS DEPOSIT – VERTICAL DISCONTINUITY

3.1 Model preparation

Fraction E dry sand is placed by automated dry pluviation achieving a uniform bulk density of $1540\,kg.m^{-3}$, for a total soil thickness of 200 mm. The relative density $Rd = 71\%$, $D_{10} = 192\,\mu m$ and $D_{60} = 525\,\mu m$, the sand friction angle is $34.6°$.

The model casing is formed from a 1.7 mm diameter solid copper core. Six model casings, split into two series of three, are positioned inside the model during sand pluviation (Fig. 2a).

For both series of tests, the lower end of the casings located 50 mm above the trap door. Series A casing end are offset 25 mm over the displacing base, while for series B, the casings are offset 50 mm to the right over the non-displacing base. Casings are spaced 30 mm apart over the width of the sample.

The casing locations are intended that, series A will be intercepted by a single fault and series B will be intercepted by two faults.

In each series, the casings are set with different dip angles (α). For series A, $\alpha = 30°, 60°, 80°$ and for series B, $\alpha = 40°, 50°, 75°$.

3.2 Instrumentation

Eight LVDT's are set to record the model surface deformation. They are spread evenly along the x-axis and centered above the trap door (Fig. 2b). The trap door itself is monitored with two LVDT's to record vertical displacement and tilting along the y-axis (off plane direction). One load cell, located between the trap door and the secondary cylinder plunger records the vertical load variation on the trap door during the experiment (Fig. 1). Instrumentation set up is completed with three cameras set to capture displacement along the z-x plane.

3.3 Experimental protocol

The test is conducted following six different phases:

- phase 1: spin up by 10 G increment up to 40 G.
- phase 2: 0.1 mm trap door displacement steps (total displacement from 0.0 to 0.5 mm).
- phase 3: 0.5 mm trap door displacement steps (total displacement from 0.5 to 5.0 mm).
- phase 4: 1.0 mm trap door displacement steps (total displacement from 5.0 to 10.0 mm).
- phase 5: 2.0 mm trap door displacement steps (total displacement from 10.0 to 20.0 mm).
- phase 6: spin down to 1 G.

LVDT's and the load cell are continuously recorded while the cameras are triggered synchronously, on demand, after each displacement step.

Once unloaded, the experimental package is prepared for post-experiment dissection (Fig. 3). x-z plane cross sections are performed along the path of each casing. The sand is removed through suction.

3.4 Post experiment analysis

All the six model casings were affected by subsurface deformation albeit at different levels (Fig. 3).

Series B models, with $\alpha = 75°$, were the least affected off all the casings, since they sit outside the area of influence of the faulting; only the casing head is exposed due to ground subsidence.

Series A model casings with $\alpha = 80°$ are also intact; these casings sit in-between the two faults, the casing head is barely exposed, however it has been dragged down with the soil over the trap door.

Series B $\alpha = 80°$ model casings are intercepted by one fault network at mid-height without any noticeable deformation. Due to soil surface subsidence, the casing head is exposed at the end of the experiment.

Series A $\alpha = 60°$, is intersected by both faults. The toe is aligned with one fault line while the head is intercepted by the right hand fault. The casing is deformed, bent in the middle (Fig. 4) with the head buried under the soil surface. The casing tends to sink with the subsiding soil except at the head where the faults interception adds a rotational component resulting in casing deformation.

On series B $\alpha = 40°$ casing, left and right faults are intercepting the casing. The casing is strongly buckled downward, cracks are observed on the gypsum plaster coating and the head is sunk into the soil. Both intercepting faults are not associated with the maximum deformation (Fig. 4). The top and bottom appear not

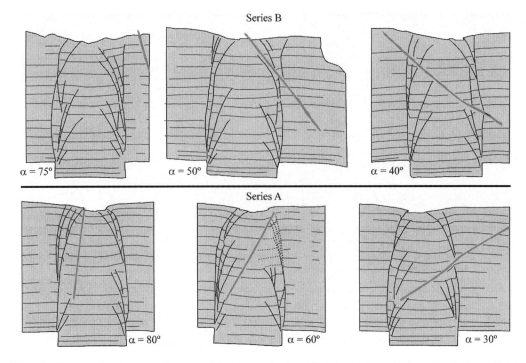

Figure 3. Post experiment cross section on model casings series A and B. Interpretation of deformation, faults position is highlighted with solid black lines, blue lines are soil deformation indicators (horizontal at the beginning of the test), and model casing position is indicated in green. Inferred zone is shown with dashed lines.

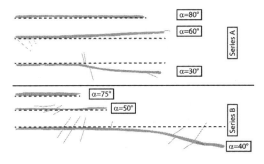

Figure 4. Compilation of post experiment model casing deformation. Thin solid grey lines are the confirmed fault positions while dashed lines are for assumed fault positions. The plot scale is different for the casing models.

to have displaced while the central section has been deformed downwards.

Finally, on series A $\alpha = 30°$ a good correlation is observed between the right fault network interception and the casing deformation. The head and toe section are sub-parallel but misaligned after the fault network interception.

3.5 Preliminary 3D numerical simulation results

Using Plaxis 3D, a numerical model was built to replicate the centrifuge test. The trap door is modelled with a local modification of the boundary conditions through imposed displacement along the z-axis and fixed displacement along the x-y axis. However, with Plaxis being finite element software (FEM), the actual displacement discontinuity in the centrifuge model is averaged over one mesh element in the numerical simulation. As a consequence, a gradient of displacement takes place within the 3 mm wide mesh element.

A full fixed boundary condition is used at the left and right basement of the trap door. Along the centrifuge box walls, a free condition allowing vertical displacement only is used. The model surface is free in all direction.

The centrifuge experimental protocol is accurately reproduced, including the increments in G and the trap doors displacements.

The sand is modelled as an elastic-plastic material with a Mohr-Coulomb failure criterion. The following parameter were used: bulk density $\rho = 1700\,\text{kg.m}^{-3}$, elasticity modulus $E' = 30\,\text{MPa}$, Poisson's ratio $v = 0.35$, effective cohesion $c' = 1\,\text{kPa}$, internal friction angle $\phi' = 35°$ and angle of dilation $\psi' = 20°$.

The FE results are compared with particle images velocimetry (PIV) plots obtained from analysis of the centrifuge experiment. Figure 5 shows the comparison after a trap door displacement of 18 mm. The simulation calculations are generally in agreement with the PIV results. The displacement contour arch shape is reproduced and the trap door boundary positions are correct. However, there is some inconsistency compared to the experimental observations; namely (1)

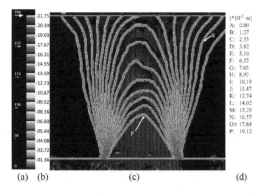

Figure 6. Sub vertical bedded rock mass with slope and displacement discontinuity prior to centrifuge testing.

Figure 7. Sub vertical bedded rock mass model post-test.

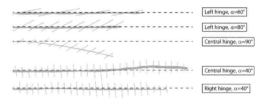

Figure 5. Centrifuge experiment PIV results compared to Plaxis 3D simulation at trap door displacement of 18 mm down. (a) Cumulated displacement from PIV in pixels, (b) PIV displacement converted in mm, (c) PIV displacement field overlaid with Plaxis 3D displacement calculation (contour line display), (d) Plaxis 3D displacement in mm.

contour line positions are mismatched; the simulation underestimates the displacement by 2 mm. (2) The simulated surface displacement is more extensive than was observed in the centrifuge experiments.

4 ROCK MASS MODEL – SLOPE AND DISPLACEMENT DISCONTINUITY

A simple model of a fracture rock mass with sub-vertical bedding was prepared and tested. In this model a basement displacement resulting in both discontinuity of both slope and displacement was introduced.

4.1 Model preparation

The model is built with gypsum plasterboard tiles (100 mm long × 200 mm wide × 9.5 mm thick) with overlapping (Fig. 6). The bedding dips at 50° to the horizontal. The rock mass model has an average total thickness of 190 mm. The plasterboard bulk density is 524 kg.m^{-3}.

Five model casings are inserted in pre-formed grooves on the front face of the rock mass.

Model casing are positioned as follows:

– Two above the left hinge with dips 60° and 80° to the right. The copper core diameter is 1.7 mm.
– Two casings above the central hinge. One with a dip angle of 90° and a copper core diameter of 0.85 mm. The second one with a dip angle of 40° to the left and a copper core of 1.7 mm.
– One casing above the right hinge with a dip angle of 40° to the left. The copper core diameter is 0.85 mm.

All model casings are 25 mm above base. Figure 6 shows the model prior to testing with the casings installed.

The experiment was carried out at 40 G, following the same experimental protocol described in §3.1.3.

Figure 8. Summary of post experiment model casing deformation. Thin solid grey line are the position and orientation of the stratigraphic joints. Display scales differ between each model casing and all casing heads are horizontally re-aligned.

4.2 Post experiment analysis

Observation of the post-test model (Fig. 7), did not reveal any evidences of any fault developing in the rock mass. A reduced amount of slippage occurred along the purely frictional stratification joints. The maximum vertical displacement recorded at the model surface is 7 mm and is located at the vertical of the left hinge suggesting the deformation was associated with slippage on the bedding planes. It is also noted that the rock mass did not follow the displacing base and was thus self-supporting.

From the five casings, only the two central ones display any deformation (Fig. 8). The vertical casing ($\alpha = 90°$) is completely arched, with the deflection going to the right, while the inclined one is deformed on his lower third with a downward deformation. The amount of deformation is similar suggesting that the casing diameter did not influence the degree of deformation.

5 DISCUSSION AND CONCLUSONS

The two basement discontinuities have induced deformation of displacement and of displacement and slope. In the former case faults were seen to propagate through the overlying sand. For the latter case no faulting in the rock mass was observed and the deformation of the overlying material was associated with sliding on bedding planes.

With or without a casing-fault interception, casing deformations were observed in both models. However, only in one casing is it clearly apparent that the maximum casing deformation is the result of an intersection fault.

From the post experiment analysis, it is only possible to observe the cumulative deformation of the models. Information of the casing deformation over time nor the chronology of the deformation (which fault is active when the casing is deforming) could not be established. Therefore uncertainties arise on the true interactions resulting in the casing deformation. The addition of numerical simulation will greatly help in understanding the interaction process between casing deformation and fault propagation.

Nevertheless this study has presented some interesting finding from which the following conclusion can be drawn:

– The mechanical interaction between an oil well casing and a fault has been investigated with simplified experimental centrifuge models.
– A trap door mechanism successfully introduced a discontinuity of displacement, resulting in the propagation of faults inside the model.
– Two scenarios have been tested: a cohesionless deposit with a basement discontinuity of displacement and a sub-vertically bedded rock mass with a basement discontinuity of slope and displacement.
– Post experiment dissections were performed to observe the interaction between the model casings and the generated faults.
– From the interpreted cross sections, casings deformations were observed. However, the relationship to the fault interception point is unclear. Only one casing exhibit a localized deformation were the casing and fault intercepted.
– As a preliminary guidance to reduce casing deformation from propagating faults it is recommended: to reduce the fault interception angle, a casing subparallel with a fault being the best scenario; reduce the number of faults intercepting the casing and having them intercept has close to the surface as possible.

REFERENCES

Bruno, M.S. 1990. Subsidence-induced well failure. *60th California Regional Meeting, Ventura, California, USA, 4-6 April 1990.* Society of Petroleum Engineers.

Bruno, M.S. 1992. Subsidence-induced well failure. *SPE Drilling Engineering* 7(02): 148–152.

Bruno, M.S. 2001. Geomechanical analysis and decision analysis for mitigating compaction related casing damage. *Annual Technical Conference and Exhibition, New Orleans, Louisiana, USA, 30 September-3 October 2001.* Society of Petroleum Engineers.

Chan, A.W. and Zoback, M.D. 2007. The Role of Hydrocarbon Production on Land Subsidence and Fault Reactivation in the Louisiana Coastal Zone. *Journal of Coastal Research* 233: 771–786.

Dusseault, M.B.; Bruno, M.S. & Barrera J. 2001. Casing shear: causes, cases, cures. *SPE Drilling & Completion* 16(2): 98–107.

He, L.; Ye, Y.; Qunyi, W.; Jianwen, Y. and Huilan, D. 2005. Challenges and countermeasures facing casing damage in Daqing oilfield. *SPE Europec/EAGE Annual Conference, Madrid, Spain, 13-16 June 2005.* Society of Petroleum Engineers.

McCauley, T.V. 1974. Planning workovers in wells with fault-damaged casing-south pass block 27 field. *Journal of Petroleum Technology* 26(07): 739–745.

Peng, S.; J. Fu, J. and J. Zhang, J. 2007. Borehole casing failure analysis in unconsolidated formations: A case study. *Journal of Petroleum Science and Engineering* 59(3-4): 226–238.

Centrifuge shaking table tests on composite caisson-piles foundation

F. Liang, Y. Jia, H. Zhang, H. Chen & M. Huang
Key Laboratory of Geotechnical and Underground Engineering of Ministry of Education, Tongji University, Shanghai, China

ABSTRACT: To study the dynamic soil-foundation-structure interaction, centrifuge shaking table tests on a composite caisson-piles foundation (CCPF) were carried out. A series of 1:50 scaled models were designed to simulate a 2 × 2 pile group connected with caisson. The infinite boundary was simulated by a laminar shear box with rubber film inside to eliminate the reflected wave. The results show that the peak bending moment of pile shaft increases gradually from pile tip to pile head for CCPF in both homogeneous and layered soils. Due to the nonlinearity of the system under strong seismic excitations, the maximum bending moments increase with the input PGA growing yet at reduced rate. Sympathetic vibration may generate in CCPF embedded in layered subgrade consisting of hard soil underlain by soft soil, which will lead to larger acceleration of superstructure and larger bending moment of piles compared with CCPF embedded in homogenous soft soil subgrade.

1 INTRODUCTION

Pile group foundations and caisson foundations are frequently utilized to support bridges crossing wide rivers and coastal waters (Dezi et al. 2012, Zafeirakos & Gerolymos 2014, Liang et al. 2017). By adding pile number, the resistance of pile group foundations to earthquake load can be increased. However, for the capacity of pile group foundation to vessel collision is insufficient, it may not be applicable in deep water area. Caisson foundations, whose capacity to vessel collision are sufficient though, the resistance to earthquake load may be insufficient for its shallow embedment depth. Many structures supported by caisson foundations were damaged seriously during strong earthquake excitations (Inagaki et al. 1996). To address the problem, the pre-construction investigation report for the highway channel connecting the mainland and Hainan Island of China proposed the composite caisson-piles foundation (CCPF). Zhong & Huang (2013, 2014) studied the seismic response of CCPF based on the Winkler model. It was concluded that adding piles beneath the caisson is of great significance to enhance the resistance against lateral dynamic loads. However, because the composite caisson-piles foundation is a relatively novel foundation type, few model tests of which the author is aware have been performed to study its behavior under seismic loads. Tests of relevance to the present research usually study the response of pile foundation or caisson foundation under earthquake load separately (Cox et al. 2014, Zhang et al. 2017).

The present paper performed dynamic centrifuge tests on composite caisson-piles foundation embedded in homogeneous clayey silt or layered soil consisting of medium sand underlain by clayey silt. The gravity level was set as 50 g. The Chi-Chi earthquake wave, which is of abundant frequency components, was employed as input motion. The CCPF model is composed of a caisson and a 2 × 2 pile group rigidly connected to the caisson. The superstructure was simplified as lumped mass and connecting columns. The accelerations of soil and superstructure, the pile bending moments are analyzed. The test results are of reference value to the design of composite caisson-piles foundation.

2 CENTRIFUGE TEST SET-UP

The dimensions and test results in this paper are presented in prototype units unless otherwise noted. The centrifuge tests were carried out with centrifuge facility TLJ-150 at Tongji University. A laminar shear box with rubber film inside was used to eliminate the reflected wave. The shear box consists of 22 hollow aluminum rings. The maximum relative displacement between the adjacent rings is 5 mm (Fig. 1).

2.1 *Soil properties*

The medium sand and clayey silt of layer σ_3 in Shanghai were utilized. The soil was compacted into the laminar shear box by layer. The relative density D_r of medium sand was 0.76. The density and moisture content of sand was 1.65 g/cm^3 and 9.0%. Figure 2 shows the grading curve for the medium sand. The clayey silt was pre-consolidated under acceleration of 50 g before the test. Table 1 shows the physical properties of clayey silt.

Figure 1. Laminar shear box.

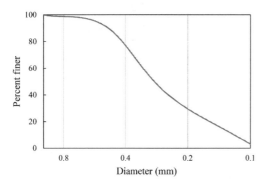

Figure 2. Grading curve for the medium sand.

Table 1. Basic properties of clayey silt.

Property	Value
Density (g/cm^3)	1.76
Water content (%)	36.2
Plastic index, I_p	12.7
Coefficient of Permeability (cm/s)	1.88×10^{-6}
Effective friction angle (degree)	27.5
Undrained shear strength, c_u (kPa)	27.0

2.2 Test model and instrumentation

The geometric similarity ratio between the model and the prototype was 1:50. The CCPF model consists of a caisson and a 2 × 2 pile group rigidly connected to the caisson. The superstructure was simplified as lumped mass and connecting columns. Figure 3 shows

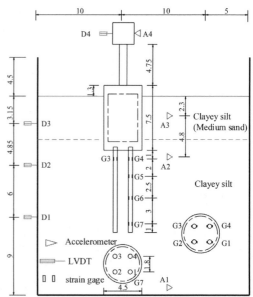

Figure 3. Elevation views (prototype scale).

Table 2. Properties of CCPF (prototype scale, 50 g).

Member	Flexural rigidity (N·m^2)	Mass (kg)	Moment of inertia (kg·m^2)
Pile	1.87×10^8 (1:50^4)		
Column	1.88×10^9 (1:50^4)		
Caisson		1220003 (1:50.5^3)	8.42×10^5 (1:50.2^5)
Structure		39062.5 (1:48.7^3)	4.07×10^4 (1:49.2^5)

Table 3. Cases arranged.

Cases	PGA (g)	Subgrade
C1	0.05	homogeneous soil
C2	0.15	homogeneous soil
C3	0.25	homogeneous soil
C4	0.05	layered soil
C5	0.15	layered soil
C6	0.25	layered soil

the schematic layout of test model along with the instrumetation in homogeneous or layered soil.

The CCPF and the superstructure model were made of aluminum alloy, of which the elastic modulus, Poisson's ratio and density were 70 GPa, 0.33 and 2.7 g/cm^3, respectively. The dimensions of piles and columns were determined according to the flexural rigidity. The dimensions of caisson and superstructure were designed according to the moment of inertia. Table 2 lists the properties of CCPF. Data in the brackets are actual scaling relationships. Table 3 lists the 6 cases arranged.

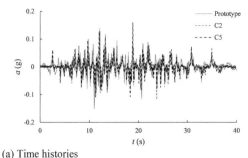

(a) Time histories

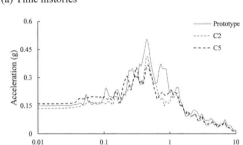

(b) Response spectra (PGA = 0.15, D = 5%)

Figure 4. Earthquake waves inputted into and output from the shaking table.

3 TEST RESULT AND ANALYSIS

The Chi-Chi earthquake wave, which represents the geological condition in coastal region of China to some extent, was employed as input motion. Figure 4 shows the response spectra of earthquake waves inputted into and output from the shaking table. Accelerations recorded are similar to the input motion, which indicates that the performance of the shaking table is reliable.

3.1 Soil acceleration

Figure 5 and Figure 6 show the output accelerations and soil accelerations at the embedment depth of caisson bottom and at the ground surface. Soil accelerations at the embedment depth of caisson bottom are similar to the corresponding output motions. However, due to the soil-caisson seismic interaction, soil acceleration at ground surface is quite different from that at the embedment depth of caisson bottom. The caisson restricted the displacement of soils surrounding the caisson. For soil accelerations at ground surface of homogeneous soil foundation, frequency components larger than 1 Hz were suppressed. While at layered soil foundation, frequency components larger than 2 Hz were suppressed and that less than 1 Hz were amplified. Soil acceleration at ground surface of layered soil foundation was larger than that in homogeneous foundation.

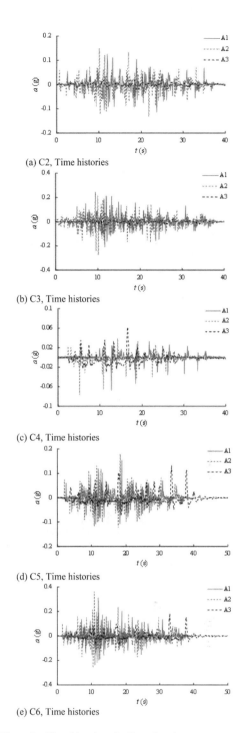

Figure 5. Time histories of soil accelerations.

3.2 Accelerations of superstructure

Figure 7 shows the accelerations of superstructure in homogeneous and layered soil foundation. The acceleration amplitude of superstructure increases with the input PGA growing. Accelerations of superstructure were larger than that at the ground surface. Compared

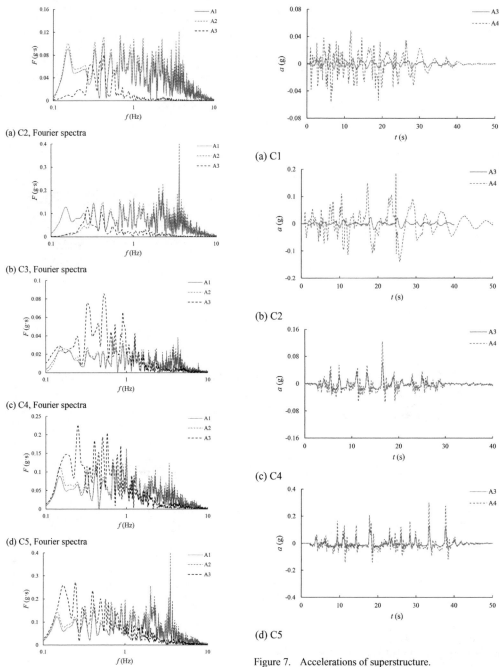

(a) C2, Fourier spectra

(b) C3, Fourier spectra

(c) C4, Fourier spectra

(d) C5, Fourier spectra

(e) C6, Fourier spectra

Figure 6. Fourier spectra of soil accelerations.

(a) C1

(b) C2

(c) C4

(d) C5

Figure 7. Accelerations of superstructure.

with accelerations of superstructure in homogeneous soil, accelerations of superstructure in layered soil subgrade were more similar to the accelerations at ground surface. Sympathetic vibration of system was generated, which may be the reason for peak accelerations of superstructure in layered soil being larger than that in homogeneous soil.

3.3 Pile bending moments

Figure 8 depicts the bending moments at each pile head. The time histories of bending moments are similar to each other. The period of bending moments in layered soil is larger than that in homogeneous soil, which is similar to the acceleration response of superstructure. The maximum bending moments of piles in layered soil were larger than that in homogeneous soil. This may be due to the generation of sympathetic

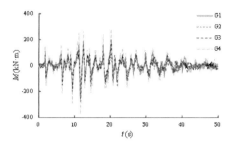

(a) C2

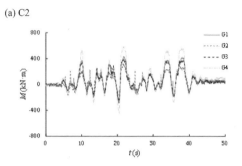

(b) C5

Figure 8. Pile bending moments at pile head.

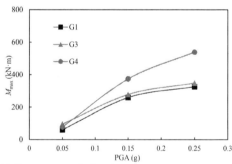

(a) Homogeneous soil subgrade

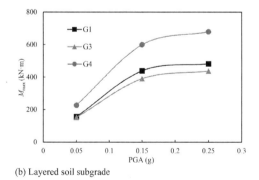

(b) Layered soil subgrade

Figure 9. Maximum bending moments at each pile head.

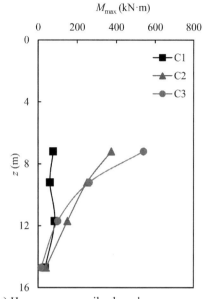

(a) Homogeneous soil subgrade

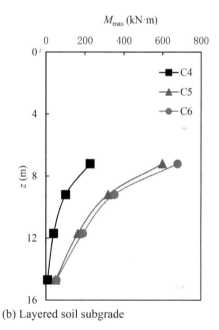

(b) Layered soil subgrade

Figure 10. Maximum bending moments along pile shaft.

vibration of the system, thus larger inertial force was transmitted from the superstructure to the foundation.

Figure 9 shows the relationship of the maximum bending moments at each pile head with the input PGA. Similar to Figure 7, the maximum bending moments of piles in layered soil were larger than that in homogeneous soil. Due to the nonlinearity of the system under strong seismic excitations, the maximum bending moments increase with the input PGA yet with reduced rate. For cases of homogeneous soil, the increasing speed of cases with PGA from 0.15 g to 0.25 g was about 30%~50% that of cases with PGA from 0.05 g to 0.15 g. For cases of layered soil, the ratio was only about 15%~20%.

Figure 10 depicts the maximum bending moments along pile shaft. The peak bending moments increases gradually from pile tip to pile head. The maximum bending moment is located at the pile head. Similar to Figure 9, the maximum bending moments increased with the input PGA yet at reduced rate. The maximum bending moments of piles in layered soil were larger than that in homogeneous soil.

4 CONCLUSIONS

Centrifuge shaking table tests at 50 g were carried out to study the seismic response of composite caisson-piles foundation (CCPF) in homogeneous and layered subgrade. The CCPF model consists of a caisson and a 2×2 pile group rigidly connected to the caisson. The superstructure was modelled by lumped mass and connecting columns. The following conclusions can be made.

(1) The peak bending moment of pile shaft increases gradually from pile tip to pile head for CCPF in both homogeneous and layered soils.
(2) Due to the nonlinearity of the system under strong seismic excitations, the maximum bending moments increase with the input PGA growing yet at reduced rate in both homogeneous and layered soils.
(3) Sympathetic vibration may generate in CCPF embedded in layered subgrade consisting of hard soil underlain by soft soil, which will lead to larger acceleration of superstructure and larger bending moment of piles compared with CCPF embedded in homogenous soft soil subgrade.

ACKNOWLEDGEMENTS

This work was supported by the National Natural Science Foundation of China (Grant No. 41672266). Financial support from the organization is gratefully acknowledged.

REFERENCES

Cox, J. A., O'Loughlin C. D., Cassidy, M., et al. 2014. Centri fuge study on the cyclic performance of caissons in sand. *International Journal of Physical Modelling in Geotechnics* 14(4): 99–115.

Dezi, F., Carbonari, S., Tombari, A., et al. 2012. Soil-structure interaction in the seismic response of an isolated three span motorway overcrossing founded on piles. *Soil Dynamics & Earthquake Engineering* 41(1): 151–163.

Inagaki, H., Iai, S., Sugano, T., et al. 1996. Performance of caisson type quay walls at Kobe Port. *Soils and Foundations* (Special issue) 119–136.

Liang, F., Jia, Y., Sun, L., et al. 2017. Seismic response of pile groups supporting long-span cable-stayed bridge subjected to multi-support excitations. *Soil Dynamics & Earthquake Engineering* 101: 182–203.

Zafeirakos, A. & Gerolymos, N. 2014. Towards a seismic ca pacity design of caisson foundations supporting bridge piers. *Soil Dynamics & Earthquake Engineering* 67: 179–197.

Zhang, L., Goh, S. H., Yi, J. 2017. A centrifuge study of the seismic response of pile-raft systems embedded in soft clay. *Géotechnique* 67(6): 479–490.

Zhong, R. & Huang, M. 2013. Winkler model for dynamic re sponse of composite caisson-piles foundations: Lateral response. *Soil Dynamics & Earthquake Engineering* 55: 182–194.

Zhong, R. & Huang, M. 2014. Winkler model for dynamic re sponse of composite caisson–piles foundations: Seismic response. *Soil Dynamics & Earthquake Engineering* 66(13): 241–251.

Dynamic behaviour of three-hinge-type precast arch culverts with various patterns of overburden in culvert longitudinal direction

Y. Miyazaki, Y. Sawamura, K. Kishida & M. Kimura
Kyoto University, Kyoto, Japan

ABSTRACT: Many three-hinge-type precast arch culverts suffered damage in the Great East Japan Earthquake (11 March 2011), which highlighted the importance of the elucidation of their seismic behaviour. The degree of damage to the culverts appeared to be closely related to the seismic wave motions in the culvert longitudinal direction. Particularly severe damage, such as mouth wall deformation and damage to arch members, seems to have been caused mainly by asymmetrical embankment loading in the culvert longitudinal direction. Thus, in order to clarify the mechanisms behind these seismic damage patterns, centrifuge tests on model culverts with three-hinge-type construction and various embankment shapes were conducted here, and the behaviour in the longitudinal direction was observed. As a result, it was clarified that the response acceleration of culverts located near a mouth wall with shallow embankment cover gets amplified and exceeds that of the surrounding embankment due to the decreased constraining effect of the embankment overburden.

1 INTRODUCTION

Hinge-type precast arch culverts (Figure 1) enable labour saving and high-quality control construction by using precast concrete arch members. According to the position of the hinge, hinge-type precast arch culverts are classified as the two-hinge type or the three-hinge type, although both types are stabilized by allowing a certain degree of movement to mobilize the passive resistance of the embankment.

In Japan, where earthquakes occur frequently, the seismic performance of precast arch culverts is closely related to the stability of the hinge, which may cause the collapse of the arch structures themselves. Therefore, the seismic behaviour in the culvert transverse direction has been investigated (e.g., Toyota & Takagai 2000) and identified as a critical issue in the section design of arch culverts. Sawamura et al. (2016a, b) conducted large shaking table tests in the culvert transverse direction (Fig. 1) on one-fifth scale models of three-hinge-type precast arch culverts, evaluating the damage morphology and ultimate state, and observed no hinge slippage before the ultimate state of the RC arch member.

On the other hand, although most damage to road embankments is correlated with ground motions in the culvert longitudinal direction (Tokida et al. 2007), research literature remains insufficient (Miyazaki et al. 2016). In the Great East Japan Earthquake (11 March 2011), the resulting damage to precast culverts appears to have been caused by strong inertial forces in the longitudinal direction (Abe and Nakamura 2014). Particularly in the old type of three-hinge arch culverts (Fig. 2), severe damage occurred, such as deformation of the mouth wall, numerous cracks in the arch

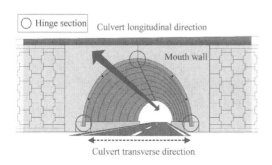

Figure 1. Three-hinge-type precast arch culverts installed in embankment.

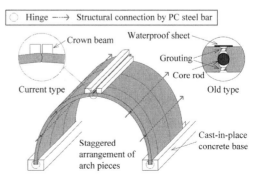

Figure 2. Schematic view of three-hinge arch culvert.

members, and patterned chipping of the foundation, as shown in Figure 3. These damaged culverts had one point in common, namely, they had either shallow soil cover or an asymmetrical embankment load in the culvert longitudinal direction due to the embankment slope.

Therefore, the aim of this study is to clarify the influence of embankment shape patterns on the seismic behaviour and soil-structure interaction in the culvert longitudinal direction. Dynamic centrifuge tests on model three-hinge-type arch culverts were carried out for varied embankment geometries.

Chipping
Damaged foundation

Deformation of mouth wall

Figure 3. Disaster examples from Great East Japan Earthquake (11 March 2011).

2 DYNAMIC CENTRIFUGE MODEL TESTS

2.1 Experimental outline

Centrifuge shaking table tests were conducted under a gravitational acceleration of 50 G using the geotechnical centrifuge device at Kyoto University's Disaster Prevention Research Institute (DPRI). A soil chamber, 340 mm (H) × 450 mm (W) × 300 mm (D), was employed. Figures 4 and 5 show schematic drawings of the model embankments including arch culverts. Three-hinge-type arch culverts were modelled with a length of 28.8 m and constructed on a 5-m-deep layer of soil. In the experiment, a half-length section, 14.4 m in length, was modelled due to the limited dimensions of the soil chamber. The measurement items were as follows: horizontal acceleration of the ground, culvert and wall, horizontal displacement of the wall, and earth pressure acting on the wall. Tests were conducted for four cases, each having a different embankment shape, as depicted in Figures 4 and 5. The experimental parameters were the soil cover thickness and the distance between the mouth and the embankment toe. All other factors were kept constant. Due to concerns of the boundary condition

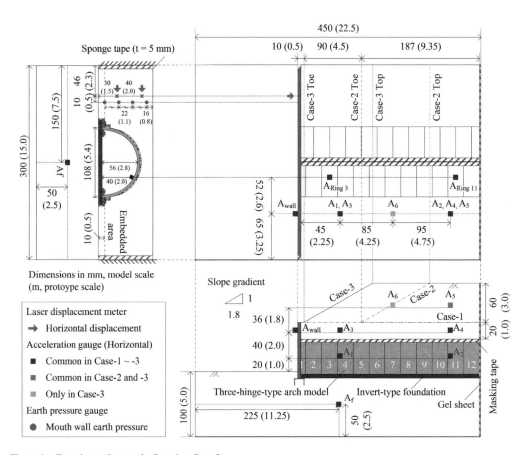

Figure 4. Experimental set-up in Case-1 ~ Case-3.

of the back of the embankment, a 2-mm-wide gel sheet (compressive stress of 10% is 0.07 N/mm²) was applied after the preliminary experiment using a rigid chamber. The gel sheet was set on the side wall of the chamber in the vertical direction against the shaking direction in order to decrease the influence of the reflected wave.

2.2 Three-hinge-type arch model

Figure 6 shows a schematic drawing of the three-hinge-type arch culvert model. The arch culvert model was designed based on the modern type (Fig. 3), which has an invert foundation. The arch model member was made of aluminium and its thickness was adjusted to match the bending stiffness of a real RC member. Each arch member was arranged in a staggered distribution as in the actual construction method. The structural connection of the arch culverts was modelled by masking tape instead of the crown beam (Fig. 3). The joints of each arch culvert were covered with polypropylene sheets to prevent the intrusion of sand.

2.3 Mouth wall model

Generally, mouth walls of three-hinge-type arch culverts are constructed as perpendicular reinforced soil walls. The wall structure is different near the mouth versus the rest of the reinforced soil wall (Fig. 1). The wall near the culvert mouth is composed of two large concrete panels, which are joined by grouting. The other part of the wall is constructed as a typical reinforced earth wall. The wall in this experiment represents only the integrated wall near the culvert. Figure 7 shows the mouth wall model. A 5-mm-thick acryl panel was used for the wall model.

Aluminium plates, 0.1 mm in thickness and 10 mm in width, were employed for the reinforcing members. The foundation for the wall model was made of an aluminium angle plate with a thickness of 1 mm and a depth of embedment of 0.5 m in the prototype scale.

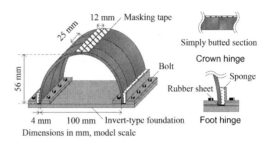

Figure 6. Three-hinge arch culvert model.

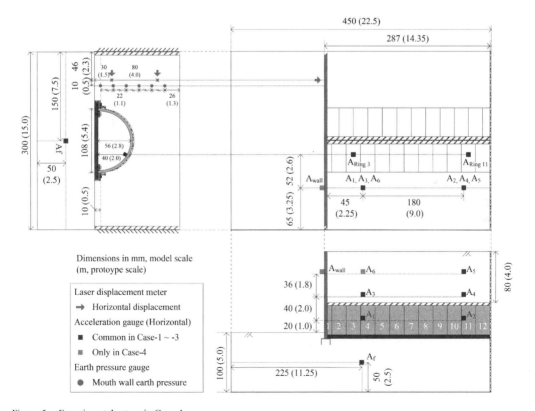

Figure 5. Experimental set-up in Case-4.

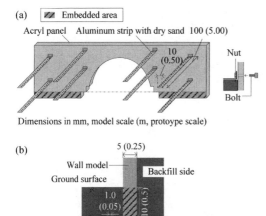

Figure 7. Schematic view of mouth wall model: (a) Structure of mouth wall model and (b) Structure of embedded area.

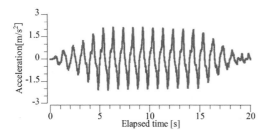

Figure 8. Input wave at STEP 5 (Maximum acceleration is 2.5 m/s^2).

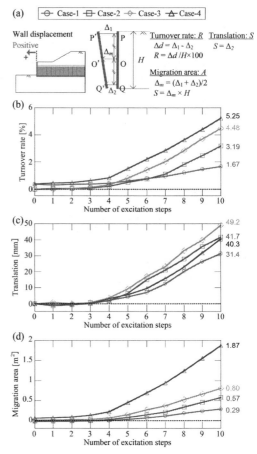

Figure 9. Deflection mode of mouth wall model: (a) Definition of physical quantity, (b) Transition of turnover rate, (c) Transition of translation, and (d) Transition of migration area.

2.4 Model ground and input wave

Hinge-type arch culverts must be built to specific standards and applied to both embankment and foundation soils. Therefore, the model ground here was made by compacting wet Edosaki sand to $Dc = 92\%$ and $w = 17.8\% (= w_{opt}$ of Edosaki sand).

The wave was input by shaking in steps to focus on the changes in the displacement of the mouth wall and the response acceleration. A continuous tapered 1-Hz wave with 20 cycles of sine waves was applied. It was applied 10 times, from STEP 1 to 10, with a gradual increase of 0.5 m/s^2 per step. Figure 8 shows the input wave of STEP 5 as an example.

3 SEISMIC RESPONSE DUE TO ASYMMETRICAL OVERBURDEN IN CULVERT LONGITUDINAL DIRECTION

3.1 Deflection mode of mouth wall

In Figure 9, the transition of the wall turnover rate, the translation, and the migration area are plotted over all the excitation steps. The definition of each physical quantity is as follows. The turnover rate is the value of the difference ($\Delta_d = \Delta_1 - \Delta_2$) between the lateral displacements of the upper portion (Δ_1) and the lower portion (Δ_2) divided by the wall height (H). The translation is equal to Δ_2. The migration area is the product of the lateral displacement at the center of the wall ($\Delta_m = (\Delta_1 + \Delta_2)/2$)) and the wall height ($H$). The figure shows the following relations:

Turnover rate: Case-4 > Case-3 > Case-2 > Case-1

Translation: Case-3 > Case-4 ≒ Case-2 > Case-1

Migration area: Case-4 > Case-3 > Case-2 > Case-1

The migration area is the path area of the wall model during excitation, which can be considered as an expression of the amount of embankment deformation. The length of the reinforcing members is constant across all experimental cases, which may result in decreased deformation in Case-4. However, the migration area increased proportionally to the gross weight of the embankment model.

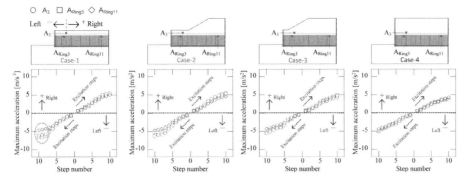

Figure 10. Translation of maximum response acceleration of A_3, A_{Ring3}, and A_{Ring11} over all excitation steps.

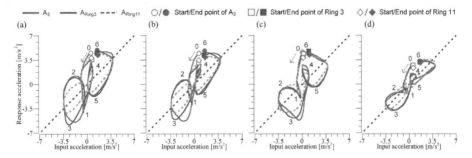

Figure 11. Hysteresis curves of response acceleration of A_3, A_{Ring3}, and A_{Ring11}: (a) Case-1, (b) Case-2, (c) Case-3, and (d) Case-4.

3.2 Seismic response of culverts and ground

Figure 10 shows the transition of the maximum response acceleration at A_3, A_{Ring3}, and A_{Ring11} over all excitation steps. The maximum response acceleration is given by the average of the negative and positive peak values from $t = 10.00 \sim 20.00$ s. In Figure 10, the maximum acceleration in the left and right directions is plotted. According to the figure, the amplification of the response acceleration and the difference between A_{Ring3} and A_{Ring11} decreased more for cases with greater embankment cover. This is because an increase in soil cover strengthens the confining pressure acting on the culverts, which reduces the amplification of the response acceleration. On the other hand, in Case-1, the response acceleration of A_{Ring3} exceeded that of A_{Ring11} as well as that of A_3 at STEPS 8, 9, and 10.

As is shown in Figure 10, the acceleration relation among A_3, A_{Ring3}, and A_{Ring11} changed at STEP 8. To explain this relation, the hysteresis curve of the response acceleration is depicted in Figure 11 from $t = 13.50$ to 14.48 s. In the figure, the start and end points of the hysteresis curves are plotted.

The mutual relationship among A_3, A_{Ring3}, and A_{Ring11} in each experimental case is visible at peak response acceleration No. 3, at which the leftwards response acceleration is maximum. In Case-1, the response acceleration of A_{Ring3} exceeded that of A_{Ring11}. On the other hand, in Case-2 and Case-3, the response acceleration of A_{Ring3} was less than that of A_{Ring11}, and the difference between A_{Ring3} and A_{Ring11} was small. Compared with the curve shape of Case-1, the curve shapes of A_3, A_{Ring3}, and A_{Ring11} converge in Case-2 and Case-3. Moreover, the curve shapes of A_3, A_{Ring3}, and A_{Ring11} are almost coincident with each other in Case-4.

From the above results, the following can be concluded. In Case-1, where the overburden thickness is a constant 1.0 m in the culvert longitudinal direction, the seismic response exceeds that of the surrounding ground, and the highest amplification is observed at the mouth. In Case-2 and Case-3, a difference in acceleration values is seen between Ring 3 and Ring 11, but the difference decreases with increasing soil cover for the culverts. On the other hand, in Case-4, where the overburden thickness is a constant 4.0 m in the longitudinal direction, integrated behaviour of the culverts and the surrounding ground is shown due to the large confining stress.

3.3 Embankment model after excitation

Figure 12 shows the embankment model surface after excitation with marked lines on the cracked areas. As seen in the figure, longitudinal cracks were observed in Case-1 $\sim$ Case-3 near the small soil cover area at the mouth. The cracks run along the arch crown and seem to have been caused by tensile forces in the culvert transverse direction. The compressive deformation on the sides of the culverts is thought to have created tensile force on the ground surface in the crown area. On

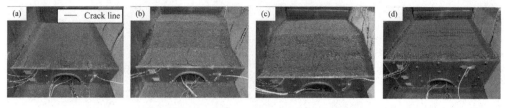

Figure 12. Embankment surface after experiment: (a) Case-1, (b) Case-2, (c) Case-3, and (d) Case-4.

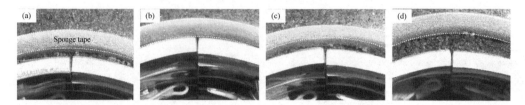

Figure 13. Crown hinge after experiment: (a) Case-1, (b) Case-2, (c) Case-3, and (d) Case-4.

the other hand, in Case-4, large cracks occurred in the culvert transverse direction in the central area of the embankment surface. This is because large deformation of the mouth wall caused tension on the surface of the embankment. While the enlargement of the earth cover has the advantage of restraining the seismic behaviour of culverts in the longitudinal direction, the larger transverse deformation of the embankment surface should be considered.

Figure 13 shows the mouth of the culverts after excitation. In this experiment, the hinge was designed as a simply butted section. That is why this hinge model would collapse more easily than an actual knuckle hinge. However, as seen in the figures, even when the embankment has large deformation, no slippage of the crown hinge is observed.

4 CONCLUSIONS

In this study, dynamic centrifuge model tests were conducted focusing on the influence of the embankment shape on the seismic performance of three-hinge arch culverts in the longitudinal direction. The following conclusions can be drawn from the results of this study:

1) The deformation of an embankment in the culvert longitudinal direction increases proportionally to the gross weight of the soil.
2) The seismic behaviour of culverts in the longitudinal direction is closely related to the degree of overburden.
3) Shallow soil cover, such as 1.0 m, allows the response acceleration of culverts to be amplified and to exceed that of the surrounding soil at the mouth. Conversely, deep soil cover, of more than 4.0 m, causes the culverts to respond as an integrated body with the surrounding soil.
4) In the experiment in which the entire overburden was more than 1.0 m and in which the hinge was a simply butted section, no slippage of the hinge portion was observed during repeated excitation in the longitudinal direction.

ACKNOWLEDGEMENTS

This work was supported by the Research Foundation on Disaster Prevention of Express Highway by Nexco-Affiliated Companies.

REFERENCES

Abe, T. and Nakamura, M. 2014. *The use of and the caution in the application of the culvert constructed by large precast element in the expressway construction*, The Foundation Engineering & Equipment, Vol. 42, No. 4, pp. 8–11. (in Japanese)

Miyazaki, Y., Sawamura, Y., Kishida, K. and Kimura, M. 2017. Evaluation of dynamic behavior of embankment with precast arch culverts considering connecting condition of culverts in culvert longitudinal direction, Japanese Geotechnical Society Special Publication, Vol. 5, No. 2, pp. 95–100. http://doi.org/10.3208/jgssp.v05.020

Sawamura, Y., Ishihara, H., Kishida, K., and Kimura, M. 2016a. Experimental Study on Damage Morphology and Critical State of Three-hinge Precast Arch Culvert through Shaking Table Tests, Procedia Engineering, Advances in Transportation Geotechnics III, Vol. 143, pp. 522–529, 2016. http://dx.doi.org/10.1016/j.proeng.2016.06.066

Sawamura, Y., Ishihara, H., Kishida, K., and Kimura, M. 2016b. Evaluation of Damage Morphology in Three-Hinge Precast Arch Culvert Based on Shaking Table Tests and Numerical Analyses, Proc. of the 8th Young Geotechnical Engineering Conference, pp. 221–226, Astana, Kazakhstan.

Tokida, K., Oda, K., Nabeshima, Y. and Egawa, Y. 2007. Damage level of road infrastructure and road traffic performance in the mid Niigata prefecture earthquake of 2004, Structural Engineering/Earthquake Engineering, Vol. 24, No. 1, pp. 51–61.

Toyota, H. and Itoh, T. 2000. Effects of Shaking Conditions and Material Properties on Dynamic Behavior of Terre Armee Foundation and 3-Hinge Arch, Proc. of Japan Society of Civil Engineers, No. 666/III-53, pp. 279–289. (in Japanese)

Dynamic centrifuge model tests on sliding base isolation systems leveraging buoyancy

N. Nigorikawa, Y. Asaka & M. Hasebe
Shimizu Corporation, Tokyo, Japan

ABSTRACT: Two simple seismic isolation systems that leverage buoyancy or excess pore water pressure generated due to liquefaction have been devised for spread foundation structures. Through dynamic centrifuge model tests, it is shown experimentally that a sliding isolation effect arises during seismic loading if buoyancy force of appropriate magnitude acts on the underside of the foundation. It is confirmed that seismic isolation performance tends to improve as the buoyancy-to-weight ratio increases. Performance with buoyancy-to-weight ratios of 90% or more is particularly good, and the maximum response acceleration of a structure can be reduced to below 100 cm/s² against a sinusoidal wave with a maximum acceleration of 500 cm/s².

1 INTRODUCTION

In recent years, the adoption of base-isolation systems in buildings is increasing as people become more conscious of disaster prevention and mitigation of earthquake damage. So there is a need to develop base-isolation methods effective for buildings with various uses as well as different building classifications, building shapes and ground conditions. A widely used method of base isolation is to increase the natural period of the upper structure by installing isolators and dampers between the base structure and the supporting ground. However, such systems are difficult to design for effective functioning in soft ground, so a seismic-isolation technology covering this target is required.

Recently, a seismic-isolation system known as the "partially floating structural system" has come into practical use (Saruta et al., 2007, Nakamura et al., 2011). In this system, the weight of the structure is partially supported by buoyancy and base-isolators such as rubber bearings are to support the remaining weight. With approximately half of the upper structure weight being supported by buoyancy, the base-isolators can be miniaturized and the natural period of the upper structure increased as compared with conventional structural systems, leading to better seismic isolation performance. During The 2011 off the Pacific Coast of Tohoku Earthquake, the system demonstrated good seismic-isolation performance in structures constructed on soft foundations.

This paper describes two new seismic-isolation concepts derived from the partially floating structural system. The major feature of these new approaches is reducing the frictional resistance of the underside of a spread foundation by buoyancy, yielding a simpler structural system that does not require base-isolators. The results of dynamic centrifugal model tests aimed at validating the seismic-isolation performance of the proposed system and confirming its effectiveness are then presented.

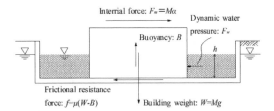

Figure 1. Schematic diagram of forces acting in the proposed system leveraging buoyancy of water.

2 CONCEPT OF PROPOSED SEISMIC ISOLATION SYSTEM

A schematic diagram of the forces acting in the proposed system is shown in Figure 1. A pit is constructed by excavating the ground, which has a high groundwater level, and a spread foundation structure is built in there. Water (or fluid) is maintained in the pit at the same level as the groundwater. The constant contact force acting on the foundation bottom surface of the structure is only that obtained by subtracting the buoyancy B from the structure weight W. When an earthquake occurs, an inertial force F_m due to the mass of the structure and the dynamic water pressure F_w act as a horizontal force on the structure. The frictional resistance force acting on the underside of the foundation f is given as

$$f = \mu(W - B) = wA \quad (1)$$

in which μ is the coefficient of friction. If f is greater than $F_m + F_w$, slip occurs between the ground and the foundation bottom.

The buoyancy B acting on the foundation bottom is written by

$$B = \rho g h A \quad (2)$$

where A is the area of the foundation bottom, h is the draft, ρ is the density of water and g is gravitational acceleration. Note that Eq. (2) is established under the assumption that there is no variation in cross section of the submerged part of the building.

The structure weight W is written by the following equation using the structure weight per unit area w:

$$W = Mg = wA \quad (3)$$

where M is the mass of the structure. Substituting Eqs. (2) and (3) into Eq (1), the frictional resistance force acting on the underside of the foundation f is given as

$$f = \mu\left(1 - \frac{\rho g h}{w}\right)Mg = \mu' Mg \quad (4)$$

in which μ' is the apparent coefficient of friction. The value of μ' varies depending on the draft and the structure weight per unit area. That is, if the structure can tolerate a certain amount of slip and residual displacement, by choosing appropriate values for these two parameters h and w and controlling the apparent friction coefficient μ', it is theoretically possible to intentionally allow sliding of the foundation and obtain a seismic isolation effect when a seismic force exceeds a certain size.

3 DYNAMIC CENTRIFUGE MODEL TESTS

3.1 Seismic isolation system leveraging buoyancy of water

In order to investigate the relationship between the frictional resistance of the foundation and seismic isolation performance, dynamic centrifuge model tests were conducted on models, taking the buoyancy acting on the structural model as an experiment parameter. The model and experimental data were prepared according to the scaling law in centrifuge modelling.

3.1.1 Test conditions

An outline of the dynamic centrifuge model tests is shown Figure 2. The scale of the model is 1/30 and the tests were performed under centrifugal acceleration of 30 g. The structural model was made by shaping aluminium into a box shape with an open top, measuring 120 mm in length, 240 mm in width and 70 mm in height with a wall thickness of 10 mm. Mean contact pressure during tests was 19.2 kPa under the centrifugal acceleration of 30 g. Figure 3 shows the grain size distributions of the samples used for the model ground. No. 3 silica sand was used as the base layer, and it was tamped into a dense layer with a thickness of 30 mm and a relative density of 90%. The base layer was de-aired under vacuum and saturated with silicon oil with a kinematic viscosity of $3.0 \times 10^{-5} m^2/sec$ (30 times

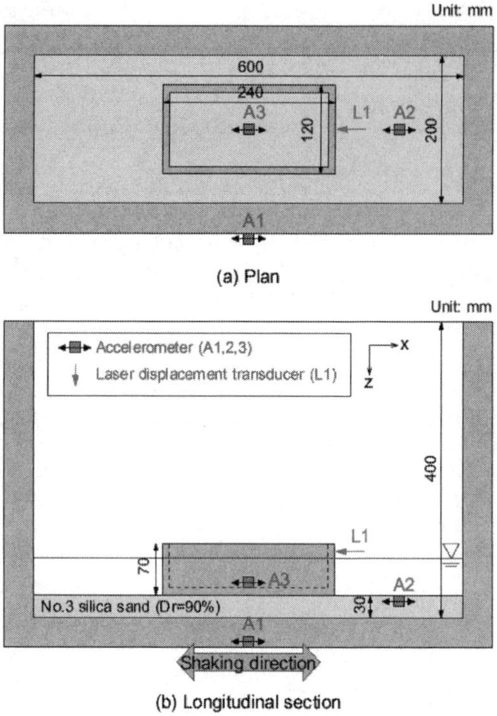

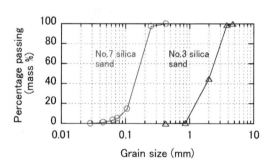

Figure 2. Outline of dynamic centrifuge model tests.

Figure 3. Grain size distributions of samples.

greater than that of water). Horizontal accelerometers (A1-3) were placed on the shaking table, the underside of the structural model and the surface of the base layer. A laser displacement transducer (L1) measured the horizontal displacement of the structural model.

The model was horizontally and sinusoidally shaken under a centrifugal acceleration of 30 g. The input motions were all regular waves with a constant frequency. Amplitude was increased linearly in each test up to a certain maximum value. The excitation frequency was 60 Hz (or, in prototype terms, 2 Hz). That is, excitation was the same in all cases except that the peak amplitude was varied.

In the discussion that follows, experimental results are presented using the proportion of buoyancy occupied in ground contact pressure by structure weight (hereinafter, referred to as "buoyancy-to-weight ratio

Table 1. Experimental cases.

Case no.	Buoyancy to weight ratio (%)	Abs. value of max. accel. (cm/s²)			Residual disp. (L1) mm
		Shaking table (A1)	Bearing stratum (A2)	Structure (A3)	
Case1-1	20	22	22	24	0
Case1-2	20	57	59	61	0
Case1-3	20	73	77	79	−1
Case1-4	20	89	95	96	0
Case1-5	20	111	116	122	−1
Case1-6	20	143	147	156	0
Case1-7	20	193	217	226	−1
Case1-8	20	261	501	371	−2
Case1-9	20	367	584	600	−6
Case2-1	75	21	21	22	0
Case2-2	75	56	59	85	1
Case2-3	75	72	77	122	−1
Case2-4	75	91	94	190	−3
Case2-5	75	111	119	225	−3
Case2-6	75	148	167	246	−2
Case2-7	75	201	230	248	−15
Case2-8	75	278	503	272	65
Case2-9	75	335	558	263	212
Case3-1	90	22	23	30	0
Case3-2	90	63	66	62	1
Case3-3	90	87	92	63	−10
Case3-4	90	87	91	59	8
Case3-5	90	107	112	58	−40
Case3-6	90	131	139	54	−132
Case3-7	90	169	181	68	124
Case3-8	90	233	264	73	97
Case3-9	90	331	478	93	340

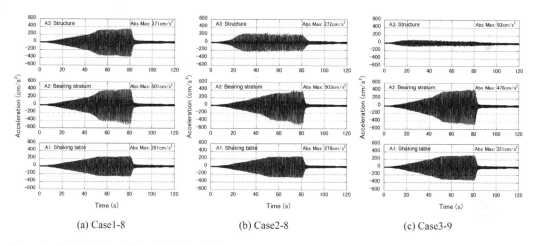

(a) Case1-8   (b) Case2-8   (c) Case3-9

Figure 4. Time histories of Cases 1-8, 2-8 and 3-9.

B/W") as an evaluation index. The behaviour of the experimental cases shown in Table 1 is examined. Buoyancy-to-weight ratio B/W is written by the following equation:

$$B/W = \frac{\rho g h}{w} \quad (5)$$

3.1.2 Test results
In the following paragraphs, all the test results are presented at prototype scale. Time histories of accelerations in Cases 1-8, 2-8 and 3-9, with maximum input acceleration set at approximately 300 cm/s², are shown in Figure 4. In all three cases, the ground responses (A2) are amplified against the input. On the other hand, the structure responses (A3) differ according to buoyancy-to-weight ratio. There is almost no reduction in amplitude at a buoyancy-to-weight ratio of 20% (Case 1-8). The structure response at a buoyancy-to-weight ratio of 75% (Case 2-8) has an upper limit of approximately 250 cm/s² after 25

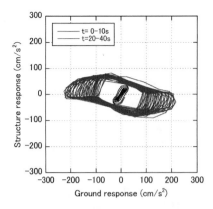

Figure 5. Orbit of response acceleration in Case 3-9.

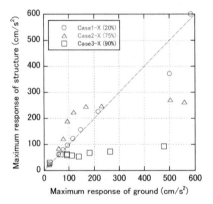

Figure 6. Orbit of response acceleration in Case 3-9.

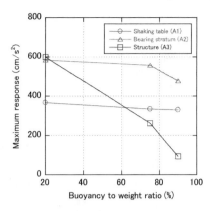

Figure 7. Relationship between buoyancy-to-weight ratio and maximum structure response.

seconds and is reduced as compared with the ground response. Furthermore the reduction in amplitude at a buoyancy-to-weight ratio of 90% (Case 3-9) is very strong, and it was confirmed that the upper limit is approximately 90 cm/s² after 20 seconds.

Figure 5 shows the orbit of response acceleration in Case 3-9, with ground response plotted on the X axis and structure response on the Y axis. In the period between 0 and 10 seconds, the ground and structure show almost the same response. After that the response becomes orbital indicating that there is a phase shift between the ground response and the structure response. This behaviour means that sliding occurs at the base of the structural model. That is, the experiments confirm that by controlling the buoyancy acting on the structural model, it is possible to intentionally allow sliding at the underside of the foundation during an earthquake and thereby obtain an isolation effect.

Figure 6 shows the relationship between maximum responses of ground and structure arranged by buoyancy-to-weight ratio for all cases listed Table 1. The solid line on the graph represents the one-to-one relationship line. In Cases 1-X, with the buoyancy-to-weight ratio set to 20%, the structure response is not amplified against ground response and closely follows this line. In Cases 2-X, with the buoyancy-to-weight ratio set to 70%, amplification of the structure response is observed when the maximum structure response exceeds 50 cm/s². Then, when it exceeds 200 cm/s², sliding occurs and a response ceiling is reached. Finally, in Cases 3-X, with the buoyancy-to-weight ratio set to 90%, no amplification of structure response is observed and the maximum structure response remains almost unchanged with an upper limit of approximately 60 cm/s².

Figure 7 shows the relationship between buoyancy-to-weight ratio against maximum structure response in the experimental cases with maximum applied excitation (Case 1-9, Case 2-9 and Case 3-9). Structure response tends to decrease as buoyancy-to-weight ratio increases. The response was approximately 100 cm/s² at a buoyancy-to-weight ratio of 90%. From these experiments, it is confirmed that the maximum response acceleration of a structure with a sliding base isolation system that leverages water buoyancy can be suppressed to approximately 100 cm/s², therefore offering a good seismic isolation effect.

3.2 Seismic isolation system leveraging liquefaction

In the previous section, it was shown that the sliding base isolation effect can be obtained between foundation and the supporting ground by leveraging buoyancy. However, achieving a buoyancy-to-weight ratio of 90% or more would require an excavation of great depth, depending on the structure to be protected, and concerns such as construction cost arise. Furthermore, such a high degree of buoyancy acting on a structure at all times would be disadvantageous in terms of foundation stability against horizontal external forces other than earthquakes.

Figure 8 shows a new seismic isolation system proposed here as a way to solve these problems. As before, the system is based on buoyancy. In this case it consists of a low-density saturated sand layer with a relative density of 30-50% in the space between the structure and a side wall. The saturated sand is intentionally allowed to liquefy during an earthquake, reducing the

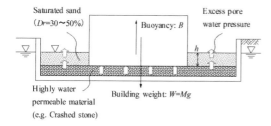

Figure 8. Schematic diagram of forces acting in the proposed system leveraging liquefaction.

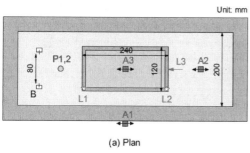

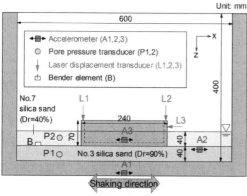

Figure 9. Outline of dynamic centrifuge model tests.

ground contact pressure due to the normal floating effect. The supporting ground is highly permeable and must be compacted sufficiently so as not to liquefy. Then, when strong ground motion due to an earthquake occurs, the excess pore water pressure developed in the ground layers acts on the underside of the foundation, allowing easy sliding between the structure and the supporting ground. This enhances the seismic isolation effect. By adopting this design, it is possible to reduce the draft required to obtain a predetermined level of buoyancy, so the amount of ground excavation is reduced and construction can be completed at reduced cost in a shorter period.

In order to verify the seismic isolation performance of the proposed system, a series of dynamic centrifugal model tests was conducted. The model and experimental data were prepared according to the scaling law in centrifuge modelling.

3.2.1 Test conditions

An outline of the dynamic centrifuge model tests is shown Figure 9. The scale of the model is again 1/30 and the tests were performed under a centrifugal acceleration of 30 g. No. 3 silica sand was used as the base layer. The low-density sand layer between the structure and the side wall was formed by pouring No. 7 silica sand to obtain a relative density of 40%. Bender elements (B) were placed in the low-density sand layer to measure S-wave velocities. The S-wave velocity V_s in the low-density sand layer before cyclic loading was approximately 90 m/s under the 30 g field. Preparation of the base layer was carried out in the same way as in the experiments in the previous section. As indicated in Figure 9, horizontal accelerometers (A1-3) were placed on the shaking table, the underside of the structural model and the surface of the base layer, respectively. Two pore pressure transducers (P1-2) were placed in the low-density layer and the base layer, respectively. Two laser displacement transducers (L1-2) measured settlement throughout the tests at two points on the structural model, while (L-3) measured the horizontal displacement of the structural model.

In this system, the buoyancy acting on the structural model under normal conditions is only the water pressure corresponding to the draft. During cyclic loading, the excess pore water pressure u_e generated in the low density layer propagates through the base layer, is added to the buoyancy, and acts on the underside of the foundation. That is, the buoyancy-to-weight ratio B/W can be written as the following equation:

$$B/W = \frac{\rho g h + u_e}{w} \qquad (6)$$

The model was horizontally and sinusoidally shaken under a centrifugal acceleration of 30 g. The maximum acceleration of the excitation was 7.0 g (or in prototype terms, 230 cm/s²) and the excitation frequency was 60 Hz (or in prototype terms, 2Hz).

In the section that follows, the behaviour of the experiment cases shown in Table 2 during this cyclic loading is examined.

3.2.2 Test results

In the following paragraphs, all the test results are presented at prototype scale. Time histories of accelerations and pore water pressure are shown in Figure 10. The response accelerations of the base layer and structure as the excess pore water pressure rises (between 5 and 15 seconds) are comparable in all cases. In Case 5, with buoyancy-to-weight ratio set to 90%, the structure response at the point when the excess pore water pressure reaches the upper limit (at 15 seconds) is approximately 100 to 120 cm/s², then it falls immediately to approximately 50 cm/s². The behaviour is similar in Case 6, with buoyancy-to-weight ratio set to 100%, and the response acceleration falls to 20 cm/s² once the excess pore water pressure reaches the upper limit. Beyond the point where the excess pore water pressure reaches the upper limit (after 15 seconds),

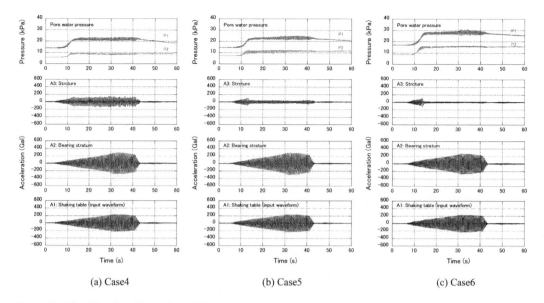

Figure 10. Time histories of Cases 4, 5 and 6.

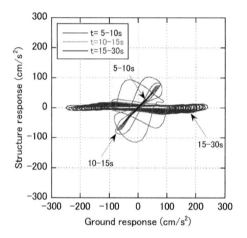

Figure 11. Orbit of response acceleration between 5 and 30 seconds in Case 6.

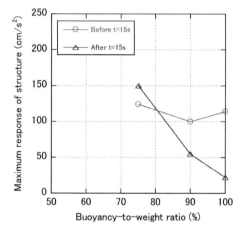

Figure 12. Relationship between buoyancy-to-weight ratio and maximum structure response.

structure response in all cases transitions to a constant upper limit.

Figure 11 shows the orbit of response acceleration in Case 6, with ground response plotted on the X axis and structure response on the Y axis. As the excess pore water pressure begins to rise (between 5 and 10 seconds), the ground and structure show almost the same response; that is, they respond in the same phase. As the excess pore water pressure approaches the upper limit (between 10 and 15 seconds), the phase of the two responses begins to diverge until they become out of phase, during which time the maximum value of the structure response is observed. It is deduced that the apparent rigidity of the low-density sand layer falls as the effective stress decreases and the maximum response acceleration occurs in a state where the structure responds readily to the input waveform.

Thereafter, as the pore water pressure reaches the upper limit (after 15 seconds), slip occurs at underside of the foundation and little vibration is transmitted from the base layer to the structure. Figure 12 shows the relationship between buoyancy-to-weight ratio and structure response for three experimental cases (Cases 4, 5 and 6). The response values in Figure 12 are plotted by dividing the maximum value into before and after the excess pore water pressure reaches the upper limit. Although the acceleration of approximately 100 to 120 cm/s^2 is transmitted to the structure as the excess pore water pressure rises, it is confirmed that once the excess pore water pressure reaches the upper limit the response tends to be lower as buoyancy-to-weight ratio is increased.

Residual displacements measured at the left and right ends of the model structure are shown in Table 2.

Table 2. Experimental cases.

Case No.	Buoyancy to weight ratio (%)		Maximum acceleration (cm/s^2)				Residual displacement (mm)			Inclined angle (rad)
	Before shaking	During shaking	Shaking table (A1)	Bearing stratum (A2)	Structure (A3)		Vertical (L1)	Vertical (L2)	Horizontal (L3)	
					Before t=15s	After t=15s				
Case4	40	75	228	280	124	150	4	7	−48	0.0004
Case5	45	90	233	294	100	55	18	15	1	0.0004
Case6	50	100	231	274	114	22	2	4	−20	0.0003

Maximum settlement is 18 mm in prototype scale and the inclination angle is maximum 0.4×10^{-3} rad. These figures indicate that the probability of structure failure resulting from differential settlement of the structure can be considered low (AIJ, 2001). Further, it was observed that the residual horizontal displacement is within 50 mm.

These experiments confirm the validity of the proposed seismic isolation system, in which sliding takes place only at the time of an earthquake as a result of leveraging excess pore water pressure generated within a low density layer.

4 CONCLUSIONS

Two simple seismic isolation systems were proposed for spread foundation structures that leverage buoyancy or excess pore water pressure generated due to liquefaction and verified their seismic isolation performance by dynamic centrifuge model tests. The results are as follows;

1. It is shown experimentally that sliding isolation effect arises during seismic loading if buoyancy force of appropriate magnitude acts on the underside of the foundation. It is confirmed that seismic isolation performance tends to improve as the buoyancy-to-weight ratio increases.
2. The possibility of achieving a seismic isolation system in which sliding occurs only at the time of an earthquake by constructing a low-density saturated sand layer in the space between the structure and a side wall was demonstrated.

REFERENCES

Architectural Institute of Japan, AIJ. 2001. Evaluation of Settlement. Recommendations for Design of Building Foundations: 150-154.

Nakamura, Y. Hanzawa, T., Hasebe, M., Okada, K., Kaneko, M. & Saruta, M. 2011. Report on the effects of seismic isolation methods from the 2011 Tohoku-Pacific Earthquake. *Seismic Isolation and Protection Systems* 2(1): 57-74.

Saruta, M., Ohyama, T., Nozu, T., Hasebe, M., Hori, T., Tsuchiya, H. & Murota, N. 2007. Application of a partially-floating seismic isolation system. *Proceeding of the 10th world conference on seismic isolation, Energy Dissipation and Active Vibrations Control of Structures, Istanbul, Turkey.*

Centrifugal model tests on static and seismic stability of landfills with high water level

B. Zhu, J.C. Li, L.J. Wang & Y.M. Chen
MOE Key Laboratory of Soft Soils and Geoenvironmental Engineering, Institute of Geotechnical Engineering, Zhejiang University, Hangzhou, China

ABSTRACT: A new type of synthetic Municipal Solid Waste (MSW) was developed, which exhibited engineering characteristics similar to those of real MSW in China. A series of centrifuge model tests on the stability of geosynthetics-lined landfills under the conditions of rising water levels and earthquake were carried out respectively. In the condition of rising water levels, the landfills tended to fail with a global slide along the liner when the water level reached the critical value. The ratio of critical water level and landfill height was about 0.8. When an extreme earthquake occurred on the landfill, a crack was developed on the crest and eventually penetrated the landfill as the water level raised. The earthquake increased the pore water pressure, and decreased the critical water level by 10–30% compared with the static condition. These results provide a valuable reference for stability control of landfills with high water level.

1 INTRODUCTION

In China, MSW landfills generally have high leachate levels resulting from the high water content of MSWs and the relatively low operational level of these landfills. Recent researches indicated that the failures of landfills were typically related to a build-up of the water level on the bottoms of the landfills (Kavazanjian et al. 2001). At the same time, China is an earthquake prone region, earthquake-induced landslide events occur frequently, which trigger serious disasters. Surface cracks induced by earthquakes have a significant impact on the instability of landfills.

Studies on instability of MSW landfills induced by rising water levels and earthquake has been limited to analytical and numerical approaches. It is desirable to perform effective model tests to evaluate different failure modes of MSW landfills and to verify existing theoretical approaches. Geotechnical centrifuge modelling exhibits unique advantages in the simulation of geotechnical infrastructure failure. Thusyanthan et al. (2006a, b) mixed Irish moss peat, kaolin clay and silica sand to produce a model MSW that primarily matched the unit weight, compressibility, and shear characteristics of real MSW, and they performed two sets of centrifuge tests to analyse both the static and seismic responses of a landfill. High garbage contents and high water contents make the physical properties of MSW in China quite different. In this study, using a new synthetic MSW, centrifuge model tests were performed to investigate the failure mechanism of MSW landfills induced by a rising water levels and earthquakes.

2 MODEL TEST SETUP

2.1 Test apparatus and water level control

The model tests were performed using the ZJU-400 centrifuge at Zhejiang University in Hangzhou, China (Chen et al. 2010; Chen et al. 2011). The maximum capacity and acceleration of this beam type centrifuge are $400\,g.\text{ton}$ and $150\,g$, respectively. The effective arm radius of the centrifuge is 4.5 m. An inflight uniaxial eletro-hydraulic shaking table was used to simulate seismic excitation. The shaking table has vibration frequencies ranging from 10–200 Hz. Its payload capacity is 500 kg, and its maximum lateral displacement and acceleration are 0.006 m and $40\,g$, respectively.

A rigid container was used to prepare the landfill model for static tests, with inner dimension of 1.0 m (length) × 0.4 m (width) × 1.0 m (height), and a front window made of Perspex for direct observation of the experiment. A water level control system was designed as shown in Figure 1(a). The flow pump was type BT-300E/153Y with a flow range of 0.001–2.24 L/min, and its digital drive was capable of real-time control and of regulating the flow rate. The water level in the water supply cavity was controlled by the flow pump, and the water in the cavity seeped into the landfill through the porous drainage plate during the test. A non-woven geotextile ($400\,\text{g/m}^2$) inverted filter prevented the local scouring of the MSW that might be induced by a high flow rate. The overflow water from the slope toe flowed into the reservoir through the drain pipe.

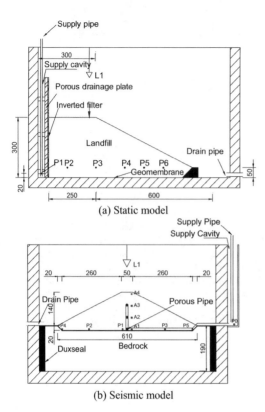

Figure 1. Layout of test setup and water level control system (unit: mm).

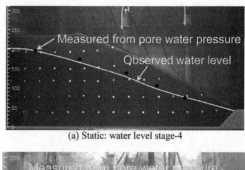

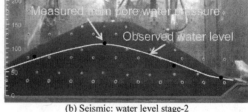

Figure 2. Water level control in the tests (unit: mm).

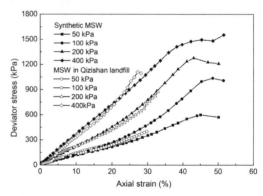

Figure 3. The CD test results of the synthetic MSWs compared with those of MSW in the Suzhou Qizishan landfill.

Another rigid container (0.77 m × 0.4 m × 0.53 m) was used to prepare the landfill model for seismic tests (see Figure 1(b)), with similar water level control system. Water was supplied to the landfill model through a flexible pipe, which was buried under the landfill and connected to three porous pipes with different height. These porous pipes were made of plexiglass, which has similar density with the synthetic MSW used in these tests and were laid perpendicular to the front Perspex window of the container. All porous pipes were enveloped with non-woven geotextile inverted filter.

During the tests, pore water pressure in the landfill was measured while the water level was raised. The measured pore water pressures at the end of each stage were converted to measured hydrostatic head and plotted on the picture of the mode landfill. As shown in Figure 2, the measured hydrostatic head was close to the observed water level.

2.2 *Materials and model preparation*

A new synthetic MSW was manufactured depending on the engineering characteristics of real MSWs of Suzhou Qizishan Landfill in China (Chen et al. 2017). Changbaishan peat, Fujian silica sand, and kaolin clay were selected as the components of the synthetic MSW, and were mixed homogeneously in varying proportions with a certain water content. A series of tests were conducted to determine both the physical and mechanical properties of the mixtures. According to the test results, the proportion of compositions of the synthetic MSW was finally determined mainly according to the stress-strain relationships. The stress-strain relationships of the synthetic MSWs are consistent with those of the real MSWs in China and both exhibit significant strain-hardening characteristics (see Figure 3). The shear strength parameters were $c_d = 22$ kPa and $\phi_d = 26.4°$ under the shear strain of 20%.

The synthetic MSW is approximately equivalent to the real MSW in China with regards to the total unit weight (9 kN/m^3), void ratio (1.6), water content (45%), compressibility (2.62 MPa^{-1}), permeability (4.4 × 10^{-6} m/s), stress-strain relationship, and shear strength, and is appropriate for centrifuge model test studies of landfill stability and deformation.

In this study, the model landfill was simplified to a homogeneous synthetic MSW landfill with a model geomembrane barrier at the bottom. No interim or

Table 1. Test programme.

Test	Slope ratio	Height (m)	Water level stage (m)	Shaking stage (g)
1	1:2	20	3.3,6.7,10,13.3	-
2	1:1	20	3.3,6.7,10,13.3	-
3	1:2	8	2,4.3	0.13,0.23,0.32,0.44
4	1:1	8	2,4.3	0.13,0.23,0.32,0.44

final cover was used. Two typical centrifuge models for static and seismic tests are shown in Figure 1. The synthetic MSW was filled on the geomembrane in 0.05 m-thick layers, and each layer was compacted to produce the expected compacted unit weight. Then the synthetic MSW was carefully excavated to form a designated slope. To model the weakest interface under the landfill, a smooth HDPE geomembrane with thickness of 0.1 mm was used as model geomembrane in the tests. The frictional angles of the peak and residual shear strength for the interfaces of the synthetic MSW and the geomembrane were 17.7° and 13.8°, which were similar to those of the shear strength for the interfaces between the liner materials in the Suzhou Qizishan landfill (Chen et al. 2017).

2.3 Test programme

As shown in Table 1, a total of four model tests were performed using the synthetic MSW. Test 1 and 2 were performed at 66.7-gravities (66.7 g) to reveal the developing processes and mechanisms of MSW landfill failures that were induced by rising water levels. The influences of the slope ratio on the stability of the landfill were also investigated. The centrifuge was spinning up to designed gravity after the installation of the model landfill, and water was supplied when the settlement development of the landfill was stabilized. During the tests, the water level in the supply cavity was elevated in four stages, increasing to 0.05 m, 0.1 m, 0.15 m, and 0.2 m in 1 minute at a constant flow rate of 1 L/min. The centrifuge operated for an extra 20 minutes after the water level in the supplying cavity reached the expected height for each water supply stage. Finally, the centrifuge was spun down when significant failure of the landfill occurred.

Test 3 and 4 were designed to understand the seismic response of the landfill with different water level and its influence on the stability of the landfill. A 25 mm thick piece of mouldable Duxseal was placed on each side of the container to reduce reflecting incident stress waves. The Taft wave was adopted in the tests, which was recorded in the earthquake happened in Kern of California in 1952, with a primary period of 0.5 s, belonging to distant earthquake. The water level in the supply cavity was elevated in two stages (with water level ratio of 0.25 and 0.53 respectively under the crest of the model landfill) when the settlement development of the landfill was stabilised at the designed gravity. In each water level stage, the seismic excitations process was divided into four shaking

(a) Shake at water level stage 1

(b) Shake at water level stage 2

Figure 4. Displacement vevtors of the MSW after the excitations (unit: mm).

stages based on the acceleration amplitudes from weak to strong. There was enough interval between two shaking stages until the excess pore pressure dissipated entirely. After all the excitations were applied, the next water level stage was supplied. Finally, the centrifuge was spun down when a significant failure of the landfill occurred.

3 TEST RESULTS

3.1 Deformation of the landfill

In these tests, vertical displacement of landfill top was measured by a laser displacement transducer, and it generally took approximately 1 hour to stabilise the settlement of model landfills with a total settlement of up to 20%. In the static tests, during the rising of the water level, horizontal displacement of the landfill under the slope was observed (Chen et al. 2017).

Figure 4 shows the displacement vectors of the MSW after the earthquake with an acceleration amplitudes of 0.44 g. At water level stage-1 (water level ratio of 0.25), the deformation of the landfill was mostly vertically downward and the MSW got denser after the excitations. While at water level stage-2 (water level ratio of 0.53), significant horizontal displacement occurred under the slopes. The settlement on the crest of the landfill induced by the earthquake was measured by a laser displacement transducer L1, as shown in Figure 5, and it increased with the acceleration amplitudes of the earthquake.

3.2 Amplification of acceleration through MSW

The acceleration signals recorded during all shaking phases show amplification from base to crest. Figure 6 shows the acceleration signals during test 3 under earthquake with PGA of 0.44 g. All acceleration signals are given in prototype scale. There was

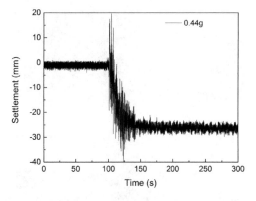

Figure 5. Settlement of the landfill during the earthquake.

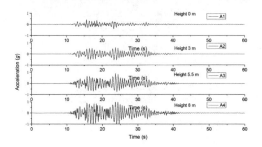

Figure 6. Amplification of acceleration through MSW.

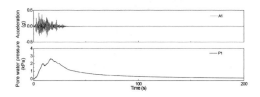

Figure 7. Pore water pressure response during shaking.

a significant increase of the amplification coefficient from the base to top of the landfill.

3.3 Pore water pressure response during the earthquake

Figure 7 shows the pore water pressure response during the earthquake with amplitude of 0.44 g in water level stage-1 in test 3. The pore water pressure in the landfill increased rapidly once earthquake loading was applied. And the pore water pressure reached peak value rapidly before the end of earthquake and then dissipated gradually. The increment in pore water pressure increased as the amplitude of the earthquake increased.

3.4 Critical water level

In test 1, the vertical displacement of the landfill surface developed slightly as the water level increased, but it changed dramatically when the water level reached

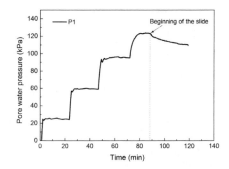

(a) Pore water pressure measured during the water supplying process

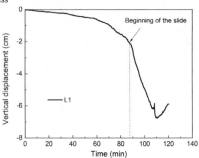

(b) Vertical displacement measured during the water supplying process

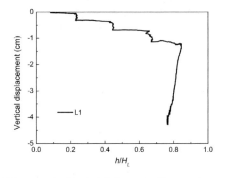

(c) Measurements of vertical displacement with respect to water level ratios

Figure 8. Determination of the critical water level in test 1.

critical value h_{max}, as shown in Figure 8(b). At the critical water level, landfill failure occurred with a sudden decrease in the water level (see Figure 8(a)) and a sudden increase in the vertical displacement. The measured data from the laser displacement transducers on the landfill surface with respect to water level ratio h/H_L is shown in Figure 8(c), where h is the water level in the supply cavity and H_L is the height of the landfill at the laser displacement transducer L1 before the water was supplied. When the failure occurred, the water level ratio reached the maximum value h_{max}/H_L, which was the critical water level ratio.

The ratio of the critical water level for the landfill with a slope ratio of 1:2 and 1:1 under static condition

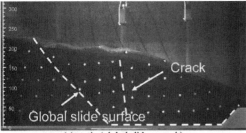
(a) static (global slide→crack)

(b) seismic (crack on shoulder→crack on slope)

Figure 9. Failure modes.

was 0.85 and 0.81 respectively, while it decreased to 0.66 and 0.61 after the application of the earthquake. The earthquake promoted the development of horizontal displacement of the landfill, especially at high water level. And this could resulted in cracks inside the landfill, thus reducing the stability of the landfill.

3.5 *Landfill failure modes*

In static tests, a large horizontal displacement was first observed, which induced a crack at the bottom of the landfill. This crack developed into a slide surface, and then a clear global slide occurred. With the sliding MSW mass continuously moving forward, another significant penetrating crack formed. The landfill failure generally began with a global slide along the liner (see Figure 9(a)) when the water level reached the critical value h_{max}. Once the global slide occurred, the water level inside the landfill decreased due to a large deformation in the landfill.

During the seismic tests, no global slide occurred along the liner. When the large earthquake was applied at high water level, a crack on the shoulder of the landfill was observed. As the water level continued rise, the crack on the slope shoulder developed and eventually penetrated the landfill when the critical water level was attained, forming a primary penetrating crack followed by large horizontal displacement. At the meanwhile, several cracks occurred on the slope, as shown in Figure 9(b).

Considering the actual sliding surface and the water level, the safety factor of the landfill calculated here was close to 1.0 when the landfill failure occurred, by slightly modifying the double-wedge analysis approach of Qian et al. (2003; 2004). And it coincided with the test result. These results provide good references for the water level and stability control of the landfill.

4 CONCLUSIONS AND SUGGESTIONS

A new synthetic MSW was developed to simulate real MSW in China. Several centrifuge model tests were performed on the stability of landfills with rising water levels under static and seismic conditions, and the failure processes were reproduced. The influences of water level and earthquake on the stability of the landfill were investigated. The following conclusions can be drawn.

1. Significant horizontal displacement inside the landfill is induced during the rising of the water level. At a low water level, the deformation of the landfill was mostly vertically downward and the MSW got denser after the earthquake, while significant horizontal displacement occurred at a high water level.
2. Landfill failure occurs when a critical water level is attained. The earthquake promoted the development of horizontal displacement of the landfill at high water level, and resulted in a decrease of the critical water level.
3. For static condition, the landfill failure generally began with a global slide along the liner when the water level reached the critical value. But for seismic condition, a primary penetrating crack occurred with several small cracks on the slope when the landfill failure.

These results provide good references for the water level and stability control of the landfill for static and seismic conditions.

REFERENCES

Chen, Y.M., Han, C., Ling, D.S., Kong, L.G., & Zhou, Y.G. 2011. Development of geotechnical centrifuge ZJU400 and performance assessment of its shaking table system. *Chinese Journal of Geotechnical Engineering*, 33(12): 1887–1894. (in Chinese)

Chen, Y.M., Kong, L.G., Zhou, Y.G., Jiang, J.Q., & Tang, X.W. 2010. Development of a large geotechnical centrifuge at Zhejiang University. *Proceedings of the 7th international conference on physical modelling in geotechnics* (ICPMG), Zurich, Switzerland, 28 June–July, 2010.

Chen, Y.M., Li, J.C., Yang, C.B., Zhu, B., & Zhan, L.T. 2017. Centrifuge modeling of MSW landfill failures induced by rising water levels. *Canadian Geotechnical Journal*. 2017, 54(12): 1739–1751.

Kavazanjian, E., Beech, J.F. & Matasovic, N. 2001. Discussion: municipal solid waste slope failure. I: Waste and foundation soil properties. *Journal of Geotechnical and Geoenvironmental Engineering*, 127: 812–815.

Thusyanthan, N.I., Madabhushi, S.P.G. & Singh. S. 2006a. Centrifuge modeling of solid waste landfill systems-Part 1: Development of a model municipal solid waste. *Geotechnical Testing Journal*, 29(3): 217–222.

Thusyanthan, N.I., Madabhushi, S.P.G., & Singh. S. 2006b. Centrifuge modeling of solid waste landfill systems-Part 2: Centrifuge testing of model waste. *Geotechnical Testing Journal*, 29(3): 223–229.

15. Seismic – liquefaction

Partial drainage during earthquake-induced liquefaction

O. Adamidis
ETH Zürich, Switzerland

G.S.P. Madabhushi
University of Cambridge, UK

ABSTRACT: Earthquake-induced liquefaction is typically viewed as an undrained phenomenon, at least in engineering practice. Undrained element tests form the core of knowledge built around it. However, there is evidence to suggest that partial drainage could take place even during an earthquake. In this paper two dynamic centrifuge tests are presented, in which a part of the soil was enclosed in a flexible chamber that prevented fluid inflow and outflow. In one test outward lateral expansion of the chamber was allowed whereas in the other it was not. The evolution of pore pressures during the imposed seismic events showed that the hypothesis of undrained behaviour was inappropriate, as fluid flow and void redistribution took place, both inside and out of the chambers. The proximity of the boundaries defined the time frame for excess pore pressure dissipation. The capacity for lateral expansion controlled the displacement response of the chambers.

1 INTRODUCTION

Earthquake-induced liquefaction is primarily understood through undrained element test results. Undrained element tests have facilitated the creation of frameworks for the estimation of the onset and the consequences of liquefaction (Ishihara 1993). However, their validity remains linked to that of the undrained hypothesis, which assumes that the effects of drainage can be neglected, at least for the duration of an earthquake.

The validity of the undrained hypothesis has been brought into question by researchers that have tried to produce a general formulation for liquefaction. According to Goren et al. (2010), two mechanisms primarily contribute to liquefaction in the field: pore volume compaction and fluid flow. Though undrained element tests capture the first mechanism, they are unable of capturing the second.

Nevertheless, the effects of fluid flow on a soil element can be significant. Vaid and Eliadorani (1998) examined how fluid inflow or outflow altered soil response in triaxial tests. They found that a sample can easily reach instability due to fluid inflow. Later, Vaid and Eliadorani (2000) showed that fluid flow can control the direction of effective stress increments and the deformation response of a sample.

Consequently, fluid migration during an earthquake could significantly affect soil response and jeopardise estimations that are based on the undrained hypothesis. Here, in order to investigate the effects of fluid flow during an earthquake, two dynamic centrifuge tests were performed. Centrifuge testing allows both pore volume compaction and fluid flow to be modelled accurately (Lakeland et al. 2014). As a result, it does not suffer from the shortcomings of undrained tabletop experiments. In the centrifuge tests presented below, the boundary conditions for different parts of the liquefiable sand were significantly altered. The hypothesis of undrained co-seismic behaviour was in all cases deemed inappropriate. The boundaries controlled the evolution of excess pore pressures.

2 EXPERIMENTAL METHODS

The use of centrifuge modelling in geotechnics is well documented (Madabhushi 2014, for instance). Its primary advantages over element tests include the accurate modelling of stress and hydraulic gradients as well as pore fluid flow.

Two centrifuge tests are presented here, designed to provide insight into the effects of drainage on the dynamic response of liquefiable sand. To achieve this goal, drainage was restricted for a part of the soil, which was enclosed within a chamber. The chamber was placed on one side of the model, while the other side allowed the behaviour expected at the free-field to be captured. A sketch of the two tests, named OA2 and OA3, is given in Figure 1. This figure also depicts the instruments used, which included miniature piezoelectric accelerometers, pore pressure transducers (PPTs), and linear variable displacement transducers (LVDTs).

A laminar box was used for the experiments, described in detail by Brennan et al. (2006). The box consisted of 25 sections which were separated by roller

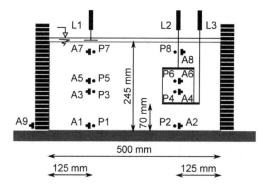

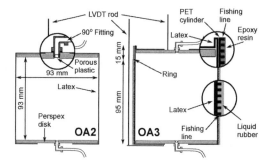

Figure 1. Cross-section of the models for centrifuge experiments OA2 and OA3. Transducers labelled A measured acceleration, P measured pore pressure, and L measured displacement. Transducer L3 was in place only for test OA3. Dimensions are in model scale.

Table 1. Properties of Hostun sand.

e_{max}	e_{min}	G_S	ϕ_{crit}	k (mm/s)
1.010	0.555	2.65	33°	1

bearings and could easily move relative to one another. The use of a laminar box limited as much as possible the boundary effects on the liquefiable sand. A thin layer of latex between the sections and the sand ensured the pore fluid remained contained.

The sand layers were prepared by air pluviation of Hostun sand. The sand was poured using the automatic sand pourer described by Madabhushi et al. (2006) with a targeted relative density of 40%. Basic properties of this sand are included in table 1, after Mitrani (2006) and Haigh et al. (2012).

A different cylindrical chamber was installed for each test. Sketches of the chambers are shown in Figure 2. The base and the top of each chamber consisted of Perspex disks, 5 mm in thickness and 93 mm in diameter (Figs. 3a, b). At the centre of the disks, 90° fittings were placed, to which 6 mm diameter pipes were connected. Thin cylinders of porous plastic placed in the fittings prevented sand from escaping into the pipes, while allowing fluid flow. The periphery of the chamber was different for each test. However, in both cases the chambers were designed so that the soil within the chamber would shear along with its surrounding soil.

For test OA2, the periphery of the chamber was flexible and could expand. It was made using a latex sheet of 0.3 mm thickness, which was wrapped around a PET cylinder of the same diameter as the base disks. The sheet was sealed to itself using two beads of aquarium silicone. Once the cylindrical periphery was created, it was sealed to the base, using two beads of aquarium silicone (Fig. 3c). Hostun sand grains were stuck to the surfaces that would come in contact with the soil (Fig. 3e). The chamber was placed in the laminar box once the sand surface had reached the right level. During sand pouring, a thin PET cylinder placed right

Figure 2. Sketches of the chambers, as used for each test. Dimensions in model scale.

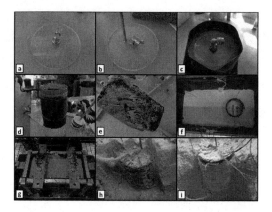

Figure 3. Preparation of the chambers: (a) the bottom disc of the chambers, (b) the top disc of the chambers, (c) the latex periphery of OA2, before being sealed to the bottom disc, (d) the periphery of the chamber of OA3 before sticking sand grains on it, (e) the periphery of the chamber of OA2 with sand on it, (f) the chamber of OA3 during sand pouring, (g) the model of OA3 mounted on the centrifuge, (h) the chamber of OA2 during post-test excavation, (i) the chamber of OA3 during post-test excavation.

outside of the chamber held the latex sheet in place. Once the chamber was full, the top cap was placed and the latex was sealed to it, again using two beads of aquarium silicone. After 24 hrs the silicone was dry, the supporting PET cylinder was removed, and the sand pouring continued.

For test OA3, the goal was to prohibit lateral expansion of the chamber's periphery. The design was inspired by simple shear test containers. Since the chamber had to remain light so as not to sink during liquefaction, the options of using stiff rings around the latex membrane or wrapping it with wire were not available. Instead, it was decided to wrap fishing line around a latex sheet, so as to prevent it from expanding without adding significantly to its weight. Initially, a latex sheet of 0.3 mm thickness was used to form a cylinder. Then, a thin layer of Tylon latex liquid rubber was applied and before it dried, fishing line Maxima Chameleon (18 lbs strength, 0.4 mm thickness) was wrapped around the chamber, leaving no gaps. Afterwards, a second coat of liquid rubber

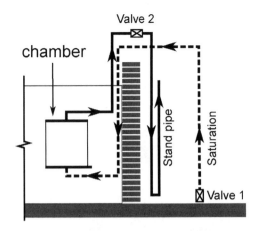

Figure 4. Sketch of the chamber, the pipes, and the valves layout.

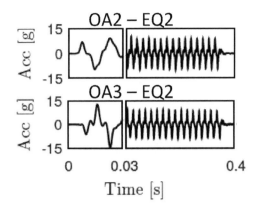

Figure 5. Input acceleration time histories. The initial cycles are shown in more detail. Dimensions in model scale.

was applied and Hostun sand grains were stuck on both sides of the periphery. This barrier corresponded to the first 95 mm of the chamber's height, as seen in Figure 2. Above that, a 15 mm tall ring was placed, within which the top cap could move. This ring was made so that upwards movement of the cap could take place without requiring the periphery to stretch. The ring consisted of a PET cylinder, reinforced with fishing line and slow-setting epoxy resin. A thin bead of aquarium silicone connected it to the rest of the chamber (Fig. 3d). As in OA2, the chamber was placed in the box once the sand was at the right level and then sand pouring continued until the chamber was full. An instance at which instruments were placed in the chamber is presented in Figure 3f. Once the chamber was full, the latex sheet was sealed to the cap with two beads of aquarium silicone. The latex sheet was folded as shown in Figure 2, allowing the top disk to move without stretching the periphery. A thin layer of grease between the ring and the latex facilitated the movement of the cap.

An aqueous solution of hydroxypropyl methylcellulose, prepared as described in Adamidis and Madabhushi (2015) was used to saturate the model. A solution of high viscosity was necessary to overcome the inconsistency between the scaling laws of dynamic and seepage time in centrifuge modelling (Madabhushi 2014, for instance). The targeted viscosity of the pore fluid was 50 cSt, since the applied centrifugal acceleration was 50 g. However, due to the sensitivity of the solutions to temperature, the actual viscosities obtained during tests OA2 and OA3 were 42 cSt and 37 cSt respectively. The saturation of the model was performed under vacuum, by placing the laminar box inside an airtight container, and was controlled via the computer system described in Stringer and Madabhushi (2009). In order to saturate the sand within the chambers, a system of pipes was in place, as shown in Figure 4. The outflow pipe formed a standpipe that ended at the same height as the fluid surface within the laminar box while spinning inside the centrifuge, ensuring no differences in hydrostatic pressure inside and outside of the chamber. During saturation, both valves of the system were open. Once saturation finished and fluid was seen coming out of the top pipe of the chamber, valves 1 and 2 were closed. The valves remained closed until the experiments started. During spinning up of the centrifuge, valve 2 was opened. Before the earthquakes were applied, valve 2 was closed. Valve 2 was opened again after each earthquake, once excess pore pressures at the free-field soil column had dissipated.

The experiments were performed on the Turner Beam Centrifuge, at the Schofield Centre of the University of Cambridge. The earthquakes applied were sinusoidal, pseudo-harmonic motions, generated using the stored angular momentum (SAM) actuator described by Madabhushi et al. (1998). The targeted duration was 0.4 s (20 s in prototype scale) and the targeted frequency was 50 Hz (1 Hz in prototype scale). The amplitude increased with each subsequent earthquake. The first earthquake of both tests was not strong enough to cause full liquefaction of the modelled sand layer. Consequently, only results from the second, stronger earthquake are presented here. Results from the first earthquake are included in Adamidis and Madabhushi (2017). The time histories of the input motions for both tests are shown in Figure 5, in model scale.

3 RESULTS

Time histories of excess pore pressures during the second earthquake of each test are presented in Figure 6. The excess pore pressures recorded inside the chamber (PPTs P4 and P6) and at similar depths of the free field (P3 and P5) are depicted. It should be noted that P6 was placed slightly higher than P5 during test OA3. The time axis is split into three parts of different scale. Within each part, time advances linearly between the two end values. Plots are separated at the endpoints of each time window by double lines. The first time window corresponds to the build up of excess pore

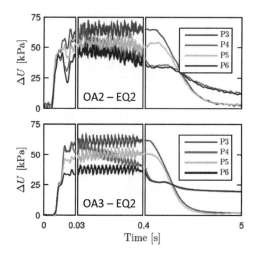
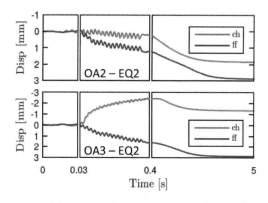

Figure 6. Time histories of excess pore pressures inside and outside of the chambers, during the second earthquake of each test. All traces are in model scale.

Figure 7. Time histories of chamber (ch) displacement and free-field settlement (ff) during the second earthquake of each test. For test OA2, chamber displacement corresponds to the trace of L2 (Fig. 1), whereas for test OA3 it corresponds to the change in height of the chamber. Negative displacement corresponds to an increase in height of the chamber. All traces are in model scale.

pressures. The second time window contains the co-seismic response. The third presents the post-seismic response.

Figure 7 presents displacements, at the free-field and at the chambers. The settlement of the free-field (ff), as recorded by LVDT L1 (Fig. 1) is included, with positive values corresponding to settlement. Chamber displacement (ch) is also included. For test OA2, this displacement corresponds to the trace of LVDT L2. For test OA3, it corresponds to the change in height of the chamber, calculated using LVDTs L2 and L3. Displacements recorded by L2 were significantly larger than those recorded by L3. Positive displacement for the chambers signifies a reduction in height.

3.1 Excess pore pressure build up

Excess pore water pressures increased quickly, leading to full liquefaction within the first two cycles of each motion. The initial rate of excess pore pressure generation was similar inside and outside of the chamber for both tests. No significant displacements were recorded before the sand liquefied completely.

The input acceleration was closer to a harmonic trace for test OA2 than for OA3, resulting in large input accelerations that were maintained for longer within the half-cycles of OA2 (Fig. 5). The application of acceleration for longer resulted in increased shearing in OA2, to which the sand responded by dilating. This can be seen by the sharp drops in excess pore pressure that were observed inside the chamber of OA2 for the first time window of Figure 6. In the free-field, fluid flow stemming from the soil beneath P3 and P5 limited the drop of excess pore pressure due to shearing. Inside the chamber, the proximity of the lower boundary allowed the drop observed to manifest. An undrained element test, where no fluid flow can exist, would be expected to produce similar results, linking increased shearing to drops in pore pressure. However, this result seems to not be appropriate for the field, where fluid flow can significantly limit drops in pore pressure. The importance of drainage as a function of the loading timescale and the proximity of the boundaries was examined in further detail in Adamidis and Madabhushi (2017).

3.2 Co-seismic response

In both tests presented, full liquefaction was reached inside and outside of the chamber. However, excess pore pressures inside the chamber started dropping fast, even as the earthquake continued. Placed near the base of the chamber, P4 recorded pore pressures that dropped faster than for P6, where increased excess pore pressures were maintained for longer, until the pressure of P4 reduced enough to match it. For test OA2, this point was reached during the earthquake, whereas for test OA3, it was reached slightly after the end of the earthquake. Once the pressures of P4 and P6 were matched, they kept decreasing in the same way until a certain value was reached, typically after the end of the earthquake.

The way pore pressures dissipated within the chamber points to void redistribution and the formation of a water film below the cap of the chamber. Research on the formation of water films due to liquefaction focuses on liquefiable layers capped with other, low permeability layers. Malvick et al. (2006) describe failure in such cases as a form of instability, where fluid is expelled from the lower part of the sand, which contracts, and is absorbed by the upper part of the sand, which dilates and loosens. If the water expelled by the lower part of the sand is more than the fluid that can be absorbed by the upper part, then instability occurs. Kokusho (2003) points out that water films can form beneath the layer of lower permeability just after the onset of liquefaction, due to rapid fluid flow. Then, excess pore pressures below the water film dissipate until they match its level of pressure. This behaviour is equivalent to a case of consolidation of liquefied sand where the top boundary is required to maintain an increased level of pressure, as opposed to the typical case of atmospheric pressure. This behaviour corresponds well to the tests presented here and has also been observed in the case of non-homogeneous soil deposits (Maharjan and Takahashi 2013, Maharjan and Takahashi 2014). During the second earthquake, excess pore pressures dissipated until, after the end of the earthquake, they reached a value of about 35 kPa for test OA2 and 28 kPa for test OA3. These values corresponded to the pressure that a fluid film formed under the top cap would have, as they were equal to the initial (pre-earthquake) vertical effective stress at the level of the cap of the chamber. Lower pressure was recorded for test OA3, as that chamber was slightly taller and its cap was closer to the surface (Fig. 2). It can be concluded that localised void redistribution and flow of pore fluid inside the chamber took place, within a time scale that was shorter than the duration of the earthquake.

In the free-field, the soil remained liquefied, retaining its elevated excess pore pressures. This behaviour is often interpreted as undrained, assuming that excess pore pressures within a soil element do not drop because fluid outflow is insignificant. Considering the behaviour observed inside the chamber and the settlement shown in Figure 7, the undrained assumption seems inappropriate. It is more likely that soil elements of the free-field maintained their excess pore pressures not because undrained behaviour prohibited fluid outflow but because the boundaries of the free-field were far enough to allow adequate inflow to sustain the elevated pressures.

The displacement response was different for each test, despite the similarities in the excess pore pressure measurements. In test OA2, the cap of the chamber moved downwards. The position of the base of the chamber was measured during placement and during excavation after the test. It was found to have moved downwards by less than 1 mm, despite the application of three earthquakes. Therefore, the height of the chamber in test OA2 is expected to have slightly decreased during the event examined here. In test OA3, the height of the chamber increased. The only difference between the two tests was in the construction of the chambers, hence in the lateral boundaries for each chamber.

Before explaining the displacement response, two points need to be highlighted. Firstly, it should be noted that some rocking of the soil mass inside the laminar box occurred during the earthquakes. This resulted in slightly larger horizontal soil displacements at the centre of the box, compared to its edges. Secondly, the top cap of the chamber was slightly heavier than sand, resulting in an increase in vertical stress of about 2 kPa for the soil inside the chamber, as compared to the soil of the free-field. This extra pressure was too small to have any visible effect before the soil liquefied completely.

In test OA2, an overall gradual reduction in the height of the chamber is expected to have occurred. Since no drainage out of the chamber was allowed, the volume inside the chamber had to remain practically constant. Therefore, the decrease of the chamber's height had to be due to the gradual extension of the flexible membrane along its periphery. This response was likely related to the increased weight of the chamber's cap.

Contrary to the chamber of OA2, the chamber of OA3 responded by increasing in height during the earthquake. This chamber could not expand due to the presence of the fishing line. Along any horizontal section of this chamber, the perimeter could not change in length. Any deviation from the initial circular shape would result in a reduction of the enclosed surface due to the isoperimetric inequality, which states that for a shape of a certain perimeter, a circle contains the maximum area. Since the volume inside the sealed chamber had to remain constant, deviations from the original cylindrical shape would necessarily result in an increase of the chamber's height. It is hypothesised that due to the rocking of the soil mass in the box, larger horizontal displacements on one side of the chamber distorted its cylindrical shape and led to the recorded displacements. This effect was not pronounced in OA2, where the periphery could expand. On the contrary, it was prominent in OA3, overshadowing the effect of the weight of the cap.

3.3 Post-seismic response

After the end of the earthquakes, the excess pore pressures in the free field gradually dissipated as the liquefied sand reconsolidated. The reconsolidation of the free-field of these tests was discussed in detail in Adamidis and Madabhushi (2016). There, it was shown that the post-earthquake excess pore pressure dissipation can be described by a consolidation equation, as long as soil properties are defined as stress-dependent.

Inside the chambers, excess pore pressures had already dropped due to fluid flow. After the earthquake, they quickly matched the top boundary condition, which demanded that pore pressures in the chamber were the same as the pressure of a fluid film below the top cap. This value of pore pressure was maintained for as long as the soil surrounding the chamber at the level of its cap remained fully liquefied. This time period corresponds to the plateau of excess pore pressures for P4 and P6 in Figure 6. While the soil outside of the chamber in OA2 was liquefied, the slightly increased weight of the cap forced the flexible membrane directly under it to expand laterally. Indeed, the top cap of the chamber moved downwards during the beginning of reconsolidation (Fig. 7). A return towards a more cylindrical shape at the end of the earthquake, when the rocking of the soil mass stopped, could have contributed to this downwards movement.

The free-field solidification front arrived at the level of the chamber's cap roughly when the pressure traces of P3 and P5 in Figure 6 crossed the traces of P4 and P6 (Florin and Ivanov 1961, for instance). After that point, dissipation of excess pore pressures began outside of the chambers, at the level of the cap and also excess pore pressures inside the chambers started dropping, faster for the expandable chamber of OA2 and slower for the chamber of OA3, whose periphery could not expand. When excess pore pressures outside of the chamber were dropping, so was total horizontal stress, as less horizontal than vertical effective stress is expected to have been regained. As a result, when excess pore pressure dissipation started outside of the membrane of the chamber in OA2, at the level of the cap, the total horizontal stress outside of the membrane started reducing and the pressure of the fluid film could no longer be maintained. The resulting drop in excess pore pressures inside the chamber likely reflected the reduction in total horizontal stress outside of the membrane. In test OA3, the soil just below the cap of the chamber was surrounded by a relatively stiff ring which could develop hoop stresses and allow differences in total horizontal stress on either of its sides (Fig. 2). As a result, excess pore pressures inside the chamber did not have to drop as much as in OA2, although total horizontal stresses outside of the chamber likely reduced in a similar way for the two tests.

4 CONCLUSIONS

In this paper, the effect of drainage was investigated experimentally using dynamic centrifuge testing. In the two tests presented, a part of the soil was enclosed within a chamber which prohibited pore fluid from flowing into or out of it. Lateral expansion of the periphery of the chamber was allowed in one test and restricted in the other.

It was concluded that undrained behaviour was not a realistic assumption for the time scale of an earthquake, in or out of the chambers. Void redistribution and fluid flow occurred both in the free-field, resulting in surface settlement, and inside the chamber, leading to excess pore pressure dissipation.

The proximity of the boundaries defined the time frame for the evolution of excess pore pressures. In the free-field, where boundaries were further away, excess pore pressures were maintained during the earthquake. Inside the chambers, the proximity of the boundaries led to excess pore pressures dropping quickly.

Finally, the different lateral boundaries of each chamber affected deformations significantly. The chamber with the expandable periphery reduced in height during the earthquake due to the weight of the top cap. On the contrary, the chamber whose periphery could not expand increased in height due to the distortion of its original cylindrical shape.

REFERENCES

Adamidis, O. & Madabhushi, G. S. P. 2015. Use of viscous pore fluids in dynamic centrifuge modelling. *Int. J. of Physical Modelling in Geotechnics* 15(3), 141–149.

Adamidis, O. & Madabhushi, G. S. P. 2016. Post-liquefaction reconsolidation of sand. *Proceedings of the Royal Society A* 472(2186), 20150745.

Adamidis, O. & Madabhushi, S. P. G. 2017. Experimental investigation of drainage during earthquake-induced liquefaction. *Géotechnique*, Available ahead of print.

Brennan, A. J., Madabhushi, S. P. G., & Houghton, N. E. 2006. Comparing laminar and equivalent shear beam (ESB) containers for dynamic centrifuge modelling. In *6th ICPMG*, Volume 1-2, pp. 171–176. Taylor & Francis.

Florin, V. A. & Ivanov, P. L. 1961. Liquefaction of Saturated Sandy Soils. In *The 5th International Conference on Soil Mechanics and Foundation Engineering*, pp. 107–111.

Goren, L., Aharonov, E., Sparks, D., & Toussaint, R. 2010. Pore pressure evolution in deforming granular material: A general formulation and the infinitely stiff approximation. *J. of Geophysical Research* 115(B9), B09216.

Haigh, S., Eadington, J., & Madabhushi, S. P. 2012. Permeability and stiffness of sands at very low effective stresses. *Géotechnique* 62(1), 69–75.

Ishihara, K. 1993. Liquefaction and flow failure during earthquakes. *Géotechnique* 43(3), 351–451.

Kokusho, T. 2003. Current state of research on flow failure considering void redistribution in liquefied deposits. *Soil Dyn. and Earthq. Eng.* 23(7), 585–603.

Lakeland, D. L., Rechenmacher, A., & Ghanem, R. 2014. Towards a complete model of soil liquefaction: the importance of fluid flow and grain motion. *Proceedings of the Royal Society A* 470(2165), 20130453–20130453.

Madabhushi, S. P. G. 2014. *Centrifuge Modelling for Civil Engineers*. Taylor & Francis, London.

Madabhushi, S. P. G., Houghton, N. E., & Haigh, S. K. 2006. A new automatic sand pourer for model preparation at University of Cambridge. In *6th ICPMG*, Volume 1-2, pp. 217–222.

Madabhushi, S. P. G., Schofield, A. N., & Lesley, S. 1998. A new Stored Angular Momentum (SAM) based earthquake actuator. In *Centrifuge '98*, pp. 111–116.

Maharjan, M. & Takahashi, A. 2013. Centrifuge model tests on liquefaction-induced settlement and pore water migration in non-homogeneous soil deposits. *Soil Dynamics and Earthquake Engineering* 55, 161–169.

Maharjan, M. & Takahashi, A. 2014. Liquefaction-induced deformation of earthen embankments on non-homogeneous soil deposits under sequential ground motions. *Soil Dynamics and Earthquake Engineering* 66, 113–124.

Malvick, E. J., Kutter, B. L., Boulanger, R. W., & Kulasingam, R. 2006. Shear Localization Due to Liquefaction-Induced Void Redistribution in a Layered Infinite Slope. *J. of Geotech. and Geoenv. Eng.* 132(10), 1293–1303.

Mitrani, H. 2006. *Liquefaction Remediation Techniques for Existing Buildings*. Ph. D. thesis, University of Cambridge, Cambridge, UK.

Stringer, M. E. & Madabhushi, S. P. G. 2009. Novel Computer-Controlled Saturation of Dynamic Centrifuge Models Using High Viscosity Fluids. *Geotech. Testing J.* 32(6), 102435.

Vaid, Y. & Eliadorani, A. 2000. Undrained and drained (?) stress-strain response. *Canadian Geotech. Journal* 37(5), 1126–1130.

Vaid, Y. P. & Eliadorani, A. 1998. Instability and liquefaction of granular soils under undrained and partially drained states. *Canadian Geotech. Journal* 35(6), 1053–1062.

Centrifuge modelling of site response and liquefaction using a 2D laminar box and biaxial dynamic base excitation

O. El Shafee, J. Lawler & T. Abdoun
Rensselaer Polytechnic Institute, Troy, New York, USA

ABSTRACT: A series of centrifuge tests with base shaking were conducted on the 150 g-ton Centrifuge at Rensselaer Polytechnic Institute to Study the effect of biaxial base excitation on sand deposits, and to calibrate and assess the performance of the newly commissioned 2D shaker. The study used biaxial base shaking on loose and medium dense sand deposits. Two centrifuge models of 32 and 26 cm-thick, level, Nevada sand deposits, were built in 2D laminar box and subjected to base excitation inflight at 25g to simulate 8 and 6.5 m soil stratum in the field. The models were subjected to uniaxial and biaxial base shakes using artificial and real earthquake records. Several configurations of soil models were calibrated, including dry and saturated models of various densities, using pore fluid with viscosity 25 times higher than water, which was used to simulate water saturated soil deposits in the field. It was found that the acceleration amplitude increases as the base shake propagate through the soil with noticeable difference between uniaxial (1D) and biaxial (2D) models, and that the shaker is capable of applying a variety the base excitations successfully with minimal differences compared to the targeted input motion.

1 INTRODUCTION

The ability to understand and predict soil response during earthquakes is an important aspect in the design and management of soil systems. Numerous studies were conducted over the last four decades to analyze and quantify the response of soil deposits when subjected to base excitation. Motions complexity progressed from single frequency sine excitation, to variable amplitude and finally multi-frequency content. The vast majority of studies using physical modeling were conducted uniaxially, and only few used simple biaxial shaking. Examples for biaxial experimental research can be found in the work done by Ng et al. (2003) studying granite embankments, Su and Li (2006) studying soil-pile interaction, and Su and Li (2008) studied level sites. All these studies used the centrifuge facility at Hong Kong University. While, Su (2012) used shake table in Shenzhen University to test piles embedded in soil. Most of the past research studying soils subjected to biaxial base excitation was conducted using earthquake shakers, which is an essential part of centrifuge testing for geotechnical earthquake applications. The common type of shakers installed in geotechnical centrifuge facilities are one-dimensional (1D) shakers, for example Madabhushi (1996), and Chazelas et al. (2008). The two-dimensional "biaxial" (2D) horizontal shakers are not very common in centrifuge facilities. According to the available literature, the centrifuge facilities that have 2D shakers are Hong Kong University Ng et al. (2001), USA Sasanakul et al. (2014), and currently under development in Korea Kim et al. (2006). These 2D shakers can produce motions in two horizontal directions (X and Y directions). They operate with minimum interaction between the two directions and minimum interference with the centrifuge machine.

Rensselaer Polytechnic Institute has developed a biaxial earthquake simulator for use on RPI's 150 G-ton geotechnical centrifuge. The RPI 2D shaker was commissioned in 2004 and has been recently upgraded with a new and powerful control system. The earthquake simulator provides two independently controlled horizontal components, utilizing advanced control algorithms for precise calibration of desired excitations. Once the desired motion is calibrated, it can be reproduced identically for use on future models with varied parameters and construction details. The work presented in this paper provides illustration of the features and capabilities of the RPI 2D shaker which is the main tool in studying soil response under biaxial shaking. Using the shaker to calibrate and create library of shakes to be used in following tests. These capabilities are tested by applying series of synthetic biaxial motions. The way these motions were designed is explained, along with the method used to produce equivalent uniaxial motions needed for comparison. The used calibration models were equipped with accelerometers inside the soil. The records from these accelerometers are used to compare the acceleration amplification inside the soil in biaxial shaking versus uniaxial shaking.

2 EQUIPMENT

2.1 2D shaker

The ES-Biax18 earthquake simulator at RPI is an integrated centrifuge platform/shaker assembly. Platform dimensions of 1000 × 660 mm include mounting points for several rigid containers, as well as 1D and 2D laminar containers. The shaker operates in two horizontal prototype directions, and is capable of supporting a model mass of 250 kg at centrifugal acceleration of 100 g. Three servo-hydraulic actuators drive the shaker as shown in Figure 1a, capable of 49 kN of force per direction, and a peak displacement of 12 mm. The table is supported by elastomeric bearings providing a mechanical system with consistent stiffness and damping over the operating range. The platform is non-resonant within the frequency range of the system, allowing precise control of the model and servo hydraulic system for fast and consistent tuning. The earthquake simulator is a powerful tool requiring a versatile and capable control system. The controller utilizes a single accelerometer per actuator which can be located in or outside the soil, and utilizes an algorithm to iteratively correct the motion. The user can tune the motion to replicate the desired time history, Use of the coupled algorithm supplies X-component information to the Y control algorithm for active cancellation of directional crosstalk. The shaker is capable of replicating completely different time histories in both prototype directions, storing the output, and replicating the identical motion in the future. This consistency is of critical importance when validating subtle changes in a series of soil models. In addition, the motions can be scaled to different acceleration levels, allowing the researcher to incrementally increase the acceleration before the model is significantly disturbed. More information about the shaker control system can be found in the work done by Sasanakul et al. (2014).

2.2 2D laminar box

The RPI 2D laminar container is designed to minimize the effects of the container boundaries and to simulate the way large expanses of soil would shear in the natural environment. The laminar box is capable of moving freely in three degrees of freedom (X, Y and θ). This provides more realistic simulation to half space soil condition in which soil is assumed not to face rigid boundaries till infinite distance from the studied soil region. This 2D laminar container is constructed from 45 stacked twelve-sided (dodecagonal) lightweight aluminum alloy rings. Each ring is 8.9 mm (0.35 inches) in height with a 594 mm (23.4 inches) mean inside diameter, and are separated from the rings above and below by 120 roller bearings per layer, specifically designed to permit translation in the two horizontal directions with minimal frictional resistance. Relative displacements of up to 2.5 mm (0.1 inch) between each layer can be achieved. The low friction between rings allows independent shear planes to

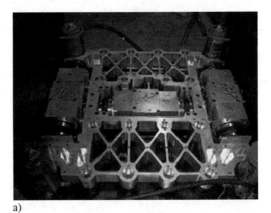

a)

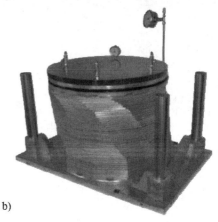

b)

Figure 1. a) RPI 2D shaker, b) RPI 2D laminar container.

develop between the layers of tested soil. The container is used to capture the response of soil systems under dynamic loading. The laminar container is shown in Figure 1b.

3 MATERIAL AND PREPARATION

One test was conducted in order to calibrate and create a library of shaking profiles. The tested model of level site deposit consisted of a Nevada 120 sand deposit. The height of the deposit was 26 cm which represent prototype depth of 6.5 m at 25 g. The model was built with relative density 45% which represents loose deposit. Nevada 120 is a medium-fine size grain sand with D_{50} of 0.15 mm. The sand grains are mainly angular and uncontaminated by clay, loam, iron compounds, or other foreign substances. The maximum and minimum void ratios of the sand are 0.763 and 0.562, respectively.

A rubber membrane is used to contain the soil model to prevent leaking of soil and water through the gaps between the rings of the 2D laminar box. The membrane is cylindrical homemade out of thin rubber (0.15 mm); this membrane is both flexible and strong. A detailed procedure is explained in El-Shafee (2016).

After the membrane is put inside the laminar box the soil model is built using dry pluviation method in lifts of 2cm. The soil layers are pluviated using a small drop height (2–3 cms) with no tamping at all. After each layer is placed and the relative density is achieved, the sensors at that level are placed. After the model is built, the laminar container is sealed and mounted on the centrifuge platform.

The saturation fluid used in the current study is "viscous fluid". The purpose of using a saturation fluid that has higher viscosity than water is to replicate the model sand permeability when spinning at high g-levels. The viscous fluid used in this study is made using Dow Chemical "Methocel LV K100 Premium CR", which is water-soluble Methylcellulose and Hydroxypropyl Methylcellulose polymers. A 1.4% concentration results in a fluid with a 25cP viscosity. The preparation steps can be found in El-Shafee (2016). Saturation took place on the shaker platform to minimize disturbance to the soil deposit. Saturation is done by applying vacuum (28 inch mercury), then injecting CO_2 to replace air in the voids. Vacuum pressure is reapplied followed by slowly saturating the model. After the saturation process finishes, the pump and all the hoses are detached. Then, the clear Plexiglas cover, top circular frame and the outer membrane with its clamps should be removed. Following that, the top restraining frame and the cross bar are mounted. Finally, the sensors are hooked to the data acquisition system (DAQ) on the centrifuge basket. Once all the sensors response is checked, the model is ready for spin-up.

4 TESTING PROCEDURE

4.1 Instrumentation

The tested models represent a level site and were equipped with accelerometers, pore pressure transducers and LVDTs. These sensors were mounted inside and outside the model at three elevations at depths of 1, 3 and 5 meters measured from the model surface. Each elevation has the following sensors: i) Two accelerometers to record the soil motion in the two horizontal X and Y directions placed equidistant from the model center. ii) Two pore pressure transducers are placed in each elevation to measure the excess pressures generated during the test. The shaking platform is equipped with accelerometers oriented in the X, Y, and Z directions. On the surface of the model there is one vertical LVDT mounted to measure the surface settlement of the model.

4.2 Input motions

Biaxial/Two Directional (2D) Shaking: The centrifuge soil model was subjected to a series of biaxial base excitations. The input consisted of synthetic motions with 1, 2, 3 Hz prototype dominant frequencies and varying amplitude. The biaxial input motions were

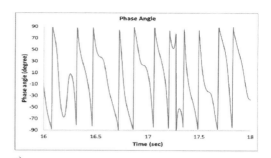

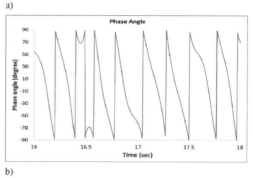

Figure 2. a) Shake A phase angle, b) Shake B phase angle.

obtained by varying the phase angle θ of the y and x accelerations ($\theta = \arctan(a_y/a_x)$), to give phase angle time history equivalent to real earthquake records. Figure 2 shows the generated phase angle time history for shakes A and B respectively. Although the used shakes are synthetic motions, the phase angle has some randomness and not far from the typical earthquake recorded phase angle time history. The variations of this angle show a pattern consistent with that of real earthquake records.

Several iterations were done to come with these biaxial motion profiles. The first trial is described in the work done by El-Shafee et al. (2014), in which a dry model was tested using biaxial shakes with sharp and abrupt changes in the acceleration profiles. Following this came the motions used in test CT1. Smoother motions were developed for the profiles used in the current test. The XY acceleration plots for shakes A and B are shown in Figure 3. The path of XY acceleration is acting on a plane and not just a uniaxial shake with a bias like the previous work mentioned in the introduction. Shake A profile has the shape of five headed (pentagon) star while shake B has an oval shaped profile.

One Directional (1D) Shaking: The uniaxial shakes used in the test have the same profile of the X component in the biaxial shakes. The uniaxial input motion is designed to have equivalent energy to the biaxial shaking. This is achieved using arias intensity so the uniaxial shake has the same input energy as the two components of the biaxial (2D) shake combined. The Arias intensity defined by Arias (1970) is the total energy per unit weight stored by a set of simple oscillators evenly spaced in frequency. The Arias

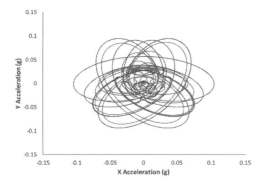

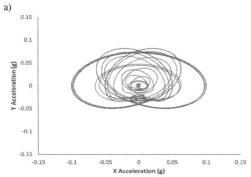

Figure 3. a) Shake A input acceleration, b) Shake B input acceleration.

intensity for ground motion in the x-direction (I_{xx}) can be written as:

$$I_{xx} = \frac{\pi}{2g} \int_0^{T_d} a(t)^2 \, dt \quad (1)$$

where $a(t)$ = The acceleration time history in the x-direction in (m/s²), g = The acceleration due to gravity in (m/s²), T_d = The total duration of motion in seconds.

The arias intensity in y-direction can be calculated similarly using the same quantities in equation (1) in the y-direction. To compare the biaxial base shake to the equivalent uniaxial base shake, the arias intensity of the uniaxial shake is equal to the resultant of the two components of the Arias intensity of the biaxial shake in x and y directions and can be written as:

$$I_{1D} = \frac{\pi}{2g} \int_0^{T_d} (a_x(t)^2 + a_y(t)^2) \, dt \quad (2)$$

where I_{1D} = Arias intensity for the 1D base shake, $a_x(t)$ = The acceleration time history in the x-direction in (m/s²), $a_y(t)$ = The acceleration time history in the y-direction in (m/s²).

Using the equation 1 and 2, the uniaxial equivalent energy shake was found to be 40% larger in amplitude. The uniaxial shake used was obtained by increasing the X component acceleration by about 40%.

5 RESULTS

5.1 Compliance of shaker

The results obtained from the test showed that the 2D shaker was able to apply the input motion to the base of the 2D laminar box. The recorded base shakes are very close to the input accelerations profiles. This was confirmed by comparing the Arias intensity of the two motions, the difference between the input motion and applied motion was in the range of 1 to 6%. In this paper samples of the recorded base accelerations will be presented.

The sequence of applying the shakes to the base of the model was Shake A in different amplitudes (0.01, 0.06, 0.1, 0.2 and 0.3g) followed by the equivalent uniaxial shake, then Shake B with the same amplitudes. Several trials were done to get an accurate record for mentioned amplitudes. For the sake of consistency with the dry test results from Shake A and B with 0.1g amplitude will be discussed in this paper. Shake A results show that the recorded XY acceleration profile is very close to the input profile with a very slight rotation as shown in Figure 4a. Arias intensity time history presented in Figure 4b shows that the recorded shake is almost identical to the input shake with 1% difference. Shake B results show similar trend to Shake A, the recorded XY acceleration profile is very close to the input profile, with Arias intensity 1% difference.

5.2 Acceleration amplification inside the model

The recorded soil response in the model consists of accelerations, excess pore water pressure (PWP) and surface settlement. Due to the preshaking of the model to achieve the desired input motion the model was subjected to large surface settlement before calibrating the targeted input motions. The model was built in a loose state ($D_r = 45\%$) but by the time the first motion was calibrated the model densified ($D_r \approx 60\%$). This densification made the generated excess PWP during the calibrated shakes insignificant and the model didn't liquefy. Because of that, the recorded accelerations inside the model gave a similar behavior to the dry test with the base acceleration amplitude increasing as the wave approaches the model surface. The acceleration amplification is compared for the directions of the two components of the biaxial shake, as well for the equivalent uniaxial shake. For the latter case the comparison is based on amplification ratio not the amplitude.

The recorded accelerations inside the model for Shake A show that for the biaxial shake the amplification in X and Y directions of the biaxial shake is almost the same at all depths, with slightly larger amplification in X direction (around 5% in average) as shown in the amplification profile in Figure 5c. For the uniaxial shake it was found that the uniaxial amplification is smaller than the biaxial amplification at all depths by about 25–30%. The accelerations time history is shown in Figures 5a and 5b. For Shake B, the results show that the amplification in X direction is larger than

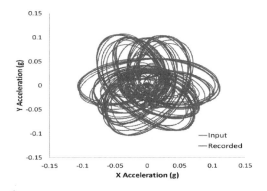

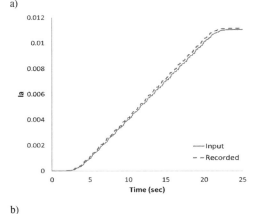

Figure 4. Shake A a) XY acceleration, b) Arias intensity time history.

amplification in Y direction at all depths by about 25% in average. The uniaxial shake results show that the amplification values lie in between X and Y amplification for the biaxial shake, with about 7% larger than Y direction and 17% smaller than X direction. These results show that although the uniaxial shaking have the same exact input energy as biaxial shaking it produces different soil behavior. This finding requires deeper analysis than just comparing recorded accelerations inside the model. This point is covered in more details in the work done by El-Shafee et al. (2016).

6 SUMMARY AND CONCLUSIONS

The experiment discussed in that paper had three main purposes: 1) Testing and assessing the behavior of the 2D shaker. 2) Calibrating biaxial and uniaxial shakes to be used in further testing with mastering building and saturation techniques. 3) Studying the effect of biaxial base shaking on level sand deposits.

In general the 2D shaker was able to apply the input biaxial shakes to the soil model successfully. Preliminary test made it clear that abrupt changes in the shake loops will decrease the efficiency of the shaker in accurately applying the input shake to the model. In the calibration tests the shakes were modified for the

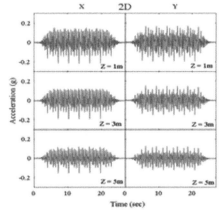

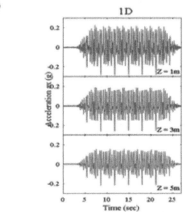

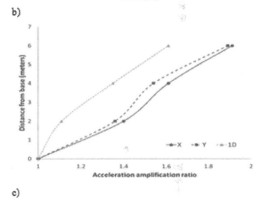

Figure 5. Shake A sample recorded accelerations at different depths a) Biaxial acceleration time history b) Uniaxial acceleration time history c) Amplification profile.

actual envelope of the shaker. Smoother loops were used and the recorded base shakes were almost identical to the input motion profiles. The phase angle change helped in creating realistic motions, and the arias intensity ensured that the energy content of the biaxial and uniaxial shakes is the same.

The calibration tests conducted helped in documenting model preparation and building procedure, and creating a data base of different profile biaxial and equivalent uniaxial shakes at various amplitudes. This shakes library will help in conducting tests without disturbing the model with any calibration or pre-shaking events. In the fully instrumented models the recorded shake from the database will be applied directly to the virgin model.

The experiment was also conducted to study the effect of biaxial base shaking on sand deposits. It have shown that the acceleration in both directions is amplified as it propagates upward in the soil as expected. The more important finding is that the amplification in biaxial shaking is not the same in X and Y directions. This means that biaxial shaking create anisotropy in the soil. Also, it was found that the amplification in uniaxial shaking is less than the amplification in biaxial shaking although both shakes have the same input energy. This finding indicates that biaxial shaking produces different response from the equivalent uniaxial shakes. The later finding needs further investigation which is not the scope of the calibration experiments discussed in this paper.

REFERENCES

Arias A. 1970 A measure of earthquake intensity, In Seismic Design for Nuclear Power Plants, Hansen RJ (ed.). MIT Press: Cambridge, MA, 1970; 438–483.

Chazelas, J.L., Escoffier, S., Garnier, J., Thorel, L. & Rault, G. 2008. Original technologies for proven performances for the new LCPC earthquake simulator Bulletin of Earthquake Engineering 6: 723–728.

El-Shafee O., Abdoun T. & Zeghal M. 2016 Centrifuge modeling and analysis of site liquefaction subjected to biaxial dynamic excitations", Geotechnique, 67, No. 3, 260–271. DOI:10.1680/jgeot.16.P.049.

El-Shafee O., Spari M., Abdoun T., & Zeghal, M. 2014. Analysis of the response of a centrifuge model of a level site subjected to biaxial base excitation. Geotechnical Special Publication, n 234 GSP, p 1081-1090, 2014, Geo-Congress 2014 Technical Papers: Geo-Characterization and Modeling for Sustainability – Proceedings of the 2014 Congress.

Kim D.S., Cho G.C. & Kim N.R. 2006. Development of KOCED geotechnical centrifuge facility at KAIST. Physical Modeling in Geotechnics. Ng, Zhang & Wang (eds), London: Taylor & Francis Group: 147–150.

Madabhushi, S.P.G. 1996. Preliminary centrifuge tests using the stored angular momentum (SAM) earthquake actuator – phase I. CUED Report, Cambridge University.

Ng C.W.W., Van Laak P.A., Tang W.H., Li X.S., & Shen C.K. 2001. The Hong Kong geotechnical centrifuge and its unique capabilities. Sino-Geotechnics, Taiwan, (83): 5–12.

Ng C.W.W., Li X.S., Van Laak P.A. & Hou, D.Y.J. 2003. Centrifuge modeling of loose fill embankment subjected to uni-axial and bi-axial earthquakes J. Soil Dynamics & Earthquake Engrg., Vol. 24: 305–318.

Sasanakul I., Abdoun T., Tessari A. & Lawler J. 2014. RPI In-flight two directional earthquake simulator Physical Modelling in Geotechnics - Proceedings of the 8th International Conference on Physical Modelling in Geotechnics 2014, ICPMG 2014, v 1, p 259–264.

Su D. & Li X.S. 2006. Centrifuge modeling of pile foundation under multi-directional earthquake loading Physical Modelling in Geotechnics-6th ICPMG, Vol. 1: 1049–1055.

Su D. & Li X.S. 2008. Impact of multidirectional shaking on liquefaction potential of level sand deposits J. Geotechnique, Vol. 58 (4): 259–267.

Su D. 2012. Resistance of short, stiff piles to multidirectional lateral loadings, J. Geotechnical Testing, Vol. 35 (2): 1–17.

Experimental simulation of the effect of preshaking on liquefaction of sandy soils

W. El-Sekelly
Mansoura University, Mansoura, Egypt

T. Abdoun & R. Dobry
Rensselaer Polytechnic Institute, Troy, USA

S. Thevanayagam
State University of NY, Buffalo, USA

ABSTRACT: The effect of preshaking on the liquefaction behaviour of sandy soils is studied in this paper. The research incorporated data from centrifuge and full-scale experiments. State of the art tools and sensors were used in the tests in order to capture as precisely as possible the response of the soil and connect it to the field. The tests simulated the effect of several decades of earthquake events on a 5-6 m uniform clean sand horizontal deposit, including both events that liquefied the deposit and others that did not liquefy it. The centrifuge model subjected to shaking events of types A, B, C and D, while the full-scale model to types A, B and C. An Event A represents a mild to moderate earthquake shaking; an Event B represents a mildly strong earthquake shaking; and Events C and D represent strong to very strong extensive liquefaction shaking in the field. The results of the experiments showed that the combination of mild/moderate (Events A) to mildly strong (Events B) shakings resulted in a significant increase in liquefaction resistance of the deposits over time. However, the occurrence of extensive liquefaction, caused by an Event C or D, resulted in a dramatic immediate reduction in liquefaction resistance.

1 INTRODUCTION

Several reported observations after earthquakes suggest that natural sands appear to liquefy less than nearby artificial fills in highly seismic areas such as California and Japan (Pyke 2003, Ishihara et al. 2011, Cox et al. 2013, Dobry et al. 2015). This has been variously attributed to geologic age and/or seismic preshaking by previous earthquakes (e. g., Arango et al. 2000, Heidari & Andrus 2012, Hayati & Andrus 2008, 2009, Dobry et al. 2015). In the specific case of the Imperial Valley of California, Dobry et al. (2015) concluded that the increased liquefaction resistance of the geologically very young silty sand deposits there, was due to intense preshaking rather than geologic age.

However, some of the same researchers studying the field evidence, have also concluded that if a field deposit is subjected to extensive liquefaction, the soil liquefaction resistance may be decreased rather than increased. On the basis of field observations in South Carolina, Heidari & Andrus (2012) found that full liquefaction can undo the beneficial effects of geologic age on both liquefaction resistance and shear wave velocity of the soil (V_s). The 200,000-year old Ten Mill Hill sand beds had low values of penetration resistance and V_s and were subjected to extensive liquefaction in the 1886 Charleston earthquake, behaving as a geotechnically young deposit despite its old geologic age (and presumably also despite the numerous earthquakes affecting these sand deposits since deposition). Heidari & Andrus (2012) concluded that whenever liquefaction occurs, the clock is reset in terms of both liquefaction resistance and V_s, with both parameters decreasing back to the values they had immediately after deposition.

This research focuses on the effect of preshaking rather than geologic age, and specifically the complex relation in the field between preshaking seismic events that increase the liquefaction resistance of young deposits, and liquefying events that may decrease it partially or totally.

In order to test that effect, an experimental program, using the RPI geotechnical centrifuge and the University at Buffalo (UB) full-scale laminar container (6m tall), was adopted.

The experimental work performed both at RPI and UB involved simulating several decades to several centuries of earthquake events applied to a uniform horizontal soil deposit about 5-6 m thick, including events that liquefied the deposit and others that did not liquefy it. State of the art tools and sensors were used at both laboratories in order to capture the behaviour of the soil as precisely as possible and connect it to the field. These tools include, but are not limited

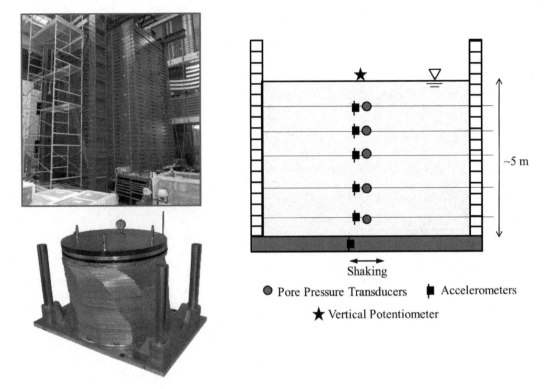

Figure 1. Large scale and centrifuge laminar containers used in this research along with typical instrumentation setup. The depth of the full-scale model was about 5 m while that of the centrifuge model was about 6 m.

to, accelerometers, LVDTs, pore pressure transducers, and laminar container (El-Sekelly 2014).

2 EXPERIMENTAL PROGRAM

2.1 Centrifuge experiment

The centrifuge model test was conducted in a 2-D laminar container at Rensselaer Polytechnic Institute's (RPI) centrifuge facility. The container has the ability to deform in two directions and to rotate about its central vertical axis. Figure 1 shows a schematic of the laminar container along with the general configuration of the sensors used. The container has an inside diameter of about 60 cm and a changing height depending on the height of the soil model. As shown in the figure, the models were instrumented with pore pressure transducers, accelerometers, and one vertical potentiometer at the ground surface.

The preparation procedures of the centrifuge experiment were according to the standard techniques used for saturated tests in the geotechnical centrifuge facility at Rensselaer Polytechnic Institute (Abdoun et al. 2013). In these procedures, the dry sand is pluviated in the box at the desired void ratio; carbon dioxide is then introduced to the model in order to replace the air; and, finally, viscous fluid is introduced by percolation for 12 h under vacuum to fully saturate the sand deposit (Abdoun et al. 2013). More details about the technique used for model construction can be found in Gonzalez (2008) and Abdoun et al. (2013).

The centrifuge Experiment was saturated with a viscous fluid having 25 times the viscosity of water and the test was performed at 25 g to yield the same prototype permeability of the full-scale experiment discussed in the following subsection. Details about the permeability of the soils used and the procedures used to calculate the prototype permeability can be found in Gonzalez (2008).

2.2 Full-scale experiment

The full-scale Experiment was performed in the full-scale laminar container at the University at Buffalo (UB), shown in Figure 1. The internal dimensions of the container are 2.8 m width, 5.0 m length and 6.1 m height. The full-scale model was instrumented with a variety of sensors to measure accelerations, pore pressures, and vertical and horizontal deformations (Fig. 1).

The deposition method used was a simulation of hydraulic fill. In this method, the sand is transferred in slurry form and pumped into the laminar container. This method allows sand grains to slowly settle down through water simulating alluvial deposition of sand in natural and man-made water bodies, such as rivers, lakes, etc. (Thevanayagam et al. 2009).

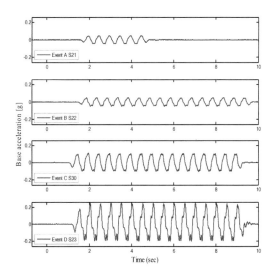

Figure 2. Target base accelerations for Events A, B, C, and D (El-Sekelly 2014).

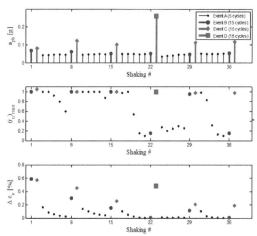

Figure 3. Histories of maximum excess pore pressure ratios and the permanent vertical strain, $\Delta \varepsilon v$, of the 37 shakings throughout the centrifuge experiment: (a) shaking sequence; (b) maximum pore pressure ratio for the whole deposit, $(r_u)_{max}$; and (c) the overall permanent vertical strain, $\Delta \epsilon_v$, of the deposit (El-Sekelly 2014).

2.3 Sand used in experiments

The sand used is clean Ottawa F#55. Ottawa F#55 sand is clean sand characterized by being fine and uniform. It has a specific gravity of 2.67, and characterized by D10 = 0.161 mm and D50 = 0.23 mm (Abdoun et al. 2013). The minimum and maximum densities are $\rho_{min} = 1475$ kg/m^3 and $\rho_{max} = 1720$ kg/m^3. The initial relative density of both the large scale and centrifuge experiments were about 40%.

3 SHAKINGS TYPES

The shaking sequences of the two experiments were somewhat different, as shown in the following section. However, in both experiments, an Event A was defined as 5 sinusoidal cycles of a peak base acceleration, $a_{pb} \approx$ 0.035-0.045 g; an Event B was defined as 15 sinusoidal cycles of a peak base acceleration, $a_{pb} \approx$ 0.04-0.05 g; and an Event C or D was defined as 15 sinusoidal cycles of a peak base acceleration, $a_{pb} \approx$ 0.1-0.25 g, all in prototype units (Fig. 2). The prototype frequency in all cases was 2 Hz. The 15-cycle duration of the Events B, C and D corresponds approximately to an earthquake of moment magnitude, $M_w \approx 7.5$; while the 5-cycle duration of the Events A correspond to $M_w \approx 6$ (Idriss and Boulanger 2008).

4 EXPERIMENTAL SIMULATION AND RESULTS

4.1 Centrifuge experiment

In this centrifuge experiment, a 6m prototype depth model of uniform saturated clean Ottawa sand was subjected to a total of 37 base shakings of different intensities and durations. The sequence of the shaking is shown in Figure 3a. It must be noted that the pore pressure was allowed to fully dissipate after each shaking event before the start of the following event.

Figure 3b shows the maximum excess pore pressure ratios, $(r_u)_{max}$, for Events A, B, C and D throughout the experiment. The $(r_u)_{max}$ typically occurred at the shallowest transducer and was generally lower at deeper locations in the model (see El-Sekelly 2014 for more details).

The behaviour can be summarized as follows:

- Events C and D always liquefied most of the deposit, as indicated by $(r_u)_{max} \approx 1$ in each Event C and D.
- Events A liquefy the deposit in the beginning, as indicated by $(r_u)_{max} \approx 1$. The liquefaction resistance gradually increases when the deposit is subjected to more and more preshaking, as indicated by lower and lower $(r_u)_{max}$.
- The behaviour is totally reset after application of Events C, as indicated by the abrupt jump of $(r_u)_{max}$ for Events A immediately after an Event C. It must be noted that the three Events A immediately after the Event D (S24-S26) had 20-25 % lower base acceleration than the rest of Events A. This helps explain why the $(r_u)_{max}$ of Events A was very low immediately after the Event D, compared to Events A that had occurred immediately after Events C.
- For Events B, the soil model experienced high excess pore pressures $((r_u)_{max} = 0.8$-$1)$ the first three times an Event B happened. The fourth time that an Event B happened (S22), the whole deposit experienced very low excess pore pressures $((r_u)_{max} = 0.2)$. After applying the extensive liquefaction shaking Event D, once again the soil deposit liquefied due to shaking Event B S29. However, in the next Events B, again the whole deposit experienced very low excess pore pressure $((r_u)_{max} = 0.2)$.

Figure 3c shows the permanent vertical strain, $\Delta\epsilon_v$, due to all shaking events. The behaviour of $\Delta\epsilon_v$ is characterized by the following trends: (i) $\Delta\epsilon_v$ decreases monotonically with each new Event B and C; (ii) $\Delta\epsilon_v$ decreases monotonically with each new Event A within each 5-Event A sequence; (iii) $\Delta\epsilon_v$ jumps to a higher value for Events A when there is an Event B and C in between; and (iv) this jump in $\Delta\epsilon_v$ is cancelled rapidly by the subsequent Events A, with the net result being a significance decrease in the $\Delta\epsilon_v$ for Events A between the beginning and the end of the 37-shaking experiment.

4.2 Full-scale experiment

In this full-scale experiment, a 5 m model of uniform clean Ottawa sand was subjected to a total of 51 base shakings of different intensities and durations. The sequence of the shaking is shown in Figure 4a. Figure 4b shows the maximum excess pore pressure ratios, $(r_u)_{max}$, for Events A, B, and C throughout the experiment. The behavior can be summarized as follows:

- Events C always liquefied most of the deposit, as indicated by $(r_u)_{max} \approx 1$ in each Event C.
- Events A liquefied the deposit in the beginning, as indicated by $(r_u)_{max} \approx 1$. The liquefaction resistance gradually increased as the deposit was subjected to more and more preshaking, as indicated by lower and lower $(r_u)_{max}$.
- The behaviour was totally reset after application of Events C, as indicated by the abrupt jump of $(r_u)_{max}$ of Events A immediately after an Event C.

For shaking Events B, the soil model experienced high excess pore pressure $((r_u)_{max} = 1)$, the first three times an Event B happened. The fourth time an Event B happened (S23), the whole deposit experienced very low excess pore pressures $((r_u)_{max} \approx 0.15)$. Afterwards, the $(r_u)_{max}$ maintained a very low value and gradually decreased even further, never returning back to the high value it had the first three times.

Figure 4c shows the permanent vertical strain, $\Delta\epsilon_v$, due to all shaking events. The behaviour of $\Delta\epsilon_v$ is characterized by the following trends: (i) $\Delta\epsilon_v$ decreases monotonically with each new sequence of Events B and C; (ii) $\Delta\epsilon_v$ decreases monotonically with each new Event A within each 5-Event A sequence; (iii) $\Delta\epsilon_v$ jumps to a higher value for Events A when there is a sequence of Events B and C in between; and (iv) this jump in $\Delta\epsilon_v$ is cancelled rapidly by the subsequent Events A, with the net result being a significance decrease in $\Delta\epsilon_v$ for Events A between the beginning and the end of the 51-shaking experiment.

5 CONCLUSION

The paper presents the summary of results for a full-scale and a centrifuge liquefaction experiments. In both experiments, the model was subjected to tens of shaking events of different magnitudes and durations

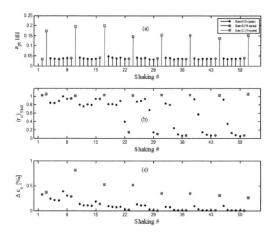

Figure 4. histories of maximum excess pore pressure ratios and the permanent vertical strain, $\Delta\varepsilon v$, of the 51 shakings throughout the full-scale experiment: (a) shaking sequence; (b) maximum pore pressure ratio for the whole deposit, (ru)max; and (c) the overall permanent vertical strain, $\Delta\varepsilon v$, of the deposit (El-Sekelly 2014).

to examine the effect of previous preshaking history on the liquefaction resistance of sand. The conclusions obtained from both experiment were similar as follows:

- There was a significant increase in the resistance of the clean sand deposits to liquefaction associated with both Events A and B, as the deposit was subjected to more and more preshaking.
- The occurrence of extensive liquefaction (Events C or D) resulted is a significant reduction in the liquefaction resistance of the deposit, to a level that was sometimes less than that of the young deposit before it was subjected to preshaking. However, it took only two to three additional Events A to undo the damage done by the Event C or D, and bring the liquefaction resistance of the deposit to where it was before the Event C or D occurred.

It must be noted that the conclusions of this research may be limited to the chosen shaking sequences and may not be general for all the cases. More research is being performed to generalize the conclusions

REFERENCES

Abdoun, T., Gonzalez, M. A., Thevanayagam, S., Dobry, R., Elgamal, A., Zeghal, M., Mercado, V. M., & El Shamy, U. 2013. Centrifuge and large scale modeling of seismic pore pressures in sands: a cyclic strain interpretation. *Journal of Geotechnical and Geoenvironmental Engineering* 139(8): 1215-1234.

Arango, I., Lewis, M.R., & Kramer, C. 2000. Updated liquefaction potential analysis eliminates foundation retrofitting of two critical structures. *Soil Dynamics and Earthquake Engineering* 20(1-4): 17-25.

Cox, B. R., Boulanger, R. W., Tokimatsu, K., Wood, C., Abe, A., Ashford, S., Donahue, J., Ishihara, K., Kayen, R., Katsumata, K., Kishida, T, Kokusho, T., Mason, B., Moss, R.,

Stewart, J., Tohyama & K., Zekkos, D. 2013. Liquefaction at strong motion stations in the 2011 great east japan earthquake with an emphasis in Urayasu city. *Earthquake Spectra* 29(S1): S55-S80.

Dobry, R., Abdoun, T., Stokoe, K., II, Moss, R., Hatton, M., & El Ganainy, H. 2015. Liquefaction potential of recent fills versus natural sands located in high-seismicity regions using shear-wave velocity. *Journal of Geotechnical and Geoenvironmental Engineering* 141(3).

El-Sekelly W. 2014. The effect of seismic pre-shaking history on the liquefaction resistance of granular soil deposits. *Ph.D. thesis* Rensselaer Polytechnic Institute, Troy, NY.

Gonzalez, M.A. 2008. Centrifuge modeling of pile foundation response to liquefaction and lateral spreading: study of sand permeability and compressibility effects using scaled sand techniques. *PhD Thesis* Rensselaer Polytechnic Institute, Troy, NY.

Hayati, H. & Andrus, R.D. 2008. Liquefaction potential map of Charleston, South Carolina based on the 1886 earthquake. *Journal of Geotechnical and Geoenvironmental Engineering* 134(6): 815-828.

Hayati, H. & Andrus, R. D. 2009. Updated liquefaction resistance correction factors for aged sands. *Journal of Geotechnical and Geoenvironmental Engineering* 135(11): 1683-1692.

Heidari, T. & Andrus, R. D. 2012. Liquefaction potential assessment of pleistocene beach sands near Charleston, South Carolina. *Journal of Geotechnical and Geoenvironmental Engineering* 138(10): 1196-1208.

Ishihara, K. Araki, K. & Bradley, B. A. 2011. Characteristics of liquefaction-induced damage in the 2011 Great East Japan earthquake. *Proc. Int. Conf. on Geotech. for Sustainable Development (Keynote Lecture), October 6-7, Hanoi, Vietnam* 1-22.

Pyke, R. 2003. Discussion of 'Liquefaction resistance of soils: summary report from the 1996 NCEER and 1998 NCEER/NSF workshops on evaluation of liquefaction resistance of soils,' by Youd, T. L. et al. (2001). *Journal of Geotechnical and Geoenvironmental Engineering* 129(3): 283-284.

Thevanayagam, S., Kanagalingam, T., Reinhorn, A., Tharmendhira, R., Dobry, R., Pitman, M., Abdoun, T., Elgamal, A., Zeghal, M., Ecemis, N. & El Shamy, U. 2009. Laminar box system for 1-g physical modeling of liquefaction and lateral spreading. *Geotech. Testing J.* 32(5): 438-449.

Dynamic centrifuge testing to assess liquefaction potential

G. Fasano, E. Bilotta & A. Flora
University of Napoli Federico II, Italy

V. Fioravante & D. Giretti
ISMGEO, University of Ferrara, Seriate (BG), Italy

C.G. Lai & A.G. Özcebe
University of Pavia, Italy

ABSTRACT: A set of centrifuge tests has been carried out at ISMGEO (Italy) laboratory on models of a liquefiable soil. A natural sand from the Emilia-Romagna region in Italy was used in the tests, in order to reproduce typical ground conditions where liquefaction occurred during the seismic sequence of 2012. The models were instrumented with miniaturised accelerometers and with pore pressure and displacement transducers. Spectrum-compatible acceleration time histories were applied at the base of the model. In this way triggering of the liquefaction was detected and post-liquefaction settlements were evaluated. The paper describes with the tests carried out on free-field models. Further tests are currently ongoing to assess the seismic response of simple model structures lying on liquefiable ground. The testing programme, funded within the H2020 research project LIQUEFACT, is aimed at an experimental verification of ground improvement techniques used to mitigate the liquefaction susceptibility of fully saturated loose sands.

1 INTRODUCTION

In seismic regions, soil liquefaction has been often one of the most significant causes of damage to structures during recent earthquakes, e.g. 2012 Emilia (northern Italy), 2011 Tohoku Oki (Japan) and particularly 2010–2011 Canterbury-Christchurch (New Zealand), where about half of the €25 billion loss was directly caused by soil liquefaction. Earthquake-induced liquefaction disasters have now been included among the most important induced effects of natural hazards on human environment. Enhancement of liquefaction risk assessment is needed to protect life and public safety and to mitigate economic, environmental, and societal impacts of liquefaction in a cost-effective manner (National Academies of Sciences, Engineering, and Medicine, 2016). In this framework physical modelling assumes relevance since it allows a number of different scenarios of hazard to be analysed in well-defined geotechnical conditions.

This work focuses on centrifuge tests carried out on free-field models of liquefiable ground, as part of a larger experimental campaign: further tests on models including simple structures are currently underway. The experimental programme, funded within the H2020 research project LIQUEFACT, is aimed at verifying the effectiveness of a number of ground improvement techniques to mitigate the liquefaction susceptibility of fully saturated loose sands.

2 BACKGROUND

Liquefaction under cyclic loading may occur in saturated soil, due to the increase of pore-water pressure. It consists in a significant loss of shear strength and stiffness of the soil that may result in instability of existing structures or trigger settlement, tilting or lateral spreading. The main aspects of the mechanics of soil liquefaction have been traditionally investigated via laboratory tests, study of case histories and physical modelling. It is worth mentioning the validation exercise VELACS (Verification of Liquefaction Analyses by Centrifuge Studies), that was carried out about 25 years ago by a number of centrifuge laboratories in UK and US. The experimental results showed significant variability. Nevertheless, they allowed a fruitful comparison of numerical blind predictions by several teams. The results indicated that numerical modelling at that time was quite challenging due to boundary problems involving soil liquefaction (Manzari et al. 2014).

More recently, the LEAP project (Liquefaction Experiments and Analysis Project) has been launched to produce a set of high quality centrifuge test data that will be used to assess the capabilities of currently available computational tools (Zeghal et al. 2014).

Nowadays it is largely accepted that a combined use of physical and numerical modelling is necessary to predict the effect of earthquake-induced liquefaction on buildings and other subsystems of civil infrastructures. The experimental activity carried out within the LIQUEFACT project follows this stream.

3 EXPERIMENTAL PROGRAMME

The set of centrifuge tests that will be shortly completed within LIQUEFACT is currently underway at ISMGEO (Italy) laboratory to assess the effectiveness of two selected ground treatment techniques against liquefaction: drainage and induced partial saturation (IPS).

ISMGEO centrifuge has an arm of 3 m and a capacity of 240 g-tonnes (maximum payload 400 kg, max acceleration 600 g). In 2010 it was equipped with a 1 degree-of-freedom shaking table. Two hydraulic actuators fire input signals up to 1 MHz at 100 g. The peak velocity of the shaking table is 0.9 m/s and the peak displacement 6.35 mm.

The LIQUEFACT testing programme includes free-field modelling to compare the response of untreated and treated ground. This paper refers only to this subset of tests on two untreated models.

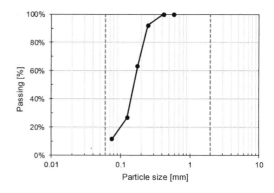

Figure 1. Grain size distribution of Pieve di Cento sand.

Figure 2. ESB container at ISMGEO laboratory.

3.1 Material

A natural sand from the Emilia-Romagna region in Italy was used in the tests to reproduce typical ground conditions where liquefaction occurred during the 2012 seismic sequence. The sand was retrieved from the field trial site of the project. This is located along the Reno river close to the town of Pieve di Cento near Bologna. This is a silica sand ($G_S = 2.69$, $\gamma_{d,min} = 12.25\,kN/m^3$, $\gamma_{d,max} = 15.75\,kN/m^3$) with a coefficient of uniformity $U \cong 1.8$ (Figure 1). The sand has been characterised under both static and cyclic loading conditions (Mele et al., unpubl.).

Model M1_S2 was made of clean Pieve di Cento sand, that is, after eliminating the finest fraction, passing at the ASTM no.200 sieve. Model M1_S3 was made using Pieve di Cento sand as it is (Fig. 1).

Table 1. Characteristics of ISMGEO ESB container.

Number of frames (–)	12
Number of rubber layers (–)	11
Frame mass (kg)	3.4
Height (mm)	337
Internal width (mm)	250
Internal length (mm)	750
Frame thickness (mm)	25
Frame width (mm)	40
Rubber layer thickness (mm)	3.36
Container weight (kg)	110 (incl. base)

3.2 Centrifuge container

An ESB (Equivalent Shear Beam, Zeng & Schofield 1996) container has been purposely manufactured by ISMGEO laboratory to be used during dynamic tests (Fig. 2). This consists of 12 rectangular aluminum frames, each 25 mm thick, separated by 11 rubber layers (shear modulus, $G_{r,0} = 1.3\,MPa$; tensile strength $\sigma_f = 4\,MPa$, tensile strain at failure, $\varepsilon_f = 250\%$), each 3.36 mm thick. The whole height of the container is 337 mm. The main characteristics of the container are shown in Table 1.

The lowest frame is connected to the base of the container having a size of 1000 mm × 495 mm and a thickness of 40 mm. The outer edge of the base is manufactured to fit the bearings on the shake table, fix to the rotating arm. The box and the table are connected on flight through a crank system. The inner part of the base is grooved to allow saturation and drainage of the model.

3.3 Preparation of physical models

The soil models were reconstituted by air pluviation of dry sand at a target void ratio.

A latex membrane between the soil and the container guarantees water-tightness along the vertical sides. The membrane is fixed to the bottom and the top of the frame stack.

A flexible aluminium mesh was inserted between the soil and the membrane along the short sides of the box. It was connected to the bottom of the stack to improve shear stress transmission at the side boundaries of the soil layer during shaking.

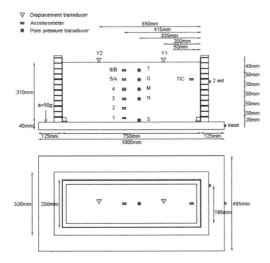

Figure 3. Instrumentation placed in Model M1_S2 to monitor the response of the physical model to ground shaking.

Table 2. Main features of input ground motions.

T_r (years)	ID (–)	M_W (–)	R_{ep} (km)	S_F (–)	Source file (–)	Model PGA (g)	f_p (Hz)
475	GM17	6.1	97.0	1.65	KiKnet EW2	5.0	500
975	GM23	5.9	10.1	2.39	ESM ACC	5.3	200
2475	GM34	6.9	28.6	0.59	NGA AT2	11.3	125

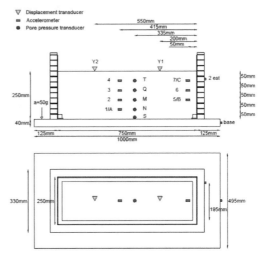

Figure 4. Instrumentation placed in Model M1_S3 to monitor the response of the physical model to ground shaking.

During soil pluviation the models were instrumented with miniaturised accelerometers and with pore pressure transducers deployed at several depths as shown in Figures 3 (model M1_S2) and 5 (model M1_S3).

Displacement transducers (potentiometers) were located at the ground surface. In this way triggering of the liquefaction was detected and post liquefaction settlements were evaluated. After deposition, the soil layer was saturated with a pore fluid with scaled viscosity. A hydraulic gradient was imposed between the bottom and the top of the layer using a vacuum system. The achievement of complete saturation was controlled by measuring the volume of fluid accumulating in the box and comparing it with the volume of voids.

3.4 Input ground motions

The time histories of acceleration were obtained through a two-step procedure aimed at reproducing the expected ground motion at a depth of 15 m below the ground surface at the prototype scale (Özcebe et al. – unpub.). These records were then applied at the base of the physical models.

In the first step, spectrum-compatible rock outcrop motions were selected from accredited international strong-motion databases for return periods of 475, 975, and 2475 years. These records were then deconvolved at the roof of seismic bedrock (in Pieve di Cento) and propagated through a soil profile in linear-equivalent ground response analyses up to a depth of 15 m below the ground surface. Finally, the ground motions were applied at the base of the centrifuge model after appropriate time scaling.

The geotechnical model required to perform ground response analysis at Pieve di Cento was defined based on the available geotechnical data at the site.

Among the 21 calculated time series, 3 were finally selected whose main characteristics are shown in Table 2. The signals were lastly corrected to take into account the transfer function of the dynamic actuator.

Each model was excited using 3 input motions. A time sufficient to the complete dissipation of the excess pore pressure was allowed between the subsequent application of the various ground motions.

4 RESULTS

A selection of results of the tests is shown in this section. In Figures 5 and 6 the acceleration time history recorded at the base and corresponding to the input signals, is plotted together with the pore pressure ratio R_u, as calculated from the recorded excess pore pressure at the top of the PPT array (T), in models M1_S2 and M1_S3 respectively and for the 3 ground motions. The results clearly show that the accumulation of pore pressures is larger for ground motion of larger amplitude. On the other hand, this accumulation drops during the last run of the series due to sand densification. However, it should be remarked that liquefaction (i.e. $R_u = 1$) was never achieved during the tests.

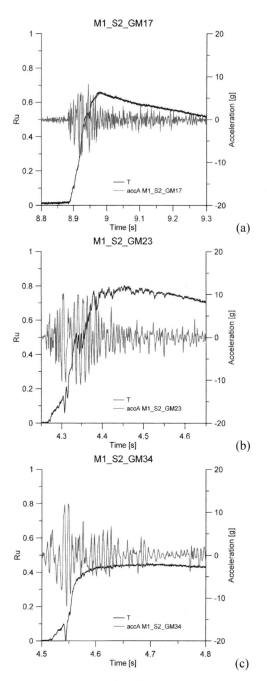

Figure 5. Acceleration time histories at the base and excess pore pressure build-up (T) recorded during the excitation of the shake table in Model M1_S2 for 3 ground motions (model scale).

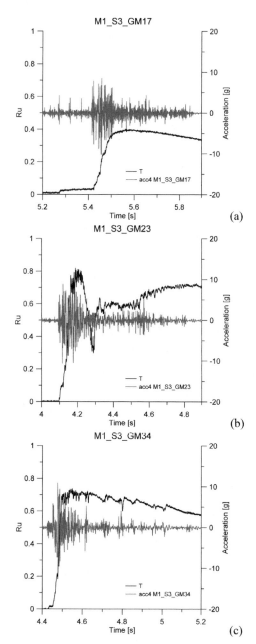

Figure 6. Acceleration time histories at the base and excess pore pressure build-up (T) recorded during the excitation of the shake table in Model M1_S3 for 3 ground motions (model scale).

Figures 7 and 8 illustrate similar plots of the accumulation of settlement at the ground surface, as measured by Y1 and Y2 potentiometers. In most cases, the displacements measured by the two transducers are the same at the end of the shaking. This indicates a uniform settlement distribution along the ground surface.

Permanent settlement after complete dissipation of excess pore pressure indicates densification of the sandy layer after ground shaking (post-seismic settlement).

Figure 9 shows a comparison of acceleration time histories recorded along the frame (2est) and within the sandy layer (A) during the 3 excitations of Model M1_S2 (Figs 10a, b, c). The good agreement among

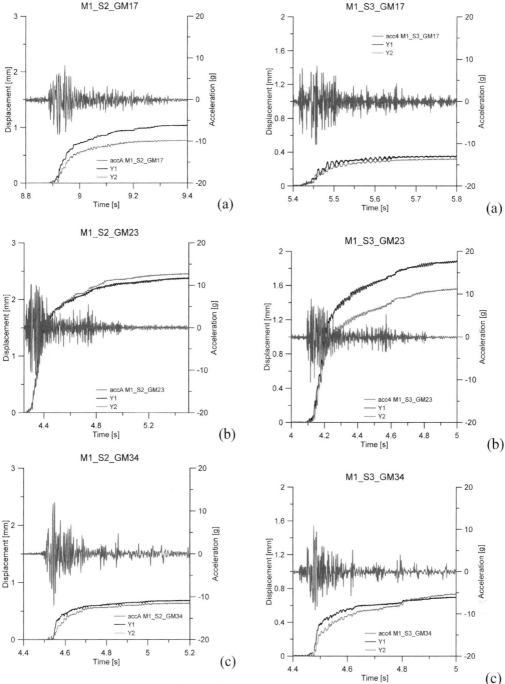

Figure 7. Acceleration time histories within the soil (A) and measured settlement at surface (Y1, Y2) recorded during three shakings in Model M1_S2 for 3 ground motions.

the time histories recorded at the same elevation along the boundary and within the domain indicates that the ESB container is performing well in mitigating the boundary effects. Only a small amplification of motion is observed towards the end of the inner recorded signal.

Figure 8. Acceleration time histories within the soil (4) and measured settlement at surface (Y1, Y2) recorded during three shakings in Model M1_S3 for 3 ground motions.

5 CONCLUDING REMARKS

This paper discussed the results of centrifuge dynamic tests on two physical model of fully saturated sand.

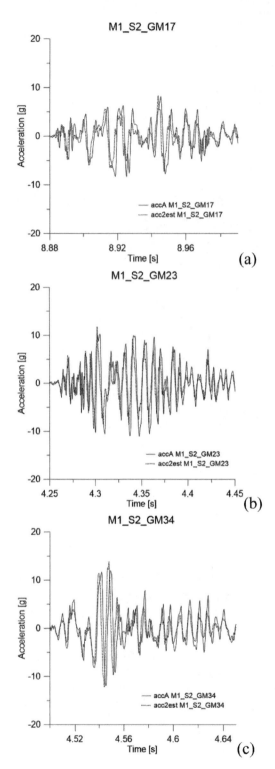

Figure 9. Acceleration time histories on the frame (2est) and within the sandy layer (A) recorded during the excitation of the shake table in Model M1_S2 for 3 ground motions.

These are preliminary tests of a more comprehensive experimental campaign that is carried out on a natural sand retrieved at a site that experienced liquefaction phenomena during the 2012 Emilia-Romagna earthquake in Northern Italy.

Considering the variability of fine content at the site, two different physical models were prepared: one using the sand retrieved at the site, the other by eliminating the fine content prior to the preparation of the model. The time series that were applied at the base of the model were obtained taking into account the expected seismic hazard at the site where the sand was retrieved. Although the pore pressure build-up was clearly observed during all excitations especially in the model with clean sand, true liquefaction was never achieved in any of these two models. Further tests are currently underway both under free-field conditions and in the presence of a building model.

ACKNOWLEDGEMENTS

This work has been carried out within the LIQUEFACT project. This project has received funding from the European Union's Horizon 2020 research and innovation programme under grant agreement No 700748. The support of F. Bozzoni (Eucentre), E. Zuccolo and A. Famà (University of Pavia) in defining the geotechnical model and the reference ground motions at Pieve di Cento is gratefully acknowledged by the authors.

REFERENCES

Manzari, M.T., Kutter, B.L., Zeghal, M., Iai, S., Tobita, T., Madabhushi, S.P.G., Haigh, S.K., Mejia, L., Gutierrez, D.A., Armstrong, R.J., Sharp, M.K., Chen, Y.M., Zhou, Y.G., 2014. LEAP Projects: Concept and Challenges, *Proc. 4th Int. Conf. on Geotechnical Engineering for Disaster mitigation and Rehabilitation* (4th GEDMAR), 16–18 September, 2014, Kyoto, Japan.

Mele L. (unpubl.) WP4 Report on the comparison of soil response before and after ground treatment at the case study pilot site. Deliverable LIQUEFACT project.

National Academies of Sciences, Engineering, and Medicine, 2016. SoAP in the Assessment of Earthquake-Induced Soil Liquefaction and Its Consequences. Washington, DC: The National Academies Press. https://doi.org/10.17226/23474

Ozcebe, A.G., Lai, C.G., Zuccolo, E., Bozzoni, F. & Famà, A. (unpubl.). Definition of shake table motions. WP4 Internal Report. LIQUEFACT project.

Zeng, X., Schofield, A.N. 1996. Design and performance of an equivalent-shear-beam container for earthquake centrifuge modelling.

Zeghal, M.M., Manzari, T., Kutter, B.L. & Abdoun, T. 2014. LEAP: selected data for class C calibrations and class A validations. Proc. 4th Int. Conf. on Geotechnical Engineering for Disaster mitigation and Rehabilitation (4th GEDMAR), 16–18 September, 2014, Kyoto, Japan.

Experimental and computational study on effects of permeability on liquefaction

H. Funahara & N. Tomita
Taisei Corporation, Yokohama, Japan

ABSTRACT: Two cases of centrifuge tests with base shaking were conducted to study the effect of soil permeability on liquefaction due to a long duration seismic excitation. It was confirmed that reduced permeability increases liquefaction extent and elongates the dissipation duration of the excess pore water pressure. Numerical simulations using an effective stress FE method were also conducted, and the same tendencies were observed. Through a series of parametric numerical studies using different permeability, it was predicted that gravelly ground might not liquefy with the same input motion and the dissipation duration of liquefied non-plastic silty ground could be more than 24 hours.

1 INTRODUCTION

Due to the 2011 Tohoku earthquake in Japan, extensive liquefaction occurred in reclaimed lands of the Tokyo Bay area that were far from the epicentre (more than 300 km). The recorded maximum accelerations at the ground surface in the area were not very large (about 1.5 m/s^2), but the duration of the seismic motion was relatively long (more than five minutes). The liquefaction observed may have been triggered by a large number of shearing cycles caused by this long duration. (Funahara & Ishizaki 2012).

In seismic design practice in Japan, some gigantic earthquakes such as the Nankai Trough earthquake and the Sagami Trough earthquake are more frequently being taken into consideration and it is becoming increasingly important to evaluate design earthquake ground motions of those earthquakes considering the effect of liquefaction. A practical methodology to evaluate design earthquake ground motions that consider the effect of liquefaction is an effective stress dynamic analysis method. Such numerical methods used in practice should be verified through comparison with observation of real phenomena, but there is no record of pore water pressure that builds up gradually by a long duration earthquake.

Therefore, the authors tried to conduct dynamic shaking table testing in a centrifugal field in order to reproduce the liquefaction caused by a long duration earthquake and observed the pore water pressure and ground accelerations. Additionally, there was a particular focus on the effect of soil permeability on this liquefaction. The observed behaviour was simulated using an effective stress analysis program that is employed to evaluate design earthquake ground motions that consider the effect of excess pore water pressure. After verification of the numerical method, additional parametric computations were conducted changing the permeability.

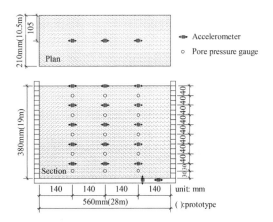

Figure 1. Test model and sensor distribution.

2 DYNAMIC CENTRIFUGE TEST

2.1 Outline of physical model

Figure 1 shows the schematic view of the test model and sensor distributions. The modelling scale is 1/50 and the applied centrifuge acceleration is 50 g. The following quantities are all presented in the prototype scale. Toyoura sand was air-pluviated in a laminar shear box and saturated by silicone oil. The targeted relative density was 60% and the model height was 19m. Test cases are shown in Table 1.

The test parameter is the dynamic viscosity of the silicone oil. The dynamic viscosity of the oil in Case 1 is 50 times that of water, which means that the model

Table 1. Test cases.

	Case 1	Case 2
Relative density	58.77%	58.94%
Dynamic viscosity of silicone oil	50 mm²/s	500 mm²/s

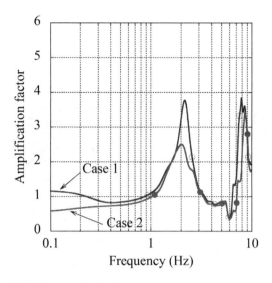

Figure 2. Transfer function.

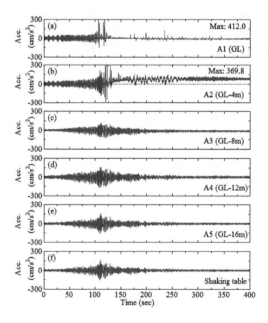

Figure 3. Acceleration time histories of Case 1.

corresponds to the prototype of clean sand saturated by water. The dynamic viscosity of oil in Case 2 is 500 times of that of water, which means that the model corresponds to the prototype of finer sand or silty sand saturated by water which has 1/10 permeability compared to clean sand.

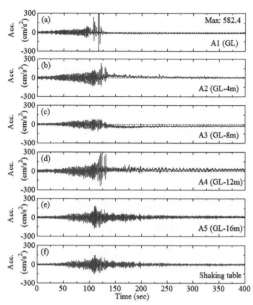

Figure 4. Acceleration time histories of Case 2.

2.2 Input acceleration

Input acceleration was created based on an acceleration record observed at the depth of 37 m at the Ooi site by PARI during the Tohoku earthquake (Port and Airport Research Institute). The observed acceleration (Maximum value: 0.659 m/s²) was input into a total stress soil column model (37 m height) for a site in Tokyo bay area (Funahara & Ishizaki 2012) and an amplified response at the depth of 19m was extracted as an input motion to the shaking table.

2.3 Natural frequency of the ground

The natural frequency of the test model was evaluated based on small-level shaking tests. Figure 2 shows the acceleration amplification factor of ground surface versus shaking table. The 1st natural frequency of the ground is about 2 Hz (0.5 seconds) in both Case1 and Case2.

2.4 Acceleration response during seismic excitation

Figure 3 and Figure 4 show the acceleration time histories observed in Case1 and Case2, respectively. Figure 5 shows the comparison of excess pore water pressure over time and Figure 6 shows the comparison of excess pore water pressure ratio over time.

Amplitude of the acceleration at ground surface and at the depth of 4m in Case1 decreased at around 120 seconds except for several spiking peaks considered to be caused by cyclic mobility (Figure 3, a, b). This timing corresponds to the timing that the pore water pressure ratio at the depth of 2m and 6m in Case1 reached almost 1.0 as shown in Figure 6.

These mean that about 6m of subsurface liquefied in Case1. The excess pore water pressure decreased

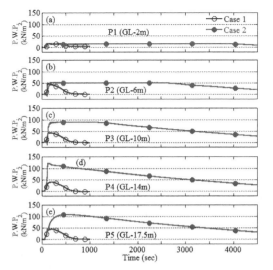

Figure 5. Time histories of excess pore water pressure.

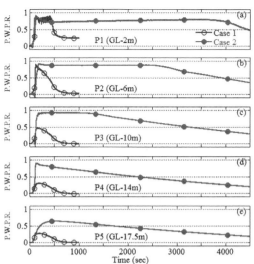

Figure 6. Time histories of excess pore water pressure ratio.

rapidly and dissipated in about 10 minutes. In Case2, the excess pore water pressure in deeper area built up much more and the excess pore water pressure at all depths remained high for a much longer duration compared to Case 1. The excess pore water ratio at the depth of 14 m reached almost 1.0 and the acceleration amplitude at the depth of 12 m decreased at around 120 seconds except for spiking peaks. It is clear that the thickness of the liquefied subsurface was much larger in Case2. Because of the lower permeability, dissipation of the excess pore water pressure took a much longer duration. Even after 75 minutes, the excess pore water pressure still remained.

3 DYNAMIC EFFECTIVE STRESS ANALYSIS

3.1 Outline of numerical model

In order to simulate the centrifuge testing and conduct parametric study, an effective stress FEM code was employed. The platform of the FEM is DIANA-J2 that is based on Biot's two phase governing equation (Zienkiewicz & Shiomi 1984) and the incorporated constitutive model for soil skeleton is Stress-Density model (Cubrinovski & Ishihara 1998).

Figure 7 shows the outline of the numerical model for the prototype subsurface of 19m depth. It is a one dimensional soil column model that consists of 38 plane strain elements. Each left-side node is tied to the corresponding right-side node to achieve the periodic boundary condition. The boundary condition at the bottom of the model is a fixed condition and the observed acceleration time history at the shaking table was input at the fixed bottom. After dynamic analysis for 400 seconds, static consolidation analysis was conducted to evaluate the pore water pressure dissipation.

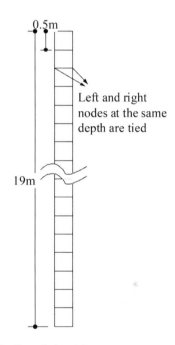

Figure 7. Numerical model.

3.2 Liquefaction strength

In order to evaluate the liquefaction strength of the physical model, a cyclic undrained triaxial shear test was conducted. The specimen was prepared by air pluviation method and a target relative density of 60% was achieved. Two kinds of initial effective confining stress, 30 kN/m^2 and 100 kN/m^2, were employed in the element tests.

The model parameters were determined through element test simulations performed aiming to reproduce the liquefaction strength curve obtained

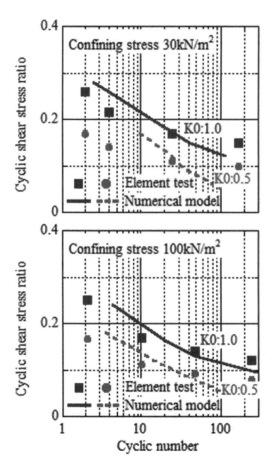

Figure 8. Element test simulation.

from the cyclic undrained shear tests. The result obtained from the test was the ratio of the cyclic shear strength to the mean effective stress. On the other hand, when modelling the natural soil deposits, the equivalent field liquefaction strength ratio, i.e. the ratio of the cyclic shear strength to the effective overburden pressure, could be employed as a target. In that case, if the coefficient of lateral earth pressure at rest (K_0) is assumed as 0.5, the mean effective stress becomes 2/3 of the effective overburden pressure (Yoshimi 1991). Then the equivalent field liquefaction strength ratio becomes 2/3 of the liquefaction strength ratio of the laboratory test. In this study, those two kinds of strength ratios are employed as target strength curves for the element test simulations and the effects on the simulation of the centrifuge testing are examined. Figure 8 shows the comparison between the target curve from the laboratory tests and the simulated curve from the constitutive model. Table 2 shows the key parameters determined through the fitting. Other parameters are shown in the reference (Cubrinovski & Ishihara 1998). Because of the difficulty to fit the whole curve, the main target was the strength ratio around the effective cyclic number (20 ~ 40) of the Tohoku earthquake (Arai 2012).

Table 2. Model parameters.

	K_0:1.0	K_0:0.5
Void ratio, e	0.726	
Dilatancy parameter, Sc	0.00358 ~ 0.00802	0.00296 ~ 0.00321

3.3 Permeability

In the centrifuge tests, as a pore fluid, we chose silicone oil whose dynamic viscosity is 50 times or 500 times that of water. As the tests were conducted in the centrifugal field of 50 g, the former model (Case 1) has the same permeability as water in clean sand in prototype. The latter one (Case 2) has lower permeability by one order, therefore, it could represent silty sand in prototype. As the employed effective stress FEM code is based on the two phase governing equation, the coefficient of permeability is an input parameter. Therefore, the coefficient of permeability of Toyoura sand is used as the input parameter for Case 1 (Hatanaka et al. 1995). The coefficient of permeability for Case 2 is 1/10 that for Case 1. Additionally, parametric computations were conducted setting the coefficient of permeability for gravel, silt and undrained condition. Table 3 shows the computational cases.

3.4 Simulation of excess pore water pressure

Figure 9 shows the comparison of excess pore water pressures between the centrifuge tests and the effective stress FEM analyses. When using the liquefaction strength from the laboratory test and the permeability of Toyoura sand (Case 1-1), the extent of liquefaction for each depth and the dissipation rate were well reproduced. On the other hand, when using the liquefaction strength from the laboratory test and the lower permeability (Case 2-1), the tendency that the amount of pore water pressure buildup in deeper areas was relatively large and the tendency that it took a longer duration for the pore water pressure to dissipate were both reproduced. But the extent of the pore water pressure buildup was much larger. It seems difficult to explain the observed large difference of the maximum excess pore water pressure in deeper areas only using the permeability. Difference in the density of the prepared physical models could be a result of the difference in the water head during the saturation process. In both cases using the equivalent field liquefaction strength considering K_0, the whole ground was completely liquefied and this is not consistent with the observed results. Therefore, it could be said that the equivalent field strength based on K_0 is an underestimation in these cases.

3.5 Acceleration at ground surface

Figure 10 and Figure 11 show the comparisons of the time history and response spectrum of the acceleration at the ground surface between the computation and the observation. The sudden decrease of acceleration level at around 120 seconds due to liquefaction is

Table 3. Computational cases.

Case	Target	K_0	Permeability (m/s)	Soil
Case1-1	Simulation	1.0	1.3×10^{-4}	Clean Toyoura sand
Case1-2		0.5	1.3×10^{-4}	Clean Toyoura sand
Case2-1		1.0	1.3×10^{-5}	
Case2-2		0.5	1.3×10^{-5}	
Case-A	Parametric study	1.0	1.0×10^{-2}	Gravel
Case-B		1.0	6.8×10^{-7}	Silt
Case-C		1.0		Undrained

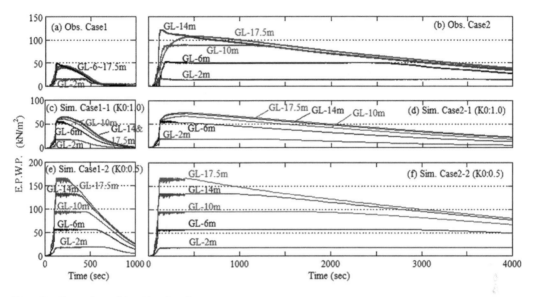

Figure 9. Comparison of time histories of excess pore water pressure.

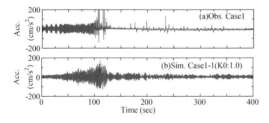

Figure 10. Comparison of acceleration time histories.

well reproduced except for the spiked response in the observation which could be considered to be the consequence of the cyclic mobility. The response spectrum suggests that the computed acceleration is conservative as a design input motion except for the range of the period shorter than 0.2 seconds and for the range of the period longer than 4.0 seconds.

3.6 Parametric study on the effect of permeability

The computed excess pore water pressures using three different permeabilities corresponding to gravel, silt and undrained condition are shown in Figure 12. The simulation results for the centrifuge tests using the liquefaction strength from the laboratory tests are also

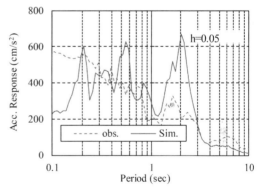

Figure 11. Comparison of acceleration response spectrum.

plotted in the same graph. The horizontal axis for time is in the logarithmic scale. The excess pore water pressure in a clean sand such as Toyoura sand dissipates in only a fraction of an hour. In contrast the excess pore water pressure in a soil which has lower permeability by 1/10 does not finish dissipation in one hour. When using the permeability of silt which was liquefied in the 2000 western Tottori prefecture earthquake (Yasuda et al. 2003), the dissipation does not finish

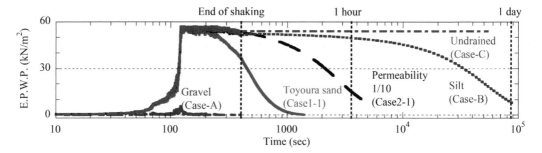

Figure 12. Permeability and dissipation time.

in one day (Case B). This is consistent with the fact that the small sand boil (weak water spring) was still observed at the surface of liquefied silty reclaimed land after more than 24 hours of the earthquake occurrence (Ishiga et al. 2001). When using the very large permeability of gravel (Case A), the ground did not liquefy at all. This is because the dissipation is faster than the excess pore water build up. When the undrained condition is applied (Case C), the built-up pore water pressure does not decrease at all.

4 CONCLUSIONS

Liquefaction due to a long duration earthquake was reproduced utilizing a centrifuge testing facility. In the centrifuge tests, the effect of permeability was examined. It was confirmed that the dissipation of the excess pore water pressure takes a longer duration if the permeability is lower, and as a consequence of this, the excess pore water pressure accumulates more easily.

Next, numerical simulations were conducted and the effect of liquefaction strength and permeability were studied. The reproduction of the liquefied depth in the test was not perfect, but the tendency that lower permeability produces a thicker liquefied zone and a longer dissipation duration was confirmed.

REFERENCES

Arai, H. 2012. Number of Equivalent Cyclic Shear and Effective Duration for Liquefaction at Tokyo Bay Area during the 2011 Tohoku Pacific Earthquake, *Summaries of Technical Papers of Annual Meeting, No. 21040*

Cubrinovski, M. & Ishihara, K. 1998. State Concept and Modified Elastoplasticity for Sand Modelling, *Soils and Foundations* 38(4): 213-225,

Funahara, H., & Ishizaki, S. 2012. Effective Stress Analyses of Two Sites with Different Extent of Liquefaction during the 2011 off the Pacific coast of Tohoku Earthquake, *47th Japan National Conference on Geotechnical Engineering*, JGS, 1583-1586.

Hatanaka et al., 1995, Permeability tests of gravel using a triaxial and large-scale permeability testing equipment, *50th JSCE annual meeting* 112-113.

Ishiga, H., et al., 2001, The boiling sands liquefied by the Western Tottori Earthquake 2000, San'in district, Japan, *Investigation report on the Western Tottori Earthquake 2000, Shimane university*.

Port and Airport Research Institute, Web page of the Strong Motion Earthquake Observation in Japanese Ports, http://www.mlit.go.jp/kowan/kyosin/eq.htm

Yasuda, S. et al., 2003, Report of special project for earthquake disaster mitigation in urban areas, 361.

Yoshimi, Y. 1991. *Liquefaction of Sandy Ground (Second Ed.)*, Gihodo

Zienkiewicz, O. C. & Shiomi, T. 1984. Dynamic behaviour of saturated porous media. *Int. Journal for Numerical and Analytical Methods in Geomechanics.* 8(1): 71-96.

The importance of vertical accelerations in liquefied soils

F.E. Hughes & S.P.G. Madabhushi
Department of Engineering, University of Cambridge, Cambridge, UK

ABSTRACT: Liquefaction-induced settlement is continuing to be a cause of significant damage in earthquakes across the world. Research is currently ongoing at the University of Cambridge to investigate the effect of the presence of a basement structure on a building sited on liquefiable soil. The inclusion of basements can provide uplift forces during the liquefied period thereby reducing overall settlement of the structures. During the dynamic centrifuge tests conducted, oscillations of excess pore pressure ratio (r_u) with peak values notably greater than one were recorded. The pore pressures generated resulted in the soil body no longer being in vertical equilibrium, and the resultant vertical accelerations were calculated using the data obtained from the pore pressure transducers. These were found to be in close agreement to the vertical accelerations measured by piezoelectric accelerometers orientated vertically in the soil body. Whilst liquefaction was found to cause horizontal accelerations to be progressively attenuated, vertical accelerations were found to be amplified. It should become common practice to measure vertical accelerations in the soil body in liquefaction problems being investigated using dynamic centrifuge testing.

1 INTRODUCTION

1.1 Motivation

Earthquake induced liquefaction is continuing to cause significant damage during earthquake events across the world, particularly in the built environment. Physical model testing, notably dynamic centrifuge testing, is an important tool being used to investigate ways to mitigate liquefaction induced damage. Dynamic centrifuge testing is currently being undertaken at the Schofield Centre at the University of Cambridge to investigate whether basement storeys can be used to reduce liquefaction induced settlement of structures. Vertical forces during the co-seismic period are an important aspect of this research (Hughes & Madabhushi 2017).

Oscillations of excess pore pressure ratio ($r_u(t)$) with peak values notably greater than one have been repeatedly observed. These have been investigated and are presented in this paper.

1.2 Background

Terzaghi's effective stress principle (Equation 1) can be used to calculate the proportion of an applied stress (σ) carried by the structure of soil particles (σ') and the proportion carried by the pore fluid (u) (Terzaghi 1943). This principle is reliant on the soil body being in equilibrium, and is hence only applicable under these conditions.

$$\sigma = \sigma' + u \tag{1}$$

Full liquefaction is traditionally defined as having occurred when the excess pore pressure ($u_{ex}(t)$) generated at a location becomes equal to the value of the vertical effective stress (σ'_{v0}) before shaking began (Equation 2) (Seed & Lee 1966). This results in a state of zero vertical effective stress (Equation 3). This definition assumes that there is vertical equilibrium in the soil body.

$$u_{ex}(t) = \sigma'_{v0} \tag{2}$$

$$\sigma'_v(t) = \sigma'_{v0} - u_{ex}(t) = 0 \tag{3}$$

Excess pore pressure ratio ($r_u(t)$) is defined as the ratio of the generated excess pore pressure divided by the initial vertical effective stress immediately before shaking commences (Equation 4). This becomes equal to one when liquefaction occurs (Equation 2).

$$r_u(t) = \frac{u_{ex}(t)}{\sigma'_{v0}} \tag{4}$$

This definition of excess pore pressure ratio is therefore also reliant on the soil body being in vertical equilibrium, resulting in the following range of possible values for $r_u(t)$:

$$0 \leq r_u(t) \leq 1 \tag{5}$$

In a number of dynamic centrifuge model experiments, excess pore pressure ratios greater than one have been observed. In most of these cases no comment is made regarding the values being greater than one (Liu and Dobry 1997, Tobita et al. 2012, Zeybek and Madabhushi 2016). In some cases it is attributed to the settlement of the pore pressure transducer and/or a rise in the water-table elevation (Fiegel and Kutter 1994, Hayden et al. 2015). The tests discussed in this paper have also observed $r_u(t)$ values greater than one. Experimental data obtained from pore pressure transducers and piezoelectric accelerometers will be compared, and the findings will be discussed.

2 DYNAMIC CENTRIFUGE MODELLING

A series of dynamic centrifuge experiments are currently being undertaken at the University of Cambridge, using the 10 m diameter beam centrifuge at the Schofield Centre (Schofield 1980). Results from one of those centrifuge tests will be presented here.

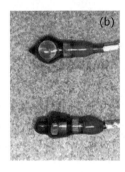

Figure 1. Instrumentation placement during model preparation. (a) Pore pressure transducer positioned perpendicular to the direction of shaking. (b) Piezoelectric accelerometers positioned both vertically (upper) and horizontally (lower).

The test presented here was conducted in a rigid container with a Perspex window which allows particle image velocimetry to be conducted. A layer of Duxseal at both ends of the the container limited the effect of the rigid boundaries (Steedman and Madabhushi 1991). A liquefiable layer of loose Hostun HN31 sand with 43% relative density was prepared by air pluviation using an automatic sand pourer. Arrays of instruments were placed underneath the structure and in the far-field. Druck pore pressure transducers (PPTs) were used (Figure 1a), which were calibrated using an air pressure chamber at 25 kPa pressure intervals. DJ Birchall (now DJB Instruments UK Ltd) A/23 piezoelectric accelerometers were positioned to measure both vertical and horizontal accelerations (Figure 1b). It is not common practice to measure vertical accelerations in the soil body in liquefaction problems being investigated using dynamic centrifuge testing. This was instigated after oscillations of excess pore pressure ratio ($r_u(t)$) with peak values notably greater than one were recorded in previous dynamic centrifuge tests conducted by the authors. The accelerometers were calibrated using a Brüel and Kjær calibrator which excited the instrument with a sinusoidal input acceleration with an amplitude of $\pm 10\ ms^{-2}$. Solartron DC50 linear variable displacement transducers (LVDTs) were used to measure vertical displacement. The locations of the instruments discussed in this paper are shown in Figure 2.

The tests were conducted at 60 g. The model was saturated with a high viscosity aqueous solution of hydroxypropyl methylcellulose with a viscosity of 60 cSt. This was done using CAM-sat, an automated, pressure controlled system (Stringer and Madabhushi 2010). A servo-hydraulic earthquake actuator was used to generate one dimensional input motions (Madabhushi et al. 2012). The horizontal sinusoidal base

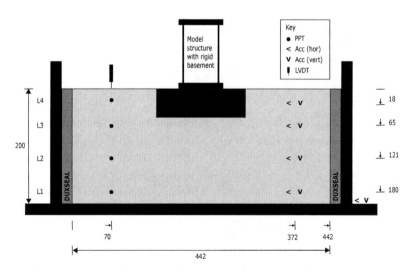

Figure 2. Centrifuge model layout and location of instruments discussed in this paper. Dimensions in mm, model scale. Locations of where instruments placed during model preparation. Piezoelectric accelerometers (Acc) were positioned both horizontally (hor) and vertically (vert). Pore pressure transducers (PPTs) were positioned at the same depths on the other side of the model. Linear variable displacement transducers (LVDTs) were used to measure vertical displacement.

shaking of the event discussed in this paper had the characteristics listed in Table 1.

3 DATA ANALYSIS

Data obtained during the centrifuge test was logged at a frequency of 6 kHz and was filtered using a low pass, eighth order Butterworth filter with a cut off frequency of 300 Hz.

3.1 Unit weight of sand

The unit weight of the uniform sand layer was calculated using the mass of sand poured and the volume it filled. Settlement during preparation, transportation and swing up were measured and used to update the unit weight of the sand. The pre shaking saturated unit weight of the sand was 19.1 kNm^{-3}.

3.2 Instrument location

The location of where each instrument was placed during the preparation of the model was measured to ±0.5 mm accuracy. It was assumed that the sand layer settled uniformly during the model preparation, transportation, loading and swing up, and that the instruments were displaced proportionally to surface settlement.

The difference between pore pressure measurements pre shaking and post excess pore pressure dissipation (Δu_{hy}) can be used to calculate the vertical displacement of PPTs during shaking and are shown in Table 2. It is not possible to track exactly when this displacement occurred during the shaking. It will be assumed that all instrument displacement occurred at the start of shaking, since positive excess pore pressures are built up during the first few cycles of shaking, causing a reduction in strength of the soil body. For the period before PPT displacement occurs, this will produce an underestimate of $r_u(t)$ for transducers that settle relative to the soil body during shaking but an overestimate for transducers that displace upwards relative to the soil body during shaking.

3.3 Initial vertical effective stress

The initial vertical effective stress (σ'_{v0}) was calculated by multiplying the depth of the PPT below the sand surface by the buoyant unit weight of the soil, and were calculated to be 92.1, 62.2, 33.2 and 10.3 kPa at PPT locations 1 to 4 respectively.

3.4 Excess pore pressure

Since it has been assumed that all instrument displacement occurs at the beginning of shaking, the excess pore pressures generated during shaking were calculated by subtracting the post shaking hydrostatic pore pressure from the pore pressure measurements obtained from the transducers during shaking.

4 RESULTS

All the results presented here are in prototype scale.

4.1 Horizontal accelerations

The input horizontal base shaking had a peak acceleration of 0.39 g in the first cycle, before reducing to 0.27 g for the subsequent cycles. Horizontal accelerations are transmitted through the soil body due to S_h wave propagation. After the initial cycle of acceleration, horizontal accelerations were attenuated at all depths in the soil body (Figure 3), with attenuation increasing with decreasing depth. This is a normal observation in liquefaction problems being investigated using dynamic centrifuge testing. The transmission of the S_h waves decreases as the soil liquefies and loses its shear strength.

4.2 Excess pore pressure ratios

The horizontal base shaking caused the generation of positive excess pore pressures as the tendency for the loose soil skeleton to contract is restricted by the pore fluid. When the excess pore pressure ratio ($r_u(t)$) reached a value of one, indicating that the excess pore pressure generated had become equal to the initial vertical effective stress (σ'_{v0}), full liquefaction is said to have occurred (Figure 4). The development of liquefaction was the cause of attenuation of the horizontal accelerations transmitted through the soil body (Figure 3).

Oscillations in $r_u(t)$ occurred with the same frequency as the horizontal base shaking. The peak values of $r_u(t)$ recorded were far greater than one. Whilst this phenomena has been observed previously, it has not been thoroughly investigated before.

For the shallowest two locations, the reduction in peak values of the excess pore pressure ratio to a constant maximum after 11 cycles implies that the

Table 1. Characteristics of base shaking (prototype scale).

PGA (g)	Freq (Hz)	No of cycles	T_{5-95} (sec)
0.37	1	20	17.7

Table 2. Instrument displacement during shaking (model scale) calculated from the difference between pore pressure measurements pre shaking and post excess pore pressure dissipation (Δu_{hy}). Arrows indicate the direction of displacement.

Instrument layer	Δu_{hy} (kPa)	Disp (mm)
4	2.90	4.9 ↑
3	4.33	7.4 ↑
2	1.18	2.0 ↑
1	−0.479	0.81 ↓

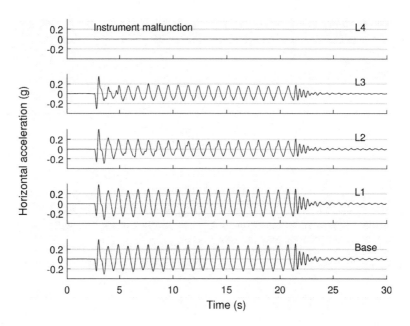

Figure 3. Horizontal accelerations (prototype scale). Comparison of horizontal accelerations measured in the far-field with the horizontal input shaking at the base of the model container.

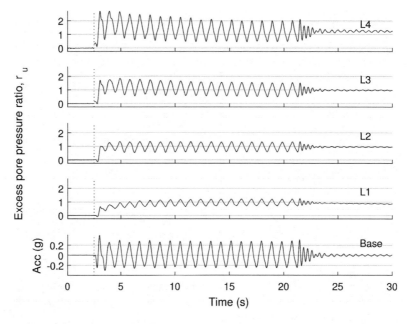

Figure 4. Variation of excess pore pressure ratio (r_u) during shaking, measured in the far-field. The vertical dashed line indicates where shaking commenced and where it is assumed that instrument displacement occurred.

instruments gradually displaced vertically during the first 11 cycles of shaking, and then remained in the same location for the remaining duration of shaking.

4.3 Vertical accelerations

The vertical accelerations transmitted to the base of the model container by the servo-hydraulic actuator were of the same frequency but much smaller magnitude than the horizontal accelerations (Figure 5). The peak vertical acceleration measured at the base of the model was 0.072 g, over and above the centrifugal acceleration.

Vertical accelerations are transmitted through the soil body due to P wave propagation. In contrast to the horizontal accelerations, the vertical accelerations

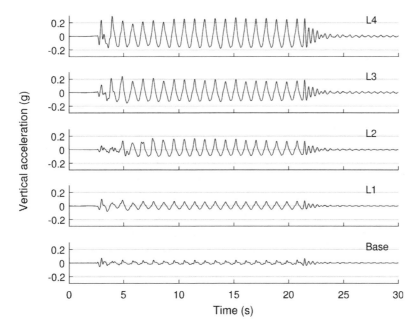

Figure 5. Vertical accelerations, over and above the centrifugal acceleration (prototype scale). Comparison of vertical accelerations measured in the far-field with the vertical shaking transmitted to the base of the model container.

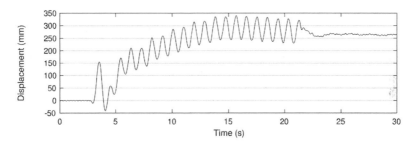

Figure 6. Co-seismic vertical displacement in the far-field (prototype scale). Heave occurred due to settlement of the structure.

were found to be amplified in the liquefied soil body (Figure 5), with amplification increasing with decreasing depth. Consequently, large vertical accelerations were measured close to the surface of the liquefied soil layer. The accelerometer located closest to the soil surface (L4) recorded vertical accelerations of approximately 0.25 g over and above the centrifugal acceleration. Accelerations of amplitude 0.25 g at a frequency of 1 Hz equate to a displacement amplitude of 62 mm. This is in close agreement with the displacements measured by the LVDT in the far-field, shown in Figure 6. Heave occurred in the far-field due to settlement of the structure.

5 DISCUSSION

Excess pore pressure ratios significantly exceeding one indicate that vertical equilibrium is not maintained during shaking. Pore pressure measurements recorded during the centrifuge test were used to calculate the net force and resulting vertical acceleration at the instrument location, independent of the measurements made using the vertically orientated accelerometers. The following methodology was used.

5.1 Vertical forces

The vertical forces were assessed at each instrument location, for a unit horizontal cross sectional area (A) (Figure 7). The downwards vertical force F_1 was due to the weight of the sand and pore fluid above the PPT.

$$F_1 = A[\gamma_{sat} \times z] = A \times \sigma_v = A[\sigma'_{v0} + u_{hy}] \quad (6)$$

Starting from Terzaghi's effective stress principle, the upward vertical force F_2 is the sum of the vertical effective stress, the hydrostatic pore water pressure and the excess pore water pressure generated. However, it is not assumed that vertical equilibrium is maintained.

$$F_2 = A[\sigma'_v(t) + u_{hy} + u_{ex}(t)] \quad (7)$$

Newton's second law is used to calculate the vertical acceleration caused by the resultant vertical force. It is assumed that the mass being accelerated is the mass of the soil and pore fluid above the location of the instrument.

$$a(t) = \frac{F_2 - F_1}{m} = \frac{A}{m}[\sigma'_v(t) + u_{ex}(t) - \sigma'_{v0}] \quad (8)$$

$$a(t) = \frac{g}{\gamma_{sat} \times z}[\sigma'_v(t) + u_{ex}(t) - \sigma'_{v0}] \quad (9)$$

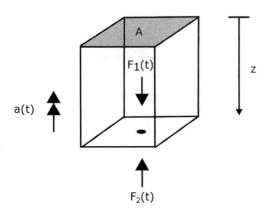

Figure 7. Vertical forces and resulting vertical acceleration a(t) acting on a volume of saturated sand, depth z and cross sectional area A, above a PPT located at depth z below the ground surface.

The generated excess pore pressure ($u_{ex}(t)$) can be measured in the centrifuge model and the initial vertical effective stress (σ'_{v0}) can be calculated using the model configuration immediately prior to the earthquake shaking. The following assumptions are made about the vertical effective stress during shaking ($\sigma'_v(t)$):

Case 1: when $u_{ex}(t) = 0$. The vertical effective stress in the soil body is unchanged ($\sigma'_v(t) = \sigma'_{v0}$) therefore upward acceleration is zero.

Case 2: when $0 < u_{ex}(t) \leq \sigma'_{v0}$. Using Terzaghi's effective stress principle, it is assumed that an increase in $u_{ex}(t)$ is equal and opposite to change in $\sigma'_v(t)$. Upward acceleration is zero.

Case 3: when $u_{ex}(t) > \sigma'_{v0}$. Since $u_{ex}(t) > \sigma'_{v0}$, it is assumed that $\sigma'_v(t) = 0$, since a negative vertical effective stress would imply that the soil particles were in tension.

$$a(t) = \frac{g}{\gamma_{sat} \times z}[u_{ex}(t) - \sigma'_{v0}] \quad (10)$$

$$a(t) = \frac{g}{\gamma_{sat} \times z}[\sigma'_{v0}[r_u(t) - 1]] \quad (11)$$

The above methodology was used to calculate the vertical accelerations at the locations of the PPTs (Figure 8). These calculations used only the pore pressures measured by the PPTs, the unit weight of the sand and the depth of the instruments.

It has been found that the vertical accelerations calculated from the pore pressure measurements using the methodology detailed above are in good agreement

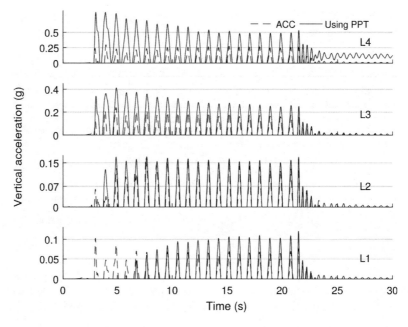

Figure 8. Vertical accelerations (prototype scale) calculated using pore pressures measured using PPTs (solid line) compared to those measured using vertically positioned piezoelectric accelerometers (dashed line). Note: different y axis scales have been used for clarity.

with the vertical accelerations measured using the vertical piezoelectric accelerometers positioned at the same depths in the centrifuge model, as seen in Figure 8. Further, the agreement increases with increasing depth. The cause of oscillations in excess pore pressure ratios with peak values exceeding one are therefore due to vertical accelerations in the soil body. Displacement of instrumentation during shaking also has an effect on excess pore pressure ratios, causing a residual change in hydrostatic pore pressure. This is anticipate to be of less significance given the agreement in the accelerations calculated.

6 CONCLUSIONS

The P waves causing the transmission of vertical accelerations are not attenuated following the onset of liquefaction, an effect observed with S_h waves transmitting horizontal accelerations. In contrast, vertical accelerations are amplified in the liquefied soil body. Amplification increases as depth below the surface of the liquefied soil decreases. It should become common practice to measure vertical accelerations in the soil body in liquefaction problems being investigated using dynamic centrifuge testing.

The vertical accelerations transmitted to and amplified through the soil body during liquefaction caused by earthquake shaking result in the soil body not being in vertical equilibrium. These vertical accelerations are the cause of oscillations of excess pore pressure ratio with peak values notably greater than one.

Definitions of liquefaction which rely on vertical equilibrium in the soil body should be updated to account for the soil body not remaining in vertical equilibrium during shaking.

REFERENCES

Fiegel, G. L. & B. L. Kutter (1994). Liquefaction Mechnism for Layered Soils. *Journal of Geotechnical Engineering 120*(4), 737–755.

Hayden, C. P., J. D. Zupan, J. D. Bray, J. D. Allmond, & B. L. Kutter (2015). Centrifuge Tests of Adjacent Mat-Supported Buildings Affected by Liquefaction. *Journal of Geotechnical and Geoenvironmental Engineering 141*(3), 04014118.

Hughes, F. & S. Madabhushi (2017). Control of liquefaction induced settlement of buildings using basement structures. In *Proceedings of the 3rd International Conference on Performance-based Design in Earthquake Geotechnical Engineering*, Vancouver, Canada.

Liu, L. & R. Dobry (1997). Seismic Response of Shallow Foundation on Liquefiable Sand. *Journal of Geotechnical and Geoenvironmental Engineering 123*(6), 557–567.

Madabhushi, S., S. Haigh, N. Houghton, & E. Gould (2012). Development of a servo-hydraulic earthquake actuator for the Cambridge Turner beam centrifuge. *International Journal of Physical Modelling in Geotechnics 12*(2), 77–88.

Schofield, A. N. (1980). Cambridge Geotechnical Centrifuge Operations. *Géotechnique 30*(3), 227–268.

Seed, H. B. & K. L. Lee (1966). Liquefaction of Saturated Sands During Cyclic Loading. *Journal of the Soil Mechanics and Foundations Division 92*(6), 105–134.

Steedman, R. S. & S. P. G. Madabhushi (1991). Wave propagation in sand medium. In *Proceedings of the 4th International Conference on Seismic Zonation*, Stanford, California.

Stringer, M. E. & S. P. G. Madabhushi (2010). Improving model quality through computer controlled saturation. In *7th International Conference on Physical Modelling in Geotechnics*, Zurich, Switzerland, pp. 171–176.

Terzaghi, K. (1943). *Theoretical Soil Mechanics*. John Wiley and Sons.

Tobita, T., G.-c. Kang, & S. Iai (2012). Estimation of Liquefaction-Induced Manhole Uplift Displacements and Trench-Backfill Settlements. *Journal of Geotechnical and Geoenvironmental Engineering 138*(4), 491–499.

Zeybek, A. & S. P. G. Madabhushi (2016). Centrifuge testing to evaluate the liquefaction response of air-injected partially saturated soils beneath shallow foundations. *Bulletin of Earthquake Engineering 15*(1), 339–356.

The effects of waveform of input motions on soil liquefaction by centrifuge modelling

W.Y. Hung, T.W. Liao, L.M. Hu & J.X. Huang
Department of Civil Engineering, National Central University, Taiwan, (R.O.C)

ABSTRACT: In recent decades, the numerical method is widely used in predictions of geotechnical phenomena and evaluations on geo-hazards. However, it has not proved its preciseness and applicability on reality. As that consequence, many centrifuge modelling laboratories has been participating the international collaboration projects, the Liquefaction Experiments and Analysis Projects (LEAP), which aims to calibrate and validate the centrifuge modelling and numerical modelling. In LEAP 2017, the goals are to perform the sufficient number of experiments to characterize the median responses and its uncertainty of a specific saturated sandy slope ground under specified input motions. The main input motion is a tapered sine wave. In this study, two input motions were conducted in one model which follow the experiment procedures required by LEAP 2017. The maintaining input motions are sine wave motion and negative sine wave motion, respectively, to discuss the interactions between the sloping model and the waveforms.

1 INTRODUCTION

Since 1960s, soil liquefaction is an important and popular research topic all over the world. The physical modelling and numerical modelling are commonly used to simulate and investigate the behaviour of soil liquefaction. Although the numerical modelling is developed in such many decades, there is no reliable and practical generally accepted process for validating such numerical models (Kutter et al. 2017). In 2013, the Liquefaction Experiments and Analysis Project (LEAP) aims to provide the experimental data of centrifuge modelling tests to validate and compare with numerical simulations. Several centrifuge facilities from USA, UK, China, Japan, Korea and Taiwan have already contributed great efforts in LEAP. The main objective of LEAP-2017 is to discuss the influences of uncertainties on lateral spreading of liquefiable soils in centrifuge modelling.

Hung et al. (2017) conducted a series of tests to discuss the influence between container boundary and the dip direction of the slope on soil liquefaction, the arrangement of the model was equal to the studies of LEAP-2017 as 5-degree-inclined saturated soil deposit, and using two kinds of container were rigid container and laminar container, as part of the result, it is difficult to realize the soil behaviour induced by different waveforms of base input motions. Therefore, two centrifuge model tests were conducted at the centrifuge model laboratory (NCU) as shown at Figure 1. The following sections describe the equipment, sand materials, model preparation and saturation, and testing procedures in this study. Finally, the effects of different input motions are mainly discussed in terms of the seismic responses towards acceleration as well as the excess pore water pressure.

2 TEST EQUIPMENT AND MATERIALS

The NCU centrifuge has nominal radius of 3 meters and a one dimensional servo-hydraulic controlled shaker is installed on its swing basket. The shaker can operate up to an acceleration of 80 g with maximum nominal force of 53.4 kN and maximum table displacement is ±6.4 mm. A rigid container which is composed by aluminium plates, shown in Figure 2, was used to contain the model. The inner dimensions of container are 767 mm (L) × 355 mm (W) × 400 mm (H).

The Ottawa sand which can be considered as whole-grain silica sand was used to prepare the sand bed of models. Figure 3 shows its grain-size distribution curve, it is classified as SP (poorly graded sand) in Unified Soil Classification System (USCS) with specific gravity of 2.66, mean size (D_{50}) of 0.203 mm, maximum dry unit weight of 17.3 kN/m^3 and minimum dry unit weight of 14.4 kN/m^3.

3 MODEL ARRANGEMENT AND TEST PROCEDURE

3.1 Model arrangement

The arrangements of models follow the experiment procedures of LEAP 2017. The model is a saturated gentle slope model with inclined angle of 5 degrees indicated in Figure 4(a). The target weight of sand

Figure 1. The geotechnical centrifuge of National Central University.

Figure 2. The aluminum rigid container.

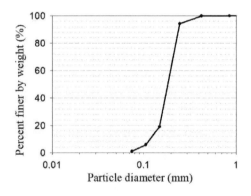

Figure 3. Grain size distribution of Ottawa sand.

bed is 16.20 kN/m³ which is corresponding to relative density of 65 %. The thickness of slope at upslope is 4.875 m in prototype (187.5 mm in model type) and at downslope is 3.125 m in prototype (120.2 mm in model type). The surface of sand bed is a curvature as shown in Figure 4(b). It relates to the rotational radius from sandy ground surface to the rotation centre of centrifuge, and it is 2.714 m for this study.

Two models were tests under acceleration field of 26 g to simulate the same prototype with the other models in LEAP. Accelerometers are attached to the shaker, top of container and embedded in the model

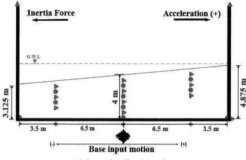

(a) Longitudinal section

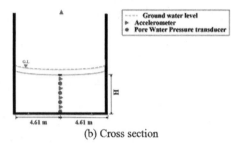

(b) Cross section

Figure 4. The arrangement of centrifuge model.

by three arrays. 10 Pore water pressure transducers are put at bottom of container and installed between every two accelerometers, as shown in Figure 4.

In this study, the positive acceleration means the shaking table moves toward right side (positive x-axis direction). Thus, the slope dip direction is toward left side (negative x-axis direction), and the inertial force caused by positive acceleration is toward negative x-axis direction.

3.2 Model preparation

The models were prepared by air pluviation. The flow rate of sand mass and drop height are the two main factors to the dry unit weight. The flow rate of sand mass would be controlled approximately 2.5 kg/minute for the sand passing the No. 16 sieve screen. Figure 5 shows the relationship between drop height and relative density under the specific flow rate, thus a constant drop height of 0.5 m was used. The pluviation process was interrupted for embedding the transducers at the specified elevations as shown in Figure 4. After pluviation, two rails with 5 degrees slope were placed on the top of container, and a curved plastic plate were suspended on the rails. The curved ground surface was sketched by moving the plastic plate along the longitude of container. Then, the dry sand model was completed.

Before saturation, the model was filled with CO_2 from the bottom to drive the air moving upward. An acrylic plate was used to tightly cover the container during the sand bed saturation process. CO_2 was then simultaneously and continuously vacuumed out from the inside to the outside of the container. At the same

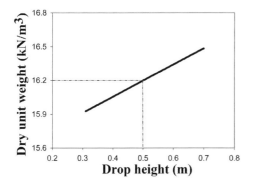

Figure 5. The relationship between dry unit weight and drop height during depositing process.

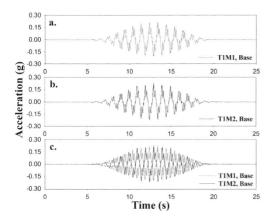

Figure 6. Time histories of base (a) Test 1-Motion 1, (b) Test 1-Motion 2, (c) Superimposition of two input motions.

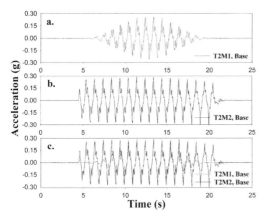

Figure 7. Time histories of base (a) Test 2-Motion 1, (b) Test 2-Motion 2, (c) Superimposition of two input motions.

time, a deionized, de-aired water-methylcellulose solution was carefully dripped into the container to saturate the sand bed until the fluid level rose to the pre-determined elevation. The viscosity of methylcellulose liquid is 26 times of water to simulate the prototype permeability at 26 g artificial acceleration field.

Table 1. Properties of base input motions.

	Pre-shaking	Test 1		Test 2	
		Main-shaking			
		Motion 1	Motion 2	Motion 1	Motion 2
Frequency (Hz)	3	1	1	1	1
Cycles	1	16	16	16	16
Amplitude (g)	0.1	0.22	0.22	0.26	0.26
Sine wave	Yes	Yes	No	Yes	Yes
Negative sine wave	No	No	Yes	No	No
Tapered shape	No	Yes	Yes	Yes	No

3.3 Test procedure

The model is fixed on the platform and the centrifuge is accelerated gradually up to 26 g. Then, a pre-shaking technique (Lee et al. 2014) is used to detect the shear wave velocity along depth by accelerometers. Two centrifuge modelling tests were conducted and subjected to different base shakings to investigate the effects of input motion waveforms on inclined saturated soil deposit.

Table 1 summarizes the properties of all input motions in prototype scale. There are two main-shakings in each test, Motion 1 and Motion 2, and they are all 1 Hz with 16 cycles. In Test 1, Motion 1 is tapered sine wave as shown in Figure 6(a), Motion 2 is negative tapered sine wave as shown in Figuree 6(b). The maximum amplitudes are about 0.22 g. Motion 1 and Motion 2 are superimposed and shown in Figure 6(c), which indicates that the phase difference is 180 degrees. In Test 2, Motion 1 is tapered sine wave as shown in Figure 7(a), Motion 2 is sinusoidal waves as shown in Figure 7(b). The maximum amplitudes are about 0.26 g. Motion 1 and Motion 2 are superimposed and shown in Figure 7(c), it can be seen that there is no phase difference for each peak but the waveforms are different.

4 TEST RESULTS

4.1 Seismic response of saturated sandy slope ground

During shaking, the time histories of acceleration and excess pore water pressure are recorded. Figure 8 is the time histories of excess pore pressure at central array for Test 1-Motion 1 (T1M1, left figures) and Test 2-Motion 1 (T2M1, right figures). The difference between two events is the peak acceleration of input motion, 0.22 g for T1M1 and 0.26 g for T2M1, respectively. The horizontal dash lines indicate the initial effective vertical stress at the depth corresponding to the elevation of censors. The soil states the initial liquefaction when the excess pore water pressure equals to the effective vertical stress. And it fully liquefies if excess pore water pressure keeps the value for certain period after shaking.

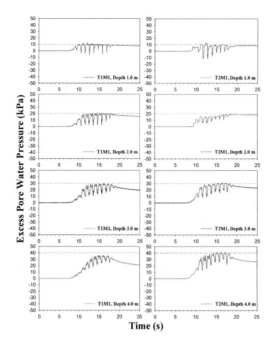

Figure 8. Time histories of acceleration at central array of Test 1-Motion 1 (T1M1) and Test 2-Motion 1 (T2M1).

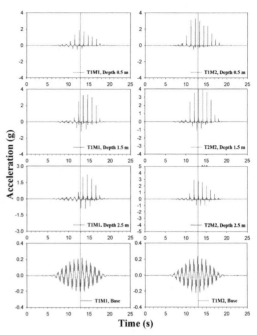

Figure 10. Time histories of acceleration at central array of Test 1-Motion 1 (T1M1) and Test 1-Motion 2 (T1M2).

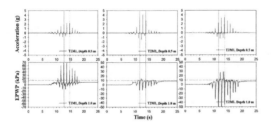

Figure 9. Time histories of acceleration and pore pressure at the shallow ground layer of downslope (left), central (middle) and upslope (right) of Test 2-Motion 1 (T2M1).

For such a 5 degree-inclined sandy ground with depth of 4 m at centre, the soil initially liquefies at depth of 3.0 m when it subjected to 0.22 g (T1M1) base input motion. After shaking, the generated excess pore water pressure dissipates immediately from bottom toward ground surface. Thus the dissipation rate at bottom layer is faster than that at shallow layer. The excess pore water pressure of soil at 1.0 m-depth keeps its value after shaking resulting to full liquefaction state. On the other hand, when a larger input base motion applies to the same model such as T2M1 with amplitude of 0.26 g, whole soil deposit liquefies and soil within 2.0 m keeps fully liquefaction state. It is reasonable to have a serious liquefaction condition when sandy soil deposit subjected to a larger base shaking.

Figure 9 is the time histories of acceleration and excess pore water pressure of T2M1 at the sallow depth of upslope, central and downslope. The slope dip direction is toward the direction of negative base acceleration that is opposite to the direction of the first half cycle of time history of base acceleration. Therefore the shear stress induced by inertial force resulting from the first half of time history of base acceleration coincides with the slope dip direction. The base acceleration causes the soil elements to be sheared, which leads to the cyclic stress paths induced by base shaking reaching the phase transformation line on the part of positive shear stress in the stress space. Once the stress path reaches the phase transformation line the shear behaviour of sand would change from contractive to dilative, which causes a drastic change in the mechanism of pore water pressure generation, from producing an increase in pore water pressure to a decrease in pore water pressure. Consequently the lower excess pore water pressure occurred in the phase of positive acceleration, leading to an increase of effective stress and resulting in the increase of shear wave velocity, indicating by vertical lines in Figure 9.

As compare with the responses at the shallow ground layer of upslope, central and downslope, the dilate behaviour by inertial force outward the slope is more obviously at upslope resulting to larger spike of acceleration.

4.2 The effect of positive and negative tapered sine wave

In Test 1, a positive and a negative tapered sine motions are input to the model in sequence (T1M1 and T1M2).

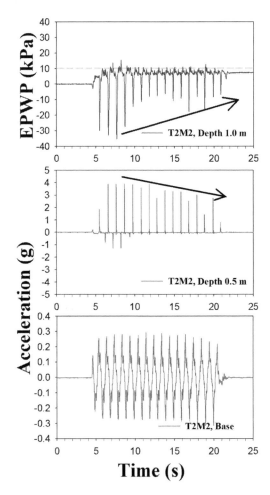

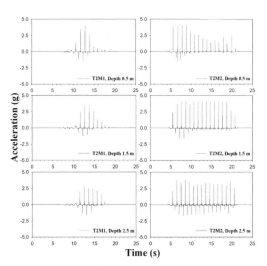

Figure 12. Time histories of acceleration at central array of Test 2-Motion 1 (T2M1) and Test 2-Motion 2 (T2M2).

4.3 The effects of tapered sine wave and sinusoidal wave

Figure 11 shows the excess pore water pressure history at depth of 1.0 m, and the acceleration histories of base and 0.5 m-depth for T2M2. It is mentioned before that a threshold base input acceleration of 0.14 g would lead to the appearance of spike. The amplitude of first half cycle of T2M2 is about 0.14 g and the spike is also observed. The excess pore water pressure accumulates very fast and reaches the effective vertical stress within the first three cycles. The ground surface soil liquefies and moves downward during shaking and changing the slope angle from 5.00 degrees to 3.49 degrees. Therefore, the dilatancy effect by inertial force during shaking reduces leading to the reduction of reverse peak of excess pore water pressure with elapsed time and the decrease of spike of positive acceleration. Figure 12 is the time histories of acceleration at central array of T2M1 and T2M2, where T2M1 is the tapered sine wave and T2M2 the sinusoidal wave. The reduction of spike for acceleration at 0.5 m-depth of central array is also observed.

For slope ground, the inertial force during shaking causes the dilatant behaviour of soil, leading to non-symmetric acceleration response and prominent spike of acceleration. A higher confining pressure coming from surrounding soil and the increase of soil density resulting from the first main shaking would cause the shear dilatancy at deeper place (such as the depth of 2.5 m) in T2M2, which is similar as the shear behaviour of dense sand. The peak values of negative acceleration increase and become more uniform for all the cycles.

Figure 11. Time histories of acceleration and excess pore pressure at upslope array of Test 2-Motion 2 (T2M2).

Figure 10 shows the acceleration response of centre slope along depth. The two vertical lines indicated the 13th second of time history during shaking of two events. Input motions of T1M1 and T1M2 have 180 degrees phase difference leading to that it is the wave trough for T1M1 and the wave peak for T1M2 at the 13th second. The acceleration time histories are observed with spikes emerging in the direction of positive acceleration. Therefore, significant spikes occur when the direction of inertial force is the same as that of slope dip. It would also lead to severe one-way movement if there is structure on the slope.

Focused on the positive acceleration time history, the amplitudes of every cycle of tapered sine wave are different with elapsed time. It seems that the spike appears while base input acceleration is larger than 0.14 g leading to enough shear stain toward to dip direction. This behaviour would affect by the properties of slope ground, soil and input motion. Also the negative spike appears especially at certain depth meaning that the enough larger inertial force inward to slope could cause the dilatancy of soil.

5 CONCLUSIONS

Two centrifuge modelling tests were conducted to simulate a 5-degree-inclined 4.0 m-depth saturated sandy ground subjecting to different input motions. The effects of amplitude and waveform of input motion on seismic response of acceleration and excess pore water pressure are discussed. From the tests, several results could be concluded.

1. The liquefaction of flat ground would usually lead to symmetric spike of positive and negative acceleration. For slope ground, the inertial force during shaking causes the dilatant behaviour, especially for the soil close to ground surface, leading to non-symmetric acceleration response and prominent spike of acceleration. The shear dilatancy is observed at deeper place leading to an increasing and uniform peak values of negative acceleration.
2. Significant spikes would occur when the direction of inertial force is the same as that of slope dip. It would also lead to severe one-way movement if there is structure on the slope.
3. A threshold of 0.14 g of positive base input acceleration is obtained. It would lead to enough shear stain toward to dip direction and cause the spikes of acceleration.
4. For rigid boundary, the dilate behaviour by inertial force outward the slope is more obviously at upslope resulting to larger spike of acceleration.
5. A deeper soil liquefaction condition would occur and a longer dissipation time when sandy soil deposit subjected to a larger base shaking.
6. The time of initial liquefaction of soil affects by the waveform of base input motion. It usually occurs within the first three cycles for uniform sinusoidal input wave with enough amplitude.

ACKNOWLEDGEMENTS

The authors would like to express our gratitude for the financial support from the Ministry of Science and Technology, Taiwan (R.O.C) (MOST 106-2628-E-008 -004 -MY3) and the technical support from the Experimental Center of Civil Engineering, National Central University and National Center for Research on Earthquake Engineering. These supports have made this study and future research possible and efficient.

REFERENCES

Elgamal A.W., Dobry R., Parra E., Yang Z. 1998. Soil Dilaiton and Shear Deformations during Liquefaction. *Fourth International Conference on Case Histories in Geotechnical Engineering 3(24): 1238–1259.*

Hung, W.Y., Lee, C.J., Hu, L.M. 2018. Study of the effects of container boundary and slope on soil liquefaction by centrifuge modeling. *Soil Dynamics and Earthquake Engineering*. (Article in press)

Kutter B.L., Carey T.J., Hashimoto T., Zeghal M., Abdoun T., Kokkali P., Madabhushi G., Haigh S.K., Burali F., Madabhu-shi S., Hung W.Y., Lee C.J., Cheng H.C., Iai S., Tobita T., Ashino T., Ren J., Zhou Y.G., Chen Y.M., Sun Z.B., Manzari M.T. 2017. EAP-GWU-2015 experiment specifications, results, and comparisons. *Soil Dynamics and Earthquake Engineering*. (Article in press)

Lee C.J., Wang C.R., Wei Y.C., Hung W.Y. 2012. Evolution of the shear wave velocity during shaking modeled in centrifuge shaking table tests. *Bulletin of Earthquake Engineering 10:* 401–420.

Manzari M.T., Kutter B.L., Zeghal M., Iai S., Tobita T., Madabhushi S.P.G, Haigh S.K., Mejia L., Gutierrez D.A., Armstrong R.J. 2014. LEAP projects: Concept and challenges, *Taylor & Francis Group*: 109–116.

Taboada-Urtuzuasteguia V.M., Martinez-Ramirezb G., Abdoun T. 2002. Centrifuge modeling of seismic behavior of a slope in liquefiable soil. *Soil Dynamics and Earthquake Engineering 22*: 1043–1049.

Tobita T, Manzari M.T., Ozutsumi O., Ueda K., Uzuoka R., Iai S. 2014. Benchmark centrifuge tests and analyses of liquefaction-induced lateral spreading during earthquake, *Taylor & Francis Group*: 127–182.

Horizontal subgrade reaction of piles in liquefiable ground

S. Imamura
Nishimatsu Construction Co. Ltd., Japan

ABSTRACT: With the purpose of elucidating the horizontal subgrade reaction of piles in liquefied ground, this study conducted centrifuge model tests, which were a combination of shaking tests of a model ground and alternating load tests of a foundation constructed with four combined piles. In these tests, several types of seismic waves were applied to a model ground to examine the influence of preliminary shaking, depth, input waves, excess pore water pressure ratio, horizontal displacement, and ground water level on the horizontal subgrade reaction of piles. All test results were then collected to construct an estimating equation of the coefficient of the subgrade reaction continuing on to the non-linear area of the ground, and a relational equation of the subgrade reaction and displacement by using a hyperbolic function.

1 INTRODUCTION

The earthquake of March 2011 off the Pacific Coast of Tohoku brought liquefaction and fluidization phenomena across a wide stretch of the Kanto region, inflicting immense damage such as tilting and sinking of many buildings and civil engineering structures; and unevenness, depression, and wavelike deformation of ground surface. Based on the experience with numerous natural disasters such as the Hyogo Prefecture Nambu Earthquake of 1995, Japan has seen repeated changes in the concepts behind the design methods for structural foundations. As the international design standardization of foundation structure is underway mainly in Europe (Simpson & Driscoll 1998), focus of design concepts has shifted from the legacy allowable stress design method to the limit state design method and the performance-based design in preparation for an internationally-harmonized foundation design code (Honjo & Kusakabe 2002). In considering the issue of lateral resistance of piles as a future model for foundation design, it is necessary to understand the interaction between pile and ground in succession, from infinitesimal displacement handled under the allowable stress method to large displacement that reaches the ultimate limit state required under plastic analysis. For this purpose, it is necessary to introduce nonlinearity into the relationship between the horizontal subgrade reaction p and the horizontal displacement y of piles. In addition, it is necessary to quantitatively elucidate the subgrade reaction by dynamic response to validate the performance of a pile foundation during an earthquake in which the occurrence of a variety of limit states is a concern, and to establish an economical aseismic design.

In this study, centrifuge model tests were performed using the alternating load test equipment developed by the authors (Imamura & Fujii 2002) for the purpose of elucidating the horizontal subgrade reaction of piles during liquefaction and during the process of dissipation of excess pore water pressure. This report discusses the effects of the applied waves and the relatively thin non-liquefied layer located in the liquefied ground surface layers on the estimation of the coefficient of the pile subgrade reaction.

2 CENTRIFUGE MODEL TESTS

2.1 Centrifuge facilities

The centrifuge equipment (effective radius 3.05 m) owned by Chuo University was used for the centrifuge model tests. The centrifuge equipment owned by Nishimatsu Construction (effective radius of 3.80 m) was used for tests on irregular waves requiring higher shaking levels. A laminar shear box (internal dimensions: width 600 mm, depth 250 mm, maximum level length 300 mm) developed by Fujii (Fujii 1988) was also used.

2.2 The alternating load test system

Figure 1 shows the comprehensive overview of the alternating lateral load test system with a foundation with four combined piles. See previous reports (Imamura & Fujii 2002; Imamura 2010) for the details about the alternating lateral load test system, the physical properties of the soil material (Grade No. 8 silica sand), strain measurement of the piles, and the calculation method for the coefficient of subgrade reaction.

2.3 Modelling of piles

The tests were carried out in a centrifugal field of 40 g using a 1/40-scale model. Copper pipes with an outside diameter of 12 mm and a wall thickness of 0.8 mm were used for the model piles to simulate steel piles with a nominal $\phi 508$ mm and a wall thickness of 12.7 mm.

For the measurement of bending strain, strain gauges were affixed at six levels shown in Figure 1, with one gauge in the direction of the shaking, and another in the opposite direction for each level.

To match the piles with the symbol of the load cell shown in the Figure 1, we will call the piles on the load motor side the P-piles, and those on the opposite side the N-piles. The pile foundation had a casting structure where four piles including a bending strain pile were combined with a centre spacing of 30 mm (1.2 m when converted to the actual size). To firmly secure the top and bottom edges of the pile to the footing and the bottom of the laminar shear box using a dedicated fixation tool, a duralumin flanged boss was inserted into the bottom edge of each pile and a flanged and tapered boss was instead into the top edge of each pile for reinforcement of the secured sections, as shown in Figure 2. Having been processed with high precision for contact with the inside wall of the pipe with no gap, these bosses were fixed inside a pipe with an adhesive before waterproofing.

2.4 Preparation and test series

Using the water sedimentation method, together with Grade No. 8 silica sand and degassed water, a model ground of saturated sand was built in the laminar shear box, in which a pile foundation had been placed. This model ground had a depth of 280 mm and a relative density of 60%. The bending rigidity of the model piles roughly matches that of the actual steel piles with a 500 mm outside diameter and a 12.7 mm wall thickness ($EI_p = 1.19 \times 10^5$ kNm^2). Four of these piles were combined with the centre spacing of 30 mm (1.2 m actual), and the top and bottom edges of the piles were rigidly combined to the footing and the bottom surface of the laminar shear box.

The piezometers and accelerometers installed on sensors were positioned at certain locations and were strained using fishing lines and springs, thereby maintaining the measuring positions before and after the shaking.

The tests were performed in four series as shown in Table 1, with a total of ten cases. In all the shaking tests, only the model ground was shaken with the footing fixed in a centrifugal field of 40 g. Immediately after the shaking, a lateral alternating load test on the pile head was performed. Further to the regular wave series, which integrated PF-1 through 3 and QF-1 and QF-2, mentioned in a previous report (Imamura 2010), we

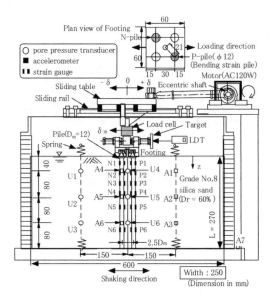

Figure 1. Comprehensive overview of the alternating lateral load testing system.

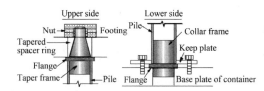

Figure 2. Fixing method for the pile head and pile edge.

Table 1. Test cases (10 Cases).

Test series	Test code	NL$_p^{*1}$ (m)	δ_m^{*2} (mm)	δ_p^{*3} (mm)	δ_p/D_p	Shaking condition in prototype scale	$I_{E(p)}^{*4}$ (m^2/sec^3)	Alternating load test condition in model scale
PF	PF-0	0	2	±80	0.17	None	0	3 cycles, f$_m$ = 1 Hz, v$_m$ = 13 mm/sec
	PF-1	0	2	±80	0.17	3.0 Hz, 140 Gal, 85 waves	29.4	4 cycles, f$_m$ = 1 Hz, v$_m$ = 13 mm/sec
	PF-2	0	4	±160	0.33	3.0 Hz, 140 Gal, 81 waves	27.2	4 cycles, f$_m$ = 1 Hz, v$_m$ = 25 mm/sec
	PF-3	0	6	±240	0.50	3.0 Hz, 150 Gal, 92 waves	34.8	4 cycles, f$_m$ = 1 Hz, v$_m$ = 38 mm/sec
QF	QF-1	0	5	±200	0.42	3.0 Hz, 110 Gal, 20 waves	2.47	1 cycle, f$_m$ = 1 Hz, v$_m$ = 31 mm/sec
	QF-2	0	5	±200	0.42	3.0 Hz, 147 Gal, 50 waves	14.3	2 cycles, f$_m$ = 2 Hz, v$_m$ = 63 mm/sec
RF	RF-1	0	5	±200	0.42	Hyogoken-numbu earthquake, 250 Gal	4.75	2 cycles, f$_m$ = 2 Hz, v$_m$ = 63 mm/sec
	RF-2	0	5	±200	0.42	Miyagi-oki earthquake, 250 Gal	9.63	2 cycles, f$_m$ = 2 Hz, v$_m$ = 63 mm/sec
GF	GF-2	0.48	5	±200	0.42	3.0 Hz, 147 Gal, 50 waves	14.3	2 cycles, f$_m$ = 2 Hz, v$_m$ = 63 mm/sec
	GF-3	1.44	5	±200	0.42	3.0 Hz, 147 Gal, 50 waves	14.3	2 cycles, f$_m$ = 2 Hz, v$_m$ = 63 mm/sec

*1 NL: Thickness of non-liquefiable sand, *2 δ_m: Lateral displacement at loading point (in model scale),
*3 δ_p: Lateral displacement at loading point (in Prototype scale), *4 I_E: Acceleration power (Hata et al. 2009).
(Where: a: Input acceleration (m/sec^2), T: Continuous time in Earthquake Time History (sec).) $I_E = \int_0^T a^2 dt$

examined the RF series, in which irregular waves (the Hyogo Prefecture Nambu Earthquake and the Miyagi-oki Earthquake) were incorporated, and the GF series, which used two thickness variants of the non-liquefied layer (GF2: 0.48 m, GF3: 1.44 m). In the PF series of the previous report (Imamura 2010) three regular waves with different shaking energies were applied to the model ground. To examine the effects of the horizontal displacement, varying values of 2, 4, and 6 mm were used as the maximum displacement δm at the loading point in the alternating load tests. Cases with no shaking were also tested. In the QF series of the previous report (Imamura 2010) the irregular waves from the Hyogo Prefecture Nambu Earthquake (RF-1) and the Earthquake off the Pacific Coast of Tohoku (RF-2), both of which were incorporated in the RF series, were converted into equivalent regular waves QF-1 and QF-2. In the alternating load tests for the QF, RF, and GF series, the maximum displacement δ m was set to 5 mm (δ/D = 0.5).

To understand the difference between the regular and irregular waves, two irregular waves shown in Figure 3 and Figure 4 (from the Hyogo Prefecture Nambu Earthquake and the Miyagi-oki Earthquake) were incorporated in the RF series. The maximum acceleration observed in the Earthquake off the Miyagi-oki Earthquake was approximately 250 Gal. To compare this earthquake with the Hyogo Prefecture Nambu Earthquake, which had the maximum acceleration of 824 Gal (the Kobe Marine Meteorological Observatory), the waveform of the Hyogo earthquake was compressed into 250 Gal. One wave with a waveform exceeding 300 Gal was used in the actual test.

In the GF series, varying thicknesses (0.48 m and 1.44 m) of the non-liquefied layer were used to examine the effects of the difference in the thickness of relatively thin non-liquefied layers on the horizontal resistance of the piles.

Five items were measured: bending strain of the pile; response acceleration; pore water pressure; horizontal load of the pile head; and horizontal displacement. Values used in the test results discussed later herein are those converted to the actual sizes.

2.5 Estimation method for the subgrade reaction and displacement of piles

The estimation of the subgrade reaction and displacement of the piles was calculated based on the measurements of the bending strain of the piles. The estimation method was based on the basic equation of the elastic beam. The bending strain of the piles measured in the tests was used to obtain the bending moment of each level. The distribution of the bending moment was approximated using a quantic equation. Differential and integral calculations were applied repeatedly to this approximate equation to obtain the shearing force (P), subgrade reaction (p), and flexible volume (y) for each second. The value of 0 was assumed as the boundary condition for the angle of deflection θ and displacement δ.

2.6 Quality management of the model ground using preliminary shaking

While the density of a model ground built using the water sedimentation method is largely uniform from the bottom of the laminar shear box to the surface, such ground with a uniform density across all layers (actual layer thickness of 11.2 m) is rare. Rather, it may be better to consider that many natural soil deposits have received shear histories from earthquakes of slight to moderate strength in the course of the depositional process over long periods of time, with the deeper sections of the ground having greater density. From this viewpoint, Imamura & Fujii (2002) aimed to bring the intensity distribution of a model ground closer to that of a natural soil deposit by proposing preliminary shaking, which repeatedly provides regular waves of a lower acceleration level (50 Gal) to a model ground made with a uniform density. This report measured the shear wave velocity Vs produced by cone penetration tests and pulse waves to examine the changes in the strength characteristics of the model ground before and after the preliminary shaking.

The report revealed that repeated application of preliminary shaking with the acceleration of 40 Gal, followed additional application of preliminary shaking with the acceleration of 80 Gal, increased the cone penetration resistance value q_c in the level depth direction, approximating the actual measurement data for natural soil deposits. It was inferred that a method

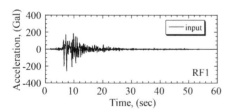

Figure 3. Input wave (RF-1).

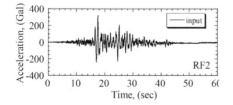

Figure 4. Input wave (RF-2).

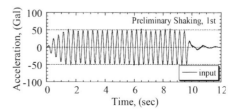

Figure 5. Preliminary shaking input waves.

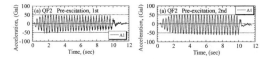

Figure 6. Time history of response acceleration (First time, Second time).

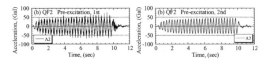

Figure 7. Time history of response acceleration (First time, Second time).

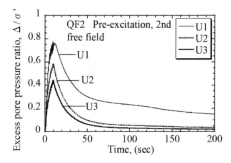

Figure 9. Time history of E.P.W.P. (2nd).

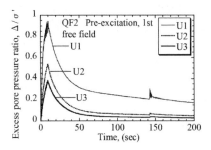

Figure 8. Time history of E.P.W.P. (1st).

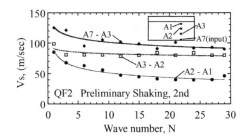

Figure 10. Correlation of frequency and seismic velocity.

like the one used in this test where piles are placed in the ground before dispersing sand in the laminar shear box tended to make the density non-uniform around the piles. Preliminary shaking was then introduced to ensure the planar uniformity including the areas around the piles and to minimize human-induced errors during the production of the model ground.

2.6.1 Response of the model ground to preliminary shaking

In the tests, sine waves (3 Hz, 50 Gal and 25 waves when converted to actual waves) shown in Figure 5 were applied as preliminary shaking to assess the dynamic response characteristics of the model ground and consider the number of preliminary waves given to the model ground. Converted to the acceleration scale of irregular waves such as actual seismic waves, the scale of this preliminary shaking is roughly equivalent to that of moderate waves with a maximum acceleration of 100 Gal.

Figure 6 and Figure 7 show the response acceleration of the adjacent ground with one preliminary shaking wave and two waves respectively, at both A1 (1.6 m deep) and A2 (4.8 m deep) levels, while.

Figure 8 and Figure 9 show the time histories of excess pore water pressure. These figures indicate that the wave shape of the acceleration response for the second preliminary wave was smoother than that of the first wave. They also indicate that the increase in the ratio of peak-time pore water pressure caused by the application of shaking was approximately 0.95, 0.55, and 0.38 as measured by the hydraulic gauges at Levels U1, U2, and U3 (in the order of closeness to the ground surface), respectively. The upward tendency of the pore water pressure caused by the application of the second preliminary shaking was the same as that of the first preliminary shaking across the board. However, the peak-time pore water pressure of the second preliminary shaking for U2 and U3 was an increase of 10% and 16%, respectively, while U1 measured pore water pressure of a mere 80% of that of the 1st shaking. While this report does not include the data on the third preliminary wave, the occurrences of the acceleration response and excess pore water pressure were approximately the same as those of the 2nd wave. These findings infer the assumption that the resultant values were caused by the fact that the preliminary shaking caused the acceleration sensors and pore water pressure gauges adapt to the adjacent ground.

2.6.2 The effects of preliminary shaking as seen in acoustic wave velocity measurement

In terms of preliminary shaking, the wave shapes of the acceleration response in the ground were captured relatively well as seen in Figure 6 and Figure 7. The zero-cross method was used to obtain the arrival times of some specific waves observed. The acoustic wave velocity Vs was then obtained by dividing the time difference by the distance between accelerometers.

Figure 10 shows the correlation of the number of waves for the second preliminary shaking with the acoustic wave velocity Vs. The figure indicates that the Vs rapidly decreases up to the tenth wave after the onset of shaking, remaining at roughly the same velocity thereafter. The initial Vs values of preliminary shaking were 125 m/s at the bottom of the ground, 99 m/s in the mid-section, and 85 m/s in the top section of the ground. From the eleventh wave

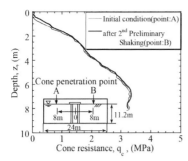

Figure 11. Cone penetration test results after preliminary shaking.

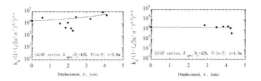

Figure 12. Correlation of horizontal displacement and the normalised coefficient of subgrade reaction.

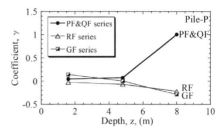

Figure 13. Correlation of the depth of subgrade reaction and the coefficient γ by series.

on, where the Vs roughly stayed the same, the values were approximately 90 m/s, 80 m/s, and 46 m/s in the order of the bottom, mid-section, and the top, which were 72%, 81%, and 54% of the initial values, respectively. Although our tests included preliminary shaking with the acceleration of 20 Gal, no increase in Vs was observed in the level depth direction. Almost no difference was observed between 20 Gal and 50 Gal regarding the degree of adaptation of the ground to the piles before and after the application of shaking, and regarding the results of the alternating load tests on the piles conducted afterwards. In light of the fact that the electromagnetic shake table (Fujii 1988) owned by Chuo University, which was used in these tests, could ensure stable input, the acceleration level was set to 50 Gal. And cone penetration test result was approximately the same as those of the 1st wave. as seen in Figure 11 These tests verified that the application of sine waves with the acceleration of approximately 50 Gal (3 Hz and 25 waves when converted to actual waves) as preliminary shaking didn't greatly disturb the initial model ground that had been built with a uniform density. This preliminary shaking may well improve the adaptation of the ground to the main pile foundation, which was built in at the time the model ground was built, and to the various measurement sensors, enabling the control over human-induced impact during the tests. For this reason, two waves of preliminary shaking with the acceleration of 50 Gal were applied in these tests, before the alternating load tests.

3 CENTRIFUGE MODEL TEST

3.1 Impact of input waves and the thickness of the non-liquefied layer

Figure 12 follows a previous report (Imamura 2010) and shows the correlation of the horizontal displacement at the level depth of 4.8 m in the irregular wave series (RF_{Series}) and the non-liquefied layer series (GF_{Series}), to the coefficient of the subgrade reaction normalized by dividing the coefficient of the subgrade reaction by $(1-(\Delta u/\sigma')^{0.3})^{0.6}$ (reduction coefficient at the level depth of 4.8 m). For a regular wave, it was verified that a normalized coefficient of the subgrade reaction tended to decrease as the horizontal displacement increased, regardless of the size of the excess pore water pressure ratio. Based on Figure 12, the correlation between the horizontal displacement and the normalized coefficient of the subgrade reaction can be interpreted as unambiguous. However, the value increased as the horizontal displacement increased in the RF series, while it showed a tendency to gradually decrease in the GF series. In the regular wave series, the correlation of the horizontal displacement on one hand, to which the initial coefficient k_{h0} of the subgrade reaction (derived from static alternating load tests(PF-0)) was added, and the normalized coefficient of the subgrade reaction on the other, could be represented by the following formula (1).

The modelling process of relationship (formula (1)) between the coefficient of horizontal subgrade reaction and displacement based on the test results is discussed in a previous report (Imamura 2010).

$$k_h = \left(1-(\Delta u/\sigma')^\alpha\right)^\beta k_{ho}\left(\frac{1}{1+\gamma|y/y_1|}\right) \quad (1)$$

where: α, β, and γ: experimental constants. y: horizontal displacement (cm), y_1: horizontal displacement used in obtaining the initial coefficient k_{h0} of the subgrade reaction (1% of pile diameter: 0.48 cm).

Figure 13 shows the results of using the least-squares method to compare the values of the coefficient γ at different depths. The γ value obtained was different for each series. The tendency to increase or decrease also differed depending on the depth. Such changes in the coefficient may be attributed to the input waves with different frequency characteristics and the varying groundwater levels, which change the degree of liquefaction in the depth level direction. And this may in turn lead to differences in the duration of liquefaction and in the degree of recovery of the effective stress, impacting the distribution characteristics of the section force of the piles.

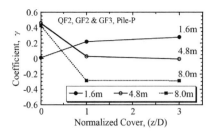

Figure 14. Correlation of the normalised non-liquefied layer thickness and the coefficient.

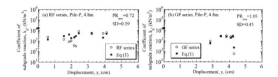

Figure 15. Comparison of test values and estimated coefficient of the subgrade reaction at the depth of 4.8 m.

Figure 14 shows the correlation of the non-liquefied layer thickness z/D_p normalized by the pile diameter D_p and the coefficient γ of the GF series. QF-2, which was added under the same conditions as the GF, is also shown here. This figure shows that the positivity and negativity of the coefficient γ is reversed depending on the presence of a non-liquefied layer or the difference in the depth. Like the approximated curve represented by the expression (1) in Figure 12, this positive-negative reversal of the coefficient γ means the increase/decrease tendency of the coefficient of the subgrade reaction that accompanies the increase in the horizontal displacement. In other words, this reversal suggests that the distribution of the section force of the piles changes depending on the presence of a non-liquefied layer, impacting the subgrade reaction displacement curve for each depth level as well. However, Figure 14 does not indicate that different thicknesses of a non-liquefied layer don't change the coefficient γ. This may be because the non-liquefied layer softened due to the excess pore water from the liquefied ground during the liquefaction process, and the ground did not receive much impact from the thickness of the non-liquefied layer at the ground water levels of 0.48 m and 1.44 m ($=3D_p$). It was verified that the use of the coefficient γ for the thickness of each non-liquefied layer would improve the estimation of the coefficient of the subgrade reaction by raising the precision of individual cases. However, in light of the fact that a difference in the coefficient γ does not significantly impact the estimation accuracy, and that the data size is also small, the GF series (GF-2 and GF-3) were integrated to obtain the coefficient γ of the estimating equation in this report.

4 COMPARISON WITH THE RESULTS OF ALTERNATING LOAD TESTS

Figure 15 shows a comparison, represented by the RF & GF series (at the depth of 4.8 m), between the estimated coefficient of the subgrade reaction in the formula (1) and the test values. Some average PR_{ave} values of the estimation ratio indicated in the figure (=actual measurement value/estimated value) exceeded 1.0 at the depths of 4.8 m, with the estimated accuracy slightly declining. And the value of the standard deviation SD was less than 1 at the depth of 4.8 m, with a good estimated accuracy. Both series produced similar results at different depths, although the behaviour of the coefficient of the subgrade reaction in relation to the horizontal displacement differed depending on the depth, ending up in results different from those of the regular wave series. This finding also indicates that differences in the coefficient γ affect the estimated error of the coefficient of the subgrade reaction. For this reason, it is necessary to construct an estimating equation with consideration for the depth, input waves, and the ground structure.

5 CONCLUSION

The following summarises the conclusion arrived at in this study.

1) When 25 regular waves of 3 Hz and 50 Gal were applied as preliminary shaking to a model ground constructed with a uniform density, application of the preliminary shaking had almost no impact on the tests afterwards. Instead, it improved the adaptation of the ground to the measurement sensors buried in the ground and enabled control over human-induced influence during the tests.
2) The coefficient of the subgrade reaction is influenced by the depth, excess pore water pressure ratio, and horizontal displacement. It was verified that the coefficient was influenced by input waves and ground water levels.

REFERENCES

Fujii, N. 1991. Development of an electromagnetic centrifuge earthquake simulator, In Ko H. Y. & Mclean F. G. (eds), *Proceedings of Centrifuge 91, Boulder*, Balkema, 351–354.

Hata, Y., Ichi, K., Kana S., Tsuchida, T. & Imamura, T. 2009. An Approach the Earthquake Motion Determination Based On Microtremor Measurement Along The Expressway, *Journal of JSCE*, F, Vol. 65, No. 4, pp. 52–541.

Honjo, Y. & Kusakabe, O. 2002. Proposal of a Comprehensive Foundation Design Code: Geocode 21 ver. 2, *Proceedings of IWS Kamakura*, 95–106.

Imamura, S. & Fujii, N. 2002. Observed dynamic characteristics of liquefying sand in a centrifuge, *Proc. of Physical Modeling in Geotechnics*, 195–200.

Imamura, S. 2010. Characteristics of horizontal subgrade reaction of pile from alternating load tests in liquefiable sand, *Proc. of Physical Modeling in Geotechnics*, 1403–1408.

Simpson, B. & Driscoll, R. 1998. Eurocod 7-a commentary. Construction Research Communications Ltd., Watford, Herts.

Experimental investigation of pore pressure and acceleration development in static liquefaction induced failures in submerged slopes

A. Maghsoudloo, A. Askarinejad, R.R. de Jager, F. Molenkamp & M.A. Hicks
Section of Geo-Engineering, Delft University of Technology, Delft, The Netherlands

ABSTRACT: Pore pressure development in loose sand layers can trigger instability and failure of earth structures. Static liquefaction is one of the most common consequences of instability, wherein pore water pressure increments under static loads lead to a decrease in the effective stresses and partial or complete liquefaction of the soil matrix. However, there is an ambiguity in the definition of static liquefaction, linking it to a number of fundamental features such as soil properties, effective stress levels and stability of particle arrangements in the soil matrix in undrained element tests. The results of static liquefaction model slope experiments show two consecutive phases of behaviour. First, a gradual change in the excess pore pressure, and subsequently the occurrence of both very fast accelerations and simultaneous excess pore pressure generation which characterize the onset of instability. The experiments are performed in a unique large scale testing device; a so-called static liquefaction tank.

1 INTRODUCTION

The initiation mechanism in the instability of loose sands in subaqueous slopes and submarine mass movements is one of the most challenging research topics due to its complexity (Molenkamp 1989, Molenkamp 1999, Md. Mizanur & Lo 2011, Lade & Yamamuro 2011, Chu et al. 2015). Various triggering mechanisms have been proposed for submarine flow slide case histories in the literature. Steepening of the slope due to sedimentation or scouring, tectonic tilts, tidal forces, earthquakes and manmade activities (construction equipment or pipeline installation) are some examples of the triggering factors (Lade & Yamamuro 2011). Moreover, the instability can be triggered by a combination of triggering elements in the field. Thus far, static liquefaction induced failure is reported to be one of the main manifestations of the instability in loose saturated granular materials (Silvis & de Groot 1995, Kramer 1988, Lade 1992, Lade & Yamamuro 2011, Jefferies & Been 2015, De Jager et al. 2017). Unlike the onset of instability which occurs at small strains, static liquefaction is known for its large deformations and rapid mass movements (Askarinejad et al. 2015). In static liquefaction, pore water pressure increments lead to significant decrease in the effective stresses and the soil matrix behaves like a liquid at the flow failure state. Instability and its consequences are still important causes for concern in analysing the stability of loose submarine slopes. However, proper experimental data is insufficient for evaluation of the available hypotheses. There are limited number of published studies on physical modelling of static liquefaction induced failures in offshore and onshore landslides (Eckersley 1990, Take et al. 2015, De Jager et al. 2017). The effect of fundamental features such as soil properties (e.g. dilatancy, particle size and shape), boundary conditions (e.g. drained and undrained conditions), effective stress levels and stability of particle arrangements on instability and liquefaction were investigated using element laboratory testing of granular materials by several researchers (Chu et al. 2003, Daouadji et al. 2009, Lade & Yamamuro 2011, Monkul et al. 2011, Dong et al. 2015, Jefferies & Been 2015). However, element tests are limited in simulating the conditions of the soil element with a realistic approach.

A unique experimental facility in the soil mechanics laboratory at Delft University of Technology is utilized to perform model slope experiments of flow failures. Preliminary results demonstrate that the facility produces consistent and reproducible liquefaction flow slides (De Jager et al. 2017). The specific objective of this study was to investigate the changes of pore water pressure and accelerations at the onset of instability. A set of new electronic sensors is developed for monitoring the pore pressures and accelerations at various points within the loose sand layer. A consistent transitional phase from drained to undrained (/partially undrained) state is observed in the recorded data. This phase is concluded to be the triggering factor of the initiation of deformation at the locations of the sensors.

2 PHYSICAL MODELING

2.1 *Material characterization*

A uniform, very fine silica sand is selected as the main material of the experiments. The sand, so called

Table 1. Material properties of Geba sand.

Parameter	Value
Permeability (m/s) (void ratio, e = 0.94)	4.2E–5
Cohesion (kPa)	0.0
φ'res (deg) (e = 0.94 & mean effective stress < 20 kPa)	36
Minimum void ratio*	0.64
Maximum void ratio*	1.07
Fines content (Silt)	4%

According to Japanese standard

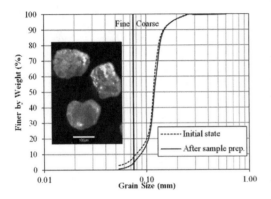

Figure 1. Grain size distribution before and after sample preparation and representative particle shapes of Geba sand.

Geba sand, is characterized by a set of conventional soil mechanics experiments. Table 1 summarizes the mechanical properties of the soil and Figure 1 illustrates the grain size distribution as well as examples of the particle shapes.

Geba sand contains about 8% of silt size particles and, after sample preparation procedure, some of the silt particles are washed away and this values is reduced to 4%. The grains are categorized as subrounded particles with a roundness coefficient of 0.77 ± 0.01, where coefficients of 0 and 1 represent fully angular and fully rounded particles, respectively (Wadell 1932).

2.2 Model slope setup, sample preparation and instrumentation

The model slope experiments are performed in an inclinable tank (5 m by 2 m by 2 m) called Liquefaction Tank (LT) (De Jager et al. 2017). Figure 2 illustrates the structure of the LT and an example of three stages in one of the experiments. The slope failure is initiated by tilting the tank at several low inclination rates (e.g. 0.1, 0.01 deg/sec) using a computer controlled hydraulic jack. The experiments reported in this paper are the ones with the tilting rate of 0.1 deg/sec. In the tilting stage, a 0.5 m thick, loosely packed horizontal sand layer will act like a steepening slope up to the point of instability. A fluidization technique is used to

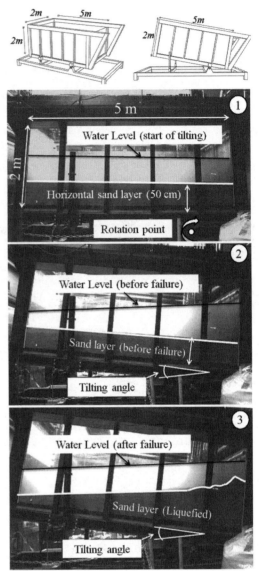

Figure 2. Schematic view of the LT and one example experiment involving triggering of the failure by tilting.

prepare reproducible loose layers with a relative density of approximately 30%. An upward flow of water is applied to the sand layer and fluidizes the sample in the fluidization phase. This phase is continued until the sample is sufficiently fluidized. Sand layer thickness expands during the fluidization. Final state of fluidization is reached when the drag forces are equal to the gravitational forces and the soil matrix reaches a fluid-like state and the sand layer thickness does not expand anymore (De Jager & Molenkam 2012, De Jager et al. 2017). Segregation is an inevitable phenomenon in the fluidized sand samples within the scale of the LT. However, selecting a uniform sand (Cu = 1.1) minimizes the possible heterogeneity and the same sample

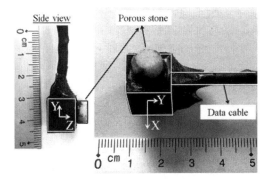

Figure 3. Mobile pore pressure and acceleration sensor (MS).

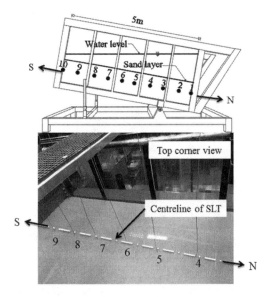

Figure 4. Installation pattern of MS sensors.

preparation method maintains it at the same range for all the experiments.

The LT is a highly-instrumented device. De Jager et al. (2017) presents a detailed explanation of the available fixed sensors installed on the LT. A set of mobile acceleration and pore water pressure sensors (MS) has been developed in addition to the stationary sensors (pore pressure, total pressure, acceleration and temperature sensors) explained by De Jager et al. (2017). Figure 3 shows a schematic view of the mobile pore pressure and acceleration sensors. Two separate micro sensors are embedded in the MS. One is an accelerometer and the other one is a pressure sensor. Therefore, the MS can record the local pore pressure and accelerations in three perpendicular directions at a specific point in the sand layer. Z-direction is normal to the circular surface of the porous stone, Y-direction is parallel to the data cable and X-direction is perpendicular to both Y and Z. The pore pressure sensors embedded in the mobile sensors can measure up to 50 kPa with the resolution of 0.15 kPa. The accelerometers can record the acceleration of 2 g ($+/-$ 1 g) with the resolution of 0.005 g (m/s^2).

Figure 4 illustrates the pattern of installation of the mobile sensors in the reported experiments. 10 sensors are suspended above the centerline of the LT at equal intervals (of about 50 cm). The length of the data cables are arranged so that they can reach the middle of the sand layer. The sensors are installed above the tank before the fluidization stage and after fluidization is terminated, they settle down along with the soil particles. This method ensures a minimum disturbance in the loose sand sample due to installation of the sensors. It should be noted that, since the sensors are suspended by flexible data cables, their location may not be exactly in the middle of the sand layer. The sampling frequencies of 1, 10 and 100 Hz are selected for recording the data files. Selection of low sampling rates at the beginning of experiments reduces the size of data files. The frequency is increased in three steps and reaches its highest value a few seconds before the failure (based on the experience of the operator).

3 TEST RESULTS AND DICUSSION

3.1 Pore water pressure

The water level in the LT remains static and horizontal during the tilting stage at the selected low tilting rates (e.g 0.1 deg/sec in this study). The volume of the water will remain constant because of the closed boundaries of the liquefaction tank. As shown in Figure 5, the pore pressure distributions recorded by several mobile sensors are all in the hydrostatic condition before the initiation of the instability. The lack of excess pore water pressure generation can be explained by hydraulic properties of the sand relative to the rate of tilting. Therefore, this implies that the hydraulic conductivity of the Geba sand is high enough to maintain drained hydrostatic condition during the tilting. It is worthy of note that, due to the resolution range of 0.1 kPa, possible excess pore pressure generation less than this range cannot be recorded with the current data acquisition system.

The mobile sensors which provide the most undisturbed data at the onset of instability are selected for further analysis. MS-01 was one of the sensors which recorded the instability prior to other sensors. Figure 6 shows one example of the pore pressure data recorded by MS-01, for the whole range of motion (0 to 10 degrees). This graph includes the abrupt jump in pore water pressures at the time of failure. The sudden jump in pore pressure, which quickly propagates through the whole soil matrix, is recorded by all of the installed pressure sensors with minor time differences.

3.2 Accelerations

The other helpful data obtained from the mobile sensors is the change in the local accelerations in three

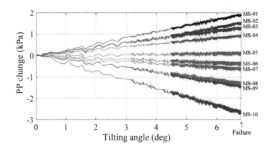

Figure 5. Pore pressure (PP) change recorded by mobile sensors during tilting of LT, before failure.

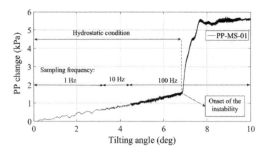

Figure 6. Pore pressure change recorded by MS-01 for the whole range of motion in the model test.

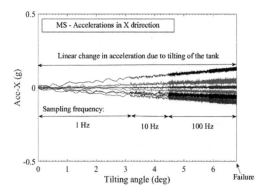

Figure 7. Acceleration in X-direction of the mobile sensors before the instability.

perpendicular direction. Figure 7 illustrates the accelerations in X-direction (see Figure 3) prior to initiation of the failure.

Figure 7 shows a linear change in the acceleration in the X-direction before the failure. This recorded linear trend is because the sensors are stable in the sand body and are rotating throughout the tilting stage. Depending on the initial orientation of the sensors, the accelerations will linearly increase or decrease due to gradual rotation. Figure 8 presents one example of the recorded acceleration for the whole range of tilting of the tank. This range includes the onset of instability and initiation of the rapid deformations at the time of failure.

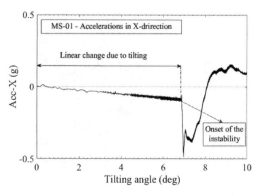

Figure 8. Acceleration in X-direction of the mobile sensors.

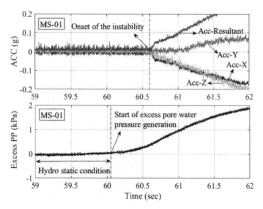

Figure 9. Acceleration and pore pressure jumps (MS-01).

3.3 Onset of instability

Pore pressure and acceleration jumps are the best measure to characterize the initiation of instability and subsequent liquefaction type of failure, since visual observation of failure cannot be very accurate for the current scale of the model test.

Recorded pore pressures and accelerations of the mobile sensors are compared to investigate the onset of instability in more detail. Figure 9 shows one example of this comparison recorded by MS-01. The most remarkable observation obtained from the data was a consistent time difference between the recorded jumps in pore pressure (excess pore pressure) and accelerations. The pore water pressures started the jump earlier than the accelerations in almost all the records.

The experiment was repeated four times with various installation patterns of the mobile sensors for some of the cases, to investigate the reproducibility of the results. The time difference between two consecutive phases of pore pressure and acceleration jumps was consistently repeated in all the experiments. Figure 10 presents the results of two of the experiments with the same installation pattern. The Tests 1 and 2 have had the same relative density and rate of tilting. The recorded jumps have started at about 6 and 6.9 degrees for the Tests 1 and 2, respectively. The observed difference in the time and location of the initiation of failure

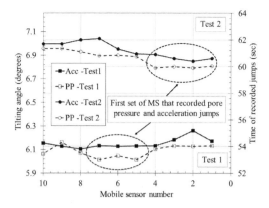

Figure 10. Comparison of the phase shift in recorded jumps of accelerations and pore pressures in two experiments with same sensor installation pattern.

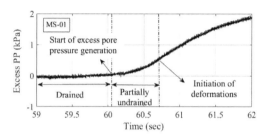

Figure 11. Transition from drained to partially undrained condition and triggering of instability.

could be explained by the possible heterogeneity in the soil layer. The mobile sensors that have recorded the first jumps in pore pressure and accelerations provide the most reliable data. The other sensors could somehow be affected by the occurring turbulence and arising deformations in their vicinity. For example, sensors number 5, 6 and 7 in Test 1 and 1, 2, 3 and 4 in Test 2 provide the most undisturbed data.

Figure 11 demonstrates a transition from the drained phase to a partially undrained condition before the initiation of instability. This transitional partially undrained phase can support the idea of triggering the instability with a change in the excess pore pressure dissipation rate. The tendency of increments in pore water pressure seems to be continued after the initiation of local instability, up to the point where the effective stress becomes very close to zero and the liquefaction takes place. The ratio of generated excess pore water pressure to effective vertical stress (pore pressure coefficient, r_u) has been calculated to be about 0.3 at the onset of instability and larger than one after the instability due to the major jump in the excess pore pressure.

Observed stable trends of change in pore water pressure and accelerations in drained condition before the failure confirms the fact that the initiation of the instability requires a trigger. Strong evidence of the triggering were clearly captured by the mobile sensors. However, these observations can best fit to the behavior of the exact locations of sensors and for further extrapolation and generalization of the onset of the instability in the LT additional experiments with various sensor installation patterns must be performed.

4 CONCLUSIONS

This study is designed to investigate pore pressure and acceleration development at the time of instability and subsequent liquefaction induced failure in submerged slope model tests. The experiments are performed in a unique instrumented device in the laboratory of soil mechanics at Delft University of Technology. The main focus of this paper is to study the triggering mechanism and the onset of failure in loosely packed sand in submerged slopes.

A set of advanced mobile pore pressure and acceleration sensors has been developed for the experiments that can record the hydro-mechanical responses at predefined points of the sand layer to increase in shearing. The sensors successfully captured increases in the pore water pressure and motion acceleration of particles at the onset of instability. The results revealed two consecutive phases of behaviour prior to and at the time of failure. First, a gradual change in the excess pore pressure that is likely to be a consequence of some soil contraction that occurs at a point in the vicinity of the sensors. The subsequent phase was the occurrence of both very fast deformations and simultaneous major excess pore pressure generation which characterizes the onset of instability and the subsequent liquefaction type of failure.

The results are in a good agreement with the published hypotheses on necessity of the triggering mechanisms for initiation of instability in granular materials (Jefferies et al. 2012, Yamamuro & Lade 2011, De Jager et al. 2017). However, these findings somewhat contradict some of the available hypotheses in the literature that state that the generation of the excess pore pressure occurs in the shear zones following the initiation of the deformations and progressive failure state (Eckersley 1990).

ACKNOWLEDGEMENTS

The authors gratefully acknowledge the support of the first author by Rijkswaterstaat (Ministry of Infrastructure and the Environment in the Netherlands). Additional resources were provided by Delft University of Technology.

REFERENCES

Askarinejad, A., Beck, A. & Springman, S.M. 2015. Scaling law of static liquefaction mechanism in geo-centrifuge and corresponding hydro-mechanical characterisation of an unsaturated silty sand having a viscous pore fluid. *Canadian Geotechnical Journal*, 52: 1–13.

Chu, J., Leroueil, S. & Leong, W.K. 2003. Unstable behaviour of sand and its implication for slope instability. *Canadian Geotechnical Journal*, 40(5): 873–885.

Chu, J., Wanatowski, D., Loke, W.L. & Leong, W.K. 2015. Pre-failure instability of sand under dilatancy rate controlled conditions. *Soils and Foundations*, 55(2): 414–424.

Daouadji, A., AlGali, H., Darve, F. & Zeghloul, A. 2009. Instability in granular materials: experimental evidence of diffuse mode of failure for loose sands. *Journal of Engineering Mechanics*, 136(5): 575–588.

De Jager, R.R. & Molenkamp, F. 2012. Fluidization system for liquefaction tank. In Eurofuge 2012, Delft, The Netherlands, April 23-24, 2012. Delft University of Tech-nology and Deltares.

De Jager, R.R., Maghsoudloo, A., Askarinejad, A. & Molenkamp, F. 2017. Preliminary results of instrumented laboratory flow slides. *Procedia Engineering*, 175: 212–219.

Dong, Q., Xu, C., Cai, Y., Juang, H., Wang, J., Yang, Z. & Gu, C. 2015. Drained instability in loose granular material. *International Journal of Geomechanics*, 16(2): 04015043.

Eckersley, D. 1990. Instrumented laboratory flowslides. *Geotechnique*, 40(3): 489–502.

Jefferies, M., Been, K. & Olivera, R. 2012. Discussion of "Evaluation of static liquefaction potential of silty sand slopes" Appearing in the Canadian Geotechnical Journal, 48 (2): 247–264. *Canadian Geotechnical Journal*, 49(6): 746–750.

Jefferies, M. & Been, K., 2015. *Soil liquefaction: a critical state approach*. CRC press.

Kramer, S.L. 1988. Triggering of liquefaction flow slides in coastal soil deposits. *Engineering Geology*, 26(1): 17–31.

Lade, P.V., 1992. Static instability and liquefaction of loose fine sandy slopes. *Journal of Geotechnical Engineering*, 118(1): 51–71.

Lade, P.V. & Yamamuro, J.A. 2011. Evaluation of static liquefaction potential of silty sand slopes. *Canadian Geotechnical Journal*, 48(2): 247–264.

Md. Mizanur, R. & Lo, S.R. 2011. Predicting the onset of static liquefaction of loose sand with fines. *Journal of Geotechnical and Geoenvironmental Engineering*, 138(8): 1037–1041.

Molenkamp, F. 1989. Liquefaction as an instability. *In Proceedings Int. Conf. on Soil Mechanics and Foundation Engineering (ICSMFE)*: 157–163.

Molenkamp, F. 1999. Collapse due to static liquefaction analysed using large deformation elasto-visco-plastic dynamics. *In: Proc. NUMOG 7, Numerical Models in Geomechanics, Graz*, Balkema: 527–53.

Monkul, M.M., Yamamuro, J.A. & Lade, P.V. 2011. Failure, instability, and the second work increment in loose silty sand. *Canadian Geotechnical Journal*, 48(6): 943–955.

Silvis, F. & De Groot, M., 1995. Flow slides in the Netherlands: experience and engineering practice. *Canadian geotechnical journal*, 32(6): 1086–1092.

Take, W.A., Beddoe, R.A., Davoodi-Bilesavar, R. & Phillips, R. 2015. Effect of antecedent groundwater conditions on the triggering of static liquefaction landslides. *Landslides*, 12(3): 469–479.

Wadell, H. 1932. Volume, shape, and roundness of rock particles. *The Journal of Geology*, 40(5): 443–451.

Centrifuge modelling of earthquake-induced liquefaction on footings built on improved ground

A.S.P.S. Marques & P.A.L.F. Coelho
Department of Civil Engineering, University of Coimbra, Coimbra, Portugal

S.K. Haigh & G.S.P. Madabhushi
Department of Engineering, University of Cambridge, Cambridge, UK

ABSTRACT: Earthquake-induced soil liquefaction is a major cause of damage that can occur as a result of seismic events, particularly in saturated deposits of cohesionless soils in seismically active regions. Shallow foundations, which are often used in structures built on soils with these characteristics, are particularly susceptible to seismic liquefaction. A centrifuge test was performed to assess the performance of innovative mitigation techniques to enhance the performance of shallow foundations built on liquefiable ground, using narrow densified zones with and without high capacity vertical drains. This paper presents some preliminary results, including the excess pore pressures developed right below the shallow foundations as well as the settlements and vertical motions taking place during the seismic simulation. Based on the major features of the performance observed, which elucidate the advantages of centrifuge modelling as a liquefaction research tool, some potential limitations of the particular test carried out are discussed.

1 INTRODUCTION

Earthquakes have been historically perceived as one of the most destructive natural hazards. Natural disasters frequently represent traumatic, economy-challenging and life-threatening situations. However, earthquakes differ from other disasters as these often correspond to severe and uncontrollable calamites, occurring without any previous warning and leaving little room for pre-event evacuation procedures or short-term measures to minimise injuries and losses. Besides, earthquakes have the increased difficulty of not having an exact ending point, exhibiting several aftershocks that can continue over months, or even years in some cases. In the recent 2010/2011 Christchurch earthquake events, for example, over 12,500 aftershocks were documented (CPAG 2014). These aftershocks lead to additional injuries and damage, as well as long-term society distress, inducing people to keep reliving the traumatic experience for a long time, which was particularly serious with children.

Earthquake-induced soil liquefaction is a major cause of damage that occurs as a result of earthquake shaking of saturated deposits of cohesionless soils in seismically active regions. This phenomenon has proved to be a serious threat to modern societies, which are increasingly vulnerable to its serious effects, namely as a result of their strong dependence on sophisticated and widespread infrastructure that is built in liquefaction-prone areas. This is exacerbated by the fact that the problem is more and more likely to cause indirect repercussion in non-seismically active countries too, due to fast growing globalisation of the world economy. For instance, earthquake-induced damage in Taiwan can severely affect Western companies while the world economy would be threatened if earthquake-induced damage affected oil production capacity in Saudi Arabia, Venezuela or Iran. Moreover, as shown by the 2004 Sumatra earthquake, losses and human misery due to earthquakes can have worldwide repercussions due to the mobility of modern civilization and the power of contemporary media. Last but not the least, many multinational companies have working places in several locations all over the world that may be prone to seismic liquefaction effects. This may result in a significant amplification of the consequences around the globe.

Shallow foundations are often used in structures built on soils with characteristics susceptible to seismic liquefaction effects. In fact, these are often located in inundated areas near river banks, lakes or sea shores, and built on alluvial deposits formed by loose cohesionless materials, their susceptibility to the devastating effects of liquefaction being often extremely high. The damaging nature of this phenomenon is frequently increased by the fact that areas geologically susceptible to liquefaction, like river floodplains and shorelines, are often extremely populated.

Clarification of the issues related to liquefaction-induced settlements on shallow foundations is a fundamental requirement to better understanding and prediction of the behaviour of these structures built on liquefiable ground and to develop innovative liquefaction resistance measures offering improved cost-benefit ratios and that are more compatible with densely populated or overpriced urban areas. These may include currently used techniques with optimised properties and/or hybrid techniques involving a mix of current techniques. In view of the objective uncertainties regarding the interpretation of field case studies, the liquefaction performance of shallow foundations has been modelled in centrifuge and large-scale shaking table experiments over the years (Liu & Dobry 1997; Kawasaki et al. 1998; Adalier et al. 2003; Dashti et al. 2010; Marques et al. 2012; Bertalot 2013; Marques et al. 2015). Overall, the ability of centrifuge models to simulate soil behaviour under realistic effective stress levels has been perceived as a distinctive advantage for experimental research in this field, being particularly valuable for developing innovative mitigation measures and for defining the mechanisms governing the complex soil-structure behaviour observed in liquefaction-related phenomena.

2 EXPERIMENTAL RESEARCH TOOLS AND MATERIALS

2.1 Characteristics of the dynamic centrifuge model

Centrifuge modelling is based on the principle of creating scaled-down models of geotechnical structures in order to depict the behaviour of a prototype model, which corresponds to a non-scaled simplified representation of the real problem. However, the behaviour of soils has been established to be highly non-linear and stress-dependent, and hence true prototype behaviour can only be observed in a model under stress and strain conditions similar to the prototype. A geotechnical centrifuge enables to recreate the same stress and strain levels within the scaled model by testing a 1:N scale model at N times earth's gravity, created by centrifugal force. The similarity of stress, strain and material behaviour ensures the realism of model behaviour as long as the boundary conditions are correctly implemented. The principles of centrifuge modelling are covered in detail by Madabhushi (2015).

This paper presents some preliminary results of a dynamic centrifuge test conducted at the Schofield Centre facilities in Cambridge, UK, to investigate the response of shallow foundations susceptible to seismic liquefaction. The model foundations were created in order to simulate shallow foundations resting on a submerged layer of loose sand, using unusual liquefaction resistance measures in order to mitigate the effects of earthquake-induced liquefaction, which often tend to be unacceptable when no ground improvement is carried out. The dynamic centrifuge test was carried out using the 10-m diameter Turner Beam Centrifuge available at the Schofield Centre, which is described in detail by Schofield (1980).

The model tested simulates two identical shallow foundations placed on top of two different liquefaction

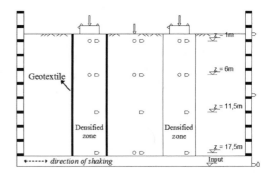

LEGEND:
▷ Horizontal Accelerometer (Acc)
◊ Vertical Accelerometer (Acc)
○ Pore Pressure Transducer (PPT)
△ Vertical Microelectromechanical Systems (MEMS)
◁ Horizontal Microelectromechanical Systems (MEMS)
⇊ Linear Variable Displacement Transducer (LVDT)
z - depth below ground level

Figure 1. Layout of the centrifuge model.

Table 1. Properties of Hostun sand (Stringer 2008).

Property	Value
Angle of repose (°)	33
D_{10} (mm)	0.286
D_{50} (mm)	0.424
Uniformity coefficient	1.59
e_{min}	0.555
e_{max}	1.067
G_s	2.65

mitigation measures, as shown in Figure 1. The prototype profile consists of an 18-m deep Hostun sand layer placed at a relative density of about 50% surrounding the narrow densified zones under the foundations, which are built with the same sand material but with a higher relative density (approximately 80%). In one of the cases, the densified zone is surrounded by supplementary vertical drains. Hostun sand is susceptible to liquefaction, as the grading curve lies well within the liquefaction susceptibility curves proposed by Tsuchida (1970), as shown in Figure 2. This material is a reference sand for many French geotechnical laboratories and it is also widely studied by the international geotechnical community (Doanh et al. 2010). Soil properties are presented in Table 1 and described in detail by Flavigny et al. (1990).

The preparation of the model is challenging, as different materials and sand densities had to be used. More details on the method implemented can

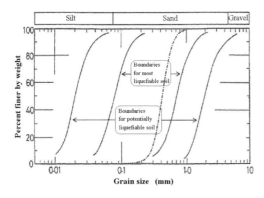

Figure 2. Particle size distribution curve for Hostun sand, superimposed on liquefaction susceptibility curves, after Tsuchida (1970) – adapted from Stringer (2008).

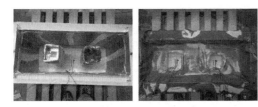

Figure 3. Process of creating dense blocks in the model soil deposit using the air dry pluviation technique.

be found in Marques et. al. (2015). The setup and instrumentation used are shown schematically in Figure 1. The model was instrumented with piezoelectric accelerometers (Acc), pore pressure transducers (PPTs) and linear variable displacement transducers (LVDTs). The shallow foundations were instrumented with micro electro-mechanical systems accelerometers and LVDTs, as shown in Figure 1. Accelerations in the soil, structures and outside the laminar box, excess pore water pressure in the free field and in the densified zones under the shallow foundations, and vertical displacement of the soil and shallow foundations were measured during the test. Figure 3 shows different model preparation stages for the creation of both densified columns of sand, showing the placement of the instruments inside the densified blocks of dense. This process had to be carefully done so that the instruments cables did not interfere too much with the pouring process of sand into the narrow dense blocks.

The model structures used to simulate the shallow foundations had a simple design, consisting of solid blocks of square plan area (3 × 3 m) with no superstructure and with a height of 1.225 m. The bearing pressure transmitted throughout the foundation base equals 95 kPa. A high-capacity vertical drain was simulated through a specific geotextile, rigid enough to avoid squeezing of the vertical drainage paths once the final horizontal stresses are applied to the model (Figure 4).

Figure 4. Geotextile used to simulate high-capacity vertical drains.

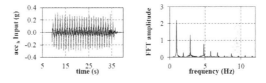

Figure 5. Time history and FFTs of the horizontal seismic motion applied to the model.

The model was excited through the Stored Angular Momentum (SAM) actuator (Madabhushi et al., 1998), which is a simple and reliable mechanical actuator that uses the energy stored in a pair of flywheels to generate the input motion. Despite not being able to reproduce real seismic actions, it generates nearly sinusoidal horizontal acceleration motions of chosen duration and amplitude, which is considered valuable for fundamental research on earthquake effects, as it avoids the difficulties introduced by more complex dynamic loading. At prototype scale, the 1-D earthquake simulation lasts about 25 s, imposes maximum peak horizontal accelerations close to 0.3 g and has a predominant frequency of 1 Hz. The time histories and FFTs of the horizontal input motion (Figure 5) show that the seismic simulation is not totally single-frequency but overall matches the desired characteristics of the seismic loading. The long earthquake duration aims at intensifying liquefaction effects and facilitating model behaviour analysis.

2.2 Model container

The centrifuge model was prepared inside an equivalent shear beam (ESB) container, which establishes the physical boundaries of the model. Consequently, the boundary conditions are extremely important in the results observed, and so, the container selected to perform the experiment had to be carefully chosen.

In the field, most structures rest on laterally unbounded soil (semi-infinite half-space). However, when performing a centrifuge test, the model is laterally limited by the presence of the container end walls, introducing an element of dissimilarity with prototype conditions. Zeng & Schofield (1996) and Brennan (2003) present a detailed evaluation concerning to the main disturbances introduced by the model container on the soil deposit, with respect to geotechnical earthquake modelling.

In order to study soil liquefaction during earthquakes, two major methods of model containers are currently in use in the Schofield Centre facilities: laminar boxes and equivalent shear beams. The ESB container (Schofield & Zeng, 1992) was the one chosen for this study. In fact, if a laminar box provides ideal boundary conditions for fully liquefied soil, its response in the pre-liquefaction phase is non-realistic and may affect the triggering of liquefaction in the soil model, introducing dissimilarity from prototype conditions. The main advantage in using laminar boxes is due to their ability to reproduce large lateral deformations occurring in liquefied soil. However, this characteristic is significant in lateral spreading ground conditions, where significant cumulative displacement may be generated during shaking, and this research is mainly on the effects of shallow foundations resting on level soil deposits. Consequently, no cumulative lateral displacement is expected as a consequence of earthquake shaking.

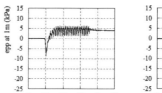
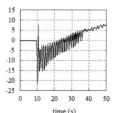

Figure 6. Co-seismic EPP generations at 1 m deep under the centre of both footings.

3 PRELIMINARY TEST RESULTS

3.1 Co-seismic excess pore pressure

Co-seismic excess pore pressure (EPP) generations measured at 1 m depth under the centre of each footing are presented in Figure 6. It is clear that the seismic liquefaction mitigation techniques used largely influences the EPP response.

Results show that when the soil under the footing is densified without the additional use of vertical drainage paths, co-seismic dilation effect are developed, as negative EPP values are achieved during almost the entire seismic motion. In fact, co-seismic EPP captured right under the footing where only densification was used as a liquefaction mitigation technique never exceeds the initial value of the free-field vertical effective stress, besides inducing negative EPP once the first couple of cycles take place. This suggests that momentary shear-stress variations occur at shallow ground levels in the dense sand just beneath the footing, whose effects are kept visible during the entire earthquake duration.

On the other hand, by using a geotextile to simulate the vertical drains response, the data clearly shows that the EPP immediately increases and reaches its maximum value after the first couple of cycles once the earthquake takes place. This suggests that the presence of the drains may mitigate the dilation effects caused by shearing the densified soil underneath the footing.

Figure 6 also suggests that right after the end of the earthquake, the EPP dissipation process seems very different in both situations analysed. In fact, it seems that the EPP starts to slightly decrease as the seismic motion ends if vertical drains are added to a narrow densified zone, while without their use the EPP shows an increasing path, probably as a result of the phenomenon of EPP migration that comes from the free-field towards the soil column under the footing. Long-term EPP measurements would be required before more definite conclusions could be established.

3.2 Improvement of the observed structural settlements

Figure 7 shows the co-and post-seismic settlements measured for both shallow foundations. The data clearly shows the great reduction in the total settlements of the footings if vertical drains are added to narrow densified zones. In fact, footing's total settlement is reduced by about 75%, which is a significant value to be taken into consideration. Actually, the absolute values of the settlements obtained by using this hybrid mitigation technique present tolerable values for field applications, especially if the extended duration of the seismic simulation is taken into account.

With respect to the co-seismic settlements obtained for both structures, it is obvious the large difference observed between both mitigation techniques used. The fact that co-seismic settlements are significantly higher for the structure resting merely on a densified block of sand might justify the co-seismic EPP measurements observed in Figure 6. By settling significantly more, the embedment and punching of the structure through the soil deposit might significantly increase soil's shear stress and consequently induce the

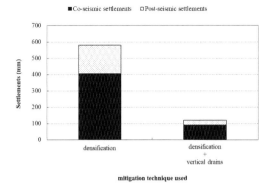

Figure 7. Absolute settlements suffered by the shallow foundations using the different mitigations techniques.

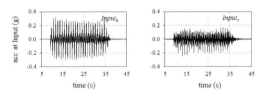

Figure 8. Horizontal and vertical input motions imposed to the base of the model.

process of dilation leading to the development of negative EPP, as typically observed in dense sand. On the contrary, for the hybrid technique the structure settles considerably less during the earthquake, which might not induce dilation in the soil, resulting in almost immediate positive EPP values.

Another interesting aspect that can be obtained from Figure 7 is that the percentages of shallow foundation's co-and post-seismic settlements in relation to the total settlements are relatively close for both mitigation techniques, corresponding to approximately 70–75% and 25–30% of the total, respectively.

3.3 Limitations of the centrifuge model

The centrifuge model tested showed a possible unexpected constraint, whose effects must be addressed. Besides the planned horizontal input motion transmitted through the soil at the base of the model, an unexpected vertical input motion was also registered (Figure 8). Even if this vertical input component is significantly smaller than the horizontal one, vertical accelerations are transmitted through the soil and measured on both footings, mostly due to this unplanned vertical motion (Figure 9).

When using vertical drains combined with densification, the vertical motions detected in the footing exceed those where only densification is used. This suggests that the overall soil's stiffness degradation by using this hybrid mitigation technique may be smaller than when just a full-depth densified zone under the footing, thus facilitating the propagation.

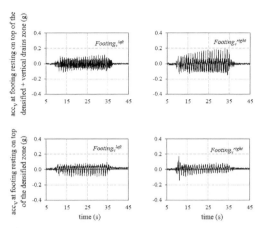

Figure 9. Vertical accelerations measured in both footings.

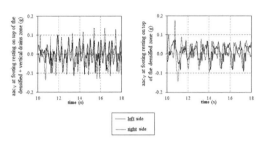

Figure 10. Comparison of the vertical accelerations measured in each footing.

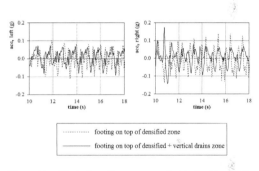

Figure 11. Vertical motions measured at both sides of the two structures.

On the other hand, by analysing the vertical motions that reach each footing (Figure 10), it is visible that the left and right sides are in phase with each other, suggesting that the footings are not subjected to any individual rocking mechanism. However, structures' vertical motions are not in phase with each other in both left and right sides, which implies that rocking of the all centrifuge model may be occurring. This is particularly visible in Figure 11, which compares the vertical accelerations measured at both sides of the two structures on a slight period of time. Even if this rocking mechanism apparently present in the model was not expected, the major conclusions are still valid as the effects on the model foundations are similar and do not exceed the effect of the EPP generated by the horizontal shaking, which govern the settlements.

4 CONCLUSIONS

A centrifuge model experiment was carried out to evaluate the performance of innovative earthquake-induced liquefaction mitigation techniques. Narrow densified zones with and without geotextiles simulating vertical drains were tested to assess the behaviour of shallow foundations built on top of different liquefaction mitigation measures.

Some of the preliminary results presented, particularly when comparing the settlements obtained using both techniques and the co-seismic EPP developed under each footing, demonstrate the ability of centrifuge modelling to depict with apparent realism the complex mechanisms governing the performance. It is clear that a significant reduction on the settlements suffered by the footing is observed when adding vertical drains around the densified zone. Moreover, the presence of the drains affects the excess-pore-pressure generation and dissipation under the footing, which further enlightens the complex mechanisms governing the performance of the hybrid liquefaction mitigation measure tested. The fact that the centrifuge experiment suggests that some rocking of the entire centrifuge model may be occurring, as structures' vertical motions are not in phase with each other in both left and right sides, may be investigated further.

Overall, the results obtained encourage further research in the field taking advantage of the unique benefits of centrifuge modelling.

ACKNOWLEDGMENTS

This research was funded by the European Community's 7th Framework Programme [FP7/2007-2013] for access to the Turner Beam Centrifuge, Cambridge, UK, under Grant Agreement n° 227887. The first author would also like to thank FCT – Fundação para a Ciência e Tecnologia, Portugal – for sponsoring her PhD programme (grant no. SFRH/BD/73170/2010).

REFERENCES

Adalier, K., Elgamal, A., Menese, J. & Baez, J.I. 2003. Stone columns as liquefaction countermeasure in non-plastic silty soils. *Journal of Soil Dynamics and Earthquake Engineering*, 23(7): 571–584.

Bertalot, D. 2013. Seismic behaviour of shallow foundations on layered liquefiable soils. PhD thesis, University of Dundee, Scotland.

Brennan, A. J. 2003. Vertical drains as a countermeasure to earthquake induced soil liquefaction. Ph.D. Thesis, Cambridge University, UK.

CPAG 2014. Children and the Canterbury Earthquakes. Child Poverty Action Group Inc., New Zealand.

Dashti, S., Bray, J.D., Pestana, J.M., Riemer, M.R. & Wilson, D. 2010. Mechanisms of Seismically-Induced Settlement of Buildings with Shallow Foundations on Liquefiable Soil. *Journal of Geotechnical and Geoenvironmental Engineering*, ASCE, 136(1): 151–164.

Doanh, T., Dubujet, P. & Touron, G. 2010. Exploring the undrained induced anisotropy of Hostun RF loose sand. *Acta Geotechnic*. 10.1007/s11440-010-0128-x, 239–256.

Flavigny, E., Desrues, J. & Palayer, B. 1990. Le sable d'Hostun "RF". *Revue Française de Géotechnique*, 53, 67–69.

Kawasaki, K. Sakai, T., Yasuda, S. & Satoh, M. 1998. Earthquake-induced settlement of an isolated footing for power transmission tower. *Proc. Centrifuge 98*, Tokyo, Balkema, pp271–2.

Liu, L. & Dobry, R. 1997. Seismic response of shallow foundation on liquefiable sand. *J Journal of Geotechnical and Geoenvironmental Engineering*, 123(6): 557–566.

Madabhushi, S.P.G. 2015. *Centrifuge modelling for Civil engineers*. Book by CRC Press.

Madabhushi, S.P.G., Schofield, A.N. & Lesley, S. 1998. A new Stored Angular Momentum Earthquake Actuator. *Proc. Int. Conf. Centrifuge '98*, Tokyo, Japan: 111–116.

Marques, A.S., Coelho, P.A.L.F., Cilingir, U., Haigh, S.K. & Madabhushi, S.P.G. 2012. Centrifuge modelling of liquefaction-induced effects on shallow foundations with different bearing pressures. *Proc. of 2nd Eurofuge conference on Physical Modelling in Geotechnics*. Delft, Netherlands.

Marques, A.S., Coelho, P.A.L.F., Cilingir, U., Haigh, S.K. & Madabhushi, S.P.G. 2015. Centrifuge Modelling of the Behaviour of Shallow Foundations in Liquefiable Ground. *Proc. of 6th International Conference on Earthquake Geotechnical Engineering*, Christchurch, New Zealand.

Schofield, A.N. & Zeng, X. 1992. Design and performance of an equivalent-shear-beam (ESB) container for earthquake centrifuge modelling. Technical Report, CUED/D-SOILS/TR245, University of Cambridge, UK.

Schofield, A.N. 1980. Cambridge geotechnical centrifuge operations. *Geotechnique*, 30(3): 227–268.

Stringer, M.E. 2008. Pile Shaft Friction in Liquefaction. First Year Report, University of Cambridge, UK.

Tsuchida, H. 1970. Prediction and countermeasure against liquefaction the liquefaction in sand deposits. In Abstract of the Seminar, Port and Harbour Research Institute.

Zeng, X. & Schofield, A.N. 1996. Design and performance of an Equivalent Shear Beam (ESB) model container for earthquake centrifuge modelling. *Geotechnique*, 46(1): 83–102.

Investigating the effect of layering on the formation of sand boils in 1 g shaking table tests

S. Miles, J. Still & M. Stringer
Department of Civil and Natural Resources Engineering, University of Canterbury, Christchurch, New Zealand

ABSTRACT: Sand boils are formed by the upwards flow of pore water following strong ground motions, and their presence is typically taken to indicate the occurrence (or non-occurrence) of liquefaction under the earthquake loading. However, it is possible for liquefaction to occur without sand boils being observed, and this is relatively common in centrifuge or 1 g shake table experiments. In this paper, the authors present the results from a study using a 1-g shaking table which aims to create sand boil features in the lab. Through this study, the mechanisms occurring in the development of sandboils are explored, as well as some of the key conditions required for these features to be observed.

1 INTRODUCTION

Following major earthquakes, it is common for reconnaissance teams to record evidence of liquefaction, based on the presence of sand boil features (i.e. Stringer et al. 2017 recorded the location of liquefaction following the 2016 Kaikoura Earthquake). These observations can later be supplemented with in-depth ground characterization to establish the "critical layer" (i.e. the layer where the ejecta material came from) in areas where the level of ground shaking is well understood. With this knowledge, case history databases can be formed, and are an important part of the semi-empirical methods being used to predict whether liquefaction is likely at a given site in the future (i.e. Boulanger & Idriss 2015). However, it is understood that while a sand boil is often associated with liquefaction, it is possible for liquefaction to occur without these features being created at the ground surface, though the particular soil profiles in which sand boils will/will not form is not well quantified or understood.

Sand boils form in the field as a result of the upwards flow of pore water, which occurs as a result of the high excess pore pressures that can exist following strong ground shaking (i.e. Towhata 2008). If the upwards flowing fluid reaches the ground surface with sufficient velocity, then material from the liquefied soil layer can be transported to the ground surface, where it is rapidly deposited as the flow velocity decreases.

Based on field observations, Ishihara (1985) linked the thickness of the liquefiable layer to the thickness of an overlying non-liquefiable crust during an earthquake of particular magnitude. More recently, Cubrinovski et al. (2017) discussed that during the Canterbury Earthquake Sequence, sites where the ground profile comprised relatively thick deposits of clean sand tended to show evidence of liquefaction, while those sites where the soil profile was heavily interlayered with sands and silts tended to show no surface evidence of liquefaction (i.e. no sand boils). In this latter (numerical) study, the importance of the fine grained layers was demonstrated as an impediment to the upwards flow of pore water following strong ground motions.

Despite their prevalence in the field, sand boils are often not apparent in laboratory experiments where researchers interested in liquefaction phenomena often favour deposits of uniform clean sand extending to the surface of the soil profile. A number of experimental studies (i.e. Huishan & Taiping 1984, Elgamal et al. 1989, Fiegel & Kutter 1994, Butterfield & Bolton 2003, Brennan et al. 2008) have shown that sand boils can form in a laboratory setting when a low permeability layer (i.e. silt or clay) impedes the upwards seepage of water after liquefaction, leading to the development of water films beneath the low permeability layer. Sand boils typically developed when the silt or clay layer ruptures, and fluid and soil rapidly travel to the surface. Kokusho (1999) further demonstrated that water films can develop in situations where even relatively small differences in hydraulic conductivity exist (i.e. between a fine sand and a silt). The aforementioned studies have demonstrated that sand boils can be reproduced in a laboratory setting as well as providing valuable insights into the mechanisms leading to their development. However, there remains a lack of research into the parameters which can affect the severity and extent of the formation of sand boils. In this paper, we present results from a preliminary series of tests which investigated the effect that the thickness of individual soil layers may have on the development of sand boils in a small-scale experiment.

Table 1. Summary of tests carried out.

Test ID	Sand	No. layers	D_r (avg)
BS01	Blenheim sand	10	30
BS02	Blenheim sand	1	39
BS04	Blenheim sand	1	45
BS05	Blenheim sand	5	37
BS06	Blenheim sand	10	41
BSNF01	Blenheim sand (fines removed)	10	24
NB01	New Brighton sand	10	31

2 EXPERIMENTAL SETUP

Small scale models were created in the laboratory using a clear plastic container, with internal plan dimensions of 254 mm × 151 mm and an approximate height of 190 mm. Models were prepared by wet pluviation; The container was filled with tap water and then dry sand was poured through a sieve onto the surface of the water. Construction of the models took place in a number of lifts (1, 5 or 10), to create layering within the models, and the fine grained particles were allowed to settle out prior to commencing the next sand layer. The mass of sand in each lift was kept constant, with a total of 10 kg of sand being used in each test.

Pluviation took place directly on the 1-g shaking table, so that no disturbance would take place between the end of construction and the beginning of testing. Once all layers were completed, the water level was slowly lowered (by reverse seepage) so that the water table was co-incident with the top of the soil. A summary of tests which are described in this paper is shown in Table 1.

The sand used in most of the tests described in this paper is liquefaction ejecta material which was recovered from Lansdowne Park in Blenheim, New Zealand, following the 2016 Kaikoura earthquake. The ejecta (denoted BS) is a medium sand, with a small proportion of fines (Fines content, FC, defined as percentage passing a 75 μm sieve). Key properties of this specific ejecta material is as follows: $D_{10} = 0.080$ mm, $D_{50} = 0.213$ mm, $D_{60} = 0.232$ mm, FC = 8%; $e_{min} = 0.573$, $e_{max} = 1.111$. BSNF01 was carried out with this same sand, after sieving through a 75um sieve, while test NB01 was carried out with New Brighton Sand ($D_{10} = 0.154$, $D_{50} = 0.218$, $D_{60} = 0.234$, FC = 0%, $e_{min} = 0.603$, $e_{max} = 0.991$). It should be noted that the New Brighton sand is more uniform in gradation than the liquefaction ejecta material used in the other tests.

The preparation of the models by wet pluviation offered a number of advantages in the context of these experiments. Firstly, as noted by Huishan & Taiping (1984), the segregation of material replicates the natural gradation of coarse to fine particles observed in the field, while at the same time creating sub-layers of fine grained material which could potentially act as a barrier to seepage following the onset of liquefaction during the experiments. The wet pluviation process

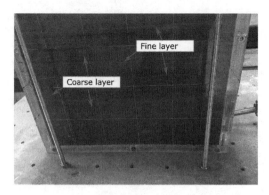

Figure 1. Layering from wet pluviation (BS06).

then simplifies the saturation of these interlayered soil profiles. If similar models were created by air pluvation, it may be difficult to achieve a good degree of saturation without essentially fluidizing the model due to the difficulty flowing water through silts, and the tendency to trap air in coarser layers located below finer ones.

It is important to recognize that average soil properties (i.e. size distributions and limiting void ratios) have been determined for the sand used in the experiment. However, the sand segregates during the pluviation process, so that no particular "soil element" will have the average properties previously determined. In particular, the average relative densities of the models have been determined, which might be significantly different to those which exist say in the coarse fraction of a particular layer. It is however the case that wet pluviation tends to create very loose soil layers, in which a particular soil tends to have a consistent relative density. It is assumed that the different constituent parts of the soil profile therefore have similar fabrics and densities across tests where similar overall relative densities have been measured.

Accelerations during the experiments were recorded on the side of the model container using a MEMS accelerometer. Pore pressures were measured using a small-range pressure transducer connected to small diameter pipe, which was located in the bottom of the container (miniaturized transducers were not available at the time of testing) as shown in Figure 2. Note that during the shaking, the pore pressure transducer was placed on the ground next to the shake table. The pipe and void within the pressure transducer were filled with water to improve the measurement of pressures during the testing. It was observed that the pore pressure did not return to their pre-shaking levels after excess pore water pressures (generated during the strong shaking) had dissipated. It is assumed that this is not a true behaviour, but a result of the relatively long pipes between the measurement point in the model, and the pressure transducer (potentially due to an air bubble in the tubes).

An example of a model prepared for testing is shown in Figure 2, showing the key elements in the test set-up.

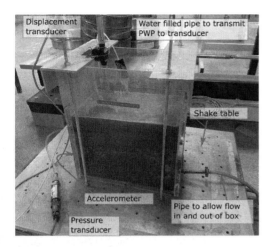

Figure 2. Typical model set up.

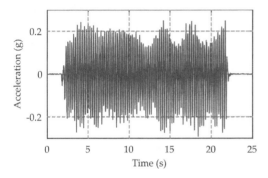

Figure 3. Typical accelerations applied to the models.

The shake-table used in the experiments operates with an offset cam, so that the ground motions are approximately sinusoidal in nature. In each test, the shaking table was activated for a period of approximately 20 s, after which it was deactivated, and the table would then stop shaking over the next few cycles. An example of the accelerations applied during test BS06 are shown in Figure 3, where peak accelerations of approximately 0.2 g were delivered at 3.5 Hz for 20 s. As shown, there is some variation in the peak accelerations with time, and during shaking, the motions tend to have a torsional component. When the acceleration records are closely examined, it is also apparent that the motion has quite a strong component at the first harmonic (i.e. 7 Hz).

After the ground shaking had been stopped, the model was observed for several minutes to allow any excess pore water pressures to dissipate. Following this, the settlement of the model during the shaking was measured and the ground water table was gradually lowered (settlement during the shaking led to the presence of water above the soil surface). The model surface was then closely inspected for the presence of sand boils, and the number of features (and approximate size) noted.

Figure 4. Sand boils observed after test BS01.

2.1 Scaling

While it is possible to re-create the phenomenon of sand boils in a controlled laboratory experiment, it is thought that results from small-scale 1-g experiments are not readily scaled to represent larger prototypes (i.e. Brennan et al. 2008). The 1-g experiments presented in this paper therefore attempt to investigate the effects of laying on a qualitative basis only.

3 RESULTS

Pore pressures developed rapidly in the first few cycles of strong shaking during each test, with liquefaction being triggered at various depths in the model. The pore water pressures were however not high enough to indicate liquefaction through the whole depth of the model, though in many tests the pore pressures reached some limiting value which was sustained for the majority of shaking. During the shaking (and shortly after), some air bubbles appeared through the top surface of the soil. The presence of the air bubbles indicates that some air was being trapped inside the model during the pluviation process (i.e. in some cases, air bubbles adhered to the soil grains as they sank through the water), but were released during the strong ground shaking. This release of air bubbles potentially creates a preferential path through which fluid might flow upwards and therefore provide the "seed" for a sand boil. However, it is important to note that these air bubbles were observed in many tests, while sand boils were not. It is also important that while sand boils sometimes formed at the location where air bubbles escaped, there were also cases where sand boils formed in locations where there had not been air bubbles, and also cases where air bubbles had been observed, but there was no sand boil after the pore pressures stopped dissipating.

Following the end of the ground shaking, the excess pore water pressures tended to dissipate relatively quickly (within 60 seconds of the end of shaking) and led to the development of vertical settlements of the soil.

Sand boils formed in a number of experiments performed with the Blenheim sand. The ejecta features

Table 2. Observations of sand boils during tests.

Test ID	No. layers	Sand Boils		
		Small	Med	Large
BS01	10	1	8	1
BS02	1	0	0	0
BS04	1	0	0	0
BS05	5	3	4	0
BS06	10	4	4	0
BSNF01	10	1	0	0
NB01	10	0	0	0

ranged in size, with the largest having a diameter of approximately 5 cm. The features formed at both the boundaries of the box, and also towards the center of the box. The sandboils which were identified after test BS01 are highlighted in Figure 4. The sandboils were grouped according to the diameter of the feature, and the number of each size of sandboil is summarized in Table 2. When the clean, New Brighton sand was used in the models, no sand boils were observed.

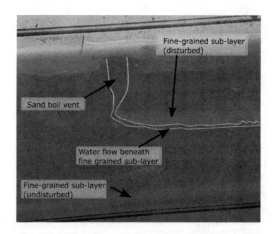

Figure 5. Disruption of fine grained sub-layers and water films below fine material.

4 DISCUSSION

4.1 Mechanisms observed in the tests

4.1.1 Segregation of the ejecta material

The formation of sandboils during each test was apparent from a visual inspection of the model surface, but it was also possible to observe the formation of these sand boils in the minutes which followed the end of shaking through the side windows. Video footage revealed a number of interesting features pertaining to the formation process. First was that near the surface, it was noticeable that the upwards flow of water was carrying material up and out of the hole as expected, but that at the same time, larger particles were also circulating downwards, both at the sides and middle (at different times) of the sand boil flow. This highlights a key issue that when ejecta is observed on the soil surface, it is not necessarily representative of the particle size distribution of the liquefied layers. Larger particles may or may not initially be carried to the surface, but as the hydraulic gradients driving the flow begin to reduce, then larger particles in the ground will no longer be carried up and out of the hole. By extension, the opposite is true – fine grained particles may be washed away from the main sand boil, suggesting that the liquefied layer is cleaner than reality.

4.1.2 Importance of fine grained layers

In the models constructed with 10 lifts, it was observed that the layers of fine grained material were being disrupted by the upwards seepage of pore water in the deeper layers as well as the layers at the surface (Fig. 5). This occurs as a result of the upward drainage of excess pore water pressure, and shows that even at depth, the upwards seepage is causing damage, and consequent preferential flow paths through layers of fine-grained material, leading to concentrated flows. It was further observed that once a deeper layer had been disrupted, then the fine grained layers above that point also became disrupted. It is assumed that this occurs due to a local increase in excess pore water pressure in the area of a break in the fine grained material. This would result in slightly larger pressures acting in the region directly above the break and hence contribute to the preferential breaks in the fine grained material above the initial breakage. It should be noted that this mechanism could equally occur from the top down, with pore fluid initially breaking through the fine grained material in the top layer. As the pore water rapidly escapes through this pathway, the excess pore water pressures will be slightly lowered in this area relative to the rest of the layer. This creates a larger differential pressure across the fine grained material directly below, leading to the pore fluid preferentially breaking through the lower layers at this same location.

4.1.3 Water films

It was noticed that where sand boils formed at the box edge (and therefore they could be observed) that upwards seeping pore water would gather directly below the fine grained material (shown in Fig. 5). In itself, this is unsurprising and has been noted in centrifuge tests investigating lateral spreading (i.e. Malvick et al. 1992 among others). This effect was surprising in the models with Blenheim sand constructed in 10 layers, where the thickness of the layers of fine grained material was very small indeed. Rather than simply break through the thin layer of fines, the water moved laterally until it reached a location where a break had already occurred. It is however important to note that the scale of the problem is very important, and it is likely that the number of points where water breaks through the fines layers is governed by: the strength and thickness of the layers as well as permeability of the sandy layers (i.e. how easily can fluid flow to other breakage points) and the hydraulic gradients which are driving the flow.

4.2 Effect of number and composition of layers

The results summarized in Table 2 shows the overall number of features and approximate quantity of liquefaction ejecta reduced with the number of soil layers, when the Blenheim sand ejecta was used. It is also shown that when a completely clean, and more uniform sand (i.e. New Brighton sand) was used, then no sand boils formed, even when the model was constructed in 10 layers. When considering a uniform clean sand, the number of layers becomes relatively unimportant, since there will be little to no segregation of material.

The rate of dissipation of excess pore water pressure was greatest in the test with the New Brighton Sand, and began during the shaking itself. This implies that the overall flow rate of water out of the model was the largest in this model, yet no sand boils developed. This suggests that the presence of some fine grained material is key to whether or not sand boils will form. As discussed previously, the fine grained material creates a barrier to the upwards flow of water so that the flow concentrates in a few locations where it manages to break through the fine layer. In these locations, the flow rate will be much higher, resulting in sand grains being carried upwards to form the sand boil features.

The results from the Blenheim Sand ejecta however point to the importance of the thickness of individual fine grained layers. As discussed by Cubrinovski et al. (2017), the presence of fine grained layers can be a significant impediment to the upwards flow of water, and hence the formation of sand boils. As the number of layers in the different models with Blenheim sand reduces, the thickness of the individual layers (and hence the thickness of the sub-layers with fine grained material) will increase. It was noted that in all of the tests, some air bubbles were observed to appear through the top of the model, both during and after shaking. These bubbles would be expected to create temporary paths in the finer grained material through which pore fluid could more easily flow. However, given the reduction in number and size of ejecta features, it appears that these flow paths were not sustained in models constructed with fewer layers. This is surprising since the role of the fine grained layers would typically be to impede/block the upwards flow of pore water (i.e. Cubrinovski et al. 2017). Assuming rapid upwards seepage in the coarser parts of the layer, it would be expected that in models with fewer layers, there would be greater hydraulic gradients and escaping pore fluid available to develop liquefaction features. However, inspection of Figure 5 shows that in a model with many layers, the fine grained material is very thin, and when the fluid breaks through, the sub-layer of fine grained material is completely destroyed (i.e. fines are dispersed upwards into the sand layer above). It is likely that when the layers are thicker, then the sub-layer of fine material does not get completely destroyed and can therefore help slow down the upwards flow of water such that material cannot be carried upwards to develop a sand boil.

5 CONCLUSIONS

In this experimental study, the authors were able to create sand boils on a small 1 g shaking table using ejecta material recovered after the 2016 Kaikoura Earthquake from Blenheim, New Zealand.

The formation of sand boils in the model tests appeared to be very sensitive to the presence of fine grained layers which were created as a result of the wet pluviation deposition. The presence of a very thin layer of fine grained material acted to concentrate the upwards flow of pore water after the strong ground shaking such that sand grains could be transported out of the ground. However, the thickness of the fine grained material plays a second role in reducing the rate of seepage, so that where the layers of fine grained material were thicker, the upwards seepage of pore water was impeded and as a consequence was unable to develop the concentrated flow paths required to create sand boils; instead the dissipation of excess pore water pressures took place more slowly, and no sand boils developed in the model.

ACKNOWLEDGEMENTS

The authors would like to thank the technical staff at the University of Canterbury for their support during the course of the project. Additionally, the authors thank Davidson Engineering for collecting and making available the ejecta material in Blenheim.

REFERENCES

Boulanger, R.W. & Idriss, I.M. 2015. CPT-based liquefaction triggering procedure. *Journal of Geotechnical and Geoenvironmental Engineering*, 142(2): 1–11.

Brennan, A.J., Moran, D. & Richie, N. 2008. Observations on sand boils from simple model tests. *Proc. Geotechnical Earthquake Engineering and Soil Dynamics IV*, ASCE GSP 181, Sacramento, USA.

Butterfield, K.J. & Bolton, M.D. 2003. Modelling pore fluid migration in layered, liquefied soils. *Proc 2003 Pacific Conference on Earthquake Engineering*. Paper 131.

Cubrinovski, M., Ntritsos, N. & Rhodes, A. 2017. System response of liquefiable soils. *Proc. PBDIII. 3rd International Conference on Performance Based Design*. Vancouver, 16-19 July 2017.

Elgamal, A., Dobry, R. & Adalier, K. 1989. Study of effect of clay layers on liquefaction of sand deposits using small scale models. *Proc. of 2nd US-Japan Workshop on Liquefaction Large Ground Deformation and Their Effects on Lifelines*, T. D. O'Rourke and M. Hamada (eds), Buffalo, USA 233–245.

Fiegel, G.L. & Kutter, B.L. 1994. Liquefaction mechanism for layered soils. *Journal of Geotechnical Engineering*. Vol 120(4) 737–755.

Kokusho, T. 1999. Water film in liquefied sand and its effect on lateral spread. *Journal of Geotechnical and Geoenvironmental Engineering*. Vol 125(10): 817–825.

Huishan, L & Taiping, 1984. Liquefaction potential of saturated sand deposits underlying foundation of structure. *Proc 8th World Conf. on Earthquake Engineering*, San Francisco, USA.(3): 199–206.

Ishihara, K. 1985. Stability of natural deposits during earthquakes. *Proceedings of the 11th International Conference on Soil Mechanics and Foundation Engineering*, San Francisco, 1: 321–376.

Malvick, E.J., Kulasingam, R., Kutter, B.L. & Boulanger, R.W. 2002. Void Redistribution and Localized Shear Strains in Slopes During Liquefaction. *Proceedings, International Conference on Physical Modelling in Geotechnics*, St. Johns, Canada, 495–500.

Stringer, M., Bastin, S., McGann, C., Cappallaro, C., El Kortbawi, M., McMahon, R., Wotherspoon, L., Green, R., Aricheta, J., Davis, R., McGlynn, L., Hargraves, S., Van Ballegooy, S., Cubrinovski, M., Bradley, B., Bellagamba, X., Foster, K., Lai, C., Ashfield, D., Baki, A., Zekkos, A., Lee, R. & Ntritsos, N. 2017. Geotechnical aspects of the 2016 Kaikōura Earthquake on the South Island of New Zealand. *Bulletin of the New Zealand Society for Earthquake Engineering*, 50(2): 117–141.

Towhata, I. 2008. *Geotechnical Earthquake Engineering*. Springer-Verlag Berlin Heidelberg.

Centrifuge modelling of mitigation-soil-structure-interaction on layered liquefiable soil deposits with a silt cap

B. Paramasivam, S. Dashti, A.B. Liel & J.C. Olarte
University of Colorado, Boulder, Colorado, USA

ABSTRACT: This paper discusses the results of three centrifuge experiments on 3-story inelastic structures founded on layered liquefiable soil deposits with and without a thin silt cap. For two of the model structures on different soils, liquefaction was mitigated with prefabricated vertical drains (PVDs) around their perimeter. Test results indicate that use of PVDs reduced the duration of large excess pore pressures in the soil, notably reducing settlement on the soil profile without a silt cap, but the PVDs amplified the seismic demand on the foundation and superstructure. In the case with the silt cap, void redistribution and strain localization limited the seismic demand on the foundation, but amplified transient foundation rotations and lateral deformations in the superstructure with or without PVDs. The presence of thin silt cap impeded drainage out of the lower sand layers and made it difficult for the PVDs to prevent shear localization below the sand-silt interface.

1 INTRODUCTION

Prefabricated vertical drains (PVDs), or earthquake drains, are often used in practice to mitigate the liquefaction hazard. PVDs are generally made of hollow perforated plastic pipes with an internal diameter ranging from about 75 to 200 mm (Rollins et al. 2003; Howell et al. 2012). The perforated plastic pipes are typically wrapped with geotextile to avoid clogging from the transportation of fines from the surrounding soil. PVDs are cheaper and faster to install than stone columns.

PVDs can limit the generation of excess pore pressures and reduce the likelihood of liquefaction triggering, by enhancing drainage with minimum shear reinforcing effects (Rollins et al. 2003; Howell et al. 2012). The effectiveness of PVDs in reducing the likelihood of liquefaction triggering in sandy soils has previously been investigated through full-scale field tests and scaled dynamic centrifuge tests (Rollins et al. 2003; Howell et al. 2012). These studies showed the successful performance of PVDs in terms of speeding the dissipation of excess pore pressures within their zone of influence, which consequently reduced vertical settlements and lateral displacements of slopes. However, the previous physical model studies were performed in the absence of structures. The influence of this mitigation technique has not yet been experimentally evaluated on soil-structure-interaction (SSI) and building's performance, for a range of soil conditions and layering.

Paramasivam et al. (2017) recently performed two centrifuge tests to evaluate the effects of PVDs on the seismic performance of 3- and 9-story, potentially-inelastic, shallow-founded structures on layered liquefiable soil deposits made of different types of clean sand with different relative densities. Test results showed that the use of PVDs around the perimeter of the footing could expedite the dissipation of excess pore pressures in the underlying (foundation) soil and, hence, generally reduced permanent foundation settlements. However, they amplified the seismic demand on the foundation and superstructure relative to the unmitigated cases. The influence of PVDs on foundation's permanent rotation and structure's permanent flexural drift strongly depended on structures' force-deformation behavior and dynamic characteristics in relation to ground motion properties.

The previous centrifuge tests involved a liquefiable soil deposit (loose, saturated Ottawa sand) that was overlain by a free draining layer (coarse, dense, saturated, Monterey sand). However, in the field, granular deposits are often stratified with sub-layers of low permeability soils. A sharp contrast in permeability could slow down drainage out of the liquefiable deposit and influence soil-structure interaction, foundation and structure's response, and the effectiveness of PVDs. These effects were not studied in previous experiments.

This paper presents preliminary results from a series of centrifuge experiments that modeled a 3-story, potentially-inelastic structure on layered, liquefiable soil deposits with and without a thin silt cap. PVDs were installed around the foundation of one structure for each soil profile. The influence of a silt cap was evaluated on mitigation-soil-structure-interaction and building performance in terms of excess pore pressure generation, foundation settlement, tilt, acceleration, and deformation patterns at the beam-column connections. Experimental results presented in this paper aim to provide insight into the influence of vertical variations in soil permeability on the effectiveness of PVDs, considering the soil-foundation-structure system holistically.

2 CENTRIFUGE EXPERIMENTAL SETUP

A series of three dynamic centrifuge tests were performed using the 5.5m-radius, 400 g-ton centrifuge facility at the University of Colorado Boulder (CU). Figure 1 shows the schematic of the test series on two different soil profiles. All tests used the same model Structure A that represented a 3-story moment resisting steel frame building on a stiff mat foundation (Olarte et al. 2017). In the first two tests (here, referred to as Tests A_{UM} and A_{DR}), the seismic response of Structure A was evaluated when unmitigated (A_{UM}) and when surrounded by PVDs or drains (A_{DR}) on a layered liquefiable soil deposit that had a free draining surface layer (Profile-1). This baseline soil profile included a 10m-thick dense layer of Ottawa sand (F-65) overlain by a 6m-thick loose layer of the same Ottawa sand, which was in turn overlain by a 2m-thick layer of dense Monterey 0/30 sand (see Figure 1). In the third test (Test $A_{UM-DR-Silt}$), both unmitigated and mitigated structures were tested simultaneously on a layered soil deposit, in which the top 2m-thick layer of Monterey sand was replaced with a 0.5m-thick layer of silica silt (Sil-Co-Sil 102) overlain by a 1.5m-thick layer of dense Monterey sand (Profile-2).

The model specimens were constructed in a flexible-shear-beam (FSB) container of length 968mm, width 376mm, and depth 304mm in model scale dimensions made of aluminum and rubber (Olarte et al. 2017), and spun to 70g of centrifugal acceleration. An automated sand pluviating machine was used to dry pluviate different layers of sand at the target relative density. For the silt layer, dry silica silt was first sieved uniformly over the entire area of the container from a constant height, and then compacted statically with a surcharge pressure of 5kPa. The properties of different soil layers used in this study were described by Olarte et al. (2017) and Badanagki et al. (2017). A solution of hydroxyl propyl methylcellulose (HPMC) with a kinematic viscosity 70 times greater than that of water was used as the pore fluid to satisfy the dynamic scaling laws. Initially, the soil specimen was flushed with CO_2 to replace the air voids. Then, the model container and the HPMC fluid tank were kept under a vacuum pressure of 70 kPa. The vacuum pressure on the fluid tank was reduced through a computer-controlled setup to initiate the flow into the model container, until the saturation was completed (Olarte et al. 2017).

The 3-story Structure A was modeled in centrifuge as a simplified, scaled, 3DOF structure (Figure 1). The design details and properties of the target prototype and the scaled model structure were described by

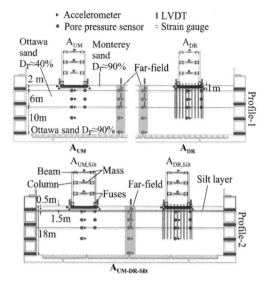

Figure 1. Centrifuge test setup and instrumentation layout of three tests. All units are in prototype scale unless otherwise specified. Tests A_{UM} and A_{DR} had another structure in the container, which is not of interest here.

Olarte et al. (2018). Maximum moments and potential nonlinearity in the structure were designed to concentrate at reduced sections (fuses) located at beam ends and column bases, as shown in Figure 1. Beams, columns, and lumped masses were constructed of steel, and the replaceable fuses were made of nickel to achieve the desired stiffness and strength in the simplified structure. The structure was founded on a stiff mat foundation embedded to a depth of 1m from the soil surface. The foundation was made of aluminum due to similarity of its unit weight with reinforced concrete.

As shown in Figure 1, the liquefiable soil under Structures A_{DR} and $A_{DR,silt}$ was mitigated with PVDs. The PVDs were installed around the foundation of the structure at a center-to-center spacing of 1.2m in prototype units. The design, construction, and properties of PVDs were detailed by Olarte et al. (2017).

All the model specimens were subject to a series of earthquake motions at the container base in flight using the servo-hydraulic shaking table. In this paper, for brevity, the results are shown only for one motion, Kobe-L, with an achieved PGA = 0.3–0.37g; Mean period, $T_m = 0.9$s; and Arias Intensity, I_a, = 1.6–1.9 m/s. Kobe-L was the first significant motion. Figure 2 shows the acceleration and Arias intensity time histories of the Kobe-L motion measured at the base of the container during all three tests presented in this paper.

3 PRELIMINARY RESULTS AND OBSERVATIONS

3.1 Excess pore pressures in the near-field

The influence of a thin silt cap on SSI, building response, and the effectiveness of PVDs was first evaluated by comparing excess pore pressures (Δu) in the soil below the center and edge of the foundation in different tests (Figure 3). Initial values of vertical effective stress (σ'_v) at each location are also shown in the same plot, to identify the occurrence of liquefaction (when Δu approached σ'_v, indicating liquefaction).

Test results showed that soil beneath Structure A_{UM} on Profile-1 exhibited the expected response of generating excess pore pressure during shaking, followed by rapid drainage after shaking ceased. The soil beneath Structure $A_{UM,Silt}$ on Profile-2 showed a similar response during shaking, but a notably slower dissipation after shaking. In fact, due to upward flow from lower elevations, and a slow rate of dissipation through the thin silt layer, the peak Δu at the top of the liquefiable layer, under both the center and edge of structure, was measured slightly after shaking. In addition, the peak Δu value under the edge of Structure $A_{UM,Silt}$ at the top of the liquefiable layer was greater than σ'_v at the corresponding location for this structure. This was likely due to the formation of a water lens around the edge of this structure below the sand-silt interface. However, that liquefaction (defined as $\Delta u = \sigma'_v$) was never observed under the center of the footing for the conditions evaluated in this study.

Use of PVDs around Structures A_{DR} and $A_{DR,Silt}$ limited the magnitude and duration of excess pore pressures in the underlying soil, particularly around the edges of the foundation (within the drains' radius of influence). The presence of silt cap beneath the Structure $A_{DR,Silt}$ increased the drainage demand on PVDs and resulted in peak Δu values that were slightly greater than those below Structure A_{DR}. PVDs, in general, expedited the dissipation of the excess pore pressures which helped limit the shear strains in the soil induced by SSI-induced ratcheting, and reduced the formation of water lenses around under Structure $A_{DR,Silt}$.

3.2 Foundation response

Figure 4 shows the time histories of settlement and foundation rotation of unmitigated and mitigated structures on both soil profiles, together with the corresponding base accelerations. The far-field settlements recorded at the top of the liquefiable soil layer in all three tests are also included in the plot for comparison. Settlements in the far-field were mainly caused by volumetric mechanisms (Dashti et al. 2010): 1) settlement due to partial drainage ($\varepsilon_{p\text{-}DR}$); 2) sedimentation ($\varepsilon_{p\text{-}SED}$); and 3) reconsolidation ($\varepsilon_{p\text{-}CON}$). The far-field settlements recorded in the first two tests (A_{UM} and A_{DR}) on Profile-1 showed minor differences, due to seismic interaction between the soil in the far-field and that under the structures with different mitigation techniques. Far-field settlements measured on Profile-1 were significantly greater than those Profile-2. The presence of a silt cap near the surface trapped or slowed down the drainage of excess pore water pressures, approaching a more undrained cyclic loading

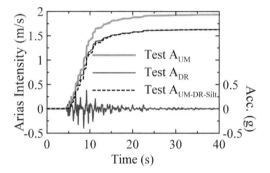

Figure 2. Comparison of container base accelerations recorded during the three tests. Note that acceleration time history is shown only for Test A_{DR}.

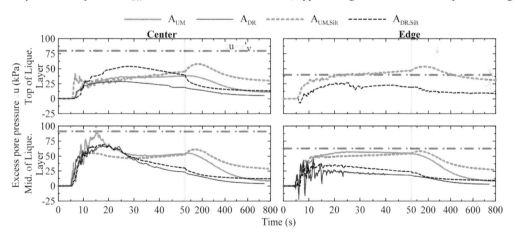

Figure 3. Excess pore pressure response under the center and edge of structures measured at elevations at middle and top of liquefiable layer. The pore pressure sensor beneath the edge of structures A_{UM} and A_{DR} at the elevation of top of liquefiable layer malfunctioned. Initial vertical effective stresses beneath the center and edge of the structure were approximated analytically based on Boussinesq (1883).

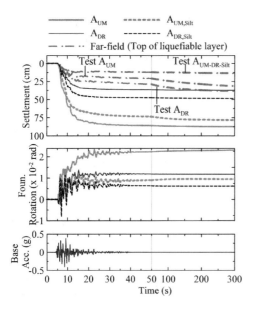

Figure 4. Time histories of foundation settlement and rotation of all four structures and the corresponding base accelerations measured in Test A_{DR}.

condition. This limited the contribution of volumetric strains due to partial drainage ($\varepsilon_{p\text{-}DR}$) and reduced the total far-field settlements in Test $A_{UM\text{-}DR\text{-}Silt}$.

Settlements recorded on the foundation of all structures were greater than those in the far-field, due to the presence of additional deviatoric deformations (partial bearing capacity loss, $\varepsilon_{q\text{-}BC}$, and soil-structure interaction induced building ratcheting, $\varepsilon_{q\text{-}SSI}$), described by Dashti et al. (2010). Settlement of Structure A_{UM} (Profile-1) was 17% greater than $A_{UM,Silt}$ (Profile-2.). Generation of larger excess pore pressures within the liquefiable layer beneath the center of Structure A_{UM} led to significant strength and stiffness loss, which amplified shear type deformations ($\varepsilon_{q\text{-}BC}$ and $\varepsilon_{q\text{-}SSI}$) and volumetric strains due to partial drainage ($\varepsilon_{p\text{-}DR}$) in Profile-1. As in the far-field, the presence of a silt cap under Structure $A_{UM,Silt}$ impeded drainage during shaking and restricted volumetric strains due to partial drainage ($\varepsilon_{p\text{-}DR}$). In this case, the accumulation of excess pore pressures led to void redistribution and shear localization below the sand-silt interface. However, these latter effects were limited during the relatively low intensity Kobe-L motion, leading to Structure $A_{UM,Silt}$ on Profile-2 settling less compared to the same structure on Profile-1. Structure $A_{UM,Silt}$ continued to settle for about 200s after shaking, due to a slower dissipation of excess pore pressures through silt, affecting different shear and volumetric mechanisms of deformation.

PVDs around the perimeter of Structures A_{DR} and $A_{DR,Silt}$ on both soil profiles limited net excess pore pressures underneath both buildings, which helped reduce deviatoric strains caused by strength loss, but amplified volumetric strains due to partial drainage ($\varepsilon_{p\text{-}DR}$). The net effect was a reduction in structure's settlement compared to the unmitigated cases for both soil profiles. Use of PVDs around Structure A_{DR} reduced settlements by approximately 57% compared to A_{UM} on Profile-1. The same degree of reduction was not observed on Profile-2. The significant drainage demand imposed on PVDs below the sand-silt interface on Profile-2 slightly reduced the drains' effectiveness in dissipating pore pressures and reducing foundation settlement. The presence of PVDs also reduced the contribution of post-shaking settlements.

Figure 4 compares the time histories of foundation rotation among different structures during the Kobe-L motion. Larger excess pore pressures and strength loss in the soil below Structure A_{UM} amplified its permanent rotation on Profile-1 compared to the other cases. PVDs surrounding Structure A_{DR} expedited drainage after shaking and helped reduce the structure's permanent rotation by approximately 50% compared to the unmitigated case (A_{UM}). However, Structure A_{DR} on Profile-1 rotated more than both the unmitigated and mitigated structures on Profile-2 ($A_{UM,Silt}$ and $A_{DR,Silt}$). The presence of silt in Profile-2 localized strength loss to the depth immediately beneath the sand-silt interface in Profile-2. This thinner zone experiencing a large degree of strength loss helped reduce the accelerations propagating to the foundation and hence, the seismic demand on the foundation and SSI-induced building ratcheting. However, structures on Profile-2 experienced a greater degree of transient foundation rotation (particularly in 5-15 s) compared to structures on Profile-1. Use of PVDs slightly reduced the permanent rotation of Structure $A_{DR,Silt}$ on Profile-2 compared to $A_{UM,Silt}$.

Figure 5 shows the time-frequency spectra (Stockwell spectra) of foundation's transverse accelerations recorded during Kobe-L. The figure shows that the amplitude and frequency content of foundation accelerations varied among structures based on the properties of the underlying soil and presence of PVDs. The content of Kobe-L did not coincide with the building's primary modes. The presence of a silt cap generally reduced the amplitude of foundation transverse accelerations, and tended to slightly shift the content to lower frequencies. Use of PVDs amplified the accelerations on both profiles (particularly Profile-1) by reducing the duration of large excess pore pressures and increasing soil stiffness. In addition, accelerations on Structure A_{DR} contained more energy around frequencies of 0.4 to 1.1 Hz, whereas the large degree of pore pressure generation and softening below Structure A_{UM} damped out frequencies greater than about 0.8Hz. Use of PVDs around Structure $A_{DR,Silt}$ slightly amplified foundation accelerations compared to $A_{UM,Silt}$. For both structures on Profile-2, foundation intensities were concentrated in frequencies in the range of 0.4-1.0 Hz.

3.3 *Response of the superstructure*

Figure 5 includes the time-frequency spectra of roof transverse accelerations during the Kobe-L motion.

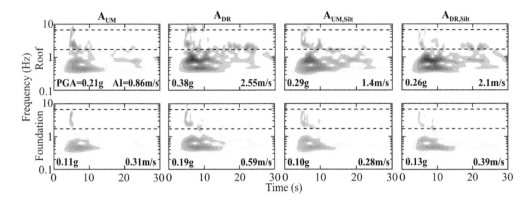

Figure 5. Time-frequency spectra of transverse accelerations measured on the foundation and roof of the structures. Colors ranging light to dark indicate shaking intensity from low to high. Peak acceleration and Arias Intensity of respective time histories are also reported. The dashed lines correspond to the natural frequency of the first two modes of the building under fixed-based conditions.

Significant loss of strength and stiffness in the soil below Structure A_{UM} on Profile-1 reduced the acceleration demand on the structure. Structure $A_{UM,Silt}$ on Profile-2 experienced greater roof transverse accelerations compared to A_{UM}. This was due to the influence of larger transient rotations on the foundation with the introduction of a silt cap (see Figure 4). Use of PVDs around both Structures A_{DR} and $A_{DR,Silt}$ amplified the roof transverse accelerations compared to their unmitigated counterparts for both soil profiles, due to the greater demand on their foundations. Peak roof accelerations on Structures A_{DR}, $A_{UM,Silt}$, and $A_{DR,Silt}$ had more content at frequencies slightly below the building's first mode, which affected the building's performance.

Figure 6 shows the profiles of bending strain and lateral displacement along the height of different structures. Peak bending strains in compression and tension were recorded on beam and column fuses by strain gauges. Transient lateral displacement profiles were obtained along the height of the structure by double integrating the accelerations recorded on the foundation and lumped masses. Peak strains recorded on beam and column fuses occurred at the same time, and lateral displacement profiles were obtained at the corresponding time.

Strain values from beam and column fuses were well below their yield limit (0.097%), indicating that none of the structures yielded during this motion. As expected, PVDs surrounding Structure A_{DR} on Profile-1 transferred a greater seismic demand and amplified transient strains and lateral displacements experienced in the beams compared to the unmitigated Structure A_{UM}. Significant soil softening and permanent settlements below Structure A_{UM} reduced flexural deformations on the superstructure (as reflected in the strains on beam fuses), while the structure's large permanent rigid-body rotation induced additional bending strains on column fuses due to P-Δ effects. Peak strains (distribution and magnitude) were

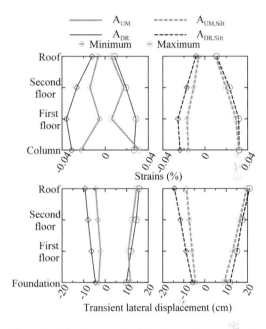

Figure 6. Distribution of peak bending strains and lateral displacement along the height of structures during Kobe-L. Strains were recorded at the beam and column fuses. Strains are positive in tension.

similar between Structures A_{DR}, $A_{UM,Silt}$, and $A_{DR,Silt}$. However, lateral displacement on Structures $A_{UM,Silt}$ and $A_{DR,Silt}$ (Profile-2) were slightly greater than Structure A_{DR} (Profile-1). The presence of a silt cap, regardless of whether drains were employed or not, appeared to amplify transient rocking (see Figure 4) on the foundation that amplified transient flexural strains and deformations in Structure $A_{UM,Silt}$. Use of drains again slightly amplified deformations on Structure $A_{DR,Silt}$ compared to $A_{UM,Silt}$ on Profile-2, but to a much lesser extent.

4 CONCLUSIONS

This paper described results of three centrifuge tests that investigated the seismic response of 3-story inelastic structures founded on a layered liquefiable deposit including sandy soils (Profile-1) and a similar profile including a thin silt cap (Profile-2). Two of the structures on both soil profiles were surrounded by PVDs to expedite drainage.

Use of PVDs around Structure A_{DR} on Profile-1 limited permanent foundation settlement and rotation compared to unmitigated Structure A_{UM}, but it amplified transverse accelerations on the foundation and therefore, the strains and deflections in the superstructure. The presence of a thin silt cap impeded drainage from the lower sand layers, leading to void redistribution and strain localization below the sand-silt interface. This effect increased the drainage demand on the PVDs around Structure $A_{DR,Silt}$, reducing their effectiveness in limiting foundation's permanent settlement and rotation compared to unmitigated Structure $A_{UM,Silt}$. Shear localization at the sand-silt interface helped reduce transverse accelerations on the foundation of both unmitigated and mitigated structures on Profile-2, but it allowed significant transient rocking when compared to Profile-1. In this way, the presence of a silt cap further increased roof transverse accelerations and flexural deformations on both unmitigated and mitigated structures.

The presented results show the important influence of vertical variations in soil's permeability on SSI and on the effectiveness of drains, which should be considered in design. In particular, PVDs may be less effective in improving the overall performance of the system when a silt cap is present.

ACKNOWLEDGMENTS

This material is based upon work supported by the National Science Foundation (NSF) under Grant No. 1362696. Any opinions, findings, and conclusions or recommendations expressed in this material are those of the author(s) and do not necessarily reflect the views of the NSF.

REFERENCES

Badanagki, M., Dashti, S. & Kirkwood, P. 2017. An experimental study of the influence of dense granular columns on the performance of level and gently sloping liquefiable sites. *Journal of Geotechnical and Geoenvironmental Engineering.* (Under review)

Boussinesq, J. 1883. Application des Potentials á L'Étude de L'Équilibre et du Mouvement desSolides Élastiques, Gauthier-Villars, Paris.

Dashti, S., Bray, J., Pestana, J., Riemer, M. & Wilson, D. 2010a. Mechanisms of seismically induced settlement of buildings with shallow foundations on liquefiable soil. *Journal of Geotechnical and Geoenvironmental Engineering,* 136(1): 151–164.

Howell, R., Rathje, E.M., Kamai, R. & Boulanger, R. 2012. Centrifuge modeling of prefabricated vertical drains for liquefaction remediation, *Journal of Geotechnical and Geoenvironmental Engineering,* 138(3): 262–271.

Olarte, J.C., Paramasivam, B., Dashti, S., Liel, A. & Zanin, J. 2017a. Centrifuge modeling of mitigation-soil-foundation-structure interaction on liquefiable ground, *Soil Dynamics and Earthquake Engineering,* 97: 304–323.

Olarte, J.C., Dashti, S. & Liel, A. 2018. Can ground densification improve seismic performance of inelastic structures on liquefiable soils, *Earthquake Engineering and Structural Dynamics,* https://doi.org/10.1002/eqe.3012.

Rollins, K.M., Anderson, J.K.S., McCain, A.K. and Goughnour, R.R. 2003. Vertical composite drains for mitigating liquefaction hazard, *13th international offshore and polar engineering conference,* Hawaii, USA: 498–505.

Paramasivam, B., Dashti, S. & Liel, A.B. 2017. Influence of prefabricated vertical drains on the seismic response of structures founded on liquefiable soil deposits. *Journal of Geotechnical and Geoenvironmental Engineering.* (Under review).

Centrifuge modelling of the effects of soil liquefiability on the seismic response of low-rise structures

S. Qi & J.A. Knappett
School of Science and Engineering, University of Dundee, UK

ABSTRACT: Earthquake-induced soil liquefaction can generate significant damage to low-rise structures, as evidenced in the 2010–2011 Canterbury Earthquake Sequence in New Zealand. In this paper, the structural response of low-rise structures on medium dense granular soils of different permeability (but both nominally liquefiable) was investigated using dynamic centrifuge modelling. In the tests, a series of consecutive motions from the 2010–2011 Canterbury Earthquake Sequence was considered, followed by a long duration 'double-pulse' motion from the 2011 Tohoku Earthquake which can potentially apply large inertial loads after liquefaction has been triggered. It was observed that the lower permeability test reached full liquefaction at shallow depth during shaking, while soil of higher permeability was only comparable in response in the first earthquake; in subsequent strong aftershocks excess pore water pressures were substantially reduced. The structural response of higher permeability soil was 10–45% larger due to the increased motion transmission ability of the soil after the initial earthquake. The structure on the higher permeability soil did, however, show reduced post-earthquake tilt in all motions tested. These results suggest that popular liquefaction triggering analyses may be limited in their ability to properly estimate the hazard posed to structures on nominally liquefiable soil when estimating resistance to subsequent motions (aftershocks).

1 INTRODUCTION

During earthquakes, the effect of soil liquefaction can cause very damaging effects in low-rise structures. An increase in excess pore water pressure (EPWP) can occur across a range of particle size distributions and relative densities in granular soils, which can result in different amounts of liquefaction in soils which are all nominally liquefiable. Earthquake induced liquefaction generally damages structures as a result of a loss of effective overburden stress within soil with EPWP generation, resulting in excessive settlement and tilting of structures with shallow foundations, as evidenced in 2010–2011 Canterbury Earthquake sequences (e.g. Cubrinovski et al. 2012).

The consequence of liquefaction induced damage on foundation can depend on a series of uncertainties (e.g. earthquake loading, site condition and the superstructure). There is lack of analytical methods cooperating effects of deviatoric and volumetric settlements together to evaluate foundation settlements when soil softens (Dashti et al. 2010). Physical modelling using centrifuge is a way of studying effects of liquefaction on building performance. Previous studies have generally considered settlement of shallow foundations alone or as a rigid structure with a centred mass representing the structure on liquefiable soil (Liu & Dobry 1997, Dashti et al. 2010, Bertalot & Brennan 2015) or as a more representative flexible structure with stiffness and mass but on non-liquefiable sand (e.g. Knappett et al. 2015). This paper aims to bring these two effects together, considering a wider structural response (co-seismic structural acceleration, inter-storey sway and drift, within a multi-degree of freedom structural system), alongside the foundation response in terms of post-earthquake settlement and tilt of a two-storey structure with strip foundations on soils of different permeability but similar pre-earthquake stiffness and strength. Through the comparison of high and low permeability subsoil, and the application of multiple earthquakes of different acceleration magnitudes it will be possible to examine: (i) the effect of the degree of liquefaction occurring; and (ii) the influence of pre-shaking and aftershocks on the linked structural and geotechnical performance of the structure in liquefiable soil.

Two dynamic centrifuge tests are presented here, consisting of the same single two-storey structure with separated strip foundations on two different permeability sands of similar relative density (i.e. so the foundations in each case have the same static factor of safety). The input ground motions considered in the two tests are a re-ordered sequence of motions from the Canterbury Earthquake Series of 2010–2011 followed by a long duration 'double-pulse' record from the 2011 Tohoku Earthquake.

2 CENTRIFUGE MODELLING

The two centrifuge tests were conducted using a model scale of 1:40 and tested at 40-g using the 3.5m radius geotechnical centrifuge at the University of Dundee. A full description of centrifuge scaling can be found in Muir Wood (2004).

2.1 Model structure

The structural model was designed to represent a two-storey, single bay, steel moment resisting frame with concrete slabs sitting on separated concrete strip foundations which is a typical damaging building type in Christchurch residential area (Cubrinovski et al. 2012). Square aluminium alloy rods were used to form individual model columns, while aluminium alloy plates were used to create the floor slabs, with additional steel plates bolted to these to allow the floor mass to be varied in later adjacent structure tests (though this feature is not used within the tests described herein). The strip foundations were also made of aluminium plates due to the close similarity in unit weight between this material and reinforced concrete. The foundation width used provides a static factor of safety of 3 against bearing failure, accounting for the total self-weight and a 3.5 kPa extra loading on each storey (applying a bearing pressure of 50 kPa on each footing). The model structure is shown in Figure 1.

The fundamental natural period of the steel frame at prototype scale was targeted using Equation 1:

$$T_n = 0.1N \qquad (1)$$

where N is the number of stories of the structure ($N = 2$ and $T_n \approx 0.2$ s here).

The mass of each floor (all in prototype) was determined based on a 3.6 m × 3.6 m × 0.5 m concrete slab (where the masses of each floor are the same, i.e. $M_1 = M_2$). The equivalent stiffness of the structure in the fundamental mode was then determined by combining Equation 1 and Equation 2, setting the columns of each storey to have the same to have the same stiffness, i.e. $K_1 = K_2$, and selecting the closest available steel Universal Column size to provide an appropriate amount of bending stiffness EI:

$$T_n = 2\pi \sqrt{\frac{M_{eq}}{K_{eq}}} \qquad (2)$$

where

$$M_{eq} = M_1 \overline{y_1}^2 + M_2 \overline{y_2}^2 \qquad (3)$$

$$K_{eq} = K_1 (\overline{y_1})^2 + K_2 (\overline{y_2} - \overline{y_1})^2 \qquad (4)$$

The normalised modal coordinates associated with the fundamental mode were $\overline{y_1} = 0.45$ and $\overline{y_2} = 0.89$, based on an eigenvalue analysis for the two-storey structure with equal stiffness and mass at each storey. The final natural period of the two-storey building was 0.21s. A summary of properties at prototype is shown in Table 1.

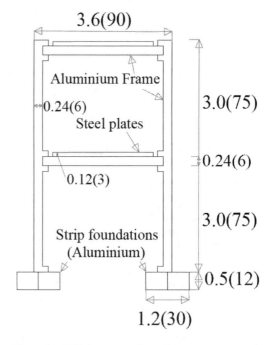

Figure 1. Model structure: dimensions at prototype scale are shown in m; dimensions at model scale are given in mm in brackets.

Table 1. Section properties (prototype scale).

Element	Description	Property
Concrete slab	3.6m×3.6m×0.5m (C25 concrete)	$M_1 = M_2 = 16.5 \times 10^3$ kg
Steel column	203×203×86 UC (3m storey height)	$EI = 20.9 \times 10^6$ Nm^2 $K_1 = K_2 = 37.1 \times 10^6$ N/m
Concrete strip foundations	1.2m×4.8m×0.5m (C25)	$M_f = 7.3 \times 10^3$ kg/strip

2.2 Model preparation and soil properties

8 m deep deposits of dry HST95 Congleton silica sand layer ($D_r = 55\%$–60%) were initially air-pluviated into an equivalent shear beam (ESB) container in ach test and saturated using water or hydroxyl-propyl methylcellulose (HPMC) pore fluid (providing a factor of 40 difference in permeability). The soil with HPMC provides a soil of comparatively lower permeability, while the water saturated soil represents a soil of comparatively higher permeability. Further information regarding the ESB container can be found in Bertalot, (2013). The container was designed for normal use at 50-g in the centrifuge, so does not represent a perfect boundary for the 40-g tests considered here. However, any unwanted boundary effects were minimised by placing the structure close to the centre of the container and far from the end walls (see Coelho et al. 2003). Physical properties of the HST95 sand are listed in Table 2 after (Lauder 2010).

Table 2. Physical properties of HST95 Congleton sand (Lauder 2010).

Property	Value
Specific gravity, G_s	2.63
D_{10}: mm	0.09
C_u (uniformity) and C_z (curvature)	1.9 and 1.06
Maximum void ratio, e_{max}	0.769
Maximum void ratio, e_{min}	0.467

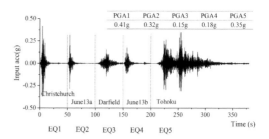

Figure 3. Input acceleration time history and peak ground acceleration (PGA).

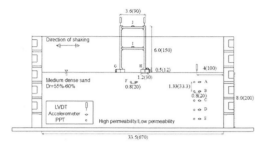

Figure 2. Layout of centrifuge test: dimensions in prototype are shown in m; dimensions in model scale are shown in mm in brackets.

The soil model was instrumented during pluviation with accelerometers and pore pressure transducers in five layers as shown in Figure 2 to measure the EPWP generation and accelerations in the free-field (full-depth, points A-E) and beneath the structures (at the same depth as free-field instrumentation, i.e. point F). Saturation using vacuum technique to reduce air bubbles in the soil in this specific case is not possible due to the presence of weakly glued latex membrane placed on the internal walls of the model container. This was instead achieved by allowing de-aired fluid to enter the model under a constant gravitational head through the bottom of the ESB container at a relatively low flow rate until it reached 2 mm above the model surface. Based on the slow speed of saturation (Bertalot 2013) and the volume of fluid entering the model it was possible to achieve conditions close to full saturation.

After loading the ESB container with saturated soil onto the centrifuge, the instrumented structure was placed carefully on the surface of the soil to be as level as possible; any initial tilt was measured using a clinometer to provide a baseline for subsequent measurements of structural rotation. MEMS Accelerometers were attached to the structures on each floor and at the foundations to measure the vertical and horizontal accelerations on each foundation and each storey (see Figure 2 for positions). Dynamic inter-storey sway and drift data were determined through careful filtering and integration of the accelerometer data. Two linear variable differential transformers (LVDTs) fixed on an overhead gantry were placed on top of model structure to measure average settlement and global rotation (tilt).

2.3 Dynamic excitation

Following spin-up, a re-ordered sequence of motions from the Canterbury Earthquake Sequence 2010–2011 (motions recorded at the Christchurch Botanical Gardens Station from Pacific Earthquake Engineering Research database) followed by a long duration 'double-pulse' motion rom the 2011 Tohoku Earthquake (recorded at Ishinomaki Station from National Research Institute for Earth Science and Disaster Resilience). These were applied in sequence using the Actidyn QS67-2 servo-hydraulic earthquake simulator (EQS) at the University of Dundee. The motions were filtered using an eighth order Butterworth filter with a band pass between 2.3–7.5 Hz (at prototype scale). The filtered time histories of input acceleration and peak ground acceleration (PGA) values are shown in Figure 3.

It was decided to apply the Christchurch Earthquake (February 2011) first as this was observed to have induced the most severe liquefaction-induced damage in the Canterbury Earthquake Sequence, and by occurring first, the initial state of the soil is fully known in the tests. Three subsequent motions from the same station ('June13a' from 2011, Darfield from 2010 and 'June13b', also from 2011) were then applied in order of expected lower EPWP generation, providing strong aftershocks generating different amounts of liquefaction. The last applied Tohoku motion was intended to be strong enough and with sufficient duration to fully re-liquefy the soil, even if the previous motions caused significant densification of the soil and reduction in liquefaction potential. This motion is also of particular interest as there are two distinct pulses of high PGA. Therefore, while in the initial Christchurch motion the early strong shaking will generate liquefaction, but thereafter apply smaller inertial demands to the structure, the final Tohoku motion can liquefy the soil and then apply a strong pulse in the liquefied state, which may be a more detrimental extreme loading case for the structure.

3 RESULTS

Results for EPWP generation and accelerations in the free-field and beneath the structure, structural response and foundation deformation will be discussed in this section. Some key performance indicators are summarised as follows: (1) excess pore pressure ratio in the free-field and beneath the structure; (2) peak storey acceleration at storey 1 (point I); (3) peak cyclic sway at storey 1; (4) peak cyclic inter-storey drift (i.e. where dynamic rocking induced displacements have been subtracted from sway data) across storey 1; (5) post-earthquake settlement and (6) structural tilt. Structural response quantities focus on storey 1 as a two-storey structure with uniform mass and stiffness distribution with height will see the largest structural deformation (inter-storey drift) within the first storey. This point is also close to the centre of mass of the structure. All data presented in this section are at prototype scale, unless otherwise stated.

3.1 Excess pore water pressure generation

Pore pressure transducers were located in the free-field and beneath the structure (Fig. 2). All EPWP readings were continuously recorded from before the start of earthquake shaking until the EPWP had fully dissipated. For the lower permeability test this required a relatively long recording time (4 minutes at model scale) compared to the higher permeability test. The readings have been corrected according to the placed initial position and instrument displacement during saturation (inferred from the static pore water pressures observed during spin-up) and between earthquake motions (based on any final static offset in the EPWP measurement after the EPWP had fully dissipated, i.e. $dr_u/dt = 0$).

Excess pore pressure ratio r_u is the maximum increase in EPWP divided by the in-situ effective vertical stress at the same depth. Figure 4 shows a comparison of free-field r_u with depth (round markers with black or grey line connected) and r_u beneath the structure (square markers without line connected) in the two tests. A reduction in r_u with depth can be generally observed in all EQs in the high permeability test (grey line) and in the aftershock motions EQ2-4 in the low permeability test (black line).

In the first earthquake, both soils show similar values of r_u down to 5 m depth. However, in all subsequent aftershocks, the higher permeability soil generally experienced much lower generation of EPWP due to the increased ability of the soil to rapidly dissipate EPWP as it is generated in a soil that is continually densifying during post-earthquake reconsolidation. This case only reaches full liquefaction near the surface in EQ1, and partial EPWP generation afterwards; however, the lower permeability soil reaches full liquefaction at essentially all depths in EQ1, and is fully re-liquefied at all depths in EQ5. The difference in EPWP resulted in greater soil softening in the lower permeability soil, which could be seen in the accelerations transmitted to the ground surface and on to the structure (Fig. 5).

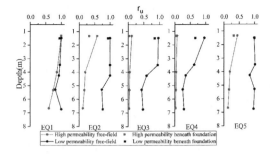

Figure 4. Excess pore pressure ratio (r_u) along depth.

By comparing square markers to free-field values in Figure 4, the EPWP generation beneath the foundation is generally smaller (or equal in stronger earthquakes) than the free-field value, which has previously been observed by Bertalot & Brennan (2015) for foundation-only models and attributed to the foundation bearing pressure increasing the confining effective stresses and inhibiting liquefaction.

3.2 Structural response

All accelerations were filtered through a high pass zero phase-shift filter. Examples of acceleration time history are shown in Figure 5. The acceleration transmitted to the top free-field accelerometer is shown in Figure 5(a). By comparing the black and grey overlap, it can be observed that less acceleration was transmitted to the top in the low permeability soil due to the much higher EPWP. Storey1 acceleration in Figure 5(b) also lower in the low permeability case due to soil softening caused by liquefaction.

Inter-storey sway across the first storey ('sway01') is the horizontal displacement of storey 1 relative to that of the foundations during shaking. The displacement was derived by high pass filtering and double integration of the accelerometer data at point G, H and I in Figure 2. The relative sway between storey 1 and the foundations was then derived by subtracting the average displacement of point H and I from point G.

The sway data is a combination of pure horizontal structural deformation (inter-storey drift, i.e. relative lateral movement between the two ends of the columns due to bending) and dynamic rocking induced displacement. Inter-storey drift (here 'Drift01') is a better indicator of the likelihood of structural damage to the building frame or any infill/curtain walling systems, as it is a direct measure of structural distortion. Dynamic rocking was derived through a high pass filtering of the rotation data (described in Section 3.3) to obtain only the dynamic shaking component, excluding the permanent rotation caused by bending deformation.

Figure 6 shows the peak sway, inter-storey drift and storey acceleration during each earthquake at or across storey 1. A general reduced structural dynamic response (by 10%–45%) is observed in the lower permeability case. This is consistent with the more

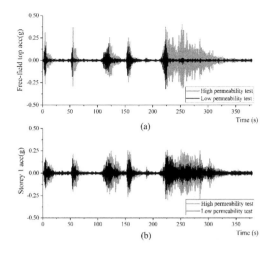

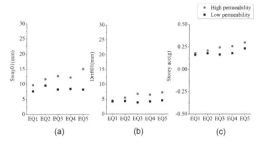

Figure 6. (a) Inter-storey sway01; (b) inter-storey drift01; (c) Storey acceleration.

Figure 5. (a) Free-field top acceleration time history; (b) Storey 1 acceleration time history.

extensive soil softening caused by greater EPWP generation in the lower permeability test. The weaker soil can transmit less acceleration so less inertial force was transferred to the structure resulting in this reduced response. It can also be seen that this reduction is much larger for sway (which includes the peak dynamic rotation) compared to inter-storey drift (which does not). This suggests, counter-intuitively, that there is greater rocking/dynamic rotation in the soil with lower EPWP, which dissipates more of the input energy to the structure, resulting in lower drift. This may be because the increased inertial actions in the structure out-weigh the potential for increased cyclic foundation vertical movement in soil with higher EPWP. The peak vertical acceleration of the foundations and the induced vertical displacement here are shown in Figure 7, indicating the dynamic rocking of the foundations during shaking in each case which support this conclusion.

It is also noticeable that similar dynamic response occurs in both soils in EQ1 for all structural measures, and the differences between responses become apparent in the subsequent earthquakes (aftershocks). This is consistent with both models exhibiting similar EPWP distribution in EQ1 (Fig. 4) and illustrates that differences in soil liquefiability may mainly complicate understanding of the response in aftershocks, i.e. in post-earthquake assessment of the resilience of an affected structure to future events.

3.3 Earthquake induced settlement and tilt

Settlement and tilt of a structure are measurements of the foundation performance during earthquakes. Large values would potentially influence the post-earthquake serviceability of a structure, even if the superstructure is largely undamaged. Both of these pieces of data were derived from synthetic LVDT data, because the LVDT alone cannot represent the shaking induced vertical displacement accurately. Figure 8 is an example

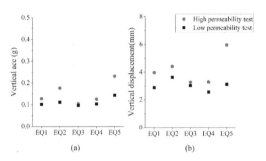

Figure 7. (a) Peak vertical acceleration on the foundation; (b) Peak vertical displacement on the foundation.

of synthetic LVDT data showing part of the time history in the final Tohoku motion (EQ5). The LVDT data was first low-pass filtered so that only the monotonic component was left in the data (Fig. 8(a)); then the double integrated vertical accelerometer data for the instruments mounted on the foundations, representing the vertical dynamic component (Fig. 8(b)) are added. The final synthetic LVDT data was used in interpreting settlement, (through averaging to values of the two foundations) and rotation (from the difference in measurements across the foundation).

Figure 9(a) shows greater post-earthquake relative settlement in the high permeability case (except for the last EQ), possibly caused by the greater acceleration transfer induced dynamic rocking and inertial force due to the lower EPWP generation at depth discussed in 3.1 and 3.2. This greater stamping effect in the higher permeability test appears to result in larger settlements than the sinking caused by high EPWP in the low permeability case, which is a surprising observation.

The higher permeability case in Figure 9(b) does, however, rotate relatively less compared to the low permeability case, and rotation is perhaps more critical to post-earthquake serviceability than gross settlement. This suggests that structural performance in liquefiable soil is a trade-off between larger structural demand but better foundation performance, or vice-versa, particularly in strong aftershocks.

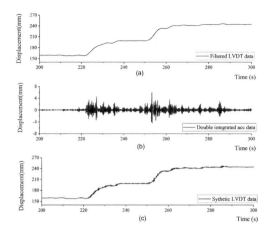

Figure 8. (a) Low-pass filtered LVDT data; (b) Double integrated acceleration data; (c) Synthetic (combined) LVDT data.

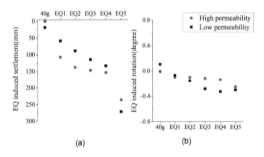

Figure 9. (a) Cumulative earthquake-induced settlement; (b) Cumulative earthquake-induced rotation.

4 CONCLUSIONS

This paper has investigated how soil permeability may influence the seismic response of low-rise structures for soils which are nominally liquefiable. A two-storey true multi-degree of freedom model was used in centrifuge tests so that both structural demand and foundation performance could be evaluated and compared, during a sequence of strong motions.

It was shown that in the first/initial earthquakes that both soils showed similar EPWP generation and structural demand. However, in subsequent strong aftershocks, soils of higher permeability exhibited much lower EPWP generation at depth, resulting in increased structural demand (by 10–45%, depending on the motion) and increased settlement due to greater foundation 'stamping'; however, post-earthquake rotation was reduced significantly.

These results indicate that even if the potential for liquefaction can be established through a liquefaction triggering analysis, it may remain difficult to estimate the consequences of an earthquake (particularly an aftershock) and therefore estimate damage potential, as the structural and foundation response depends on the EPWP that is developed.

REFERENCES

Bertalot, D. 2013. Foundations on layered liquefiable soils. PhD thesis, University of Dundee, Dundee, UK.

Bertalot, D. & Brennan, A.J. 2015. foundations Influence of initial stress distribution on liquefaction-induced settlement of shallow foundations. *Géotechnique*, 65(5), 418–428. doi: 10.1680/geot.SIP.15.P.002.

Coelho, P., Haigh, S.K & Madabhushi, S.G. Boundary effects in dynamic centrifuge modelling of liquefaction in sand deposits. *Proceedings of the 16th ASCE engineering mechanics conference*, University of Washington, Seattle, USA, 2003: 1–12.

Cubrinovski, M., Henderson, D. & Bradley, B. 2012. Liquefaction impacts in residential areas in the 2010–2011 Christchurch earthquakes. *Proceedings of International Symposium on Engineering Lessons Learned from the 2011 Great East Japan Earthquake*, 811–824.

Dashti, S., Bray, J.D., Pestana, J.M., Riemer, M.R. & Wilson, D. 2010. Centrifuge Testing to Evaluate and Mitigate Liquefaction-Induced Building Settlement Mechanisms. *Journal of Geotechnical and Geoenvironmental Engineering* 136(7): 918–929.

Dashti, S., Bray, J.D., Pestana, J.M., Riemer, M.R. & Wilson, D. 2010. Mechanisms of seismically-induced settlement of buildings with shallow foundations on liquefiable soil. *J. Geotech. Geoenviron. Engng.*, ASCE, 136(1): 151–164.

Knappett, J.A., Madden, P. & Caucis, K. 2015. Seismic structure–soil–structure interaction between pairs of adjacent building structures, *Géotechnique*, 65(5), pp. 429–441.

Lauder, K. 2010. The performance of pipeline ploughs The performance of pipeline ploughs. PhD thesis, University of Dundee, Dundee, UK.

Liu, L. & Dobry, R., 1997. Seismic response of shallow foundation on liquefiable sand. *Journal of Geotechnical and Geoenvironmental Engineering*, ASCE, 123(6), 557–567.

Muir Wood, D. 2004. Geotechnical modelling. London, UK: Spon Press, Taylor and Francis.

Liquefaction behaviour focusing on pore water inflow into unsaturated surface layer

Y. Takada
Graduate School of Engineering, Kyoto University, Kyoto, Japan

K. Ueda & S. Iai
Disaster Prevention Research Institute, Kyoto University, Kyoto, Japan

T. Mikami
Maeda Corporation, Tokyo, Japan

ABSTRACT: It has been pointed out that the effect of partially saturated surface layer cannot be ignored when ground is liquefied. In this research, a series of centrifuge tests was made to clarify the effect of pore water inflow into a partially saturated surface layer during liquefaction. Four tests were carried out by changing groundwater level and pore fluid type (water or viscous fluid). The extent of the expansion of saturated area during and after shaking was different among the cases. Comparison between liquefied and non-liquefied cases indicated that there were obvious differences: a part of the unsaturated layer changed into a saturated one due to the seepage. In addition to the tests, three phase numerical analyses which take the effect of pore air pressure into consideration were performed for the centrifuge tests. The simulation well captured the change of the degree of saturation and the response of excess pore water.

1 INTRODUCTION

Liquefaction of the ground is caused when the sandy ground beneath the ground water level was subjected to the cyclic shear stress under undrained condition. In 2011 off the Pacific coast of Tohoku earthquake, a part of Urayasu city was liquefied due to the main shock and aftershock. It is indicated that partially saturated surface layer was changed into the saturated one after the main shock, which enhanced the liquefaction damage during the aftershock. The key to solve this mechanism is the seepage effect between partially and fully saturated layers during and after liquefaction.

In order to clarify this behaviour, many researchers have studied. Uzuoka et al. (2001) discussed the effect of the unsaturated seepage characteristics. The numerical results showed that the seepage to partially saturated soil caused the delayed dissipation of excess pore water pressure in the subsurface liquefied soil. Yoshikawa & Noda (2013) have shown the soil-water-air coupled finite deformation analysis and its application to an unsaturated embankment on a clayey ground. This simulation result indicated the change of the degree of saturation, the excess pore water pressure and air pressure. However, there have been few attempts to consider the seepage into an unsaturated surface layer due to liquefaction.

In this research, a series of centrifuge tests and two-dimensional effective stress analyses are made to clarify the effects of an unsaturated surface layer on liquefaction behaviour during and after earthquake.

The change of the excess pore water pressure and the degree of saturation is focused in this study. In the centrifuge experiment, the degree of saturation is monitored by a soil-water moisture sensor. Numerical analysis is used to simulate the experimental results and to evaluate the distribution of the degree of saturation.

2 CENTRIFUGE TESTS

2.1 Geotechnical centrifuge

In this study, the geotechnical centrifuge at the Disaster Prevention Research Institute (DPRI), Kyoto University, was used. The rotation radius, defined as the length from the rotation axis of the arm to the centre of the model, is 2.5 ± 0.05 m. The maximum centrifugal accelerations are 200G for static tests and 50G when using a shaking table.

A rigid container (450 mm long × 150 mm wide × 300 mm deep) was used in this study. The front face of the container was transparent so that the sand layers are directly observed. In this model, six accelerometers (produced by SSK, Co., Ltd., A6H-50), six pore water pressure transducers (produced by SSK, Co., Ltd., P306A-2) and one moisture sensor (produced by Campbell Scientific, Co., Ltd., C-CS650) were used (Fig. 1). Soil moisture sensor was put in the unsaturated layer to measure the change of the degree of saturation. The rod length is 300 mm (Fig. 2).

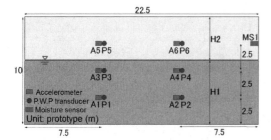

Figure 1. Schematic view of experimental setup.

Figure 2. Soil moisture sensor.

Table 1. Properties of Silica Sand No. 7.

Maximum void ratio e_{max}	1.2
Minimum void ratio e_{min}	0.7
Mean particle size D_{50}	0.13
Coefficient of uniformity Uc	1.9
Specific gravity Gs	2.62

2.2 Model setting

The procedure for the construction of the model is shown below:

(1) The model ground (height: 20 cm in model scale) was made by Silica Sand No. 7 (Tab. 1) with air pluviation in all cases. The relative density was set to 50%.
(2) The model was fully saturated under a vacuum (close to −0.1 MPa) after injecting carbon dioxide.
(3) A given amount of pore fluid was drained from the bottom in the gravitational field (at 1G).

2.3 Test cases

Test cases are summarized in Table 2. Four tests were carried out by changing groundwater level and pore fluid type (water or viscous fluid). H1 and H2 in Table 2 indicate the thickness of saturated and unsaturated layers, respectively, as shown in Figure 1. In the centrifugal model tests, considering both the fluid flow and dynamic motion, a viscous fluid (e.g. metolose solution) is generally used as a pore fluid. However, in the case of unsaturated soils, the suction level is

Table 2. Test cases.

	Ground (H1: H2)	pore fluid	input wave
Case 1	6m: 4m	water	300gal 30s
Case 2	7m: 3m		
Case 3	6m: 4m	metolose	300gal 40s
Case 4	7m: 3m		400gal 60s

affected by the metolose solution whose surface tension is less than that of water (Okamura et al., 2013, Higo et al., 2015). Because both water and metolose have merits and demerits, two kinds of pore fluid were used. Note that the location of the moisture sensor was constant in all cases.

2.4 Results

Figure 3 shows the excess pore water pressure (E.P.W.P) response of Cases 1~4. In Cases 1 and 2, E.P.W.P quickly dissipated within about 30s. The maximum E.P.W.P was low and did not reach the initial overburden pressure, which indicated that the ground was not liquefied. This must be due to the use of water as pore fluid. In Case 3, the maximum E.P.W.P (P1, P3) reached 40~60% of the initial overburden pressure, and the dissipation speed of E.P.W.P (P1, P3) was slower compared to the two cases described above. However, the ground was not liquefied. Finally in Case 4, the saturated layer was liquefied because P1 and P3 responses reached the initial overburden pressure. Taking a look at the response after shaking, it took long time for E.P.W.P to dissipate and did not completely reach 0 kPa. It is highly possible that the pore water transducer was subsided due to the liquefaction.

The Time history of degree of saturation is given in Figure 4. The measured time is only for 1,000s due to the sensor's capacity. The change of the degree of saturation during and after shaking was totally different: In Case 1, no change was observed. This is because the E.P.W.P was too small. In Case 2, the rate of saturation increase was large during shaking even if the amount of E.P.W.P was much the same as Case 1. This is because the moisture sensor was located near the G.W. level. In Case 3, the tendency for the change of the degree of saturation was different from the cases described above. The increase of saturation during shaking was not seen, while the degree of saturation increased after shaking. It is noted that an increase of saturation during shaking might be seen if the moisture sensor was located near the G.W. level. However, this fact was not confirmed yet. At last in case 4, the change of the degree of saturation was occurred not only during shaking, but also after shaking. The degree of saturation began to decrease from 400 s (Sr: 84%→83%). It may be attributed to a stop of the water supply from the saturated layer into the unsaturated one as shown in the decrease of pore pressure response at P5 in Case 4 (Figure 3). In this way, when the ground below the water level was liquefied, the degree of saturation

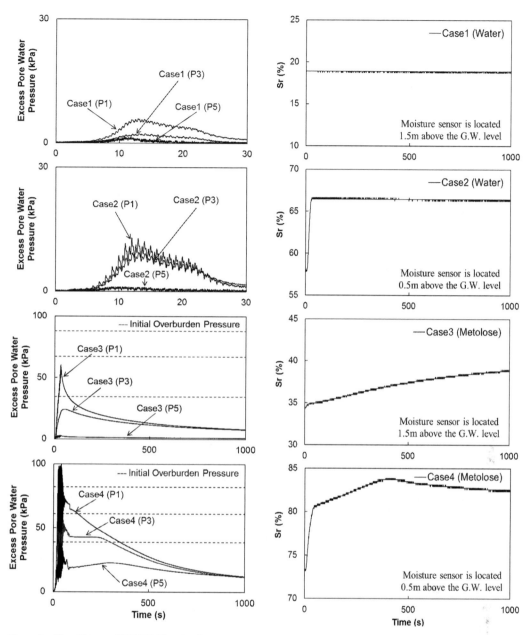

Figure 3. Time history of E.P.W.P (Case 1~4).

Figure 4. Time history of degree of saturation.

of unsaturated layer above the water level increases largely and continues to rise up for a long time.

3 EFFECTIVE STRESS ANALYSIS

3.1 Overview

In addition to the centrifuge model tests, a series of 2D effective stress analyses was conducted by using a 2D effective stress analysis program, FLIP (Finite element analysis of Liquefaction Program), based on the finite element method. In this program, a strain space multiple mechanism model (Iai et al., 2011), called Cocktail glass model, is used as a constitutive model to consider the nonlinearity of soils. The model consists of a multitude of simple shear mechanisms with each oriented in an arbitrary direction, and can describe the behaviour of granular materials under complicated loading paths, including the effect of rotation of principal stress axes.

In order to carry out seismic response analyses of unsaturated ground, three phase numerical analyses which take the effect of pore air pressure into consideration were performed. The governing equations are

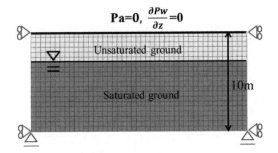

Figure 5. Mesh used in the analysis and boundary condition.

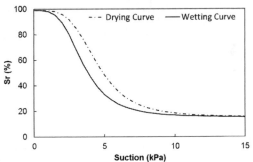

Figure 6. Soil-water characteristic curve of Silica No. 7.

Table 3. Analysis parameters.

Initial shear modulus G_0, (kPa) at Pa	5.46×10^4
Initial Bulk modulus K, (kPa) at Pa	1.42×10^5
Mean effective confining stress Pa, (kPa)	73.5
Coefficient of confining stress dependency m	0.5
Internal friction angle ϕ_f, (deg.)	35.99
Phase transformation angle ϕ_p, (deg.)	28
Cohesion c, (kPa)	0
Maximum dumping coefficient H_{max}	0.24
Coefficient of permeability of water k_f, (m/s)	1.46×10^{-4}
Coefficient of permeability of air k_a, (m/s)	1.0×10^{-2}
Mass density ρ, (kN/m³)	2.092
Porosity n	0.487

as follows:

$$div\sigma + \rho g = \rho \ddot{u} \quad (1)$$

$$n\chi(\dot{\rho}_a - \dot{\rho}_f) + nS_r\dot{\rho}_f/K_f + S_r div\dot{u} = -div[k^f(-grad\rho_f + \rho_f g - \rho_f \ddot{u})] \quad (2)$$

$$-n\chi(\dot{\rho}_a - \dot{\rho}_f) + n(1-S_r)\dot{\rho}_a/K_a + (1-S_r)div\dot{u} = -div[k^a(-grad\rho_a + \rho_a g - \rho_a \ddot{u})] \quad (3)$$

where σ is total stress, ρ is overall density of three phase material, **g** is gravitational acceleration, **u** is the displacements of the solid matrix, n is void ratio, χ is specific soil water content, ρ_f is the pore water density, p_f is pore water pressure, p_a is pore air pressure, S_r is the degree of saturation, K_f is the bulk modulus of pore water, $\mathbf{k}^f$ is the coefficient of permeability, ρ_a is the pore air density, $\mathbf{k}^f$ is the coefficient of permeability of air, K_a is the bulk modulus of pore air.

3.2 Setup

As shown in Figure 5, the ground (saturated layer + unsaturated layer) is numerically modeled for the FEM analysis, and boundary condition is given. This analysis has three steps below. First, initial steady state flow analysis is conducted to calculate the distribution of saturation at first. In this phase, the proper saturation is determined. Calculated saturation is saved for next phase of analysis. The second phase of the analysis is initial gravity analysis that provides initial condition for dynamic analysis, and the saturation distribution remains the same as the first phase. This phase corresponds to the physical consolidation of the ground. Following the initial gravity analysis, dynamic analysis is conducted by using the achieved input wave in the centrifuge test, and gravity correction due to the change of saturation is added.

The parameters used for analysis are shown in Table 3. The soil-water characteristic curve of Silica Sand No.7 is shown in Figure 6. This graph is derived from the result of Toyoura sand whose particle size distribution is similar to Silica Sand No. 7.

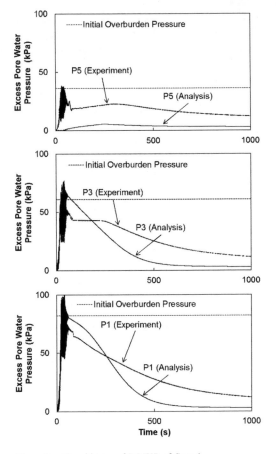

Figure 7. Time history of E.P.W.P of Case 4.

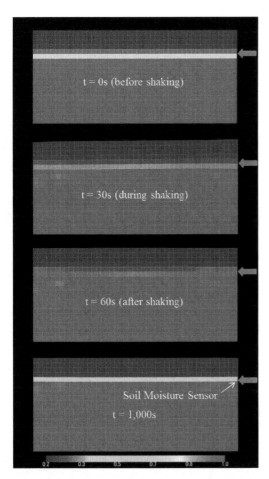

Figure 8. Distribution of saturation in Case 4.

3.3 Results

The time history of the response E.P.W.P of Case 4 is shown in Figure 7. The simulated responses in the saturated layer (i.e. P1, P3) are comparable with the experimental result. Although the dissipation speed is slightly different from the experimental results, the rate of increase and the maximum of the response E.P.W.P are well captured. However, some difference can be recognized between the analysis and experiment in the unsaturated layer (i.e. P5). The difference may be caused by an installation mistake. It is supposed that the pore water transducer was installed lower than the target height shown in Figure 1.

Figure 8 shows the distribution of saturation in Case 4 during and after shaking. It is seen that the degree of saturation in the unsaturated layer just above the ground water table (or just below the moisture sensor) increases during and after shaking (78% at 0 s, 82% at 30 s, 90% at 60 s and 97% at 1,000 s). This phenomenon well captured the experimental results. The red zone, which indicates saturated zone, seems to be expanded. It is highly possible that this mechanism is related to the compression of the pore air in unsaturated layer.

4 CONCLUSION

To investigate the pore water inflow into partially saturated layer due to liquefaction, a series of centrifuge tests and two dimension effective stress analyses were conducted in this research. In the centrifuge tests, two points were confirmed. First, when the saturated ground was liquefied, the pore water in it infiltrated into the partially saturated layer and caused the change of the degree of saturation. Second, the change of the degree of saturation in the unsaturated layer is dependent on the excess pore water pressure and the ground water level. In addition to the tests, three phase numerical analyses which take the effect of pore air pressure into consideration were performed for the centrifuge tests. Based on the governing equation of unsaturated soil by utilizing the soil-water characteristics curve and coefficient of permeability of air, the simulation well captured the response of excess pore water pressure and the degree of saturation. The numerical analysis also showed the distribution of the degree of saturation. According to this result, a part of the partially saturated layer, which is located close to the saturated ground, changed into the saturated one. The past study revealed that the increase of the degree of saturation could cause the decrease of shear strength. Therefore, it is important to consider the seepage in to partially saturated layer when liquefaction potential is evaluated for ground including partially saturated layer.

REFERENCES

Higo Y., Chung-Won Lee, Doi T., Kinigawa T., Kimura M., Kimoto S. and Oka F. 2015. Study of dynamic stability of unsaturated embankments with different water contents by centrifugal model tests. *Soils and Foundations* 55(1): 112-126.

Iai, S., Matsunaga, Y., and Kameoka, T. 1992. Strain space plasticity model for cyclic mobility. *Soils and Foundations* 32(2): 1-15.

Iai, S., Tobita, T., Ozutsumi, O. and Ueda, K.. 2011. Dilatancy of granular materials in a strain space multiple mechanism model. *International Journal for Numerical and Analytical Methods in Geomechanics* 35(3): 360-392.

Okamura M., Tamamura S. and Yamamoto R. 2013. Seismic stability of embankments subjected to pre-deformation due to foundation consolidation. *Soils and Foundation* 53(1): 11-22.

Towhata, I. and Ishihara, K.. 1985. Modeling Soil Behavior Under Principal Stress Axes Rotation. *Proc. of 5th International Conf. on Num. Methods in Geomechanics. Nagoya*, Vol.1: 523-530.

Uzuoka, R., Kubo, T., Yashima, A. and Zhang, F. 2001. Liquefaction analysis with seepage to partially saturated layer. *JSCE* 694/III-25: 153-163 (in Japanese).

Yasuda, S., Harada, K. and Ishikawa, K. 2012. Damage to structures in Chiba Prefectures during the 2011 Tohoku-Pacific Ocean Earthquake. *Japanese Geotechnical Journal* 7(1): 103-115 (in Japanese).

Yoshikawa, T. and Noda, T. 2013. Soil-water-air coupled finite deformation simulation of an unsaturated soil structure during construction and during/after a seismic motion. *25th Annual Conference of JSCE Chubu Branch*, 23-28 (in Japanese).

16. Dams and embankments

Performance of single piles in riverbank clay slopes subject to repetitive tidal cycles

U. Ahmed & D.E.L. Ong
Research Centre for Sustainable Technologies, Swinburne University of Technology Sarawak Campus, Sarawak, Malaysia

C.F. Leung
Department of Civil and Environmental Engineering, National University of Singapore, Singapore

ABSTRACT: This paper presents the performance of a single pile embedded in a soft riverbank clay slope subject to repetitive tidal cycles in a series of 50-g centrifuge model tests. The slope movements were triggered by 5 m tidal cycles at a 3.5 day interval. In-flight T-bar tests were also carried out periodically to study the strength development of the clay slope. Upon interpretation, the clay slope and the embedded pile showed complex time-dependent behaviour. Over the course of the tidal cycles, it was observed that the pile bending moment initially increased but subsequently reduced over time. In contrast, the pile head deflection would steadily increase over time unabatedly while the slope creeps towards the river. This research has shown that piles installed in soft riverbank clay slope subject to repeated tidal fluctuations over time should be designed for strength in short-term undrained condition and for serviceability in long-term condition.

1 INTRODUCTION

Cyclic loading conditions defined in terms of non-static repetitive soil loading conditions are observed in a variety of geotechnical situations. A common situation where quasi-cyclic behaviour is observed is that of cyclic fluctuation of water levels. This condition is observed when the soil body is influenced by external conditions such as infiltrations due to a storm, seasonal freezing and thawing of ice, reservoir impounding/drawdown and seasonal/tidal variation in river water levels. In the latter case, the influence and effect on the soil, especially sloping ground, can be detrimental to foundations embedded in affected areas. Several authors have reported such cases of structural distress due to fluctuation in water level where the continuing fluctuations result in the cumulative lateral movement of the soil (Sullivan 1972, Ting & Tan 1997; Thomas & Margeson 2004; Johansson & Edeskär 2014; Lee et al. 2014). These reports show a certain susceptibility of soft clay slopes with low undrained shear strength to repetitive lateral soil movements.

Most of the studies to date focus on the short-term behaviour of slopes due to a particular case of drawdown or rising water level. An instrumentation program carried out by Lee et al. (2014) showed the typical behaviour of clay slope subject to tidal fluctuation. The lateral movement was closely related to the external water level where during low tides, maximum lateral movements occurred. Over a period of time, the continuing fluctuations result in accumulated movement of the slope. More relevant and significant work has been carried out by Wong et al. (2016). Physical models of a river bank with a slope of 1V:3H was used to simulate a tide fluctuation of 5 m in 7 days with piles installed at various locations along the crest. The results showed that there was an initially large movement during the reconsolidation period of testing, resulting in large bending moments before the commencement of water level fluctuations. Repetitive cycles showed that the soil movement and the pile head movement crept asymptotically.

Table 1. Test configurations.

Test	Pile Locations	Free Length, A mm	Embedded Length, B mm	Distance from toe, C mm
1	Toe	100	200	0
2	Mid slope	50	250	150
3	Crest	0	300	300

2 CENTRIFUGE TESTS ON MODEL RIVERBANK SLOPES

A series of centrifuge model tests were carried out at the National University of Singapore Geotechnical Centrifuge Laboratory. The testing program (see Table 1) consisted of a kaolin clay slope (1V:3H)

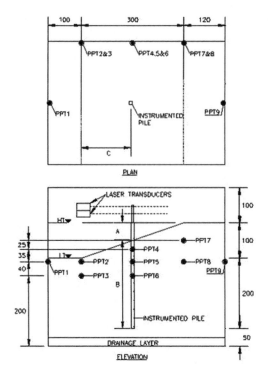

Figure 1. Test setup in mm (model scale).

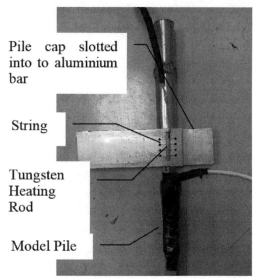

Figure 2. Pile cap and aluminium attachment used for in-flight installation of pile.

subjected to repetitive water level fluctuations of 5 m in 3.5 days (prototype). Model piles were installed after 20 cycles of fluctuations at different locations. Three additional tests were carried out to monitor the shear strength and pore pressure developments at different locations of the slope before the tests and after 50 and 100 cycles of fluctuation.

2.1 *Model setup and test*

The model was prepared in a 520 mm by 450 mm container with an internal standpipe (Wong et al. 2016). The water height in the container could be controlled by moving a water tank fixed to the back panel of the container via an actuator. The tank was connected to the internal standpipe with screw holes cut at every 25 mm.

Kaolin clay was mixed with water in mass ratio of 1:1.2 and simultaneously de-aired for four hours. The properties of kaolin clay used are given in 2.2. The container was heavily greased while the front Perspex window was covered in polystyrene film and vacuum greased. The bolt holes in the drainage standpipe were screwed and the bottom drainage valve was opened.

Fully de-aired Druck PDCR81 and TML KPE-PB models of pore pressure transducers (PPT) were attached to the container walls. A 20 mm thick sand layer sandwiched in geotextile was placed at the bottom for drainage. The mixed slurry was carefully poured into the container underwater and loaded with 50 kg dead weights for 3 days. The container and sample were then spun at 50 g for 12 hours and the pore pressure and surface settlement which was monitored until the soil reached 95% consolidation.

After consolidation, the front Perspex was removed and multi coloured flock powder rained on to the kaolin front face before fixing the Perspex back onto the container. Aluminium guides made of 5 mm thick plates were then gently pushed into the soil so that the desired slope inclination could be achieved. The soil above the guides was carefully scraped off.

The bolt plugs in the drainage standpipe above the soil level were then removed and water allowed to fill up to the crest height. The container was then placed in the centrifuge and the two Wenglor (CP24MHT80) non-contact laser transducers with an effective measurement range of 40–120 mm were installed. An aluminium cap was slotted into an aluminium bar and held together by a piece of cotton string. This prevented lateral deflection of the pile during installation and also enabled the pile cap to be used to measure pile head movements. A tungsten heating rod was also installed so that by supplying a voltage to the heating rod, the string could be burned off and the aluminium cap could be released. Additionally, the pile could be screwed onto the pile cap (see Figure 2). In order to carry out in-flight T-bar tests at different locations, the loading frame consisted of a rotary motor attached to a movable platform. A hollow square aluminium pile of 10 mm was used in the tests. The pile was fixed with 11 strain gauges at 25 mm intervals in bending configuration along its length. The pile height was 300 mm with bending stiffness EI of 2.15×10^5 kNm^2 which represent a 600 mm diameter concrete bored pile.

After setup of the model slope and the loading frame, the water level in the container was allowed to stabilise with the water tank at the back. The sample was then spun for 8 hours to reconsolidate. During the

Table 2. Properties of remoulded Kaolin Clay specimens. (Jiahui 2013).

Properties	Mean Value
Compression index	0.236
Swelling index	0.040
Specific gravity	2.637
Plastic limit	35.3%
Liquid Limit	76.5%
Plasticity Index	41.2%
%Clay	96.9%
%Silt	3.1%
%Sand	0%
Permeability (m/s) 100 kPa	5.33×10^{-8}

initial spin up, the changes in the radial acceleration caused the water level to be higher at the crest.

After reconsolidation, the water tank was allowed to fluctuate by 100 mm within two minute cycles. Using the scaling law of $1/N^2$ (where N is the scaling factor) for consolidation dependent time and $1/N$ for length, this can be scaled to 5 m of fluctuation within 3.5 days. After 20 cycles of fluctuation, the tank was maintained at a high position (high tide situation) so that the pile could be installed. The rate of pile installation was determined based on the dimensionless velocity group proposed by Finnie (2003) defined as:

$$\frac{vB}{C_v} \quad (1)$$

where v = velocity, B = appropriate length dimension and C_v = coefficient of consolidation. Using a velocity of 3 mm/s, pile width of 10 mm and C_v of 40 m²/year, the value of $vB/C_v = 24$ which falls within the velocity region where undrained conditions prevail.

The pile was driven in and once the pile reached the required position, the tungsten heating element was powered on which burned the string attaching the pile to the aluminium bar. The bar was then removed and a waiting period of 20 minutes was allowed for any soil movements to stabilise. Additional 80 cycles of fluctuations were carried out before test completion and the bending moment and pore pressures were monitored.

2.2 Shear strength characterisation

Three tests were carried out to observe the shear strength development over successive cycles and to determine the residual shear strength. Inflight T-bar tests (Stewart & Randolph 1991) were carried out at the crest, mid-slope and toe locations. The T-bar was inserted and extracted into the soil at a constant rate of 3 mm/s. Initial T-bar tests carried out before the water level fluctuation showed that there was 1 m of over consolidated crust over the slope. The strength gain over depth was higher for the toe (5.89 kPa/m) and lower at the mid-slope (3.9 kPa/m) and crest (3.66 kPa/m) (see Figure 3). This could be attributed to the fact that

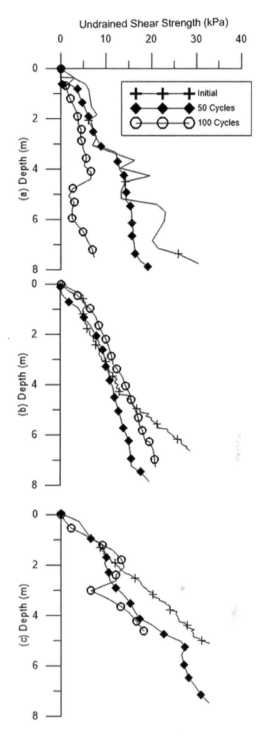

Figure 3. Development of shear strength over 100 cycles of water fluctuation at (a) Crest, (b) Mid-slope and (c) Toe.

the pre-overburden pressure was higher at the slope toe and reduces at the crest. This was due to the slope preparation procedure where at the toe 110 mm (5.5 m in prototype) of soil was scraped while at the crest

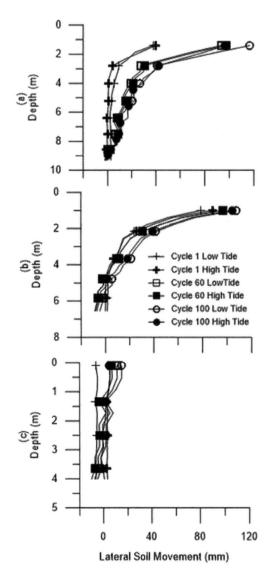

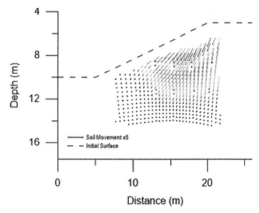

Figure 5. Soil movement profile after 100 cycles of fluctuation (prototype scale).

2.3 Soil movement

The soil movements were analysed via Particle Image Velocimetry. A SJCAM 5000 model camera was used to capture high-definition videos at a frame rate of 24 frames per second. High-resolution images were extracted for the required durations at 1-second intervals and the relevant images selected for PIV processing. GeoPIV-RG (Stanier & White 2013) was used to process the images. The results showed that the largest soil movements occur near the surface of the slope crest. Figure 4 shows the variation of soil movement with time for three locations along the slope. During the drawdown of water level, the soil moved towards the toe and during the subsequent rise of water level, the movement tend to reverse towards the crest. A clear slip plane was observed to have developed during the first drawdown and the subsequent cycles (Figure 5).

During the initial drawdown, the mid-slope tend to achieve the highest lateral soil movement while the crest lateral movement was comparatively low. Over time, the cumulative movement at both the crest and mid-slope reached 120 mm. Subsequently, the toe tend to show minimal movement as the slip plane did not reach the toe location.

Figure 4. Lateral soil movement profiles at (a) Crest, (b) Mid-slope and (c) Toe.

only 10 mm (0.5 m in protype) was scraped. Unless otherwise stated hereafter, all dimensions stated are in prototype scale.

The results showed that with consecutive cycles, the depth of over consolidated crust significantly reduced for the crest and mid-slope locations. In the case of the crest, the 2 m thick crust was unidentifiable after 100 cycles as was the case for the 1 m thick crust at mid-slope locations. Alternatively, the toe developed a 2 m thick layer over the 100 cycles which was over consolidated, possibly due to repetitive loading and unloading by water. This is of significance as over consolidated clay tends to develop negative pore pressures when sheared which can increase the effective soil strength thus reducing the bending moments.

2.4 Generation and dissipation of pore water pressures

Pore pressures were monitored via PPT's at locations shown in Figure 1. An additional PPT 1 (not shown) was used to monitor the water level inside the container. Figure 6 shows the variation of the pore pressures with fluctuation cycles. Similar to the soil movement, the pore pressure with each cycle of drawdown and rising water levels showed a decrease and an increase of the water level. It was observed that the pore pressures at the crest decreased slightly (negative excess pore pressures are generated) while pore pressures at the toe and mid-slope showed more significant but similar behaviour. The pore pressure in a slope undergoing water level fluctuation was influenced by

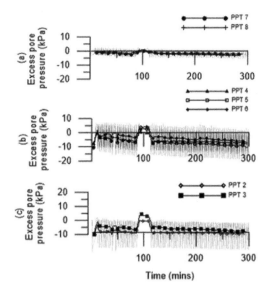

Figure 6. Excess pore water pressure variations over time.

(1) groundwater head and the seepage flow forces and (2) the cyclic shearing of the soil body resulting in a decrease in pore water pressure and increase in effective stresses. The soil shearing across the pile could also influence the pore pressures but since the PPT's were located 7.5 m away from the pile, this effect was not observed in the various tests. Comparing the data from the PPT at different depths of each location, it was apparent that within the shearing zone of the slip circle and closer to the surface, the excess pore pressures generated was higher. In the case of mid-slope, PPT 4 which was closest to the surface showed an average excess pore pressure of 10 kPa while PPT 6 which was closer to the slip plane but yet further from the surface, shows an average of 3.5 kPa at the end of the test. This suggested that the pore pressure generated was due to the cyclic shearing of the over consolidated surface crust.

Clays have low permeability and the contribution from seepage may be lower than that from shearing. Since the drawdown was relatively rapid, pore pressures developed in over consolidated soil did not have a chance to effectively dissipate. This was especially evident in the initial drawdown where pore pressures dramatically dropped but did not attain the initial values.

2.5 Soil-pile interactions

Pile bending moment profiles are shown in Figure 7 at specific intervals after pile installation. The behaviour at the three locations showed different trends in the short-term and long-term. The highest bending moment at first cycle occurred at the mid-slope (120 kNm) while the toe showed the lowest (71 kNm). In the case of mid-slope, the bending moments tend to decrease over time with a final maximum value of 36 kNm. Alternatively, the crest showed

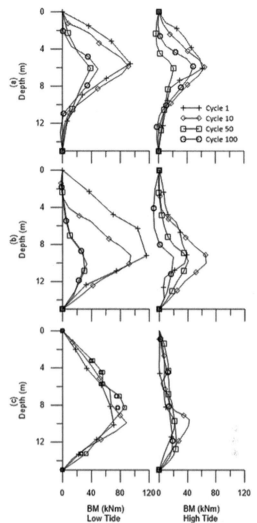

Figure 7. Bending moment profiles at specific days after pile installation at (a) Crest, (b) Mid-slope and (c) Toe.

an increase in the bending moment for the first few cycles, reaching a peak of 94 kNm after four cycles and subsequently started to decrease over time to a maximum value of 40 kNm. The marked difference in the behaviour at the crest and other locations could be attributed to the slope soil movement after a period of stabilisation where the crest locations showed a smaller movement compared to the mid-slope, but reached a final value as that of the crest in the long-term.

It was interesting also to note that at the mid-slopes and the toe, the bending moment did not reach zero at the surface. This could be explained by looking at the induced bending moments at the toe where soil movement was small but the moments tend to be significant. The two forces acting on the pile at any location would be the lateral soil movement and the drag from water flow. Since the lateral soil movement was small at the toe location, this suggested that there was a drag

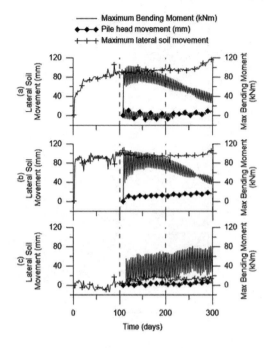

Figure 8. Maximum pile bending moment, pile head movement and maximum soil lateral movements at (a) Crest, (b) Mid-slope and (c) Toe.

force being applied on the free length of the pile which contributes to the pile head movement. Large movements of water levels especially in flooding cases, has shown to have developed such drag forces on piles of bridge abutments (Liu et al. 2007; Wang et al. 2015) but further research is required to quantify the forces involved.

Figure 8 shows that the soil movement and the pile head movement recorded continuously increasing trend. The developed slip circle could result in a weakened region around the slip plane, which could allow the back and forth movement of the soil within the slip circle with a cumulative movement towards the direction of the river.

3 CONCLUSIONS

The results from the series of tests carried out to simulate the effect of river bank soil movements on pile behaviour shows that the influence of repetitive loading on a soft soil slope could be detrimental to embedded foundations in the long-term. The changes in soil strength during successive cycles of fluctuation plays an important role in the short-term and long-term behaviour. While piles need to be designed for the larger bending moments in the short-term, particularly following its installation, long-term serviceability issues also need to be considered as they could lead to potential failure.

REFERENCES

Finnie, I.M.S. 2003. *Performance of shallow foundations in calcareous soils, phd thesis*. Perth, University of Western Australia.

Jiahui, H. 2013. *Cylic and post cyclic behaviour of soft clays, phd thesis*. Singapore, National University of Singapore.

Johansson, J.M.A. & Edeskär, T. 2014. Effects of external water-level fluctuations on slope stability. *Electronic Journal of Geotechnical Engineering*, 19 K, 2437–2463.

Lee, L.J., Kaniraj, S.R. & Taib, S.N.L. 2014. Effect of tidal fluctuation on ground movement and pore water pressure. *Tunneling and Underground Construction*, 35–44.

Liu, S.X., Li, Y.C. & Li, G.W. 2007. Wave current forces on the pile group of base foundation for the east sea bridge, china. *Journal of Hydrodynamics, Ser. B*, 19, 661–670.

Stanier, S.A. & White, D.J. 2013. Improved image-based deformation measurement for the centrifuge environment. *Geotechnical Testing Journal*, 6, 915–927.

Stewart, D.P. & Randolph, M.F. 1991. A new site investigation tool for the centrifuge. *Proc. Int. Conf. On Centrifuge Modelling, Centrifuge 91*. Rotterdam, Netherlands, Balkema.

Sullivan, R.A. 1972. Behaviour of wharf by effected of river fluctuations. *Journal of the Soil Mechanics and Foundation Division*, 98, 939–954.

Thomas, J.E. & Margeson, G.R. 2004. Impacts of sediment mounds under pile supported wharf structures. *Ports 2001*, 27612, 1–11.

Ting, W. & Tan, Y.K. 1997. The movement of a wharf structure subject to fluctuation of water level. *14th International Conference on Soil Mechanics and Foundation Engineering*. Rotterdam, Netherlands, Balkema.

Wang, Y.H., Zou, Y.S., Xu, L.Q. & Luo, Z. 2015. Analysis of water flow pressure on bridge piers considering the impact effect. *Mathematical Problems in Engineering*, 2015, 1–8.

Wong, S.T.Y., Ong, D.E.L., Leung, C.F., Arulrajah, A., Evans, R. & Disfani, M.M. 2016. Centrifuge model study of pile subjected to tidal induced soil movement. *Asiafuge. Second Asian Conference on Physical Modelling in Geotechnics*. Shanghai, China.

Load transfer mechanism of reinforced piled embankments

M.S.S. Almeida
Federal University of Rio de Janeiro, COPPE, Rio de Janeiro, RJ, Brazil

D.F. Fagundes
School of Engineering, Federal University of Rio Grande – FURG, Rio Grande, RS, Brazil

M.C.F. Almeida
Federal University of Rio de Janeiro, POLI, Rio de Janeiro, RJ, Brazil

D.A. Hartmann
Federal University of Pampa, Unipampa, Alegrete, RS, Brazil

R. Girout, L. Thorel & M. Blanc
IFSTTAR, GERS, GMG, Bouguenais, France

ABSTRACT: A series of centrifuge tests was performed at the IFSTTAR centrifuge on piled embankments with basal geosynthetic reinforcement. The objectives of these studies were to assess the influence of pile spacing, embankment height, pile cap size and number of layers of geosynthetic and stiffness on the load transfer mechanism. Preliminary tests studied the influence of the level of pretension on the geosynthetic as well as the number of layers of geosynthetic, one or two adequately spaced. The measurements of the forces on the piles made it possible to assess the load transfer mechanisms, and then 100% efficiency was achieved in most tests performed.

1 INTRODUCTION

The load transfer mechanisms of piled embankment reinforced with a basal geosynthetic platform are shown in Figure 1(a). The arching effect is defined as the part of the embankment load directly transferred to the piles, and the remaining total load not transferred by arching effect is the vertical stress applied on the subsoil and the basal Geosynthetic Reinforcement (GR). The GR in tension allows the transference of this remainder of the load back to the piles. This mechanism is called the membrane effect and its magnitude depends on both the support provided by the soil beneath the geosynthetic layer and on the GR stiffness (Van Eekelen et al. 2013). All these mechanisms are strictly dependent on the area ratio values $\alpha = \pi \cdot d^2 / 4s^2$, where d is the pile diameter, or the cap diameter if there is one, and s is the pile spacing (Fig. 1b). In the present paper, the load transfer efficiency E is the result of the arching and the membrane effect combined. The efficiency E is the ability of the embankment to transfer the load F to the piles, as defined by:

$$E = \frac{F}{(\gamma H + w_s)s^2} \quad (1)$$

A Mobile Tray Device (MTD), used in the present study, was especially developed to study in a centrifuge the behaviour of piled embankments subjected an imposed settlement of the soft ground (Rault et al. 2010).

The aim of this paper is to analyse the influence of different types of geometries and geosynthetic stiffness on the load transfer mechanisms for embankments with area ratios α in the range 2% up to 20%. With this purpose centrifuge tests were performed, combining two pile diameters d, three pile spacings s, six embankment heights H and two geosynthetic stiffness J. The influences of the level of pretension on the geosynthetic as well as the number of geosynthetics used were also assessed.

2 CENTRIFUGE MODELLING OF PILED EMBANKMENTS

2.1 Materials and methods

The scaled models were tested in the IFSTTAR centrifuge at a g-level N equal to 20, but data and results are presented in prototype values.

The granular soil representing the embankment material is placed on a perforated steel mobile tray,

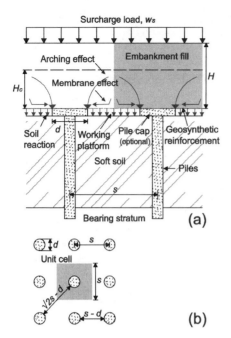

Figure 1. Schematic representation of load transfer mechanisms in piled embankments reinforced with geosynthetics.

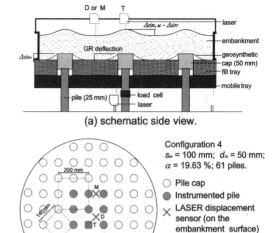

Figure 2. Schematic view of the mobile tray with locations of the instrumentation for CF 4, $s_m = 100$ mm, $d_m = 50$ mm; (a) cross section; (b) top view.

as shown in Figure 2. The settlement of the working platform is simulated by the vertical displacement of the tray $\Delta\omega_m$ controlled by two LASER displacement sensors (Fig. 2b). When the tray descends (Fig. 2a), the load transfer mechanisms shown in Figure 1 are triggered.

The circular tray (inner diameter 900 mm) is perforated with 61 holes, through which piles can be inserted (Fig. 2b). Some holes can be closed to simulate different mesh configurations. The mesh configurations used two pile diameters d, 0.5 m and 1.0 m in prototype scale; and three values of pile spacing s, 2.0 m, 2.8 m and 4.0 m in prototype scale.

Nine central piles (Fig. 2b) were instrumented with load cells to measure the total force F transferred to the piles. Mean values of F (scatter less than 10%) were used for the efficiency analysis. Three LASER displacement sensors were placed on the top of the embankment to measure the settlement at the surface of the embankment (Fig. 2), but these measurements are outside the scope of this paper.

The geosynthetic layer placed between the mobile tray and the embankment is shown in Figure 2(a). Two different woven polypropylene geosynthetic materials, Geolon® PP-25 and Geolon® PP-60, were used to study the influence of the reinforcement tensile stiffness on the membrane effect. The representative stiffness of GR PP25 and PP60 in the prototype are $J1 = 3.86$ MN/m and $J2 = 16.8$ MN/m, respectively.

2.2 Experimental program

The experimental programme consisted of tests with five geometrical configurations (CF) and two geosynthetic reinforcements with different tensile stiffness, as listed in Table 1. These configurations (Fig. 2) combined three pile spacings (s) and two pile diameters (d) with three different area ratios (α). Configurations 1 and 2 had the same area ratio (4.9%) but different values of s and d. Embankment heights evaluated varied from 0.7 m to 7.2 m in prototype scale.

2.3 Tests to assess geosynthetic pretension and number of geosynthetic layers

Some tests were performed on geosynthetics with and without pretension in configurations CF0 and CF1. The pretension level was chosen based on its maximum tensile strength, and arbitrary values of 0.2 and 1% were chosen in order to maintain the deformations close to those observed in the field. The influence of one and two layers of geosynthetics was also investigated. The two geosynthetic layers had spacing of 10 mm in model scale, hence 20 cm in prototype scale, about one standard layer of compacted soil.

3 TEST RESULTS

3.1 Influence of geosynthetic pretension and number of geosynthetic layers on efficiency

Efficiency curves were plotted relative to the tray displacement (Δ_w) normalized by the pile diameter d. To verify the pretension influence on efficiency four plots

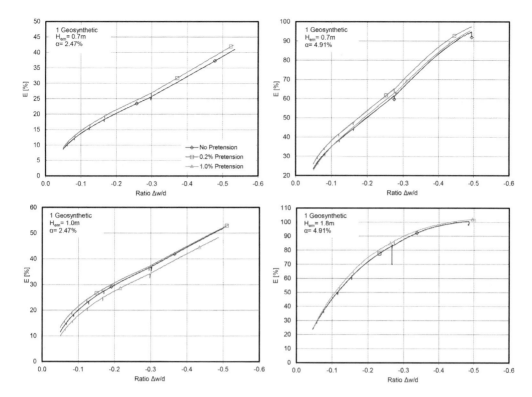

Figure 3. Influence of reinforcement pretension on efficiency.

Table 1. Test configurations.

Configuration	s (m)	d (m)	α (%)
CF0	2.8	0.5	2.47
CF1	2.0	0.5	4.9
CF2	4.0	1.0	4.9
CF3	2.8	1.0	9.8
CF4	2.0	1.0	19.6

with different embankment heights and cover ratios are shown in Figure 3. All plots are related to tests with one reinforcement layer.

From these plots, it can be concluded that the geosynthetic pretension had little to no influence on the load transfer efficiency. This may be linked to the small pretension values acting upon the geosynthetic, as one could expect that by applying a greater pretension from the beginning of the test should lead to greater values of efficiency.

The addition of a geosynthetic layer to piled embankments is a step that should further improve its efficiency, as it helps to develop a membrane effect, redirecting the embankment weight not supported by the arching effect to the piles. The influence of the number of geosynthetic reinforcements on efficiency is analyzed in Figure 4 by means of four plots in which embankment heights and cover ratios vary.

From these graphs it is clear that the inclusion of a geosynthetic layer leads to larger efficiency values for all embankment heights and cover ratios studied here. For the highest cover ratio efficiency reaches 100% at larger tray displacements. A general increase in efficiency of about 30% is noticed for most cases, except for the case with the highest cover ratio, which presented an increase of about 80%.

It is interesting to notice the development of the membrane and arching effects on the graphs, especially for small embankments. With no geosynthetic the efficiency starts to increase as arching develops, but as the tray reaches a certain value the arching ceases and the efficiency drops. This problem is clearly amended when a geosynthetic layer is added.

As for the use of a second reinforcement layer, its influence is not as marked as the inclusion of the first layer. There is some influence however, and it is more important for tests with lower cover ratios.

3.2 Efficiency E versus basal embankment settlement

Figure 5 presents typical data of load efficiency E versus basal embankment settlement $\Delta\omega$. Fig. 5(a) shows the results of the tests performed with the embankment height $H = 1.0, 1.8, 3.2$ m for the three configurations with the same d and the geosynthetic stiffness $J1$. Figure 5(b) shows the tests performed with the embankment height $H = 1.8$ m and geosynthetic stiffness $J2$ for the four configurations studied. Based on Fig. 5 it is seen that for most configurations (e.g. CF 1,

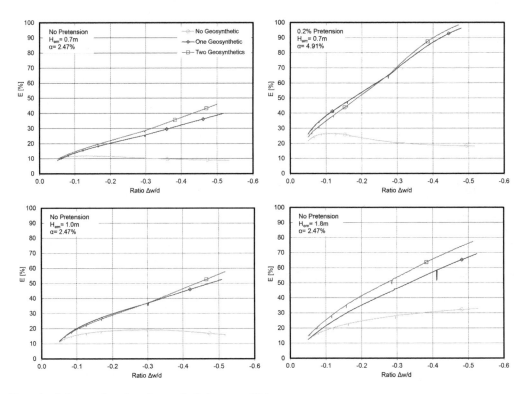

Figure 4. Influence of number of geosynthetic layers on efficiency.

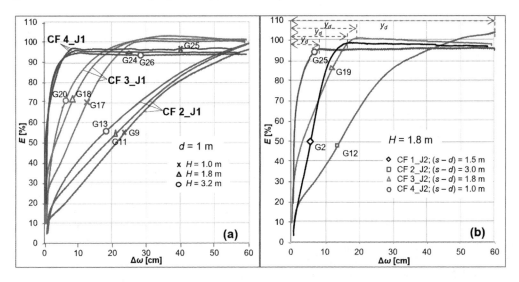

Figure 5. Load efficiency E versus $\Delta\omega$ for the tests conducted with: (a) $J1$ and $H = 1.0$, 1.8 and 3.2 m for configurations with $d = 1.0$ m (b) $J2$ and the same value of H (1.8 m) for all configurations.

2 and 3), the value of the load efficiency E increased, reached a peak value, and then a plateau with a constant value while the basal embankment settlement increased continuously. However, the tests with CF 2 did not reach a post peak plateau.

For these tests, which have the largest clear span between piles ($s - d = 3.0$ m), it seems that the mobile tray did not go down enough so that a constant value of efficiency could not be reached. In Fig. 5 all the tests resulted in efficiency values close to 100% (E_{max}).

The more the basal embankment settlement increases, the more the geosynthetic deforms and is placed under tension, thus the efficiency values increase until the entire available load is transferred to

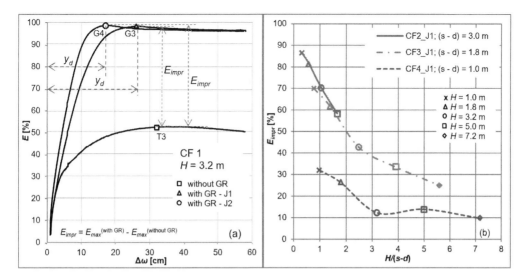

Figure 6. Differences in the load efficiency, between tests conducted with and without geosynthetic, E_{impr}.

the piles (E approximately 100%). The basal embankment settlement $\Delta\omega$, at which E_{max} was reached, is the main difference between the tests. As E_{max} is reached the vertical stress applied on the subsoil has decreased to zero, i.e., no stress is applied on the soft soil. In other words, the geosynthetic is no longer in contact with the tray and the GR maximum deflection (y_d) has been found as indicated in Figure 5(b).

Values of y_d increased with the increasing clear span between the piles $(s-d)$. Figure 5(b) shows that the value of y_d increases slightly as H increases. The tests performed with the stiffer reinforcement J2 for the same configurations resulted in similar behaviours, as shown in Figure 5(b), but the values of E_{max} were obtained at smaller values of basal embankment settlements $\Delta\omega$, i.e., at smaller values of y_d. For tests with the same embankment height and pile spacing CF1_J2 and CF4_J2 (Fig. 5b), the larger pile cap increases the arching mechanisms (Fagundes et al. 2015 and Girout et al., 2016) decreasing the vertical stress on the reinforcement geosynthetic and, consequently, the magnitude of the maximum deflection. Comparing tests with $\alpha=4.9\%$ in Figure 5(b) (CF1_J2 and CF2_J2), the y_d increases with an increasing clear span $(s-d)$.

3.3 Improvement of efficiency by reinforcing with geosynthetics

Fagundes et al. (2015) and Girout et al. (2016) presented results of non-reinforced centrifuge tests with the same geometrical configurations (s, d, H, α), soil characteristics and experimental procedures adopted in the present paper (see T3 in Fig. 6(a)). These results are used here to evaluate the influences of the geosynthetic and the GR stiffness on load efficiency and differential settlements. Fig. 6(a) presents the differences in load efficiency for tests with and without geosynthetic, $E_{impr}=E_{max}^{(with\ GR)}-E_{max}^{(without\ GR)}$, for the test with configuration 1 and $H=3.2$ m. In Fig. 6(b) the E_{impr} is plotted against the embankment height normalised by $(s-d)$ for the configurations with the same diameter d and geosynthetic $J1$.

Figure 6 shows that the geosynthetic reinforcement always improves efficiency, as also observed by Chen et al. (2008), Abusharar et al. (2009), Blanc et al., (2013, 2014) and Fagundes et al. (2017). As observed in Fig. 6(a), the efficiency E increases with the increase of the basal embankment settlement $\Delta\omega$ in the same proportion for the tests with and without GR, for values of $\Delta\omega<3.5$ cm. However, E_{max} is reached at lower basal settlements for tests with GR than for those without GR. It must be noticed that the differences observed on the results with and without GR may not only be due to the membrane effect but also to the fact that the presence of the GR may improve arching within the embankment (Van Eekelen et al., 2013).

The efficiency improvement E_{impr} is evidently influenced by the clear span $(s-d)$ and the embankment height H (Fig. 6(b)). This improvement in efficiency increases for a larger clear span and a lower embankment height, i.e., when the arching mechanism are less effective, $H<H_{arch}$. It was observed for the tests performed with the same configuration that the influence of the GR stiffness on the E_{impr} was negligible. However, as mentioned before, E_{max} was reached at lower basal settlements for tests with $J2$ (Fig. 5b and Fig. 6a).

4 DIFFERENTIAL SURFACE SETTLEMENT

The settlements at the embankment base and embankment surface are key factors in understanding the behaviour of piled embankments. Figure 7 presents photos, at the end of the tests (after a decrease in g), of the embankment surface for a test without differential

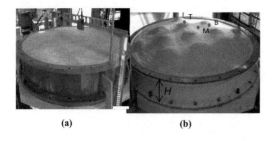

(a) (b)

Figure 7. Surface view of the embankment at the end of the tests: (a) Test with no differential settlement: CF3, $H = 3.2$ m and $J1$; (b) Tests with differential settlement at the top of the embankment: CF2, $H = 1.8$ m and $J1$.

settlement (Figure 7a) and for a test of low embankment where differential settlements were observed (Figure 7b). The commonly required serviceability state condition is zero differential surface settlement.

The test results indicated that surface differential settlement decreases with increases of both embankment height and area ratio (or decrease in pile spacing). Thicker embankments showed negligible surface differential settlement Δu. The surface differential settlements were more dependent on the relationship between the clear span and the embankment height (H_c) than on the presence of the GR.

The stiffness of the reinforcement reduced the magnitude of Δu but did not affect the critical height. The stiffer reinforcement $J2$ leads to a greater reduction in the Δu than the less stiff reinforcement $J1$. The present series of tests showed that the ratio of the normalized height of embankment by the clear span $H/(s-d)$ corresponding to the critical height (embankment height for zero settlements) is around 2.1 but lower values of $H/(s-d)$ may result in zero surface settlement depending on the relevant area ratio.

5 CONCLUSIONS

This paper presents a study of load transfer mechanisms and deformation of piled embankment with basal GR based on the results of twenty-eight centrifuge tests. A wide range of geometric parameters (H, s and d) and geosynthetic stiffnesses J were varied, covering typical values used in practice.

Preliminary tests analyzed the number of geosynthetic layers (none, one or two) and pretension levels (0%, 0.2% and 1%) of the maximum tensile strength of the geosynthetics. The results indicated that geosynthetic pretension levels showed little influence on the efficiency. It was also observed that the inclusion of a geosynthetic layer improves the efficiency, especially for the lower embankment height adopted here, 0.70 m prototype scale. However, the inclusion of a second reinforcement layer did not improve the efficiency as much as the inclusion of a single layer, and therefore it is not recommended based on the present tests program.

The experimental results showed that the membrane effect combined with soil arching provides values of efficiency regarding load transfer to the piles close to 100% for most tests. The efficiency increased with tray displacement ($\Delta \omega$) until reaching the 100% efficiency plateau. The improvement in efficiency by adding a GR layer, E_{impr}, increases the less effective the arching mechanism is, e.g., $H < H_{arch}$ and the larger the pile mesh is, E_{impr} being influenced more by the height and clear span than by the stiffness of GR. Therefore, the application of GR is of greater importance than the stiffness of the GR in the improvement of the load transfer mechanisms in piled embankments.

REFERENCES

Abusharar, S.W., Zeng, J.J., Chen, B.G., Yin, J.H., 2009. A simplified method for analysis of a piled embankment reinforced with geosynthetics. Geotextiles and Geomembranes 27, pp 39–52.

Blanc, M., Rault, G., Thorel, L., Almeida, M.S.S., 2013. Centrifuge investigation of load transfer mechanisms in a granular mattress above a rigid inclusions network. Geotextiles and Geomembranes, 36: 92–105.

Blanc, M., Thorel, L., Girout, R., Almeida, M.S.S., 2014. Geosynthetic reinforcement of a granular load transfer platform above rigid inclusions: Comparison between centrifuge testing and analytical modelling. Geosynth. Int. 21 (1): 37–52.

Chen, Y.M., Cao, W.P., Chen, R.P., 2008. An experimental investigation of soil arching within basal reinforced and unreinforced piled embankments. Geotextiles and Geomembranes 26, 164–174.

Fagundes, D.F., Almeida, M.S.S. Girout, R, Thorel, L. and Blanc, M., 2015. Behaviour of piled embankment without reinforcement. Proc. Inst. Civ. Engs – Geotechnical Engineering 168 (6): 514–525.

Fagundes, D.F., Almeida, M.S.S. Thorel, L. and Blanc, M., 2016. Load transfer mechanism and deformation of reinforced piled embankments Geotextiles and Geomembranes 45 (2017) 1–10.

Girout, R., Blanc, M., Thorel, L., Fagundes, D. F., Almeida, M. S. 2016. Arching and Deformation in a Piled Embankment: Centrifuge Tests Compared to Analytical Calculations. Journal of Geotechnical and Geoenvironmental Engineering, 04016069.

Rault, G., Thorel, L., Néel, A., Buttigieg, S., Derkx, F., Six, G., Okyay, U., 2010. Mobile tray for simulation of 3D load transfer in pile-supported earth platforms. In 7th ICPMG Int. Conf. on Physical Modelling in Geotechnics (Springman, Laue, and Seward (eds)). Taylor & Francis, Zurich, pp. 261–266.

Van Eekelen, S.J.M., Bezuijen, A., Van Tol, A.F., 2013. An analytical model for arching in piled embankments. Geotextiles and Geomembranes, 39, 78–102.

Experiments for a coarse sand barrier as a measure against backwards erosion piping

A. Bezuijen
Ghent University, Ghent, Belgium
Deltares, Delft, The Netherlands

E. Rosenbrand & V.M. van Beek
Deltares, Delft, The Netherlands

K. Vandenboer
Ghent University, Ghent, Belgium

ABSTRACT: Backward erosion piping can occur when an unfiltered exit is present on the downstream side of a levee and seepage forces are sufficient to transport sand grains. A pipe forms, which progresses upstream, in the opposite direction of flow, below the levee. Small-scale model tests have shown that a coarse sand barrier can be effective to prevent piping. This paper describes some of these tests. These tests are part of a research programme that also includes medium-scale, large-scale tests and numerical calculations. The small-scale tests confirm that the coarse sand barrier is very effective at this scale and offers possibilities for prediction at larger scale. This paper describes some practical issues related to the physical modeling of the barrier and how these affect the results. In addition, scale effects are discussed.

1 INTRODUCTION

1.1 Backward erosion piping

Backward erosion piping is described in Vandenboer et al., (2015): Backward erosion piping is an important failure mechanism for cohesive water-retaining structures founded on a sandy aquifer. A local disruption of the downstream top layer leads to concentrated seepage flow towards the opening. This entails high local hydraulic gradients causing upward forces on the sand grains, which may result in the onset of erosion at that particular location (pipe initiation). The erosion process continues in the upstream direction, resulting in the formation of shallow pipes in the sand layer (pipe progression). These pipes do not collapse because of the bridging nature of the overlying cohesive material.

The term 'backward erosion piping' refers to the growth direction of the pipes, which is opposite to the flow direction, i.e. from downstream to upstream. Eventually, the pipe forms a direct connection between upstream and downstream, which leads to a facilitated water transport and to the action of accelerated erosion. The pipe finally reaches unbridgeable dimensions resulting in a (partial) collapse (Fig. 1).

1.2 The coarse sand barrier

The idea of the coarse sand barrier (CSB) is to stop the backward erosion process. This is realized by filling

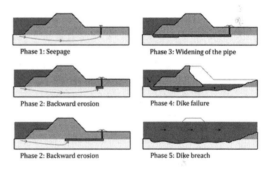

Figure 1. Backward erosion piping (Van Beek, 2011).

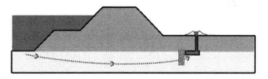

Figure 2. Principle of the coarse sand barrier.

a trench of limited width and depth with coarse sand, which is less erodible than the surrounding fine sand. The filled trench is covered with cohesive material. In this way, the seepage underneath the dike will hardly increase, while the material is more resistant to erosion (van Beek et al., 2015), see Figure 2.

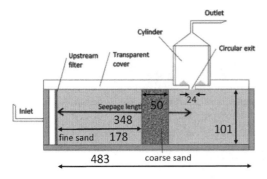

Figure 3. Set-up of the small-scale experiments. The width is 300 mm. Dimensions in mm.

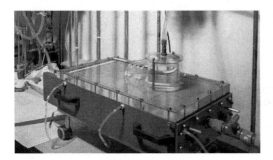

Figure 4. Setup with horizontal container.

Figure 5. Set-up with the narrow high box.

1.3 Research programme

Backward erosion piping is considered the most important failure mechanism for the levees in the Netherlands (Van Beek, 2015) and also is mentioned as a dangerous failure mechanism in the USA along the Mississippi (USACE, 1956), in China (Yao et al., 2009, Cao, 1994) and in Italy (García Martínez et al., 2016). The CSB is potentially a cost-effective intervention to prevent piping, based on natural materials. Therefore, a research programme has been initiated to investigate the potential of this method. This research programme consists of small-scale model tests, medium-scale model test, nearly full-scale model tests and numerical simulation (Koe-lewijn et al., 2017). This paper describes the first phase: some of the small-scale model tests.

2 SMALL-SCALE MODEL TESTS

2.1 Set-up

Two set-ups have been used for investigation of the CSB at small scale: a flat box (L×W×D = 483×300× 101 mm) for which a sketch of the set-up is shown in Figure 3 and a narrow higher box (L×W×D = 483×101×300 mm), which is shown in Figure 5. The general concept and seepage length are the same for both set-ups.

The flat box, consisting of a PVC container with an acrylate lid, is filled with sand, see also Figure 4. The acrylate lid has an outlet hole of 24 mm diameter with a cylinder to collect the sand that is removed from the container by erosion. Two experiments have been conducted with the narrow high box, consisting entirely of acrylate, to investigate the effect of scale on the results, from which one is presented in this paper. The set-up with this box is quite comparable to what is shown in Figure 3, however the thickness of the sand layer is increased from 101 mm to 300 mm and the width is decreased from 300 to 101 mm.

Standpipes were installed at different locations on the lid and on the bottom of the container to measure the hydraulic head. The locations in the lid of the flat box are shown in Figure 6. In the narrow high box a similar pattern of capillaries was installed, but without number 16 and 19.

2.2 Test process

During the experiments, a constant head difference is maintained between the inlet and outlet of the set-up. If there is no sand transport, the head is increased after 5 minutes. In case of sand transport, the head is kept constant for 5 minutes after the transport has stopped. When the sand transport stops, the head difference is increased again. The head difference at which the pipe grows into the barrier is considered as one of the critical gradients, since the barrier is damaged at this point. The tests were continued past this point in

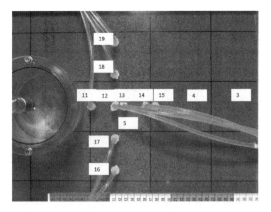

Figure 6. Standpipes connected to the model and numbering used. Locations with only a number mean that the standpipe is connected there with the bottom of the container. Upstream at right side (in all other figures on the left side).

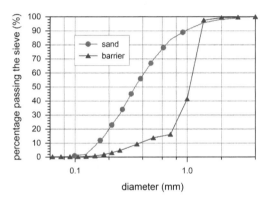

Figure 7. Grain size distribution of the sand and the barrier material.

Table 1. Parameters of the sand used.

Parameter	Barrier	Fine sand
Density grains (kg/m^3)	2650	2650
min. void ratio (–)	0.48	0.43
max. void ratio (–)	0.7	0.7
permeability (m/s)	$2.6\ 10^{-3}$	$3\ 10^{-4}$
d_{50} (mm)	1.06	0.342

order to also capture the head difference at which the pipe progresses through the barrier. When an erosion pipe has crossed the CSB and the pipe in the fine sand progresses upstream the test was stopped.

2.3 Model preparation

This paper presents the results of 3 tests with one type of barrier sand and one type of finer background sand. The depth and relative density of the barrier were varied. The background sand used in the experiments had an average grainsize of 0.342 mm, The average grainsize of the coarse barrier sand was 1.05 mm. The grainsize distribution of both the background sand and barrier sand are shown in Figure 7.

The barrier sand needs to be significant coarser than the background sand, because then the larger grains will only erode at larger gradients. However, if the ratio between the grainsizes is too large, the fine sand will penetrate into the barrier and decrease its permeability. The sand upstream of the barrier had a relative density of 95%. The density of the barrier varied in the experiments. The sand downstream of the barrier had a relative density of around 60%, except for the test in which the barrier had a lower relative density. Then the relative density was 40%. Other properties are mentioned in Table 1.

To prepare the sand model, the container was tilted 90 degrees in a way that the inlet pointed downwards, the cover that is on the top in this turned position is removed and the first sand layer is rained in (the part left from the coarse sand in Figure 3). The sand is placed underwater in thin layers (thickness 0.05 m) in case the highest density has to be reached (relative density 95%) and it was rained homogeneously over the whole depth when the relative density was lower than 65%. It was observed that some fines are in suspension and the suspension is filtered off in order to prevent the settling of fines in the layers. This part is densified to the desired relative density. Densification for relative densities up to 65% was done by dropping the box over a few centimeters and in this way applying a shockwave through the sand and the box. This method has shown to result in a homogeneous density throughout the sample (Van der Poel and Schenkeveld, 1998). In case a higher relative density was needed, the different layers were densified by tamping the sand under water. Then the coarse sand of the CSB is applied and densified until the necessary density.

For tests where the barrier did not continue all the way to the bottom of the model, both the sand below the barrier (when the box is returned in the testing position) and the sand of the barrier were placed at the same time by keeping a custom-fitted dividing plate between the barrier and the sand. After placing the materials this was withdrawn and the materials were densified.

On top of that, the last sand layer is applied (the sand layer on the right handside in Figure 3). This last sand layer has a lower density compared to the other layers. This is necessary, because densification of that layer would also densify the other layers. After this procedure, the cover is mounted again, the container is turned back 90 degrees, the standpipes are connected and the set-up is ready for testing.

One test was performed without a coarse sand barrier. This test was used to be able to compare the results with and without barrier.

3 RESULTS

3.1 Typical course of test

At a relatively low head difference of 6 cm, the sand starts to erode and can be noticed in the cylinder of

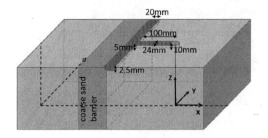

Figure 8. Sketch of the development of erosion pipes during a test. The dimensions are indicative only.

top of the set-up. Increasing the head further leads to progression of the pipe, which grows from the opening in the acrylate plate to the barrier. With increasing head difference, this pipe widens at the barrier and forms a T-shape as sketched in Figure 8 and described by Negrinelli et al. (2016). Further increasing of the head difference leads to progression of the pipe into the sand barrier. Subsequently as the head is further increased, several other pipes form into the barrier and the pipes in the barrier lengthen. Generally, one or two pipes lengthen most in the barrier and reach the upstream end. Then a further increment in head is typically required before the pipe continues to erode in the upstream sand. From that point, the pipe progresses to the upstream end without further increments in head difference.

3.2 Results horizontal container

Test 191 can be seen as a reference test: the coarse sand barrier depth is over the whole depth of the container, as indicated in Figure 3 and it is densified to a relative density of 90%. In Test 192, the barrier is only applied over the upper 30 mm, but with a comparable relative density. In Test 193, the barrier is applied over the whole depth but the relative density is only 40%. Test 194 is the test without barrier.

Failure occurred in Test 191 and 192 at the same difference of piezometric head (560 mm) and at a higher head difference for test 193 (670 mm). Figure 10 shows that the head difference between standpipes 3 and 4 is the same for Test 191 and 192, since the same fine sand is used, this means that also the discharge must be comparable. This is the case, see Table 2. The head difference and discharge are larger for Test 193 (with the barrier with a lower density). We noticed a filter cake forming upstream of the barrier. The fines that formed this filter cake were fine sand grains that migrated from the upstream sand probably resulting in sand with a larger permeability upstream. In all tests, the drop in piezometric head is small for the standpipes 11, 12 and 13. Indicating that the flow resistance in the erosion pipes is small.

The critical head difference for the tests with barrier was taken as the head difference where the erosion pipe starts to grow into the barrier. This head difference can be compared with the head difference that leads to ongoing erosion in the test without barrier, Test 194.

Table 2. Results of 5 tests. Critical head difference (ΔH) and corresponding discharge. See also text.

Test	ΔH mm	Q cc/min	Description
191	560	917.6	Reference test with barrier
192	560	987.2	Shallow barrier (30 mm height)
193	670	1185.6	Non compacted barrier
194	125	359	Test without barrier
198	350	635	Narrow high box setup

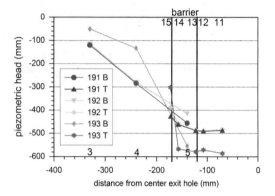

Figure 9. Piezometric head measured on top (T) and bottom (B) of the model container in 3 different tests at critical head for the barrier. Numbers indicate the standpipes as shown in Figure 6. See also text.

It is clear that the critical head difference for Test 194 is much lower than for the other tests.

Figure 9 shows the piezometric head measured at various positions for Test 191, 192 and 193 at the moment a pipe grew into the barrier. The piezometric head was taken zero at the inlet. The total head difference and discharge at that moment for these 5 tests is presented in Table 2. Compared with the critical head of 120 mm measured in tests without a barrier, it is clear that the barrier has a large influence.

In the tests shown in Figure 9 most head loss is over the fine sand section upstream from the barrier (between x = −400 and −175 mm). When the pipes have formed as indicated in Figure 8, the head loss over the barrier is relatively small, because the barrier consists of permeable sand and the pipes caused by erosion are also permeable.

3.3 Results narrow high box

The test with the narrow high box presented here is shown in Figure 10. In this test, there was a 30 mm deep barrier with a width of 50 mm over the full width of the container. See Figure 10. The relative densities were comparable to those in Test 191.

The measured critical head and discharge at the critical head (when the pipe grows into the barrier) are significantly lower than for the other tests with a barrier. See Table 2. The ratio between the discharge and the head difference is the same as in Test 192, which

Figure 10. Top view of the narrow high box setup and side view of detail showing the barrier, indicated with the white dashed line, Test 198.

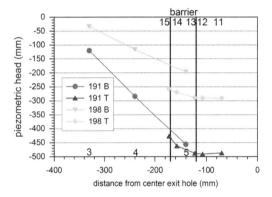

Figure 11. Piezometric head as function of x critical head difference (the moment the pipe grew into the barrier). Test 198 with the narrow high box compared with Test 191 in the flat box.

means that the permeabilities of the used sands are comparable. The course of the piezometric head as a function of the position at the moment the pipe grew into the barrier is shown in Figure 11. The measured piezometric head is compared with the head measured in Test 191, the 'reference test' also at the moment the pipe grew into the barrier. It is clear that the difference between the piezometric head measured at the top of the container (T) and measured at the bottom (B) is larger in the narrow high box test and that the total head difference is lower.

4 ANALYSING THE RESULTS

4.1 Clogging as a function of density of the barrier

Remarkable is the pressure drop at the upstream side of the barrier measured at the top site of the barrier during Test 193. There is a significant head loss between the standpipes 15 and 14 although the distance between these two standpipes is only 15 mm. The most likely explanation is clogging of the barrier sand by a fine fraction of the sand upstream. In this test, the barrier sand has a low relative density, which means that the

dimension of the pores between the grains is larger than for densified sand. Consequently, the fine sand can penetrate between the larger grains, creating an area with a rather low permeability. This means that without densification of the coarse sand barrier. The fine sand may block the pores in the barrier leading to an area with a permeability that is even lower than the permeability in the fine sand, which explains this high drop in piezometric head.

4.2 Flow regime in Test 191 and consequences

For Test 191, with a densified barrier over the full height of the container, the configuration of the erosion pipes is as sketched in Figure 8 just before an erosion pipe will grow into the CSB. The largest drop in piezometric head is over the fine sand, as shown in Figure 9. The piezometric head in the barrier is constant at the inlet and nearly constant in all locations of the barrier and close to the value of the piezometric head in the outflow area.

This qualitative description of the flow explains why such a barrier is quite effective to prevent erosion. The sand upstream of the barrier is still fine sand with a relatively low permeability, thus the discharge through the barrier and consequently the gradient in the barrier are relatively small (compared to the situation that all sand was replaced with the coarser barrier material). However, the removal of the coarse particles from the CSB requires a relatively high gradient. Furthermore, the erosion pipe along the barrier at the downstream side of the barrier causes a relative large outflow area over the whole width of the barrier. The flow lines will not concentrate to one pipe tip, as will happen without a barrier. With less flow concentration to one point, the flow velocities will be lower and again a higher average head difference across the sand bed is necessary to start the erosion in the CSB. This explains the significant difference in head drop at the critical gradient for the tests with and without barrier as shown in Table 2. With CSB the head drop is 2.8 to 5.4 times higher than without CSB.

4.3 Setup with narrow high box

The description of the flow through the setup as presented in the previous section is also the reason for the tests with the narrow high box. It is reasonable to assume that for this box the depth of the pipe parallel to the barrier is the same. However, the height of the soil sample is now 3 times larger and the width 3 times smaller. This means that flow resistance of the setup, which is to a large extend determined by fine sand upstream of the barrier, is more or less the same for the normal position and the narrow high box tests. However, the outflow area of the barrier, which is mainly to the erosion pipe parallel to the barrier, is now 3 times less and consequently the flow velocity in the outflow area of the barrier is roughly 3 times higher. It is reasonable to assume that in both setups erosion of the barrier starts when the flow velocity in the outflow area exceeds a certain threshold and becomes high enough to transport the grains from the barrier. This could mean that the erosion in the barrier of the narrow high box starts at a 3 times lower head difference. In Test 198 there was only a shallow barrier of 30 mm. This means that the flow situation is a bit more complicated and the flow through the pipes is not simply 3 times higher. However, as can be seen in Table 2, both the discharge and the head difference over the set-up are considerably smaller in case of the narrow high box. A factor $560/350 = 1.6$ was found in critical head between Tests 191 and 198.

5 SCALING

The consequence of the erosion mechanism as described above is that scaling of the tests to what can be expected in prototype is not straightforward. At a larger scale, the barrier will be deeper, which means that for the same hydraulic gradient there will be more water flowing to the barrier. Consequently, the flow from the CSB to the erosion pipe will be larger at a larger scale and the barrier may be less effective in field scale conditions than in the small-scale model tests described in this paper. This was already proven to some extent in the experiments with the narrow high box. To test this further, experiments are foreseen at different scales. Medium scale tests in a container with all dimensions 4 times larger than the dimensions of the small-scale tests described here and large-scale tests in a large wave flume of Deltares with a seepage length of 15 m (Koelewijn et al., 2017).

6 CONCLUSIONS

Small-scale model tests on the influence of a coarse sand barrier (CSB) on backward erosion piping have been performed as a part of a larger research project. Based on the 5 experiments described here, the conclusions are drawn:

– The CSB is very effective in small-scale tests. The stability against backward erosion piping for the sand tested in these tests increases with 2.8 until 5.4 depending on the configuration.
– The obtained critical head depends on the configuration. This may indicate that the results also depend on the scale of the model. Consequently, tests at different and larger scale are necessary to see what the increase in stability in the field is.
– Since it is not possible to perform field tests on a scale 1:1 in most situations, it is also necessary to develop numerical tools to extrapolate the results to the field situation.
– Results depend on details, as was shown with the differences in the results between Test 191 and 193, where only the density of the barrier is different and this leads to a low permeable filter cake in the upstream part of the barrier. This means that for all tests, regardless the scale, careful and reproducible model preparation is essential and that in the field the contractor should pay attention to the details.

ACKNOWLEDGEMENTS

The authors acknowledge F.M. Schenkeveld and J. Terwindt for the careful preparation and execution of the experiments and their efforts improving the setup.

REFERENCES

Cao D. 1994. Countermeasures for seepage erosion of Yangtze River main dikes. *Yangtze River* 25(1): 25–30.
García Martínez, M.F., Gragnano, C.G., Gottardi, G., Marchi, M., Tonni, L. and Rosso, A. 2016. Analysis of Underseepage Phenomena of River Po Embankments. *Procedia Engineering*, 158, 338–343.
Koelewijn A., Van Beek V.M., Förster U. and Bezuijen A., 2017. The development of a coarse sand barrier as an effective measure against piping underneath dikes. Proc 19th ICSMGE, Seoul, South Korea.
Negrinelli G., Van Beek, V.M., Ranzi R. 2016. Experimental and numerical investigation of backward erosion piping in heterogeneous sands, Scour and Erosion – Harris, Whitehouse & Moxon (Eds) © 2016 Taylor & Francis Group, London, ISBN 978-1-138-02979-8.
Van Beek, V.M., H. Knoeff, and H. Sellmeijer 2011. Observations on the process of backward erosion piping in small-, medium- and full-scale experiments. European Journal of Environmental and Civil Engineering 15(8):1115–1137.
Van Beek, V.M., Koelewijn A.R., Negrinelli, G. and Förster, U. 2015. A coarse sand barrier as an effective piping measure (in Dutch), Geotechniek 19 September:4–7.
Van Beek, V.M. 2015. Backward erosion piping, initiation and progression, PhD-thesis, TU Delft. ISBN 978-94-6259-940-6, http://repository.tudelft.nl/
Vandenboer K., Beek van V.M., Bezuijen A. 2015. 3D character of backward erosion piping: Small-scale experiments. Scour and Erosion – Cheng, Draper & An (Eds) © 2015 Taylor & Francis Group, London, 978-1-138-02732-9.
Van der Poel, J.T. & Schenkeveld, F.M. 1998. A preparation technique for very homogenous sand models and CPT research. In Kimura & Kusakabe (eds), Proceedings of the International Conference Centrifuge 98: 149–154. Balkema, Rotterdam.
USACE 1956. Investigation of underseepage and its control, Lower Mississippi river Levees. Technical memorandum No. 3-424, Volume 1, Waterways Experiment Station, Vicksburg, Mississippi.
Yao, Q., Xie, J., Sun, D., Zhao, J. 2009. Data collection of dike breach cases of China. Sino-Dutch Cooperation Project Report. China Institute of Water Resources and Hydropower Research.

Load transfer mechanism of piled embankments: Centrifuge tests versus analytical models

M. Blanc & L. Thorel
IFSTTAR, GERS, GMG, Bouguenais, France

R. Girout & M.S.S. Almeida
COPPE-UFRJ-Federal University of Rio de Janeiro, Rio de Janeiro, Brazil

D.F. Fagundes
School of Engineering-FURG, Federal University of Rio Grande, Rio Grande-RS, Brazil

ABSTRACT: Measured loads in centrifuge tests on piled embankments were compared with values calculated by using three analytical arching models. The Zaeske and van Eekelen models use sets of arches that give the direction of the load transfer and result in an increasing load on the soft soil with increasing embankment thickness or surcharge load, which matches the measurements. The Svanø model, however, assumes that the load on the soft soil remains constant with increasing embankment thickness or surcharge load, thereby being transferred directly to the piles. Therefore, the Zaeske and van Eekelen models match the measurements better than the Svanø model. The van Eekelen model slightly overestimates the pile efficiency for relatively shallow embankments, but matches the measurements better than the Zaeske model for increasing embankment thickness or surcharge load.

1 INTRODUCTION

Soft soils may be reinforced by using a network of rigid piles supporting a part of the load applied on the soft soil surface. The granular mattress with thickness H (Figure 1) is located above the soft soil and inside this mattress shearing and arching occurs, which increases the load transfer towards the piles. The piles network is characterised by the area replacement ratio α which is the proportion of pile area in a unit cell, the pile diameter d and the spacing between the pile axes s.

The ratio of the average load F on pile head to the total load Q consisting of the mattress weight and any additional surcharge q_0 exerted on a square mesh is called the load transfer efficiency E_F.

The behaviour of the granular mattress for the conventional piled embankment and with geosynthetic reinforcement has been investigated using different scales: 1g models (Van Eekelen et al. 2012a, b), centrifuge tests (Ellis and Aslam 2009; Blanc et al. 2013; Blanc et al. 2014) and large-scale tests (Chen et al. 2010; Briançon and Simon 2012; Xing et al. 2014).

Several analytical models have been proposed to estimate the load transferred towards the pile network. Among the arching models, Van Eekelen et al. (2013) distinguished the calculation based on (i) equilibrium of a granular mattress volumetric element (Hewlett and Randolph 1988, Van Eekelen et al. 2013, Zaeske

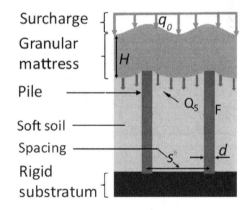

Figure 1. Soft soil reinforcement using a network of piles.

2001) and on (ii) rigid arches, where the load on the pile comes from some granular mattress volumetric shape (Svanø et al. 2000). The shape of the volume and arches is discussed further below.

In this study, the first objective was to provide a better understanding of the influence of the piled embankment parameters on the arching using additional tests corresponding to: (i) a larger range of H/s ratios, (ii) an area replacement ratio α ranging from 4.91% to 19.63% and (iii) different values of the surcharge q_0 applied at the surface of the

Table 1. Arching analytical models studied.

Arching model	Schematic representation	Analytical formulation
Svanø		$E_F = F/(s^2 \cdot (\gamma_d H + q_0))$ $F = \left[\dfrac{s^3 - B^3}{6} \beta \gamma_d + \gamma_d \cdot s^2 (H - \beta \cdot (s - B)/2)\right] + q_0 \cdot s^2$ if $H > 0.5 \cdot (s - B) \cdot \beta$ with $\tan(\theta) = \beta$ else $F = s^2 \cdot (\gamma_d H + q_0) - \left[\gamma_d s^2 H - \beta \gamma_d / 6 \cdot ((B + 2H/\beta)^3 - B^3) + q_0 \cdot (s - B) \cdot (s - B - 2H/\beta)\right]$
Zaeske		$E_F = 1 - \dfrac{1-\alpha}{\gamma_d s^2 H} \cdot \lambda_1^\chi \cdot \left(\gamma_d + \dfrac{q_0}{H}\right) \times \left[H(\lambda_1 + h_g^2 \lambda_2)^{-\chi} + h_g \cdot \left((\lambda_1 + \dfrac{h_g^2 \lambda_2}{4})^{-\chi} - (\lambda_1 + h_g^2 \lambda_2)^{-\chi}\right)\right]$ with $\lambda_1 = (s\sqrt{2} - B)^2/8, \lambda_2 = ((s\sqrt{2})^2 + 2B(s\sqrt{2}) - B^2)/(4s^2)$ $\chi = B(K_P - 1)/(s\sqrt{2} \cdot \lambda_2)$ and $h_g = \min(H, s\sqrt{2}/2)$
Van Eekelen (vE)		$E_F = \dfrac{F}{(\gamma_d H + q_0)s^2}$ where $F = (\gamma_d H + q_0)s^2 - F_{GRsquare} - F_{GRstrip}$ where $F_{GRsquare}$: load exerted on the square and $F_{GRstrip}$: load exerted on the strip

mattress. As the second objective, the physical modelling results obtained for the load transfer were then compared to the most commonly used analytical models quoted previously (those based on the equilibrium of a granular volumetric part and rigid arches).

2 ARCHING MODELS

Three analytical arching models (Girout et al. 2016) are used here to predict the efficiency E_F of the load transferred to the pile (Table 1).

The first model (Svanø et al. 2000), which is a rigid arching model, considers that the load transfer results in the piles supporting the weight of the mattress edge (3D case). Its advantage is to be a simple model, proposed that the friction of the granular material is taken into account by the coefficient β, where $2.5 < \beta < 3.5$. Le Hello and Villard (2009) proposed to use $\beta = 1/\tan(0.5\phi)$.

The second model (Zaeske, 2001) is based on the lower bound theorem of the plasticity theory. In the arch zone where stress redistribution takes place, the lower and upper spans of the arching zone are not concentric.

The last arching model, which is also an equilibrium model, is based on the assumption that stress arches occur between the piles and are responsible for the load transfer. The last modification of this kind of model is proposed by Van Eekelen et al. (2015). The load is supposed to be transferred towards the piles via a series of concentric arches occurring between the piles on the entire strip between two piles and on the square between the edges of four piles. The model of Van Eekelen was initially developed for piled embankments reinforced with Geosynthetic.

3 EXPERIMENTAL SET-UP

3.1 Mobile tray device

Load transfer towards end-bearing piles in the granular mattress has already been studied using a centrifuged small-scale device ($N = 20$) called the mobile tray device (MTD) (Blanc et al. 2013 and Girout et al. 2014). In this model, the settlement of a soft soil is simulated by the downward displacement $\Delta \omega$ of a tray around a network of metallic piles rather than by using a soft material. Before each test starts, the pile and the tray surface are at the same level. For each test, a sand surface layer (or a rigid foam) is normally added on the tray under the granular mattress (Blanc et al., 2013; de Freitas Fagundes et al., 2015). The interface between the granular material and the tray is described as smooth for this test (and rough in the others). A ring placed on this mobile tray is then filled with sand, containing the granular mattress. During the test, as the tray displaces downwards, the piles punch the mattress. There is no soft soil below the tray around the piles.

The spacing $s^{(m)}$ between model pile axes may be 100 mm, 141 mm or 200 mm and the pile diameter may be $d^{(m)} = 25$ mm or 50 mm. Thus the area replacement ratio α is equal either to 1.23%, 2.45%, 4.91% ($d^{(m)} = 25$ mm) or 4.91%, 9.82% or 19.63% ($d^{(m)} = 50$ mm). The thickness of the granular mattress is related to the pile spacing and diameter using the ratio H/H_{arch} where H_{arch} is $(s - d)/2$ in accordance with a semi-spherical arch. A homogeneous surcharge q_0 can also be applied at the surface of the mattress by filling a tank reservoir, the bottom of which is a soft membrane. Alternatively, tests can be made without surcharge by removing the tank, which makes it possible to increase the mattress thickness up to 360 mm.

Table 2. Geometries of tests configurations.

Test Conf.	H (m)	d (m)	α (%)	s (m)
T1 to T6	0.7 / 1.0 / 1.8 3.2 / 5.0 / 7.2	0.5	4.91	2.0
T7 to T11	1.0 / 1.8 / 3.2 / 5.0 / 7.2	1.0	4.91	4.0
T12 to T16	1.0 / 1.8 / 3.2 / 5.0 / 7.2	1.0	9.82	2.8
T17 to T21	1.0 / 1.8 / 3.2 / 5.0 / 7.2	1.0	19.6	2.0

From this point onwards, every symbol will be default on the prototype scale.

For each test configuration, the nine central piles monitor the load transfer. The measured load distribution (load-transfer efficiency and the stress acting on the tray) was compared to values calculated using three analytical models.

3.2 Test chronology

The small-scale centrifuge device is placed inside the centrifuge basket. Hostun sand mix (Blanc et al. (2013) is then poured manually. The mean density value obtained for these tests is $\rho_d = 1.62$ kg/m³ (where the dry unit weight $\gamma_d = 15.9$ kN/m³) corresponding to a relative density D_r of 70%.

The next steps of the physical modelling process are: (i) increasing the g-level to 20, (ii) adding water if a surcharge load is used (at a rate of 2.6 kPa/min), and (iii) prescribing the downward movement of the tray.

3.3 Typical test results

Experimental parametric studies are based on 21 test configurations (H, s, d) (Table 2). Some test configurations were performed with different surcharge q_0.

The load transfer efficiency E_F is plotted as a function of the ratio $\Delta\omega/H$ for test configuration T1 to T4 with $q_0 = 80$ kPa (Figure 2). The displacement $\Delta\omega$ is normalised by the thickness to allow comparison of different thicknesses, the spacing being the same for these tests ($s = 2.0$ m). Before any displacement of the tray, the load on the piles should only be due to the weight of the mattress and of the surcharge exerted on its surface, thus the initial value of efficiency should be equal to α (here, 4.91%). Nevertheless, the bending of the tray due to the surcharge and the g-increase initially induces a load transfer towards the rigid piles before the displacement is prescribed. Hence, the efficiency E_F starts from a value higher than α, with subsequent increases of E_F with the tray displacement $\Delta\omega$.

For the cases $H = 0.7$ m, 1.0 m and 1.8, E_F reaches a peak, then decreases to a residual value varying with the thickness. This softening could be explained in a similar way to the case of drained triaxial tests: for very high displacement prescribed to its bottom, the mattress reaches a critical state with large shear deformation.

For the cases $H = 3.2$ m, there is no obvious load transfer decrease because it was not possible to prescribe higher displacement $\Delta\omega$ with this configuration

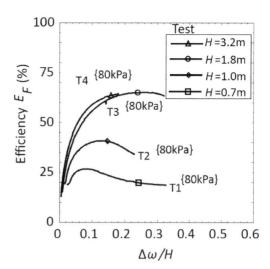

Figure 2. Typical experimental results.

without damaging the soft membrane of the water tank. In the real case, the settlement of the soft soil would stop by itself while the load transfer toward the piles increases. To analyse the load transfer, the value of the efficiency corresponding to the peak of the load transfer is defined as efficiency E_F.

4 ANALYTICAL VS. PHYSICAL MODELLING

Measured load transfer results are compared to the analytical ones in the present section. Each chosen analytical formulation (noted in the text as the 'Svanø', 'Zaeske' or 'vE' models) corresponds to a load transfer mechanism. In order to take into account the frictional angle of the granular material, we take $\beta = 2.9$ corresponding to $\varphi = 38°$.

4.1 Mattress thickness

The influence of the mattress thickness on the load transfer is studied using the tests with H ranging from 1.8 m to 7.2 m, without surcharge and considering the results with 1.0 m pile diameter (Figure 3). It illustrates the comparison between the analytical and experimental results, both for the case $\alpha = 4.91\%$ ($s = 4.0$ m) and $\alpha = 19.63\%$ ($s = 2.0$ m). The experimental results are plotted with a circular marker, except when noted.

The experimental results shown in Figure 3 for $s = 4.0$ m indicate that the measured efficiency E_F increases with H up to 55% (for this geometrical parameters set). The 'vE' overestimates the efficiency when H is lower then 3 m, but good agreement was found for thick embankments. On the other hand, the 'Zaeske' model is in relatively close agreement with the experimental results. The 'Svanø' model seems to be in agreement with the measured results here. For $s = 2.0$ m the experimental results show an increase of E_F according to the thickness and this may reach a maximum value. For each thickness, the efficiency is higher than for the equivalent

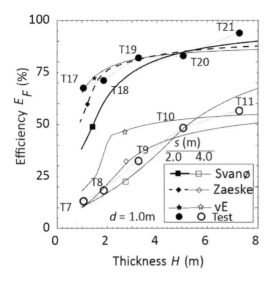

Figure 3. Efficiency E_F versus H.

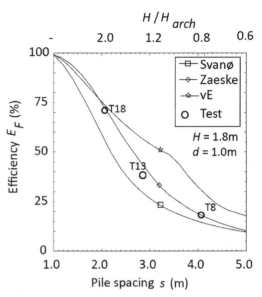

Figure 4. Efficiency E_F versus s.

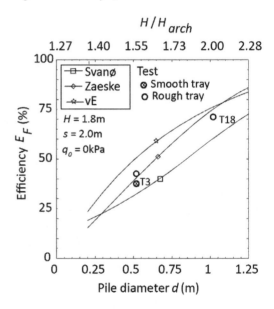

Figure 5. Efficiency E_F versus d.

thickness with $\alpha = 4.91\%$ because more arches occur in the embankment. All analytical formulations tend to an asymptotic value close to 90%. The 'Zaeske' model gives close agreement with the results of the mobil tray device (MTD) for the entire range of H. The 'vE' model slightly overestimates the pile efficiency for relatively thin embankments, but matches the measurements better than the 'Zaeske' model for increasing embankment thickness.

4.2 Pile spacing

Figure 4 compares measured and calculated E_F values for s ranging from 1.0 m to 5.0 m for $H = 1.8$ m and $d = 1.0$ m. This thickness has been chosen in order to calculate both a thin embankment case (i.e. $H/H_{arch} < 1$) and a thick embankment case. E_F decreases if the spacing s increases. All of the analytical results follow this trend. The 'Zaeske' values are the closest to the three available experimental results. The 'vE' model overestimated the load transfer. It seems that the differences between the 'vE' model and the experimental results increases with increasing spacing and reaches its maximum value for the ratio H/H_{arch} equal to 1. This model uses, as a limit between full and partial arching, the condition H greater than according to the concept proposed by Hewlett and Randolph (1988). This corresponds to a singular point in the analytical formulation which explains that the gap is maximal for $s = 3.4$ m. The 'Svanø' model shows a decrease in the efficiency until reaching a minimum value close to that obtained by the 'vE' and 'Zaeske' models.

4.3 Pile diameter

Figure 5 shows the measured and calculated efficiency for the area replacement ratio varying with d, $s = 2.0$ m (and $H = 1.8$ m). The efficiency E_F increases with d because, for increasing diameter, a decreasing part of the granular mattress exerted its weight on the soft soil surface. The values of the efficiency predicted by 'Zaeske' and 'vE' are equal for $d = 1.0$ m. The configurations modelled experimentally are close to the 'Zaeske' model. With the configuration of the test for $d = 0.5$ m, a second value is obtained with smooth interface. The efficiency is lower but the difference is small.

4.4 Area replacement ratio

Figure 6 compares calculations and experimental results with a constant value of $\alpha = 4.91\%$, considering large ($d = 1.0$ m) and small ($d = 0.5$ m) pile diameter.

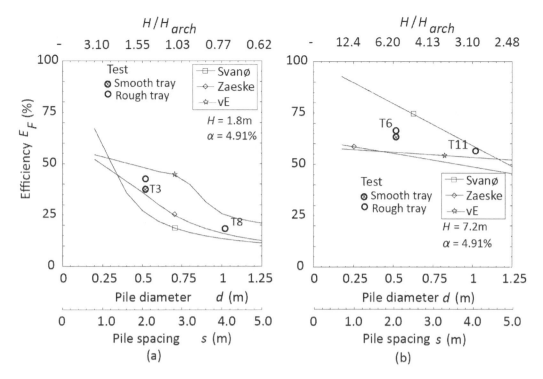

Figure 6. Efficiency E_F versus d & s ($\alpha = 4.91\%$) – (a) $H = 1.8$ m; (b) $H = 7.2$ m.

and for both mattress heights: 1.8 m (Figure 6.a) and 7.2 m (Figure 6.b).

Considering the case $H = 1.8$ m, the measured efficiency is quite close to the 'Zaeske' results. The 'Svanø' strongly depends on the pile diameter. For a large pile diameter, the 'Svanø' model is quite close to the experimental results. The calculated efficiency predicted by the 'vE' model is higher than other analytical results and experimental values, especially for a ratio $0.8 < H/H_{arch} < 1.4$ (Figure 6.a). Nevertheless, the 'vE' model shows the same trend as the physical modelling results.

For a thicker mattress (Figure 6.b), the 'vE' and 'Zaeske' models are more conservative than the experimental results. The 'vE' model gives similar results to the 'Zaeske' model. The 'Svanø' model overestimates the efficiency for smaller pile diameter, considering that the thickness is much greater for the critical thickness for smaller spacing s.

For a constant area replacement ratio, the efficiency decreases if the spacing increases. Arches occur in the mattress between the piles as a function of the thickness of the mattress. There are more arches if the total number of piles is higher. The difference between two networks of piles is limited because the ratio H/H_{arch} is always greater than 2.

As a conclusion, the efficiency increases with the number of pile elements replacing the soft soil. The load transfer is higher for a large number of piles, even if the pile diameter is relatively small. Analytical formulations based on the arching effect are relatively close to the experimental results, especially

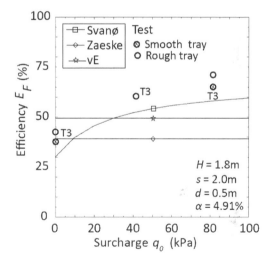

Figure 7. Efficiency E_F versus q_0.

the model of Zaeske et al. (2001). The choice of the pile configuration (s, d) is negligible compared to the mattress thickness if this is large enough (and with a minimum α).

4.5 Surcharge

Figure 7 compares the influence of the surcharge load q_0 for low cover ratios considering a mattress thickness $H = 1.8$ m (T3). The experimental E_F increases with the surcharge and reaches an asymptotic value close to 70% for $\alpha = 4.91\%$. The value for $q_0 = 80$ kPa represents about twice the value of the efficiency without surcharge. The 'Svanø' model matches the experimental results relatively well (considering that the thickness H is lower than H_{arch}). The impact of the surcharge q_0 in the load transfer efficiency in accordance with the analytical models based on arching effect is nil, as previously observed by Van Eekelen et al. (2013). The 'vE' and 'Zaeske' model calculation do not take q_0 into account for the efficiency because E_F is obtained initially for the case without surcharge ($q_0 = 0$ kPa), then E_F is multiplied by a factor.

5 CONCLUSIONS

A set of centrifuge tests on piled embankments was performed to investigate the influence of a number of parameters affecting its performance: the granular mattress thickness H, the pile spacing s, the pile diameter d and the surcharge load q_0. The purposes of this study were: (i) to gain more insight into the influence of different parameters on the arching and surface settlement, and (ii) to determine the performance of three analytical arching models.

The parametric studies showed that:

– Load transfer towards the piles and the pile efficiency E_F increase with H, d, q_0 and decrease with increasing s;
– Arching improves with the area replacement ratio, decreasing of the pile spacing being more efficient than increasing the pile dimension;
– Arching increases for higher values of the surcharge load at the embankment surface.

Three analytical arching models were selected for comparison with the measured efficiency E_F: the constant shape arching model of Svanø et al. (2000), and the models of Zaeske (2001) and Van Eekelen et al. (2013) which are limit equilibrium models. To conclude, the 'Zaeske' and 'van Eekelen' models match the measurements better than the 'Svanø' model. The 'van Eekelen' model slightly overestimates the pile efficiency for relatively thin embankments, but matches the measurements better than the 'Zaeske' model for increasing embankment thickness or surcharge load.

REFERENCES

Blanc, M., Rault, G., Thorel, L. and Almeida, M.S.S. 2013. Centrifuge investigation of load transfer mechanisms in a granular mattress above a rigid inclusions network. *Geotext. Geomembr.*, 36, 92–105.

Blanc, M., Thorel, L., Girout, R. and Almeida, M. S. S. 2014. Geosynthetic reinforcement of a granular load transfer platform above rigid inclusions: comparison between centrifuge testing and analytical modelling. *Geosynth. Int.*, 21(1), 37–52.

Briançon, L. and Simon, B. 2012. Performance of pile-supported embankment over soft soil: full-scale experiment. *J. Geotech. Geoenviron.*, 138(4), 551–561

Chen, R., Xu, Z., Chen, Y., Ling, D., and Zhu, B. (2010). Field test on a piled embankment over soft ground. *J. Geotech. Geoenviron. Eng.*,136(6), 777–785.

Ellis, E. and Aslam, R. 2009. Arching in piled embankments: comparison of centrifuge tests and predictive methods – part 1 of 2. *Ground Eng.*, 42(6), 28–31.

de Freitas Fagundes, D., de Almeida, M.S.S., Girout, R., Blanc, M. et Thorel, L 2015. Behaviour of piled embankment without reinforcement. *Proceedings of the Institution of Civil Engineers – Geotechnical Engineering*, 168(6), 514–525.

Girout, R., Blanc, M., Dias, D. and Thorel, L. 2014. Numerical analysis of a geosynthetic-reinforced piled load transfer platform – validation on centrifuge test. *Geotext. Geomembr.*, 42(5), 525–539.

Girout, R., Blanc, M., Thorel, L., Fagundes, D.F. and Almeida, M.S.S. 2016. Arching and Deformation in a Piled Embankment: Centrifuge Tests Compared to Analytical Calculations. *J. Geotech. Geoenviron. Eng.*, 142.(12),

Hewlett, W. and Randolph, M. A. 1988. Analysis of piled embankments. *Ground Eng.*, 21(3), 12–18.

Le Hello, B., and Villard, P. 2009. Embankments reinforced by piles and geosynthetics - Numerical and experimental studies dealing with the transfer of load on the soil embankment. *Eng. Geology*, 106(1-2), 78–91.

Svanø, G., Ilstad, T., Eiksund, G. and Watn, A. A. 2000. Alternative calculation principle for design of piled embankments with base reinforcement. *Proc., 4th Int. Conf. Ground Improvement Geosystem*, Finnish Geotechnical Society, Helsinki, Finland.

Van Eekelen, S. J. M., A. Bezuijen, H. J. Lodder, et A. F. van Tol. 2012a. Model experiments on piled embankments. Part I. *Geotext. and Geomembr.*, 32(1), 69–81.

Van Eekelen, S. J. M., A. Bezuijen, H. J. Lodder, et A. F. van Tol. 2012b. Model experiments on piled embankments. Part II. *Geotext. and Geomembr.*, 32(1), 82–94.

Van Eekelen, S. J. M., Bezuijen, A. and van Tol, A. F. 2013. An analytical model for arching in piled embankments. *Geotext. Geomembr.*, 39(1), 78–102.

Van Eekelen, S. J. M., Bezuijen, A. and van Tol. A. F. 2015. Validation of analytical models for the design of basal reinforced piled embankments. *Geotext. and Geomembr.*, 43(1), 56–81.

Xing, H., Zhang, Z., Liu, H. and Wei, H. 2014. Large-scale tests of pile-supported earth platform with and without geogrid. *Geotext. Geomembr.*, 42(6), 586–598.

Zaeske, D. 2001. Zur Wirkungsweise von unbewehrten und bewehrten mineralischen Tragschichten über pfahlartigen Gründungselementen. [To see the effect of non-reinforced and proven mineral base courses over pale like Gründungselemnte]. *In Geotechnical engineering. Univ Kassel.* (in German).

Physical model testing to evaluate erosion quantity and pattern

M. Kamalzare
Civil Engineering Department, California State Polytechnic, Pomona, USA

T.F. Zimmie
Civil & Environmental Engineering Department, Rensselaer Polytechnic Institute, Troy, USA

ABSTRACT: The objective of this research is to develop tools that would improve the understanding of the process of levee failure due to erosion and reduce the risk of failure. Detailed results of the project have been presented else-where. This paper deals with one part of the research, the determination of the amounts and rates of sediment transport due to the soil erosion. Hydraulic erosion is a complicated phenomenon and depends on many different parameters. This project dealt with the study of surface erosion of levees, earth dams and embankments. To improve design criteria of these structures, development and verification of realistic computer models that can simulate the erosion process is necessary. In this research, a large number of physical levee erosion tests were performed at 1-g and at high g's using a geotechnical centrifuge. The erosion was modeled physically in detail. A Kinect device was used to scan and evaluate the volume of eroded soil. The variation of the shape of the channels as a function of time along with the amount of eroded soil was measured using implementing this method.

1 INTRODUCTION

Most geotechnical physical modeling facilities feature a system of image capture and analysis in order to provide measurements of deformation fields. This paper aims to describe a new technique developed for identifying and characterizing the soil hydraulic erosion process of levees, earthen dams, embankments, and similar structures due to overtopping incidents. Evaluations have been done for levee overtopping erosion tests performed at 1-*g* and higher *g* levels (centrifuge) conditions but are also applicable in other experiments with different environments. The new technique includes various advances beyond the state of the art described previously. With the use of a new and improved apparatus and software system, improved measurement performance was achieved. This method provides the capability to measure quantities that were not easy and sometimes impossible to measure previously. This method results in a step forward in measurement utility: small particle-scale transportation features can be detected, allowing soil deformation and erosion patterns to be quantified as a function of time. The proposed methodologies can also be used for other experiments.

During recent decades, there have been continuing efforts to study dam-break hydraulics, including numerical and experimental investigations of dam-break flows and the potential damage caused by the flows. This is understandable since real-time field measurements are extremely difficult to obtain. In fact, numerical simulations play an increasingly important role for dam-break flow problems. The erosion process is dependent on several parameters such as soil fines and clay size content, plasticity, and dispersivity, compaction water content, density and degree of saturation, clay mineralogy, and possibly the presence of cementing materials such as iron oxides. Considering all these parameters, it is very difficult to understand and numerically model the erosion phenomena. Validation of these numerical simulations is also a very challenging task. In order to provide associated guidance for more rational designs, there is a need for methods that can visualize and evaluate different phases of erosion in laboratory experiments.

Some studies have focused on soil classification. Using an Erosion Function Apparatus (EFA), Briaud et al. (2008) investigated the erodibility of several types of soil. The soils are classified into different categories of erodibility based on degree of compaction, erosion rate, water velocity and hydraulic shear stress. Dean et al. (2010) applied experimental steady-state results for different ranges of overtopping. Laboratory results consisting of velocities and durations for acceptable land side levee erosion due to steady flows were examined to determine the physical basis for the erosion. Three bases are examined: (1) velocity above a threshold value, (2) shear stress above a threshold value, and (3) work done on the landside of the levee above a threshold value. The work basis provides the best agreement with the data and a threshold work value, and a work index representing the summation of the product of work above the threshold and time were developed.

Recent research has used effective methods to investigate levee erodibility. Kamalzare et al. (2011) per-formed a number of laboratory tests on levees with different geometries, and investigated the effects of different parameters on levee erodibility. The rilling process occurring on the landside slope was studied, and erosion effects occurring on the waterside slope were not considered. The durations of various erosion phases were measured, but no data was collected regarding the quantity of eroded soil. Stanier and White (2013) describe a new apparatus and techniques for performing deformation measurements using particle image velocimetry in the centrifuge environment. The new system includes camera, lighting, and control equipment that facilitates image capture at least 30 times faster than that in legacy systems. Methods for optimizing the addition of artificial seeding on the exposed plane of a geotechnical model were used. These techniques ensured that the precision of the deformation calculations was optimized even in models with multiple soil layers. An example application of a flat footing penetrating sand overlying clay was used to illustrate the performance of the equipment and the artificial seeding optimization technique. Analyses highlight not only the benefits of the new technology, but also the need for carefully optimized experimental procedures to maximize the measurement precision. Dong and Selvaduria (2006) presented a color visualization-based image processing technique for the quantitative determination of a chemical dye concentration in a fluid-saturated porous column composed of glass beads. In this image processing technique, an image filter is designed by taking into account the porous structure of the medium and color characteristics of both the fluid and the solid particles to extract the color representation of the dye solution in pore space, which enables the image quantification. A comparison of experimental results with analytical and numerical simulations illustrates the efficiency and accuracy of the image processing method for determining the chemical concentrations in the porous medium.

Lo et al. (2010) proposed a new water-based transparent material called "Aquabeads" for modeling flow in natural soils. Three types of this material were used to model miscible and multiphase flow transport process in layered soil systems. An optical system was set up to trace flow movements in a two-dimensional (2D) physical model of a soil profile, and analyzed using digital image processing to define images of 2D concentration profiles in the model. Model surfactant flushing tests were conducted using a layered soil system and two contaminants, mineral oil and motor oil, in order to illustrate the feasibility of using this water-based polymer to visualize geoenvironmental con-amination problems. Because a transparent soil was used, the optical systems allow for visualizing surfactant flushing. The study demonstrates that Aquabeads are suitable for modeling multiphase flow, particularly in educational settings. Kamalzare et al. (2013a and b), Kamazare et al. (2012a and b), Holmes et al. (2011), Stetzle et al. (2011), Yu et al. (2009), and Xiao et al. (2009) also apply different numerical techniques to study and simulate various aspects of the erosion phenomena.

Although much work has been done to simulate erosion in the field of computer graphics, there has been limited validation. This is mostly because of limitations of current laboratory measurement methods. A primary objective of this research was to find a methodology to validate computer simulations by laboratory experimentation. Therefore, in this research, laboratory tests using model levees have been performed to improve the computer simulations of levee and embankment erosion. To evaluate the effects of water flow on real levees, some centrifuge tests were also performed simulating full-scale prototype levees and embankments. A new visualization methodology has been introduced that not only provides 3D images of erosion channels but also the capability to validate the quantity of the erosion and the evolution of erosion channels as a function of time.

2 CONVENTIONAL VISUALIZATION METHODS

Techniques for the measurement of deformations and soil transport in geotechnical models have developed significantly in recent years. Early studies by Butterfield et al. (1970) and Andrawes and Butterfield (1973) reported the use of stereo photogrammetry, in which individual particle movements as seen in stereo pair photographs, were measured manually. The recent introduction of digital technology has removed the need for painstaking manual film measurements. The technique of PIV (also known as digital image correlation) has widely applied across many branches of engineering (Raffel et al. 2007; Pan et al. 2009; Sutton et al. 2009). The use of PIV and photogrammetry to measure soil displacement in small-scale physical models has led to a significant increase in measurement accuracy and precision relative to previously utilized techniques (White et al. 2003). Furthermore, the number of measurement points available in the analysis process has become a function of discretization (patch size) rather than the number of identifiable features (e.g., target markers) on an exposed plane of the model (Stanier and White, 2013). The other conventional and widely used visualization method for both outdoor large scale and indoor small-scale experiments is laser scanning. Three-dimensional object scanning allows enhancement of the design process, speeds up and reduces data collection errors, saves time and money, and thus makes it an attractive alternative to traditional data collection techniques. 3D scanning is also used for mobile mapping, surveying, scanning of buildings and building interiors, and in archaeology. Lidar is a remote sensing technology that measures distance by illuminating a target with a laser and analyzing the reflected light. Lidar was developed in the early 1960s, shortly after the invention of the laser, and combined laser's focused imaging with radar's ability to calculate distances by measuring the time for the signal to return (Goyer and Watson, 1963). Lidar uses ultraviolet, visible, or near infrared light to image objects and can be used with a wide range of targets, including non-metallic objects, rocks, rain, chemical compounds, aerosols, clouds and even single molecules.

In our research, the initial and final surface geometries of the model levee for different overtopping erosion experiments were recorded using a three-dimensional Laser Range Scanner (Lidar) and were analyzed using the data structure developed in this study. The acquired surface data from each test could then be used to further visualize the final results of the physical overtopping simulations. The Laser Range Scanner rotated through a user specified angle and, using a single laser beam, conducted a scan of the surface at each incremental rotation within the range of rotation. Each incremental movement was characterized by a new pulse of the laser beam that collected data based on features in surface elevation or geometry of the object of interest at that specific position being scanned. The Laser Range Scanner used in this research was a Leica 30 HDS 3000, by Leica Geosystems HDS, LLC. Layer surface data was collected in the form of a point cloud via a three-dimensional laser range scanner. This 3D point data was then run through a data preparation script that, for each scan, registered the points. It was then aligned to a regular grid in the XY plane, retaining the height values of the points. The details of the calculation and analyses performed on the Lidar data has been presented in Kamalzare et al. 2016.

Figure 1 shows a schematic drawing of elevation and plan view of a typical experimental setup. Different drainage systems were used based on the water flow and the type of the soil (Kamalzare et al. 2013a). The grain size distribution of the soil was determined ac-cording to ASTM D6913-04(2009). The soil is classified as "SC" according to the Unified Soil Classification System (USCS). Other characteristics of the soil are presented in Table 1.

The collected scan data, which was processed digitally, yielded the visualized representation of the data shown in Figure 2.

During the course of this research a computer simulation was also developed to model hydraulic soil erosion (Kamalzare, 2013 and Stuetzle, 2012). To model the levee system, the high-resolution particle-based Lagrangian method based on Smoothed Particle Hydrodynamics (SPH) was used.

As this paper is more focused on the physical model testing of the erosion, the details of the computer model have not been presented here. However, more information regarding the computer model can be found in Kamalzare, 2013.

3 A VISUALIZATION METHOD TO MEASURE EROSION QUANTITY AND EVOLUTION

Because of the limitation of current experimental methods, it is quite difficult if not impossible to

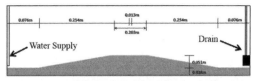

Figure 1. Schematic drawing of elevation and plan view of a typical experimental setup.

Table 1. Soil characteristics.

Property	Numerical value
D_{10} (mm)	0.074
D_{30} (mm)	0.11
D_{60} (mm)	0.19
Coefficient of uniformity	2.57
Coefficient of curvature	0.86
Liquid limit	17
Plastic limit	11

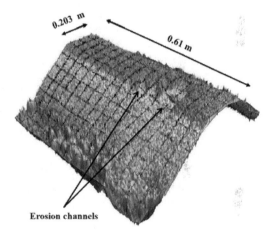

Figure 2. Visualization of an overtopping erosion experiment.

measure the exact amount of transported soil and the erosion evolution during hydraulic erosion experiments. However, the process of channel formation during an overtopping experiment was recorded with a Kinect sensor. The Kinect is a low-cost sensor, and enabled the measurement of the rate of soil erosion, which if done at all, usually requires expensive equipment. Unlike a regular camera or even a high-speed camera, the Kinect sensor records additional information regarding elevations and depths of different

Table 2. Technical details of the Kinect sensor.

Property	Description
Sensor	Color and depth-sensing lenses
	Voice microphone array
	Tilt motor for sensor adjustment
Field of View	Horizontal field of view: 57 degrees
	Vertical field of view: 43 degrees
	Physical tilt range: 27 degrees
	Depth sensor range: 1.2 m - 3.5 m
Data Streams	320×240 16-bit depth at 30 frames/sec
	640×480 32-bit color at 30 frames/sec
	16-kHz, 24-bit mono pulse code
Skeletal	Tracks up to 6 people,
Tracking	including 2 active players
System	Tracks 20 joints per active player

parts of the channels. The Kinect sensor is part of a Microsoft gaming system, and due to the economics of mass production, can be typically purchased for about $100.00. The technical capabilities of the Kinect system can be duplicated in a custom design, but the cost will be several thousand dollars. A brief introduction about the Kinect sensor (Xbox 360, by Microsoft) is presented below, followed by details of the methodology. This paper and associated research is in neither an endorsement nor promotion for the Kinect system.

Table 2 summarizes the technical properties of the Kinect device.

Considering the dimensions of the modeled levees and the Kinect viewing angle, a minimum distance of 1.3 m was required to visualize the whole model. This distance was measured from the highest point of the modeled levee to the Kinect, and is consistent with the practical ranging limit of 1.2–3.5 m for the depth sensor. In 1-g experiments, the Kinect was connected to a vertical metal bar, and secured to a table placed in the middle of the room. An important consideration for setting up the Kinect prior to experiments is to place the sensor on a stable surface in a location where it will not fall or be struck during use. It should also be accurately leveled; otherwise there will be errors in the recorded depth, and a complicated and time-consuming process is needed to cancel this error.

The Kinect was also used in centrifuge experiments to investigate soil erosion. In centrifuge experiments forces increase with increased g levels, and materials can be subjected to large forces compared to 1-g tests. Some of the challenges were to ensure that the Kinect would function at high g loads and not move during the centrifuge tests. A frame and an extended arm were constructed to secure the Kinect during the centrifuge experiments. The arm attached the Kinect frame to the centrifuge beam. The Kinect was placed as close as possible to the center of the centrifuge to minimize the centrifugal loads.

4 RESULTS

The process of channel formation during an overtopping experiment was recorded with the Kinect. Unlike

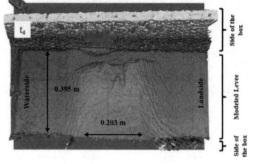

Figure 3. Kinect visualization of progression of erosion for a 1-g overtopping experiment (plan view) $t_1 < t_4$.

a regular camera or even a high-speed camera the Kinect records additional information regarding elevations and depths of different parts of the channels. Differing depths can be presented using different colors, enabling one to follow the channel formation process and investigate the erosion process. Figure 3 show the results of the Kinect visualization at the time of breach as an example. Similar images can be produced for other times periods. The sides of the experiment box and different parts of the levee can be seen. The progression of overtopping during the experiment has been clearly recorded and shown in these figures. As the measured depth increases, the coloration from brown proceeds to green. The water overtopped the levee from the waterside, which is on the left, to the landside, which is on the right. Some small channels propagate on the crest of the levee, and eventually the primary channel forms and the levee breaches. In addition to the colors, there are contour lines on each figure that show elevations.

These results can be useful for obtaining an understanding of channel propagation and are used for verifying predicted depths of the channels obtained by digital simulations; however, measurements of the quantity of erosion are required, that is the volume of eroded soil as a function of time. The Kinect sensor meshes the surface of the levee to a rectangular grid, and records the depth of each cell at different time steps. Theoretically it should be easy to find the differences between recorded values of the frames and compute the amount of eroded soil (Fig. 4), but there are several challenges to completing this task. Each frame contains information about the location

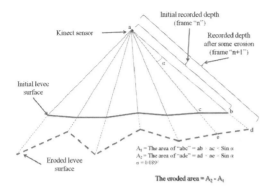

Figure 4. Theoretical analyses of Kinect recorded data.

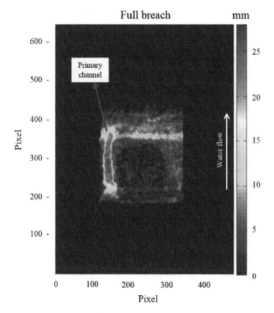

Figure 5. Results of analyses on the measured depths for a centrifuge experiment at full breach (plan view).

and depth of 307200 cells. For each experiment, about 5000 to 8000 frames were recorded, resulting in a large amount of data. Obviously, it is necessary to use a computer code to handle the data.

Another challenge is noise in the data. Some noise can be explained because of movement of the water during the experiment, and consequently reflection. The nature of this type of noise is random, and random noise is challenging since it is not similar to regular noise that can be easily filtered. Fig. 5 show the results of the measured depths for the same centrifuge experiment after a noise reduction process (Kamalzare et al. 2016) at the same time (full breach).

5 CONCLUSIONS

The following specific conclusions can be drawn from the study:

1. The data produced over the course of this research validated the behavior predicted by the numerical models. Although the process of channel formation during an overtopping experiment was recorded with conventional methods, the Kinect sensor was used as a new methodology to measure and evaluate soil transportation in the erosion experiments.
2. The new visualization methodology not only makes it feasible to record the precise duration of different erosion phases, but also to measure erosion quantity and calculate the volume of eroded soil in an erosion experiment. To the authors' knowledge, this is the first time that the methodology has been applied to the study of soil erosion.
3. Unlike a regular camera or even a high-speed camera, the Kinect records additional information regarding elevations and depths of different parts of the channels. The variations of shape and volume of the eroded channels could be measured as a function of time during the experiments, and the rate of sediment transport can be calculated.
4. Centrifuge tests were also performed to simulate real (prototype) size levees, and the application of this methodology was also studied in centrifuge testing.
5. The Kinect functioned well in 1-g experiments and at high g levels in the centrifuge tests. It is believed this is the first use of a Kinect device in centrifuge experiments. The application of the Kinect sensor in other laboratory experiments was also presented.

REFERENCES

Andrawes, K. Z. and Butterfield, R., 1973. The Measurement of Planar Displacements of Sand Grains, *Geotechnique*, Vol. 23(5), pp. 571-576.

ASTM D6913-04(2009)e1, Standard test methods for particle-size distribution (gradation) of soils using sieve analysis. *ASTM International,* West Conshohocken, PA.

Briaud, J. L., Chen, H. C., Govindasamy, A. V., and Storesund, R., 2008. Levee erosion by overtopping in New Orleans during the Katrina hurricane. *Journal of Geotechnical and Geoenvironmental Engineering ASCE*, Vol. 134, No. 5, pp. 618-632.

Dean, R. G., Rosati, J. D., Walton, T. L., and Edge, B. L., 2010. Erosional equivalences of levees steady and intermittent wave overtopping. *Ocean Engineering*, Vol. 37, No. 1, pp. 104-113.

Dong, W. and Selvadurai, APS., 2006. Image processing technique for determining the concentration of a chemical in a fluid-saturated porous medium. *Geotechnical Testing Journal*, Vol. 29, No. 5.

Goyer, G. G. and Watson, R., 1963. The laser and its application to meteorology. *Bulletin of the American Meteorological Society*, Vol. 44, No. 9, pp. 564-575.

Holmes, D. W., Williams, J. R., and Tilke, P., 2011. Smooth particle hydrodynamics simulations of low Reynolds number flows through porous media. *International Journal for Numerical and Analytical Methods in Geomechanics*, Vol. 35, No. 4, pp. 419-437.

Kamalzare, M., Zimmie, T. F., Cutler, B., and Franklin, W. R., 2016. New visualization method to evaluate erosion quantity and pattern. *Geotechnical Testing Journal*, Vol. 39, No. 3, pp. 431-446.

Kamalzare, M., 2013. Investigation of levee failure due to overtopping and validation of erosion evolution and quantity. *Ph.D. dissertation*, Department of Civil and Environmental Engineering, Rensselaer Polytechnic Institute, Troy, NY.

Kamalzare, M., Han, T. S., McMullan, M., Stuetzle, C., Zimmie, T., Cutler, B., Franklin, W. R., 2013a. Computer simulation of levee erosion and overtopping. *Proc. Geo-Congress 2013: Stability and Performance of Slopes and Embankments III*, San Diego, CA, pp. 1851-1860.

Kamalzare, M., Zimmie, T. F., Han, T. S., McMullan, M., Cutler, B., and Franklin, W. R., 2013b. Computer Simulation of Levee's Erosion and Overtopping. *Proceedings of the 18th International Conference on Soil Mechanics and Geotechnical Engineering (ICSMGE)*, Paris, France.

Kamalzare, M., Stuetzle, C., Chen, Z., Zimmie, T.F., Cutler, B., and Franklin, W. R., 2012a. Validation of erosion modeling: physical and numerical. *Geo-Congress 2012*, Oakland, CA, pp. 710-719.

Kamalzare, M., Zimmie, T.F., Stuetzle, C., Cutler, B., and Franklin, W. R., 2012b. Computer simulation of levee's erosion and overtopping. *XII International Symposium on Environmental Geotechnology, Energy and Global Sustainable Development,* International Society for Environmental Geotechnology, Los Angeles, CA, pp. 264-273.

Kamalzare, M., Chen, Z., Stuetzle, C., Cutler, B., Franklin, W. R., and Zimmie, T. F., 2011. Computer simulation of overtopping of levees. *14th Pan-American Conference on Soil Mechanics and Geotechnical Engineering (64th Canadian Geotechnical Conference)*. Toronto, Ontario, Canada.

Lo, H.Ch., Tabe, K., Iskander, M., Yoon, S.H., 2010. A transparent water-based polymer for simulating multiphase flow. *Geotechnical Testing Journal*, Vol. 33, No. 1.

Pan, B., Qian, K., Xie, H., and Asundi, A., 2009. Two-dimensional digital image correlation for in-plane displacement and strain measurement: A review. *Meas. Sci. Technol.*, Vol. 20(6), 62001.

Raffel, M., Willert, C., Wereley, S., and Kompenhaus, J., 2007. *Particle Image Velocimetry -A Practical Guide*, Springer- Verlag, Berlin.

Stanier, S. A. and White, D. J., 2013. Improved image-based deformation measurement in the centrifuge environment. *Geotechnical Testing Journal*, Vol. 36, No. 6.

Stuetzle, C., 2012. Representation and generation of terrain using mathematical modeling. *Ph.D. dissertation*, Department of Computer Science, Rensselaer Polytechnic Institute, Troy, NY.

Stuetzle, C., Cutler, B., Chen, Z., Franklin, W. R., Kamalzare, M., and Zimmie, T. F., 2011. Measuring terrain distances through extracted channel networks. *19th ACM SIGSPATIAL International Conference on Advances in Geographic Information Systems*, Chicago, IL, Vol. 3 (3), pp. 21-26.

Sutton, M. A., Oreu, J.-J., and Schreier, H. W., 2009. Image correlation for shape, motion and deformation measurements, *Spring, Science and Business Media*, New York.

White, D. J., Take, W., and Bolton, M., 2003. Soil deformation measurement using particle image velocimetry (PIV) and photogrammetry. *Geotechnique*, Vol. 53(7), pp. 619–631.

Xiao, H., Huang, W., and Tao, J., 2009. Numerical modeling of wave overtopping a levee during hurricane Katrina. *Computers and Fluids*, Vol. 38, No. 5, pp. 991-996.

Yu, M., Deng, Y., Qin, L., Wang, D. and Chen, Y., 2009. Numerical simulation of levee breach flows under complex boundary conditions. *Journal of Hydrodynamics*, Vol. 21, No. 5, pp. 633-639.

Physical modelling of large dams for seismic performance evaluation

N.R. Kim & S.B. Jo
K-water Research Institute, Daejeon, Korea

ABSTRACT: The performance and safety of large embankment dams subjected to strong earthquake ground motion is one of the major risks in new construction projects. The design and performance of the structure is generally verified by numerical methods in most of engineering practice. Reduced scale physical modelling is often carried out in design and management stages for the same purpose where numerical methods cannot provide sufficient information. In general, physical model tests are designed and carried out according to rigorous scaling principles. However, it is sometimes difficult to follow these principles due to various limitations; equipment, boundary conditions, characteristics of prototype structures, etc. This paper will review the limitations of physical modelling methods and its effects in seismic performance evaluation of large embankment dams. It will also introduce alternative modelling and testing methods to overcome the limitations. A series of tests with a reduced scale embankment dam model are carried out to examine the limitations and to verify the applicability of alternative modelling method.

1 INTRODUCTION

Seismic safety is one of the greatest risks in design and constructions of large dams. Number of embankment dams have been damaged due to strong earthquakes for last decades, and the deteriorated safety performance of the structure may result in tragic consequences. Therefore, evaluation of seismic performance is an important issue in engineering practice of dam safety management.

Various methods are employed in order to evaluate safety performance or to validate design parameters of large earth systems. Numerical methods are widely adopted in engineering practice since the engineers can easily model entire structure and simulate various loading conditions. While we are able to model complicated structural and material characteristics, and to simulate various and more realistic loading conditions by numerical simulation with great advances of technology, reduced scale physical modelling is still useful and conducted in both research and practice. It is carried out in order to simulate complicated conditions such as boundary conditions or effect of interfaces between different materials, and to validate the result of numerical simulation.

There have been cases of physical modelling of embankment dams to evaluate seismic response characteristics such as amplification, deformation and/or settlement, liquefaction susceptibility, slope stability, etc. With development of in-flight shaking tables for centrifuge, various seismic tests for embankment dams were conducted by a number of researchers, especially for embankments sitting on a liquefiable foundation (Ng et al. 2004; Sharp & Adalier 2006; Peiris et al. 2008). More recently, seismic response and amplification characteristics of rock-fill dams were investigated by a series of centrifuge tests (Kim et al., 2011). In these cases, physical modelling and dynamic centrifuge tests were carried out to understand the general behaviour subjected to seismic loading, and the information could be utilized to understand the global behaviour of prototype scale structures.

The prototype dams of previous research are relatively small. In reality, failure of large dam has great risks to public safety, and most of practical cases which geotechnical engineers are asked to evaluate the seismic performance by physical modelling deal with large dams. However, there are limitations in physical modelling of large prototype structures in engineering practice; mainly difficulties in large scaling ratio of the structure, limited centrifugal acceleration, performance of loading system such as earthquake simulator, etc. Hence the limitations have to be resolved and alternative modelling strategies have to be applied in order to simulate the behaviour of a specific large prototype structure in practical problems.

In this paper, various limitations in modelling of large embankments such as scaling problems, boundary conditions and performance characteristics of testing machines are reviewed. Alternative modelling principles to overcome these limitations are introduced. Finally, case studies utilizing the alternative modelling strategy in engineering practice are introduced and its applicability is discussed.

2 LIMITATIONS AND DIFFICULTIES IN MODELLING OF A LARGE EMBANKMENT FOR SEISMIC TESTS

2.1 Dimensions and modelling scale

Physical modelling is often utilized in engineering practice for various purposes. It is mainly applied to validate a conceptual design approach or to evaluate the performance of a certain structure subjected to design loads. The experiments are generally designed considering various factors and there can be flexibility in the scale especially when the objective of the experiment is to understand the structural behaviour. However, the scale must be precisely reserved if the tests are conducted for performance evaluation of an existing structure or for verification of a design.

Embankment dams which requires a physical modelling study for performance evaluation in engineering practice are generally huge earth structures as parts of large development projects. For example, we may design a test to evaluate seismic performance of an embankment dam which is 100 m high and 300 m long in river direction. The entire structure should be considered in the reduced scale model to simulate structural response and to understand global behaviour. If the model is built in a 1.5 m long and 0.5 m high container, the scale will be 1:200. In general, the test should be carried out at 200 g considering the scaling principles and the shaking acceleration will proportionally increase with the scale. However, the experiment is only feasible with a testing system with unrealistic large capacity.

In some cases, the test can be alternatively conducted at low g-level and the prototype scale behaviour is inferred from the testing results. If the test is conducted at 50 g using the model mentioned above, it is considered as simulating 25 m high embankment dam. Even though the typical behaviour of a certain type of structure can be understood, it is difficult to confirm the performance of prototype structure with this limited condition.

The limitations of centrifuge modelling are raised in other types of problems such as design of foundation systems, nuclear power projects, natural hazard, etc. Therefore, advanced modelling strategies are required to overcome the limitations, e.g. generalised scaling relations suggested by Iai et al. (2005), and the applicability of new scaling relations should be understood prior to experiment.

2.2 Boundary conditions

Effects of boundary conditions of the container are often important issues in physical modelling, especially in earthquake problems. In 2D modelling of an embankment, the rigidity of the container's end wall does not affect the seismic response, and rigid containers can be recommended (Kim et al., 2010). To verify the effect of flexibility of the container, an embankment model has been prepared in an equivalent shear beam (ESB) container and the amplification at various

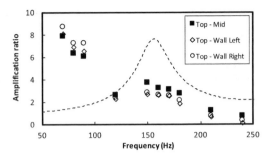

Figure 1. Frequency responses and amplifications measured at top of embankment model and the effect of frequency response of ESB container.

frequencies has been evaluated at 20g centrifuge shaking tests. The embankment model was prepared using gravelly sand material, and the contact between the embankment and side wall was lubricated using thin rubber sheets and grease. Figure 1 shows the crest-to-base amplification ratio at three different locations; one in the middle of the embankment, Top-Mid, and two close to container walls on each side, Top-Wall Left and Top-Wall Right. The amplification ratio was evaluated by measuring responses corresponding to sinusoidal input motions varying frequencies. Greatest amplification was detected at 70 to 80 Hz range which is equivalent to the natural frequency of the ESB container, which means the response of the container highly affects at its natural frequency range. The natural frequency of the embankment is around 150 Hz according to analytical solution by Dakoulas & Gazetas (1985) which is presented by dotted line, and it could be verified by experiment. It should be noted that there could be misinterpretation if the response of other testing system can affect the measurement.

In Figure 1, the amplification ratio measured close to the container wall is greater when the excitation frequency is close to the natural frequency of the container, while it is smaller at the middle point when the frequency is close to the natural frequency of the embankment. It is due to interaction between the model and the container wall even though the contact was lubricated. In reality, the embankment in narrow valley tends to be stiffer due to 3D valley effect, and the fundamental period of the embankment model evaluated by dynamic centrifuge test can be shorter. Ideal 2D behaviour cannot be achieved hence this effect has to be understood in interpretations.

2.3 Materials

One of the greatest advantage of physical modelling is that the testing model is prepared using the same soil material hence the uncertainties of soil characteristics and its effect on structural behaviour can be resolved in the testing results and interpretation. Even though the exactly same material cannot be used in the modelling, alternative material with equivalent engineering parameters, which plays a key role in structural behaviour can be adopted.

In large geotechnical construction projects, rubbles are frequently dumped to construct massive embankment or used as filling material. Large embankment dams are generally constructed as rockfill type and the effective diameter of largest particle is often bigger than 1m. The engineering properties of the material are important because it determines structural safety.

In physical modelling of large rockfill dams, the model can't be constructed using the same material and alternative soil material should be used, which will be non-cohesive and the largest particle size will not affect the shear behaviour of the embankment.

To simulate seismic response and to evaluate the safety of embankment by physical modelling, similitude of deformation and strength characteristics of the material should be satisfied. However, the behaviour of rockfill materials cannot be replicated by using sandy soil. For example, unit weight of dumped rockfill material is higher than compacted gravelly sand but relative density of the sand is higher. This inconsistency affects various parameters of soil behaviour; stiffness, strength, dilatancy, etc. It is difficult to prepare a physical model which completely reflects the prototype condition by using different material, and the difference should be considered in design and interpretation of the tests.

2.4 Earthquake input motions

The frequency content of the input motion is as important as the response characteristics of the structural system in simulation of seismic behaviour. Resonance of the system induces large amplification and the structural system will be more vulnerable to earthquake.

In design of physical modelling, time (or frequency) should be scaled for input motion. At the same time, the frequency contents of the input motion must be modified or filtered considering the frequency response of the testing system. For example, the reliable linear frequency band of state-of-the-art hydraulic actuator systems for shakers is from 20 to 400 Hz or even narrower, and the contents out of this range in input motion cannot be reliably simulated or can cause unexpected resonance of the testing system.

Figure 2 shows the effect of band-pass filtering on earthquake input motion. The signal becomes smoother than the original record and the peak acceleration reduces. Moreover, the spectral acceleration in both short and long period range drastically reduce, hence the structural response subjected to this modified input motion will be different from the prototype conditions.

In design of experiments, the natural period of the structure should be compared with the frequency band of the shaking table especially when the performance is evaluated using any earthquake record as input motion. This inconsistency can happen for modelling of either a huge and rigid system or a small and flexible system. If the system is either too rigid or flexible so that the shaking table cannot replicate the

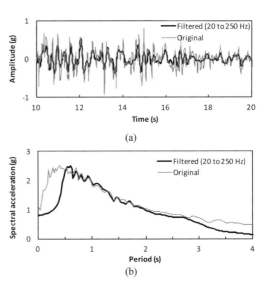

Figure 2. The effect of band-pass filtering on input signal; (a) time domain signal becomes smooth and the peak acceleration reduces. (b) spectral acceleration decreases at both short period and long period due to band-pass filtering. The frequency band is determined considering testing g-level (50g) and linear frequency band of the actuator.

input motion corresponding to the natural period of the system, the test can't simulate appropriate frequency response of the system and the amplification can be underestimated.

3 ALTERNATIVE APPROACHES

Basically, the reduced scale physical model for centrifuge testing is prepared according to the geometry of the prototype structure using equivalent materials. The main limitation in modelling of large embankment dams, especially for earthquake simulations, is that the test cannot be conducted at acceleration level corresponding to prototype-to-model scale, and there are other minor physical limitations. Alternative modelling and testing strategies to overcome these limitations are introduced.

3.1 Generalised scaling relations

The behaviour of soil is highly affected by self-weight stress, and the basic concept of centrifuge modelling is to replicate the stress distribution in the reduced scale model so that the behaviour of full scale prototype structure can be estimated from the test results. When a dynamic test is conducted at a certain g-level lower than general centrifuge testing case, the testing results will not directly reflect the response characteristics of prototype structure. However, the prototype structural response may be inferred from the physical model by analysing the effect of different stress condition on the global behaviour and utilizing appropriate scaling relations. For example, the natural period of a geotechnical system can be calculated as

$$T_n = c \frac{H}{V_S} \quad (1)$$

where c is a constant, T_n and H are the natural period and the height of the system respectively, and V_S is average shear wave velocity of the construction material, which represents the stiffness of the system. The stiffness of a certain type of soil is mainly determined by effective stress conditions hence the scaling relations of natural period between prototype and testing conditions can be easily derived. The same scale can be applied to time and various scaling relations can be derived by similar manner.

Iai et al. (2005) suggested generalised scaling relations, a special scaling principles which can be applied where ordinary scaling principles for centrifuge test cannot be directly applied because of either the testing machine or the largeness of prototype structure. The technique basically combines scaling principles for 1g shaking table test and Ng dynamic centrifuge test. In this method, if we define the scale between virtual intermediate prototype and model as η and that between virtual prototype and full scale prototype as μ, every scale factor for physical quantities in a test is derived by combination of η and μ.

The scaling relations derived by dimensional analysis can be validated by various method. Modelling of models (Ko, 1994) is the most common way to validate either scaling relations or modelling technology in geotechnical centrifuge modelling, and the idea can be also applied for generalised scaling. Figure 3 shows the concept of generalised scaling relations by indicating the role of each linear scaling factor, i.e. η and μ, and the alternative modelling of models can be also explained. In general cases, any points on an equivalent modelling line represent the same prototype condition, and the scaling principles or modelling technology can be verified by comparing the testing results from different model size/g-level conditions.

For generalised scaling, g-level can be reduced considering various testing limitations (A to A_1 or A_2) and the testing conditions will represent virtual prototypes (B_1 or B_2 from A_1 or A_2 respectively). Finally, the reduction can be compensated by additional scaling factor, μ_2 for B_2 to B, and μ_1 for B_1 to B. In this concept, a single model can be used to validate generalised scaling principles by varying g-level. At each g-level, μ should also change in verification of scaling relations, i.e. $\eta_1\mu_1 = \eta_2\mu_2$, and the results predict the same structural behaviour of the prototype. Ideally, any point in this domain can represent the same prototype scale by adjusting η and μ ($\eta\mu = N$).

3.2 Scaling of loading conditions and interpretation

The frequency contents of input motion as well as response characteristics of the structure determines the seismic response and performance. While

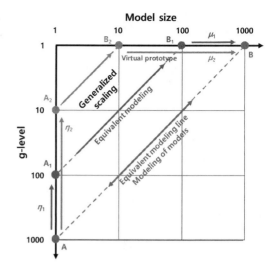

Figure 3. Modelling of models process to verify generalised scaling relations.

construction of physical model only requires scaling of linear dimensions, testing conditions should follow a consistent scaling method. When a seismic test using a 1:200 scaled embankment model is conducted at 50g for example, generalised scaling should be applied in scaling of input motion for both time and amplitude; $\eta = 50$ and $\mu = 4$. The scaling relation should be also applied on interpretation of testing result even though scaling on the physical quantities are not considered in design stage.

The frequency content of the input motion is limited by the performance of hydraulic actuator, the frequency band of filtered input motion should be reviewed prior to shaking tests. The frequency band in prototype scale generally gets narrower when the prototype to model scale is larger, and the cut-off frequency or period determined by performance of earthquake simulator and the scale can be easily calculated.

For example, the limitation for above scaling case can be calculated using generalised scaling factor for time $(\mu\mu_\varepsilon)^{0.5}\eta$. Calculating the minimum cut-off period of input motion for centrifuge test, it is 0.66sec for ordinary centrifuge scaling, $\eta=200$ and $\mu=1$, and it becomes 0.47sec for applying generalised scaling, $\eta=50$ and $\mu=4$. When short period input motion is required to evaluate the frequency response of the structure, generalised scaling has advantage in preserving short period input motion.

4 APPLICATIONS

Seismic responses of large embankment dams have been evaluated by centrifuge tests. The models and testing procedures were designed by generalised scaling relations. The details of two testing cases are introduces in this chapter.

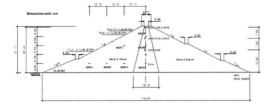

Figure 4. Design of testing model for 41 m high ECRD.

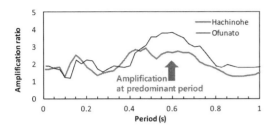

Figure 5. Amplification at crest for each input motion.

4.1 A rockfill dam model

Seismic responses of an earth cored rockfill dam (ECRD) subjected various earthquake input motions have been evaluated. The prototype embankment is 45 m high and 300 m long in up-/down-stream direction, and the testing model was designed by 1:135 scale considering the limitations of container dimensions. Generalised scaling factors for linear dimensions were selected as $\eta = 45$ and $\mu = 3$, hence the tests were conducted at 45 g. Number of transducers (e.g. accelerometers, displacement transducers) were installed to measure the response and displacement, and the details are given in Figure 4. In this case, two types of earthquake records (Hachinohe and Ofunato) were applied as input motions and various PGA were applied to investigate the effect of intensity.

The acceleration responses at each location were analysed, and the fundamental period of the embankment were calculated to validate the scaling relations. It was calculated using ratio of response spectra between base motion and crest (Okamoto, 1984). Figure 5 shows the ratios of response spectra for two different input motions. These two results show similar frequency responses of the embankment, and the predominant period from the experiment is close to the period calculated by earthquake monitoring data.

4.2 AG dam case

AG dam is a 131 m high rockfill dam with inclined impervious core zone located in a seismically active region in East Asia. The site is close to an active seismic fault, and the PGA for the MCE level determined by PSHA was 0.94 g. The physical modelling has been carried out to verify the seismic safety of the structure and to compare with the results from numerical simulation.

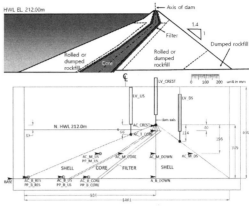

Figure 6. Design of reduced-scale model of AG dam and instrumentation layout.

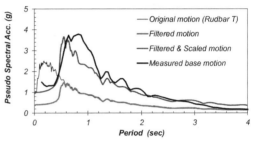

Figure 7. Frequency filtering effect on input motion for centrifuge test.

The model representing 131 m embankment was constructed in a 1460 mm long rigid container, which results in 1:350 scale and generalised scaling was applied. Large horizontal shaking force from the earthquake simulator proportional to centrifugal acceleration is required hence the centrifugal acceleration has to be limited. Therefore, η for generalised scaling was determined as 30 considering maximum force capacity of the hydraulic actuators, and it determined other scaling factors. Figure 6 shows drawings of the prototype structure and model design.

Figure 7 compares response spectra of original input motion and one modified for actual tests considering the frequency response of testing system. The cut-off high frequency of testing system and prototype to model scale determine the shortest period of response spectrum, hence short period (high frequency) component of the input motion, shorter than 0.55 sec, had to be eliminated by filtering process. If ordinary scaling for centrifuge modelling is applied, the cut-off period in the response spectra becomes higher and there is an advantage to preserve broad range of frequency components for the tests.

The response of the embankment could be evaluated by centrifuge test, and the results shows similar frequency response characteristics with numerical modelling. Even though the response or deformation couldn't be directly matched with numerical simulation due to limitations such as material, boundary conditions, large prototype-to-model scale, the testing results showed reasonable response and amplification characteristics and the full-scale prototype structure could be sufficiently inferred.

The application cases in this study only focused on the response characteristics of embankments. If the effect of pore fluid plays a key role in the behaviour of the system, e.g. liquefaction of foundation or embankment, appropriate scaling should be applied on the fluid; generally scaling of viscosity to overcome time-scaling conflict between dynamic time and dissipation phenomena.

5 CONCLUSIONS

The limitations in physical modelling of large embankment dams for experimental evaluation of seismic performance are reviewed and alternative scaling and modelling strategies are introduced. The scaling principles must be thoroughly reviewed for appropriate modelling of structural behaviour in design stage, and alternative scaling relations can be derived considering stress conditions in testing model and its effect on structural response characteristics. The generalised scaling concept can be experimentally validated by modified modelling of models process. The same scaling relations should be applied in loading conditions as well as interpretation of testing results. The limitations of testing equipment and its effect on structural behaviour should also be considered in design process of experiments.

Finally, two physical modelling cases of large embankment dams by generalised scaling were introduced. Frequency responses characteristics of the embankments could be reliably evaluated despite that the input motions have to be modified considering the frequency response of testing equipment. Even though this method cannot ideally duplicate the behaviour of full scale prototype structure, it can be sufficiently inferred from the physical modelling if key mode of response can be simulated. This modelling strategy may be applied in various geotechnical problems once the applicability of the scaling relations is verified.

REFERENCES

Dakoulas, P. & Gazetas, G. 1985. A class of inhomogeneous shear models for seismic response of dams and embankments. Soil Dyn Earthq Eng. 4(4): 166–182.

Iai S, Tobita T. & Nakahara T. 2005. Generalised scaling relations for dynamic centrifuge tests. Géotechnique. 55(5): 355–362.

Kim, M. K., Lee, S. H., Choo, Y. W., & Kim, D. S. 2011. Seismic behaviors of earth-core and concrete-faced rock-fill dams by dynamic centrifuge tests. Soil Dyn Earthq Eng. 31(11): 1579–1593.

Ko, H.Y. 1988. Summary of the state-of-the-art in centrifuge model testing, Centrifuges in Soil Mechanics, Craig, James and Schofield Eds. Balkema. Rotterdam. pp. 11–18.

Ng, C.W.W., Li, X.S., Van Laak, P.A. & Hou, D.Y.J. 2004. Centrifuge modeling of loose fill embankment subjected to uni-axial and bi-axial earthquake. Soil Dyn Earthq Eng. 24(4): 305–318.

Okamoto, S. 1984. Introduction to Earthquake Engineering. 2nd Edition. University of Tokyo Press. pp. 466–477.

Peiris, L.M.N., Madabhush, S.P.G. & Schofield, A.N. 2008. Centrifuge modeling of rock-fill embankments on deep loose saturated sand deposits subjected to earthquakes. J Geotech Geoenviron. 134(9): 1364–1374.

Sharp, M.K. & Adalier, K. 2006. Seismic response of earth dam with varying depth of liquefiable foundation layer. Soil Dyn Earthq Eng. 26(11): 1028–1037.

Centrifuge model tests on levees subjected to flooding

R.K. Saran & B.V.S. Viswanadham
Department of Civil Engineering, Indian Institute of Technology Bombay, Mumbai, India

ABSTRACT: The objective of the paper is to examine the deformation behaviour and stability of levees subjected to flooding through centrifuge model tests at 30g in a large beam centrifuge available at IIT Bombay. Three levee sections were modeled, i.e. i) levee without any drainage layer (or clogged drain), ii) levee with horizontal drainage layer and iii) levee with chimney drainage layer. The flood was induced using a custom developed and calibrated in-flight flood simulator. At the onset of flood and subsequent seepage, pore water pressures within levee section were measured using pore water transducers (PPTs). Digital image analysis was employed to trace displacement vectors, and downstream slope face movements at the onset of flooding during centrifuge tests. Levee section without any horizontal drain or clogged drain experienced a catastrophic failure. In comparison, the levee sections with horizontal drainage layer or chimney drain were found to be stable at the onset of flooding. Further, seepage and stability analyses were carried out and compared with centrifuge test results.

1 INTRODUCTION

1.1 General

In recent times, the frequent occurrence of the flood has been witnessed in the Indian subcontinent due to several natural causes (i.e. flash flood, course correction in the river and irregular monsoon). It not only destroys the levees, earthen embankment, and public utilities but also disturbs the socio-economic fabric of the nation. For the prevention of the disaster, Intergovernmental Panel on Climate Change IPCC (2013) recommended the improvisation and strengthening of levees, earthen embankment.

The main cause behind the failure in a levee is the development of high pore water pressure. It leads to decrease in the effective stress within the levee, which in turn may be translated in catastrophic slope failure on the downstream side. US army Corps of Engineers USACE (2000) suggests the provision of horizontal drainage (HD) layer and chimney drainage layer (CD) to dissipate the pore water pressure, as shown in Figure 1. Similarly, USACE (2000) also renders the guidelines for selecting drainage material and determining the minimum thickness of drainage layer (i.e. 600 mm).

Several researchers adopted numerical (Wang & Castay, 2012), analytical and 1-g physical modeling (Jia et al. 2009) and N-g physical modeling (Cargill & Ko 1983, Steedman & Sharp 2011; Saran & Viswanadham 2013) for examining the performance of levee subjected to seepage. Among these, the centrifuge (Ng) modeling is a state of the art method to investigate the performance of levees because it facilitates the simulation of real stress levels in the model

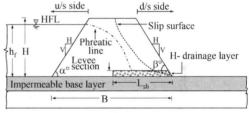

a) Levee with horizontal drainage layer

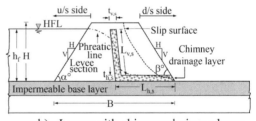

b) Levee with chimney drainage layer

Figure 1. Schematic diagram of typical levee sections.

due to elevated gravity level. The elevated gravity level can be achieved with the help of centrifugal acceleration by rotating a model in a horizontal plane about a vertical axis. However, in this approach, it is necessary to follow certain scaling principles while using the centrifuge modeling.

From the literature, it was found that very few researchers conducted centrifuge model tests to study the performance of levees subjected to flooding. Further, a comparative study on the levee with HD and CD

Table 1. Summary of scaling factors used in the present study.

Properties	Unit	Prototype	Model
Levee parameters			
Levee height (H)	m	1	1/N
U/s and d/s slope of levee (α, β)	(°)	1	1
Length of HD (L_{sh})	m	1	1/N
Length of CD ($L_{h,s}$, $L_{v,s}$)	m	1	1/N
Thickness of HD (t_{sh})	m	1	1/N
Thickness of CD ($t_{h,s}$, $t_{v,s}$)	m	1	1/N
Flood parameters			
Flood level (h_f)	m	1	1/N
Flood duration (t_{sf})	Days	1	$1/N^2$
Flood rate (r_f)	m/day	1	N
Soil parameters			
Cohesion (c)	kPa	1	1
Angle of internal friction (ϕ)	(°)	1	1
Unit weight of soil (γ)	kN/m^3	1	N
Seepage parameters			
Pore water pressure (u)	kPa	1	1
Seepage time (t_{sf})	Days	1	$1/N^2$
Coefficient of permeability (k)	m/sec	1	N

layer is also not present. Hence, in the present study, an attempt has been made to evaluate the performance of levee section with HD and CD layer. For this purpose, three centrifuge model tests were conducted. They are i) levee without any drainage layer (or clogged drain), ii) levee with a horizontal drainage (HD) layer and iii) levee with chimney drainage (CD) layer. The results were compared in terms of pore water pressure, elevation of the phreatic surface, deformation vectors, and face movements. Further, seepage and stability analyses were also carried out on levee section analogous to centrifuge tests and results were compared with the centrifuge test results.

2 SCALING FACTORS

As discussed earlier, centrifuge modeling requires the adoption of scaling relations for several parameters. The scaling relations should be considered for the properties of the model soils, parameters contributing to the performance of the levee section and the simulation of the flooding phenomenon. Table 1 summarises the scaling relations used in this study.

Figure 1 demonstrates the schematic diagram of levees with horizontal and chimney drainage layer. Several parameters related to levee geometry such as levee height (H), upstream and downstream slope angle (α° & β°) were scaled down in the model as per the scaling relations mentioned in Table 1.

2.1 Scaling relation for model materials

The soil, used in the construction of the levee model, was similar to the soil used in prototype levee. Hence, several properties (c and ϕ°) remains same as prototype values. The seepage through levee is influenced by pore water pressure (u), seepage time (t_{sf}). The pore water pressure remains identical in both model and prototype, whereas the seepage time (t_{sf}) in the model is reduced by N^2 times as compared to prototype. It facilitates the simulation of long-term seepage within a short span of time.

2.2 Scaling relations for simulation of flooding

Important parameters pertaining to the flooding phenomenon are flood rate (r_f), flood level (h_f), and flood duration (t_{sf}). In order to achieve identical hydraulic boundary conditions and subsequent seepage in the model and prototype, the simulated flood level (h_f), flood rate (r_f) and flood duration (t_{sf}) in the model were kept 1/N, N and N^2 times of the prototype, respectively.

3 CENTRIFUGE MODEL TESTS ON LEVEES SUBJECTED TO FLOODING

3.1 Test programme

In this study, three centrifuge tests were performed at 30 gravities in a 4.5 m radius large beam centrifuge available at IIT Bombay. They are i) levee without any drainage layer (S-1), ii) levee with HD layer (S-5) and iii) levee with CD layer (S-6). Table 2 summarises the details of parameters of levee models. The levee model, having a height (H) of 7.2 m, u/s slope inclination (α°) of 1.5V:1H and d/s slope inclination (β°) of 1.5V:1H, was constructed using the scaling factors according to 30 g. The HD layer, having a thickness (t_{sh}) of 0.6 m, was extended from d/s toe to the mid point of the bottom of the levee. Similarly, the horizontal portion of CD layer was extended from d/s toe to the mid-point of the bottom of the levee, whereas vertical portion was extended up to the elevation of HFL in the upstream side. The thickness of CD layer ($t_{s,h}$ and $t_{s,v}$) and angle of inclination of the vertical portion (δ°) were kept equal to 0.6 m and 6°, respectively.

3.2 Material properties of model soil

In this study, two types of soils were used. They are a) silty sand for construction of levee section and b) sand for the construction of drainage layer. The silty sand was formulated by mixing the sand and Kaolin in 4:1 ratio. The maximum dry unit weight $\gamma_{d,max}$ and optimum moisture content (OMC) of model soil were 19.75 kN/m^3 and 10.5%, respectively. The coefficient of permeability of model soil, moist-compacted at $\gamma_{d,max}$ and OMC, is 8×10^{-7} m/s.

The model soil (moist-compacted at $\gamma_{d,max}$ and OMC and saturated) was found to have an effective cohesion of 5 kN/m^2 and an effective angle of internal friction of 35° obtained by conducting consolidated undrained triaxial tests.

The sand used in the present study, having an effective particle size $d_{10} = 0.10$ mm, is classified as SP according to Unified Soil Classification System. The sand was found to have a maximum void ratio of 0.89

Table 2. Details of centrifuge model tests.

Test	H	α° & β°	Drainage layer Type	Length	Thickness
S-1	7200	56.3	–a	–a	–a
S-5	7200	56.3	HD	$L_{sh} = 7050$	$t_{sh} = 600$
S-6	7200	56.3	CD	$L_{h,s} = 7050$ $L_{v,s} = 6600$	$t_{h,s} = 600$ $t_{v,s} = 600$

–aNot applicable; H = height of Levee; HD = Horizontal drainage layer; CD = Chimney drainage layer; All dimensions are in mm and reported in prototype dimensions.

Figure 2. Perspective view of model test package.

and minimum void ratio of 0.58. The coefficient of permeability of sand placed at 85% relative density is 1.49×10^{-4} m/s. The cohesionless sand was found to have an angle of internal friction of 36° obtained by conducting direct shear tests.

3.3 Test procedure

The levee model was constructed in the strong box having internal dimensions of 760 mm × 410 mm × 200 mm. The base layer was constructed by compacting 20 mm thick Kaolin layer above the 30 mm thick silty sand layer. The pore pressure transducers (PPTs) were placed above the base layers at the geometrical points corresponding to left crest, mid-point of levee crest, right crest and toe region. As shown in the Figure 2, levee was constructed by compacting the levee soil, at OMC and $\gamma_{d,max}$, in 30 mm thick horizontal layers. The geometry of the levee model (i.e. height of levee and the angles of inclinations of u/s and d/s slopes of levee) was achieved with the help of wooden formwork, which was removed upon completion of model preparation. Food dye was placed at a constant interval to trace the seepage of water within the levee. Further, the L-shaped plastic markers were also placed within the horizontal soil layer at a horizontal and vertical spacing of 20 mm and 30 mm, respectively.

For the construction of drainage layer, sand was compacted at 85% relative density with the help of pluviation method. The end of drainage layer on d/s side was wrapped by a thin geotextile having anchorage length of 20 mm, in order to prevent the washing of sand particle due to seepage.

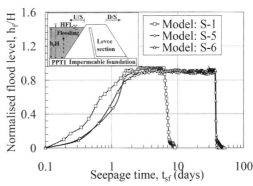

Figure 3. Variations in flood function with t_{sf}.

After the construction of the model, the model test package was mounted on the centrifuge basket. During the test, the 30 g was achieved by gradually increasing the angular speed of the centrifuge. Upon achieving at 30 g, the model was maintained for 5 minutes to attain equilibrium within the model. Further, the flood was triggered by switching on the In-flight Flood and Drawdown Simulator (IFDS) from the control room. The flood generated by IFDS was monitored by both a digital camera and PPT placed in u/s.

4 ANALYSIS AND INTERPRETATION OF RESULTS

This section reports the results of the centrifuge tests, which are expressed in term of variations in pore water pressure, elevation phreatic surface and face movement of d/s slope with seepage time. Further, the image analysis was also employed to plot the displacement vectors at the penultimate state.

4.1 Simulation of flooding

The flood was simulated by triggering the IFDS set-up upon attaining 30 gravities. The simulated flood function can be expressed in terms of normalised flood level (h_f/H) and seepage time (t_{sf}), as shown in Figure 3. The average values of flood rate (r_f), maintained for all the centrifuge tests, was 4.5 m/day. The Maximum elevation of flood level, which is also termed as High Flood Level (HFL), was kept equal to 6.6 m by maintaining the free board of 0.6 m.

4.2 Variation in pore water pressure

Figures 4a-4c depict the variation in pore water pressure within the levee section with the seepage time (t_{sf}). The values of pore water pressure measured at PPT4 ($u_{d/s}$) and PPT5 (u_{toe}) can be used to compare the performance of drainage layer.

The maximum values of u at PPT4 ($u_{d/s,max}$) in levee models S-1, S-5 and S-6 were found to be 51.3 kPa, 11.77 kPa and 5.20 kPa, respectively. Similarly, the maximum values of u at PPT5 ($u_{toe,max}$) in levee models

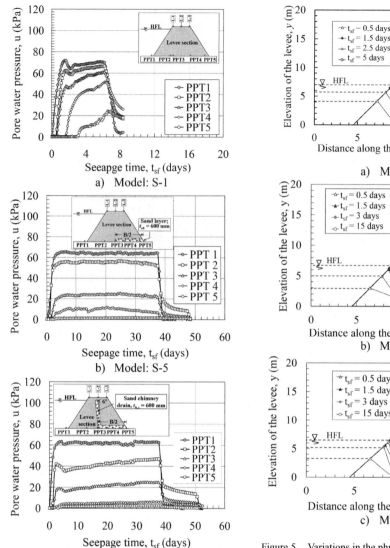

Figure 4. Variations in u with t_{sf}.

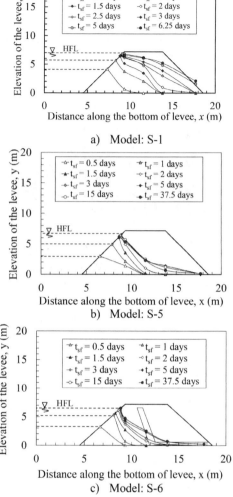

Figure 5. Variations in the phreatic surface with t_{sf}.

S-1, S-5 and S-6 were found to be 24.22 kPa, 3.32 kPa and 1.52 kPa, respectively. High values of $u_{d/s}$ and u_{toe} in S-1 highlight the need for an efficient and unclogged drainage layer to improve the stability of levee in the downstream side. Further, the negligible value of $u_{d/s}$ and u_{toe} in S-6, as compared to the S-5, indicates towards the better performance of chimney drain as compared to the horizontal drain, which can be attributed to the increased drainage face provided by a vertical portion of chimney drain in model S-6.

4.3 Elevation of phreatic surfaces

The elevation of phreatic surface indicates the extent to which a levee section is saturated due to seepage of water within the levee. The higher elevation can lead to the slope failure in the d/s side of the levee. Figures 5a-5c show the variation in elevation of phreatic surfaces with t_{sf} for all the centrifuge tests. In the case of model S-1, it can be noted that phreatic approached to the downstream slope face at the penultimate stage. On the contrary, phreatic surface in model S-5 and S-6 remained at a significantly low elevation. It concludes that the placement of HD and CD layer is an effective way to dissipate the pore water pressure and impart the stability to the levee.

4.4 Image analysis

The images captured by digital camera were analysed with the help of ImageJ (ImageJ, 2012), a Java based software, to determine the displacement vectors and slope face movement (s_f). The movement of movable marker was traced with respect to permanent markers in order to evaluate the s_f. The levee model with

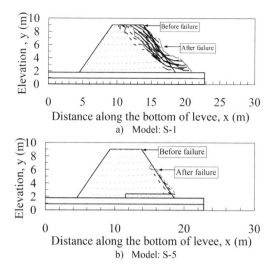

Figure 6. Displacement vectors at penultimate stage.

chimney drain (S-6) did not undergo any significant deformation even after the seepage time of 37.5 days. Hence, in this section, the displacement vectors and slope face movements (s_f) are plotted only for levee models S-1 and S-5.

4.4.1 Displacement vectors

Figures 6a-6b demonstrate the displacement vectors of the levee section at the t_{sf} of 6.17 days and 30.66 days. It can be noticed that levee without any drainage layer experienced a catastrophic failure, whereas the levee with HD layer was subjected to a minor localised failure in the downstream side. However, the levee with CD layer, model S-6, remained intact even after the t_{sf} of 37.5 days.

4.4.2 Face movement

Figures 7a-7b depict the variation in normalised slope face movements (S_f/H) with the normalised levee height (z/H) and seepage time (t_{sf}). Here, the normalised slope face movements (S_f/H) and normalised levee height (z/H) are expressed as the ratio of slope face movement (S_f) and elevation of a geometrical point on d/s slope (z) with the height of levee (H), respectively. It can be noted that the levee without any drainage layer experienced catastrophic failure on the downstream side, which is evident from $s_{f,max}/H$ of 0.576 at t_{sf} of 6.17 days. On the other hand, localised failure in levee with a horizontal drain was observed, which is clear from $s_{f,max}/H$ of 0.06 at t_{sf} of 30.66 days.

5 SEEPAGE AND STABILITY ANALYSIS OF CENTRIFUGE MODEL TESTS

The coupled flow-deformation analyses on levee models, analogous to the centrifuge tests, were conducted by using a finite element based software Plaxis-2D (Plaxis, 2012). It includes the seepage, deformation

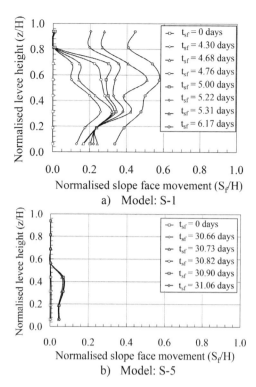

Figure 7. Variations in S_f/H with z/h and t_{sf}.

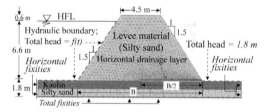

Figure 8. Schematic diagram of numerical scheme.

and stability analyses of levees. For this purpose, the geometry of levee sections, hydraulic boundaries, and stress boundaries were kept similar to the centrifuge tests. Figure 8 presents the numerical scheme adopted in this study. The flood was simulated in the form of flood function shown in Fig. 3.

Figures 9a-9c depict the pore water pressure contours at the steady state seepage. It was observed that the values of $u_{toe,max}$ in models S-1, S-5 and S-6 were found to be 20.52 kPa, 3.02 and 1.8 kPa, respectively. The drastic reduction in $u_{toe,max}$ in S-5 and S-6 reflects the efficient dissipation of pore water pressure by HD and CD layer. Furthermore, Table 3 summarises the results of coupled flow-deformation analyses.

Similarly, the maximum normalised face movement of downstream slope in model S-5 and S-6 were found to reduce to 0.04 and 0.01, respectively, as compared to the $S_{f,max}/H$ of 0.42 in the case of model S-1. Similarly, the minimum factor of safety, for models S-5 and S-6, is also found to increase to the 1.09 and 1.1

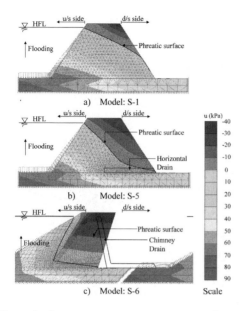

Figure 9. Pore water pressure contours at steady state seepage obtained from Plaxis-2D.

a) Model: S-1
b) Model: S-5
c) Model: S-6

2. The values of $u_{toe,max}$ in levee models with HD and CD was found to be 3.32 and 1.52 kPa, respectively. They are significantly less as compared to the levee without any drainage layer, wherein the $u_{toe,max}$ was found to be equal to 24.22 kPa. This highlights the efficacy of HD and CD layer in dissipating the pore water pressure.

3. The results of numerical models are comparable to the centrifuge tests. The dissipation of pore water pressure helped in improving the deformation and stability behavior of levees. It is evident from the values of $s_{f,max}/H$ obtained in S-1, S-5 and S-6, which were found to be 0.42, 0.04 and 0.01 kPa, respectively. Further, the minimum values of FOS for levee models S-1, S-5 and S-6 were found to be 0.92, 1.09 and 1.1, respectively.

In this study, the efficacy of horizontal and chimney sand drainage layer in dissipating the pore water pressure and enhancing the overall stability to the levee section was examined. However, the future scope of study involves the exploring use of the alternative materials in place of conventionally used sand, as drainage material, which can withstand sustained and long-term seepage.

Table 3. Summary of tests results.

	N-g tests			Plaxis-2D		
Parameter	S-1	S-5	S-6	S-1	S-5	S-6
Drainage	–[a]	HD	CD	–[a]	HD	CD
$t_{failure}$ (days)	4.30	30.66	–[b]	5.20	30.66	–[b]
$u_{d/s,max}$ (kPa)	51.30	11.77	5.20	45.3	11.77	5.20
$u_{toe,max}$ (kPa)	24.22	3.32	1.52	20.52	3.02	1.80
$S_{f,max}/H$	0.576	0.069	–[b]	0.42	0.04	0.01
FOS	–[a]	–[a]	–[a]	0.92	1.09	1.10

–[a] Not applicable; –[b] Failure did not occur even after 37.5 days; FOS = Minimum factor of safety.

as compared to the FOS of 0.92 in the case of model S-1. Hence, it can be concluded that the reduction in pore water pressure near the toe region was also reflected in the deformation and stability behavior of the levee.

6 CONCLUSIONS

In the present study, centrifuge model tests were carried out on levees with and without drainage layer (horizontal or chimney) subjected to flooding having flood rate (r_f) of 4.5 m/days. Based on the analysis and interpretation of results, the following conclusions can be drawn.

1. A levee without any drainage layer experienced catastrophic slope failure in downstream side. On the other hand, a levee with horizontal drainage layer experienced only localised failure in downstream side slope. Furthermore, levee with chimney could maintain stability in d/s slope without undergoing any notable deformation.

REFERENCES

Cargill, K.W. & Ko, H.Y. 1983. Centrifugal Modeling of Transient Water Flow. *Journal of Geotechnical Engineering, ASCE*, 109(3): 281–300.

ImageJ User guide. 2012. Version 1.45s, National Institutes of Health. USA.

Intergovernmental Panel on Climate Change (IPCC). 2013. Climate Change – The Physical Science Basis, Contribution to the Fifth Assessment Report of the IPCC, UNEP, New York, USA.

Jia, G.W., Zhan, T.L.T., Chen, Y.M. & Fredlund, D. G. 2009. Performance of a Large-scale Slope Model Subjected to Rising and Lowering Water Levels. *Engineering Geology* 106(1-2): 92–103.

PLAXIS 2012. Reference manual: PLAXIS 2D – version 9.0, R.B.J., Bringkgreve, E., Engin, W.M., Swolf (Eds.), Delft University of Technology, Delft, Netherlands.

Saran, R.K. & Viswanadham, B.V.S. 2013. Use of geocomposites as an internal drain in levee subjected to seepage: centrifuge model study. Bhatia & Boyle, (eds.), *Geosynthetics*-2013, Long Beach, CA, USA: 807–817.

Steedman, R.S. & Sharp, M.K., 2011. Physical Modelling Analysis of the New Orleans Levee Breaches. *Geotechnical Engineering* 164(6): 353–372.

U.S. Army Corps of Engineers (UACE). 2000. Engineering Manual for design and construction of levees, Department of the Army, U.S. Army Corps of Engineers Washington, DC.

Wang, X. & Castay, M. 2012. Failure Analysis of the Breached Levee at the 17th Street Canal in New Orleans during Hurricane Katrina. *Canadian Geotechnical Journal*, 49(7): 812–834.

Centrifuge model test of vacuum consolidation on soft clay combined with embankment loading

S. Shiraga
Kinjo Rubber Corporation, Osaka, Japan

G. Hasegawa, Y. Sawamura & M. Kimura
Kyoto University, Kyoto, Japan

ABSTRACT: Vacuum consolidation is a ground improvement method for soft clays which decreases pore water pressure by applying vacuum pressure through installed drains to promote consolidation. Simultaneous application of static embankment loading during vacuum consolidation is a desirable and promising ground improvement method, and has seen increased application in recent years. However, there are some unclear aspects of the behavior of pore water pressure during vacuum consolidation. It is important to predict the distribution of water pressure during combined vacuum consolidation and loading for evaluating embankment stability. In this study, a centrifugal test was conducted at centrifugal acceleration 50 g and the combined ground improvement was reproduced using an air cylinder and a loading plate for loading and a drain material for vacuuming. The tests revealed that the negative pressure was greater in areas closer to the drain, and that the increase in water pressure during embankment loading was suppressed in regions.

1 INTRODUCTION

Figure 1 shows an outline of the vacuum consolidation method. Vacuum consolidation involves applying vacuum pressure through installed drains using vacuum pumps to consolidate soft clay (Chu et al. 2008, Chai et al. 2008). Figure 2 shows the procedure of vacuum consolidation, separated into four stages:

1. Application of vacuum pressure after installing drains in the clay foundation and setting pumps
2. Vacuum pressure propagates to drains
3. Negative pressure gradually propagates from the drains
4. Vacuum pressure in the surrounding soil reaches steady state at a value close to that of the drains

Vacuum consolidation stabilizes soft soils without shear deformation, reducing the lateral displacement during embankment (Chai et al. 2006 and Hayashi et al. 2002). In practice, embankment stability is evaluated by the degree of consolidation based on the 1D solution by Barron (1948). Many 2D FEM analyses have been conducted to evaluate the effect of vacuum consolidation during embankment loading (Nguyen et al. 2015 and Tashiro et al. 2015). However, there are few studies considering three dimensional consolidation behavior around the drain. It is important to comprehend the three dimensional distribution of pore water pressure for predicting the consolidation process during combined use, especially when evaluating stability during embankment

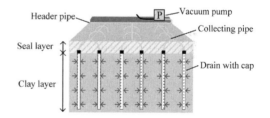

Figure 1. Outline of vacuum consolidation method.

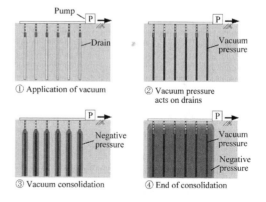

Figure 2. Propagation process of negative pressure.

loading. Centrifuge model tests were conducted to comprehend the distribution of water pressure during vacuum consolidation combined with embankment loading.

2 CENTRIFUGE MODELING TEST

2.1 Cases of centrifuge modeling

Centrifuge tests were conducted for two cases at 50 g acceleration. We modeled embankment loading only (Case-1) and vacuum consolidation combined with embankment loading (Case-2). In this paper, Case-1 and Case-2 are compared based on observed pore pressure measurements and progression of the consolidation process.

2.2 Model ground

Figure 3 shows the schematic of Case-2. The conditions of Case-1 were the same as Case-2, but without a drain. All experiments were conducted on Fujinomori clay. Figure 4 shows e-logp relationship of Fujinomori clay as determined by oedometer tests and Table 1 shows some basic index properties of Fujinomori clay. Silicon grease was applied to the walls of the box to minimize the friction between the soil and box during the test. First, the water content of the clay slurry was adjusted to 1.5 times the liquid limit ($w_L = 56.4\%$). Next, the slurry was poured into a rigid box after 2 hours of degassing and mixing. Then, self-weight consolidation of the sample and loading plate was conducted at 50 g centrifugal acceleration following centrifugal self-weight consolidation of the slurry. The clay sample had a layer thickness of 54 mm and a surface area of 240 × 240 mm in model scale, and a layer thickness of 2.7 m and surface area 12.0 × 12.0 m in prototype scale.

Silica sand was used as a drainage layer overlying the clay. This sand layer was made by water pluviation with a void ratio of 0.6. The sand layer had a layer thickness of 10 mm in model size and 0.5 m in prototype size.

Layers of water and oil were laid on the soil surface after preparing the clay and sand layers. The water layer prevents the ground for drying and the thin oil layer prevents the water from evaporating.

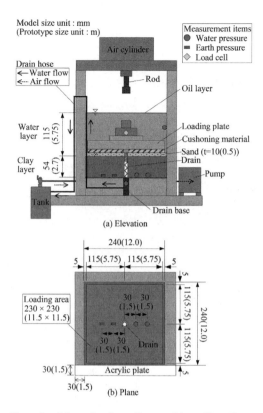

Figure 3. Schematic of centrifuge model test (Case-2).

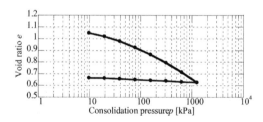

Figure 4. e-logp relationship of Fujinomori clay.

2.3 Modeling of vacuum consolidation

A model drain is installed at the center of ground and vacuum consolidation is simulated by vacuuming inside the drain with a pump. The drain is connected to a pump via a water storage tank through a drain hose. The pump draws out water inside the drain model and creates a pressure difference between the drain and the surrounding clay. This pressure difference causes water flow into the drain from the clay as consolidation progresses. The water flowing into the drain model is discharged to the tank.

Figure 5 shows (a) model drain and (b) core material of the water collecting section. The drain was composed of a base, water collecting section, drain hose, and a pore-water pressure gauge used to measure the vacuum pressure acting on the drain. An aluminum pipe with an outer diameter of 12 mm and an inner diameter of 10 mm was used for the core material. The pipe has many holes of 2 mm diameter. The aluminum pipe is used as a core material to prevent deformation of the core material by restraint pressure during the experiment. The nonwoven fabric filter covers the core material of the water collecting section and keeps clay from entering the drain. Nonwoven fabric is used in actual sites for the same reason. Glue is applied to the upper 15 mm of the nonwoven fabric filter to make it impermeable to water. This is to prevent water on the ground from being drained through the sand layer by the drain whose inside is vacuumed.

2.4 Modeling of embankment loading

Embankment loading was simulated by the air cylinder and loading plate shown in Figure 6. An air cylinder pushes out the rod with compressed air, whose pressure

Table 1. Properties of Fujinomori clay.

Item	Value
Unit volume weight γ_t (kN/m³)	15.7
Compression index C_c	0.32
Consolidation coefficient c_v (cm²/day)	520
Permeability at the center of the clay before starting the test k (m/s)	5.1×10^{-9}
Slope of $e - \log k\lambda_k$	0.15
Void ratio at the center of the clay before starting the test e_0	0.9
Plasticity index I_p (%)	31.3

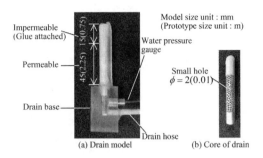

Figure 5. (a) Model drain and (b) Core material of the water collecting section.

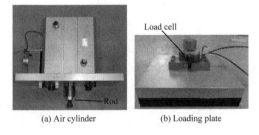

Figure 6. (a) Air cylinder and (b) Loading plate.

is adjusted loading the steel plate as shown in Figure 2. The loading plate has a thickness of 25 mm and a surface area of 230 × 230 mm and NR sponge rubber (10 mm thickness, Young's modulus E = 500 kPa) is affixed to the plate and used as a cushioning material between the ground and the plate. It has previously been confirmed that uniform loading on the sand layer can be applied using NR sponge rubber as a cushioning material. In Case-2, a hole of 20 mm diameter was drilled in the center of the loading plate to prevent loading on the drain and equally apply the embankment load to the ground.

The load acting on the loading plate by the air cylinder was measured by a load cell attached to the upper surface of the plate. By controlling the increase of the load on the plate, the loading rate of an actual embankment construction was reproduced. Also, by keeping the load constant during consolidation the persistent embankment load was reproduced.

2.5 Testing procedure

The experiment procedure for each case is shown below. Embankment loading in the tests aimed to simulate an embankment construction with height 3 m at a construction speed of 5 cm/day by an air cylinder and a loading plate.

Case-1

- Set the rigid box in centrifuge machine and apply 50 g to the box
- Consolidate until the excess pore water pressure in the clay layer dissipates
- Apply embankment loading

Case-2

- Set the rigid box in centrifuge machine and apply 50 g to the box
- Consolidate until the excess pore water pressure in the clay layer dissipates
- Apply vacuum pressure
- Apply embankment loading about 50 days (prototype scale) after the start of vacuuming

3 TEST RESULTS

The following results are written in prototype scale and the values on the vertical axis of the following graphs show the change of the pressure relative to the initial pressure at the time when 50 g was loaded to the box and excess water pressure dissipated. Due to evaporation of water and discharge through the drain, the water level in the box decreased throughout the tests. Accordingly, vertical pressure and excess pore water pressure are corrected using the change of the water level in the box.

3.1 Case-1: Embankment loading on soft ground

3.1.1 Change of water level in the box
Figure 7 shows the water pressure at 3.0 m depth. Water pressure at this point decreased by about 1 kPa from the start of this test and the change of water level in the box was minimal.

3.1.2 Embankment load
Figure 8 shows change of bearing pressure acting on the ground through the loading plate with time. The pressure is calculated by dividing the load measured in the load cell by the area of the loading plate (11.5 m × 11.5 m). A pressure of 55 kPa was applied on the ground over 55 days after the start of loading and the pressure increased at a rate of 1.00 kPa/day during embankment construction. After the end of loading, a pressure of 54~55 kPa was maintained during consolidation. This loading simulated construction

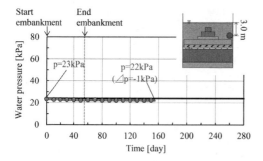

Figure 7. Water pressure at 3.0 m depth (Case-1).

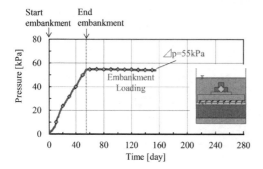

Figure 8. Bearing pressure acting on the ground through the loading plate (Case-1).

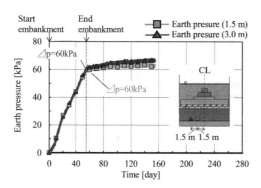

Figure 9. Vertical stress at the bottom of the clay layer (Case-1).

of an embankment with height 3.06 m at a construction speed 5.6 cm/day.

3.1.3 Vertical stress

Figure 9 shows change of vertical stress at the bottom of the clay layer with time. There are two measurement points (1.5 m away and 3.0 m away from the center of the drain). Vertical stress increased by 60 kPa at a rate of 1.01 kPa/day at both points. It is thought that even loading was simulated.

3.1.4 Pore-water pressure

Figure 10 shows the change of pore water pressure with time. There are two measure points (1.5 m away and 3.0 m away from the center of the drain). Yoshimura &

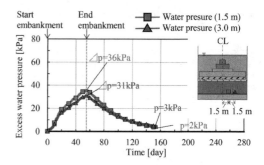

Figure 10. Pore-water pressure at the bottom of clay layer (Case-1).

Tohda (1997) confirmed the friction between the side wall of the box and the ground in centrifuge tests. In this test, silicon grease was applied to the walls of the box. However, excess pressure increased during embankment loading by 36 kPa at 1.5 m away from the center and 31 kPa at 3.0 m away, resulting in a difference of 5 kPa. This difference suggests that there was still some friction. The loss due to friction between the side wall of the box and the soil is smaller and water pressure tends to increase more in areas closer to the center of the ground. As a result, it is thought that the difference in the increase of water pressure happened during embankment loading. The loading condition can be described as almost one-dimensional loading under drainage condition.

3.2 Case-2: Vacuum consolidation combined with embankment loading

3.2.1 Change of water level in the box
Figure 11 shows the water pressure at 3.0 m depth. Water level started decreasing from the start of vacuum application and kept decreasing until the end of the test. The cause seems to be drainage from the model drain to the storage tank.

3.2.2 Negative pressure acting on the drain
Figure 12 shows negative pressure acting on the drain. The vacuum pressure increased to −43 kPa when vacuuming is started, but decreased with time until −30 kPa at the end of the test. The drain hose connecting the drain and the pump passed through a position higher than the water surface. A decrease in the water level made larger a difference in elevation between the hose and the water surface and the head loss increased, causing a decrease in vacuum pressure.

3.2.3 Embankment load
Figure 12 shows the change of pressure acting on the ground by the loading plate with time. The pressure is calculated by dividing the load measured in the load cell by the area of the loading plate (11.5 m × 11.5 m). A pressure of 56 kPa was applied on the ground in 55 days after the start of loading and the pressure increased at a rate of 0.93 kPa/day.

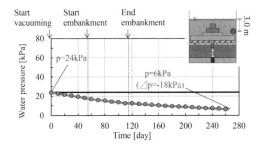

Figure 11. Water pressure at 3.0 m depth (Case-2).

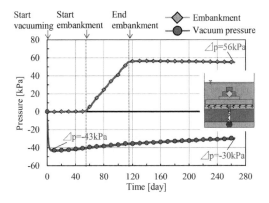

Figure 12. Pressure acting on the ground and negative pressure acting on the drain (Case-2).

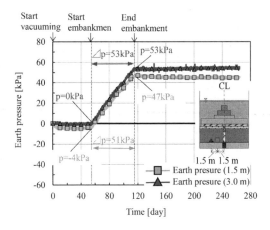

Figure 13. Vertical stress at the bottom of clay layer (Case-2).

After the end of loading, a constant pressure of 54~55 kPa was kept during consolidation. This loading simulated construction of an embankment with height 3.06 m, construction speed 5.6 cm/day.

3.2.4 Vertical stress

Figure 13 shows the change of vertical stress at the bottom of clay layer with time. There are two measurement points (1.5 m away and 3.0 m away from the center of the ground). Vertical stress increased during embankment loading by 51 kPa at 1.5 m away from the center and 53 kPa at 3.0 m away. Increase of vertical stress became almost the same with the pressure acting on the ground (55 kPa).

3.2.5 Pore-water pressure

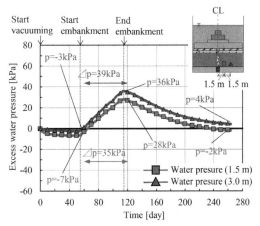

Figure 14. Pore water pressure at the bottom of clay layer (Case-2).

Figure 14 shows the change of pore water pressure at the bottom of clay layer with time. There are two measurement points (1.5 m away and 3.0 m away from the center of the ground). Excess pore water pressure decreased from the start of vacuuming. Excess pore water pressure decreased by 7 kPa at a point 1.5 m away and 3 kPa at a point 3.0 m away from the center and the decrease of pore water pressure at both points stabilized 55 days before the start of embankment loading. It was confirmed that during vacuum consolidation, the magnitude of negative pressure was greater in areas closer to the drain. Notably, the magnitude of the negative pressure in the clay was smaller than the vacuum pressure acting on the drain. In practice, multiple drains are installed in the foundation and the negative pressure acting on each drain is almost the same as the negative pressure in the soil at the end of the vacuum consolidation due to the influence of nearby drains. However, in this experiment only one drain material was installed. Additionally, some vacuum pressure is presumed to escape through the ground surface. We aim to evaluate the causes of this discrepancy further in future studies.

Excess water pressure increased from the start of loading. Excess pressure increased to 28 kPa (increment 35 kPa) at a point 1.5 m away from the center and 36 kPa (increment 39 kPa) at a point 3.0 m away from the center. The results of Case-1 show that the increase of pore water pressure during embankment loading was greater in regions closer to the drain due to friction between the side wall and the soil. But in Case-2, the increase in water pressure during the embankment loading was smaller in regions closer to the drain installed at the center of the ground. This result indicates that suppression of the increase of pore water pressure is greater in regions closer to the drain material during vacuum consolidation combined with the embankment loading. Excess water pressure dissipated at the same rate at the both points during consolidation period. Excess pressure decreased to $-2\,\text{kPa}$ at a point $1.5\,\text{m}$ away from the center and $4\,\text{kPa}$ at a point $3.0\,\text{m}$ away from the center at the end of the test (165 days from the end of loading).

Closer to the drain, greater propagation of negative pore water pressures was observed, resulting in greater suppression of the pore water pressure increase during embankment loading. It was confirmed that the behavior of the pore water pressure varies depending on the distance from the drain.

4 CONCLUSION

In this research, centrifuge model tests were conducted to comprehend the pore water pressure distribution in soft clay subjected to vacuum consolidation combined with the embankment loading. The tests confirmed the progression of negative pore water pressures in the clay, and provided insight into the interaction of vacuum consolidation and embankment loading and its effects on pore water pressure. Two centrifuge tests were carried out: simple embankment loading on soft clay (Case-1) and vacuum consolidation combined with embankment loading (Case-2). A summary of results and findings obtained are shown below.

1. It was confirmed that the uniform loading on the ground was achieved using a loading device consisting of an air cylinder and loading plate with the loading speed controlled.
2. By comparing water pressure measurements within the drain and the sample clay, it was confirmed that vacuum consolidation can be modeled using a vacuum pump and an aluminum drain wrapped with a nonwoven fabric filter with a vacuum pressure in the drain base.
3. It was confirmed that the change in pore water pressure during vacuuming is greater in regions closer to the drain. Consequently, the pore water pressure increase due to embankment loading was suppressed in these regions, demonstrating that the pore water pressure profile varies with distance to the drain.

Moving forward, FEM analysis of this experiment will be conducted and analysis under conditions close to actual site conditions where multiple drain materials are installed will be conducted.

REFERENCES

Barron, R.A. 1948. Consolidation of fine-grained soils by drain wells. *Trans. ASCE* 113: 718–742.

Chai, JC. Hayashi, S. & Carter, JP. 2006. Vacuum consolidation and its combination with embankment loading. *Canadian Geotechnical Journal* 43(10): 177–184.

Chai, JC. Miura, N. & Bergado, DT. 2008 Preloading clayey deposit by vacuum pressure with cap-drain: analyses versus performance. *Geotextiles and Geomembranes* 26(3): 220–230.

Chu, J. Yan, S. & Indraranata, B. 2008. Vacuum preloading techniques – recent developments and applications. In Milind V. Khire, Akram N. Alshawabkeh & Krishna R. Reddy (eds), *Geosustainability and Geohazard Mitigation*; Proc. GeoCongress., New Orleans, 9-12 March 2008. : 391–398.

Hayashi, H. Nishikawa, J. & Sawai, K. 2002. Improvement effect of vacuum consolidation and prefabricated vertical drains in peat ground. In CI-Premier (eds.), *Ground Improvement Techniques*; Proc. intern. Conf., Kuala Lumpur, 26-28 March 2002. : 391–398.

Nguyen, HS. Tashiro, M. Inagaki, M & Noda, T. 2015. Simulation and evaluation of improvement effects by vertical drains/vacuum consolidation on peat ground under embankment loading based on a macro-element method with water absorption and discharge functions. *Soils and Foundations* 55(5): 1044–1057.

Tashiro, M. Nguyen, HS. Inagaki, M. Yamada S & Noda, T. 2015. Simulation of large-scale deformation of ultra-soft peaty ground under test embankment loading and investigation of effective countermeasures against residual settlement and failure. *Soils and Foundations* 55(2): 343–358.

Yoshimura, H. & Tohda, J. 1997. Fe elastic analysis on mechanical behavior of buried flexible pipes measured in centrifuge model test. *Proc. of Japan Society of Civil Engineers* No. 596/III-43: 175-188. (in Japanese).

17. Geohazards

Effects of viscosity in granular flows simulated in a centrifugal acceleration field

Miguel Cabrera
Departamento de Ingeniería Civil y Ambiental, Universidad de los Andes, Bogota, Colombia

Patrick Kailey
Department of Civil and Natural Resources Engineering, University of Canterbury, Canterbury, New Zealand

Elisabeth T. Bowman
Department of Civil and Structural Engineering, The University of Sheffield, Sheffield, UK

Wei Wu
Institute of Geotechnical Engineering, University of Natural Resources and Life Sciences, Vienna, Austria

ABSTRACT: Centrifuge modelling is increasingly employed as an experimental tool for the simulation of dry and wet granular flows, providing good agreement in the representation of highly stress-dependent materials. However, the scaling relations of the interstitial fluid remain a challenging task, focusing on the matching of model and prototype processes. In this paper we present four different techniques for the simulation of the interstitial fluid (i.e. water W, water-glycerine W-G, water-methylcellulose W-M, and water-kaolin W-K). The fluid is mixed with a granular material, flowing down an inclined plane in a model tested in a drum or beam centrifuge, respectively. The granular flow behaviour developed in each case is compared in terms of the flow height, velocity and flow regimes, according to dimensional scaling. We note that in both centrifuge configurations, different flow regimes, such as inertial, collisional, or macro-viscous, can be observed, depending on the careful selection of particle size, fluid viscosity, flow rate and slope dimensions.

1 INTRODUCTION

A common characteristic of mass flows like debris flows, rock avalanches, and mudflows is that gravity is the main driving force, defining the packing density during flow and the intensity and duration of the main interactions between particles and their surrounding media (i.e., particle-particle, particle-fluid, and fluid-fluid). At the same time, gravity delimits the occurrence of phase separation, inverse segregation, and mass consolidation (Gray and Ancey 2009, Johnson et al. 2012, Leonardi et al. 2015).

In the exploration of the physics of granular flows, it is important to account for the role and scaling of gravity in laboratory-scale models, and its influence on the selection of the interstitial fluid. With this motivation in mind, we present four different techniques for the simulation of the interstitial fluid (i.e., water, glycerine, water-methyl cellulose, and water-kaolin) and compare the resultant behaviour in a centrifugal acceleration field.

The most common laboratory test for the study of granular flows is the inclined plane. The inclined plane shares many similarities with free-surface flows in most geophysical processes, being characterized by a mass flowing over a sloping bed with occasional lateral confinement. This simple configuration has been extensively researched experimentally (Iverson 1997, Ancey 2001, Sanvitale and Bowman 2017), and has helped to develop an understanding of the physics and formulate constitutive laws for their numerical simulation (Savage and Hutter 1989, Iverson and Denlinger 2001, Jop et al. 2006).

The research of granular flows in a geotechnical centrifuge was initially reported in 2006 in the study of granular flows down an incline (Vallejo et al. 2006), and in the study of creeping granular motion in a rotating drum (Arndt et al. 2006). Continuing research has broadened the model flow configurations, testing granular flows down curved channels (Bowman et al. 2010, Imre et al. 2010) and studying the kinematics involved during the emptying of a silo (Dorbolo et al.

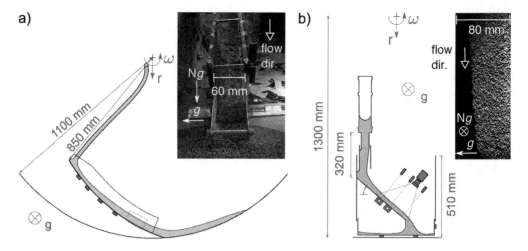

Figure 1. Experimental models sketch. a) Drum centrifuge model, and b) Beam centrifuge model. Insets: a) photography after test, and b) photography during test.

2013, Mathews 2013). In this paper, we present experiments performed in a curved channel tested in a drum centrifuge (Kailey 2013), and in an inclined channel tested in a beam centrifuge (Cabrera 2016).

In a debris flow, the combined action of particle and fluid forces act simultaneously. In this system, particle interactions are strongly linked with the fluid drag through the porous media, meaning that there is no strong distinction between seepage flow and the overall flowing mass. However, in an experimental model, the decision of scaling the interstitial fluid viscosity η becomes a challenging task, being necessary to balance the viscous shear resistance with the pore-pressure diffusion (Iverson 2015). In such a scaled system, the controlling parameters are the flow length L, flow height H, permeability k, granular medium stiffness E, particle and fluid density ρ_f and ρ_p, and fluid viscosity η. In a centrifugal acceleration field, inertial effects and diffusional effects scale differently over the same time period, being controlled by a factor of N^{-1} and N^{-2}, respectively (Garnier et al. 2007). In this paper we present two strategies for the scaling of the viscous pore fluid in the simulation of particle-fluid flows in a geotechnical centrifuge.

2 METHODS

2.1 Experimental models

2.1.1 Drum centrifuge model

The ETH Zurich drum centrifuge has a nominal working radius of 1.1 m and a maximum payload of two tonnes at 440 g. For the experiments described here, conducted at 40 g, a slurry of the selected fluid and solids was poured down the central axis of the centrifuge (Bowman et al. 2010) to run via a feed tube to the head of the 60 mm wide channel (see Fig. 1.a). The flow accelerated then decelerated down the curved channel at a decreasing slope from 36 to 12 degrees until it reached the bed formed by the drum itself, whereupon it became unconfined and then to a halt as a fan. PPTs in the base of the channel recorded the passing of the flow front while a small high speed camera was mounted to view through the transparent sidewall at a position 420 mm from the head of the channel (280 mm from the exit). The particles were comprised of silty-sand mixtures. Three different masses of solids were used in the experiments (1.0, 1.75, and 2.5 kg) with three initial moisture contents (33%, 36% and 39%) with three different fluids-water, a water-methylcellulose mixture and a water-glycerine mixture. The two latter fluids were designed to have a kinematic viscosity of 0.04 Pa s at the expected mean shear rate of the flow, although the water-methyl cellulose mixture was non-Newtonian (see Fig. 2).

2.1.2 Beam centrifuge model

Experiments were performed in the beam centrifuge (Trio-Tech 1231) at the University of Natural Resources and Life Sciences, Vienna (BOKU). The centrifuge has a nominal radius of 1.3 m, maximum payload capacity of 0.9 tonnes at 100 g, and can hold maximum model dimensions of 540 mm wide, 560 mm long, and 560 mm high.

For this series of experiments, a cylindrical silo was filled with 1.0 kg of particle-fluid mixture. Once the desired angular velocity was reached, the mixture was released from the silo through a corrugated flexible tube towards an inclined plate of 400 mm long and 80 mm wide, covered by a 2 mm thick rubber sheet, confined between a 15 mm thick aluminium wall and a 30 mm thick Plexiglass GS window (see Fig. 1.b). At the end of the inclined plate the flowing mixture fell into a collection box 480 mm long, 76 mm wide, and 80 mm high (Cabrera and Wu 2017b).

In this series of experiments, the granular material employed was composed of glass beads mixed with water or a kaolin-water dispersion with its viscosity controlled by the kaolin concentration. All experiments were performed at a fixed inclination of 30 degrees and equivalent centrifugal accelerations of 10, 15 and 20 g. Additionally, 1 g experiments were performed outside the centrifuge, providing a reference behaviour of the particle-fluid mixture flowing down the inclined plate.

2.2 Materials

Two main simulation strategies are adopted for the scaling of the viscous pore fluid in a particle-fluid flow in a geotechnical centrifuge. The first simulation strategy, adopted in (Kailey 2013), increases η by a factor proportional to N or shifts down the particle grading curve by $N^{1/2}$ scaling. By increasing η or decreasing the mean particle size D_{50}, the seepage flow is expected to be slower and therefore move synchronously with the solid phase. The second strategy, used in (Cabrera 2016), tests the effects of keeping a constant fluid viscosity under variable Ng levels versus scaling the fluid viscosity proportional to N, relating the changes on the mixture flow regime. On the first strategy the input mass was varied, while on the second the effective acceleration Ng was varied.

The experiments here presented are performed over three sets of granular materials, with mean particle size ranging from fine to coarse sand, mixed with an interstitial viscous fluid (see Table 1). Coarse and fine sandy soil is employed in the experiments in the drum centrifuge, with $D_{50}=0.30$ mm and $D_{50}=0.05$ mm, respectively. Monodisperse 1.45 mm diameter glass beads of are employed for the experiments in the beam centrifuge.

The viscous fluid employed in experiments present a Newtonian-like behaviour for water (denoted W) and the water-glycerine dispersion (W-G), and a non-Newtonian behaviour for the water-methylcellulose (W-M) and the water-kaolin (W-K) dispersions (see Fig. 2). For comparison purposes, all fluid dispersions are characterized as a Bingham fluid with a linear function between shear stresses τ and shear strain ratio $\dot{\gamma}$ with slope equal to the fluid dynamic viscosity η. For the experiments in the drum centrifuge, η was kept constant and proportional to the scale factor of $N = 40$ for the W-M and W-G dispersions (Kailey 2013). For the experiments in the beam centrifuge, the fluid viscosity was varied by controlling the concentration of kaolin in the W-K dispersion (Cabrera 2016).

3 RESULTS

Table 1 presents a resume of the experiments performed and the direct measurements of flow height and flow velocity. For the experiments in the drum centrifuge the flow height remained almost constant for the three discharge conditions, while the flow velocity increases as a function of the input mass. The mean flow height for the W-M and W-G experiments was $\bar{h} = 14.4$ mm, with the W-G experiments having

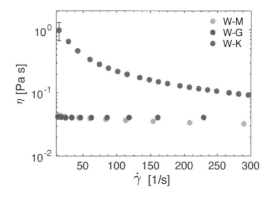

Figure 2. Rheological characterization of the viscous fluids employed in this experiments. The mixture with glycerine presents a nearly Newtonian behaviour, while the mixtures with methyl-cellulose and kaolin present a shear-thickening behaviour.

slightly thicker flows, while the mean flow height for the W mixtures was of $\bar{h} = 6.0$ mm. Moreover, the W mixtures ($\bar{u} = 2.3$ m/s) were overall faster than the W-M and W-G mixtures, with ($\bar{u} = 1.1$ m/s) and ($\bar{u} = 0.5$ m/s), respectively. The last is an indication of how the higher viscosity fluids lead to higher shear resistance at the margins of the flows.

Similarly to the drum experiments, the particle-fluid mixtures tested in the beam centrifuge do not present extreme variations with respect to flow height ($\bar{h} = 11.9$ mm). However, a clear relation is observed between fluid viscosity and the acceleration level, presenting higher flow velocities for the W mixtures ($\bar{u} = 4.4$ m/s) and lower flow velocities for the W-K mixtures with the scaled viscosity proportional to N ($\bar{u} = 1.8$ m/s). A similar relation is studied on dry granular flows in centrifugal acceleration field (Cabrera and Wu 2017a). An important remark is that by scaling η by a proportional factor of N, deposition of material over the inclined plane is observed.

Both experiments point to the relevance of the Non-Newtonian nature for the W-M and W-K mixtures, in the cases where the shear imposed by the particles does not overcomes the viscous behaviour in this range.

4 DISCUSSION

By adding a viscous fluid to the flow of glass beads, the fluid tends to change the time scale at which the interactions occur. The Froude number Fr allows to relate the intensity and frequency of this interactions as a function of the flow velocity and the driving acceleration (see Eq. 1).

$$\text{Fr} = \frac{u}{\sqrt{Ngh\cos\zeta}} = \frac{\text{flow velocity}}{\text{gravity wave speed}} \quad (1)$$

It is notable that the flows on the beam are all supercritical (Fr > 1) while those on the drum are subcritical (Fr < 1), apart from flows in which water is the pore

Table 1. Experimental parameters and direct measurements. Nomenclature: (W-M) water methyl-cellulose mixture, (W-G) water glycerine mixture, (W-K) water kaolin mixture, (W°) water mixtures in the drum centrifuge, and (W*) water mixtures in the beam centrifuge.

Test id	Fluid type	N [–]	D_{50} [mm]	Viscosity η [Pa s]	Mass [kg]	Height [mm]	Velocity [m/s]
Drum centrifuge model (Kailey 2013)							
T-9	W-M	40	0.30	0.04	1.00	13.0	0.60
T-7	W-M	40	0.30	0.04	1.75	14.5	1.40
T-10	W-M	40	0.30	0.04	2.50	13.3	1.32
T-15	W-G	40	0.30	0.04	1.00	15.0	0.35
T-14	W-G	40	0.30	0.04	1.75	15.4	0.52
T-20	W-G	40	0.30	0.04	2.50	15.4	0.60
T-21	W°	40	0.05	0.001	1.00	5.4	2.06
T-24	W°	40	0.05	0.001	1.75	6.5	2.50
Beam centrifuge model (Cabrera 2016)							
30d_gb-kw_10g_01	W-K	10	1.45	0.02	1.07	7.7	3.19
30d_gb-kw_10g_01	W-K	10	1.45	0.02	1.07	10.2	3.10
30d_gb-kw_15g_01	W-K	15	1.45	0.02	1.07	-	-
30d_gb-kw_20g_01	W-K	20	1.45	0.03	1.07	10.9	4.27
30d_gb-kw_10g_01	W-K	10	1.45	0.17	1.11	10.6	1.79
30d_gb-kw_15g_01	W-K	15	1.45	0.28	1.00	10.9	1.85
30d_gb-kw_20g_01	W-K	20	1.45	0.40	1.00	9.0	1.89
30d_gb-w_10g_01	W*	10	1.45	0.001	1.05	14.4	3.70
30d_gb-w_20g_01	W*	20	1.45	0.001	1.05	15.9	4.45
30d_gb-w_15g_01	W*	15	1.45	0.001	1.05	16.1	4.97
30d_gb-kw_1g_01	W-K	1	1.45	0.02	1.07	9.3	0.63
30d_gb-kw_1g_01	W-K	1	1.45	0.02	1.07	9.7	0.60
35d_gbWater_1g_01	W*	1	1.45	0.00	0.99	5.2	1.23

fluid. In most debris flow experiments at 1 g and full scale field flows, Fr has been found to vary between 0.5 and 6 (Cui et al. 2015), so with the exception of the highly viscous W-G (glycerin mixture) flows, the majority of tests fall into this range. The experiments on the drum show that the mass of material has an influence on the maximum flow speed, but less so on the flow height. This may be due to the tube delivery of the material to the head of the channel, which may have restricted the flow-rate at that point. Nevertheless, the results show that Fr increases with total flow quantity.

Bagnold and Savage numbers (Ba and Sa, respectively) provide a dimensionless measure of the role of viscous fluid interactions in a flowing granular mass (see Eqs. 2 and 3). Bagnold performed experiments on a shear cell filled with a neutrally buoyant particle-fluid mixture (Bagnold 1954), defining a transition region between viscous dominated flows (macroviscous regime) and grain inertia dominated flows (40 < Ba < 450). Moreover, Savage and Hutter found experimentally that for Sa > 0.1 single grain collisions control the flow dynamics of the flowing mixture (Savage and Hutter 1989), and for Sa < 0.1 long lasting contacts dominate the flow.

$$\text{Ba} = \sqrt{\frac{\phi^{1/3}}{\phi_c^{1/3} - \phi^{1/3}}} \frac{\rho_p d^2 \dot\gamma}{\eta} = \frac{\text{inertial grain stress}}{\text{viscous shear}} \quad (2)$$

$$\text{Sa} = \frac{\dot\gamma^2 \rho_p d}{(\rho_p - \rho_f) N g h} = \frac{\text{grain collisions}}{\text{frictional contacts}} \quad (3)$$

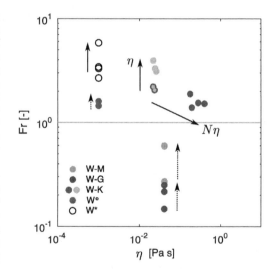

Figure 3. Froude number variation as a function of fluid viscosity η. The arrows point to the increase in mass flowing through the model (dotted line) or the increase in the centrifugal acceleration (continuous line).

Overall, the experiments performed in the drum centrifuge model fall in the Bagnold macroviscous and Savage frictional regimes, and the experiments performed in the beam centrifuge model fall in the Savage collisional regime starting in the Bagnold transition and moving towards the Bagnold inertial and Bagnold macroviscous as a function of η. Fig. 4 presents two

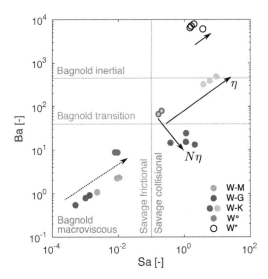
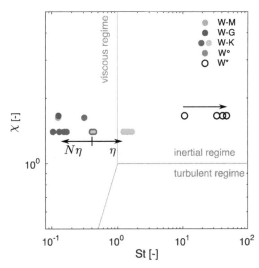

Figure 4. Phase diagram in the Bagnold (Ba) and Savage (Sa) plane (the plane is plotted in a logarithmic scale). The arrows point to the increase in mass flowing through the model (dotted line) or the increase in the centrifugal acceleration (continuous line).

Figure 5. Phase diagram as a function of the Stokes number St and the density ratio χ (the plane is plotted in a logarithmic scale). The arrows point to the increase of the centrifugal acceleration in the model.

dominating trends: i) by increasing the mass flow discharge or the centrifugal acceleration Ng, the flows move towards a collisional regime, and ii) by increasing simultaneously η and N, the increase in viscosity overcomes the increase in the flow inertia towards the macroviscous regime. Interestingly, the effect of the mean particle size employed in both experiments is reflected in the differentiation between the Savage regimes, acknowledging the permeability decrease for the drum centrifuge experiments, in which the grain size distribution included fines.

Courrech du Pont et al. (2003) presented an analysis on the time scales of the motion of a sphere immersed in a fluid driven by gravity. In their analysis, the Stokes number St (see Eq. 4) and the density ratio χ (see Eq. 5) were found to govern the grain dynamics in an elementary case of a falling particle along its own diameter d. St and χ relate the characteristic time of a particle flow controlled by the fluid viscosity t_η, with the characteristic time of a particle in free-fall t_{ff}, and the characteristic time of a particle driven by the turbulence of the fluid t_t.

$$\mathrm{St} = \frac{t_\eta}{t_{\mathrm{ff}}} = \frac{1}{18\sqrt{2}} \frac{\sqrt{\rho_p d^3 (\rho_p - \rho_f) Ng \sin\zeta}}{\eta} \quad (4)$$

$$\chi = \frac{t_\mathrm{t}}{t_{\mathrm{ff}}} = \sqrt{\frac{\rho_p}{\rho_f}} \quad (5)$$

Fig. 5 presents the experimental results in the St $- \chi$ plane. The experiments mixed with the W-K dispersion successfully present the effect of scaling the viscosity of the fluid phase proportional to the acceleration in the model (Ng). In them, by increasing Ng in a mixture with constant η_0, the flows move towards a regime controlled by the particle interactions (free-fall regime). Alternatively, by scaling the viscosity by N and increasing Ng in the model, the drag of the fluid phase becomes evident, resulting in the deposition of material over the inclined plate. When the viscosity of the fluid phase is not scaled, an increase in Ng moves the granular flows towards larger St values, becoming more inertia dominated. Moreover, because of the nature of St and χ there is no effect on increasing the mass flow through the model, reaffirming the observation that the particular experiments undertaken in the drum centrifuge are dominated by the fluid viscosity.

5 CONCLUSIONS

Two experimental models in a drum and a beam centrifuge were used to test granular flows mixed with a viscous fluid flowing down an inclined channel. Flow height and flow velocity were directly measured for different granular materials (fine and coarse sand, and glass beads) and different viscosity fluids (water, water-glycerine, water-methylcellulose, and water-kaolin). Two simulation strategies were used to approach the scaling of viscous fluids in a kinematic system in a centrifuge model: i) increase the viscosity η by a factor proportional to N or shift down the particle grading curve by $N^{1/2}$ scaling, and ii) keep a constant fluid viscosity under variable Ng levels and compare it against flows with fluid viscosity proportional to N. For the first strategy the input mass m was varied, while for the second, the effective acceleration Ng was varied. For experiments performed under a common η, the flow height remained nearly unaffected by m or Ng, but the flow velocity was found to be proportional to m or Ng.

These separate suites of experiments show that different strategies can be used to study different granular flow behaviours and regimes within both beam and drum centrifuges. First, it is possible to vary viscosity using a number of different pore fluids or additives such as a kaolin, resulting in either Newtonian or non-Newtonian behaviour. Second, it is possible to change the particle size and size distribution to alter both fluid-particle and particle-particle interactions. Use of non-dimensional scaling is useful to quantify the flow regime developed, although should be used with care in the case of well-graded granular materials, as a result of uncertainties over the appropriate length scale (Sanvitale and Bowman 2017).

REFERENCES

Ancey, C. (2001). Dry granular flows down an inclined channel: Experimental investigations on the frictional-collisional regime. *Phys. Rev. E 65*(011304).

Arndt, T., A. Brucks, J. Ottino, & R. Lueptow (2006). Creeping granular motion under variable gravity levels. *Phys. Rev. E 74*(031307).

Bagnold, R. (1954). Experiments on a gravity-free dispersion of large solid spheres in a newtonian fluid under shear. In *Proceedings of the Royal Society London*, Volume 225, pp. 49–63.

Bowman, E., J. Laue, & S. Springman (2010). Experimental modelling of debris flow behaviour using a geotechnical centrifuge. *Canadian Geotechnical Journal 47*(7), 742–762.

Cabrera, M. & W. Wu (2017a). Experimental modelling of free-surface dry granular flows under a centrifugal acceleration field. *Granular Matter*. in press.

Cabrera, M. & W. Wu (2017b). Scale model for mass flows down an inclined plane in a geotechnical centrifuge. *Geotechnical Testing Journal 40*(4), 719–730.

Cabrera, M. (2016). *Experimental modelling of granular flows in rotating frames*. Ph. D. thesis, University of Natural Resources and Life Sciences, Vienna.

Courrech du Pont, S., P. Gondret, B. Perrin, & M. Rabaud (2003). Granular avalanches in fluids. *Physical Review Letters 90*(4), 044301–044304.

Cui, P., C. Zeng, & Y. Lei (2015). Experimental analysis on the impact force of viscous debris flow. *Earth Surface Processes and Landforms 40*(12), 1644–1655. ESP-14-0196.R3.

Dorbolo, S., L. Maquet, M. Brandenbourger, F. Ludewig, G. Lumay, H. Caps, N. Vandewalle, S. Rondia, M. Mlard, J. van Loon, A. Dowson, & S. Vincent-Bonnieu (2013). Influence of the gravity on the discharge of a silo. *Granular Matter 15*, 263–273.

Garnier, J., C. Gaudin, S. Springman, P. Culligan, D. Goodings, D. Konig, B. Kutter, R. Phillips, M. Randolph, & L. Thorel (2007). Catalogue of scaling laws and similitude questions in geotechnical centrifuge modelling. *International Journal of Physical Modelling in Geotechnics 7*(3), 1–23.

Gray, J. M. N. T. & C. Ancey (2009, 6). Segregation, recirculation and deposition of coarse particles near two-dimensional avalanche fronts. *Journal of Fluid Mechanics 629*, 387.

Imre, B., J. Laue, & S. Springman (2010). Fractal fragmentation of rocks within sturzstroms: insight derived from physical experiments within the eth geotechnical drum centrifuge. *Granular Matter 12*, 267–285.

Iverson, R. (1997). The physics of debris flows. *Reviews of Geophysics 35*(3), 245–296.

Iverson, R. (2015). Scaling and design of landslide and debris-flow experiments. *Geomorphology*.

Iverson, R. & R. Denlinger (2001). Flow of variably fluidized granular masses across three-dimensional terrain: 1. coulomb mixture theory. *Journal of Geophysical Research: Solid Earth 106*(B1), 537–552.

Johnson, C., B. Kokelaar, R. Iverson, M. Logan, R. LaHusen, & J. Gray (2012, March). Grain-size segregation and levee formation in geophysical mass flows. *J. Geophys. Res. 117*, Issue F1.

Jop, P., Y. Forterre, & O. Pouliquen (2006, June). A constitutive law for dense granular flows. *Nature 441*(7094), 727–30.

Kailey, P. (2013). *Debris flows in New Zealand alpine catchments*. Ph. D. thesis, University of Canterbury February.

Leonardi, A., M. Cabrera, F. Wittel, R. Kaitna, M. Mendoza, W. Wu, & H. Herrmann (2015, Nov). Granular-front formation in free-surface flow of concentrated suspensions. *Phys. Rev. E 92*, 052204.

Mathews, J. (2013, September). *Investigation of granular flow using silo centrifuge models*. Ph. D. thesis, University of Natural Resources and Life Sciences, Vienna.

Sanvitale, N. & E. T. Bowman (2017). Visualization of dominant stress-transfer mechanisms in experimental debris flows of different particle-size distribution. *Canadian Geotechnical Journal 54*(2), 258–269.

Savage, S. & K. Hutter (1989). The motion of a finite mass of granular material down a rough incline. *J. Fluid Mech. 199*, 177–215.

Vallejo, L., N. Estrada, A. Taboada, B. Caicedo, & J. Silva (2006, July). Numerical and physical modeling of granular flow. In C. Ng, Y. Wang, and L. Zhang (Eds.), *Physical Modelling in Geotechnics*, pp. –. Taylor & Francis.

Using pipe deflection to detect sinkhole development

E.P. Kearsley, S.W. Jacobsz & H. Louw
University of Pretoria, Pretoria, South Africa

ABSTRACT: In South Africa leaking water distribution networks contribute to the formation of sinkholes in soil underlain by dolomite areas. By detecting the position of leaks in pipe networks as soon and as accurately as possible, maintenance action can be taken and the formation of sinkholes prevented. Local changes in the support conditions of a pipeline, due to leaks, would result in localized deformation of the pipeline. The detection of strain concentrations in a pipe network could thus potentially be used as an early warning system for sinkhole detection. In this paper centrifuge trapdoor experiments are described investigating the effect of void formation on the deformation and strain development in semi-flexible pipes. The effect of void size and cavity migration for different soil conditions and relative pipe stiffnesses are discussed.

1 INTRODUCTION

Sinkholes regularly occur in dolomitic ground in South Africa. The dolomitic bedrock is mostly covered by a layer of heterogeneous material comprising of transported material, blocks of chert and dolomitic rock fragments, ranging in particle size from clays and sands to gravel. In the relatively dry climate this highly variable overburden material normally occurs as an unsaturated, cohesive layer spanning over cavities and gulleys that developed in the dolomite bedrock over geological time. When overburden spans over a stable void, equilibrium has to be disturbed before a sinkhole will form. Infiltrating water, resulting in a loss of strength of the arching soil, is seen as one of the main causes of sinkhole formation. It has, in fact, been reported by Buttrick and Van Schalwyk (1995) that 98.9% of all new sinkholes in the Tshwane area are triggered by leaking water services. It is regularly mentioned in the media that up to a third of potable water distributed in South Africa is lost from the system and this creates a serious risk for sinkhole formation in developments built on dolomitic soils.

Before the leak is detected, and as the sinkhole develops due to leakage, the leak is likely to be made worse by the ground movement associated with the development of the sinkhole, resulting in a large volume of water being lost. Perhaps the biggest problem with water lost from the distribution system is that the presence and location of leaks are not easily detected before a very large volume of water had been lost so that remedial action is normally only taken very late. A research project funded by the South African Water Research Commission (WRC) is currently being undertaken by researchers in the Department of Civil Engineering at the University of Pretoria to develop an early warning system, where changes in pipeline temperature and strain can be used as early indicators of leaks, causing permanent subsurface deformations.

By detecting the position of leaks in pipe networks as soon, and as accurately as possible, maintenance action can be taken and the formation of sinkholes prevented. Local changes in the support conditions of a pipeline, due to leaks, would result in localized deformation of the pipeline. The detection of strain concentrations in a pipe network could thus potentially be used as an early warning system for sinkhole detection.

The intention is to detect sinkholes by means of instrumented pipelines while the soil above the cavity still has the capacity to bridge the growing cavity underneath. Before any full scale sections are instrumented, scale model testing is underway to determine not only the effect of the buried pipeline on the soil behaviour during sinkhole formation, but also establish zones of influence and factors affecting the behaviour. As the sinkholes will not be visible on the surface, it is essential to quantify the behavioural trends experimentally, thus making it possible to interpret results obtained from actual leaking pipelines.

To ensure realistic soil behaviour, centrifuge models were used in the experimental study. In this paper centrifuge trapdoor experiments are described investigating the effect of void formation on the deformation and strain development in semi-flexible pipes. The effect of void size and cavity migration for different soil conditions and relative pipe stiffnesses are discussed.

2 BACKGROUND

The effect of permanent ground displacement on buried pipelines has been extensively investigated by various research groups using centrifuge models with either split-box (O'Rourke et al., 2005) or tunnel induced movement (Vorster et al., 2005a & 2005b; Klar et al., 2005; Marshall et al., 2010). Tunnelling beneath buried pipelines results in bending of the pipeline and this bending behaviour of the pipe is governed primarily by both the relative stiffness of the pipe compared to the soil, and the characteristics of the shape of the greenfield settlement trough. These relationships are expressed in terms of the normalized parameters M_n and R as indicated in Equation 1 and Equation 2 (Marshall et al., 2010):

$$M_n(x) = \frac{M(x)i^2}{EI\,S_{max}} \quad (1)$$

$$R = \frac{EI}{E_s r_0 i^3} \quad (2)$$

where $M(x)$ = bending moment function in terms of distance x along the pipeline, i = distance from the tunnel centreline to the inflection point of the greenfield settlement trough, EI = bending stiffness of the pipe, S_{max} = maximum greenfield settlement, E_s = Young's modulus of the soil and r_0 = outer radius.

The term R is a "rigidity" parameter and in tunnelling tests it was found that for values of less than about 0.1, the soil-pipe interaction was negligible and the bending moments in pipelines can be determined directly from greenfield settlements, indicating that an infinitely flexible pipeline will perfectly follow the greenfield displacement profile. When R is greater than about 5, the soil-pipe interaction is critical and the greenfield settlement trough less important. An extremely rigid pipeline will experience the greenfield displacement as a localized disturbance. The greenfield soil displacement at the level of the pipe can be described by a Gaussian curve given by Equation 3:

$$S_v(x) = S_{max}\,exp[-0.5\left(\frac{x}{i}\right)^2] \quad (3)$$

Researchers investigating the effect of tunnelling on pipelines (Vorster et al., 2005b) found that the relative stiffness of the pipe increased in comparison to that of the soil (increase in R-value) as the volume loss increased. Volume loss was however limited to less than 10%. The tunnel volume loss was achieved by deflating a membrane around a cylinder. In all of the permanent ground displacement tests the induced displacement was limited, and although the relative stiffness between the pipe and the soil was varied, tests were mostly conducted using fine, dry silica sand.

Jacobsz (2016) studied cavity propagation using centrifuge trapdoor experiments using moist and dry, medium and fine grained sands at loose or dense consistencies. He concluded that in dense sand the width of a sinkhole appearing at the surface was dependent only upon the width over which support was removed at depth. Test results indicated that sinkholes would tend to propagate to the surface in a vertical chimney-like fashion. Sinkholes in dolomitic areas would however not occur in silica sand. The overburden normally arch between rock pinnacles and the zone of influence would not necessarily propagate realistically from a flat surface as arching over the cavity would be limited by the lack of horizontal restraint provided by a metal strongbox floor.

Physical testing is required to determine whether the effect of sinkhole propagation on pipeline deformation could be accurately predicted using the normalized parameters developed by researchers such as Vorster et al. (2015), Klar et al. (2015) and Marshall et al. (2010) to express the bending behaviour of buried pipes above tunnels. The intension of the tests described in this paper was to determine the effect of sinkhole propagation on pipelines, taking into account the possible effect of cohesive soil and sloping rock pinnacles acting as horizontal restraint for arching soil.

3 EXPERIMENTAL SETUP

Large-displacement trapdoor experiments were carried out in a loading frame that allows the upward propagation of the zone of influence to be studied in two-dimensional plane-strain conditions. The model set-up is illustrated in Figure 1. The model comprises of a frame built from aluminium alloy channel sections that contains a volume measuring 500 mm wide up to 450 mm high. The thickness of the soil body was 76 mm.

The front of the sample is contained by a 20 mm thick safety glass panel which allows the effect of lowering the 50 mm wide trapdoor to be observed throughout the sample using a high resolution digital camera. The trapdoor, located in the base of the frame, could be lowered during experiments by extracting water from the piston using a second piston attached to a linear drive powered by a stepper motor. The stepper motor is controlled from the centrifuge control room, thus allowing precise control over the lowering of the

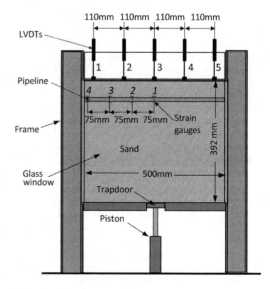

Figure 1. Test frame.

Figure 2. Test setup containing silica sand.

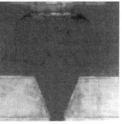

Figure 3. Test setup containing chert material (a) Arching of void (b) After sinkhole formation.

trapdoor during tests. A photo of the test setup can be seen in Figure 2.

Two tests were conducted using a fine silica sand, with a median particle size of $62 \mu m$ and specific gravity of $2\,670\,kg/m^3$, that was pluviated into the model to obtain a density of approximately $1\,540\,kg/m^3$ (50% relative density). Based on oedometer test results, an E-value of 40MPa was used for the sand. The first of these tests was used to determine the greenfield displacement profile caused by cavity propagation without any pipeline. In the second test a 500 mm long aluminium pipe with an outer diameter of 12.72 mm and a wall thickness of 0.76 mm was placed to have a prototype invert depth of 1.08 m. The pipeline was instrumented with a total of eight strain gauges configured in four half bridges as shown in Figure 1. Each half bridge comprised a gauge on the pipe crown with an opposite gauge on the pipe invert, forming a beam-type configuration measuring bending strain. These tests were carried out at an acceleration of 30G. When viewed in cross-section, the presence of the pipe will result in settlement varying over the thickness of the model. Therefore, assuming plane-strain conditions, when viewing the model from the front (e.g. Figure 1), is an approximation. The thickness of the model was intentionally chosen to be narrow (76 mm) to minimise this error.

Two more tests were conducted using a silty-clay containing chert gravel. The median particle size of the silty clay was $25\,\mu m$ with 8% particles smaller than $2\,\mu m$, a specific gravity of $2\,670\,kg/m^3$ and a compacted density of approximately $1\,900\,kg/m^3$ at a moisture content of 10%. The Young's modulus, as determined from oedometer tests, was about 20MPa. This material is a typical overburden material in dolomitic areas surrounding Pretoria and is referred to as "chert material" in this paper. These tests were conducted to establish whether the behaviour observed in sand holds true for cohesive materials containing moisture.

To make cavity propagation from the base more realistic rock pinnacles slopes were modelled by placing timber blocks sloping at 25° (measured from the vertical) adjacent to the trapdoor, thus allowing the overburden to arch over the void developing at the trapdoor by exerting horizontal forces against the model pinnacle side slopes. Due to the cohesive nature of the soil, sinkhole propagation only occurred at an acceleration of 50G. To aid the simulation of a sinkhole in the chert material provision was made to introduce water into the model. Once the trapdoor settlement stopped (see Figure 3(a)) water was released at a flow rate of 1.2 ml/s at the soil surface to trigger the formation of the sinkhole (see Figure 3(b)). Without the presence of additional water, the modelled sinkhole would not have formed due to the strength of the overburden. The surface deformation was recorded using a set of 5 LVDT's placed at 110 mm intervals on the soil surface, while an additional set of LVDT's were used to measure the actual pipeline deformation. Particle Image Velocimetry (PIV) was conducted on photos using 32 pixel patched spaced 32 pixels apart. This spacing of patches resulted in sets of 75 patches at each level in the sample.

4 RESULTS

The effect of cavity propagation on the greenfield settlement profile was established by slowly lowering the trapdoor and plotting the surface movement as a function of trapdoor movement as can be seen in Figure 4. The position of the LVDTs can be seen in Figure 1 and the trapdoor movement is expressed as a percentage of the total depth of the model. From these graphs it can be seen that the trapdoor movement immediately caused significant surface settlement in the sand.

The cavity propagated to the surface in a chimney-like fashion. Although LVDT 3 (placed above the centre of the trapdoor) indicate that the chimney broke the surface at a trapdoor movement of about 8% (approximately 31.5 mm), limited surface settlement was observed beyond the width of the trapdoor. In contrast, the chert model showed little surface settlement and not only the surface above the trapdoor, but a large area around it started settling. Although both these graphs indicate that surface settlement took place as soon as the trapdoor opened, the surface settlement in the moist cohesive material is insignificant and would not be useable as an indicator of subsurface cavity propagation.

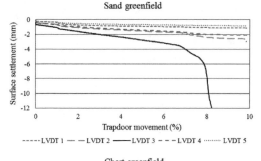
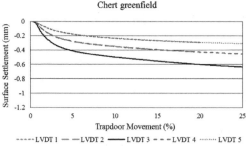

Figure 4. Surface settlement caused by trapdoor movement.

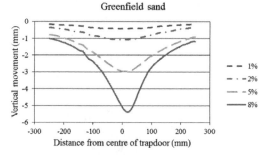
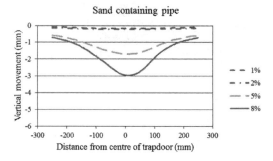

Figure 5. Surface settlement comparison for sand.

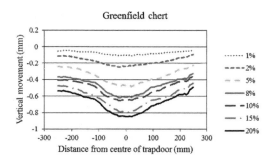
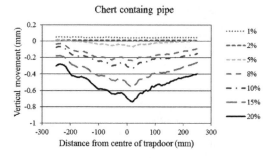

Figure 6. Surface settlement comparison for chert.

The effect of buried pipes on surface soil settlement can best be seen by comparing the greenfield surface profile, at specific trapdoor settlements, with the surface profiles of the models containing pipes. The profiles were drawn using the vertical movement of the PIV patches on the soil surface and the results can be seen in Figure 5 for sand and Figure 6 for chert, respectively. As expected, the sand settlement trough follows a Gaussian shape and the inclusion of the pipe reduces the depth of the trough, but increases the width of the settlement trough. The PIV results for the moist, cohesive chert material are more variable as a result of some particles sticking to the window causing irregular observed movement and wild vectors.

The chert material shows much less surface settlement than the sand and the settlement curves cannot be presented by Gaussian fits. There are significant similarities in the behavioural trends observed for the two material types. The inclusion of a pipe in the cohesive material results in a significant reduction in settlement at small trapdoor movements. The effect of the pipe on the surface settlement reduces as the cavity volume increases, and the difference between the greenfields settlement and the settlement in the model containing the pipe reduces with increased trapdoor settlement. This difference can be expressed as relative settlement (greenfield surface settlement – pipe surface settlement), indicated in Figure 7, where a negative relative settlement indicates that the inclusion of a pipe resulted in reduced model surface settlement. At small trapdoor movements (small cavities) the sand settlement is reduced by the inclusion of the pipe. It is only when the pipe prevents the cavity from propagating to the surface that the soil on top of the pipe stays in position, because the pipe is bridging the gap, resulting in increased pressure on the sand shoulders adjacent to the cavity propagation chimney, that the vertical movement adjacent to the chimney exceeds the greenfields settlement. The fact that the pipe affects movement of the soil means that the pipe will exert pressure on the soil surrounding it, which would cause bending moments in the pipe, resulting in stresses and strains

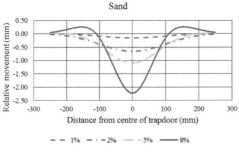

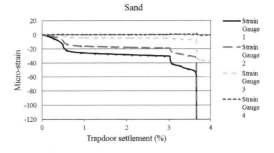

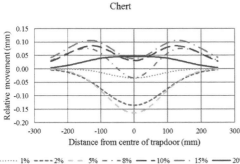

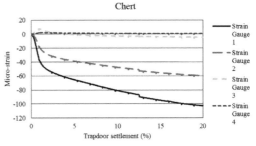

Figure 8. Pipe strain development.

Figure 7. Relative surface settlement for pipes.

which could be used as indicators of subsurface cavity propagation.

The trend observed for the chert material is similar to that observed for sand, where the inclusion of the pipeline reduced vertical movement at limited trapdoor movement (up to 5%), but increased surface movement for large trapdoor movements.

The effect of differential settlement on the pipes can be assessed by taking the strain gauge readings before sinkhole formation into account. From Figure 8 it can be seen that the strain development in the pipes differ significantly. Limited change in strain took place in the sand prior to sinkhole formation, while the strain in the pipe in moist cohesive material increased continuously while the volume of extracted material increased. This indicates that localized changes in pipeline strain could be used as indicators of cavity development.

As the aim of the experiment was to determine whether sinkhole propagation, caused by leaking water pipes could be detected by measuring the longitudinal strain in pipes, the chert material centrifuge models were provided with a means to introduce water at the soil surface to model the effect of a leaking pipe. The effect of water introduction can be seen in Figure 9 where a noticeable change in pipe strains was observed as soon as the wetting surface started moving into the soil. It was interesting to note that the strain in the pipe reduced directly above the trap door when the soil became saturated (strain gauge 1), while the strain in the adjacent pipe section increased.

The effect of water leaking through the soil was immediately noticeable from the strain recorded at three of the strain gauge bridges, giving an indication of the strain influence sphere. As soon as the wetting

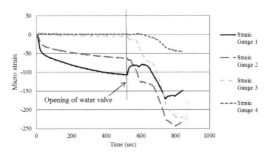

Figure 9. Strain caused by sinkhole propagation.

front reached the crown of the arch that was bridging the cavity formed by lowering of the trapdoor, the arch collapsed, resulting in sinkhole propagation.

The onset of sinkhole propagation coincided with an increase in strain in the area above the arch (approximately equal to the trapdoor width). The recording of strain was stopped when the sinkhole reached the model surface.

Pipeline maintenance programs can in future use localized strain development in instrumented pipelines for detecting leak and subsurface cavity formation. Fibre optic strain measurement could be a feasible means to conduct practical strain measurement along the length of a pipeline for this purpose.

5 CONCLUSIONS

The onset of cavity propagation results in small surface settlements in both dry sand and moist cohesive chert. This settlement is however not of a magnitude so that the surface settlement can be used as indicator of subsurface cavity propagation. The observed settlement in chert material is also significantly less than that observed in dry sand for similar trapdoor settlements.

Although the surface settlement curve for sand at small trapdoor displacements follow a Gaussian distribution, the deflected shape of moist, cohesive chert material cannot be modelled using a Gaussian function.

The cavity, caused by the trapdoor, propagates to the surface in a chimney-like fashion, resulting in a sinkhole with a surface width similar to that of the trapdoor. Although the moist cohesive material bridges the cavity caused by the trapdoor movement, surface settlement still takes place, indicating that it would be possible to use localized pipe movement as indicator of subsoil movement or sinkhole propagation.

The inclusion of a pipe in the model resulted in reduced surface settlement at small trapdoor movements for both sand and chert material models. At larger trapdoor movements, surface settlement adjacent to the cavity propagation chimney increased, indicating that the pipeline caused additional vertical stresses on the remaining soil when a cavity developed below a section of the pipe.

There was no significant increase in strain for increased trapdoor movement prior to sinkhole formation for the pipeline in sand. In chert material the pipeline strain increased with increased trapdoor movement, indicating that differential pipeline strain could in future, possibly be used as early warning system for both sinkhole formation and leak detection.

REFERENCES

Buttrick, D. & Van Schalwyk, A. 1995. The method of scenario supposition for stability evaluation of sites on dolomitic land in South Africa. *Journal of the South African Institution of Civil Engineering.* Fourth quarter 1995: 9–14.

Jacobsz, S.W. 2016. Trapdoor experiments studying cavity propagation. *Proceedings of the 1st Southern African Geotechnical conference, Sun City, South Africa, 5–6 May 2016.* CRC Press, Netherlands.

Klar, A., Vorster, T.E.B., Soga, K. & Mair, R.J. 2005. Soil-pipe interaction due to tunneling: comparison between Winkler andelastic continuum solutions. *Géotechnique 55(6)*: 461–466.

Marshall, A.M., Klar, A. & Mair, R.J. 2010. Tunneling beneath buried pipes: View of soil strain and its effect on pipeline behaviour. *Journal of Geotechnical and Geoenvironmental Engineering, December* 2010: 1664–1672.

O'Rourke, M., Gadicherla, V. & Abdoun, T. 2005. Centrifuge modelling of PGD response to buried pipe, *Earthquake Engineering and Engineering Vibration* 4(1): 69–73.

Vorster, T.E.B., Klar, A., Soga, K. & Mair, R.J. 2005. Estimating the effect of tunneling on existing pipelines. *Journal of Geotechnical and Geoenvironmental Engineering, November* 2005: 1399–1410.

Vorster, T.E.B., Mair, R.J., Soga, K. & Klar, A. 2005. Centrifuge modelling of the effect of tunnelling on buried pipelines: Mechanisms observed. *Geotechnical aspects of underground construction in soft ground. Proceedings of the 5th International conference of TC28 of the ISSMGE, Netherlands, 15–17 June 2005.* London: Taylor & Francis.

White, D., Take, W. & Bolton, M. 2003. Soil deformation measurement using particle image velocimetry (PIV) and photogrammetry. *Géotechnique,* 53(7): 619–631.

Model tests to simulate formation and expansion of subsurface cavities

R. Kuwano
Institute of Industrial Science, The University of Tokyo, Japan

R. Sera
Geo Search Co. Ltd., Japan

Y. Ohara
Graduate School, The University of Tokyo, Japan

ABSTRACT: Sinkholes or cave-in's of the ground often occur in urban roads. The complicated underground situation as well as the necessity of urgent restoration do not usually allow full investigation of the real cause. The detailed mechanism of the phenomenon has not been, therefore, well understood. A cave-in is usually initiated by the formation of a cavity in the ground. It is possible that the hidden cavity expands to eventually cause apparently sudden collapse. In this study, a series of model tests simulating the process of subsurface cavity formation/expansion was conducted, using a small soil chamber having an opening in a base plate. A cavity was formed above the opening at the base, after the water with soil flow out. It was found that a cavity and loosened ground above the cavity can extend rapidly upward when the ground consists of poorly graded sand, especially when it is fully saturated.

1 INTRODUCTION

An old deteriorated sewer pipe, when it is damaged, may eventually cause local subside or a cave-in in the road. Recently, the number of pipes which are older than the service life time of 50 years, has been rapidly increasing. Accordingly, incidents of a road cave-in are found to be more frequent in urban areas. In spite of the significance, full investigation of the road cave-in is often skipped as the urgent road restoration is usually priotised. It is difficult to clearly identify the real cause by the fact that eventual cave-in is likely to occur long after the initial formation of a cavity in soil. The congested underground lifeline structures also make it complicated.

Kuwano et al. (2006) conducted a survey to obtain basic information on how the damaged sewer pipes were related to the collapses of road, which occurred from 2001 to 2003. The survey was performed by sending questionnaires and interviewing local government officers in seven cities where the sewerage system had started more than 30 years ago and the management and maintenance of old sewer pipes are likely to be the concerned issues. It was found that even small gaps or cracks could lead to road cave-in and the rainfall appears to be one of the most important factors. Based on the survey results, Mukunoki et al. (2009) and Kuwano et al. (2010a, 2010b) performed a series of model tests to investigate how a cavity initially forms in soil and how it progresses up to the ground surface.

In this study, the typical process of cavity formation/expansion in poorly graded sand was re-evaluated and the potential size of cavity is roughly quantified.

2 APPARATUS AND TEST PROCEDURE

2.1 Apparatus

A test apparatus used in this study is schematically shown in Figure 1. Model ground of 300 mm wide, 80 mm long and 200 mm high was made in a small soil chamber having an opening of 5 mm in a base plate. Water was supplied to the model ground from a water tank connected to sides or bottom of the ground. Water level in the model ground was controlled by the height of a water tank. Soil flowed out of the opening with water. Pore water pressure transducers were placed at the bottom of both sides to monitor ground water level. Surface ground settlements were measured using non-contact displacement transducers.

2.2 Tested materials and model ground

Uniform silica sand having mean grain size of 0.4 mm was used for the model ground. The maximum and minimum void ratios were 1.04 and 0.66 respectively. Dry sand was gently placed by a scoop in the soil chamber and colored sand layer was put in front at intervals of 2.5 cm. Relative density of the model ground was about 50% ($\rho_d \fallingdotseq 1.4\,\text{g/cm}^3$). A photo of apparatus and model ground setup is shown in Figure 2.

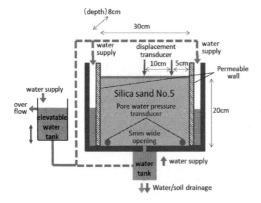

Figure 1. Schematic figure of test apparatus.

Figure 2. Apparatus and model ground setup.

2.3 Test procedure and test case

After setting up of the model ground, water was supplied to the ground. Two conditions were considered for the flow of water; i) water flows in and out of the base opening, ii) water was supplied through side walls and drained out of the opening. Considering the real situation, when the soil outlet exists above the ground water level the condition can be the former and when the soil outlet is below the ground water level, it is similar to the latter condition.

Non-woven cloth was put in the opening as a plug to prevent soil grains from flowing out in the preparation stage. When the water table was stabilized in the model ground, the plug was removed to start testing. Deformation of the ground was observed, with the measurement of surface displacement, amount of drained water and soil. Water was supplied/drained several times from the bottom opening for Test B. Tests were continued until the surface collapsed for Test A, B and D, while, the penetration resistance was measured above the cavity formed in the Test C. Test cases are presented in Table 1.

Table 1. Test condition.

Test case	Specified water level	Water supply/ drain cycle	Surface collapse	Penetration test
A	20 cm	–	yes	–
B	10 cm	4 times	yes	–
C	10 cm	–	–	done
D	10 cm	–	yes	–

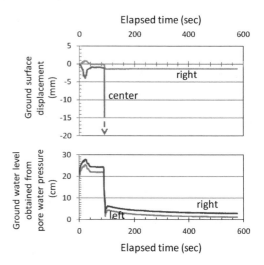

Figure 3. Surface settlements, and ground water levels in Test A.

3 CAVITY FORMATION IN THE GROUND

3.1 Soil loss below the ground water level

Transition of surface settlements, ground water levels obtained from pore water pressure measurement for Test A are plotted with time in Figure 3. Deformation of the ground is presented in Figure 4.

As soon as the plug was taken out, soil started to flow out of the opening. Deformation of the soil immediately spread to the above without forming a clear cavity and reached the surface. The settlement at the center was recorded to increase and decrease at around 20 second in Figure 3. It seems to be affected by the water level above the depressed ground. Area of significant deformation was limited to a band of approximately 5 cm. Almost all the soil of deformed area flowed out in a short period and at the end of the test, about 1400 g soil was lost to form a hole.

3.2 Cavity formation in/under unsaturated layer

Surface settlements, ground water levels and amount of drained water and soil for Test B are shown in Figure 5 with some photos in Figure 6. Water was supplied from bottom and initially ground water level was set to 10 cm from the bottom. As soon as the plug was taken out, water quickly drained and soil below the initial ground water level was dragged down with water/soil drainage. A cavity of about 5 cm wide emerged at

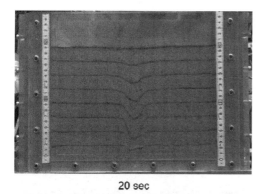

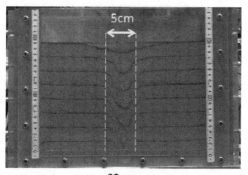

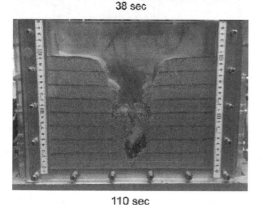

Figure 4. Deformation of the ground and cavity formation in Test A.

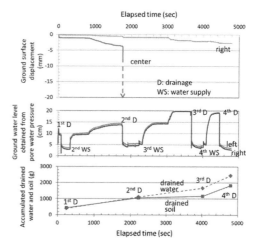

Figure 5. Surface settlements, ground water levels and amount of drained water/soil in Test B.

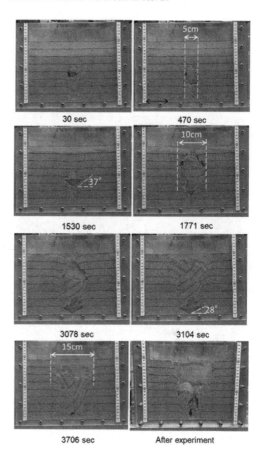

Figure 6. Deformation of the ground and cavity formation in Test B.

height of 10 cm as a result. A surface settlement of about 1 mm was observed at the center. After the 1st soil/water drainage was completed, the 2nd cycle of water supply started. Water penetrated into the cavity. Inside the cavity, soil under the water slid down towards the center along the slope formed in both sides. Angle of the slope was 37° and seemed to approach to the angle of repose of the material. The width of the cavity of fan-like shape became larger and when it became about 10 cm, the soil above lost the stability. Then during 3rd and 4th water supply/drain cycle, the deformation of the ground expanded to the ground surface. Finally a hole of 15 cm wide appeared. 1823 g of soil, which is 27.5% of initial amount, was drained out in total.

3.3 Penetration resistance of soil above the cavity

Surface settlements, ground water levels and amount of drained water and soil for Test C are shown in

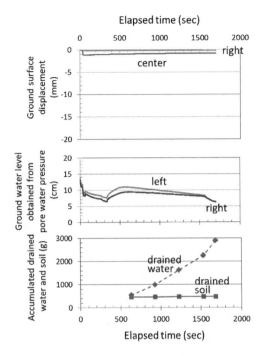

Figure 7. Surface settlements, ground water levels and amount of drained water/soil in Test C.

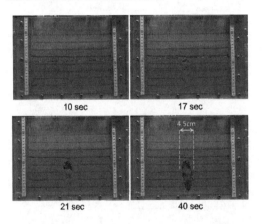

Figure 8. Deformation of the ground and cavity formation in Test C.

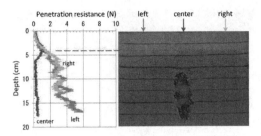

Figure 9. Penetration resistance above the cavity in Test C.

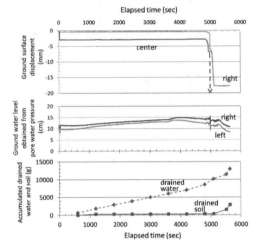

Figure 10. Surface settlements, ground water levels and amount of drained water/soil in Test D.

Figure 7 with some photos in Figure 8. Water was supplied from both sides and ground water level was kept at 10 cm.

Similar behavior to the 1st cycle of Test B was observed in Test C. A cavity of 4.5 cm wide was formed within 40 seconds below the initial ground water level as shown in Figure 8. Although the water was continuously supplied from both sides, it drained from the opening without seeping into the cavity. The cavity was never filled with water and the water level at the center was zero. The cavity was not expanded and soil was not drained any more.

Penetration resistance was measured at three locations as shown in Figure 9, using a rod of 3 mm diameter having a cone shape head. Penetration resistances at left, right and center showed similar trend up to 4 cm from the surface, while that at the center significantly reduced below.

3.4 Expansion of cavity and collapse of surface

Surface settlements, ground water levels and amount of drained water and soil for Test D are shown in Figure 10 with some photos in Figure 11. Water was supplied from both sides and ground water level was at 10 cm.

A cavity of about 5 cm wide was initially formed under the ground water level, accompanied by the surface settlement of about 3 mm. Once the water in the cavity was drained, no significant soil outflow occurred. The ground water level was raised to 15 cm, then the width of cavity increased. The soil above the cavity started to sink and finally collapsed into the cavity.

Accumulated amount of drained soil and water in four tests are compared in Figure 12. Soil below the ground water level is the potential area of outflowed soil. When the water drained quickly out of the opening, vertically elongated cavity formed. Dry weight of drained soil at such a state was about 433 g, 463 g, and 300 g for Test B after the 1st drain, Test C and Test D at 1200 seconds respectively. Assuming the cavity is fan-shape with the radius of 10 cm and chord of

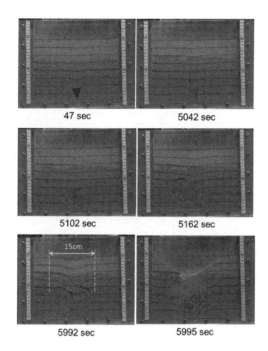

Figure 11. Deformation of the ground and cavity formation in Test D.

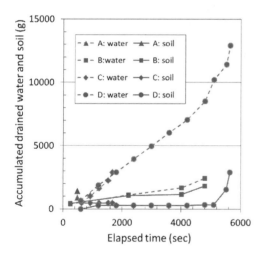

Figure 12. Amount of drained water and soil.

5 cm, the volume of the cavity is about 200 cm^3 and the equivalent dry soil weight is 280 g, which is reasonable estimation compared to the measured values as indicated in Figure 13. The width of the initial cavity may be affected by various factors such as permeability and mechanical properties of soil, ground water level, size of opening, and etc. Only the water drained at constant rate without significant soil outflow as shown in Test C and D in Figure 12.

Ground arching can be developed when the width of the cavity is small enough compared to the length of covered soil above the cavity as schematically shown

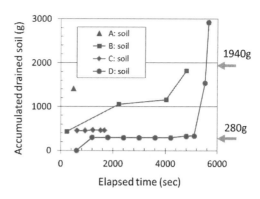

Figure 13. Comparison of drained soil amount.

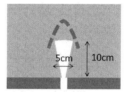

When water quickly drains

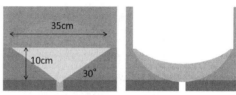

When water remains in a cavity

Figure 14. Development of a cavity.

in Figure 14. In such a case, the cavity can sustain without surface collapse.

In the later stage of Test B and Test D, when the water penetrated into the cavity by the cycle of water supply or the rise of surrounding water level, angle of fan-shape becomes wider as the soil starts to slide down to the opening along the slope formed both sides above the opening. If sufficient amount of water was supplied and the cavity was filled with water, the angle of slope would be the angle of repose of the tested material. Assuming that the angle of repose of the tested sand is 30°, the maximum width of the cavity developed above the opening can be about 35 cm, as indicated in Figure 14. Soil mass above the cavity cannot be sustained and it would collapse, since arching effect is not developed under such condition. Total amount of drained soil can be 1940 g or more, as compared to the measured value in Figure 13. If the collapsed soil mass above the cavity also flowed out in succession, total amount of drained soil would be 6680 g.

4 CONCLUSIONS

The sinkhole or cave-in phenomenon was simulated in a sand model ground and the fundamental mechanism of subsurface cavity formation and development was investigated. In case the outlet of soil leakage exists in uniform sand, followings were observed.

- Soil under the ground water level can be the potential area of cavity. Soil flowed out with water and when the water drained quickly, vertically elongated shape of cavity developed, while when the water level maintained in the cavity, the cavity developed wider and bigger.
- The shape of the cavity was roughly fan-shape. If the cavity was filled with water, the angle of slope developed above the opening could be angle of repose of tested sand. Estimated values of cavity size approximately agreed with the measured values of drained amount of soil.
- When the width of cavity was small enough compared to the covered soil length, a cavity was stable as the ground arching effect seemed to work, while when the cavity became wider and the ceiling of the cavity became shallow, the soil mass above the cavity lost the stability and the surface collapse eventually occurred.
- Surface settlements above the cavity were not significant until the moment of surface collapse, since the soil above the cavity seemed to be supported by the ground arching.

It is implied that the subsurface cavity can be initiated from the soil outflow and the ground water level is the location of cavity ceiling. If the water remains in the cavity, it would grow wider. The maximum width and height of the potential cavity are estimated from angle of repose of sand and the distance between soil outlet and ground water level. When the cavity becomes wide and shallow, ground arching effect is lost and surface collapse would happen.

REFERENCES

Kuwano, R., Horii, T., Kohashi, H. & Yamauchi, K. 2006. Defects of Sewer Pipes Causing Cave-in's in the Road, *Proc. 5th International symposium on new technologies for urban safety of mega cities in Asia, USMCA, Phuket, November 2006.*

Kuwano, R., Sato, M. & Sera, R. 2010. Study on the detection of underground cavity and ground loosening for the prevention of ground cave-in accident, *Japanese Geotechnical Journal*, 5(2): 219-229 (in Japanese).

Kuwano, R., Horii, T., Yamauchi, K. & Kohashi, H. 2010. Formation of subsurface cavity and loosening due to defected sewer pipe, *Japanese Geotechnical Journal*, 5(2): 349-361 (in Japanese).

Mukunoki, T., Kumano, N., Otani, J. & Kuwano, R. 2009. Visualization of Three Dimensional Failure in Sand due to Water Inflow and Soil Drainage from Defected Underground Pipe using X-ray CT, *Soils and Foundations*, 49(6): 959-968.

Centrifuge modelling of a pipeline subjected to soil mass movements

J.R.M.S. Oliveira, K.I. Rammah, P.C. Trejo, M.S.S. Almeida & M.C.F. Almeida
Federal University of Rio de Janeiro, Rio de Janeiro, Brazil

ABSTRACT: This paper presents the results of a series of physical modelling tests carried out on a geotechnical centrifuge to investigate the influence of a soil mass movement on the behavior of a model pipeline buried in Roncador oil field marine clay. The soil mass movement was modelled using a vertical plate that pushed the soil towards the pipe. The test setup allowed the measurement of deformations and forces developed in the pipe during the consolidation of the soil and due to the soil movement. The results indicate that the force transmitted to the pipeline decreases with the increase of the pipe freedom to displace when subjected to the soil loading. Additionally, the force on the pipe increases as the embedment ratio increases. The tests results were also compared with different analytical approaches documented in the literature in order to understand the behavior of buried pipelines under similar loading conditions.

1 INTRODUCTION

Onshore and offshore oil and gas pipelines are buried in order to provide stability, thermal insulation and mechanical protection from the surrounding environment. In many cases, these pipelines can be subjected to external surcharge loads such as those caused by soil mass movements. These loads can cause significant deformation and hence, failure of the pipeline structure. This is the case of many pipelines crossing marine soil areas in the south and southeast Brazil. Therefore, reliable estimates of the induced stresses within the pipeline and thus, its deformation, are essential to provide the design engineer with the relevant information necessary for a safe design of these structures.

Problems of interactions of pipelines buried in clayey soils have been investigated by a number of researchers, both analytically and numerically. Some of these studies have yielded guidelines intended to develop design provisions for evaluating the integrity of buried pipes for a range of applied loads (ASCE, 2005). O'Rourke (2005) discuss some of the possible loading conditions that could induce stresses and hence, deformations of a buried pipe. These loading conditions could be due to earthquakes, soil sliding and excavations adjacent to the zone in which the pipe is buried. They carried out tests using large-scale models and developed analytical solutions that describe the three-dimensional distribution of deformation along the pipeline.

Sahdi et al. (2014) investigated the impact forces of submarine slides in pipelines through a set of centrifuge tests where the resistance of a model pipe moving horizontally within a soil mass was measured for different values of the degree of consolidation of the kaolin and the pipe velocity. These authors propose a hybrid approach equation to calculate de lateral resistance based on a bearing capacity (NH) component and an inertial drag component. For low velocities and high strength values, the bearing capacity component dominates, and NH = 7. On the other hand, for high velocities and low strength values, the inertial component dominates and NH increases significantly.

2 TESTED SOIL AND TEST CONCEPT

The soil used in the present study was the Roncador oil field marine clay obtained by Kullenberg samplers. The liquid limit (w_L) and the plastic index (I_p) for the adopted soil were 82% and 59%, respectively. The coefficient of consolidation $c_v = 1.5 \times 10^{-8}$ m^2/s, specific gravity $G_s = 2.61$, specific weight $\gamma = 15.8$ kN/m3, effective friction angle $\phi' = 24.8°$, Young's modulus $E_u = 12.9$ MPa and shear modulus $G = 4.3$ MPa. The soil was placed into the centrifuge box using the technique of lumps, adopted by a number of studies carried out by many authors (e.g., Rammah et al., 2014).

The objective of the centrifuge tests is to evaluate the forces developed on a pipeline buried in clayey soil subjected to a soil slide. It is expected that the pipeline will be subjected to flexural loading and hence, deflections, due to the potential forces induced by the soil mass moving towards the pipeline. Figure 1 shows a schematic diagram that describes the problem.

A pipeline model was designed to simulate a prototype pipeline in terms of flexural stiffness, which governs the stress-deformation behavior of the pipeline structure subjected to distributed load associated, in this case, with a soil mass moving towards the pipe (Rammah et al., 2014). Normally, this soil mass loading occurs in a portion of the pipeline, which is buried

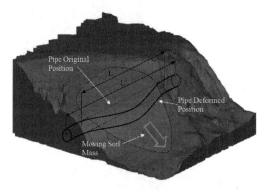

Figure 1. Schematic view of a pipeline subjected to soil mass movements.

in the part of the slope that is in movement. The dimensions (including diameter, wall thickness and length) of the pipeline model were calculated as described below. The calculations presented in this section has the sole objective of finding the pipe dimensions in model scale which better represent the real dimensions of the prototype pipeline. Therefore, the mathematical approximation described ahead will not be used to assess the forces on the pipe. Those forces were measured directly by means of strain gages, which were previously calibrated outside the centrifuge with increasing loads, as will be described later on the paper.

The force acting on the pipeline is defined as

$$F_p = K_p \cdot \delta_p \qquad (1)$$

where F, δ and K are the pipeline force, displacement and stiffness, respectively. The subscripts m and p refer to the model and prototype, respectively.

Applying centrifuge-scaling factors to convert from model to prototype conditions gives

$$F_m = K_m \cdot \delta_m \qquad (2)$$

Thus, the relation between the prototype and model stiffness has to comply with the equation below:

$$K_p = K_m \cdot N \qquad (3)$$

where N is the g-level.

Figure 2 describes the structural system of the pipeline model with an influenced length equal to L, which is the distance between two assumed points of zero moment (rotational joints) across the moving soil mass width (in this case L = embankment width). The two extremes of null rotation (fixed-end points) are away from the rotational joint and separated by a distance L' greater than L. It should be noted that the advantage of adopting the rotational joints was to simulate a real pipeline that has a length L', i.e., longer than L.

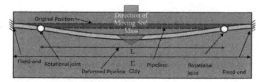

Figure 2. The structural system adopted for the pipe model (plan view).

Table 1. Parameters of pipeline prototype and model.

Parameter	Prototype	Model
Material	Steel	Aluminum
Young's modulus (E)	2.00×10^8 kPa	0.65×10^8 kPa
Length (L)	9.0 m	180 mm
External diameter (D)	0.457 m	9.0 mm
Wall thickness (t)	12.7 mm	1.0 mm

2.1 Pipeline model I (3D-pipe)

For the case of a simply supported beam subjected to uniform distributed load, the stiffness is

$$K = \frac{48 \cdot E \cdot I}{L^3} \qquad (4)$$

where E is the elastic modulus, I is the moment of inertia and L is the pipe length. Table 1 summarizes the parameters for both the prototype and the centrifuge model of the pipeline. Using the parameters presented in Table 1 and Equation (4) to calculate K_p and K_m, N can be estimated by Equation (3) as presented in Equation (5) where the target value of the g-level on the centrifuge was found as $N = 50$.

$$N = \frac{K_p}{K_m} = \frac{5760}{115} = 50 \qquad (5)$$

It should be noted that the centrifuge model simulates the portion of the pipe that has a distance L between the two points of zero moments, and therefore, the simply-supported beam model adopted in the above analytical approach is considered the closest estimation of the relationship between the stiffness of the prototype and the model of the pipeline.

In order to assess the maximum flexural moment acting on the pipe, and hence the maximum deformation, a full Wheatstone bridge was installed at the centre of the pipe using four Vishay strain gauges of the type MM-CEA-125UN-350. The four gauges were glued onto the outer face of the pipe and were offset by 90 degrees; this configuration of the strain gauges allowed the measurement of the forces acting in both the vertical and the horizontal directions.

The instrumented pipe, designated here as 3D-pipe, was mounted inside the centrifuge box using the two rotational adapters shown in Figure 3 and then was calibrated by applying a uniformly distributed load (q) along the span of the pipe and acquiring the reading of the strain gages associated with the bending moments.

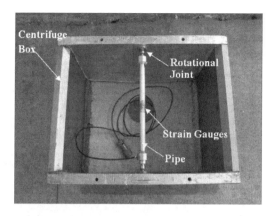

Figure 3. Instrumented 3D-pipe mounted inside a centrifuge box (empty).

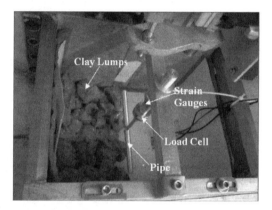

Figure 4. View of the centrifuge box with the 2D-pipe set up and some clay lumps.

2.2 Pipeline model II (2D-pipe)

The designed 3D-pipe has both ends attached to the centrifuge box, restraining the pipe displacements at these ends, giving a three-dimensional approach to the analysis. In that way, another pipe with free ends, which represents a bi-dimensional approach, was necessary to allow a behavior comparison between both. Thus, another pipe model arrangement, here designated as 2D-pipe, with the same length and diameter of the 3D-pipe, was attached to an instrumented rod capable of measuring the vertical and horizontal forces which the pipe is subjected to, by means of a load cell and strain-gauges full bridge, respectively. Figure 4 shows an inside view of the centrifuge box with the 2D-pipe set up mounted on during the clay lumps placing.

2.3 Soil movement procedure

After the design of the pipes, a special procedure had to be developed to create a controlled moving soil mass passing around the pipe body. A moving plate was used to generate this soil movement in a way that the necessary amount of force could be measured by the plate and compared with the force transmitted to the pipe. Thus, a strain gauge instrumented plate with a thickness of 6 mm was designed to be introduced in the soil and drag a section of this material towards the pipe, as shown in Figure 5, where L_i is the initial distance between the moving plate and the front of the pipe ($6D = 54$ mm), and L_f is the final distance ($2D = 18$ mm). In addition, the moving plate penetration depth is always kept as a function of the pipe burial depth ($c + H$). The plate was instrumented with a full Wheatstone bridge and the force is derived from the bending moment.

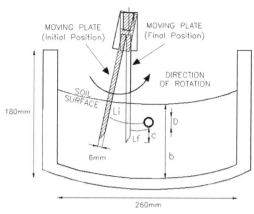

Figure 5. Schematic view of the tests with 2-D and 3D pipes.

3 CENTRIFUGE TEST STAGES AND PROGRAMME

The preparation of the centrifuge test was carried out following the sequence below:

1. Installing the instrumented pipe in its position inside the strongbox;
2. Installing two pore-pressure transducers for monitoring the development and dissipation of pore water pressure during consolidation and testing phases;
3. The soil was placed in the centrifuge strongbox using the technique of clay lumps;
4. Consolidating the clay soil at a g-level of 50 for a period of 18 hours during which the pore-water pressure, the forces developed in the pipe and the settlement of the sample were acquired. This period was previously confirmed as sufficient to reach full consolidation of the clay layer. The settlement of the soil model was measured using a laser transducer. After all tests, cylindrical samples were taken from the soil to water content and specific weight control;
5. Installing the instrumented plate in position to the required depth below the soil surface as shown in Figure 5, and then initiating the test by moving the plate towards the pipe for a distance of $2.5D$.
6. Assessing the undrained shear strength (S_u) by carrying out a T-bar penetrometer test on the clay layer that underwent the consolidation phase;

Table 2. Centrifuge tests programme.

3-D Pipe Model		2-D Pipe Model	
Test	H/D	Test	H/D
3D-1	200%	2D-1	200%
3D-2	111%	2D-2	111%
3D-3	177%	2D-3	144%
3D-4	167%	2D-4	211%
3D-5	200%	2D-5	100%
3D-6	111%	2D-6	100%
3D-7	244%	2D-7	139%

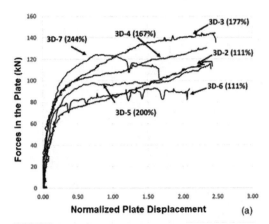
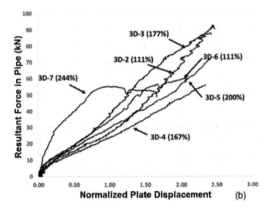

Figure 6. Force versus normalized plate displacement curves for the 3D-pipe tests (a) Plate; (b) Pipe.

A total of 14 centrifuge tests were carried out, each including the stages mentioned above. Table 2 summarizes the tests programme where the first set of 7 tests corresponds to the 3D-pipe, and the second set of 7 tests is related to the 2D-pipe. In both sets of tests the embedment ratio (H/D) varies, always taking into account that H/D is calculated considering that H is the height measured from the clay surface up to the bottom of the pipe (i.e., the soil cover above the pipe's crest + the pipe diameter).

To determine the strength profile, a 5mm diameter T-bar penetrometers was used over the course of the 14 T-bar tests conducted in the centrifuge.

4 TEST RESULTS

4.1 Forces versus displacements

The force versus displacement curves in prototype scale regarding the moving plate and the 3D-pipe are plotted in Figure 6 and the same set of curves related to the moving plate and the 2D-pipe are plotted in Figure 7. In these curves the plate displacement is normalized by the pipe diameter. The forces developed in the moving plate are composite mainly by the passive forces ahead of the pipe and also by some side friction forces. The latter are considered negligible when compared to the former due to the magnitude of the passive forces and also due to the small thickness of the plate.

It can be observed that the plate force versus displacement curves (Figs 6a and 7a) show a bilinear behavior with a steeper initial phase, probably associated with the full mobilization of the undrained shear strength along the slip surface, followed by a gentler second phase after a breakout point, where the soil mass is kept in movement towards the pipe.

The development of forces on the pipe (Figs 6b and 7b) shows almost linearly increasing behavior, indicating that soil strength has not been fully mobilized during the test. However, reach fully soil strength mobilization was not the intention, but comparing forces on the plate and on the pipe at specific displacements in order to estimate the amount of effort that is transmitted to the pipe.

Some influence of the shaft that holds the rigid pipe is expected in horizontal forces measurement for H/D values between 100% and 244%. However this effect it was not taken into account, as previous calculations show influences less than 4% in a worst case scenario.

Due to some variation in the undrained strength of each sample, the curves do not show a clear increase in resistance with H/D, indicating that a normalization procedure regarding forces data is necessary.

4.2 Pipe forces versus plate forces

In order to minimize as possible the influence of the plate proximity on the pipe measurements, which is an important issue on these tests, all forces comparisons were taken at $0.5D$ plate. This distance was considered sufficient to mobilize full soil strength at the plate, still being far enough $(5, 5D)$ from the pipe to avoid plate influence.

Considering a normalized displacement of $0.5D$, which is close to the breakout resistance in the plate, the values of forces in the plate and forces in the pipe were normalized for the 3D-pipe and for the 2D-pipe. The normalization adopted in both cases is presented in Equation (6), where F is the force, S_u is the undrained shear strength at the pipe depth, D is the diameter of

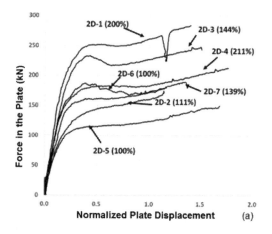

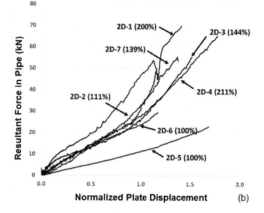

Figure 7. Force versus normalized plate displacement curves for the 2D-pipe tests (a) Plate; (b) Pipe.

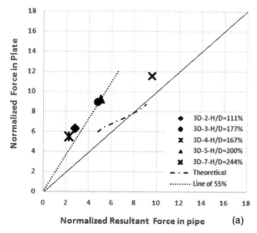

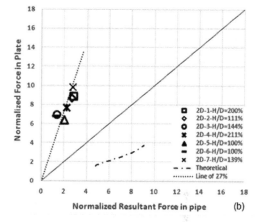

Figure 8. Normalized force in plate versus normalized resultant force in pipe for: a) 3D-pipe and; b) 2D-pipe tests.

the pipe or the burial depth of the plate and L is the length of the pipe or the plate.

$$\bar{F} = \frac{F}{S_u \cdot D \cdot L} \quad (6)$$

The classic Rankine theoretical upper bound solution for the normal component of the passive normalized force in a rough retaining wall for undrained conditions can be considered as a theoretical approach to the normalized force in plate (Equation 7).

$$N' = \frac{\frac{1}{2}\gamma \cdot h^2 + 2S_u h \left(1 + \frac{S_w}{2S_u}\right)}{S_u L_p} \quad (7)$$

where S_w is the maximum mobilized shear strength in pipe-soil interface, adopted as half the undrained shear strength value, L_p is the length of the plate.

The normalized resultant force in the pipe can be assumed as the bearing capacity factor of a cylindrical shallow foundation. Oliveira et al. (2010) proposed Equation (8) on the variation of the T-bar bearing factor for shallow depths (Almeida et al., 2013). This solution can also be considered as a theoretical approach to the normalized resultant force in 3D and 2D-pipe, respectively.

$$N_b = 0.0053\left(\frac{H}{D}\right)^6 - 0.1102\left(\frac{H}{D}\right)^5 + 0.9079\left(\frac{H}{D}\right)^4 - 3.7002\left(\frac{H}{D}\right)^3 - 7.2509\left(\frac{H}{D}\right)^2 - 3.9168\left(\frac{H}{D}\right)^2 + 5.3519 \quad (8)$$

Figure 8 presents the normalized force in plate against the normalized resultant force in pipe for both 3D and 2D-pipes. Bisector lines were plotted as a reference of equal values of the normalized force in plate and the normalized force in pipe. The theoretical curves lay below the bisector lines, that is below the 2D and 3D-pipe curves. In both cases, the data is located above the bisector line, meaning that the forces on the plate are greater than the forces on the pipe. This means that just part of the energy associated with the soil mass movement generated by the plate is transmitted to the pipe.

The region where the data are located in Figure 8 indicates that the ratios between the values measured in the pipe and in the plate are approximately 55% for the

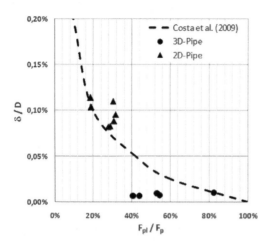

Figure 9. Variation of the ratio between the force in the plate and the force in the pipe against the normalized displacement.

3D-pipe and just 27% for the 2D-pipe. This might be associated with the degree of constraint in both cases. Normalising the pipe displacement (δ), relative to its original position, by the pipe diameter (D), the 3D-pipe has a mean maximum measured normalized displacement of $\delta/D = 7.7 \times 10^{-3}$%, while the 2D-pipe has an average measured value of $\delta/D = 9.66 \times 10^{-2}$%, i.e., almost 13 times greater than the 3D-pipe. It means that the force transmitting process loses efficiency as the pipe displacement increases due to the supporting system constraint.

This behavior has similarity with Rankine earth pressure theory, where the soil expansion is directly related to the active state and decrease in earth pressure forces, while the soil compression is related to the passive state and increase in earth pressure forces. When the pipe deforms, the relative movement between the pipe and the soil allows some stress relief in the soil mass behind the pipe due to an active state mobilization, and thus a decrease in force induction.

The smaller is the pipe moving restrain, the smaller is the soil pressure against the pipe, and, therefore, the lower the forces on the pipe. On the other hand, the bigger the pipe restrains the closer to the passive state and the bigger the soil pressure on the pipe. In that way, changes in pipe design can be proposed in order to increase pipe flexibility and thus decrease force induction.

Figure 9 shows the ratio between the force in plate (F_{pl}) and the force in the pipe (F_p) against the maximum normalized displacement in the pipe. Normalized 3D pipe displacements appear very close to each other because they are much smaller than those from the 2D pipe tests.

Costa et al. (2009) studied the variation in the vertical force over a deep active trapdoor in sandy soils where, the bigger is the trapdoor displacement, the smaller are the forces. This behavior is related to Rankine's earth pressure conditions, and although tested for sands, it is applicable for other types of soil. The 3D and 2D-pipes data compares well with Costa et al. (2009) and the variation in normalized displacement can be associated with different degrees of Rankine's earth pressure condition, leading to the conclusion that the arching phenomena is responsible for the decrease in the pipe force.

5 CONCLUSIONS

The normalized resultant force in the pipe against the normalized force in the plate showed that the 3D-pipe, with small displacement, comprises bigger stress levels than the rigid pipe, with bigger displacement.

When comparing the ratio between the force on the pipe and force on the plate with the normalized pipe displacement, it becomes evident that the bigger is the moving flexibility, the smaller are the force levels on the pipe due to the soil mass. On the other hand, the bigger are the pipe restraints, the bigger are the forces that affect the pipe.

These results can be used to improve the design analysis of real situations such as pipelines crossing areas subjected to soil mass movements, although further investigations are necessary to validate the proposed equations for other types of soils.

REFERENCES

Almeida, M.S.S., Oliveira, J.R.M.S., Rammah, K.I. & Trejo, P.C. 2013. Investigation of bearing capacity factor of T-bar penetrometer at shallow depths in clayey soils. *J. of Geo-engineering Sciences* 1: 1-12. DOI 10.3233/JGS-13005

ASCE 2005. Guidelines for the design of buried steel pipe. In: A.L. Alliance (ed.), *American Society of Civil Engineers*.

Costa Y.D.J., Zornberg J.G., Bueno B.S. & Costa C.L. 2009. Failure Mechanism in sand over a deep active trapdoor. *J. of Geotech. and Geoenviron. Eng.* 135(11): 1741–1753.

Oliveira, J.R.M.S., Almeida, M.S.S., Almeida, M.C.F. & Borges, R.G. 2010. Physical Modeling of Lateral Clay-Pipe *Interaction. J. Geotech. Geoenviron. Eng.*, 136(7): 950–956. DOI 10.1061/(ASCE)GT.1943-5606.0000311

O'rourke, T.D. 2005. Soil-Structure interaction under extreme loading conditions. In: T.A.M. University (ed.), *The Thirteenth Spencer J. Buchanan Lecture*, 9-35. November 2005, USA.

Rammah, K.I, Oliveira, J.R.M.S, Almeida, M.C.F, Almeida, M.S.S. & Borges, R.G. 2014. Centrifuge modelling of a buried pipeline below an embankment. *Int. J. of Physical Modelling in Geotechnics* 14(4): 116–127. HTTP://DX.DOI.ORG/10.1680/IJPMG.14.0

Sahdi, F., Gaudin, C., White D.J., Boylan N. & Randolph, M.F. 2014. Centrifuge modelling of active slide-pipeline loading in soft clay. *Géotecnique* 64(1): 16-27 HTTP://DX.DOI.ORG/10.1680/GEOT.12.P.191

Effects of earthquake motion on subsurface cavities

R. Sera & M. Ota
GEO SEARCH Co. Ltd, Japan

R. Kuwano
Institute of Industrial Science, The University of Tokyo, Japan

ABSTRACT: After a large earthquake, road function may be lost due to collapses of road, which makes emergency situations difficult. Earthquakes frequently cause invisible sub-surface cavities and some of them become apparent as cave-ins or subsidence. Based on Japanese governmental emergency surveys, essential characteristics of cavity caused by earthquake were identified. In this study, a series of shaking model experiments was conducted to investigate effects of earthquake motion on the subsurface cavities. A cavity was formed in a model ground in a small model chamber, and then earthquake motion was applied using a shaking table. It was observed that when arch action decline, a cavity caused a cave-in with shear cracks, subsidence and loosening by earthquake motion.

1 INTRODUCTION

Large-scale earthquakes, such as the 2011 Great East Japan Earthquake and the 2016 Kumamoto Earthquakes, occurred in recent years in Japan. After these disasters, road functions were lost partially due to cave-in problems (Fig. 1), which made emergency rescue operation difficult. Earthquakes frequently cause invisible sub-surface cavities and some of them become apparent as cave-ins or subsidence.

Since the 1990s, several Japanese public administrations have carried out road cavity surveys for the maintenance of their roads as a countermeasure to road cave-in accidents. Such regular surveys focus on cavities due to damage of aged buried infrastructures, influenced by high-density underground development and due to fluctuation of groundwater level. In recent years, emergency surveys were conducted after large earthquakes in order to ensure disaster recovery activities. Based on these emergency surveys, some valuable actual conditions reported (Abe et al. 2007, Koike et al. 2012, Agatsuma et al. 2014, and Okamoto et.al 2017) and a study on the mechanism of cavity development caused by earthquakes has been carried out by GEO SEARCH and The University of Tokyo in Japan (Sera et al. 2013, Kuwano et al. 2013 and Sera et al. 2015). In this study, essential characteristics of cavity caused by earthquake were investigated. For example, the potential of cavity occurrence increases due to earthquakes, especially in the regions recorded $I_{jma} = 5+$ or greater (I_{jma} stands for Japan Meteorological Agency Seismic Intensity Scale). It was found that the cavities were formed in the vicinities of buried items or underground structures.

Figure 1. Photos of road cave-in and cavities taken at earthquake recovery working by GEO SEARCH.

This contribution aims to understand effects of earthquake motion on the development of subsurface cavities. A series of shaking model experiments was conducted and deformation characteristics of a cavity was discussed.

2 EXPERIMENTAL PROCEDURE

2.1 Experimental cases

In this study, four cases of experiments were conducted to observe the process of cavity formation and the behavior of cavity under earthquake motion. They are categorized roughly into two types according to method of cavity formation: formed by melting an ice block in model ground (hereinafter referred to as I-Cavity-Model) and formed by water supply and drainage that simulated more realistic occurrence of cavity under subsurface (hereinafter referred to as W-Cavity-Model). In addition, the cavities were formed by other ways in each method. The explanations of each cases are listed in Table 1.

Table 1. Experimental cases.

Case	Method of formation of a cavity
I-Cavity-Model-1	By melting an ice block in model ground at a room temperature for 24 hours or longer
I-Cavity-Model-2	By melting an ice block in model ground rapidly by blowing hot air
W-Cavity-Model-3	By water supply and drainage with two cycles
W-Cavity-Model-4	By water supply and drainage with three cycles

Figure 3. A rigid box on the shaking table.

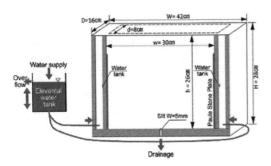

Figure 2. Schematic illustration of soil rigid.

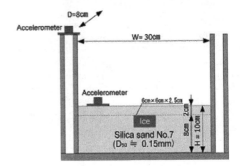

Figure 4. Schematic illustration of I-Cavity-Model.

2.2 Apparatus

These series of experiments were conducted as focused on realistic occurrence of cavity under road subsurface. Owing to actual cavities have been found at a few decimeter depth around buried utilities which are factors of cavity occurrence, the particle size of sand and slit size for drainage are selected similar to actual condition. The size of model ground was made about one-tenth of prototype. A model ground was prepared in a rigid box. The dimension of it was 42 cm in width, 16 cm in depth and 28 cm in height as shown in Figure 2. In addition, two accelerometers and three video cameras were used to observe the behavior of the cavity and surrounding soils. In cases of W-Cavity-Model, water was supplied from water tanks at both sides which was connected to elevatable water tank. The water head of experiments were controlled to be 5–10 cm from the bottom of rigid box. To maintain the hydraulic gradient constant, the water was drained from a 5 mm-wide slit at the bottom of the rigid box. After a cavity was formed in the model ground, the rigid box was fixed to a shaking table as shown in Figure 3.

2.3 Model ground

2.3.1 I-Cavity-Model

In both cases of I-Cavity-Model, silica sand No.7 which has a mean particle diameter of 0.15 mm was used. Its hydraulic conductivity was 3.8×10^{-3} cm/s and the relative density of these model grounds was adjusted close to 50%. The dimension of model ground was 30 cm in width, 8 cm in depth and 10 cm in height. They were filled by the air-pluviation method. An ice block was buried in the ground having dimensions of $6 \times 6 \times 2.5$ cm at the depth of 2 cm. One side of ice block faced the front panel of the rigid box to observe a cavity, and the other side was kept off from the behind panel to observe the water infiltrating into surrounding soils. Block-colored sand was put in front of the ground at depth of 2cm to observe the ground deformation around the surface of ice block as shown in Figure 4.

2.3.2 W-Cavity-Model

In both cases of W-Cavity-Model, silica sand No. 6 which has a mean particle diameter of 0.29 mm was used. The dimension of model ground was 30 cm in width, 8 cm in depth and 20 cm in height. The purpose of using silica sand No. 6 in W-Cavity Model is to make it easier to create a cavity by increasing permeability. The relative density of these model ground also was adjusted close to 50% and they were filled by the air-pluviation method. Besides a box as a buried structure was buried against side wall having dimensions of $10 \times 8 \times 10$ cm. The surface of box was sealed by gum elastic to fix in the ground. Colored sand was put in front of the ground every 2.5 cm depth to observe of the ground deformation as shown in Figure 5.

2.4 Earthquake motion

For all the experiments, an earthquake intensity equivalent to $I_{jma} = 7$ was adopted, and it corresponds to a

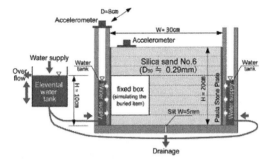

Figure 5. Schematic illustration of W-Cavity-Model.

Table 2. Specification of earthquake motion.

Condition	Value
Direction	Horizontal
Frequency of sinusoidal motion	5 Hz
Acceleration value	1,200 cm/s^2
Duration of shaking per 1 cycle	30 s

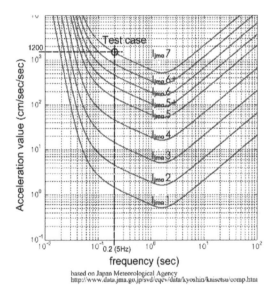

based on Japan Meteorological Agency
http://www.data.jma.go.jp/svd/eqev/data/kyoshin/kaisetsu/comp.htm

Figure 6. Determination of earthquake intensity according to Japan Meteorological Agency's chart.

Table 3. Duration of water supply in W-Cavity-Model.

Experimental case	Duration of WS (with as WH)*		
	Cycle-1	Cycle-2	Cycle-3
W-Cavity-Model-3	30 min (5 cm)	15 min (10 cm)	–
W-Cavity-Model-4	30 min (5 cm)	30 min (10 cm)	11 min (10 cm)

* WS stands for water supply; WH stands for water head.

Table 4. Additional experiment after 1st earthquake motion.

Additional experiment after 1st earthquake motion	WS/WD*	EM*	Penetration test
I-Cavity-Model-1	–	–	–
I-Cavity-Model-2	–	–	–
W-Cavity-Model-3	2 cycles	3 cycles	yes
W-Cavity-Model-4	–	–	yes

*WS/WD stands for water supply and drainage; EM stands for earthquake motion.

serious earthquake. Details of the earthquake motion is tabulated in Table 2.

Figure 6 shows the relationship between acceleration and frequency of earthquake motion used to determine I_{jma} according to Japan Meteorological Agency. Referring to Table 2, the earthquake intensity considered in the present study corresponds to $I_{jma} = 7$.

2.5 Cavity formation

2.5.1 I-Cavity-Model
As shown in Table 1, a cavity was formed by melting an ice block by two ways in experiments of I-Cavity-Model. In I-Cavity-Model-1, the ice was melted at a room temperature for 24 hours or longer. In I-Cavity-Model-2, the ice was melted rapidly by blowing hot air. As a result, there was a different degree of saturation in the surrounding soils in these two cases.

2.5.2 W-Cavity-Model
In cases of W-Cavity-Model, a cavity was formed by water supply and drainage that simulated more realistic occurrence of cavity under road subsurface. The experiments which water was supplied from side tanks and then sand and water drained allowed by opening the slit were carried out. W-Cavity-Model-3 was carried out by two cycles water supply and drainage while W-Cavity-4 was imposed three cycles. The duration of water supply and the water head considered in each cases are shown in Table 3.

2.5.3 Additional experiments and penetration resistance
After each cavity was formed, the dynamic earthquake motion was applied to the model grounds. Except W-Cavity-Model-3, the earthquake motion was applied one cycle for 30s. In W-Cavity-Model-3, after 1st cycle earthquake motion, additional experiments were carried out to observe model ground. Besides in cases of W-Cavity-Model, after each series of experiments, the water in the model ground was drained for a period of more than 24 hours, and penetration resistances were measured at more than seven different locations using a needle having a diameter of 3 mm. The experiments of each case after 1st earthquake motion are shown in Table 4.

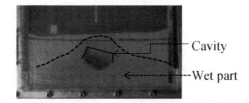

Figure 7. I-Cavity-Model-1; cavity formed by melting ice cube.

Figure 8. I-Cavity-Model-1; beginning and after of boring.

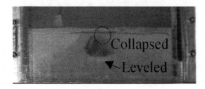

Figure 9. I-Cavity-Model-1; after earthquake motion.

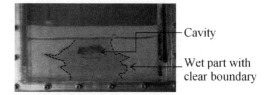

Figure 10. I-Cavity-Model-2; formed by melting ice cube as blowing hot air drastically.

Figure 11. I-Cavity-Model-2; collapsed by earthquake motion.

3 CHARACTERIZATION OF INITIAL CAVITY AND OBSERVATIONS UNDER EARTHQUAKE MOTIONS

3.1 I-Cavity-Model

3.1.1 I-Cavity-Model-1

In I-Cavity-Model-1 case, after melting ice block at a room temperature for 24 hours or longer in sand, a cavity was formed. From this cavity observation, the roof of cavity has an arch effect with horn-shaped and the surrounding soils were wet equally as shown in Figure 7. The region of wet sand expanded evenly from the cavity with fuzzy boundary of dry part.

To confirm the effectiveness of arching, a stick was penetrated into the model ground. When the stick reached around the top of cavity, a greater resistance was confirmed. When the stick was removed, a small hole with the diameter of the stick was formed. Some sand around the hole collapsed and the shape of cavity was changed from horn-shape to arch-shape similar to the shape of wet area as shown in Figure 8. This indicates that the cavity was maintained due to an action of arching.

In this case, one cycle of earthquake motion experiment was carried out. During earthquake motion, it was observed inside the cavity that a little sand collapse from the right shoulder and the bottom was leveled, but the arch of cavity had been maintained. The cavity did not expand and the surface of model ground did not exhibit subsidence as shown in Figure 9.

3.1.2 I-Cavity-Model-2

In I-Cavity-Model-2 case, the ice was melted rapidly by blowing hot air. From this cavity observation, the roof of cavity has an arch effect with a little twist horn-shape and the surrounding soils were mixed wet and dry unequally. The area of wet sand expanded from the cavity unevenly with clear boundary of dry part as shown in Figure 10. It is presumed that the degree of saturation of sand was not uniform based on the shape of wet area.

When the model ground was vibrated by earthquake motion, the cavity collapsed in just 3s, referring to Figure 11. At 2s (10 pulses), a vertical crack upwards occurred from the left shoulder of the cavity. At 3s (15 pulses), second crack also occurred from the right shoulder and the cavity was filled with collapsed sand. After one cycle for 30s, the surface of model ground subsided by 2 mm.

The result of I-Cavity-Model experiments led to a new understanding of arch action under earthquake motion with different condition of surrounding soils. It is suggested that when the sand around a cavity is wet equally, the arch action is effective maintained even under earthquake motion.

3.2 W-Cavity-Model

3.2.1 W-Cavity-Model-3

In W-Cavity-Model-3 case, a cavity was formed by 2-cycle water supply and drainage that simulated realistic occurrence of cavity under road subsurface. The observation of process of forming a cavity is as described below and shown in Figure 12. In saturated condition, deformation of model ground started at lowest layer with moving along a streamline. During 2nd drainage, the deformation expanded across the sixth layer. After 3rd drainage, an arch was created at 5 cm depth as horizontal crack opened downward and width of streamline was expanded.

After one cycle of the earthquake motion was applied, the arch located at 5 cm depth was still formed but became bigger due to collapse of sand mass inside

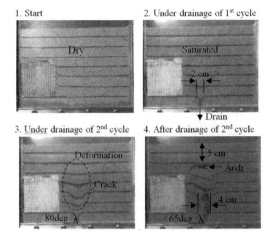

Figure 12. W-Cavity-Model-3; formed by 2-cycle water supply and drainage.

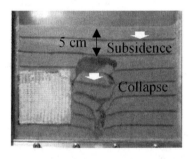

Figure 13. W-Cavity-Model-3; after 1st earthquake motion.

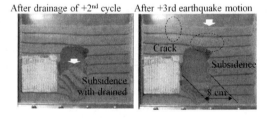

Figure 14. W-Cavity-Model-3; additional experiments.

cavity. And subsidence of ground surface was observed at the right part (Figure 13).

After 1st earthquake motion, several additional experiments were carried out with 2-cycle of water supply and drainage and 3-cycle of earthquake motion. After 2nd drainage, it was observed that the arch was maintained but the sand inside cavity settled down when drained. After three following experiments of earthquake motion, although the entire model ground deformed downwards progressively as subsidence increased and some cracks appeared at ground surface, but the arch was still maintained as shown in Figure 14.

As a result of a penetration experiment, referring to Figure 15, the maximum resistance of arch was

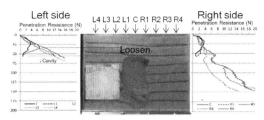

Figure 15. W-Cavity-Model-3; penetration resistance.

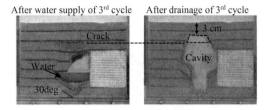

Figure 16. W-Cavity-Model-4; formed by 3-cycle water supply and drainage.

measured at 2.5 cm depth above the arch. Further penetration exhibit a loosened region just above the arch. A loosen area has confirmed as lower resistance was measured.

3.2.2 W-Cavity-Model-4

In W-Cavity-Model-4 case, a shallower cavity was formed as shown in Figure 16. The experiments of forming a cavity consisted of 3-cycle water supply and drainage with longer time of water supply than W-Cavity-Model-3 case. Because the first shallowest horizontal crack that becomes the arch afterward, occurred at 5 cm depth at 2nd drainage same as W-Cavity-Model-3. The 2nd water supply time was extended continuously and one additional cycle of experiment was carried out to expand the cavity. After 2nd drainage, a 5cm square-shaped cavity mounted the buried box. Under 3rd water supply, the ground had fluid in the cavity with forming an angle of repose at lower side. Finally, a horizontal crack occurred at 3 cm depth and an arch formed with collapse of sand mass.

After 10s from the start of the model ground was vibrated by earthquake motion, a sand mass collapsed from inside the arch and caused a shear crack on right side of the arch. At 29s, the shear cracks opened and the arch collapsed shortly. Besides the left side of the model ground leaned towards center a little as shown in Figure 17.

The result of penetration experiment is shown in Figure 18. Three loosened areas were confirmed along lines L4, L2 and R1. Lines of L2 and R1 were located beside the cave-in, but L4 which recorded the lowest value equal to zero was located opposite of the buried box with leaning a little towards center. On the other hand, in W-Cavity-Model-3, it is recognized that because the arching created above the cavity was effective, the model ground did not lean and no loosened area was observed except just above the cavity.

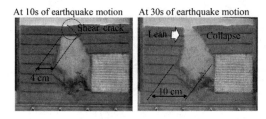

Figure 17. W-Cavity-Model-4; under earthquake moon.

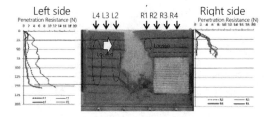

Figure 18. W-Cavity-Model-4; penetration resistance.

4 CONCLUSIONS

The influence of the earthquake motion on sub-surface cavity was examined using a sand model ground. The earthquake motion considered was equivalent to $I_{jma} = 7$. Four types of experiments were conducted; those conditions were different in degree of saturation of the sand and thickness of soil layer above the cavity.

- In I-Cavity-Model-1 case as the sand around cavity was wet uniformly, the earthquake motion did not mark large deformation with cavity and surrounding soils in the model ground.
- From the observation of I-Cavity-Model-2 case, it is presumed that the degree of saturation of sand was not uniform based on the shape of wet area. When the model ground was vibrated by earthquake motion, the cavity collapsed in just 3s, after earthquake motion. The result of I-Cavity-Model experiments led to a new understanding of arch action under earthquake motion with different condition of surrounding soils that when the sand around a cavity is wet equally, the arch action is effective maintained even under earthquake motion.
- In W-Cavity-Model-3 case as the width of arch of the cavity was 8 cm and thickness of soil layer above the cavity was 5 cm, subsidence of surface occurred by the earthquake motion. The collapsed sand as high porosity into the cavity was compacted by the earthquake motion, but arch of cavity did not move.
- In W-Cavity-Model-4 case as the width of arch of the cavity was 4 cm and thickness of soil layer above the cavity was 3 cm, shear cracks occurred on the shoulder of the cavity in 10 s of earthquake motion, and 20 s after that, the arch of cavity collapsed. Left side sand leaned a little.

Through those experiments, it was observed that when arching is created, a sub-surface cavity will not be deformed easily by horizontal earthquake motion. However, when arch action is lost, a cavity caused a cave-in with shear cracks, subsidence and loosening by earthquake motion. In conclusion, arch action of cavity is important to resist against horizontal earthquake motion, and suction of soil and thickness of sand layer above the cavity are also important.

REFERENCES

Abe, T., Saika, M., Kusakabe, T., Kichikawa, S. & Fujii, K. 2007. Management of the road cave-in risk after a large earthquake using subsurface cavity survey technology. 23rd World Road Congress of the World Road Association (PIARC).

Agatsuma, K., Tobita, Y., Sera, R., Hironaka, Y., Amari, N. & Konno, C. 2014. Characteristics of subsurface cavities in Miyagi pref. caused by the Great East Japan Earthquake – incidence rate trend analysis-. The Japanese Geotechnical Society special symposium overcome the Great East Japan Earthquake (in Japanese).

Koike, Y. & Sera, R. 2012. Basic Consideration on Occurrence of Sinkhole under Pavement – General Characteristics. 47th Geotechnical Symposium (in Japanese).

Kuwano, R., Kuwano, J., Taira, S., Sera, R., & Koike, Y. 2013. Model tests simulating sub-surface cavities formed in the liquefied ground. New Technologies for Urban Safety of Mega Cities in Asia 2013.

Okamoto, J., Matsukuma, T. & Hamazaki, T. 2017. Sub-surface cavity survey as advanced initiatives for early restoration of damaged road by earthquake. Japan Society of Civil Engineers 2017 Annual Meeting (in Japanese).

Sera, R., Koike, Y., Nakamura, H., Kuwano, R. & Kuwano, J. 2013. Survey of sub-surface cavities in the liquefied ground caused by the Great East Japan Earthquake. New Technologies for Urban Safety of Mega Cities in Asia 2013.

Sera, R., Ota, M., Kuwano, R., Horiuchi, Y. & Kikuchi, T. 2015. A study of the impact of earthquake on expansion of sub-surface cavity – Model test report 1. 50th Japan National Conference on Geotechnical Engineering (in Japanese).

The earthquake motion condition produced by Japan Meterological Agency; http://www.data.jma.go.jp/svd/eqev/data/kyoshin/kaisetsu/comp.htm

Preliminary study of debris flow impact force on a circular pillar

A.L. Yifru, R.N. Pradhan, S. Nordal & V. Thakur
Department of Civil and Environmental Engineering, Geotechnical Division,
Norwegian University of Science and Technology (NTNU), Trondheim, Norway

ABSTRACT: This paper presents results from a series of preliminary tests in a flume model, where impact forces on a circular pillar from debris flow is measured. The flume model is 0.3 m wide and 9 m long. It comprises of a run-out channel and a deposition area. A 75 mm diameter hollow circular steel pillar equiped with a load cell is placed at the end of the run-out channel, perpendicular to the flow. Nine tests were conducted with different volumes of debris which resulted in varying discharges. In all the tests, the debris flow material that is a mixture of soil and water, was made to have solids concentration C_s of 60% by volume. Despite some scatter in the data, the results suggest that the impact force on the pillar increases linearly with the debris flow discharge. The recorded impact forces are also compared and discussed in light of the analytical solutions suggested in the literature.

1 INTRODUCTION

Debris flows are natural hazards that occur in regions with mountainous terrain posing threats to infrastructures and population. Countries like Norway which has coastal roads at the foot of mountains are prone to such hazards. Closed roads incur additional travel cost and time since large detours may be needed. Moreover, damages result in additional expenses for maintenance of roads and road structures.

Different aspects of debris flow behaviours have been studied around the world, such as: in hazard mapping e.g. Sandersen (1997), Fischer et al. (2012), Meyer et al. (2014), in its countermeasures e.g. Van-Dine (1996), Jakob and Hungr (2005), Mizuyama (2008), and in estimating its impact force e.g. Proske et al. (2011), Bugnion et al. (2012), Vagnon and Segalini (2016).

In designing countermeasures such as check dams, deflection walls, or underpasses, one need to consider the impact force of debris flow. Number of flume experiments dealing with the impact process of debris flow on rigid barriers have been reported in literature e.g. Zanuttigh and Lamberti (2006), Wendeler et al. (2006), Tiberghien et al. (2007), Huang et al. (2007), Ishikawa et al. (2008), Hübl et al. (2009), Armanini et al. (2011), Proske et al. (2011), Vagnon and Segalini (2016), Song et al. (2017). The impact force was compared with empirical and analytical formulas in relation to the debris flows characteristics like flow height and flow velocity. Such formulas are reported in literature e.g Hungr et al. (1984), VanDine (1996), Rickenmann (1999).

It is worth mentioning that the majority of experimental studies are conducted using dry granular materials. In Norway, debris flows are mostly water-triggered. Therefore, experimental studies with solid-fluid mixtures are of particular interests: such as in Iverson (1997) and Iverson et al. (2010). Also, separate calculation of the slurry impact pressure and the particle load force of debris flows are of interest.

Several studies were focused on impact force against impassable barriers and more knowledge on passable structures is needed in designing bridge columns or underpass countermeasures. Therefore, this study is designed to investigate impact force of debris flow on circular pillar using a laboratory flume model.

2 METHODOLOGY

It is not always easy and economical to address debris flow problems in its natural scale or using field tests. Small-scale flume tests, alternatively, are found to be the most commonly preferred approaches to study different behaviours of debris flow. Representativeness and the reliability of the outcome from such flume tests depend on the scaling law. This has been addressed in the literature e.g. Hübl et al. (2009), Bowman et al. (2010), Iverson (2015).

The flume model used in this study has a scale ratio of 1:20. It has been extensively used to model debris flow with a reasonable kinematic and dynamic similarities e.g. Le et al. (2016), Laache (2016).

2.1 The flume model

The flume model used in this study is 9 m long and has two parts: run-out channel and deposition area. The run-out channel has two parts with inclinations of 23° and 14° slopes and is 0.3 m wide. The deposition area is 3.6 m long and 2.5 m wide with 2° inclination. The whole model set-up and placement of the instruments are illustrated in Figure 1.

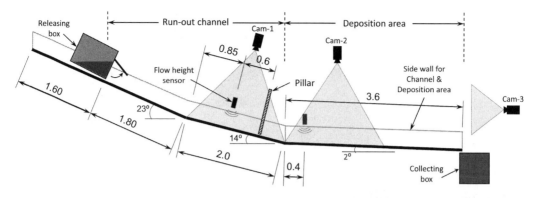

Figure 1. The flume model profile view and its instrumentation (modified after (Yifru et al. 2018)) (All dimensions in metres).

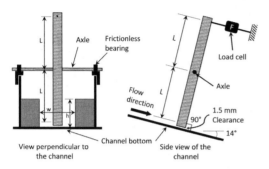

Figure 2. Configuration of the pillar and the load cell. Here $L = 0.45$ m, $w = 0.3$ m, and $h = 0.22$ m.

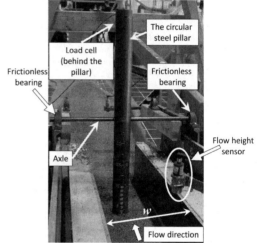

Figure 3. Photo of the pillar configuration.

The flume model is instrumented with two flow height sensors and a rigid circular steel pillar with a load cell at its top. The type of load cell used is S2M Force Transducer where as that of the flow height sensors is MIC+35/IU/TC Ultrasound. The two flow height sensors are placed at mid-section of the run-out channel and the deposition area according to the dimensions given in Figure 1.

Three video cameras were used to capture the flow behaviours (Fig. 1) in which Cam-1 was placed above the run-out channel, Cam-2 was placed above the deposition area and Cam-3 was placed at the front of the flume model. Cam-1 is used to record the flow behaviour in the run-out channel and debris interaction with the pillar. The other two cameras are used to record the overall flow behaviour and the deposition pattern.

Two identical wooden boxes (crates) with size 0.9 × 0.6 m × 0.8 m are used for containing the debris flow material. One is used for releasing the debris during a test and the other is used for collecting the debris after a test.

2.1.1 Set-up and working principle of the force measuring pillar

The fundamental working principle of the force measuring pillar is the concept of torque. Torque is a product of force and perpendicular distance (so called lever-arm) from some rotation point. The set-up of the pillar, as shown in Figure 2, is positioned to have a perpendicular impact direction and a perpendicular load cell attachment with the pillar. The load cell was attached to the top of the pillar, at a distance $L = 0.45$ m from the centre of the pillar equal to the expected point of impact at the bottom. The pillar is hinged on an axle attached at its mid section where the axle is secured by two frictionless bearing on either sides. In order to avoid any friction against the flow bed, the pillar has a 1.5 mm clearance at its bottom (see Fig. 2). So, this equal lever-arm configuration allows the application of principle of torque to record the forces from the bottom section at the top of the pillar. The force measuring pillar was calibrated prior to the flume tests. In addition, a photo of the pillar configuration is presented in Figure 3.

2.1.2 Boundary conditions

In this study, the model is made to simulate a well-developed debris flow in a defined sloped channel and deposition area. The entire run-out channel has a constant 0.3 m width. The side walls and the flow bed of the flume model are made to be smooth. The triggering and initiation of the debris flow are not considered in

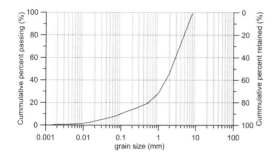

Figure 4. Grain size distribution of the debris material.

Table 1. Summary of the recorded results.

Test Name	Q [m³/s]	h_{Max} [mm]	v [m/s]	F_{Max} [N]	F_R [N]
T1	0.0134	16.54	2.70	13.78	2.0
T2	0.0154	20.00	2.56	17.79	3.0
T3	0.0156	19.25	2.70	16.16	2.0
T4	0.0160	21.36	2.50	21.23	3.5
T5	0.0173	23.65	2.44	17.14	2.5
T6	0.0200	30.61	2.17	26.45	7.0
T7	0.0247	32.05	2.56	29.25	4.0
T8	0.0264	35.14	2.50	28.32	4.5
T9	0.0278	31.52	2.94	28.17	8.0

the simulation. There was, as well, no entrainment or resistance to the flow from the bed or the side walls. The characteristics of the surfaces and rheology of the debris material are assumed to be uniform throughout the repetition of the tests.

2.2 The test material

The test material is a crushed-aggregate sand from Vassfjell, Norway. The grain size distribution (GSD) of the material used for the testing is shown on Figure 4 and has $d_{max} = 8$ mm, $d_{50} = 2.2$ mm, and $d_{10} = 0.1$ mm. The coefficient of uniformity $C_u = 26$, which suggests that the material is well-graded. The particle density of the test material is $\rho_s = 2.72$ g/cm³. During the repetition of tests, this GSD was maintained with out any significant variation.

2.3 Test procedure

The experiments were conducted using different discharges of debris with solid-water mix having solids concentration C_s of 60% by volume. A total of 9 experiments were conducted in such a way that they resulted in different flow discharges by varying the total release volume.

To conduct a test, measured mass of the test material and water are added in the releasing box and mixed. Then the box is lifted and placed at the top of the run-out channel (Fig. 1). Before opening the box and start an experiment, the debris flow material is thoroughly mixed using a hand mixer for about a minute. While keeping the debris in suspension, the box is made open. This procedure simulates the flow of a well developed debris flow in a defined channel. The released debris then accelerates through the run-out channel to hit the pillar and accumulates in the deposition area.

During a test, force on pillar, flow height before and after the pillar are recorded starting from the releasing of the mass until the flow stops. A custom made data acquisition program written using LabVIEW 2016 software is used to acquire the measurements from the force transducer and the flow height sensors. The continuous data measurement was acquired at a rate of 50 Hz for both flow height and force. In the mean time, the entire test procedure was being recorded using the three video cameras at a rate of 60 fps.

3 RESULTS

The recorded forces and flow heights of the nine tests (T1–T9) are presented in Table 1. The resulting average discharges Q are found to vary from 13 litre/second to 28 litre/second. These discharges are calculated by computing the flow velocity v just before hitting the pillar, maximum flow height h_{Max}, and the total channel width w. Here the h_{Max} of the debris flow, which is recorded at 0.6 m before the pillar, is considered as the maximum impact height. The impact velocity v was computed from the video recorded using Cam-1 (Fig. 1) by tracking the debris flow front.

The recorded forces with time are presented in Figure 5 in three groups according to the calculated Q. In these plots, the bold lines represent the moving-average of the recorded data that are drawn with a light gray lines. Here, some of the plots show values below zero in their initial parts which are results of noise and induced vibrations.

In Figure 6, the recorded force and flow height are plotted together. It shows the characteristic values, i.e. F_{Max}, h_{Max}, and F_R that are given in Table 1. Here F_R is the residual force or simply the static force at $v = 0$ m/s. This force corresponds to a force resulting from passive earth pressure of the accumulated soil after the debris flow has stopped. In the same figure, there is a 0.27-second lag time between F_{Max} and h_{Max}. This is a result of the 0.6 m gap between locations of the pillar and the flow height sensor.

In Figures 5 and 6, the initial time $t = 0$ sec refers to the time at which the data acquisition program is made to start recording immediately prior to releasing of the material.

4 DISCUSSION

According to the result presented in Table 1, different analyses were conducted to assess and find correlations between F_{Max} and the other flow parameters. Comparing F_{Max} with Q gave an interesting correlation and is presented in Figure 7. The F_{Max} is positively correlated with Q with $R^2 = 0.8329$. This positive correlation can be expressed by the following trend line equation:

$$\bar{F}_{Max} = \alpha Q \qquad (1)$$

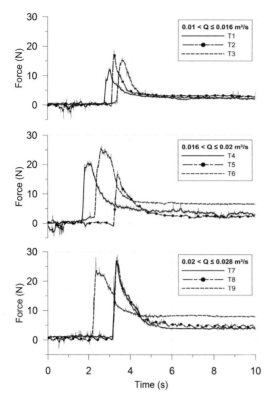

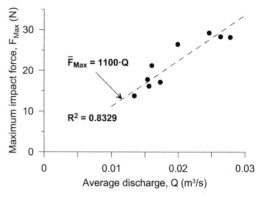

Figure 7. Correlation between maximum impact force and flow discharge.

F_r (i.e. $v/\sqrt{gh}$) with the normalized impact pressure (Proske et al. 2011).

$$F_{hd1} = \rho A v^2 sin\beta \qquad (2)$$

$$F_{hd2} = 5\rho v^{0.8}(gh)^{0.6} A \qquad (3)$$

where β is the flow angle with respect to the structure; ρ is the bulk density of the flowing debris; v is velocity of flow; A is area of impact; g is gravitational acceleration; and h is the impact height which is considered as h_{Max} for this study. In addition, $\beta = 90°$ as the flow direction is perpendicular to the pillar (see Fig. 2) which result in $sin\beta$ equals unity.

A mixed model is also presented in Arattano and Franzi (2003) and Vagnon and Segalini (2016) by considering both static and dynamic contribution of the flow. The general equation for this mixed model of impact force is given by:

$$F_{mixed} = \frac{1}{2}\gamma h A + \rho v^2 A \qquad (4)$$

where γ is the bulk unit weight of the debris which is equal to ρg.

Figure 5. The recorded impact and residual forces on the pillar for all the tests grouped by ranges of discharge.

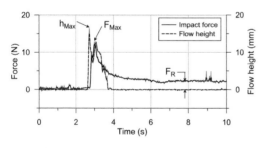

Figure 6. The recorded force and upstream flow height for T1.

where α is the multiplying factor and $\bar{F}_{Max}$ is average maximum impact force. From Figure 7 and Equation 1, α is found to be 1100 Ns/m³. Here it is worth noting that one may expect different α with different set-up of the flume model i.e. with change of slope, width and length of run-out channel.

This study has also compared the recorded F_{Max} with the available analytical hydro-dynamic and static models reported in literatures. The first hydro-dynamic model equation is calculated using the momentum quation (Hungr et al. 1984) and given by Equation 2. The second hydro-dynamic model equation is proposed by Hübl and Holzinger (2003) and given by Equation 3. This relation is obtained by normalizing measured impact pressure with the hydro-dynamic pressure, and relating the Froude number

Figure 8a shows the recorded impact forces along with the above analytical impact force equations. The equations are plotted by considering the maximum and minimum impact heights recorded on the pillar. Almost all, except one, of the recorded F_{Max} values are encompassed by the minimum and maximum curves of F_{hd2} given by Equation 3. In addition, this model gives a closer fit to the individual recorded values than the others when plotted with their respective impact height and velocity. In fact, there is no general trend between the F_{Max} and v alone. This can be seen by the relatively small variation in flow velocity when compared with the widely varying impact force records.

When comparing Equation 2 with Equation 4, the mixed model incorporates static contribution of the flowing debris, of which the effect could not be seen on Figure 8a. This is because of the flow height which was found relatively thin for the static contribution when compared with the dynamic term.

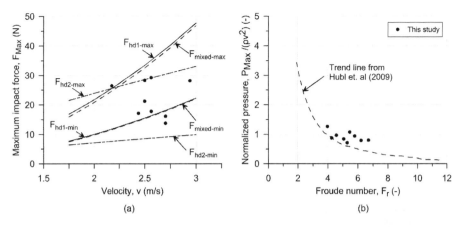

Figure 8. (a) Maximum impact force compared with hydro-dynamic and mixed impact force estimation models and (b) Plot showing normalized maximum pressure against Froude number F_r comparing it with other miniaturized laboratory test impact force results (Adopted from (Hübl et al. 2009)).

In order to compare the results from the flume model to debris flows in nature, similitude for all dimensionless quantities shall be attained for both. It is not possible to achieve full dynamic similarity between down-sized experimental debris flows model and natural debris flow if the viscosity of water kept constant over both scales. In other words, both Froude and Reynolds similitudes cannot be satisfied simultaneously. This dynamic similitude can be partly approached pragmatically using a hydrodynamic scaling that can be achieved by considering the most dominant forces of open channel flow (Choi et al. 2015). This will result in F_r which Hübl et al. (2009) and Armanini et al. (2011) identified as a key dimensionless similarity factor to scale debris flow impact on structures.

Debris flows in nature usually have velocity v between 5 and 10 m/s, and flow height h around 1 m (that can be up to 3 m for very large debris flows). This can give F_r up to around 3. It is, however, demonstrated among others by Fink et al. (1981) and Costa (1984) that debris flow in nature can assume higher F_r value where debris flows travel at higher velocity on steep terrain.

Interestingly, the recorded data shows some agreement with the normalized hydro-dynamic relation when F_r is employed (see Fig. 8b). In this plot, the recorded maximum pressure P_{Max} (i.e $P_{Max} = F_{Max}/A$) is normalized by the respective hydrodynamic factor, ρv^2, to obtain a dimensionless peak pressure. The F_r obtained in this study i.e. 4 – 7 fall with in the range of other flume model tests. In general, the results suggest that the impact force best relates to combinations of flow parameters, instead of only one parameter like flow velocity.

Ideally, the F_r similarity should have been maintained with the field observation that does not usually exceed 3 (Hübl et al. 2009, Wendeler and Volkwein 2015). On the other hand, the F_r in this test is relatively high. This is because the flow bed roughness was smooth as well as the fluid viscosity was not scaled.

The desired F_r can be achieved by increasing the flow bed roughness of the flume model and/or increasing the clay fraction of the material (Iverson et al. 2010, Holmes 2018). This will be considered in upcoming researches.

5 CONCLUSION

This paper presented a preliminary flume model study to measure impact force of a moving debris on a rigid circular steel pillar. It was found that the impact force is positively correlated with flow discharge which is a function of the flow velocity and the flow height. No clear trend was found when the recorded force was compared with the velocity alone. However, the hydrodynamic models seem to fit better when compared in light of both flow velocity and height, which means with the Froude number F_r. Despite some scatter in the data, the results are promising. More comprehensive tests shall be conducted at various C_s and discharges to verify empirical and analytical approaches for use in assessing impact forces on pillars.

ACKNOWLEDGEMENT

This study was supported financially by the Norwegian Public Roads Administration (SVV) E39 ferry-free highway project. The authors would like to acknowledge the Klima2050 project. The authors would also like to acknowledge Harald Norem and Petter Fornes for their discussions regarding testing and model set-ups. The authors are grateful to Einar Husby, Espen Andersen, Karl Ivar Volden Kvisvik, Per A. Østensen, Frank Stæhli who have participated in building and instrumenting the flume model.

REFERENCES

Arattano, M. & L. Franzi (2003). On the evaluation of debris flows dynamics by means of mathematical models. *Natural Hazards and Earth System Science* 3(6), 539–544.

Armanini, A., M. Larcher, & M. Odorizzi (2011). Dynamic impact of a debris flow front against a vertical wall. In *Proceedings of the 5th International Conference on Debris-Flow Hazards Mitigation: Mechanics, Prediction and Assessment, Padua, Italy*, pp. 1041–1049.

Bowman, E. T., J. Laue, B. Imre, & S. M. Springman (2010). Experimental modelling of debris flow behaviour using a geotechnical centrifuge. *Canadian geotechnical journal 47*(7), 742–762.

Bugnion, L., A. Bötticher, & C. Wendeler (2012). Large scale field testing of hill slope debris flows resulting in the design of flexible protection barriers. In G. Koboltschnig, J. Hübl, and J. Braun (Eds.), *Proceedings of 12th Interpraevent, 23-26 April 2012*, Grenoble, France, pp. 59–66.

Choi, C. E., C. W. W. Ng, S. C. H. Au-Yeung, & G. R. Goodwin (2015). Froude characteristics of both dense granular and water flows in flume modelling. *Landslides 12*(6), 1197–1206.

Costa, J. E. (1984). Physical geomorphology of debris flows. In J. Costa and F. P.J. (Eds.), *Developments and applications of geomorphology*, pp. 268–317. Berlin, Heidelberg: Springer.

Fink, J. H., M. C. Malin, R. E. D'Alli, & R. Greeley (1981). Rheological properties of mudflows associated with the spring 1980 eruptions of mount st. helens volcano, washington. *Geophysical research letters 8*(1), 43–46.

Fischer, L., L. Rubensdotter, K. Sletten, K. Stalsberg, C. Melchiorre, P. Horton, & M. Jaboyedoff (2012). Debris flow modeling for susceptibility mapping at regional to national scale in norway. In *Proceedings of the 11th International and 2nd North American Symposium on Landslides*, pp. 3–8.

Holmes, J. L. (2018). An assessment of experimental debris-flow scaling relationships. Master's thesis, Dunham University, Dunham, England.

Huang, H.-P., K.-C. Yang, & S.-W. Lai (2007). Impact force of debris flow on filter dam. *Momentum 9*(2).

Hübl, J. & G. Holzinger (2003). Entwicklung von grundlagen zur dimensionierung kronenoffener bauwerke für die geschiebebewirtschaftung in wildbächen: Kleinmaßstäbliche modellversuche zur wirkung von murbrechern. *WLS Report 50*.

Hübl, J., J. Suda, D. Proske, R. Kaitna, & C. Scheidl (2009). Debris flow impact estimation. In *Proceedings of the 11th international symposium on water management and hydraulic engineering, Ohrid, Macedonia*, pp. 1–5.

Hungr, O., G. C. Morgan, & R. Kellerhals (1984). Quantitative-analysis of debris torrent hazards for design of remedial measures. *Canadian Geotechnical Journal 21*(4), 663–677.

Ishikawa, N., R. Inoue, K. Hayashi, Y. Hasegawa, & T. Mizuyama (2008). Experimental approach on measurement of impulsive fluid force using debris flow model. In M. Mikos, J. Hübl, and G. Koboltschnig (Eds.), *11th International Interpraevent congress, 26-30 May 2008*, Volume 1, Durnbirn, Vorarlberg, Austria, pp. 343–354.

Iverson, R. M. (1997). The physics of debris flows. *Reviews of geophysics 35*(3), 245–296.

Iverson, R. M. (2015). Scaling and design of landslide and debris-flow experiments. *Geomorphology 244*, 9–20.

Iverson, R. M., M. Logan, R. G. LaHusen, & M. Berti (2010). The perfect debris flow? aggregated results from 28 large-scale experiments. *Journal of Geophysical Research: Earth Surface 115*(F3).

Jakob, M. & O. Hungr (2005). *Debris-flow hazards and related phenomena*. Berlin, New York: Springer.

Laache, E. (2016). Model testing of the drainage screen type debris flow breaker. [master's thesis], Norwegian University of Science and Technology (NTNU), Trondheim, Norway.

Le, T. M. H., S. O. Christensen, A. Watn, L. F. Christiansen, A. Emdal, & H. Norem (2016). Effects of deflection wall on run-up height of debris flow. In S. Aversa, L. Cascini, L. Picarelli, and C. Scavia (Eds.), *Landslides and Engineered Slopes. Experience, Theory and Practice*, pp. 1237–1244. CRC Press.

Meyer, N. K., W. Schwanghart, O. Korup, B. Romstad, & B. Etzelmller (2014). Estimating the topographic predictability of debris flows. *Geomorphology 207*, 114–125.

Mizuyama, T. (2008). Structural countermeasures for debris flow disasters. *International Journal of Erosion Control Engineering 1*(2), 38–43.

Proske, D., J. Suda, & J. Hübl (2011). Debris flow impact estimation for breakers. *Georisk 5*(2), 143–155.

Rickenmann, D. (1999). Empirical relationships for debris flows. *Natural Hazards 19*(1), 47–77.

Sandersen, F. (1997). The influence of meteorological factors on the initiation of debris flows in norway. *European Paleoclimate and Man 12*, 321–332.

Song, D., C. Choi, C. Ng, & G. Zhou (2017). Geophysical flows impacting a flexible barrier: effects of solid-fluid interaction. *Landslides*, 1–12.

Tiberghien, D., D. Laigle, M. Naaim, E. Thibert, & F. Ousset (2007). Experimental investigations of interaction between mudflow and an obstacle. In *Debris-flow hazards mitigation: mechanics, prediction and assessment*, Rotterdam. Millpress.

Vagnon, F. & A. Segalini (2016). Debris flow impact estimation on a rigid barrier. *Natural Hazards and Earth System Sciences 16*(7), 1691–1697.

VanDine, D. (1996). Debris flow control structures for forest engineering. *Res. Br., BC Min. For., Victoria, BC, Work. Pap. 8/1996*.

Wendeler, C., B. McArdell, D. Rickenmann, A. Volkwein, A. Roth, & M. Denk (2006). Field testing and numerical modeling of flexible debris flow barriers. In *Proceedings of international conference on physical modelling in geotechnics*, Hong Kong, China, pp. 4–6.

Wendeler, C. & A. Volkwein (2015). Laboratory tests for the optimization of mesh size for flexible debris-flow barriers. *Natural Hazards & Earth System Sciences Discussions 3*(3).

Yifru, A. L., E. Laache, H. Norem, S. Nordal, & V. Thakur (2018). Laboratory investigation of performance of a screen type debris-flow countermeasure. *Submitted to: HKIE Transactions - Special Issue on Landslides and Debris Flow 25*(2).

Zanuttigh, B. & A. Lamberti (2006). Experimental analysis of the impact of dry avalanches on structures and implication for debris flows. *Journal of Hydraulic Research 44*(4), 522–534.

18. Slopes

Centrifuge modelling of earth slopes subjected to change in water content

P. Aggarwal & R. Singla
Department of Civil Engineering, National Institute of Technology, Kurukshetra, India

A. Juneja
Department of Civil Engineering, Indian Institute of Technology Bombay, Mumbai, India

ABSTRACT: In the present study stability of slopes prepared using residual soil, compacted in dry state and was examined in different water content condition using a small beam centrifuge. Slopes were prepared in a small strong perspex box by tamping. A perforated water tank was placed behind the slopes to permit seepage of water through the soil mass. Provision was made to collect the water above the base plate. The slope angle was varied from 60 to 75° and the slope was examined in different water content conditions. Digital images were captured at fixed intervals and analysed. The result from the parametric studies shows the quantitative data points of Stability number at different gravity scale. It was observed that the relationships obtained in this study, are comparable to the Taylor's Stability number.

1 INTRODUCTION

Cross-Section of an embankment is usually of trapezoidal shape and comprises of slope structure. To calculate its stability, it is necessary to understand its side slope stability. The factors which affect the side slope stability include the physical properties of the soil, external loads and water infiltration (Terzaghi, 1950). Its key work is to support entire structure and the loads. Support to the embankment is given by its slope filled soil. Slopes are generally characterized by non-uniform stress field (Picarelli 2000). Bishop's modified method, Janbu's generalised procedure of slices and force equilibrium are the few methods which are used in the stability analysis. Failures in unsaturated residual soil slopes generally occur during the wet season. During this period, the shear strength is reduced and the pressure applied by the soil is increased. Shimada et al. (1995), Fredlund and Rahardjo (1994), Oeberg (1995) and Alonso et al. (1995) studied the effect of infiltration on stability of dry slopes.

Residual soils collected from west coast of India were used in this study. These soils have traditionally been formed by chemical weathering of laterites and basalt rocks bedded deep below the ground surface. Stability of the embankment made of the above soils was modelled using a small beam centrifuge. A prototype physical model similar to the one developed by Bucky (1931) was used at a reduced scale. Centrifuge modelling helps to draw relationship between stability number and angle of slope. The failure event in nature cannot be stimulated on the laboratory floor at 1 g. Reduce scale model is relatively inexpensive and favourable process to understand changes in stress and soil deformation in the prototype. In the model, the linear dimensions of the embankment were reduced by n-g, were n is the centrifuge acceleration and g is the earth's gravity. In accordance of centrifuge scaling laws, n-times greater than model was achieved by increasing the acceleration by n.

Table 1. Material properties of residual soil.

Soil Properties	Magnitude
Natural Moisture Content (%)	21
Specific Gravity	2.75
Liquid Limit (&)	49
Plastic Limit (%)	29
Shrinkage Limit (%)	19.7
Plasticity Index (%)	20
Soil Classification	ML
Maximum Dry Unit weight (kN/m^3)	14.5
Optimum Moisture Content (%)	24
UCS (kPa)	35.6

1.1 Experimental setup

The analysis of slope stability was performed using centrifuge model. Prior to the generation of the physical model, properties of soil were investigated since these results were considered to be helpful in further analysis. The soil properties are listed in Table 1. Figure 1 shows the grain size distribution of the soil sample.

Scanning Electron Microscope is helpful in developing electron digital images. The images analysis was done at different magnifications. Figure 2 shows the

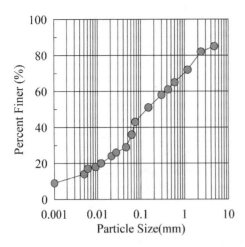

Figure 1. Particle size distribution of soil.

Figure 2. SEM image of soil.

SEM image of soil. The image was taken at CRNTS IIT Bombay.

1.2 Test program for the small model tests

The modelling was performed on balance beam small centrifuge having arm radius as 0.49 m. When driving unit stimulates the centrifugal movement, the angular velocity attained stimulates a particular Gravity Scale (N) called Scaling Factor. The scaling factor attained helps to calculate the critical height of prototype.

Driving unit consist of universal 0.5HP motor, with a swinging bucket on both side of arm. The maximum payload can be taken is 0.02 kN. Speed Regulator helps to initiate the in-flight condition and helps to achieve the failure of slopes. Digi-Strobe called stroboscope helps to appear cyclic moving objects in a slow moving string. It's as synchronized as the sample arrives in front of transparent screen of lid, the light flashes and makes box appear. It yields 50-50000 flashes per min.

Physical modelling includes strong box of Perspex material, containing bed slopes with seepage tank. The internal dimensions of tank were 160 mm × 1130 mm × 60 mm. Perforated Seepage tank attached adjacent the tank. Holes were arranged in a single central line with gauge distance of 2 mm and of 1.5mm diameter. Seepage flow was controlled by applying geotextile sheet on inner face of tank. Sheet makes water to flow all around the face of tank uniformly. The small scale model of slope was housed in strong box. The slope model consisted of a homogeneous residual soil slope. Different geometrical dimensional slope were prepared as listed in table 2. The analysis consists of different seepage conditions, by perforated seepage tank, placed behind the slope built. For generating in-flight condition slope was prepared and corresponding pay load was measured.

The following experimental program focused mainly on the performance of the slope model when the seepage was performed. The primary purpose of the experiment was to probe the critical height of slope at different slope angles. The modes of failure of slopes were seen in on toe, slope surface i.e. due to erosion, hair line cracks on slope surface. Before preparing soil slope model, the weight of soil was calculated by unit weight of sample and the volume of desired slope. The soil was tamped in different layers. Free board of minimum 1mm was provided. The soil cuboidal block was prepared and cutting was performed according to desired marked slope. As the sample was dry, sample was handled and trimmed gently.

The following test programme was adopted in the study.

1. Dry soil slope was prepared and was stimulated without seepage. Slope was checked till failure at every constant interval of angular velocity.
2. Dry soil slope was prepared and was stimulated without seepage. Slope was checked directly to the failing angular velocity.
3. Dry soil slope was prepared and was stimulated with seepage. Slope was checked till failure at every constant interval of angular velocity.
4. Dry soil slope was prepared and was stimulated with seepage. Slope was checked directly to the failing angular velocity.

Table 2 summarises the geometry of the models under different test conditions.

Table 2. Geometrics of models under different test conditions.

Test no.	Slope angle (degree)	Model height (m)	Base Length (m)
1	60	0.08	0.46
2	65	0.08	0.37
3	70	0.08	0.29
4	75	0.08	0.21

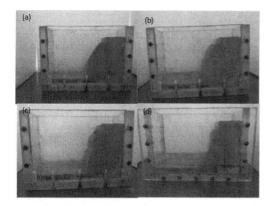

Figure 3. Slope condition for Dry sample with slope angle of (a) 60°; (b) 65°; (d) 70°; and (d) 75° (Condition 1).

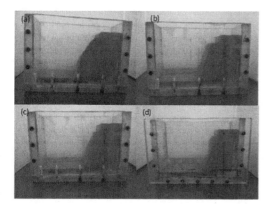

Figure 4. Slope condition for Dry sample with slope angle of (a) 60°; (b) 65°; (d) 70°; and (d) 75° (Condition 2).

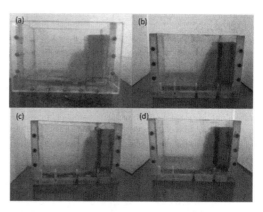

Figure 5. Slope condition for Dry sample with slope angle of (a) 60°; (b) 65°; (d) 70°; and (d) 75° (Condition 3).

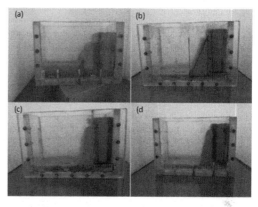

Figure 6. Slope condition for Dry sample with slope angle of (a) 60°; (b) 65°; (d) 70°; and (d) 75° (Condition 4).

1.2.1 Condition 1

After the sample preparation for each slope, counter balance was measured and both were placed in-flight setup. In this case the centrifuge was operated at incremental angular velocity. At every constant increment of angular velocity the centrifuge was slowed down. The box was removed and condition of slope was checked. Digital images were captured immediately after the removal of strong box. The box was then replaced in the centrifuge for the higher g level. The increment process was continued until the slope movement was recorded in the digital images. Figure 3a–d shows the slope conditions for condition 1.

1.2.2 Condition 2

The test conditions are similar to above condition, but in this case the centrifuge was operated directly at a speed near to the failing angular velocity. To check the exact failing point Stroboscope is placed over transparent screen of centrifuge.

Stroboscope is electronic devices which emits rapid flashes of light as sample passes over it. This was helpful to view sample. Figure 4a–d shows the slope condition for condition 2.

1.2.3 Condition 3

Centrifuge modelling is helpful in checking different condition of slopes. In this condition slopes were designed at different angles and condition of controlled seepage was introduced. Seepage was equally distributed along the face with the help of geotextile. Slopes were placed in centrifuge modelling and same procedure of stopping and removing of box was done. In this, seepage movement was also checked. As the angular velocity increase water starts its movement from toe to crest side of slope. Every constant interval of angular velocity images were produced and failing angular velocity was recorded. Figure 5a–d shows the slope condition in condition 3.

1.2.4 Condition 4

After checking the centrifuge at every incremental angular velocity for activated seepage conditions, in this condition direct increment of angular velocity to failing value was done. This process consists of image processing at starting and final time of setup. Figure 6a–d shows the slope condition for condition 4.

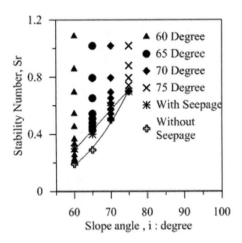

Figure 7. Comparing incremental loading – unloading conditions with and without seepage.

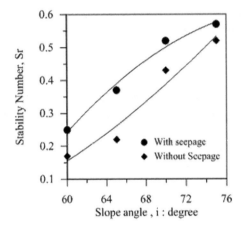

Figure 8. Comparison of direct centrifuge modelling with and without seepage.

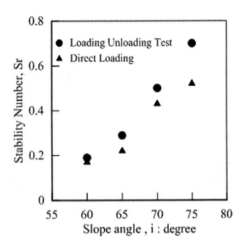

Figure 9. Comparison of without seepage centrifuge modelling.

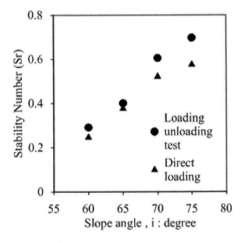

Figure 10. Comparison of with seepage centrifuge modelling.

Condition 4 comprises of incremental loading and unloading. At every constant interval of angular velocity, the condition of slope was checked.

Failure points at direct incremental condition are shown in Fig. 7. Graph shows variation of stability number with slope angles. Failure points with and without seepage conditions are compared.

2 CENTRIFUGE TEST RESULTS

Soil slopes ranging from 60° to 75° were tested under with and without seepage conditions. Table 2 summarise the geometric of models. Fig. 7 shows the relations between Stability number (Sn) and angle of slope under with and without seepage condition during incremental loading and unloading.

Fig. 8 shows the summarised failure points from Fig. 7. From the results it is observed that as the angular velocity increases stability number decreases. Stability number was computed using relation given below

$$Sn = \frac{C_u}{F * \gamma * H} \tag{1}$$

where F = 1 (factor of safety).

Figure 9 shows the comparison between direct incremental and loading-unloading incremental condition. In this condition seepage condition wasn't stimulated. The figure shows the variation of stability number along different slope angles.

Figure 10 shows the comparison between direct incremental and loading-unloading incremental

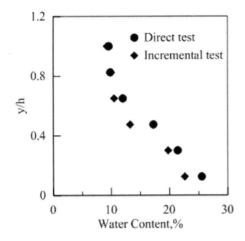

Figure 11. Comparison between Incremental and direct test at slope angle 60°.

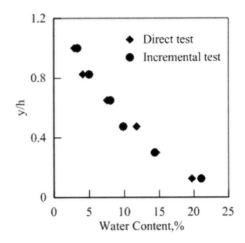

Figure 13. Comparison between Increment and direct test at slope angle 70°.

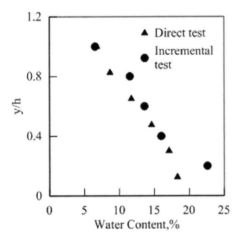

Figure 12. Comparison between Increment and direct test at slope angle 65°.

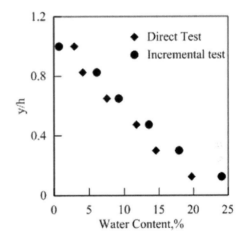

Figure 14. Comparison between Increment and direct test at slope angle 75°.

condition. In this comparison seepage was stimulated. Graph shows variation of stability number along different slope angles.

Test were performed on 60°, 65°, 70° and 75° angles at dry compacted soil. Proposed model Height of soil slope was 80 mm. Graphs depicts that water content varies along the height from w = 3% at top and w = 24% at bottom portion. So we can expect variation of apparent cohesion along the height of model. For peak strength, UU triaxial and unconfined compression strength Test were performed.

In the seepage analysis, water content along height was checked.

Figures 11 to 14 show variation in water content along the slope height under direct and incremental loading-unloading condition at different slope angles.

In phase 2, slopes were prepared at 65° at Optimum Moisture Content. Total 6 model test series with different slope height were formulated. Table 3 shows the detained geometry of the models tested at OMC. Water content along the slope height was calculated at every constant interval. Fig. 15 shows the variation in

Table 3. Summary of tests.

Test no.	Slope angle (degree)	Model height (m)	Base Length (m)
1	65	0.40	0.18
2	65	0.45	0.21
3	65	0.50	0.23
4	65	0.55	0.25
5	65	0.70	0.32
6	65	0.75	0.35

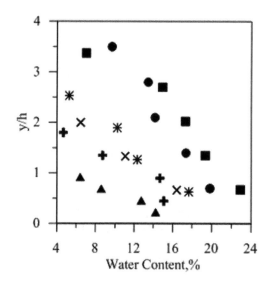

Figure 15. Variation of water content along slope height in OMC soil conditions.

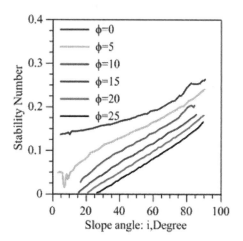

Figure 16. Taylor's stability chart (1948).

water content in each test model. In every slope, soil at top crest has low water content (1 to 3%) whereas soil at toe has water content varies from (14 to 23%).

The results were compared with Taylor stability graph (1948) in Figure 16.

3 CONCLUSION

Residual soil with its high permeability and silty clay shows unusual characteristics. The water flow in soil very easily and causes much variation along the slope height. In every test, slopes water content along the height increases from apex to bottom. Fig. 11–14 and Fig. 15 show variation of water along the slope height in different slope conditions. In either of the case, water content at toe was less than Optimum Moisture Content ($S < 1$). These finding imply following

- In case of OMC prepared slope, as the slope height increases, saturation at toe side has reached up to 100%. In case of dry soil slopes, failing angular velocity is more for without seepage condition in both direct and incremental loading-unloading condition.

REFERENCES

Cai, F. and Ugai, K. 2004. Numerical Analysis of rainfall effects on slope stability, International Journal of Geomechanics, 4(2): 69–78.

Fredlund, D. G. and Rahardjo, H. 1993. Soil mechanics for unsaturated soils. John Wiley & Sons, New York.

Grkceoglu, C and Aksoy, H 1996. Landslide susceptibility mapping of the slopes in the residual soils of the Mengen region (Turkey) by deterministic stability analyses and image processing techniques, Engineering Geology 44: 147–161.

Juneja, A., Chatterjee, D. and Kumar, R 2012. Embankment failure in residual soils at Nivsar, Ratnagiri, International Journal of Geo engineering Case Histories, 2(3): 229–251.

Kim, J. Jeong, S. Park, S. and Sharma, J. 2004. Influence of rainfall-induced wetting on the stability of slopes in weathered soils, Engineering Geology 75: 251–262.

Ling; H. I., Wu; M.H., Leshchinsky, d., and Leshchinsky, B 2009. Centrifuge Modelling of Slope Instability, Journal of Geotechnical and Geoenvironmental Engineering, ASCE, 135: 758–767.

Ng, C.W.W. and Shi. Q. 1997. A numerical investigation of the stability of unsaturated soil slopes subjected to transient seepage, Computers and Geotechnics, 22(1): 1–28.

Tan, L.Y. Lee and Sivadass, T. 2008. Parametric study of residual soil slope stability, ICCBT, E-04: 33–42.

Tohari, A. Nishigaki, M. and Komatsu. M. 2007. Laboratory Rainfall-Induced Slope Failure with Moisture Content Measurement, Journal of Geotechnical and Geoenvironmental Engineering ASCE, 133: 575–587.

Vishwanathan, B.V.S and Mahajan, R.R. 2007. Centrifuge model tests on geotextile-reinforced slopes, Geosynthetics International, 14(6).

Centrifuge and numerical modelling of static liquefaction of fine sandy slopes

A. Askarinejad, W. Zhang, M. de Boorder & J. van der Zon
Faculty of Civil Engineering and Geosciences, Section of Geo-Engineering, Delft University of Technology, Delft, The Netherlands

ABSTRACT: A series of centrifuge tests on fully saturated slope samples was performed to investigate the unstable behaviour of statically liquefied slopes. The samples were saturated with a viscous fluid to satisfy the time scaling factors of generation and dissipation of pore pressures. The pore pressures were monitored inside the slopes. The behaviour of these models under increasing g-level were also simulated using the Finite Element Method (FEM). The 2D plane strain FE simulations indicate that the stress states of the soil elements of a slope fall inside the instability zone. However, the centrifuge tests showed that a global flow slide failure is unlikely to occur if a triggering mechanism is applied to a limited extent of the soil body due to the radial dissipation of the excess pore pressures. The results are applicable to offshore pile driving projects.

1 INTRODUCTION

Submarine landslides in loose sand deposits seriously damage many flood barriers in deltaic areas all over the world. For instance, more than 1000 flow-slides have been damaging the dikes in the south-western part of the Netherlands during the last two centuries. This number is expected to increase in near future as a result of the predicted global rise in seawater level and also the increase in the extreme meteorological events. However, very few reliable field measurements and laboratory observations are available due to technical difficulties and high observational costs. Therefore, the factors controlling the domino type chain of micro-collapses (Askarinejad et al. 2014; Maghsoudloo et al. 2017) leading to a major event are not fully understood. A series of centrifuge tests on loose fine sand slopes was performed to study the triggering mechanisms of flow slides. These tests were simulated using the Finite Element Method.

2 CENTRIFUGE TESTS

A very fine, uniform silica sand is used in the experiments of this study. A set of tests has been performed to characterise the geotechnical properties of this sand (Table 1).

The centrifuge experiments were conducted using the beam centrifuge at TU Delft. A viscous fluid with the viscosity 10 times higher than that of water (Askarinejad et al. 2017) was used to saturate and submerge the slopes in order to unify the time factors of generation and dissipation of pore pressures due to collapse of micro-pores as the triggering mechanism of static liquefaction (Askarinejad et al. 2014).

In total 4 centrifuge tests were performed (Table 2). The loading on the samples was conducted by controlled increase of acceleration. The dimensions of the strong box used in these tests were 270 (L) × 150 (H) × 135 (W) mm³. Several batches of sand saturated with the viscous fluid were prepared and kept in vacuum for 24 hours to ensure full saturation. The centrifuge models were prepared by wet pluviation of the saturated sand below a 20 mm layer of viscous fluid up to a height of 124 mm in the strong box. The extra sand was very carefully dredged out of the box to form

Table 1. Geotechnical properties of the soil (after De Jager et al. 2017).

Parameter	Values
D_{50} (mm)	0.11
$\varphi'_{residual}$ (°)	34
Angle of repose (°)	34
Permeability (m/s)	4.2E–5
Min-Max void ratio	0.64–1.07
Particle shape	Sub-rounded

Table 2. Summary of the centrifuge tests.

Test no.	Initial Slope angle	Final slope angle
1	33°	26°
2a	29°	28°
2b	28°	Local failure
2c	28°	Local failure

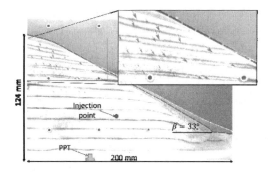

Figure 1. Pre- and post-failure profiles (dark and light dashed-lines, respectively) of the slope surface. The arrows in the crest show the movement of the soil.

a slope. The relative density of samples was measured to be about 5%. Miniature Pore Pressure Transducers (PPT) were installed at the inner base of the box. The pore pressure was locally increased artificially by injecting viscous fluid using a syringe in Tests 2b and 2c. The inner diameter of the injection pipe was 3 mm. These tests were performed on the same sample as that of the Test 2a. Particle Image Velocimetry (PIV) was used to determine the movement field of sand.

3 TEST RESULTS

The centrifugal acceleration was increased in steps of 10 g at a rate of 0.15 g/s. The slopes of the Tests 1 and 2 showed failures at less than 10 g while the centrifuge was accelerating and resulted in milder slope angles of 26° and 28°, respectively. These failures could be visualised by small settlements in the crest of the slope (Figure 1). The high viscosity of the submerging fluid has potentially prohibited further extension of the failure.

The excess pore fluid pressures built up with increase of the g-level (Figure 2) indicating a partially drained condition. The rate of excess pore pressure change increased and peaked about 33 seconds after the failure onset in Test 1. Similar observation was also reported by Eckersley (1990). He also mentioned that excess pore pressures are generated during and after, rather than before movement. However, a slight triggering mechanism was needed to mobiles the failure and this was not detected by the PPTs.

The results indicate that the slope of Test 1 failed slightly earlier and the excess pore fluid pressure (EPFP) kept on increasing with increasing centrifugal acceleration. However, both tests show that the EPFP attained a local peak after the failure. The slope of Test 1 was slightly steeper and marginally looser than that of Test 2. These factors have resulted in larger runout extent and milder failure slope angles. Moreover, the slope of Test 2 indicates a dilative behaviour as the centrifugal acceleration increased. The changes in EPFP are shown in Figure 3.

The internal mechanism leading to static liquefaction can be explained by the collapse of saturated voids,

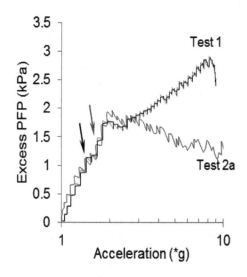

Figure 2. Evolution of excess pore fluid pressure during the increase in g-level for Tests 1 and 2a. The arrows show the failure points.

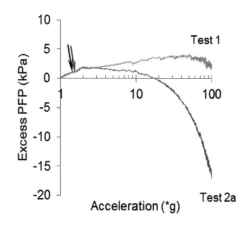

Figure 3. Evolution of excess pore fluid pressure during the increase in g-level for Tests 1 and 2a. The arrows show the failure points.

which results in local and abrupt increase of the pore pressure (Take et al., 2004). The locally increased pore pressure reduces the effective stress, and hence the shear strength, and can trigger larger movements in the soil mass. To simulate the initiation of failure mechanism, pore water pressure was locally increased using an injection system in the slope of Tests 2b and 2c at 10 g in order to trigger the chain of the domino-type collapse events. The location of the injection was selected as the most critical point in a liquefiable sandy slope based on the zonation that Lade (1992) suggested according to simplified limit equilibrium analysis (Figure 1). The increase of the pore pressure was applied in the middle of the slope surface next to the transparent longitudinal wall at a depth of 50 mm.

The pore pressure increase was conducted in two phases of injection at rates of 0.31 mL/s (low) and

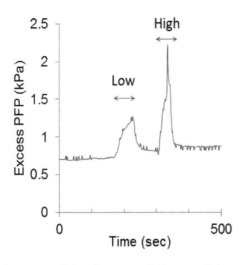

Figure 4. Evolution of excess pore fluid pressure during two phases of injection at low and high injection rates.

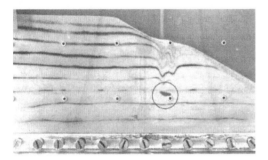

Figure 5. Local failure around the injection point in Test 2c.

0.5 mL/s (high) for 30 sec and 19 sec, respectively. The results indicate that the pore pressure increased by a value of 0.4 and 1.4 kPa for the two phases (Figure 4). Although a local failure occurred close to the outlet, this failure did not lead to a major collapse of the slope (Figure 5). The major collapse of the slope is prohibited because of the radial dissipation of the EPFP. The same observation has been made during the offshore pile driving projects in loose sand slopes (Lamens et al., 2018).

4 NUMERICAL SIMULATIONS USING UBCSAND CONSTITUTIVE MODEL

The Finite Element software Plaxis 2D is used to investigate soil behaviour in the processes of rising gravity and elevating of pore pressure in loose sand slopes. The soil behaviour was described by UBCSAND constitutive model which was developed for prediction of sand liquefaction behaviour (Puebla et al. 1997).

There are 15 parameters required for UBCSAND model. The parameters were calibrated based on a drained triaxial shear test on sand samples with a relative density (D_r) of 50% (Figure 6 and Figure 7). Given

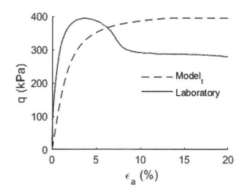

Figure 6. Shear stress vs. axial strain of drained triaxial tests at cell pressures of 100 kPa.

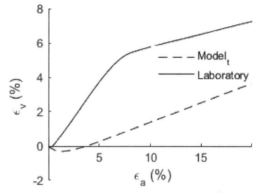

Figure 7. Volumetric strain vs. axial strain of drained triaxial tests at cell pressures of 100 kPa.

the sand models in centrifuge tests had D_r around 15%, the already calibrated parameters were extrapolated based on the relative density difference between two tested conditions empirically.

The material property parameter φ_{cv} (constant volume friction angle) was determined as 34° (Chavez Abril 2017). The Cohesion and Tension Cut-off were both given as 0 kPa. Furthermore, the peak friction angle (φ_p), the elastic bulk modulus number (K_B^e), the plastic shear modulus numbers (K_G^P), the failure ratio (R_f) were gained by curve fitting with the drained triaxial test. The elastic shear modulus number (K_G^e) was calculated from;

$$K_G^e = K_B^e \frac{3(1-2\nu)}{2(1+\nu)} \quad (1)$$

where the Poison ratio (ν) was taken as 0.3. Moreover, the indices of the elastic bulk modulus (m_e), the elastic shear modulus (n_e) and the plastic shear modulus (n_p) were all assumed as default values of 0.5, 0.5, 0.5, respectively. Additionally, the densification factor and the post liquefaction factor were assigned as 1.0 and 1.0, respectively, since the cyclic load is out of the scope of this study. The SPT value, $N1_{60}$, was estimated from the corresponding D_r value (in the unit of percent), based on the following empirical equation (Petalas & Galavi 2013).

$$N1_{60} = \left(\frac{D_r}{15}\right)^2 \quad (2)$$

It is well known that soil is density dependent material. Hence the parameters calibrated from drained triaxial shear test should be modified in order to be suitable for the centrifuge test sample. Maeda and Miura (1999) performed triaxial tests on about 80 sand materials and presented a relationship between the void ratio extent ($e_{max} - e_{min}$) and Young's moduli at various relative densities:

$$\frac{\Delta \log E'_{50}}{\Delta D_r} = 0.7(e_{max} - e_{min}) + 0.27 \quad (3)$$

The range of void ratio for the sand that was used in the centrifuge tests is 0.37. Therefore:

$$\frac{E'_{50}(D_r = 50\%)}{E'_{50}(D_r = 15\%)} = e^{0.186} \quad (4)$$

Furthermore, the bulk modulus (K) is proportional to Young's modulus and is defined as equation (6) in UBCSAND constitutive model (Petalas and Galavi, 2013),

$$K = \frac{E'_{50}}{2(1+v)} \quad (5)$$

$$K = K_B^e P_A \left(\frac{p}{P_{ref}}\right)^{me} \quad (6)$$

where, the reference stress level (P_{ref}) is given as the 100 kPa (the same as the atmosphere pressure, P_A) and p is the mean normal stress. It can be inferred that K_B^e and K_G^e (equation (1)) of Sand samples at $D_r = 15\%$ and 50% should follow the relationship shown in equation (4).

Considering that the loose sample would show contractive behaviour, the peak friction angle was assigned as 33°. The constant volume friction angle was kept as 34°, since it is a stress and density independent parameter. Tsegaye (2010) suggested that, for contractive soil, the plastic shear modulus number can be estimated from the following equation:

$$K_G^P = -72 K_B^e \eta_{il} \left(\frac{\sin\varphi_p}{(6+\eta_{il})\sin\varphi_p - 3\eta_{il} R_f}\right)^2 \frac{3\eta_{il} - (6+\eta_{il})\sin\varphi_{cv}}{16 + (6+\eta_{il})\sin\varphi_{cv}} \quad (7)$$

where, η_{il} is the inclination of the instability line (IL). Based on the loose fine sands results performed by

Table 3. Calibrated UBCSAND parameters for Geba Sand models with relative densities of 50% and 15%.

Parameters	Triaxial test ($D_r = 50\%$)	Centrifuge tests ($D_r = 15\%$)
φ_{cv} (°)	34	34
φ_p (°)	41.4	33
c (kPa)	0	0
K_B^e (–)	418.6	347.6
K_G^e (–)	193.2	160.4
K_G^p (–)	500	176.7
m_e (–)	0.5 (default)	0.5
n_e (–)	0.5 (default)	0.5
n_p (–)	0.5 (default)	0.5
R_f (–)	0.98 (default)	0.98
Fac_{hard} (–)	1.0 (default)	1.0
$N1_{60}$ (–)	11.1	1.0
Fac_{post} (–)	1.0 (default)	1.0
P_A (kPa)	100	100

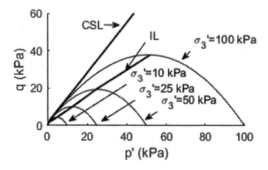

Figure 8. Effective stress diagram of numerical undrained triaxial tests at confining stresses of 10 kPa, 25 kPa, 50 kPa and 100 kPa.

Sladen et al. (1985), Chavez Abril (2017) made an expectation that the ratio of φ_{cv} and the slope angle of instability line (φ_{il}) should be around 1.77. He found η_{il} a value of 0.74 when the φ_{cv} of Sand was 34°. The input parameters of the UBCSAND model for the drained triaxial test and the centrifuge test are presented in Table 3.

Undrained triaxial tests at various cell pressures were conducted numerically using the UBCSAND parameters modified for the centrifuge test sample. The effective stress paths are presented on p'-q diagram in Figure 8. Thereafter, the critical state line (CSL) is plotted based on the value of φ_{cv} which gives the M value of 1.37 in this space. The IL is plotted by connecting the peak points of the effective stress paths. It is indicated that the IL of undrained triaxial tests at high confining stresses should intersect the origin of p'-q space (Yamamuro and Lade, 1997, Bopp, 1994). It is observed that the inclination of IL varies with cell pressure conditions. The inclinations of the segments IL between stress paths of cell pressures from 100 kPa to 50 kPa, from 50 kPa to 25 kPa and from 25 kPa to 10 kPa are 0.73, 0.68 and 0.77, respectively.

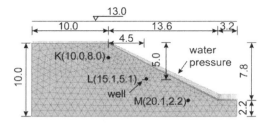

Figure 9. Mesh distribution of the Plaxis model with three analysed points and the injection point (unit: cm).

5 NUMERICAL SIMULATION OF SUBMERGED SLOPE

The numerical FE model has been performed in plain strain conditions with geometries tested in centrifuge. A Coarseness factor of 0.5 was used to generate the mesh for the model (Figure 9).

The *Gravity Loading* was selected as the calculation type in the initial phase to initialize stress distribution of the slope. The process of increasing gravity was simulated by adding a new calculation phase giving $\sum M_{weight}$ a value of 100 in PLAXIS. In order to obtain drained condition, the drainage type of UBC-SAND constitutive model was taken as Drained and the calculation type was provided with Consolidation. Furthermore, the ground flow conditions of the left, right and bottom boundaries were set to be closed (impervious). However, the ground flow conditions of the slope surfaces were set to be Seepage to allow fluid flow through slope surface.

After reaching 100 g condition, the slope was stable. The effective stress paths of three slope-parallel points (Figure 9) are compared with the CSL and IL in Figure 10. The CSL and IL were obtained from undrained triaxial simulations (Figure 8). It can be seen that the effective stress conditions of all the points rose linearly with the increase of gravity condition. Moreover, at the initial stage, all effective stress states located lower than IL, however, after a certain of value of gravity, they passed over IL and arrived at the instability zone bounded by CSL and IL. Therefore, it can be concluded that the fact that the effective stress condition located within instability zone itself does not mean liquefaction.

Thereafter, the fluid injection test was investigated. A *well structure* was constructed with a length of 4 mm in (Figure 9). The Infiltration option was selected to simulate the injection of fluid. The calculation was performed using the *Plastic calculation type*. In this calculation phase, the drainage type of the soil was assigned as Undrained, and the permeability in x and y directions were both taken as 7.5×10^{-5} m/s (Krapfenbauer, 2016). It should be noted that the fluid injection in the centrifuge test was at one point, while, in this Plaxis model, the elevated pore pressure was along the slope in the out-of-plan direction (Plane strain condition).

During the infiltration of fluid, the slope became instable. The points K, L and M reached at excess pore

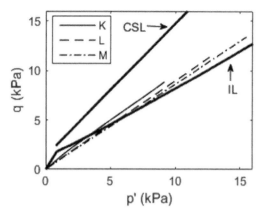

Figure 10. Effective stress paths of points K, L and M with increasing of gravity.

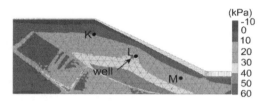

Figure 11. Excessive pore pressure distribution in the process of fluid injection (pressure = positive).

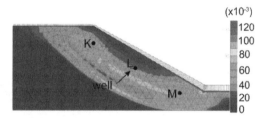

Figure 12. Incremental deviatoric strain at the moment of failure.

pressures of 9.5 kPa, 32.5 kPa and 19.3 kPa, respectively, which yield excess pore pressure ratios (the ratio of excess pore pressure and the initial vertical effective stress) of 0.65, 1.7 and 0.97, respectively. The location of point L was very close to the well which might cause the excess pore pressure ratio at point L relatively higher than those of positions K and M. The excess pore pressure ratios at the failure of submerged slopes were varying from 0.54 to 1.0 measured by Zhang et al. (2015), 0.7 presented by L'Heureux et al. (2013) and 0.9 suggested by Kvalstad et al. (2005) and Masson et al. (2006).

The excess pore pressure contour (Figure 11) show that the elevated pore pressure tended to dissipate through the slope surfaces while accumulated in the slope since the ground flow conditions of model boundaries were impervious. The highest pore pressure zone shows the accumulation of pore pressure and the failure surface position which can be recognized

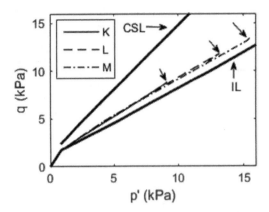

Figure 13. Effective stress paths of points K, L and M during the infiltration of fluid under drained condition (arrow heads show the starting effective stress states).

also from the incremental deviatoric strain contour (Figure 12). It can be inferred that the accumulated pore pressure dissipated through the slip surface.

The change of the effective stress states at K, L and M are presented in Figure 13. The stress states dropped linearly and appeared to end at a same stress state ($p' = 1.20$ kPa and $q = 2.07$ kPa) nearly intersecting the instability line.

6 CONCLUSIONS

This paper demonstrates centrifuge modelling and FE modelling of underwater slopes composed of loose fine sand. The following conclusions can be drawn from the physical and numerical modelling:

1. The results of the centrifuge tests show that the measureable pore pressures occur during and after the onset of static liquefaction, although micro collapses might have resulted in the general failure of the slope.
2. The extent of the local increase in pore pressure is important in triggering a flow slide.
3. The effective stress state located within the instability is the necessary but not the sufficient condition for static liquefaction;
4. Excess pore pressure tends to accumulate around the failure surface and may liquefy the soil above the failure surface.

REFERENCES

Askarinejad, A., Beck, A. & Springman, S. M. 2014. Scaling law of static liquefaction mechanism in geocentrifuge and corresponding hydromechanical characterization of an unsaturated silty sand having a viscous pore fluid. *Canadian Geotechnical Journal*, 52, 708–720.

Askarinejad, A., Sitanggang, A.P.B. & Schenkeveld, F.M. Effect of pore fluid on the cyclic behavior of laterally loaded offshore piles modelled in centrifuge. *19th International Conference on Soil Mechanics and Geotechnical Engineering*, 2017 Seoul, South Korea. 905–910.

Bopp, P.A. 1994. The effect of initial relative density on instability and behavior of granular materials at high pressures. *PhD Thesis, University of California*.

Chavez Abril, M.A. 2017. Numerical simulations of static liquefaction in submerged slopes. *MSc Thesis, TU Delft*.

De Jager, R.R., Maghsoudloo, A., Askarinejad, A. & Molenkamp, F. 2017. Preliminary results of instrumented laboratory flow slides. *1st International Conference on the Material Point Method*. Delft, The Netherlands: Elsevier Ltd.

Eckersley, J.D. 1990. Instrumented laboratory flowslides. *Géotechnique*, 40, 489–502.

Krapfenbauer, C. 2016. Experimental investigation of static liquefaction in submarine slopes *MSc Thesis, TU Delft*.

Kvalstad, T.J., Andresen, L., Forsberg, C.F., Berg, K., Bryn, P. & Wangen, M. 2005. The Storegga slide: evaluation of triggering sources and slide mechanics. *Marine and Petroleum Geology*, 22, 245–256.

L'heureux, J., Vanneste, M., Rise, L., Brendryen, J., Forsberg, C., Nadim, F., Longva, O., Chand, S., Kvalstad, T. & Haflidason, H. 2013. Stability, mobility and failure mechanism for landslides at the upper continental slope off Vesterålen, Norway. *Marine Geology*, 346, 192–207.

Lamens, P., Askarinejad, A., Sluijsmans, R. W. & Feddema, A. 2018. Ground response during offshore pile driving in a sandy slope. *Geotechniqe*, Under review.

Maeda, K. & Miura, K. 1999. Relative density dependency of mechanical properties of sands. *Soils and Foundations*, 39, 69–79.

Maghsoudloo, A., Galavi, V., Hicks, M. & Askarinejad, A. 2017. Finite element simulation of static liquefaction of submerged sand slopes using a multilaminate model. *19th International Conference on Soil Mechanics and Geotechnical Engineering*. Seoul.

Masson, D., Harbitz, C., Wynn, R., Pedersen, G. & Løvholt, F. 2006. Submarine landslides: processes, triggers and hazard prediction. *Philosophical Transactions of the Royal Society of London A: Mathematical, Physical and Engineering Sciences*, 364, 2009–2039.

Petalas, A. & Galavi, V. 2013. Plaxis Liquefaction Model UBC3DPLM. *PLAXIS Report*.

Puebla, H., Byrne, P.M. & Phillips, R. 1997. Analysis of CANLEX liquefaction embankments: prototype and centrifuge models. *Canadian Geotechnical Journal*, 34, 641–657.

Sladen, J., D'hollander, R. & Krahn, J. 1985. The liquefaction of sands, a collapse surface approach. *Canadian Geotechnical Journal*, 22, 564–578.

Take, W.A., Bolton, M.D., Wong, P.C. P. & Yeung, F.J. 2004. Evaluation of landslide triggering mechanisms in model fill slopes. *Landslides*, 1, 173–184.

Tsegaye, A. 2010. Plaxis liquefaction model. *external report. PLAXIS knowledge base: www.plaxis.nl*.

Yamamuro, J.A. & Lade, P.V. 1997. Instability of granular materials at high pressures. *Soils and Foundations*, 37, 41–52.

Zhang, J., Lin, H. & Wang, K. 2015. Centrifuge modeling and analysis of submarine landslides triggered by elevated pore pressure. *Ocean Engineering*, 109, 419–429.

Modelling of MSW landfill slope failure

Y.J. Hou, X.D. Zhang, J.H. Liang & C.H. Jia
China Institute of Water Resources and Hydropower Research (IWHR), Beijing, China

R. Peng
Beijing Municipal Construction Co. Ltd, China

C. Wang
State Nuclear Electric Power Planning Design & Research Institute Co. Ltd, Beijing, China

ABSTRACT: Large portion of municipal solid waste (MSW) has to be stored in valleys or landfills around cities in China. There are always potential risks of slope failure for those solid wastes. This paper presents the centrifuge model tests to simulate the various MSW slopes on horizontal ground and inclined ground. Artificial MSW materials with different mix ratio were adopted to simulate different store ages of waste. Models with different section layouts were tested to failure respectively by using a newly developed rotating container in the swing basket, to increase the slope angle until model collapse. The test results show that the old age landfill slope is generally stable than the middle age one with the same boundary condition, and for the landfill with leachate inside, slope failure can be easily activated when the slope toe loss its strength at certain leachate level. A simple method is presented in this paper to estimate the safety reservation of the model slope based on slope stability analysis together with the slope failure information from centrifuge model tests.

1 INTRODUCTION

The investigation on 31 major cities in China shows that the increasing rate of solid waste is over 10% in the past decade, with accumulated waste of 7 billion tons stored around cities, according to the Annual Report on Environment Development of China (Yang et al. 2011). This tendency has been the same in recent years and about one-fourth of the Cities in China do not have suitable areas to set up MSW landfills, which add more risks to the existing landfills. The influence of MSW to the environment includes air pollution, ground water pollution, soil pollution, and potential diseases spread by mosquitos and insects. Sometimes at extreme environment conditions, such as heavy rain and earthquake, when higher leachate level is inside the landfill, slope failure or flow slide may be generated in the landfill (Zhan et al. 2010, Matasovic et al. 1998).

The compositions of MSW are very complex with its physical and mechanical properties varying with the cities and locations. Its density is not only related with the compositions and water content, but also changing with time and buried depth. The water content of MSW is directly related with the original material types, local climate, transporting method, leachate collecting and drainage system under the landfill. Therefore a big variety of material density is detected by some researches (Geoffrey 2008, Jessberger & Kockel 1993), with the density of $0.3\,g/cm^3$ to $1.4\,g/cm^3$ for different kinds of MSW. There void ratios vary from 40% to 52% depending on the compositions and compaction method. Usually there is a lot of leachate existing in MSW, owning to self-weight and loading, or seepage water from outside, such as rainfall, irrigation and ground water in the landfill. Most of the MSW in China comes from kitchen with higher potential of leachate production. The leachate levels in some landfills become higher gradually when their drainage systems fail, which is a great threatening to the stability of landfill slope.

It is essential to understand the characteristics of the MSW before making a model simulation. There are mainly three methods to study its properties, such as in-situ test, sampling and laboratory test and back analysis from field loading test (Koerner et al. 2000). Mccreanor & Reinhart (2000) suggested that the mechanical parameters of MSW may be deducted from stress back analyses for the design application. However there are still many uncertainties about the performance of landfill with MSW.

The MSW material is different from common geotechnical materials, but its mechanical behavior is also closely related with the gravity. Therefore the centrifuge simulation rules are also applicable to MSW material. Artificial waste materials were often adopted for centrifuge modeling.

For centrifuge modeling, the original landfill or slope can be tested by reducing its dimension by scale N, putting the model with container into the centrifuge, accelerating the model to Ng (g is gravity acceleration) so that the stress level inside the model is the same as that in prototype. The deformation and failure process of landfill slope can be monitored by transducers and cameras preinstalled in the model container. Thusyanthan & Madabhushi (2006) had conducted a series centrifuge model tests on the static and dynamic performance of MSW, presented the simulating method and slope deformation result. However it's not easy to design a model at the beginning, to make it collapsed at certain g level in order to capture the slope failure phenomenon during centrifuge test.

This paper presents the research results on the landfill slope deformation and failure with different storing ages and design sections. A rotating container was developed to fit the swing basket of IWHR centrifuge. With the help of the rotating container, every model slope can be tested to failure by increasing the slope angle during centrifuge flight.

Figure 1. IWHR 450 g-ton geotechnical centrifuge.

Figure 2. Rotating container with driving facilities.

2 TEST FACILITIES AND MATERIAL

2.1 Centrifuge and rotating container

The tests were conducted with IWHR 450 g-ton Centrifuge (Fig. 1) which has been put into operation since 1991. Its maximum design acceleration is 300 g. The maximum radius of the centrifuge is 5.03 m. Inner size of the swing basket is 1500 mm × 1000 mm × 1500 mm. A rotating container with driving facilities was developed as shown in Figure 2. It can be operated under centrifuge acceleration of 50 g. The inner size of the rotating container is 568 mm × 342 mm × 308 mm (L × W × H). Its maximum inclination is 75° with different rotating rates. A constant rotating rate of 1.4°/min was adopted for all the model tests in this research.

2.2 Model material

Original MSW is hard to use directly for centrifuge model test because the model material is almost unrepeatable and harmful for operators. Some researchers designed their own methods to simulate MSW. Thusyanthan (2006) used the mixture of peat, E-grade kaolin clay and fraction-E fine sand. After tests on its unit weight, compressibility and shear strength, the results show that the characteristics of the model waste are match well with those reported for real MSW. Zhu et al. (2012) presented a kind of model MSW in China, which is a mixture of turf-soil, standard fine sand and Kaolin clay. By adjusting the propositions of above materials, different storing ages of the model MSW can be simulated, as given in Table 1.

These simulated materials in Table 1 were also adopted for the research in this paper. The physical and mechanical characteristics of the model MSW were tested in IWHR soil lab and listed in Table 2.

Table 1. Material proportions of model MSW.

Type of MSW	Middle age	Old age
Storing age / Years	3.3~6.6	6.6~10
Buried depth / m	10~20	20~40
Turf-soil : fine sand : kaolin clay	0.54:1:1	0.23:1:1

Table 2. Physical and mechanical characteristics.

Sample	Water content %	Dry density g/cm^3	Seepage coefficient cm/s	Strength parameters	
				C_d kPa	Φ_d °
Old age MSW	43	0.84	–	0.8	18.6
		0.98	–	10.7	22.8
		1.12	–	19.0~21.5	23.4
Middle age MSW	47	0.54	1.3*10^{-3}	0	26.2
		0.61	1.4*10^{-4}	0	28.0
		0.81	7.2*10^{-5}	10.9~12.4	28.5

Generally for the middle age MSW, its density is relatively lower and water content higher, and the density gradually increased along the age, water content reduces in the old age. The model MSW is classified as highly compressible soil based on the compression test results, with medium to high seepage coefficient.

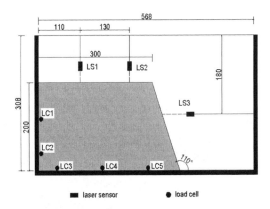

Figure 3. Configuration of model type A.

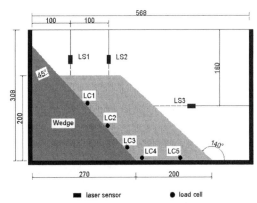

Figure 4. Configuration of model type B.

Table 3. Illustration of the tested models.

No.	Model name	Illustration
1	AM70	Middle age MSW slope angle 70°
2	AO70	Old age MSW slope angle 70°
3	BM40	Middle age MSW Inclined lining slope 45° Slope angle 40°
4	BM60	Middle age MSW Inclined lining slope 45° Slope angle 60°
5	BM70	Middle age MSW Inclined lining slope 45° Slope angle 70°
6	CO70	Old age MSW slope angle 70° with water inside the container

3 SLOPE FAILURE TESTS

3.1 Model preparation

Three types of model sections and layouts were designed in the research, with total of six models. All the models had the height of 200 mm and were tested at 40 g, to simulate the prototype MSW slope with 8 m high. The original average dry density is 0.78 g/cm³ for middle age waste and 0.98 g/cm³ for old age waste, and the water contents are 47% for middle age and 43% for old age waste respectively.

Model type A has two models (AM70 and AO70), using middle age MSW and old age MSW respectively for model compaction before cutting the slope into an angle of 70°. The model section of type A is shown in Figure 3. Several load cells were installed at the bottom of the model. There were laser sensors LS1 and LS2 on the top of the model to measure the settlement of the top surface, and LS3 arranged horizontally facing the slope inside the container to measure the horizontal deformation of the slope. A high speed camera facing the transparent side of the model container was installed, so that the deformation of the model section can be recorded for later PIV analysis.

There were three models for type B, mainly concentrate on different slopes for middle age MSW. Their slope angles are 40°, 60° and 70° respectively, labeled as model BM40, BM60 and BM70. For simulating the MSW stored in a mountain area with inclined slopes on the ground. A concrete wedge with the slope angle of 45° under the model landfill was designed as shown in Figure 4. A thin layer of geomembrane with a thickness of 1mm was stickered to the container bottom surface and the wedge slope for model type B, with fine sand glued on the membrane surface to simulate the friction effect coming from the composite lining of the landfill. The friction angle of the model lining is about 25°.

For each model test, after the centrifuge acceleration reached 40 g and model deformation did not increase obviously, the rotating container was operated to increase the slope angle until the model failed. The failure information or data was recorded before stop the rotating container and the centrifuge.

The third model type C has a model CO70 designed to observe the leachate influence on the old storing age model slope. The model section and profile of CO70 is the same as model type A in Figure 3. Distilled water was adopted instead of real leachate. Before centrifuge rotation, the water was filled into the model container slowly until the water height reached 50 mm, which was 1/4 of the model height. All the models for comparison are illustrated in Table 3.

3.2 Slope safety reservation

Sometimes it is necessary to know the instability potential of the designed model slope. The safety reservation S_r for centrifuge model slope is defined in this paper, based on the slope failure information from the centrifuge test. In practice and slope stability analysis, the calculated safety factor should be $F_f \leq 1$ when the slope fails. The model slope section and boundary condition at the failure moment in centrifuge test can be adopted for back analysis to get the material coherent strength c_f at failure when $F_f \approx 1$, using the

Table 4. Model conditions at failure.

Model	Test acc. g	Model dry density after compaction in centrifuge g/cm³	Strength for calculation		Slope angle at failure °
			C_f kPa	Φ_d °	
AM70	40	0.81	11.0	28.5	83
AO70	40	1.12	19.0	23.4	88
BM40	40	0.81	10.7	28.5	66
BM60	40	0.81	10.0	28.5	72.8
BM70	40	0.81	10.5	28.5	79.6
CO70	33.5	1.10	17.0	23.4	70

Table 5. Safety reservations of designed model slopes.

Model	Safety factor at failure F_f	Initial safety factor F_i	Safety reservation S_r %
AM70	0.99	1.30	31
AO70	1.00	1.39	39
BM40	0.99	1.79	80
BM60	0.99	1.25	26
BM70	0.98	1.13	15
CO70	1.04	1.44	40
	With water	No water	

parameters from soil tests, in which only c_f can be adjusted in its reasonable range of a material for stability analysis. Therefore the safety factor F_i of the initial model slope can be obtained by the calculation with the same parameters as used for back analysis. The safety reservation may be defined as $S_r = (F_i - F_f) \times 100\%$ represented by percentage.

For example, Model AM70 failed when the final slope angle reached 83°. With the knowing section right before the slope collapse, using the slope stability analyses software STAB with the material initial parameters in Table 2, slightly adjusting the coherent strength at failure to get c_f, the safety factor of the model at failure can be calculated with the result of $F_f = 0.99$. Based on the same parameters and c_f, the initial safety factor of the model slope before rotating the container is $F_i = 1.30$ by calculation. Therefore the safety reservation of the model slope is $S_r = (1.30 - 0.99) \times 100\% = 31\%$.

For the stability analysis in this study, the model scale is converted into prototype scale, with the consideration of the settlement at 40 g for each model test. The initial average model density was estimated based on the measured section profile at 40 g before the container rotation, and its appropriate strength parameters were adopted for trial analysis. The initial dry density for each model after compaction at 40 g is also provided in Table 4, suppose the model material is uniform and its water content is kept original. Each model section right before its failure moment was recorded and used for back analysis to get the coherent strength at failure c_f. Based on the information in Table 4, the

(a) Model AM70

(b) Model AO70

Figure 5. Top views of cracks and slope failure.

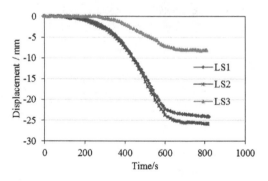

Figure 6. AM70 displacement before container rotation.

safety reservations for all designed model slopes in this study are listed in Table 5.

3.3 *Test results and interpretation*

The test result comparison of model AM70 and AO70 shows that old age landfill slope is more stable than the middle age one with the same section, based on the safety reservation estimation. The S_r of the old age model is 8% higher than that of the middle age one, which shows a tendency that the landfill becomes stable along with the time.

From the test results of model BM40, BM60 and BM70, it shows that more sliding deformation may happen along the inclined ground slope if the inclined angle is greater than the landfill slope, based on the test result from model BM40. The higher the slope angle, the smaller the safety reservation as given in Table 5. Compare model AM70 and BM70, for the model BM70 with landfill situating on the inclined ground, its safety reservation is 15% less than that of the model AM70.

The preliminary deformation of the middle age model AM70 from 1 g to 40 g was given in Figure 6,

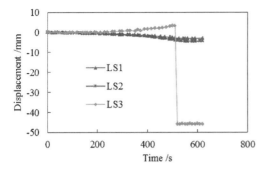

Figure 7. AM70 displacement after container rotation.

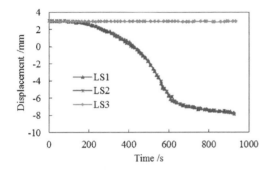

Figure 8. AO70 displacement before container rotation.

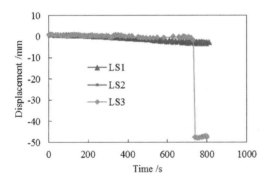

Figure 9. AO70 displacement after container rotation.

about 24~25 mm settlement on the top surface of the model, 8 mm in horizontal towards left on the slope at the position of LS3. However the slope remained stable without failure. After the model settlement became stable, the container was rotated at a constant rate of 1.4°/min until the slope collapsed with a sudden settlement as shown in Figure 7. The failure happened 510s after the container rotating corresponding to a rotating degree of 13° based the rated rotating speed. Figure 8 and Figure 9 shows the model AO70 deformation before and after the container rotating respectively.

Based on the measured deformation of the model at certain points, the model dry density after settlement can be estimated with approximately 0.81 g/cm^3 for middle age material and 1.12 g/cm^3 for old age material.

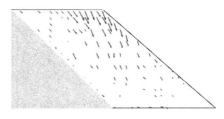
(a) Before container rotating

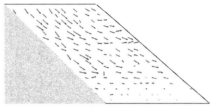
(b) Right before the slope failure

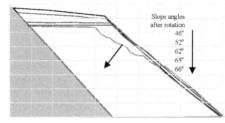
(c) Slope profile at different rotation stages

Figure 10. Deformation of model BM40 by photogrammetry.

Test on model BM40 shown that the landfill surface settled for 10 mm at centrifuge acceleration of 40 g. A relative movement along the inclined lining was observed by the camera facing the model section. However the slope did not fail at centrifuge acceleration of 40 g before rotating the container. The model section displacement distribution pattern is given in Figure 10(a). Figure 10(b) shows the section displacement distribution right before the slope failure when the slope angle became $\phi = 65°$. The slope profiles at different rotation stages are present in Figure 10(c) when the model slope collapsed at $\phi = 66°$, which was identified and analyzed based on the video images.

For model CO70 to simulate the leachate inside the model, the model slope collapsed with obvious cracks on the top when the centrifuge acceleration reached 33.5 g (Fig. 11), corresponding to prototype slope height of 6.7 m, and the water level height of 1.8 m. The video images during the test show that the model waste material at the slope toe was collapsed first by losing its effective strength under water. Safety factor F by calculation is 1.04 at the failure moment, and is 1.44 for the same model without leachate. Therefore the safety reservation of model CO70 is reduced 40% when the leachate level is about 1/4 height of the slope. It also explains that the MSW slope with high leachate level is much easier to get failure.

(a) Initial model section (b) Monitoring sensors

(c) Section view of collapse (d) Top view of slope collapse

Figure 11. Photos of Model CO70 with water inside.

4 CONCLUSIONS

The MSW slope failure phenomenon was observed by centrifuge model tests on simulated MSW material. The rotating container developed in IWHR made it possible that all the models be tested to failure. A simple method is presented in this paper to give a general estimation of the safety condition quantitatively for a designed model slope, by calculating the safety reservation of a model slope based on its failure information from centrifuge test.

The test result shows that the old age landfill slope is generally stable than the middle age one with the same boundary conditions. The safety reservation for the old age MSW is about 8% higher than that of the middle age. For the landfill on inclined ground with geomembrane lining, the failure of the waste slope tends to move along the contact area when the ground inclination is greater than the MSW slope angle.

For the landfill with the height of 6.7 m in prototype and slope angle of 70°, when the leachate level reaches 1/4 height of the slope, it activated the slope failure based on the centrifuge test in this paper. Compare with the same model without water, its safety reservation reduced about 40%, indicating that the instability of MSW slope will increase a lot under high leachate level. It will be effective to deduct the leachate outside the landfill and lower the water level inside to prevent cracks and slope failure.

ACKNOWLEDGEMENTS

The authors appreciate the professional work of Ms. Bian J.H., Ms. Song X.H. and other technicians in IWHR soil test lab and centrifuge center, for assisting the research work in this paper. This research was supported by the National Basic Research Program (No. 2012CB719803) and the National Science Foundation (No. 51679003).

REFERENCES

Blight, G. 2008. Slope failures in municipal solid waste dumps and landfills: a review. Waste Management & Research 26(5): 448–463.

Matasovic, N., Kavazanjian, E. & Anderson, R. L. 1998. Performance of Solid Waste Landfills in Earthquakes. Earthquake Spectra, 14(2): 319–334.

Jessberger, H L. & Kockel, R. 1993. Determination and assessment of the mechanical properties of waste materials. Proc., Sardinia 93, 4th Int. Landfill Symp., San Margherita di Pula, CISA, Cagliari, Italy: 1383–1392.

Koerner, R. M. & Soong, T. Y. 2000. Leachate in landfills: the stability issues. Geotextiles and Geomembranes 18: 293–309.

Pen, R., Hou, Y. J., Zhan, L.T. & Yao, Y.P. 2016. Back-Analyses of landfill instability induced by high water level: case study of Shenzhen Landfill. International Journal of Environmental Research and Public Health 13: 126–138.

Thusyanthan, N I. & Madabhushi, S P G. 2006. Centrifuge modeling of solid waste landfill systems—Part 1: Development of a model municipal solid waste. Geotechnical Testing Journal 29(3): 217–222.

Yang, D.P., et al. 2011 Annual report on environment development of China (in Chinese), Social Sciences Academic Press (China).

Zhu, B., Yang, C.B. & Wang, L. 2012. Solid waste configuration and centrifugal model test on the deformation of landfill (in Chinese). The 1st China National Symposium on Multi-field Coupled Problems of Geo-materials and Geo-environmental Engineering: 425–432.

Zhan, L.T., Guan R.D. & Chen Y.M. 2010. Monitoring and back analysis of slope failure process at a landfill, Chinese Journal of Rock Mechanics and Engineering, 29(8): 1697–1705.

Effects of plant removal on slope hydrology and stability

V. Kamchoom
*Faculty of Engineering, King Mongkut's Institute of Technology Ladkrabang, Bangkok, Thailand,
(formerly, Department of Civil and Environmental Engineering, Hong Kong University of Science
and Technology, Hong Kong SAR)*

A.K. Leung
*Department of Civil and Environmental Engineering, Hong Kong University of Science and Technology,
Hong Kong SAR
(formerly, Division of Civil Engineering, University of Dundee, Dundee)*

ABSTRACT: Removal of vegetation in close vicinity to transportation corridors would have an impact on slope stability due to the loss of hydro-mechanical root reinforcement. Fallowed slope would be subjected to accelerated episodic movements and shallow failure, causing potential ultimate and serviceability limit state problems. This study investigates the effects of plant removal on slope hydrology and stability, through centrifuge and numerical modelling. This study uses artificial roots that are capable of simulating the effects of transpiration in-flight and have mechanical properties closely representative to real roots. Model slopes before and after plant removal (i.e. with and without considering transpiration effects, respectively) were subjected to identical intense rainfall in the centrifuge. Removing all vegetation on the slope lost all the beneficial effects of transpiration-induced suction that would otherwise be present without plant removal. Relying on mechanical reinforcements alone are insufficient to maintain slope stability. Preliminary parametric analyses reveal that to maintain slope stability, plant removal should not be conducted beyond two-third from the slope crest. Even removing only one-fourth of plants from the slope toe would significantly reduce factor-of-safety of slope and temporary support of the slope is needed.

1 INTRODUCTION

Vegetation in close vicinity to transportation assets (i.e., embankments or cuttings) can potentially cause vehicular disruptions and accidents. The overhanging plant branches and leaves can obstruct the road and railway, hence limiting the driver's sight to other vehicles on the road, traffic control devices, pedestrian and wildlife (Berger 2005). Local transport agency is often required to provide regular maintenance to prune or remove the vegetation so that vehicles, signs and pedestrian path can be visible by the road's users.

Most of the road and railway were built either on an embankment or next to a cut slope. Previous research has shown that the presence of vegetation has beneficial effects to slope stability. Plant roots behave like a structural element for reinforcing the soil (Wu et al. 1988, Barker 1999). On the other hand, plant transpiration would induce matric suction. This suction would affect soil hydraulic conductivity (hence rainfall infiltration; Leung et al. 2015, 2017b) and increase soil shear strength, both favourable to slope stability (Ng et al. 2016, Leung et al. 2017a). In short term, plant removal would eliminate the positive contributions from transpiration (Briggs et al. 2013, Smerthurst et al. 2016), leaving only root reinforcement for temporary slope stabilisation. In longer term, the roots would decay and lose strength (O'Loughlin 1974, Watson et al. 1999), hence reducing slope stability. It is important to incorporate soil-root interaction theories and engineering principles when making decisions of plant removal adjacent to transportation corridors, in addition to solely based on operational and logistical constraints.

This study uses centrifuge modelling technique to experimentally quantify the effects of plant removal on slope hydrology and stability. Artificial roots were used to simulate the root hydro-mechanical reinforcements effects in the centrifuge. By controlling suction during a test, any changes in soil hydrology and stability before and after plant removal were studied. Finite-element seepage-stability models were developed and validated for subsequent investigation of the effects of progressive plant removal and removal patterns on slope stability.

2 CENTRIFUGE MODELLING

2.1 Test plan

Two centrifuge tests were performed at 15 g using the geotechnical beam centrifuge at HKUST. Figure 1 shows a typical centrifuge model package of a 1:15

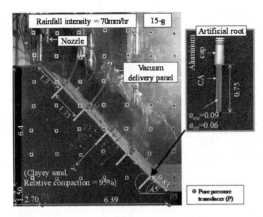

Figure 1. Typical centrifuge model package of the root supported slopes (Ng et al. 2016).

Table 1. Summary of soil properties and the input parameters used for the finite element seepage-stability analyses.

Parameter		Value
Bulk unit weight (γ_t)		20 kN/m^3
Specific gravity (G_s)		2.59
Maximum dry density		1890 kg/m^3
Optimum moisture content		15.10%
Sand content ($\leq$4.760 mm)		56.8
Fine content ($\leq$0.074 mm)		43.2
Effective cohesion (c')		0 kPa
Critical-state friction angle (f_{cr}')		37.4 degrees
Dilation angle (ψ)		5 degrees
Young's modulus (E)		35 MPa
Poisson's ratio (ν)		0.26
Saturated hydraulic conductivity		1×10^{-7} m/s
Air-entry value (AEV)		1 kPa
Saturated water content (Θs)		Drying 41, Wetting 30%
Residual water content (Θr)		Drying 15, Wetting 15%
Fitting parameters for van Genuchten (1980)	α	Drying 0.5, Wetting 1
	n	1.8
	m	0.44

scaled model slope supported by arrays of artificial taproots. Note that unless stated otherwise, dimensions shown in each figure and the corresponding discussion are expressed in prototype scale. The first test, denoted as CT (as reported in Ng et al. 2016), was to study the slope stability before plant removal, meaning that the slope was supported by both root mechanical reinforcement and transpiration-induced suction. The second test, denoted as CN (new test in this study), was a repetition of the first test, but without simulating transpiration in all roots. This represents a condition where all plants on the slope were removed, relying only on the mechanical root reinforcement for slope stability.

2.2 Artificial roots

In order to simultaneously simulate both the effects of transpiration and mechanical reinforcement of plant roots in the centrifuge, the artificial root developed by Ng et al. (2014) and Kamchoom et al. (2015) was adopted in this study. The artificial root was made of cellulose acetate (CA), which is connected to a vacuum delivery panel for applying a vacuum pressure to each root. The applied vacuum would reduce water pressure inside the artificial root, creating a hydraulic gradient to remove the moisture of the surrounding soil. Because the CA has high air-entry value (AEV; up to 100 kPa), the hydraulic connection between surrounding soil and the roots can be maintained. The flow of moisture from soil to the roots hence induces matric suction, simulating the effects of plant transpiration. The scaling factor of matric suction is 1.0 (Taylor 1995).

Besides being a suitable material for modelling the effects of plant transpiration, CA also has scaled mechanical properties that are suitable for modelling mechanical root reinforcement (Kamchoom et al. 2014). The scaling factors of axial rigidity (EA) and flexural rigidity (EI) are $1/N^2$ and $1/N^4$, respectively (where N is the gravitational acceleration applied in the centrifuge). Hence, for the geometry and material of the artificial root considered, the prototype of EA and EI are about 290 kPa.m^2 and 0.2 kPa.m^4, respectively.

2.3 Soil type

Each slope model was made of completely decomposed granite (CDG; clayey sand (CL)) after 2-mm sieved following ASTM (2011) D2487-11. A pair of drying and wetting soil water retention curves (SWRCs) of the CDG was measured using a pressure-plate apparatus. Each SWRC was fitted with the equation proposed by van Genuchten (1980). Drying and wetting soil hydraulic conductivity functions (SHCF) were estimated using a combination of the prediction equation proposed by van Genuchten (1980) and saturated hydraulic conductivity k_s measured from a series of falling-head tests. The soil properties are summarised in Table 1.

2.4 Model configuration and preparation

Identical slope geometry (Fig. 1) was adopted for both Tests CT and CN. The CDG was compacted by moist tamping at the RC of 95% (i.e. 1777 kg/m^3) and gravimetric water content of 15.1%. A total of 15 roots were arranged in a square pattern. The bottom and side boundaries of each model slope were set to be impermeable. This would allow any water table to be formed due to seepage in both tests. The crest and the slope surface were left exposed for evaporation and rainfall infiltration. In order to minimise any suction induced by evaporation during testing, a wind cover was placed over the centrifuge model box. A rainfall simulation system was adopted (see Fig. 1) to apply and control a rainfall event during each test. Each model slope was subjected to an intense rainfall event with a constant intensity of 70 mm/h for 8 hours. More details of this test setup can be found in Ng et al. (2016).

2.5 Instrumentation

Pore-water pressure (PWP) changes during testing were monitored by five pore pressure transducers (PPTs; Druck PDCR-81; see Fig. 1). Surface runoff during the simulation of rainfall was measured by a collection frame (340 mm in length × 340 mm in width × 100 mm in height) that was located at a horizontal distance of 4.11 m away from the slope toe. The interfaces between the soil, the frame and the strongbox were sealed so that only runoff water that was not infiltrated into the slope would be collected in the frame. To identify any slope failure mechanism, the displacement field of each slope was captured using high-speed and high-resolution cameras. The measured displacement field was subsequently interpreted by the particle image velocimetry (PIV) technique coupled with a close-range photogrammetry correction.

2.6 Test procedure

Each centrifuge test was carried out in five stages. In the first stage, the centrifuge model was spun up, until the centrifugal acceleration reached 15 g. The second stage was to maintain the 15 g acceleration for the soil mass to consolidate so as to dissipate any excess PWP from the first stage. When all PPTs showed negligible change in PWP within 24 h in prototype, in the third stage, the effects of transpiration were simulated by applying an identical vacuum pressure to all artificial roots. The vacuum was kept constant, again until all PPTs showed negligible changes. In the fourth stage, a rainfall event was applied through the nozzles, with a constant intensity of 70 mm/h for 8 h. After that, rainfall was ceased and the model was span down to 1 g in the last stage.

3 NUMERICAL MODELLING

Seepage-stability analyses, denoted as T and N, were to back-analyse the slope hydrology and stability observed from Tests CT and CN, respectively. The validated numerical models were then used for subsequent parametric studies, which were concerned on the effects of progressive plant removal and removal patterns (i.e., starting from the slope crest or toe) on slope stability.

Figure 2 shows the finite element mesh using SEEP/W (Geo-Slope Int. 2009a) for all transient seepage analyses. Each analysis adopted Darcy-Richard's equation to describe transient seepage in unsaturated soils. The equation assumes that any PWP change would not induce soil volume change. Previous study (Chiu & Ng 2012) has shown that compacted CDG exhibited negligible volume change when suction is less than 40 kPa.

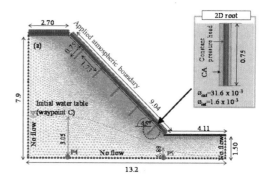

Figure 2. Typical finite element mesh of the root supported slopes.

The model setup and procedure in all numerical models were identical to the centrifuge tests. The initial water table was identified based on the readings from P4 and P5. In an attempt to consider the effects of hydraulic hysteresis, the drying SWRC and SHCF were used during any simulation of root transpiration before rainfall (i.e. third stage). For the simulation of the rainfall event (i.e. fourth stage), the wetting SWRC and SHCF were inputted. The input parameters for each SWRC and SHCF are listed in Table 1.

Each artificial root (see inset; Fig. 2a) was modelled by creating a material that has an AEV of 100 kPa and saturated hydraulic conductivity of 2×10^{-6} m/s. In order to model three-dimensional (3D) transpiration process by each root in a two-dimensional (2D) model, the root internal diameter was adjusted so that the total water volume flow in this 2D model is equal to that in the 3D circular structure.

Minimum factor of safety (FOS_{min}) corresponding to a critical slip surface was calculated by the strength reduction method (SRM; Dawson et al. 1999, Griffiths and Lane 1999) using SIGMA/W (Geo-Slope Int. 2009b). The CDG was assumed to obey the extended Mohr-Coulomb failure criterion (Vanapalli et al. 1996). Each artificial root was modelled as a beam element to capture both the elastic axial and bending responses. The interface element surrounding each root was modelled by a perfectly-plastic model, following the Coulomb's friction law. All required parameters are listed in Table 1 In each stability analysis, the PWP fields computed at the end of second, third and fourth stages from the seepage analysis were inputted to determine FOS_{min}.

4 PLANT REMOVAL EFFECTS ON SLOPE HYDROLOGY AND STABILITY

Figure 3 shows the measured and back-analysed PWP profiles in the two tests. In both tests, the initial PWP profiles aligned with a hydrostatic line. Before plant removal when transpiration was present, (Fig. 3a), PWP at 0.3 and 0.6 m depths decreased substantially by almost 10 kPa. On the contrary, for the plant

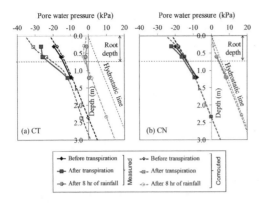

Figure 3. Measured and computed PWP profiles of root supported slopes (a) before and (b) after plant removal.

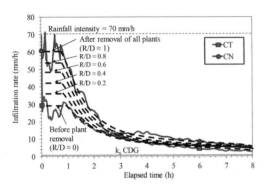

Figure 4. Measured variations in infiltration rate with time and the fitted curves used for seepage analyses.

removal case, the PWP at all depths increased only slightly (i.e., less than 2 kPa), contributed mainly by soil surface evaporation. After rainfall, it can be seen that the vegetated slope retained some suctions (up to 3 kPa) within the root zone (Fig. 3a), while only a slight positive PWP of about 1 kPa was built up at 1.2 m depth. After plant removal, no suction was retained at the end of the rainfall event (Fig. 3b). Instead, significant positive PWP (i.e., up to 10 kPa) was built up between 1–1.5 m depth, leading to almost a hydrostatic distribution. In general, good agreements of PWP profiles are obtained between the measurements and simulations. Thus, the numerical modelling procedures adopted captured the physical modelling of the effects of plant transpiration well.

Figure 4 shows the variations in the water infiltration rate during rainfall. Both tests showed an exponential reduction in infiltration. However, the case without plant removal (i.e. when transpiration effect was present) had substantially lower infiltration rate than the plant removal case. This is because of the reduction of PWP by the transpiration before rainfall (Fig. 3a), hence causing substantial reduction of soil hydraulic conductivity and infiltration rate during the subsequent rainfall. The measured variations of infiltration curve from each test was imposed on

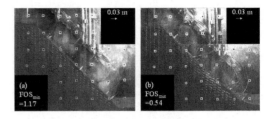

Figure 5. Measured displacement fields after 8 h of rainfall of root supported slopes (a) before and (b) after plant removal.

the top and slopping face of the corresponding slopes for back-analyses.

Because of suction retained within the root zone for the case without plant removal (Fig. 3a), the slope displaced uniformly in a translational manner parallel to the slope (Fig. 5a). The FOS_{min} is found to be higher than 1.0, consistent with the centrifuge observation that the slope remained stable. For the slope considering plant removal (i.e., supported by root reinforcement alone), the large positive PWP of up to 10 kPa built up (Fig. 3b) caused substantial slope movement (Fig. 5b). The displacement was accumulated near slope toe, leading to slope failure after rainfall. This is consistent with the FOS_{min} value of 0.54. It is important to reveal from both the measurements and simulations that the absence of even a small amount of transpiration-induced suction due to short-term plant removal (i.e. before root decay and root mechanical reinforcement is still effective) would cause slope instability as FOS_{min} was almost halved. Engineering intervention is required to provide short-term slope stability when plant removal has to be done.

5 EFFECTS OF PLANT REMOVAL PATTERNS ON SLOPE STABILITY

The validated numerical seepage and stability models were used to investigate the effects of progressive plant removal on slope stability, either starting from the slope crest (namely top-down removal) or the slope toe (namely bottom-up removal). The removal pattern is quantified by a normalized distance, R/D, where R is the removal distance either from the slope crest or toe and D is the length of the slope face. For each removal pattern, "short-term" condition is considered when plant transpiration has lost but root mechanical reinforcement exists (i.e. root has not decayed yet). In terms of modelling, this means that all root transpiration was not simulated (i.e., no vacuum pressure was applied) and the structural element remains for simulating mechanical reinforcement. "Long-term" condition refers to the condition in which plants have lost both mechanical reinforcements (full root decay) and hydrological reinforcements. In this case, all the structure elements were deactivated. As plant is removed (i.e. R/D increases), the rate of rainfall infiltration is assumed to increase proportionally from the lower- to the upper-bound infiltration curves shown in Figure 4.

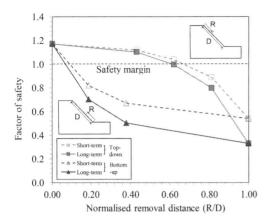

Figure 6. Short- and long-term stability of root supported slopes subjected to top-down and bottom-up plant removal.

Following the top-down removal pattern, the slope remained stable (i.e. $FOS_{min} > 1.0$) even top-half of the plants have been removed, under either short- or long-term conditions. Interestingly, the difference of short- and long-term FOS_{min} becomes prominent only when more than half of the vegetation is removed from the slope (i.e. $R/D > 0.5$). Recent field data presented by Smerthurst et al. (2015) showed that after removing the vegetation up to two-third of the slope height from the crest (equivalent to $R/D = 0.6$ in the simulation), no slope failure was reported during the monitoring period. Based on the preliminary numerical analysis conducted in this study and the limited field data, the general recommendation of vegetation clearance for not more than two-thirds of the slope height (Briggs et al. 2003) seems to be reasonable. As expected, in longer term when root mechanical reinforcement is not available, the rate of FOS_{min} drop with R/D is faster (Fig. 6). This means that plant hydro-mechanical reinforcement at the bottom-half of the slope is required to maintain slope stability.

On the contrary, following the bottom-up pattern, removing the plants even only the lower one-fourth of the slope would significantly reduce the FOS_{min} by almost half, consequently leading to slope failure. This agrees with the centrifuge observation where most of slope displacement was mobilised near the slope toe (Fig. 5). It is thus crucial to provide immediate support to the slope if any plants in the bottom part of the slope have to be removed.

6 CONCLUSIONS

This paper uses centrifuge modelling technique to quantify the effects of plant removal on slope hydrology and stability. The plant hydro-mechanical contributions to slope stability were simulated by using artificial roots. These model roots have mechanical properties close to real roots and they are capable of simulating the effects of plant transpiration and induced matric suction in the centrifuge. Model slopes before and after plant removal were subjected to an intense rainfall event. The observed slope behaviour during an intense rainfall was back-analysed through a series of finite-element seepage-stability analyses to improve the understanding of plant effects on slope hydrology and stability.

The results from centrifuge tests revealed that plant removal has pronounced effects on slope stability. Before plant removal, considerable amount of transpiration-induced suction was retained within the root zone after rainfall, maintaining the slope stability with FOS_{min} higher than 1.0. After removing all plants, due to the loss of the benefit of plant transpiration, much larger positive PWP was built up and a higher infiltration rate were measured, causing a significant drop of FOS_{min} by more than 50%. Slope movement was localised near the slope toe, consequently leading to slope failures. Preliminary parametric study about plant removal pattern highlights that it is important to not to remove more than two-third of plants from the slope crest to maintain the slope stability. The analysis also reveals that even removing only one-fourth of plants from the slope toe would significantly reduce the slope stability by almost half (in terms of FOS_{min}). Temporary support of slope is needed if any plants near the slope toe area have to be removed.

ACKNOWLEDGEMENT

The first author would like to acknowledge the Hong Kong-Scotland Partners in Post-Doctoral Research (S-HKUST601/15) provided by the Research Grants Council of Hong Kong SAR and the Scottish Government and travel grant from King Mongkut's Institute of Technology Ladkrabang. The second author also acknowledges the funding provided by the EU FP7 Marie Curie Career Integration Grant (CIG) under the project BioEPIC Slope as well as the Engineering and Physical Sciences Research Council (EPSRC), UK (EP/N03287X/1).

REFERENCES

ASTM. 2011. Standard practice for classification of soils for engineering purposes (Unified Soil Classification System). ASTM standard D2487-11. ASTM International, West Conshohocken, Pa.

Barker, D.H. 1999. Live pole stabilisation in the tropics. Ground and Water Bioengineering for Erosion Control and Slope Stabilisation, 301–308.

Berger, R.L. 2005. Integrated roadside vegetation management. Transportation Research Board., Washington, DC

Briggs, K.M., Smethurst, J.A., Powrie, W., O'Brien, A. S., & Butcher, D.J.E. (2013) Managing the extent of tree removal from railway earthwork slopes. Ecological engineering, 61, 690–696.

Chiu, C.F., & Ng, C.W.W. 2012. Coupled water retention and shrinkage properties of a compacted silt under isotropic and deviatoric stress paths. Canadian Geotechnical Journal, 49(8): 928–938.

Geo-Slope International Ltd. 2009a. Seepage Modeling with SEEP/W 2007, An Engineering Methodology, Fourth Edition.

Geo-Slope International Ltd. 2009b. Stress-Deformation Modeling with SIGMA/W 2007, An Engineering Methodology, Fourth Edition.

Kamchoom, V., Leung, A.K. & Ng, C.W.W. 2014. Effects of root geometry and transpiration on pull-out resistance. Géotechnique Letters. 4(4), 330–336.

Kamchoom, V., Leung, A.K. & Ng, C.W.W. 2015. A new artificial root system to simulate the effects of transpiration-induced suction and root reinforcement, 15th Asian Regional Conference on Soil Mechanics and Geotechnical Engineering, Kumamoto University, Fukuoka, Japan, 9-13 November 2015: 236–240.

Leung, A.K., Garg, A., & Ng, C.W.W. 2015. Effects of plant roots on soil-water retention and induced suction in vegetated soil. Engineering Geology, 193(1):183–197.

Leung, A.K., Kamchoom, V. & Ng, C.W.W. 2017a. Influence of root-induced soil suction and root geometry on slope stability: a centrifuge study. Canadian Geotechnical Journal. 54(3): 291–303.

Leung, A.K., Boldrin, D., Liang, T., Wu, Z., Kamchoom, V., & Bengough, A.G. 2017b. Plant age effects on soil infiltration rate during early plant establishment. Geotechnique. DOI: 10.1680/jgeot.17.T.037.

Ng, C.W.W., Kamchoom, V. & Leung, A.K. (2016) Centrifuge modelling of the effects of root geometry on the transpiration-induced suction and stability of vegetated slopes. Landslides. 13(5), 925–938.

Ng, C.W.W., Leung, A.K., Kamchoom, V. & Garg, A. 2014. A novel root system for simulating transpiration-induced soil suction in centrifuge. Geotechnical Testing Journal, ASTM. 37(5), 733–747.

Ng, C.W.W., Leung, A.K., Yu R. & Kamchoom, V. 2017. Hydrological Effects of Live Poles on Transient Seepage in an Unsaturated Soil Slope: Centrifuge and Numerical Study. Journal of Geotechnical and Geoenvironmental Engineering, ASCE. 143(3).

O'Loughlin, C.L. 1974. A study of tree root strength deterioration following clearfelling. Canadian Journal of Forest Research, 4(1), 107–113.

Smethurst, J., Briggs, K., Powrie, W., Ridley, A. & Butcher, D. 2015. Mechanical and hydrological impacts of tree removal on a clay fill railway embankment. Geotechnique, 65 (11), 869–882.

Smethurst, J.A., Clarke, D., & Powrie, W. 2006. Seasonal changes in pore water pressure in a grass covered cut slope in London Clay. Géotechnique, 56(8), 523–537.

Taylor, R.N. 1995. Geotechnical centrifuge technology. CRC Press.

Van Genuchten, MT. 1980. A closed-form equation for predicting the hydraulic conductivity of unsaturated soils. Soil Science Society of America Journal, 44(5): 892–898.

Vanapalli, S.K., Fredlund, D.G., Pufahl, D.E. & Clifton, A.W. 1996. Model for the prediction of shear strength with respect to soil suction. Canadian Geotechnical Journal, 33(3):379–392.

Watson, A., Phillips, C. & Marden, M., 1999. Root strength, growth, and rates of decay: root reinforcement changes of two tree species and their contribution to slope stability. Plant and Soil 217, 39–47.

Wu, T.H., Beal, P.E. & Lan, C. 1988. In-situ shear test of soil-root systems. Journal of Geotechnical and Geoenvironmental Engineering, ASCE. 114(12): 1376–1394.

Centrifuge model test on deformation and failure of slopes under wetting-drying cycles

F. Luo & G. Zhang
State Key Laboratory of Hydroscience and Engineering, Tsinghua University, China

ABSTRACT: Slopes near the reservoirs and rivers usually experience frequent wetting-drying cycles due to water variation and induced new deformation and even failure of the slope. Centrifuge model tests were conducted to investigate the deformation and failure behavior of slopes under wetting-drying cycles. The test procedure contained two wetting-drying cycles with a wetting process and a drying process in each cycle. The results showed that the water variation induced deformation of the slope with a more significant influence on the upper part of the slope. The drying process induced deformation within a limited zone, mainly in the upper part of the slope and close to the slope surface. The wetting-drying cycles caused a significant progressive slope failure that developed downward during the second water drawdown. The failure mechanism was revealed via the deformation localization analysis on the basis of measurement of full-field displacement of the slope. It is the accumulation of deformation localization during the wetting-drying cycles that caused the ultimate slope failure. On the other hand, the local failure led to more significant deformation localization near it.

1 INTRODUCTION

Slopes near the reservoirs and rivers usually experience frequent wetting-drying cycles due to water level variation. The soil properties of slopes could be significantly influenced by the wetting and drying cycles. This leads to new deformation and even failure of the slope (Smethurst et al. 2012). The safety level of slopes should be reasonably analyzed with the application of wetting-drying cycles.

Centrifuge model tests have been widely used to simulate the deformation and failure behavior of the slopes considering different conditions (Hudacsek et al. 2009; Sommers and Viswanadham 2009; Li et al. 2011; Wang and Zhang 2014). Few centrifuge model tests were employed to investigate the effect of the wetting-drying cycles on the behaviors of the slopes.

This paper conducted centrifuge model tests to observe the deformation and failure processes of a slope during the wetting and drying cycles. The deformation behavior and failure mechanism of the slope were investigated on the basis of the full-field displacement measurement results.

2 TESTS

2.1 Devices

The centrifuge model tests were conducted using the centrifuge at Tsinghua University with a maximum acceleration of 250 g and a capacity of 50 g-ton.

Figure 1. Photograph of the wetting-drying devices.

The model container with a length of 600 mm, a width of 200 mm and a height of 500 mm was used in the tests (Fig. 1). On the lateral side of the model container was a transparent plexiglass, through which the deformation and failure of the slope can be observed during the tests.

A water variation simulator was used to simulate the change of water level during the centrifuge model tests (Luo and Zhang 2016). The raise and drop of the water level were realized by controlling the inlet magnetic valve and the outlet magnetic valve, respectively. A heating equipment containing four heating tubes with a length of 150 mm and a power of 300 W was used to dehydrate the slope surface in the process of drying (Fig. 1). The heating tubes were fixed on the model

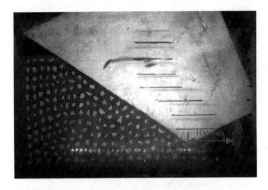

Figure 2. Photograph of the slope model.

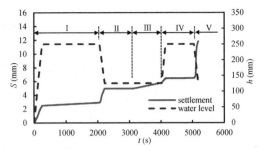

Figure 3. Histories of the settlement of point A (location shown in Figure 5) and water level during the whole test. S, settlement; h, the water level.

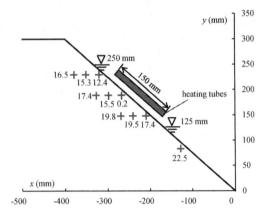

Figure 4. Distribution of the water content in preliminary test.

container and parallel to the slope surface with an equal space. The distance between the slope surface and the heating tube was 30 mm.

2.2 Model

A type of silty clay was used for model preparation. According to the GB/T 50123-1999 standard methods of China, the specific gravity of the soil was 2.7, and the plastic limit and liquid limit of the soil was 15.5% and 33.5%, respectively.

In the model preparation process, the soil was compacted to a dry density of 1.65 g/cm^3 layer by layer with an initial water content of 18% in the model container. The thicknesses of each layer was 50 mm. The height of the slope was 300 mm. The surplus soil was removed to achieve a predetermined gradient of 0.75:1 (Vertical: Horizontal) (Fig. 2). The width of the top of the slope was 100 mm. White particles were embedded on one side of the slope model to reach a higher grey differences for the convenience of image-based measurement. The permeability coefficient of the soil was approximately 10^{-5} cm/s.

An extra soil layer was set at the bottom of the slope and 100 mm gap was provided between the side of the model container and the toe of the slope. Silicone oil was smeared between the interface of the slope and the model container to reduce friction. Those measures could significantly reduce turbulence and friction of the model container on the slope.

2.3 Procedure

The wetting-drying cycles were carried out at 50 g level. The application of the centrifugal acceleration level was a five-step increase. More specifically, the acceleration level was raised up from 1 g to 10 g, 20 g, 30 g, 40 g, 50 g. There were pauses lasting for 5 min between steps so that the gravity-induced deformation of the slope stabilized at the g level.

Figure 3 shows the histories of water level that were controlled during the centrifuge model tests. It can be seen that the slope experienced two wetting-drying cycles. In the first cycle, the water level was increased from 0 to 250 mm and maintained for 30 minutes after the slope deformation stabilized. Then the water level was decreased from 250 mm to 125 mm and after 15 minutes later, the heating equipment was turned on to dehydrate the slope for 15 minutes. The second wetting process was started right after the first wetting-drying cycle. The water level was increased from 125 mm to 250 mm and maintained for 15 minutes so that the slope deformation due to impoundment stabilized. In the second drying process, the water level was decreased from 250 mm to 125 mm resulting in the slope failure. Thus, the test was ended.

The whole wetting-drying process was divided into five stages (Fig. 4). Stages I and IV were the first and second wetting processes. Stages II and V were the first and second water drawdown processes. Stage III was the heating process in the first wetting-drying cycle.

2.4 Measurements

Pictures and videos were captured using an image-capture and displacement-measurement system during the tests (Zhang et al. 2009). These pictures and videos were dealt with to obtain an image series to determine a full-field displacements of the slope.

The water level was measured using a pore pressure transducer installed at the bottom of slope model.

For the sake of a clearer understanding, a Cartesian coordinate was set up with the origin at the toe of the model slope (Fig. 4). The positive directions of the x-axis and y-axis were leftwards and upwards, respectively. All the dimensions referred in this paper are in the dimension of model. The prototype dimension was acquired by amplifying the dimension of model with the g level, 50.

A preliminary test was conducted to obtain the distribution of the water content at the end of a wetting-drying cycle. The test ended after a wetting-drying cycle. The water content of several points on the slope were measured right after the test ended (Fig. 4). It can be seen that the water content was smaller near the slope surface in the same height. The smallest water content, 0.2%, appeared at the middle part of the water variation area. All the water contents in the drying area were smaller than the one under the water level of 125 mm, 22.5%. The results indicated the effectiveness of the drying device.

3 OBSERVATION

3.1 Slope failure

Figure 3 shows the history of settlement of a typical point on the slope during the test. It can be seen that the settlement increased in each stage, that is to say the water level rise and drop processes both caused deformation of the slope. In stage V, the settlement exhibited a rapid increase and indicated that there was a slope failure. Figure 5(a) shows the photograph of the slope after failure at the end of the test. The slip surface was portrayed in the Figure 5(b) according to the photograph. The slip surface ran from the top to the bottom of the slope, dividing the slope into two parts, namely the sliding body and the base body.

3.2 Deformation due to water variation

Figure 6 shows the histories of settlements in different stages of water variation at several points on the slope. The locations of the points were shown in Figure 5(b), namely Points A, B and C. Points A and C were located near the slope surface and in the ultimate sliding body, while Point B was on the base body with the same height as the point C. The settlements of all points increased as the water level increased and decreased in Stage I, II, V. This indicated that the water variation led to new deformation of the slope. The settlement of Point A was larger than those of Points B and C in stages I and II. This showed that the water variation had a more significant influence on the upper part of the slope than the lower part. In stage V where the slope failed, the settlements of Points A and C increased rapidly to 6.2 mm and 4.6 mm, respectively. Whereas, the settlement of Point B, located on the base body, exhibited a small increase in Stage V.

Figure 7 compares the settlement increments of Point A during Stages I and IV where the first and the second impoundment processes occurred. It should be

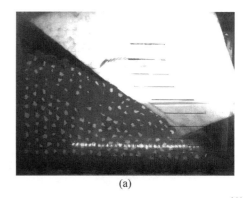

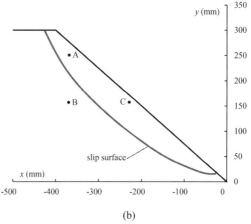

Figure 5. Photograph and schematic view of slope failure. (a) Photograph; (b) structural view.

pointed out that the settlement increment was determined from the water level of 125 mm to 250 mm for ease of comparison between the two wetting processes. The increase of settlement reached approximately 0.6 mm in Stage I and was significantly larger than the maximum settlement increment in Stage IV, approximately 0.15 mm. This result indicated that the first wetting process had a more significant impact on the deformation of the slope than the second wetting process under the same water variation conditions. This implied that the slope properties was changed by the first wetting-drying cycle, which induced different deformation response during the second wetting process.

3.3 Deformation due to drying

Figure 8 shows the vectors of displacement that was induced by drying process during Stage III. Generally speaking, the drying process mainly caused displacement with a direction to the vertical. It can be seen that the drying process induced deformation within a limited zone. This zone was in the upper part of the slope and close to the slope surface. Specifically, the displacement increased with increasing altitude. The largest displacement appeared on the highest and

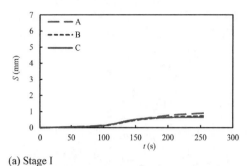

(a) Stage I

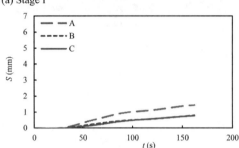

(b) Stage II

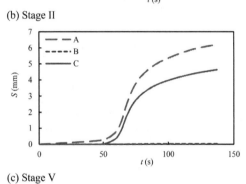

(c) Stage V

Figure 6. Histories of settlements of representative points (location shown in Figure 5(b)) in stage I, II, V. (a) Stage I; (b) stage II; (c) stage V. S, settlement.

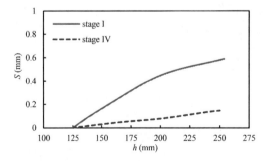

Figure 7. Histories of settlements of point A (location shown in Figure 5(b)) in stage I and IV. S, settlement; h, the water level.

most superficial part of the slope with a maximum vertical displacement of 0.66 mm and a maximum horizontal displacement of 0.34 mm. The displacements of the slope under the water level of 125 mm were very

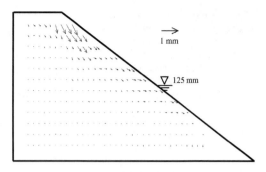

Figure 8. Distribution of displacement vectors of stage III.

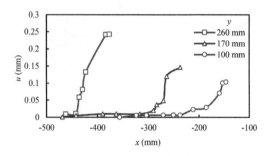

Figure 9. Distributions of the horizontal displacements of stage III at three heights. u, the horizontal displacement.

small, which indicated the drying process had a negligible effect on the slope water. This finding was further illustrated by the observations from the distributions of the horizontal displacement (Fig. 9).

4 PROGRESSIVE FAILURE ANALYSIS

4.1 Failure process

Point couple analysis has been successfully applied to the determination of the failure process of slopes under different loading conditions (Li et al. 2011; Zhang et al. 2015). The point couple is defined as two adjacent points with one on the base body and the other on the slip body. The relative displacement of the two points can be decomposed to the tangential direction of the slip surface and the normal direction. Figure 10 shows the histories of relative displacement of four point couples in Stage V. It can be seen that the tangential relative displacements were larger than the normal relative displacements for all point couples during the drawdown process. This indicated that the shear failure occurred in the slope and the tangential relative displacement can be used to analyze the failure process. The tangential relative displacement increased slowly at the beginning of the drawdown process, and then grew rapidly from a certain moment. The inflection point was considered the moment when the local failure occurred, as outlined using the dashed line in Fig. 10. In this way, the failure process of the slope can be determined by quantifying the local failure moment of each point couple alongside the slip surface.

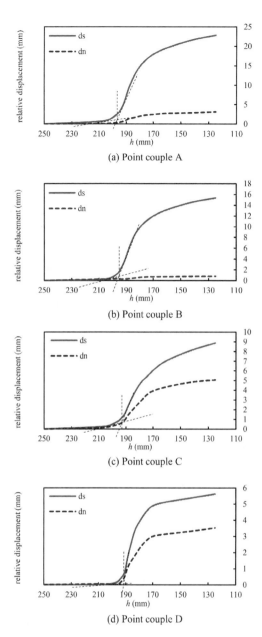

(a) Point couple A

(b) Point couple B

(c) Point couple C

(d) Point couple D

Figure 10. The relative displacement histories of point couples (locations shown in Fig. 11) in stage V. ds, the tangential relative displacement; dn, the normal relative displacement; h, the water level.

Figure 11 marked the water levels at failure moments in the brackets behind the corresponding point couples. It can be seen that the local failure occurred from the top to the bottom as the water level deceased.

4.2 Failure mechanism

Figure 12 shows the distributions of the horizontal displacement of the slope at three heights during State V. It

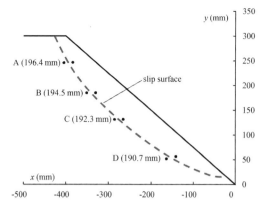

Figure 11. Failure process of the slip surface in stage V.

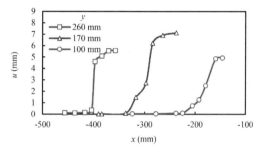

Figure 12. Distributions of the horizontal displacements of stage V at three heights. u, the horizontal displacement.

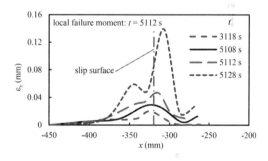

Figure 13. Horizontal strain distributions of the height of 170 mm at representative moments. ε_x, the horizontal strain.

can be seen that there was an area where the horizontal displacement exhibited an evident change on the distribution curve. This implied that the deformation localization occurred in this area.

Figure 13 shows a horizontal distribution of the horizontal strain at representative moments during the wetting-drying cycles. It can be seen that there was a peak value on the distribution curve and the horizontal strains were larger around this peak area. It indicated that obvious deformation localization occurred around the peak. It can be observed that the local failure occurred around the location of the peak area of the horizontal strain. This finding illustrated that the deformation localization gradually developed during the

1141

wetting and drying cycles and ultimately caused the local failure. On the other hand, the local failure led to more significant deformation localization near it. As the horizontal strain increased during the wetting-drying cycles from the beginning, it can be concluded that the wetting-drying cycles induced the deformation localization, and it is the accumulation of deformation localization during the wetting-drying cycles that caused the ultimate slope failure.

5 CONCLUSIONS

The deformation and failure behavior of a slope under wetting-drying cycles was investigated using centrifuge model tests. The failure mechanism was revealed via the deformation localization analysis on the basis of measurement of full-field displacement of the slope. The main conclusions are as follows:

1. The water variation induced deformation of the slope with a more significant influence on the upper part of the slope. The wetting-drying cycle changed the behavior of the slope. This change resulted in less deformation induced by the subsequent wetting process.
2. The drying process induced deformation within a limited zone, mainly in the upper part of the slope and close to the slope surface. The drying process had a negligible effect on the slope underwater.
3. The wetting-drying cycles caused a significant progressive slope failure during the second water drawdown process. The slip surface gradually developed from the top to the bottom of the slope.
4. The wetting-drying cycles induced the deformation localization, and it is the accumulation of deformation localization during the wetting-drying cycles that caused the ultimate slope failure. On the other hand, the local failure led to more significant deformation localization near it.

ACKNOWLEDGEMENTS

The study is supported by the National Natural Science Foundation of China (No. 51479096), Tsinghua University Initiative Scientific Research Program (No. 20161080105, 20171080348) and the China southern power grid co., LTD. Technology project (No. 060200KK52160004).

REFERENCES

Hudacsek P, Bransby MF, Hallett PD, Bengough AG. 2009. Centrifuge modelling of climatic effects on clay embankments. *Proc ICE-Eng Sustainability* 162 (2): 91–100.

Li M, Zhang G, Zhang JM, Lee CF. 2011 Centrifuge model tests on a cohesive soil slope under excavation conditions. *Soils and Foundations* 51(5): 801–812.

Luo Fangyue, Zhang Ga. 2016. Progressive failure behavior of cohesive soil slopes under water drawdown conditions. *Environmental Earth Sciences*, 75 (11): 973.

Smethurst JA, Clarke D, Powrie W. 2012. Factors controlling the seasonal variation in soil water content and pore water pressures within a lightly vegetated clay slope. *Geotechnique* 62 (5): 429–446.

Sommers AN, Viswanadham BVS. 2009. Centrifuge model tests on the behavior of strip footing on geotextile-reinforced slopes. *Geotextiles and Geomembranes* 27(6): 497–505.

Wang LP, Zhang G. 2014. Centrifuge model test study on pile reinforcement behavior of cohesive soil slopes under earthquake conditions. *Landslides* 11(2): 213–223.

Zhang G, Hu Y, Wang LP. 2015. Behaviour and mechanism of failure process of soil slopes. *Environmental Earth Sciences* 73(4): 1701–1713.

Zhang G, Hu Y, Zhang JM. 2009. New image-analysis-based displacement-measurement system for geotechnical centrifuge modeling tests. *Measurement*, 42(1): 87–96.

Centrifuge model studies of the soil slope under freezing and thawing processes

C. Zhang, Z.Y. Cai, Y.H. Huang & G.M. Xu
Nanjing Hydraulic Research Institute, Nanjing, China

ABSTRACT: A tentatively experimental campaign has been realized to study the effect of the freeze-thaw processes on the soil slope by centrifugal modelling. The heat transfer has been achieved by the heat-exchange plate at 30×g and one of tests has paved with copper on model surface. During 600 mins in model time, the output temperature of the heat-exchange plate is from –40~25°C. The temperature changes in models, frozen depth in freezing period as well as the deformation of measuring points on the slope surface are obtained. The results of the tests show the temperature changes on model surface are far less than that of heat-exchange plate. The heat transfer process and testing method of slope model are also discussed.

1 INTRODUCTION

Frost heave and thaw-induced settlement are significant engineering problems to be solved in seasonally frozen regions. More than any other engineered facility, canals, roadbeds are the more at risk for damage due to frost heave and thaw settlement, by cycles of heave and thaw is detrimental to their function. As society develops, the necessity of developing cold regions becomes inevitable and with it a demand for cold regions engineering. However, it must be deal with climatic extremes and where existing technology is often stretched to the limit. Practical experience in many areas is still quite limited and while field tests have been undertaken, the costs are often prohibitive, hazardous working conditions and the time scales very long is freezing or thawing behaviour is of interest.

During the construction of centrifuge models, it is normal practice to ensure that at a microscopic level the model soil is identical to the prototype soil in order that the stress-strain constitutive behaviour is the same in model and prototype (Taylor, 1995). In centrifuge model, a reduced-scale model of lineal dimensions N^2 times smaller is used to simulate the full-scale problem under an acceleration N times the gravity. However, compared with other fields of geotechnical centrifugal modelling, the potential of using centrifuges to investigate cold regions engineering problem has still not been widely explored. Few attempts have been made to investigate the evolution of deformation on slopes during freezing and thawing except thaw-induced slope instability of high cut slope in periglacial regions (Bommer et al., 2012; Harris et al., 2008) and permafrost warming in rock slopes (Davies et al., 2003).

Centrifuge modelling of such problems is an attractive proposition. With this aim in view, the objective of this research, therefore, is to improve the research of the frost damage on soil slopes in seasonal frozen ground by centrifugal modelling technique and propose some suitable test methods. In addition, the testing procedures and results need to be discussed.

2 METHODOLOGY

2.1 Model soil

The model soil consists of clay from an area of significant frost damage to local canals in the North Xinjiang, China. The relevant properties of this clay are provided in Table 2. The soil coefficient of hydraulic conductivity is 4.2×10^{-7} cm/s. A series of frost heave tests show the water content had a strong impact on the degree of frost heave (Cai and Wu, 2014).

2.2 Model procedures

A small-scale model have prepared in the insulation package to simulate freezing-thawing effects occurring in 3m in slope height. The inner wall is made of aviation organic glass, and has inner dimensions of 450×350×275 mm³ (length × width × height). Using PTFE foam as the insulate martials which has a thermal conductivity of 0.025W/m.k. The model was prepared with initial water contents of 13.5% and hit-solid with maximum dry density of 1.89 g/cm³. After hit-solid process, the slope was excavated to its final grade

Table 1. Characteristics of soil used.

Grain size (%)			w_L (%)	w_P (%)	ρ_{dmax} (g/cm³)	w_{opt} (%)
>0.075 mm	0.075~0.005 mm	<0.05 mm				
17.9	62.1	20	29.1	15	1.89	13.5

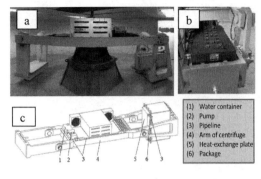

Figure 1. Image of centrifugal facility.

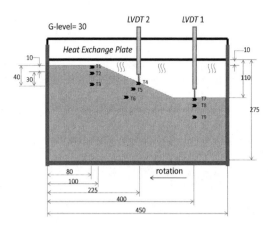

Figure 2. Centrifuge modeling of model and arrangement of sensors (unit in mm).

of 1 (V):2.5 (H). In order to obtain frost heave and thawing obviously, water was added to reach 15.5% (saturation is 97%). After model preparation, a plastic film was attached to the model surface for 1∼2 days. Vaseline was smeared around the models in order to not restrain the movement of the slope as it deformed. the distance of slope top and bottom to the top of box is 10 mm and 110 mm, respectively.

2.3 Device

The centrifuge is used for study at Nanjing Hydraulic Research Institute (Fig. 1a), the capacity of the centrifuge is 60 g.ton.

A suitable device which provides the cold-thermal boundary is the premise of this test. The heat-exchange plate is fixed on the top of the inner wall (Fig. 1b). Due to semiconductor refrigeration, which is based on the "Peltier effect", a uniform plate temperature above the model can be maintained. This technology was first used by Chen (Chen, et al. 1993; Chen, et al. 1999) and the cooling condition can be converted into warming-up condition by changing the DC current direction. The nominal refrigerate capacity of heat-exchange plate is 3600 watts, which can provide −40°C to 25°C on the plate surface.

By connecting this to four high-pressure water pumps, the circulating water, as a cooling medium, can be continuously delivered from water container to the heat-exchange plate. During the heat exchange process, the circulating water was re-cooling by corresponding air-cooled radiator where fixed on the arm (Fig. 1c).

2.4 Experimental campaign

Thermistors and LVDTs are prepared to measure the model temperature and surface displacement during the test. As presented in Figure 2, this measure profile contains 9 thermistors and two LVDTs located where the long side of package is 100 mm.

After the preparation phase, two tests are carried out in order to examine freezing and thawing processes on slope model over a longer, slower cooling/thawing sequence. The Test 1 subjected the temperature boundary from initial temperature to −40°C in freezing period through 300 mins in model time. Similarly, after the freezing period, the plate temperature is increased up to 25°C through 300 mins in model time. Model boundary temperature durations are calculated as $\tau_{model} = \tau_{prototype}/N^2$ (where N is model scale). With a model scale of 30, 300 mins is equivalent to one year of prototype. The centrifuge would always keep rotating in the process of testing at model scale is 30.

In Test 2, all the conditions are the same as that in Test 1 except for laying a copper plate on the surface of model (the initial temperature of model is nearly identical to Test 1). The thermal conductivity of using copper is 115 W/m.k. The idea of laying metal material on model surface had been proposed by Taylor (1995), also aiming at investigating the difference of heat transfer in models.

3 EXPERIMENTAL RESULTS

3.1 Temperature change

Figure 3 shows temperature of heat exchange plate and every thermistor in soils in Test 1 and Test 2. Note that in the freezing period, the system of heat exchange plate enters the self-tuning stage and then approaches the target temperature of −40°C after the temperature drops impetuously from initial temperature to nearly −30°C. Similar process also appears in thawing period (from 20°C to 30°C).

By taking the data of thermistors in soil during the Test 1, the soil temperatures have been depicted in Figure 4. It is found that the soil surfaces have the greater temperature change, yet the temperature difference between the bottom, central slope and slope top during the test is obvious.

On the other hand, the characteristics of temperature changes in thawing period were showed different pattern, and the ultimate temperature did not reach to the initial level. The initial temperature, minimum temperature and maximum temperature of all thermistors in model are listed in Table 2. The sensitivity

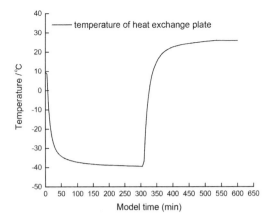

Figure 3. Centrifuge modeling of model and arrangement of sensors (unit in mm).

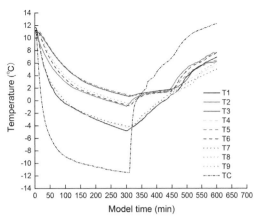

Figure 5. Temperature change in Test 2.

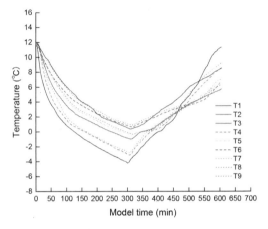

Figure 4. Temperature change in Test 1.

Table 2. Initial and extreme temperature of thermistors in test 1.

Thermistor number	Temperature (°C)		
	Initial	Minimum	Maximum
T1	12.47	−4.1	11.34
T2	12.16	−0.81	5.72
T3	11.96	0.49	8.53
T4	11.95	−3.15	6.39
T5	12.06	−0.18	7.02
T6	11.91	0.9	8.49
T7	11.87	−2.68	6.64
T8	12.16	−0.15	7.66
T9	11.86	1.15	9.2

Table 3. Initial and extreme temperature of thermistors in test 1.

Thermistor number	Temperature (°C)		
	Initial	Minimum	Maximum
T1	11.68	−4.73	6.96
T2	11.21	−0.76	7.67
T3	11.01	0.7	6.2
T4	11.03	−4.33	5.78
T5	11.22	0.76	6.44
T6	11.3	−0.84	7.68
T7	11.05	−4.12	5.03
T8	11.3	−0.51	7.71
T9	10.99	0.68	6.05

of temperature change at the top surface is obviously higher than other locations.

The temperature of thermistors (another thermistor TC was added to the copper before testing) in Test 2 are presented in Figure 5, and the initial temperature, minimum temperature and maximum temperature of all thermistors in model are listed in Table 3. As expected, the surface temperatures, in freezing period, are changed almost simultaneously, and the temperature differences are reduced, including the same depth in soil. Additionally, the final temperatures of all points where under the surface of 10 mm in slope approach to negative temperature. However, the temperature on model surface is much lower than that of the cooper and heat exchange plate (May be due to a small gap between the model and the copper plate, or the thermistors on surface are covered by the soil).

Compared with the Test 1, the surface temperature differences in thawing period in Test 2 are reduced but the temperature of the top surface is lower than the previous. Other, all thermistors in soils are slowly growth in 305 min~450 min, then the growth rate have improved until test is terminated.

3.2 Deformation of slope surface

Figure 6 is a plot of change on surface elevation in response to freezing and thawing of soil in 2 tests. Both the value of LVDTs decreased in the first 5 mins, the settlement of copper plate may be related to the rotation of the centrifuge.

In the freeing period in test 1, the heave eventually occurs 17 mins later after the temperature reached to minus degree on bottom surface (T7), the same as it

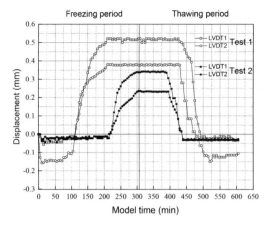

Figure 6. Deformation of slope surface in tests.

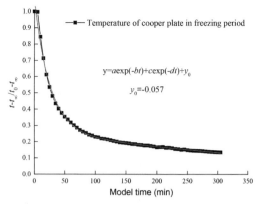

Figure 7. Fixing curves of Copper plate's temperature in freezing period in Test 2.

occurs 15 mins in test 2. From 114 mins to 160 mins in model time, the heave rate in Test 2 is 0.01 mm/min. However, the rate decreased to 0.002 mm /min from 160 mins to 210 mins. This heave mode almost occurs in central slope. Note that at the end of freezing period, even the temperature in soils dropped continuously, heave in models reached near steady state. This phenomenon maybe related to the reduced flow of water to the freezing zone in close system and its subsequent termination. The ultimate heave in bottom (0.56 mm at model scale, 16.8 mm at full scale) is greater than central slope (0.47 mm at model scale, 14.1 mm at full scale), where including initial settlement. Similarly, the rule of temperature change in test 1 but the moment of initiate freezing is postponed, so that the heave rate decreased until the end of freezing period.

The surface displacements in the thawing period are similar to in the freezing period but the slow deformation process is not obvious and duration of process are slightly shorter than the frost heave. The total settlement in bottom and central slope in test 2 are 0.58 mm and 0.54 mm at model scale, respectively. Homologous 0.23 mm in bottom and 0.34 mm in central slope are in test 1. However, both settlement rates are dropped drastically than the frost heaven, relatively the settlement process in test 1 with a short time.

4 ANALYSIS AND DISSCUSSION

4.1 Heat transfer in tests

Indirect heat transfer involves the mounting of heat exchange plate that relies on conduction, convection, and radiation (Taylor, 1995). As mentioned above, the form of temperature boundary is output by PID self-turning controller, more attention should be paid to the amplitude of the temperature in this model using this method. In Test 1, although temperature boundary changes slowly in self-turning stage, the temperature of model surface decreased continuously due to convective transfer, which formed spontaneously in circumstances where upper temperature is lower than model surface (Holman, 1986), even in the centrifugal field (Taylor, 1995). This process is governed by the Rayleigh number; the heat flux can be calculated by Nusselt number where the model surface is horizontal (Taylor 1995; Chen et al. 1999), which is related to the power of using device, the distance from surface to the plate and acceleration level. However, both the horizontal and inclined surfaces in the model (where calculation strategies mentioned above are not applicable) are related to the gap between plate and surface in terms of test results.

In Test 2, there is a significant change of the processes of heat transfer on the model surface by laying a copper plate. It can be considered that the copper plate indirectly provides a relatively stable temperature boundary for the model surface. In the freezing period, due to high thermal conductivity and thickness, the temperature of copper plate dropped quickly. The average temperature of plate $T_{p\text{-ave}}$ and copper plate $T_{c\text{-ave}}$ is $-35°C$ and $-8°C$, respectively, then the reference temperature can be given: $t\infty = T_{p\text{-ave}} - T_{c\text{-ave}}/2 = -13.5°C$. Now this test introduces the excess temperature as: $\theta = t - t_\infty$. Thus, the temperature of copper plate with initial temperature t_0 can be depicted by exponential function:

$$\frac{t-t_\infty}{t_0-t_\infty} = a\exp(-bt) + c\exp(-dt) + y_0 \qquad (1)$$

where a, b, c and d are fitting parameters, $y_0 = -0.057$. The fitting results presented in Fig. 7 show that the ultimate temperature can reach to $-14.9°C$ in circumstances where the device continually works. It can be seen that the copper plate has played a medium role that makes the convective boundary the conductive boundary which drastically dropped stage and relatively steady-state stage. This temperature mode is similar to the previous research like Han and Goodings (2006).

Under the condition of small temperature differences on the model surface, the heat flux can be calculated by Rayleigh number and Nusselt number under the premise of the distance from surface to the plate can be considered to have the same. With reference temperature −13.5°C, the parameters of air for ideal gas including conductivity (k), kinematic viscosity (v) and coefficient thermal expansion(α) are $k \approx 2.2 \times 10^{-6}$ W/m.k, $v \approx 10.8 \times 10^{-6}$ m²/s, $\alpha \approx 14.9 \times 10^{-6}$ m²/s respectively. With the distance is 10 mm, the heat flux due to pure conduction can be given by $q_{cond} = kT/H \approx 59.4$ W/m², and Rayleigh number $R_{ah} \approx NgH^3T/\alpha v = 3.72 \times 10^6$, according to Nusselt number:

$$Nu = q_{conv}/q_{cond} = C(R_{ah})^m \qquad (2)$$

where q_{conv} is heat flux by convection, C and m are correlation (Taylor, 1995), when $R_{ah} > 3.2 \times 10^5$, $C = 0.061$, $m = 0.33$. Hence $q_{conv} = C(R_{ah})^m = 533.8$ w/m².

In thawing period, since the surface temperature of plate is lower than the plate, the convection heat transfer cannot take place. In this situation, the heat transfer relies on conduction and radiation. The calculation of the magnitude of heat transfer occurring if both surface can be considered as black bodies is given by:

$$q_{rad} = \sigma(T_p - T_s)F \qquad (3)$$

where the $\sigma = 56.7 \times 10^{-9}$ W/m²k⁴, and the F is the transmittance coefficient which depends on the nature of the transmitting and receiving surfaces ($0 < F < 1$), taking $T_p = 20.1°C = 293.1K$, $T_s = 6.6°C = 279.6K$. (The average temperature of plate and copper plate), then $q_{rad} = 72$ W/m².

A closer examination of the analysis reveals that the effect of the convection is more significant than conduction and radiation under the condition of indirect heat transfer in freezing period. However, for the slope model, the above analysis is only applicable to the space between plate and model surface under the premise of using copper plate. In other words, the copper plate plays the role of the slope equivalent to "a horizontal plane", reducing the unnecessary temperature difference on slope surface in small scale model test (example like in Test 1).

In Test 2, for the process of heat transfer from copper to model surface is dominated by conduction. Normally the temperature of the surface of model should be consistent with the copper during the test. The reasons for unobserved consistency temperature of copper and model surface in Test 2 are not clear. May be due to a small gap between the model and the copper plate, and if the thermistor on surface is covered by the soil.

4.2 Freezing process of the slope model

The frozen depth is an important index of the freezing level of model. Deficient sensor results, however, cannot be used directly to obtain the continuous temperature data along the depth. On the other hand, an interesting note is that the temperature difference, in Test 2, does not change after 10 mm below surface

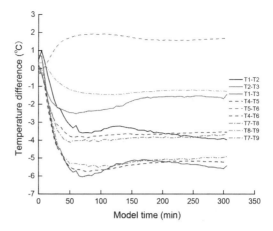

Figure 8. The temperature difference during the freezing period in Test 2.

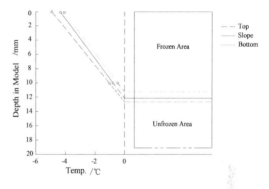

Figure 9. Ultimate frozen depth of model in Test 2.

reaches the negative temperature (in Fig. 8). Under the premise of the temperature in soil is transferred by a certain gradient, the ultimate frozen depth can be estimated approximately. At the model time on 305 min (the connection of two points are prolonged to the zero temperature). The ultimate frozen depth can be given as (Fig. 9): 14 mm on top, 12.7 mm on central slope and 11.7 mm on bottom. According to the N scale factor, where in prototype are 42 cm, 38.1 cm and 35.1 cm, respectively.

The ultimate heave of the bottom is greater than that of the central slope. The reason is that the directions of LVDT on central slope are not consistent with directions of the freezing front. Where the soil slope canals, damage usually occurs at near the slope bottom which accounts for 33% of length of slope lining (Wang, et al. 2012), as well as the largest displacement site. The necessary condition of this phenomenon is the sticking behaviour between lining and soils. However, due to uniform temperature distribution, the ultimate heave, in their tests, on the surface of the bottom is not so different from that of the central slope under the premise of converting the measured value to the normal displacement without pavement lining. Temperature gradient is the main factor influencing the heave under certain conditions of water content in slope model.

5 SUMMARY AND SUGGESTIONS

Slope model with freezing-thawing process have been tested using geotechnical centrifuge at $30 \times g$ in order to determine the temperature distribution in model and deformation of model surface, by using heat-exchange plate.

A significant difference has been analysed as regards the performances of heat transfer in use of copper plate. The copper plate can reduce the temperature difference on the surface of the model, which is beneficial to the simulation of "low slope project" in cold regions. It is also convenient for analysing the heat transfer process. However, restricted by present testing condition, it is unable to form an objective conclusion about the relationship between the degree of frost damage on soil slope and water content, temperature boundary and soil properties. Other, no further discussion about heat transfer in thawing period, due to the testing results of the rule is not significant. This work needs to be improved in the future.

According to the tentatively set of testing results, some of the ways or advises for further researches are drawn as follows:

1. The uniformity of heat transfer to a slope surface is seen as the primary problem in modeling. Considerations relating to equipment design like transporting "cold" to the slope surface directly by ancillary equipment, or change the location of heat exchange plate that the cold face can be "blown" laterally across the slope surface are possible.
2. A sufficient rate of heat transfer is the premise to achieve experimental results. In other case the achievement of temperature on slope below the final desired is identified as necessary, improving the power of heat exchange plate, and compiling the relationship between the temperature of plate and slope surface, under the distance of theirs and acceleration levels may be an advantage to a particular modelling requirement.
3. As mentioned, continuously thermistor can acquire the temperature distribution with depth and the location of frozen front. Theoretically, continuous temperature in soil can be achieved by Fibre Bragg Grating (FBG), and the sensors can be miniaturized according to monitoring objects. The influence of cold environment and electromagnetic on the degree of testing precision are needed to be concerned.
4. Due to limited measuring points, the formation of deformation on slope model during centrifuge tests will not presented at a global view. Visualization in package design (lateral view) can be concerned on the premise of preferable insulation. On the other hand, Suggestion that the study of thawing problem can be carried out independently, analysed the stability of the slope by ultimate strength, water content, earth pressure, etc. By pre-freezing on 1g ground and warm it up by heat exchange plate during melting period, "uncoupled model preparation" can be concerned in thawing problem. The PE film can be applied on the surface of the model to eliminate the influence of frost when prefreezing.

ACKNOWLEDGEMENTS

This work was supported by the National key research and development program (2017YFC0405100). Technical demonstration project of China Ministry of water Resource (SF-201704), and the National Natural Science Foundation of China (51709185). Special thanks to the NHRI Geotechnical centrifuge team for its technical support and assistance during the centrifuge experimental campaign.

REFERENCES

Bommer C., Fitze P., Schneider H. Thaw-Consolidation Effects on the Stability of Alpine Talus Slopes in Permafrost, 2012. Permafrost and Periglacial Processes. 23: 267–276.

Cai, Z.Y., Wu, Z.Q., Huang, Y.H., et al. 2014. Influence of water and salt contents on strength of frozen soils. Chinese Journal of Geotechnical Engineering 36(9): 1580–1586. (In Chinese)

Chen X.S., Schofield A.N., Smith C.C. 1993. Preliminary tests of heave and settlement of soils undergoing one cycle of freeze-thaw in a closed system on a small centrifuge. Proceeding. 6th International. Conference Permafrost. 2: 1070–1072.

Chen X.S., Pu J.L. 1999. Centrifuge modelling test of soil freezing heave. Journal of China Coal Society. 6(24): 615–619. (In Chinese)

Davies, M.C.R. Hamza, O., Harris, C.. 2003. Physical modelling of permafrost warming in rock slopes. Proceedings of the 8th International Conference on Permafrost, Zurich. Lisse, Netherlands, Balkema. 169–173.

Han S.J., Goodings D.J. 2006. Practical model of frost heave in clay. Journal of geotechnical and geoenvironmental engineering. 132(1): 92–101.

Harris, C., Smith, J.S., Davies, M.C.R., et al. 2008. An investigation of periglacial slope stability in relation to soil properties based on physical modelling in the geotechnical centrifuge. Geomorphology. 93(3): 437–459.

Holman J.P. 1986. Heat Transfer. Singapore: McGraw-Hill book Company.

Huang Y.H., Cai Z.Y., et al., 2015. Development of centrifugal model test facility for frost-heave of channels. Chinese Journal of Geotechnical Engineering. 37(4): 615–621. (In Chinese)

Taylor, R.N. 1995. Geotechnical Centrifuge Technology. Glasgow, UK. Blackie Academic and Professional.

Wang Z.Z. 2004. Establishment and application of mechanics models of frost heaving damage of concrete lining trapezoidal open canal. Transactions of Chinese society of agricultural engineering. 20(3), 24–29. (In Chinese)

An experimental and numerical study of pipe behaviour in triggered sandy slope failures

W. Zhang, Z. Gng & A. Askarinejad
Faculty of Civil Engineering and Geosciences, Delft University of Technology, Delft, The Netherlands

ABSTRACT: The soil-pipe interaction in an artificial landslide condition has been investigated. Each small-scale slope instability was triggered by loading a rigid strip footing on the crest of the slope. Five scenarios with different assigned pipe positions have been designed to study the soil-pipe interaction. The failure mechanisms and pipe movements were observed with the aid of PIV analysis. Furthermore, two dimensional finite element slope models in the plane-strain condition were built to simulate the experimental tests and to obtain the maximum external stresses imposed to the pipes caused by the slope failures. Results indicate that the modification of the Audibert & Nyman (1977) method with a geometrical factor can be used to estimate the maximum normal stress on a pipe buried in a slope.

1 INTRODUCTION

Buried pipelines have been widely used for transporting energy material, such as oil and gas, all over the world. Due to the long transportation distance between extraction sites to refinement plants, buried pipelines may pass through various geological conditions. Large ground movement, such as slope failure, is reported as one of the four major causes of pipeline failures (Daiyan et al. 2011). As the acting stresses on a pipeline induced by a slope failure are non-uniformly distributed, it is a challenge to analyse the soil-pipeline interaction triggered by landslides. External forces acting on pipelines induced by the moving of soil mass are the main reason of deflection and rupture of pipelines. Therefore, it is paramount to evaluate the external stresses on the pipe induced by the moving soil mass in order to decrease the possibility of pipe break and to mitigate the potential environmental and social effects.

Physical modelling is a widely accepted approach to investigate engineering problems. Many experimental model tests have been conducted to investigate the behaviour of laterally moving pipelines. The soil-pipe interaction is normally studied in terms of the maximum acting stress/force and the lateral displacement at the maximum stress/force. Audibert & Nyman (1977) developed a small-scale apparatus to study the pipe-soil interaction. In each test, a buried pipe was pulled by a hydraulic jack, and the applied horizontal load and the pipe displacement were measured. Audibert & Nyman (1977) suggested that the ultimate soil resistance can be expressed as:

$$q_u = \gamma H N_q \quad (1)$$

where γ = soil unit weight; H = pipe buried depth (soil surface to the centre of pipe); N_q = bearing capacity factor (Hansen 1961). They also presented relationships between the ratio of y_u/H_b and pipe diameter (D) for both loose and dense sand, where, y_u is the displacement at ultimate soil resistance, H_b is the embedment depth from soil surface to the bottom of pipe.

By performing pipe pulling tests, Trautmann & O'Rourke (1985) studied the influence of soil density and H/D ratio on the ultimate soil resistance. They proposed a formula to predict the ultimate soil resistance, which had a format the same as Equation 1, but the bearing capacity factor was proposed by Ovesen (1964). Based on the observation, y_u was suggested to be $0.13H$, $0.08H$ and $0.03H$ for loose, medium and dense soil respectively. Liu et al. (2015) designed lateral pull-out tests to investigate the resultant forces on the pipelines buried both in sand and soft clay samples. A formula was proposed by taking into account the coefficients of Rankin's passive and active earth pressure.

Based on the literature review, most researchers studied the soil-pipeline interaction by using horizontal loading systems in the laboratory (Audibert & Nyman 1977; Liu et al. 2015; Trautmann & O'Rourke 1985). However, this artificially pulling method may not be correctly extrapolated to the field condition when a landslide is happening. In the above-mentioned tests, each pipe was pulled artificially in a predetermined direction. While during a landslide, the pipe should be pushed by the failure soil mass and could move nonlinearly.

A series of slope model tests were designed and slope failures were triggered in this study. The objective of this study is to investigate the soil-pipe interaction in the landslide condition. In this study, failure mechanisms of the slopes considering the different locations of the pipe were investigated by conducting small-scale model tests. Moreover, Finite Element Method (FEM) was used to simulate the laboratory tests by using PLAXIS 2D (Version 2015.02).

2 PHYSICAL MODELLING

2.1 Soil characterisation

The yellow-white Merwede river sand (also called Delft Centrifuge Sand) from the Netherlands was used to perform the tests. The soil properties are shown in Table 1. Direct shear tests at three normal stress conditions were conducted on samples with relative densities of around 75%. The shear stress-horizontal displacement curves are illustrated in in Figure 1.

2.2 Test setup

In each test, slope samples were prepared in the strong box (Figure 2) by using the travelling pluviation method (Presti et al. 1992; Pozo et al. 2016). The strongbox had inner dimensions of 268 mm in length, 133 mm in width and 150 mm in height. The two transparent walls were made of plexiglass in order to monitor the soil and pipe movements using the Particle Image Velocimetry (PIV) analysis method (Stanier et al. 2015; Askarinejad et al. 2015). Artificial landslides were induced by penetrating a displacement-controlled footing on the crest. The footing had dimensions of 50 mm × 133 mm (width and length, respectively). All the tests were conducted in the plane-strain condition.

3 NUMERICAL MODELLING

All numerical models had the same geometries as those of samples tested in the laboratory. The soil behaviour was described using Mohr-Coulomb constitutive model. Although sand is regarded as cohesionless soil, it is suggested to set $c > 0.2$ kPa for sand material (PLAXIS 2015). Parameters were calibrated by comparing the bearing capacities and the slip surfaces obtained from PLAXIS models to those of laboratory tests (Table 2).

A rigid plate element and a prescribed displacement were placed on the crest to model the displacement-controlled rigid footing. The pipe properties were taken according to the experimental pipe. The parameters of the footing and pipe are listed in Table 3.

Interfaces between structure elements and soil were created. In order to reduce the strain concentration effect adjacent to the right end of the footing, extended interfaces were constructed, as shown in Figure 3. R_i for the footing was 0.5. Considering the shape effect of the pipe, $R_i = 0.1$ was applied for the steel pipe in this study. The numerical models were carried out under the plane-strain condition. 15-node triangular elements were selected for the mesh generation process.

Table 1. Properties of Delft Centrifuge Sand.

Property	Value
Particle sizes, $D_{10}, D_{30}, D_{50}, D_{60}$: mm	0.74, 0.85, 0.92, 0.98
Specific gravity, G_s (–)	2.647
Minimum void ratio (–)	0.52
Maximum void ratio (–)	0.72

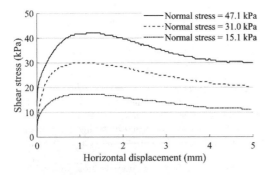

Figure 1. Shear stress-displacement curves of direct shear tests at various normal stresses.

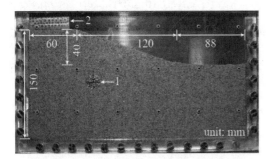

Figure 2. Test Environment: 1) Steel pipe; 2) Footing.

Table 2. Soil parameters in PLAXIS.

	Units	Value
Unsaturated unit weight γ	kN/m^3	16.86
Young's modulus E	MPa	1.6
Poisson's ratio ν		0.3
Cohesion c	kPa	0.3
Friction angle φ'	(°)	43
Dilatancy angle ψ	(°)	15

Table 3. Structure elements' parameters in PLAXIS.

	Footing	Pipe
Normal stiffness (EA, kN/m)	5.00E+06	9.70E+04
Flexural rigidity (EI, kN m^2 /m)	8.50E+03	2.00E−03
Weight (kN/m/m)	0	0.4
Poisson's ratio (ν)	0.3	0.3

4 RESULTS & DISCUSSION

This section reports both laboratory and numerical simulation results. It total, 6 types of tests were conducted including one test without the pipe (P0) and five tests with the pipes located at different burial positions (P1 to P5, Figure 3 and Table 4). The ratio of the embedment (distance from the soil surface to the pipe centre, H, in Figure 3) over the pipe diameter (D) of each test is illustrated in Table 4.

4.1 Failure surface

For a cohesionless soil slope with a footing on the crest, it is commonly believed that the region of the slipping zone consists of three parts starting from the footing bottom towards to the slope toe as shown in Figure 4. Zone I is regarded as an elastic wedge, and zone II is a radial slip zone which is followed by a rectilinear failure zone. (Graham et al. 1988, Raj & Bharathi 2013).

The maximum incremental shear strain plots, analysed by PIV method, at three penetration depths of P0 and P3, are demonstrated in Figure 5. The slip surfaces can be noticed by observing the development of maximum incremental shear strains during the loading process. For the test P0, with the increase of the penetration depth (D_p), a shear strain band became more and more noticeable. When D_p was about 8 mm, the shear band became the failure surface (FS0). While, for the test P3 when the pipe was placed deeper than FS0, some shear strains occurred below the pipe at an early stage of penetrating. With the subsequent loading, a shear band joining the left-bottom corner of the footing, the bottom of the pipe and the slope toe was getting significant. When the penetration depth reached at about 6.1 mm, the slip surface (FS3) was fully developed.

It can be noticed that the failure surfaces of all other tests passed through the bottom of the pipe (Figure 6) also. This phenomenon is in accordance with the observation of Feng et al. (2015) who conducted a large scale slope failure test. Since the pipe was hollow, for the same volume, the weight of the pipe was less than that of the soil mass. Furthermore, the sand-pipe interface was not as strong as that of sand-sand interface. It was therefore the pipe was a "weak point" in the slope. In the case of test P2, two failure surfaces were observed. In practice, once a landslide is happening it may be strong enough to cause fracture of the adjacent buried pipe, therefore, only the first slope failure was analysed in this study.

Figure 7 shows the results of incremental deviatoric strain plots of PLAXIS models. It can be seen that all the failure surfaces formed below the corresponding pipes. The failure surfaces obtained from both experimental and numerical models are plotted in Figure 8 and Figure 9. It can be inferred that the existence of the buried pipes influenced the development of the slip bands. Figure 8 illustrates that, for the condition that the pipes had the same horizontal distance to the crest, the deeper buried depth of the pipe the deeper the failure surface formed. It is observed that the failure surfaces of P4 and P5 were longer and deeper than FS0. For the condition that the pipes had the same altitude, the further the pipe was buried to the slope crest the further the failure surface ended (Figure 9). However, this might be only true when a pipe was placed within the slope and lower than the slope toe.

The inclinations of the linear segments (CD in Figure 4) of all the failure surfaces for experimental and numerical models are listed in Table 5. For a dense-sand specimen with $D_{50} = 0.92$, the thickness of the shear band was estimated to be around 11 D_{50}. (Alshibli & Sture 1999), which was 10.1 mm in this study. Due to this thickness, the obtained inclinations of the failure surfaces might be not very accurate, but still they are useful to evaluate the influence of the existence of the pipes. The linear segments of the failure surfaces became steeper with increasing pipe burial

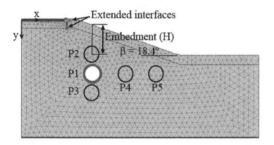

Figure 3. Plan view of numerical slope model of P1 and other pipe positions.

Table 4. Test scenarios and corresponding bearing capacity factors.

Tests	D_r %	Pipe positions			N_q (Hansen 1961)	
		x (mm)	y (mm)	H/D	$\varphi' = 37°$	$\varphi' = 43°C$
P0	78.1	–	–	–	–	–
P1	74.1	80	60	2.96	14.0	23.9
P2	72.6	80	40	1.85	12.1	20.6
P3	74.4	80	80	4.07	15.7	27.1
P4	73.1	120	60	2.22	12.7	21.7
P5	75.8	160	60	1.48	11.4	19.4

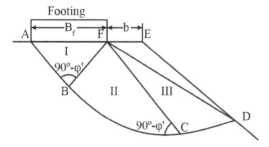

Figure 4. Theoretical slip surface shape when a foundation is on the crest.

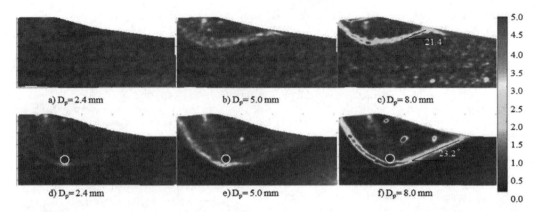

Figure 5. Development of maximum incremental shear strains plots of P0 (a), b) and c)) and P3 (d), e) and f), D_p is the penetration depth).

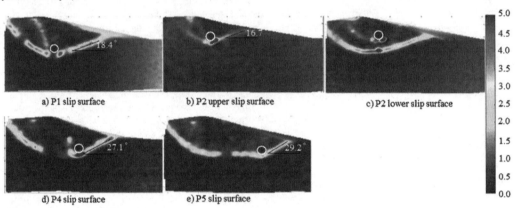

Figure 6. Maximum incremental shear strains plots of P1, P2, P4 and P5.

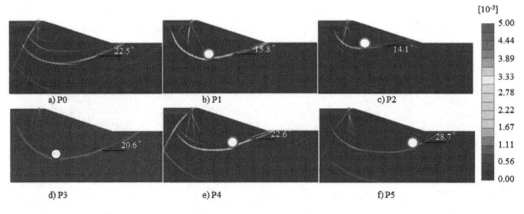

Figure 7. Incremental deviatoric strain plots of FEM models.

depth. The same trend was noticed when the horizontal distance between the pipe and toe became shorter.

4.2 External stresses acting on the pipe

The maximum external stresses acting on the pipe of numerical models are summarized in Table 6. The maximum acting stresses became higher as the pipe positions became deeper (P2, P1 and P3). This can be explained by considering the change of passive soil zone. As a result of deepening of failure surface, the soil zone right to the pipe became larger which should cause the increase of the passive soil resistance. Therefore, as a result of locating closer to the slope toe, the resultant maximum normal stresses of P4 and P5

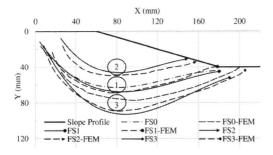

Figure 8. Failure Surfaces of pipe position 0, 1, 2(upper) and 3.

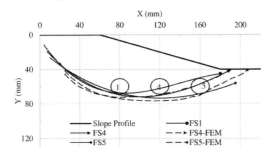

Figure 9. Failure Surfaces of pipe position 1, 4 and 5.

Table 5. Inclination of the linear part of failure surfaces.

	P0	P1	P2	P3	P4	P5
Experimental tests	21.4°	19.3°	16.7°	23.2°	27.1°	29.7°
PLAXIS simulation	22.5°	15.0°	14.1°	20.6°	22.8°	28.7°

Table 6. Maximum normal stresses on pipes (unit: kPa).

	P1	P2	P3	P4	P5
PLAXIS (small scale)	12.53	7.45	19.14	7.35	8.64
Estimation* (small scale)	21.52	11.56	33.66	14.64	8.72
Estimation** (small scale)	12.27	6.59	19.12	8.34	–
Anticipation (prototype, $\varphi' = 37°$)	716.4	386.1	1107.1	488.8	510.8

* Estimated based on Audibert & Nyman (1977) method.
** Estimated based on Zhang & Askarinejad (2018) method.

very close to the slope toe (Figure 7f), the effect of the geometric difference can be neglected.

$$\alpha = \frac{\tan(45° - \varphi'/2)}{\tan\beta + \tan(45° - \varphi'/2)} \quad (2)$$

4.3 Estimation of external stresses in prototype condition

In practice, if a pipe is buried in a slope which is geometrically 100 times larger than the tests, the maximum external stress that will be acting on the pipe can be estimated from the above-mentioned method. The stress condition in prototype condition can be roughly regarded as 100 times higher than that of the laboratory test condition. According the results of direct shear tests (Figure 1), the soil friction angle and the dilatancy angle for the filed soil can be selected as 37° and 6°, respectively. Hence, the maximum external stress on pipes for all positions in prototype can be predicted by using N_q at $\varphi' = 37°$ (Table 4) and $\alpha = 0.60$. The anticipations are presented in Table 6.

were smaller than P1. However, P5 had a larger maximum normal stress comparing to P4 which might be attributed to the fact that the failure surface of P5 ended up further comparing to that of P4.

The external stresses estimated based on Equation 1 are illustrated in Table 6 using N_q (when $\varphi' = 43°$) listed in Table 4. It should be noticed that Audibert & Nyman (1977) conducted their tests by artificially pulling pipes in the flat soil surface condition which was different from the practical condition where pipe movements would be induced by landslides. The difference between these conditions can explain the fact that the estimated acting stresses are approximately 1.5 times higher than those obtained from FEM except for P5. Zhang & Askarinejad (2018) modified the method of Audibert & Nyman (1977) with a factor (Equation 2) which can be calculated according to the geometry difference between a flat ground and a slope ground, where β is the slope angle. By multiplying the results from Equation 1 with α, which is 0.57 for $\varphi' = 43°$, the results show an excellent agreement with numerical results. In the case of test P5, as the pipe was

5 CONCLUSIONS

In this paper, the buried pipe behaviour in a sliding slope was studied through experimental tests. The influence on the failure surface development caused by the presence of the buried pipe was discussed. Experimental tests were simulated numerically by using a FE software. Furthermore, the maximum induced external forces on the pipes from the finite element models were compared with those from two analytical methods. The conclusions are as follows:

1. The slip surfaces tended to pass through the bottoms of the pipes.
2. The external stresses acting on the pipe depended on the passive soil resistance. In the slope condition, these stresses can be estimated from Audibert & Nyman (1977) method by multiplying a geometrical factor which depends on the slope inclination and the soil friction angle.

The above conclusions are made based on the condition that the H/D ratio of the pipe is less than or equal to 4.0 and the pipe is located within the slope.

REFERENCES

Alshibli, K.A. & Sture, S. 1999. Sand shear band thickness measurements by digital imaging techniques. *Journal of computing in civil engineering* 13: 103–109.

Askarinejad, A., Beck, A. & Springman, S.M. 2015. Scaling law of static liquefaction mechanism in geocentrifuge and corresponding hydromechanical characterization of an unsaturated silty sand having a viscous pore fluid. *Canadian Geotechnical Journal* 52:1-13, doi:10.1139/cgj-2014-0237.

Audibert, J.M. & Nyman, K.J. 1977. Soil restraint against horizontal motion of pipes. *Journal of the Geotechnical Engineering Division* 103: 1119–1142.

Daiyan, N., Kenny, S., Phillips, R. & Popescu, R. 2011. Investigating pipeline–soil interaction under axial–lateral relative movements in sand. *Canadian Geotechnical Journal* 48: 1683–1695.

Feng, W., Huang, R., Liu, J., Xu, X. & Luo, M. 2015. Large-scale field trial to explore landslide and pipeline interaction. *Soils and Foundations* 55: 1466–1473.

Graham, J., Andrews, M. & Shields, D. 1988. Stress characteristics for shallow footings in cohesionless slopes. *Canadian Geotechnical Journal* 25: 238–249.

Hansen, J.B. 1961. The ultimate resistance of rigid piles against transversal forces. *Bulletin* 12: 5–9.

Liu, R., Guo, S. & Yan, S. 2015. Study on the Lateral Soil Resistance Acting on the Buried Pipeline. *Journal of Coastal Research* 73: 391–398.

Ovesen, N.K. (1964) *Anchor Slabs: Calculation Methods and Model Tests,* Bulletin 16, Danish Geotechnical Institute.

Plaxis, B. 2015. Plaxis material models manual. *Delft University of Technology & PLAXIS bv,* The Netherlands

Pozo, C., Gng, Z. & Askarinejad, A. 2016. Evaluation of Soft Boundary Effects (SBE) on the Behaviour of a Shallow Foundation. *3rd European Conference on Physical Modelling in Geotechnics (EUROFUGE 2016)* France: Nantes

Presti, D.C.L., Pedroni, S. & Crippa, V. 1992. Maximum dry density of cohesionless soils by pluviation and by ASTM D 4253-83: a comparative study. *Geotechnical Testing Journal* 15: 180–189.

Raj, D. & Bharathi, M. 2013. Bearing Capacity of Shallow Foundation on Slope: A Review. *Proc. GGWUIP* India: Ludhiana

Stanier, S.A., Blaber, J., Take, W.A. & White, D. 2015. Improved image-based deformation measurement for geotechnical applications. *Canadian Geotechnical Journal* 53: 727–739.

Trautmann, C.H. & O'rourke, T.D. 1985. Lateral force-displacement response of buried pipe. *Journal of Geotechnical Engineering* 111: 1077-1092.

Zhang, W. & Askarinejad, A. 2018. Behaviour of Buried Pipes in Unstable Sandy Slopes. *Landslides.* Under Review.

19. Ground improvement

Investigation of nailed slope behaviour during excavation by N*g* centrifuge physical model tests

A. Akoochakian, M. Moradi & A. Kavand
School of Civil Engineering, College of Engineering, University of Tehran, Iran

ABSTRACT: Nailing method, is one of the common methods for stabilizing slopes. In this method, however, if the displacement of the soil is more than the tolerable limits, irreplaceable damage will occur. In this study, a vertical trench reinforced with nails was tested in a geotechnical centrifuge device under acceleration of 30g where the settlement of the strip footing placed nearby the trench was monitored during the excavation. The displacement of the nailed wall during excavation and loading was recoded using continuous photography as well. The results show that decreasing spacing between the nails in a certain area decreases soil settlement at top of the slope. Also, the reduction in settlement due to excavation in the region above the nails is not noticeable when moving away from the slope edge. In this region, the settlement is almost uniform and with increasing the distance of footing from the trench edge, horizontal displacement of nailed wall decreases. This reduction in horizontal displacement is also observed with decreasing the spacing between the nails.

1 INTRODUCTION

Due to increase in human needs for construction of building structures and execution of civil projects, deep excavations are necessary where buildings may be constructed in the proximity of slopes or trenches. Collapse and failure of these trenches result in financial damage and loss of life in different parts of the world. Lateral pressure required for occurrence of these failures can develop by the weight of the slope and possible surcharges on the ground nearby the excavation. These surcharges can include soil above horizontal level at excavation edge, nearby buildings, loads due to using nearby roads, etc. In order to prevent failure of unstable trenches, different stabilization methods can be employed. The structures used to stabilize the unstable trench are called retaining structures. Soil nailing operation means stabilization of steep slopes by placing steel bars close to each other called nails in a slope or excavation, during excavation from top to bottom. Nailing is used for passive consolidation (without creating pre-tension) in the ground and this is done by installing steel bars (nails). Finally, these nails are injected with cement grout to enhance load transfer. By continuation of construction which takes place towards the bottom, a shotcrete layer is also executed on the excavation surface to create continuity between the nails. Nailing technique is commonly used for consolidating slopes and excavations in which the earth removing is done from top to bottom.

Austrian engineers used nailing for the first time at the beginning of 1960s to stabilize the rock slopes in tunnels. This method was later used by German and French engineers for stabilization of soil slopes. In Germany, for the first time, a comprehensive research was done on nailed walls in Karlsruhe University and also in Boer building company between 1975-1981. After that, laboratory tests including scaled centrifuge models were done. Shen et al. (1982) in University of California were the first who successfully modeled nailed structure in the centrifuge (Shen et al. 1989). Centrifuge modeling was then continued in other research centers.

The purpose of present study is to investigate the effect of nail configuration and surcharge distance from the edge of nailed slope on displacement of the nailed wall at different stages of excavation and loading. For this purpose, 8 physical model tests were done using geotechnical centrifuge equipment of University of Tehran. During these tests, the behavior of the models such as wall displacement and ground settlement were monitored using various electronic transducers as well as continuous photography technique.

2 MODEL DIMENSIONS AND EXPERIMENT SET-UP

Developing a laboratory equipment for using in centrifuge has some limitations including weight and size of the equipment, which is totally related to centrifuge specifications. Besides, it should have sufficient strength and stability against the forces induced by high acceleration in centrifuge. It should have good performance, minimum weight, an appropriate size to be accommodated in the centrifuge basket and lastly, its dimensions should be appropriate according to shape and size of the modeled footing.

In current experiments, the length of the soil box was selected based on the footing model width, B.

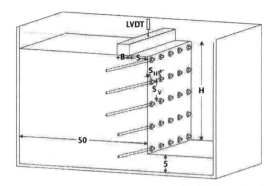

Figure 1. Schematic view of the soil box used in current study.

Table 1. Characteristics of Firooz Kooh sand no. 161.

Soil Type	G_s	e_{max}	e_{min}	D_{50}	C_u	C_c	φ
161 Firooz Kooh	2.658	0.874	0.574	0.32	1.87	0.88	34
Toyora	2.65	0.977	0.597	0.17	–	–	
Sengenyama	2.72	0.911	0.55	0.27	–	–	

Table 2. Geometric and mechanical characteristics of the nails.

Property	
Yield stress (kN/m^2)	100×10^3
Yield strain	100×10^{-6}
Modulus of Elasticity (kN/m^2)	55×10^6
Area (mm^2)	12.57

In other words, it was selected long enough so that the slipping wedges in limiting state can form completely without reaching the box wall. In this regard, some researchers have pointed out that the ratio of the box length to the width of a strip footing should be larger than 10 to minimize the boundary effects (Barghi Khezrloo 2013).

The height of the soil box must be selected so that its rigid bottom does not affect the model behavior. In this regard, if the depth of the soil layer is more than 2B from the bottom of footing, it will not affect the measured load bearing capacity of the footing (Pfeifle et al. 1979). However, some researchers have suggested minimum value of 4B for soil box height.

In this study, based on the size of centrifuge basket and the aforementioned considerations, we used a box with internal dimensions of 0.5 m in height, 0.8 m in length and 0.4 m in width. In one of the lateral sides of the box, a transparent window made of plexi-glass with thickness of 40 mm was provided which allows taking photos during the experiments. It should be noted that the lateral deformation of the transparent side is negligible and does not affect the plane strain behavior of the model. The experiments were conducted under the centrifuge acceleration of 30 g (the geometrical scale factor is 30); therefore, based on the scaling similitude rules, a trench of 9 m high was tested. The height of the model nailed wall was 300 mm. Also, the length of the model in all experiments was 500 mm and its width was 400 mm. To reduce the boundary effect, a depth of 50 mm was accommodated between bottom of the wall and bottom of the box as shown in Figure 1.

2.1 *Specifications of the soil*

The soil used in physical model was Firooz Kooh sand no. 161. This sand is widely used as the standard soil for physical model testing in Iran (Bahadori et al. 2008). This sand has a uniform gradation similar to Japanese Toyora sand (Saber Mahani 2008). Physical and mechanical characteristics of this sand is presented in Table 1 in comparison with two other types of standard sand. In this table, Gs is the specific gravity of solid soil particles, e_{max} & e_{min} are the maximum and the minimum void ratios, D_{50} is the average particle size, C_u is the coefficient of uniformity, C_c is the coefficient of curvature and φ is the friction angle. Based on ASTM D2187 (2004), the soil used in this study is a medium sized sand which is designated as SP according to USCS. The model ground was constructed under controlled conditions based on the wet tamping method. The target relative density of the model ground was 60%.

2.2 *Specifications of the footings*

The footings used in physical models were made of Aluminum with dimensions of 500 mm × 398 mm (W×L) and 750 mm × 398 mm (W×L). In order to avoid unwanted deflections in the footings, the vertical loading was uniformly applied on the footings through a rectangular steel bar having the same size as the footing.

2.3 *Specifications of the nails*

Considering the index properties of the soil (i.e. $D_{50} = 0.32$ mm), minimum nail diameter is 4.8 mm and 11.2 mm, based on Ovesen (1975) and Miligan and Tei et al. (1993) criteria, respectively. In this study, aluminum nails with 5 mm outside diameter and 3 mm inside diameter were used due to practical limitations and ease of nail placement in the model ground. Mechanical and geometric characteristics of Aluminum nails are presented in Table 2. The lengths of the nails were 200 mm and 150 mm. The head of each nail was threaded for installing a nut at its contact point to the wall facing. For continuity of movement between the facing and the nail, a plastic washer was used between the nut and the facing. To properly simulate the injection area around the nails, a thin layer of sand was glued to their outer surface.

2.4 *Specifications of the facing*

In order to avoid local failure in soil slopes, using a facing is necessary. Based on the results of previous studies (Sharifinejad 2012), the facing used in this

Table 3. Mechanical properties of Polycarbonate facing.

Property	
specific gravity (G_s)	1.19
Tensile strength (kg/cm^2)	760
Modulus of elasticity (kN/m^2)	30000
Flexural strength (kg/cm^2)	1050

study was a 1.0 mm thick Polycarbonate sheet. Characteristics of the facing are shown in Table 3.

2.5 Instrumentation

Linear variable differential transformer (LVDT) with 10 cm course and 0.05 mm accuracy, made by GEFRAN company was used for measuring soil settlement below the footing. To measure the applied forces, a 15-ton Zemic load cell was used. Moreover, to record the deformation of the model during centrifuge spin, a Canon G7 camera was used. Finally, in order to observe the failure surface in the soil, thin layers of colored sand with 50 mm spacing were created in the model.

2.6 Excavation simulation

For simulation of the excavation and the corresponding earth pressures, the space in front of the wall was filled with water. To keep water in front of the wall, two-layer resilient nylons were used which had enough strength in high acceleration. In order to simulate the excavation stages, water was discharged according to the induced pressure at each excavation stage at the target acceleration level of 30 g. An electric 0.5-inch valve was used for this purpose.

2.7 Data acquisition

In order to record the data provided by the transducers, the internal data logger of centrifuge equipment was used. This data logger provides the possibility of using either slipping connections or a wireless communication system. In this study slipping connections were used for data acquisition.

2.8 Physical model preparation

The model ground was prepared according to the wet tamping method in which the soil was carefully compacted in 2.5 cm thick layers. After construction of the entire soil slope, a Polycarbonate facing was placed in front of the vertical face of the slope and the nails were inserted into the soil at predetermined locations. After installing all nails, the Nylon water container was placed in front of the wall and then the test box was placed inside the centrifuge basket. Finally, the water container in front of the wall was filled with water.

3 RESULTS AND DISCUSSION

In the present study, 8 physical model tests were conducted (Table 4). In following figures horizontal

Figure 2. Inserting the nails into the trench.

Figure 3. The physical model in the centrifuge before testing.

displacement of the wall in different conditions during excavation is shown. It is obvious from Figure 4a that when the nail spacing is 80 mm, as the distance of surcharge application point from the slope edge increases, the horizontal displacement of wall becomes significantly smaller. In case of nail spacing equal to 50 mm (Fig. 4b), the same behaviour is observed, however, the horizontal displacement of the wall is not that much sensitive to the distance of surcharge from the slope edge. Figure 5 shows horizontal displacement of the wall for a 4.5 m excavation depth for two different nail spacing. As expected, with decreasing the nail spacing, horizontal displacement of the wall decreases significantly. The horizontal displacement of top and middle of the wall for different excavation depths are shown in Figure 6. In all graphs, it is observed that with increasing the excavation depth, horizontal displacement of the wall increases. Jacobsz & Phalanndwa (2011) presented a graph similar to the above mentioned diagram based on their experiments (Fig. 7). It can be observed that the results of the present study are in general agreement with the graph presented by them. However, the horizontal displacement of the wall in the current experiments is a little higher than the values obtained by these researchers which can be attributed to the differences in configuration and length of the nails in these two experiments.

Regarding the soil settlement below the footing, the following results were obtained:

By comparing the results of tests no. 2 and no. 3, we realized that with getting further away from the slope edge, the ground settlement due to excavation becomes smaller. As Figure 8a shows, in test no. 2 (S/B = 1.5), the slope failed when the excavation depth was approximately 5.7 m. In test no. 3, by placing the footing further from the slope edge, the height of wall can be increased by 1.0 m. For example, as shown in Figure 8a, in case of S/B = 3 the soil fails as the excavation depth reaches 6.6 m. Moreover,

Table 4. Details of the experiments

Experiment no	Footing width (B:cm)	Distance of footing from trench edge (S:cm)	Trench height (H:cm)	Vertical and horizontal distance of nails (cm)	Facing thickness (t:cm)	Excavation height (h:cm)
1	5	21.5	25	–	0.1	–
2	5	7.5	30	8	0.1	19
3	5	14.5	30	8	0.1	23
4	7.5	7.5	30	5	0.1	25
5	7.5	15	30	5	0.1	30
6	7.5	22.5	30	5	0.1	30
7	7.5	7.5	30	5	0.1	30
8	7.5	15	30	5	0.1	30

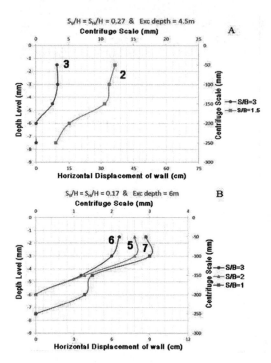

Figure 4. Horizontal displacement of wall for different conditions, (a) 4.5 m excavation depth in tests no.2 and no.3, (b) excavation depth of 6 m in tests no. 5, no. 6, and no. 7.

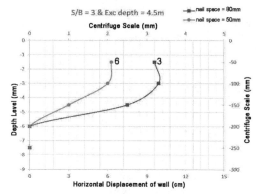

Figure 5. Horizontal displacement of wall in 4.5 m depth of excavation in tests no. 3 and no. 6.

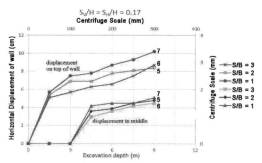

Figure 6. Horizontal displacement on top and middle of wall in tests no. 5, no. 6 and no. 7.

the slope of curve in case of S/B = 3 is smaller than that for S/B = 1.5 which means that with moving away from the slope edge the effect of excavation on ground settlement reduces. Figure 8b shows the ground settlement curves for $S_h/H = S_y/H = 0.17$. As seen in this figure, during the excavation, no failure occurred. For S/B = 1 and S/B = 2, the two curves are almost the same while for S/B = 3, the footing experiences less settlement. The trends also indicate that with moving away from the slope edge, the ground settlement under the footing decreases. However, this reduction in settlement is not noticeable in the region above the nails where the ground experiences an almost uniform settlement. Referring to Figure 9, one can observe that for S/B = 3, the recorded ground settlements for the nail spacing of 80 mm and 50 mm are nearly the same before the occurrence of slope failure. Therefore, it can be inferred again that decreasing the nail spacing causes reduction in ground settlement only in the zone above the nails while with moving away from this area, the effect of nail spacing on ground settlement becomes smaller. In Figure 10, the values of ground settlement versus the distance from the slope edge are presented for two different excavation depths. Based on this figure, when the nail spacing is 50 mm, the increase of S/B value causes a very small reduction in observed ground settlement. However, for a nail spacing of 80 mm, this reduction is significant. Another noteworthy point is that for S/B = 3, the two curves

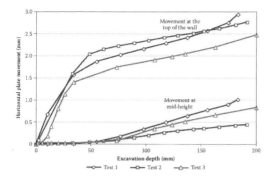

Figure 7. Horizontal displacement of a nailed wall in different excavation depths, Jacobsz and Phalanndwa (2011).

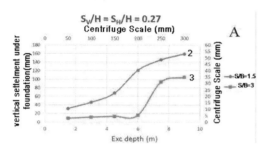

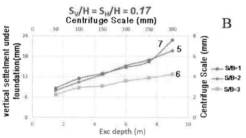

Figure 8. Settlement of footing during excavation for different tests, (a) Tests no. 2 and 3 (b) tests no. 5, 6, 7.

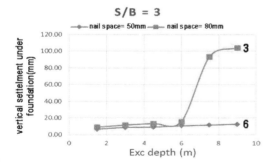

Figure 9. Footing settlement during excavation for tests no. 3 and no. 6.

are close to each other. This indicates that by moving away from the slope edge and passing the nail region, the effects of the nail spacing and the excavation depth on ground settlement become negligible.

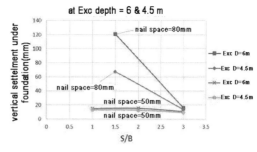

Figure 10. Footing settlement for excavation depth of 4.5 m and 6.0 m.

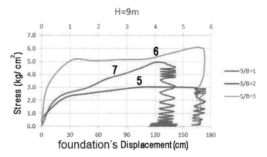

Figure 11. Stress – Settlement curves of soil in tests no. 5, 6, 7.

Lazarte et al. (2003) proposed relations for estimating vertical and horizontal settlement of nailed slopes. They proposed the value of H/333 (H is the wall height) as the maximum expected vertical settlement for fine granular soils. This value is equal to approximately 1 mm for our physical models which is higher than the observed values during the tests. Lazarte et al. (2003) proposed following equation for estimation of wall displacements:

$$D_{DEF}/H = C \times (1 - \tan \alpha) \quad (1)$$

in which C is 0.7 for fine soils, α is the wall angle which is zero in this study and H is the wall height. Therefore, we have:

$$D_{DEF} = 0.7 \times 300 = 210 \text{ mm} \quad (2)$$

The result of current experiments shows that for S/B = 3, the effect of excavation depth and the nail configuration on ground settlement is minimal as significant settlements are not observed in the physical model. However, this is not the case for closer distance to the slope edge. For a width of 75 mm, S is equal to 225 mm which is very close to the value predicted by Equation 1.

In Figure 11, the stress-settlement curves of the soil beneath the footings are drawn for a slope height of 9 m and for different S/B values. It is observed that for S/B = 1, the ground fails after experiencing a stress level of approximately 5 kg/cm² since the footing is above the nail region. Besides, it should be noted that the curve for S/B = 2 is below the curve for S/B = 1.

This is because the nails in the uppermost elevation have 200 mm length, therefore for S/B = 2, the outer edge of the footing is located right above the tip of the nails where the soil structure may be weakened during the nail installation. For S/B = 3, the footing is sufficiently far from the slope edge and the soil beneath the footing shows a significant bearing capacity. Moreover, considering the initial slope of the curve for S/B = 3, it can be inferred that the soil shows a stiffer behavior in this case compared to the two previous cases.

Naimifar et al. (2013) conducted a numerical study on 10 case studies in Iran and proposed failure surfaces for nearby buildings of nailed slopes. In their study, it was found that at a distance approximately equal to the nail length measured from the excavation edge at the ground surface, the degree of damage is the highest. They called this distance the "critical distance". This result is confirmed by the finding of current experiments.

4 CONCLUSION

Present study investigates the effect of surcharge distance from the nailed slope edge as well as that of the nails configuration on displacement of nailed wall in different stages of excavation using centrifuge physical modeling. The most important results from the experiments can be highlighted as below:

– As the distance of surcharge from the slope edge increases, the horizontal displacement of nailed wall due to excavation becomes smaller.
– By decreasing the nails spacing, horizontal displacement of the nailed wall decreases during the excavation.
– By decreasing the nails spacing, effect of distance of surcharge from the slope edge on horizontal displacement of the wall becomes significantly smaller.
– With moving away from the slope edge, reduction of ground settlement due to excavation is not significant in an area above the nails where the ground experiences a nearly uniform settlement.
– By decreasing the nails spacing, the ground settlements become more uniform.
– Decreasing the nails spacing in a specific region, reduces the ground settlement, however, by moving away from the slope edge, the effect of this parameter on the ground settlement becomes negligible.

REFERENCES

ASTM D2187-94, 2004. Standard Test Methods for Physical and Chemical Properties of Particulate Ion-Exchange Resins, ASTM International, West Conshohocken, PA.

Bahadori, H., Ghalandarzadeh, A. & Towhata, I. 2008. Effect of non-plastic silt on the anisotropic behavior of sand. *Soils and Foundations* 48(4): 531–545.

Barghi Khezrloo, A. 2013. Investigation load bearing capacity of shallow footings at the edge of geosynthetic reinforces slopes. *Masters Thesis* University of Tehran (in Persian).

Jacobsz, S.W. & Phalanndwa, T.S. 2011. Observed axial loads in soil nails. *Proceedings, 15th African Regional Conference on Soil Mechanics and Geotechnical Engineering, Mozambique.*

Lazarte, C.A, Elias, V., Espinoza, R.D. & Sabatini, P.J. 2003. Geotechnical Engineering Circular No. 7 Soil nail walls. *Publication FHWA0-IF-03-017.* Federal Highway Administration, Washington D.C.

Naimifar, I., Yasrobi, S.S., Fakher, A. & Golshani, A. 2016. Performance analysis of deep soil nail walls based on excavation-induced damage. *Sharif Journal of Civil Engineering* 31.2(4.2): 123-131 (in Persian).

Ovesen, N.K. 1975. Centrifugal Testing Applied to Bearing Capacity Problems of Footings on Sand. *Geotechnique* 25(2): 394–401.

Pfeifle T.W. & Das B.M. 1979. Model tests for bearing capacity in sand. *ASCE Journal of Geotechnical and Geoenvironmental Engineering* 105(GT9): 1112.

Saber Mahani, M. 2008. Investigating deformation modes and seismic response of reinforced soil walls with 1g shake table. *Ph.D. Thesis* University of Tehran (in Persian).

Sharifinejad, S. 2012. Centrifuge modeling for investigation of possibility of safe excavation with nailing method and investigating its behavior. *Masters Thesis* University of Tehran (in Persian).

Shen, C.K. Bang, S. Kim, Y.S. & Mitchell, J.F. 1982. Centrifuge Modelling of lateral Earth support *ASCE Journal of Geotechnical Engineering Division* 108(GT9): 1150-1164.

Tei, K., Taylor, R.N. & Milligan, W.E. 1993. Centrifuge model test on nailed soil slopes. *Soils and Foundations* 38(2): 165–177.

Relative contribution of drainage capacity of stone columns as a countermeasure against liquefaction

E. Apostolou & A.J. Brennan
Division of Civil Engineering, University of Dundee, Dundee, UK

J. Wehr
Erfurt University of Applied Sciences, Germany

ABSTRACT: Soil liquefaction is the complete or partial loss of the strength of soil, the transition of sand into the state of very low stiffness followed by a period of reconsolidation as temporarily elevated excess pore pressures dissipate. The phenomenon appears mostly during strong earthquakes with high shaking duration, in loose soil profiles. Thus, soil liquefaction mitigation methods are required for constructions projects founded on soils with high liquefiability. Compared to other liquefaction mitigation methods, stone columns offer three basic advantages. The column installation leads to the densification of the surrounding soil, the potential for dissipation of excess pore water pressure is enhanced by the presence of the higher permeability column and the overall soil stiffness increases. However, the relative contribution of each of these benefits is unclear. The purpose of this work is to separate out the relative contribution of drainage capacity via centrifuge modelling. Appropriate modelling considerations for a small-scale stone column system are addressed, especially particle size. Centrifuge tests are carried out to record the excess pore pressures in a stone column reinforced liquefiable loose soil during different earthquake motions. Further testing is taken place for comparison, with unreinforced soil samples or reinforced with modelled vertical drains in loose soil state. The results show that host soil characteristics are the key factor of the excess pore pressure dissipation of all models.

1 SOIL LIQUEFACTION PHENOMENON

1.1 Liquefaction and effects on soils

Liquefaction is the phenomenon when soil loses a large percentage of its shear strength and is transformed into a new, dense sediment with lower stiffness. The loss of the shear strength can be achieved by any type of monotonic or cyclic loading, such as rapid, vibratory, wave or shock loading, or an increase in ground-water pressure. Thus, the loading that causes the loss of the shear resistance and the increase of pore pressure ratio, makes the soil flow and displace. When the soil mass comes to rest a new denser soil sediment is formed (Florin & Ivanov 1961). Although liquefaction can occur in different types of soil, the phenomenon is usually observed in sands (Castro 1977).

Many soil profiles can be defined as liquefiable, as there are many different properties which characterise a liquefiable soil. Generally, soils susceptive to liquefaction are saturated, uniform grain sized loose soils with rounded particles- where friction between them is small. Thus, fine and medium-grained sands are the most common liquefiable soil types. However, there are other soil categories, less susceptible to the phenomenon, but potentially liquefiable, such as coarse silty soils and fine gravels in loose state. The range of

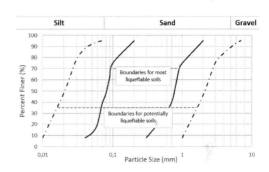

Figure 1. Range of grain size distribution curve for liquefiable soils (replotting following Ishihara 1985).

grain size distribution curves for soils with higher and lower liquefaction potential are shown in Figure 1.

A number of effects are felt when soil liquefies, many examples of which were seen as a result of the Christchurch earthquake in New Zealand, on 21 February 2011. The phenomenon was widely expanded on the street network and in various positions inside the residential area of Christchurch (EEFIT 2011). Clearly there is a need for effective ground improvement in order to mitigate or eliminate such events.

1.2 Stone columns for mitigation

There are various methods for the limitation of liquefaction effects on soil. Stone column installation method is a technique used for improving soil conditions since 1950. Generally, the method includes the penetration of a cylindrical vibrating poker to the soil profile until the desired depth and the filling of the created hole with coarse gravel particles. Thus, when the poker is extracted a stone column is formed in the soil layer.

Stone columns work in three ways. First of all, the introduction of stiff, gravel columns in the ground increases the overall stiffness and shear strength of the area, thus permitting greater design loadings of structures. Second, the installation procedure of stone columns leads to the densification of the surrounding soil. The effect is even more significant in non-cohesive, loose soils, such as would be susceptible to liquefaction. Thus, the soil profile after the column construction is denser, leading to smaller surface settlements during an earthquake event. Third, stone columns have relatively high permeability, so should work as drains during shaking, resulting to the rapid dissipation of excess pore water pressures. The main question on this is how permeable the stone column is and if its drainage capacity can be compared with mitigation methods which focus only on drainage.

2 STONE COLUMN CENTRIFUGE MODELLING

2.1 Scaling issues in centrifuge modelling

Centrifuge modelling is an advanced physical modelling technique for testing small scale geotechnical engineering models in the enhanced gravity field of a centrifuge. The main principle of testing in centrifuge is the equivalence between the used small-scale models and the full-scale prototype via well-established scaling laws. In order to retain the stresses of a small model in full size levels, the centrifuge increases the strength of the gravity field as many as times as the ratio difference between the model and the prototype dimensions. The outcome of modelling is the stress and soil behaviour similarity between the model and the field structure (Schofield 1980). In geotechnical centrifuge engineering, the aforementioned principle is still valid, while the soil used in models is usually the same as in prototype, so that the mechanical properties of the material are identical in the two cases. This is based on the principle that macroscopic properties are more important that particle-scale effects, as is common in other non-particle soil models e.g. finite elements. However, for stone column testing which is described in this paper, conventional modelling is not possible, as prototype aggregate is too large in diameter for the model column. Thus, stone columns must be filled with a finer material scaled down in particle size. For consistency then, the surrounding soil must be scaled down by the same rule. At the same time, the fine surrounding material should be capable of

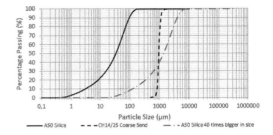

Figure 2. Particle size distribution curve of coarse silt-graded A50 Silica, same curve multiplied by forty times and of CH14/25 coarse sand (replotting following Apostolou et al. 2016).

liquefaction, as it represents a liquefiable soil. Particle scaling has an added benefit in that, water may be used as pore fluid rather than high-viscosity alternatives that are conventionally used in liquefaction modelling to correctly model seepage timescales (Stewart et al. 1998). The proposed model is a coarse silt liquefiable layer with sand in the model stone columns and water as the pore fluid (Apostolou et al. 2016). The centrifuge test series were performed in the geotechnical centrifuge of Dundee University. The characteristics of Dundee University geotechnical centrifuge and the earthquake simulator were described by (Brennan et al. 2014). The centrifuge box used for the models consists of rectangular aluminium frames, with rubber layers among them, for sealing purposes and it allows the model saturation from the bottom to the top through suitable holes at the base plate and a carved grid of 2 mm deep channels on the bottom of the box (Bertalot 2013).

2.2 Coarse silt-graded A50 Silica properties and liquefiability

The most common liquefiable soil type is sand. However, since sand was too coarse for the stone column centrifuge test and silt was chosen as substitute, the silty material had to be proved capable of liquefaction. A length scale factor of 40 was selected for the centrifuge test, necessitating a test acceleration level of 40 g, to balance the requirements of a high g level for modelling larger prototype deposits and enabling the earthquake actuator to deliver shaking in its operational frequency range of 40–400 Hz (Brennan et al. 2014), balanced against the desire for a smaller scale factor to minimise any discrepancies caused by particle scaling. As a result, the chosen material should be equivalent to 1/40th scale coarse sand in terms of particle size. Thus, coarse silt-graded A50 Silica Flour was selected. The particle size distribution curve, which was determined via a laser diffraction technique, by using the Malvern Mastersizer 2000 Particle Size Analyser, proved that the median grain size of the silt D_{50} was 33.81 μm, representing 1.35 mm at 1/40th scale, which is within the required coarse sand range (BSI 2014). In Figure 2, the particle size distribution curves of A50 Silica and CH14/25 coarse

Figure 3. Comparison between particle grains of A50 Silica Flour (left), zoomed in 3000 times and coarse sand (right), zoomed in 35 times (Apostolou et al. 2016).

Table 1. Soil properties of coarse silt-graded A50 Silica Flour and CH14/25 Coarse Silica Sand (Apostolou et al. 2016).

Soil properties	A50 silica flour	CH14/25 silica flour
e_{max}	1.385	0.882
e_{min}	0.612	0.623
G_s	2.65	2.65
φ_{peak}	$28.0° \sim 32.0°$ ($e_o = 1.08$)	$36.0°$
k (m/sec) for Loose State	1.58E-04	
k (m/sec) for Dense State	1.31E-04	3.07E-03

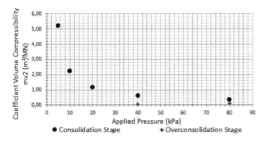

Figure 4. Coefficient volume compressibility m_v of A50 Silica Flour in loose state for normally consolidated and over consolidated data.

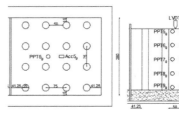

Figure 5. Typical instrumentation ordinance for various depths in centrifuge model.

sand are shown and compared. There is more to soil particles than size, however. The scanning electron microscope in Dundee University was used to inspect particle shape. The silt grain particles, zoomed in 3000 times, proved to be more angular in shape than a comparable coarse sand with its more rounded particles and contained significant characteristics of a platy nature (Fig. 3). Thus, the mechanical characteristics and, as a result, the liquefiability of A50 Silica may be affected and further laboratory testing was needed.

Apostolou et al. 2016 report a series of tests characterizing the A50 silica flour as summarized in Table 1. They also demonstrated that the material is able to liquefy when expected under shaking motion on the centrifuge. Since then, further measurements have been taken of, as the coefficient volume compressibility m_v of coarse silt-graded A50 Silica Flour using an oedometer, as shown in Figure 4. Coarse sand will represent the granular material inside the model columns. The characteristics of the chosen CH14/25 coarse sand are also given in Table 1.

2.3 Construction procedure of centrifuge models

The purpose of the paper is to separate out the relative contribution of drainage capacity of stone columns, by comparing centrifuge model response of stone columns, earthquake drains and unreinforced sample.

Three tests are presented in this paper: a saturated, unreinforced coarse silt sample in loose state and two similar loose samples reinforced with stone columns and earthquake drains respectively. The stone columns and drains are arranged in grid as shown in Figure 5. The first stage of model construction was to sieve a dense, 40 mm thick layer of fine HST95 sand at the bottom of the box using a BS 600 μm sieve. The sand layer was necessary for keeping the tubes that formed the stone columns and the plastic cores of the earthquake drains stable for reinforced models and for letting water to spread over the whole sample more easily during saturation.

After that, the silt was sieved in the centrifuge box through a BS sieve with a diameter of 1.18 mm, with a target relative density Dr of 40% (initial void ratio e_o 1.08). Soil sieving was paused at certain positions to allow instrumentation installation in various positions throughout the sample, including accelerometers (ACC), linear variable differential transformers (LVDT) and pore pressure transducers (PPT). The instrumentation ordinance in a typical test sample is given schematically in Figure 5. The soil was sieved up to a depth of 225 mm, which corresponds to a soil profile depth of nine metres in prototype scale. After that, all models were saturated with de-aired water.

For the stone column cases, a group of sixteen aluminium tubes were firstly pushed down and placed vertically in the aforementioned HST95 sand layer. The tubes were closed on top to avoid silt flow into them. After that, A50 Silica Flour was sieved up to the desired height, as described above. After sieving, CH14/25 coarse Silica sand was poured into the tubes through a funnel in three equal layers. The desired height of the column was equal to the surrounded soil height and each pouring layer was one third of it. Among the layers, the poured sand was compacted, by tamping a metal rod on it 12 times throughout the

Figure 6. Stone column model preparation.

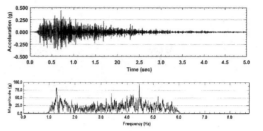

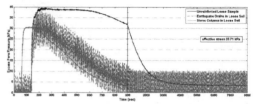

Figure 8. Acceleration time history and frequency graphs for Maule motion in prototype scale.

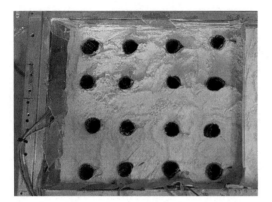

Figure 9. Excess pore pressure difference between unreinforced, earthquake drain and stone column sample in loose state for Maule earthquake.

Figure 7. Earthquake drain model preparation.

whole sand surface, so that it was compacted evenly. After all tubes were filled with the coarse material, they were gently removed and stone columns were formed (Fig. 6). The final height of stone columns was 225 mm and the diameter was 25 mm, which correspond in prototype scale to a stone column of nine metres in height with a diameter of one metre.

Finally, in reference to the earthquake vertical drains, a group of cylindrical cores were created by flexible plastic mesh. The core diameter was 25 mm and the clear height 250 mm in model scale, or one metre and ten metres in prototype scale respectively. The cylinders were later covered by geotextile of high permeability (2.8×10^{-4} m). The material used was VD27 type geotextile, fabricated by Geotechnics BV. The foundation length of the drains was 15 mm and the clear length 225 mm, identical to the stone column tests, so that the results could be comparable. Similar to stone column tests, the drains were pushed down in the fine sand layer, and afterwards coarse silt was sieved in loose state, as described above (Fig. 7).

2.4 Input earthquake motions

Three earthquake events were used for the centrifuge test series. All motions were applied as a sequence to the same sample in each centrifuge test, rather than a fresh sample each time, for reasons of economy.

There is some evidence that the results of an earthquake motion applied to a liquefiable soil sample do not greatly affect the pore pressure response measured in subsequent earthquake events (Coelho et al. 2006). The motion sequence though, for the specific test series was chosen carefully, so that the effect of each earthquake to the following one on the samples is minimised. The acceleration time history and the frequency graph of the Maule earthquake is shown in Figure 8. In this paper, the results of Maule earthquake are presented and discussed.

3 CENTRIFUGE TEST RESULTS

All centrifuge models showed a clear excess pore pressure increase during shaking followed by dissipation after the earthquake motion finished. As expected, dissipation time was different for each pore pressure transducer, as it was affected by the reinforcement soil type, the transducer depth and the applied earthquake event as well as position in the model that has a strong influence on the start of dissipation (Brennan & Madabhushi 2002). Centrifuge data extracted from representative pore pressure graphs for similar depths (PPT7 in Figure 5, 112.5 mm from surface in model scale or 4.5 metres in prototype scale) for the Maule earthquake motion are shown in Figure 9. As identified by Brennan & Madabhushi (2002), the dissipation has two elements, the time at which dissipation starts, and the rate at which is occurs when it does begin. It is observed that at this depth dissipation starts more quickly and the inclination is steeper for the reinforced loose soil samples. Moreover, the peak pore pressure value is clearly greater for the unreinforced model.

The rate of dissipation during reconsolidation may be quantified through the parameter c_v, the coefficient of consolidation. The main advantage of using c_v is that it takes into account both the soil compressibility and the model permeability for each centrifuge

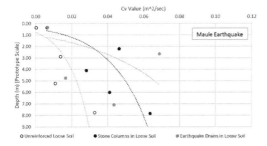

Figure 10. c_v factor comparison among unreinforced loose model and stone column and earthquake drains in loose state for Maule earthquake.

test (Brennan & Madabhushi 2011). It also enables the comparison of samples even if the soil did not liquefy, so it gives reliable results for all depths. c_v is extracted from excess pore pressure transducer graphs, defined by a range of pore pressure values starting at the start of dissipation time until the time when pore pressure values reach half the value of the effective stress in each case. This happens in order to keep the initial conditions the same for all tests and the only difference to be the reinforcement soil type. In depths where the half value of effective stress was much higher than the pore pressure values and there was no common point with the pore pressure curve, the c_v value for the specific depth of the specific test was ignored, so the presented graphs refer to a standard initial condition for c_v definition. The c_v factor values for the three centrifuge loose models for various depths are shown in Figure 10 for the Maule earthquake.

In the unreinforced centrifuge sample, the dissipation rate is much lower compared to the rest of the models, confirming the pore pressure transducers response difference among the tests. At the same time, stone columns and earthquake drain installation in loose soil provide similar excess pore pressure dissipation results. As expected, dissipation rate is lower at surface layers, where effective stress is negligible. For the unreinforced sample, c_v is zero, as excess pore pressure reaches effective stress value and remains constant during shaking, leading to the conclusion that full liquefaction occurs at this depth. This can be also confirmed by the large surface settlements which were observed at the model due to the soil contractive behaviour during shearing. For greater depths however, the c_v factor becomes higher. On the other hand, it is greatly increased for the reinforced soil cases. For the presented motion, c_v factor values are clearly lower for the unreinforced sample, approximately three times lower for the layer below the surface and two times lower for the deeper layers.

Slightly surprisingly, the two reinforcement soil types show similar excess pore pressure response, despite the much greater permeability of the drains. This can be explained by the consideration that the host soil remains the same in each case. Both drains and stone columns are much more permeable than the host soil, so it is likely that the dissipation through the host soil is the weak link, controlling the process. So, as long as the applied reinforcement is permeable enough to maintain zero excess pore pressure inside its internal volume, then the dissipation is practically controlled by the permeability characteristics of the host soil. It should be underlined that the two reinforcement types were compared for same host soil conditions. However, stone column installation in the field leads to soil densification, unlike earthquake drains. Thus, in real field conditions, stone columns are expected to perform better, because of the additional densification of the soil and, as shown above, comparable drainage contributions to highly permeable earthquake drains.

4 CONCLUSIONS

Stone columns and earthquake drains in loose soil state allowed a quicker excess pore pressure dissipation compared to the unreinforced soil in all depths. However, the dissipation rate was very similar between the two reinforced soil types, as differences of c_v factor values were negligible. Thus, it can be concluded that as long as the reinforcement applied in the soil profile is permeable enough to keep excess pore pressure inside its internal volume zero, then the pore pressure dissipation is practically controlled by the permeability characteristics of the surrounding soil. In the aforementioned centrifuge test series, stone columns and earthquake drains proved to be permeable enough to achieve this performance. In conclusion, stone column mitigation method proved to be capable of increasing the excess pore pressure dissipation rate during shaking two or three times more compared to the unreinforced soil profile for extreme seismic events. Results showed similar behaviour between columns and drains for loose soil state. Stone columns are anticipated however to be more effective, due to the soil densification achieved during the installation process.

ACKNOWLEDGEMENTS

The first author is supported by funding from Keller Holding GmbH and Durham Bequest at University of Dundee, which is acknowledged with thanks. Special thanks are also given to Geotechnics B.V. and especially to Mr. R.M. Bodamèr, for providing the geotextile material for the earthquake drain models used at the centrifuge test.

REFERENCES

Apostolou, E., Brennan, A.J. & Wehr, J. 2016. Liquefaction characteristics of coarse silt-graded A50 silica flour. *1st International Conference on Natural Hazards & Infrastructure. Chania, Greece.*

Bertalot, D. 2013. Seismic behaviour of shallow foundations on layered liquefiable soils. University of Dundee.

Brennan, A.J. & Madabhushi, S.P.G. 2002. Effectiveness of vertical drains in mitigation of liquefaction. *Soil Dynamics and Earthquake Engineering* 22(9–12): 1059–1065.

Brennan, A., Knappett, J., Bertalot, D., Loli, M., Anastasopoulos, I. & Brown, M. 2014. Dynamic centrifuge modelling facilities at the University of Dundee and their application to studying seismic case histories. *Proceedings of the 8th International Conference on Physical Modelling in Geotechnics 2014 (ICPMG2014). Perth, Australia: CRC Press, pp. 227–233. doi: 10.1201/b16200-25.*

Brennan, A.J. & Madabhushi, S.P.G. 2011. Measurement of coefficient of consolidation during reconsolidation of liquefied sand. *Geotechnical Testing Journal* 34(2).

BSI. 2014. BS EN ISO 17892-1:2014 Geotechnical investigation and testing – Laboratory testing of soil Part 1: Determination of water content. *British Standards Institution.*

Castro, G. & P.S.J. 1977. Factors affecting liquefaction and cyclic mobility. *Journal of the Geotechnical Engineering Division, Proceedings of the American Society of Civil Engineers* 103(6): 501–515.

Coelho, P.A.L.F., Haigh, S.K. & Madabhushi, S.P.G. 2006. Effects of successive earthquakes on saturated deposits of sand. *Physical Modelling in Geotechnics – 6th ICPMG '06. Hong Kong: Taylor & Francis/Balkema.*

EEFIT. 2011. The Christchurch, New Zealand earthquake of 22 February 2011. *Field Investigation Report, EEFIT, UK.*

Florin, V.A. & Ivanov, P.L. 1961. Liquefaction of saturated sandy soils. *Proceedings in 5th International Conference on Soil Mechanic Foundation Engineering. Paris,* pp. 107–111.

Ishihara, K. 1985. Stability of natural deposits during earthquakes. *Proc. 11th International Conference on Soil Mechanics and Foundation Engineering. Rotterdam,* pp. 321–376.

Schofield, A. N. 1980. Cambridge Geotechnical Centrifuge Operations. *Géotechnique* 30(3): 227–268.

Stewart, D., Chen, Y.-R. & Kutter, B. 1998. Experience with the use of methylcellulose as a viscous pore fluid in centrifuge models. *Geotechnical Testing Journal. ASTM International* 21(4): 365–369.

Observed deformations in geosynthetic-reinforced granular soils subjected to voids

T.S. da Silva
Department of Engineering, University of Cambridge, Cambridge, UK

M.Z.E.B. Elshafie
Laing O'Rourke Senior Lecturer in Construction Engineering and Technology, University of Cambridge, Cambridge, UK
Qatar University, Doha, Qatar

ABSTRACT: Geosynthetic-reinforced soils are used in the design and construction of road and railway embankments over geotechnically challenging areas with high potential for void formation in order to limit the surface deflection over the void. Validation of the mechanisms by which deformation at the geosynthetic level is transferred to the surface is lacking in existing research and design methodology. Hence, centrifuge modelling of the behaviour of basal-reinforced granular fills over voids was conducted using a trapdoor model to simulate the void formation. A plane-strain model was used with a transparent window to allow the observations of soil deformations using Particle Image Velocimetry. Analysis of the soil displacement contours for tests conducted with different soil thicknesses allowed a description of the deformation behaviour to be defined in terms of the surface settlement profile, the shape of the zone of subsidence, and the expansion in the soil.

1 INTRODUCTION

Geosynthetic-reinforced soils are used in road and railway infrastructure over geotechnically challenging areas where the potential for the formation of voids, localised subsidence or differential settlement exist. Examples include sinkholes, and collapse of mines or underground infrastructure such as tunnels. The reinforcement is used as a preventative measure to support the soil and infrastructure above the void using the tensile strength generated by the deflection of the geosynthetic. The deformation is transferred through the soil to the surface where it could have a critical impact on the infrastructure serviceability.

The current assumptions regarding the way this deformation at the geosynthetic level is propagated through the soil to the surface vary widely in existing design methodologies. The study of geosynthetic-reinforced soils over voids is important in order to quantify the benefit of the reinforcement in reducing the extent and magnitude of soil deformations in response to void formation, and hence provide for an appropriate design. Physical model tests of the behaviour of basal-reinforced granular fills over voids were conducted in a geotechnical centrifuge to investigate these mechanisms and contribute to the understanding of this behaviour.

2 EXISTING DESIGN METHODOLOGY

There are three primary analytical design methods used in practice for the evaluation of geosynthetic-reinforced soils above voids; these are:

- British Standard BS8006: 'Code of practice for strengthened reinforced soils and other fills' (BS8006 2010);
- RAFAEL: a design method developed from the experimental results of the French research program (Blivet et al. 2002, Villard et al. 2000); and
- EBGEO: German design guideline (EBGEO 2010).

The standard design procedure in addition to the assumptions, and limitations of the different methods are presented in the paragraphs that follow.

2.1 Design procedure

The geometry considered in the design problem is presented in Figure 1, where:

H	Total height of soil fill
d_s	Surface deflection
B_s/D_s	Width/diameter of trough at surface
B/D	Width/diameter of a void
d	Geosynthetic deflection
θ_D	Angle of draw of subsidence zone

The design procedure is as follows:

1. The limiting surface settlement (d_s), or surface settlement ratio (d_s/B_s or d_s/D_s) is specified.
2. Based on a known void size, and assumed failure model through the soil, the resulting trough width or diameter is calculated.
3. Considering expansion in the soil, and a relationship between the volume loss at the soil surface and at the geosynthetic level, the geosynthetic deflection is calculated.
4. The geosynthetic deflection ratio (d/B or d/D) is used to calculate the geosynthetic strain.
5. The load applied to the geosynthetic is determined independently by a consideration of the arching in the soil (where relevant); this is combined with the calculated strain to determine the required tensile strength of the geosynthetic.

Soil expansion is taken into account using a coefficient of expansion, C_e. This is the ratio of the final volume of the soil in the subsidence zone to its initial volume, and is related to volumetric strain as follows:

$$C_e = \frac{V_f}{V_i} = \frac{V_i + \Delta V}{V_i} = 1 + \frac{\Delta V}{V_i} = 1 + \varepsilon_V \quad (1)$$

2.2 Assumptions

The design guidelines all assume that both the geosynthetic deflection profile and the surface deflection profile are parabolas (plane strain) or paraboloids of revolution (circular voids). The geosynthetic deflection is calculated using Equation 2 for circular voids assuming cylindrical failure mechanisms. This is adapted to Equation 3 for plane-strain problems.

$$d = d_s + 2H(C_e - 1) \quad (2)$$

$$d = d_s + (4/3)H(C_e - 1) \quad (3)$$

The assumption of the failure models (shape of the zone of subsidence) and the expansion of the soil vary widely in the design codes; these are summarised in Table 1. The effects of the assumptions on the size of the subsidence zone and the geosynthetic deflection (as a result of the expansion in the soil using the average value of the recommended range of C_e), are shown schematically in Figure 2.

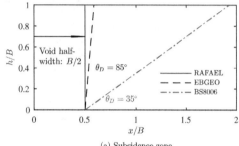

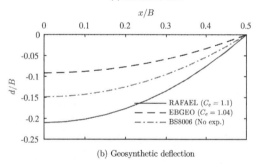

(a) Subsidence zone

(b) Geosynthetic deflection

Figure 2. Effects of the assumptions of the design codes, example with infinitely long void, $d_s/B_s = 0.01$, $H/B = 1$, and $\phi = 35°$.

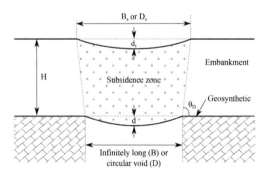

Figure 1. General geometry of design problem considered.

Table 1. Characteristics and assumptions of primary analytical design methods.

Design code	Description	Failure model	Soil expansion and surface settlement
BS8006	British Standard	Funnel: $\theta_D = \phi_{fill}$	Equal parabola between soil and geosynthetic settlements, no soil expansion: $dB = d_s B_s$
RAFAEL*	French design recommendations	Cylindrical collapse	Paraboloids of revolution; $C_e = 1.05 - 1.15$ d calculated from Equation 2
EBGEO*	German design guideline	$H/D < 1$: Cylindrical collapse $1 < H/D < 3$: Funnel: $\theta_D = 85°$ $H/D > 3$: Complete arch	Paraboloids of revolution; $C_e = 1.03 - 1.05$ d calculated from Equation 2 No surface settlement

* Developed for use with circular voids.

From the figure it is evident that, given the same design problem geometry and constraints, the varying assumptions have a significant impact on the zone of soil that is influenced by the deformation of the geosynthetic into the void, and the geosynthetic deflection required to induce a certain settlement at the soil surface resulting in the requirement for entirely different reinforcement strengths.

2.3 Limitations

Limited research has been conducted into the fill behaviour and characterisation of the volumetric expansion and soil surface settlement profile. The EBGEO and RAFAEL design guidelines are based on recommendations from a small number of tests with circular voids and shallow H/D ratios. These observations can therefore not readily be applied to thicker fills, or infinitely long voids. In addition, proper observation and validation of the funnel failure model adopted by BS8006 has not been made. The limitations of the codes are as follows:

1. All of the guidelines use a parabolic profile at the soil surface; research conducted by Potts et al. (2008) showed that a Gaussian distribution may be more appropriate, and would mean that the assumptions made in the design codes are unconservative as a result of the steeper slopes of the Gaussian distribution.
2. The width of the surface trough varies from the void width to a wide funnel based on the internal angle of friction. If the predicted surface trough is too wide, this may be unconservative in the implications of degree of differential settlement; conversely, if it is too narrow, structures outside the predicted zone of influence that would be expected to be safe may be impacted by differential settlement.
3. Expansion in the soil is taken into consideration in some codes and not others, and is applied as a uniform constant to the whole deforming soil body. Additionally, no indication is given to estimate C_e from soil properties. It is expected that the surface settlement trough volume is less than that at the geosynthetic due to shearing in the fill which induces dilation and expansion. This expansion is, however, not expected to be uniformly distributed across the soil, and would be confined to zones with the highest deformation.

2.4 Research objectives

It is important to ensure that the shape of the subsidence zone and surface settlement profile, and the expansion behaviour is well understood to ensure that the design procedure allows an accurate prediction of performance. The work presented in this paper is an attempt to address the uncertainties mentioned above.

3 METHODOLOGY

3.1 Centrifuge package

Centrifuge modelling was conducted in the 10 m balanced beam centrifuge at the University of Cambridge (Schofield 1980) at a gravitational acceleration of $N = 40$. Plane strain testing with the inclusion of a transparent boundary was chosen for the ability to obtain optical measurement of soil displacements using Particle Image Velocimetry (PIV) (Stanier et al. 2015).

The void was simulated by means of a rectangular trapdoor which was lowered between two rigid abutments. The trapdoor was controlled by a system of hydraulic cylinders with a linear actuator. More details about the test set-up and trapdoor mechanism can be found in da Silva et al. (2016). An annotated photograph of the centrifuge package with the trapdoor is shown in Figure 3.

3.2 Granular soil

Dry Hostun sand (HN31), a fine grained siliceous sand from France, was used for the granular material. A mix of approximately 5% dyed blue sand was included to create more contrast allowing better tracking of strains and deformations by the PIV analysis.

The soil has an average particle size, d_{50}, of 0.356 mm; using the trapdoor width as the critical model dimension, $B/d_{50} \approx 150$, and particle size effects are therefore not expected to influence the observed soil behaviour (Iglesia et al. 2011). The measured shear strength from shearbox tests gave peak and critical angles of friction of 45° and 35° respectively. The sand model was prepared at a minimum relative density of 85% by using air pluviation in the automatic sand pourer as described by Zhao et al. (2006); this creates a consistent and uniform relative density of the soil.

3.3 Model geosynthetic

A model geosynthetic was selected based on scaling of the tensile strength-strain behaviour as presented by

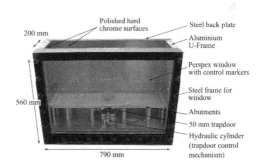

Figure 3. Plane strain centrifuge package with trapdoor to simulate voids and Perspex window for image capture of soil displacements.

Table 2. Model geosynthetic properties.

Property	Units	Model	Prototype
2% secant stiffness	kN/m	20.27	811
5% secant stiffness	kN/m	20.18	807

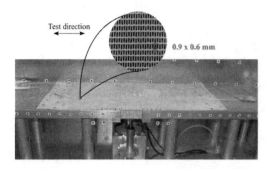

Figure 4. Model geosynthetic installed at the base of the centrifuge package.

Springman et al. (1992) and Viswanadham and König (2004). In the behaviour of geosynthetic-reinforced soils above voids, the dominant geosynthetic property influencing the behaviour is the material stiffness. The scaling laws required that the stiffness is reduced by a factor of N from the prototype (full-scale) to the model material.

A curtain fabric was used as the model geosynthetic-reinforcement. Cyclic testing was conducted to determine the material stiffness; the curtain showed a repeatable stiffness and linear stress-strain response in the expected deformation range of 5% strain. The stiffness is shown in Table 2; the equivalent prototype stiffness of $\approx$800 kN/m is similar to the the stiffness of prototype materials used in this application in industry.

A photograph of the model geosynthetic showing a close-up of the fabric structure, and the final installation in the centrifuge package is shown in Figure 4. The material was laid flat on across the trapdoor prior to the sandpouring; long extensions into the abutment area were included to ensure that it was sufficiently anchored adjacent to the void.

3.4 Test procedure

Three centrifuge tests are reported in this paper; the ratio of the soil height to void width, H/B, was varied in the three tests. Ratios of 1, 2 and 3 were adopted (i.e. soil depths of 50, 100 and 150 mm respectively); these ratios were selected as these are the boundaries indicated in EBGEO between different modes of behaviour, and would allow comparison of the deformation mechanism between thin and thick fills.

4 RESULTS & DISCUSSION

The soil deformation results from the PIV were used to plot the surface and sub-surface soil settlement

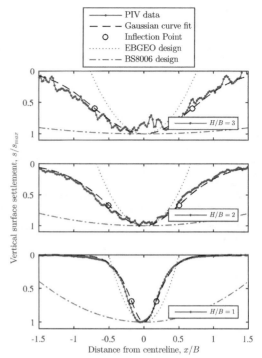

Figure 5. Surface settlement profiles obtained from PIV data; compared to design models and Gaussian curve-fit.

profiles. These results were used to characterise the soil behaviour by considering three areas: (1) the surface settlement profile, (2) the shape of the subsidence zone, and (3) the expansion in the soil. The results and discussion for each of these areas is presented in the paragraphs that follow.

4.1 Surface settlement profile

The surface settlement profile determined from the PIV data is plotted in Figure 5 with the vertical settlement normalised by the maximum settlement at the centreline, s_{max}, and the distance from the centreline normalised by the trapdoor width, B. The EBGEO and BS8006 design profiles were fitted through s_{max} as a comparison to the observed profile; the RAFAEL profile is similar to that of EBGEO (see Table 1), and has been excluded for clarity. The results show that neither model matches the correct shape or extent of the surface settlement trough.

The Gaussian distribution, as is well used in characterising the settlement curve above tunnels (e.g. Peck 1969), was used to fit the observed data. The Gaussian distribution equation is shown in Equation 4; the curve fitting results are also plotted in Figure 5. This shows the much better shape and extent of this curve to describe the surface settlement.

$$s_v(x) = s_{max}\, e^{-0.5(x^2/i^2)} \tag{4}$$

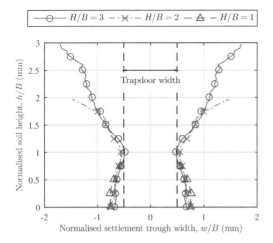
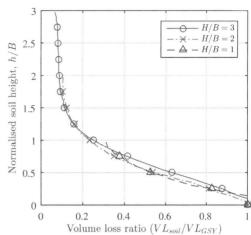

Figure 6. Calculated trough width from the settlement curves as a function of soil height showing a consistent shape of the zone of subsidence between the three tests conducted.

Figure 7. Normalised soil volume loss as a function of height above the trapdoor.

$s_v(x)$ Vertical settlement of the soil at distance x from the centreline;
s_{max} Maximum vertical settlement at centreline
i Horizontal distance measured from s_{max} to inflection point of the curve
x Horizontal distance measured from the centre of the trough

4.2 Shape of the zone of subsidence

The shape of the zone of subsidence was determined by fitting the Gaussian curve to the observed soil deformation through the height of the soil, calculating the width of the trough, w, and plotting this as a function of the soil height above the trapdoor, h. The width of the trough is calculated from Equation 5; the results from this procedure are presented in Figure 6.

$$w = 2i\sqrt{2\pi} \qquad (5)$$

From this figure, it is observed that a consistent deformation mechanism exists between the tests, with the zone of subsidence following the same shape, with a slight variation as the soil surface is reached. The use of a simple vertical or funnel with a specified angle of draw is insufficient in describing the observed zone of subsidence. The shape appears to follow an approximately vertical failure zone slightly wider than the void width up to $h/B = 1$, and then follow a funnel towards the soil surface with an observed $\theta_D \approx 60°$.

4.3 Expansion

The volume of the soil settlement trough, VL_{soil}, was calculated from the PIV data through the height of the soil. This was plotted as a function of the volume displaced by the geosynthetic as it deflects into the void, VL_{GSY}, also calculated from the PIV; the result is shown in Figure 7. This shows that the volume loss ratio is a function of the height above the base of the soil; at large heights the ratio appears to stabilise indicating that no more soil expansion is occurring in the upper regions of soil. This confirms the expectations that, firstly, there is a significant expansion in the soil that needs to be accounted for, and secondly, the most volumetric expansion occurs in the regions where there is the most deformation.

The coefficient of expansion, C_e, was calculated from the volume loss data assuming the formation of a vertical failure zone equal to the trapdoor width throughout the soil (Equation 6); the results are presented in Figure 8, and the recommended values by the EBGEO and RAFAEL design guidelines are highlighted. Above $h/B = 1$ where the subsidence zone widens, the calculated C_e-value will be an upper estimate as the actual initial volume will be bigger due to the funnel mechanism.

$$C_e = 1 + \frac{\Delta V}{V_i} = 1 + \frac{VL_{GSY} - VL_{soil}}{Bh} \qquad (6)$$

Figure 8 shows that the use of a single, uniform coefficient of expansion to describe the relationship between the soil settlement and the geosynthetic deflection is inappropriate. Given a similar subsidence zone profile (Figure 6) and volume loss relationship (Figure 7), the calculated C_e varies between the tests. The calculated value is within the recommended ranges, but varies with height and is not a unique number for a given geometry. A more consistent model can be described using the volume loss ratio than using the coefficient of expansion.

5 CONCLUSION

Understanding the embankment behaviour is crucial in the design of geosynthetic-reinforced fills above voids, as the design is contingent on limiting the

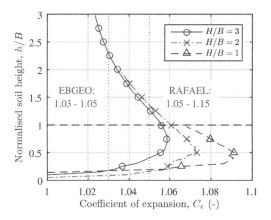

Figure 8. Calculated coefficient of expansion, C_e, using the observed volume loss and assuming a vertical failure zone.

surface differential deflection. This requires understanding of the surface settlement profile, the shape of the zone of subsidence, and the soil expansion. Current design codes have large variations in the assumptions of the soil behaviour which imposes limitations in the appropriate design of these systems. A series of centrifuge tests were conducted to visualise the deformation mechanism from physical models. The following conclusions can be drawn from this study:

- The surface settlement profile is better described by a Gaussian curve than a parabolic curve; this is more pronounced for shallow fills than deeper fills.
- The width of the observed surface settlement trough is wider than predicted by EBGEO and RAFAEL, but narrower than predicted by BS8006.
- The vertical failure mechanism used by EBGEO ($H/B < 1$) and RAFAEL which was developed for circular voids matched the observation made in the plane strain tests conducted up to $h/B = 1$.
- A funnel-shaped zone of subsidence is insufficient to describe the observed shape for $H/B > 1$; a better description would include a vertical zone up to $h/B = 1$ with a funnel above it.
- θ_D for the funnel is in between the recommended values by EBGEO and BS8006.
- Soil expansion is observed through the soil height, stabilising to an approximately constant value in the upper regions of the fill.
- A single, unique, coefficient of expansion is insufficient to describe the soil behaviour; a better model would be to predict the volume loss ratio which showed the same behaviour between tests.

These observations confirmed some of the expected behaviour highlighted in the limitations to the current design methods. A set of results has been provided which can inform further research and investigation into the most accurate designs and prediction of behaviour, allowing efficient designs based on thorough understanding of the soil and geosynthetic behaviour.

ACKNOWLEDGEMENTS

The centrifuge tests reported in this paper were conducted at the Schofield Centre at the University of Cambridge. Thanks are due to the technicians and staff for their assistance without which this testing would not have been possible.

The first author would like to acknowledge the Gates Cambridge Trust for funding to conduct this research.

REFERENCES

Blivet, J., J. Gourc, P. Villard, H. Giraud, M. Khay, & A. Morbois (2002). Design method for geosynthetic as reinforcement for embankment subjected to localized subsidence. In *Proceedings of the Seventh International Conference on Geosynthetics*, Volume 1, France, pp. 341–344.

BS8006 (2010). Code of practice for strengthened/reinforced soils and other fills. Technical report, British Standards Institution, London.

da Silva, T., M. Elshafie, & G. Madabhushi (2016). Centrifuge modelling of arching in granular soils. In *Proceedings of the 3rd European Conference on Physical Modelling in Geotechnics (EuroFuge 2016)*, Nantes, France, pp. 301–306.

EBGEO (2010). *Recommendations for Design and Analysis of Earth Structures Using Geosynthetic Reinforcements-EBGEO* (2nd ed.). Ernst & Sohn.

Iglesia, G. R., H. H. Einstein, & R. V. Whitman (2011). Validation of centrifuge model scaling for soil systems via trapdoor tests. *Journal of Geotechnical and Geoenvironmental Engineering 137*(11), 1075–1089.

Peck, R. B. (1969). Deep excavations and tunnelling in soft ground. In *7th International Conference on Soil Mechanics and Foundation Engineering*, Volume 1, Mexico City, Mexico, pp. 225–290.

Potts, V., L. Zdravkovic, & N. Dixon (2008). Assessment of BS8006: 1995 design method for reinforced fill layers above voids. In *Proceedings of the 4th European Geosynthetics Conference*, Edinburgh, UK.

Schofield, A. N. (1980). Cambridge geotechnical centrifuge operations. *Geotechnique 30*(3), 227–268.

Springman, S., M. Bolton, J. Sharma, & S. Balachandran (1992). Modelling and instrumentation of a geotextile in the geotechnical centrifuge. In *Proc. Int. Symp. on Earth Reinforcement Practice, Kyushu*, Volume 167, pp. 172.

Stanier, S., J. Blaber, W. A. Take, & D. White (2015). Improved image-based deformation measurement for geotechnical applications. *Canadian Geotechnical Journal 53*(5), 727–739.

Villard, P., J. Gourc, & H. Giraud (2000). A geosynthetic reinforcement solution to prevent the formation of localized sinkholes. *Canadian Geotechnical Journal 37*(5), 987–999.

Viswanadham, B. & D. König (2004). Studies on scaling and instrumentation of a geogrid. *Geotextiles and Geomembranes 22*(5), 307–328.

Zhao, Y., K. Gafar, M. Elshafie, A. Deeks, J. Knappett, & S. Madabhushi (2006). Calibration and use of a new automatic sand pourer. In *Proceedings of the Sixth International Conference on Physical Modelling in Geotechnics*, pp. 265–270. Taylor & Francis Group.

Analytical design approach for the self-regulating interactive membrane foundation based on centrifuge-model tests and numerical simulations

O. Detert
HUESKER Synthetic GmbH, Gescher, Germany

D. König & T. Schanz
Ruhr-Universität Bochum, Bochum, Germany

ABSTRACT: An innovative foundation system for embankments on soft soils has been analyzed at the Ruhr-Universität Bochum, Germany, in cooperation with HUESKER Synthetic GmbH, Germany. To get a deeper insight to the complex interactive system behavior centrifuge model tests and numerical simulations have been conducted. Based on the outcome of these investigations an analytical design approach has been developed, which considers the complex interactive behavior of the system. The analytical design approach has been applied to system configurations investigated in former studies but not used for the development of the approach. The comparison of the predicted and measured or calculated system behavior demonstrates the applicability of the analytical design approach. The procedure is an example of how results of physical model tests support the development of innovative solutions for practical engineering.

1 INTRODUCTION

The construction of embankments on very soft soils, e.g. for transportation, breakwaters or stockpiles, is a challenge due to the low shear strength, low permeability, high compressibility and high water content of the subsoil. The surcharge load imposed by the embankment can not only result in a local or total loss of stability, but also in unacceptable settlements and horizontal deformations or thrust, which could endanger nearby structures. To overcome these problems, different construction methods are applicable. However, each method has its limitations, which can be related to the thickness of the soft soil layer, height respectively load of the embankment, time and cost constrains as well as ecological or technical reasons.

An innovative foundation system for embankments on soft soils has been analyzed at the Ruhr-Universität Bochum, Germany, in cooperation with HUESKER Synthetic GmbH, Germany. The foundation system (Fig. 1) consists of two parallel vertical walls, which are installed into the soft subsoil and connected at ground level via a horizontal geosynthetic reinforcement which acts as a tension membrane (from here on referred to as tension membrane). The tension membrane is assumed to cover the whole area in-between the vertical walls. The vertical walls may end within the soft soil layer or reach further down into a firm layer. The soft soil beneath the embankment is therefore confined by the vertical and horizontal elements.

The embankment is constructed on top of the tension membrane. The load from the embankment over the soft soil generates a significant horizontal pressure increase onto the vertical walls which provokes outward movements. These movements are restricted by the tension membrane. At the same time an additional tension force is mobilized within the tension membrane due to deflections beneath the embankment. This additional tension force may lead to a

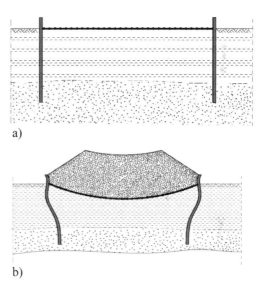

Figure 1. Undeformed and unloaded system (a); deformed (exaggerated) and loaded system (b).

further restriction of the outward movements. The system ensures global stability, controls the horizontal deformation and reduces the horizontal thrust in the subsoil on either side of the embankment, e.g. to protect nearby structures. The complex interactive system behavior has been analyzed by centrifuge model tests and numerical simulations.

The basic ideas of the system are on the one hand to confine the soft soil by the vertical and horizontal elements to prevent excessive lateral deformation or even extrusion of the soft soil, which results in reduced vertical deformation. On the other hand, the self-regulating mechanism of the system, where each load increment provokes an increased pressure on the vertical walls and therefore a further outward deformation, which in return results in a larger strain of the tension membrane and consequently a higher restraining anchor force to counteract the outward movements.

The complex interactive system behavior depending on various parameters has been analyzed by centrifuge model tests and numerical simulations. Based on the results of the centrifuge model test and the numerical investigations, an analytical design approach has been developed, which considers the complex interactive behavior of the system.

Within the paper the centrifuge model test set-up, test procedure and results, as well as the findings of the numerical investigations and the concept of the analytical design approach, derived from the results of the before mentioned analyses will be presented. The benefit of the results of the physical modelling for the set up of the design approach will be discussed.

2 ANALYSES OF THE LOAD BEARING AND DEFORMATION BEHAVIOUR

2.1 Physical and numerical analysis of the system

For a sound design of the system, it is important to know and understand the stress and strain evolution in the system over time. Due to the complex and time dependent interaction and the multitude of influencing parameters, a comprehensive numerical parametric study has been conducted for the system analysis. For the validation of the numerical model, measurement data are required to demonstrate its capability of reproducing the main mechanisms of the self-regulating foundation system. With a series of centrifuge model tests, some principal configurations of the system were analyzed, before a systematic investigation by numerical simulations started.

2.2 Centrifuge model tests

The centrifuge model tests were done in the beam centrifuge of the Department of Foundation Engineering, Soil and Rock Mechanics, at the Ruhr-Universität Bochum in Germany. A detailed description of the beam centrifuge can be found in Jessberger & Güttler (1988).

The tests were done on a model which represents an embankment of 10 m height founded on 10 m thick soft soil layer. The model was constructed at a scale of 1:50 and consequently accelerated in the centrifuge to $ng = 50g$. The stress field in the centrifuge model is therefore equivalent to the stress field of the real scale system set-up (prototype) due to the elevated acceleration field of 50 g. This is important to reproduce the correct stress-dependent behavior of the soils. The structural elements, such as sheet pile walls and tension membrane, were scaled according to validated scaling laws (Jessberger 1992, Springman et al. 1992, Schürmann 1997, Viswanadham & König 2004, Garnier & Gaudin 2007). So the bending stiffness of the sheet pile wall was reduced in the model compared to the prototype by n^3 and the tensile modulus of the tension membrane by n.

Only half of the foundation system requires model simulation, due to the symmetry of the system. The embankment is constructed in three stages by means of an in-flight refillable and moveable sand hopper. Each construction stage is followed by a consolidation phase of about 1 hour.

The centrifuge models were extensively instrumented to measure stresses, pore water pressure, deformations and bending moments of the sheet pile wall. A detailed description of the centrifuge test set-up and execution can be found in Detert et al. (2012). The centrifuge model dimensions are shown in Figure 2.

2.3 Numerical simulations

The numerical simulations have been executed with the program Plaxis 2D using the Hardening Soil model to describe the mechanical behavior of the soft soil (kaolin) and of the dam material (sand). According to Potts & Zdravkovic (2008) and van Eekelen at al. (2011) the influence of the friction between soil and membrane is not significant for the behavior of a dam placed on a membrane. Due to this a perfect bound between membrane and soil is simulated by interface elements. In the first step the numerical model was validated based on the results of the centrifuge model tests. With the validated numerical model, the stresses and strains within the system could be investigated and the load bearing behavior analyzed. The dominating parameter on the system behavior have been determined by global sensitivity analyses; their quantitative influence on the system behavior was evaluated subsequently by parametric studies and can be represented in design charts (Detert et al. 2016).

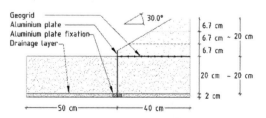

Figure 2. Centrifuge model dimensions.

3 RESULTS

3.1 Arching mechanism in the embankment body

The results of the centrifuge model tests are presented at model scale. In Figure 3a, the total vertical pressure over time measured in the drainage layer beneath the soft subsoil and embankment can be seen. The three construction phases of the embankment are clearly shown by the strong increase of the total vertical pressure. During the consolidation phases, a decrease of the total vertical pressure $\Delta\sigma_v$ as indicated in Figure 3b over time is observed.

While in the first consolidation phase, the decrease of the total vertical pressure is only 4 kPa; a much stronger decrease can be observed in the second and third consolidation phase. The evaluation of all executed centrifuge model tests shows a clear relation between the embankment height and the decrease of the total vertical pressure (Fig. 4). At the same time, an increasing outward movement of the wall is observed during the consolidation phases, although the excess pore water pressure, and therefore the pressure on the vertical walls, does decrease (Fig. 5).

By means of the numerical simulations, the principal stresses in the embankment body before and after consolidation can be analyzed. As shown in Figure 6a, immediately after the construction phase of the embankment, the principal stresses are nearly vertical and horizontal. In the embankment slope a minor rotation occurs due to spreading forces. After consolidation a clear rotation of the principal stresses in the embankment body can be seen in Figure 6b. The rotation of the principal stresses occur due to a load transfer from the middle of the embankment towards the embankment slopes. Due to the settlement of the embankment during the consolidation phase, the friction between the soil particles in the embankment body is mobilized and an arching mechanism develops.

A more detailed analysis of this mechanism shows that the load redistribution stabilizes the system,

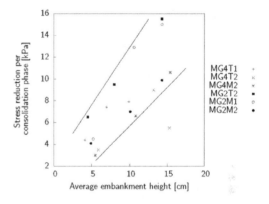

Figure 4. Decrease of total vertical pressure over average embankment height (results of different combination between tensile stiffness of the geotextile and thickness of model wall; MG for MembranGründung; 4 for 4 mm and 2 for 2 mm model wall thickness; T for raw material Polyester, M for raw material Polyvinylalcohol, last number either test no. 1 or test no. 2).

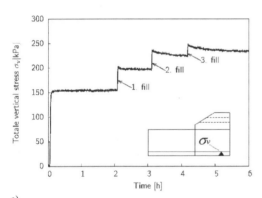

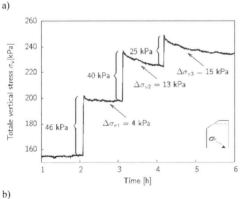

Figure 3. Total vertical pressure over time measured below the soft subsoil and the embankment over time for the full range (a) and limited range to analyze the different load steps (b).

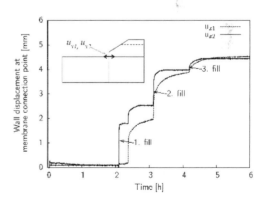

Figure 5. Horizontal wall displacement over time measured at the connection point of the tension membrane to the vertical wall (positive values represent an outward movement).

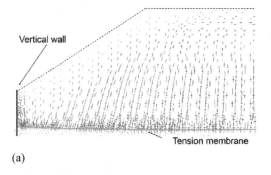

(a)

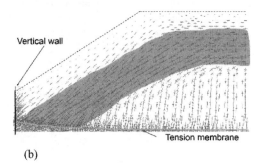

(b)

Figure 6. Principal stresses within the embankment body before (left side) and after consolidation (right side). Used soil parameters for soft soil are $\phi' = 21.6°$, $c' = 5.5\,\text{kN/m}^2$, $E_{50,100\,\text{kN/m}^2} = 4100\,\text{kN/m}^2$, $E_{oed,100\,\text{kN/m}^2} = 3000\,\text{kN/m}^2$, $E_{ur,100\,\text{kN/m}^2} = 10{,}000\,\text{kN/m}^2$, $m = 0{,}75$ and for the dam material $\phi' = 35°$, $c' = 1\,\text{kN/m}^2$, $E_{50,100\,\text{kN/m}^2} = 60{,}000\,\text{kN/m}^2$, $E_{oed,100\,\text{kN/m}^2} = 60{,}000\,\text{kN/m}^2$, $E_{ur,100\,\text{kN/m}^2} = 350{,}000\,\text{kN/m}^2$, $m = 0{,}5$.

since a rotational failure mechanism in the subsoil (Fig. 7), which develops in the transition zone between embankment slope and crest towards the vertical wall at the slope toe, is retained by the arch. The zone where this rotational failure comes "up" is the zone where the load transfer arch props on to the subsoil. Due to the beginning rotational failure, the subsoil "presses" into the embankment and creates a zone of higher stiffness, which in turn attracts the load from the load transfer mechanism.

4 DESIGN APPROACH

4.1 Concept

The conducted investigations have shown that two main loading conditions do occur within the system.

The maximum loading on to the sheet pile walls occur directly after the placement of the embankment material. At this stage the whole weight of the additional material is carried by the pore water pressure, so that excessive pore water pressure is generated. This excess pore water pressure is acting on to the walls.

During the consolidation phase the excess pore water pressure decrease and the embankment load is

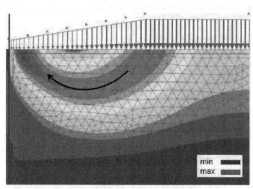

Figure 7. Rotational movement (most probable failure mechanism) within the soft soil beneath the embankment.

carried by the shear strength of the subsoil. This leads to a reduction of the pressure onto the sheet pile walls. At the same time the embankment settles and further tensile forces within the membrane are activated. Due to the observed arching mechanism in the embankment body and the rotational movement of the soft sub soil beneath the slope of the embankment the loading of the sheet pile wall in the upper zone does increase.

Based on this observation there two main design situations: directly after material placement and after the consolidation phase.

For design purposes the system is split into two subsystems (Fig. 8). Subsystem 1 is to calculate the stresses and deformation of the sheet pile wall and subsystem 2 to calculate the maximum tensile force of the tension membrane as well as the connection forces of the tension membrane to sheet pile wall. The last force is the connection between the two subsystems.

4.2 Factors

Within in the numerical investigations a global sensitivity analysis has been conducted to determine the dominating parameter of the system on the load-bearing behavior. Those parameters are

– Embankment height and weight
– Length of the sheet pile wall extension above ground (protrusion)
– Tensile modulus of the tension membrane
– Soft soil stiffness
– Bending stiffness of the sheet pile wall

The maximum stresses and tensile forces have been observed for a ratio between embankment height and base width of 0.25. Based on comprehensive numerical parameter studies design charts have been developed to determine the membrane forces as well as the acting forces on the sheet pile wall (Detert 2016). The design charts are based on nine standard configurations of the system. Adaption factors can be derived from the design charts for further configuration within a certain range to determine the loads on the sheet pile wall or the forces within the membrane.

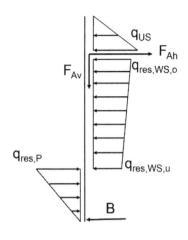

a)

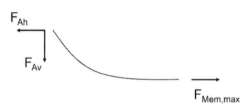

b)

Figure 8. The system is split into two subsystems for design purpose. Subsystem 1 is to calculate the stresses and deformation of the sheet pile wall and subsystem 2 to determine the forces within the membrane. (q_{us} resulting loading on the sheet wall protrusion; F_{av}/F_{ah} vertical and horizontal connection forces of the tension membrane; $q_{res,WS,o}/q_{res,WS,u}$ resulting loading on the sheet wall within the soft soil layer at the top and bottom; $q_{res,P}$ resulting loading on the sheet pile wall within in the firm foundation layer; B substitution force).

The tensile force of the tension membrane is derived as follows:

$$F_i = F_{0.25,i} \cdot A_{geo,i} \cdot A_{\gamma,i} \cdot A_{E\_oed,i} \cdot A_{J,i}$$

with

F   Tensile force within the membrane
i   vertical, horizontal connection force or maximum tensile force
$F_{0.25}$ Tensile force of standard configuration (ratio embankment height to base width of 0.25)
A_{geo} Adaption factor for deviating geometries
A_γ Adaption factor for densities of the embankment fill material greater 17 kN/m³
A_{E_oed} Adaption factor for soft soil stiffness smaller 3000 kN/m²
A_j Adaption factor for tensile moduli of the tension membrane smaller the 50,000 kN/m

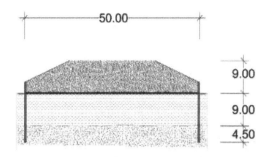

Figure 9. System configuration for comparative calculation between the developed approach using the design charts and numerical methods ($\gamma = 22$ kN/m³, $J = 20{,}000$ kN/m, $E_{oed} = 2500$ kN/m²).

Table 1. Deviation of calculation results using design charts and numerical methods before and after consolidation.

F_{ah} [kN/m]		F_{max} [kN/m]		M_{field} [kN/m]	
before	after	before	after	before	after
8%	7%	3%	0%	8%	12%

The determination of the other loadings onto the sheet pile wall follows an analogue procedure. Those diagrams have been developed for undrained and drained conditions, which allows a check for both cases. Whereas the force in the membrane can be directly determined by the use of the design charts, a calculation has to be carried out, to calculate the stresses within the sheet pile wall.

4.3 Example

Figure 9 shows a system which has been calculated based on the design charts and with numerical methods. The configuration of this system has not been used before to derive the design charts.

Table 1 shows the deviation between the both design approaches. Figure 10 shows the bending moment over the height of the sheet pile calculated with both methods.

The results of table 1 comparing the forces as well as the comparison of the bending moments in figure 10 demonstrate the very good fit between the calculation methods.

5 CONCLUSION

The complex load bearing behavior has been analyzed by means of sophisticated centrifuge model tests and comprehensive numerical studies. The results of the model tests helped to understand the principal system behavior and to validate the numerical simulations. Based on the conducted tests and simulations an analytical design approach, including design charts,

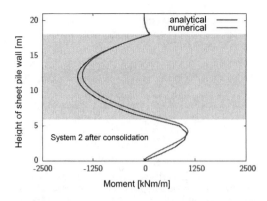

Figure 10. Bending moment of the sheet pile wall over the height. The grey area reflects the soft soil layer.

has been developed. The comparison of the results between the developed design approach and numerical simulation has shown very good agreement.

The physical modeling of the system by means of centrifuge model tests and the observed time and geometry depending behavior has been fundamental for the general understanding of the system behavior and also for the validation of the numerical simulation. This research is a further example how results of physical model tests supports the development of innovative solutions for practical engineering.

REFERENCES

Detert, O., König, D. & Schanz, T. 2012. Centrifuge modeling of an adaptive foundation system for embankments on soft soils, *Proceedings of Eurofuge 2012*, Delft, Netherlands.

Detert, O, Schanz, T, Alexiew, D. & König, D. 2016. Zum Tragverhalten der selbstregulierenden interaktiven Membrangründung, *Bautechnik* 93(9).

Detert, O. 2016. Analyse einer selbstregulierenden interaktiven Membrangründung auf Schüttkörper auf geringtragfähigen Böden, *Dissertation*, Ruhr-Universität Bochum.

Garnier, J. & Gaudin, C. 2007. Physical modelling in geotechnics – catalogue of scaling laws and similitude questions in centrifuge modelling, *Technical report*, ISSMGE TC2.

Jessberger, H.L. & Güttler, U. 1988. Bochum geotechnical centrifuge, *Int. Conf. Centrifuge88*, Paris, Balkema, pp. 37–44.

Jessberger, H.L. 1992. Praxisbezogene Anwendung der Zentrifugen-Modelltechnik in Grundbau, Tunnel- und Schachtbau und Umwelttechnik, *Geotechnik*, Sonderheft, pp. 21–35.

Potts, VJ. & Zdravkovic, L. 2008. Assessment of BS8006:1995 design method for reinforced fill layers above voids. EuroGeo4, *Fourth European Geosynthetcis Conference*, Edinburgh, UK, CD volume, paper ID: 116.

Springman, S., Bolton, M. Sharam, J. & Balachandran, S. 1992. Modelling and instrumentation of a geotextile in the geotechnical centrifuge, *Earth Reinforcement Practice*, pp. 167–172.

Schürmann, A, 1997. Zum Erddruck auf unverankerte flexible Verbauwände, *Dissertation*, Ruhr- Universität Bochum.

van Eekelen, S., Bezuijen, A., Lodder, H. & van Tol, A. 2011. Model experiments on piled embankments. Part I. *Geotextiles and Geomembranes*, pp. 69–81.

Viswanadham, V. & König, D. 2004. Studies on scaling and instrumentation of a Geogrid, *Geotextiles and Geomembranes*, pp. 307–328.

# Earthquake-induced liquefaction mitigation under existing buildings using drains

S. García-Torres & G.S.P. Madabhushi
Schofield Centre, Department of Engineering, University of Cambridge, Cambridge, UK

ABSTRACT: Earthquake-induced liquefaction causes extensive damage to infrastructure. Soil liquefaction-induced effects can account for a significant proportion of damage such as settlement of shallow foundations in saturated soils. Several mitigation measures for liquefaction phenomenon have been previously studied, such as the use of vertical drains. They are used as a method to dissipate excess of pore pressure generated during an earthquake and their efficiency has been previously examined. However, no studies have been focused on the effectiveness of vertical drains below existing structures after a seismic event. In this study, a dynamic centrifuge test was performed to address this issue. Under foundation, vertical drains containing coarse high permeable sand were placed in a square arrangement. The same test was performed for a similar foundation without drains to facilitate direct comparison. The effectiveness of vertical drains under structures was evaluated by comparing the settlement of the foundation with and without drains after ground shaking. In this paper, the reduction in the settlement of structures and the excess pore pressure in the presence of drains are presented.

1 INTRODUCTION

Earthquakes are natural hazards with high socioeconomic impact that represent a constant risk for our society. From 1990 to 2010, around 3,000 damaging earthquakes have occurred (Daniell et al. 2011) and until 2010, earthquakes have caused 8.5 million fatalities around the globe. Furthermore, the economic losses due to earthquakes reached $2.1 trillion in 2000 (converted to dollars from HNDECI 2011). Landslides, liquefaction, fault rupture, tsunamis or fires are some of the secondary effects that a seismic event can cause with severe consequences (Daniell et al. 2011). Earthquake-induced liquefaction is a latent phenomenon that generates extensive damage to infrastructure. Soil liquefaction-induced effects such as settlement of shallow foundations or rotation of buildings can cause significant damage to civil infrastructure and has acquired relevance in geotechnical research over the years. Two of the most representative liquefaction events were registered in the earthquakes of Niigata in Japan (1964) and Alaska in USA where failures of buildings as consequence of liquefaction were the principal concern. Since then, liquefaction consequences began gaining importance as a research topic and as a theme that required study. Although, research advances have been elaborated over the last decades related to liquefaction effects, it becomes essential to understand the consequences of these phenomena under existing buildings. While it is not feasible to predict earthquakes, it is necessary to study possible solutions to reduce the damage of its effects using engineering tools and studying which ones yield the best results in solving the problem. Damage reduction measures under existing and new buildings such as vertical drains have shown positive improvements on liquefiable soils. Techniques used as mitigation measures for structural damage after liquefaction phenomenon are being studied, showing effectiveness in the way they improve ground conditions and prevent damage. Previous research work has been carried out using vertical drains as a drainage measure in free field, showing positive results (Brennan & Madabhushi 2002). However, the behaviour of vertical drains under a more realistic context i.e. underneath existing buildings has not been analysed yet. Consequently, the objective of this work was to analyse the behaviour of soil below an existing foundation with vertical drains and without them, considering generation of pore pressure in the soil due to earthquakes.

2 MITIGATION TECHNIQUES

2.1 *Mitigation techniques against liquefaction*

Effects of liquefaction involves settlement, rotation of buildings and lateral spreading of the soil (Haigh et al. 2000). Settlement of buildings is considered one of the most dangerous effects of liquefaction and for this reason different remediation techniques against this effect on infrastructure have been developed over the last decades. According to Yasuda (2007), constructions have been greatly influenced by countermeasures techniques after the large earthquake of Niigata in 1964. Several damages were registered after this seismic event; however, the effectiveness of using countermeasure methods against liquefaction was corroborated in certain places with compaction improvement that remained undamaged (Watanabe 1966, Ohsaki 1970). According to Brennan (2004), ground improvement strategies used in order to avoid liquefaction damage are: replacement, densification, drainage, dewatering, grouting and shear strain limitation. Engineers and technicians are still working on new methods to improve ground behaviour during liquefaction; nevertheless, the use of the majority of them can be expensive (Baez 1997).

2.2 Drains as countermeasure method

The effectiveness of remediation techniques were evaluated during recent years, especially in densification and drainage methods. Effective and rapid dissipation of excess pore pressures in the soil during liquefaction is an optimal alternative used to reduce damage in structures. The use of vertical drains as a countermeasure technique has as principal objective to relieve the excess pore pressure generated during the shaking before they reach high values that can finally cause damage and loss to infrastructures (Brennan & Madabhushi 2006).

Gravel drains have been used for many years as the best method to mitigate liquefaction effects (Marinucci et al. 2008). Previous studies focused on gravel drain performance as a system to mitigate damage were elaborated by Seed & Booker (1977). These become the point of reference for following research in this field. Numerical and physical modelling of vertical drains during the following years were performed based on equations proposed by Seed & Booker (1977), in which the consolidation equation with an additional term of pore pressure generation obtained from the curve of Seed et al. (1976) are considered. Moreover, many experimental studies have been carried out with the objective of analysing the presence of drains in saturated liquefiable sand. Gravel drain performance has been evaluated using large scale models in shaking table tests. In the work elaborated by Iai et al. (1988) a scheme of the necessary procedure to obtain effective spacing between drains was proposed including an analysis of the effectiveness of using Seed & Booker (1977) equation. Sasaki et al. (1982) analysed gravel drains on sandy soil using cyclic tests and finite element analysis at the same time for the generation and dissipation of pore pressures. Moreover, centrifuge modelling was used to analyse easy drainage techniques such as stone columns in silty sand (Adalier et al. 2003) and concluded that drains act efficiently in relieving excess pore pressure. Adalier et al. (1998) worked with stone columns in order to enable drainage and dissipation of pore pressure and they established that vertical deformation decreased by a significant magnitude and that lateral spreading reduced completely in certain cases in the presence of these columns. Furthermore, centrifuge tests were carried out during the last decade to study the behaviour of soil with vertical drains. Brennan & Madabhushi (2002) explained vertical drains performance in liquefied soil by using centrifuge tests focusing on the analysis of flow front arrival times in a soil with specific drains arrangement.

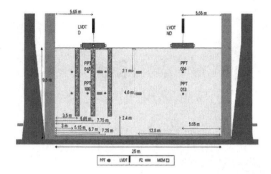

Figure 1. Centrifuge model layout.

3 CENTRIFUGE TEST

In this study, a dynamic centrifuge test was performed using a 1:50 scale model to analyse the behaviour of soil below a foundation with drains and without drains. The bearing pressure exerted by the foundation was 50 kPa. Figure 1 shows a cross section of the model with dimensions in prototype scale. The test was performed using the Turner Beam Centrifuge of University of Cambridge (Madabhushi 2014). Two types of sand materials were used, loose Hostun sand in the free field and Fraction B ($k = 0.7 \times 10^{-2}$ m/s. (Brennan 2004)) coarse material inside the drains. Layers of Hostun sand (Dr = 35%, e = 0.875) were poured by air pluviation using an automatic sand pourer (Madabhushi et al. 2006). Vertical drains have a diameter of 12 mm and a height of 142 mm, this means that in prototype scale drains have a diameter of 0.6 m and a depth of 7.1 m. The laminar box container was used for this test as it allows free deformation of the soil (Brennan et al. 2006). The Servo-hydraulic actuator was used to generate the sinusoidal motions (Madabhushi 2014). Four types of instruments were placed in the model, pore pressure transducers (PPTs), linear variable displacement transducers (LVDTs), microelectromechanical system accelerometers (MEMs) and piezoelectric accelerometers (PZs). Location of instruments can be seen in the layout of Figure 1. After the sand pouring, saturation of the sand stratum was executed using solution of hydroxypropyl methylcellulose (HPMC) and operated by the Schofield CAM-Sat system (Stringer & Madabhushi, 2009).

3.1 Model construction

The model considers a stratum of soil with a square arrangement of drains on one side and without drains

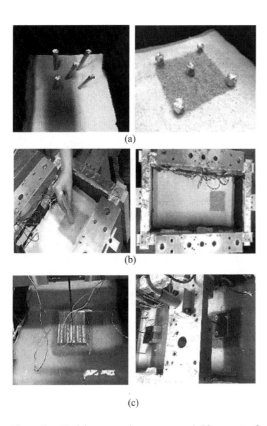

Figure 2. Model construction sequence a) Placement of plastic tube drains b) Tubes withdrawal c) Placement of foundations in the soil.

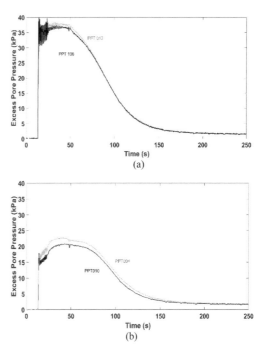

Figure 3. Excess pore pressure in the soil; (a) under foundation for soil with drains (PPT 106) and no drains (PPT 013) for a depth of 4.6 m below the surface and (b) under foundation for soil with drains (PPT 010) and no drains (PPT 004) for a depth of 2.1 m. below the surface.

in the other. Figure 2 shows the complete construction sequence. After pouring the first layer of Hostun sand, plastic tubes with a diameter of 12 mm. were stood at 48 mm from the base of the box in the half right area of the container (Fig. 2a). Drains in the square arrangement were spaced at a distance of 60 mm from mid-point of the central drain to the corner drains, considering a spacing ratio of $a/b = 0.2$ where b is the half distance between drains centre points and a a drain radius. These tubes were sealed with aluminium tape at the top and Hostun sand continued pouring until the required height. Then, the tubes were filled with fraction B sand and tamped to obtain uniformity. A layer of 5 mm of fraction B sand was poured manually over the area enclosed by the drains and the plastic tubes were carefully removed following a vertical direction to avoid unwanted disturbance to the adjacent soil (Figure 2b). Finally, after the saturation stage, foundations blocks were placed over the sand surface in both sides of the stratum (Figure 2c). The foundations were sufficiently far apart so that the influence of one on the other was considered to be minimal. These blocks were designed to apply the required bearing pressure of 50 kPa at 50 g.

3.2 Results and discussion

The model was tested at 50g and three earthquakes were carried out with peak input motion of 0.11 g, 0.31 g, 0.38 g. All earthquakes show similar behaviour of pore pressure generation and dissipation, for this reason the analysis of results is concentrated on Earthquake 3 (EQ3), which is the biggest earthquake with a duration of 10.7 seconds and a frequency of 1 Hz in prototype scale. This earthquake is shown in Figure 5b.

Time histories of excess pore pressures under the central axis of the foundations for EQ3 are shown in Fig. 3 for the region with vertical drains and for the region without them. Graphs present excess pore pressure behaviour for all the shaking time including dissipation stage for depths of 2.1 m and 4.6 m (prototype scale) below the soil surface. It is possible to see an influence in the generation of pore pressures, showing peak values during the first cycle of shaking in both depths and regions. However, it can be seen that the values of excess pore pressure for depth of 2.1 m are about the half of excess pore pressure at deeper depth (4.6 m) as they are closer to the surface. In the soil with drains, for a depth of 4.6 m, excess pore pressure reaches a value around 36 kPa while in the soil without drains excess pore pressure reaches a peak value of 38 kPa in the central point (Fig. 3a). For the other depth (Fig. 3b), approximately the same difference of 2 kPa of excess pore pressure is presented between both regions of the

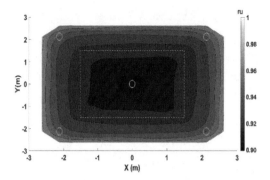

Figure 4. Contours of excess pore pressure through horizontal plane at the start time of dissipation, depth 2.1 m below the surface.

stratum. Comparing performances under foundations, a lower rate of pore pressure generation is shown in the presence of vertical drains.

After the end of the earthquake, approximately ten seconds later, dissipation of pore pressures starts from the base of the stratum in the area of soil with drains as can be seen in Figure 3 for both depths. For PPT 106 and PPT010, which are the closest instruments to the internal central drain of the arrangement, dissipation starts at $t = 47$ and $t = 68$ seconds respectively, earlier than the dissipation times of the four perimeter drains. Viewing in a horizontal plane, for the initiation time of dissipation at a depth of 2.1 m, lower values of pore pressure ($r_u = 0.90$) are recorded in the central part and higher values close to the perimeter drains due to the constant arrival of fluid coming from more distant places. This performance can be seen in Figure 4 in which excess pore pressure contours were plotted together with the foundation and drains, represented by a dotted square and circles respectively. This explains that the internal drain is acting as a unit cell, enclosing a specific area in which the drain is responsible for the flow (Brennan 2004). For this same start time of dissipation, excess pore pressure under the foundation (PPT013 and PPT004) in the soil without drains presents higher values and starts to dissipate few seconds later ($t = 50$ and $t = 72$ seconds). The arrangement of drains is working effectively in the dissipation of pore pressures.

In addition, it was possible to verify liquefaction state of the soil due to the curved profile presented in the model surface after the test. Complete acceleration decoupling occurs almost instantaneously as a consequence of excess pore pressure. The high level of excess pore pressure induced softening in the soil, reducing acceleration magnitudes and resulting in a disassociation between input motion (Fig. 5b) and soil acceleration recorded at the mid-depth of the stratum (Fig. 5a). These results illustrate that the effect of the square arrangement of drains with permeability of 0.7×10^{-2} m/s located under a foundation of 50 kPa is not working to prevent liquefaction effectively in all depths. However, the objective of the experiment was

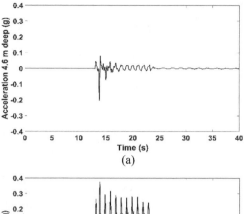

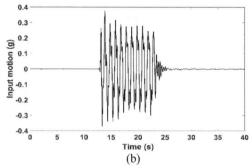

Figure 5. (a) Acceleration (g) in the soil at mid-depth of the stratum and (b) Input motion (g).

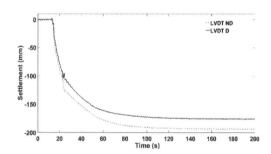

Figure 6. Settlement of foundations in soil with vertical drains (LVDT D) and without vertical drains (LVDT ND).

not to prevent liquefaction but to evaluate pore pressure behaviour using a specific drain arrangement and load foundation.

Settlement time histories obtained from LVDTS instruments located over foundations in soil with vertical and no vertical drains are shown in Figure 6. LVDTS were collocated above both foundations at the same level. During the shaking, settlement in the soil with drains (LVDT D) was around 100 mm and reached a ultimate settlement of 178 mm following dissipation of excess pore pressures. On the other hand (LVDT ND), foundation settlement during the shaking extended to a peak of 120 mm and to a total of 193 mm finishing the earthquake.

Although total settlement between both areas of the stratum do not have big differences in magnitude, it

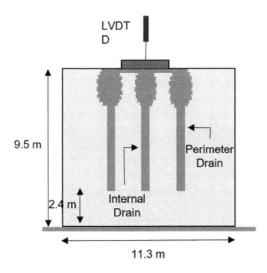

Figure 7. Profile view of vertical drains deformation under foundation during liquefaction.

is possible to state clearly that there is some reduction of settlement of foundations in the presence of vertical drains. Even though one might expect a larger reduction in the settlements due to the larger vertical stiffness of the drains, this did not materialise, as seen in the results above. With the onset of liquefaction, the soil surrounding the drains was unable to support the drain material. Thus, drains suffered deformation by widening in the upper zone, generating a greater area of instability and the settlement of the structure in this area (see Fig. 7). Further analysis is required on this specific issue in order to improve the stability of vertical drains.

4 CONCLUSIONS

Control or prevention of liquefied ground effects on infrastructures using countermeasure techniques is a relevant topic to study. Vertical drains as a method to reduce liquefaction-induced effects has been studied in this work by using dynamic centrifuge modelling to evaluate behaviour of soil with and without drains under foundations. The arrangement of drains analysed in this work were not acting to completely avoid liquefaction; however, behaviour of vertical drains in the soil shows positive results in terms of settlement and excess pore pressure generation. Soil with vertical drains presents lower generation rate of excess spore pressure in comparison with the soil without drains. In addition, soil with drains starts dissipation faster in this area of the stratum, verifying the effectiveness of using this technique.

Moreover, advantages in the effect of drains can be explained by settlement results that show that soil with drains suffer less settlement than a soil without drains. At the same time, settlement in the soil with drains could not be fully avoided as vertical drains lose stability due to free field liquefied soil. Further work is required to consider more realistic scenarios including increase of foundation loads, and also more detailed analysis related to stability of drains.

ACKNOWLEDGMENTS

First author is supported by funding of the Peruvian Council of Science, Technology and Technological Innovation-CONCYTEC.

REFERENCES

Adalier, K., Elgamal, A.W., & Martin, G.R. 1998. Foundation liquefaction countermeasures for earth embankments. *Journal of Geotechnical and Geoenvironmental Engineering*, 124(6): 500–517.

Adalier, K., Elgamal, A., Meneses, J., & Baez, J.I. 2003. Stone columns as liquefaction countermeasure in non-plastic silty soils. *Soil Dynamics and Earthquake Engineering*, 23(7): 571–584.

Baez, S.J. 1997. A design model for the reduction of soil liquefaction by vibro-stone columns.

Brennan, A.J., & Madabhushi, S.P.G. 2002. Effectiveness of vertical drains in mitigation of liquefaction. *Soil Dynamics and Earthquake Engineering*, 22(9): 1059–1065.

Brennan, A. 2004. Vertical Drains as a countermeasure to earthquake-induced soil liquefaction, from Cambridge.pdf. *University of Cambridge*.

Brennan, A.J., & Madabhushi, S.P.G. 2006. Liquefaction remediation by vertical drains with varying penetration depths. *Soil Dynamics and Earthquake Engineering*, 26(5): 469–475.

Brennan, A.J., Madabhushi, S.P.G., & Houghton, N.E. 2006. Comparing laminar and equivalent shear beam (ESB) containers for dynamic centrifuge modelling. *In Physical Modelling in Geotechnics, Proceedings of the 6th International Conference ICPMG* (6): 171–176.

Daniell, J.E., Khazai, B., Wenzel, F., & Vervaeck, A. 2011. The CATDAT damaging earthquakes database. *Natural Hazards and Earth System Sciences*, 11(8): 2235.

Haigh, S. K., Madabhushi, G.S.P., Soga, K., Taji, Y., & Shamoto, Y. 2000. Lateral spreading during centrifuge model earthquakes. In *ISRM International Symposium*. International Society for Rock Mechanics.

Iai, S., Koizumi, K., Noda, S., & Tsuchida, H. 1988. 2. Large Scale Model Tests and Analyses of Gravel Drains. Un'yushō Kōwan Gijutsu Kenkyūjo.

Madabhushi, S.P.G., Houghton, N.E., & Haigh, S.K. 2006. A new automatic sand pourer for model preparation at University of Cambridge. *In Proceedings of the 6th International Conference on Physical Modelling in Geotechnics*: 217–222. Taylor & Francis Group, London, UK.

Madabhushi, G. 2014. Centrifuge modelling for civil engineers. CRC Press.

Marinucci, A., Rathje, E., Kano, S., Kamai, R., Conlee, C., Howell, R. & Gallagher, P. 2008. Centrifuge testing of prefabricated vertical drains for liquefaction remediation. In *Geotechnical earthquake engineering and soil dynamics IV*: 1–10.

Ohsaki, Y. 1970. Effects of sand compaction on liquefaction during the Tokachioki earthquake. *Soils and Foundations*, 10(2): 112–128.

Sasaki, Y., & Taniguchi, E. 1982. Shaking table tests on gravel drains to prevent liquefaction of sand deposits. *Soils and Foundations*, 22(3): 1–14.

Seed, H.B., Martin, P.P., & Lysmer, J. 1976. Pore-water pressure changes during soil liquefaction. *Journal of Geotechnical and Geoenvironmental Engineering*, 102 (Proc. Paper# 12074).

Seed, H.B., & Booker, J.R. 1977. Stabilization of potentially liquefiable sand deposits using gravel drains. *Journal of Geotechnical and Geoenvironmental Engineering*, 103 (ASCE 13050).

Stringer, M.E. & Madabhushi, S.P.G. 2009. Novel computer controlled saturation of dynamic centrifuge models using high viscosity fluids, *ASTM Geotechnical Testing Journal*, 32(6): 53–59.

Watanabe, T. 1966. Damage to oil refinery plants and a building on compacted ground by the Niigata earthquake and their restoration. *Soils and foundations*, 6(2): 86–99.

Yasuda, S. 2007. Remediation methods against liquefaction which can be applied to existing structures. *Earthquake geotechnical engineering*: 385–406.

Deformation behaviour research of an artificial island by centrifuge modelling test

X.W. Gu, Z.Y. Cai, G.M. Xu & G.F. Ren
Nanjing Hydraulic Research Institute, China

ABSTRACT: A centrifuge modelling test is performed to study the deformation behaviour of western artificial island of the HZMB (Hong Kong, Zhuhai and Macao Bridge). The island is designed for tunnels and constructed on soft ground improved by sand compaction piles (SCP) using lattice type steel plate piles. The test includes 2 phases, the first investigates stability of the island from its formation till tunnel pit excavation, and the second studies the long-term deformation development of the island. According to the results, major settlement occurs for the sloped embankment of artificial island from its formation till tunnel pit excavation, the pit and the cofferdam structure are stable during excavation. Also, the prediction is made of post-construction deformation for the artificial island as well as the tunnel.

1 INTRODUCTION

Since the 1980s, the transportation passages among Hong Kong, Macao and Chinese mainland, have witnessed the obvious progress in their construction so as to ensure and promote the mutual development of economy in Hong Kong and the Pearl River Delta. However, the transportation connection between Hong Kong and the western areas of the Pearl River has always been weak. Therefore, approved by the Chinese Central Government, Guangdong Province, Hong Kong and Macao have constructed the HZMB, connecting the Shisanshi Bay of Dayushan in Hong Kong, Mingzhu in Macao and Gongbei in Zhuhai.

According to the overall layout of principal engineering of the HZMB and artificial islands for tunnels, the artificial islands are divided into two maritime ones, that is, eastern island and western island. The western island is the object of this experimental study. It is close to Zhuhai. Its eastern end is connected with the tunnel and its western end is connected with the approach bridge of the Qingzhou Bridge. The western island is basically elliptic in plan (Fig. 1).

The western artificial island is 625 m in length and 183 m in width. The island wall employs the structure of sloped embankment with riprap. The enclosure structure for the foundation pit directly employs the cellular steel sheet pile cofferdam as retaining structure and anti-seepage and water sealing measure for the foundation pit. For the foundation improvement by use of the SCP method, the sand piles in the muddy soil layer have a replacement rate of 30%, and those in the silty clay and clay have a replacement rate of 12%. The main construction sequences include the important time nodes of the insertion of the cellular steel sheet piles, formation of SCP composite foundation,

Figure 1. Impression of western artificial island.

dewatering and excavation of the foundation as well as the final backfill.

However, the improvement efficiency of composite foundation has been a hot issue and one of the difficulties in the geotechnical engineering field. Especially the stability and long term deformation (post-construction settlement) of the composite foundation need to be deeply and systematically studied.

A typical section of the western artificial island is selected to carry out a centrifuge modelling test. The study will investigate the following: (1) the observed lateral displacement of the cellular steel sheet piles when the foundation pit for the tunnel is excavated to the depth of -12.5 m; (2) the total settlement and deformation characteristics of the sloped embankment during the construction period simultaneously; (3) prediction of the post-construction settlement deformation of the tunnel foundation and the embankment after the project completion.

Figure 2. NHRI – 400 gt large geotechnical centrifuge.

The test is carried out at the NHRI – 400 gt centrifuge (Fig. 2). The centrifuge's maximum payload is 2 tons at up to 200 g, that means the facility's capacity is 400 g-ton. The machine's largest radius is 5.5 m. Details of the facility were stated by Dou & Jing (1994).

2 MODEL DESIGN

The determination of model length scale is the first step for the model design. There are 2 factors for the model length scale: one is the dimension of the model container for the model, which is dependent on the geotechnical centrifuges and the box conditions; the other is the prototype range simulated by the model. The simulated range by the model should be large enough, and the standards to be determined are as follows: if instability failure occurs in the artificial island, the existence of model box boundaries should not stop the instability failure phenomenon from producing and developing freely. Therefore, the centrifugal model tests employ the plane strain model box with a net dimension of 1200 mm (length) × 400 mm (width) × 800 mm (height). Owing to the plane symmetry of the artificial islands, the area of one side along the central line is selected as the simulating range. Accordingly, along the width direction only 120 m is simulated. Along the depth direction, the bottom elevation to be simulated is −65 m and the top elevation of the cofferdam is +8 m, thus, the total vertical height to be simulated is 73 m.

After a comprehensive consideration according to the above conditions, the selected similarity scale for the model length is 100, that is, $n = 100$.

Figures 3 and 4 show the model setup with excavated tunnel pit and with the tunnel present respectively. They show different phases of the model test, stated later.

3 MODEL MEASUREMENTS

The most important measurement items in the model test are settlements and horizontal displacements of the artificial island foundation during various stages

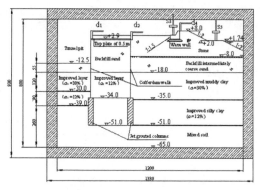

Figure 3. Model with excavated tunnel pit (unit: m for elevations, mm for others).

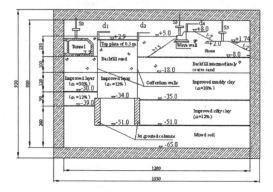

Figure 4. Model with the tunnel set up (Elevation unit: m; other unit: mm).

as well as the lateral displacements of the cellular steel sheet piles.

As shown in Figures 3 and 4, the settlements of island surface (s0, s3 and s5) and the lateral displacements of cellular steel sheet piles (d1, d2 and d4) are measured by 6 laser displacement gauges.

The displacements of model island profiles are measured by the PIV technology (White, Take & Bolton, 2003), which is the most advanced technology for analysing images. A PIV displacement field image measuring system is newly installed. The ES11000 CCD camera is 11000000 pixels, the resolution is 4000×2672, and the sampling frequency is up to 5 frames/s. In the test, layout of measurements by the PIV technology is shown in Figure 5. Through the correlation analysis of 2 neighbouring images, the displacement of soil elements and displacement vector fields of the whole section are obtained.

4 MODEL PREPARATION

The preparation of model foundation is based on the properties of various prototype soil strata (Table 1) and model similarity scales. Different preparation methods are required with regard to different kinds of soil layers. As for the non-cohesive sandy soil layer, its dry

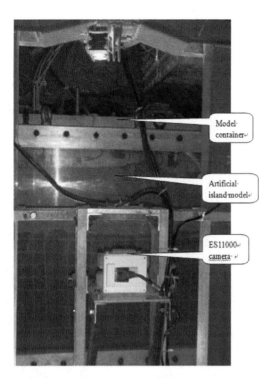

Figure 5. Layout of measurements by use of PIV technology.

initial strength is estimated by the following formula: $(1 + a_s(\lambda - 1))$, in which a_s is replacement ratio, i.e., 0.30, and λ is ratio of stress between SCP and soil, ranging from 2 to 4. If λ is given 3, the ratio of strength increase is 1.6, and the value of in-situ strength of the improved muddy clay must be up to 44 kPa.

In the present model test, the key structures are the cellular steel sheet piles. During the process of dewatering and excavation of the foundation pit, they mainly bear the differences of earth pressures and water pressures between the inner and outer sides, that is, action of bending moment. Therefore, it is designed according to the similarity principles of the flexural rigidity. If the model preparation selects non-prototype materials, the wall thickness should be designed based on the transformation of relevant module. In the bending plane, that is, in the vertical profile of the cellular steel sheet pile cofferdam, it should satisfy $E_p I_p = n^3 E_m I_m$, and along the width direction, that is, along the circumference direction of the cellular steel sheet pile cofferdam, the components are prepared according to the model length scale, i.e., $b_p = b_m n$.

The equivalent wall thickness of the cellular steel sheet piles for the model is calculated according to the following equation:

$$\delta_m = \frac{\delta_p}{n} \sqrt[3]{\frac{E_p}{E_m}} \qquad (1)$$

where δ is the wall thickness of the cellular steel sheet piles; E is the elastic modulus. Through calculation, the model cellular steel sheet piles for the final design are made of aluminium alloy plates 5 mm in thickness, 369 mm in length and 390 mm in width, which is 10 mm smaller than that of the model box. The height along the bending direction is 220 mm, and the net height is 210 mm. The net spacing between division plates is 210 mm.

The water level at the outer side of the artificial islands to be simulated in the model tests corresponds to the elevation of the prototype +1.74 m.

5 TEST PROCEDURES

To simulate the two working conditions: island formation after the foundation treatment and excavation elevation of the foundation pit of −12.5 m, steady operation of 4∼5 years after the project completion. The test procedures are as follows:

(1) Preparation of the mixed layer with the elevation of −51 m ∼ −65 m for model natural foundation using remoulded soils;
(2) Simulation of preparation silty clay improvement layer after SCP treatment with the elevation of −35 m ∼ −51 m (replacement rate of SCP = 12%);
(3) Simulation of preparation muddy clay improvement layer after SCP treatment with the elevation of −18 m ∼ −25 m (replacement rate of SCP = 30%);

density or relative density should strictly agree with that of the prototype; as for the cohesive soft soil layer, its in-situ undrained shear strength is controlled to agree with that of the prototype, and the stage preloading drained consolidation method should be employed for the preparation of model soils; as for the composite foundation with sand compaction piles, its equivalent strength and compressibility should agree with those of the prototype.

The soil strata of western artificial island from top to bottom mainly include: stone cushion, backfill medium/coarse sand layer, muddy clay improvement layer (SCP $A_s = 30\%$), silty clay improvement layer (SCP $A_s = 12\%$) and mixed layer of silty fine sand and coarse gravelly sand. The model preparation refers to the typical section of the western artificial island. The thickness is proportionally reduced according to the geometrical scales. As for the layers of the stone cushion and the backfill medium/coarse sand, their dry density is controlled to be in agreement with that of the prototype; as for other soil layers, their in-situ undrained strength is controlled to be in agreement with that of the prototype.

It should be noted that the in-situ strength of soils (Table 1) is calculated by using the empirical fitting formula: $s_u = -(12.1 + H)/0.53$, in which H is the value of Elevation. Concerning the increase in strength due to SCP improvement, only the improved muddy clay is modelled in the test according to its increased strength. And the ratio of its increased strength to the

Table 1. Physical and mechanical properties of main soil strata.

Name of soil stratum	Thickness of soil stratum t/m	Moisture content w/%	Unit weight γ/kN/m^3	In-situ strength s_u/kPa	Compression index C_c	Permeability coefficient k/cm/s
Stone cushion	16(+8~−8)	/	20	/	/	3.0E−4
Backfill medium/coarse sand layer	10(−8~−18)	/	14.4*	/	/	5.0E−3
Improved muddy clay ($A_s = 30\%$)	17(−18~−35)	60	16.9	27 (46) *	/	/
Improved silty clay ($A_s = 12\%$)	16(−35~−51)	46	17.2	58	0.27	/
Mixed layer	14(−51~−65)	33	18.3	/	0.22	/

Notes: 1: * stands for the dry density; 2: value in bracket being increased in-situ strength through SCP improvement.

(4) Simulation of preparation of medium/coarse sand layer for replacement with the elevation of −8 m ∼ −18 m;
(5) Insertion of model cellular steel sheet piles;
(6) Simulation of preparation of island embankment with stone blocks and backfill of medium/coarse sand behind the cellular steel sheet piles;
(7) Simulation of excavation of the foundation pit, the excavation surface corresponds to elevation of −12.5 m;

After completion of the model construction, the centrifuge starts and accelerates to the design acceleration of 100 g by stepless adjustment and stops after 1 hour.

(8) Simulation of backfill of the foundation pit and arrangement of the tunnel, medium/coarse sand is employed to fill the spacing between the tunnel and steel sheet piles and the upper part of the tunnel till the design elevation (+5.0 m);
(9) Owing to the occurrence of settlement on the island embankment surface at the first time of operation, crushed stone is employed to place the island embankment till the original design elevation (+8.0 m).

The centrifuge starts again and directly accelerates to the design acceleration of 100 g and stops after 4 hours so as to simulate the operation of 4∼5 years after the projection completion.

6 TEST RESULTS

6.1 Deformation behaviour of artificial island from its formation till tunnel pit excavation

As shown in Figure 3, the model with excavated tunnel pit is moved onto the platform of centrifuge and brought into flight of 100 g in steps. Each step is 20 g. On arrival of 100 g, one hour is taken for model of soil to be stressed under self-weight. The surface settlement and lateral movement of structure are measured and given on prototype scale in the plot of displacement versus time in Figure 6, in which lateral movement pointing to tunnel pit is accounted to positive, and downward vertical, settlement, denotes positive.

It can be seen from the plot that the lateral movement of two walls of cofferdam structure at their top, d1 and d2, are of positive sign and of identical value.

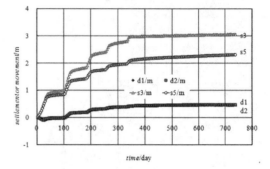

Figure 6. Displacement versus time (Tunnel pit excavated to El. −12.5 m).

Therefore, the cofferdam structure of cellular steel sheet piles moves toward tunnel pit due to excavation.

It can be also seen from Figure 6 that the lateral movement of cofferdam structure at the top continues to develop with a decreasing rate after being excavated to El. −12.5 m but that the value of lateral movement soon reached a stable one, indicating that the tunnel pit and the cofferdam structure is of good stability during excavation.

Following the plots of Figure 6, two walls of the cofferdam structure of cellular steel sheet piles move toward the tunnel pit by 0.46 m at top. And according to analysis result of PIV measurement (Figure 7), the cofferdam structure of cellular steel sheet piles move toward the tunnel pit by 0.30 m at its bottom. As a result, a difference of about 0.16 m of lateral movement is formed during excavation.

The plots of Figure 6 shows that a great amount of settlement of the artificial island occurs from formation of embankment till excavation of tunnel pit, mainly resulted from consolidation of the underlain compressible layers. The settlement develops with time at a decreasing rate and approaches to a constant value, showing that the sloped embankment with riprap is stable at the end of construction.

Following the plots of Figure 6, the total settlement of the embankment from its formation till excavation is of about 2.31 m∼3.06 m, which are listed in Table 2 later.

Figure 7 shows that the soils enclosed by cellular steel sheet piles and their adjacent soils move toward tunnel pit while the soils outside embankment moves a

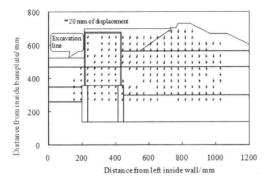

Figure 7. Map of vectors of soil movement.

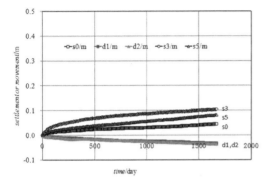

Figure 8. Displacement versus time after the tunnel is set up.

little toward seaside during their settlement. It is found that on the whole, the direction of displacement experience by artificial island is downward, indicating no possibility of sliding at the end of construction.

6.2 Post-construction deformation behaviour of artificial island after its completion

As shown in Figure 4, the model with the tunnel set up is moved onto the platform of centrifuge once again and brought into flight of 100 g to be stressed under self-weight. Because of the NHRI – 400 gt centrifuge is not in good situation and is about to be maintained and upgraded thoroughly, the second flight lasted only for four hours, simulating a term of about 4.6 years of prototype time after the completion of artificial island. The change of the surface settlement of embankment and tunnel and the lateral movement of cofferdam structure are depicted in the plots of displacement versus time in Figure 8. It can be noticed that the sign of readings is negative of lateral movement of two walls of cofferdam structure, indicating that the upper wall parts moves away from tunnel pit because of the presence of tunnel structure and backfilled sand.

It can be seen that the magnitude of lateral movement of the upper walls develops gently at a reducing rate with prototype time. During 4.6 years, the movement amounts to 0.062 m at d1 and 0.044 m at d2, respectively.

Table 2. Total settlement at the end of construction and predicted settlement of post construction.

Gauging point		s0/m	s3/m	s5/m
Total settlement at the end of construction		/	3.06	2.31
Predicted settlement of post construction	2 years	0.027	0.078	0.045
	4 years	0.039	0.097	0.073
	6 years	0.045	0.117	0.086
	8 years	0.051	0.131	0.101
	10 years	0.057	0.143	0.113
	12 years	0.062	0.153	0.125
	14 years	0.066	0.163	0.136
	50 years	0.119	0.268	0.271
	120 years	0.178	0.377	0.434

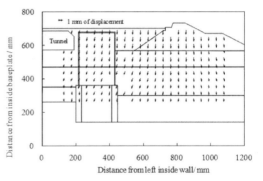

Figure 9. Map of vectors of soil movement.

It can be also seen from the plots in Figure 8 that the magnitude of settlement of the embankment and the tunnel develops gradually at a reducing rate with prototype time. During 4.6 years, the settlement amounts to 0.103 m at s3, 0.081 m at s5, and 0.044 m at s0, respectively.

Since it is hard to simulate post-construction such as 50 years even 120 years of operation by running centrifuge without a stop, the prediction of post-construction settlement cannot be made reliably. But it can be anticipated that the settlements will come to their steady values as long as the artificial island is put into use long enough. If the power curves are used to fit the plots of settlement versus time obtained, it can be roughly estimated in a extrapolation method that the embankment of artificial island and the tunnel structure will have 0.38 m~0.43 m and 0.18 m of post-construction settlement in 120 years, respectively (Table 2).

It is noted in Table 2 that the magnitude of the total settlement at s5 is less than that at s3, and that it is the case for the predicted post-construction settlement at s3 over a period of less than 46 years. This is attributed to the boundary effect of model container, which brings about a hysteresis of development of compression for the adjacent soils.

As shown in Figure 9, a map of vectors of soil movement on the cross-section is acquired by means of PIV measurement technology for the second stage of centrifuge flight. It can be seen that on the whole, the direction of displacement experience by artificial island is downward during 4.6 years of post-construction. It is noticed that the soils adjacent to the cofferdam structure still move slightly toward tunnel pit while the soils outside embankment moves a little toward seaside during their settlement.

7 SUMMARY

Two working conditions of artificial island are simulated by means of centrifuge model test, one case being excavation of tunnel pit and the other case being post-construction operation after the placement of tunnel. Following points are drawn from the test results.

(1) Two walls of the cofferdam structure of cellular steel sheet piles move toward the tunnel pit by 0.46 m at the top and 0.30 m at the bottom due to the excavation, so the difference of lateral movement of the cofferdam structure is about 0.16 m between the top and the bottom. It is shown that the tunnel pit and the cofferdam structure are stable during excavation since the displacement developed can come to a steady value.

(2) Major settlement occurs for the sloped embankment of artificial island from its formation till excavation. The part of artificial island adjacent to the cellular steel sheet piles moves laterally towards tunnel pit and the outside part of artificial island moves laterally towards seaside when they settle. The total settlement of the embankment from its formation till excavation is of about 2.31 m~3.06 m.

(3) The upper part of two walls of the cofferdam structure of cellular steel sheet piles move away from tunnel pit by 0.044 m~0.062 m after placement of tunnel structure and backfilling of middle coarse sand. New settlement of 0.08 m~0.10 m occurs for the sloped embankment of artificial island during the 4.6 years of operation modelled. The displacement of embankment and underlain soil layers is composed of settlement and movement of small magnitude, whose direction depends on the position of the part of embankment.

(4) From the settlement curve, settlement develops with time at a reducing rate. Confined by capability of geotechnical centrifuge modelling test, an accurate prediction of post-construction settlement can be made by combining with other physical simulation measures and analytical methods with more precision. It is estimated by the power fitting curve that the embankment of artificial island will have a post-construction settlement of 0.38 m~0.43 m in 120 years, and the tunnel will experience 0.18 m of settlement in 120 years.

REFERENCES

Y. Dou & P. Jing. 1994. Development of NHRI – 400 gt geotechnical centrifuge. Leung, Lee & Tan (eds), *Centrifuge 94*: 69–74. Rotterdam: Balkema.

D. J. White, W. A. Take & M. D. Bolton. 2003. Soil deformation measurement using particle image velocimetry (PIV) and photogrammetry. *Géotechnique* 53, No. 7, 619–631.

Z. Y. Cai. 2009. Key technical study on foundation improvement of artificial islands. Nanjing, China.

Effect of lateral confining condition of behaviour of confined-reinforced earth

H.M. Hung & J. Kuwano
Department of Civil Engineering, Saitama University, Saitama, Japan

ABSTRACT: Earthquakes often induce differential settlement between bridges and their approach embankments. Consequently, vehicles cannot pass the step-wise settlement. The confined-reinforced earth (CRE) method, has been proposed to make it possible for vehicles, especially emergency vehicles to pass road surfaces roughened. In this study, the effect of the length of geogrids on the behavior of the CRE subjected to differential settlement was investigated. The results showed that the CRE surface settlement distribution and tensile strain along the geogrids were significantly affected by the geogrid length. CRE with long geogrids is recommended to use in practice.

1 INTRODUCTION

A previous study has shown that earthquakes commonly induce differential settlement between bridge abutments and approaches due to slope movement or grainslip (Siddharthan & El-Gamal 1996). Consequently, vehicles cannot pass the unevenness created by the earthquakes. Geosynthetic reinforced soil has been widely used to mitigate such damage from earthquakes because of its high seismic resistance (Koseki 2012; Kuwano et al. 2014). Furthermore, important aspects of reinforced-soil are its potential to reduce differential settlement (Miao et al. 2014; Monley and Wu 1993) and its ability to be used as bridges over cavities (Briancon & Villard 2008; Poorooshasb, 2002). The confined-reinforced earth (CRE) method, has been proposed to make it possible for vehicles, especially emergency vehicles, which are needed to approach affected areas as soon as possible, to pass road surfaces roughened by earthquakes. The CRE method employs geogrid layers, confining steel tie rods, and granular soil applied to subgrade layers under pavements of roads. In this method, reinforced soil is confined by prestressed tie rods as shown in Figure 1. Prestress is introduced into the tie rods, contributing to the integrity of reinforced soil that has high stiffness (Tatsuoka et al. 1997). As a result, deformation of the CRE can be reduced when the CRE is subjected to traffic loading (Uchimura et al. 2003, Shinoda et al. 2003).

A full-scale field test has previously been carried out to compare the behavior of road structures prepared using the CRE method and a conventional method (without using geosynthetics) when subjected to a differential settlement of 550 mm. The results of the trials have shown that the road with CRE would allow a vehicle to pass, whereas the other would not, because of a significant crack in the pavement (Ohta et al. 2013).

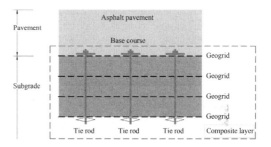

Figure 1. Confined-reinforced earth.

Kuwano et al. (2013) have also proved that the CRE method effectively improves the stiffness of the subgrade layer and can reduce a step-wise deformation of road pavement under differential settlement.

In reinforced soil, the behavior of reinforced soil is strongly affected by embedded geogrid length (Moraci & Recalcati 2006; Teixeira et al. 2007). They pointed out that an increase in specimen length leads to increasing pullout resistance, increased initial stiffness and increased displacement at peak pullout resistance. In retaining walls, previous studies have shown that the length of reinforced soil significantly affects reinforced soil behavior when the ratio of reinforcement length to height of retaining walls is between 0.5 and 0.7 (Rowe & Ho 1998). In fact, 100% reduction in displacement is caused by an increase in length-height (L/H) ratio from 0.5 to 0.7.

In CRE method, the effect of geogrid length on the behavior of CRE has not yet been studied. In this study, the effect of geogrid length was studied to investigate the behavior of the CRE subjected to differential settlement. A physical model was proposed to simulate embedded geogrids, which were replaced by their pullout resistance. Consequently, large model laboratory tests can be done in laboratories where the space is usually limited.

2 CONFINED-REINFORCED EARTH TEST

2.1 Materials

The geogrid (Adeam HG-200) used in this study is made of high density polyethylene with an opening size of 26 × 28 mm (longitudinal × transversal). The longitudinal rib is reinforced by aramid fibers. The geogrid has a tensile strength of 200 × 5 kN/m (longitudinal × transversal) with a rupture strain of 4.5%. The granular soil used in the model was dry Toyoura sand (uniform sand) with a specific gravity $\gamma_s = 2.645$; $e_{max} = 0.973$; $e_{min} = 0.609$; $D_{50} = 0.2$ mm.

2.2 Model test

An application of CRE was proposed to reduce differential settlement between a bridge and its embankment as shown in Figure 2. It is noted that confining tie rods should be used in reinforced soil zone where it is bent to improve its stiffness. To investigate the effect of geogrid length of CRE, the geogrid length must be changed. Three types of model experiments are often used, including small-scale model laboratory, large-scale model laboratory, and full-scale field tests (Yi et al. 2015). Model laboratory tests are preferred to be conducted as large as possible because these approach to their realistic conditions. Therefore, the reliability of the experiments increases. However, large-scale tests are difficult to be performed in laboratories where the space is usually limited.

To overcome this issue, embedded geogrid length of reinforced soil (Figure 2) was simulated by its pullout resistance. Therefore, a soil box in which the CRE was constructed can be shortened and the pullout resistance of the embedded geogrids is applied at the boundary of the soil box to model a larger scale test. Therefore, instead of constructing a large physical model test, a smaller size of the physical model test applied boundary conditions can be used to simulate the larger scale model test. Consequently, pullout tests must be done first to get the load-displacement behavior. The relationship between overburden and pullout resistance was then established. As overburden on each geogrid was different, the corresponding pullout resistance was different. It was then calculated based on the length of an embedded geogrid. It is noted that pullout resistance increases linearly with geogrid length (Teixeira et al. 2007). This resistance was connected to the geogrids at the boundary of the soil box to simulate its embedded geogrid.

The pullout resistance, was simulated by a system attached at the boundary of the CRE at one side of the soil box. The friction at the soil-geogrid interfaces was modelled using the Coulomb friction law, including elastic and plastic components (Villard & Briancon

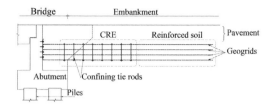

Figure 2. Application for bridge and its embankment.

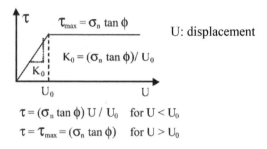

Figure 3. Coulomb friction law (Villard & Briancon 2008).

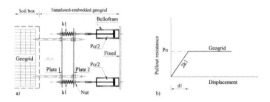

Figure 4. a) Arrangement of springs and bellofram and b) mechanism of system.

2008), as shown in Figure 3. To simulate this friction law, the system consisted of springs and belloframs (air cylinders) that were connected in series as illustrated in Figure 4. In this study, the embedded geogrid length outside the soil box, L_R, was set at 0, 0.54 m, and ∞ (the right end of the geogrid was horizontally fixed to simulate long geogrid with large resistance so that the geogrids were not pulled out).

The model test setup is shown in Figure 5. CRE was made in a soil box with inner dimensions of length × width × height = 1200 mm × 400 mm × 800 mm. The soil box includes front and back transparent Plexiglass plates used to enable visualization of the models during the testing. The composite layer was supported by two bottom plates, Plate 1 and Plate 2. Plate 1 was fixed on a frame (to simulate an abutment of a bridge) while the other plate was supported by jacks so that it could move down to simulate settlement of an embankment. Four geogrid layers (G1, G2, G3, and G4) and three sand layers with a thickness of 95 mm were constructed. The total thickness of the CRE was 300 mm. The lowest geogrid (G4) was placed directly onto the plates. The tensile strains of the geogrids G1, G2, G3, and G4 were measured at (A)

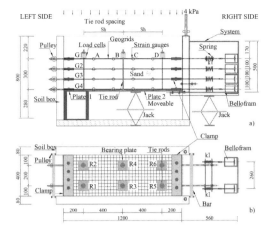

Figure 5. Model test, (a) elevation view, (b) plan view.

300 mm, (B) 500 mm, (C) 700 mm, and (D) 900 mm from the left to the right of the soil box in Figure 5.

Two strain gauges were attached in pairs on both sides of each measuring point, i.e., on the upper and lower surfaces of the geogrid. The tie rods were set with spacing of 400 mm in the longitudinal direction and 200 mm in the transversal direction. They were placed at 200 mm, 600 mm and 1000 mm from the left to right of the soil box. Overburden on the CRE was 4 kPa to simulate weight of a pavement. Therefore, the overburden on the geogrids (G1-4) were 4, 5.5, 7, and 8.5 kPa, respectively (weight of each sand layer was 1.5 kPa). These values were smaller than those used in the field because the size of the CRE model in the laboratory was smaller. It should be noted that in order to keep the sand from leaking when the CRE deformed, the sand layers were wrapped in a thin low strength geotextile.

The pullout tests were carried out to determine the pullout resistance in the soil box with dimensions of $300 \times 202 \times 400$ mm = width × height × length. The plastic components would be estimated corresponding to the overburden of the model test based on the relationship between the pullout resistance and the overburden.

The elastic parameter E50, which is the modulus at 50% of the peak strength and commonly used in Soil Mechanics, was estimated for each overburden. The plastic parameter was determined based on the balance of energy, of which friction energy = pullout resistance × displacement must be the same in the model and reality. The elasticity for each overburden was different. However, for simplicity, in the model, an average elasticity was estimated for all overburden. The average elastic component, k, was 6.5 kN/mm. This value was very large and springs with this value were not available in the market. It is noted that the load-displacement of a geogrid depends on the length of geogrid and overburden (Moraci & Recalcati 2006). Therefore, it changes depending on specific conditions. In this study, the spring constant taken for the system is 0.62 kN/mm.

Table 1. Test program.

Cases	Resistance Length, L_R, (m)	Tie rod spacing, (cm)	Overburden (kPa)
1	0 (Free)	40	4
2	0.54	40	4
3	∞ (Fixed)	40	4

L_R: length of geogrid was simulated by pullout resistance of Belloframs.

2.3 Test program

Three cases of tests performed for the current study are shown in Table 2. The geogrid length of the reinforced soil, which was outside the soil box, was changed to investigate the effect of geogrid length. It simulated by pullout resistance corresponded to the lengths of embedded geogrid of 0, 0.54 m, and ∞ (horizontally fixed). It should be noted that in Case 1 and 3 were extreme cases. The geogrids in Case 1 were considered to be very short while in Case 3 were long enough so that the geogrids was not pulled out.

2.4 Test procedure

A detailed description of the experimental procedure is as follows. In the first step, the lowest geogrid layer (G4) and six tie rods, R1 – R6 were placed on the bottom Plates 1 and 2. Then, the 95 mm – thick sand layer was made by air pluviation technique as proposed by (Miura and Toki 1982). The sand was poured from a height ranging between 25 cm and 55 cm using multiple sieve layers to maintain a constant relative density of 80% (dense sand), and leveled. Miura and Toki (1982) have noted that the effect of sand drop height is very small and this was checked and confirmed in this study. The other geogrid layers (G3, G2, and G1) and upper sand layers were prepared in a similar way. After the placement of the top geogrid (G1), the clamps to hold the geogrids were connected to the left side of the soil box through pulleys, and connected to the springs of the system at right side (Fig. 5).

The tie rods were then preloaded to 3 kN, a pressure high enough for effective CRE, but still below failure level (Tatsuoka et al. 1997). Then, the boundary conditions were applied at both sides of the CRE, i.e. horizontally fixed at the left and pulled out at the right of the tie rods. It is should be noted that the pullout resistance increased from G1 to G4 due to increased overburden on the geogrids. The overburden used steel balls with a diameter of 10 mm in bags of 10 kg was applied next. In the last step, Plate 2 was lowered by the jacks to induce differential settlement, Sv, with respect to Plate 1. Plate 2 in Case 3 was lowered until it was detached from the CRE layer, the settlement in the other cases was the same as Case 3, to compare their behavior. The deformation of Case 3 was expected to be the minimum because it was horizontally fixed at both of its ends.

The deformation of the CRE was measured using AutoCAD tool based on a series of photo, which were taken at the initial and final stage of the settlement. Strains of the geogrids and force in the tie rods, which were measured by strain gauges and load cells, respectively, were recorded by the data recorder during the tests.

3 RESULTS AND DISCUSSION

3.1 *Deformation*

A typical deformation of the CRE subjected to differential settlement is shown in Figure 6 which indicates the deformation of Case 2 at 9 cm. It should be noted that the surface settlement no longer increased for the increase in Sv more than 9 cm in Case 3 (both ends of the geogrids were horizontally fixed), therefore, to compare with the other cases, the settlement, Sv = 9.0 cm was applied for all cases. It can be seen from the figure that the bending deformation mode, which caused the rotation of the CRE, can be observed through the rotation of the tie rods. The rotation of the middle tie rods, R3 and R4 was at the maximum while the minimum was seen at R1 and R2 which were on the fixed plate, Plate 1 (1.75^0 compared to 0.45^0 in Case 3, Figure 6).

Surface settlement of the CRE is the most important to evaluate reducing differential settlement of this method. As can be seen from Figure 7, the surface settlement distribution decreased with increasing resistance length, L_R, (embedded geogrid length outside the soil box). The maximum surface settlement was observed at Case 1 (Free case) while the minimum was seen at Case 3 (Fixed case). This is due to the right-end boundary condition that restrained lateral movement of the CRE. Therefore, the CRE with long geogrids showed better performance than the CRE with short geogrids.

3.2 *Strain in geogrids*

Strain changes along the geogrids with the increase of settlement, S_v, were monitored at four locations (A, B, C, and D) as shown in Figure 8. A positive strain value indicates the tensile deformation of the geogrid while a negative one means compression.

In general, as can be seen from Figure 8, the tensilel strain increased with increasing geogrid length. The maximum tensile strain in each geogrid of Case 3 was larger compared to that of the other cases except G1. This is because the geogrids in Case 3 were horizontally fixed at both their ends, their strains developed significantly when they supported vertical load due to the membrane effect of geogrid (Holtz et al. 1998) like a hammock. However, the maximum tensile strain of G1 of Case 2 and 3 was not much different. This is due to the right-end boundary condition of Case 3 that restrained the bending of the CRE at the section between Plate 1 and 2. The maximum tensile strain in

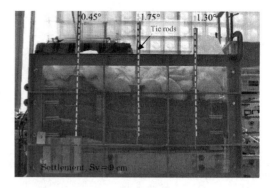

Figure 6. CRE deformation (Case 2).

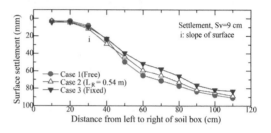

Figure 7. Surface settlement of CRE.

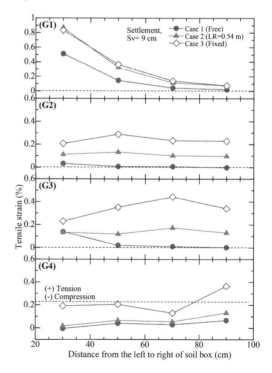

Figure 8. Distributions of tensile strain along geogrids.

the geogrids was seen in G1 at A (G1 A) in all cases. This is because of deformation mode of CRE, which was bent when the CRE was subjected to the differential settlement. As a result, at the top geogrid, G1 A showed the largest tension. The highest tensile strain

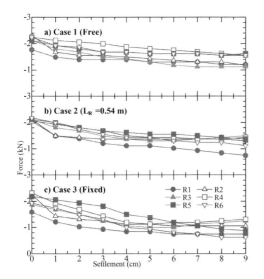

Figure 9. Force in tie rods.

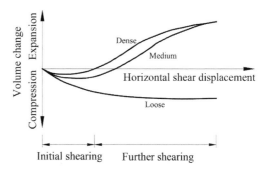

Figure 10. Direct shear test results in sands (Das 1999).

reached approximately 0.9% in Case 2 and 3, and 0.5% in Case 1 at G1 A. The tensile strain in G2 and G3 near the right end of the CRE of Case 1, which were not restrained by pullout resistance at the right end, was equal to almost zero due to the right-end-movement of the CRE.

It was also found in Case 2 and 3 (as considered the case of long geogrids) the location of the peak strain shifted from point A in the top geogrid of G1 to point D in the bottom geogrid G4, i.e. the location of the highest tensile strain shifted from the left to the right with depth. This is also because the mode of bending deformation of the CRE, causing clockwise rotation of the tie rods as illustrated in Figure 6. This shows a good agreement with the results of the full-scale field test (Ishigaki et al. 2012; Ohta et al. 2013).

By contrast, in Case 1 (as considered a case of short geogrid) the strain showed both tension and small compression, the small compression was at G4 A. This is because the mode of deformation in those cases was rather step-wise than a hammock and compressional deformation dominated at the bottom geogrid when the bottom of the tie rods R3 and R4 (middle tie rods) moved further left than in Case 2 and 3. However, the compression was very small. The significant difference between cases with long and short geogrids is that large tension developed in all geogrids in the cases with the long geogrids to support the vertical load whereas large tension was only observed at G1 in the cases with the short geogrid.

3.3 Force in tie rods

The forces in the six tie rods, R1–R6 were measured by the load cells while settlement of Plate 2 increased as shown in Figure 9. The tie rods were preloaded to 3 kN. However, the preload then decreased gradually before the increase of settlement because of stress relaxation of the soil.

It is seen from the Figure 9 that a decrease in the force in the tie rods of all cases subjected to the differential settlement was observed in general. An increase in R3 and R4 was seen slight after the decrease. This is probably due to dilatancy of the dense sand confined between the geogrids in tension after certain amount of shear strain in Case 2 and 3. When dense sand is sheared, it initially shows volume decrease within the small strain range. However, the dense sand then shows volume increase with an increase in shear strain (Das 1999) as shown in Figure 10. Furthermore, one more reason for the increase in force is that the differential settlement between the top and bottom due to the soil arching (Huang et al. 2015; Poorooshasb 2002). Therefore, if a tie rod was arranged into this zone, the tension of the tie rods would increase.

4 CONCLUSIONS

A series of laboratory model tests were carried out to investigate the effect of the length of geogrids on the behavior of confined-reinforced earth (CRE). The following conclusions were obtained:

1. The surface settlement distribution decreases with increasing geogrid length (embedded geogrid). CRE with long geogrids is recommended to be used in practice.
2. Tensile strain increases with an increase in geogrid length. The maximum tensile strain is seen at the top geogrid, near the structure (i.e. G1 A in this study). In case of long geogrids, the peak tensile strain shifts from the left to right with depth. This is due to the fact that the CRE behaved like a beam.
3. The force in all tie rods decreases due to the differential settlement in general. A slight recovery is seen at the middle tie rods, which are on the Plate 2, close to Plate 1.

ACKNOWLEDGEMENTS

This research project has been sponsored by JSPS KAKENHI Grant Number 16H04406, Monbukagakusho Scholarship (MEXT) and Maeda Kosen Co., Ltd.

REFERENCES

Briancon, L. & Villard, P. 2008. Design of geosynthetic-reinforced platforms spanning localized sinkholes. Geotextiles and Geomembranes, 26, 416–428.

Das, B.M. 1999. Shallow foundation: bearing capacity and settlement, Florida, USA: CRC Press.

Holtz, R., Christopher, B. & Berg, R. 1998. Geosynthetic Design and Construction Guideline., p.460.

Huang, J., Le, V., Bin-Saphique, S. & Papagiannakis, A.T. 2015. Experimental and numerical study of geosynthetic reinforced soil over a channel. Geotextiles and Geomembranes, 43(5), 382–392.

Ishigaki, T. Tatta, N., Kawasaki, H., Tachibana, S., Kuwano, J. & Ohta, H. 2012. Experimental investigation on reinforcement mechanism of the confined-reinforced earth method. In Proc. of 67th JSCE Annual Meeting, CD-ROM, V-369. (in Japanese).

Koseki, J. 2012. Use of geosynthetics to improve seismic performance of earth structures. Geotextiles and Geomembranes, 34, 51–68.

Kuwano, J., Miyata, Y. & Koseki, J. 2014. Performance of reinforced soil walls during the 2011 Tohoku earthquake. Geosynthetics International, 21(3), 179–196.

Kuwano, J., Tachibana, S., Ishigaki, T. & Tatta, N. 2013. Confined reinforced subgrade to reduce differetial settlement of road pavement.pdf. Proceedings of 5th KGS-JGS Geotechnical Engineering workshop, Seoul, Korea, 149–154.

Miao, L., Wang, F., Han, J. & Lv, W. 2014. Benefits of geosynthetic reinforcement in widening of embankments subjected to foundation differential settlement. Geosynthetics International, 21(5), 321–332.

Miura, S. & Toki, S. 1982. A Sample Preparation Method and Its Effect on Static and Cyclic Deformation-Strength Properties of Sand. Soils and Foundations, 22(1), 61–77.

Monley, G. & Wu, J.T. 1993. Tensile reinforcement effects on bridge-approach settlement., 119(4), 749–762.

Moraci, N. & Recalcati, P. 2006. Factors affecting the pullout behaviour of extruded geogrids embedded in a compacted granular soil. Geotextiles and Geomembranes, 24(4), 220–242.

Ohta, H., T Ishigaki & Tatta. N. 2013. Retrofit technique for asphalt concrete after seismic damage. In Proceedings of the 18th international conference on soil mechanics and geotechnical engineering. Paris, 1333–1336.

Poorooshasb, H.B. 2002. Subsidence Evaluation of Geotextile Reinforced Gravel Mats Bridging a Sinkhole. Geosynthetics International, 9(3), 259–282.

Rowe, R.K. & Ho, S.K. 1998. Horizontal deformation in reinforced soil walls. Canadian Geotechnical Journal, 35(1), 312–327.

Siddharthan, R. & El-Gamal, M. 1996. Earthquake-induced ground settlements of bridge abutment fills. Proceeding of ASCE National Convention, Washington, D.C., 100–123.

Shinoda, M., Uchimura, T. & Tatsuoka, F. 2003. Increasing the stiffness of mechanically reinforced backfill by preloading and prestressing. Soils and Foundations, 43(1), 75–92.

Tatsuoka, F., Uchimura, T. & Tateyama, M. 1997. Preloaded and Prestressed reinforced soil. Soils and Foundations, 37(3), 79–94.

Teixeira, S.H.C. et al. 2007. Pullout Resistance of Individual Longitudinal and Transverse Geogrid Ribs. Journal of Geotechnical and Geoenvironmental Engineering, (January), 37–50.

Uchimura, T., Tateyama, M., Koga, T. & Tatsuoka, F. 2003. Performance of a preloaded-prestressed geogrid-reinforced soil pier for a railway bridge. Soils and Foundations, 43(6), 155–171.

Villard, P. & Briancon, L. 2008. Design of geosynthetic reinforcements for platforms subjected to localized sinkholes. Canadian Geotechnical Journal, 209, 196–209.

Yi, Y., Menles, D., Nassiri, S. & Bayat, A. 2015. On the compressibility of tire-derived aggregate: comparison of results from laboratory and field tests. Canadian Geotechnical Journal, 52(4), 442–458.

An experimental study on the effects of enhanced drainage for liquefaction mitigation in dense urban environments

P.B. Kirkwood & S. Dashti
University of Colorado Boulder, Boulder, USA

ABSTRACT: Structures founded on liquefiable soils are vulnerable to large rotation and settlement during seismic events. Expressions for settlement calculation during design are based on free-field conditions. Consequently, they do not capture soil-structure interactions, often resulting in unconservative and inaccurate settlement predictions. To reduce the probability of excessive settlement or rotation, mitigation measures such as soil drains are becoming increasingly popular. Experimental studies have shown that appropriate use of soil drains usually reduces permanent settlement and rotation of the structure but increases the seismic demand. In dense urban environments, where the region of drain influence extends under adjacent structures, there is concern that improved soil drainage may have a detrimental effect on the performance of unmitigated neighbouring structures. A series of centrifuge tests, conducted at the University of Colorado Boulder, investigated the performance of neighbouring structures, and the influence of soil drains when installed around just one structure. Results show serious detrimental consequences for unmitigated structures located next to a structure protected by soil drains, which should be considered in design when in dense urban settings.

1 INTRODUCTION

Structures founded on liquefiable deposits often experience unacceptable settlement and rotation during earthquakes. During the Christchurch earthquake series (2010-2011), despite rigorously enforced seismic building codes, much of the city was severely damaged as a result of soil liquefaction (Cubrinovski, 2013). The risk of similarly widespread damage is significant for major population centres on the West coasts of North and South America, Japan, Indonesia, and elsewhere. An important challenge for geotechnical earthquake engineers is to establish design methodologies, which ensure structures founded on liquefiable deposits remain serviceable after major earthquakes.

At present, it is possible for engineers to estimate, with reasonable accuracy, the settlement of free-field sites using empirical procedures. However, predicting the settlement of structures remains problematic. Bertalot et al. (2013) published correlations for liquefaction induced structural settlement as a function of foundation width, thickness of the liquefiable layer and bearing pressure. However, it is unreasonable to assume that these relationships, developed from empirical correlations, will remain accurate in more general circumstances. For example, structures that have been mitigated to reduce the liquefaction hazard, or closely spaced structures that interact with one another through the soil, are expected to behave differently. Structure-soil-structure interaction (SSSI) is known to affect the settlement and rotation response of structures on liquefiable deposits (Hayden et al., 2015; Kirkwood & Dashti, 2017a, b & c). In such cases, it is not yet possible to accurately predict the settlement (or rotation) of structures founded on liquefiable deposits.

Kirkwood and Dashti (2017a) showed that overlapping effective stress bulbs between neighbouring structures can lead to differential soil stiffness between opposite sides of a structure, which along with the differences in soil confinement compared to an isolated structure, affect the magnitude of both volumetric and deviatoric soil strains. In general, SSSI was observed to decrease structural settlements compared to an isolated structure but increase its permanent rotation. To mitigate the increase in rotation, various strategies might be employed such as soil densification, structural walls, and soil drains. This study focuses on the efficacy of liquefaction mitigation with soil drains in dense urban environments.

Various published studies demonstrate the viability of soil drains for reducing settlement of isolated, shallow founded structures on liquefiable deposits (Paramasivam et al. 2017; Hausler & Sitar 2001). In general, the reduction in liquefaction-induced settlement is accompanied by an increase in seismic demand and superstructure damage potential. In this study, attention is devoted to the effects of soil drains on settlement, rotation, and structural damage potential of both the mitigated structure and an adjacent unmitigated structure.

Four centrifuge tests were conducted, each involving a pair of closely spaced structures founded on a layered liquefiable deposit. A comparison is presented of the response of two similar 3-story structures, with and without prefabricated vertical drains (PVDs) around one structure. Also presented is a more general case of two dissimilar adjacent structures, one of 3 stories, the second of 9. In each case, for both similar and dissimilar structure pairs, the effect of mitigating just one structure was severe for the adjacent unmitigated structure in terms of increased permanent rotation and damage potential. The results demonstrate the importance of holistic liquefaction mitigation design rather than considering individual structures in isolation.

2 EXPERIMENTAL PROCEDURE

The four centrifuge tests presented herein were performed at the University of Colorado Boulder (CUB) Geotechnical Centrifuge Facility under a radial acceleration of 70 g. The models were prepared in the flexible shear beam (FSB), each using the same soil profile shown in Figure 1. The layered soil profile was fabricated using the automated sand pourer (ASP) at CUB.

From the surface, down were layers of dense Monterey sand, loose Ottawa, and dense Ottawa sand with initial relative densities of 88, 36 and 90%. Each model also contained two structures. Structure A was a 3-story moment resisting steel frame structure with representative bearing pressure, height/width ratio, and natural frequencies. This structure was placed on a 1 m-thick mat foundation. Structure B was representative of a 9-story structure, modelled as an equivalent 2-story structure owing to limitations on overhead space in the centrifuge. The structure was constructed on a stiff basement embedded to 3m depth. Important structural properties for Structures A and B are documented in Table 1. To enable non-linearity and damage, both structures incorporated fuses at the column-foundation intersection, and at column-beam joints. Fuses were instrumented with strain gauges to monitor demand on the superstructure during shaking.

Figure 1 shows the location of structures within each model. Models AA and AA$_{DR}$ each contained two Structure A's with a horizontal separation of 3m. Models BA and BA$_{DR}$ contained a Structure B located 3m South of Structure A. The subscript "DR" denotes models in which PVDs were placed around the perimeter of the Northmost Structure A. Save for the presence, or otherwise, of PVDs, Models AA and AA$_{DR}$, BA and BA$_{DR}$ were identical in terms of building placement and instrumentation. To differentiate between the structures in each model, further subscripts are employed. AA$_{DR,S}$ is the South structure from Model AA$_{DR}$, while BA$_A$ is the Structure A within Model BA.

In each model, the structures were instrumented with accelerometers, strain gauges, and displacement sensors (LVDTs) as shown in Figure 1. Beneath the structure and in the far-field (mid-way between the structures and the container wall), pore pressure transducers (PPTs) and accelerometers were installed in addition to vertical LVDTs at different depths.

Following dry preparation using the ASP each model was saturated with hydroxypropyl methylcellulose (HPMC) prepared to 70 cSt, to satisfy scaling laws at 70 g. Saturation was conducted using the automated

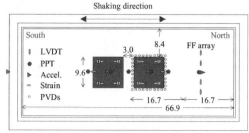

(a) AA$_{DR}$ & BA$_{DR}$ (plan view)

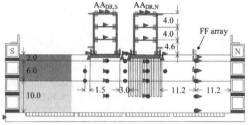

(b) AA$_{DR}$ (side view)

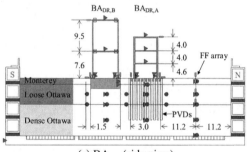

(c) BA$_{DR}$ (side view)

Figure 1. Model layout for tests AA$_{DR}$ and BA$_{DR}$. Tests AA & BA had identical layout, but without PVDs. (Dimensions in prototype meters.)

Table 1. Structure properties.

	Structure A	Structure B
Bearing pressure (kPa)	77	195
Modal frequencies (Hz)[1]	1.7; 6.7; 16.7	0.35; 1.41
Damping ratio (%)[2]	5	12
Footing dimensions (m^2)	9.56 × 9.56	9.56 × 9.56
Embedment (m)	1	3
Height/ Width (m/m)	1.8	2.5
Fuse yield moment (kNm)	1,470	655
Fuse plastic moment (kNm)	7,640	3,400

[1]Measured via hammer impact test under fixed base conditions.

facility at CUB which draws HPMC up from the base of the model under differential vacuum. Following saturation, the model was accelerated to 70 g and subjected to a series of 1D earthquake motions.

3 KEY EXPERIMENTAL RESULTS

The first significant motion (sufficiently intense to cause measurable settlement and a rise in excess pore pressures) applied to the base of each model was a scaled and filtered version of the accelerations recorded at the Takatori field station during the Kobe, 1995 earthquake in Japan. Figure 2 shows the variation in acceleration, Arias Intensity (I_a), and 5% damped spectral acceleration (S_a) from each test and the mean from across all tests. The acceleration variation in time and frequency domains was small. Therefore, it is reasonable to compare directly the performance of structures from each model.

3.1 Response in the far-field

During the Kobe motion, significant excess pore pressures were generated, and large far-field surface settlements were observed. These are presented alongside the base acceleration in Figure 3. The reader is advised to note the change in time scale at 40s.

The different models showed similar responses in terms surface settlement and excess pore pressure at mid-depth of the loose Ottawa sand layer. This demonstrates reasonable similarity in the effect of the structures on the far-field between tests.

3.2 Response of the structures

Structural settlement arises from combined volumetric and deviatoric soil strains beneath the structure (Dashti et al., 2010). Rotation develops in response to differences in the magnitude of soil strains across the width of the foundation, and the response of the structure to inertial moments.

To explain the relative importance of deviatoric and volumetric mechanisms of deformation on the settlement-rotation response of the structures, it helps to consider the driving stresses in the soil for each mechanism. Figures 4a, b and c show respectively, the vertical effective stress (σ'_z) beneath the structures, horizontal shear stresses (τ_{xz}) beneath the center of the structures in the direction of shaking, and horizontal shear stresses midway between the two structures perpendicular to the direction of shaking (τ_{yz}). The σ'_z was calculated based on the analytical solution for stresses in an elastic half space, while static shear stresses were calculated for elastic soil with Poisson's ratio of 0.3 using a 3D finite element simulation. The origin in Figure 4 is located at the centre of the container at the soil surface.

The static effective normal stresses are higher beneath BA models compared to AA models. This impacts the soil stiffness, which depends also on the

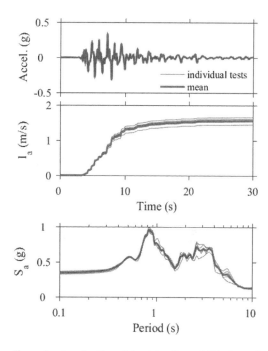

Figure 2. Acceleration, Arias intensity and 5% damped spectral acceleration at the container base.

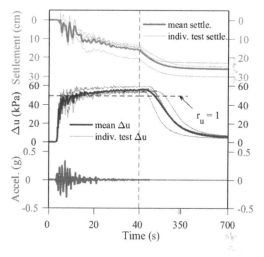

Figure 3. Surface settlement, excess pore pressure at mid-depth of the loose Ottawa sand, and base acceleration in the far-field from all tests.

excess pore pressure rise over time beneath the structures. It is later shown that this affected both the settlement-rotation response and the demand on the superstructure. Also, important to note, is non-uniform σ'_z beneath the structures. Figure 4a shows higher σ'_z between the structures (in all models) compared to beneath their outside edge. This implies that the soil between the structures will exhibit a stiffer response than soil beneath the outside edges of the structures. Such behavior would encourage rotation of adjacent structures away from one another.

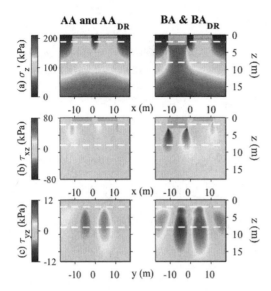

Figure 4. Static stresses beneath the structures in each model: (a) vertical effective stress, (b) horizontal shear stresses beneath the centreline of the structures in the direction of shaking, (c) horizontal shear stresses midway between the structures perpendicular to the shaking direction.

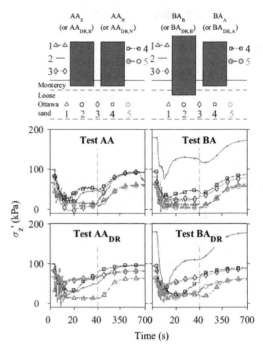

Figure 5. Vertical effective stress as a function of time at mid-depth within the loose Ottawa sand layer.

Similarly, shear stresses τ_{xz} shown in Figure 4b cancel through superposition between the structures. Therefore, there is a larger driving force (and reduced confinement) for the ejection of softened sediment from beneath the outside edge of each structure compared to the inside edge. Again, this encourages the structures to rotate away from one another. When PVDs are positioned around the North structure in the model, the stiffness of soil between the structures is likely to be higher than that beneath the outside edge of the unmitigated (South) structure owing to lower excess pore pressures, amplifying its rotation potential.

The tendency of adjacent structures to rotate away from one another is reduced by deviatoric strains driven by τ_{yz}, which is strongest between the structures compared to near the outside edge and therefore encourages the rotation of the structure towards one another. However, with enhanced drainage, the soil between the structures may not soften significantly and the importance of this mechanism in determining the net rotation of the structures may be reduced. This is investigated in Figure 5, which presents the temporal variation in vertical effective stress beneath the structures in each model at different locations at mid-depth of the loose Ottawa sand layer. The calculation of σ'_z over time assumes the initial vertical effective stress less the excess pore pressure measured at the indicated location. As such, changes in effective stress due to relative movement between the structure and PPT, or due to dynamic loading are not accounted for.

Consider the σ'_z beneath structures AA compared to structures AA_{DR}. Beneath the edges of the drained structure, $AA_{DR,N}$ (locations 3 and 5), the reduction in vertical effective stress is relatively small, σ'_z remains substantially greater than beneath the corresponding structure in Test AA (AA_N). The PVDs increased the average soil stiffness, thus reducing net volumetric strains (the reduction in sedimentation was greater than the increase in strain due to partial drainage). Deviatoric soil strains were also reduced by the PVDs, which by limiting the magnitude of excess pore pressures, preserved the shear strength of the soil. The net result, shown in Figure 6, was a reduction in the settlement of $AA_{DR,N}$. Similarly, the settlement of $BA_{DR,A}$ was substantially less than that of Structure BA_A for the same reasons.

The effect of enhanced drainage on an adjacent, unmitigated structure is a cause for major concern. Compare the σ'_z response beneath structures AA_S, without PVDs, and Structure $AA_{DR,S}$ which had PVDs placed around the adjacent structure. The vertical effective stress beneath $AA_{DR,S}$ at location 3 (visible as blue diamonds, partially hidden behind the near identical response at location 5) was substantially higher than at location 1 (beneath the South edge of the structure). By comparison, the σ'_z at locations 1 and 3 beneath Structure AA_S were reasonably similar. The difference in the soil stiffness owing to the presence of PVDs around an adjacent structure had a major impact on the settlement-rotation response. Large differential soil stiffness beneath $AA_{DR,S}$ resulted in differential volumetric soil strains within the foundation soil. This promoted counter-clockwise rotation of the structure (away from the neighbouring, drained structure). Further, high shear strength reduced deviatoric soil strains

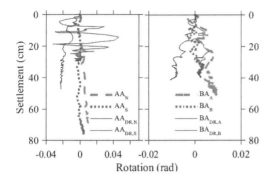

Figure 6. Settlement-rotation response for all structures.

(under the action of both τ_{xz} and τ_{yz}) from between structures AA$_{DR}$. Meanwhile, larger deviatoric soil strains occurred in the softer soil at location 1. The net result was a significant increase in permanent rotation owing to the PVDs around the neighbouring structure.

Similar behaviour was experienced by Structure BA$_{DR,B}$ compared to Structure BA$_B$. For any two neighbouring structures where one is mitigated by soil drains, and the region of influence of the drains extends beneath the neighbouring structure, the rotation of the unmitigated structure is likely to be more severe (in the event of significant soil softening beneath the unmitigated structure) than a similar case without soil drains. For this reason, in dense urban environments, the design of soil drainage for liquefaction mitigation should be conducted in a holistic manner that considers the response of all structures.

A final important point to draw from Figure 6 is the difference in SSSI effects on Structure A. When located next to an identical structure A, the final settlement was 50% greater than when next to Structure B with greater confinement and embedment. However, the rotation was similar. The principal cause of reduction in settlement is thought to be increased soil stiffness (due to higher σ'_z as shown in Figure 4). However, further research is required to show if this behaviour is repeated in all similar scenarios.

3.3 Demand on the superstructures

In the previous section, it was shown that PVDs located around a structure reduced liquefaction induced settlements, but at the expense of increased permanent rotation of the neighbouring structure. One of the causes of reduced settlement was increased soil stiffness resulting from a reduction in net excess pore pressures. A side effect of increased soil stiffness is decreased damping, resulting in increased seismic energy incident to the structures.

Figure 7 shows, as a measure of the seismic demand on the superstructure, the maximum and minimum strains recorded on structural fuses at each level of the different structures. Each strain gauge is represented by a single point, while the mean maximum and minimum strains are represented by lines connecting each story.

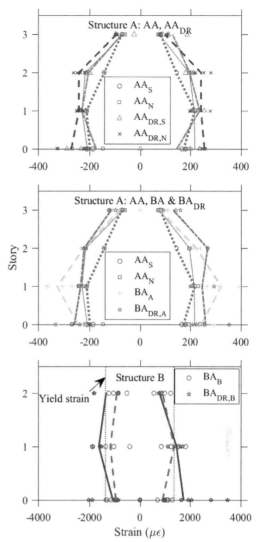

Figure 7. Maximum and minimum strains recorded on each structural fuse.

For Structure B, drains paced around the adjacent structure caused significant increases in the measured strains resulting in widespread plastic behaviour. The increase in strains measured on BA$_{DR,B}$ was largely due to P-Δ effects arising from increased rotation (see Fig. 6).

Similarly, the strains measured on both structures in Model AA$_{DR}$ were greater than those for the corresponding structures in Model AA. This shows that enhancing drainage increased the structural damage potential for both the mitigated and the adjacent structure.

However, similar trends were not observed for structures BA$_A$ and BA$_{DR,A}$. The strain on the top two stories of BA$_{DR,A}$ increased owing to enhanced drainage, but strains on the lower two stories were little affected or reduced.

4 CONCLUDING REMARKS

Four centrifuge tests were conducted on adjacent structures founded on a layered liquefiable deposit. Two tests compared the response of equivalent 3-story structures with and without PVDs around one of the structures. A similar comparison was presented for adjacent structures of 3 and 9 stories both with and without PVDs around the 3-story structure. In all cases, the settlement and permanent rotation of the mitigated structure was reduced. However, the adjacent structure paid a heavy cost. Permanent rotation of the adjacent unmitigated structure increased dramatically owing to the use of PVDs around its neighbour. The demand on the superstructure was also observed to increase with the inclusion of PVDs around a neighbouring structure. For the case of $BA_{DR,B}$, this led to widespread plastic behaviour.

This study illustrates the need for holistic design of soil drains. In dense urban environments, it is not acceptable to design mitigation for a single structure without considering the potential consequences for neighbouring structures.

ACKNOWLEDGEMENTS

This research was supported by the U.S. National Science Foundation (NSF) through grant number 145431. The opinions or findings presented in this paper are those of the authors and do not necessarily reflect the views of the NSF.

REFERENCES

Bertalot, Brennan & Villalobos. 2013. Influence of bearing pressure on liquefaction-induced settlement of shallow foundations. *Géotechnique*, 63(5): 391–399.

Cubrinovski, M. 2013. Liquefaction-Induced Damage in the 2010-2011 Christchurch (New Zealand) Earthquakes, *Proc. 7th Int. Conf. Case Histories in Geotech. Eng.* Chicago.

Dashti, Bray, Pestana, Riemer & Wilson. 2010. Mechanisms of Seismically Induced Settlement of Buildings with Shallow Foundations on Liquefiable Soil. *J. Geotech. & Geoenv. Eng.* 136(1): 151–164.

Hausler and Sitar. 2001. Performance of Soil Improvement Techniques in Earthquakes. *4th Int. Conf. on Recent Advances in Geotech. Earthquake Eng. and Soil Dyn.* Rolla.

Hayden, Zupan, Bray, Allmond, & Kutter. 2015. 'Centrifuge Tests of Adjacent Mat-Supported Buildings Affected by Liquefaction', *J. Geotech. & Geoenv. Eng.*, 141(3).

Kirkwood, P. and Dashti, S. 2017a (in press). A Centrifuge Study of Seismic Structure-Soil-Structure Interaction on Liquefiable Ground and Implications for Design in Dense Urban Area. *Earthquake Spectra*.

Kirkwood & Dashti. 2017b (under review). Influence of Prefabricated Vertical Drains on the Seismic Performance of Similar Neighboring Structures Founded on Liquefiable Deposits. *Géotechnique*.

Kirkwood & Dashti. 2017c (under review). Considerations for the Mitigation of Earthquake-Induced Soil Liquefaction in Urban Environments. *J. Geotech. & Geoenv. Eng.*

Olarte, Paramasivam, Dashti & Liel. 2017. Centrifuge Modeling of Mitigation-Soil-Foundation-Structure Interaction on Liquefiable Ground. *Soil Dyn. & Eq Eng.* 97: 304–323.

Paramasivam, Dashti & Liel. 2017 (under review). Influence of Prefabricated Vertical Drains on the Seismic Performance of Inelastic Structures Founded on Liquefiable Soils. *J. Geotech. & Geoenv. Eng.*

Influence of tamper shape on dynamic compaction of granular soil

S. Kundu & B.V.S. Viswanadham
Department of Civil Engineering, Indian Institute of Technology Bombay, India

ABSTRACT: Dynamic compaction (DC) has gained significant popularity in the last few decades as an effective ground improvement technique for densification of loose granular soils. The objective of the paper is to examine the influence of tamper geometry on the response of underlying deposits for improving the efficiency of DC process in the field. An experimental set-up is designed and developed for the above purpose, and the advantages of the actuator over existing test-setups are discussed. Physical model tests are subsequently carried out with the developed actuator at earth's gravity on loose dry sandy soil using circular and conical based tampers. The degree and depth of improvement due to DC is assessed in each case using Geo-PIV on images captured during experimentation. The pattern of propagation of energy waves though the soil mass plotted in terms of displacement vectors and contours varied considerably depending on tamper shape, and for a given drop height and tamper mass, the use of conical tamper resulted in larger extent of influence zone (of the order of 16%) as compared to flat based tamper. Further, the volume of crater induced by conical tamper was observed to be 18% more than that of conventional flat based tamper.

1 INTRODUCTION

In recent years, dynamic compaction (DC) has evolved as an economically viable method of ground remediation for loose cohesionless soils. The technique involves repeated dropping of a heavy tamper mass from a determined height on an impact surface, wherein the underlying soil gets compacted by virtue of transfer of kinetic energy. The method is widely employed in various civil engineering projects, including design of foundations on poor subgrades, construction of highways, airports and densification of landfills. In principle, efficacy of DC in the field necessitates controlled transmissions of dynamic stresses to the ground surface, induced by the impact of steel or concrete tampers. The transmitted energy propagates in the form of shear waves and compression waves into the underlying soil strata, thereby rearranging the soil particles into a denser state. Hence, the shape of tamper mass inducing the energy waves has a considerable influence on the improvement depth and crater geometry obtained after DC.

To date, there is no discrete information reported in literature regarding the relative contribution of compression waves, shear waves and Rayleigh waves as possible constituents of energy generated during DC. However, as conventional tampers used in the field possess a flat base similar to shallow foundation, it may be assumed that the partition of dynamic compaction energy is similar to that experienced by shallow foundations subjected to steady-state energy waves [Van Impe (1989)]. Thus, based on the observations of Miller and Pursey (1955) related to shallow foundations undergoing steady-state vibrations, it can be concluded that nearly 67%, 27% and 6% of the generated energy is transmitted by Rayleigh waves, compression waves and shear waves respectively. The numbers hereby suggest that only a small fraction of the total compaction energy is transmitted through shear waves, which is predominantly responsible for densification during DC in view of shear displacements and associated re-arrangement of soil particles [Feng et al. (2000)]. A possibility of increasing the shear wave energy generated during DC is by modifying the base of tamper mass used in the process.

Laboratory DC tests using traditional flat based-tampers were reported by researchers like Orrje & Broms (1970), Ellis (1986) and Poran & Rodriguez (1992). Further, previous studies related to variable tamper geometry in connection with DC have been based on displaced volume at the surface, expressed in terms of crater depth and associated heave [Mullins et al. (2000), Feng et al. (2000), Arslan et al. (2007) and Ghazavi & Niazipour (2010)]. These studies did not address the densification at depth, which becomes crucial while designing foundations in problematic soils. In view of the above mentioned research gaps, an attempt has been made in the present paper to investigate the influence of tamper shape on the response of granular soil subjected to DC. Laboratory tests were conducted in this regard on sand deposit using a conical tamper along with a conventional circular flat-based tamper. Details of the model test package developed, materials and methodology involved are discussed in subsequent sections.

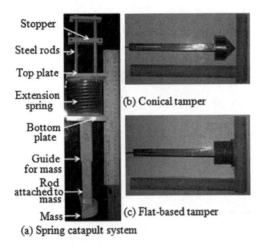

Figure 1. Components of developed DC set-up.

2 DETAILS OF MODEL TEST PACKAGE

2.1 Description of set-up and tampers used

Earlier researchers like Hajialilue-Bonab and Rezaei (2009) and Bonab and Zare (2014) have simulated low energy dynamic compaction in the laboratory by making circular holes on the gravity centre of each quadrant of the tampers, and inserting cylindrical rods passing through them into the underlying soil. The embedded steel rods impart additional strength to the soil mass, on account of which, the actual gain in strength due to DC cannot be compared with that observed in the field. Secondly, these set-ups are designed for dropping the tamper by considerable drop heights of the magnitude of (2–3) m, thereby making the overall set-up costly and cumbersome, and impossible to use where restrictions in height exist.

In order to overcome these drawbacks, a metallic spring was introduced in the present set-up [as shown in Fig. 1(a)] to contribute a major share to the overall energy required to be imparted to the soil surface by virtue of its stiffness. The additional potential energy of the spring contributes a major amount to the total energy required to be transmitted to the soil mass. This, in turn, reduces the height of fall of the tamper, making the developed set-up compact and robust. Detailed description of the DC simulator developed, and its advantages over existing set-ups are discussed in Kundu and Viswanadham (2015).

The conical and flat-based tampers made of mild steel used in the present study are presented in Fig. 1(b)-1(c) respectively. Considering the front plane (perspex sheet) of the model container as the plane of symmetry, half of the tampers were modelled in this case. For the sake of comparison, the mass, diameter and height of fall of the tampers were kept identical as 1.75 kg [0.875*2 kg], 0.075 m and 1 m respectively during the experiments. A cone angle of 90° was adopted in the present case, in view of the findings reported by Feng et al. (2000), wherein, a 60° conical tamper was found to penetrate excessively into the soil sample, whereas, higher angles of 120° resulted in comparatively smaller craters.

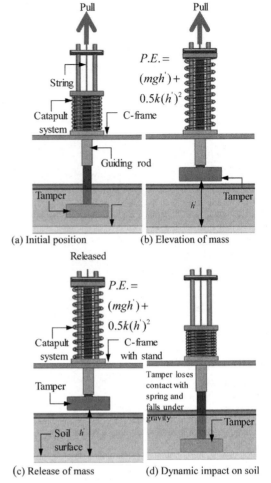

Figure 2. Working mechanism of actuator.

2.2 Working mechanism

Figures 2(a)-2(d) depict the working mechanism of the developed DC test set-up for performing 1g small-scale model tests. As shown in Fig. 2(a), a steel string is attached at the end of the rod used for holding the tamper. As the string is pulled up by a distance h' vertically, the rod touches the movable top plate of the spring and pushes it upwards. This causes the spring to extend by the same distance h', which the tamper traverses as shown in Fig. 2(b). After attaining h', the steel wire is released [refer Fig. 2(c)], causing the tamper along with square hollow rod to fall under the combined force of gravity (mgh') and with the force induced by the extended spring ($0.5kh'^2$). Just before impact, the mass losses contact with the spring, falls under gravity and hit the soil surface as shown in Fig. 2 (d). The tamper thus derives a small fraction of its total required energy from gravitational potential energy (mgh'), whereas, the major fraction is derived from the potential energy stored in the spring ($0.5kh'^2$), thereby reducing the actual drop height to h' in the process.

2.3 Soil properties

The sand used in the present study was found to completely pass through BS 36 sieve (0.425 mm) and retained in BS 200 sieve (0.075 mm). The soil is classified as SP according to Unified Soil Classification System (USCS) [$d_{10} = 0.101$ mm; $d_{50} = 0.191$ mm, $C_u = 2.065$ and $C_c = 1.117$]. The specific gravity was 2.654, whereas, the maximum and minimum void ratio were 0.94 and 0.63 respectively. The unit weight of the soil was reported to be 14 kN/m^3 at a relative density (RD) of 35%. In addition, shear strength parameters were determined for the model soil by conducting CU triaxial tests at 35% RD, and were found to be 0 kPa (c') and 32° (ϕ) respectively. The model soil chosen in the study thus represents the typical characteristics of loose granular deposit commonly encountered in the field on which DC technique is applied.

(a) Conical tamper (b) Flat-based tamper

Figure 3. Front-view of deformed soil with conical and circular tamper [At the end of 15th impact].

(a) Conical tamper (b) Flat-based tamper

Figure 4. Extent of crater formation after 15th impact.

3 TEST PROCEDURE

Low-energy DC tests were conducted in the laboratory in a model container having 720 mm length, 450 mm breadth and 410 mm height internally. The front wall of the container consists of a thick Perspex sheet for enabling proper view of the front elevation of soil model during testing. Low energy compaction process was chosen in this case in order to overcome the problems related to instrumentation and data acquisition involved in capturing high-energy impact during laboratory 1 g testing. An air pluviation technique was adopted for preparing loose dry sand specimens with an initial thickness of 320 mm. The flow rate and drop height of sand particles were carefully adjusted in order to ensure uniformity in density at the time of model preparation. In the beginning, pre-weighed sand quantities were poured into a conical container having an adjustable valve at the base. At a preset position, the valve was partially opened, thereby allowing the sand particles to flow through the open area freely under gravity. A constant drop height was maintained by lifting the conical container at regular intervals corresponding to the rate of increase in the thickness of the sand specimen.

Model tests were conducted by raising and dropping the circular and conical steel tampers 15 times on the surface of the sand deposit in each test. Selection of 15 drops was made in accordance with the observations during preliminary tests, where no substantial increase in crater depth was observed beyond this point. After each impact on the soil surface, a digital image was captured of the deformed soil using a Canon Powershot digital camera having an image resolution of 16 megapixels. In addition, the front elevation of the model was illuminated with the help of LED light assembly placed uniformly along the periphery of the container to enable clear visualization of the model during successive impacts.

4 RESULTS AND DISCUSSION

Figure 3 presents the front view of the deformed soil surface captured at the final stage of tamping for conical [Fig. 3(a)] and flat-based tampers [Fig. 3(c)]. The vertical and radial extent of the crater formed in each case as captured during post-investigation are shown in Figs. 4(a)-4(b), which depict a larger zone of influence in case of conical tampers.

Using GeoPIV software [White et al. (2003)], displacement vectors were obtained from the images captured during tests from the plane of the perspex sheet, corresponding to various stages of tamping. The images were processed as per the procedure outlined in White et al. (2003) in order to generate displacement vector fields for both conical and flat-based tampers, as shown in Figs. 5(a)-5(b) respectively. During analysis, irregular wild vectors arising due to excessive soil deformations were encountered. The resultant vectors in image space were subsequently converted to object space measurements and plotted against normalized depth (Y/D) and normalized distance from tamper centre (X/D) [D: tamper diameter]. As can be seen from Fig. 5, irrespective of the tamper geometry, the soil mass beneath the tamper gets compressed due to dynamic impact, and is displaced sideways and down the depth of soil stratum. In either case, no heaving of soil was observed above the ground surface. Further, it is evident from Figs. 5(a)-5(b), that the pattern of propagation of energy waves though the soil mass varied considerably depending on tamper shape, and all parameters remaining same, the conical tamper could

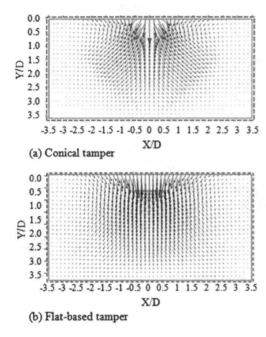

(a) Conical tamper

(b) Flat-based tamper

Figure 5. Displacement vectors obtained by Geo-PIV [Vector magnification = 1.0].

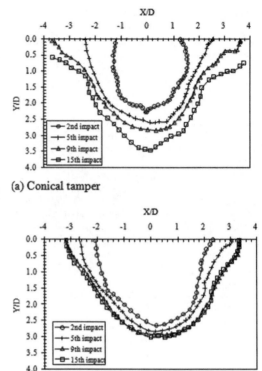

(a) Conical tamper

(b) Flat-based tamper

Figure 6. Displacement contours at different drops.

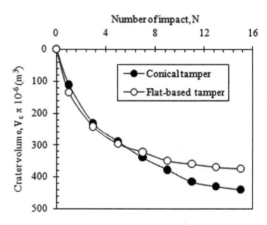

Figure 7. Crater volume induced by various tampers.

penetrate further and affected a greater quantum of the soil than the flat tamper.

The displacement vectors shown in Fig. 5 were subsequently used for calculating displacement contours using the method described in Hajialilue-Bonab & Rezaei (2009). Figures 6(a)-6(b) present the contours considering a displacement of 0.02 D (1.5 mm) as the influence boundary plotted against normalized depth (Y/D) and normalized distance from tamper centre (X/D) for different impacts for both conical and flat-based tampers. As visible from Figs. 6(a)-6(b), the movement of the soil in horizontal and vertical direction was initially more for the flat based tamper, up to almost the 5th impact. However, the final depth of improvement after 15th impact was higher by 16% for the conical tamper, and the influence zone affected measured about 3.5D in vertical extent and 4D in horizontal extent in contrast to flat-based tampers, wherein the vertical and lateral boundaries of influence zone were restricted to 3D in either case. This may be attributed to higher degree of shearing of soil particles in the case of conical tampers, resulting in greater improvement with DC. Figure 6 thus gives an understanding about the densification at depth, which has not been addressed before with tamper shape.

An alternative way of expressing the efficiency of DC is by quantifying the crater volume (V_c) formed due to impact. It is basically the difference between the depressed soil volume and the heave volume. As no heave was observed in this case, the soil volume displaced due to DC effectively represents the volume of crater formed. The V_c values for both the tampers is presented in Fig. 7, which highlights the fact that initially, up to the 5th impact, the flat-based tamper induced slightly higher crater volume as compared to the conical tamper, which is in line with earlier observations made with respect to displacement contours shown in Fig. 6. However, the final crater volume induced by the conical tamper is 18% higher than that induced by the conventional flat-based circular tamper. This becomes important and useful in the case of densification of municipal solid waste (MSW) landfills, where use of conical tampers are recommended for reducing the volume of wastes accumulated with

Table 1. Summary of test results.

Parameter	T1	T2
Tamper shape	Conical	Flat-based
Vertical influence, Y/D	3.5D	3.0D
Radial influence, X/D	4.0D	3.0D
Crater volume, V_c/D^3	1.04	0.90

Note: Tamper mass, diameter and height of fall were kept identical as 1.75 kg, 0.075 m and 1 m respectively in tests T1 and T2.

time. The results hereby obtained for both the tampers are summarized in Table 1.

5 CONCLUSIONS

In the present paper, an attempt has been made to evaluate the influence of tamper shape on the efficiency of dynamic compaction process. Two types of tampers, namely, a conventional flat based tamper and a conical tamper (less frequently adopted in the field) were used to conduct laboratory investigations related to low-energy DC process. An innovative DC simulator was developed and used in the present study having a specialized spring-catapult system, by virtue of which, it possesses distinct advantages over set-ups developed by earlier researchers. The mass, radius and height of fall were kept constant in each of the two tampers used, and a total of 15 blows were imparted to the soil surface in each case, beyond which, no significant improvement was monitored. The images captured at various stages of tamping were processed by Geo-PIV software, and the displacement vectors and contours of 0.02 D (1.5 mm) were plotted to get an understanding of the improvement due to DC with depth of soil stratum.

The Geo-PIV analysis results indicated that the pattern of propagation of energy waves though the soil mass varied considerably depending on tamper shape, and all parameters remaining constant, the conical tamper penetrated further and affected a greater quantity of soil than the flat-based tamper. The displacement contours revealed that the movement of soil in horizontal and vertical direction was initially more for the flat based tamper, although the final depth of improvement was higher by 16% for the conical tamper. Further, the zone of influence affected by the conical tamper measured about 3.5D in vertical extent and 4D in horizontal extent in contrast to flat-based tampers, wherein the vertical and lateral boundaries of influence zone were restricted to 3D in either case, thereby highlighting the efficiency of conical tampers over flat-based tampers. In addition, the crater volume formed due to impact was quantified in both the cases, and the volume induced by the conical tamper was observed to be 18% higher than that induced by the conventional flat-based circular tamper. The study suggests the advantage of using conical tampers over circular flat-based ones for improving the over-all efficiency of DC process. However, the effect of varying tamper shapes on adjoining compaction sites can be investigated for an enhanced understanding of the results. In addition, further parametric studies need to be conducted, and the effect of variable soil type and in-situ moisture content of the soil needs to be addressed before the results can be generalized.

REFERENCES

Arslan, H., Baykal, G. & Ertas, O., 2007, Influence of tamper weight shape on dynamic compaction, *Ground Improvement*, 11(2), 61–66.

Bonab, M.H. & Zare, F.S., 2014, Investigation on tamping spacing in dynamic compaction using model tests, *Ground Improvement*, 167(3), 219–231.

Ellis, G. W. 1986, Dynamic consolidation of fly ash, *Proceedings of International Symposium on Environmental Geotechnology*, pp. 564–573.

Feng, T.W., Chen, K.H., Su, Y.T. & Shi, Y.C., 2000, Laboratory investigation of the efficiency of conical based tampers for dynamic compaction, *Geotechnique*, 50(2), 667–674.

Ghazavi, M. & Niazipour, M., 2010, An experimental setup for the investigation of tamper geometry effects, *Proceedings of the 7th International Conference on Physical Modelling in Geotechnics, Zurich*, S. Springman, J. Laue and L. Seward (Eds.). A.A. Balkema (Pub.), Vol. 1, pp. 235–238.

Hajialilue-Bonab, M. & Rezaei, A.H., 2009, Physical Modelling of low energy dynamic compaction, *International Journal of Physical Modelling in Geotechnics*, 9(3), 21–32.

Kundu, S. & Viswanadham, B.V.S., 2015, Studies to evaluate the impact of tamper on the depth of improvement in dynamic compaction, *Japanese Geotechnical Society Special Publication*, 2(59), 2033–2037.

Miller, G. F. & Pursey, H., 1955, On the partition of energy between elastic waves in a semi-infinite solid, *Proceedings of the Royal Society*, London, pp. 233, 55–69.

Mullins, G., Gunaratne, M., Stinnette, P. & Thilakasiri, S., 2000, Prediction of Dynamic Compaction Tamper Penetration, *Soils and Foundations*, Japanese Geotechnical Society, 40(5), 91–97.

Orrje, O. & Broms, B., 1970, "Strength and deformation properties of soils as determined by a falling weight", *Swedish Geotechnical Institute Proceedings*, No. 23, pp. 1–25.

Poran, C. J. & Rodriguez, J. A., 1992, "Design of dynamic compaction", *Canadian Geotechnical Journal*, 29(5), 796–802.

Van Impe, W. F. 1989, "Soil improvement techniques and their evolution", *Rotterdam: Balkema*.

White, D.J., Take, W.A., and Bolton, M.D., 2003, "Soil deformation measurement using particle image velocimetry (PIV) and photogrammetry", *Geotechnique*, 53(7), 619–631.

Behaviour of geogrid reinforced soil walls with marginal backfills with and without chimney drain in a geotechnical centrifuge

J. Mamaghanian & H.R. Razeghi
School of Civil Engineering, Iran University of Science and Technology, Tehran, Iran

B.V.S. Viswanadham & C.H.S.G. Manikumar
Department of Civil Engineering, IIT Bombay, Mumbai, India

ABSTRACT: This paper is aimed to study the effect of chimney drain on the performance of geogrid reinforced soil walls with marginal backfills. Three centrifuge model tests were performed using a 4.5 m radius beam centrifuge facility available at IIT Bombay at 40 g. All models were subjected to seepage simulating rising ground water condition. Marginal soil type, the wall height, reinforcement length, and spacing and facing type was kept constant. Two types of chimney drain were used in this study including chimney nonwoven geotextile drain and chimney sand drain. Pore water pressure and displacement transducers were used to monitor the wall behaviour during centrifuge tests. Moreover, image analysis technique was used to obtain the displacement field of the models. The geogrid reinforced soil wall without chimney drain experienced global stability failure soon after inducing seepage. Application of chimney geotextile layer could not improve the wall behaviour but caused a delay on the occurrence of global stability failure. On the other hand, the application of chimney sand drain improved the wall behaviour effectively.

1 INTRODUCTION

The conventional design guidelines of Mechanically Stabilized Earth (MSE) walls have stringent recommendations in selecting backfill materials. FHWA (Elias, 2009) limits backfill with maximum fines of 15% and plasticity index of 6% for MSE walls. Therefore this type of backfill generally accounts 50–75% of the total construction costs (Koerner & Koerner 2011). The use of soils with relatively high percentage of fines, here termed as marginal soil, can save 20–30% of the total wall cost (Christopher & Stuglis 2005).

However, it should be noticed that the most failures in reinforced soil structures were mainly due to the use of marginal soil (Mitchell & Zornberg 1995; Yoo & Jun, 2006; Koerner & Koerner 2013). Therefore, several concerns have be taken into account in the use of marginal soils as backfill. First, the low permeability of marginal soils results in generation of pore water pressure due to infiltration. Second, wetting of the soil can lead to decrease in soil stiffness and strength as well as decrease in interface strength of soil-reinforcement and third, wetting of the soil intensifies its creep deformations.

Therefore, attempts has been made to find viable solutions to alleviate problems imposed by substandard backfills. The main conclusion drawn from the literature is related to the provision of adequate external or internal drainage systems. The use of nonwoven geotextiles in dissipating pore water pressures generated in steep clay slopes have been reported by Tatsuoka & Yamauchi (1986). Provision of chimney drain out of reinforced soil zone is a good example of external drainage system, suggested by NCMA (2009) in the case of marginal soils with fines up to 35%. A distinct example of internal drainage system is the use of geocomposite layers with dual function of reinforcement and drainage in the reinforced soil zone (Raisinghani & Viswanadham 2011; Kang et al. 2014).

The lack of proper drainage system is the main reason in most of reinforced soil wall failures with marginal backfill. Moreover, the knowledge behind the effect of proper external drainage is limited. Therefore, this paper was primarily aimed to study and compare the effects of chimney nonwoven geotextile drain and chimney sand drain systems on the geogrid reinforced soil wall behaviour. In this paper the deformation behaviour of three geogrid reinforced soil walls through centrifuge model tests at 40 g are presented. The backfill soil was taken as marginal soil and the wall facing was selected as precast panel facing which is one of common facing types in the practice. In order to simulate the rising water condition due to heavy rainfall the centrifuge models were subjected to seepage condition. The marginal soil type, the wall geometry and the layout of reinforcements were kept constant in this research. The research does not include the effect of soil and reinforcement creep behaviour on the long-term performance of geosynthetic reinforced soil walls.

2 SCALING CONSIDERATIONS

In centrifuge modelling of geogrid reinforced soil walls, the dimensions of wall geometry and layout of geogrid layers are scaled down by a factor N. Then centrifuge acceleration is taken as N times of the gravity acceleration (g) to reach identical stress field in both model and prototype. Moreover, the scaling consideration of the properties of soil, geogrid and facing panels as well as the interaction between geogrid-soil, soil-facing and facing-geogridshould be taken into account. In order to simulate seepage condition in a geotechnical centrifuge, the dimensional analysis of seepage flow, permeability of backfill and transmissivity of geotextile drain layer are required. Details of these scaling considerations were discussed by Raisinghani & Viswanadham (2011) and Viswanadham et al. (2017).

3 MODEL MATERIALS

3.1 Model soil

The model marginal soil in this study was silty sand and classified as SMaccording to the USCS (Unified Soil Classification System). This model soil comprised of locally available fine sand and commercially kaolin in the ratio of 4:1 by dry weight. The soil has a liquid limit of 18.4% and is non-plastic. Maximum dry unit weight ($\gamma_{d, max}$) of the soil is 19.42 kN/m^3 at an optimum moisture content (OMC) of 9% (standard Proctor compaction test). Coefficient of permeability of the model soil compacted at its $\gamma_{d, max}$ and OMC was measured as 1.25×10^{-6} m/s from falling head test. The effective shear strength parameters of 5 kPa and 30.9° were obtained for cohesion and internal friction of the soil respectively using CU triaxial compression test.

3.2 Model geosynthetics

Two types of geosynthetic materials were used in the present study including geogrid (GG) as reinforcement and nonwoven geotextile (NGT) as chimney geotextile drain. These materials were modeled based on the scaling consideration of geosynthetic reinforced soil walls in centrifuge under seepage condition. A fine model geogrid material was selected to meet both tensile stiffness and soil-geogrid interaction requirements (Viswanadham & König 2004). Wide width tensile strength tests were performed on model geogrid according to ASTM-D4595 (2011) at a strain rate of 10 mm/min to get their tensile load-strain behaviour. An ultimate tensile load of 1.01 kN/m at an ultimate strain of 51.6% was found for geogrid material. The secant stiffness of geogrid material up to 5% strain was 6.42 kN/m. The transmissivity of nonwoven geotextile material was determined based on the procedure outlined in ASTM-D6574 (2013). Accordingly, a transmissivity of 28.2×10^{-6} m^2/s was measured at normal stress of 20 kPa for geotextile layer. In this study, the selection of model geogrid was such a way to induce failure in geogrid reinforced soil wall without chimney drain at the onset of seepage.

3.3 Model facing

Precast panel concrete was selected for wall facing in this study. In order to select a proper material to model this type of facing in a geotechnical centrifuge flexural rigidity scaling regarding to the effect of thickness and elastic modulus of facing elements should be considered simultaneously. Moreover the facing material should have lightweight and low water absorption. Therefore marine plywood sheets was found as a proper material to serve these requirements to some extent. In model dimensions, thickness of facing panel was taken as 10 mm. Accordingly, for making facing panels three layers of marine plywood with 3 mm thickness were stuck together with the help of a special wood glue. The details of facing panel scaling as well as its preparation and installation were discussed by Viswanadham et al. (2017).

The test set-up consisted primarily of six panels constituting the facing element, connecting six geogrid layers at a vertical spacing of 40 mm (in model scale). To ensure proper contact between facing panels and the backfill, a very thin layer of grade II sand was spread on the backside of each panel and held in place with the help of epoxy resin. Further, a special geogrid-facing anchorage system was custom fabricated by employing a hinged-type connection as outlined in Viswanadham et al. (2017) for ensuring adequate load transfer between the geogrid layers and the panel facing. The model soil wall was constructed in six layers of 40 mm thickness (in model scale) by compacting the soil at the respective $\gamma_{d, max}$ and OMC as evaluated during laboratory tests. During model preparation, adequate care was taken to ensure simulation of plane strain conditions by minimizing interface friction, on account of which, the sides of the container in contact with the soil wall were covered with thin flexible polyethylene sheet strips of 100 mm width lubricated with a thin layer of white petroleum grease.

4 CENTRIFUGE TEST SETUP AND TESTING PROGRAM

Centrifuge model tests were performed using 4.5 m radius large beam centrifuge available at Indian Institute of Technology Bombay (IIT Bombay), India. Figure 1 shows schematic of model test package used for modelling of geosynthetic reinforced soil walls at 40 g. Digital image analysis technique was used to find the displacement field of the wall models with the help of permanent and discrete markers (Figure 1). A special seepage simulator setup was used in this study to induce raising ground water condition in the wall models (Raisinghani & Viswanadham 2011). Three Linear Variable Differential Transformers (LVDTs) were used at the top surface of the wall models to get surface settlement profile (Figure 2a). Moreover four miniature

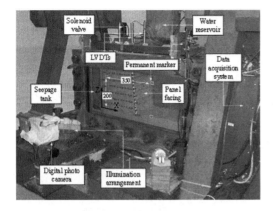

Figure 1. Perspective view of a centrifuge model on swinging basket (Model M2; Dimensions are in mm).

Pore water Pressure Transducers (PPTs) were used to depict the flow of water inside the reinforced soil wall (Figure 2a).

In this paper, the results of three centrifuge tests were reported and discussed with an aim to study the effect of chimney nonwoven geotextile drain and chimney sand drain on the performance of geogrid reinforced soil walls with marginal backfills. The wall height (H = 250 mm), facing inclination ($\beta = 84°$ with horizontal), reinforcement length ($L_r/H = 0.8$) and spacing ($S_v/H = 0.160$) was kept constant in all models. The facing and marginal soil type was also the same in all models. All centrifuge models were subjected to 40 gravity loading. This gravity level was selected based on primary experiences in order to impose negligible deformation on the wall models before applying seepage.

Considering Figure 2, model M1 was reinforced with six layers of geogrid (GG) without chimney drain. In model M2 six layers of geogrid (GG) were used and a chimney nonwoven geotextile drain was also provided. In model M3, chimney sand drain was applied and other parameters were kept constant. The thickness of chimney sand drain was 15 mm (600 mm in prototype dimension) and grade II quartz sand was used to construct it. This sand has particle size ranging 0.4–1 mm and coefficient of permeability of 1.54×10^{-4} m/s (constant head test on sand at 85% relative density). Further detailed description of the construction procedure adopted were discussed by Viswanadham et al. (2017).

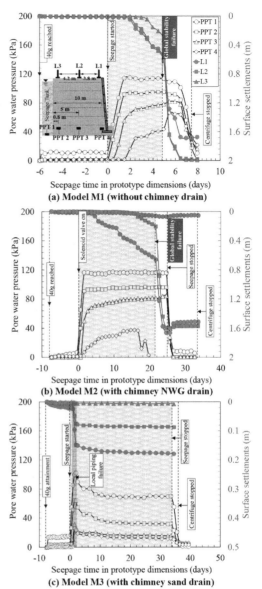

Figure 2. Measured variation of pore water pressure within the backfill and top surface settlements during centrifuge test at 40 g.

5 RESULTS AND DISCUSSION

5.1 Pore water pressure and surface settlements

The process of pore water pressure development within the reinforced soil wall for models M1, M2 and M3 with seepage time in days (in prototype dimensions) are shown in Figures 3a-c respectively. The inset window placed in Figure 3a shows, PPT1 was placed inside the seepage tank, while PPT2, PPT3 and PPT4 were placed horizontally at 0.8 m, 5 m and 10 m (corresponding prototype dimensions at 40 g) distance from the left soil boundary of the model. Figure 3 covers the time when 40 g attained to the end of the test (when centrifuge stopped). The time zero day indicates the onset of seepage and before that the value of time is negative, which indicates the behaviour of model before the onset of seepage at 40 g. The duration of seepage was highlighted with grey background. Moreover Figures 2a-2c illustrate the measured top surface settlements with seepage time in days as a result of pore water pressure development within the model walls. The variation of measured settlements with seepage

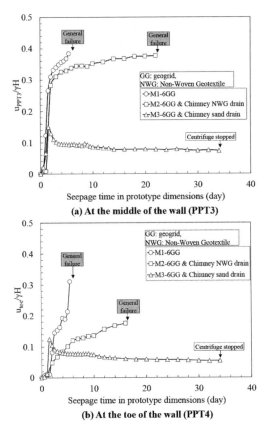

Figure 3. Comparison of normalized pore water pressure variations in the wall models at the onset of seepage.

time is also illustrated in Figures 2a-2c. As shown at the inset window of Figure 2a, LVDT L1 was placed at the crest of the wall, while LVDTs L2 and L3 were put at 3.8 m and 8 m distance far from the crest.

At first sight, it can be clearly seen that the surface settlements increases with seepage time, as a result of pore water pressure development. Model M1 without any chimney drain, experienced catastrophic failure after 5.07 days of seepage (Figure 2a). A maximum pore water pressure of 78 kPa and 43 kPa was recorded by PPT3 and PPT4 respectively at the onset of failure. The failure was proceeded with the formation of five tension cracks during various stages of seepage. The development of high pore water pressures within the soil in the absence of any drainage system was the main reason for the failure.

In model M2, the application of nonwoven geotextile as chimney drain did not improve the wall behaviour effectively and high pore water pressures were measured through the reinforced soil wall (Figure 2b). Nevertheless, in comparison with model M1 a delay was found in pore water pressure development. Therefore, catastrophic failure was seen after 21.7 days of seepage with 16.63 days delay compared with model M1. Model M3 with application of chimney sand drain experienced the very low values of pore water pressure indicating the effectiveness of chimney sand drain. Maximum pore water pressures of 17 kPa and 13 kPa were recorded at the middle and at the toe of the wall. However, during early times of seepage a large amount of water drained through chimney drain resulted in occurrence of local piping after 1.19 days of seepage. This phenomenon can be found in Figure 2c while a rapid rise and down was seen in the graphs

The normalised value of maximum settlement at the crest ($S_{c,max}$) with respect to the height of the wall ($S_{c,max}/H$) was equal to 0.169 and 0.151 for models M1 and M2 respectively. Model M3 with was found to be stable and experienced marginal settlements even after applying seepage. The value of $S_{c,max}/H$ was equal to 0.018 in model M3. For better understanding the function of chimney drain system, the normalized pore water pressure at the middle and at the toe of the wall was compared for all models. The values of pore water pressure at the middle (PPT3) and at the toe (PPT4) of the wall were normalized with the product of bulk unit weight of the soil, γ, and height of the wall, H and termed as $u_{PPT3}/\gamma H$ and $u_{toe}/\gamma H$.

Figures 3a-3b show the variations of $u_{PPT3}/\gamma H$ and $u_{toe}/\gamma H$ with seepage time in days. The non-effectiveness of application of chimney NWG drain in model M2 in comparison with model M1 (without any chimney drain), can be seen in Figure 3, in which the development of pore water pressure was not controlled in model M2. By contrast, the values of $u_{PPT3}/\gamma H$ and $u_{toe}/\gamma H$ for model M3 with chimney sand drain were observed to be 78% and 65% lower than the value registered for model M1 respectively. Interestingly the better comparison can be found in Figure 4 with phreatic surfaces plotted with PPT readings within the wall models during the penultimate stage of the centrifuge tests.

In Figure 2a horizontal dashed lines indicate the location of reinforcement layers in the model. At first glance, the higher location of phreatic surface in models M1 and M2 could be noticed. The great depletion of phreatic surface was registered for model M3 with chimney sand drain application. This is attributed to the high drainage capability of chimney sand drain system.

5.2 *Digital image analysis of centrifuge models*

Digital image analysis (DIA) technique has been used by different researchers in centrifuge modeling of reinforced soil walls and slopes (Viswanadham & Mahajan 2007; Zornberg & Arriaga 2003). Digital image analysis was performed on photographs captured from front view of the model geosynthetic reinforced soil walls during flight at 40 g. Image analysis was conducted using ImageJ software (ImageJ, 2012).

Figures 5a-5b indicate the displacement field of model geosynthetic reinforced soil walls M1 and M3 from the moment when seepage induced to the penultimate stage of the test in prototype dimensions. Figure 5a represents the large surface settlement and facing deformation occurred in model M1 without

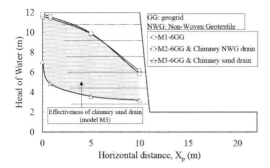

Figure 4. Development of phreatic surfaces within geogrid reinforced soil walls with and without chimney drain system during penultimate stage of centrifuge test.

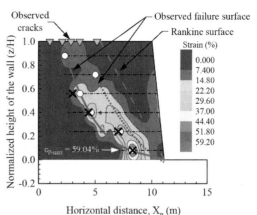

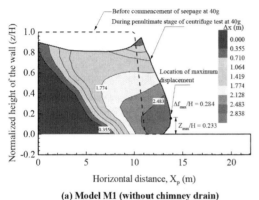

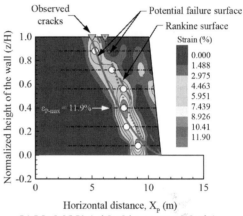

Figure 6. Strain fields of geogrid layers deduced from image analysis during penultimate stage of centrifuge test.

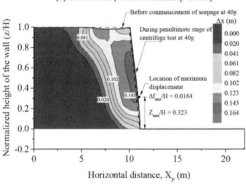

Figure 5. Displacement fields obtained by digital image analysis during penultimate stage of centrifuge test.

chimney drain system as a result of catastrophic failure. The overview of failure surface is obvious in Figure 5a. In comparison, the surface settlements and facing deformations were observed to be controlled effectively in model M3 with chimney sand drain (Figure 5b). The value of maximum facing displacement during penultimate stage of the test was normalized with respect to the height of the wall ($\Delta f_{max}/H$) and presented in Figure 5. The height of point of maximum facing displacement from the base layer of the wall was also mentioned in the figure with respect to the height of the wall (Z_{max}/H).

The magnitude of strain along the reinforcement layers were calculated for each selected image using the coordinates of discrete plastic markers glued on geosynthetic layers and facing panels (Figure 1). To obtain the strain, the relative displacement between two consecutive markers was divided by initial distance between them (Viswanadham and Mahajan, 2007). The values of calculated strains were considered at the initial location of each discrete marker along geogrid layers and after that the strain contours were plotted using Origin 9 software.

Figures 6a-6b indicate the strain field in reinforced soil zone for penultimate stage of centrifuge models of M1 and M3. The locations of peak strains along reinforcement layers were plotted using circle markers with white fill and the value of maximum peak strain (ε_{p-max-}) was shown in Figures 6a-6b. The tension cracks at the top surface of the wall were also located using triangle markers with grey fill with respect to the chronological order. Moreover the observed potential failure surfaces in models M1 and M3 along with the Rankine failure surface were plotted in each Figures 6a-6b.

In the case of model M1, the amount of peak strain was observed to be maximum at the bottom half portion of the wall confirming the profile of face movement in Figure 6a. The value of maximum peak strain ($\varepsilon_{p\text{-max}}$) reached its highest value at 59% at the onset of failure being greater than the ultimate strain of model geogrid and indicating the rupture occurred at geogrid layer. Moreover the rupture was seen at the bottom four geogrid layers during post-investigation of model M1 being compatible with the location of peak strain at these geogrid layers. In comparison with model M1, the amount of strain in geogrid layers was found to be limited in model M3 a result of chimney sand drain. Model M3 was observed to experience the maximum peak strain value of 11.6% at third reinforcement layer from the bottom. In model M3, the location of potential failure surface is observed to close to Rankine failure surface.

6 CONCLUSIONS

The geogrid reinforced soil wall without chimney drain experienced general failure soon after applying seepage. Application of chimney nonwoven geotextile drain could not improve the wall behaviour but caused a delay on the occurrence of general failure.

On the other hand, the application of chimney sand drain improved the wall behaviour effectively and decreased pore water pressure not only at the wall toe but also at mid-distance from toe of the wall and resulted in enhancing the wall behaviour. The values of pore water pressure at the middle and at the toe of the wall were observed to be 78% and 65% lower than the value registered for wall without chimney sand drain application. More studies are warranted to consider the effect of transmissivity of chimney non-woven geotextile drainage layer as well as thickness of chimney sand drainage layer.

REFERENCES

ASTM-D4595 2011. Standard Test Method for Tensile Properties of Geotextiles by the Wide-Width Strip Method. West Conshohocken, PA, USA. *ASTM International*.

ASTM-D6574 2013. Standard Test Method for Determining the (In-plane) Hydraulic Transmissivity of a Geosynthetic by Radial Flow. West Conshohocken, PA, USA. *ASTM International*.

Christopher, B.R. & Stuglis, R.P. 2005. Low permeable backfill soils in geosynthetic reinforced soil walls: state of the practice in North America. *Proceedings of North American Geosynthetics Conference (NAGS)*, 2005 Las Vegas, Nevada, USA, GRI-19,14-16.

Elias, V.C., Barry R. & Berg, R. 2009. Mechanically Stabilized Earth Walls and Reinforced Slopes Design and Construction Guidelines, *FHWA*.

ImageJ. 2012. User guide, version 1.45 s. *US National Institutes of Health, Bethesda*.

Kang, Y., Nam, B., Zornberg, J.G. & Cho, Y.H. 2014. Pullout resistance of geogrid reinforcement with in-plane drainage capacity in cohesive soil. *KSCE Journal of Civil Engineering*, 19(3): 1–9.

Koerner, R.M. & Koerner, G.R. 2011. The Importance of Drainage Control for Geosynthetic Reinforced Mechanically Stabilized Earth Walls. *Journal of GeoEngineering*, 6(1): 3–13.

Koerner, R.M. & Koerner, G.R. 2013. A data base, statistics and recommendations regarding 171 failed geosynthetic reinforced mechanically stabilized earth (MSE) walls. *Geotextiles and Geomembranes*, 40(5): 20–27.

Mitchell, J.K. & Zornberg, J. 1995. Reinforced soil structures with poorly draining backfills, Part II: Case histories and applications. *Geosynthetics International*, 2(1): 265–307.

NCMA. 2009. In: Collin, J. (Ed.), Design manual for segmental retaining walls, third ed., Herndon, Virginia, USA, *National Concrete Masonry Association*.

Raisinghani, D.V. & Viswanadham, B.V.S. 2011. Centrifuge model study on low permeable slope reinforced by hybrid geosynthetics. *Geotextiles and Geomembranes*, 29(6): 567–580.

Tatsuoka, F. & Yamauchi, H. 1986. A reinforcing method for steep clay slopes using non-woven geotextile. *Geotextiles and Geomembranes*, 4(3-4): 241–268.

Viswanadham, B., Razeghi, H.R., Mamaghanian, J. & Manikumar, CH. 2017. Centrifuge model study on geogrid reinforced soil walls with marginal backfills with and withou't chimney sand drain. *Geotextiles and Geomembranes*, 45(5): 430–446.

Viswanadham, B.V.S. & König, D. 2004. Studies on scaling and instrumentation of a geogrid. *Geotextiles and Geomembranes*, 22(5): 307–328.

Viswanadham, B.V.S. & Mahajan, R.R. 2007. Centrifuge model tests on geotextile-reinforced slopes. *Geosynthetics International*, 14(6): 365–379.

Yoo, C. & Jung, H. 2006. Case History of Geosynthetic Reinforced Segmental Retaining Wall Failure. *Journal of Geotechnical and Geoenvironmental Engineering*, ASCE, 132(12): 1538–1548.

Zornberg, J. & Arriaga, F. 2003. Strain Distribution within Geosynthetic-Reinforced Slopes. *Journal of Geotechnical and Geoenvironmental Engineering*, ASCE, 129(1): 32–45.

Centrifuge model tests on effect of inclined foundation on stability of column type deep mixing improved ground

S. Matsuda, M. Momoi & M. Kitazume
Tokyo Institute of Technology, Tokyo, Japan

ABSTRACT: A column type deep mixing improved ground has frequently been used to increase stability of embankment constructed on a soft ground. A lot of model tests and numerical analyses have been carried out to study the behaviour and failure pattern of the improved ground subjected to embankment load, where the shear failure of the deep mixing columns are studied in the internal stability and the sliding, overturning and bearing capacity failures of the improved ground are long studied in the external stability. Almost all studies investigate the stability of the improved ground sitting on a horizontal foundation. However the stability of the improved ground sitting on an inclined foundation has not well been studied yet. In this study, the internal stability of DM columns on an inclined foundation was investigated by centrifuge model tests.

1 INTRODUCTION

Soft soil deposits are often encountered in many construction projects, where large ground settlement and stability failure are anticipated. Accordingly, a large number of soil improvement techniques have been developed in order to provide reinforcement of these soft soil deposits. The Deep Mixing Method (DMM), a deep in-situ soil stabilization technique using cement as a binder, has often been applied to improve soft soils (Coastal Development Institute of Technology, 2002; Kitazume and Terashi, 2013).

A special deep mixing machine used to treat soft soil in-situ is basically composed of several mixing shafts and blades and a system supplying binder (Figure 1).

By one operation, a column shaped stabilized soil is constructed in the ground. Group column type improvement, where many columns are constructed in rows with rectangular or triangular arrangements, has been extensively applied to foundation of embankment and lightweight structures. A design procedure for the group column type DM improved ground has been established in Japan mainly for application of embankment foundation (Public Work Research Center, 2004). Two major failure patterns are assumed in the design procedure as shown in Figure 2: external and internal stabilities. In the external stability, the possibility of sliding, overturning and bearing capacity failures are evaluated, in which the DM columns and the clay between them are assumed to behave as a whole without any failure in the columns. In the internal stability analysis, rupture breaking failure is evaluated by a slip circle analysis, in which the shear failure of DM columns is assumed. Obviously, the improved ground could fail by one of various failure patterns depending on the ground and loading conditions. Each failure pattern is characterized by a particular failure envelope in

Figure 1. Deep mixing machines for on land construction.

a loading plane. It is reasonable that the ground should fail by one of the failure patterns that gives a minimum capacity under certain condition.

The ground behaviour of the column type improved ground under embankment has been investigated by many model tests (e.g. Akamoto & Miyake, 1989; Kitazume & Maruyama, 2006, 2007; Nguyen *et al.*, 2017a, 2017b), the numerical analysis (e.g. Han *et al.*, 2005; Filz & Navin, 2006; Adams *et al.*, 2009; Nguyen *et al.*, 2015) and the design calculation (Kitazume, 2008).

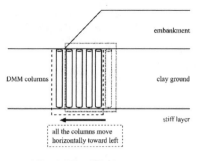

(a) External stability (sliding failure)

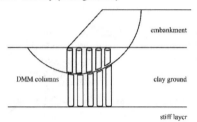

(b) Internal stability (rupture breaking failure)

Figure 2. Assumed failure patterns of DM improved ground in the current design method.

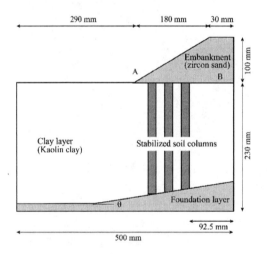

Figure 3. Model ground (Case I.S-2).

However almost all researches focused on the DM columns sitting on a horizontal foundation, and little research focused on the DM columns sitting on an inclined foundation. Though the behaviour and failure pattern of DM columns on an inclined foundation may be different from those on a horizontal foundation, the current design procedure based on the horizontal foundation has been simply applied to the inclined foundation in some case histories.

The authors conducted a research project on the failure mechanism and stability of group column type DM improved ground on an inclined foundation. The external stability of the ground was investigated by the centrifuge model tests and FEM analyses (Toshinari et al., 2017).

In this study, the internal stability of DM columns on an inclined foundation layer was investigated by a series of centrifuge model tests. In the model tests, a soft Kaolin clay ground and column type DM improved ground were modelled, where the inclination of foundation was changed as 0, 10 and 30 deg. The model improved ground was subjected to the embankment load to cause ground failure under 50 G centrifugal acceleration. The effect of the inclination of foundation layer on the improved ground behaviour and failure pattern were investigated in detail.

In this manuscript, the effect of the inclination of foundation on the behaviour and failure of the DM improved ground are discussed in detail.

2 TEST CONDITIONS

2.1 Model ground preparation

A series of model tests was carried out in the TIT Mark III Centrifuge in order to simulate the prototype stress conditions (Takemura et al., 1999). A rectangular model container was used for the tests the inside dimensions of which are 150 mm in width, 500 mm in length and 362 mm in depth. Its front side is made of an acrylic plate for visual observation during test. One of the model grounds is exemplified in Figure 3, which consists of Kaolin clay layer, model DM columns, inclined foundation and embankment. Though an infinite inclined foundation would be simulated in the model test, an inclined foundation was constructed under the embankment area due to the small size container. The FEM analyses were carried out to investigate the effect of the bent foundation on the ground behaviour and revealed it could be negligible.

In the preparation of the model ground, a horizontal foundation was made on the bottom of the container by compacting silica No. 3 sand. Kaolin clay slurry, specific gravity, $Gs = 2.61$, liquid limit, $w_l = 77.5\%$ and plastic limit, $w_p = 30.3\%$, with a water content of 100% was poured on the layer and one dimensionally consolidated at the consolidation pressure of 200 kPa to make a rectangular shaped clay layer with the uniform strength distribution along the depth. As the consolidation pressure was larger than the overburden pressure at 50 G, the ground was over-consolidated condition in the centrifuge. The thickness of clay layer was 200 mm in Case I.S-1 and 225 mm in Cases I.S-2 and I.S-3 so that the clay thickness at the most front column was the same of 200 mm throughout the test series. In Cases I.S-2 and I.S-3, inclined foundation cases, the front acrylic window was disassembled and a part of the clay layer was excavated and filled with silica No. 3 sand to make an inclined foundation layer. The optical markers were placed on the side surface of the clay ground for measuring the deformation of model ground (Figure 4). Detailed displacement of the model ground was obtained by the Particle Tracking

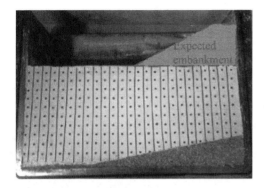

Figure 4. Model ground with embankment (Case I.S-2).

(a) Removing clay in a thin-wall tube.

(b) Installation of model DM column.

Figure 5. Preparation of model improved ground.

Velocimetry technique (PTV) (Dracos, 1996; Murthy, 2013).

The acrylic window was assembled again, and the model DM columns were installed into the clay ground. The DM columns are constructed by injecting cement slurry into the ground and mixing together in the field. However, in the model tests, a thinwall tube with an outer diameter of 20 mm was penetrated into the clay ground. The clay inside the tube was then carefully removed using a tiny auger to make hole (Figure 5(a)), and a model DM column, the cement stabilized columns manufactured in advance, was inserted into the hole after removing the tube, as shown in Figure 5(b). The length of the front most DM column was constant as 200 mm throughout the test series, and the column length was changed according to the inclination of foundation. Though DM columns were usually key into a foundation layer in actual construction to assure their tight connection in the fix type improved ground (Kitazume & Terashi, 2008). However, in this study the column toe was designed standing on the foundation layer without any penetration for ease of model ground preparation, as shown in Figure 3. This procedure was repeated to produce the improved ground with 3 rows and 4 lines columns in a square pattern with an interval of 37.5 mm. The improvement area ratio, a_s, defined as the ratio of sectional area of DM column to the hypothetical cylindrical area was constant of 0.28.

The model ground thus prepared was subjected to the 50 G centrifugal acceleration. Soon after reaching 50 G, the embankment loading was carried out, while the ground was under the over-consolidated condition. An in-flight sand hopper was used to construct an embankment during the flight, in which Zircon sand having a large specific gravity of 4.66 was poured from about 170 mm height on the clay ground in order to cause the model ground failure in the undrained condition. The embankment pressure was measured at the centre line of embankment by earth pressure gauges installed on the clay ground surface. Three model tests were carried out changing the inclination of foundation: a horizontal foundation layer (Case I.S-1), inclined foundation layer with 10 deg. (Case I.S-2) and 30 deg. (Case I.S-3).

2.2 Property of model ground

Figure 6(a) shows the water content distribution along over depth, which were measured at the completion of the consolidation. Though there is scatter in the measured data, they show almost constant water content of about 65 % along over depth and throughout the test cases. The undrained shear strength was also measured at the completion of the consolidation by a hand vane apparatus, and shown in Figure 6(b) along the depth. The figure shows almost uniform shear strength of about 40 kPa along the depth throughout the test cases. These test results can confirm that the uniform clay ground could be made throughout the test cases.

2.3 Property of cement DM columns

Cement stabilized soil columns were produced by mixing Kaolin clay of 160 % in initial water content and ordinal Portland cement of cement content of 10 %. The soil and cement mixture was poured into a acrylic tube with inner diameter of 20 mm. As the stabilized soil strength was influenced by many factors, sufficient number of the columns were manufactured at a time to minimize the strength difference during the test series. Unconfined compression test was conducted on the sample trimmed 20 mm in diameter and 40 mm in height from the stabilized model columns. The target unconfined compressive strength was 500 kPa, while the obtained strength was 465 kPa (Case I.S-1),

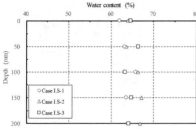
(a) Water content distribution along depth.

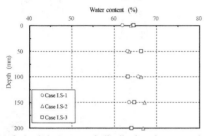
(b) Undrained shear strength distribution along over depth.

Figure 6. Property of model clay ground.

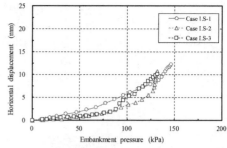
(a) Horizontal displacement and embankment pressure.

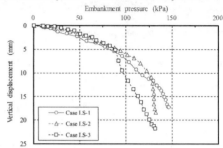
(b) Vertical displacement and embankment pressure.

Figure 7. Displacements and embankment pressure.

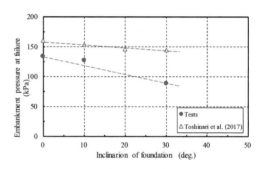

Figure 8. Embankment pressure at yield and inclination of foundation.

485 kPa (Case I.S-2) and 505 kPa (Case I.S-3), while the mean strength of 485 kPa was slightly smaller than the target strength.

3 TEST RESULTS AND DISCUSSIONS

3.1 Displacement of embankment

The embankment was constructed on the model ground at the 50 G within short time to simulate the undrained loading condition. Figure 7(a) shows the relationship between the horizontal displacement at the embankment toe (point A in Figure 3) and embankment pressure. In Case I.S-1, the horizontal foundation, the horizontal displacement was increased almost linearly with the embankment pressure but rapidly when the embankment pressure exceeded about 130 kPa. Similar phenomenon can be seen in Cases I.S-2 and I.S-3, an inclined foundation, in which the horizontal displacement was increased rapidly at about 130 kPa for Case I.S-2 and about 90 kPa for Case I.S-3.

The relationship between the vertical displacement at the ground surface (point B in Figure 3) and embankment pressure is shown in Figure 7(b). In Case I.S-1, the vertical displacement was increased gradually with the embankment pressure and then rapidly when the embankment pressure exceeded about 130 kPa. This phenomenon is consistent with that in the horizontal displacement as shown in Figure 7(a). In Cases I.S-2 and I.S-3, the vertical displacement was increased gradually beginning but increased rapidly when the embankment pressure exceeded about 90 kPa for Case I.S-2 and about 130 kPa for Case I.S-3, which are consistent with those in the horizontal displacement.

3.2 Embankment pressure at failure and inclination of foundation

The ground failure is defined when the vertical and horizontal displacements were increased rapidly with the embankment pressure. The embankment pressure at failure thus obtained is shown in Figure 8 along the inclination of foundation. It can be seen that the embankment pressure at failure decreases almost linearly with the inclination. The failure pressure at the inclination of 30 deg. is about 70% of that of the horizontal foundation case.

In the figure, the relationship for the external stability is plotted together (Toshinari et al., 2017), in which the model ground material and preparation are the same as this study except the column material. The model column for the external stability was made by acrylic pipe to avoid any columns' failure

(Toshinari et al., 2017). The embankment pressure made by acrylic pipe to avoid any columns' failure (Toshinari et al., 2017). The embankment pressure at failure for the external stability is decreased slightly with inclination of foundation. By comparing them, the embankment pressure at failure for the internal stability shows considerably larger decrease with the inclination than those for the external stability.

3.3 Deformation of model ground

The model ground deformation after the embankment loading is shown in Figure 9. In Case I.S-1, Figure 9.(a), large ground settlement took place beneath the embankment, while large horizontal displacement and ground heaving took place beneath the embankment toe and ground surface respectively. The large slip circle can be found, which is illustrated by a red line. In Case I.S-2, Figure 9(b), large slip failure with a sharp bend beneath the embankment can be seen. In Case I.S-3, Figure 9(c), large ground settlement took place beneath the embankment. A slip failure can be seen, which consisted a linear slip line perpendicular to the inclined foundation and circle shape failure. The slip line and circle is a little bit smaller than that in Case I.S-1.

3.4 Horizontal displacement distribution

The horizontal displacement distribution along the depth at embankment toe was measured and shown in Figure 5 at the various embankment pressures. In Case I.S-1, horizontal foundation, Figure 10(a), a small linear horizontal displacement was observed at the embankment pressure of 60 kPa. The horizontal displacement was increased rapidly with the embankment pressure, particularly at the shallow to middle depth, while the displacement at the bottom of column was negligible. When the embankment pressure was increased to 147 kPa, quite large horizontal displacement of about 180 mm can be seen at the middle depth. In Case I.S-2, an inclined foundation of 10 deg., Figure 10(b), small displacement can be seen at the embankment pressure of 60 kPa, which is similar to the horizontal foundation case, Case I.S-1. With the increase of embankment pressure, the horizontal displacement was increased rapidly along the depth, particularly at the shallow to middle depth. The maximum displacement was found at the depth of about 145 mm at the embankment pressure of 132 kPa, which was little shallower than that in Case I.S-2. In Case I.S-3, inclination of foundation of 30 deg., Figure 10(c), similar phenomenon can be seen, where the horizontal displacement was increased with the increase of the embankment pressure. The maximum displacement was found at the depth of about 70 mm at the embankment pressure of 131 kPa.

After the embankment loading test, the clay ground was excavated to observe the DM columns' failure in detail. The deformation and bending failure points of the most front DM column are plotted together in Figure 10. It is found that the bending failures took place

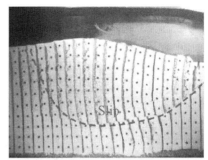

(a) Case I.S-1

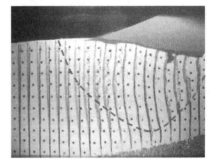

(b) Case I.S-2.

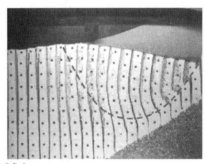

(c) Case I.S-3.

Figure 9. Ground deformation after embankment loading.

at various depths in all three columns irrespective of the inclination of foundation and the deformation of DM column was almost coincided with the horizontal displacement profile at the embankment toe.

4 CONCLUSIONS

In this study, a series of centrifuge model tests and FEM analyses were carried out to investigate the effect of inclination of foundation on the deformation and internal stability of the column type DM improved ground subjected to the embankment loading.

(1) The failure of DM improved ground was induced by the failure of DM columns. The embankment pressure at failure decreased almost linearly with the inclination of foundation.
(2) A large slip circle shape ground deformation was found in the flat foundation case, while a slip

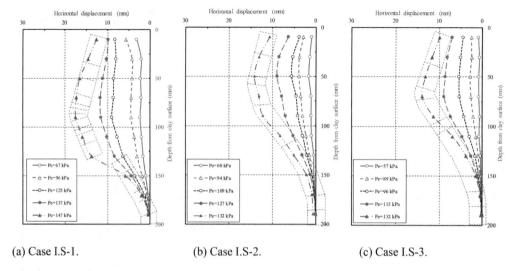

(a) Case I.S-1.   (b) Case I.S-2.   (c) Case I.S-3.

Figure 10. Horizontal displacement distribution along depth.

failure can be seen, which consisted a linear slip line perpendicular to the inclined foundation and circle shape failure.
(3) The DM columns were failed by the bending failure mode instead of the shear failure mode irrespective of the inclination of foundation layer.
(4) The large amount of horizontal displacement took place along the depth in the flat foundation case, while large horizontal displacement took place locally in the inclined foundation cases.

REFERENCES

Adams, T., Filz, G. & Navin, M. 2009. Stability of embankments and levees on deep-mixed foundations. *Proc. of the International Symposium on Deep Mixing and Admixture Stabilization.*

Akamoto, H. & Miyake, M. 1989. Deformation characteristics of the improved ground by a group of treated soil columns, *Proc. of the 24th Annual Conference of the Japanese Society of Soil Mechanics and Foundation Engineering.* (in Japanese).

Coastal Development Institute of Technology. 2002. *The Deep Mixing Method – Principle, Design and Construction-.* A. A. Balkema Publishers. 123p.

Dracos, T., 1996. Particle tracking velocimetry. In T. Dracos, ed. *Three-dimensional velocity and vorticity measuring and image analysis techniques.* Springer, pp. 155–256.

Filz, G. M. & Navin, M. P. 2006. Stability of column-supported embankments. *Final contract report, VTRC 06-CR13, Virginia Transportation Research Council.*

Han, J., Parsons, R. L., Huang, J. & Sheth, A. R. 2005. Factors of safety against deep-seated failure of embankments over deep mixed columns. *Proc. of the International Conference on Deep Mixing - Best Practice and Recent Advances,* Stockholm. pp. 231-236.

Kitazume, M. 2008. Stability of group column type DM improved ground under embankment loading. *Report of the Port and Airport Research Institute.* Vol. 47. No. 1. pp. 1-53.

Kitazume, M. & Maruyama, K. 2006. External stability of group column type deep mixing improved ground under embankment loading. *Soils and Foundations.* Vol. 46. No. 3. pp. 323-340.

Kitazume, M. & Maruyama, K. 2007. Internal stability of group column type deep mixing improved ground under embankment loading. *Soils and Foundations.* Vol. 47. No. 3. pp. 437-455.

Kitazume, M. & Terashi, M., 2013. *The deep mixing method,* CRC Press, Taylor & Francis Group, 410p.

Murthy, T. G., 2013. Evolution of deformation fields in 1-g model tests of footings on sand. *In ICPMG2014 -Physical Modelling in Geotechnics,* CRC Press, pp. 653–657.

Nguyen, B. T., Takeyama, T. and Kitazume, M., 2015, Numerical analyses on the failure of deep mixing columns reinforced by a shallow mixing layer, *Proc. of the 16th Asian Regional Conference on Soil Mechanics and Geotechnical Engineering, Fukuoka 2015.*

Nguyen, B. T., Takeyama, T. & Kitazume, M. 2017a, Using a shallow mixing layer to reinforce deep mixing columns during embankment construction, *Proc. of the 5th International Grouting Conference, Grouting 2017.*

Nguyen, B. T., Takeyama, T. and Kitazume, M. 2017b, Study on failure mechanisms of the deep mixing columns reinforced by a shallow mixing layer, *Proc. of the 19th International Conference on Soil Mechanics and Geotechnical Engineering, Seoul 2017.*

Public Works Research Center. 2004. *Technical Manual on Deep Mixing Method for On Land Works.* 334p. (in Japanese).

Takemura, J., Kondoh, M., Esaki, T., Kouda, M. & Kusakabe, O. 1999. Centrifuge model tests on double propped wall excavation in soft clay. *Soils and Foundations.* Vol. 39. No. 3. pp. 75–87. doi:10.3208/sandf.39.3_75.

Toshinari, Y., Matsuda, S., Kitazume, M. & Nguyen, B. 2017. Centrifuge model studies on external stability of column type deep mixing ground on inclined foundation, *Proc. of the 5th International Grouting Conference, Grouting 2017.*

Large-scale physical model GRS walls: Evaluation of the combined effects of facing stiffness and toe resistance on performance

S.H. Mirmoradi & M. Ehrlich
Federal University of Rio de Janeiro, Rio de Janeiro, Brazil

ABSTRACT: Two large-scale geosynthetic-reinforced soil (GRS) walls were constructed at the COPPE/UFRJ Geotechnical Laboratory to evaluate the effect of facing type and toe resistance on the performance of the walls. The walls, identified as Walls 1 and 2, were similar apart from the facing type. Walls with block and wrap facing were considered in this study. After the end of construction, a surcharge was applied to the entire surface of the top of the walls up to 100 kPa. During construction and surcharge application, the toe of Wall 1 (block-face wall) was restricted. After applying the surcharge up to 100 kPa, in Wall 1, the surcharge was kept constant at 100 kPa while the toe of the wall was gradually released. In the wall with wrap facing (Wall 2), there was no toe restriction during surcharge application. The walls were well-instrumented in order to monitor the values of the reinforcement load, toe load, and horizontal facing displacement. The results clearly call attention to the importance of the combined effect of facing stiffness and toe resistance on the behaviour of GRS walls.

1 INTRODUCTION

Geosynthetic reinforced soil (GRS) walls are widely used throughout the world because of the many advantages over other wall types, such as low cost, simple construction, and the ability to accommodate deformation (e.g., Fannin & Hermann 1990; Rowe & Ho 1993; Kazimierowicz-Frankowska 2005; Yoo & Kim 2008; Yang et al. 2009).

In recent decades, several studies have been carried out to evaluate controlling factors on the behavior of GRS walls; the facing of a reinforced soil structure and toe restriction are two of those factors (e.g., Tatsuoka 1993; Bathurst et al. 2006; Huang et al. 2010; Leshchinsky & Vahedifard 2012; Ehrlich & Mirmoradi 2013; Mirmoradi & Ehrlich 2014, 2016).

Tatsuoka (1993) showed that for a stiffer facing, the earth pressure acting on the back of the face increases. Furthermore, Bathurst et al. (2006) reported the higher wall deformation and reinforcements loads for the flexible wrapped-face wall than the modular block-face wall.

Regarding the toe resistance, the studies indicated the sensitivity of the reinforcement loads to the toe resistance. Decrease of toe resistance leads to an increase in reinforcement load (Huang et al. 2010; Leshchinsky & Vahedifard 2012; Mirmoradi & Ehrlich 2015, 2016; Mirmoradi et al. 2016). Nevertheless, Mirmoradi & Ehrlich (2017a) using numerical analyses showed that for vertical reinforced soil walls with segmental block facing, the combined effect of the facing and reinforcement stiffness, wall height, and toe resistance on the distribution of the maximum reinforcement load with depth may be limited to ~4 m above the base of the wall.

In the current study, the combined effect of facing stiffness and toe resistance is explicitly evaluated using data from two large-scale physical models — block and wrap-faced GRS walls — constructed at the COPPE/UFRJ Geotechnical Laboratory (Mirmoradi 2015). The toe of the block-faced wall was restricted during the construction and surcharge application. After the end of loading, the toe of this wall was gradually released. While for the wrapped-facing wall, no toe restriction was applied during the test as the toe restraint does not affect the behaviour of the walls with low facing stiffness (e.g., Ehrlich & Mirmoradi 2013; Mirmoradi & Ehrlich 2014, 2015, 2017a).

2 EXPERIMENTAL STUDY

2.1 Test characteristics and material used

A series of well-instrumented physical model walls were constructed at the COPPE/UFRJ Laboratory of Physical Models (Mirmoradi 2015). The results of two of these walls are used in this paper to evaluate the combined effect of facing stiffness and toe resistance on the behaviour of GRS walls. The walls described here are identified as Walls 1 and 2 with block and wrapped-facing, respectively.

Cross-sections of physical model walls are shown in Figure 1. The height of each physical model wall was 1.2 m. The length and vertical spacing of the geogrid were 2.2 and 0.4 m, respectively. A flexible

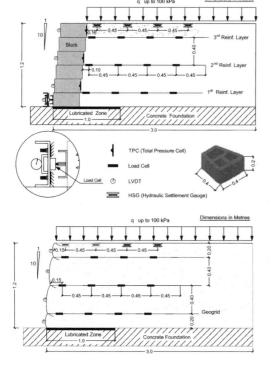

Figure 1. Cross-sectional views of (a) block and (b) wrap-faced walls.

Figure 2. Views of (a) Wall 1 and (b) Wall 2 at the end of construction.

polyester geogrid was used as reinforcement. Based on the mechanical properties provided by the company, the tensile stiffness modulus of the reinforcement, J_r, is equal to 917 kN/m. Three layers of reinforcement were installed along the height of the wall, placed at 0.2, 0.6, and 1.0 m above the foundation.

The backfill material consisted of well-graded sand, composed of crushed quartz powder with a significant amount of fines (19% <0.075 mm), $D_{50} = 0.25$ mm, curvature coefficient $C_c = 1$, uniformity coefficient $C_u = 8.9$, and plasticity index PI equal to zero. A light vibrating plate was used for the compaction operation. The equivalent static load of the compactor was determined through Kyowa accelerometers installed in the compactor's body, and a value of 8 kPa was achieved. The soil unit weight after compaction was 21 kN/m^3. The soil peak friction angles, considering the measured unit weight, were determined by triaxial and plane-strain tests as 42° and 50°, respectively (Ehrlich et al. 2012; Ehrlich & Mirmoradi 2013; Mirmoradi & Ehrlich 2017b).

The differences in the walls are related to facing type. In Wall 1, precast blocks were used for the wall facing. Wall 2 was similar to Wall 1, however wrapped facing was used. Figure 2 shows views of the walls with block and wrap facing.

The toe of the block-faced wall was restricted during construction and surcharge application. Figure 1 also shows a schematic view of the procedure used to guarantee the toe restraint of the walls. Lateral movement of the toe was restricted by a steel beam that was fixed to the concrete U-shaped wall box using two bolts on each side of the beam.

2.2 Construction sequence

The construction of the model was performed in six soil layers, 0.2 m thick and placed dry. The sequence of construction of the walls was developed in two stages per layer of soil: (1) soil placement and (2) compaction of the placed backfill using the vibrating plate. During construction of the wrap-faced wall, the face of the wall was restricted using wood fixed to the U-shaped concrete wall. The wooden support was removed after the end of construction.

A 1.0 m-wide zone at the bottom of the walls was lubricated through a sandwich of rubber sheets and Teflon grease. To reduce the effect of lateral friction at the interface between the backfill soil and the concrete wall, PVC sheets were installed on all lateral faces of the wall that comprises the U-shaped concrete box of the model. In addition, a thin layer of Teflon grease covered by plastic sheets was used in order to assure a plane-strain condition during the tests.

Three layers of reinforcement were installed along the height of the wall, placed at: 0.2 m (first layer), 0.6 m (second layer), and 1.0 m (third layer) above the foundation. Each reinforcement layer was longitudinally divided into three sections, and only the 0.5 m reinforcement placed at the centre of the wall was instrumented (see Figure 3).

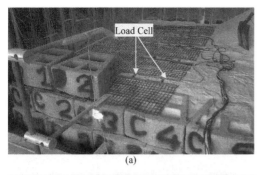

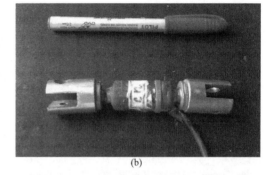

Figure 3. Views of (a) Wall 1 and (b) Wall 2 during reinforcement placement.

After the end of construction, a surcharge up to 100 kPa was applied over the entire surface of the backfill soil using an air bag. Then, for the block faced wall (Wall 1), the surcharge was kept constant at 100 kPa. In the meantime, the toe of the wall was gradually released to the free base condition (0.5 mm release in each step) using the load cells shown in Figure 4a.

2.3 Instrumentation

The walls were instrumented to monitor the values of the reinforcement load, toe load, horizontal facing displacement, and vertical displacement at the top of the walls.

Reinforcement loads were monitored using load cells installed at four points along the reinforcement (i.e. two load cells at each point). The load cells allowed for reinforcement load monitoring without the need to determine the reinforcement stress-strain curves, which are time dependent. The load cells were also capable of counterbalancing the temperature effects and bending moments and were strong enough to resist the stress induced during the operation of the compaction equipment (Ehrlich et al. 2012; Ehrlich & Mirmoradi 2013; Mirmoradi et al. 2016). Figure 4b shows a view of the load cell used to measure reinforcement load.

The horizontal displacements of the wall face were monitored by LVDTs. The horizontal facing displacements were measured at the second (0.3 m height), fourth (0.7 m height), and sixth (1.1 m height) layers.

Figure 4. Views of the instruments; (a) load cell to measure reinforcement load, (b) load cell to measure toe load, (c) hydraulic settlement gauge.

For the block-faced wall, the toe load was measured using the load cells installed on the steel beam fixed to the concrete U-shaped wall box. The load cells were placed between the aforementioned steel beam and another steel beam installed on the blocks of the first layer. As stated earlier, a 1.0-m-wide zone at the bottom of the walls, that included the base of the block facing, was lubricated. Thus, the toe was free and the restriction of lateral movements was guaranteed through the load cells; the toe load was measured using these load cells (see Fig. 4a).

A hydraulic settlement gauge (HSG) was used for monitoring the vertical displacements (see Figure 4c). This hydraulic settlement gage (HSG) consists of an acrylic settlement cell filled with mercury connected to a plastic tube, also filled with mercury, which is monitored by a pressure transducer. Any settlement or heave in the settlement cell can thus be related to the readings in the pressure transducer. For both walls, the monitoring points, HSG 1, HSG 2, HSG 3, and HSG 4, were located at four different distances from the back of the facing: 0.15, 0.6, 1.05, and 1.50 m, respectively. These instruments were installed at the top of the walls and monitored the settlement after the end of construction.

3 RESULTS

This paper presents and discusses the values of the reinforcement loads and horizontal facing displacements measured through the load cells and LVDTs in the walls, respectively.

3.1 Reinforcement load

Figure 5 shows the variation in the sum of maximum reinforcement loads, ΣT_{max}, during surcharge application for the walls. The figure indicates that ΣT_{max} is lower for the block-faced wall than for the wrap-faced wall. This is attributed to the combined effect of toe resistance and facing stiffness on the magnitude of mobilised tension along the reinforcements as discussed by Ehrlich & Mirmoradi (2013) and Mirmoradi & Ehrlich (2016). However, this discrepancy is decreased during release of the toe.

Figure 6 shows the variation of ΣT_{max} during toe release. In this figure, a point corresponding to the surcharge value equal to 100 kPa is shown to represent the value of ΣT_{max} for the wrap-faced wall. As explained earlier for the block-faced wall (Wall 1), after applying the surcharge up to 100 kPa, the surcharge was kept constant at 100 kPa while the toe of the wall was released.

In Figure 6, although the facing stiffness of the block and wrap-faced walls is different, after the toe release of the block-faced wall, ΣT_{max} for both the wrap and block-faced walls were similar. This supports the results presented by Ehrlich & Mirmoradi (2013), Mirmoradi & Ehrlich (2014; 2015; 2016; 2017a) which numerically and experimentally considered the combined effect of toe resistance and facing stiffness on the behaviour of GRS walls.

3.2 Horizontal facing displacement

For the wrap-faced wall, the LVDTs were installed after removing the wooden support fixed on the U-shaped concrete wall and therefore the recorded displacement values represent the magnitude of the horizontal facing displacement due to the surcharge application, only. For the block-faced wall the lateral movements were measured during and after construction.

Figure 7 shows the average of the horizontal facing displacement increments, ΔH_{ave}, during surcharge application for both of the walls. As expected, the value

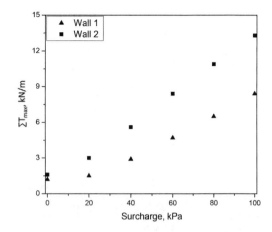

Figure 5. Sum of maximum reinforcement load versus surcharge application.

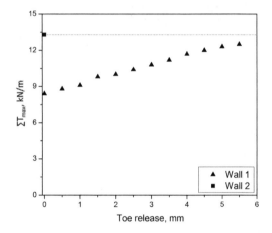

Figure 6. Sum of maximum reinforcement load versus toe release.

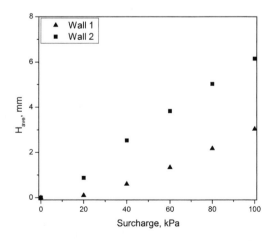

Figure 7. Average of the horizontal facing displacement increment versus surcharge application.

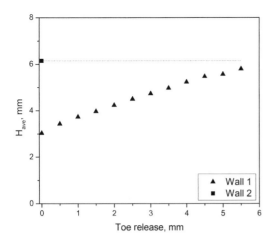

Figure 8. Average of the horizontal facing displacement increment versus toe release.

of ΔH_{ave} is higher for the wrap-faced wall. Note that, in the comparison presented in this figure the lateral movements of the block-faced wall are related to the toe restriction condition.

Figure 8 indicates the average of the horizontal facing displacement increments during toe release. In this figure, a point corresponding to the surcharge value equal to 100 kPa is shown to represent the value of ΔH_{ave} in the wrap-faced wall. The figure shows that in Wall 1, the ΔH_{ave} increases and at the end of toe release, a very similar value of ΔH_{ave} is observed for both walls, irrespective of the facing type.

4 CONCLUSIONS

Two large-scale physical models — block and wrap-faced geosynthetic-reinforced soil (GRS) walls — were constructed at the COPPE/UFRJ Geotechnical Laboratory (Mirmoradi 2015). The walls were identified as Walls 1 and 2, and were similar except for facing type. In Walls 1 and 2, precast block and wrap facing were used for the wall facing, respectively. After the end of construction, a surcharge up to 100 kPa was applied to the entire surface of the top of the walls. During the construction and surcharge application, the toe of Wall 1 (block-faced wall) was restricted. After applying the surcharge up to 100 kPa to Wall 1, the surcharge was kept constant at 100 kPa while the toe of the wall was gradually released.

As demonstrated and discussed in the previous physical and numerical studies performed by the authors (e.g., Ehrlich & Mirmoradi 2013; Mirmoradi & Ehrlich 2014, 2015, 2017a) for walls with low facing stiffness that may correctly represent the wrapped-facing wall, the toe restraint does not affect the behaviour of the walls. It means that no matter the toe condition, in Wall 2, the results would be similar.

The results show, as expected, that the values of the reinforcement load and horizontal facing displacement during surcharge application are greater for the wall with wrap facing than for the block-faced wall. Nevertheless, due to the toe release in the block-faced wall, the magnitude of the reinforcement load and horizontal facing displacement increased and at the end of toe release similar values for the sum of reinforcement loads and horizontal facing displacements were observed.

This supports the results of the numerical analyses presented by Ehrlich & Mirmoradi (2013). They stated that "variation in facing stiffness may play an effective role in affecting the magnitude of ΣT_{max} when the base of the face is restricted. This means that the effect of facing on the magnitude of ΣT_{max} is not solely associated with facing stiffness, but also with the mobilised shear stress at the interface of the base of the facing column and foundation soil." Note that after toe release both walls yielded similar values of ΣT_{max}, which means that for the condition without toe resistance, the facing type (facing stiffness) did not significantly affect the value of ΣT_{max}.

It is important to notice that in this study, the combined effect of the facing stiffness and toe restraint was fundamentally evaluated considering two large-scale physical model GRS walls. As shown by Mirmoradi & Ehrlich (2015, 2017a), the combined effect of the facing stiffness and toe restraint substantially depends on the height of the reinforced soil walls and when the height of the wall increases, the relative effect of the facing stiffness and toe restraint on the reinforcement load would significantly decrease and this effect is limited to approximately 4 m above the base of the walls.

ACKNOWLEDGMENTS

The authors greatly appreciate the funding for this study provided by the Brazilian Research Council, CNPq, and the Brazilian Federal Agency for Support and Evaluation of Graduate Education, CAPES. We also thank Flavio Montez and Andre Estevao Ferreira da Silva from the HUESKER Company for their support, as well as Cid Almeida Dieguez for help with experiments.

REFERENCES

Bathurst, R.J., Vlachopoulos, N., Walters, D.L., Burgess, P.G. & Allen, T.M. 2006. The influence of facing stiffness on the performance of two geosynthetic reinforced soil retaining walls. *Canadian Geotechnical Journal* 43: 1225–1237.

Ehrlich, M., Mirmoradi, S.H. & Saramago, R.P. 2012. Evaluation of the effect of compaction on the behaviour of geosynthetic-reinforced soil walls. *J. Geotextile Geomembr.* 34: 108–115.

Ehrlich, M. & Mirmoradi, S.H. 2013. Evaluation of the effects of facing stiffness and toe resistance on the behavior of GRS walls. *J. Geotextile Geomembr.* 40(Oct): 28–36.

Fannin, R.J. & Hermann, S. 1990. Performance data for a sloped reinforced soil wall. *Canadian Geotechnical Journal* 27 (5): 676–686.

Huang, B., Bathurst, R.J., Hatami, K. & Allen, T.M. 2010. Influence of toe restraint on reinforced soil segmental walls. *Canadian Geotechnical Journal* 47(8): 885–904.

Kazimierowicz-Frankowska, K. 2005. A case study of a geosynthetic reinforced wall with wrap-around facing. *J. Geotextile Geomembr.* 23(1): 107–116.

Leshchinsky, D. & Vahedifard, F. 2012. Impact of toe resistance in reinforced masonry block walls: design Dilemma. *Journal of Geotechnical and Geoenvironmental Engineering, ASCE* 138(2): 236–240.

Mirmoradi, S.H. 2015. Evaluation of the behaviour of reinforced soil walls under working stress conditions. *Ph.D. Thesis. COPPE/UFRJ*, Rio de Janeiro, BR: 340.

Mirmoradi, S. H. & Ehrlich, M. 2014. Geosynthetic reinforced soil walls: experimental and numerical evaluation of the combined effects of facing stiffness and toe resistance on performance. *Proc., 10th Int. Conf. on Geosynthetics, International Geosynthetics Society (IGS)*, Berlin, Germany.

Mirmoradi, S.H. & Ehrlich, M. 2015. Numerical evaluation of the behavior of GRS walls with segmental block facing under working stress conditions. *ASCE J. Geotech. Geoenviron. Eng.* 141(3), 04014109.

Mirmoradi, S.H., Ehrlich, M., & Dieguez, C. 2016. Evaluation of the combined effect of toe resistance and facing inclination on the behaviour of GRS walls. *J. Geotextile Geomembr.* 44(3): 287–294.

Mirmoradi, S. H. & Ehrlich, M., 2016. Evaluation of the effect of toe restraint on GRS walls. *Transportation Geotechnics, SI: Geosynthetics in Tpt,* 8: 35–44.

Mirmoradi, S.H. & Ehrlich, M., 2017a. Effects of facing, reinforcement stiffness, toe resistance, and height on reinforced walls. *J. Geotextile Geomembr.* 45(1): 67–76.

Mirmoradi, S.H. & Ehrlich, M. 2017b. Experimental evaluation of the effects of surcharge width and location on GRS walls. *International Journal of Physical Modelling in Geotechnics*, doi.org/10.1680/jphmg.16.00074.

Rowe, R.K. & Ho, S.K., 1993. Keynote lecture: A review of the behaviour of reinforced soil walls. Earth Reinforcement Practice, *Proc. Int. Symposium on Earth Reinforcement Practice*, Vol. 2, Kyushu University, Fukuoka, Japan, H. Ochia et al. eds., Balkema, Rotterdam, NL: 801–830.

Tatsuoka, F. 1993. Keynote lecture: Roles of facing rigidity in soil reinforcing. In Y. Ochiai, S. Hayashi, J. Otani (Eds.) *Earth Reinforcement Practice*. Balkema, Rotterdam, NL: 831–870.

Yang, G., Zhang, B., Lv, P. & Zhou, Q. 2009. Behavior of geogrid reinforced soil retaining wall with concrete-rigid facing. *J. Geotextile Geomembr.* 27: 350–356.

Yoo, C. & Kim, S.B. 2008. Performance of a two-tier geosynthetic reinforced segmental retaining wall under a surcharge load: full-scale load test and 3D finite element analysis. *J. Geotextile Geomembr.* 26 (6): 460–472.

Deep vibration compaction of sand using mini vibrator

S. Nagula, P. Mayanja & J. Grabe
Institute of Geotechnical Engineering and Construction Management, Technical University Hamburg, Hamburg, Germany

ABSTRACT: The deep vibration compaction method includes densification of loose sands by means of shear deformation processes imparted by horizontal vibrations of vibrator probe. Model tests are conducted using a mini vibrator in order to study the physical processes the granular material undergoes during deep vibration compaction. Tests were performed in test rig of 1 m diameter and 1.1 m height. The mini vibrator is driven into the sand layer in the test rig at a frequency of 45 Hz, after reaching the required depth, the vibrator was set into the compaction frequency. The effect of compaction frequency on the degree and extent of compaction is studied. Compaction is carried out in frequency range of 15 Hz to 60 Hz. The effectiveness of the compaction is evaluated by means to mini cone penetration test (CPT). CPT values before compaction and after compaction are used to evaluate the effectiveness of the compaction.

1 INTRODUCTION

The deep vibration compaction method is an established ground improvement technique for granular materials. This technique is used to improve the properties of loose to medium dense granular soils by compacting deep layers of the soil and therefore reducing settlements and increasing the vertical bearing capacity of foundations. Additionally, it also aids in liquefaction mitigation. The process primarily involves a vibrator probe being installed in the loose granular media and the horizontal vibrations of the vibrator then initiate shear deformation processes in the surrounding soil eventually leading to a denser state of soil mass.

The process on deep vibration compaction mainly works on field experience and requires further scientific insight in order to optimize the process. Numerical simulations of the process can help one understand the various physical processes the soil undergoes during the compaction process and hence help optimize the process. The validation of such numerical models is of utmost importance before such models can be applied for predictive and optimization algorithms. Field measurement data for such complex processes from sensors installed in field are not only difficult to obtain but also prone to errors. Model tests under controlled lab conditions can not only provide ideal data for the validation of the numerical model but also help understand in detail the various physical processes granular materials undergo during deep vibration compaction. A prototype mini vibrator was fabricated in order to conduct deep vibration compaction model tests in sands.

2 DEEP VIBRATION COMPACTION

2.1 Concept

The deep vibration compaction technique comprises mainly of a vibrator probe generally made of steel with diameters ranging around 0.3 m to 0.5 m. The vibrator consists of an eccentric rotating mass. The rotation of these eccentric masses around the vertical axis generates horizontal vibrations which imparts shear deformations in the soil. The rearrangement of the non-cohesive soil particles is facilitated hence leading to a denser state. The frequency of rotation of the vibrator also referred to as compaction frequency, is an important parameter in order to maximize the degree of compaction (Massarsch & Fellenius 2005). The deep vibrator compactor typically functions between 15 to 60 Hz, which allows a maximum deflection of 0.003 to 0.021 m at the toe of the vibrator. This range of frequency ensures maximum compaction of granular materials on field and is based on field experience and field monitoring during compaction (Massarsch & Fellenius 2005). The weight of the eccentric masses and the eccentricity determine amount of force the vibrator can impart to the surrounding soil which generally lies in the range of 150 to 700 kN (Fellin 2000).

2.2 Methodology

The vibrator is penetrated to the required depth under its dead Load and in certain cases with aid of drilling fluid and vertical vibrations. The vibrator is pin jointed to a stay tube which of same diameter as the vibrator.

After reaching the desired depth, the horizontal vibrations are initiated which leads to the compaction of the surrounding soil. The extent of compaction can be observed in an area of radius 0.6 m to 1.75 m (Fellin 2000, Witt 2009) around the vibrator. The vibrator is retraced back in steps of 0.3 m to 1 m of depth. Shear waves imparted to the surrounding soil leads to compaction. The soil from the adjacent areas moves into the void spaces, which can be observed in terms of surface settlements.

3 MODEL LAWS

The model vibrator design is based on the S-vibrator commonly used for deep vibration compaction. The scaling laws used in the work are based on Wood (2004). The linear dimensions of the vibrator has been scaled by a factor of $\lambda = 13.15$. The force that can be generated by the vibrator has been scaled by a factor of λ^3. Time scale factor of $\lambda^{1/2}$ is considered. The stress state in the sand in the test tank is considered under 1g. In order to reduce error due to hydromechnical modelling errors, tests are conducted in dry sands, rather than in saturated sands where scaling of time would play an important role. On prototype level, compaction is generally executed in saturated granular materials. The effect of vibration is not observed immediately due to increase in pore water pressures but is a gradually observed with the drainage of the excess pore water pressure. Hence the outcome of the final compaction in both saturated granular material and dry granular material can be compared at qualitative level and hence the model tests in dry sands would lead to good insight on the compaction process at primary level.

4 MINI VIBRATOR

A prototype vibrator was designed as shown in Figure 1. The mini vibrator consists of a steel tube of 0.038 m diameter and has a length of 0.213 m which is pin jointed to a stay tube of similar diameter and a length of 0.3 m as shown in Figure 1. It has pointed tip and wings at the tip of the vibrator in order to aid the penetration of the vibrator in the soil. It consists of eccentric masses made of brass weighing 0.675 kg located near the tip of the vibrator as shown in Figure 2. The eccentric masses maintain a constant eccentricity of 7.213 mm from the central vertical axis of the vibrator. The masses are connected to a motor through an internal shaft. The motor is EC-max 30 from Maxon and is used in the working range of 25 Hz to 66 Hz. The data from the motor in terms of the functioning frequency can be recorded on the data logger.

5 MODEL TEST

5.1 Test setup and materials used

Model tests are conducted in test tank with dry sand. The steel cylindrical test tank measures 1 m in diameter

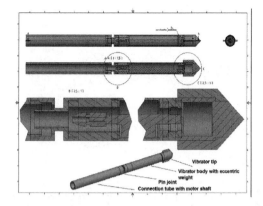

Figure 1. Design of mini vibrator with stay tube.

Figure 2. Eccentric mass of vibrator.

and 1.1 m in height as shown in Figure 3. The test stand is equipped with an automated sand filling system. The sand is stored in an overhead tank which can be moved up vertically at a controlled rate while the sand is deposited from the overhead tank into the test tank through a funnel. The rate of the movement of the overhead tank is controlled as per requirement based on trial and error in order to ensure that sand of fairly uniform density is deposited in the test tank. The test tank frame consists of a connection rig through which the CPT can be connected and driven into the soil at the required rate with aid of the motor attached to the test frame. Pressure sensors are installed at the boundaries of the test tank in order to record the stresses the soil is subjected to at the boundaries.

Hamburger sand consisting of nearly spherical grains whose physical properties are well established is used for the tests. The basic physical properties of the Hamburger sand are as tabulated in Table 1.

5.2 Procedure

Trials were made in order to determine the rate at which the storage tank should be raised in order to have a sand deposit of uniform density. Once the rate was identified the same procedure was used to deposit the sand in order to avoid discrepancies between various tests. The overhead tank was raised at a uniform rate (obtained by trail and error) in order to ensure that sand of uniform density was deposited in the test tank. Sand was deposited in the test tank in as loosest state as

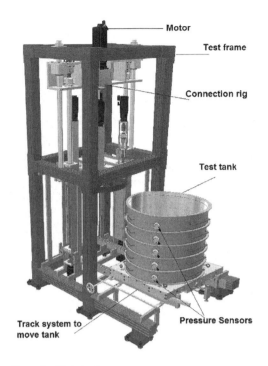

Figure 3. Test stand.

Table 1. Basic physical properties of Hamburger sand.

Property	Value
e_{min}	0.526
e_{max}	0.813
Friction Angle	32°
Density max	1,736 kg/m^3
Density min	1,462 kg/m^3
Particle Size	1–2 mm

possible at an average relative density of 0.2. The sand surface was smoothened in order to ensure a horizontal surface.

The process of compaction using the mini vibrator involved two primary processes. First being the penetration of the vibrator into the sand and the second part being the horizontal oscillations of the vibrator at required depth. It was observed that the vibrator could easily be penetrated into the sand at higher frequency of around 45 Hz. Hence in all the tests the vibrator was inserted into the sand at a frequency of 45 Hz. Tests were conducted at different compaction frequencies ranging from 15 to 60 Hz. This range was chosen in accordance to the frequency range used on field for deep vibration compaction as stated earlier. The mini vibrator was penetrated into the sand layer at the center of the test tank. In order to ensure uniformity, the vibrator was penetrated to a depth of 0.423 m from top of the sand layer. After reaching the required depth, the frequency of the compacter was set to the required compaction frequency. After compaction the vibrator was again withdrawn at a frequency of 45 Hz.

Figure 4. Compaction trough created due to compaction by vibrator.

In order to access the success of compaction, CPT tests are performed on the sand before and after the compaction. CPT-15 of 36 mm diameter was mounted on the test frame and was driven into the sand at a rate of 20 mm/s. CPT results are recorded at the center of the test tank and at radial distances of 70 mm, 140 mm and 250 mm from the center. For each of the test the CPT was driven into the soil before compaction and after the compaction at these four locations. Calibration was done in order to convert the CPT test data into tip resistance.

6 RESULTS

6.1 Compaction by mini vibrator

Figure 4 depicts the trough created around the vibrator due to compaction. The shear waves generated by tie vibrator leads to the compaction of sand at the penetration depth. The soil from adjacent regions will move into the resulting void spaces. This movement can be noticed as a settlement of the ground surface as shown in Figure 4.

6.2 Effectiveness of compaction

The effectiveness of the mini vibrator to compact the loose sand was studied in terms of the CPT test values before and after vibration. The degree of compaction can be indirectly associated with the CPT resistance. As the density of the sand increases with compaction the CPT values are bound to increase. Hence the CPT tip resistance is used as an indicator of degree of compaction. It is observed in Figure 5 that there is an increase in the CPT values before and after compaction indicating that the mini vibrator lead to the compaction of the loose sand. The CPT values show some discrepancy as they approach the base of the test tank probably due to boundary effects.

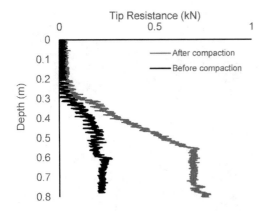

Figure 5. CPT values before and after compaction at 60 Hz.

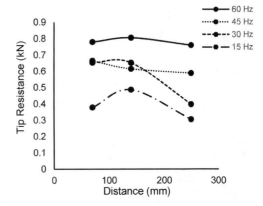

Figure 6. CPT values at various radial distances.

6.3 Extent of compaction

Compaction profiles show formation of three zones around the vibrator: a zone of limited compaction around the vibrator, zone of maximum compaction and zone of no compaction at remote distances from the vibrator (Arnold et al. 2009). CPT tests were performed at three radial distances from the center of the compaction point in order the study the extent of compaction. Figure 6 depicts the CPT values for various frequencies at various radial distances. The degree of compaction increases with frequency. But it can also be observed that for a larger extent of compaction higher frequencies are more suitable. It can be observed in Figure 6 that minimum compaction is observed at 70 mm and maximum compaction is obtained at a distance of 140 mm. It can also be observed that extent of compaction is greater at higher frequencies than at lower frequencies where the CPT values drop after 140 mm.

6.4 Degree of compaction

The density of the sand achieved depends predominantly on the vibration frequency. There is an optimum frequency at which both the extent and degree of compaction achieved is maximum (Massarsch & Fellenius 2005). It is observed in Figure 5 that the extent of compaction is maximized at higher frequencies of 45 and 60 Hz. It can also be observed that maximum values of CPT resistances are recorded for 60 Hz compaction. In Figure 7 it can be observed that maximum density of sand at a radial distance of 250 mm is achieved when the vibrator runs at a frequency of 60 Hz and as the frequency reduces, the degree of compaction also reduces. The degree of compaction at 250 mm for frequencies of 15 and 30 Hz is the similar due to the larger radial distance where the effect of effect of compaction by lower frequency is hardly perceivable.

6.5 Geophone measurement

During the compaction phase, maximum energy should be transferred from the vibrator to the sand in order to have an effective compaction. This is

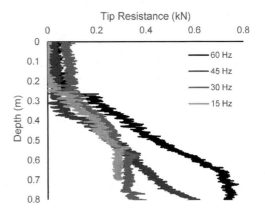

Figure 7. CPT values at various frequencies at 250 mm radial distance.

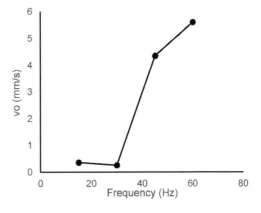

Figure 8. Maximum RMS value of particle velocity amplitude from measure frequency response.

achieved when the vibrator vibrates in resonance with the soil. Resonance between the vibrator-soil-system leads to an increase in ground vibrations (Massarsch & Fellenius 2005) and hence measurement of the ground vibrations can provide a useful insight on the compaction process. In order to study the effect of frequency on the ground vibrations, geophone is installed on top of the sand layer at a distance of 250 mm from the center of the compaction point in the test tank. Figure 8. describes the increase in the particle velocity amplitude with frequency increase. It can be observed that maximum particle velocity and the degree of compaction at 250 mm radial distance (Fig. 7) for 30 and 15 Hz is comparable. The geophone data serves as a preliminary source of resonance frequency investigation (Wersäll et al. 2015). It should be noted that geophone data is biased due to reflex vibrations from the test tank. Various alternatives are being considered to be executed in the tests in future.

7 CONCLUSIONS

Model test with mini vibrator can serve as an effective mean to understand the physical processes granular materials undergo during deep vibration compaction. Frequency at which the compaction is executed predominantly governs both the degree and extent of compaction. The CPT results indicate that there is substantial effect on the results nearing boundary due to boundary effect. The pressure sensor data is being processed in order to understand the magnitude of the effect. Future work aims at the installation of accelerometer at the tip of the vibrator to measure the phase angle between acceleration and direction of movement of the vibrator in order to investigate the idea of Fellin (2000). Numerical simulations of the deep vibration compaction process is to be validated against the model test with the mini vibrator.

REFERENCES

Arnold, M. & Herle, I. 2009. Comparison of vibrocompaction methods by numerical simulations. *International Journal of Numerical and analytical methods in Geomechanics* 33: 1823–1838.

Fellin, W. 2000. *Rütteldurckverdichtung als plastodynamisches Problem*. Insitute of Geotechnics and Tunneling, University of Innsbruck, Advances in Geotechnical Engineering and Tunneling, Heft 2.

Massarsch, K.R. & Fellenius, B.H. 2005. Deep vibratory compaction of granular soils. In B. Indranatna & C. Jian, (eds) *Ground Improvement-Case Histories* 1:633–658.

Wersäll, C., Larsson, S., Rydén, N., & Nordfelt, I. 2015. Frequency Variable Surface Compaction of Sand Using Rotating Mass Oscillators. *Geotechnical Testing Journal*, 38(2): 198–207.

Witt, K.J. 2009. *Grundbau-Taschenbuch, Teil 2: Geotechnische Verfahren*. Berlin: Ernst & Sohn.

Wood, D.M. 2004. *Geotechnical Modelling (Applied Geotechnics)*. London and New York: Spon Press.

Dynamic centrifuge tests on nailed slope with facing plates

S. Nakamoto
Kajima Corporation, Tokyo, Japan

N. Iwasa
Nippon Steel & Sumikin Metal Products Co. Ltd., Japan

J. Takemura
Tokyo Institute of Technology, Tokyo, Japan

ABSTRACT: Rock bolts with individual facing plates is one of the soil nailing methods against relatively shallow sliding failure, which can be applied to natural slopes without cutting trees. In this research, dynamic centrifuge tests were conducted to study the reinforcement mechanism of the rock bolt with individual facing plates and the effects of facing plate width, rock bolt spacing, and rock bolt inclination angle on the seismic stability of the reinforced slope. From the test results, the facing plate resistance working well against the seismic load was confirmed. For weak slopes, plate bearing failure and local surface failure are critical failure mechanisms for the reinforced slopes, which cause reduction of plate resistance and leads to large deformation and slip failure of the slope. Downward inclined installation of the rock bolt from the normal to the slope surface can effectively prevent the movement and secure the slope stability with less mobilised resistance.

1 INTRODUCTION

Rock bolts with facing plates is one of soil nailing methods against relatively shallow sliding failure. Rock bolts are inserted with grout in the slope, and stiff individual facing plates are attached at the bolt head to cover the slope surface with a certain preloading. Compared to the nail without facing or with wall facing, this method can be applied on loose natural slopes without cutting trees (Figure 1). To study the reinforcing effect and the resistance force mobilisation of soil nails without facing protection works or with wall facing, several researches have been conducted by numerical analyses (Jewell & Pedley (1992)), laboratory tests (McGown & Andrawes (1977)), centrifuge model tests (Tei et al. (1998)) and site monitoring (Lin et al. (2016)). While few studies have been done on the stability of nailed slope with individual facing plate (Hayashi et al. (1986)). And to the best of authors' knowledge, the reinforcing effects of nailing with individual facing plate under dynamic conditions have not been studied well. In this study, dynamic centrifuge tests were conducted on reinforced and unreinforced slopes with rock bolt and individual facing plates to study its reinforcement effect with different plate size, rock bolt spacing and rock bolt installation angle. From the test results, the reinforcing effect of this method, and the failure mechanism of the nailed slope is discussed.

2 CENTRIFUGE MODEL TESTS

2.1 Test setup

For the rock bolts installed in the immobile layer, three critical conditions, which cause slope failure, can be considered, a) local and bearing failure in the mobile layer, b) pull-out failure of the rock bolt grouted in the immobile layer, and c) tensile failure of the bolt as shown in Figure 2. Among the three conditions,

Figure 1. Soil slope reinforced by rock bolts with individual facing plates.

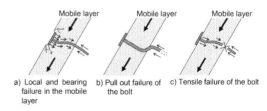

Figure 2. Three critical conditions of reinforced slope by rock-bolt with individual facing plate.

the first one is focused in this study. A mobile layer was modelled with relatively shallow soft soils with constant thickness resting on a hard immobile layer to which the rock bolts with enough tensile strength are fixed.

Using an aluminum made rigid container with inner dimensions of 600 mm length, 250 mm breadth and 400 mm depth, the conceptual model was made as shown in Figure 3. An aluminum made rigid triangle box with 45 degrees' slope was fixed at the bottom of the container. Silica sand was glued to the base of the slope to create rough surface and 28 tapered holes with 80 mm or 160 mm spacing were provided to accommodate the model rock bolts. The model rock bolts were 3 mm diameter brass screw rods with plastic resin pasted on the surface to simulate grouting. The rock bolts were installed in a rectangular alignment in the compacted mobile layer with 120 mm thickness. Model facing plates, 2 mm thick aluminum square plates, were then attached at the end of bolts by nuts with miniature load cells.

Parameters investigated in this study are facing plate width (B), spacing of the rock bolts (S) and rock bolt installation angle (α) as shown in Table 1 and Figure 4.

2.2 Model slope

The mobile layer of the model slope was made of Edosaki sand with optimum water content of 14.5% and compaction degree of 85%. For comparison, a slope model of mobile layer with 95% degree of compaction (CaseN-95) was also conducted without reinforcement. Physical and mechanical properties of

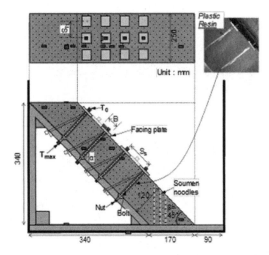

Figure 3. Model setup and test parameters.

Table 1. Model test conditions [prototype scale in bracket].

Test	Compaction degree D_c	Facing plate width B	Rock bolt spacing Sloping direction S_s	Horizontal direction S_h	Covering ratio $B^2/(S_s \times S_h)$	Installation Angle α	Number of shaking	Max input acceleration amplitude a_{inp}
CaseN-95	95%	—	—	—	0	—	4	12.5 g [490 gal]
CaseN-85	85%	—	—	—	0	—	2	9.3 g [370 gal]
CaseR-S_n-2	85%	20 mm [0.5 m]	80 mm [2 m]	82.5 mm [2.06 m]	0.06	90°	3	12.9 g [515 gal]
CaseR-M_n-2	85%	30 mm [0.75 m]	80 mm [2 m]	82.5 mm [2.06 m]	0.14	90°	5	19.0 g [760 gal]
CaseR-L_n-4	85%	40 mm [1.0 m]	160 mm [4 m]	125 mm [3.13 m]	0.08	90°	5	9.5 g [380 gal]
CaseR-M_i-2	85%	30 mm [0.75 m]	80 mm [2 m]	82.5 mm [2.06 m]	0.14	70°	5	18.5 g [740 gal]

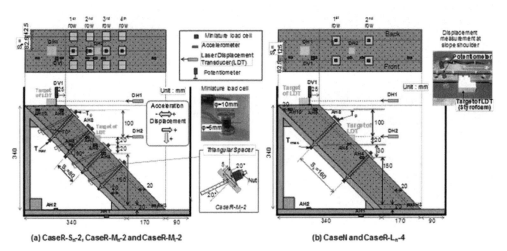

Figure 4. Details of reinforcements and instrumentations.

the sand used for making the model slope are shown in Table 2. The strength parameters shown in the table were obtained from triaxial compression tests under the drained condition with confining stresses from 25 to 100 kPa. The results of the triaxial compression tests are shown in Figure 5. Plate loading tests were conducted on the Edosaki sand mobile layer and the load-displacement relations obtained from those tests are shown in Figure 6.

Table 2. Physical and mechanical properties of Edosaki-sand used for the model slope.

Specific gravity:	Gs	2.72
Particle size	D_{10}	0.012 mm
	D_{30}	0.16 mm
	D_{50}	0.27 mm
	D_{60}	0.34 mm
Bulk density:	ρ_t (85%)	1.76 g/cm³
Bulk density:	ρ_t (95%)	1.94 g/cm³
Water content	w (=w_{opt})	14.5%
Degree of compaction**	D_c	85%, 95%
Friction angle*:	φ' ($D_c=85\%$)	37°
	φ' ($D_c=95\%$)	41°
Cohesion:	c' ($D_c=85\%$)	3.3 kPa
	c' ($D_c=95\%$)	12.8 kPa

*Drained triaxial compression ($\sigma_3 = 25$–100 kPa)
**JIS A 1210(JGS, 2009)

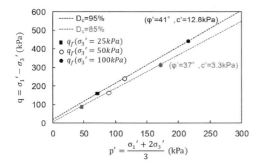

Figure 5. Relationships between the mean principal stress p and the ultimate deviator stress q obtained from drained triaxial compression tests.

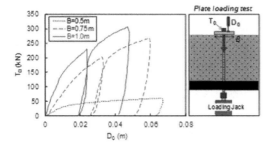

Figure 6. Vertical displacement D_0 and bearing resistance force T_0 of the facing plate measured in the plate loading tests under 25 g.

2.3 Model preparation and test procedures

In the preparation of the mobile layer, Edosaki-sand was compacted horizontally layer by layer from the bottom of the container. After the compaction, the front and back walls were detached and then the compacted soil was cut to the slope with 45 degrees' inclination angle using a template. In the cases with reinforcement, 3 mm holes were augured in the normal direction, except of one model, to the slope surface to the tapered holes in the base and then model rock bolts were inserted in the holes. After installing the rock bolts and placing udon-noodles on the front slope surface for the visual observation of the slope displacement, the walls were attached to the model container.

Before the shaking tests, 40 g centrifugal acceleration was once applied to the model slope and then stopped as the preloading process to secure certain facing plate loads. After applying the preload, the centrifugal acceleration was again increased up to 25 g, which is called G-up process in these tests. Figure 7 shows the facing plate resistance forces measured by small load cells before the shaking test. From the figure, it can be confirmed that the facing plate resistance force at each location is very similar in all the cases except of CaseR-L_n-4 with larger rock bolt spacing. Under 25 g, a several input motions were applied to the model as shown in Figure 8. Firstly, small amplitude white noise wave was inputted and then 50 Hz (which is equivalent to 2 Hz in the prototype scale) tapered sinusoidal waves were applied with the amplitudes, which were increased waves by waves until a large deformation observed on the slope. The maximum amplitudes of the sinusoidal waves applied to the slope are summarised in Table 1. In Figure 6, Figure 7 and the following section, the results are shown in prototype scales. And the records of the acceleration in this study were filtered by cutting out frequencies of greater than 10 Hz in prototype scales.

3 TEST RESULTS AND DISCUSSIONS

3.1 Slope deformation and failure characteristics

Figure 9 shows the input acceleration amplitude (a_{inp}) and cumulative vertical displacement at the slope

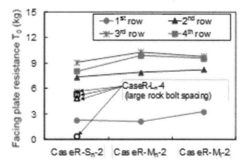

Figure 7. Measured facing plate resistance forces before 1st shaking (T_{0i-1}) and the shaking when the failure (T_{0i-f}).

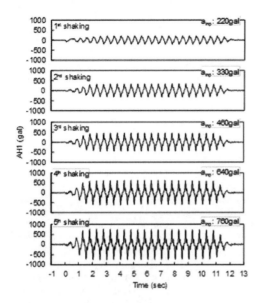

Figure 8. Example of input waves, CaseR-M_n-2.

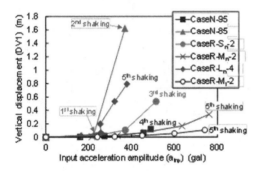

Figure 9. Relationships between input acceleration amplitude (AH1) and slope shoulder vertical displacement (DV1).

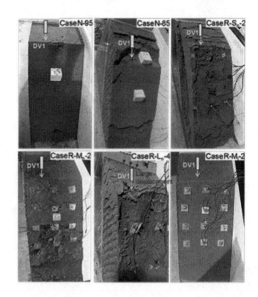

Figure 10. Failures of the slope surface observed after the tests.

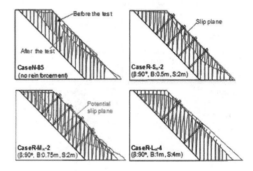

Figure 11. Deformations of the slope observed after the tests.

shoulder (DV1) relationships measured after each shaking. For no-reinforcement case with $D_c = 85\%$ mobile layer (CaseN-85), a clear failure occurred in the slope and the slope shoulder displacement increased sharply in the 2nd shaking step. While in the case with $D_c = 95\%$ mobile layer (CaseN-95), no clear slope deformation could be found even when relatively large motion was applied. From these results, $D_c = 85\%$ was chosen for the mobile layer of soil in the reinforcement cases to clarify the difference in the different reinforcement conditions. The displacements caused by the shaking could be greatly decreased by the reinforcement, and the displacements tends to be smaller with the larger facing plate for the same rock-bolt spacing. However, even though the plate covering ratio was larger, in CaseR-L_n-4 than CaseR-S_n-2 (Table 1), the larger displacement took place at the smaller acceleration in the former than the latter. From the smallest displacement in CaseR-M_i-2 ($\beta = 70°$), a significant effect of bolt installation angle in preventing slope deformation can be confirmed.

Slope top views and deformations observed after the tests are shown in Figures 10 and 11 respectively. A crack was shown at the center of the slope in the case with $D_c = 95\%$ mobile layer (CaseN-95), what could be because of the deformation at the top of the slope. However, no clear slope failure and slip line could be found in this case. On the other hand, large deformation with clear slip lines took place in the case with $D_c = 85\%$ mobile layer without reinforcement (CaseN-85) and the case with small size facing plate (CaseR-S_n-2). On the slope surface of CaseR-S_n-2, the bearing failure of the plate and local surface sliding between the plates were observed. In CaseR-M_n-2 and CaseR-L_n-4, although no clear slip plane occurred, large deformations with deep potential slip plane were observed. The location of the plane reached to the near the slope toe in the former case, while it was limited in the upper slope in the latter. Both cases showed local surface failures between the plates, but the locations were at the lower part of the slope in CaseR-M_n-2 and near the slope shoulder in CaseR-L_n-4, which could cause the difference of the locations of

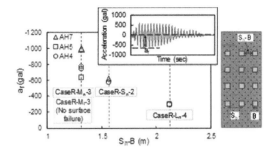

Figure 12. Maximum response acceleration at the slope surface before the slope surface failure occur.

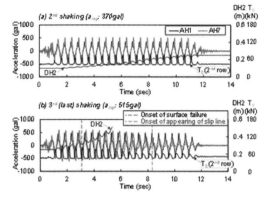

Figure 13. Time histories of the input accelerations (AH1) and surface response accelerations (AH5, AH7), horizontal displacement at the slope center (DH2), and facing plate resistance (T_0): (a) CaseN-85 and (b) CaseR-S_n-2.

potential slip plane. By contrast, no clear local failure and bearing failure occurred in CaseR-M_i-2 even the maximum input acceleration was larger than those of the most other cases. Figure 12 shows the relationships between the uncovered width (S_h-B), which is the width of the area without facing, and the maximum amplitudes of the response accelerations of the slope surface at the time of surface failure. As shown in the Figure 4, the acceleration to the valley side direction is taken as positive, the negative acceleration means the inertia force acting to the valley side as driving force of slope sliding. The maximum amplitudes of response accelerations at the slope surface has a clear relationship with the uncovered width. As the uncovered width increases, the critical acceleration causing the surface failure decreases. This is a main reason why CaseR-L_n-4 with coving ratio of 0.08 was less stable against the shaking than CaseR-S_n-2 with that of 0.06. While in CaseR-M_i-2 with downward inclined nail ($\alpha = 70°$), though similar response accelerations were observed, no surface failure occurred.

3.2 Resistance force mobilisation on the reinforced slope

Figure 13 shows the time histories of AH1, AH7 and DH2 with facing plate resistance force (T_0) measured at the 2nd row of the reinforcement during the 2nd shaking in CaseR-S_n-2 with smallest facing plate. In this case, a clear slip surface occurred during the 3rd shaking. And in the 2nd shaking, neither clear slope failure nor sharp increase of the horizontal displacement could be observed. As shown in the figure, some residual resistance force has been developed due to the previous steps. Additional facing plate resistance was mobilised in limited time of each cycle against the inertia force acting to the valley direction. The mobilised resistance gradually increased with increasing displacement amplitude. Initiations of the local surface failure and slip plane could be identified from the video inspection during the 3rd shaking. The onsets of the detected surface failure and slip failure are indicated in Figure 13. The plate resistance started decreasing at the time of surface failure and then the slope acceleration attenuated just before the appearance of slip plane. These observations imply that the gradual increase of plate resistance could prevent the large slope deformation, but the local surface failure leads the reduction of plate resistance, resulting the slope failure. The same behavior also could be found in CaseR-M_n-2 during the last shaking, and in CaseR-L_n-4 during the 3rd shaking when the slope surface failure was observed (Nakamoto et al. 2017), though the input acceleration in the former was much larger than the latter. And the maximum decrease of the facing plate resistance reduction ratio, which is the ratio of the peak of facing plate resistance before surface failure to that after surface failure ($T_{0peak-after}/T_{0peak-before}$) in CaseR-$S_n$-2, CaseR-$M_n$-2, CaseR-$L_n$-4 is 0.61, 0.71 and 0.65 respectively. Maximum value of the facing plate resistance and additional facing plate resistance developed against the shaking (T_{0max} and ΔT_0) are plotted to the amplitude of the input acceleration at AH1 in Figure 14-(a) and Figure 14-(b), and the residual facing plate resistance forces measured at the facing plates at 2nd rows after each shaking are plotted to the residual vertical displacement measured at the slope shoulder, DV1, in Figure 14-(c). In CaseR-S_n-2, the maximum facing plate resistance during the 3rd shaking was much larger than the bearing resistance obtained from the static loading test (Figure 6), which implying that large facing plate penetration could induce the bearing failure of the facing plate and the local failure between the facing plates. On the hand, in CaseR-L_n-4, the observed facing plate resistance in this case was much smaller than those observed in the static loading, while the time of facing plate resistance reduction corresponds to the occurrence of the local surface failure between the facing plates. The additional facing plate resistances (ΔT_0) increased with increasing facing plate width. No clear inclement of residual facing plate resistance could be found in the large range of displacement in CaseR-S_n-2, which also implying the facing plate bearing resistances were very close to the ultimate bearing capacity before the shaking for the small facing plate. In CaseR-L_n-4, the facing plate resistance did not increase at the mid upper part, which could be also due to the serious local failure between

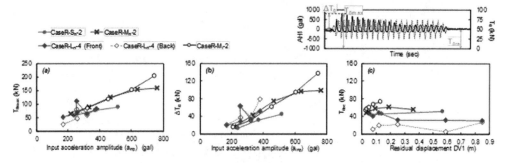

Figure 14. a_{inp}-T_{0max}, a_{inp}-ΔT_0 and DV1-T_{0re} relationships (facing plate at 2nd row).

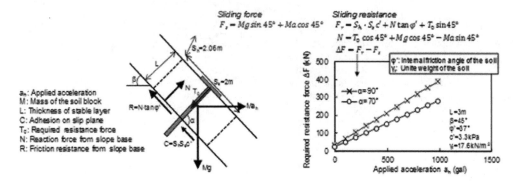

Figure 15. Simple calculation of required resistance of the facing plate with different nail inclination angle against the applied acceleration using infinite slope assumption.

the facing plates as shown in Figure 10. The residual resistances at the same displacement were larger in CaseR-M_i-2 than CaseR-M_n-2, which also indicating the effective mobilisation of the resistance by relatively small slope displacement for the inclined rock bolts than those installed normal to the slope surface.

The effect of the angle of inclination can be captured by a simple analysis on the reinforced infinite slope as shown in Figure 15. The necessary resistance force for keeping the conditions for the equilibrium of the forces ($\Delta F = F_r - F_s$) is smaller for the case with $\alpha = 70°$ than $\alpha = 90°$, which also can be found in Figure 14-(b). This small resistance force requires a small facing plate penetration for stabilising the slope, which also reduces the crack formation from the plate edge and the local failure initiated from the crack.

4 CONCLUSIONS

In this study, a series of dynamic centrifuge model test was conducted on nailed slope with individual facing plates to study the failure mechanism of the slope and the effects of the facing plate width, rock bolt spacing and rock bolt installation angle. From the results, it could be found that the facing plate resistance well works against the seismic load, which suppresses the slope deformation during the shaking and the generated resistance increases with increasing facing plate size. Plate bearing failure and local surface failure are critical failure mechanisms for the reinforced slope with individual facing plates that occur when the plate is small or uncovered area is large respectively. These surface failures cause reduction of plate resistance and leads to large deformation and slip failure of the slope. Downward inclined installation of the nailing from the normal to the slope surface can effectively reduce the movement and secure the slope stability with less mobilised resistance.

REFERENCES

Hayashi, S., Ochiai, H., Tayama, S. & Sakai, A. 1986. Effect of top plates on mechanism of soil reinforcement of cut-off slope with steel bars. *Journal of JSCE*. 367(3): 62–70.

Jewell, A. & Pedley, M.J. 1992. Analysis for soil reinforcement with bending stiffness. *Journal of Geotechnical Engineering*. 118(10): 1505–1528.

Lin, P., Bathurst, R. & Liu, J. 2016. Statistical Evaluation of the FHWA Simplified Method and Modifications for Pre-dicting Soil Nail Loads. *Journal of Geotechnical and Geo-environmental Engineering*.

McGown, A. & Andrawes, K.Z. 1977. The influence of non-woven fabric inclusions on the stress strain behavior of a soil mass. *Proceeding of International Conference on the Use of Fabrics in Geotechnics:* 161–166.

Nakamoto, S., Iwasa, N. & Takemura, J. 2017. Effects of nails and facing plates on seismic slope response and failure. *Géotechnique Letters*. 7(2): 136–145.

Tei, K., Taylor, N. & Minnigan, G.W.E. 1998. Centrifuge model tests of nailed soil slopes. *Soils and Foundations*. 38(2): 165–177.

Influence of slope inclination on the performance of slopes with and without soil-nails subjected to seepage: A centrifuge study

V.M. Rotte
Institute of Infrastructure Technology Research and Management, Ahmedabad, India

B.V.S. Viswanadham
Indian Institute of Technology Bombay, Powai, Mumbai, India

ABSTRACT: This paper presents a study on influence of slope inclination on the performance of slopes with and without soil-nails subjected to seepage. Centrifuge model tests were performed on 2V:1H (63.43°) and 5V:1H (79°) slopes with and without soil-nails at 30 gravities. All model slopes were instrumented and movable markers were digitised in order to get displacement vectors with rising ground water table. The model slope (unreinforced) with inclination of 63.43° was observed to be stable at 30 g till when the seeped water reached to toe of the slope. However, model slope (unreinforced) with inclination of 79° was noticed to fail as soon as it reached 30 g, before being subjected to seepage. Soil-nailed model slope inclined at 63.43° registered crest settlement of 3.13 mm. In comparison, model soil-nailed slope inclined at 79° experienced a crest settlement of 9.8 mm. The observed centrifuge model tests were found to match well with results of the Finite Element analysis.

1 INTRODUCTION

The most important forces that cause instability in existing slopes are the force of gravity and the force of ground water seepage. The force of gravity accelerates movement of an unsupported soil mass in downward and outward directions to obtain a more stable condition. It increases with an increase in slope inclination and that affects the stability and behaviour of a slope. Seepage-induced slope failures are highly catastrophic and cause geo-hazards leading to heavy loss of human life and property. Hence, researches attempting to design and construct effective engineering measures for strengthening slopes have significant relevance. Soil nailing is one of the in-situ reinforcement techniques that has been used widely for the past five decades to stabilise existing slopes and to retain excavations. Stability of a soil-nailed slope is closely related with the geometry of slope and seepage force.

Several researchers studied the effect of slope inclination on the behaviour of slopes with and without soil-nails, under various loading conditions, through numerical, full-scale and small-scale (at normal and high gravity) studies. Fan & Luo (2008) and Rotte et al. (2011) used numerical/analytical approach to find out an optimal nail inclination of a soil-nailed slope considering various slope inclinations, back slope inclinations and nail inclinations. The nail inclination with horizontal which fulfils both the requirements of safety and economy of a soil-nailed slope is termed as optimal nail inclination (α_{opt}) of the considered slope.

It was observed to decrease with an increase in slope inclination for horizontal backslope. However, for a constant slope inclination, it was noted to increase with an increase in back slope inclination. Mittal & Biswas (2006) reported that the factor of safety of a soil-nailed slope increases with a decrease in slope inclination. Rawat et al. (2013) performed a series of 1-g model tests on soil-nailed slopes ($\beta = 30°$, 45° and 60°) subjected to surcharge loading to study the influence of slope inclination on the performance of soil-nailed slopes. For all slopes ($\beta = 30°, 45°$ and $60°$) tested with soil-nails, the stability and settlement profiles of soil-nailed slopes were noted to be influenced by slope inclination.

Aminfar (1998) conducted centrifuge tests at 20 gravities on soil-nailed slopes of 375 mm high with $\beta = 45°, 60°$, and $70°$. It was reported that the magnitudes of maximum axial force in soil-nail, lateral deformation and surface settlement of slopes were observed to be influenced strongly by the slope inclination. Morgan (2002) studied the performance of soil-nailed slopes of 300 mm height with a flexible facing at 20 g to find out the effect of slope inclination on the performance of soil-nailed slopes. It was noted that an increase in slope inclination mobilise soil-nail capacity completely. The deformation behaviour of soil-nailed slopes was noticed to be influenced by seepage force developed within the slope (Morgan 2002; Ng et al. 2006; Ming 2008; Deepa & Viswanadham 2009). Soil-nailed slope without slope facing failed to resist local failure at the toe of the slope when subjected to seepage (Tei et al. 1998; Deepa & Viswanadham 2009; Viswanadham & Deepa 2010). Centrifuge studies on soil-nailed slopes subjected to seepage reported by Tei et al. (1998) and Rotte & Viswanadham (2014) indicated that stiffness of the slope facing influences deformation behaviour of the slope, significantly.

Based on the literature review, it was found that very few researchers have attempted to understand the influence of slope inclination on the performance of slopes with and without soil-nails in a centrifuge. With the exception of Aminfar (1998), Tei et al. (1998) and Morgan (2002), the slope inclination with the horizontal was not varied in the centrifuge. Some of the investigators (Morgan 2002; Ng et al. 2006; Ming 2008; Deepa & Viswanadham 2009; Viswanadham & Deepa 2010; Rotte & Viswanadham 2013, 2014) have studied the behaviour of soil-nailed slopes at the onset of seepage in a centrifuge. However, except for Morgan (2002), the study related to effect of slope inclination on the behaviour of soil-nailed slopes subjected to seepage is unaddressed.

Since the stability and deformation behaviour of slopes with and without soil-nails are influenced by prototype stress levels, full-scale model studies or centrifuge model studies can provide more realistic information when compared to small-scale physical model testing at normal gravity. Hence, centrifuge modelling (CM) technique was used in the present study to address the influence of slope inclination on the behaviour of slopes with and without soil-nails when subjected to seepage.

2 CENTRIFUGE MODEL TESTS ON SLOPES WITH AND WITHOUT SOIL-NAILS

2.1 Centrifuge equipment

The 4.5 m radius large beam centrifuge facility available at the Indian Institute of Technology, Bombay was used to perform centrifuge tests on slopes with and without soil-nails subjected to seepage. With the help of an on-board central processing unit (CPU), LAN connections and embedded signal conditioning and filter cards data was continuously acquired from various transducers. The scaling considerations of parameters related to soil, slope, soil-nails and slope facing need to be fulfilled and are discussed in detail by Rotte & Viswanadham (2013).

2.2 Model materials

2.2.1 Model soil
Model soil was composed of a mixture of sand and kaolin in the ratio of 4:1 by dry weight. According to USCS, the model soil was classified as silty sand (SM) and it was found to have a specific gravity of 2.63. The maximum dry unit weight ($\gamma_{d,max}$) and optimum moisture content (OMC) of the model soil were found to be 19.1 kN/m^3 and 9.5% respectively (standard Proctor compaction). Undrained shear strength parameters of the model soil c_u and ϕ_u were obtained as 12.5 kN/m^2 and 28° and effective shear strength parameters c' and ϕ' were found to be 10 kN/m^2 and 32° (consolidated undrained triaxial test). The coefficient of permeability (k) of the soil was measured as 1.54×10^{-6} m/s through falling head tests on sample moist-compacted at $\gamma_{d,max}$ and OMC.

2.2.2 Model soil-nail
Based on the scaling considerations, thin aluminium tube having outer diameter (D_o) of 6 mm and having thickness of 0.5 mm was selected as model soil-nail and used. Interface behaviour between grout and surrounding soil (usually observed in the field) was simulated by applying a 1.25 mm thick layer of epoxy covered with standard sand (Grade I) on the surface of aluminum tube. R_{int} is the strength reduction factor between standard sand (Grade I) smeared aluminium plate and model soil. This was used as one of the input parameters for finite element analysis (FEA) of soil-nailed slopes and was found to be 0.83 through modified saturated direct shear tests. A 6 mm thick acrylic sheet of 20 mm × 20 mm size was selected to model the nail head and Key pins were used to connect nail head and the soil-nail.

2.2.3 Model slope facing
Performance of a soil-nailed slope without facing is significantly affected (Rotte & Viswanadham 2013, 2014). In the present study, an adapted fiber blended plaster of Paris facing (FbPoP) was used to model facing of soil-nailed slopes. In order to develop FbPoP facing, plaster of Paris powder was blended with 0.5% (by dry weight) of polyester fibers uniformly. Model FbPoP facing having thickness of 5 mm was prepared by mixing fiber blended poP powder with a water content of 60% and allowed to set and harden for 72 hours. Holes of 9 mm diameter were predrilled at desired locations with nail spacing of 70 mm × 70 mm to insert soil-nails. Custom prepared FbPoP facing was assessed for its engineering properties in the laboratory and it was found to have a unit weight of 12.1 kN/m^3, unconfined compressive strength (q_u) of 4,450 kPa, Cohesion (c_u) of 2,225 kPa and Modulus of Elasticity (E_f) of 320,000 kPa.

2.3 Test set up, programme and procedure

Figure 1 shows a perspective view of model soil-nailed slope (VBS 12) along with sensors and other accessories used in the present study. A strong box with internal dimensions of 760 mm × 200 mm × 410 mm was used for making slope models at normal gravity. A thin layer of white petroleum grease was applied on the front and rear walls of the strong box before placing thin flexible polythene sheet strips to reduce the friction between the inner walls and soil as well as to approximate plane-strain conditions. In order to induce seepage during the centrifuge model test at 30 gravities (g), a fabricated custom designed seepage simulator was used. The details and working of

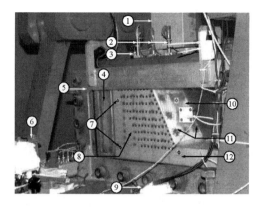

Figure 1. Photographs of a soil-nailed slope model: 1) Water reservoir; 2) LVDTs; 3) Solenoid valve; 4) Seepage Simulator tank; 5) Strong box; 6) Digital Camera; 7) Permanent Markers; 8) Soil-nailed slope; 9) Illumination arrangement; 10) Model Slope facing; 11) Model soil-nails; 12) Base layer.

seepage simulator unit can be obtained from Rotte & Viswanadham (2013).

A grid (350 mm × 210 mm) of permanent markers was positioned on the internal side of front Perspex sheet at pre-determined locations (Figure 1). The 70 mm thick base layer was prepared using model soil itself, moist-compacted at its maximum dry unit weight ($\gamma_{d,max}$) and optimum moisture content (OMC). After construction of the base layer, the model slope of 240 mm high with predetermined slope inclination was erected in layers of 30 mm thick using model soil moist-compacted at its $\gamma_{d,max}$ and OMC. In order to track movements of the slope models, movable plastic markers were embedded (2 cm from the slope face) within the front elevation of the slope as well as along the slope facing. The detailed procedure of construction of slope model and insertion of nail with facing can be attained by Rotte & Viswanadham (2013). In order to measure pore water pressure (PWP) within the slope during centrifuge tests, one PPT was placed within the seepage simulator tank and four PPTs were placed within the base layer. The surface settlements of the slope were measured with the help of LVDTs.

All the slopes with and without soil-nails were tested at 30 g, keeping the model soil type, slope height (H), and back slope inclination as constant. At 30 g, the solenoid valve connected to the water reservoir (Figure 1) was opened to permit water flow into the model slope through the seepage tank.

A digital camera was used for capturing images at every 30 second intervals to register the physical deformations of the slope model during the centrifuge test. All data from LVDTs and PPTs were recorded at 1 second intervals using an onboard computer data acquisition system.

3 ANALYSIS OF RESULTS AND DISCUSSION

In order to investigate the influence of slope inclination on the performance of soil-nailed slopes, two slope

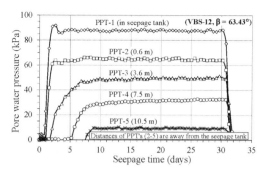

Figure 2. Variation of pore water pressure with seepage time.

inclinations of 63.43° and 79° were adopted. Four centrifuge tests (VBS-1, VBS-2, VBS-10 and VBS-12) were performed. Of these, VBS-1 ($\beta = 63.43°$) and VBS-2 ($\beta = 79°$) were unreinforced slopes whereas VBS-10 ($\beta = 79°$) and VBS-12 ($\beta = 63.43°$) were reinforced with soil-nails inclined at 25° along with square nail layout [$S_v/H = S_h/H = 0.29$] and FbPoP facing. The results are expressed in terms of measured pore water pressure (PWP) profiles, surface settlements and facing movements at the onset of seepage in the prototype. The status of front elevations and top surfaces of slopes, with and without soil-nails, and observed during post investigation of model slopes were also considered to study the effect of slope inclination on the performance of slopes subjected to seepage. Figure 2 depicts the variation of measured pore water pressures within the soil slope at the onset of seepage (along with the measured water pressure in seepage tank) for model VBS-12. The steady state seepage conditions could be achieved during the centrifuge test (Figure 2). With the help of measured pore water pressures within the slope, heads of water (i.e. phreatic surfaces) could be attained during various stages of seepage. Figure 3 presents the measured phreatic surfaces at various seepage times within the soil-nailed slope with FbPop facing (VBS-12) in the prototype.

Figure 4 presents the variation of crest settlements with seepage time for slope models VBS-1, VBS-2, VBS-10 and VBS-12. The model VBS-1, with slope inclination of 63.43°, was stable at 30 g up to a time till the seeped water not reached to toe of the slope. Subsequently, it was observed to fail due to built up of pore water pressure at the toe of the slope at seepage time equal to 5.1 days. The model VBS-2, with steeper slope inclination of 79°, was just stable before reaching 30g and failed as soon as it reached at 30 g, prior to being subjected to seepage. The soil-nailed slope models, with slope inclinations of 79° (VBS-10) and 63.43° (VBS-12), were noticed to exhibit maximum crest settlements of 0.294 m and 0.094 m respectively. As can be clearly seen, the model VBS-10, having slope inclination of 79°, settled more as compared to model VBS-12 having slope inclination of 63.43°.

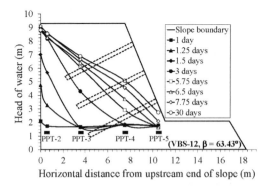

Figure 3. Developed phreatic surfaces at various seepage times.

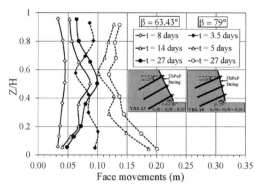

Figure 5. Lateral displacements at the onset of seepage.

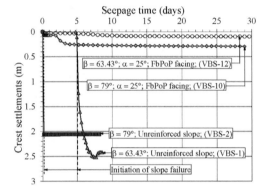

Figure 4. Variation of crest settlements with seepage time.

Figure 5 depicts face movements with normalised slope height (Z/H) for soil-nailed slopes with inclinations of 79° (VBS-10) and 63.43° (VBS-12) at various seepage times. As can be noted from Figure 5, the face movements increase with an increase in seepage time (days) for both the reinforced slopes (VBS-10 and VBS-12). For model slope with $\beta = 79°$ (VBS-10), the maximum face movements (measured for bottommost face marker) of 0.096 m, 0.186 m and 0.2 m were observed at seepage times (t) = 3.5 days, 5 days and 27 days respectively. However, soil-nailed slope having $\beta = 63.43°$ (VBS-12) experienced face movements (measured for bottommost face marker) equal to 0.033 m, 0.041 m and 0.048 m at seepage times of 8 days, 14 days and 27 days respectively. On comparing the results of models VBS-10 and VBS-12, Model VBS-12 ($\beta = 63.43°$) gave better performance than model VBS-10 ($\beta = 79°$).

The status of front elevation of slope models during the centrifuge tests (without seepage and at the onset of seepage) is presented in Figures 6a, b for unreinforced slopes (Models: VBS-1 and VBS-2) for various $u/\gamma H$ values. In this paper, $u/\gamma H$ is defined as the ratio of PWP measured by PPT3 at a distance of 245 mm from toe of the slope to the product of bulk unit weight of the soil and slope height. The model VBS-2 ($\beta = 79°$) was just stable before reaching 30 g. The top portion of

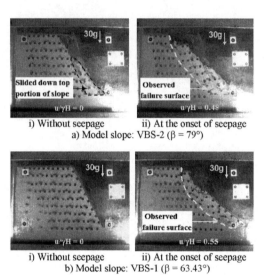

Figure 6. Status of front elevation of soil-nailed slope models.

the model VBS-2 ($\beta = 79°$) was noticed to fail immediately after reaching 30 g, showing a catastrophic failure of the slope when subjected to seepage ($u/\gamma H = 0.48$ as can be seen from Figure 6a). However, model VBS-1 ($\beta = 63.43°$) was observed to stable at 30 g and experienced catastrophic failure at $u/\gamma H = 0.55$ (Figure 6b). In addition, the developed failure surface was found to be away from the slope face for model VBS-2 ($\beta = 79°$) as compared to model VBS-1 ($\beta = 63.43°$).

Figures 7a, b show the status of top surfaces of the soil-nailed slopes observed during post-test investigations of model slopes (VBS-10 and VBS-12). Model VBS-10 ($\beta = 79°$) was observed to experience major multiple cracks on its top surface at $u/\gamma H$ of 0.49 (Figure 7a). However, for model VBS-12 ($\beta = 63.43°$), only a single major crack was noted at $u/\gamma H$ of 0.51 (Figure 7b). It was noticed that maximum top surface crack width for model VBS-10 ($\beta = 79°$) was 0.06 m and 0.045 m for model VBS-12 ($\beta = 63.43°$).

a) [Model: VBS-10; β = 79°]   b) [Model: VBS-12; β = 63.43°]

Figure 7. Status of top surface of soil-nailed slope models.

a) [Model: VBS-10; β = 79°]   b) [Model: VBS-12; β = 63.43°]

Figure 8. Deformation profiles for different slope inclinations.

a) [Model: VBS-10; β = 79°]   b) [Model: VBS-12; β = 63.43°]

Figure 9. Variation in nail forces for different slope inclination.

4 FINITE ELEMENT ANALYSIS (FEA)

A finite element analyses (FEA) of slopes ($\beta = 63.43°$ and $\beta = 79°$) with and without soil-nails subjected to seepage was performed in this study using PLAXIS 2D (PLAXIS, 2010), a geotechnical finite element code. The present study was simplified as a plane strain problem. Boundary conditions were fixed similar to those in centrifuge model tests in prototype dimensions. Six-noded triangular iso-parametric elements were used to model soil and Mohr–Coulomb constitutive model was selected to model stress–strain behaviour of soil. In FEA, soil-nails and slope facing were modelled as a rectangular shaped structural element 'Plate', considering elastic-perfectly-plastic behaviour with a limiting tensile force (T_p) and maximum bending moment (M_p). The analysis was carried out in three phases: (a) Gravity loading, (b) Plastic analysis and (c) Safety analysis. The detailed procedure followed for FEA of slopes was discussed in detail by Rotte & Viswanadham (2014).

Slope geometry influences the stability of soil-nailed slopes. To verify the significance of slope inclination on the performance of a slope with and without soil-nails, two slope inclinations (β) of 79° and 63.43° (with H = 7.2 m, $S_v/H = S_h/H = 0.29$; $\alpha = 25°$ and $t_f = 30$ mm) were used in the FEA. Effect of slope inclination on the performance of soil-nailed slopes was studied in terms of factor of safety (FOS), combined deformation of soil-nailed slopes in vertical as well as horizontal directions (D_{xy}) and distribution of nail forces. For unreinforced slopes ($\beta = 79°$ and $\beta = 63.43°$), the calculated safety factors were observed to be less than unity when subjected to seepage. Figures 8-9 show the deformation profiles and the variation in nail forces, respectively, for soil-nailed slopes $\beta = 79°$ and $\beta = 63.43°$. Soil-nailed slope with $\beta = 79°$ was found to experience maximum deformation (D_{xy}) of 0.347 m and showed FOS equal to 1.04, as seen in Figure 8a. However, for a soil-nailed slope with $\beta = 63.43°$, maximum deformation (D_{xy}) was 0.183 m and FOS was equal to 1.45 (Figure 8b). It was observed that as disturbing forces increased, the resisting forces developed in soil-nails were also noted to increase for ensuring stability to the slope. The total forces developed in soil-nails (i.e. summation of forces measured in top, middle and bottom soil-nails) for $\beta = 79°$ is 97.7 kN/m (Figure 9a) while for $\beta = 63.43°$ it is 58.7 kN/m (Figure 9b). However, FOS is on the higher side for relatively flatter slope inclination. It

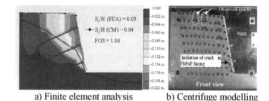

a) Finite element analysis   b) Centrifuge modelling

Figure 10. Comparison of results of soil-nailed slopes obtained from centrifuge tests and FEA [Model: VBS-10].

confirms that for steeper slope inclination the deformation and developed nail forces are greater than the relatively flatter slope inclination.

Comparison of observed performance of soil-nailed slope model VBS-10 ($\beta = 79°$) in centrifuge test along with FEA results is shown in Figure 10. In centrifuge model tests, crest settlement normalised with slope height (S_c/H) was observed to be 0.04 m for the soil-nail reinforced slope ($\beta = 79°$) with FbPoP facing, whereas S_c/H obtained from FEA was 0.03 m (Figure 10a). In addition, cracks at the top surface of the slope and a longitudinal crack in FbPoP facing along the bottom row of soil-nails were observed in centrifuge model tests (Figure 10b). The factor of safety of the soil-nailed slope with $\beta = 79°$ was 1.04 (Figure 10a). Both the results from centrifuge model tests and FEA show that the performance of soil-nailed slope with $\beta = 63.43°$ is superior to the soil-nailed slope with $\beta = 79°$.

5 CONCLUSIONS

Based on the results, analyses and interpretation of centrifuge tests and FEA on slopes with and without soil-nails, the following conclusions are drawn:

a. Presence of water table reduces shear strength of soil and thus affects the stability of slopes, with and without soil-nails, considerably. The unreinforced slope models with slope inclination of 63.43° and 79° turned into unstable slopes and exhibited catastrophic failure when subjected to seepage. The deformations of soil-nailed slopes ($\beta = 63.43°$ and 79°) were observed to increase with an increase in water level within the slope.

b. The soil-nailed slope with $\beta = 63.43°$ showed lower crest settlement value of 0.013 times the slope height as compared to 0.04 times slope height recorded for soil-nailed model slope with inclination of 79°. This confirms that for identical nail parameters and facing material type relatively flatter slope inclination improves the deformation behaviour of soil-nailed slopes.

c. The summation of forces developed in soil-nails for $\beta = 79°$ is 97.7 kN/m and 58.7 kN/m for $\beta = 63.43°$. Though the developed nail forces for a slope with $\beta = 63.43°$ are less than forces developed in soil-nails for a slope with $\beta = 79°$, the slope with $\beta = 63.43°$ was found to be more stable than the slope with $\beta = 79°$. Thus, it can be concluded that relatively flat slope is stable as compared to the steep slope and hence the nail forces required to maintain the stability of a soil-nailed slope are less. The trend of observed deformation behaviour of soil-nailed slopes obtained from FEA was found to match well with the physically observed centrifuge test results.

d. The soil nailing technique is effective in improving the stability and performance of existing slopes subjected to seepage. Natural slopes, rail/road side slopes, retaining walls, canal slopes, Levees, etc. can be strengthened effectively by soil-nailing technique.

ACKNOWLEDGEMENTS

The authors thank staff at the National Geotechnical Centrifuge Facility (IIT Bombay, India) for their active support in the course of the present study.

REFERENCES

Aminfar, M.H. 1998. Centrifuge modelling of stabilisation of slopes using the technique of soil nailing. *Doctoral Thesis*, Cardiff School of Engineering, University of Wales, UK.

Deepa, V. & Viswanadham, B.V.S. 2009. Centrifuge model tests on soil-nailed slopes subjected to seepage. *Ground Improvement Journal*, 162(3): 133–144.

Fan, C.C. & Luo, J.H. 2008. Numerical study on the optimum layout of soil-nailed slopes. *Computer and Geotechnics* 35(4): 585–599.

Ming, C.Y. 2008. Centrifuge and three dimensional numerical modelling of CDG filled slopes reinforced with different nail inclinations. *M.Phil. dissertation*, Hong Kong University of Science and Technology, Hong Kong.

Mittal, S. & Biswas, A.K. 2006. River bank erosion control by soil nailing. *Geotechnical and Geological Engineering*, 24(6): 1821–1833.

Morgan, N. 2002. The influence of variation in effective stress on the serviceability of soil-nailed slopes. *Ph.D. Thesis*, University of Dundee, UK.

Ng C.W.W., Zhou R.Z.B. & Zhang M. 2006. The effects of soil-nails in a dense steep slope subjected to rising groundwater. C.W.W. Ng, L.M. Zhang & Y.H. Wang (eds.), *Proc., 6th Int. Conference on Physical Modelling in Geotechnics 2006*, Vol. 1, 397–401, Taylor & Francis, London.

PLAXIS 2010. Plaxis 2D Reference Manual, Version 10. Delft, Netherlands.

Rawat, S., Zodinpuii, R., Manna, B. & Sharma, K.G. 2014. Investigation on failure mechanism of nailed soil slopes under surcharge loading: testing and analysis. *Geomechanics and Geoengineering: An International Journal*, 9(1), 18–35.

Rotte, V.M., Viswanadham, B.V.S. & Chourasia, D. 2011. Influence of slope geometry and nail parameters on the performance of soil-nailed slopes. *International Journal of Geotechnical Engineering*, 5(3), 267–281.

Rotte, V.M. & Viswanadham, B.V.S. 2013. Influence of nail inclination and facing material type on soil-nailed slopes. *Ground Improvement J.*, 166(2): 86–107.

Rotte, V.M. & Viswanadham, B.V.S. 2014. Centrifuge and numerical model studies on the behaviour of soil-nailed slopes with and without slope facing. W. Ding & X. Li (eds.), *Geo-Shanghai 2014*, Special Publication-242, 581–591, ASCE, Shanghai, China.

Tei, K., Taylor, R.N. & Milligan, G.W.E. 1998. Centrifuge model tests of nailed soil slopes. *Soils and Foundation*, 38 (2): 165–177.

Viswanadham, B.V.S. & Deepa, V. 2010. Evaluation of performance of soil-nailed slopes subjected to seepage in a centrifuge. S. Springman, J. Laue & L. Seward (eds.), *Proc., 7th Int. Conference on Physical Modelling in Geotechnics 2010*, 1151–1156, Taylor & Francis, London.

Performance of soil-nailed wall with three-dimensional geometry: Centrifuge study

M. Sabermahani
School of Civil Engineering, Iran University of Science and Technology, Tehran, Iran

M. Moradi & A. Pooresmaeili
School of Civil Engineering, University of Tehran, Tehran, Iran

ABSTRACT: For an irregular shaped excavation, walls deformation pattern is greatly affected by the concave and convex location. The vicinity of convex corners is a problematic zone, and causes problems in everyday practice. In this paper, a series of centrifuge model tests is conducted to investigate the impact of convex corners on the deformation of a soil-nailed walls, and to evaluate the influence of soil-nail layout on wall behavior. In this regard, an excavation with equivalent convex geometry, and different soil-nail layouts were considered, and results were compared with two-dimensional plane strain model. In each test, the centrifugal acceleration was increased gradually to the 40g-level. The observations indicate that deformation pattern of convex corners is thoroughly affected by soil nail horizontal inclination angle, Furthermore, the results show that wall facing plays an effective role in controlling the wall deformations with three-dimensional geometry.

1 INTRODUCTION

Due to urban population growth, the need for construction of high-rise buildings and related deep excavations has become increased. There are several engineering problems related to the complexity and diversity of excavations geometry in urban environments. For an irregular shaped excavation, the deformation pattern of the walls is greatly affected by the concave and convex location. In comparison to the most concave locations, which are beneficial for the excavation stability, the convex areas usually yield a complex stress state during the excavation of corner sides. Previous studies such as Ou & Shiau (1998), Lee et al. (1998) and Finno et al. (2007) indicated that the wall deformation and ground surface settlement near the concave locations are smaller than other locations due to the arching effect.

Therefore, the vicinity of convex corners in excavation areas is an especially problematic zone, and causes numerous problems in everyday practice. However, due to complexity and time-consuming of three-dimensional analysis, the excavation design is usually simplified into plane-strain problem. This simplification leads to conservative design in concave locations, and also will cause an unsafe situation in convex locations. The spatial effects of convex corners have been studied by researchers such as Zhao et al. (2015), Zhang et al. (2011) and Wu & Tu (2007) using numerical simulation method. The results showed that the deformation pattern of convex areas has many differences from the deformation in the middle positions and, the corner effects should be seriously considered to restrict the deformation of reinforced walls.

Soil-nailing is one of the soil reinforcement techniques that can be used to stabilise the excavation walls effectively. However, the behavior of soil-nailed walls with three-dimensional geometry is not completely discussed in the literature and, the experimental studies are still rather scanty for better understanding of the actual behavior of this problems.

Centrifuge model tests have been widely employed to investigate the performance of the soil-nailed slopes subjected to different conditions. Davies & Morgan (2005) performed a series of centrifuge model tests to investigate the serviceability performance of soil-nailed slopes under both short and long term conditions. It was reported that the nail loads can increase significantly when an earth structure is subjected to a reduction in effective stress resulting from an increase in pore water pressure. In addition, the cyclic changes in effective stress can led to an increase in reinforcement loads resulting from accumulated deformations, moreover, Zhou et al. (2006) has also reported the further mobilisation of axial nail forces after increasing the ground water level.

Chan (2008) carried out four centrifuge model tests to investigate the effect of nail inclination on the stability of soil-nailed slopes. It was observed that with an increase in nail inclinations, the measured settlements of model crest also increased. Rotte & Viswanadham (2012) conducted centrifuge model tests to examine the stability and the deformation of soil-nailed slopes subjected to seepage conditions. Results indicated that the soil-nails and slope facing have significant effects on lowering the crest settlement and, the variation of nail horizontal spacing can greatly affected the stability of soil-nailed slopes as compared to the nail vertical spacing.

In this paper, four centrifuge model tests was conducted to investigate the impact of convex corner on deformation of a soil-nailed wall, and to evaluate the influence of soil-nail layout on wall behavior. The results can be used to determine the soil-nails horizontal splaying in convex corners to achieve better performance of soil-nailed systems.

2 MODELING EQUIPMENT AND SCALING CONSIDERATION

Geotechnical centrifuge at Tehran University, with nominal radius of 2.7 m and maximum acceleration of 130 g, is used in the present study. The basket provides a 0.8 m wide by 1 m long platform which has a capacity to carry a payload of up to 1.5 ton under 100 g acceleration. The model container used for the tests was made of 10 mm thick stainless steel plates with inner dimensions of 800 mm length, 570 mm width, and 500 mm height. The 40 mm thick transparent plexi-glass sheet is placed at one side of container to facilitate capturing images of the model during centrifuge tests.

In order to simulate the prototype condition of soil-nailed wall, all significant dimensions and mechanical properties of model materials including soil, soil-nailed and facing should be considered based on scaling rules, which are outlined in Table 1.

3 SCHEMES

The centrifuge tests were designed to study the behavior of right angled convex corner, considering two distinct soil-nail pattern. The nails were inserted to the wall perpendicular to the wall facing, or with the angle of 45 degrees which will be parallel to bisector of convex angle that are abbreviated as PDD and PRR, respectively. Soil-nail execution in the form of PDD pattern is highly adopted in practice, nails may need to be splayed on plan view to avoid manholes and other obstructions, and to avoid convex corners due to interference with adjacent nails.

Table 1. Scaling laws for centrifuge testing.

Parameter	Model/Prototype	Dimensions
Length	1/N	L
Strain	1	1
Stress	1	$ML^{-1}T^{-2}$
Axial rigidity	$1/N^2$	MLT^{-2}
Flexural rigidity	$1/N^4$	ML^3T^{-2}

Note: N, Scaling factor. L, Length. M, Mass. T, Time.

The plan geometry of centrifuge model tests is indicated in Figure 1. The size of convex corner was chosen based on the frequently observed cases in urban areas. Thin transparent Perspex sheets were used inside container to minimise the boundary effects.

The reinforcement was designed to prevent the wall instability in order to record the wall displacements. Cross sections of test models are the same, which are shown in Figure 2. Thus, the three tests were conducted with the same convex geometry to investigate the effect of nail splaying on wall stability and deformation. It should be noted the sides facing (two solid sheets) in the test number 3 are well stuck together at the corner tip which represents "rigid facing connection". On the other hand, there is no appropriate connection between sides facing at the corner tip in the other convex geometries which represent "flexible facing connection".

At last, a plane strain model was tested to compare the results of 2D and 3D wall geometry to evaluate the impact of convex corner on deformation of a soil-nailed wall.

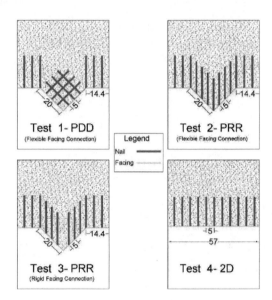

Figure 1. Geometry of centrifuge model tests on plan view (dimensions are in cm).

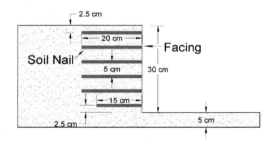

Figure 2. Cross sections of test models.

4 MODEL MATERIAL

Firoozkouh-161 sand was selected for this study. It is produced by crushing of parent rocks of the Firoozkouh region. The soil mainly consists of golden angular grains. From a mineralogical view, 90% by weight of the soil mass is of feldspar and chert and the remaining is of quartz and calcite origin (Farahmand et al., 2016).

Friction angle of Firoozkouh-161 sand was 32°. It had a specific gravity of 2.66 and, maximum dry unit weight of $18.75 \, kN/m^3$. The average grain diameter (D_{50}) was 0.3 mm. The detail specification of Firoozkouh-161 sand compared with other well-known sandy soils is presented in Table 2 (Rojhani et al., 2012).

Aluminum tubes having 5 mm outer diameter and 1 mm thickness were used to simulate the nails of reinforced wall. The yield strength and the elastic modulus of adopted soil-nails model were 100 MPa and 69 GPa, respectively. The axial stiffness (EA) of soil-nails model was equivalent to the 40 mm steel bar with surrounding grout of 3.5 cm thickness as a prototype nail at a centrifugal acceleration of 23 g. Two different nail lengths of 200 mm and 150 mm were used in the models. Distribution of soil nail lengths in elevation was similar in all the tests (Figure 2). In order to simulate the interface behavior of soil-nails model, a 0.4 mm thick layer of epoxy mixed with sand was applied on the outer side of the tubes.

The facing of all four tests was made of 1 mm thick polycarbonate solid sheets. The tensile strength, flexural strength and the elastic modulus of adopted facing model were 75 MPa, 100 MPa and 3 GPa, respectively.

Aluminum tubes were connected to the facing through desired steel nuts. Plastic washers with 19 mm diameter and 4 mm thickness were used between nuts and facing to improve load-transfer mechanisms from wall facing to the soil-nails.

Table 2. Firoozkouh-161 sand properties compared with other sands.

Sand type	Firoozkouh-161	Toyoura	Sengenyama
Specific gravity, Gs	2.65	2.65	2.72
Maximum void ratio, e_{max}	0.894	0.977	0.911
Minimum void ratio, e_{min}	0.622	0.597	0.55
Average particle size, D_{50} (mm)	0.16	0.17	0.27
Fine content, FC (%)	1	0	2.3
Coefficient of uniformity, Cu	1.27	1.54	2.15
Coefficient of curvature, Cc	0.96	1.25	1.21

5 TEST SET UP AND PREPARATION

A base soil layer of 50 mm thickness was prepared in the container. Prefabricated 50 mm thick expanded polystyrene sheets with a desired shape were placed on the base layer as a temporary support to construct a wall in the form of right angled convex corner and then, wall facing with pre-punched holes was attached to the temporary support.

Soil with moisture content of 5% was prepared and placed inside the container (in front of temporary support) in 2.5 cm layers that were then compacted using steel hammer to achieve 60% soil relative density. Fourteen soil layers were compacted to get wall surface level. Then, upper part of polystyrene sheets was removed to drive the first row of nails in the wall through pre-punched holes of wall facing. Nails were driven through the wall facing with the help of a mallet hammer. Plastic washer and steel nut were then placed at the nail head. In order to integrate the components, plastic washers of nails head were stuck to the facing by appropriate glue. After that, second polystyrene sheet was removed to execute next row of nails. This process will be continued, step by step, to get to the bottom of excavation.

It should be noted that the nails of either side of convex corner in the case of PDD pattern were inserted to the wall with 10 mm offset to avoid interference. In the case of PRR pattern, a little gap will be remained between head nail and facing, therefore, clayey soil was used as a filling material to increase the contact surface of head nail and facing (Fig. 3).

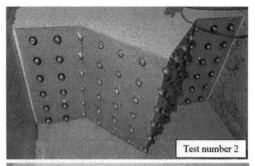

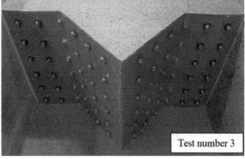

Figure 3. Model preparation of centrifuge test models.

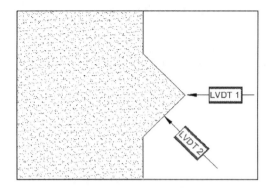

Figure 4. Measurement of horizontal displacement of reinforced convex corner.

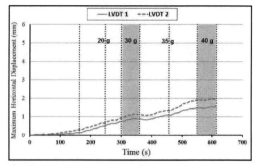

Figure 6. Cumulative horizontal displacements of test 2.

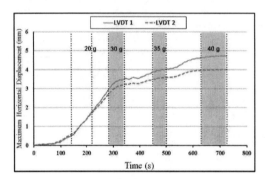

Figure 5. Cumulative horizontal displacements of test 1.

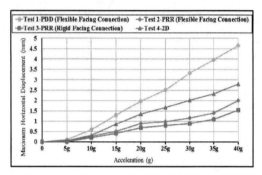

Figure 7. Maximum horizontal displacement of test models.

The soil-nailed models were instrumented by linearly variable displacement transducers (LVDT) to monitor wall deformation during the test. In each test, the centrifugal acceleration was increased gradually to the 40 g-level. Two LVDTs were placed at the top of the wall (2 cm below the soil surface) to measure horizontal displacement of the soil-nailed wall during the increase in centrifuge acceleration (Fig. 4).

6 RESULTS AND DISCUSSION

The cumulative horizontal displacements of test number 1 and 2 are shown in Figure 5 and 6, respectively. It can be observed that the maximum displacement of test number 1 with nail PDD pattern has occurred at the corner tip (LVDT 1) while, the maximum displacement of test number 2 with nail PRR pattern was located in the middle of corner sides (LVDT 2). The results indicated that the deformation pattern of soil-nailed walls is highly affected by soil nail horizontal splaying and, due to axial resistance of soil nails, the highest level of efficacy against wall deformation is achieved in the direction of nail axis.

The maximum horizontal displacement of test models is plotted in Figure 7 with respect to centrifuge acceleration. The graph indicated that the maximum horizontal displacement of convex corner with PDD nail pattern is significantly higher than others.

The stability of a soil-nailed system is often considered in terms of two zones divided (idealistically) by a potential failure plane (Phear et al., 2005), active and resistant (passive) zones. The resistant zone behind the potential failure plane contains the length of the soil nail that can develop pullout resistance. Therefore, the nail length in the resistant zone has an important role in the stability of soil-nailed walls. The probable active wedge of centrifuge soil-nailed models are plotted in Figure 8 which indicates that a significant proportion of nail length of PRR pattern compared with PDD pattern is located in the resistant zone. Hence, the stability contribution of soil nails with PRR pattern is relatively greater than PDD pattern. It should be noted that the total length of some nails adjacent to the corner tip with PDD pattern may be located in the active zone and offer no resistance to the wall instability. As a result, the PRR pattern indicates better performance in controlling horizontal deformation of soil-nailed walls.

According to Figure 7, the maximum horizontal displacement of convex corner with nail PRR pattern is lower than plane strain model. Two main factors, reinforcement density and role of the wall facing, can be

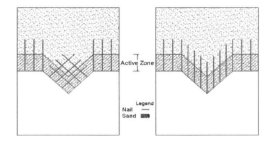

Figure 8. Probable active wedge of reinforced wall.

considered to describe the behavior of convex corner with nail PRR pattern compared to plane strain model.

Soil nail spacing of two mentioned models is equal in the direction of the excavation line as shown in Figure 1. Therefore, the horizontal nail spacing in the convex area is less than nail spacing in the plane strain model, leading to increased density of reinforcement in convex area which is beneficial in controlling the wall displacements. This issue is one of the objective of current study to investigate how horizontal nail spacing of 2D analysis can be used in such 3D locations to improve wall stability.

Role of the wall facing is another effective factor in controlling the wall displacements. Based on geometry condition of such areas, soil displacements are restrained by mobilsation of tensile forces in the sides facing. On the other hand, due to soil confinement in the convex areas (between the sides facing), a significant proportion of active soil pressure can be transferred to the wall facing. As shown in Figure 9, the resultant force of soil pressure (F) has been transferred to the wall facing in the form of tensile forces (F′). The proportion of soil pressure transferred to the facing varies depending on the facing stiffness and appropriate connections of the sides facing.

Deformation reduction of test number 3 (Fig. 7), which represents the rigid connection of the sides facings is another reason to prove the effective role of the wall facing to control the deformation of 3D soil-nailed walls in the form of convex corners.

As a result, the PRR pattern of soil nails showed a better performance to control the horizontal deformation of reinforced walls in the form of convex corners. According to reference manual of FHWA (Lazarte et al., 2015), nails splaying on plan view can possibly improve stability at internal corners. Test results of current study reveal that nails splaying can certainly improve stability at external corners. As it is mentioned before, 2D design computer programs do not account for the splay angle and, three-dimensional analysis is less utilised in practical works. The results indicated that the required nail spacing in the plane strain analysis can be adopted in convex areas that nails execution in the form of PRR pattern (increased density of reinforcement) and rigid connection of sides facing (effective role of the wall facing to retain soil pressure) can possibly lead to reduction of wall horizontal deformation compared to plane strain model.

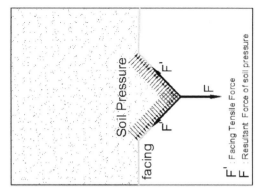

Figure 9. Role of the facing to retain soil pressure.

7 CONCLUSIONS

The centrifuge model tests were performed to investigate the effects of convex corner on deformations of soil-nailed walls. In this paper, the convex corner with specific three-dimensional geometry was considered to also compare the results of the different soil-nail layouts. In overall, two distinct soil-nail layout were considered which are abbreviated as PDD and PRR. Based on the measurement results of centrifuge model tests, the main conclusions can be drawn as follows:

1. The deformation pattern of soil-nailed walls is highly affected by soil nail horizontal splaying and, due to axial resistance of soil nails, the highest level of efficacy against wall deformation is achieved in the direction of nail axis.
2. The PRR pattern indicates better performance in controlling horizontal deformation of soil-nailed walls. In the other hand, nails splaying on plan view should be determined considering the nail length in soil resisted zone behind the wall.
3. The wall facing plays an influential role in controlling the wall deformations with three-dimensional geometry.
4. The required nail spacing in the plane strain analysis can be adopted in convex areas considering nails splaying on plan view (PRR pattern) and rigid connection of sides facing, which can possibly lead to reduction of wall horizontal deformation compared to plane strain model.

REFERENCES

Chan, Y.M. 2008. Centrifuge and three dimensional numerical modelling of CDG filled slopes reinforced with different nail inclinations. PhD Thesis, Hong Kong University of Science and Technology.

Davies, M. & Morgan, N. The influence of the variation of effective stress on the serviceability of soil nailed slopes. *Proceedings of the international conference on soil mechanics and geotechnical engineering, 2005. A. Balkema, 1335.*

Farahmand, K., Lashkari, A. & Ghalandarzadeh, A. 2016. Firoozkuh sand: introduction of a benchmark for geomechanical studies. *Iranian Journal of Science and Technology, Transactions of Civil Engineering,* 40, 133–148.

Finno, R.J., Blackburn, J.T. & Roboski, J.F. 2007. Three-dimensional effects for supported excavations in clay. *Journal of Geotechnical and Geoenvironmental Engineering,* 133, 30–36.

Lazarte, C.A., Robinson, H., Gómez, J.E., Baxter, A., Cadden, A. & Berg, R. 2015. Soil Nail Walls Reference Manual.

Lee, F.H., Yong, K.Y., Quan, K.C. & Chee, K.T. 1998. Effect of corners in strutted excavations: Field monitoring and case histories. *Journal of Geotechnical and Geoenvironmental Engineering,* 124, 339–349.

Ou, C.Y. & Shiau, B.Y. 1998. Analysis of the corner effect on excavation behaviors. *Canadian Geotechnical Journal,* 35, 532–540.

Phear, A., Dew, C., Ozsoy, B., Wharmby, N., Judge, J. & Barley, A. 2005. *Soil nailing-best practice guidance.*

Rojhani, M., Moradi, M., Galandarzadeh, A. & Takada, S. 2012. Centrifuge modeling of buried continuous pipelines subjected to reverse faulting. *Canadian Geotechnical Journal,* 49, 659–670.

Rotte, V. & Viswanadham, B. 2012. Performance of 2V: 1H slopes with and without soil-nails subjected to seepage: Centrifuge study. *GeoCongress 2012: State of the Art and Practice in Geotechnical Engineering.*

Wu, Z.M. & Tu, Y.M. 2007. Space effect of soil-nailing excavation protection. *Yantu Lixuem(Rock and Soil Mechanics),* 28, 2178–2182.

Zhang, M., Wang, X.H., Yang, G.C. & Wang, Y. 2011. Numerical investigation of the convex effect on the behavior of crossing excavations. *Journal of Zhejiang University-Science A,* 12, 747–757.

Zhao, W., Chen, C., Li, S. & Pang, Y. 2015. Researches on the Influence on Neighboring Buildings by Concave and Convex Location Effect of Excavations in Soft Soil Area. *Journal of Intelligent & Robotic Systems,* 79, 351.

Zhou, R., Ng, C., Zhang, M., Pun, W., Shiu, Y. & Chang, G. The effects of soil nails in a dense steep slope subjected to rising groundwater. *6th International Conference on Physical Modelling in Geotechnics,* 2006. 397–402.

Behaviour of geogrid-reinforced aggregate layer overlaying poorly graded sand under cyclic loading

A.A. Soe, J. Kuwano, I. Akram, T. Kogure & H. Kanai
Department of Civil and Environmental Engineering, Saitama University, Japan

ABSTRACT: Deformation behavior of geogrid-reinforced aggregate layer was investigated through laboratory model tests. Unpaved road was modelled by using silica sand and aggregate. Cyclic loading tests were performed, considering the effects of aggregate thickness and pressure intensities. Aggregate thickness had minimal influence on surface settlement, except under 550 kPa pressure. However, this was significant on the sand layer settlement since under 200 kPa. Sand layer settlement highly contributed to surface settlement in 10 cm thick aggregate layer under all pressure levels. This contribution became small with increase in aggregate thickness. However, loss of aggregate thickness increased with increase in thickness. This loss was reduced with the inclusion of geogrid, especially when geogrid was inside 20 cm thick aggregate layer. With geogrid inclusion, surface settlement was reduced in the respective cases, by 45% in 10 cm, and 25% in 20 cm thicknesses.

1 INTRODUCTION

Flexible pavement generally includes asphalt, base course and/or subbase, and subgrade layers. Among them, base course is the main structural layer to transfer the wheel load to the subgrade layer. For the existing poor subgrade, thick base course is generally required. To reduce this problem, geosynthetics have been introduced and used in the road construction. Previously, geotextile was used in improving flexible pavement performance (Al-Qadi et al. 1994). The benefits are measured in terms of traffic benefit ratio (TBR) and base course reduction (BCR) (Zornberg 2012). Later, geogird has been developed and used in flexible pavement as a reinforcement layer. The improvement behavior due to geosynthetic reinforcement has been studied and reported by many researchers (Anderson 2006, Al-Qadi et al. 1994, Schuettpelz et al. 2009). However, it is still necessary to examine this behavior because this may vary depending on the geosynthetic type, soil properties and loading conditions. In general, the deformation behavior was investigated under the constant pressure, corresponding to standard axle load. In reality, wheel load may vary depending on the type of vehicle. This varying load characteristics may affect on the deformation behavior of geosynthetic-reinforced layer. According to Dawson and Kolisoja (2006), rutting in flexible pavement can occur due to the combination of different modes, including compaction, inadequate shear strength in aggregate, shear deformation within subgrade layer and particle damage. In this study, this deformation behavior is investigated for the geogrid-reinforced layer, considering aggregate layer thickness and pressure intensities. Surface and subgrade settlements are separately analyzed to investigate the deformation behavior. In addition, benefits due to geogrid inclusion are quantified by means of reductions in surface settlement and aggregate layer deterioration.

2 EXPERIMENTAL PROGRAM

The unbound base course layer overlaying soft subgrade was modelled by using the well-graded aggregate and poorly graded sand. In this study, six cyclic loading tests were performed in a large container, with three unreinforced tests and three geogrid-reinforced tests. The base layer thicknesses were 10 cm, 20 cm and 30 cm in the unreinforced tests and 10 cm and 20 cm in the geogrid-reinforced tests. Subgrade layer thickness was also changed in each test. These thicknesses are presented in Table.1. In each test, cyclic pressure was applied in steps: 100 kPa, 200 kPa, 300 kPa, 400 kPa and 550 kPa. 550 kPa was selected because this pressure corresponds to the standard axle load (80 kN). Each pressure level was allowed for 500 cycles, except under 550 kPa pressure. This pressure was applied up to 1000 cycles to check the performance difference with load cycles.

3 MATERIALS

3.1 Model subgrade and base course

The subgrade layer was modelled by using uniform silica sand, which is commercially available in Japan. This sand has the following index properties: the specific gravity, $G_s = 2.675$, the average particle size, $D_{50} = 0.488$ mm, coefficient of uniformity, $C_u = 1.98$,

Table 1. Thicknesses of subgrade and base course layers in each test.

Test ID	Subgrade cm	Base course cm	Geogrid
10X	62	10	No
20X	53	20	No
30X	42	30	No
10G	64	10	At bottom
20G	52	20	At bottom
20G-mid	51	20	At middle

coefficient of curvature, $C_c = 0.943$, maximum dry density, $\rho_{max} = 1.628$ g/cm^3, and minimum dry density, $\rho_{min} = 1.338$ g/cm^3. According to the unified soil classification system (USCS), this sand can be regarded as poorly graded sand (SP). The particle size distribution curve of this sand is shown in Figure 1.

The base course layer was prepared by using the crushed-stone aggregate, (commercial name: C40), which is usually used for the base course layer in pavement construction in Japan. This aggregate has the average particle size, $D_{50} = 14.793$ mm, uniformity coefficient $C_u = 4.609$ and curvature coefficient $C_c = 1.579$. This aggregate can be considered as well-graded aggregate according to USCS. The particle size distribution curve is shown in Figure 1.

3.2 Reinforcement material

For the reinforcement, the multiaxial geogrid TX160, with triangular aperture, was selected for this study. This geogrid has been recently developed by Tensar. The properties of this geogrid are: radial secant stiffness at 0.5% strain = 390 kN/m; pitch length on each side of triangle = 40 mm; junction thickness = 3.1 mm and weight = 0.22 kg/m^2 (Tensar, 2013).

4 EXPERIMENTAL SETUP AND TESTING

In this model test program, tests were performed in the large container (1 m × 1 m × 0.8 m). The walls of the container are supported by using the rigid steel frames. A rigid circular steel plate, diameter = 17.5 cm, was used to simulate the repeated wheel load. This plate size was selected so that the size ratio of container and plate was large enough (100/17.5 ≈ 6.0) to minimize the boundary interference (Yetimoglu et al. 1994, Alawaji 2001). In each test, the total depth of model soil layers was prepared to be larger than 70 cm, as presented in Table 1. For this depth, the boundary influence would be minimal according to Boussinesq Stress Distribution Analysis.

To model the poor subgrade layer, sand raining method was adopted in accordance with the previous study (Miura & Toki 1982). A large raining box was developed to cover the plan area of container. By using this raining method, the density of sand layer was achieved, ranging from 1.60 g/cm^3 to 1.64 g/cm^3.

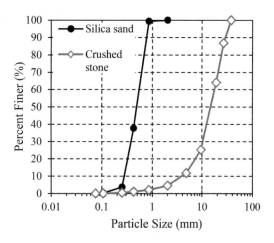

Figure 1. Particle size distribution curves of silica sand and crushed stone.

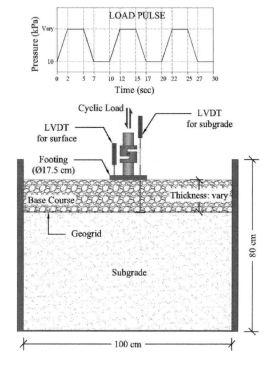

Figure 2. Schematic view of test layout.

The bearing capacity of this subgrade layer is 110 kPa, determined by using monotonic loading test.

As seen in Figure 2, a small aluminum tube was set directly above a small plastic plate, so that the sand layer settlement could be monitored through the loading plate and the aggregate layer. Then, the base course aggregate layer was prepared in layers by the manual compaction (approximately 1.31 N-cm/cm^3). This low compaction energy was applied through a wooden plate to minimize the breakage of materials and disturbance to subgrade layer. After compaction, the average thickness of 5 cm was achieved in each layer, for which 80 kg aggregate was used (average density = 1.6 g/cm^3). The same procedure was continued to get the desired aggregate thickness in each test.

After the preparation of sand and aggregate layers, a linear variable displacement transducer was set on the loading plate to measure the surface settlement, while another transducer was used to monitor sand layer settlement directly under the plate. The seating pressure of 10 kPa was applied before loading started. To simulate the trafficked loading, the cyclic loading with trapezoidal load pulse was applied with the frequency of 0.1 Hz, maximum available value in the loading system. This load pulse was chosen so that the deformations under maximum and minimum pressures would be easily recorded in each load cycle. The load magnitude and settlements were recorded by the data acquisition system at every 1 second.

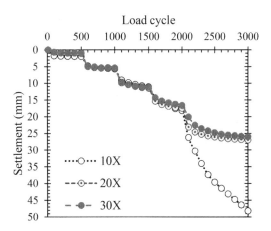

Figure 3. Surface settlement of the unreinforced aggregate layer.

5 RESULTS AND DISCUSSION

5.1 Effect of aggregate thickness on unreinforced layer overlaying poorly graded sand

Three tests were performed with different aggregate thicknesses to investigate the influence of aggregate layer thickness on the deformation behavior of unreinforced aggregate layer. The aggregate surface settlements and sand layer settlements are presented in Figure 3 and 4.

As seen in Figure 3, the differences in surface settlements were not obvious until the end of 400 kPa pressure (2000 cycles). Under 550 kPa pressure, the surface settlement of 10X increased rapidly with the increasing load cycles, while the settlements of 20X and 30X were almost same. At the end of the test, the surface settlements were 48.2 mm, 26.97 mm and 25.98 mm in 10X, 20X and 30X respectively. The significant reduction of 21.23 mm was achieved when the aggregate thickness was increased from 10 cm to 20 cm. However, this reduction was only about 1 mm when aggregate thickness was further increased to 30 cm. Hence, it was realized that the influence of aggregate thickness was insignificant on the surface settlement under the given pressure levels when the thickness was greater than 20 cm.

Although differences were not observed in the surface settlement, significant differences were noticed in sand layer settlement as seen in Figure 4. Since the bearing capacity of this layer is 110 kPa, this layer will not support the vertical stress larger than this level. Because of the thinner aggregate layer in 10X, this aggregate layer may not be able to redistribute the applied pressure wide enough on the sand layer. As a result, the intensity of vertical stress on sand layer would rapidly increase with increase in pressure. Under 550 kPa, hence, sand layer settlement was progressive with load cycles due to bearing failure.

Although surface settlements of 20X and 30X were almost same, their respective sand layer settlements

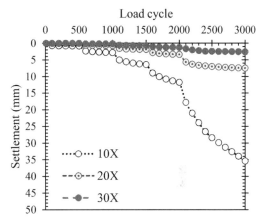

Figure 4. Sand layer settlement underlaying the unreinforced aggregate layer.

were different. This difference became more obvious with the increasing pressure after 200 kPa (1000 cycles). This was because the applied vertical stresses might have been distributed to the wider area on the sand layer with increasing aggregate layer thickness. In addition, the compacted well-graded aggregate layer may have higher modulus than that of underlaying sand layer. According to the layered elastic theory (Burmister 1958), vertical stress will decrease with increase in modular ratio and depth (thickness). Hence, sand layer settlement became smaller with increase in aggregate layer thickness.

With increase in aggregate layer thickness, sand layer settlement was significantly reduced by about 28 mm in 20X test and 33 mm in 30X test, comparing with 10X. However, this reduction was only 5 mm when aggregate thickness increased from 20 cm to 30 cm. Though there was no obvious difference in surface settlement until 400 kPa, differences in sand layer settlement were noticed among the tests since under 200 kPa. This implied that settlement in aggregate layer also increased with increase in thickness.

This might have been resulted from both compaction and lateral flow of particles under cyclic loading.

5.2 Effect of aggregate thickness on geogrid-reinforced layer overlaying poorly graded sand

Three tests were performed with the inclusion of geogrid, presented in Table 1. The surface settlements of these tests are shown in Figure 5. As seen in this figure, the surface settlements were almost same under 100 kPa through 400 kPa (2000 cycles). The differences were started notice under 550 kPa pressure.

When aggregate thickness increased from 10 cm (10G) to 20 cm (20G), the reduction in the surface settlement was only 6.38 mm at the end of the test. This reduction was more than 21 mm in case of the unreinforced tests for the same increment of aggregate thickness. This implies that the influence of aggregate layer thickness becomes less on the surface settlement with the inclusion of geogrid reinforcement. This was because vertical stresses might have been reduced with the inclusion of geogrid in 10G test. Hence, under 550 kPa, the rate of surface settlement became slow in 10G test with increasing load cycles, compared to 10X. On the other hand, the stress on sand layer might have been already reduced by the aggregate layer in 20G test, according to the layered elastic theory (Burmister 1958). Hence, stress reduction by geogrid would be small in this test, compared to 10G test. Comparing with the unreinforced tests, geogrid reinforcement reduced the surface settlement from 48.2 mm to 26.62 mm (≈ 45%) in 10 cm aggregate thickness, and from 26.97 mm to 20.24 mm (≈25%) in 20 cm case. This implied that the effect of geogrid became insignificant with increase in the thickness of aggregate layer when geogrid was placed at the bottom of that layer. To investigate the effectiveness of geogrid reinforcement 20G-mid test was performed by placing the geogrid at the mid-depth of 20 cm aggregate thickness. From Figure 5, a small improvement was observed at the end of 400 kPa and under 550 kPa pressures. Reduction in surface settlement increased from 6.38 mm (20G) to 9.05 mm (20G-mid) by changing the geogrid position. Since geogrid was placed inside the aggregate layer, geogrid would provide the more confinement effect by interlocking the aggregate particles on both sides of geogrid. This would make the aggregate layer stiffer and result in better performance (≈35% settlement reduction, compared to 20X).

As seen in Figure 6, like in the unreinforced tests, the obvious difference was noticed since under 200 kPa pressure. Comparing with the surface settlement shown in Figure 5, the sand layer settlement of 10G test was about 20 mm at the end of the test while its surface settlement was 26.62 mm. This implied that the surface settlement in 10G test was mainly contributed by sand layer settlement. On the other hand, the sand layer settlements of 20G and 20G-mid tests were 4.54 mm and 5.82 mm while their surface settlements were 20.24 mm and 17.57 mm respectively. Since the surface settlements in all cases were almost

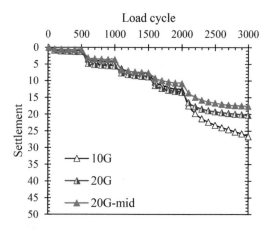

Figure 5. Surface settlement of geogrid-reinforced aggregate layer.

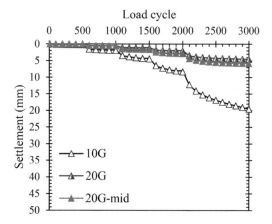

Figure 6. Sand layer settlement underlaying the geogridreinforced aggregate layer.

same until 400 kPa pressure, as shown in Figure 5, the surface settlements of 20G and 20G-mid tests were mainly contributed by the aggregate layer, rather than by the sand layer settlement.

Interestingly, the sand layer settlement of 20G-mid was slightly greater than that of 20G test under 550 kPa pressure, though the surface settlement of 20G-mid test was smaller than that of 20G test. This was because the aggregate deterioration potential was reduced by means of interlocking effect from both sides of geogrid in 20G-mid test. By this action, the reduction in surface settlement was achieved in this test. However, the stress increase would be expected on the sand layer because geogrid was inside the aggregate layer rather than directly on the sand layer. This might cause a slight increase in sand layer settlement. Schuettpelz et al. (2009) reported that shear stresses may propagate below reinforcement when a geogrid is secured too close to the surface. Since the different in sand layer settlement was small between 20G and 20G-mid tests, it was realized that the sand layer settlement was mainly influenced by the aggregate layer

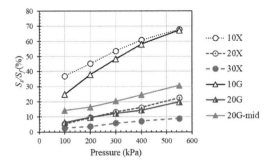

Figure 7. Percent contribution of sand layer settlement.

thickness, rather than geogrid reinforcement position which had noticeable influence on the aggregate surface settlement, as shown in Figure 5.

5.3 Percent contribution of sand layer settlement on the surface settlement of aggregate layer

From the aggregate surface settlement (S_f) and sand layer settlement (S_s), the contribution percentage of sand layer settlement to the surface settlement, S_s/S_f, was calculated for each test to distinguish how much the surface settlement was influenced by the underlaying sand layer. The average contribution percentages under respective pressure levels are presented in Figure 7.

As seen in Figure 7, the contribution percentages, S_s/S_f, of thin aggregate tests (10X & 10G) were significantly higher than those of thick aggregate tests under the given pressure level. In addition, these contribution values increased more rapidly with increasing pressure in 10 cm thick aggregate tests than 20 cm ones, followed by 30 cm test (30X). Since the surface settlements were almost same until the end of 400 kPa, as presented in Figure 3 for unreinforced layers and Figure 5 for geogrid-reinforced layers, it was realized that these similar characteristics were mainly resulted from the sand layer settlements in thin aggregate tests and from the aggregate layer in thick ones. Because the aggregate layer was manually compacted in each test, the compacting energy might not be high enough to achieve the maximum compaction state of aggregate layer. Hence, the layer might have been densified under the initial loading. After it had reached to its densest state, the lateral flow might have been initiated in the aggregate layer. This lateral flow would also depend on the frictional resistance of aggregate particles. In case of thin aggregate layer, the rate of sand layer settlement was higher than that of lateral flow due to higher stress intensity on the sand layer. Compared to the unreinforced 10X test, geogrid-reinforced 10G test initially showed the smaller amount of S_s/S_f values. However, these contribution values became close to each other with increasing pressure. This was because the geogrid reduced the stress level on the sand layer in 10G test under the initial loading, resulting in the smaller S_s/S_f percentage. However, with increasing pressure, geogrid might have been deformed and tension membrane effect would occur in geogrid. This bending deformation would also allow more settlement in sand layer, resulting in the higher S_s/S_f values.

For thicker aggregate layers, S_s/S_f values were obviously small because of lower stress intensity on sand layer. Interestingly, the values of 20X and 20G were almost same, regardless of reinforcement condition. This implied that reinforcement mechanisms of geogrid might not be fully mobilized under the initial loading, due to larger thickness. The slightly smaller values were noticed after 300 kPa pressure stage in 20G test, implying reinforcement mechanism was mobilized. For 20G-mid test, the contribution percentages were larger than those of counterpart ones: 20X and 20G. Because geogrid was inside the aggregate layer, lateral movement of aggregate particles might have been reduced by means of confinement. As a result, the surface settlement due to lateral flow became smaller. However, this would cause high stress propagation to the lower sand layer, resulting in larger sand layer settlement and higher S_s/S_f values.

As seen in Figure 7, S_s/S_f value increases with different rates depending on aggregate thickness, with increasing pressure. The increasing rate of S_s/S_f becomes slow when the aggregate layer is thicker, regardless of reinforcement condition. On the other hand, this rate is not affected by the geogrid position because the trends are roughly parallel among 20X, 20G and 20G-mid tests. Hence, the influence of aggregate thickness is more obvious on the S_s/S_f percentage than geogrid position.

5.4 Accumulative loss in aggregate layer thickness (h_l) under each pressure level

The accumulative loss of aggregate layer thickness (h_l) was calculated, considering the surface settlement (S_f) and the sand layer settlement (S_s). The h_l is the summation of the rate differences between S_f and S_s in each load cycle under given pressure level. The formula of h_l is shown in equation (1).

$$h_l = \sum (\Delta S_f - \Delta S_s) \qquad (1)$$

where h_l = accumulated loss in aggregate thickness; ΔS_f = rate of surface settlement in each cycle; and ΔS_s = rate of sand layer settlement in each cycle. For each pressure level, the rate differences were summed up for 500 load cycles, and then, this sum was added to the value obtained under previous pressure level. In this way, the incremental loss in aggregate layer thickness was calculated and presented in Figure 8.

As seen in Figure 8, the h_l values were significantly larger with increasing pressure in the thick layers 20X, 30X and 20G. Though h_l values in 10X test were smaller than those of 20X, the respective values of 20X and 30X tests were almost same. This revealed that a constant amount of h_l could be expected for the aggregate layer thickness larger than 20 cm under the given test conditions.

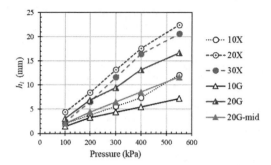

Figure 8. Accumulative loss in aggregate thickness.

The improvement due to reinforcement in 20G test was observed under 300 kPa and afterward. Significant improvement was noticed in 20G-mid test since under 200 kPa pressure. Despite having the same thickness of 20 cm, h_l could be reduced in 20G-mid test, by an approximately half of h_l in 20X, and about one-third in 20G. This was because the effective confinement was achieved in 20G-mid test. For 10G test, the benefit of geogrid inclusion was noticed under 300 kPa and following pressures, same as the 20G test. Hence, for the bottom reinforcement of aggregate layer, a certain amount of deformation will be needed to fully mobilize the reinforcement action. Here, this was about 3 mm for 10G test (≈3% of thickness) and nearly 7 mm for 20G test (≈3.5% of thickness).

6 CONCLUSIONS

Deformation behavior of geogrid-reinforced aggregate layer was studied by conducting six cyclic loading tests, considering aggregate layer thickness and pressure intensities. Aggregate thickness has minimal effect on surface settlement until 400 kPa (2000 cycles) in all cases. The differences were noticed under 550 kPa pressure. The improvement due to reinforcement became small with increase in the thickness, 45% settlement reduction in 10 cm and 25% in 20 cm cases. This reduction increases to 35% in 20G-mid case. The aggregate thickness has obvious influence on the sand layer settlement since under 200 kPa. From S_s/S_f values and h_l behavior, surface settlement is considerably contributed by the sand layer in 10 cm cases and by the aggregate deterioration in 20 cm and 30 cm cases.

This deterioration can be reduced by placing geogrid inside the aggregate layer. From the current study, it was realized that surface settlement is attributed to the different combinations of aggregate deterioration and sand layer settlement, depending on the aggregate layer thickness and geogrid reinforcement. Compared with the real situations such as compaction energy and wheel load condition, the deformation behavior may be different. However, it can be excepted to achieve the potential improvements due to geogrid reinforcement.

ACKNOWLEDGEMENTS

This research was collaborated with Nippo Corporation and Mitsui Chemicals Industrial Products Ltd.

REFERENCES

Alawaji, H. A. 2001. Settlement and bearing capacity of geogrid-reinforced sand over collapsible soil. *Geotextiles and Geomembranes* 19: 75-88.

Al-Qadi, I. L., Brandon, T. L., Valentine, R. J., Lacina, B. A., & Smith, T. E. 1994. Laboratory evaluation of geosynthetic-reinforced pavement sections. *Transportation Research Record* 1439(1439): 25-31.

Anderson, R. P. 2006. Geogrid Separation. *Proceedings of International Conference on New Developments in Geoenvironmental and Geotechnical Engineering. Incheon, Republic of Korea.*

Burmister, D. M. 1958. Evaluation of pavement systems of the WASHO road test by layered system methods. *Highway Research Board Bulletin* (177): 26-54.

Dawson, A. & Kolisoja, P. 2006. Managing rutting in low volume roads, Executive summary. Northern Periphery, EU.: Roadex III Project.

Miura, S. & Toki, S. (1982). A sample preparation method and its effect on static and cyclic deformation-strength properties of sand. *Soils and Foundations* 22(1): 61-77.

Schuettpelz, C., Fratta, D. & Edill, T. B. 2009. Evaluation of the zone of influence and stiffness improvement from geogrid reinforcement in granular materials. *Transportation Research Record* (2116): 76-84.

Tensar. 2013. Tensar TriAx® Stabilization Geogrid Technical Data – TX160. Blackburn, UK: Tensar International Limited.

Yetimoglu, T., Wu, J. T. & Saglamer, A. 1994. Bearing capacity of rectangular footings on geogrid-reinforced sand. *Journal of Geotechnical Engineering* 120(12): 2083-2099.

Zornberg, J. G. 2012. Geosynthetic-reinforced pavement systems. *5th European Geosynthetics Congress. Valencia.*

Physical modelling of compaction grouting injection using a transparent soil

D. Takano & Y. Morikawa
Port and Airport Research Institute, Yokosuka, Japan

Y. Miyata & H. Nonoyama
National Defense Academy, Yokosuka, Japan

R.J. Bathurst
Royal Military College of Canada, Kingston, Canada

ABSTRACT: Compaction grouting (CPG) is an in-situ grout injection technique used to improve the liquefaction resistance of loose sandy ground by densification and increasing lateral confining pressure. The present study investigates ground response during grout injection using a transparent granular soil and a natural sand soil. The injection process was simulated using models placed in a geotechnical centrifuge. The experimental results relating injection pressure to injection depth are shown to be in good qualitative agreement with field data. The results of models using transparent granular soil are shown to be in good agreement with the same models using standard opaque silica sand. The experimental methodology to simulate CPG and to observe grout volumes and ground movements during injection holds promise to better understand the mechanisms of compaction grouting and to improve its efficiency in the field.

1 INTRODUCTION

Compaction grouting (CPG) is an in-situ compaction technique that uses grout injection to improve the liquefaction resistance of loose sandy ground (e.g. Boulanger and Hayden 1995, Miller and Roycroft 2004). The advantage of CPG is its small footprint, mobility, and less vibration and noise compared to competing ground improvement techniques. Therefore, CPG is attractive for ground treatment near or beneath existing structures. Figure 1 illustrates the basic concept of CPG. Viscous grout is injected through an injection pipe into the loose ground which leads to the surrounding ground being compacted. Nevertheless, unexpected ground heave or damage to existing structures has been reported. These problems are caused partly by variability in ground conditions and are not fully understood. Experience plays a major role in the success of ground improvement using CPG.

Nishimura et al. (2012) investigated ground response due to grout injection with special focus on stress changes due to grout injection. Takano et al. (2013) studied the changes in ground density due to compaction grouting using X-ray tomography. From these studies, mechanisms of densification and increases in lateral confining pressure were evaluated. To optimize densification and increase lateral confinement, systematic studies of changes in ground conditions and interaction with existing structures are required. This is the motivation for the current study.

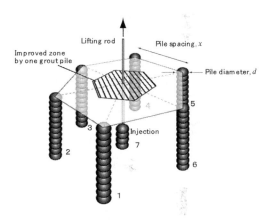

Figure 1. Illustration of compaction grouting.

Transparent granular soils are a powerful tool to visualize ground behavior at model scale (Iskander et al. 2015). Transparent granular soil has been used for geotechnical centrifuge model testing by Black (2015). An important requirement of a candidate surrogate (transparent) soil is that its mechanical behavior must conform to that expected for natural soils. Ezzein & Bathurst (2011) conducted a series of conventional geotechnical laboratory tests using transparent soil comprised of fused silica particles saturated with mineral oil and water and confirmed that this material possessed mechanical properties that could be expected for a natural sand soil. The same type of transparent soil is used in the current study. However, there are only a limited number of studies in which the results of physical modeling using transparent soil has been compared with sand models using opaque natural sand and field data.

The present study is focused on visualization of ground behavior due to compaction grouting by using transparent granular soil. The injection process was simulated using a surrogate grout injected through a rod into a transparent soil in one test and anFN opaque silica sand in the other test. The test specimens were contained in a box mounted on a large geotechnical centrifuge. The injection pressures recorded from the centrifuge tests were compared with field data. Finally, ground movements due to grout injection were recorded using marker particles embedded in the transparent soil.

2 TEST SETUP

2.1 Experimental methodology

A beam-type geotechnical centrifuge with effective radius of 3.8 m located at the Port and Airport Research Institute was employed. The centrifugal acceleration for each test was 30 g. Figure 2 shows schematics of the test arrangements. Case 1 is the model with transparent soil and Case 2 is the model with Soma silica sand #5. The model ground for Case 1 consisted of three layers. The loose sand layer with relative density 50% was confined between dense layers. Colored transparent particles were embedded in the model ground as targets for Particle Tracking Velocimetry (PTV) measurements.

In Case 1, two grout piles were installed simultaneously in the loose layer by injecting grout from the bottom-up through the injection rods which were retracted as injection proceeded.

In Case 2, the model ground consisted of a single loose sand layer. A single grout pile was installed within a hexagonal cylinder with diameter of 116 mm (3.5 m at prototype scale). The rigid PVC wall of the soil container simulated the improved volume of surrounding ground by grout piles that were installed earlier.

In both cases, saturation of the model ground was carried out after the model ground was subjected to the target centrifugal acceleration of 30 g.

2.2 Materials

The materials for the model ground were crushed fused silica particles saturated with mineral oil (Case 1) and Soma Silica Sand #5 saturated with water (Case 2). The pore fluid for transparent soil was a mixture of two different mineral oils having different refractive index. The volume mixture of the fluid was adjusted to match the refractive index of the silica. Table 1 shows mechanical properties of ground materials. Crushed

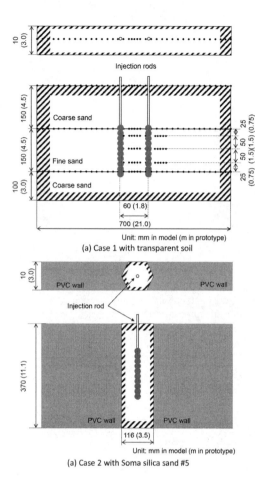

Figure 2. Vertical and horizontal cross-section of model ground for: (a) Case 1 (transparent soil); and (b) Case 2 (Soma Silica Sand #5).

Table 1. Properties of transparent soil and Soma Silica Sand #5.

	Transparent silica sand	Soma silica sand #5
Specific gravity, ρ_s (g/cm³)	2.21	2.65
Maximum void ratio, e_{max}	1.018	1.115
Minimum void ratio, e_{min}	0.707	0.710
Friction angle, ϕ' (degree)	41	36
Mean grain size, D_{50} (mm)	1.2	0.35
Coefficient of uniformity, U_c	1.0	1.5

silica particles have larger particle size than the Soma silica sand.

Figure 3 shows relationship between deviator stress and mean effective stress studied from drained triaxial compression tests for crushed silica particles saturated with oil or water, and Soma silica sand saturated with water. It can be seen that the granular materials used in this study have similar frictional shear strength.

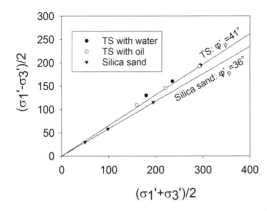

Figure 3. Shear strength envelope for Transparent soil (TS) saturated with mineral oil and water, and Soma sand saturated with water.

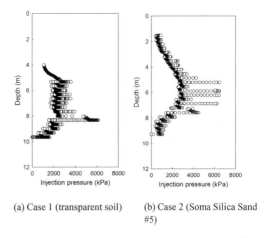

(a) Case 1 (transparent soil)   (b) Case 2 (Soma Silica Sand #5)

Figure 4. Injection pressure-depth relationship at prototype scale.

Table 2. Mixture of injected grout.

	Case 1	Case 2
Soma Silica Sand #5 (%)	–	25
Kawasaki clay (%)	57	37
Portland cement (%)	7	8
Water (%)	36	30

Table 2 shows the mixture components for the simulated grout used for Case 1 and Case 2. In each model test, the grout was injected through a small tube attached to the top of the injection rod. The grout must be sufficiently non-viscous to avoid plugging.

3 RESULTS

The injection procedure involved staged injection from the bottom-up. The injection volume was controlled to create a grout pile that was 0.7 m in diameter at prototype scale. The injection and injection pipe retracting rates were selected to match typical field practice.

The injection pressure is plotted against depth for both test cases in Fig. 4. The injection pressure increases immediately after the start of the experiment but repeatedly increases and decreases at each injection and pipe retraction step. Peak injection pressures occur over depths of around 5–8 m. Furthermore, when the injection depth becomes shallow, the injection pressure generally decreases; this phenomenon was observed in both tests. It can be seen that there are injection cycles in which the injection pressure becomes extremely large at depths of 5–8 m, especially for Case 2. It should be noted that when a relatively large sand particle is present in the grout in a field installation, the tip can become plugged.

The inner diameter of the injection rod in this investigation was scaled to match the inner diameter of 50 mm at prototype scale used in practice (CDIT 2007). Similar plugging behavior to that observed in the field was observed for Case 1 with transparent soil.

When the injection pressure increased rapidly, the grout was compressed at the rod tip without being discharged, and plugging at the tip occurred. The plugged grout in the rod was suddenly discharged when injection stopped and the rod was lifted above the loose layer and ground stress decreased.

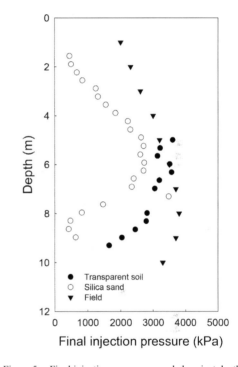

Figure 5. Final injection pressure recorded against depth.

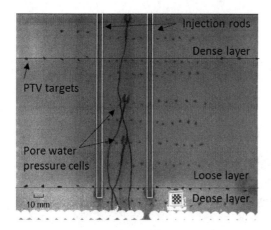

Figure 6. Transparent model ground before injection (Case 1).

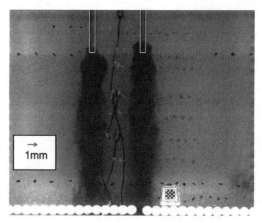

Figure 7. Transparent model ground after injection showing displacement vectors deduced from embedded targets (Case 1).

Figure 5 shows the final grout pressure recorded against depth in Case 1 and Case 2. The injection pressure due to plugging was excluded from the plot and the maximum injection pressure at each depth was taken as the final injection pressure at each depth. In addition to the results for Case 1 and Case 2, the distribution of injection pressure with depth obtained from actual CPG injection in the field is shown in Fig. 5. These measured data were obtained at the site of liquefaction countermeasure construction beneath a runway at Niigata Airport. As shown in Fig. 5, when the injection depth is less than 4 m, the injection pressure-depth relationship is linear in both experimental cases and in the field case; at greater depths the final injection pressure is non-linear with depth. In the centrifuge experiments, the final injection pressure shows a peak near the depth of 6 m, whereas in the field, the final injection pressure appears at around a depth of 8 m. The decrease in injection pressure at a depth deeper than 8 m is not as pronounced in the field test compared to the model tests. These results suggest that the response of the ground varies depending on the depth of the injection grout. However, the trends in results for the transparent soil and silica sand tests, and the field data are judged to be good agreement, and the peak final pressures are in reasonable quantitative agreement.

Figure 6 shows an annotated photograph of the model ground for Case 1 after ground saturation. The 16 bit gray scale image was taken by a high-speed camera placed in front of the model container. The embedded target particles are clearly visible. There is no difference in gray scale between dense and loose layers.

Figure 7 shows the model ground after injection was complete. Displacement vectors deduced from target particle measurements using PTV analysis are superimposed on the image. The magnitude of the displacement vector was set to 10 times larger than the actual value. The targets just beyond the right side grout pile show that ground displacement has occurred to a distance of twice the pile diameter. In addition, when the injection depth is deep, the horizontal direction of ground movement is pronounced, while at shallow depth, ground heave is the predominant ground movement. Figure 8 shows ground heave near the injection rod recorded against injection depth in Case2. The ground heave was recorded by Linear Variable Differential Transformer (LVDT) displacement sensor. The ground heave gradually occurs from the injection depth around 6–7 m. If the injection depth is shallower than 4.5 m, the ground heave is relatively large. This feature indicates that the different deformation mode such as cavity expansion mode with deep injection and uplift deformation mode with shallow injection occurred. These results suggest that the same response of the ground was observed from the models made of silica sand and the transparent soil.

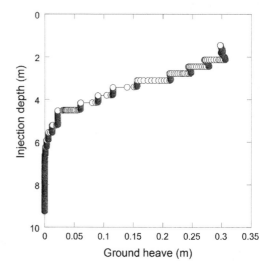

Figure 8. Deformation of ground surface against injection depth.

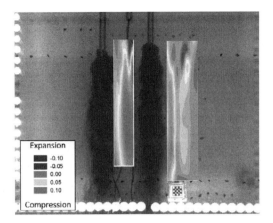

Figure 9. Distribution of horizontal principal strains.

Figure 9 shows horizontal principal strains calculated from target displacement vectors. From the strain distribution, the compression zone around the grout pile is clearly detectable. It is particularly noticeable over the lower part of the grout pile where the compressive strain is about 10%. Between the two grout piles, the strain distribution confirms that the compression region extends to the center between the piles.

4 CONCLUSIONS

The injection process of compaction grouting was simulated using models placed in a geotechnical centrifuge. Transparent granular soil and a natural silica sand were used for the model ground. Final peak grouting pressures for both cases were in good agreement with injection pressures recorded from a field site. Ground movements due to grout injection were easily visualized using the transparent soil.

REFERENCES

Black, J. A. 2015. Centrifuge Modelling With Transparent Soil and Laser Aided Imaging, *ASTM Geotechnical Testing Journal*, Vol. 38, No. 5, 2015, pp. 631–644.

Boulanger R.W. & Hayden R.F. 1995. Aspects of compaction grouting of liquefiable soil. *Journal of Geotechnical Engineering, ASCE*, 121 (12), 844–855.

CDIT (Coastal Development Institute of Technology), 2007, Manual of Compaction grouting method as a countermeasure for liquefaction (in Japanese).

Ezzein, F. & Bathurst, R.J. 2011. A transparent sand for geotechnical laboratory modeling. *ASTM Geotechnical Testing Journal* 34(6): 590–601.

Iskander, M, Bathurst, R.J. & Omidvar, M. 2015. Past, present and future of physical modeling with transparent soils. *ASTM Geotechnical Testing Journal* 38(5): 557–573

Miller E.A. & Roycroft G.A. 2004. Compaction grouting test program for liquefaction control, *Journal of Geotechnical and Geoenvironmental Engineering, ASCE*, 130 (4), 355–361.

Nishimura S., Takehana K., Morikawa Y. & Takahashi H. 2012. Experimental study of stress changes due to compaction grouting. *Soils and Foundations*, 51 (6), 1037–1049.

Takano, D. Nishimura, S. Takehana, K., Morikawa, Y. & Takahashi, H. 2013. Effect of compaction grouting as a countermeasure against liquefaction, JGS Geotechnical Engineering Journal, Vol. 8, No. 1, pp. 81–95 (in Japanese).

Centrifuge modelling of remediation of liquefaction-induced pipeline uplift using model root systems

K. Wang, A.J. Brennan, J.A. Knappett & S. Robinson
School of Science and Engineering, University of Dundee, Dundee, UK

A.G. Bengough
The James Hutton Institute, Invergowrie, UK
University of Dundee, Dundee, UK

ABSTRACT: Buried pipelines are susceptible to floatation within liquefiable soil after earthquakes. When soil liquefies, its shear strength is significantly reduced due to generation of excess pore pressure. A buried pipeline within such soil can then uplift due to a combination of (i) an upwards pore pressure gradient across the pipe and (ii) the resisting force contributed from soil shear strength being significantly reduced. Roots have been confirmed to increase the shear resistance of soil, so they can potentially be used as a new countermeasure against pipeline uplift, in locations where there is no above-ground infrastructure (i.e. where uplift is most likely). Three centrifuge tests have been conducted in this study to evaluate this potential. One was performed as a benchmark and the other two included one of two overlying model shallow root systems (either fibrous roots only, or fibrous and large structural roots) respectively. The results show that roots can be used as a remediation method against pipeline uplift induced by soil liquefaction. Model fibrous roots were shown to reduce uplift displacements by 15% while the model system consisting of both large structural and fibrous roots further reduced uplift to approximately 28%.

1 INTRODUCTION

Pipelines can suffer severe damage from strong earthquakes. One of the possible damage patterns is pipeline uplift within the liquefiable soil. In the field, this phenomenon has been observed in numerous earthquakes including: 1989 earthquake of Loma Prieta (O'Rourke et al. 1991), 1995 earthquake of Kobe (Shinozuka 1999), 2004 earthquake of Chuetsu (Yasuda & Kiku 2006), and recent 2011 earthquake of Tohoku (Chian & Tokimatsu 2012).

One of the possible remediation methods, in locations without overlying infrastructure (e.g. roads or buildings) is to increase the strength of overlying soil above the pipeline. The reinforcing effect of roots on soil has been recognised by its ability to increase slope stability (Gray & Leiser 1982, Coppin & Richards 1990; Gray & Sotir 1996) and many studies have been conducted for quantifying the contribution of roots (Wu, 2013), including using geotechnical centrifuge modelling with real live plants (Sonnenberg et al. 2010) and 3D printed analogue root models (Liang et al. 2016). To the best of the authors' knowledge, there has been no previous study about effects of roots on remediation of soil liquefaction. Nevertheless, research on fibre reinforcement, which has been used to simulate root reinforcement in laboratory tests, could assist our understanding of how this might work. Fibres were first found to increase soil liquefaction resistance by Noorany & Uzdavines (1989) using cyclic triaxial tests and this result has been verified by other researchers through similar element tests (Krishnaswamy & Isaac 1994; Boominathan & Hari 2002; Noorzad & Amini 2014; Pasha et al. 2016). Based on the geotechnical centrifuge modelling conducted by Wang & Brennan (2014, 2015), fibres can increase soil stiffness and limit significant de-formation caused by soil liquefaction.

As a pioneering study, three centrifuge tests were conducted to investigate the potential of shallow root systems to remediate the pipeline uplift caused by soil liquefaction. A benchmark test was performed first to demonstrate the uplift of the pipeline within liquefiable soil during and after a sequence of ground motions. Synthetic fibres were then introduced into the overlying soil above the pipeline in the subsequent test to mimic the mechanic effects of a purely fibrous root system on limiting the uplift of the pipeline. Three-dimensional models of large structural roots were placed together with fibres in the final test to investigate an alternative shrub-type root system on reducing the uplift of the pipeline.

2 CENTRIFUGE MODELLING

2.1 Apparatus and instruments

All tests were performed using the Actidyn C67-2 geotechnical centrifuge at the University of Dundee

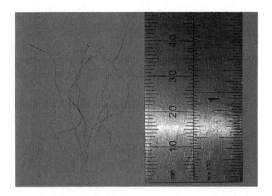

Figure 1. Loksand™ fibres.

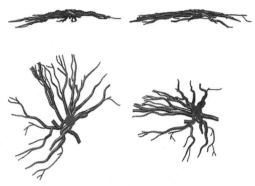

Figure 2. Modified design of architecture of large structural roots.

under 30 g. Input ground motions were simulated using the Actidyn Q67-2 earthquake simulator mounted on the centrifuge. More details about the centrifuge and the earthquake simulator can be found in Bertalot (2013) and Brennan et al. (2014). Models were prepared in an equivalent shear beam (ESB) container with internal dimensions of $674 \times 312 \times 280$ mm, which was described by Bertalot (2013) in detail. Accelerometers (ACCs), pore pressure transducers (PPTs), linear variable transducers (LVDTs) and draw wire transducers (DWs) were used for relevant measurements.

2.2 Model materials

The sand used in the models was HST 95 Congleton sand. It is a uniform fine silica sand with a mean particle size $D_{50} = 0.13$ mm, an effective size $D_{10} = 0.1$ mm, a coefficient of uniformity $C_u = 2.25$, a coefficient of curvature $C_c = 1.36$ and a specific gravi-ty $G_s = 2.63$. The maximum and minimum void ratios are $e_{max} = 0.795$ and $e_{min} = 0.463$ respectively. The fine fibrous roots were modelled by synthetic fibres with the commercial name Loksand™ (Figure 1). Their nominal length and diameter are 35 mm and 0.1 mm respectively. The specific gravity of this synthetic material is 0.91 and the tensile strength is 200MPa. All property data of Loksand™ was provided by the manufacturer.

The pipe model was made from a hollow Polyvinyl Chloride (PVC) plastic tube with closed ends on both sides. The length of the pipe model was 255 mm, the outer diameter was 40 mm and the mass was 231.95 g.

The *uPrint SE* Acrylonitrile Butadiene Styrene (ABS) prototyper (known as a 3D printer) at the University of Dundee was used to construct the scaled model of the large structural roots with representative complex architecture. The tensile strength of the ABS is around 17 MPa. More details about the ABS and the procedures of 3D printing root models can be found in the work of Liang et al. (2014).

The prototype of the root architecture used was from the structural roots of Arctostaphylos pungens (a chaparral shrub) for which detailed root architecture was available from Wu et al. (2014). When scaling

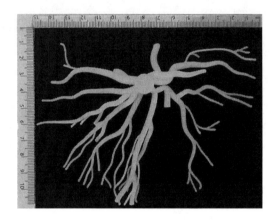

Figure 3. ABS model of large structural roots from 3D printer.

the prototype down by 1:30 for centrifuge modelling according to the scaling law, diameters of the model are comparable to those of the Loksand™. The model 3D printed models hence would lose their functionality as larger structural roots in the centrifuge tests. The prototype was therefore reasonably modified by increasing diameter six times while maintaining length as the original. The architecture adopted, reconstructed by AutoCAD, is shown in Figure 2 and the 3D printed model is shown in Figure 3. The maximum depth of model of the large structural roots was 20 mm.

Methylcellulose solution with 30 times viscosity of water was used as the pore fluid instead of water. This is to resolve the disparity between the scaling laws for the time of diffusion processes and a dynamic event (Madabhushi 2014).

2.3 Model preparation

There are three centrifuge models presented in this study. Their schematic profiles are shown in Figure 4 and Figure 5. The first model (PU1) represents the benchmark condition in which there was no remediation method applied to limit the uplift of pipe induced

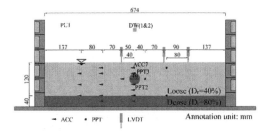

Figure 4. Centrifuge model and instrument distribution of PU1.

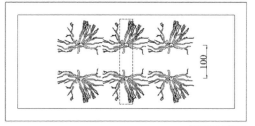

Figure 6. Plan distribution of large structural root models in PU3.

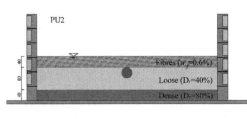

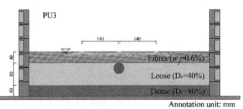

Figure 5. Schematic profiles of PU2 and PU3.

by soil liquefaction (Figure 4). The second and third models (PU2 and PU3) represent the conditions in which fibrous roots and fibrous with large structural roots were applied in the 40 mm (1.2 m in prototype) soil layer above the pipeline (Figure 5). In each case the normalised cover-depth of the pipe was one diameter (1.2 m). Units used in this subsection are at model scale while those in other sections are at prototype scale (unless otherwise stated).

Layered sand models with relative densities (D_r) of 40% overlying a lower layer of 80% were prepared by dry pluviation using a spot pluviator and slot pluviator respectively. The reinforced layers with fibrous roots in model PU2 and the combined root system in model PU3 were also prepared by dry pluviation using the slot pluviator. Fibre content (w_f) was 0.6% by mass relative to that of sand in the corresponding layer. The equivalent fibre content in volume is 1%, which is within the range of root content in volume found in the field (Bengough 2012) Further details about the dry pluviation method used in this study can be found in the work of Wang & Brennan (2014). ACCs, PPTs and the pipe model and the structural roots model were placed at the pre-determined locations during the dry pluviation, and LVDTs and DWs were installed after completion of dry pluviation. Instrument distribution was identical in all models – this is shown using model PU1 as an example in Figure 4. The plan distribution of large structural root models and pipeline model is shown in Figure 6. The roots were placed in this way to ensure symmetrical behaviour at either end of the pipe, so as to avoid twisting out of plane and jamming the pipe between the walls. Securing supports were then placed to avoid the accidental movement of pipe model before centrifuge testing. After completion of these procedures, models were saturated with viscous methylcellulose solution.

2.4 Test programme

After completing the preparation, the model was loaded onto the earthquake simulator. The centrifuge was then spun up at intervals of 10 g until reaching the desired g-level (30 g in this study). A succession of three input motions (EQ1, EQ2 and EQ3) were then simulated. There was an adequate time interval between each motion to allow the completion of excess pore pressure dissipation within the model. This was confirmed by observations of the PPTs. The three input ground motions were ramped sinusoidal motions having the same properties (frequency content and duration) except for the acceleration amplitudes (Figure 7). The time histories of the input motions can be described by the following equations and parameters of three input motions are summarised in Table 1.

$$A(t) = \begin{cases} \dfrac{t}{nT} A_0 \sin(\omega t) & 0 \leq t < nT \\ A_0 \sin(\omega t) & nT \leq t < (N-n)T \\ \dfrac{NT-t}{nT} A_0 \sin(\omega t) & (N-n)T \leq t \leq NT \end{cases}$$
(1a)

$$\omega = \dfrac{2\pi}{T}$$
(2b)

where A = amplitude of acceleration; A_0 = maximum amplitude; t = time; T = period of motion; n = number of ramped motion cycles; and N = total number of motion cycles.

Table 1. Properties of input motions (prototype scale).

Input motion ID	A_0 (g)	T(s)	N	n
EQ1	0.045	0.5	28	9
EQ2	0.100			
EQ3	0.210			

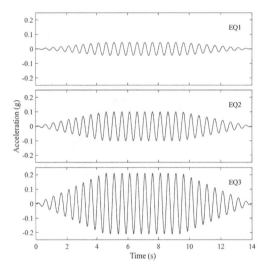

Figure 7. Time histories of input ground motions.

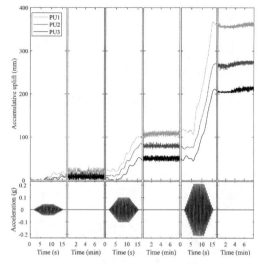

Figure 8. Time histories of accumulative uplift of pipeline.

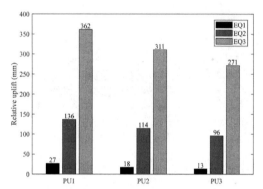

Figure 9. Relative uplift of pipeline to the ground.

3 TEST RESULTS

3.1 Uplift response of pipeline

Based on the time histories of pipeline uplifts shown in Figure 8 (derived from the average of direct measurements of DW1 and DW2), the uplift predominantly occurred co-seismically. Ultimate accumulative uplift displacements increase with the intensity of the ground motions. Around 70% of the total accumulative uplift was attributed to that in the final ground motion event regardless of the reinforcement condition. Uplift displacements were effectively reduced by introducing root systems in model PU2 and model PU3. The relative uplift displacements, which are the measured uplift displacements of pipeline relative to the settled ground surface, in each ground motion event are shown in Figure 9. The total relative 525 mm pipeline uplift in model PU1 was limited to 443 mm in model PU2 and 380 mm in model PU3 (around 15% and 28% of pipeline uplift were inhibited by the two root systems, respectively). Thus, the large structural roots contributed around 13% to limiting pipeline uplift when acting with the fibrous roots. More specifically, 33% and 52% of relative pipeline uplift displacements were reduced in EQ1, 16% and 29% in EQ2, and 14% and 25% in EQ3 in model PU2 and model PU3 respectively. Reduction of the relative pipeline uplift attributed to root systems increases with the pipeline uplift potential (ground motion intensity) when considering value rather than the percentage. In EQ1 event, the relative pipeline uplift in model PU1 was 27 mm, and 9 mm and 14 mm of uplift displacements were reduced in model PU2 and model PU3 respectively. In EQ3 event, such reduction increased to 51 mm and 91 mm in model PU2 and model PU3 while the relative uplift displacement in model PU1 was 362 mm.

3.2 Dynamic response of soil above pipeline

Displacements in the EQ3 event played a dominant role in all models and due to limitations of paper length, results only in the EQ3 event are shown in this and the following subsections.

As shown in Figure 10, the accelerations recorded by ACC7 within the cover soil above the pipeline were significantly de-amplified compared with the original EQ3 input motion. The peak acceleration was reduced to around 0.1g in model PU1 and model PU2, and to

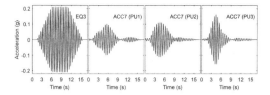

Figure 10. Acceleration responses of cover soil in EQ3 event.

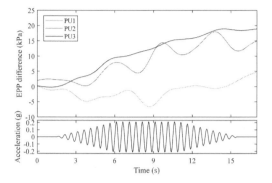

Figure 11. Excess pore pressure difference between PPT2 and PPT3 in EQ3 event.

around 0.15 g in model PU3. It indicates soil softening caused by liquefaction, even though the soil was reinforced with root systems. The overlying soil in model PU3 was less softened than those in the other tests. It is also suggested that soil in model PU2 was less softened than that of model PU1. The dynamic responses of overlying soil in the three tests support the conclusion that the root systems increase the shear strength of soil, which is beneficial to limit the uplift of the pipeline.

3.3 Excess pore pressure generation under pipeline

The time histories of excess pore pressure difference between the measurement of PPT3 (at the invert of the pipeline) and that of PPT2 (at the crown of the pipeline) in the EQ3 event are shown in Figure 11.

The differences indicate the uplift force due to excess pore pressure in each model. The excess pore pressure differences in model PU2 and model PU3 were larger than that in PU1. This is mainly caused because pipeline uplift was effectively reduced by introducing root systems in overlying soil above the pipeline. When a pipeline uplifts, a cavity is formed beneath it (Stone & Newson 2006). The formation of the cavity induces a negative change in pore pressure and therefore reduces the excess pore pressure generated by the cyclic loading. The excess pore pressure at the crown of the pipeline, however, remains much less affected. Therefore, the more the pipeline uplifts, the less excess pore pressure induced uplift force will generate. Model root systems did not reduce excess pore pressure induced uplift force.

4 DISCUSSION

Pipeline uplift in liquefied soil is due to the buoyant force being greater than the resistance against it. This resistance consists of the weight of pipeline, the weight of overlying soil and the shear resistance of soil (Chian & Madabhushi 2013). In this study, the pipeline model was identical and the weight of overlying soil was not significantly increased by introducing the root systems. The excess pore pressure induced uplift force was also not reduced by introducing root systems. Therefore, improving overlying soil shear strength should be the main contribution to limiting pipeline uplift. The overlying soil was less softened when reinforced with root systems based on their dynamic response. More soil resistance is mobilized when the more the pipeline attempts to uplift. This is because the roots only become effective in mobilising soil strength when larger monotonic strains are induced. This interpretation is based on previous research on fibre-reinforced sand suggesting that a certain threshold strain level is required to mobilise the interlocking of soil and fibres to increase the shear strength of soil (Li & Zornberg 2013; Wang & Brennan 2015).

5 CONCLUSIONS

This study described centrifuge modelling investigating the use of root systems to remediate soil liquefaction induced uplift of buried pipelines. Introducing root systems in the overlying (cover) soil above the pipeline can reduce the pipeline uplift at various intensity levels of input motions. The fibrous root system reduced uplift by 15% of the total relative uplift after the three ground motion events, and the reduction increased to 28% when the root system also contained larger structural roots. Model root systems did not reduce the uplift force induced by excess pore pressure. The improvements appeared to be provided through the increased shear strength of the cover soil.

Root systems are probably beneficial in reducing pipeline uplift induced by soil liquefaction. The model root system including fibrous with large structural roots appears to be more effective than that only includes fibrous roots.

Applying root systems to cover soil above the pipeline is a promising low-cost method for limiting pipeline uplift induced by soil liquefaction in an urban area.

ACKNOWLEDGEMENT

The authors would like to acknowledge the financial support from the Leverhulme Trust (Grant no. RPG-2015-091). The authors would like to thank Ms. Yuan Wu and Dr. Xihong Cui at the Beijing Normal University for providing original data for constructing the model of structural roots. The authors also would like to express their gratitude to Gary Callon, Mark

Truswell and Grant Kydd at the University of Dundee for their assistance in printing the 3D root models and operating centrifuge tests. The Loksand™ fibres were kindly supplied by Drake Extrusion Ltd (UK).

REFERENCES

Bengough, A.G. 2012. Water Dynamics of the Root Zone: Rhizosphere Biophysics and Its Control on Soil Hydrology. *Vadose Zone Journal* 11(2). doi:10.2136/vzj2011.0111

Bertalot, D. 2013. Seismic behaviour of shallow foundations on layered liquefiable soils. *PhD Thesis*. University of Dundee. Dundee, UK.

Boominathan, A. & Hari, S. 2002. Liquefaction strength of fly ash reinforced with randomly distributed fibers. *Soil Dynamics and Earthquake Engineering* 22(9): 1027–1033.

Brennan, A.J., Knappett, J., Bertalot, D., Loli, M., Anastasopoulos, I. & Brown, M.J. 2014. Dynamic centrifuge modelling facilities at the University of Dundee and their application to studying seismic case histories. Gaudin & White (eds.), *Physical Modelling in Geotechnics – Proceedings of the 8th International Conference on Physical Modelling in Geotechnics 2014, ICPMG 2014, Perth Australia, 14–17 January 2014*.

Chian, S. & Tokimatsu, K. 2012. Floatation of underground structures during the M w 9.0 Tōhoku Earthquake of 11th March 2011. *Proceedings of the 15th World Conference on Earthquake Engineering, Lisbon, Portugal, 24–28 September 2012*.

Chian, S.C. & Madabhushi, S.P.G. 2013. Remediation against floatation of underground structures. *Proceedings of the Institution of Civil Engineers – Ground Improvement* 166(3): 155–167.

Coppin, N.J. & Richards, I.G. 1990. *Use of vegetation in civil engineering*. London: Construction Industry Research and Information Association London.

Gray, D.H. & Leiser, A.T. 1982. *Biotechnical slope protection and erosion control*. New York: Van Nostrand Reinhold Company Inc.

Gray, D.H. & Sotir, R.B. 1996. *Biotechnical and soil bioengineering slope stabilization: a practical guide for erosion control*. Hoboken: John Wiley & Sons.

Krishnaswamy, N. & Isaac, N.T. 1994. Liquefaction potential of reinforced sand. *Geotextiles and Geomembranes* 13(1): 23–41.

Li, C. & Zornberg, J.G. 2013. Mobilization of Reinforcement Forces in Fiber-Reinforced Soil. *Journal of Geotechnical and Geoenvironmental Engineering* 139(1): 107–115.

Liang, T., Bengough, A., Knappett, J., Muirwood, D., Loades, K. & Hallett, P. 2016. Realistic scaling of plant root systems for centrifuge modelling of root-reinforced slopes. *4th International Conference on Soil Bio- and Eco-Engineering-'The Use of Vegetation to Improve Slope Stability', Sydney, Australia, 11–14 July 2016*.

Liang, T., Knappett, J.A. & Bengough, A.G. 2014. Scale modelling of plant root systems using 3-D printing. Gaudin & White (eds.), *Physical Modelling in Geotechnics – Proceedings of the 8th International Conference on Physical Modelling in Geotechnics 2014, ICPMG 2014, Perth, Australia 14–17 January 2014*.

Madabhushi, G. 2014. *Centrifuge modelling for civil engineers*. Boca Raton: CRC Press.

Pasha, S.M.K., Hazarika, H., Bahadori, H. & Chaudhary, B. 2016. Dynamic behaviour of saturated sandy soil reinforced with non-woven polypropylene fibre. *International Journal of Geotechnical Engineering*: 1–12.

Noorany, I. & Uzdavines, M. 1989. Dynamic behavior of saturated sand reinforced with geosynthetic fabrics. *Proceedings of Geosynthetics '89, San Diego USA, 21–23 February 1989*.

Noorzad, R. & Fardad Amini, P. 2014. Liquefaction resistance of Babolsar sand reinforced with randomly distributed fibers under cyclic loading. *Soil Dynamics and Earthquake Engineering* 66: 281–292.

O'Rourke, T. D., Stewart, H. E., Gowdy, T. E. & Pease, J. W. 1991. Lifeline and geotechnical aspects of the 1989 Loma Prieta earthquake. *1991 – Second International Conference on Recent Advances in Geotechnical Earthquake Engineering & Soil Dynamics, Rolla, Missouri, USA, 11–15 March 1991*.

Shinozuka, M. 1999. *The Hanshin-Awaji Earthquake of January 17, 1995: Performance of Lifelines*. Darby: DIANE Publishing Company.

Sonnenberg, R., Bransby, M. F., Hallett, P. D., Bengough, A. G., Mickovski, S. B. & Davies, M. C. R. 2010. Centrifuge modelling of soil slopes reinforced with vegetation. *Canadian Geotechnical Journal* 47(12): 1415–1430.

Stone, K. & Newson, T. 2006. Uplift resistance of buried pipelines: An investigation of scale effects in model tests. Ng, Wang & Zhang (eds.), *Physical Modelling in Geotechnics-Proceedings of the Sixth International Conference on Physical Modelling in Geotechnics, 6th ICPMG '06, Hong Kong, 4–6 August 2006*.

Wang, K. & Brennan, A.J. 2014. Centrifuge modelling of saturated fibre-reinforced sand. Gaudin & White (eds.), *Physical Modelling in Geotechnics – Proceedings of the 8th International Conference on Physical Modelling in Geotechnics 2014, ICPMG 2014, Perth, Australia, 14–17 January 2014*.

Wang, K. & Brennan, A.J. 2015. Centrifuge modelling of fibre-reinforcement as a liquefaction countermeasure for quay wall backfill. *6th International Conference on Earthquake Geotechnical Engineering, Christchurch, New Zealand, 1–4 November 2015*.

Wu, T.H. 2013. Root reinforcement of soil: review of analytical models, test results, and applications to design. *Canadian Geotechnical Journal* 50(3): 259–274.

Wu, Y., Guo, L., Cui, X., Chen, J., Cao, X. & Lin, H. 2014. Ground-penetrating radar-based automatic reconstruction of three-dimensional coarse root system architecture. *Plant and Soil* 383(1): 155–172.

Yasuda, S. & Kiku, H. 2006. Uplift of sewage manholes and pipes during the 2004 Niigataken-chuetsu earthquake. *Soils and Foundations* 46(6): 885–894.

Comparative study of consolidation behaviour of differently-treated mature fine tailings specimens through centrifuge modelling

G. Zambrano-Narvaez, Y. Wang & R.J. Chalaturnyk
Department of Civil and Environmental Engineering, University of Alberta, Edmonton, Canada

ABSTRACT: The oil sands industry in Western Canada is actively testing new treatment techniques to increase the consolidation rate of Mature Fine Tailings (MFT). It is extremely time consuming to physically evaluate the long-term response of new treatment methods with conventional settling column tests or field pilot tests. Centrifuge modelling considerably shortens the consolidation process, while maintains the stress similarity between a downscaled model and the prototype. A series of centrifuge modelling tests were conducted in the University of Alberta to compare the long-term consolidation behavior of differently-treated MFT specimens. The centrifuge models shared the same prototype scale and consolidation time. Results revealed the long-term divergence of consolidation behavior in terms of settlement, geotechnical properties evolvement and porewater pressure response as the result of different treatment methods. The research proves that centrifuge modelling is an effective and highly-efficient experimental method in obtaining the long-term behavior of MFT consolidation of new amendments.

1 INTRODUCTION

Oil sands operations in northern Alberta generate a significant volume of mature fine tailings (MFT). The fluid fine tailings (slurry) are conventionally discharged into ponds. After reaching 30% solids content in few years, called MFT, the dewater rate becomes very low; therefore, the MFT disposal demands long-term commitment and large land space (FTFC, 195). New treatment techniques are being developed to increase the dewatering rate, including the use of chemical additives. It is extremely time consuming to evaluate the long-term behavior of these methods, especially for comparative purposes, as conventional settling column tests (1-g) or field pilot tests would take years to produce meaningful results.

A geotechnical centrifuge can be used for scale modeling of any large nonlinear problem for which gravity is the primary driving force, including the self-weight consolidation of MFT. Research indicated that the self-weight consolidation of MFT can be modelled in a shorter time frame, as centrifugal force replicates important stresses that govern the consolidation process (Sorta, 2015). This paper presents the method and results of a series of centrifuge tests of MFT self-weight consolidation that verify the effectiveness of using centrifuge modelling in the area of treatment method evaluation.

2 EXPERIMENTAL SYSTEM & SETUP

2.1 Centrifuge

The Geotechnical Centrifuge Research Facility (Geo-CERF) at the University of Alberta operates a 2-m radius geotechnical beam centrifuge, which can model a high-gravity environment up to 150 times Earth's gravity acceleration (Zambrano-Narvaez & Chalaturnyk, 2014).

The fundamental principle of centrifuge modelling is the stress similarity between the prototype and the centrifuge model. Scaling factors for stress, strain, time, dimensions and density are used to design the appropriate centrifuge operation. Primary self-weight consolidation is governed by the scaling laws as shown in Table 1, where the N values is the centrifuge acceleration level (N times Earth's gravity acceleration).

Table 1. Theoretical scaling laws for self-weight consolidation modelling.

Property	Model	Prototype
Acceleration	1	1
Length	1	N
Time	1	N^2
Stress/Strain	1	1

Additional details of scaling laws between models and prototypes for self-weigh consolidation phenomena can be found in several publications, for example Croce et al. (1985), Taylor (1995), and Garnier et al. (2007).

2.2 The consolidation cell & porewater pressure transducers

Figure 1a shows the test setup on the centrifuge test platform. The consolidation cell constructed for this study is made of a transparent Plexiglas cylinder of 177 mm internal diameter, 300 mm length and a wall thickness of 13 mm and a supporting aluminum structure. The transparent wall allows for readings of material height during the test. Settlement-time relationship curve is a clear indication of the progress of consolidation and can be used to determine important consolidation parameters to quantitatively assess the consolidation characteristics. Rulers with millimeter graduations were glued with epoxy on the internal wall of the Plexiglas cylinder. The movement of mudline (solids/water interface) was monitored during centrifuge tests through a high-resolution in-flight camera (IDS, 2448 × 2048 pixel, gigabit ethernet uEye RE model, CINEGON 1·8/4·8 CMPCT RUGGEDZD lenses model).

At the base of the cell, there is a disk-shaped porous stone which is connected to a bottom drainage port. The bottom drainage port remained closed during the test. All the tests conducted were carried out under one-way drainage conditions, with drainage towards the top of the model. The consolidation cell was covered during the test to eliminate the evaporation of the cap water layer.

At the side of the Plexiglas cylinder aligns five circular openings for miniature porewater pressure transducers (EPRB1-NU33-3-35B). Figure 1.b presents the vertical configuration of the five porewater pressure sensors. Porewater pressure was monitored during the consolidation and used to estimate excess porewater pressure, also an indicator of consolidation progress.

2.3 T-bar penetration device

A penetration device (T-bar) was used to estimate the undrained shear strength of the consolidated specimens. The device is installed on top of the consolidation cell. It has a T-shaped penetration head with strain gauges attached. The T-bar was driven by a stepper motor at a constant speed (1 mm/s) and penetrated the sample while loading force data from the sensor head were measured. The undrained shear strength is determined via the following empirical relationship, after Stewart & Randolph (1994):

$$s_u = \frac{Pd}{N_b} \quad (1)$$

where P is the force per unit length acting on the bar, d is the diameter of the bar, and N_b is the bar factor. An intermediate value of 10.5 for N_b was used

Figure 1. a) Test setup on centrifuge platform; b) cross-section of cell and pore pressure transducer locations.

as recommended by Randolph & Houlsby (1984) for general use.

2.4 Experimental procedure

Test specimens were collected in 20L pails from a production facility where the MFT were amended. During the model preparation, the specimen in the pail was divided into 5 vertical layers of equal thickness and placed into sub-containers. In order to reduce shearing, the samples were handled with a scoop. Then the 5 layers of sample were transferred to the consolidation cell in the same order, in order to maintain the density profile that had already developed inside the pail. The specimen was trimmed inside the consolidation cell to match the required model height. In the meantime, several samples were taken from the pail and measured for initial geotechnical index properties, including void ratio, solids content, water content and degree of saturation.

Then the consolidation cell was loaded on the centrifuge platform and the centrifuge was started at target speed for 48 hours. During the spinning, the settlement of MFT was pictured at regular intervals through the in-flight camera. Water cap thickness was also tracked and combined with porewater pressure measurements for the calculation of excess porewater pressure. Near the end of the test, the T-bar penetration test was conducted.

After completing each centrifuge test, the consolidated specimen was sampled by thin-wall glass tubes. The tubes were inserted gently to the bottom of the specimen. Each tube of sample was carefully sliced into pockets of equal thickness for index properties measurement.

3 COMPARATIVE TEST MATRIX

3.1 Treated MFT sample

Four tests were conducted to compare the self-weight consolidation behaviors of differently-treated MFT specimens under the same prototype conditions. The specimens were amended MFT with the same parent material. They were treated different concentration of polymer flocculants and coagulant, and shearing conditions. The treatment techniques were confidential and are therefore labeled as Specimen A, B, C and D.

The clay-size fraction of the parent MFT mainly consist of kaoline (80%) and illite (15%). The MFT consisted of fine sized material ranging from 98% to 99% (defined as <45 μm). The MFT liquid limits varied from 50% to 60% and the plasticity indexes from 24% to 30% (FTFC, 1995; Miller et al., 2010).

3.2 Test matrix

All specimens were tested under 80 times Earth's gravity acceleration, with an initial sample height of 12.5 cm and were spun for 48 hours in centrifuge. Per centrifuge scaling laws, the models represented the same prototype height of 10 m and prototype time of 35 years.

The test matrix offers a comparative perspective to evaluate the behavior of different specimens. With equal sampling condition, prototype thickness, and consolidation time, the treatment techniques can be effectively analyzed in terms of the up-scaled and long-term prototype behavior.

4 RESULTS

4.1 Consolidation curves

Based on MFT consolidation curves and scaling laws, prototype settlement curves are generated and shown in Figure 2. Specimen A, Specimen B and Specimen C have similar prototype settlements between 2.87 m and 2.95 m. Specimen D has a smaller prototype settlement of 2.28 m. Specimen B and C have overlapping curves, indicating an identical evolvement of consolidation over time. Specimen A reaches the end of primary consolidation phase (∼7 years) faster than Specimen B and Specimen C samples (∼24 years). Specimen D

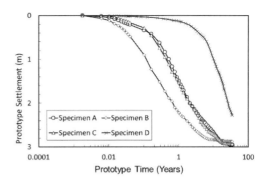

Figure 2. Prototype settlement-time relationship for treated MFT Tests.

MFT is experiencing primary consolidation through the end of the prototype time.

Specimen B has the normalized settlement of 34%. Specimen D has normalized settlement of 23%. Specimen A and Specimen C have the same normalized settlement of 31% and 33% respectively.

4.2 Pore pressure dissipation

The dissipation of excess porewater pressure is a clear indication of the progress of consolidation. Excess porewater pressure is interpreted as the difference between measured porewater pressure and hydrostatic pressure. For each centrifuge test, two pressure transducers measured porewater pressure changes in-flight. Hydrostatic pressure for each transducer location was determined with a separate spin with the consolidation cell filled with water of the same height. Combining measured porewater pressure and hydrostatic water pressure, excess pore-water pressure dissipation curves were plotted for the specimens and shown in Figure 3.

Transducer P1 is located at 4 cm above the bottom of specimen. Specimen B shows the highest initial rate of excess pore-water pressure dissipation, followed by Specimen C and Specimen A. Specimen D does not reach a stable stage for pore-water pressure at P1 location compared to other specimens and sensor location. The stable curve near the end of the dissipation curve signals the end of the primary consolidation stage. Temperature effect are observed in some tests and account for the upwards trend in some sensors. The response of excess pore-water pressure for all tests synchronizes with the respective settlement behavior shown in Figure 2, that Specimen B shows the quickest consolidation and Specimen D MFT doesn't finish primary consolidation during test period.

4.3 Index properties

Figure 4 presents the solids content and water content profiles before and after centrifuge spinning. Initially, Specimen D has the highest solids content and Specimen A has the lowest solids content (only measured at

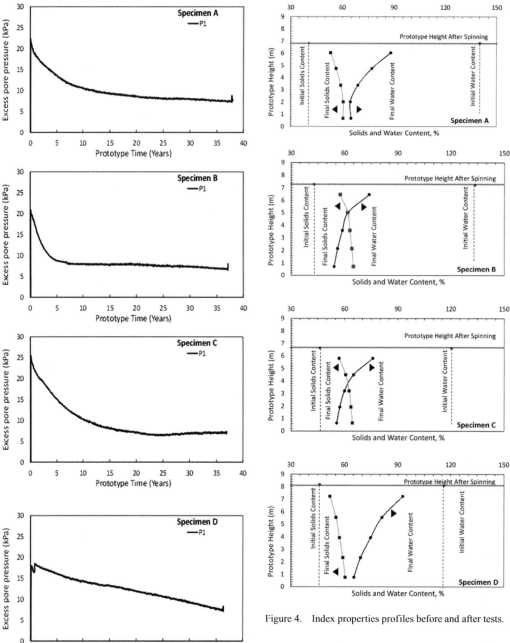

Figure 3. Excess porewater pressure dissipation with time.

the surface of the sample). After 35 years of proto type consolidation, Specimen D has, in contrary, the lowest solids content, which implies the smallest change in solids content of all the materials. The other three materials show comparable increase in solids content. Specimen B and Specimen C show very close initial solids content and identical solids content after centrifuge spinning. Other index parameters could not be reported due to a confidentially agreement, but the ability to track the changes of these properties, mainly Atterberg limit, void ratio, unit weight, flocculant and coagulant rations, etc., in the prototype time of years using the geotechnical centrifuge has proven an effective tool to assess the overall performance of the different treated mine tailings specimens.

Figure 4. Index properties profiles before and after tests.

4.4 *Undrained shear strength*

Figure 5 presents the shear strength profiles obtained through T-bar penetration test conducted at the end

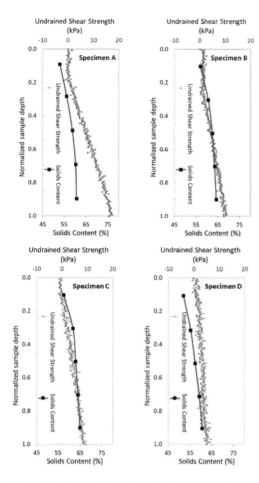

Figure 5. Undrained shear strength profiles measured by at the end of tests.

of each centrifuge test. The measured solids content data are also marked on the profiles. The developed shear strength for each material coincides well with the solids content profile. At the bottom layer of the material, Specimen A developed the largest undrained strength of 77 kPa and Specimen D has the smallest of 61 kPa.

Undrained shear strength is a direct indicator in terms of the reclamation capability of tailings ponds. The T-bar penetration test was conducted while the centrifuge was in flight. The obtain profiles represent prototype shear strength distribution at the end of the 35 years prototype consolidation.

5 CONCLUSIONS

This paper presents a testing time of 48 hours per model that reproduces similar effective stress of a prototype with 35 years of self-weight consolidation behavior. This makes the geotechnical centrifuge an important and attractive tool for physical modelling self-weight consolidation behavior of slurry in a cost-effective way and saving time.

The alternative to reproduce the self-weight consolidation performance of multiple years of multiple MFT materials using conventional testing methods at 1 g requires settling column tests and numerical models based on laboratory derived consolidation parameters. These methods and tests (e.g. large strain consolidation tests) are well accepted and suitable in quantifying the consolidation parameters and understanding of the self-weight consolidation behavior of slurries. However, these tests will require many months if no years to complete and calibrate with field observations.

The oil sands industry is actively developing new MFT treatment techniques. The use of geotechnical beam centrifuge in this research area can speed up the evaluation process for new techniques. The research conducted in GeoCERF proves that centrifuge modelling is an effective and highly-efficient experimental technique in obtaining the long-term behavior of MFT consolidation. The effectiveness of each new MFT amendment can be experimentally assessed in a matter of days instead of years of field study. On-board instrumentations enable the measurement and analysis of multiple consolidation effects and processes, including settlement, pore pressure response and shear strength development. The change of material index properties can also be obtained.

Centrifuge modelling is an effective method in assessing the performance of new amended MFT products. It can also be used to verify or complement results from conventional methods such as field test, lab column test or numerical simulation. The cost-effective and time-saving aspects of centrifuge modelling provide advanced tool in the world of MFT treatment and deposition technologies.

ACKNOWLEDGEMENT

The GeoCERF center was funded by CFI and AEAE. The authors acknowledge the advice and financial support from Shell International Limited and Shell Technology Center Calgary.

REFERENCES

Croce, P., Pane, V., Znidarcic, D., Ko, H.Y., Olsen, H.W. & Schiffman, R.L. 1985. Evaluation of consolidation theories by centrifuge modelling. *In Proceedings of the International Conference on Applications of Centrifuge Modelling to Geotechnical Design.* Balkema, Rotterdam, the Netherlands, pp. 380–401.

FTFC (Fine Tailings Fundamental Consortium) 1995. Advance in Oil Sands Tailings Research. FTFC, Oil Sands and Research Division, Alberta Department of Energy, Edmonton, Alberta, Canada.

Garnier, J., Gaudin, C., Springman, S.M., Culligan, P.J., Goddings, D., Konig, D., Kutter, B., Phillips, R., Randolph, M.F. & Thorel, L. 2007. Catalogue of scaling laws and similitude questions in geotechnical Centrifuge Modelling. *International Journal of Physical Modelling in Geotechnics.* 3, 2007: 01–23.

Miller, W.G., Scott, J.D. & Sego, D.C. 2010. Influence of the extraction process on the characteristics of oil sands fine tailings. *CIM Journal*.12, 2010: 93–112.

Randolph, M. F. & Houlsby, G. T. 1984. The limiting pressure on a circular pile loaded laterally in cohesive soil. Cambridge University Engineering Department.

Sorta, A. R. 2014. Centrifuge Modelling of Oils Sands Consolidation (Doctoral dissertation), University of Alberta.

Stewart, D. P.& Randolph, M. F. 1994. T-bar penetration testing in soft clay. *Journal of Geotechnical Engineering* 120.12 1994: 2230–2235.

Taylor, R. N. 1995. Geotechnical Centrifuge Technology. London: Blackie Academic.

Zambrano-Narvaez, G. & Chalaturnyk, R. 2014. The New GeoREF Geotechnical Beam Centrifuge at the University of Alberta, Canada, *Proceeding of the 8th International conference in Physical Modelling in Geotechnics*, Perth, Australia, pp. 163–167.

Physical modelling and monitoring of the subgrade on weak foundation and its reinforcing with geosynthetics

A.A. Zaytsev, Y.K. Frolovsky & A.V. Gorlov
Russian University of Transport RUT(MIIT), Moscow, Russia

A.V. Petryaev & V.V. Ganchits
Sankt-Petersburg State University of Railway Engineering (PGUPS), Saint-Petersburg, Russia

ABSTRACT: Nowadays in the Russian Federation there is a problem of the construction and exploitation of the subgrade on the weak foundation, including the efficiency assessing of subgrade stability assessing methods on the weak silty soils. Especially this problem has a significant importance in the northwestern part of Russia where the foundation soil is silty clay. The paper presents the results of physical modelling and monitoring carried out by the Russian Railway Universities of Moscow and Saint-Petersburg to assess the efficiency of several methods to reinforce the embankments foundation. There is the evaluation of the performance models for the foundation reinforce by geomattresses, filled with stone material. The simulation results showed that the embankment reinforced with geotextile has significant uniform precipitation, which can later lead to bulging of the underlying subsoil outside the contour of the slopes. The optimal solution in this case is the lightweight embankments with a core of expanded polystyrene (foam). This solution is effective to reduce the total embankment deformations on soft soils and to limit the lateral displacement.

1 INTRODUCTION

The construction of the subgrade on weak soils often lead to increased pore pressure. The effective stresses stay low as the result of undrained behavior and contractors use special stages for compaction and relaxation to improve the earthworks. The excess pore pressure dissipates during compaction and the soil shear strength is increased up to values which allow resumption construction. The main objective of this work is to estimate the viability of the decisions for the reconstruction of the approach highway embankment to the over-bridge through the Volhov River (North-Western part of Russia) and typical cross-sections for the newly-constructed railway embankment in the same region (Fig. 1).

The geology of this area was studied up to a depth of 30 m consisted of modern overburden soils, lacustrine deposits and Lontovassky set of rock of the Cambrian system (speckled and bluish clay). The lacustrine deposits are present clay silts with fluid or fluid-plastic consistency of coarse lens in saturated sands. The in with the paleo-valley of the Volhov River is filled with silts from 1 to 24 m deep.

Stabilization of the embankments on weak foundations in the highway and railway industries is provided by using berms, at high costs and big volumes of draining soils: sand, gravel, ground columns, cement mixing etc. (Yakovleva, T.G. & Ivanov, D.I. 1980, Almeida 1985, Arulrajah et al 2009, Aslam, R. & Ellis, E.A. 2010, Hayashi et al 2004, Pupatenko 1993, Petryaev et al 2015, Svatovskaya et al 2012, Vinogradov et al 2002, 2005, 2006, Zaytsev 2005, 2014). Some alternative engineering solutions include different geosynthetics: geogrids, geomattresses, geodrains and geofoams are used in the major it of them.

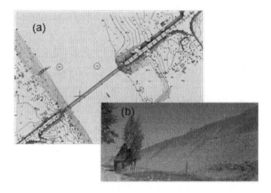

Figure 1. Bridge crossing plan The Volkhov River (a) and the embankment (b).

2 MODELLING AND MONITORING OF HIGHWAY EMBANKMENT

2.1 *The description of the prototypes, models and the characterization of the possible deformations*

A characteristic cross-section was chosen to evaluate the design solution. This reconstructed structure

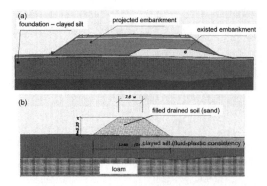

Figure 2. Scheme of the structure highway (a) and railway (b) embankments (scale factor for the modeling N = 75).

Table 1. The sample soil parameters (clayed silt).

Parameter name	Value		
	dimensionless	percent	g/cm³
Moisture	–	43	–
Plastic limit	–	36	–
Liquid limit	–	57	–
Plasticity Index	–	21	–
Density	–	–	1.71
Dry unit weight	–	–	1.20
Maximum dry unit Weight	–	–	1.67
Porosity	–	53	–
Porosity index	1.12	–	–

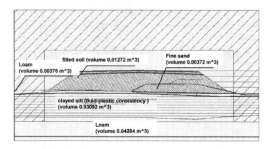

Figure 3. The view of the model preparation (for the highway embankment).

Figure 4. Stages for the model preparation.

consists of the jointed structure with the existing old-term embankment, which had a heighted of 2.4 m and a 1:2.3 slope for the left slope and 1:5 for the right, and a newly-constructed part of embankments which is elevated and filled under the existing embankment. The height of the projected newly-constructed highway prototype embankment is 6 m and the slope 1:1.5 (Figure 2 a). The prototype embankment was filled using drained fine sand and this embankment had the height of 6 m too (Figure 2 b). The first models of the embankments were prepared without any reinforcement to estimate the behavior of the embankments on these weak foundations.

Several hypotheses are preliminarily suggested to solve the problem of embankment stability. There are deformations – loss of slope stability; settlements of embankments as the results of the foundation soil uplift and settlement caused by the compression of the embankment and foundation soils. The scale factor N = 75 was chosen for the physical centrifuge modelling (Fig. 2).

2.2 The soil parameters of embankments foundation and modelling without reinforcement

Weak clay soil was used as foundation. Laboratory tests were done to estimate the main physical and mechanical parameters and these results are presented in Table 1 according to the State Standards (GOST 5180 2015, GOST 25100-2011). The clay silt of the fluid-plastic consistency was defined as the results of the lab tests. The soil moisture was increased up to 50% during the preparation of the foundation model.

The foundation soil was compacted layer by layer until it reached a density value which is equal to the maximum value of the dry unit weight of the 16.7 kN/m³ with the moisture content of was filled with 50%.

The of embankment was filled with soil (existing and newly constructed in prototype terms) after the foundation preparation (Fig. 3). Some sample for the moisture and density of soil were taken for the estimate the physical characteristics of the model soil (Fig. 4).

Laser profile-measurements (on three cross-sections) were constructed for each model after finishing of the model preparation and the same measurements were taken after centrifuge modeling.

The results of physical modeling show, that the embankment has stability, but values of displacement exceed the admissible valuation. It was decided to increase the soil moisture contest of the next experiment to 50%.

The settlements of the main top of the model at the prototype scale were equal to 0.45 m.

2.3 The results of modelling and design of reinforced highway embankments

The next model was prepared as the embankment model reinforced by vertical sand drains and geomatrasses, which were filled with gravel sand (Fig. 5).

The foundation soil was compacted layer by layer until it reached a density value equal to the maximum value of the dry unit weight of 16.7 kH/m³ at the moisture of 50% (see paragraph 2.2) Holes were the drilled into the found for the sand drains with depth of 1.05 m². These holes were filled with sand and compaction took place after moistening.

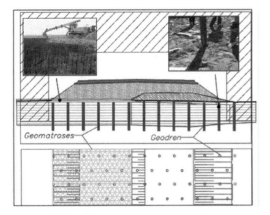

Figure 5. The scheme of the highway embankments reinforced by sand drains and geomatrasses.

Figure 6. The initial stage of the geomatresses and geo drain installation (prototype).

The model of embankments was prepared after the foundation and elements – models of the geomattress jointed together with foundation.

Then geomattress models were filled with coarse of 3 mm² and compacted. Dimensions of the models equal to scale N = 75.

For the physical modeling of the foul lane highway load on the top surface (GOST P 52748 2007) was chosen take a distributed static load of 45 kPa. The elements of the static load where made from ten separate pasteboard sections with filling sand- lead mix filling.

The settlements of the top of the model at the prototype scale were equal to 0.075 m.

The results of the modeling made it possible to recommend decisions for the prototype and to confirm stabilization of the foundation limited settlements making it possible to start the construction (Fig. 6).

2.4 Prototype data, subsidiary technical solutions (geogrids and geofoam) and monitoring of the highway embankments (the Volhov river)

To increase the foundation consolidation, the vertical drains were placed on the section of a new construction site for the highway embankment. The installation of these drains led to increased water drainage and to the decreased hydrostatic and pore pressure.

For the prototype conditions: the distance between drain centers was 0.8 m and the depth of installation ranged from 8 to 12 m.

Intensive consolidation stopped six months after the full filling and compaction of the embankment soil was completed.

In the triangular scheme of the placing, the distance between the drains was 0.80 m, and between the rows the distance was 0.70 m.

Some special technical solutions were used:

– Geomattress were used to provide a foundation bearing capacity;
– The geofoam blocks were placed above the existent embankment on the right side to reduce own weight of the embankment.

Figure 7. The geofoam blocks.

This structure of geofoam blocks was covered by geotextile. The geomembrane, with a thickness of 1 mm, was laid under the pavement to protect the blocks. The geofoam blocks were placed on the subgrade after cutting the existing pavement to a depth of 1.2 m and these blocks did not add load to the foundation (Fig. 7).

The results of the monitoring show that this decision made it possible to lay the pavement immedicable after finishing the earthworks. The monitoring of the pore pressure allowed estimation of the consolidation time for complete dissipation within 6 months (Fig. 8).

Insufficient bearing capacity of the subgrade foundation, for construction of a new embankment on weak soils, made it necessary to fill the layers of the embankment layer in stages, using the method of preliminary consolidation.

When filling a new embankment on weak soil foundations it is possible to exceed the permissible values of pore pressure, which can lead to settle ment of the foundation soil of the embankment. To prevent this, it is necessary to monitor the amount of excess pore pressure and the dimensions of the embankment.

3 PHYSICAL MODELING OF TYPICAL RAILWAY EMBANKMENTS

Successful modelling of the highway embankments during their foundation reconstruction, reinforced by geosynthetics structure, predetermined the need for

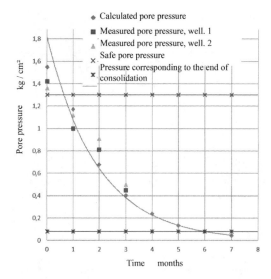

Figure 8. The monitoring of the pore pressure on the construction site.

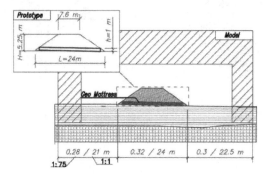

Figure 9. Schemes of model of railway embankment reinforced by geotextile mattresses.

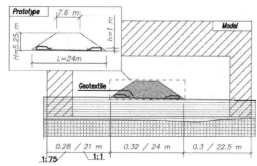

Figure 10. Schemes of model of railway embankment reinforced by geotextile.

The operating time of the centrifugal unit in the stationary mode at each stage of the simulation was 60 minutes, which equals to 7.7 months of the prototype scale. The time of simulation in stationary mode was counted from the moment of deviation of the carriage axis by an angle of 84° from the vertical, which, according to simulation experience, occurs when $\frac{3}{4}$ of the set angular rotation speed is reaches. The operating time of the centrifuge under steady-state conditions was 60 minutes. It corresponds to the operating time of the embankment for real conditions of 7.7 months.

The model of the embankment and foundation was made from field soils – loam and clayed silt. Achieve similarity between the physical and mechanical properties of the soil model and the full-scale prototype. When making the soil model, the density and moisture content of the soils were determined and matched actual conditions.

The foundation soil of the embankment model was compacted to reach its maximum density at 60 b moisture content. Due to the high moisture content required for compacting the soil under manual tamping, the soil was abandoned to give the required compaction for this case.

The soil compaction of the foundation with the fluid-plastic consistency was carried out by means of preliminary consolidation of the soil with the help of a press up to achieve a density of 1.5 ± 7 g/cm^3.

The preliminary consolidation of the foundation lasted 72 hours.

After laying and compaction of the foundation, models of geosell mattresses were laid, which were fixed to the foundation by metal J-shaped rods and fastened together.

Mattresses (mattress models) were filled with coarse sand of a 3-mm fraction and the sealing of this structure was made. The size of geosynthetics mattresses corresponded to the scale modelling of 1: 75.

The geotextile model was laid on the foundation and a layer of sandy soil was poured down with a thickness of 1 m (on the scale of the prototype). The layer was compacted and covered with geotextile model material, as shown in Figure 11.

For modelling the impact of the rolling stock on the subgrade of the mound model in a centrifugal

modelling of a typical solution for a railway embankment in new construction sites on weak foundations.

To assess the operational reliability of the design solution for strengthening the railway embankment, a characteristic cross profile was chosen.

The structure consists of the embankment, filled with draining soil (sand of medium size). The height of the new project is 6 m and slope 1: 1.5 (Fig. 9).

Two models of the subgrade were made for this modelling. To stabilize the structures, additional measures were envisaged to strengthen the embankment foundation. During the testing two variants of reinforcing were checked.

The foundation of the first model was reinforced with geotextile mattress filled by stone material (Fig. 9). The foundation of the second model was reinforced by a geotextile semi-moat structure (Fig. 10). The following centrifuge operation modes used at RUT (MIIT) centrifuge: dispersal – 6 minutes, working stationary mode – 60 minutes, centrifuge stop – 6 minutes.

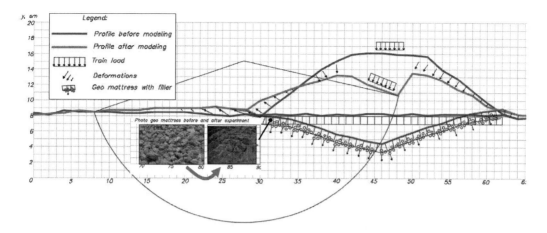

Figure 11. The results of modeling railway embankment reinforced by geomattresses.

installation, a set of sections corresponding to a static load at a maximum stress of 80 kPa was used. The static model of the train load consists of 4 separate sections.

It was found that in the model reinforced by geomattresses, the stone material of geomattresses was pushed into the foundation (Fig. 11), which could be a reason for changing the characteristics of the geomattress and, as a result, could lead to the destruction of the embankment. There was an extrusion of the foundation soil on both sides of the embankment, as seen in the photographs and the cross profile.

At the same time, in the analysis of the model reinforced by geotextile, it was found that during compaction a foundation deposit during compaction of the soil of the embankment occurred and, consequently, it caused additional compaction of the soil foundation look place. At the same time, the maximum settlement along the axis of the embankment was 1.35 m and it is possible to interpret this as a uniform settlement being realized during the construction period.

After 60 minutes of centrifugation, the embankment model showed its efficiency in the given conditions, without uneven deformations. The maximum settlement of the foundation during the construction period was 1.45 m in the prototype scale. The maximum settlement of the embankment after the modelling was 1.275 m. The outburst of the ground was observed only during the construction and compaction when the embankment height reached 2.025 m. Detailed results of the model profiling and calculation of settlements of model embankments on the scale of the model and the prototype are shown in Figure 12.

4 CONCLUSIONS

The results of the modelling showed that stability of the embankment and stability of the foundation soil are ensured both for the variant without reinforcing and for the variant with reinforcing.

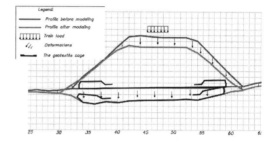

Figure 12. The results of modeling railway embankment reinforced by geotextile.

The analysis of the results of modelling and physical characteristics of soils has shown that when a weak foundation model is constructed and then the clayed soils have a soft-plastic consistency, stability of the foundation soils is provided, and the appearance of elastic-plastic deformations does not occur.

When comparis results for the Volhov highway embankment and the model settlements, before and after reinforcing corresponding (among themselves) sections it was established reveals that the values of the settlements of the model with the reinforcement of the foundation by sand piles and the geomattress turned out to be considerably less than the model settlements without reinforcement. As a result of the reinforcement of the soil foundation of the reconstructed embankment, the settlements values decreased by 50 to 83%, compared to the embankment, where the reinforcement of the foundation was not performed. This indicates a certain efficiency and for the prototype-technical solution for increasing the stiffness of the weak foundation soils by the sand piles with a cover construction from geomatress filled with stone material.

The control of the road bed construction by monitoring the pore pressure allowed it to be completed without any losses of bearing capacity of the foundation.

REFERENCES

Almeida, M.S.S. & Parry, R.H.G. 1985. Centrifuge studies of embankment foundations strengthened with granular columns. Proc. 3rd Geotech. Sem. Soil Imp., Sing.: 153–156.

Arulrajah, A., Abdullah, A., Bo, M.W. & Bouazza, A. 2009. Ground improvement techniques for railway embankments. Ground Improvement 162(1): 3–14.

Aslam, R. & Ellis, E.A. 2010. Centrifuge modelling of piled embankments. In S.M. Springman, J. Laue & L. Seward (eds), Phys. Mod. Geotech.: 1297–1302. London: T&F.

GOST 5180-2015. Soils. Laboratory methods for determination of physical characteristics.

GOST P 52748-2007 Automobile roads of the general using. Standard loads, loading systems and clearance approaches GOST 25100-2011. Soils. Classification.

Hayashi, H., Nishimoto, S. & Sawai, K. 2004. Peat ground treated by deep cement mixing with low improvement ratio. Proceedings of the International Symposium on Engineering Practice and Performance of Soft Deposits, IS-Osaka: 229–234.

Petriaev AV, 2015. Thawing railroad bed and methods of its reinforcing. Computer Methods and Recent Advances in Geomechanics. In: Proceedings of the 14th International Conference of International Association For Computer Methods and Recent Advances in Geomechanics. Kyoto, Japan, pp. 265.

Petryaev, A.A., Ganchits, V.V. 2015. The use of lightweight embankments in the construction and reconstruction of the subgrade on weak grounds. Proceedings of international scientific conference Transportation system infrastructure problems. Saint-Petersburg, Russia, September 30th–October 1st.: 70–73.

Pupatenko VV, 1993. Subgrade strength of narrow gauge railways under the influence of rolling stock. (In the context of the Sakhalin Railway.). Ph.D dissertation, St. Petersburg State Transport University, St. Petersburg, pp. 179.

Svatovskaya, L.B., Baidarashvily, M.M., Sakharova, A.S., Petryaev, A.V. 2012. Using of geomembrane is in ecoprotective aims. Transport construction No. 8, pp. 26–28.

Vinogradov, V.V., Yakovleva, T.G., Frolovsky, Y.K. & Zaitsev, A.A. 2002. Centrifugal modelling of the railway embankments with reinforcement by the various reinforced earth constructions. Proceedings of the International Conference on Physical Modelling in Geotechnics, St. John's, Newfoundland, Canada, 10–12 July: 987–991.

Vinogradov V.V., Yakovleva T.G., Frolovsky Y.K. & Zaytsev A.Al. 2005 Evaluation of slope stability of railway embankments. Proc. of the 16th International Conference on Soil Mechanics and Geotechnical Engineering, Osaka, Japan.

Vinogradov, V.V., Yakovleva, T.G., Frolovsky, Y.K., Zaitsev, A.A. 2006. Physical modelling of railway embankments on peat foundations. Proceedings of the 6th International Conference on Physical Modelling in Geotechnics—6th ICPMGE'06, Hong Kong, 4–6 August: 591–595.

Yakovleva, T.G. & Ivanov, D.I. 1980. Modelling of Stiffness and Stability of the Subgrade. Moscow: Transport.

Zaytsev, A.A. 2005. Modeling of stability of the railway embankments on the weak peat foundations, geotechnical problems on sedimentary soils in seismic region. Proceedings of the third international Central-Asian geotechnical symposium, Tadzhikistan, Dushanbe, 10–12 November: 171–174.

Zaytsev, A.A. 2014. Physical modeling embankment on peat foundation with reinforcing of the wooden piles. Physical Modelling in Geotechnics. 8th International Conference on Physical Modelling in Geotechnics (ICPMG), Perth, Australia, 13-17 January. 877-881

20. Shallow foundations

Effect of spatial variability on the behaviour of shallow foundations: Centrifuge study

L.X. Garzón
Department of Civil Engineering, Escuela Colombiana de Ingeniería, Bogotá, Colombia

B. Caicedo & M. Sánchez-Silva
Department of Civil and Environmental Engineering, University of Los Andes, Bogotá, Colombia

K.K. Phoon
Department of Civil and Environmental Engineering, National University of Singapore, Singapore, Singapore

ABSTRACT: This paper presents the results of a reduced-scale model of a vertically loaded single rigid strip footing resting on a spatially varying soil. The experimental work was carried out in a mini geotechnical centrifuge, and a micro loading system was developed. A total of 35 bearing capacity tests were performed, 30 tests resting in heterogeneous soils and 5 in homogeneous soils. The heterogeneous models match three different random fields that are defined by three coefficient of variations of the liquid limit, COV_{WL} 51%, 30% and 13%. Results show a reduction in the mean bearing capacity of the heterogeneous soils compared with the corresponding bearing capacity of the homogeneous soils that have the same mean property. Particle Image Velocimetry (PIV) analysis was made for analyzing the failure mechanisms. Results show that the inherent spatial variability of the soil properties can modify the basic form of the failure mechanism drastically.

1 INTRODUCTION

The spatial variability of soil properties is a critical factor that brings inevitable uncertainty on the design and analysis of geotechnical structures. In particular, the effects of soil heterogeneity on the bearing capacity of a shallow foundation has been widely studied numerical using random field theory to model the heterogeneity, and finite elements to compute the bearing capacity response, in conjunction with Monte Carlo Simulation approach (Fenton & Griffiths, 2002; Fenton & Griffiths, 2003; Popescu et al., 2005; Griffiths et al. 2006; Haldar & Babu, 2008; Hicks & Spencer, 2010; Huber et al., 2010; Kasama & Zen, 2011; and Al-Bittar & Soubra, 2012). The principal conclusions that can be drawn from these studies are the following: (i) the inherent spatial variability of the soil shear strength parameters modify the basic form of the failure mechanisms drastically; (ii) the average soil bearing capacity is lower when spatially variability is taken into account in comparison with the deterministic value obtained for a homogeneous soil; (iii) a critical bearing capacity value occurred when the autocorrelation length is equal to the footing width; (iv) differential settlements appear in the spatially varying soil leading to the rotation of the footing; (v) the average value of the ultimate footing load capacity is more sensitive to the variation of the horizontal autocorrelation length than to the vertical one. However, in the case of soil heterogeneity, all previous research efforts involved only numerical analysis because of the difficulty of modeling spatial variation of soil properties physically. Nonetheless, it is necessary to validate the growing volume of numerical results physically. Only one attempt to model soil heterogeneity at reduced-scale was made by Chakrabortty & Popescu, (2012), where seismic behavior and liquefaction mechanism of heterogeneous sand was studied.

This paper presents the results of a reduced-scale model of a vertically loaded single rigid strip footing of width 1.5m (prototype scale) resting on a spatially varying soil. The heterogeneous models were constructed, with variable liquid limit, will result in models having variable mechanical soil properties, e.g., undrained shear strength. The experimental work was carried out in a mini geotechnical centrifuge. For that purpose, a micro loading system that is capable of performing tests at a constant rate of strain was designed and built.

In the experimental program, three heterogeneous models that were built with a different coefficient of variations of the liquid limit, COV_{WL} 51%, 30% and 13% were tested. For each case, ten realizations were performed for a total of 30 bearing capacity tests. Also, five homogeneous models that have the same mean liquid limit of the heterogeneous models were tested. Preliminary results support the findings of the analytical research due to the average soil bearing capacity is lower when spatially variability is taken into account in comparison with the deterministic value obtained for a homogeneous soil.

The paper is organized as follows: in Section 2 we present the descriptions of the micro-loading system. In Section 3 we describe the bearing capacity test of the single rigid strip footing resting on a spatially varying soil. In Section 4 we present the results of the tests performed including the test on the homogeneous soil. Finally, in Section 5 we discuss the results and draw the conclusions of the paper.

2 MICRO LOADING SYSTEM

2.1 General description

The device has two major parts: (1) a driving motor that has a primary supporting beam, a linear actuator, and two shafts, (2) a moving platform that has a secondary supporting beam, three load cells, a profile rail guide, a footing strip, and auxiliary elements. The two parts are connected by the motor shaft, which is attached to the moving platform beam. It (the moving platform) moves vertically guided by two bushings, which shafts are joined at the principal supporting beam of the driving motor part. The beams, the auxiliary parts, and the footing strip are made of epoxy fibreglass 10 mm in thickness. This material was selected due to its high yield point and lightweight. The bushings and the shafts are of stainless steel. Figure 1 shows the two major parts of the micro-loading system. The micro-loading system can drive the footing strip at a constant rate into the clay.

2.2 Driving motor part

As described above the driving motor part has a main supporting beam of 57 mm in width and 140 mm in length. On top of the beam, a linear actuator is fixed, and two stainless steel shafts of 75 mm in length are joined at the edges. The linear actuator has a body size of 57 × 57 mm and 4.3 mm in height. It has an EAD motors Size 23 hybrid stepping motor, which internal rotating nut is made of SAE 660 bearing bronze. Figure 2 shows the sizes of the driving motor part.

2.3 Moving platform part

The moving platform is the most complicated part of the device since in this part the loading system is housed. The following description system's parts are up and down: the first element is the second supporting beam of 57 mm in width and 120 mm in length, which has at the edges two stainless steel bushings. Then, two miniature tension-compression load cells of 222.4 N, 30.73 mm in height and 24.89 mm in diameter are fixed in the inner part of the second supporting beam. Next, another supporting beam of 26 mm in width and 80 mm in length holds the bottom of the tension-compression load cells. Then, a miniature S beam load cell of 44.5

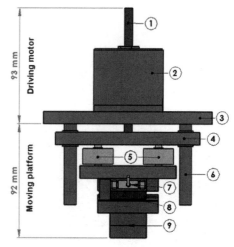

1. Actuating shaft
2. Linear actuator
3. Principal beam
4. Secondary beam
5. Vertical loads
6. Bushing and shaft
7. Horizontal load
8. Profile rail guide
9. Footing strip

Figure 1. Schema of the micro-loading system.

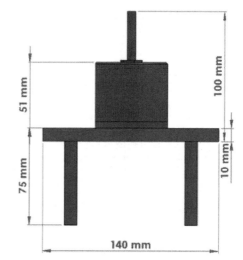

Figure 2. Schema of the driving motor part.

N is holding to an L-shaped beam, which is fixed to a profile rail guide that is attached to another beam. Finally, the footing strip of 70 mm in width and 30 mm in length is formed by two epoxy fibreglass elements to create a footing of 20 mm in depth. Figures 3 and 4 shows the moving platform part.

2.4 Software development

A computer program was developed using LabView (National Instruments Corporation, 2012) to control the micro-loading device. As mentioned above, the device was designed to perform tests at a constant rate

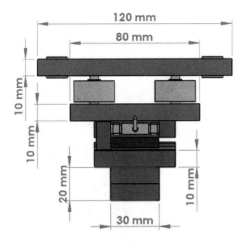

Figure 3. Schema of the moving platform front view.

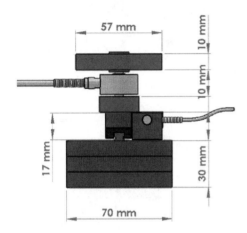

Figure 4. Schema of the moving platform lateral view.

of strain; for that reason, the program allows manual control of the displacement rate. The minimum and maximum rate of displacement are 0.01 mm/min and 5 mm/min respectively. The load and displacement data are sent to an external computer through the wireless data acquisition system mounted at the rotation centre of the mini centrifuge.

3 BEARING CAPACITY TESTS

The bearing capacity tests were performed on spatially varying soil models. These models were constructed following the procedure of a new technique to prepare reduced scale soil models with controlled variability for centrifuge modeling proposed by Garzón et al. (2015). In this technique, variability is controlled in the sense that each heterogeneous soil model is a physical representation of a random field realization. The heterogeneous soils are built by reproducing the variability in mineralogy, i.e., liquid limit and reproducing the history of field stresses by using an oedometric compression and then reproducing the field stresses in a geotechnical centrifuge.

A total of 35 bearing capacity tests were performed for this study, 30 tests resting in heterogeneous soils and 5 in homogeneous soils. The 30 heterogeneous models match three different random fields that are defined by three coefficient of variations of the liquid limit, COV_{WL} 51%, 30% and 13%; for each case, ten realizations were performed. The five homogeneous models have the same mean liquid limit of the heterogeneous models. All the tests were performed in the mini centrifuge of Universidad de los Andes, Bogotá, Colombia. Sections 3.1 and 3.2 summarized the terms in which the variability of the heterogeneous models was defined as well as a summary of the building process.

3.1 Variability of the heterogeneous models

The variability of the heterogeneous models was defined regarding the liquid limit, W_L. The soil is described by a mesh of equally sized elements for which a liquid limit, W_L, is assigned randomly according to the properties of the random field, which are the mean, μ_{WL}, the standard deviation, σ_{WL}, and the autocorrelation length, δ_{WL} of the liquid limit. Table 1 summarizes the random field properties and the number of realizations for the model. The 2-D field space for the physical model had 14 cm in length and 8 cm in height and was divided into square elements of 1 cm. The matrix decomposition technique proposed by El-Kadi & Williams (2000) was used to generate the random fields. For practicality, the continuous variation of liquid limit obtained from the matrix decomposition technique was discretized into eight values using the nearest 0.5 value. An example of a typical realization of an isotropic liquid limit random field within m = 112 divisions is shown in Figure 5.

3.2 Preparation of the heterogeneous models

The heterogeneous models were prepared following the next steps: first, a realization of the prescribed random field of liquid limit was generated numerically using the covariance matrix decomposition method as

Table 1. Random field input parameters.

Model	Constant property WL	Value1	Variable property WL	Value*	Realizations
Heterogeneous	μ_{WL}^a	157%	COV_{WL}^d	51%	10
	δ_h^b	1.5 m		30%	10
	δ_v^c	1.5 m		13%	10
Homogeneous	μ_{WL}^a	157%	COV_{WL}^d	0%	5

[1]Values are given in prototype dimensions, [2]mean liquid limit, [3]horizontal scale of fluctuation liquid limit, [4]vertical scale of fluctuation liquid limit and [5]coefficient of variation liquid limit.

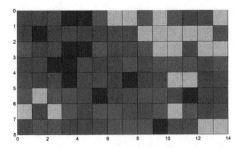

Figure 5. Random field with COV$_{WL}$ 51%.

Table 2. Bearing capacity results.

Model	COV$_{WL}$	μ qu* (kPa)	% Reduction
Heterogeneous	51%	18.4	29%
	30%	20.3	21%
	13%	22.5	13%
Homogeneous	0%	25.9	0%

*Ultimate bearing capacity.

explained in Section 3.1. Eight different soil types are reconstituted artificially so that the discrete set of liquid limit values is covered. These artificial homogeneous soils were prepared by mixing different percentages of kaolin and bentonite with a water content of about 1.5 times the liquid limit. Then, the physical (heterogeneous) model was constructed manually to match the discrete version of the random field realization. One of the eight slurries soil types fills each discrete cell in the physical model. This procedure was made at one time using a manual caulking gun and a grid from the bottom to the top of the container, completing the eight rows of the random field. This assembly of spatially heterogeneous cells, which constitutes the entire physical model, was next subjected to the consolidation process that was developed in two stages. For the first stage, the model was subjected to one-dimensional compression at a vertical stress of 30kPa. For the second stage, the models were transferred to the centrifuge and consolidated under 50g. The duration of the whole preparation process is approximate two weeks. More details of the preparation process can be found in Garzón et al. (2015). To ensure that the models behaved as a single block and not as a bricks assembly, the models were built in a single phase and the reconstituted soils were with a moisture content of 1.5 times the liquid limit. A Computerized Axial Tomography (CAT Scan) was performed showing no fissures or voids between the joints of the blocks.

3.3 Bearing capacity test setup

The bearing capacity tests were undertaken at the mini-geotechnical centrifuge of the Universidad de los Andes, Bogotá, Colombia. This mini size centrifuge is housed in a circular chamber of 1.7 m in diameter by

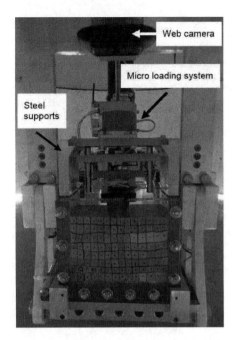

Figure 6. Micro-loading system installed in the mini centrifuge.

0.7 m in height. The mini centrifuge has a nominal radius of 56.5 cm and is capable of accelerating a 4 kg model package to 400 gravities. The container has a depth of 7 cm, a width of 14 cm and a height of 12 cm. A wireless data acquisition system mounted at the rotation center of the centrifuge send the measured data, at a specified time interval, to an external computer; as well, a webcam with LED lighting provides in-flight monitoring.

The micro-loading device was fixed to the centrifuge container using two stainless steel supports. The models surface was sliced to guarantee a horizontal surface. To analyze the failure mechanism, a grid was painted matching the cells of the random field; in the case of the homogeneous models, the grid was painted at a separation of 1 cm × 1 cm. The rigid strip footing was placed lightly touching the horizontal surface of the model. Then, the centrifuge was accelerated to 50g; next, the model was allowed to fly for one minute until the system got stabilized. Finally, the bearing capacity test was performed at a displacement rate of 0.05 mm/min. The test finished until reach a displacement of 20 mm. A web camera fixed to the centrifuge beam captured images during the test (Fig. 6). The recorded images were used to assess the strip footing displacement and failure mechanism using Particle Image Velocimetry (PIV).

4 RESULTS

4.1 Bearing capacity test

Figures 7, 8, 9 and 10 show the load-settlement curves of the five homogenous models and the

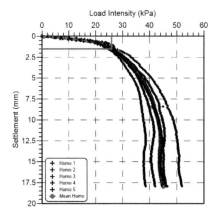

Figure 7. Load intensity-settlement curves homogenous models.

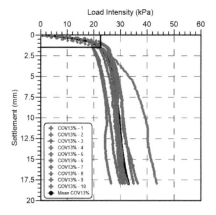

Figure 8. Load intensity-settlement curves heterogeneous models COV_{WL} 13%.

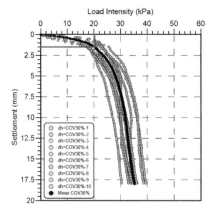

Figure 9. Load intensity-settlement curves heterogeneous models COV_{WL} 30%.

heterogeneous models with COV_{WL} of 51%, 30% and 13% respectably. For all the curves the mean was calculated and graphed; also, the ultimate bearing capacity of the tests was calculated using the criterion of 10% displacement of the foundation diameter proposed by Amar et al., (1994). Figure 11

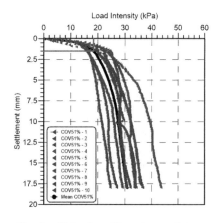

Figure 10. Load intensity-settlement curves heterogeneous models COV_{WL} 51%.

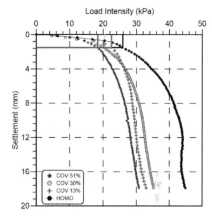

Figure 11. Mean load intensity-settlement curves of heterogeneous and homogeneous models.

shows the load-settlement curve of the mean bearing capacity of the three heterogeneous models and the homogeneous models. Table 2 summarized the results of the ultimate bearing capacity. It can be seen that there is a reduction in the mean bearing capacity of the heterogeneous models compared with the corresponding bearing capacity of the homogeneous models that have the same mean property. Likewise, as the COV_{WL} decreases the mean bearing capacity decreases approaching the value of the bearing capacity of the homogeneous soil with the same liquid limit properties.

Figures 12 and 13 show results of the displacement and failure mechanism using Particle Image Velocimetry (PIV) for two models: homogenous model Homo-2 and heterogeneous model COV_{WL} 30%-5. The graphs show the evolution of the failure mechanism as the bearing capacity test runs in four different times. In the left column of the graph, it can see the load-settlement curve, in the middle column the displacement vectors and in the right column the xy strains. A general conclusion of the failure mechanics analysis of the 30 heterogeneous soils and the five homogeneous soils is

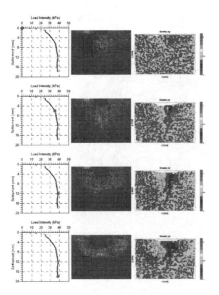

Figure 12. Evolution of the failure mechanics of homogeneous model HOMO – 2.

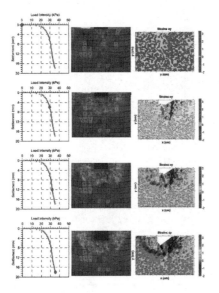

Figure 13. Evolution of the failure mechanics of heterogeneous model COV$_{WL}$ 30%-5.

that in the case of the homogeneous models symmetric shape failure was present, whereas in the case of the heterogeneous soils a non-symmetric shape of the failure mechanism is visible. This phenomenon is critical to understand since the inherent spatial variability of the soil properties is not only affecting the bearing capacity value of the soils but it is also modifying the basic form of the failure mechanics.

5 CONCLUSIONS

This paper presents the preliminary results of the bearing capacity test, conducted in a mini-centrifuge, of a single rigid strip footing of width 1.5 m (prototype scale) resting on a spatially varying soil. For that purpose, a micro loading system that allows performing bearing capacity tests with controlled displacement was development. The results show that the average bearing capacity of a spatially random soil is lower than the value obtained for a homogeneous soil for which the soil properties are equal to their mean values. Furthermore, the results show that the inherent spatial variability of the soil properties can modify the basic form of the failure mechanism drastically.

REFERENCES

Amar, S., Baguelin, F., Canepa, Y. & Frank, R. 1994. Experimental study of the settlement of shallow foundations. *In Vertical and horizontal deformations of foundations and embankments*, ASCE, (2) GSP, 40, 1602–1610.

Al-Bittar, T. & Soubra A.H. 2012. Bearing capacity of strip footings on spatially random soils using sparse polynomial chaos expansion. *International Journal for Numerical and Analytical Methods in Geomechanics*, 37, (13) 2039–2060.

Chakrabortty, P. & Popescu, R. 2012. Numerical simulation of centrifuge tests on homogeneous and heterogeneous soil models. *Computers and Geotechnics*, 41, 95–105.

El-Kadi, A.I. & Williams, S. A. 2000. Generating Two-Dimensional Fields of Autocorrelated, Normally Distributed Parameters by the Matrix Decomposition Technique. *Ground Water*, 38, (4) 523–532.

Fenton, G.A. & Griffiths, D.V. 2002. Probabilistic Foundation Settlement on a Spatially Random Soil. *Journal of Geotechnical and Geoenvironmental Engineering*, 128, (5) 381–390.

Fenton, G.A. & Griffiths, D.V. 2003. Bearing capacity prediction of spatially random c– ϕ soils. *Canadian Geotechnical Journal*, 40, (1) 54–65.

Garzón, L.X., Caicedo, B., Sánchez-Silva, M. & Phoon, K.K. 2015. Physical modeling of soil uncertainty. *International Journal of Physical Modelling in Geotechnics*, 15, (1), 19–34.

Griffiths, D. V. Fenton, G.A. & Manoharan, N. 2006. Undrained Bearing Capacity of Two-Strip Footings on Spatially Random Soil. *International Journal of Geomechanics*, 6, (6) 421–427.

Haldar, S. & Babu, S. 2008. Effect of soil spatial variability on the response of laterally loaded pile in undrained clay. *Computers and Geotechnics*, 35, 537–547.

Hicks, M.A. & Spencer, W.A. 2010. Influence of heterogeneity on the reliability and failure of a long 3D slope. *Computers and Geotechnics*, 37, 948–955.

Huber, M., 1, Hicks, M.A., Vermeer, P.A. & Moormann, C. 2010. Probabilistic calculation of differential settlement due to tunneling. *Proceedings of the 8th International Probabilistic Workshop*, 1–13.

Kasama, K. & Zen, K. 2011. Effects of Spatial Variability of Soil Property on Slope Stability. *Vulnerability, uncertainty, and risk analysis, modeling and management*. ASCE, 691–698.

Popescu. R., Deodatis. G. & Nobahar. A. 2005. Effects of random heterogeneity of soil properties on bearing capacity. *Probabilistic Engineering Mechanics*, 20, 324–341.

1g model tests of surface and embedded footings on unsaturated compacted sand

A.J. Lutenegger
University of Massachusetts, Amherst, Massachusetts, USA

M.T. Adams
Federal Highway Administration, McLean, Virginia, USA

ABSTRACT: A series of 1g model footing tests were performed on compacted sand to evaluate the settlement and bearing capacity. Compacted sand beds were prepared in a large test pit at relative densities of 35, 50 and 75% in order to evaluate footing response in both loose and dense compacted sands. The sand beds were moist and footing tests were performed at the compaction water content which represented unsaturated conditions. Square concrete footings with widths of 0.30, 0.61 and 0.91 m were tested with relative embedment ratios (D/B) of 0, 0.25, 0.5 and 1.0 to determine both scale effects and embedment effects on settlement and bearing capacity. Characteristics of the sand and results of the load tests are presented. A comparison is presented of the measured bearing capacities for the different conditions and the load-displacement behaviour of the footings is presented in simple terms to allow a prediction of settlement behaviour from sand characteristics.

1 INTRODUCTION

1.1 Shallow foundations on compacted sand

Shallow foundations are often placed on compacted coarse-grained soils at sites where near surface soils are marginal and may not support a shallow foundation directly. Simple remove and replacement technology is a common economical shallow ground improvement method with the compacted soil supporting the shallow foundation. In many cases, the compacted soil is placed in lifts and never becomes saturated. The bearing capacity and settlement behaviour of the shallow foundation is controlled by the unsaturated behaviour of the soil.

1.2 Large-scale model footing tests

A series of 1g large-scale model footing tests were performed to evaluate the bearing capacity and settlement behaviour of footings on compacted sand and included tests on both unsaturated and saturated sand compacted to different relative density (D_R). Some tests have previously been reported by Lutenegger & Adams (1998; 2003). In this paper, results of tests performed only on unsaturated sand are presented. The tests are used to illustrate the influence of relative density, relative footing embedment and footing size (scale) on the bearing capacity and load-displacement behaviour. The sizes of footings ranged from 0.30 m to 0.91 m and are considered large-scale (prototype) models as compared to small-scale (<0.30 m) models often used in laboratory studies and thus eliminate some of the scaling issues associated with small-scale model tests.

2 FOOTING TESTS

2.1 Sand characteristics and compaction

Footing load tests were conducted at the Federal Highway Administration Turner-Fairbank Highway Research Centre at McLean, Virginia. Tests were performed in a 5.5 m × 7.1 m × 6.1 m deep concrete test pit on compacted sand beds prepared at different relative densities. Sand was placed in the test pit in 0.3 m loose lifts and then compacted using a vibratory plate compactor to achieve a desired relative density. In order to achieve the desired relative density for each pit fill, the water content of the sand needed to be adjusted. This resulted in sand at different levels of saturation. In-place density tests were performed using a nuclear moisture-density gauge at several locations around the pit on each lift to verify the density and moisture achieved with each pit fill. The sand used for the testing was a subrounded, uniform fine mortar sand having a mean grain size of 0.29 mm, uniformity coefficient of 2.7 and coefficient of curvature of 1.0. The fines content is less than 2%. Minimum index unit weight of the sand is 1.41 Mg/m³ and maximum index unit weight is 1.70 Mg/m³. The specific gravity of the sand is 2.66 giving minimum and maximum

Table 1. Summary of footing tests.

Test Series	D_R (%)	W (%)	S (%)	Width (m)	D/B
95SD1	35	6.4	22.0	0.61	0
				0.61	0.25
				0.61	0.5
				0.61	1.0
97SD1	50	10.6	38.9	0.61	0
				0.61	0.25
				0.61	0.5
				0.61	1.0
100SD1	75	12.7	52.3	0.61	0
				0.61	0.25
				0.61	0.5
				0.61	1.0
95	38	5.6	19.4	0.30	0
				0.61	0
				0.91	0
95GA3	38	9.5	33.0	0.30	1
				0.61	1
				0.91	1
97SD1	50	10.6	38.9	0.30	0.5
				0.61	0.5
				0.91	0.5

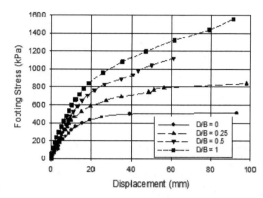

Figure 1. Typical stress-displacement curves: Series 97SD1.

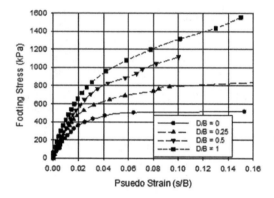

Figure 2. Load-pseudo-strain curves: Series 97SD1.

void ratios of 0.564 and 0.887. Load tests described in this paper were performed on the as compacted moist condition. After completion of each test series, the sand was removed, air dried and then replaced and compacted for the next series of tests.

2.2 Footings and load tests

Footings were constructed of reinforced concrete and had widths (B) ranging from 0.30 m to 0.91 m. Footings were placed at different depths (D) in the sand as the sand was placed to give embedment ratios (D/B) ranging from 0 to 1. Incremental load tests were performed using a hydraulic loading system with the central vertical load measured using an electronic load cell. Vertical displacement was measured at the four corners of the footing using LVDT's. Each of the footing tests was conducted so that a total settlement of approximately 10% of the footing width was achieved in the test. Table 1 gives a summary of the sand conditions and footing tests in each series of tests described in the current paper.

3 RESULTS – BEARING CAPACITY

3.1 Typical load test results

Typical load-displacement curves for one set of load tests are shown in Figures 1 and 2. When results from an actual footing load test are available, the ultimate capacity may be obtained directly from the footing performance. However, the load-displacement curve may be subject to interpretation, especially in the absence of a "plunging" failure. It is necessary to have a consistent and reasonable definition of ultimate bearing capacity in order to compare observed with predicted behaviour. For the current tests, the ultimate capacity was taken as the footing stress producing a relative displacement (s) of 10% of the footing width; i.e., $q_{ult} = q$ @ $s = 0.10B$. Table 2 gives a summary of the ultimate capacities determined for each of the footing tests. Figure 3 shows the normalised footing stress vs. pseudo-strain curves for the tests given in Figure 1.

In some cases the load tests did show a "plunging" type of failure and the load increment could not be maintained without continuous pumping of the hydraulic load ram. In general, this typically occurred for the smallest size surface footings (0.30 m) at low relative density, indicative of punching failure.

3.2 Influence of relative density

Figure 4 shows the variation of measured ultimate capacity for the same size footings in sands compacted to different relative density. The influence of relative density is pronounced between 35% and 50% relative density, but less between 50% and 75% relative density. Also, the influence of relative density between 50% and 75% becomes less as the footing embedment

Table 2. Summary of bearing capacity.

Test Series	D_R (%)	Width (m)	D/B	q_{ult} (kPa)
95SD1	35	0.61	0	240
		0.61	0.25	345
		0.61	0.5	405
		0.61	1.0	525
97SD1	50	0.61	0	508
		0.61	0.25	800
		0.61	0.5	1110
		0.61	1.0	1320
100SD1	75	0.61	0	1000
		0.61	0.25	1175
		0.61	0.5	1160
		0.61	1.0	1350
95	38	0.30	0	245
		0.61	0	300
		0.91	0	380
95GA3	38	0.30	1.0	480
		0.61	1.0	655
		0.91	1.0	770
97SD1	50	0.30	0.5	755
		0.61	0.5	1110
		0.91	0.5	1350

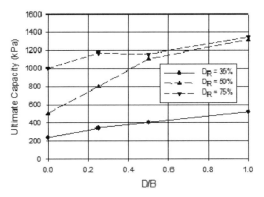

Figure 4. Ultimate capacity vs. D/B for different D_R.

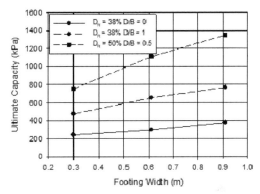

Figure 5. Influence of footing width on ultimate capacity.

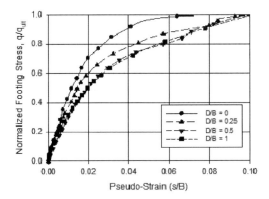

Figure 3. Normalised stress-pseudo-strain curves: Series 97SD1.

increases. Some of this difference is likely related to differences in saturation producing differences in soil matric suction.

3.3 Influence of embedment

Figure 4 also shows how the ultimate bearing capacity is increased by increasing the relative embedment of the footing, as is predicted with conventional bearing capacity theory.

3.4 Influence of footing size

Figure 5 shows the increase in bearing capacity with increase in footing width. Back-calculation of appropriate bearing capacity factors would require estimating soil suction in order to evaluate scale effects.

4 RESULTS – STRESS-DISPLACMENT

4.1 Generalised load-displacement behaviour

Using published full-scale footing load tests on sands, Mayne & Illingworth (2010) and Mayne et al. (2012) suggested that the behaviour of shallow foundations could be represented by a simple linear expression as:

$$q = r_s (s/B)^{0.5} \qquad (1)$$

where r_s is the slope of the footing stress vs. pseudo-strain curve and is equal to an empirical soil stiffness factor, reported to be in the range of about 0.5 to 5.5 (MPa) for sands. This approach using Eq. 1 was attempted to describe the stress-pseudo-strain curves for the current set of footing tests. The results are shown in the following figures. Figure 6 shows the results of the footing tests previously shown in Figures 1–3 using two different scales.

Even though the results clearly indicate the influence of footing embedment on the displacement

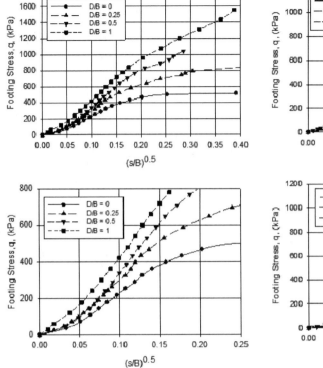

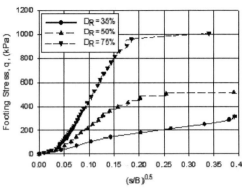

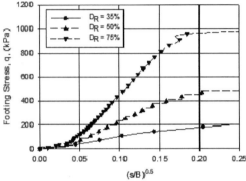

Figure 6. Footing stress vs. $(s/B)^{0.5}$: Series 97SD1.

Figure 7. Influence of relative density on footing response.

behaviour, i.e., increased embedment gives lower displacement at the same stress level, the results did not give a perfect linear relationship as in some cases the footings experienced very large displacements nearing failure conditions. Even when the data are only shown up to a pseudo-strain level representing a relative displacement of about 5%, the results show a curved behaviour. Linear correlations could be developed that would approximate the behaviour in this range giving r_s values in the range of about 2.5 to 4.9 (MPA), consistent with previous results.

4.2 Influence of relative density

The influence of relative density on the displacement behaviour is shown in Figure 7 which shows the results again presented at two scales. As the relative density increases, the stiffness increases as expected, with r_s values increasing dramatically from 0.9 to 5.3 for the same size footing with the same embedment. The range of pseudo-strain representing a stress level closer to the serviceability range for shallow foundations ($q/q_{ult} = 0.33$) is about 0.01 or 1% of the footing width. It appears that up to this level the use of Eq. 1 would certainly be appropriate.

The results shown in Figure 7 suggest that there was some initial soft behaviour in the load tests, especially at the highest relative density. This may indicate that the load system had some "slack" which does not

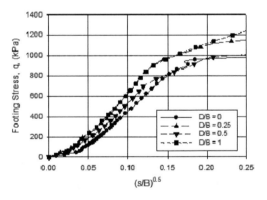

Figure 8. Footing response for test series 100SD1.

actually represent the footing response and should be adjusted for.

4.3 Influence of embedment

Results previously shown in Figure 6 demonstrated that the embedment influences the displacement behaviour. Additional results obtained from Test Series 100SD1 ($D_R = 75\%$) are shown in Figure 8. These results indicate that at higher relative density (75% vs. 50%) the embedment is less important on displacement with r_s values in the range of 5.3 to 6.6.

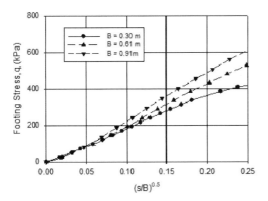

Figure 9. Footing response for test series 95GA3.

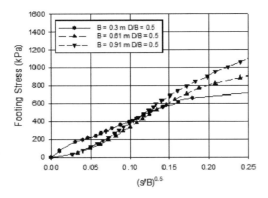

Figure 10. Footing response for test series 97SD1.

4.4 Influence of footing size

The sizes of footings used in the current tests were referred to as "model" size footings, however in some cases the width approximates sizes of footings that might be used to support individual column loads, especially for lightly loaded structures. In fact the sizes used in this work are within the range of several of the footings used in the data bases to create Eq. 1 (Mayne & Illingworth 2010; Mayne et al. 2012; Viswanath & Mayne 2013).

Figures 9 and 10 show results of load tests conducted on different size footings with the same embedment and the same relative density (Series 95GA3 and 97SD1). In both cases, the results show that there is very little difference in the footing response, i.e., footing response is independent of footing size for the same test conditions. These results support the use of normalised footing behaviour that has previously been shown applicable to shallow foundations (Lutenegger & Adams 2003).

5 DISCUSSION

Many results of laboratory scale model footing tests on both saturated and unsaturated sands have been reported in the past 50 years; even up to recent times (e.g., Vanapalli & Mohamed 2007; 2013). In general, most of these previous tests have used model footings with B less than 0.10 m. Unfortunately, many of these tests may also suffer from scale effects, which depend on both the size of the model footing and the grain size of the sand. From a practical standpoint, small-scale model tests may be used to evaluate relative trends in test results but may not necessarily give absolute values that can be applied to full-scale design without some adjustment. Larger scale models are less influence by scale and approach full-scale behaviour.

5.1 Bearing capacity

The bearing capacity of shallow foundations on unsaturated soils can include the soil matric suction (e.g., Oh & Vanapalli 2011; Vanapalli et al. 2011) which can either be measured at the time of the load tests or it may be estimated by direct measurement of the sand Soil-Water Characteristic Curve (SWCC) which will be a function of water content (saturation). Alternatively, the matric suction may be estimated using a general SWCC appropriate for the sand and test conditions. An appropriate "cohesion" term is included in the general bearing capacity equation to account for apparent cohesion resulting from the moisture and matric suction. The bearing capacity terms for footing size and embedment do not necessarily need adjustment for unsaturated conditions and the footing behaviour is similarly dependent of relative embedment as shown.

5.2 Stress-displacement

Most settlement estimates for shallow foundations on sand use some form of strain distribution model which requires an estimate of sand modulus. Alternatively, the linear model for footing response previously suggested provides a simple approach, provided that an estimate of the stiffness parameter, r_s, may be made. For full-scale footings, the value of r_s may be reflected in CPT or SPT results over the appropriate zone of soil beneath the footing. Results presented in the current tests show that for the same relative density, the sand stiffness increase as relative embedment increases, especially for lower relative density. Footing shape, although not considered in the current tests, may also be a factor.

6 SUMMARY

Results of several series of 1g model tests of shallow foundations on unsaturated compacted sands have been used to demonstrate the influence of sand relative density, footing embedment and footing size on the footing response to loading, i.e., footing stress-displacement behaviour. The results indicate that the unit bearing capacity is related to D_R, D/B and footing size. The simple linear response model proposed by others using the pseudo-strain (s/B) appears appropriate to describe the footing behaviour within the serviceability stress state. Unlike results previously reported by others, the linear model did not describe the footing response over the full range of footing loading to failure conditions. However, the Authors consider the model to be useful, since in most cases, the displacement and not the bearing capacity of shallow foundations on coarse-grained soils controls the design.

REFERENCES

Lutenegger, A.J. & Adams, M.T. 1998. Bearing capacity of footings on compacted sand. *Proceedings of the 4th International Conference on Case Histories in Geotechnical Engineering*: 1216–1224.

Lutenegger, A J. & Adams, M.T. 2003. Characteristic load-displacement curves of shallow foundations. *Proceedings of the International Conference on Shallow Foundations*, Paris, France, 2: 381–393.

Mohamed, F.M.O., Vanapalli, S.K. & Saatcioglu, M. 2012. Settlement estimation of shallow footings on saturated and unsaturated sands. *GeoCongress*, ASCE, 2552–2561.

Mayne, P.W. & Illingworth, F. 2010. Direct CPT method for footing response in sands using a database approach. *Proceedings of the 2nd International Symposium on Cone Penetration Testing*, 3: 315–322.

Mayne, P.W., Uzielli, M. & Illingworth, F., 2012. Shallow Footing Response on sands using a direct method based on cone penetration tests. *Full-Scale Testing and Foundation Design*, ASCE, 644–679.

Oh, W.T. & Vanapalli, S.K. 2011. Modelling the applied vertical stress and settlement relationship of shallow foundations in saturated and unsaturated sands. *Canadian Geotechnical Journal* 46: 1337–1355.

Vanapalli, S.K. & Mohamed, F.M.O. 2007. Bearing capacity of model footing in unsaturated soils. *Experimental Unsaturated Soil Mechanics: Proceedings in Physics.* 112: 483–493.

Vanapalli, S.K. & Mohamed, F.M.O. 2013. Bearing capacity and settlement of footings in unsaturated sands. *International Journal of Geomaterials.* 5(1): 595–604.

Vanapalli, S.K., Sun, R. & Li, X. 2011. Bearing capacity of an unsaturated sand from model footing tests. *Unsaturated Soils*, 1217–1224.

Viswanath, M.L. & Mayne, P.W. 2013. Direct SPT method for footing response in sands using a database approach. *Proceedings of the 4th International Symposium on Geotechnical and Geophysical Site Characterization*: 1131–1136.

Experimental study on the coupled effect of the vertical load and the horizontal load on the performance of piled beam-slab foundation

L. Mu, M. Huang, X. Kang & Y. Zhang
Department of Geotechnical Engineering, Key Laboratory of Geotechnical and Underground Engineering of Ministry of Education, Tongji University, Shanghai, China

ABSTRACT: With the development of the wind farms, a new type of foundation, piled beam-slab foundation, is invented to support wind turbines on land in the advantage of saving investment. Foundations for wind turbines are usually subject to loads combined of vertical loads, horizontal loads and moments. The behaviors of the foundation under vertical loads, horizontal loads and moments are designed separately in practice without considering the coupled effect of those loads for the coupled effect is unclear since now. In order to investigate the coupled effect of those load on piled beam-slab foundation, a series of model tests on behaviors of piled beam-slab foundation under vertical loads, horizontal loads and moments are carried out.

1 INTRODUCTION

In order to meet the requirement of clean energy all around the world, more and more wind farms have been constructed on land and offshore. The wind turbines on land are usually supported by pile-raft foundations and gravity foundations. Wind farms on soft ground area in China, which are usually supported by pile-raft foundation, are losing money due to the high cost of the construction. According to Sharma et al. (2014), beam & slab foundations can save great amount of steel and concrete comparing to raft foundation. In order to save investment in wind farms, a new type of foundation, piled beam-slab foundation, is invented to replace the pile-raft foundation. Only a few numerical studies (Singla 2009, He et al. 2011) and field tests (Zhang et al. 2011) have been carried out to study the performance of piled beam-slab foundation for wind turbines. In order to develop a reasonable design method for this foundation, much more work is needed to study the mechanism and the performance of the piled beam-slab foundation under the loads transferred from the wind turbines.

Many researches have been carried out to study the performances of the pile-raft foundations (Hain & Lee 1978, Kuwabara 1989, Poulos 1968, Ta and Small 1996, Horikoshi & Randolph 2015, Mendonca & Paiva 2000, Sanctis & Russo 2008 and Huang et al. 2009) which can provide a lot of useful information for design of piled beam-slab foundation. In most of these researches the pile-raft foundation are subjected to only vertical load or horizontal load. Meanwhile, the piled beam-slab foundation for wind turbines are usually subjected to vertical loads, horizontal loads and moments transferred from the structures above simultaneously. Significant couple effects of vertical loads and horizontal loads were observed in some existing researches (Anagnostopoulos & Georgiadis, 1993). According to the researches on coupled effect of vertical loads and horizontal loads on pile-raft foundation (Mu et al. 2014, Kitiyodom & Musmoto 2013), the couple effects of the loads on the raft are important factors that could influence the distribution of forces on the piles and the soils under the raft which is very important for design of piled beam-slab foundations. Due to the effect of beam, the behaviour of the piled beam-slab foundation is significantly different from the behaviour of the pile-raft foundation (Sharma et al. 2014). Thus, it is important to investigate the behaviour of piled beam-slab foundation under loads combined with vertical loads, horizontal loads and moments.

It is well recognized that full-scale field tests may provide high quality data for studying the behaviour of foundation. However, such tests have well-known limitations in term of cost, loading conditions, control mechanisms and metrical properties. To avoid these disadvantages, model tests have been used in all areas of engineering to understanding the complex behaviours of prototypes. These physical models are often advantageous in providing information and have more flexibility in comparison with full-scale field tests, particularly in desired loading conditions. Noting that the response of the models differs from that of a true prototype, therefore model scaling laws are essential in correlating the response of a model to the prototype. Comprehensive review on model scaling involving soil is provided by Rocha (1957) and Roscoe (1968). Rocha (1957) assumed that the stress and strain in the model are linearly related to those in the prototype. A lot of studies on the research and application

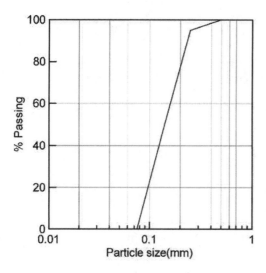

Figure 1. Grading curve of the sand.

Table 1. Properties of the sand.

Parameters	Value
Natural dry density (g/cm^3)	1.433
Maximum dry density (g/cm^3)	1.512
Minimum dry density (g/cm^3)	1.340
Relative density	57.1%
Friction angle	33°

Figure 2. Piled beam-slab foundation in situ.

Table 2. Scaling relations for the piled beam-slab foundation.

Parameters	Scaling law	Scaling factor	Dimension
Length l	S_l	$\lambda_l = 0.01$	L
Line displacement δ	$S_\delta = S_l$	0.01	L
Angle displacement φ	$S_\varphi = 1$	1	/
Strain ε	$S_\varepsilon = 1$	1	/
Elastic modulus E	$S_E = S_\sigma$	$\lambda_E = 2.3$	FL^{-2}
Stress σ	S_σ	2.3	FL^{-2}
Force F	$S_F = S_\sigma S_l^2$	2.3E−4	F
Line load q	$S_F = S_\sigma S_l$	2.3E−4	FL^{-1}
Surface load p	$S_F = S_\sigma$	2.3	FL^{-2}
Moment M	$S_F = S_\sigma S_l^3$	2.3E−5	FL

of model scaling effect were carried out during the following decades (Iai, 1989; Goit & Saitoh, 2013; Goit & Saitoh, 2014; Goit et al., 2013). The literature showed that the model tests were effective for studying the bearing mechanisms and bearing rules of pile foundations. In addition, the response of a model can be correlated to the response of the prototype by a set of scaling relations.

The present study focuses on experimental investigation on the performance of piled beam-slab foundations under coupled load. The objectives of this study were the following: (1) to describe the bearing characteristics of the piled beam-slab foundation under horizontal loads, (2) to investigate the coupled effect of vertical loads and horizontal loads on the performance of piled beam-slab foundation.

2 EXPERIMENTAL MODEL

2.1 Model Setup

The experimental models consist of soil-foundation systems cased in a steel tank with dimensions of 800 mm × 1000 mm × 1200 mm (Width × Length × Height). Homogeneous dry Shanghai fine sand is used, and the soil system is prepared by layered rolling. The grading curve of the sand is shown in Figure 1. The soil system is prepared by controlling the properties as shown in Table 1.

Figure 2 shows the piled beam-slab foundation in situ. The model of piled beam-slab foundation is prepared according to scaling relations as shown in Table 2. Aluminium pipes with an outer diameter $d_{out} = 14$ mm, inter diameter $d_{int} = 10$ mm, effective length $l_e = 440$ mm, total length $l_t = 476$ mm are used for the piles in the foundation system. The piles are shown in Figure 3. The beam-slab is pouring molded in a workshop with aluminium. The details of the beam-slab are shown in Figure 4. The piles and the beam-slab are threaded connected, providing strict restraint. Thus, all the force and deformation translations are allowed, including the vertical loads, horizontal loads, moments, vertical displacements, horizontal displacements and rotations.

The layout of the piled beam-slab foundation is shown in Figure 5.

2.2 Experimental instruments

A loading facility is developed for the test. As shown in Figure 6, the loading facility is constituted with three loading rails (in vertical, horizontal and rotation directions) and three servo motors in three directions respectively. The facility can achieve any combination of the loads in those three directions. The loading can be controlled in force or in displacement. The loading accuracy of the facility is 10 N or 0.01 mm in vertical and horizontal direction, 0.017° in rotation direction. The maximum loads are 10 kN, 5 kN and 1 kNm in vertical, horizontal and rotation direction respectively.

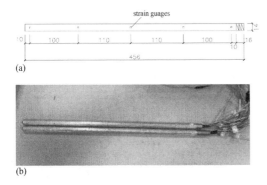

(a)

(b)

Figure 3.   Illustration of piles (mm).

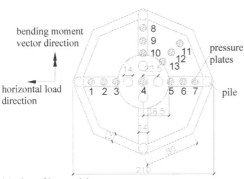

(a) plan of beam-slab

(b) vertical section of beam-slab

(c) beam-slab foundation

Figure 4.   Illustration of beam-slab (mm).

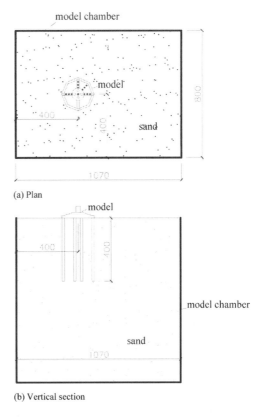

(a) Plan

(b) Vertical section

Figure 5.   Layout of piled beam-slab foundation (mm).

Figure 6.   Loading system.

The forces at the top of the foundation are measured by a multi-axis force/torque sensor made by ATI with an accuracy of 1% of indicating value. The displacements at the top of the foundation are measured by the sensors fixed in the loading facility with and accuracy of 0.001 mm in vertical and horizontal directions and 0.017° in rotation direction. The internal forces of the foundation are measured by the strain gauges attached on the piles and beams in the foundation, as shown in Figure 3 and Figure 7. The soil pressure under the raft is measured by film pressure plates, as shown in Figure 4(a).

2.3   *Loading processing*

Five loading cases are carried out herein to study the coupled effect of the vertical loads, horizontal loads

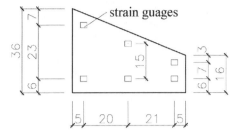

(a) strain gauges on rib beam

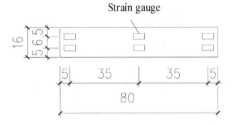

(b) strain gauges on circle beam

Figure 7. Layout of strain gauges on the beams.

Table 3. Loading cases.

	Loads		Loading speed	
Case	Vertical (kN)	Horizontal (mm)	Vertical (kN/min)	Horizontal (mm/min)
1	10	/	1	/
2	0	50	1	/
3	2.5	50	/	10
4	5	50	1	10
5	10	50	1	10

and moments on responses of piled beam-slab foundation, as shown in Table 3. The vertical loading is force controlled, while the horizontal loading is displacement controlled. The foundation is loaded slowly in order to simulate the static load. For the first case, the foundation is vertically loaded to the maximum loading capacity of the facility or until the failure of the foundation. As shown in Figure 8, the foundation did not reach the failure state when the facility is loaded to its loading capacity. Thus, for the 2nd to 5th cases, the foundation is firstly vertically loaded to 0, 0.25, 0.5 and 1 times of the loading capacity respectively. Then, the foundation is horizontally loaded to failure.

3 RESULTS

The axial forces of the piles are obtained from the measured strains through the following equation:

$$F = E_p A_p \frac{\varepsilon_+ + \varepsilon_-}{2} \quad (1)$$

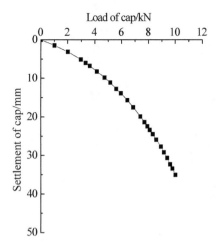

Figure 8. Q-s curve of the foundation.

The bending moments of the piles and the beams are obtained from the stains measured through the following equation:

$$M_p = \frac{E_p I_p}{D}(\varepsilon_+ + \varepsilon_-) \quad (2)$$

$$M_b = \frac{E_b I_b}{H}(\varepsilon_+ + \varepsilon_-) \quad (3)$$

where E_p is the elastic modulus of the pile, A_p is the effective section area of the pile, I_p is the moment of inertia of the pile, E_b is the elastic modulus of the beam, I_b is the moment of inertia of the beam, ε_+ and ε_- is the strain measured along the pile or the beam.

The H-y curves of the foundation under different vertical loads are shown in Figure 9. Assuming the foundation reaches its working limit state when the horizontal deformation of the foundation reaches 50 mm, the horizontal bearing capacity of the foundation is equal to the bearing load at the working limit state. As shown in Figure 10, the horizontal bearing capacity of the piled beam-slab foundation increases with the increase of the vertical load. Also does the stiffness of the foundation as shown in the Figure.

The axial forces along pile 1 are shown in Figure 11. It is common knowledge, that under this horizontal load the axial force on pile 1 should be the largest. As shown in Figure 11, the axial force along the pile is very small which is negligible when the vertical load is 0. It can be indicated that the axial force along the pile is mostly caused by the vertical load. Although the horizontal force along the pile cannot be measured directly, it is clear horizontal load does not transfer to axial forces on the piles for a piled beam-slab foundation. The horizontal load influences the distribution of the axial forces on the pile heads a bit. Figure 12 shows the distribution of the axial forces on the piles when the horizontal load on the foundation is 1300 N. It can be seen that the distribution of the axial force on the pile is more uniform when the vertical load on the foundation is larger. This is because that the piled beam-slab

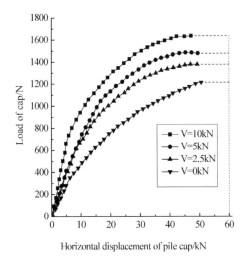

Figure 9. H-y curves.

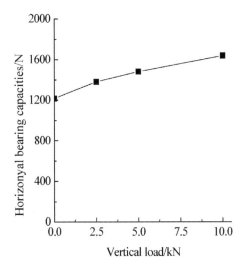

Figure 10. Relationship of horizontal bearing capacities and vertical loads.

foundation is a relative flexible foundation, the piles at the outer circle carries less load when the vertical load on the foundation is small. Figure 12 shows the ratio of the vertical load carried by the piles in different cases. It can be indicated that the pile carried less load when the vertical load on the foundation increases. And when the horizontal load increases the load carried by the piles increase. It is important for pile-raft design in practice for we usually didn't consider the influence of the horizontal load when we design the pile-raft foundation under vertical loads.

4 CONCLUSIONS

Based on the model tests on piled beam-slab foundation under combined loads of vertical loads and horizontal loads, the bearing characteristics of the

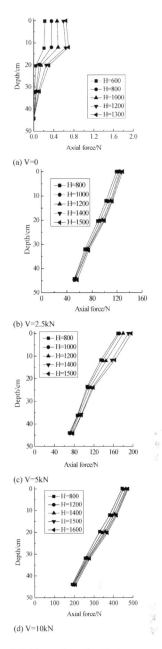

Figure 11. Axial force along the pile 1.

piled beam-slab foundation is investigated. The results shows that the presence of vertical loads increase the horizontal bearing capacity of the piled beam-slab foundation linearly. When the vertical loads is 10 kN, the horizontal bearing capacity of the foundation is increased as much as 30%. Also the pile carried less load when the vertical load on the foundation increases. Thus, it is important to consider the influence of the vertical load in practice when designing the horizontal bearing capacity of the piled beam-slab foundation.

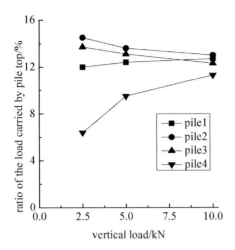

Figure 12. Distribution of axial force on the pile.

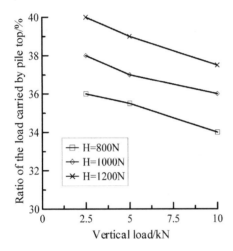

Figure 13. Ratio of the load carried by the piles.

REFERENCES

Anagnostopoulos, C. & Georgiadis M. 1993. Interaction of Axial and Lateral Pile Responses. *Journal of Geotechnical Engineering*, 119(4): 793–798.

Goit, C.S. & Saitoh, M. 2013. Model tests and numerical analyses on horizontal impedance functions of inclined single piles embedded in cohesionless soil. *Earthquake Engineering & Engineering Vibration* 12(1): 143–154.

Goit, C.S. & Saitoh, M. 2014. Model Tests on Horizontal Impedance Functions of Fixed-Head Inclined Pile Groups under Soil Nonlinearity. *Journal of Geotechnical & Geoenvironmental Engineering* 140(6): 971–984.

Goit, C.S., Saitoh, M. & Mylonakis, G. & Kawakami, H. & Oikawa, H. 2013. Model tests on horizontal pile-to-pile interaction incorporating local non-linearity and resonance effects. *Soil Dynamics and Earthquake Engineering* 48: 175–192.

Hain, S.J & Lee, I.K. 1978. The analysis of flexible raft-pile systems. *Geotechnique* 28(1): 65–83.

Horikoshi, K. & Randolph, M.F. 2015. Centrifuge modelling of piled raft foundations on clay. *Géotechnique* 46(4): 741–752.

Huang, M.S., Zhang, C.R. & Li, Z. 2009. A simplified analysis method for the influence of tunneling on grouped piles. *Tunnelling and Underground Space Technology* 24: 410–422.

Iai, S. 1989. Similitude for shaking table tests on soil-structure-fluid model in 1g gravitational field. *Soils and Foundations* 29(1): 105–118.

Jie, H.E., Tao, T. & Chen, Y. 2011. Numerical simulation and study on new type beam-slab foundation of wind turbine generator. *Yangtze River*: 43–52. (In Chinese)

Kitiyodom, P. & Matsumoto, T.A. 2003. Simplified analysis method for piled raft foundations in non-homogeneous soils. *Int. J. Numer. Anal. Meth. Geomech* 27: 85–109.

Kuwabara, F. 1989. An elastic analysis of piled raft foundations in a homogeneous soil. *Engineering Structures* 29(1): 81–92.

Mendonca, A.V. & Paiva, J.B. 2000. An elastostatic FEM/BEM analysis of vertically loaded raft foundation on piles. *Engineering Analysis with Boundary Elements* 24(3): 237–247.

Mu, L.L., Huang, M.S & Lian, K.N. 2014. Analysis of pile-raft foundations under complex loads in layered soils. *International Journal for Numerical & Analytical Methods in Geomechanics* 38(3): 256–280.

Poulos, H.G. 1968. Analysis of the settlement of pile groups. *Geotechinque* 18: 449–471.

Rocha, M. 1957. The Possibility of Solving Soil Mechanics Problem by the Use of Models. *Proceedings of 4th International Conference on Soil Mechanics*, London, 183–188.

Roscoe, K.H. 1968. Soils and model tests. *Journal of Strain Analysis for Engineering Design* 3(1): 57–64.

Sanctis, L.D. & Russo, G. 2008. Analysis and Performance of Piled Rafts Designed Using Innovative Criteria. *Journal of Geotechnical & Geoenvironmental Engineering* 134(8): 1118–1128.

Sharma, S.M., Vanza, M.G. & Mehta, D.D. 2014. Comparison of Raft foundation and Beam & Slab Raft Foundation for High Rise Building, *International Journal of Engineering Development and Research* 2(1): 571–575.

Singla, A.R. 2009. Finite Element Analysis for Punching Shear Performance of Beam-slab Type Raft Foundation. *Journal of Architecture & Civil Engineering* 26(2): 687–696.

Ta, L.D. & Small, J.C. 1996. Analysis of piled raft systems in layered soil. *International Journal for Numerical and Analytical Methods in Geomechanics* 20(1): 57–72.

Zhang, Y.J. & Mu, L.L. & Qian, J.G. & Huang, M.S. 2014. Field test of piled beam-slab foundation. *Yantu Lixue/rock & Soil Mechanics* 35(11): 3253–3258.

Determining shallow foundation stiffness in sand from centrifuge modelling

A. Pearson & P. Shepley
Department of Civil and Structural Engineering, University of Sheffield, UK

ABSTRACT: Centrifuge modelling using sands excels for large deformation ultimate limit state (ULS) problems. Stiffness remains more challenging due to the inherent particle size effects that lead to displacements being incorrectly simulated. A series of modelling of models tests has been conducted to investigate the influence of the relative size between a vertically loaded shallow foundation and the mean sand particle size on the measured stiffness. The results highlight that the number of particles within a failure mechanism directly affects the observed stiffness. The number of particles can be estimated through taking the footing width to mean particle size ratio and a linear relationship relating this ratio and the foundation stiffness.

1 INTRODUCTION

Small scale physical model testing is an essential tool in geotechnical engineering. They have facilitated myriad studies to shed light on complex, multi-faceted problems for a broad range of applications at significantly lower costs than field-scale trials. Continuous research effort has been expended to improve these physical models and increase their output. Arguably the most significant step was the implementation of physical modelling onboard geotechnical centrifuges which allowed small scale models to exhibit similar soil stresses to the field scale, such that the most appropriate soil stress-strain response was achieved throughout the physical model. Despite accruing fifty years of experience, geotechnical engineers are still doubtful over the prediction of foundation or structural stiffness using centrifuge models based on sands. Centrifuge modelling maintains an excellent reputation for simulating large strain ultimate limit state (ULS) problems, however continues to struggle with serviceability limit state (SLS) situations, despite serviceability becoming ever more important for the design of monopile foundations for wind turbines or the assessment of infrastructure stability in congested urban environments. This paper seeks to outline how model stiffness might be improved in the future by demonstrating initial experiments dedicated to the investigation of how stiffness scales onboard a geotechnical centrifuge.

2 BACKGROUND

There has been a wealth of element testing conducted to explore relative size effects in cohesionless soils. This has led to the agreement that the ratio of the mean particle size (d_{50}) to the sample size is an important controlling parameter (Omar and Sadrekarimi 2015) where, as the ratio reduces the peak shear strength increases. This result is integral for small scale physical modelers where the number of particles mobilized in a given failure mechanism is reduced when maintaining the same material in both model and prototype scenarios.

The trend is explained by the failure mechanism of granular materials being controlled by localized deformations rather than the more coherent response of cohesive soils. Previous experimental work has highlighted how these localizations manifest as shear bands (Roscoe 1970, Peters et al. 1988) which are required for a sand to reach steady state shearing and a fully mobilized ULS failure mechanism. But their formation must be understood in order to determine how stiffness may evolve through changing the particle:sample size ratio.

There is some consensus that shear bands are 8-18 particle diameters in width (Roscoe 1970, Scarpelli and Wood, Vardoulakis and Graf 1985, Oda and Kazama 1998, and Alshibi and Sture 2000). The actual thickness depends on the sand particle size, angularity, and density – all granular properties – and not the sample size being tested. Experiments by Alshibi and Sture (1999) demonstrate only a minor effect on shear band thickness due to confining pressure, reinforcing the idea that shear bands are dominated by the granular properties. Hence a shear band of the same dimension will form in both the centrifuge model and the prototype scale, but would represent something N times larger in the centrifuge model (using N as the length scaling number).

It follows that to generate a feature such as a shear band a certain absolute displacement must be imposed on the sand which would also be a function of the particle properties. This could be considered to be proportional to the width of the shear band itself. Applying this fact to the issue of particle:sample size ratios, it

Table 1. Properties for Fraction B and C Leighton Buzzard sand.

Sand	Fraction B	Fraction C
Critical friction angle (ϕ'_c)	36.6°	33.0°
Minimum voids ratio (e_{min})	0.495	0.491
Maximum voids ratio (e_{max})	0.820	0.829
Mean particle size (d_{50})	0.89 mm	0.45 mm

would suggest that the smaller the number of particles in the soil volume, the smaller the stiffness response would be.

If the same soil of mean grain size d_{50} is placed in a small model with volume l^3 where $l = 200 d_{50}$ and a larger model with volume L^3 where $L = 2000 d_{50}$, then the relative size of the shear band (assuming a shear band width of $10 d_{50}$) compared to the base lengths would be 0.05 and 0.005. Therefore the strain required to meet the steady state frictional response is smaller in the larger volume and would result in a larger stiffness being measured.

This concept will be experimentally verified in this paper.

Figure 1. Photograph of completed strongbox.

3 PHYSICAL MODELLING

A suite of modelling of models centrifuge tests was designed to investigate how foundation stiffness varied with footing width and sand particle size, with a focus on the ratio between particle size and footing width. Two sands were used with three different footing widths to ensure a range of ratios were simulated during the modelling campaign.

3.1 Sand properties

Leighton Buzzard sand Fractions B and C were used in the study. Both are rounded to sub-rounded yellow-ish sands supplied by David Ball Associates. The properties of the two sands are provided in Table 1. The friction angles were determined from direct shear box testing by the authors, whilst the minimum and maximum voids ratios were taken from Lee (1989). Sieve testing was completed to find the mean particle size.

3.2 Model geometry

The model testing was conducted using the small teaching centrifuge at the University of Sheffield (Black 2014). A photograph of a prepared strongbox is shown in Figure 1 with a detailed cross section in Figure 2. The sand body had dimensions 160 mm * 80 mm * 90 mm.

Three footing widths were used during the work, of 15 mm, 20 mm and 30 mm. Each footing was machined to the same 20 mm height and 78 mm length, ensuring an equal pressure was initially exerted on the sand body and no sand could be trapped between the

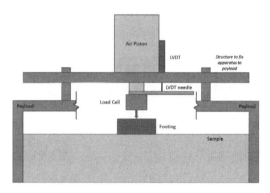

Figure 2. Cross section of experimental package.

footing and the Perspex viewing windows present on both sides of the box. Table 2 gives the properties of the footings and the B/d_{50} ratios for each configuration.

The 15 mm footings has a B/d_{50} ratio less than the limit suggested within the scaling law catalogue (Garnier et al. 2007) for ULS modelling. But given the focus of this study on the stiffness performance of the foundation, the load capacity of the equipment and the maximum footing width allowed within the strongbox, this dimension was the best option to achieve three different footing widths. This decision was justified by the consistent load-displacement results observed for the 15 mm footing in all tests, given that no footing was loaded to ULS.

3.3 Model preparation

Sand was air pluviated into the centrifuge strongbox using a manual sand pouring apparatus. The drop

Table 2. B/d₅₀ ratios for all configurations.

Sand fraction	d_{50} (mm)	B/d_{50} ratios		
		15 mm	20 mm	30 mm
Fraction B	0.89	17	22	34
Fraction C	0.45	33	45	67

Table 3. List of tests conducted.

Sand fraction	Footing width (mm)	Acceleration (g)	Repeat number
B	15	53	1
B	15	53	2*
B	15	53	3
B	20	40	1**
B	20	40	2
B	20	40	3
B	30	27	1
B	30	27	2
B	30	27	3
C	15	53	1
C	15	53	2
C	20	40	1**
C	20	40	2
C	20	40	3
C	20	40	4
C	30	27	1
C	30	27	2***
C	30	27	3

* Eccentrically applied load, resulting in twisted footing and non-linear response.
** Relative density of top 30 mm of sample was significantly greater than cohort, resulting in significantly stiffer responses.
*** Data transmission issues, resulting in unreliable data.

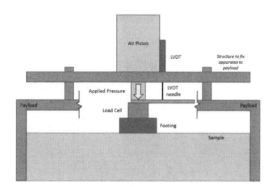

Figure 3. Cross section showing onboard instrumentation.

height varied as the strongbox was filled, but the sand density was consistently measured for every 15 mm layer poured to a model depth of 90 mm. Despite the constant pluviator height, the relative density, I_D, of the samples was controlled to $I_D = 80 \pm 5\%$ in the top 30 mm of the model depth. As this was the sand material likely to govern the vertical stiffness of the footing, this was considered acceptable.

3.4 Test procedure

The footing was loaded by a pneumatic piston positioned centrally above the model footing. For each new test run, the bottom of the piston was aligned to be just in contact with the top of the footing at 1 g. The centrifuge was then spun up to the target acceleration in a single ramp. Once at the target acceleration the footing was loaded by increasing the air pressure in the top of the piston at 900kPa/hour until reaching a maximum load of 600 N or by failing the footing.

The loading rate was selected to prevent any shock loading of the footing that might limit particle rearrangement in response to the applied forces. The loading rate was kept constant for all tests to ensure consistency in the overall soil response.

The footing response was monitored using an onboard load cell and LVDT positioned as shown in Figure 3. Additionally, images of the soil movement below the footing were recorded at 30 second intervals using a GoPro™ camera.

3.5 Tests conducted

Eighteen tests were conducted using the two sands and three footings previously mentioned. Table 3 lists

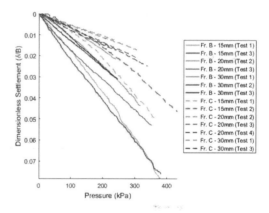

Figure 4. Applied pressure plotted against normalized settlement for all tests conducted.

the tests and the accelerations used. The accelerations were selected to ensure all models were representative of a footing width of 800 mm at prototype scale.

The test list in Table 3 displays all 18 tests conducted and highlights four anomalous results which have been removed from the study as they displayed demonstrably different load-displacement behaviors compared to the other repeats.

4 RESULTS

The load-settlement responses for the 14 representative tests are listed in Table 3 are plotted in Figure 4.

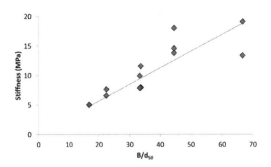

Figure 5. Footing stiffness plotted against B/d_{50}.

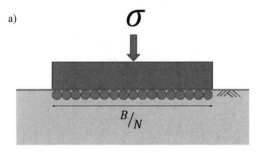

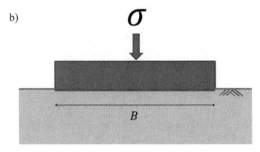

Figure 6. Idealized representation of (a) model and (b) prototype geometries.

The pressure given is the applied pressure below the footing and the settlements reported are normalized by the footing width. Due to the limited load capacity, only one footing reached complete failure and it is therefore impossible to compare the ULS response of the footings.

The secant stiffness of each footing was calculated from the results in Figure 4. To complete this, the load-normalized displacement behavior was approximated by a linear least squares regression fit to the data, with the gradient taken to represent the secant stiffness of the footing. The secant stiffness was used due to the limitations of the loading system not being able to monitor the small strains around the point of initial loading. The stiffness for the 14 footing tests shown in Figure 4 are plotted against the B/d_{50} ratio in Figure 5.

Figure 5 shows a clearly increasing trend in the stiffness results as the B/d_{50} ratio increases. A linear trend is fitted to the data with reasonable least squares regression correlation (the regression coefficient was $R^2 = 0.71$). The trend line was also forced to pass through the origin on the premise that an infinitely thin footing would penetrate into the sand without developing any load and therefore have zero stiffness.

5 DISCUSSION

The results shown in Figure 5 highlight that the number of particles in the failure mechanism has a significant effect on the measured stiffness. This observation can be considered to be caused by the number of particle contacts within the failure mechanism.

Figure 6(a) shows an idealized model footing overlying a number of particles, compared to the prototype scenario in Figure 6(b). The length scale factor, N, is applied as per traditional centrifuge modelling methods, and equal stress, σ, is applied to both footings. Assuming that the sand in both model and prototype has the same particle size, another common assumption in centrifuge modelling, then the force per particle, $F_{particle}$, is approximately the same in both situations, assuming the stress is evenly distributed through the soil.

The main difference in the two situations is the number of particles within the failure mechanism.

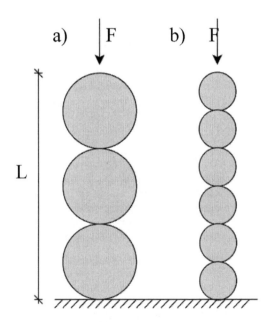

Figure 7. Diagrammatic representation of loading, F, moving through (a) Fraction B and (b) Fraction C sand particles.

Assuming the volume of affected soil is approximately twice the footing width in depth, then the number of particles in the failure mechanism increases as per N^3.

Consider the tests conducted as described in Sections 3 and 4. The largest footing was 30 mm in width and the number of particle contacts against the footing can be considered to be approximately double for the Fraction C sand compared to the Fraction B sand (the B/d_{50} ratio was 34 and 67 respectively). The linear regression fit shown in Figure 5 suggests that a doubling of the B/d_{50} ratio also leads to a doubling of the footing stiffness, but the cause of this is unknown.

5.1 Normal contact forces

One possible explanation for the observed doubling in footing stiffness is due to the particle contact properties. At small to moderate strains, it is reasonable to consider only small movements in the sand structure below the footing, allowing the consideration of a static sand structure in a hypothetical analysis. In this mode, assuming the two sands have a very uniform particle size distribution and that the footing load is equally distributed across its area, the average normal force acting at a pair of particle contacts would be the same for both materials.

Two contact models were considered to see if the particles undergoing purely normal contact would explain the observed stiffness trend. Figure 7 shows two idealized particle columns demonstrating Fraction B and Fraction C sized particles. The strain for both columns can be calculated either from a Hertzian contact model or an asperity hardness model using the simplification that the force is constant along the chain of particles.

This is not strictly the case for the sand below a footing, but given the same size of footing was used, then both sands would have the same stress profile with depth, ensuring equal loading in both examples.

The Hertzian contact model (Johnson, 1985) determines the contact force from the material stiffness, E^*, particle radius, R, and interference distance, δ via the equation:

$$F = \frac{4}{3} E^* \sqrt{R} \delta^{3/2} \quad (1)$$

Assuming the two sands have the same stiffness due to their identical mineralogy, and the same force acting normal to the contact points allows the interference for the materials to be compared, giving:

$$\delta_C = \sqrt[3]{2} \delta_B \quad (2)$$

δ_B = interference distance (Fraction B sand)
δ_C = interference distance (Fraction C sand)

This implies that Fraction C would exhibit greater strains for the same force compared to Fraction B, and therefore give a lower stiffness for the footing loading above – counter to the observed behavior.

An asperity hardness model removes the dependency on the particle radius in the calculation of particle force. Instead, the force between a particle contacts is considered to be proportional to the material hardness, H, and the interference distance:

$$F \propto H, \delta \quad (3)$$

Despite removing the particle size dependency, the asperity hardness model agrees with the Hertzian result in that it also suggests that Fraction C should give a softer response to the footing load.

Both methods predicted the opposite effect on stiffness by increasing the number of particles within the mechanism. Therefore the observed increase in stiffness must be due to a more complex system than a straightforward approach to the particle contact forces and normal displacements.

5.2 Turbulent shearing

The alternative explanation for the observed increase in stiffness with B/d_{50} ratio is that it is essential to consider how the shear bands form in the sand during loading. Theoretical work has been conducted in this regard by Muhlhaus and Vardoulakis (1987). They determined the importance of rotations at particle contacts, the Cosserat effect, during the propagation of shear bands. Their work suggests a more complex mechanism is required to understand the effects of the B/d_{50} ratio when estimating model stiffness from centrifuge modelling.

5.3 Implications

The work presented in this paper reinforces the danger of predicting moderate-strain stiffness from centrifuge modelling. The observed difference in stiffness is challenging to explain from a mechanistic perspective, but it is clear that an appreciation of particle motion during shearing will be required rather than solely displacements and interference at the particle contacts. This will require a more intimate understanding of how particles respond to initial shearing and the physical mechanism which leads to shear band formation. It is likely that as suggested at the start of the paper, the shearing distance required to form a shear band is directly related to the particle size.

The results from this study suggest a linear relationship between the stiffness of a vertically loaded footing and the B/d_{50} ratio. An alternative expression to the B/d_{50} ratio would be the estimated number of particles in the heavily loaded soil, however this is much more challenging to determine. Extrapolating the trend would suggest that by using smaller and smaller particles, an infinite stiffness could be achieved, yet this is known to not be possible. Therefore a stiffness cut-off must be reached where increasing the number of particles in the mechanism stops having an effect.

Other studies have indicated that a B/d_{50} cut-off for ultimate limit state, above which particle size effects can be ignored as the shear band is sufficiently small, can be applied in centrifuge modelling. Garnier et al. (2007) proposed a cut-off of $B/d_{50} > 35$ and later studies, Toyosawa et al. (2013), have suggested this can be $B/d_{50} > 50$ in some scenarios. This study has shown a cut-off of $B/d_{50} > 35$ is not applicable when investigating stiffness. As demonstrated, stiffness results are very sensitive and heavily dependent on the density of the mobilised volume of the sample and insufficient data is presented at $44 > B/d_{50} > 67$ to confidently assess a cut-off of $B/d_{50} > 50$ or present a separate cut-off.

The stiffness cut-off is therefore unknown and further tests conducted at $B/d_{50} > 44$ are needed.

6 CONCLUSIONS

The number of particles has a significant effect on the observed stiffness of a foundation structure. In the case of a vertically loaded shallow foundation, it can be suggested that stiffness is linearly related to the B/d_{50} ratio, but a cut-off is anticipated. Therefore caution must be taken when predicting stiffness of soil-structure interaction problems via small-scale modelling when the soil being modelled is cohesionless.

Any potential prediction of stiffness from centrifuge modelling will rely on an appreciation of the physical mechanism of shear bands form in sands. The generation of stiffness appears to be dependent on the number of particles in the loaded soil region, rather than due to the contact force induced displacements.

ACKNOWLEDGEMENTS

The authors wish to thank the tireless technical support of Alex Cargill who ensured testing was conducted to the highest standard as well as the Department of Civil and Structural Engineering and the Centre for Energy and Infrastructure Ground Research for access to the centrifuge facilities.

REFERENCES

Alshibli, A. & Sture, S. 1999. Sand shear band thickness measurements by digital imaging techniques. *Journal of Computations in Civil Engineering*, 1999, **13**(2), pp. 103–109.

Alshibli, A. & Sture, S. 2000. Shear band formation in plane strain experiments on sand. *Journal of Geotechnical and Geoenvironmental Engineering* **126**(6), pp. 495–503.

Black, J.A., 2014. Development of a small scale teaching centrifuge. *Proceedings of the 8th International Conference on Physical Modelling in Geotechnics (ICPMG2014)*, 14–18 January 2014, Perth, Australia.

Garnier, J., C. Gaudin, S.M. Springman, P.J. Culligan, D. Goodings, D. Konig, B. Kutter, R. Phillips, M.F. Randolph, & Thorel, L. 2007. Catalogue of Scaling Laws and Similitude Question in Geotechnical Centrifuge Modelling. *International Journal of Physical Modelling in Geotechnics*, **7**(3), pp. 1–23.

Johnson, K. L. 1985. *Contact Mechanics*, Cambridge University Press, Cambridge, UK.

Lee, S. Y. 1989. Centrifuge modelling of cone penetration testing in cohesionless soils. *Ph.D. dissertation*, Univ. of Cambridge, Cambridge, U.K.

Muhlhaus, H. B. & Vardoulakis, 1987. The thickness of shear bands in granular materials *Géotechnique*, **37**(3), pp. 271–283.

Oda, M. & Kazama, H. 1998. Microstructure of shear bands and its relation to the mechanisms of dilatancy and failure of dense granular soils, *Géotechnique*, **48**(4), pp. 465–481.

Omar, T. & Sadrekarimi A. 2015. Specimen size effects on behavior of loose sand in triaxial compression tests, *Canadian Geotechnical Journal*, **52**(6), pp. 732–746.

Peters, J., Lade, P., & Bro, A. 1988. Shear band formation in triaxial and plane strain tests. *Advanced Triaxial Testing of Soil and Rock, ASTM, STP 977*, R. Donoghue, R. Chaney, and M. Silver, ed., ASTM, West Conshohoken, Pa., pp. 604–627.

Roscoe, K.H., 1970. The influence of strains in soil mechanics, 10th Rankine Lecture. *Géotechnique* **20**(2), pp. 129–170.

Scarpelli, G. & Wood, D.M. 1982. Experimental observations of shear band patterns in direct shear tests. *Proc. IUTAM Conf: Deformation and Failure of Granular Materials, Delft*, pp. 473–484. Rotterdam: Balkema.

Toyosawa, T., Itoh, K,. Kikkawa, N,. Yang, J. J., Lui, F. 2013. Influence of model footing diameter and embedded depth on particle size effect in centrifugal bearing capacity tests. *Soils and Foundations*, **53**(2), pp. 349–356.

Vardoulakis, I., & Graf, B. 1985. Calibration of constitutive models for granular materials using data from biaxial experiments. *Géotechnique*, **35**(3), pp. 299–317.

The effect of soil stiffness on the undrained bearing capacity of a footing on a layered clay deposit

A. Salehi, Y. Hu & B.M. Lehane
Department of Civil, Environmental & Mining Engineering, University of Western Australia, Perth, Australia

V. Zania & S.L. Sovso
Department of Civil Engineering, Technical University of Denmark, Copenhagen, Denmark

ABSTRACT: The bearing capacity of footings on layered clays has been studied previously, but with the clays assumed as rigid plastic materials. In this study a circular footing was penetrated into a stiff-over-soft clay deposit in a centrifuge. It was found that, within a small displacement range (i.e. ~15% of footing diameter), the footing capacity did not reach its limit, as predicted by the rigid plastic analysis. Both small strain and large penetration finite element (FE) analyses were performed. The centrifuge test data were compared with the FE results from small strain and large penetration analyses. It was observed that the footing capacity on the layered clay surface is dependent on the soil stiffness.

1 INTRODUCTION

The capacity of strip footings on layered clay deposits has been studied using physical testing, rigid-plasticity and limit analysis by many researchers, such as Brown & Meyerhof (1969), Meyerhof & Hanna (1978), Michalowski (1992) and Merifield et al. (1999). Among these studies, Brown & Meyerhof (1969) and Meyerhof & Hanna (1978) also studied the bearing capacity of circular footings (with diameters, D = 50 mm & 75 mm) on layered clay deposits using a series of small scale model tests at 1 g. The ultimate capacity occurred in these experiments at a settlement of 0.05 D to 0.25 D, depending on the strength ratio of the two clay layers. They proposed empirical and semi-empirical bearing capacity factors based on these model test results. Michalowski (1992) provided limit loads for strip footings on two-layer clays using strict upper bound analysis. Merifield et al. (1999) combined limit theories (i.e. upper bound and lower bound theories) with finite element analysis and provided upper bound (UB) and lower bound (LB) solutions for the bearing capacity factors of strip footings. The UB and LB solutions were closely comparable, verifying the accuracy of the proposed solutions.

Numerical investigations of circular footings with flat and conical bases (i.e. spudcan) on layered clays were carried out using finite element analysis (Wang & Carter 2002; Liu et al. 2005, Hossain & Randolph 2010a) and centrifuge testing (Hossain & Randolph 2010b). The governing factors of the problem were identified as: the strength ratio of the two layers, the relative thickness of the upper layer (with respect to the diameter of the footing), the normalized strength of the soft clay layer, and the roughness of the footing. Small strain and large deformation analyses concluded that the bearing capacity of a surface footing was also dependent on the soil stiffness characteristics.

In the current study, a centrifuge experiment was performed to measure the load-penetration response of a circular footing (model diameter $D_m = 80$ mm) under a centrifuge acceleration of 50 g (and hence modelling a prototype diameter D = 2R of = 4 m). The footing was founded at the surface of a stiff clay layer which was underlain by a soft clay deposit. In spite of the study of 'punch-through' failures being common, most research has focused on spudcan foundations with conical bases. Research on flat circular footings has largely been conducted numerically and by limit analysis, where the focus has been on the ultimate bearing capacity. Experiments examining the behaviour of flat circular footings on layered clay stratigraphy undergoing large penetrations are scarce. Understanding the response of footings founded on a relatively thin stiff clay layer which is underlain by a soft layer has clear relevance to serviceability limit state design. For example, the ultimate bearing capacity may not be reached when the footing reaches its service design settlement.

This paper presents the results and interpretation of a test performed on a thin stiff kaolin layer overlying a soft kaolin layer, with the primary aim being to examine the effect of the stiffness of the clay layers on the bearing capacity of the footing. Small strain and large deformation finite element analyses were conducted to assist interpretation of the experimental findings.

2 EXPERIMENTAL DETAILS

2.1 Centrifuge apparatus

The footing experiments were carried out in the beam centrifuge at the Technical University of Denmark (DTU). The beam centrifuge at DTU was built in 1976 and upgraded in 1998. The length of the beam arm is 1.7 m from the rotational axis to the hinge and the U-shaped yoke is 0.93 m long, giving total radius of 2.63 m (Leth et al., 2008). The strongbox had an internal diameter of 527 mm and internal height of 495 mm.

2.2 Soil sample preparation

Commercially available kaolin, with respective liquid and plastic limits of 59% and 30%, was used to create the stiff and soft clay layers in the experiments. The properties of the kaolin are well documented and reported in Stewart (1992) and Lehane et al. (2008). The kaolin powder was mixed as slurry at a water content of 120% for a period of 12 hours under an applied vacuum. The pressure inside the mixer was approximately 15 kPa, and could be monitored via the vacuum pump connected to the lid.

The slurry was consolidated in two centrifuge strongboxes at two different maximum stress levels (50 kPa and 400 kPa) in order to obtain a soft and a stiff clay layer. A 100 mm thick sand layer with upper geofabric was initially placed at the base of one of the strongboxes (the one intended for the soft clay layer) before placement of the slurry. A geofabric was also attached on the inner walls of the strongboxes to reduce the time required for consolidation. Hydraulic presses with rigid upper platen applied the vertical pre-consolidation stresses stepwise, while the consolidation was monitored with displacement transducers.

The centrifuge experiment was conducted using the strongbox containing the soft clay layer. The stiff clay layer was removed from the other strongbox after disassembling the bottom platen. A slice of the stiff layer was cut into two equal pieces and then carefully placed on the top of the soft clay sample. Kaolin slurry was poured over the joins between the two parts and the entire sample was then ramped up to a centrifuge acceleration of 50g for a period of 24 hours. Every effort was made to ensure that no footing or penetration tests were conducted in the vicinity of the joints.

The layered sample prepared for the test reported in this paper (referred to as Test 1) had layer thicknesses for the stiff and soft clay layers as indicated in Table 1. Cone Penetration Tests (CPTs) were performed in-flight after the footing test. The cone diameter employed was 11.2 mm, which is almost 50% of the thickness of the upper stiff layer in Test 1 (see Table 1). Corrections for both shallow penetration effects and for the presence of the underlying soft layer make the interpretation of undrained strength (s_u) in the top layer from the CPT uncertain. The undrained strength of the soft clay, however, could be reasonably estimated from the CPT as steady state penetration was attained after a penetration of about three cone diameters (=33 mm) into the soft clay. Applying an N_{kt} value of 12 (Randolph & Hope, 2004) to the measured net cone resistance gave a strength of the soft clay layer as $s_{ub} = 4.5$ kPa. This strength is consistent with an over-consolidation ratio (OCR) of about 3 at the upper surface reducing to unity at 35mm below the layer's top surface.

The water content of the stiff clay was measured immediately after the footing test and allowed inference of a pre-consolidation pressure for this layer of 120 kPa which is lower than the applied pressure of 400 kPa. A possible explanation for this discrepancy is the disturbance to the stiff layer induced during its placement on the top of the soft clay layer, and/or swelling of the clay after consolidation pressure was removed. The undrained shear strength with $\sigma'_{vy} = 120$ kPa consolidation pressure at 12 mm depth in the sample was interpreted as 14.5 kPa (i.e. $\sim 0.12 \sigma'_{vy}$); this strength is in line with estimations made using hand vanes as well as with the CPT data when account is taken for the uncertainties in interpretation mentioned above.

Table 1. Test details.

	Top stiff clay	Bottom soft clay
Thickness (mm)	Ht = 24	Hb = 154
Prototype (m)	Ht = 1.2	Hb = 7.7
Strength (kPa)	sut = 14.5	sub = 4.5
H/D	0.3	1.825

2.3 Test details

The footing test was carried out as displacement controlled vertical loading of a 20 mm thick circular steel footing with a diameter (D) of 80 mm. The footing was tested from its initial location at the upper surface of the sample to a maximum displacement of 40 mm at a displacement rate of v = 0.04 mm/sec at a centrifuge acceleration of 50 g. This rate induces a normalized velocity $V = vD/c_v = 39$ which exceeds the value of 30 proposed by DeJong et al. (2013) as a minimum for undrained condition. Load was measured via a load cell connected to the centrifuge actuator and seated on a ball bearing located in a recess at the centre of the footing. The test details are listed in Table 1 along with the corresponding prototype dimensions. The test configuration is also shown in Figure 1.

3 TEST RESULTS

The measured footing penetration response is presented in Figure 2 as the reaction pressure ($q = F_v/A$, where F_v is the reaction force and A is the cross section area of the circular footing) against the footing settlement (s) normalized by the footing diameter (s/D). It is apparent that the footing capacity increases

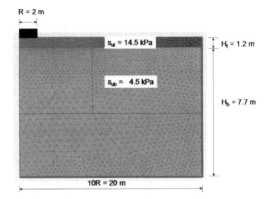

Figure 1. Mesh employed in axisymmetric FEA.

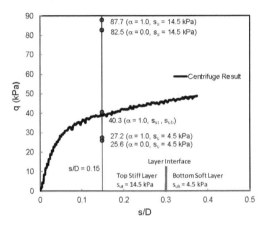

Figure 2. Footing penetration response in centrifuge test with existing analytical solutions.

Figure 3. View of the extracted footing in centrifuge test.

with increasing footing displacement, even in excess of 30% of the footing diameter (i.e. s/D = 0.3 and s > 1.2 m in prototype). There is no ultimate bearing capacity reached at s/D = 0.4, where the footing has fully penetrated through the stiff clay layer (which has a thickness of $H_t/D = 0.3$) and into the soft clay layer.

The bearing capacity of surface footing has been studied by many researchers. It can be estimated using the bearing capacity factors proposed by Martin & Randolph (2000) for a circular footing on the surface of a single clay layer. The bearing capacity factors are $N_c = 5.69$ and 6.05 for smooth (i.e. the roughness factor $\alpha = 0.0$) and rough (i.e. $\alpha = 1.0$) respectively. The calculated surface footing capacities are compared with the centrifuge data at penetration depth of s/D = 0.15, since the surcharge effect (or penetration depth effect) on the footing capacity becomes more significant with further penetration. By using these bearing capacity factors, the footing capacities (i.e. $q = N_c \times s_u$) on a single clay layer with the top stiff clay ($s_{ut} = 14.5$ kPa) or the bottom soft clay ($s_{ub} = 4.5$ kPa) in this test are displayed in Figure 2. It is apparent that the bearing capacity of the footing lies between these bounds.

Brown & Meyerhof (1969) conducted model tests to investigate the capacity of a footing with a rough base on a stiff-over-soft clay deposit. They suggested a simple shear punching failure around the footing perimeter, and proposed a formula for the footing bearing capacity factor given in Equation 1, where N_{cmax} is an upper bound for a footing on single clay layer suggested by Brown & Meyerhof (1969). Here the exact solutions of Martin & Randolph (2000), $N_{cmax} = 6.05$ for a rough footing is used instead. As the test condition is $H_t/D = 0.3$ (Table 1) in this study, the capacity of the footing on the two-layer clay is 40.3 kPa (i.e. in Equation 1, $N_c = 3.0 \times 0.3 + 6.05 \times (4.5/14.5) = 2.78$, $q = N_c \times s_{ut} = 2.78 \times 14.5 = 40.3$ kPa) for a rough footing, as the formula in Equation 1 was proposed for rough footings for a thin top stiff clay layer over a soft clay layer. Various solutions are indicated in Figure 2. It can be seen that the two-layer solution provided by Equation 1 generally matches the centrifuge measurements at s/D = 0.15. However, the footing capacity still increases with further penetration.

$$N_c = 3.0(H_t/D) + N_{cmax}(s_{ub}/s_{ut}) \quad (1)$$

where $H_t/D < 1$ and $s_{ub} < s_{ut}$.

A photo taken at the end of the test after the extraction of the footing is shown on Figure 3. The absence of any significant disturbance outside the footprint of the foundation indicates that the classical bearing capacity failure mode of a surface footing did not take place and the mode was one of punching. It is also evident from the clay on upper side of the footing that some collapse and/or flow-around failure took place.

4 NUMERICAL RESULTS

4.1 Small strain FE analysis

An axisymmetric finite element analysis of the centrifuge footing test was performed at prototype scale (see Table 1 and Figure 1) using the Plaxis finite element program (Brinkgreve et al. 2012). The clay layers were modelled as elastic-perfectly plastic material

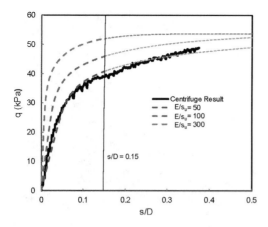

Figure 4. Comparison of small strain FE results (PLAXIS) with centrifuge test data.

with a Tresca yield criterion. The undrained strengths are given in Table 1 with a Poisson's ratio of $\nu = 0.49$ for undrained analysis. The ratio of Young's modulus (E_u) to the undrained shear strength s_u was set as $E_u/s_u = 50$, 100, and 300 to investigate the effect of the stiffness. The footing was modelled as a rigid disk, located at the surface prior to loading. The soil domain had a radius of 20 m which is equivalent to ten times the footing radius. Parametric studies showed that selected dimensions of the mesh were sufficiently large to avoid any significant boundary effects. The depth of the lower boundary was 8.9 m, which is equal to the scaled depth of the clay sample in the centrifuge strongbox (see Fig. 1). The 15-node triangular elements were used in the analysis with a refined zone near the footing. The fine elements near the footing were tested for accuracy. The vertical boundaries of the mesh were fixed in the horizontal direction and free to move vertically while the base horizontal boundary was fixed in both the horizontal and vertical directions.

Figure 4 presents a comparison of the bearing stress (q) versus normalised displacement (s/D) relationship obtained in the centrifuge test and in the small strain FE analyses, which were conducted for a range of E/s_u values. Evidently, a best fit between both sets of results is obtained for a E/s_u ratio of 50.

It can be seen that the soil stiffness affects the initial load-displacement response, but has little or no effect on the ultimate bearing (other than greater settlement is required to develop ultimate conditions at lower stiffness values). In this study the ultimate capacity is defined at a normalised footing penetration, s/D, of 0.15.

4.2 Large deformation FE analysis

Large deformation finite element (LDFE) analysis has been developed for engineering use over the last 30 years. For geotechnical engineering, the remeshing and interpolation technique with small strain (RITSS) approach is one such technique. The RITSS approach is based on small strain analysis with frequent remeshing. After each remeshing, the geometry of the soil domain is updated and the stresses and soil properties in the soil domain are interpolated from an old mesh to a new mesh. The process is repeated until a desired penetration is reached. The details of the approach can be found in Hu & Randolph (1998a). There have been many applications of this approach for offshore geotechnical problems (Randolph et al., 2008). The RITSS method has shown its advantages over the conventional large deformation methods, such as Lagrangian and Eulerian methods (Hu & Randolph 1998b, c).

The LDFE analysis employed the same configuration shown in Figure 1 with parameters given in Table 1. However only one soil stiffness ratio $E_u/s_u = 50$ was employed based on the small strain FE results in Figure 4. Six-noded triangular elements with three internal Gauss points were used (Carter & Balaam, 1990). At the soil domain boundaries, all points along the vertical sides were restrained in the horizontal direction and nodal points at the base were restrained vertically. Also the bulk unit weight of the clay was measured as 16.1 kN/m^3, hence the effective unit weight of the clay is 6.1 kN/m^3. The penetration analysis was displacement controlled. The total reaction force (F_v) acting on the footing is obtained from the LDFE analysis, and the reaction pressure is calculated as $q = F_v/A$, where A is the cross-section area of the footing.

Figure 5 presents the LDFE result together with the small strain FE result and centrifuge test result. It is evident that, at s/D values in excess of about 0.15, the rate of change of applied stress with s/D, as predicted by the LDFE analysis, provides a better match to the experimental results compared to the small strain analysis. It is also noted that, during the initial penetration (s/D < 0.1), the small strain analysis predicts a slightly lower stiffness than the centrifuge test data, despite both analyses having a E_u/s_u ratio of 50. The LDFE result captures the initial stiffness of footing response very well. Although the soil domain deformation is ignored in the small strain analysis, under the small strain range of s/D < 0.15, the small strain analysis response matches with the LDFE. Once the footing penetration becomes large (i.e. s/D > 0.15), the LDFE result follows the trend of the centrifuge data very well. Since the soil domain is updated as the footing penetrates the upper clay layer, the soil surface heave and soil surcharge load at the footing base are accounted for in the LDFE analysis. The effect of soil deformation on footing capacity in LDFE analysis is demonstrated further in the following section.

4.3 Results discussion and soil flow mechanism

The small strain FE analyses (Fig. 4), clearly illustrate the influence of soil stiffness on the predicted footing penetration response. The mesh updating performed in LDFE/RITSS analysis (Fig. 6), automatically updates the footing response and evidently is capable of matching the full load displacement response from low to

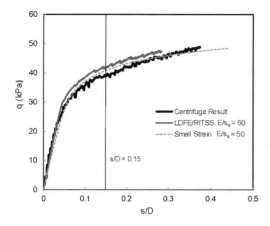

Figure 5. Comparison of LDFE/RITSS result with small strain and centrifuge test result.

very large penetrations. It is also evident comparing Figures 4 and 5 that the effect of the footing penetration is minimal at shallow penetration (s/D < 0.15), but that it shouldn't be ignored at larger penetration (s/D > 0.15).

Figure 6 presents the soil layer deformation and soil flow mechanisms during footing penetration into the two-layer clay deposit. Figure 6a shows the soil deformation at a shallow embedment of $s/D = 0.15$. It is apparent that the top stiff layer is punching into the bottom soft layer without obvious heaving at the soil surface. This matches the centrifuge observation where soil heaving is absent (see Figure 3). The punching failure matches the assumption by Brown & Meyerhof (1969) and their equation (Eq. 1) is seen to provide good predictions if capacity is defined at $s/D = 0.15$.

Once the footing reaches the clay layer interface at $s/D = 0.3$, the centrifuge experiment does not indicate a drop in footing capacity (see Figure 4). This is because, as seen in Figure 6b ($s/D = H_t/D = 0.3$), the top stiff soil is trapped underneath the footing and the trapped top stiff soil penetrates into the bottom soft clay with the footing. This means an effectively larger footing is penetrating into the bottom soft soil. The soil above the footing base act as a surcharge hence resulted in promoting an increase in the reaction pressure on the footing with increasing penetration depth.

Furthermore, parametric studies using LDFE/RITSS analyses for footings on layered soils are planned to investigate the effects of footing size, layer configuration and layer strength ratio on footing penetration response.

5 CONCLUSION

The bearing capacity of a circular surface footing on a stratigraphy comprising a stiff clay layer underlain by a soft clay layer was tested in the centrifuge. The footing continued to gain capacity even after punching through the upper stiff clay layer. Both small strain and large deformation FE analyses were conducted to assist the interpretation of the footing penetration response. It was found that the soil stiffness played an important role and the capacity corresponding to the upper clay layer could not be developed as the footing displacement led to reducing distance between its base and the underlying soft clay.

The LDFE analysis results have confirmed that a punching shear failure occurred for the footing at a shallow penetration depth. With deep penetration, the stiff soil trapped underneath the footing and the surcharge loading from the footing penetration should be taken into consideration into the bearing capacity estimation, which is otherwise underestimated.

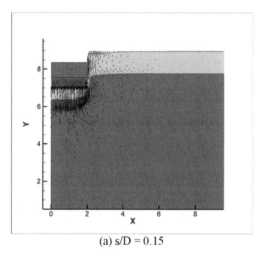

(a) s/D = 0.15

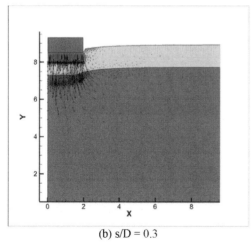

(b) s/D = 0.3

Figure 6. Soil layer deformation and soil flow mechanisms from LDFE/RITSS analysis ((a): $s/D = 0.15$, (b): $s/D = 0.3$).

REFERENCES

Brinkgreve, R.B.J. Engin, E. & Swolfs, W.M. 2012. *PLAXIS 2D 2012 Manual*. The Netherlands: PLAXIS B.V.

Brown, J. & Meyerhof, G. 1969. Experimental study of bearing capacity in layered clays. *Proc. Of the 7th int. conf. on soil mech and foundation eng.* 2: 45–51.

Carter, J.P. & Balaam, N.P. 1990. *AFENA Users' Manual*. Centre for Geotechnical Research, The University of Sydney.

DeJong, J.D., Jaeger, R.A., Boulanger, R.W., Randolph, M.F. & Wahl, D.A.J. 2013. Variable penetration rate cone testing for characterization of intermediate soils. *Geotechnical & Geophysical Site Characterization 4, Vol. 1 (Proc. ISC-4, Pernambuco), Taylor & Francis Group, London:* 25–42.

Hossain, M.S. & Randolph, M.F. 2010a. Deep-penetrating spudcan foundations on layered clays: numerical analysis. *Geotechnique* 60(3): 171–184.

Hossain, M.S. & Randolph, M.F. 2010b. Deep-penetrating spudcan foundations on layered clays: centrifuge tests. *Geotechnique* 60(3): 157–170.

Hu, Y. & Randolph, M.F. 1998a. A practical numerical approach for large deformation problems in soil. *Int. J. Numerical and Analytical Methods in Geomechanics* 22(5): 327–350.

Hu, Y. & Randolph, M.F. 1998b. Deep penetration of shallow foundations on non-homogeneous soil. *Soils and Foundations* 38(1): 241–246.

Hu, Y. & Randolph, M.F. 1998c. H-adaptive FE analysis of elasto-plastic non-homogeneous soil with large deformation. *Computers and Geotechnics* 23, (1–2): 61–83.

Lehane, B.M. Gaudin, C. Richards, D.J. & Rattley, M.J. 2008. Rate effects on the vertical uplift capacity of footings founded in clay. *Géotechnique* 58(1): 13–21.

Leth, C.T., Krogsboll, A. & Hededal, O. 2008. Centrifuge facilities at Technical University of Denmark. *Proceedings of the 15th Nordic Geotechnical Meeting. (NGM2008)*. Sandefjord Norway. Norsk Geoteknisk Forening: 335–342.

Liu, J., Hu, Y. & Kong X.J. 2005. Deep penetration of spudcan foundation into double layered soils. *China Ocean Engineering* 19(2): 309–324.

Martin, C.M. & Randolph, M.F. 2001. Applications of the lower and upper bound theorems of plasticity to collapse of circular foundations. *10th Int. Conf. of IACMAG*, Arizona, USA, 2: 1417–1428.

Merifield, R.S. Sloan, S.W. and Yu, H.S. 1999. Rigorous plasticity solutions for the bearing capacity of two-layered clays. *Geotechnique* 49(4): 471–490.

Meyerhof, G.G. & Hanna, A.M. 1978. Ultimate bearing capacity of foundations on layered soils under inclined load. *Can. Geotech. J.* 15: 565–572.

Michalowski, R.L. 1992. Bearing capacity of nonhomogeneous cohesive soils under embankments. *J. Geotech. Engrg., ASCE* 118(7): 1098–1118.

Randolph, M.F. & Hope, S. 2004. Effect of cone velocity on cone resistance and excess pore pressures. *Engineering Practice and Performance of Soft Deposits*, Osaka, Japan: 147–152.

Randolph, M.F. Wang, D. Zhou, H. & Hu, Y. 2008. Large deformation finite element analysis for offshore applications. *The 12th Int. Conf. of IACMAG*, Goa, India: 3307–3318.

Stewart, D.P. 1992. *Lateral loading of piled bridge abutments due to embankment construction*. Ph. D. Thesis. The University of Western Australia.

Wang, C.X. & Carter, J.P. 2002. Deep penetration of strip and circular footings into layered clays. *International Journal of Geomechanics* 2(2): 205–232.

# Centrifuge investigation of the cyclic loading effect on the post-cyclic monotonic performance of a single-helix anchor in sand

J.A. Schiavon & C.H.C. Tsuha
Department of Geotechnical Engineering, University of São Paulo, São Carlos, Brazil

L. Thorel
LUNAM University, IFSTTAR, Nantes, France

ABSTRACT: Screw anchor application includes supporting structures subjected to cyclic loadings, as in the case of guyed towers. In the design of transmission line tower foundations in Brazil, most calculations are based on the limiting loading condition caused by an extreme environmental monotonic loading, and, therefore, the effects of cyclic loadings on the anchor response are disregarded. In this context, centrifuge tests were carried out to evaluate cyclic and post-cyclic responses of a single-helix anchor prototype in very dense sand. The centrifuge modelling used two single-helix anchor models (one instrumented with a load cell). A sequence of up to 2000 tensile loading cycles followed by a monotonic uplift loading was performed in each test. Noticeable displacement accumulation occurred in the first 10-500 cycles depending on the loading amplitude. In the post-cyclic phase, some tests exhibited up to 9% of uplift capacity degradation although other tests exhibited up to 5% of capacity increase.

1 INTRODUCTION

Over time, several structures are subjected to different types of loadings of varying magnitude. Transmission towers, chimneys and other tall structures are submitted to a series of loading and unloading cycles due to wind action. Offshore structures face the action of waves, which can be characterized as a cyclic loading that occurs during the entire service life of the structure. Therefore, the effects of cyclic loadings on the foundation response should be taken into account.

The uplift capacity of single-helix anchors is the sum of the helix bearing resistance and the shaft resistance (Lutenegger, 2013). Thus, previous observations on the cyclic response of pile foundations and plate anchors can contribute to a better understanding of the helical anchor behaviour under cyclic loadings. However, two important aspects are responsible for a significant difference in the behaviour of helical anchors compared to conventional piles and plate anchors:

i) When the dense sand is penetrated by the helix during the anchor installation, the degree of compactness reduces in the penetrated zone around the shaft. Therefore, the shaft friction resistance is significantly affected.

ii) The sand within the cylinder circumscribed by the helices experiences increase in void ratio whereas the sand outside the cylinder becomes denser (Mitsch & Clemence, 1985). This disturbance of the lateral stresses increases the potential for a local failure surface with cylindrical shape to develop above the helical plate during anchor uplift, which characterizes the helix bearing mechanism and significantly differs from the conical failure surface observed in buried plate anchors. Perez et al. (2017) proposed a method for the estimation of the helix uplift bearing capacity assuming the development a cylindrical failure surface with the length of the shear failure zone equals to $2.5D$. The assumption of Perez et al. (2017) based on observations from centrifuge and numerical modelling.

Previous studies on helical anchors have reported that cyclic loading causes anchor performance degradation in both coarse-grain and fine-grain soils. A reduction on the horizontal stresses acting on the failure surface above the helix occurs due to the movement of the sand particles toward the gap formed below the helical plate because of the displacements (upward) that accumulate in each load cycle in tension. Consequently, a significant capacity decrease occurs (Clemence & Smithling, 1984). In the case of an expected inflow of soil into the cavity, Trofimenkov & Mariupolskii (1964) mentioned that the limiting value of the applied load cyclic conditions should be limited to between 60 and 70% of the anchor uplift capacity in one-way cyclic loading, and from 30 to 40% of the uplift capacity in two-way cyclic loading. However, the experimental tests on plate anchor models of Andreadis et al. (1981) showed that the application of repeated loads increased the stiffness of the soil-anchor system in some cases. Additionally, the results presented in Buhler & Cerato (2010) indicated that cyclic loadings improved the helical anchor uplift capacity. According to the authors, the densification of the soil above the helices during the cyclic loading is

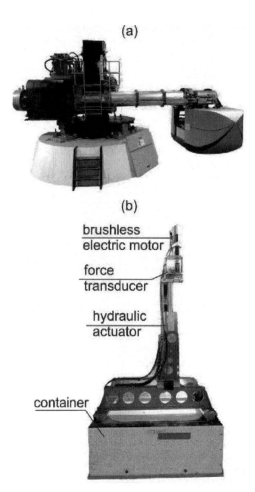

Figure 1. (a) Ifsttar beam centrifuge; (b) container.

Table 1. Characteristics of HN38 Hostun sand.

Specific gravity of the sand particles	G_s	2.64
Maximum dry unity weight (kN/m^3)	$\gamma_{d(max)}$	15.24
Minimum dry unity weight (kN/m^3)	$\gamma_{d(min)}$	11.63
Maximum void ratio	e_{max}	1.226
Minimum void ratio	e_{min}	0.699
Average grain size (mm)	d_{50}	0.12
Coefficient of uniformity	C_U	1.97

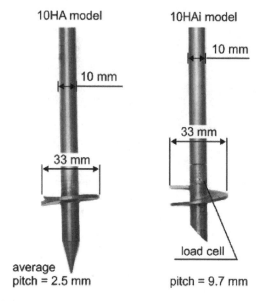

Figure 2. Helical anchor models.

the most likely effect to explain the capacity increase. Moreover, the cyclic load amplitude was observed to cause more influence on the anchor response rather than at the maximum load.

Under this scenario described above, the current investigation was performed to provide better understanding of the post-cyclic behaviour of single-helix anchors in dense sand. To this aim, a series of centrifuge model tests was conducted to evaluate the effects of different combinations of tensile cyclic loadings on the anchor post-cyclic response.

2 CENTRIFUGE MODELLING

Centrifuge modelling was conducted in the beam centrifuge at the Ifsttar's GMG Laboratory at a gravitational acceleration of $10 \times g$. The cyclic loading tests were carried out on model anchors installed in a rectangular container filled with sand of internal dimensions of 1200 mm × 800 mm × 360 mm (length × width × depth); a photograph of the centrifuge package used is shown in Figure 1.

2.1 Sand sample preparation

The centrifuge tests were performed in dry HN38 Hostun sand with relative density (D_r) varying from 94 to 99%. HN38 Hostun sand is a fine silica sand consisting of angular particles. This sand is used in other laboratory investigations, and its physical properties are well established (Flavigny et al., 1990). Table 3 lists some characteristics of HN38 Hostun sand.

The sand was backfilled using a manual air pluviation technique (slot width = 3 mm (where the sand pours from the hopper); drop height = 600 mm; horizontal speed = 18.5 cm/s; round-trip movements before re-adjusting the drop height = 4).

2.2 Helical anchor models

The helical anchor models used in the centrifuge tests are shown in Figure 2. A first experimental campaign was performed with a non-instrumented model (10HA). Posteriorly, a second anchor model was machined to attach a load cell to the shaft (10HAi). Both anchor models were fabricated with 33 mm helix diameter (D) and 10 mm shaft diameter (d). The helix and shaft dimensions were chosen to prevent undesirable particle-size effects (Schiavon et al., 2016).

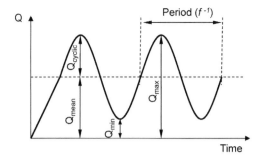

Figure 3. Cyclic loading parameters.

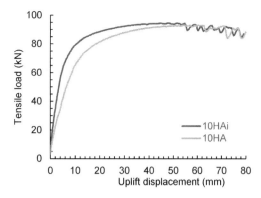

Figure 4. Monotonic load-displacement response (prototype scale).

Table 2. Load test information.

Test	Anchor model	Loading type	Q_{mean}	Q_{cyclic}	Number of Cycles N
M1	10HA	Monot.	-	-	-
M2	10HAi	Monot.	-	-	-
C1	10HA	Cyclic	$0.32Q_T$	$0.10Q_T$	2000
C2	10HA	Cyclic	$0.36Q_T$	$0.15Q_T$	2000
C3	10HA	Cyclic	$0.50Q_T$	$0.28Q_T$	1000
C4	10HAi	Cyclic	$0.41Q_T$	$0.20Q_T$	2000
C5	10HAi	Cyclic	$0.51Q_T$	$0.29Q_T$	1000
C6	10HAi	Cyclic	$0.59Q_T$	$0.34Q_T$	300
PC1	10HA	Monotonic (post-cyclic)			-
PC2	10HA	Monotonic (post-cyclic)			-
PC3	10HA	Monotonic (post-cyclic)			-
PC4	10HAi	Monotonic (post-cyclic)			-
PC5	10HAi	Monotonic (post-cyclic)			-
PC6	10HAi	Monotonic (post-cyclic)			-

2.3 Test procedure

Model installation and load tests were carried out at $10\times g$ in order to simulate an anchor prototype with $D=330$ mm and $d=100$ mm. The anchor models were installed with a constant rotation rate of 5.3 RPM up to a helix embedment depth of $7.4D$ (2.4 m in prototype scale). According to the AB Chance's (2012) guide to model specification, the helical anchors are usually installed with a rotation rate ranging from 5 to 20 RPM to provide a continuous and smooth advance into the soil. Lower rates are preferred to minimize soil disturbance. According to Perko (2009), typical rotation rates range from 10 to 30 RPM.

The vertical feed rate during installation was set to provide the optimum rate of helix penetration equal to 1 pitch per revolution. The use of a servo-controlled system ensured the installation met the requirements.

After each model installation, a tensile cyclic loading test following a sine function was conducted according to the parameters described in Figure 3. The cyclic loading characteristics of each test are presented in Table 2.

According to Wichtmann (2005), the limit between quasi-static and dynamic loading is considered to be within about 5 Hz. Therefore, the cyclic loading frequency was adopted as $f=1$ Hz in model scale.

Since the cyclic loading intensity was defined as a function of the pre-cyclic monotonic capacity (Q_T), monotonic pull-out tests were conducted on the anchor models before the cyclic tests. The anchor model installation for both monotonic tests followed the same procedures as the cyclic tests. The anchor model was pulled-out at a rate of 0.3 mm/s in model scale.

The number of cycles N established for the cyclic tests was based on results of preliminary tests and on the study of Tsuha et al. (2012) that proposed to separate the styles of cyclic response according to the number of cycles necessary to achieve cyclic failure caused either by cyclic pull-out or limiting displacement. In this cited study, when the number of cycles (N_f) to reach the limiting permanent displacement is greater than 1000, the foundation cyclic response is considered Stable. Otherwise, the cyclic response can be considered Metastable ($100 < N_f < 1000$) or Unstable ($N_f \leq 100$). In the current study, since no cyclic pull-out was observed, cyclic failure is recognized when the anchor model reaches 10%D of accumulated displacement.

A monotonic pull-out was performed after each cyclic test in order to evaluate the post-cyclic capacity degradation. The model pull-out rate was similar to the tests M1 and M2.

3 RESULTS & DISCUSSION

Figure 4 presents the monotonic load-displacement response of 10HA and 10HAi models, both installed at $7.4D$ embedment depth at $10\times g$. The monotonic capacity (Q_T) recognized as the peak tensile force exhibited similar values (93 kN and 94 kN for 10FH and 10FHi, respectively), which were observed at around 17%D of vertical displacement. For 10%D of vertical displacement, the difference between both models in measured force was 4.5% (89 kN and 93 kN, respectively for 10HA and 10HAi).

The 10FHi model showed stiffer response than 10FH model. The difference in stiffness of the two load-displacement responses may be related to the greater number of turns required for the installation of the anchor model 10FH compared to 10FHi (the anchor model 10FH was constructed with a reduced helix pitch). On the other hand, the uplift capacity seemed to have not been affected by the greater number of turns.

During the monotonic test on the model 10FHi, the load cell measurements showed inconsistent values (for details, see the Appendix H of Schiavon (2016)). Therefore, helix bearing and shaft resistance could not be distinguished in this case.

3.1 Cyclic displacement accumulation

For various types of structures, the Service Limit State dictates the design. This means that the anchor accumulated displacements during cycling may exceed a certain limit that could compromise the stability of the structure before it reaches the Ultimate Limit State. The acceptable accumulated displacement can be related to a slack in the guy cable, for example. Therefore, the current study is also focused on the displacement accumulation during cycling.

Figure 5a shows the accumulated displacement (U_{acc}) normalised by the helix diameter (D) versus the number of cycles (N) in linear scale. Although C3 and C5 tests were carried out with similar cyclic loading characteristics, C3 test exhibited greater accumulated displacements than C5 test. This observation corroborates the hypothesis that the 10HA axial stiffness was affected by the greater number of times that the helix traversed the soil when compared to the anchor model 10HAi. Moreover, the C2 test had a less intense cyclic loading than the C4 test. However, the accumulated displacements of C2 test were greater than C4 test for a number of cycles up to $N = 2000$. For a number of cycles larger than 2000, the tendency is that the C4 accumulated displacements would exceed C2 displacements.

In Figure 5b, the displacement accumulation response of test C2 exhibits a larger slope up to 100 cycles than C4 test. Witchmann (2005) observed a significant decrease of strain amplitude in the first 100 cycles in triaxial tests in dense sand specimens. During this phase, the elastic volumetric deformation decreases and considerable soil densification occurs. Following this phase, the elastic deformation remains approximately constant and represents the major portion of the total deformation for an individual cycle.

The number of cycles required for the transition between both phases seems to be related to the amplitude of the loading cycles. Silver & Seed (1971), Youd (1972) and Witchmann (2005) observed a significant increase of the densification rate with increasing shear strain amplitude. Therefore, the lower the load amplitude, the more cycles are required to complete the densification phase.

In Figures 5b and 5c, an approximately bi-linear trend in the displacement accumulation response can be identified in most tests. The transition between both trends may indicate the *locus* where the densification phase is complete. For large-amplitude cyclic loadings (e.g. tests C3 and C5), the transition occurs with less cycles compared to the tests with small-amplitude cyclic loadings (e.g. tests C1, C2 and C4).

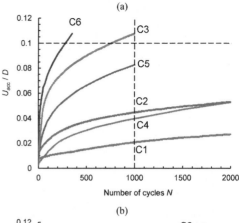

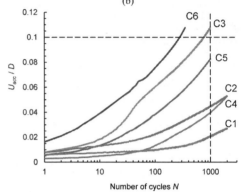

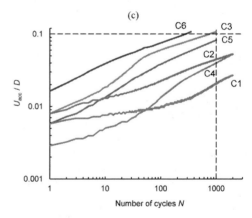

Figure 5. Normalised accumulated displacements (U_{acc}/D) with: (a) N linear scale; (b) N in log scale; (c) U_{acc}/D and N in log scale.

For the cyclic tests C3 and C6, the anchor cyclic failure occurred at approximately 700 and 250 cycles, which characterizes a Metastable cyclic response. The other tests exhibited accumulated displacements of less than 10%D after 1000 cycles, which characterizes Stable cyclic response.

After few cycles, soil densification occurred and both models exhibited increase in stiffness. For the 10HAi model, the stiffness at the end of the cycles was larger for cyclic loadings with larger Q_{cyclic} (C6).

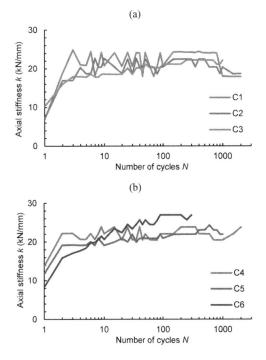
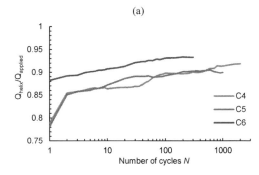
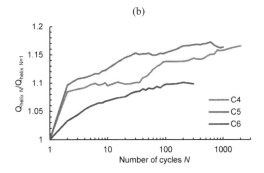

Figure 6. Axial stiffness of cyclic tests with (a) 10HA model, and (b) 10AHi model.

However, the same trend was not observed for the tests performed on the anchor model 10HA.

No sudden pull-out was observed in any cyclic test. In general, the tests presented a tendency of gradual decrease in the rate of displacement accumulation, and a consequent stiffness increase. Figure 6 shows the axial stiffness for both 10HA and 10HAi models. The initial stiffness of the 10FHi tests was 40% greater than the 10HA tests. This difference is probably related to the greater soil disturbance caused by the installation of the 10HA model, as argued before.

3.2 Helix bearing during cycling

Figure 7a shows the load measured with the load cell at a shaft section immediately above the helix (load related to the helix bearing resistance). In tests C4 and C5, the maximum load resisted by helix bearing at an individual cycle corresponded to between 78% and 88% of the applied load. The cyclic loading was observed to cause the increase of the load mobilized by the helix of up to 16% (Fig. 7b). The increase in helix bearing occurs due to the degradation of shaft friction resistance. During cycling, the disturbed soil around the shaft becomes denser, which results in volume decrease and relaxation of the horizontal stresses acting on the shaft. Consequently, the more the soil above the helix becomes denser, the more degradation of shaft friction resistance occurs.

3.3 Post-cyclic monotonic capacity

Figures 8 and 9 presents the load-displacement responses of the post-cyclic monotonic tests for 10HA

Figure 7. (a) maximum force registered with the load cell (Q_{helix}) normalised by the applied load ($Q_{applied}$); (b) maximum force registered with the load cell in each cycle ($Q_{helix\,N}$) normalised by the maximum load with the load cell at the 1st cycle ($Q_{helix\,N=1}$).

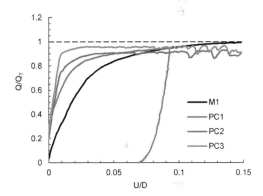

Figure 8. Normalised post-cyclic monotonic load-displacement responses for the 10HA model.

and 10HAi models, respectively. Degradation of the post-cyclic uplift capacity was observed for the three tests on 10HA model (Fig. 8). In contrast, tests PC5 and PC6 on 10HAi showed a slight increase in uplift capacity after the cyclic loadings (Fig. 9). Table 3 summarizes the results of the post-cyclic monotonic tests.

Figures 8 and 9 also show that the anchor axial stiffness increased after all cyclic loading tests. The anchors subjected to cycles of larger load amplitude presented a stiffer post-cyclic monotonic response.

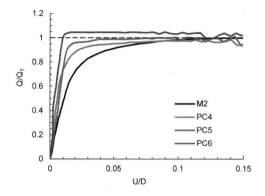

Figure 9. Normalised post-cyclic monotonic load-displacement responses for the 10HAi model.

Table 3. Summary results of the post-cyclic monotonic tests.

Test	Anchor model	Q_{mean}	Q_{cyclic}	Q_{T-PC}/Q_T	U/D for Q_{T-PC}
PC1	10HA	$0.32Q_T$	$0.10Q_T$	0.94	0.09
PC2	10HA	$0.36Q_T$	$0.15Q_T$	0.91	0.05
PC3	10HA	$0.50Q_T$	$0.28Q_T$	0.96	0.03
PC4	10HAi	$0.41Q_T$	$0.20Q_T$	0.98	0.08
PC5	10HAi	$0.51Q_T$	$0.29Q_T$	1.01	0.04
PC6	10HAi	$0.59Q_T$	$0.34Q_T$	1.05	0.02

4 CONCLUSIONS

Six tensile cyclic loading tests have been carried out on single-helix anchor models installed in very dense sand. The results of this investigation have indicated two different phases of displacement accumulation. The first phase occurred in the first 10-500 cycles with the soil experiencing densification and the anchor exhibiting increase in axial stiffness. In the second phase, the rate of displacement accumulation tended to decrease but not cease.

The cyclic loading effects on the post-cyclic capacity was different for the two anchor models tested. The anchor model with a smaller helix pitch (10HA) showed a decrease between 4% and 9% in post-cyclic monotonic capacity. On the other hand, the model anchor with greater and constant pitch (10HAi) showed an increase in post-cyclic capacity between 1% and 5%, which occurred in uplift tests performed after the large-amplitude cyclic loadings. Based on these observations, the cyclic loadings performed in this investigation can be assumed to cause a limited and negligible influence on the post-cyclic uplift capacity for the conditions and range of cyclic loading tested.

REFERENCES

AB Chance Civil Construction. 2012. *Guide to model specification: helical piles for structural support. Bulletin 01-0303, Centralia.*

Andreadis, A. et al. 1981. Embedded anchor response to uplift loading. *Journal of the Geotechnical Engineering Division* 107(GT1): 59-78.

Buhler, R.T. & Cerato, A.B. 2010. Design of dynamically wind-loaded helical piers for small wind turbines. *Journal of Performance of Constructed Facilities* 24(4): 417-426, doi: 10.1061/(ASCE)CF.1943-5509.0000119.

Clemence, S.P. & Smithling, A.P. 1984. Dynamic uplift capacity of helical anchors in sand. In E. D. Storr (ed.), *4th Australia-New Zealand Conference on Geomechanics, Proc. int. conf., Perth, 14-18 May 1984.* Austr. Geomech. Soc. and N. Zeal. Geomech. Soc.

Flavigny, E. et al. 1990. Le sable d'Hostun RF. *Revue française de géotechnique* 53(1):67-70. (in French).

Mitsch, M.P. & Clemence, S.P. 1985. The uplift capacity of helical anchors in sand. In S. P. Clemence (ed.), *Uplift behavior of anchor foundations in soil, ASCE Convention, Detroit, 24 October 1985.* ASCE.

Perko, H.A. 2009. *Helical piles: a practical guide to design and installation.* John Wiley & Sons.

Pérez Z. et al. 2017. Numerical and experimental study on the influence of installation effects on the behaviour of helical anchors in very dense sand. *Canadian Geotechnical Journal*, (ahead of print), https://doi.org/10.1139/cgj-2017-0137.

Silver, M.L. & Seed, H.B. 1971. Deformation characteristics of sands under cyclic loading. *Journal of the Soil Mechanics and Foundations Division* 97(SM8): 1081-1098.

Trofimenkov, Y.G. & Mariupolskii, L.G. 1964. Screw piles as foundations of supports and towers of transmission lines. *Soil Mechanics and Foundation Engineering* 1(4): 232-239.

Tsuha, C.H.C. et al. 2012. Behaviour of displacement piles in sand under cyclic axial loading. *Soils and Foundations* 52(3): 393-410, doi: 10.1016/j.sandf.2012.05.002.

Schiavon, J.A. 2016. Behaviour of helical anchors subjected to cyclic loadings. PhD thesis, University of São Paulo and LUNAM Université.

Schiavon, J.A. et al. 2016. Scale effect in centrifuge tests of helical anchors in sand. *International Journal of Physical Modelling in Geotechnics* 16(4): 185-196, doi: 10.1680/jphmg.15.00047.

Wichtmann, T. 2005. *Explicit accumulation model for non-cohesive soils under cyclic loading.* PhD thesis. Bochum: Ruhr University Bochum.

Youd, T.L. 1972. Compaction of sands by repeated shear straining. *Journal of the Soil Mechanics and Foundations Division* 98(SM7): 709-725.

Bearing capacity of surface and embedded foundations on a slope: Centrifuge modelling

D. Taeseri, L. Sakellariadis, R. Schindler & I. Anastasopoulos
ETH Zurich, Zurich, Switzerland

ABSTRACT: The static bearing capacity of surface and embedded foundations at the vicinity of slopes was predominantly studied analytically in the past and various well-known upper bound solutions have been proposed. However, only a few relevant 1-g or n-g physical model tests can be found in the literature, focusing on small variations of key variables, such as the slope angle α, the static safety factor F_s, the distance Δx between the crest of the slope and the foundation, the roughness μ of the footing, and the embedment ratio D/B. This paper summarizes existing analytical solutions, and provides verification through centrifuge model tests conducted at the ETH Zurich geotechnical drum centrifuge. A series of centrifuge model tests are conducted on dry sand, investigating a wide range of the aforementioned key problem variables. The results obtained through the centrifuge model tests are compared to the analytical solutions, setting the basis for further testing to address the combined M-H-V capacity of foundations in slope.

1 INTRODUCTION

The foundation system, apart from transferring the loads of the superstructure to the ground, affects significantly the overall seismic response. In order to assess the structural behaviour during an earthquake it is essential to understand the dependency between the superstructure and the soil – foundation system. Ignoring soil – structure interaction (SSI) can lead to gross errors in the estimation of the seismic response of a system.

A wide experimental research has been dedicated in the past to the SSI effect on the rocking of shallow foundations (e.g. Algie et al. 2010, Butterfield & Gottardi 1994, Drosos et al. 2012, Faccioli et al. 2001, Gajan et al. 2005, Gajan & Kutter 2008, Georgiadis & Butterfield 1988). It has to be mentioned though, that many of these tests were performed at low confining stresses. However, it is of importance to examine such behaviour under prototype stresses, taking into account also the stress – depended soil properties. The latter can be achieved either with large-scale field tests or with centrifuge experiments with increased gravitational acceleration compensating the reduced soil depth in the model.

The present study is part of a detailed centrifuge experimental study focusing on the nonlinear behaviour of foundations in inclined terrain under monotonic and cyclic loading. This paper focuses on the experimental investigation of the vertical bearing capacity of surface and embedded foundations at the crest of slope. Thereby, the key factors affecting the stability of the foundation, such as the slope angle and the embedment ratio are examined and quantified. The tests results are then verified with well-known analytical solution present in the literature. The outcome of this study contributes to the improvement of the investigation techniques and the current design methods for structures in seismically active and mountainous regions and therefore towards cost-effective design and seismic safety of engineering structures.

2 METHODOLOGY

The current work is part of an extensive experimental investigation, focusing on the response of surface and embedded foundations on level and sloped sand. The tests are carried out with the geotechnical drum centrifuge of the ETH – Zurich (Springman et al., 2001). The present paper focuses on the experimental determination of the vertical bearing capacity of surface and embedded foundation at the crest of a slope. To that end, eighteen (18) displacement–controlled vertical push tests were conducted with key variables the embedment depth D/B and inclination angle α. The load settlement curves not only allow the assessment of the ultimate bearing capacity, but also for the determination of the yielding load and the stiffness of the foundation system. Furthermore the vertical bearing capacity V_f sets the basis for the estimation of the safety factor $F_s = V_f/V$ of the system. The safety factor represents a fundamental parameter for the assessment of the response of the foundation under general loading.

Although small-scale 1g physical model tests are cost effective and simple to prepare, they are incapable of reproducing the actual prototype stresses in the soil. Geotechnical centrifuge model tests overcome these limitations by accelerating the soil model, and

Table 1. Geotechnical centrifuge scaling laws.

Parameter	Model/Prototype
Length	$1/n$
Acceleration	n
Velocity	1
Strain	1
Force	$1/n^2$
Mass	$1/n^3$
Time	$1/n$
Frequency	n

thus increasing the gravitational level inside the small-scale model by N times. Soil parameters such as the stiffness and strength are strongly stress dependent (Taylor 1995). It is therefore fundamental to be able to reproduce the correct full-scale prototype confining pressure in the soil in order to be able to evaluate correctly the response of the system.

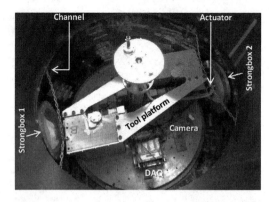

Figure 1. Configuration of the centrifuge model tests: photo of the ETH drum centrifuge with the two strongboxes mounted.

2.1 Experimental setup

The 2.2 m diameter drum centrifuge of ETH – Zurich was manufactured by Thomas Broadbent & Sons Ltd. and can reach a maximum centrifugal acceleration of 440 g at a payload of 2 tons. A total number of 36 tests were carried out at a centrifugal acceleration of 60 g and 100 g. The dimensions of the models were scaled down following the scaling law summarized in Table 1.

The drum centrifuge was equipped with two cylindrical boxes (Fig. 1) where the soil models were prepared. The model was prepared outside the centrifuge and then mounted on the drum channel.

In the first phase the soil is transferred into the soil container via air pluviation. Depending on the pluviation height h_t and the diameter d of the diffuser, the relative density D_r is controlled (Morales 2015). In order to reproduce the slope, the box was inclined with the desired slope angle of α. This simple but efficient technique allowed the preparation of an undisturbed inclined terrain without touching-up the model and creating an inhomogeneous soil density around the foundation. Figure 2 shows schematically the cylindrical strong boxes and the experimental setup utilized in the vertical pushover tests.

The displacement–controlled vertical pushover tests were conducted at a constant speed of 1.44 mm/s, using the actuator attached on the tool platform (Fig. 1). Initial experiments (Taeseri et al., 2017) showed that the response of the foundation is not affected if the velocity of load application is less than 1.5 mm/s. The actuator was armed with a load cell and was attached to the foundation with a rigidly connected square plate. To avoid boundary effects, the vertical push tests were carried out at 100 g, so that the lateral and vertical distance between the foundation and the side wall is larger than $3B$.

The model preparation occurred in two different phases: (a) outside the centrifuge; and (b) inside the centrifuge. The soil is poured inside the box (Fig. 3a)

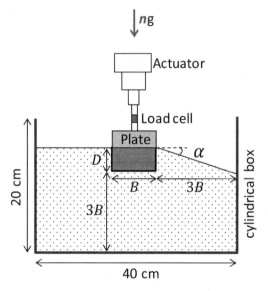

Figure 2. Configuration of the centrifuge model tests: experimental setup for vertical push tests (VP).

via air pluviation. At the desired soil layer depth the foundation is placed (Fig. 3b) and the slope is shaped with the predefined inclination angle. The model undergoes a saturation and desaturation process (Laue et al. 2002) (Fig. 3c) in order to create an apparent cohesion inside the model, allowing the rotation of the soil container by 90° and its mounting on the drum channel, keeping the sample stable until the centrifuge starts to spin (Fig. 3d). The centrifuge is then accelerated to the desired g-level (Fig. 3e) and the remaining water inside the soil is drained out. The vertical pushover test is performed using the tool platform actuator arm (Fig. 3f).

The soil used for the vertical and lateral monotonic pushover tests was the well-documented Perth Sand (Arnold 2012). Table 2 summarises the most important parameters derived from triaxial and odometer tests. Along with the aforementioned experiments, a novel

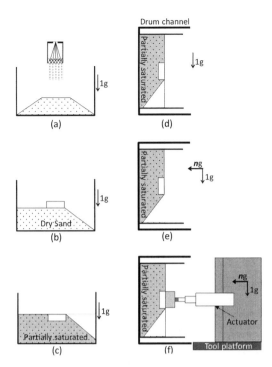

Figure 3. Model preparation: (a) dry sand pluviation outside the centrifuge; (b) structure installation and slope shaping; (c) saturation and subsequent desaturation; (d) rotation of the model by 90° – Installation on the drum channel; (e) spinning up to the desired g – level; (f) vertical push through the actuator.

Table 2. Soil parameters of the Perth sand used in the tests.

Parameters			Perth Sand
Effective particle size,	D_{10}	(mm)	0.14
Average particle size,	D_{50}	(mm)	0.23
Uniformity coefficient,	C_u	(–)	1.79
Coefficient of curvature,	C_c	(–)	1.26
Specific density,	ρ_s	(kg/m³)	2700
Dry density,	ρ_d	(kg/m³)	1700
Relative density,	D_r	(%)	80
Void ratio,	$e_{min} \ldots e_{max}$	(–)	0.50…0.75
Friction angle,	ϕ'_{max}	(°)	38
Dilatancy angle,	Ψ	(°)	8
Surface shear modulus,	G_0	(kPa)	35000
Sur. shear wave vel.	$V_{s,0}$	(m/s)	150

experimental technique was utilized for the estimation of the small strain shear modulus distribution of the Perth sand with the depth $G(z)$. The latter consisted on the combination of Spectral Analysis of Surface Waves (SASW) combined with shaking table experiments. The SASW measurement allowed the estimation of the small strain shear modulus at the surface G_0, and the shaking table experiments; its distribution with the depth. A detailed description of this technique can be found in Taeseri et al. (2017).

A generalised parabolic expression for the shear modulus distribution with the depth was proposed by Gazetas (1991):

$$G(z) = G_0(1+\varepsilon z)^{0.5} \quad (1)$$

where parameter ε defines the shape of the parabola. In order to fit the experimentally obtained parabola to the analytical formulation, the parameter was set to $\varepsilon = 2.5$ (1/m).

As summarized in Table 3 a total of nine vertical pushover tests were conducted and the examined variables were the embedment ratio D/B and the slope angle α. The width of the square shaped foundation ($B = 2.8$ m) and the relative density ($D_r = 80\%$) were kept constant throughout all the tests.

Table 3. Test matrix vertical pushover.

ID	B (m)	D (m)	D/B (-)	α (°)	D_r (%)
1	2.8	0	0	0	81
2	2.8	0	0	20	83
3	2.8	0	0	30	80
4	2.8	1.4	0.5	0	82
5	2.8	1.4	0.5	20	83
6	2.8	1.4	0.5	30	78
7	2.8	2.8	1	0	81
8	2.8	2.8	1	10	80
9	2.8	2.8	1	30	82

3 TEST RESULTS AND COMPARISON WITH EXISTING SOLUTION

The upcoming chapters are dedicated to the comparison between the physical modelling test results and well-known analytical formulations. The experimentally derived vertical bearing capacity values are compared to the solutions proposed by Graham et al. (1988). The assumption made by him represents as close as possible the characteristic of the physical model tests discussed herewith in. In fact the slope is only present on one side of the foundation and the distance between the crest of the slope and the foundation is reduced to zero.

The load settlement curves of foundations with $B = 2.8$ m and different embedment ratios D/B and slope angle α subjected to vertical push are illustrated in Figures 4 & 5 along with the analytical ultimate bearing capacity values $V_f^{Grah.}$:

$$V_f^{Grah.} = B^2(\gamma D N_q s_q d_q + 0.4\gamma B N_\gamma s_\gamma d_\gamma) + 4\gamma BD2\mu \quad (2)$$

where:
γ (kN/m³) = effective unit weight of Perth sand
B (m) = width of the square foundation
μ (–) = interface friction
d_q, d_γ (–) = deep factors
s_q, s_γ (–) = shape factors
N_q, N_γ (–) = bearing capacity factors

The first part of the equation represents the bearing capacity of surface and embedded square foundation in frictional soil and the second part considers the

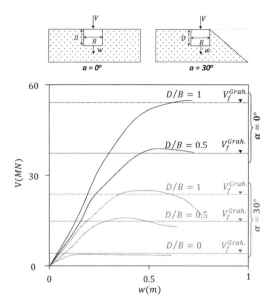
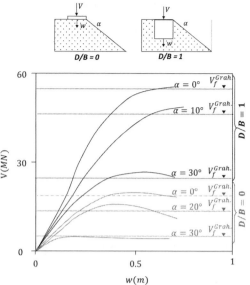

Figure 4. The effect of the embedment ratio D/B on the ultimate bearing capacity for $\alpha = 0°$ and $\alpha = 30°$.

Figure 5. The effect of the inclination angle α on the ultimate bearing capacity for D/B = 0 and D/B = 1.

mobilized vertical frictional forces acting on the lateral walls of the foundation depending on the friction μ of the soil structure interface.

For the calculation of the vertical bearing capacity V_f^{Grah}, the formulation proposed by Villalobos (2006) was considered and the values of the bearing capacity factors were determined according to Graham et al. (1988), which permitted the estimation of ultimate vertical capacity for shallow and embedded foundations at the crest of a slope in granular soil based on the method of stress characteristics. For the calculation of the capacity, the friction coefficient μ between the soil and the foundation (aluminium) was assumed equal to 0.5 (Uesugi et al. 1988).

In order to assess the impact of the embedment ratio on the ultimate bearing capacity, the response curves of the tests are represented in the same graph for a given inclination angle α and increasing embedment ratio D/B (Fig. 4). Due to the mobilized skin friction along the side walls of the embedded foundation and the larger mobilized failure body, the capacity increases with an increase of the embedment ratio. Furthermore it can be noticed that the increase of the embedment ration between 0 to 0.5 causes a larger gain in the vertical bearing capacity if compared to an increment of the embedment ratio between 0.5 and 1. In fact, in case of an inclination angle of 30°, the capacity enlarges by a factor of 5.3 from embedment ratio 0 to 0.5, but only enlarges by a factor of 1.6 from embedment ratio 0.5 to 1. The influence of the slope vanishes with increasing embedment ratio and the contribution in reduction of the capacity due to the slope becomes marginal in relative terms.

The shape of the load settlement curve (Fig. 3) suggested that a local shear failure type of mechanism occurred, which is defined by Vesic (1963) as an intermediate failure mechanism between the general and punching failure. In fact the post failure images of the tests showed that the failure mechanism developed up to the surface and the load settlement curves exhibited a softening behaviour after reaching the peak value.

On the other hand in order to assess the influence of the slope angle α on the ultimate capacity, the curves are plotted for a given embedment ratio D/B (Fig. 5). The results indicate that with an increase of the inclination angle, the bearing capacity reduces drastically. As illustrated in Figure 5, the bearing capacity is reduced by almost 60% with a slope inclination of 30°, compared to the flat case. The image analysis of the experiments outlined that the failure mechanism is activated on the weaker part of the footing, meaning on the sloped side of the foundation. In fact, the apparent "missing" portion of the soil along the inclined side facilitates the development of the failure mechanism, rendering the downslope side more vulnerable to failures. The load settlement curves also confirm this tendency. Indeed, as shown in Figure 5, the capacity of the system with $\alpha = 30°$ reaches the peak value at smaller deformation if compared to the system with $\alpha = 0°$ indicating that for the flat terrain, larger deformations are needed to fully mobilize the larger failure body.

In the case of surface and embedded foundations in flat terrain ($\alpha = 0°$) the experimentally obtained ultimate capacity are in good agreement with the solution proposed by Graham et al. (1988) ($V_f^{Grah.}$). With the presence of the slope, the experimentally and analytically obtained capacities diverge slightly from each other, with differences smaller than 13%. In fact due to the more elaborated preparation of the sloped soil model compared to the flat case, the homogeneity of the soil around the foundation is impaired causing negligible deviations between the two capacities.

4 CONCLUSIONS

The current work analysed the vertical bearing capacity of surface and embedded foundations in sand at the crest of a slope. Physical model tests were conducted with the drum centrifuge of the ETH – Zurich. The key variables of the experiments were the embedment ratio D/B and the inclination angle α. The physical modelling test results were verified against well-known existing analytical solutions. The results of the tests confirmed that with increasing slope angle the bearing capacity decreases. This effect is more pronounced with decreasing embedment ratio. Furthermore, with increasing slope angle the stiffness of the foundations system decreases for a specific embedment ratio. On the other hand, the stiffness is not affected significantly by the slope in the case of surface foundations.

All conducted vertical pushover test showed a clear increase of bearing capacity with increasing embedment ratio. Thereby the increase from embedment ratio 0 to 0.5 resulted in a higher relative increase than from 0.5 to 1. Increasing embedment ratio also resulted in an increase of the stiffness of the footing system. Summarizing, through the present centrifuge study, the effects of key variables, such as the slope angle and the embedment ratio, were clearly revealed. The outcome sets the basis for the validation of future numerical models to examine parametrically the response of such systems.

ACKNOWLEDGEMENTS

The authors would like to thank Prof. Dr. S. Springman for her valuable scientific support throughout the experimental investigation. We greatly appreciate the scientific and technical support of the MSc student S. Britschgi as well as the technicians M. Iten and E. Bleiker.

REFERENCES

Algie, T.B., Pender, M.J. and Orense, R.P., 2010. Large scale field tests of rocking foundations on an Auckland residual soil. *Soil Foundation Structure Interaction (R Orense, N Chouw, and M Pender (eds)), CRC Press/Balkema, The Netherlands*, pp. 57–65.

Arnold, A., 2012. *Tragverhalten von nicht starren Flachfundationen unter Berücksichtigung der lokalen Steifigkeitsverhältnisse* (No. 19516). vdf Hochschulverlag AG.

Butterfield, R. and Gottardi, G., 1994. A complete three-dimensional failure envelope for shallow footings on sand. *Géotechnique*, 44(1), pp. 181–184.

Drosos, V., Georgarakos, T., Loli, M., Anastasopoulos, I., Zarzouras, O. and Gazetas, G., 2012. Soil-foundation-structure interaction with mobilization of bearing capacity: Experimental study on sand. *Journal of Geotechnical and Geoenvironmental Engineering*, 138(11), pp. 1369–1386.

Faccioli, E., Paolucci, R. and Vivero, G., 2001. Investigation of seismic soil-footing interaction by large scale tests and analytical models. In Proceedings of the *4th International Conference on Recent Advances in Geotechnical Earthquake Engineering and Soil Dynamics*, San Diego, California, pp. 1–12.

Gajan, S., Kutter, B.L., Phalen, J.D., Hutchinson, T.C. and Martin, G.R., 2005. Centrifuge modeling of load-deformation behavior of rocking shallow foundations. *Soil Dynamics and Earthquake Engineering*, 25(7), pp. 773–783.

Gajan, S. and Kutter, B.L., 2008. Capacity, settlement, and energy dissipation of shallow footings subjected to rocking. *Journal of Geotechnical and Geoenvironmental Engineering*, 134(8), pp. 1129–1141.

Gazetas G., 1991. Formulas and charts for impedances of surface and embedded foundations. *Journal of Geotechnical Engineering*, ASCE, 117(9): 1363–1381.

Georgiadis, M. and Butterfield, R., 1988. Displacements of footings on sand under eccentric and inclined loads. *Canadian Geotechnical Journal*, 25(2), pp. 199–212.

Graham, J., Andrews, M. and Shields, D.H., 1988. Stress characteristics for shallow footings in cohesionless slopes. *Canadian Geotechnical Journal*, 25(2), pp. 238–249.

Laue, J., 2002. Centrifuge technology. In *Workshop on constitutive and centrifuge modeling, two extremes, Monte Verita, Rotterdam, Balkema* (pp. 75–105).

Morales, W.F., 2015. River dyke failure modeling under transient water conditions (Vol. 247). vdf Hochschulverlag AG.

Springman, S.M., Laue, J., Boyle, R., White, J. and Zweidler, A., 2001. The ETH Zurich geotechnical drum centrifuge. *International Journal of Physical Modelling in Geotechnics*, 1(1), pp.59–70.

Taeseri, D., Martakis, P., Chatzi, E., Laue, J., Anastasopoulos, I., 2017. Static and dynamic rocking stiffness of shallow footings on sand: centrifuge modelling. *International Journal of Physical Modelling in Geotechnics (in print)*

Taylor, R.N. ed., 1995. *Geotechnical centrifuge technology*. CRC Press.

Uesugi, M., Kishida, H. and Tsubakihara, Y., 1988. Behavior of sand particles in sand-steel friction. *Soils and foundations*, 28(1): 107–118.

Vesic, A.B., 1963. Bearing capacity of deep foundations in sand. *Highway research record*, (39).

Villalobos J., F.A., 2006. *Model testing of foundations for offshore wind turbines* (Doctoral dissertation, University of Oxford).

21. Deep foundations

Performance of piled raft with unequal pile lengths

R.S. Bisht & A. Juneja
Department of Civil Engineering, Indian Institute of Technology Bombay, Mumbai, India

A. Tyagi & F.H. Lee
Department of Civil and Environmental Engineering, National University of Singapore, Singapore

ABSTRACT: The role of piles in piled raft is complex as the load distribution and the reduction in settlement is not uniform across the foundation. For this reason, until recently, piles were used only to improve the serviceability and the ultimate limit state of the raft foundations. In reality, however, significant interaction takes place between the raft and the piles, and it is not uncommon to overload the piles. In order to appreciate the above effects, two issues are addressed in this paper. First, to identify the contribution of the raft and the proportion of load it transfers to the piles. The location of the piles was important in this case. Second, is to obtain the optimal length of the corner and central piles. Centrifuge model tests were conducted at 30 g on model piled raft resting on sand bed.

1 INTRODUCTION

Piled raft foundations have gained lot of attention in the last few decades. The contribution of raft is usually ignored in conventional design method by designing the piles to carry whole superstructure load. This approach results in installation of more number of piles than that are necessary under the raft. However, in reality raft carries a significant proportion of total superstructure load in a piled raft (Butterfield & Banerjee 1971; Cooke 1986; Kuwabara 1989; Yamashita et al. 2011). The raft not only contributes in load carrying capacity but also in stiffness of a piled raft (Poulos 2001). The design of piled raft foundation can be optimised by incorporating the contribution of raft.

The use of piles in these foundations as settlement reducers in piled-raft has long been recognised (Burland et al. 1977). Now emphasis is given to optimum design which can make the piled raft foundations more economical and provide better solution over the conventional pile foundation. The number of piles, their location and length are important parameters which control the optimum design of piled raft. There is of course a limit on the number of piles which can be added beneath the raft. Beyond this threshold, the piled raft behaviour would be unaffected (Poulos 2001). Likewise, for the same total pile-length, the use of a small number of long piles is observed to be more beneficial in reducing the settlement rather than the use of a large number of short length piles (Reul and Randolph 2004).

In the usual case, the loading type that is either uniformly distributed or concentrated loading on the raft is the single most important factor that decides the position and length of the piles beneath a raft. In much of the available literature, piled rafts are designed to carry uniformly distributed load, although heavy column loads are now more common in practice. This results in unequal distribution of load amongst the piles and therefore non-uniform settlement across the width of the foundation. It is therefore convenient to use piles of longer length underneath heavily loaded area of the raft which otherwise is more prone to large settlement (e.g. Katzenbach et al. 2000; Chow & Small 2005). The use of pile of dissimilar lengths can considerably improve the performance of piled raft and can also optimise the pile length (Leung et al. 2009). Likewise, the use of piles under heavy column loads in piled raft, can minimises the settlement but at the cost of increased pile load. For instance, in Phung (1993) large scale field tests, the centre pile was observed to carry higher proportion of load than the outer piles in a group of five piled raft resting on sand. Similarly, Nguyen et al. (2013) centrifuge tests showed that the pile position can considerably affect the performance of piled raft. In their tests, the total and differential settlement in a closely spaced concentrated pile group placed under heavily loaded columns were significantly less than that of uniformly spaced piles below the raft. It is therefore not surprising that longer piles are used at the central portion to control differential settlement of a flexible piled raft (Tan et al. 2005). However, there are no experimental studies that deal with improvement in performance of piled rafts with the unequal pile length despite being used in field.

In this paper, the performance of piled raft with unequal pile lengths is investigated by using a series

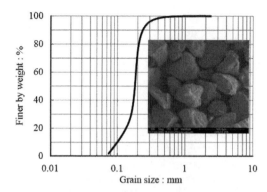

Figure 1. Grain size distribution and SEM of sand.

Table 1. Properties of Yamuna sand.

Soil Parameters		Value
Specific gravity	G_s	2.66
Angle of internal friction	ϕ	36.8°
Minimum dry unit weight	γ_{dmin}	12.4 kN/m^3
Maximum dry unit weight	γ_{dmax}	15.7 kN/m^3
Effective size	D_{10}	0.11 mm
Mean particle size	D_{50}	0.18 mm
Uniformity coefficient	C_u	2.02
Coefficient of curvature	C_c	1.019

of centrifuge tests on small piled raft. The inclusion of one-additional central pile and the effect of the uneven length of the piles on load-settlement behaviour of the foundation were investigated. The limitation of the centrifuge loading apparatus did not permit the use of large piled rafts. Because of which, the contribution of raft on load capacity and effect of number of piles on load sharing between piles and raft was examined by the installation of large model piled rafts on the laboratory floor.

Centrifuge model tests were conducted using a 4 m radius beam centrifuge available at Indian Institute of Technology Bombay. Details of this centrifuge and its specifications are reported elsewhere (Chandrasekaran 2001).

1.1 Test soil and model preparation

Centrifuge model tests were conducted in a rigid rectangular container of internal dimension 720 mm × 450 mm × 410 mm. The three side-walls of the strong box were made of solid steel whereas its front wall was made of a thick transparent Perspex sheet. Model ground was prepared using uniformly graded Yamuna sand. The sand was cleaned, washed and air-dried before the test to prepare dry homogenous sand bed. Fig. 1 shows the grain size distribution curve and SEM images of the sand. As can be seen, the sand consisted of sub-angular to sub-rounded particles. Table 1 shows the properties of the sand.

Sand beds were prepared by pluviating air-dry sand using a sand hopper. The opening of the hopper and the height of fall were maintained to obtain uniform and homogenous sand beds. Using this procedure, uniform sand beds of about 55% relative density could be prepared.

1.2 Model foundation and test procedure

An aluminium plate of size 100 mm × 100 mm and 20 mm thick was used to model the raft. Since the stiffness of raft not only depends upon its thickness but is also a function of the relative raft–soil stiffness ratio, K_{rs}. This affects the magnitude of differential settlement of the piled raft. In this case, the equation proposed by Horikoshi & Randolph (1997) was used to determine K_{rs}, that is

$$K_{rs} = 5.57 \frac{E_r}{E_s} \left(\frac{1-v_s^2}{1-v_r^2} \right) \left(\frac{B}{L} \right)^\alpha \left(\frac{t_r}{L} \right)^3 \quad (1)$$

where E_r and E_s are equal to the Young's modulus of the raft and the soil, respectively, v_r and v_s are equal to the Poisson's ratio of the raft and the soil, respectively, B and L are equal to the width and the length of the raft respectively and t_r is equal to the thickness of the raft.

Horikoshi & Randolph (1997) suggested that the value of K_{rs} for flexible raft ranges from 0.01 to 1. In this case, K_{rs} was greater than 100, that is, a rigid raft was used.

The plate was punched with five holes to attached 12 mm diameter aluminium rods to it. 120 to 180 mm long rods were used to simulate the piles under the raft. The piles were rigidly fixed to the raft with the help of nuts.

The model piled raft was centrally loaded using an in-flight electro-mechanical actuator. Figure 2 (c) and (d) shows the actuator and its component. In essence, the actuator consisted of an AC servo-motor with an encoder. The planetary gear box along with the ball screw and driving rod, permitted its operation under both load and displacement control modes. The linear travel of the ball screw was restricted with the use of two limit switches fixed at the top and the bottom of the actuator. The AC motor was controlled using a feedback system consisting of programmable logic controller (PLC) and drive housed in a panel box.

The model piled raft was installed at the centre of the container. Disturbance during the installation was kept a minimum and care was taken to prevent tilting of the raft.

Two LVDTs at the two ends of the raft helped measure the foundation settlement. Figure 3 shows the assembly on the centrifuge basket.

Six tests were performed at 30g centrifugal acceleration. Table 2 summarises the test details. Test UR was conducted on raft alone. The load-settlement curve of this test was used as a reference to determine the change in foundation capacity in the other five tests. The number of piles varied from 1 to 5. The centre pile

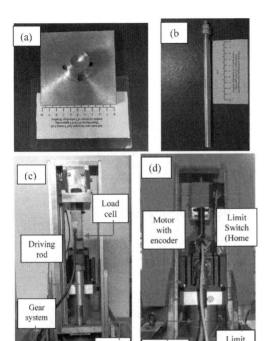

Figure 2. Test apparatus (a) Model raft; (b) Model pile; (c) and (d) Front and rear view of electro-mechanical actuator developed at IITB.

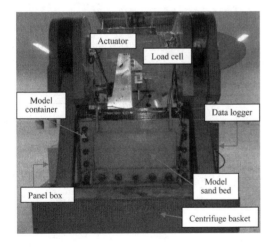

Figure 3. Model container mounted on centrifuge basket.

in tests PRCL180 was 180 mm while the remaining four piles were 120 mm in length. Spacing of 3.5 to 4 times the diameter of piles was maintained between the piles in all the tests.

2 CENTRIFUGE TEST RESULT

Figure 4 shows the effect of piles on the piled raft. In the figure, the load capacity of the piled raft was

Table 2. Test layout for centrifuge model test

Sl	Test	No. of piles	Central pile	Length of corner piles mm	Length of central pile mm
1	UR	0	N	–	–
2	PR1	1	Y	–	120
3	PR4	4	N	120	–
4	PR5	5	Y	120	120
5	PRAL180	5	Y	180	180
6	PRCL180	5	Y	120	180

* Y = yes, N = no

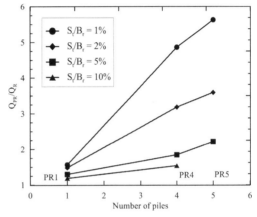

Figure 4. Effect of number of piles on performance of piled raft.

normalised by the load capacity of the raft alone, Q_{PR}/Q_R. Q_{PR}/Q_R is often referred to as the load improvement ratio.

In this case the load-settlement curve of the raft was used as the reference to help assess change in the performance of piled raft. Similarly, the total settlement of the raft was normalised by the width of the raft, S_t/B_r. The figure shows that Q_{PR}/Q_R increased with the increase in number of piles that is, the performance of foundation improved with the increase in the number of piles, as expected. Q_{PR}/Q_R was the greatest at S_t/B_r equal to 1%, with a decreasing trend when S_t/B_r increased from 2 to 10%.

This particular behaviour of the piled raft is related to the fact that, initially, the piles carry majority of the load at small loads. At this stage, the piles are able to transfer this load to depth where the soil has greater modulus and consequently the settlement is less and the ultimately stiffness is high. Later, as the load level increased, the contribution of the raft increased. This lead the raft to transmits the loads to shallow soils with small modulus. This stage resulted in an increase in settlement.

Figure 5 shows the effect of pile length on the load versus settlement of the piled raft. In the figure, the pile length was normalised by the equivalent radius of raft,

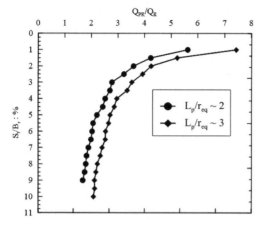

Figure 5. Effect of pile length on behaviour of piled raft.

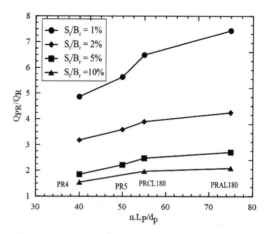

Figure 6. Effect of unequal pile length on performance of piled raft.

L_p/r_{eq}. r_{eq} was calculated by converting the square raft into an equivalent circular raft. Furthermore, Q_{PR}/Q_R is plotted against S_t/B_r for two pile lengths in tests PR5 and PRAL180. The figure shows that, Q_{PR}/Q_R exponentially increased with S_t/B_r to nearly reach a peak. Q_{PR}/Q_R also increased with the increase in L_p/r_{eq}. When the S_t/B_r was equal to 9% and L_p/r_{eq} increased from 2 to 3, there was 27% increase in Q_{PR}/Q_R. It shows that long piles with small rafts, are more useful in reducing the settlement of the piled raft foundation.

Figure 6 shows the effect of unequal pile length in the piled raft. In the figure, the sum of the lengths of all the piles was normalised by the diameter of the piles, nL_p/d_p and was plotted against Q_{PR}/Q_R. The figure shows that when S_t/B_r was 1%, the stiffness was high in test with 180 mm piles. Furthermore, when S_t/B_r increased from 2 to 10%, the difference between the tests PRCL180 and PRAL180 was reduced. For instance, when S_t/B_r was 10%, Q_{PR}/Q_R in tests PRAL180 was only 6% more than the test PRCL180 compared to the value when S_t/B_r was 1%, and Q_{PR}/Q_R was in excess of 14.5%. This variation shows a greater pile-pile interaction amongst dissimilar length as the settlement increased. The use of longer piles at central area of raft was therefore more effective in reducing the total settlement of the piled raft.

3 PILED RAFT TESTED AT 1 g

A few larger piled rafts were tested on the laboratory floor at 1 g because of the limitation of the axial capacity of the actuator. In this case, a 150 mm × 150 mm plate was used to model the raft. Piles were modelled by using 12 mm hollow aluminium rods as shown in Figure 7(a). In these tests, up to 16 piles were used. To separately determine the raft and the pile contribution, some of the piles were strain gauged to measure the axial load which was transferred from the raft to the pile. The strain gauges were attached by first machining a notch at three different locations along the length of the pile as shown in Figure 7(b). Strain gauges were then fixed at the notch locations to complete a full Wheatstone bridge circuit. The wires of the gauges were made to pass through the hollow section of the piles using 2 mm diameter holes pre-drilled through the pile surface. A thin layer of silicon rubber was then applied over the bonded strain gauges to protect the exposed surface of the gauges and to keep all the wires in intact position. The strain gauges were then covered with a thin epoxy layer to prevent them from being compressed additionally by radial stresses due to earth pressure. Sand particle were then glued to the pile surface. These instrumented piles were calibrated to convert the output voltage into load. Up to 5 instrumented piles were used in the tests Table 3 shows the test details.

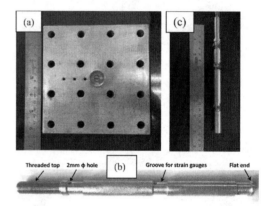

Figure 7. Model piled raft: (a) Model raft; (b) Model pile; and (c) Pile instrumented with strain gauge.

3.1 Test results

Figure 8 compare the load-settlement of the raft to the piled raft. The figure shows that even one pile altered the load-settlement results. The increase in the size of

Table 3. Test layout for 1 g model test.

Sl. No.	Test name	No. of piles
1	Raft only (GR)	–
2	Piled raft (GPR1)	1
3	Piled raft (GPR4)	4
4	Piled raft (GPR9)	9
5	Piled raft (GPR16)	16

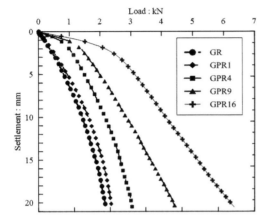

Figure 8. Comparison of load settlement behaviour of raft and piled raft.

piled raft resulted in an increase in not only the stiffness but also the load capacity of the foundation. The figure shows that settlement of the raft can effectively be controlled even with the use of a small pile group.

The load shared between the raft and the piles in a piled raft foundation is usually represented by a piled raft coefficient, α_{PR} defined as

$$\alpha_{PR} = \frac{\sum Q_{pile}}{Q_{Total}} \quad (1)$$

where $\sum Q_{pile}$ is equal to the sum of the load on the piles and Q_{Total} is equal to the total load on the piled raft.

Figure 9 shows the load shared between the piles and the raft. In the figure, α_{PR} momentarily increased immediately upon loading to ultimately decrease with the settlement of the raft. Once α_{PR} attained a constant value to reach a limit, the settlement indefinitely increased. It shows that the piles initially carry the majority of the applied load but because of the increase in the settlement, the load proportion carried by the piles also decreases. It results in additional loads shared by the raft. With further increase in the settlement, the load sharing became constant. The figure also shows the effect of number of piles on the contribution of the raft. As can be seen α_{PR} varied from 0.09 to 0.81 for piled rafts with 16 piles. It further goes to show that even with a large pile group, the contribution of the raft can be significant. The rate of increase in the

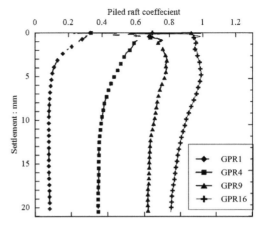

Figure 9. Load sharing in piled raft foundation.

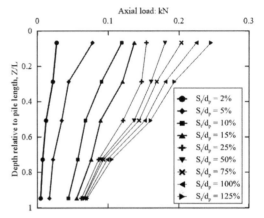

Figure 10. Axial load distribution in pile for test GPR4.

load carried by the piles decreased with the increase in the number of piles.

Figure 10 shows the axial load distribution along the length of the pile. The settlement of the raft was normalised by the diameter of pile, S_t/d_p. In the figure, the distance of the measurement point from the top of the pile was normalised by the length of the pile, Z/L. The figure shows that the pile head and the tip load increased with S_t/d_p. However, the pile-tip load beyond S_t/d_p equal to 10% remained within a narrow band. It can also be observed from the figure that the upper half portion of the piles carried more load compared to the bottom half portion. This may be due to raft-soil-pile interaction effect which is significantly more in the upper portion of the pile. Similarly, the raft-soil-pile interaction effect was dominant in upper portion of pile with the increase in S_t/B_r value.

4 CONCLUSIONS

The effect of equal and unequal pile length on the performance of piled raft was examined using centrifuge model tests at 30 g. The piled raft showed high stiffness at low settlement, which decreased as the settlement increased. The raft stiffness also found to be increased with increase in number and length of pile. The piled raft performance increased significantly with the use of unequal piles under the raft. The difference in load improvement ratio, Q_{PR}/Q_R of piled raft with unequal length pile was only 6% less compared the all long length piles at S_t/B_r of 10%. The paper also explored the contribution of raft in a piled raft using a series of model scale experiments performed on the laboratory floor. The raft contributes significantly in load carrying capacity even when a large pile group was used beneath the raft. The raft-soil pile interaction was more effective in the upper portion of pile due to which the upper half of the piles carried additional load compared to the bottom half of the piles.

REFERENCES

Burland, J.B., Broms, B.B. & De Mello, V.F.B. 1977. Behaviour of foundations and structures. *Proceedings of the 9th International Conference on Soil Mechanics and Foundation Engineering*, Tokyo, Japan, 2: 495–546.

Butterfield R. & Banerjee P.K. 1971. The problem of pile group-pile cap interaction. *Geotechnique*, 21(2): 135–142.

Chandrasekaran, V.S. 2001. Numerical and centrifuge modelling in soil structures interaction, *Indian Geotechnical Journal* 31(1): 1–59.

Chow, H.S.W. & Small, J.C. 2005. Behaviour of piled rafts with piles of different lengths and diameters under vertical loading. *Geotechnical special publications (GSP),132, ASCE Geo-Frontiers, Austin: 841–55*.

Cooke, R.W. 1986. Piled raft foundations on stiff clays- a contribution to design philosophy. *Geotechnique*, 36(2): 169–203.

Horikoshi, K. & Randolph, M.F. 1996. Centrifuge modelling of piled raft foundation on clay. *Geotechnique*, 46(4): 741–752.

Horikoshi, K. & Randolph, M.F. 1997. On the defination of raft-soil stiffness ratio for rectangular rafts. *Geotechnique*, 47(5): 1055–1061.

Katzenbach, R., Arslan, U. & Moormann, C. 2000. Piled raft foundation projects in Germany. Design Application of Raft Foundation, J.A Hemsley, Thomas Telford, 23–391.

Kuwabara, F. 1989. An elastic analysis for piled raft foundations in a homogeneous soil. *Soils and Foundations*, 28(1): 82–92.

Leung, Y.F., Klar, A. & Soga, K. 2009. Theoretical study on pile length optimization of pile groups and piled rafts. *Journal of Geotechnical and Geoenvironmental Engineering*, 136(2): 319–330.

Nguyen, D.D.C., Kim, D.S. & Jo, S.B. 2013. Settlement of piled raft with different pile arrangement schemes via centrifuge tests. *Journal of Geotechnical and Geo Environmental Engineering*, 139(10): 1690–1698.

Phung, D.L. 1993. Footings with Settlement-Reducing Piles in Non-Cohesive Soil. Ph.D Thesis. University of Technology, Göteborg, Sweden.

Poulos, H.G. 2001. Piled raft foundations: design and applications. *Geotechnique*, 51(2): 95–11.

Reul, O. & Randolph, M. F. 2004. Design strategies for piled rafts subjected to non-uniform vertical loading. *Journal of Geotechnical and Geoenvironmental Engineering*, 130(1): 1–13.

Tan, Y. C., Chow, C. M. & Gue, S. S. 2005. Piled raft with different pile length for medium-rise buildings on very soft clay. *Proc., 16th Int. Conf. Soil Mechanics and Geotechnical Engineering*, Balkema, Rotterdam, The Netherlands: 2045–2048.

Yamashita, K., Yamada, T. & Hamada, J. 2011. Investigation of settlement and load sharing on piled rafts by monitoring full-scale structures. *Soils and Foundations*, 51(3): 513–532.

Pile response during liquefaction-induced lateral spreading: 1-g shake table tests with different ground inclination

A. Ebeido, A. Elgamal & M. Zayed
Department of Structural Engineering, University of California, San Diego, La Jolla, California, USA

ABSTRACT: Past earthquakes demonstrate that lateral spreading induced by liquefaction may cause excessive movement, and significant damage to structures and their underlying pile foundation. A 1-g shake-table series of experiments was conducted to investigate the mechanism of liquefaction-induced lateral spreading effects on pile foundations in mildly inclined ground. A single steel pipe pile of 25 cm diameter was tested under earthquake excitation in two different mildly inclined soil profiles of 2 and 4 degrees. The ground stratum was built of sand at about 1.8 m in height with a 1.1 m base saturated layer and an upper 0.7 m dry crust. For each inclination, data is employed to compare and assess peak pile response and soil behaviour pre- and post-liquefaction. Soil and pile lateral displacement as well as excess pore pressure are discussed. Such a pile-ground interaction mechanism is of consequence for analyses that correlate pile bending moments to the accumulated lateral soil deformation.

1 INTRODUCTION

Lateral spreading induced by liquefaction may cause excessive movement and possible failure of pile foundations (e.g. Yasuda & Berrill 2000). This presents a complex loading situation as the soil undergoes a significant change in its dynamic properties (Finn 2015).

Case history investigations describe a wide range of damage to structures and their pile foundations (Tokimatsu & Asaka 1998). In the Kobe region, the 1995 Hyogoken-Nambu earthquake was responsible for the damage of numerous pile foundation supported structures. The observed damage indicated the contribution of both inertial and kinematic forces. Tokimatsu and Asaka (1998) describe a specific waterfront structure with deformed piles due to lateral spreading. Horizontal cracks appeared at the interface between the liquefiable layer and the underlying non-liquefiable ground. This cracking points to local plastic demands on the piles due to the significant change in soil properties at the interface between these layers.

Other investigators show the dependence of the failure mode on the relative stiffness between pile and soil (Martin & Chen 2005). They conclude that soil might eventually flow around stiffer piles whereas more flexible pile experience lower lateral loads while undergoing larger deflections. Many other field investigations have provided valuable insight and helped define the scope of necessary research (e.g. Tokimatsu et al. 2005, Ishihara & Cubrinovski 2004, Koyamada et al. 2006).

Physical modelling is a valuable resource to complement field investigations. Centrifuge experiments were conducted to study pile kinematic effects during liquefaction and lateral spreading in mildly inclined ground. Studies by Abdoun et al. (2003) and Dobry et al. (2003) in a laminar box with multi-layered soil profiles found the largest bending moment at the interface between liquefied and underlying non-liquefied soil layers. Conclusions from tests by Brandenberg et al. (2005) show that lateral load imposed by the different soil layers depends on the incremental and total relative displacement between pile and soil. Brandenberg et al. (2007) further discuss the softening of crust layer load transfer mechanisms during the associated passive soil failure.

In addition to the above, large scale one-g shake table experiments were performed. He (2005) and He et al. (2006, 2009) discuss four experiments in a mildly inclined laminar box with different single pile and pile group configurations. These experiments address the evolution of pore-water pressures, total pressures and displacements on steel pipe piles along with pile foundation response to such loading. Chang and Hutchinson (2013) tested a reinforced concrete pile in a medium size inclined laminar box and their analysis focused on local plastic demands on the pile and the failure mechanism.

Following up on the above efforts, this paper reports on the results of a 1.8 m steel pipe pile embedded in a 2-layer profile within a mildly inclined laminar soil box (at 2 and at 4 degrees) at the University of California, San Diego. The response is analysed to

Figure 1. Laminar box on the shake table at UC San Diego, a) side fish-eye lens view, and b) top view.

compare ground and pile displacements, excess pore water pressures and pile bending moments.

2 EXPERIMENTATION PROGRAMME

Figures 1 and 2 present a layout of the experiment using a medium size laminar box on the shake table at the University of California, San Diego. A single steel pipe pile in a 2 layer stratum was tested. The saturated soil stratum was constructed by sand deposition in water using Ottawa F-65 sand (Bastidas 2016).

Box inner dimensions are 3.9 m in length, 1.8 m width and 1.8 m height with 28 laminates. In two different experiments, the box was inclined at both 2° (Test 1) and 4° (Test 2) to the horizontal. Water table was 0.7 m below the downslope ground surface. As such, the bottom saturated layer is 1.1 m high and the overlying dry crust is 0.7 m. The bottom saturated layer has an estimated relative density of 50–60% and saturated density of 1900 kg/m^3. The upper dry layer is placed with a 10% water content and compacted in layers to have an estimated relative density of 100%, and bulk density of 1900 kg/m^3.

Input motion (Fig. 3) was in the form of a sinusoidal acceleration with a 2 Hz frequency and 0.30 g peak amplitude. Duration of motion for Test 1 was 34 cycles with the amplitude building up gradually in the first 10 cycles followed by a constant 0.30 g amplitude for

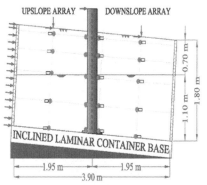

Figure 2. Schematic experiment layout with instrumentation locations.

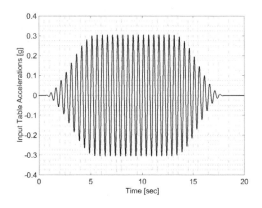

Figure 3. Input shake table acceleration for Test 1 (34 total cycles at 2 Hz). Test 2 had a slightly shorter input motion (30 total cycles).

14 cycles, then ramping down for the last 10 cycles. Test 2 had a similar input motion (2 seconds shorter) with a total shaking of 30 cycles. Motion had the same 10 cycle gradual increase and 10 cycle ramping down at the end. However, it only had 10 cycles of constant amplitude instead of 14 cycles.

2.1 Pile details

The circular steel pipe pile (Fig. 4) was 0.25 m in diameter and 3 mm in thickness. The pile was welded to a steel plate at the base then bolted to the box floor. Rotational and translational fixity were preferred for the base, however the connection flexibility allowed minimal rotation at the pile base. As such, this connection was tested and characterized to have a rotational flexibility of 200 kN.m/rad. Pile material is mild steel with 455 MPa yield strength and 2.1×10^5 MPa elastic modulus. The theoretical yield moment is calculated based on yield stress and section modulus, estimated to be 66 kN-m. Moment curvature for the section was identified using the software Opensees (Mazzoni et al. 2006). The yield curvature was identified as 0.016 rad/m.

2.2 Instrumentation

The model was instrumented with a large number of accelerometers, pore pressure sensors, soil pressure transducers, strain gauges and LVDTs (Fig. 2). Instrumentation was placed along the pile shaft and along the depth of the ground stratum. Strain gauges were densely deployed along the pile height to aid in back-calculation of the bending moment during shaking. The gauges, 40 in all, were placed on both sides of the pile at 10 cm spacing. A total of 16 displacement transducers were mounted on the laminar box exterior wall to measure lateral displacements approximately every other laminate, 2 on the soil surface to measure horizontal, and 2 to measure vertical displacements. The pile was also instrumented with transducers to measure head displacement above the ground surface. In addition, a total of 25 pore pressure transducers and 29 accelerometers were also placed on the pile.

Figure 2 presents the distribution of these sensors along the pile and soil. In general, locations chosen for soil embedment were in the middle between the pile and box boundary in an effort to reduce the influence of boundary effects on the readings. For the upslope-downslope shaking direction, instrumentation was placed approximately 1 m away from the pile. In addition, in the perpendicular direction, instrumentation was 0.45 m away from the pile. Figure 4 shows a picture of the pile with the instrumentation attached to it.

Figure 4. Steel pipe pile with attached instrumentation.

3 RESULTS

3.1 Analysis protocol

Focus is placed on system response mainly discussing excess pore-water pressure, displacements and pile bending moments. Thus, representative time histories were chosen to display this response. Bending moment was calculated based on strain gauge readings placed on the pile wall. As such, bending moment is an indicator of lateral pressures acting on the pile. For ease of analysis, the time corresponding to the peak bending moment was identified and represented as a dashed line on all time history plots.

As the box was mildly inclined, the soil started moving downslope due to liquefaction of the underlying saturated layer. In both tests, only the soil exhibited permanent displacements, while the pile head had no residual displacements.

3.2 Soil response

Input motion (Fig. 3) was a 17 second 2Hz sinusoidal wave with gradual increase and decrease in amplitude for Test 1 and 15 seconds for Test 2. Liquefaction occurred early on, approximately 5.25 seconds into the shaking phase. Representative time history of excess pore water pressure ratio (Fig. 5) clearly display this mechanism. The vertical line in these figures denotes the time corresponding to the instant of maximum pile bending moment as will be shown below, occurring at about 5.11 seconds for Test 1 (2°) and 3.12 seconds for Test 2 (4°) and is included on all time histories for ease of tracking. Maximum moment occurred before liquefaction for both tests at an excess pore pressure ratio of about $r_u = 0.95$ for Test 1 and $r_u = 0.70$ for Test 2.

3.3 Excess pore-water pressures

Figure 5 shows representative excess pore water pressure ratio response. The general trend of the pore pressure data displays rapid pore pressure build up with instantaneous reductions for both downslope and upslope readings. This shows some dilative tendency in the liquefied soil response.

Dips in excess pore pressure for Test 2 are seen to be much larger than those of Test 1 (Fig. 5). This can be attributed to influence of the increased static driving shear stress in the 4° Test 2, compared to those of 2° Test 1.

3.4 Displacements

Figure 6 shows the downslope deformation for the soil box, and single pile for Tests 1 and 2 respectively. Both tests show an increase of ground surface displacement with shaking and a resulting accumulated permanent value. Deformations started with shaking and stopped thereafter.

In light of the acting initial driving shear stress, it is seen that Test 2 incurred a significantly higher level of deformation earlier during the shaking phase (Fig. 6). Of interest as well is that the instant of peak moment on the pile coincided approximately with the same level of ground deformation in both experiments (about 30 mm).

As shaking started, the pile began to oscillate back and forth recording its highest value at the time of

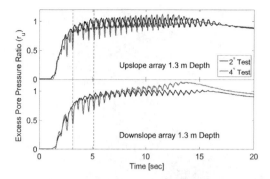

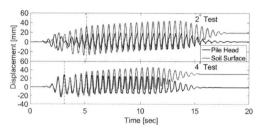

Figure 6. Displacement time histories for the 2 tests showing the soil surface and pile head.

Figure 5. Representative excess pore water pressure ratio time histories for the 2 tests at 1.30 m depth (Upslope and downslope of the pile).

maximum bending moment. Values close to maximum displacement were reached before liquefaction, oscillating thereafter around a constant but slightly lower value. Pile head displacement gradually decreased with the ramping down of the input acceleration reaching zero at the end of the shaking event.

Maximum pile head displacement was 16.6 mm and 29.2 mm for Tests 1 and 2 respectively. This shows a 76% increase in pile head displacement as the inclination changed from 2 to 4 degrees. On the other hand, accumulated ground surface displacement was 19.1 mm and 30.1 mm for Tests 1 and 2 displaying an increase of about 60%.

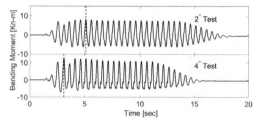

Figure 7. Representative bending moment time histories for the 2 tests at the pile base.

earlier in the higher 4° inclination test (at 3.12 s) rather than a 5.11 s for 2° experiment. This suggests that the higher inclination caused earlier mobilization of soil movement, just due to pre pressure increase, but well before the onset of liquefaction.

3.5 Pile response

Figure 7 shows representative time histories of bending moment along the pile shaft. The location chosen (Fig. 7) was at the base of the pile where the maximum values were recorded. Peaks in bending moment are seen to correspond to the excess pore pressure transient drops, denoting coincidence with the cyclic large shear strain excursions (Zeghal & Elgamal 1994).

As mentioned earlier, peak bending moment was observed early during the shaking phase, before soil liquefaction (Fig. 5). After maximum moment was reached, both tests show oscillation of the recorded value around a constant slightly lower value, gradually decreasing as the shaking was ramped down (Fig. 7). At the end of shaking, both piles rebounded to an almost zero position. As such, pile response was linear elastic.

Figures 6 and 7 demonstrate that soil keeps on moving, after maximum pile response, with no appreciable further loading as the ground displacement continues to accumulate. As such, there is no correlation between the final accumulated ground deformation and the observed peak loading on the pile. Brandenberg et al. (2005, 2007) provide further details about the involved dry crust lateral loading mechanism.

Maximum bending moment recorded for Test 1 was 8 kN-m and 13 kN-m for Test 2 as shown in Figure 7. Therefore, the difference in inclination caused the pile to incur an additional bending moment of about 60 %. Furthermore, maximum response occurred much

4 SUMMARY AND CONCLUSIONS

A large scale 1-g shaking table test series was performed on a single steel pipe pile embedded in a 2-layer soil profile. The base layer was saturated while the upper was dry and highly compacted. The pile was tested to investigate the liquefaction-induced lateral spread loading as ground inclination changed from 2 to 4 degrees. Shaking was conducted using a 2 Hz sine wave with 0.30 g peak amplitude. Results were compared for both cases. Settlement of the upper stratum and cracking upslope of the pile were observed, while a gap was formed between the pile and the downslope soil (Fig. 8). The pile represents a stiff behavior as evident from the large cyclic response then settling at zero displacement at the end. This indicates the soil flowing around the pile.

The pattern of pile response is similar in both cases as indicated by the measured bending moment and displacements. However, the increase in ground inclination caused an additional 60% in bending moment and 76% in pile head displacement. The higher locked-in driving shear stress in the 4 degree scenario triggered the crust to start moving earlier, with lower excess pore pressure ratio $r_u = 0.70$ (rather than at $r_u = 0.95$ for the 2-degree scenario). Around this time instant, the pile approached its maximum recorded response. With the subsequent increase in soil displacement, no appreciable increase in pile bending moment occurred.

Figure 8. Upslope cracking and downslope gap formation in the crust post-shaking.

In general, from this particular data set it may be concluded that: i) values approaching peak bending moment were noted quite early in the shaking phase. Increased lateral spreading due to the continued soil downslope deformation, did not result in appreciable increase in pile displacements or moments, and ii) as the driving static shear stress increases, it is likely that peak moment will occur earlier, with excess pore pressures that might be well below those corresponding to the onset of liquefaction.

ACKNOWLEDGEMENT

The presented research is supported by the California Department of Transportation with Dr. Charles Sikorsky as the Program Manager. This support is gratefully acknowledged. Testing was conducted in the UCSD Powell laboratories, with assistance graciously provided by Dr. Christopher Latham, Mr. Darren Mckay, Mr. Noah Aldrich, Mr. Abdullah Hamid and Mr. Mike Sanders.

REFERENCES

Abdoun, T., Dobry, R., O'Rourke, T.D. & Goh, S.H. 2003. Pile response to lateral spreads: centrifuge modeling. *J. of Geotechnical and Geo-environmental Engineering* 129(10), 869–878.

Bastidas, A.M. 2016. Ottawa F-65 Sand Characterization. Ph.D Dissertation, University of California, Davis.

Brandenberg, S.J.; Boulanger, R.W., Kutter, B.L. & Chang, D. 2005. Behavior of pile foundations in laterally spreading ground during centrifuge tests. *J. of Geotechnical and Geo-environmental Engineering.* 131(11), 1378–1391.

Brandenberg, S.J., Boulanger, R.W., Kutter, B.L. & Chang, D. 2007. Liquefaction-Induced Softening of Load Transfer between Pile Groups and Laterally Spreading Crusts," *J. of Geotechnical and Geo-environmental Engineering*, 133(1), 91–103.

Dobry, R., Abdoun, T., O'Rourke, T.D. & Goh, S.H. 2003. Single piles in lateral spreads: Field bending moment evaluation. *J. of Geotechnical and Geo-environmental Engineering.* 129(10), 879–889.

Chang, B. J. & Hutchinson, T.C. 2013. Experimental investigation of plastic demands in piles embedded in multi-layered liquefiable soils. *Soil Dynamics and Earthquake Engineering* 49, 146–156.

Finn, W. L. 2015. 1st Ishihara Lecture: An overview of the behavior of pile foundations in liquefiable and non-liquefiable soils during earthquake excitation. *Soil Dynamics and Earthquake Engineering.* 68, 69–77.

He, L. 2005. Liquefaction-induced lateral spreading and its effects on pile foundations. Ph.D Dissertation, University of California, San Diego.

He, L., Elgamal, A., Abdoun, T., Abe, A., Dobry, R., Meneses, J., Sato, M. & Tokimatsu, K. 2006. Lateral loads on piles due to liquefaction induced lateral spreading during one-g shake table experiments. *Proc. of the 8th U.S. National Conference on Earthquake Engineering, San Francisco, CA.*

He, L., Elgamal, A., Abdoun, T., Abe, A., Dobry, R., Hamada, M., Menses, J., Sato, M., Shantz, T., & Tokimatsu, K. 2009. Liquefaction-Induced Lateral Load on Pile in a Medium Dr Sand Layer, *J. of Earthquake Engineering.* 13:7, 916–938.

Ishihara, K. & Cubrinovski, M. 2004. Case studies on pile foundations undergoing lateral spreading in liquefied deposits. *Fifth International Conference on Case Histories in Geotechnical Engineering*, New York, NY.

Koyamada, K., Miyamoto, Y. & Tokimatsu, K. 2006. Field investigation and analysis study of damaged pile foundation during the 2003 Tokachi-Oki Earthquake. *ASCE*, 97–108.

Martin G, Chen C. (2005). Response of piles due to lateral slope movement. *Computers and Structures.* 83:588–98.

Mazzoni, S., McKenna, F. & Fenves, G.L. 2006. Open System for Earthquake Engineering Simulation User Manual, *Pacific Earthquake Engineering Research Center, University of California, Berkeley.* (http://opensees.berkel-ey.edu/OpenSees/manuals/usermanual/).

Tokimatsu, K. & Asaka, Y. 1998. Effects of Liquefaction-Induced Ground Displacements on Pile Performance in the 1995 Hyogoken-Nambu Earthquake. *Soils and Foundations, Special Issue*, pp. 163–177.

Tokimatsu, K., Suzuki, H. & Sato, M. 2005. Effects of inertial and kinematic interaction on seismic behavior of pile with embedded foundation. *Soil Dynamics and Earthquake Engineering*, No. 25, pp. 753–762.

Yasuda, S. & Berrill, J.B. 2000. Observations of the Earthquake Response of Foundations in Soil Profiles Containing Saturated Sands. *1st International Conference on Geotechnical and Geological Engineering.* Melbourne, Australia, Issue Lecture, pp. 1441–1471.

Zeghal, M. & Elgamal, A.-W. 1994. Analysis of site liquefaction using earthquake records. *J. of Geotechnical Engineering* 120(6), 996–1017.

Effect of the installation methods of piles in cohesionless soil on their axial capacity

I. El Haffar, M. Blanc & L. Thorel
IFSTTAR, Bouguenais, France

ABSTRACT: An experimental campaign has been realized to study the effect of the installation method on the axial capacity of piles. Different installation methods have been used to jack the tested piles at $1 \times g$ and $100 \times g$ in dense Fontainebleau sand. The results of the compression and tension tests carried out on these piles show important impacts of the installation method on their axial capacities. The pile total compression capacity, as well as shaft and tip resistance are analysed. Pile total tension capacities are also studied. A significant increase in the tension capacity is observed in cyclically-jacked piles unlike piles monotonically jacked at $100 \times g$.

1 INTRODUCTION

In the literature, the use and capacity of driven piles are extensively studied (Randolph et al. 1994; Jardine et al. 2005; Puech & Benzaria 2013). These studies have helped to reduce the uncertainty related to the axial capacity of driven piles. They have initiated the development of approaches and standards nowadays widely used in pile design (e.g. API, DNVGL, Eurocode 7, ICP (Jardine et al. 2005)). However, driving piles into the soil may cause a high level of noise and ground vibration as well as ground movement. As an alternative to this traditional installation method, the use of jacked piles (which can be placed using a hydraulic jack) has received increasing attention in the past few years. The possibility of jacking piles without noise and vibration is indeed more suitable for urban use and more acceptable by current European recommended limits for noise and vibration (Eurocode 3, White et al. 2002). However, compared to driven piles, jacked pile behaviour remains largely unknown and little research has been devoted to the comparison of their respective capacities (Yu et al. 2012; Yang et al. 2006).

The objective of this research, therefore, is to improve the understanding of the axial capacity evolution of jacked piles and propose a comparison with other installation methods. To achieve this, an experimental programme is conducted on $100 \times g$ centrifuged model piles. The present paper presents the findings of the detailed investigation carried out to examine static axial capacity of close-ended piles in dense sand when using different installation methods. The key feature of this analysis is the use of in-flight jacking installation techniques. This method is most representative of stress state susceptible to develop around prototype jacked piles.

2 METHODOLOGY

2.1 Centrifuge modelling

Centrifuge modelling is used here on small scale models installed in a strong gravity field allowing for the replication of stress state occurring within prototype soil.

The tests presented are carried out on 1/100 scale piles scaled at 100 times the Earth's gravity ($100 \times g$).

2.2 Model soil

The model soil consists of Fontainebleau NE34 poorly graded sand (Table 1) with a relative density of 99% obtained by air pluviation into a rectangular strongbox.

2.3 CPT tests

In order to verify the good uniformity of the sand used for this study. Several CPT ($B = 12$ mm) tests have been realized inside the sand strongboxes. Figure 1 shows the results from the CPT performed in the dense sand strongbox. Two tests have been realized in this strongbox and show clearly the existence of a good match between these tests results.

It proves that the pluviation method used during the preparation of the sand give a high level of sand uniformity.

2.4 Model pile

The model pile used is a rigid aluminium rough pile (Fig. 2-b) with dimensions $B = 18$ mm and $D = 250$ mm. The corresponding prototype pile is then 1.8 m in diameter and 25 m in embedded length. The normalized roughness R_n introduced by Uesugi & Kishida (1986) is used to define the pile roughness

Table 1. Characteristics of used sand Fontainbleau NE34.

Sand	$U_C = d_{60}/d_{10}$	d_{50} (μm)	γ_{min} (g/cm³)	γ_{max} (g/cm³)
Fontainebleau NE34	1.53	210	1.46	1.71

U_c is coefficient of uniformity (Silva 2014).
d_x is grain size at which x% of particles by weight, respectively, are smaller (Silva 2014).
γ_{min}, γ_{max} is minimum and maximum, respectively, dry unit weigh tested in the lab according to the standard (NF P 94-059).

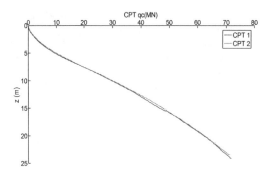

Figure 1. CPT in dense sand.

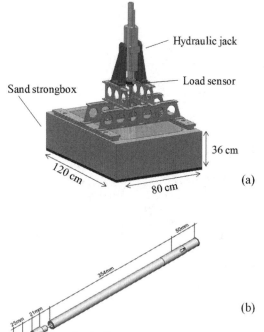

Figure 2. (a) Experimental set up (inside dimensions), (b) Model pile.

$R_n = 0.35$. Its represents the ratio of the maximum height of the pile profile R_z (NF EN ISO 4287, 1998) on the mean diameter of the sand particle D_{50}. For this model pile, R_z has been measured and is equal to 69.8 μm. The pile is instrumented using a 25 KN load sensor (XF3059 from *Measurement*) with a thickness of 21 mm placed at a distance of 25 mm from the tip. Another 25 KN load sensor (FN3070 from *FGP*) is placed between the pile head and the hydraulic jack (Figure 2-a) also measures the total bearing capacity of the pile. The pile displacement can be determined using a magnetostrictif displacement sensor (1/3000350S010–1E01 from *TWK*) which controls the displacement of the hydraulic jack.

2.5 Piles installation and experimental campaign

All the tests where realized in dry Fontainebleau sand prepared as indicated in the above sections.

Two different installation methods are compared: first, 1×g jacking used to model bored pile installation; then, in flight jacking used to represent installation effects and soil displacement occurring during jacked pile installation.

With the first method (MJP1G, Monotonic Jacked Pile at 1×g), the piles are jacked at 1×g to the desired embedded depth of 250 mm before application of the centrifuge acceleration and loading test itself (compression or tension).

With the second method, the piles are jacked in flight up to a depth of 250mm. The tests are carried out without stopping the centrifuge. Different types of jacking techniques are used:

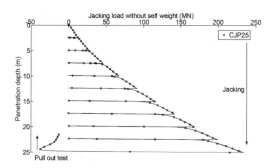

Figure 3. Jacked pile using strokes of 2.5 m.

- MJP100G (Monotonic Jacked Pile at 100×g): the piles are jacked monotonically at a speed of 0.1 mm/s before a pull out test is performed.
- CJP25 (Cyclic Jacked Pile): the piles are jacked using a series of jacking strokes, equal to 25 mm each, at the same speed than the first technique. Between each stroke, the force returns to zero. Then, the jacking pattern is repeated every 25 mm until reaching the desired 250-mm embedded length. Similarly, a pull out test is performed at the end of the jacking phase (Figure 3).
- CJP10, CJP5 and CJP2.5 follow the same procedure as CJP25 with only stroke length differences, which are 10 mm for CJP10, 5 mm for CJP5 and 2.5 mm for CJP2.5, respectively. The different tests are described in Table 2.

The spacing between the piles is 300 mm and the distance between the piles and the rigid wall of the strongbox is 250 mm in the width of the strongbox and 300 mm in it length.

3 BEARING CAPACITY ANALYSIS

3.1 Determination of ultimate capacity

Compression and/or tension tests are carried out after completion of the pile installation. The determination of compression and tension capacities from the force-displacement curve is detailed in Blanc & Thorel (2016). When no compression test is performed after in-flight jacking, the final jacking force is considered as the ultimate compression capacity of the pile.

3.2 Determination of shaft and tip resistance

As described above, all the tests are performed using model piles instrumented with a sensor placed at 2.5m

Table 2. Experimental campaign.

Test	Description
MJP1G	Monotonically jacked pile at 1×g
MJP100G	Monotonically jacked pile at 100×g
CJP25	10 Jacking strokes of 25 mm at 100×g
CJP10	25 Jacking strokes of 10 mm at 100×g
CJP5	50 Jacking strokes of 5 mm at 100×g
CJP2.5	100 Jacking strokes of 2.5 mm at 100×g

from the pile tip. Sensor results, however, cannot be used directly to deduce tip capacity and shaft friction. The sensor measures the sum of the tip capacity and the shaft resistance for the first 4.6 m at the bottom of the pile. In order to deduce shaft resistance along the entire piles, an experimental analysis method is developed in several steps:

1. Subtracting the tip sensor load ((b) on Figure 4) from head sensor load (curve (a) Fig. 4) gives only the shaft resistance for the first 20.4m of the pile ((c) in Figure 4).
2. Shifting up this curve on 4.6 m gives the first 20.4 m of the total shaft resistance ((d) in Figure 4).
3. By extending this curve up to 25m, using a 3rd order polynomial, the total shaft resistance is assumed.
4. The tip capacity ((e) in Fig. 4) is deduced from the difference between total head load and shaft resistance.

3.3 Compression test analysis

Compression force experimental results are presented in Table 3 and Figure 5.

A first comparison (Fig. 5-a) clearly shows, as expected, the existence of a significant difference in compression force between piles jacked at 1×g and piles jacked at 100×g. The compression capacity of MJP100G, for instance, is three times higher than MJP1G. This difference has been found also in the work of Ko et al. (1984) where the bearing capacity of the piles installed at 1×g was found to be 40% less than that of the piles installed at 70×g.

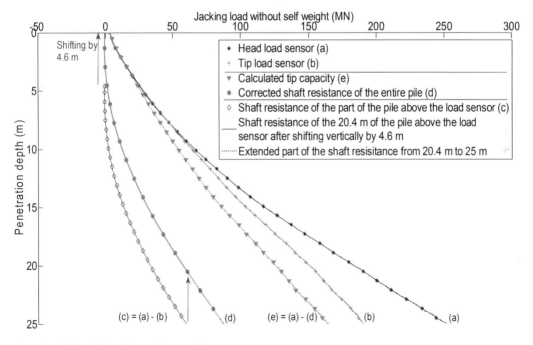

Figure 4. Shaft and tip resistance decomposition.

Table 3. Static ultimate capacities in compression and tension (prototype values).

Installation method	Compression			Tension	
	Total force [MN]	Tip capacity [MN]	Shaft resistance [MN]	Shaft resistance [MN]	Initial tension stiffness [MN/m]
MJP1G	74	–	–	−23	580
MJP100G	252	165	87	−30	309
CJP25	234	142	92	−37	341
CJP10	241	135	106	−46	452
CJP5	223	135	88	−47	474
CJP2.5	228	137	91	−50	523

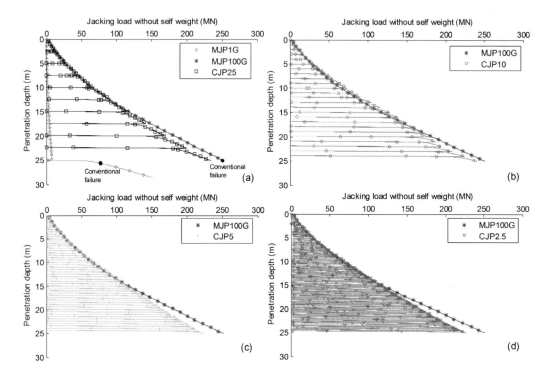

Figure 5. Jacking load without self-weight versus penetration depth for MJP100G (reference) and (a) MJP1G and CPJ25 – (b) CJP10 – (c) CJP5 – (d) CJP2.5.

A closer examination of the results reveals some differences between the piles jacked at 100×g (Figs 5-b, c, d). MJP100G has the highest compression capacity but there is neither clear trend nor clear relationship between compression capacity and stroke number. The difference between cyclically and monotonically jacked piles may be accounted for by both shaft resistance and tip capacity (Tab. 3). The tip capacity decreases as the number of jacking strokes increases. This could be related to soil densification under the tip because the one direction only displacement of the pile MJP100G may cause a higher degree of densification of the soil under the pile tip compared to the cyclically jacked piles. On the contrary, MJP100G shaft resistance is the lowest among all the piles tested at 100×g (Tab. 3).

4 PULL OUT TEST ANALYSIS

With the development of deep foundations in recent engineering projects, the use of piles, not only in compression but also in tension, is spreading. With this aim in view, the pull out capacity of the piles installed according to the methods described is studied. Results are displayed in Figure 6 where only the tension part of the load-displacement curves is plotted. The tension displacement is normalized by the pile diameter B. Moreover, the initial tension stiffnesses of the foundation are calculated for each test.

They correspond to the slopes crossing the traction displacement curve at the half of the maximum traction capacity. Main tests results are summarised in Table 3.

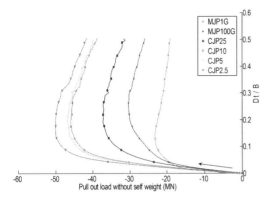

Figure 6. Pull out tests.

It should first be noted that the pull out capacity of MJP100G jacked at 100×g is 25% higher than that of MJP1G jacked at 1×g. This difference is not as high as with the compression capacity of both piles however. The difference between the pile capacity in compression and tension is not due solely to the mobilization of the tip capacity. The shaft resistance in compression is twice the tensile shaft resistance. The shaft resistance ratio (tension to compression) obtained in this study is lower than the values of 0.7–0.8 found in the literature (Schneider et al. 2008). Despite having lesser pull out capacity, it is noticed from Table 3 that the MJP1G initial stiffness is higher than the MJP100G one. For MJP1G, the g increase occurs while the pile is already installed. The soil surrounding the pile settles and rearranges giving a good contact between soil and pile. This condition is close to a cast in place pile behaviour. For MJP100G, the pull out test is realized directly after the jacking. The shearing mechanism along the shaft has to be reversed which required some displacement to be fully mobilized. The initial tension stiffness is then less rigid.

On the other hand, when it comes to the comparison of the pull out capacity of piles jacked at 100×g using different stroke protocols; the graph clearly shows that the tension capacity of the piles increases with the number of strokes (cf. Tab. 3). Such results raise questions about (1) the gain in capacity with the increase of cyclic strokes and (2) the actual pile ultimate tension capacity that must be used for design methods.

During the installation of displacement piles, the sand tends to dilate and generate high level of lateral contact stresses with the piles. The idea of the dilation is also found in the work of Lings & Dietz (2005). They observed i) in case of intermediate and rough surfaces in contact with the sand, the motion of the particles is increasingly characterized by rolling, resulting in dilation and ii) the increase in roughness and the increase in density bring about increased dilation and a resulting increase in strength. The dilation of the sand can be affected by the installation method used during the jacking of the piles. This phenomenon has been highlighted by Lehane & White (2005) where they have compared a monotonic and cyclic installation methods used during the jacking of displacement piles. They have concluded that cyclic installation creates either greater dilation during pile loading or stiffer confinement as a result of densification of the surrounding soil. So the gain of capacity with the increase of cyclic strokes can be mainly related to the increase of sand dilation during the cyclic installation in comparison with the monotonic installation. Dilation can generate an increase in the pile radial stress applied, which is directly related to the shaft resistance of the pile and to the gain observed during testing. It is also noticed that this mobilization is more rapidly in case of cyclic installation than the case monotonic installation. Table 3 shows clearly that the initial stiffness in traction of piles installed at 100×g increase with the number of strokes. The results obtained in this paper are also in good accordance with Lehane & White (2005) where they have revealed that the pull out capacity of their monotonically-installed piles was only 60% that of their cyclically-installed piles.

Another phenomenon, which is also related in second order to this gain in capacity, is observed during the emptying of the strongbox containing the sand. The sand, indeed, is crushed near the shaft and under the tip of the piles. The grain size of sand in contact with the pile varies from its initial state and may cause an increase in the friction angle between piles and sand. Similar results are found in Yang et al (2010) where the installation of displacement pile in pressurized sand produces particle breakage. These results also underline the two phenomena discussed above, i.e., the sand around the displacement pile is over-consolidated at the end of the installation and has a final void ratio substantially below its initial e_{min} value. Their conclusion is that the sand response to further static loading is highly likely to be strongly dilatant.

5 CONCLUSIONS

Model piles with an embedment depth of 250 mm and a diameter of 18 mm have been tested using geotechnical centrifuge at 100×g in order to determine axial capacity in dense sand. Different installation techniques have been compared. A significant difference has been observed as regards the performances of jacked and bored piles in compression: jacked piles have a resistance three times higher than that obtained with bored piles. The piles jacked at 100×g do not differ in compression except that the monotonically jacked piles provide a capacity approximately 8% higher than the cyclically jacked piles.

In the second part of the study, the impact of installation techniques on the pull out capacity of the piles has been studied. The most significant difference has been found for piles jacked at 100×g if they are jacked monotonically or cyclically. Despite the limitation to the study to one test for each installation method and to one studied density, the pull out capacity of the piles showed a clear tendency to increase with the increasing of the number of the installation strokes. The results show a pull out capacity gain of up to 67% with increasing cyclic installation strokes. The explanation discussed in this paper has proposed that a possible relationship may exist between the gain in the capacity and the dilation and crushing of sand usually observed in cases where rough surfaces are in contact with dense sand.

ACKNOWLEDGEMENTS

The authors wish to thank the IFSTTAR and the Region Pays de Loire for their financial support to the thesis grants, within the context of which this study has been conducted. Special thanks to the IFSTTAR Centrifuge team for its technical support and assistance during the centrifuge experimental campaign.

REFERENCES

API (American Petroleum Institute). 2011. *API RP 2GEO: geotechnical and foundation design considerations.* American Petroleum Institute, Washington, DC, USA.

Blanc, M. & Thorel, L. 2016. Effects of cyclic axial loading sequences on piles in sand. *Géotechnique Letters* 6(2): 163-167.

DNVGL-ST-0126. 2016. *Support structures for wind turbines. Appendix F pile resistance and load-displacement relationships.* 168p.

Eurocode 3 DD ENV 1993-5:1998. 2002. *Design of steel structures, Chapter 5, piling.*

Eurocode 7. 2005. *Geotechnical design – Part 1 general rules* (NF EN 1997). 145p.

Jardine, R., Chow, F., Overy, R. & Standing, J. 2005. *ICP design methods for driven piles in sands and clays.* Thomas Telford, London.

Ko, H.Y., Atkinson, R.H., Globe, G.G. & Ealy, C.D. 1984. *Centrifuge Modelling Of Piles Foundations, Analysis and Design Of Piles Foundations,* J.R. Meyer, Ed., ASCE, New York: 21-40.

Lehane, B.M. & White, D.J. 2005. Lateral stress changes and shaft friction for model displacement piles in sand. *Can. Geotech. J.* 42(4): 1039-1052.

Lings, M.L. & Dietz, M.S. 2005. The peak strength of sand-steel interfaces and the role of dilation. *Soils and Foundations* 45(6): 1-14.

NF EN ISO 4287. 1998. *Geometrical Product Specifications (GPS) - Surface texture : profile method - Terms, definitions and surface texture parameters.* 58p.

Puech, A. & Benzaria, O. 2013. Effects of installation method on the static behaviour of piles in highly overconsolidated Flanders clay. *Proceedings of TC 209 Workshop – 18th ICSMGE, Paris 4 September 2013. Design for cyclic loading: Piles and other foundations.* 69-72.

Randolph, M. F., Dolwin, J. & Beck, R. 1994. Design of driven piles in sand. *Géotechnique* 44(3): 427-448.

Schneider, J.A., Xu, Xiangtao & Lehane, B.M. 2008. Database Assessment of CPT-Based Design Methods for Axial Capacity of Driven Piles in Siliceous Sands. *J. of Geotech. Geoenviron. Engng* 134(9): 1227-1244.

Silva, M. 2014. *Experimental study of ageing and axial cyclic loading effect on shaft friction along driven piles in sands.* Ph. D. thesis. Université de Grenoble.

Uesugi, M. & Kishida, H. 1986. Frictional resistance at yield between dry sand and mild steel. *Soils and Foundations* 26(4): 139-149.

White, D., Finlay, T., Bolton, M. & Bearss, G. 2002. Press-in piling: Ground vibration and noise during pile installation. *Proceedings of the International Deep Foundations Congress. Orlando, USA. ASCE Special Publication 116,* 363-371.

Yang, J., Tham, L.G., Lee, P.K.K., Chan, S.T. & Yu, F. 2006. Behaviour of jacked and driven piles in sandy soil. *Géotechnique* 56(4): 245-259.

Yang, Z.X., Jardine, R.J., ZHU, B.T., Foray, P. & Tsuha, C.H.C. 2010. Sand grain crushing and interface shearing during displacement pile installation in sand. *Géotechnique* 60(6): 469-482.

Yu, F., Kou, H., Liu, J., Yang, Y. 2012. Jacking Installation of Displacement Piles: from Empiricism toward Scientism. *EJGE 17 ,Bund. J* . 1381-1390.

Model testing of rotary jacked open ended tubular piles in saturated non-cohesive soil

D. Frick, K.A. Schmoor, P. Gütz & M. Achmus
Institute for Geotechnical Engineering, Hannover, Germany

ABSTRACT: Conventional installation methods for displacement piles (i.e. impact-driving or vibratory-driving) induce vibrations and settlements in a zone close to the pile. Moreover they lead to noise pollution unfavourable especially in urban areas. A possible solution to these problems is the technology of rotary pile jacking. Thereby the pile is continuously pushed into the subsoil by a vertical force. For large pile dimensions a rotation is additionally applied to overcome high axial forces. This paper presents a new testing facility for axially loaded piles in saturated non-cohesive soils as well as the preparation procedure of the test sand. The facility is capable of installing and loading model piles by applying an axial load as well as a simultaneous rotary motion and enables model testing in different scales up to pile diameters of 101.6 mm and a maximum embedment length of 2.3 m. All test stages can be monitored and recorded with a data acquisition system, making it possible to study the influence of different installation parameters on the required installation forces and the resulting bearing behaviour. First results prove the efficiency of a rotation regarding the required axial installation force.

1 INTRODUCTION

Piles can generally be classified as replacement piles or displacement piles. Bored and cast-in situ piles belong to replacement piles. Thereby the soil is excavated and replaced with reinforced concrete. Driven piles, screw piles and jacked piles belong to displacement piles, where the soil is displaced by the pile within the installation process.

The installation of driven piles is usually done by impact-driving. Thereby the pile is driven into the subsoil by the application of several blows with an impact-hammer. This leads to high dynamic acceleration of the surrounding subsoil, which may cause dynamic excitation of neighbouring structures and settlements of the soil next to the pile. In contrast, jacked piles are installed into the subsoil by the application of a permanent vertical force, where for screw piles and rotary jacked piles a rotation is applied simultaneously. In case of a rotary jacked pile, two main benefits can be expected according to White et al. (2010). Firstly, the required vertical force during installation is reduced since a ratio of the shaft resistance is utilized in the direction of rotation. Secondly, the vertical stress of the soil within the pile at the pile tip is reduced. Hence, the plugging ability and therewith the tip bearing resistance of the pile is also reduced during the installation.

2 THEORETICAL BACKGROUND

By applying a movement to an open ended pile into the direction of increasing geostatic stress an arching effect within the inner soil plug can be noticed. This arching effect can be described with the stress distribution theory within silos by Janssen (1895). Equation 1, where $A = 4k_0 \tan \delta$; $k_0 = \tau_i/(\sigma_z \tan \delta)$; δ = interface friction angle and h_{plug} = height of the soil plug, describes the solution for the stress σ_z at the pile tip as function of the plug height. Thus, for a purely jacked pile, the increase in stress at the pile tip rises exponentially with increasing plug height.

$$\sigma_z = \frac{\gamma d}{A}\left(e^{A\frac{h_{plug}}{d}} - 1\right) \quad (1)$$

The application of an additional rotation θ leads to reduction of the required vertical installation force. However, a reduction of the vertical installation force is achieved due to an application of an additional torque. According to White et al. (2010) this installation method should not necessarily be a more efficient one. Anyhow, it allows overcoming high required vertical installation forces (cf. Figure 1).

The ratio of the vertical movement velocity and the movement velocity in direction of the rotation is described by the pitch according to Equation 2, where $\dot{\theta}$ = rotation speed [rad/s]; $\dot{w}$ = vertical speed [m/s]. For instance, a pitch value of 2 means that the pile has been moved twice the length in circumferential as in vertical direction.

$$p = \dot{\theta}D/2\dot{w} \quad (2)$$

Deeks (2008) executed centrifugal installation tests on a rotary jacked closed ended model pile. Within

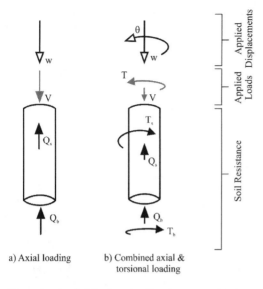

a) Axial loading b) Combined axial & torsional loading

Figure 1. Applied loads and soil resistances acting on a (a) jacked and (b) rotary jacked pile.

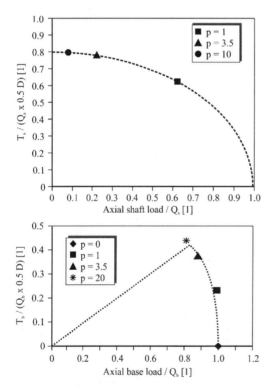

Figure 2. Failure envelopes of the shaft resistance (top) and base resistance (bottom) of a rotary jacked pile (acc. to Deeks (2008)).

these tests the axial shaft resistance Q_s and the axial base resistance Q_b as well as the shaft and base torques T_s and T_b were determined separately. Figure 2 depicts for various pitches the determined failure envelopes of the shaft resistance and the base resistance. As it can be

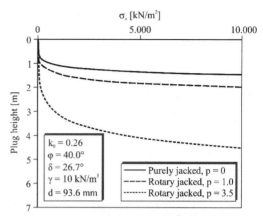

Figure 3. Relationship between plug height h_{plug} and vertical stress σ_z at the pile tip for different pitches.

seen even the base resistance is significantly reduced by the application of an additional rotation.

Regarding the vertical stress at the pile tip, Deeks (2008) suggested a modification of the parameter A for Equation 1 according to Equation 3. Thereby an additional rotation causes a decline of the resulting mean principal stress, so that the vertical stress is minimized as a function of the applied pitch.

$$A = \frac{4k_0 \tan \delta}{\left(1 + p^2\right)^{0.5}} \quad (3)$$

Based on that modification, Figure 3 elucidates the theoretical vertical stress at the pile tip for a purely jacked pile as well as for different pitches. Thereafter a plugging of the pile should be interfered by the application of a rotation.

3 NEW TESTING FACILITY

3.1 General structure

The new testing facility consists of a large sand container, a computer controlled loading frame and a high accuracy data sampling system. The sand container has an inner diameter of 2.50 m and a total height of 3.27 m (cf. Figure 4). A layer of filter gravel is placed at the bottom of the container having a thickness of 0.40 m. A geotextile is used to prevent erosion between gravel and sand. The height of the sand is 2.40 m and the water level exceeds the sand's surface by 0.32 m. The sand container is connected to a hydraulic system, which allows water to be pumped through the soil from the bottom to the top of the sand container. A drainage pipe installed in the filter gravel provides a uniform distribution of water pressure at the bottom of the sand container. An outflow at the top of the sand container gathers the surplus water, which runs in an overflow vessel providing the reservoir for the variable speed pump. To reach fully saturated sand state, de-aired water is used in the testing facility.

Table 1. Model piles and dimensions.

Model pile no.		Unit	1	2	3
Length	L	[mm]	3300	3300	3300
Outer diameter	D	[mm]	101.6	51.0	20.0
Inner diameter	d	[mm]	93.6	45.2	14.2
Wall thickness	t	[mm]	4.0	2.9	2.9

The loading frame is mounted on the sand container as shown in Figure 4. The actuator is installed on the loading frame and can be moved over the entire diameter of the sand container. A computer controls the actuator and allows the application of precise loading conditions.

For the considered problem, of course scaling effects have to be expected. In order to identify scaling effects and scaling laws, a "model family" of open-ended piles with outer diameters of 20, 51 and 101.6 mm will be investigated. The dimensions of the model piles are summarized in Table 1.

3.2 Sand parameters and preparation

The container of the testing facility contains 20 tons of medium grained, well graded silica sand. The sand was chosen to have a very low uniformity coefficient preventing the separation of particular grain fractions especially during the test preparation procedure. Sieve analyses have been performed with several samples. The envelope of the determined grain size distributions is depicted in Figure 5 and further properties are summarized in Table 2.

Before each test, the soil has to be prepared at a specific relative density under reproducible conditions. This aim is reached by first loosening the sand with an upward directed hydraulic gradient, followed by a compaction procedure. The loosening process is started with a hydraulic gradient of 0.9 which is then incrementally increased up to 1.1 in steps of 0.05. Each stage is kept for at least 15 minutes or until there is no more elevation of the sand surface visible. Finally, hydraulic failure of the sand normally occurs and the loosening process is stopped. Afterwards, the soil is compacted by a stiffened internal vibrator, which is pushed into the soil and then extracted with a constant velocity of 1.0 m/min. Every second point of a regular grid with dimensions of 25 cm is compacted leading to a staggered triangular shape. This process is supported by an upward directed hydraulic gradient of 0.8 to ensure that the internal vibrator reaches the final depth.

The relative density is controlled before each test series using a self calibrated laboratory cone penetration test (CPT) having a diameter of 30 mm. This procedure was proved by Foglia & Ibsen (2014) to be a reliable method of evaluating the relative density of non-cohesive soil in the laboratory. CPTs are usually conducted at five locations, which are equally distributed over the diameter of the sand container as shown in Figure 6. A constant displacement rate of

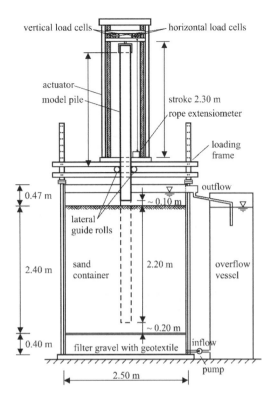

Figure 4. Model test facility.

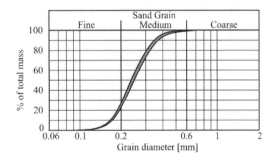

Figure 5. Grain size distribution of the sand used in the tests.

Table 2. Soil properties.

Description	Parameter	Unit	Value
Uniformity coefficient	C_u	[–]	1.5
Min. void ratio	e_{min}	[–]	0.553
Max. void ratio	e_{max}	[–]	0.873
Specific density	ρ_s	[t/m^3]	2.65

5 mm/s is used for the CPT. Figure 7 elucidates typical CPT results.

3.3 Evaluation of the relative density

To obtain a relationship between the CPT cone resistance and the relative density of the sand, the results

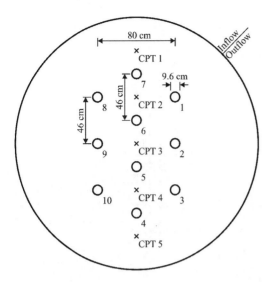

Figure 6. Position of CPT's and soil samples.

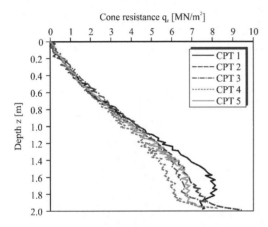

Figure 7. Typical CPT-results used for correlation of the relative density.

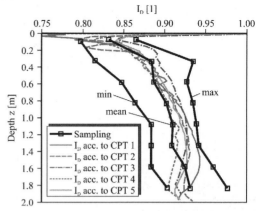

Figure 8. Relative density derived from CPTs and soil samples.

of a typical CPT test campaign were correlated with the actual relative densities. To obtain a reliable correlation, 80 soil samples were taken using sampling cylinders at eight equally spaced depths. Figure 6 shows a schematic top view of the container with the locations of the ten samples taken at each depth level as well as the location of the CPTs. The relative density of the soil samples was analysed with regard to homogeneity at each level and distribution over the extraction depth. The results are summarized in Figure 8, wherein the mean value as well as the lower and upper bound of the relative densities are depicted for the corresponding extraction depth. The scatter of relative densities must of course be considered in the evaluation of the pile installation tests. However, due to relatively small deviations, no severe falsification of test results is expected.

The measured relative densities were correlated with the CPT-profiles from Figure 6 using the equation of Jamiolkowski et al. (2003) by adjusting the parameters C_0, C_1 and C_2 to 0.043, 0.93 and 9.75, respectively ($R^2 = 0.87$). The use of the calibrated Equation 4, where q_c = measured cone tip resistance; σ_v' = vertical effective stress [kPa] and p_a = atmospheric pressure [kPa], allows the calculation of relative densities I_D for similar CPT-profiles measured in this testing facility.

$$I_D = \frac{1}{C_2} \ln\left(\frac{q_c/p_a}{C_0(\sigma_v'/p_a)^{C_1}}\right) \quad (4)$$

Figure 8 shows the results of the correlation for the exemplary CPT-profiles depicted in Figure 6. Obviously the correlation has minor accuracy for shallow depths up to approximately 200 mm, which is due to the low vertical stress at these depths. However, the estimation of the relative density is very good for greater depths.

3.4 Load application and measuring system

The testing facility is driven by two computer controlled servo motors, each equipped with a gearbox. It is capable of applying monotonic displacement-controlled vertical loads as well as a rotation to the mounted model pile. The vertical displacement rate of the actuator, realized by a spindle drive and actuated by a 400 W AC motor, is up to 5 mm/s within a stroke range of 2.3 m (cf. Figure 5). The maximum vertical force is limited to 50 kN. The rotational velocity of the model pile can be varied from 0.1 to 180°/s and is powered by the second 5.5 kW AC motor allowing a maximum torque of 1250 Nm. Furthermore the testing facility is equipped with a precise measuring and data acquisition system. For measuring the vertical displacement while performing CPTs, pile installations or pile load tests, there is a rope extensiometer which is able to measure vertical movements in the entire 2.3 m stroke range of the actuator. The axial load acting on a model pile or the CPT rod system is measured by two load cells installed directly beneath the moving part of the actuator. The load cells operate up to 50 kN vertical load, either compressive or tensile, which is essential for controlling the load being applied by the actuator while testing. The torque, resulting from the rotation of the pile during the installation process, is measured by two horizontal load cells. They are placed between the moving part of the actuator and the pile bracket system, having a defined lever arm. This allows measuring the force needed to hold the pile bracket system in place. In this way the torque acting on the pile is calculated.

3.5 Testing procedure

Before conducting model tests, the soil is prepared at uniform relative density via loosening and compaction, which is verified by laboratory CPT. Due to the large size of the sand container, multiple tests can be conducted without any impact between the individual tests after once preparing the soil. To ensure there is no mutual influence of the pile tests, a minimum distance depending on the pile diameter (minimum 5 D) is kept between the pile test positions as well as the container boundary. After positioning of the loading device, the chosen model pile is fixed to the bracket system on the actuator and the lateral guide rolls at the bottom steel plate of the actuator are attached to the pile. Thereafter the actuator is slowly moved downward until the model pile is directly above the sand's surface. Subsequently, the installation process with optional rotation and the data acquisition are started. After reaching the final depth, the installation is stopped and the plug height within the pile is measured with a plump bob through a prepared borehole at the pile head. Finally compression or tension tests can be carried out on the installed model pile.

3.6 Test campaign

A comprehensive testing campaign with varying installation parameters (penetration rates and rotational speeds) is planned to study the influence of rotary jacking on the needed installation forces as well as the bearing behaviour of piles. For reference purpose, also tests on solely jacked and additionally on driven piles will be conducted. To address scaling issues, all tests will be performed with the three different model piles of distinct dimensions (cf. Table 1). In this way, accounting for scale effects and thus an extrapolation of the results and a prediction of the behaviour of piles with greater dimensions will be possible.

4 RESULTS WITH PILES D = 101.6 MM

The entire testing program has not been conducted yet, so there will be further publications focusing on the results. This article presents measurements of four preliminary installation tests conducted with model pile no. 1. Two of the tests have been purely jacked installations with a penetration rate of 3 mm/s, while the other

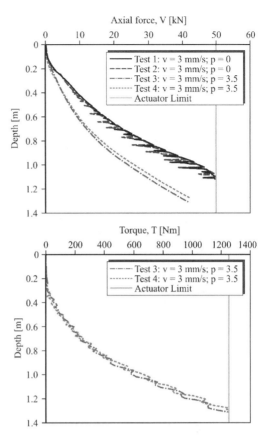

Figure 9. Results: axial force (top) and torque (bottom) for rotary jacked pile installations of model pile no. 1.

two have been rotary jacked with the same penetration rate, but with a pitch of p = 3.5. The installations have been stopped after reaching the axial force or rather torque limit of the actuator. The relative density for all tests was determined in the range of 0.85 to 0.95 with an average value of 0.94 for the region of interest from 0.2 m to 1.2 m depth.

The results of the four tests are depicted in Figure 9. The charts present the axial installation forces (top) as well as the torques in case of rotary jacking (bottom) with increasing penetration depth. Furthermore the installation depths, measured plug heights and resulting plug length ratios (PLR) are given in Table 3. Thereby the PLR describes the ratio of the soil plug height and the installed pile length.

As it can be seen from Figure 9 (top), the purely jacked piles reached a penetration depth of approximately 1.10 m for the maximum axial force of 50 kN. For the rotary jacked piles the required vertical installation force was reduced to 32 kN for the same depth. However, due to the actuator torque limit only an embedded pile length of approximately 1.30 m was reached within the tests for the rotary jacked piles. The axial force at this depth was approximately 42 kN.

With respect to the depth of 1.1 m, the ratio of the required force under simultaneous application of

Table 3. Plug measurements of the four pile tests.

Test no.	Plug height [m]	Installation depth [m]	PLR [%]
1	0.38	1.13	33.63
2	0.36	1.10	32.73
3	0.27	1.31	20.61
4	0.32	1.29	24.81

a torque and the required force for pure jacking was $32\,kN/50\,kN = 0.64$. For a depth of 0.6 m, the respective ratio was 0.56. These results might be used in the assessment of calculation methods regarding the effect of a rotation on the required jacking force.

In case of the jacked piles an irregular measurement curve for the required axial force can be seen from Figure 9 (top). This effect is caused by the irregular soil plug movement inside the pile. The soil plug at the pile tip suddenly sliced further inside the pile for a certain critical stress level which led to higher stress state inside the soil plug. Hence, the critical stress level increased with the embedded pile length.

Regarding the PLR it can be noticed from Table 3 that an additional rotation leads to an increase in the vertical stress. Hence, the PLR for the rotary jacked piles decreased approximately about 11% in comparison to the purely jacked piles. This is in contrast to the theoretical solution according to Deeks (2008) (cf. Figure 3) and shows the need for further research regarding the installation of rotary jacked piles.

5 CONCLUSIONS

The installation of piles via rotary jacking provides an alternative technique where dynamic excitation and noise emission can be avoided. Thereby, an additional rotation should lead to reduction of the required vertical force in combination with a reduction of soil plugging potential inside the pile. A new test facility for the execution of rotary jacked model pile tests was introduced. Therewith the sand preparation and evaluation was discussed in detail. According to the first performed model tests, it could be proved that an additional rotation leads to a reduction of the required vertical force. However, for the investigated boundary conditions the transfer of the required vertical force in favour of a torque was not as effective as predicted by the theoretical solution. In addition the plugging of the soil inside the pile is favoured in case of an application of an additional rotation, which is in contrast to published theory.

ACKNOWLEDGEMENT

The presented study was supported by the Federal Ministry for Economic Affairs and Energy. The authors sincerely acknowledge the support.

REFERENCES

Deeks, A.D. 2008. An investigation into the strength and stiffness of jacked piles in sand. *PhD Thesis, University of Cambridge*

Foglia A. & Ibsen L.B. 2014. Laboratory experiments of bucket foundations under cyclic loading. *DCE Technical Report No. 177. Aalborg*

Jamiolkowski M., Lo Presti D.C.F., Manassero M. 2003. Evaluation of relative density and shear strength of sands from CPT and DMT. *Geotechnical Special Publication* 119: 201–238

Janssen, H.A. 1895. Versuche über Getreidedruck in Silozellen, *Zeitschrift des Vereines deutscher Ingenieure* 39, 1045–1049

White, D.J., Deeks, A.D., Ishihara, Y. 2010. Novel Piling: Axial and Rotary Jacking. *Keynote paper, 11th Int. Conference of the Deep Foundation Institute, London*

Model tests on soil displacement effects for differently shaped piles

A.A. Ganiyu
Department of Civil Engineering and Quantity Surveying, Military Technological College, Muscat, Oman

A.S.A. Rashid & M.H. Osman
Faculty of Civil Engineering, Universiti Teknologi Malaysia, Skudai, Johor, Malaysia

W.O. Ajagbe
Department of Civil Engineering, Faculty of Technology, University of Ibadan, Ibadan, Nigeria

ABSTRACT: Transparent synthetic soil synthetic soil surrogates which permit real-time visualisation of soil continuum during testing is a novel development for geotechnical physical model tests. This paper presents model tests on pile penetration effects of differently shaped model piles subjected to axial loads in transparent synthetic soil model. Model piles, made of mortar of square, hexagonal, octagonal and circular shapes were utilised for the research. The transparent soil was made from fumed silica powder and pore fluid containing Paraffin and Technical White Oil. Soil displacement patterns were captured non-intrusively using close range photogrammetry while Particle Image Velocimetry (PIV) was employed to analyse the images. The analysed results revealed that the displacements of soil beneath the square pile aligned perfectly vertical with the edge of the pile, while it inclined with the vertical for piles of other shapes; this angle of inclination θ also varies for the variously shaped piles. This result, which depicts varying displacement patterns of soils beneath the differently shaped piles and marks distinguishable features for each shape of pile is significant. It could be used as the basis for the evolution of design charts and protocols based on shape of piles.

1 INTRODUCTION

Transparent synthetic soil surrogates which permit real-time visualisation of soil continuum during testing is a novel development to geotechnical physical model tests (Ganiyu et al., 2016, Kong et al., 2018, Xiao et al., 2017). Transparent soil is produced from synthetic aggregates and pore fluids with matching refractive indices, thus permitting complete penetration of light (Iskander et al., 2015, Kong et al., 2017). Transparent soil is an effective tool for modelling soil-structure interaction mechanisms; and its incorporation with Particle Image Velocimetry (PIV) provides an innovative technique for the visualization of failure mechanisms and the quantification of soil–structure interaction problems in geotechnical physical models non-intrusively (Omidvar and Iskander, 2017, Xiao et al., 2016, Yin et al., 2017).

PIV is an image-processing technology that computes fields of incremental displacement by comparing two successive images through a precise identification of several minute parts in image space (pixel) which are later converted to object space (in millimetres) (Take, 2015, Sui and Zheng, 2017, Rashid et al., 2017). GeoPIV is a MATLAB based PIV module developed specifically for geotechnical applications. In GeoPIV, image processing algorithms are written to apply the PIV principle to the images of soil (Stanier et al., 2015, Chen et al., 2016, Rashid et al., 2014). Internal displacements and deformations in a transparent soil slurry are measured by relating images of speckles that are generated both before and after soil deformation. This is obtained by employing GeoPIV combined with advance photogrammetry and the use of laser light source to optically slice the soil (Ganiyu, 2016, Qi et al., 2016).

The study of soil movement beneath a pile during installation is highly essential. The repositioning of soil beneath a pile subjected to dynamic forces may impact underground services such as tunnels and pipelines, adjacent foundations or archaeological remains (Hird et al., 2011). Furthermore, the design of pile foundation is mostly a function of the applied load and the resisting capacity of pile. The resistance of the pile is influenced by the state and properties of soils within the critical zone immediately surrounding the pile (Burland, 2012, Tomlinson and Woodward, 2014).

Precast concrete piles are prefabricated, displacement piles of different solid cross sections. However, little is known about the effect of the different shapes of piles when they interact with the adjoining soils. Many design equations and charts in geotechnical engineering were developed based on failure planes measured from physical model tests (Iskander and Liu,

Figure 1. Piles cast with pipes at top.

2010). The goal of this paper is to advance the understanding of soil-pile interaction for differently shaped piles through the investigation of displacement of soil underneath the piles using transparent synthetic soil and PIV.

2 MATERIAL AND TESTING

2.1 Model piles

The model piles were cast with mortar in the laboratory. The cement was Ordinary Portland Cement (OPC) while the fine aggregate was naturally occurring river sand; a mix ratio of 1:3 for cement: sand; and a water: cement (w/c) ratio of 0.45 was used to make the mortar. Four model piles of circular, square, hexagonal and octagonal shapes with overall cross sectional area of 452.4 mm^2 and length 100 mm were cast. The circular pile is 24 mm in diameter, while the length of side of the square, hexagonal and octagonal piles are 21.27, 13.2 and 9.68 mm respectively. During casting 8 mm diameter circular plastic pipes were inserted to a depth of 10 mm at the top centre of each pile, Figure 1 shows the pile cast with pipes at top centre.

The castings were demoulded after 24 hours and cured in water for 28 days. The average value of the compressive strength of the mortar cubes cast along with the model piles was 38.0 MPa. The inserted pipes were removed after the curing process; and the removal of these pipes created a hole of 8 mm diameter and 10 mm deep at the top centre of each of the piles. The holes later served as the point of insertion for the connector during penetration tests. The piles have a smooth surface finish and the surfaces were later sprayed with black Aerosol spray paint to ensure that the piles have the required black background necessary for laser light during penetration tests (Sills et al., 2017).

2.2 Transparent synthetic soil

Transparent synthetic soil was made from Fumed Silica powder (HDK–N20) as the aggregate component, while Technical White Oil ISO 15 (Grade A) and Paraffin oil (P1000) form the pore fluid. A ratio of 23% Paraffin and 77% technical white oil was adopted for the pore fluids (Stanier et al., 2014); while the aggregate was 5% of the total mass of the transparent soil. This mix gave the best quality of transparency after a series of trial mixes. Specifically, 850 g of technical white oil was mixed with 254 g of paraffin oil to form the pore fluid. Also, 0.012 g of Timiron Ultraluster MP-111 powder was added to the pore fluid to provide the contrast (texture) (Hird and Stanier, 2010). The mixture was then added to 58.2 g of fumed silica and an intense mixing was carried out using the whisk until a completely homogeneous mix was formed. The mass of Timiron represents 0.02% of the aggregate mass (Stanier et al., 2012).

The slurry formed was yet to be transparent as it contained a significant amount of air. Due to this, a vacuuming process was performed to de-air the slurry. The slurry was poured into a cylindrical chamber of diameter 90 mm and height 260 mm, made from Perspex and connected to a vacuum pump. The vacuum pump was switched on and the de-airing process took place for six hours (Iskander et al., 2002). Thereafter, testing cylinder containing the transparent soil was coupled to the consolidation frame. The arrangement was left for 24 hours and it ensures consolidation under self-weight of the transparent soil. Successive increasing load of 3.125, 6.25, 12.5, 25, 50 and 100 kPa was added and each load was maintained for a period of 24 hours (Hakhamaneshi and Black, 2016). After the peak load of 100 kPa, the loading was reduced to 50 kPa and maintained for another 24 hours. This gives an overconsolidation ratio of 2. The height of the sample was recorded before the consolidation began, prior to adding successive loadings and at the end of the consolidation test. Figure 2 shows the full arrangement of the consolidation process.

2.3 Penetration tests

After the consolidation was achieved, the set-up was decoupled again and the target markers were carefully gummed to the surface of the testing cylinder to serve as control points for the imminent penetration test. Target markers were made as stationary control points because they were needed in the image to permit a distinction between the soil and camera movements (Kelly and Black, 2012). Two columns of the target markers were spaced at 20 mm centre to centre on vertical axis, and the columns spaced at 50 mm centres.

The testing cylinder was placed directly underneath the driving unit. A driving unit that generates a constant rate of penetration of the model pile into the soil was provided for the loading test. The rate of drive unit penetration was 50 mm/minute as obtained in the calibration test. The LVDT was connected to the top, while the load cell was mounted at the lower end connected

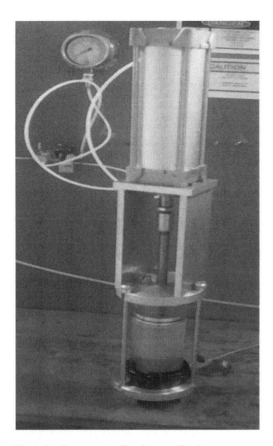

Figure 2. Transparent soil under consolidation.

Figure 3. Set-up for the pile penetration test.

to a flat tipped adaptor. Meanwhile, the adaptor has been rigidly joined with the black painted model pile via the hole at its top, with the help of gap sealant. The driving unit was carefully lowered until the base of the model pile was at the surface of the transparent soil.

Afterwards, a green diode-pumped solid-state (DPSS) laser apparatus was connected to power source and placed on the left side of the testing chamber. The laser head was adjusted until its illuminated vertical section aligned with the centreline of the pile and focused on the area directly beneath the pile. The interaction between the transparent soil and the laser light produced a distinctive laser speckle pattern. A Nikon D5100 digital camera was mounted on its tripod base and placed in front of the testing chamber; 300 mm away, and carefully set to focus the test arrangement, particularly the interface of the model pile and the transparent soil. A spirit level was used to confirm the correctness of both the vertical and horizontal alignments of the camera, and a remote control was connected to the camera. Figure 3 shows the set-up for the penetration test.

The actual penetration test was carried out in a 'dark room' situation (Qi et al., 2016, Black and Tatari, 2015). The driving unit was released to facilitate the downward movement of the pile with a simultaneous activation of the camera. The images were captured in a continuous shooting mode at an average rate of 0.8 frame per second. The values of the loads and displacements of the pile during the penetration test were recorded by the logging system. The test was terminated when the penetration of the pile reached 20 mm. The procedure was repeated for the remaining three piles of other shapes. Figure 4 shows the progressive downward movement of pile with example images taken at different times during a penetration test.

2.4 PIV analysis

The images from the penetration tests were downloaded from the camera to the computer. A total of 100 images were selected for analysis for each test; the images adequately cover the penetration process from the beginning to a convenient point proximate to the termination of each tests. GeoPIV8, a Matlab based PIV module developed by (White and Take, 2002) was employed for the analysis of the images.

3 RESULTS AND DISCUSSION

3.1 Consolidation properties of transparent soil

Figure 5 shows the time-settlement graph for the consolidation for 0 to 100 kPa for the square pile, and this is typical for all the four tests as the values closely matched one another. The steep slope between the first

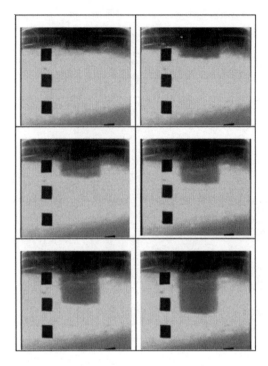

Figure 4. Progressive downward movement of pile during the penetration test.

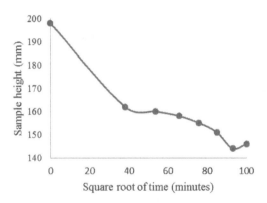

Figure 5. Consolidation plot for transparent soil.

two points is as a result of the consolidation under self-weight of the freshly prepared transparent soil. There is rapid release of pore fluid during this period; which is the first 24 hours after the soil was prepared, due to agglomeration of the aggregate components.

The coefficient of consolidation c_v was derived by using the Taylor's method for the loading range between 3.125–100 kPa. The coefficient of consolidation, c_v obtained was 0.965 m²/year. This value of c_v falls within (0.9 ± 0.2) reported by (Lehane and Gill, 2004) and (0.8–1.2) reported by (McKelvey et al., 2004) and (Sivakumar et al., 2007). The coefficient of compressibility, m_v and the hydraulic conductivity, k were obtained as 1.15 m²/MN and 2.93×10^{-9} m/s

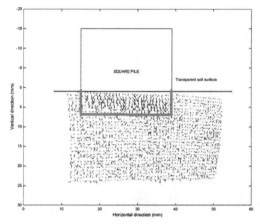

Figure 6. Displacement beneath square pile.

respectively. (McKelvey et al., 2004) reported similar coefficients of compressibility, m_v values while the hydraulic conductivity, k value is also within the range of values reported by (Lehane and Gill, 2004) and (Song et al., 2009). Thus, the transparent soil prepared and employed for this research can be categorized as a normally consolidated alluvial clay.

3.2 Pile penetration effects in transparent soil

PIV analysis was done and terminated at the vector plots stage to show the soil movements beneath the piles; the stage shows the displacements of soils at vertical directions underneath the differently shaped piles. According to previous researches on anchors (Ilamparuthi et al., 2002, Ilamparuthi and Muthukrishnaiah, 1999), and laterally loaded piles (Liu et al., 2010), it was assumed that the failure plane from soil displacement is similar to that of the shear strain field. The failure plane was delineated beneath the piles, and it was depicted by joining the points which had clear-cut vertical displacements of half the maximum vertical displacements beneath the piles. Figures 6–9 show the soil displacements underneath the square, hexagonal, octagonal and circular piles, respectively.

From the Figures, it was observed that the displacements of soil beneath the square pile align perfectly vertical with the edge of the pile, while it inclined with the vertical for the piles of other shapes. This angle of inclination θ also varies for the variously shaped piles; θ was determined to be 45° and 49.4° for the hexagonal pile; 33.7° and 41.6° for the octagonal pile; and 36.9° and 51.3° for the circular pile, respectively. It can be deduced that the behaviour manifested by the square pile is due to the flat plane of its sides/edges; while for the other shapes, it is based on the curvilinear plane of their respective sides/edges. Hence, it is concluded that the soil movements underneath an axially loaded penetrating pile is influenced by the shape of the pile, therefore, soil – pile interaction beneath differently shaped pile under axial load will be affected by the shape of the pile.

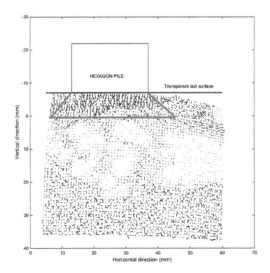

Figure 7. Displacement beneath hexagonal pile.

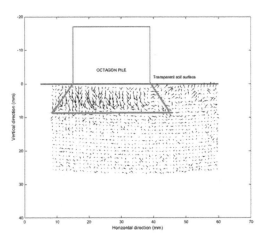

Figure 8. Displacement beneath octagonal pile.

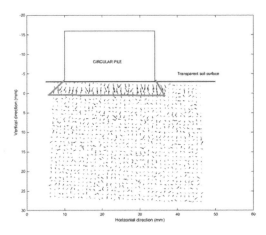

Figure 9. Displacement beneath circular pile.

4 CONCLUSIONS

Transparent soil with geotechnical properties replicating normally consolidated alluvial clay was prepared from synthetic materials. The soil enabled a real-time visualisation of the penetration of differently shaped piles installed in a unit gravity model test. In addition, the displacement effects of soil beneath the differently shaped piles were studied using close range photogrammetry while PIV was used to analyse the images taken during the penetration tests. The results revealed the differences in displacement patterns of soils beneath the differently shaped piles. It was therefore concluded that soil movements underneath an axially loaded penetrating pile can be associated with the shape of the pile. These marked distinguishable features for each shape of pile is a potential. In future, it could be the basis for the evolution of design charts and protocols based on the geometry of piles.

REFERENCES

Black, J. A. & Tatari, A. 2015. Transparent Soil to Model Thermal Processes: An Energy Pile Example. Geotechnical Testing Journal 38(5):1–13.

Burland, J. B. 2012. Behaviour of Single Piles Under Vertical Loads. In ICE Manual of Geotechnical Engineering, Institution of Civil Engineers, United Kingdom, pp. 231–246.

Chen, Z., Li, K., Omidvar, M. & Iskander, M. 2016. Guidelines for DIC in geotechnical engineering research. International Journal of Physical Modelling in Geotechnics: 1–20.

Ganiyu, A. A. 2016. Influence of Axially Loaded Shaped Pile on Geotechnical Capacity. PhD Thesis, Faculty of Civil Engineering, Universiti Teknologi Malaysia, Skudai, Johor, Malaysia.

Ganiyu, A. A., Rashid, A. S. A. & Osman, M. H. 2016. Utilisation of transparent synthetic soil surrogates in geotechnical physical models: A review. Journal of Rock Mechanics and Geotechnical Engineering 8(4):568–576.

Hakhamaneshi, M. & Black, J. A. 2016. Shear Strength of Transparent Gelita—The Effect of the Mixture Ratio, Displacement Rate, and Over-Consolidation Ratio. In Geo-Chicago 2016, pp. 443–452.

Hird, C. C., Ni, Q. & Guymer, I. 2011. Physical modelling of deformations around piling augers in clay. Géotechnique 61(11):993–999.

Hird, C. C. & Stanier, S. A. 2010. Modelling Helical Screw Piles in Clay Using a Transparent Soil. In Physical Modelling in Geotechnics, CRC Press, Zurich, Switzerland, pp. 769–774.

Ilamparuthi, K., Dickin, E. A. & Muthukrisnaiah, K. 2002. Experimental investigation of the uplift behaviour of circular plate anchors embedded in sand. Canadian Geotechnical Journal 39(3):648–664.

Ilamparuthi, K. & Muthukrishnaiah, K. 1999. Anchors in sand bed: delineation of rupture surface. Ocean Engineering 26(12):1249–1273.

Iskander, M., Bathurst, R. J. & Omidvar, M. 2015. Past, Present, and Future of Transparent Soils. Geotechnical Testing Journal 38(5):1–17.

Iskander, M. & Liu, J. Y. 2010. Spatial Deformation Measurement Using Transparent Soil. Geotechnical Testing Journal 33(4):314–321.

Iskander, M. G., Liu, J. Y. & Sadek, S. 2002. Transparent Amorphous Silica to Model Clay. Journal of Geotechnical and Geoenvironmental Engineering 128(3):262–273.

Kelly, P. & Black, J. A. 2012. Optimisation of Stone Column Design Using Transparent Soil and Particle Image Velocimetry (PIV). In ISSMGE – TC 211 International Symposium on Ground Improvement IS-GI, Brussels, Belgium, vol. III, pp. 443–452.

Kong, G., Li, H., Yang, G. & Cao, Z. 2018. Investigation on shear modulus and damping ratio of transparent soils with different pore fluids. Granular Matter 20(1).

Kong, G. Q., Li, H., Hu, Y. X., Yu, Y. X. & Xu, W. B. 2017. New Suitable Pore Fluid to Manufacture Transparent Soil. Geotechnical Testing Journal 40(4):20160163.

Lehane, B. M. & Gill, D. R. 2004. Displacement Field Induced by Penetrometer Installation in an Artificial Soil. International Journal of Physical Modelling in Geotechnics 4(1):25–36.

Liu, J., Liu, M. & Gao, H. 2010. Influence of Pile Geometry on Internal Sand Displacement around a Laterally Loaded Pile Using Transparent Soil. In GeoShangai International Conference ASCE, pp. 1–7.

Mckelvey, D., Sivakumar, V., Bell, A. & Graham, J. 2004. Modelling Vibrated Stone Columns in Soft Clay. Proceedings of the Institution of Civil Engineers-Geotechnical Engineering 157(3):137–149.

Omidvar, M. & Iskander, M. 2017. Soil Deformations during Finless Torpedo Installation. In Geotechnical Frontiers 2017, pp. 389–397.

Qi, C.-G., Iskander, M. & Omidvar, M. 2016. Soil Deformations During Casing Jacking and Extraction of Expanded-Shoe Piles, Using Model Tests. Geotechnical and Geological Engineering 35(2):809–826.

Rashid, A. S. A., Black, J. A., Mohamad, H. & Mohd Noor, N. 2014. Behavior of Weak Soils Reinforced with End-Bearing Soil-Cement Columns Formed by the Deep Mixing Method. Marine Georesources & Geotechnology 33(6):473–486.

Rashid, A. S. A., Kueh, A. B. H. & Mohamad, H. 2017. Behaviour of soft soil improved by floating soil–cement columns. International Journal of Physical Modelling in Geotechnics:1–22.

Sills, L.-a. K., Mumford, K. G. & Siemens, G. A. 2017. Quantification of Fluid Saturations in Transparent Porous Media. Vadose Zone Journal 16(2):0.

Sivakumar, V., Glynn, D., Black, J. & Mcneill, J. 2007. A Laboratory Model Study of the Performance Of Vibrated Stone Columns in Soft Clay. In 14th European Conference on Soil Mechanics and Geotechnical Engineering, Madrid, Spain, pp. 1–6.

Song, Z. H., Hu, Y. X., O'loughlin, C. & Randolph, M. F. 2009. Loss in Anchor Embedment During Plate Anchor Keying in Clay. Journal of Geotechnical and Geoenvironmental Engineering 135(10):1475–1485.

Stanier, S. A., Blaber, J., Take, W. A. & White, D. J. 2015. Improved Image-Based Deformation Measurement for Geotechnical Applications. Canadian Geotechnical Journal 53(5):727–739.

Stanier, S. A., Black, J. A. & Hird, C. C. 2012. Enhancing Accuracy and Precision of Transparent Synthetic Soil Modelling. International Journal of Physical Modelling in Geotechnics 12(4):162–175.

Stanier, S. A., Black, J. A. & Hird, C. C. 2014. Modelling Helical Screw Piles in Soft Clay and Design Implications. Proceedings of the Institution of Civil Engineers-Geotechnical Engineering 167(5):447–460.

Sui, W. & Zheng, G. 2017. An experimental investigation on slope stability under drawdown conditions using transparent soils. Bulletin of Engineering Geology and the Environment, pp. 1–9.

Take, W. A. 2015. Thirty-Sixth Canadian Geotechnical Colloquium: Advances in Visualization of Geotechnical Processes Through Digital Image Correlation. Canadian Geotechnical Journal 52(9):1199–1220.

Tomlinson, M. & Woodward, J. 2014. Pile Design and Construction Practice. 6th edn., CRC Press.

White, D. J. & Take, W. A. 2002. GeoPIV: Particle Image Velocimetry (PIV) Software for use in Geotechnical Testing, CUED/D-SOILS/TR322, University of Cambridge.

Xiao, Y., Sun, Y., Yin, F., Liu, H. & Xiang, J. 2017. Constitutive Modeling for Transparent Granular Soils. International Journal of Geomechanics 17(7):04016150.

Xiao, Y., Yin, F., Liu, H., Chu, J. & Zhang, W. 2016. Model Tests on Soil Movement during the Installation of Piles in Transparent Granular Soil. International Journal of Geomechanics, 17(4), pp. 06016027-1/7

Yin, F., Xiao, Y., Liu, H., Zhou, H. & Chu, J. 2017. Experimental Investigation on the Movement of Soil and Piles in Transparent Granular Soils. Geotechnical and Geological Engineering, pp. 1–9.

Comparison of seismic behaviour of pile foundations in two different soft clay profiles

Thejesh Kumar Garala & Gopal S.P. Madabhushi
Schofield Centre, Department of Engineering, University of Cambridge, UK

ABSTRACT: In this study, an attempt has been made to investigate seismic behaviour of friction pile foundations in two different soft clay profiles under the action of dynamic loading. A single pile and a 1 × 3 pile group (aluminium tubular model piles) were embedded into two soft clay profiles with different strengths and tested aboard centrifuge facility at the University of Cambridge. This paper discusses the preliminary results obtained from these two centrifuge experiments. A few important findings of this study are: (i) the pile foundation response changes significantly with a relatively small change in undrained shear strength of the clay, (ii) larger dynamic amplification of input motion was observed in the pile foundations during small intensity sinusoidal signals (SS) rather than during the strong intensity SS and (iii) the pile foundation continues to respond for the seismic motion though the clay fails in shear during the strong intensity SS.

1 INTRODUCTION

Soft soils are usually considered problematic because of high moisture content and low strength. They are found in almost all parts of the world, particularly coastal areas. Structures constructed over the soft soil are susceptible to large settlements, especially differential settlements, which is more troublesome. Pile foundations are found to be efficient in such soils for transferring heavy super structure loads to deep layers of stiffer soil or rock which can satisfy both bearing capacity and settlement requirements as per the standard code provisions. In earthquake prone areas, along with the static loads from super structure, these pile foundations are subjected to additional dynamic loads during earthquakes. A few studies have been conducted on amplification of bed rock acceleration in soft soils during earthquakes, earthquake induced bending moment in pile foundations and seismic behaviour of structures supported on pile foundations in soft clay (Idriss 1990, Cafe 1991, Banerjee 2009, Zhang et al. 2017). However, the seismic behaviour of pile foundations in soft clay is not fully understood, especially the influence of undrained shear strength of clay and magnitude of earthquakes on the behaviour of pile foundations during earthquakes. In this study, two centrifuge experiments (test 1 and test 2) are conducted to study the influence of undrained shear strength and intensity of earthquake on the dynamic behaviour of pile foundations. A single pile and a 1 × 3 pile group (see figure 1), made from aluminium tubes, are embedded into two soft clays with different undrained shear strength profiles and tested at 50 times the earth's gravity using the Turner beam centrifuge facility at the Schofield Centre, University of Cambridge. Also, the strength and stiffness of clay profiles are evaluated using the T-bar test (Lau 2015) and air hammer device (Ghosh & Madabhushi 2002), respectively. Though friction piles in very soft clays are not so common

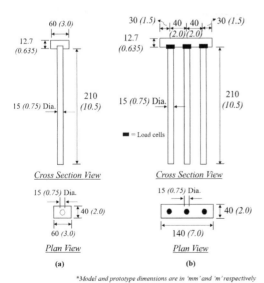

Figure 1. Dimensions of (a) single pile and (b) pile group used in the study.

in practice, these experiments are conducted as a part of series of centrifuge experiments which investigates dynamic behaviour of both friction piles and end bearing piles in soft clay. More details about this series of centrifuge experiments can be found in Garala (2016).

2 EXPERIMENTAL PROCEDURE

Centrifuge modelling is known for its field stress-strain simulation in scaled-down models of geotechnical structures. The stress-strain similarity in model and prototype is achieved by means of testing the models within the increased g-field of a large geotechnical centrifuge. Centrifuge modelling is always advantageous over field tests because of its convenient scaling, higher efficiency in data monitoring, easy repeatability, controlled conditions and less costs. The principle, mechanics, scaling laws, various types of centrifuge, advantages and disadvantages of centrifuge modelling are well covered in Schofield (1980) and Madabhushi (2014).

2.1 Centrifuge setup

The centrifuge experiments were conducted in Turner beam centrifuge at Schofield Centre, University of Cambridge. It is a 10 m diameter centrifuge which rotates about the central vertical axis with a working radius of 4.125 m. The complete design description can be found in Schofield (1980). Stored angular momentum (SAM) earthquake actuator developed by Madabhushi et al. (1998) is used in this study. Equivalent shear beam (ESB) box (Zeng & Schofield 1996) is used as a model container, which is a rectangular box (645 mm × 228 mm × 385 mm) made by stacking hollow aluminium sections with rubber layers in between the aluminium sections.

2.2 Experimental setup

2.2.1 Clay
Laboratory grade speswhite kaolin clay is widely used in different research studies at Schofield Centre for the reasons of homogeneity in mineralogy and ease of repeatability. The properties of speswhite kaolin clay used in this study can be found in Lau (2015).

2.2.2 Model pile
An aluminium tubular model pile of diameter 15 mm and thickness 1 mm was used in this study to fabricate a single pile and 1 × 3 pile group. The flexural stiffness of tested pile represents a 0.75 m diameter high strength concrete pile in prototype. The surface of the model pile is smooth and the pile head is covered with a pile cap in both single pile and pile group. Figure 1 shows the single pile and pile group along with the pile caps used in the study. In pile group, load cells are placed between the pile cap and piles to measure the load distribution among piles in the pile group during the earthquakes. The dimensions shown in figure 1 are in model scale and the values within the parenthesis represent the prototype dimensions (model and prototype dimensions are in mm and m respectively).

2.2.3 Model preparation
The speswhite kaolin clay powder was first mixed with de-aired water in 1:1.25 ratio under vacuum for two-three hours using the clay-mixer available at Schofield Centre. Prior to transferring the clay slurry into the ESB box, the inner surface of the box was coated with silicone grease to minimize friction against the clay. The box was filled with water to a depth of about 5–10 mm in order to prevent air from being entrapped within the slurry during its placement. Clay slurry was then transferred into the ESB box with utmost care. The top surface of clay was covered with a filter material and perforated top loading plate was placed over it. The clay slurry was allowed to consolidate on its self-weight for 24 hours. Though it is possible to consolidate the clay quickly at higher gravity levels using centrifuge, for operational reasons in the Schofield Centre, the clay samples are usually consolidated outside the centrifuge at normal gravity levels. In order to obtain more realistic clay profile (with some strength at surface and then increased strength with depth), a combination of consolidation under vertical stress and hydraulic consolidation by suction-induced seepage (HCSS) were used to consolidate the sample in both the tests. In HCSS, suction is applied at the bottom of clay bed and top surface is subjected to near atmospheric pressure. The working principle along with the validity of this method can be found in Robinson et al. (2003).

The ESB box filled with clay slurry was placed under a computer-controlled hydraulic press to consolidate under a vertical stress. Once the clay was consolidated for the applied vertical stress, the sample was then subjected to HCSS method by applying a suction pressure at the bottom of ESB box with certain increments. The vertical stress on top of the clay was constantly maintained even during the HCSS method of consolidation. A vertical stress of 175 kPa and suction pressure of −70 kPa was used to consolidate the clay slurry in test 1 and vertical stress of 40.5 kPa and suction pressure of −50 kPa was used to consolidate the clay slurry in test 2.

Once the clay was consolidated, sample was unloaded and taken out from the consolidation chamber. Following removal, the clay top surface was slightly trimmed to obtain levelled surface. Then, T-bar penetrometer tests were conducted at normal gravity (1g) to obtain the undrained shear strength of consolidated clay. The depths of the clay obtained after consolidation in tests 1 and 2 are 250 mm and 220 mm respectively (initial clay slurry depths are not same in both tests). The densities of the clay in tests 1 and 2 are calculated as $1704\,kg/m^3$ and $1623\,kg/m^3$ respectively. Piezo-electric accelerometers and micro-electro-mechanical systems (MEMS) accelerometers were used to measure the accelerations in clay and pile foundations respectively. A stand made out of

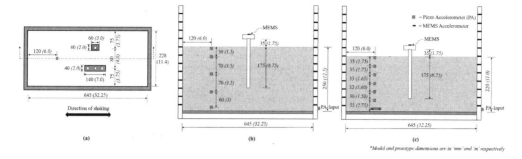

Figure 2. (a) Plan view of model in tests 1 and 2, (b) and (c) are cross-section of models in tests 1 and 2 respectively.

Poly-Tetra-Fluoro-Ethylene (PTFE) was used to hold the instruments in proper direction and exact location.

Piles were installed manually, approximately at a rate of 2–5 mm/sec in test 1 and 5–10 mm/sec in test 2 with intermittent pauses to check the verticality of pile installation. Single pile was installed first followed by the insertion of three piles of the group at same time in both the tests. Out of 210 mm length of model pile, 175 mm is embedded into the clay, keeping 35 mm above the clay surface in both the experiments even though the clay depths are different in tests 1 and 2. Figure 2 shows the plan view and cross sections of the model in both the tests.

Once all the instruments and pile foundations were placed successfully in the clay, the ESB box was moved and loaded in centrifuge testing platform. The model was swung up to 50 g and planned sinusoidal signals were fired using the SAM actuator.

3 RESULTS AND DISCUSSION

The preliminary results obtained from the two centrifuge experiments are discussed in this section. The results are presented in prototype units unless stated as model dimensions.

3.1 Strength and stiffness of clay

T-bar tests were performed at normal gravity levels on consolidated clay profiles. Figure 3 shows the undrained shear strength of clay evaluated from the T-bar in tests 1 and 2. Even though the T-bar tests are performed at 1 g, the depth units are shown in prototype scale for better understanding of the results discussed in following sections. Also, strength of the clay at 50 g may be slightly greater than the clay strength at 1 g. Increase in stroke length of the actuator in test 2 helped to assess strengths to greater depths compared to test 1. It can be seen from figure 3 that the strength of clay in test 1 is approximately four times the clay strength in test 2. The small hump nearly at 1.85 m in test 2 is because of disturbance to the driving actuator. Apart from that, it is clear from the figure that the undrained shear strength of clay is increasing with depth in both tests 1 and 2, indicating the success of HCSS consolidation.

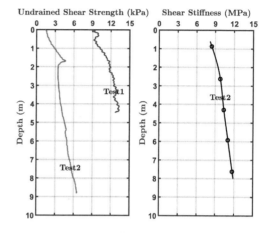

Figure 3. Undrained shear strength of clay in tests 1 and 2 (measured at 1g) and shear stiffness of clay at 50 g in test 2.

Further, air hammer tests were performed in test 2 at 50g before firing the earthquakes to calculate the shear stiffness of clay. Figure 3 shows the shear stiffness (G_o) variation with the depth of the clay. For calculating natural frequency, an average shear stiffness of 10 MPa is considered.

Natural frequency (f) of clay in test 2 can be computed from $f = \frac{1}{4D}\sqrt{\frac{G_0}{\rho}}$, where D is depth and ρ is density of clay, which gives 1.78 Hz as natural frequency. Considering clay in test 1 is relatively stiff and deeper than test 2, one can expect its natural frequency to be in the range of 2–3 Hz.

3.2 Pile foundation's response during small intensity sinusoidal signals

In both test 1 and test 2, a monofrequency sinusoidal-like signal of frequency 0.6 Hz (30 Hz at model scale) was fired to study the behaviour of piles during the small intensity sinusoidal signals. Figure 4 shows the input motion for test 1 (peak acceleration = 0.087 g) and its corresponding fast fourier transform (FFT) while figure 5 shows the input motion for test 2 (peak acceleration = 0.076 g) and its corresponding FFT. In addition to the driving frequency, the first five harmonics of the driving frequency which possess some

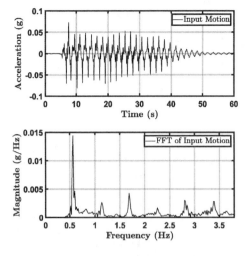

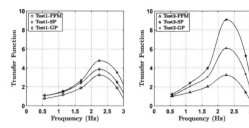

Figure 6. Transfer functions obtained from test 1 and test 2 during 0.6 Hz small intensity sinusoidal signal.

than individual pile in the pile group. Free field soil, single pile and pile group show peak value in transfer function at 2.25 Hz in both the models. The pile response shown in figure 6 is a result of both kinematic (surrounding clay) and inertial effects (pile cap weights).

Therefore, it can be concluded that a relatively small change in shear strength of clay (undrained shear strength of clay in test 1 is four times that of clay in test 2) can significantly influence the dynamic response of pile foundations during small intensity sinusoidal signals, although the free field clay behaviour itself remains more or less unchanged.

3.3 Pile foundation's response during strong intensity sinusoidal signals

If the shear stresses generated during the propagation of shear waves under strong sinusoidal signals are greater than the undrained shear strength of the clay, then the clay fails to propagate the waves to the upper layers. As a result, the free field motion can be less than the input motion. To study the behaviour of single pile and pile group under such conditions, the centrifuge models are subjected to a 1 Hz (50 Hz at model scale) monofrequency sinusoidal wave having a peak acceleration of 0.32 g and 0.24 g in tests 1 and 2 respectively.

Figure 7 shows the acceleration recorded along a vertical profile of piezo accelerometers in test 1 and attenuation of acceleration with respect to input motion can be observed at the surface. It is clear from the figure 7 that clay should have sheared somewhere in between 5.25 m and 8.75 m. Similarly, figure 8 shows the acceleration recorded along a vertical profile in test 2 and again, attenuation of acceleration with respect to input motion can be observed at the surface. The acceleration at the clay surface is almost half of the input motion. In this case, the clay should have sheared somewhere in between 6.75 m and 8.25 m.

Maximum shear stress generated because of acceleration signal propagation between 8.75 m and 5.25 m in test 1 can be computed as:

$$\tau_{max} = \rho \times a_{max} \times z \approx 18.72 \text{ kPa} \quad (1)$$

where ρ is density (1704 kg/m^3), a_{max} is maximum acceleration at a value of 0.32 g and z represents the depth of the region considered (3.5 m).

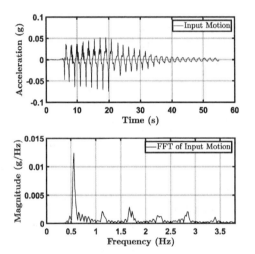

Figure 4. Input motion and corresponding FFT of 0.6 Hz small intensity sinusoidal signal in test 1.

Figure 5. Input motion and corresponding FFT of 0.6 Hz small intensity sinusoidal signal in test 2.

energy were considered for analysing clay, single pile and pile group behaviour. The FFT component of acceleration recorded at clay surface, top of single pile and top of pile group were normalized with the FFT component of input motion at all frequencies considered to obtain the transfer function. Figure 6 shows the transfer functions obtained in both tests 1 and 2. In figure 6, FFM stands for transfer function of free field motion, SP represents transfer function of single pile and GP indicates transfer function of pile group.

From figure 6, it can be observed that the clay response (i.e., FFM) did not change significantly with the reduction in strength of the clay. However, the single pile and pile group oscillate more in test 2 compared to the test 1 case. Also, single pile oscillates more than the pile group, which is to be expected considering the stiffness of single pile is smaller than that of the pile group and single pile is carrying more weight

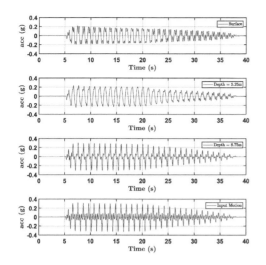

Figure 7. Acceleration recorded at different depths for 1 Hz strong intensity sinusoidal signal in test 1.

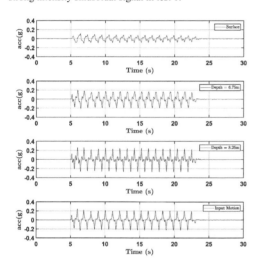

Figure 8. Acceleration recorded at different depths for 1 Hz strong intensity sinusoidal signal in test 2.

Though the undrained shear strength (c_u) at depths greater than 4.5 m is not available in test 1 (see figure 3), an approximate extrapolation indicates that c_u between 5.25 m and 8.75 m will be about 14–16 kPa. This means $\tau_{max} > c_u$, causing the clay to fail in shear and resulting in smaller acceleration at depth 5.25 m (see figure 7).

Similarly, average shear stress generated because of acceleration signal propagation between 8.25 m and 6.75 m in test 2 can be computed using (1) as:

$$\tau_{max} = 1623 \times (0.24 \times 9.81) \times 1.5 \approx 6.02 \text{ kPa}$$

It can be seen from figure 3 that the c_u is less than 6 kPa at depths less than 8 m for clay in test 2. This indicates that $\tau_{max} > c_u$ at depths between 8 m and 6.75 m resulting in shear failure of clay, followed by smaller acceleration at depth 6.75 m (see figure 8).

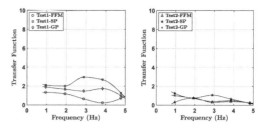

Figure 9. Transfer functions obtained from test 1 and test 2 during 1 Hz strong intensity sinusoidal signal.

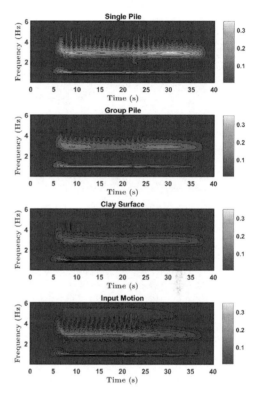

Figure 10. Continuous wavelet transforms from test 1 during 1 Hz strong intensity sinusoidal signal.

Figure 9 shows the transfer functions obtained for strong intensity sinusoidal signal in tests 1 and 2. Transfer function of value less than one for free field motion in both tests indicates that the magnitude of signal recorded at surface is less than the input motion. In test 1, the higher values of transfer function for pile foundations indicate that the piles continued to oscillate even though the clay failed in shear. Also, the magnitude of transfer functions obtained in figure 9 are less than those values shown in figure 6, which suggests that small intensity sinusoidal signal can attract larger dynamic amplification response in the pile foundations than the strong intensity sinusoidal signal, especially in very soft clays.

Figures 10 and 11 show the continuous wavelet transforms (CWTs) of input motion, signal recorded

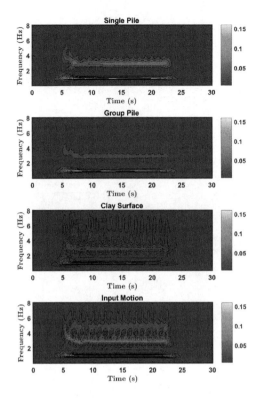

Figure 11. Continuous wavelet transforms from test 2 during 1 Hz strong intensity sinusoidal signal.

at clay surface, top of single pile and top of pile group in tests 1 and 2 respectively.

It can be seen from the figures 10 and 11 that though the clay failed to transmit the actual input motion to the surface, both single pile and pile group continued to respond. The reason for this is that the pile foundations are embedded to a depth of 8.75 m and in both the tests clay failed in shear at depths shallower than 8.75 m. Therefore, the bottom tip of pile foundations received the unattenuated motion from the bedrock and responded accordingly. Again, single pile oscillates more than the pile group because of its lower stiffness compared to the pile group and single pile is carrying more weight than individual pile in the pile group. It would be interesting to see the pile foundation's response if the clay fails in shear at depth greater than the bottom tip of the pile foundation.

4 CONCLUSIONS

The behaviour of soft clay and friction pile foundations during dynamic loading were investigated in this study. A single pile and a 1×3 pile group were tested in two different soft clay profiles by conducting two dynamic centrifuge experiments. The clay in test 1 has an undrained shear strength approximately four times that of the clay in test 2. During small intensity sinusoidal signals, it was observed that the single pile and pile group behaviour changes significantly with the change in undrained shear strength of the clay though the free-field clay response remains more or less unchanged. Further, these small intensity sinusoidal signals can generate greater dynamic amplification response in the pile foundations compared to strong intensity sinusoidal signals, especially in very soft clays. Acceleration attenuation at the surface was observed in both the tests because of local shear failure during the strong intensity sinusoidal signals. Further, pile foundations continued to oscillate even though the clay fails in shear during the strong intensity sinusoidal signal as their embedded length is greater than the depths at which the clay failed in shear.

ACKNOWLEDGEMENT

The authors would like to thank Prof. Luca de Sanctis, Parthenope University of Naples for his collaboration in this project. The first author would also like to thank the Commonwealth Scholarship Commission (CSC) and Cambridge Trust for their doctoral scholarship.

REFERENCES

Banerjee, S. (2009). *Centrifuge and numerical modelling of soft clay-pile-raft foundations subjected to seismic shaking*. PhD Thesis, National University of Singapore, Singapore.

Cafe, P. F. M. (1991). Dynamic response of a pile-supported bridge on soft soil. Master's thesis, Univ. of California, Davis.

Garala, T. K. (2016). Dynamic behaviour of pile foundations in soft clay. First Year Report, Department of Engineering, University of Cambridge.

Ghosh, B. & S. P. G. Madabhushi (2002). An efficient tool for measuring shear wave velocity in the centrifuge. In *Proc. of the international conf. on physical modelling in geotechnics*, St. Johns, Newfoundland: Balkema, pp. 119–124.

Idriss, I. M. (1990). Response of soft soil sites during earthquakes. In *Proc. H. Bolton Seed Memorial Symposium*, University of California, Berkeley, pp. 273–289.

Lau, B. H. (2015). *Cyclic behaviour of monopile foundations for offshore wind turbines in clay*. PhD Thesis, University of Cambridge.

Madabhushi, G. (2014). *Centrifuge modelling for civil engineers*. CRC Press, Taylor & Francis Group.

Madabhushi, S. P. G., A. N. Schofield, & S. Lesley (1998). A new stored angular momentum (SAM) based earthquake actuator. In *Proceedings of the International Conference Centrifuge 98*, Tokyo, Japan: Balkema, pp. 111–116.

Robinson, R. G., T. S. Tan, & F. H. Lee (2003). A comparative study of suction-induced seepage consolidation versus centrifuge consolidation. *Geotechnical Testing Journal* Vol. 26(1), 92–101.

Schofield, A. N. (1980). Cambridge geotechnical centrifuge operations. *Geotechnique* 30(3), 227–268.

Zeng, X. & A. N. Schofield (1996). Design and performance of an equivalent-shear-beam container for earthquake centrifuge modelling. *Geotechnique* 46(1), 83–102.

Zhang, L., S. H. Goh, & J. Yi (2017). A centrifuge study of the seismic response of pileraft systems embedded in soft clay. *Gotechnique* 67(6), 479–490.

Issues with centrifuge modelling of energy piles in soft clays

I. Ghaaowd & J. McCartney
University of California San Diego, La Jolla, California, USA

X. Huang
Hohai University, Gulou Qu, Nanjing Shi, Jiangsu Sheng, China

F. Saboya & S. Tibana
State University of Norte Fluminense, Rio de Janeiro, Brazil

ABSTRACT: This study focuses on an experimental evaluation of issues encountered when using a geotechnical centrifuge to evaluate the heat transfer, pore water pressure generation, volume change, and subsequent soil-structure interaction phenomena associated with energy pile operation in normally-consolidated Kaolinite clay. Although the scaled zones of influence of heat transfer and volume change due to thermal consolidation in centrifuge models may be wider than those expected in a field-scale prototype, centrifuge modelling results can still be used to validate numerical simulations performed at model scale. Further, topics such as the impact of temperature changes on the ultimate capacity of energy piles are well-suited to centrifuge testing due to the difficulty of performing such tests in the field. The energy pile investigated in this study is an aluminium cylinder whose temperature is controlled using an embedded electrical resistance heater. This paper focuses on an evaluation of the transient response of a soft clay layer during heating and cooling of the energy pile, an evaluation of the change in undrained shear strength profiles measured using a T-bar, and a comparison of the pullout capacity of the heated energy pile with that of an unheated energy pile.

1 INTRODUCTION

When a heat source such as an energy pile is embedded in a saturated clay layer, temperature changes will result in changes in pore water pressure as the rate of heating is typically faster than the rate of drainage (Ghaaowd et al. 2016). Although all soils will initially expand during undrained heating, normally consolidated soils typically exhibit permanent contraction while overconsolidated soils exhibit expansion after sustained heating (Vega & McCartney 2015). Subsequent cooling typically leads to elastic contraction. Thermally-induced changes in pore water pressure and volume may affect the soil-structure interaction of energy piles, as well as the ultimate capacity of the energy pile.

There have been several previous studies that have emphasized the importance of considering the effects of temperature changes on the behaviour of soils and associated effects on embedded structural elements like energy piles. Booker & Savvidou (1985) proposed analytical solution to predict the temperature change and thermal excess pore water pressure as a function of distance from a cylindrical heat source in a thermo-elastic soil layer. Savvidou (1988) found that centrifuge modelling of thermo-mechanical processes in soils requires consideration of thermal convection effects, and that lower permeability soils are less affected by convection. Ghaaowd et al. (2016) developed and validated a model to predict the change in pore water pressure of saturated clay as a function of depth in a soil layer during undrained heating based on the model of Campanella & Mitchell (1968). Their model indicates that the initial mean effective plays a critical role in the magnitude of pore water pressure change during undrained heating, which implies that centrifuge testing has an advantage over 1g testing in that the effective stresses are similar to those in a prototype soil layer. Takai et al. (2016) found that thermally-induced pore water pressures during undrained heating cannot be used to predict the volume change observed during drained heating, although they may be related. Maddocks & Savvidou (1984) studied the heat transfer from a hot cylinder installed in soft clay in a centrifuge test and observed changes in thermal excess pore water pressure and time-dependent contraction as a function of distance from the cylinder. Stewart & McCartney (2013) studied the soil-structure interaction in an end-bearing energy pile in unsaturated silt and found that the heat transfer does not scale, resulting in a greater zone of influence of temperature changes. They found that the centrifuge still can provide useful information on soil-structure interaction phenomena and recommended use of model-scale results in validation of numerical models. Ng et al. (2014) used a centrifuge to study the effects of cyclic heating of an aluminium energy pile in soft clay and observed permanent settlement that accumulated with each cycle. McCartney & Murphy (2017) evaluated the long-term response of a full-scale energy pile in claystone during building heat pump operation and observed a transient change in the thermal axial strains over a period of five years that was attributed to dragdown that may arise from the effect of temperature.

This study investigates the behaviour of a normally-consolidated Kaolinite clay layer during heating and cooling of an embedded aluminium energy pile. The Actidyn C61-3 centrifuge at the University of California San Diego was used to perform two tests on energy piles in separate soil layers, the first in which an energy pile was installed using jacked-in procedures, heated to a constant temperature, cooled, then pulled out at a constant rate after reaching equilibrium, and the second where the energy pile was installed similarly then pulled out at a constant rate after reaching equilibrium without heating. The results for the heated energy pile include the variations in temperature and pore water pressure generation in the clay layer, the undrained shear strength profile, as well as the pull-out capacity for heated and unheated energy piles are also compared. For brevity, the soil-structure interaction evaluation involving assessment of thermal axial strain profiles and potential dragdown effects are not presented in this paper.

Table 1. Properties of the Kaolinite clay and initial conditions of the specimen used in this study.

Liquid limit	47%
Plastic limit	28%
Plasticity index	19
Specific gravity	2.6

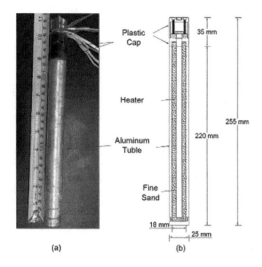

Figure 1. (a) Assembled energy pile, (b) Pile cross section with model-scale dimensions.

2 MATERIALS

2.1 Kaolinite clay

The soil used in the two experiments was commercially-available Kaolinite clay from M&M Clays Inc. of McIntyre, Georgia whose geotechnical properties of the clay are summarized in Table 1. The clay classifies as CL according to the Unified Soil Classification Scheme. An isotropic compression test indicates that the slopes of the normal compression line (λ) and the recompression line (κ) for the clay are 0.100 and 0.016, respectively. The clay specimens were formed from a slurry to reach initially normally consolidated conditions as will be described below.

2.2 Scale-model energy pile

The scale-model energy pile is a 25 mm-diameter, 255 mm-long, split-shell aluminium cylinder having a wall thickness of 3.3 mm, as shown in Figure 1. At 50 g, the model pile corresponds to a prototype pile having a diameter of 1.25 m and a length of 12.75 m. The insides of the cylinder halves were instrumented with five temperature-compensated strain gages and thermocouples at the locations shown in Figure 2. The halves are held together by screw-on top and bottom caps. An internal electrical resistance heater running the length of the energy pile is connected to the top cap, and heat is conducted from the heater to the outside of the pile through a sand fill. The top cap of the

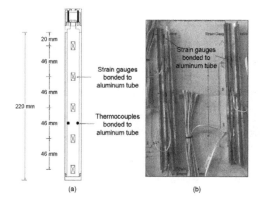

Figure 2. (a) Schematic showing the model-scale locations of strain gauges, (b) Picture of strain gauges bonded to the aluminium tube.

pile was fabricated from plastic to minimize heating of the water ponded on the soil surface, and all wiring passes through the top cap.

3 EXPERIMENTAL SETUP

A schematic of the container used in this study to evaluate the behaviour of an energy pile in normally consolidated clay is shown in Figure 3. The aluminium container consists of a base plate, a cylindrical tank, and an upper reaction plate. The base and

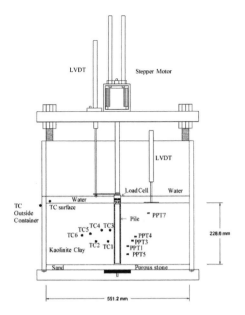

Figure 3. Cross-section of the test setup showing the energy pile, T-bar, thermocouples, and pore water pressure sensors.

Table 2. Sensor locations with respect to the container centre and soil surface

Transducer	Depth (mm)	Radius (mm)	Sensor	Depth (mm)	Radius (mm)
PPT1	160	36	TC1	142	36
PPT3	140	85	TC2	142	83
PPT4	125	53	TC3	108	28
PPT5	190	66	TC4	102	60
PPT7	38	116	TC6	121	136

infer the undrained shear strength of disturbed clay as a function of depth (Stewart & Randolph 1994).

4 EXPERIMENTAL PROCEDURES

The kaolinite clay in powder form was mixed with water in a vacuum mixer to reach a slurry having a gravimetric water content of 130%. The slurry was carefully poured into the container to avoid air inclusions. The clay specimen was drained from the bottom via a sand layer having a thickness of 20 mm thickness and from the top via a filter paper and a 50 mm-thick porous stone placed on the top of the slurry. After 24 hours of self-weight consolidation, dead-weights applying vertical stresses of 2.4, 6.3, 10.2 kPa were added in 24-hour increments. The surcharge was then increased to 23.6 kPa using a hydraulic piston and maintained for another 24 hours. Then the container was placed inside the centrifuge basket for in-flight self-weight consolidation at 50 g. This procedure was found to produce a normally-consolidated clay layer with an overconsolidated portion at its top, and is similar to the approach proposed by Cinicioglu et al. (2006).

During in-flight self-weight consolidation at 50 g, the excess pore-water pressures were monitored using five pore pressure transducers. An example of the excess pore water pressure measured by pore water pressure transducer (PPT4) is plotted against the square root of time in Figure 4. This data permits definition of the value of t_{90} using the root time method (Taylor 1948). The measured pore water pressure at the end of the consolidation was compared with the theoretical hydrostatic pore water pressure profile in in Figure 5. These results confirm that primary consolidation was reached achieved throughout the clay layer. The gravimetric water content of the clay layer with the unheated pile measured at the end of testing ranged from 38.5 to 52%, while the void ratio ranged from 1.0 to 1.35. These values can be assumed to correspond to the initial conditions for the energy pile before the heating and cooling cycle.

After consolidation, the pile was inserted into the clay at a model-scale velocity of 0.1 mm/s until the tip reached the sand layer. The pile was inserted so that the aluminium portion was embedded in the clay and the plastic top cap was above the clay surface, with a prototype-scale embedment of approximately 11 m. The pile was loaded to a prototype-scale compressive

reaction plates of the container have dimensions of 0.62 m × 0.62 m × 0.05 m. The cylindrical tank has an inside diameter of 0.55 m, a wall thickness of 16 mm, and a height of 0.47 m, and was connected to the base plate via four threaded rods atop an "O"-ring seal. The top reaction plate was connected to the top ends of the same threaded rods. The reaction plate supports stepper motors for loading the energy pile and T-bar, as well as displacement sensors.

The energy piles were installed in flight through sedimented clay layers having a thickness of 228 mm so that they would bear on a dense sand layer to simulate end-bearing boundary conditions. A stepper motor was used to insert the pile at a constant-displacement rate to simulate jacked-in conditions. The motor movement is controlled using a LabVIEW program capable of applying both constant-displacement conditions as well as constant axial head loads (i.e., to simulate the dead weight of an overlying structure). The applied axial head load was measured using a load cell. Six thermocouples and five miniature pore water pressure sensors were inserted through the container side wall into the clay layer at the locations shown in Figure 3 and Table 2.

A T-bar with a diameter of 14 mm and length of 57 mm was used to measure the undrained shear strength profiles of the clay layers containing heated and unheated energy piles. The T-bar was designed to permit model-scale penetrations up to 230 mm and was driven by a second stepper motor at a model-scale velocity of 0.2 mm/s to ensure undrained conditions during insertion and extraction. Insertion of the T-bar into the clay layer can be used to infer the undrained shear strength of the intact clay layer as a function of depth, while extraction of the T-bar can be used to

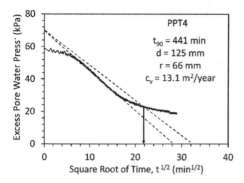

Figure 4. Excess pore water pressure measured by PPT4 versus the square root of time.

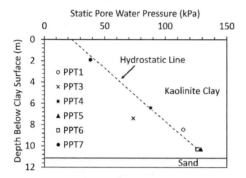

Figure 5. Pore water pressure profile after consolidation with depth in prototype-scale.

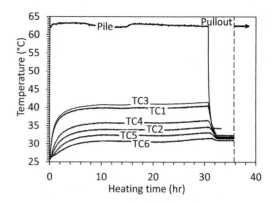

Figure 6. Temperature versus heating time at different radii.

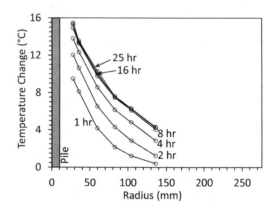

Figure 7. Temperature changes at different heating times.

load of 1445 kN, which is expected to mobilize most of the side shear resistance and mobilization of the end bearing. A control loop was used to maintain the load under constant-load conditions (i.e., free displacement conditions). When the excess pore water pressure generated due to the pile insertion was dissipated, the temperature of the heated energy pile was increased using the Watlow heat controller until reaching a maximum temperature at the pile wall of 63.4°C. Heating continued for 31 hours during centrifugation to permit stabilization of both temperature and pore water pressure. As the strains scale 1:1 in the centrifuge, the thermal expansion strains are expected to be the same in the model and prototype. After heating, the pile was cooled and was then pulled out at the same insertion speed used for installation. T-bar tests were executed in each test at a radius of 100 mm from the heated and unheated energy piles, which is far enough away to be undisturbed from the pile insertion and extraction but close enough to be influenced by temperature.

5 EXPERIMENTAL RESULTS

5.1 Temperature response

Time series of the pile and clay temperatures at different locations are shown in Figure 6, and the temperature changes versus radius at various times are shown in Figure 7.

The pile temperature increased very quickly due to the rapid response of the electrical resistance heater, but the temperature of the clay increased gradually and stabilized after 10 hours in model scale (1041 days in prototype scale). Although the heater in the pile increased in temperature from 26.0 to 63.4°C ($\Delta T = 37.4°C$), the soil only reached a maximum temperature change of 15.9°C. The temperature decreased away from the energy pile with a negligible change at the container boundary of 276 mm. When the electrical resistance heater was turned off, the clay temperature rapidly decreased to a temperature of 30°C. This is greater than the initial temperature, perhaps because the heat pulse was still moving through the clay layer and because the temperature of the centrifuge chamber increased to nearly 29°C in during the 50 hours of centrifuge heating.

5.2 Thermal excess pore water pressure

Thermal excess pore water pressures were generated in the clay immediately after the increase in temperature, as shown in Figure 8. This increase in pore water pressure occurs due to partially undrained conditions and the difference between the coefficients of thermal

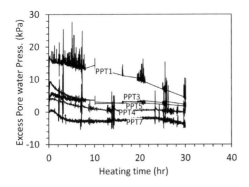

Figure 8. Thermal excess pore water pressure versus heating time at different radii and different depths.

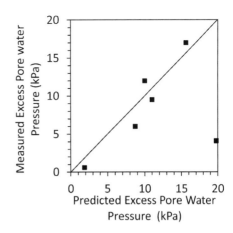

Figure 9. Measured excess pore water pressure versus predicted pore water pressure from the model of Ghaaowd et al. (2015).

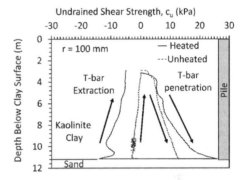

Figure 10. Undrained shear strength measured by T-bar (positive values for insertion and negative values for extraction).

expansion of the clay solid skeleton and the water (e.g., Ghaaowd et al. 2015). The pore water pressures immediately started to dissipate, with thermal consolidation finishing after approximately 21 hours. Although the data acquisition unit used for the pore water pressure sensors in this test had some noise issues that have since been rectified, the dissipation trends in the data are clear. PPT7 shows an initial increase in pore water pressure followed by a dissipation that goes below zero, which could be due to issues with the sensor or slight changes in the surface water level due evaporation induced by the elevated soil temperature. The maximum measured thermal excess pore water pressure was 18.2 kPa at PPT1, due to the combination of the depth of the sensor and the greater temperature change at this location. The thermally induced excess pore water pressure did not fully dissipate at this location by the end of heating, although they likely returned to static values after cooling.

The change in pore water pressure at the effective stress associated with the prototype depth of PPT1 predicted by the model of Ghaaowd et al. 2015) is 17 kPa, so the measured value is reasonable. Similar checks were performed for the other sensors, and the predicted and measured maximum excess pore water pressure are shown in Figure 9. In general, most of the predicted pore water pressures were above those measured by the sensors. Differences in the predicted and measured values may be due to changes in vertical position of the PPTs due to consolidation, transient changes in temperature at the different locations, and the different distances from the drainage boundaries that may result in partial drainage of the thermal excess pore water pressures.

6 T-BAR MEASUREMENTS AND INTERPRETATION

The T-bar measurements in the clay layers with heated and unheated energy piles provide further evidence as to the effects of heating and cooling on the temperature effects on the behaviour of normally consolidated clay layers. The correlations of Stewart & Randolph (1991) were used to interpret the undrained shear strength profiles from the T-bar measurements. The undrained shear strength profiles for the clay layer at a model-scale distance of 100 mm from the unheated and heated energy piles are shown in Figure 10. The undrained shear strength profiles indicate that heating of the energy pile leads to an increase in undrained shear strength of the clay layer during both insertion and extraction. Greater differences in undrained shear strength were observed deeper in the clay layer, which may be because the thermally induced excess pore water pressures were greater at these depths. The initial T-bar position is slightly below the clay surface to maximize the stroke of the T-bar test, so the undrained shear strength near the surface of the clay layer may not be well characterized.

7 PULLOUT CAPACITY

The net pullout pile capacities for the heated and unheated energy piles in clay layers having similar initial conditions are shown in Figure 11, with respect to

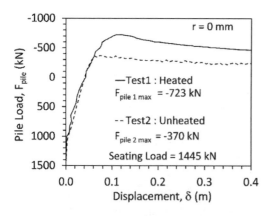

Figure 11. Pullout-displacement curves for heated and unheated energy piles in prototype scale.

the initial compressive seating load of 1445 kN. The heated energy pile had a pullout capacity of -723 kN that is 20% greater than that of the unheated energy pile that had a pullout capacity of -370 kN. This significant increase in capacity confirms that the drained heating processes leads to substantial improvements in the undrained shear strength of the soil surrounding the heated energy pile. The slopes of the pullout curves are similar, even though the aluminium energy pile experiences a significant axial strain in the direction opposite to loading during heating. This will be further investigated in the future using the axial strain measurements.

As a check on the measured capacity of the unheated energy pile, the pullout capacity was predicted using the equation of Dennis and Olson (1983):

$$Q_s = c_u \, \alpha \, A_s \, F_c \, F_L \qquad (1)$$

where the c_u is the mean undrained shear strength measured using the T-bar, α is an empirical side shear reduction capacity factor (equal to 1 for the soft clay evaluated in this study), A_s is the pile surface area, F_c is a sampling correction factor (equal to 1.0), and FL is a length correction factor (equal to 1.0). It is assumed that the pile has no negative end bearing. The predicted pullout capacity of the unheated energy pile is -362 kN, while the measured pullout capacity of the unheated pile is -370 kN.

8 CONCLUSIONS

Heated and unheated scale-model aluminium energy piles were tested to evaluate the impacts of temperature on the a normally consolidated clay layer and on the corresponding pullout capacity of the energy pile. The following specific conclusions can be drawn:

- The temperature in the clay stabilized after 10 hours of heating, which led to relatively undrained heating of the clay layer. The pore water pressure gradually dissipated after this time.
- The temperature change and the vertical effective stress are directly proportional to the thermal excess pore water pressure generation.
- The measured undrained shear strength profiles at a model-scale radius of 100 mm both the tests on heated and not heated energy piles were interpreted using from T-bar tests. The undrained shear strength of the soil surrounding the heated-cooled energy pile was higher than the soil surrounding the unheated pile, with greater increases in undrained shear strength deeper in the soil layer.
- The pullout capacity of the heated energy pile was approximately 20% greater than that for the unheated energy pile, with a similar slope for the pullout curve.

REFERENCES

Booker, J.R. & Savvidou, C. 1985. Consolidation around a point heat source. *International Journal for Numerical and Analytical Methods in Geomechanics*, (9): 173-184.

Campanella, R.G. & Mitchell, J.K. 1968. Influence of temperature variations on soil behavior. *Journal of the Soil Mechanics and Foundation Division*. 94(SM3): 709–734.

Cinicioglu, O., Znidarčić, D. & Ko, H.-Y. 2006. "A new centrifugal testing method: Descending gravity test." *Geotechnical Testing Journal*. 29(5): 1-10.

Dennis, N.D., & Olson, R. E. 1983. Axial capacity of steel pipe piles in clay. *Proc., Conf. on Geotech. Pract. In Offshore Eng.*, S. G. Wright, ed., ASCE, New York: 370–388.

Ghaaowd, I., Takai, A., Katsumi, T. & McCartney, J.S. 2016. Pore water pressure prediction for undrained heating of soils. *Environmental Geotechnics*. 4(2): 70-78.

McCartney, J.S. & Murphy, K.D. 2017. Investigation of potential dragdown/uplift effects on energy piles. *Geomechanics for Energy and the Environment*. 10(June): 21-28.

Maddocks, D.V. & Savvidou, C. 1984. The effect of the heat Transfer from a hot penetrator installed in the ocean bed. *Proc. Symp. on the Application of Centrifuge Modeling to Geotechnical Design. W.H. Craig. Manchester U.K*

Ng, C.W.W., Gunawan, A. & Laloui, L. 2014. Centrifuge modelling of energy piles subjected to heating and cooling cycles in clay. *Géotechnique Letters*, 4(4): 310-316.

Savvidou, C. 1988. Centrifuge modelling of heat transfer in soil. Proc. *Centrifuge 88*. Balkema: 583–591.

Stewart, D. & Randolph, M. 1994. T-Bar penetration testing in soft clay. *J. Geotech. Eng. Div.* 120(12): 2230-2235.

Stewart, M.A. & McCartney, J.S. 2013. Centrifuge modeling of soil-structure interaction in energy foundations. *Journal of Geotechnical and Geoenvironmental Eng*. 04013044.

Takai, A., Ghaaowd, I., Katsumi, T., & McCartney, J.S. 2016. Impact of drainage conditions on the thermal volume change of soft clay. *Proc. GeoChicago 2016*: 32-41.

Taylor, D.W. 1948. *Fundamentals of Soil Mechanics*. John Wiley and Sons, New York.

Vega, A. & McCartney, J.S. 2015. "Cyclic heating effects on thermal volume change of silt." *Environmental Geotechnics*. 2(5): 257-268.

Centrifuge modelling of non-displacement piles on a thin bearing layer overlying a clay layer

Y. Horii & T. Nagao
Taisei Corporation, Yokohama, Japan

ABSTRACT: Pile foundations constructed on thin bearing layers have been widely used in Japanese estuary regions. Although several methods for predicting the bearing capacity of the pile tip have been proposed, the relationship between the load and settlement of the pile tip, which is necessary for predicting how the foundation settle, remains under-investigated. It depends on the nonlinear characteristics and confining stress of the soil. In this study, centrifuge modeling is used to investigate the bearing behavior of non-displacement piles. Seven tests of vertical loading were conducted on model piles (fabricated at 1/50 scale with a diameter, D, of 20 mm) on a bearing layer with an effective thickness ratio, H/D, which varied from zero to eight. Based on the experimental results, a method previously proposed by RTRI (1987) for predicting the bearing capacity is validated, and a method is proposed for predicting the settlement relationship using a hyperbolic curve.

1 INTRODUCTION

Thin bearing layers widely underlie Japanese estuary regions near large cities, while some of those are thick (more than 50 meters deep). By designing piles on thin bearing layers, the pile lengths can be shortened, thereby reducing the cost of the project, construction time, and environmental impact. Piles on thin bearing layers have been applied to various civil structures and buildings in Japan (see Fig. 1).

Several studies have demonstrated that the thinner the bearing layer is, the lower its bearing capacity will be. For shallow foundations, Yamaguchi (1963) proposed the projected-area method which assumed a truncated sand block with a spread angle of 27° (a 1:0.5 vertical:horizontal ratio) below the foundation penetrating into a lower clay layer and had a base resistance equal to the bearing capacity. Hanna & Meyerhof (1980) proposed a punching shear method. More recently, alternative prediction methods which include a failure mechanism that combines the concept of the projected-area and punching-shear methods have been developed by Okamura et al. (1998), Teh (2007), and Lee et al. (2013). However, the applicability of these methods to deep foundations has not been thoroughly investigated.

For deep foundations, Meyerhof (1978) conducted tests with model piles, and based on the results, proposed a relationship between the effective thickness ratio of the bearing layer H/D (wherein H denotes the thickness of the bearing (sand) layer below the pile, and D denotes the pile diameter) overlying the weak soil and the load on the pile tip. Matsui & Oda (1991) conducted a load test on a bored cast-in-place pile on a thin bearing layer and further proposed a method to predict the bearing capacity using finite-element (FE) analyses. Railway Technology Research Institute (1987) presented a modified version of the projected-area method for deep foundations using a spread angle, θ, of 17° (a 1:0.3 vertical:horizontal ratio) and a bearing capacity in the lower layer of six times the shear strength of soil, s_u (see Fig. 2). Although this approach has been used since 1987 for the design of railway structures and buildings in Japan, its applicability has not been explicitly verified.

The load–settlement relationship for a non-displacement pile tip on a thin bearing layer is necessary for predicting the settlement of the foundation.

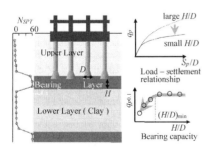

Figure 1. Schematic of piles on a thin bearing layer overlying a clay layer.

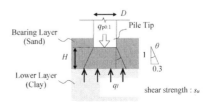

Figure 2. Modified projected-area method (RTRI 1987).

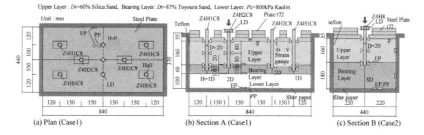

Figure 3. Schematic of the centrifuge package.

However, this characteristic has rarely been investigated except the FE-analysis conducted by Matsui & Oda (1991). Considering piles on a thick bearing layer, Yasufuku et al. (2001) proposed a method to predict the load–settlement curve using a hyperbolic function. Horii et al. (2016) recently conducted centrifuge modeling to investigate the relationship.

However, the bearing behavior of non-displacement pile tips on a thin bearing layer has been insufficiently investigated. This behavior should be experimentally tested with bearing layers of varying thicknesses, relative densities and confining stresses using models with soil materials and stress conditions similar to those of the prototype. Without this information, the applicability of various methods for predicting the bearing capacity remains unclear, and methods to predict the load–settlement relationship have not been developed.

Using centrifuge modeling, this paper investigates the bearing capacity and the load–settlement relationship of a non-displacement pile tip (approximately 40 meters depth in real scale) on a thin bearing layer overlying over-consolidated clay. The modeling used Toyoura sand and Kaolin clay for a bearing layer and a lower layer, respectively. Using test results, the prediction method for the bearing capacity and the load–settlement curve are developed.

2 CENTRIFUGE MODELING

2.1 Centrifuge modeling

The bearing behavior of a pile tip on a thin bearing layer overlying clay depends on H/D as well as the strength and stiffness values of the bearing and the lower layer. These have strong nonlinear characteristics and stress-dependency. Therefore, geotechnical centrifuge modeling is conducted in which the stress conditions and soil materials are similar to those of the prototype. Further, the load–settlement relationship for the pile tip is influenced by the pile shape and the installation method; in this study, a round non-displacement pile are modeled as they correspond to the bored pile, which is widely used in Japanese urban areas. The centrifuge with a effective radius of 2.65 m has been described in detail by Nagura et al. (1994). In this study, a centrifugal acceleration of 50g (where g denotes the gravitational acceleration) were applied to a 1/50-scale model.

Table 1. Test cases and conditions (all dimensions in mm).

Case	Pile*	H/D	Bearing Layer D_r %	Bearing Layer σ'_v** kPa	Lower Layer P_{cmax} kPa	Lower Layer OCR	Lower Layer s_u kPa
1	Z4H0C8	0	87	326	800	2.18	124
	Z4H1C8	1					
	Z4H2C8	2					
2	Z4H8	8	87	326	–	–	–

* ZaHbCc a indicates the upper level of bearing layer (approximately 40 m), b indicates the H/D ratio, and c indicates the P_{cmax} value, ** value for the upper level

2.2 Centrifuge package and model preparation

Table 1 lists the parameters of the test cases considered here. Figures 3a and 3b show schematics of the centrifuge package, which represent piles on a thin bearing layer (Case 1). A rigid box with a length of 840 mm, width of 440 mm, and height of 375 mm was used. Teflon sheets were attached to the sidewalls to reduce the wall friction. The main soil constituents (saturated and layered from top to bottom) were medium dense sand, dense sand and over-consolidated clay, here called the upper layer (160 mm thick), the bearing layer (60 mm thick) and the lower layer (approximately 100 mm thick), respectively. As shown in the schematic diagrams, 20-mm-diameter model piles were placed on the bearing layer with H/D ratio varying from zero to two. As shown in Figure 3c, in Case 2, the piles on a thick bearing layer, which has a H/D ratio of eight and which is thicker than that in Case 1, and there is no lower layer.

The basic properties of the soil materials and the effective overburden pressure of the layers are summarized in Tables 2 and 3, respectively. 72-mm-thick steel plates were placed on the ground surface to increase the effective overburden pressure during loading tests on the ground surface to 245 kPa and that on the upper portion of the bearing layer to 326 kPa; these values correspond to the soil stresses approximately 40 meters below the ground surface. The model piles were made of smooth-surface aluminum pipes with an outer diameter of 20 mm and a thickness of 2 mm (see Figs 3, 4 and Table 1). The pile tips were closed using a steel end plate. The axial forces were measured by strain gauges attached at different heights inside the pipe.

Table 2. Soil properties.

(a) Sand

Soil	D_{50} mm	G_s g/cm³	e_{max}	e_{min}
Toyoura*	0.19	2.654	0.940	0.595
Silica	0.24	2.643	1.204	0.679

* shear resistance angles of another Toyoura sand ($D_r = 85\%$) vary from 46.9 to 23.5 degrees as confining pressure increase from 25 to 6400 kPa.

(b) Clay

Soil	D_{50} mm	G_s g/cm³	w_L %	w_P %	I_p
Kaolin Clay	<0.005	2.551	78.1	31.5	47

Table 3. Effective overburden pressure on the soil.

Soil	γ kN/m³	γ' kN/m³	σ'_v kPa
Upper layer	18.3	8.5	245
Bearing layer	19.7	9.9	326
Lower layer	17.0	7.2	367

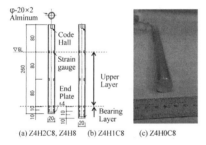

Figure 4. Model piles (dimensions in mm).

The model of Case 1 involving piles on a thin bearing layer was prepared as follows. First, the bottom drainage layer was prepared by placing filter papers into the box followed by a de-aired clay slurry with 100% water content. Preconsolidation was conducted at 1g using steel plates and hydraulic jacks with a maximum pressure of 800 kPa. The piles (Z4H0C8) were set and air-pluviated dry Toyoura sand with a relative density, D_r, of 87% was poured to reach the level of following pile tips. Second, once the piles (Z4H1C8 and Z4H2C8) were set, the remaining the Toyoura sand as mentioned, and silica sands with a D_r of 61% were poured, and the soil was subsequently saturated with water down to 20 mm below the ground surface. Finally, steel plates were set on the soil surface. The centrifuge package of Case 1 is shown as Figure 5.

The model of Case 2 with piles (Z4H8) on a thick bearing layer was prepared by the same procedure except for the preparation of the clay layer.

Figure 5. Centrifuge package (Case 2).

2.3 Test procedure

The model was mounted on the centrifuge with an electric vertical loading jack, a load cell, and laser displacement meters. Here, the jack was not fixed to the pile heads before loading, so piles freely settled with the soil consolidating in the flight. Prior to the loading tests, a centrifugal acceleration of 50g was applied to reduce the settlement of the soil surface and the negative friction on the pile. Once the increase in the settlement of the soil surface had almost converged 6 hours after applying 50g in Case 1, a load was applied to the top of the pile at a displacement rate of 0.3 mm/s. During the test, the load, the settlement of the top of the pile, the earth pressure, the pore water pressure, and the strain of the pile were monitored. After the loading test, the soil deformation was observed in a vertical section after draining the water from the box.

To characterize the lower (clay) layer of Case 1, the water content, w_n, and the shear strength, s_u, was investigated in soil samples collected from the model following the tests after the centrifuge flight. The average w_n value was 47%. Further, a linear relationship between s_u and the mean effective confining stress, σ'_c, was obtained from results of unconfined compression tests and confined triaxial compression tests after consolidated (i.e. CU, wherein $\sigma_c = 360$ kPa):

$$s_u = 0.289\sigma_c' + 36.5 \qquad (1)$$

Substituting $\sigma'_c = 306$ kPa ($\sigma'_v = 367$ kPa, as shown in Table 3, assumed $K_0 = 0.75$ because the clay was overconsolidated) into Equation 1, a s_u value of 124 kPa was calculated (see Table 1) for the lower layer in Case 1.

3 TEST RESULTS

3.1 Load-settlement relationship, soil deformation

Figure 6 shows the relationship between the load per unit area of the pile tip, q_p, obtained from the measured strain of the pile at a part 10 mm above the tip and the S_p/D ratio where S_p represents the settlement of the pile tip and was calculated by subtracting the pile shrinkage based on the measured strain from the average settlement of the top of the pile. The measurements were initialized when the vertical loading started. The results from a single test of Z4H1C8 ($H/D=1$) are shown, and two tests with each of Z4H0C8, Z4H2C8, Z4H8 ($H/D=$ 0, 2 and 8, respectively) under the same condition are included to confirm the repeatability. With $H/D=8$ (Z4H8), q_p increased with S_p/D and continued to increase in the range of $S_p/D > 0.5$. With $H/D \leq 2$ (for Z4H0C8, Z4H1C8 and Z4H2C8), q_p increased with S_p/D and reached a maximum value when $S_p/D > 0.2$, thereafter it maintains approximately the same load until $S_p/D = 0.5$ indicating an ultimate state.

The soil deformation observed after the centrifuge in each type of piles is shown in Figure 7. Here, colored soil layers were placed at a 10-mm intervals prior to the test for better visualization. The deformation of the soil was the most significant in the region below each pile tip. With $H/D=8$, a cone formed under the pile tip appeared to penetrate and expand the peripheral soil. With $H/D=2$, the deformation was spread to a deeper region than in the case of $H/D=8$. For $1 \leq H/D \leq 2$, a hollow was formed below each pile on the upper surface of the lower layer and was distributed within approximately 23 of the spread angle before the tests relative to the edge of the pile. The width of the hollow increased as H/D increased for $0 \leq H/D \leq 2$. This indicates that a punching failure occurred with a truncated sand block under the pile in the case of piles on thin bearing layers. Furthermore, the shape of the cone formed with $H/D=2$ was less defined than that formed with $H/D=8$, and it was nearly undetectable when $H/D=1$. It is concluded that as H/D decreased, the punching failure occurred before the cone was fully formed.

3.2 Bearing capacity of pile tip and stiffness in the initial stage

Figure 8 shows that q_{pmax} (the maximum of q_p) increased with H/D; the solid curve represents the analytical result of solving Equation 2 with a θ angle of 23° (1:0.43 vertical:horizontal ratio or $\tan^{-1}(0.43)$) and a bearing capacity, q_l, of six times s_u for the lower layer, as classified in the projected-area method proposed by Yamaguchi (1963). Equation 2 was found to closely represent the test results; thus, can be used to predict the bearing capacity.

$$q_{p\max} = \left(1 + 2\frac{H}{D}\tan\theta\right)^2 q_l \tag{2}$$

Figure 9 shows that as H/D increased, the bearing capacity, $q_{p0.1}$, of the pile tip when $S_p/D = 0.1$ also increased. In practical design application, $q_{p0.1}$ is used as the ultimate bearing capacity. The solid curve represents the analytical result obtained using the modified projected-area method (RTRI 1987, see Fig. 2) described in Equation 2 using a θ angle of 17° for (1:0.3 vertical:horizontal ratio or $\tan^{-1}(0.3)$) and six times s_u of q_l. The results indicate that this method can be used to adequately predict $q_{p0.1}$ from test results.

Figure 10 shows that as H/D increased, the stiffness in the initial stage, $K_{p0.005}$, also increased and reached a maximum at a sufficiently large H/D. Here, $K_{p0.005}$ denotes the secant modulus of the $q_p - S_p/D$ curve corresponding to $S_p/D = 0.005$ ($S_p = 0.1$ mm). The + markers in Figure 10 represent the analytical results obtained using Equation 3 (see Fig. 11):

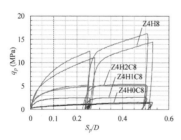

Figure 6. $q_p - S_p/D$ relationship.

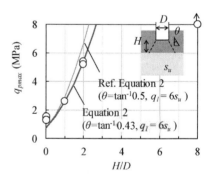

Figure 8. q_{pmax}–H/D relationship.

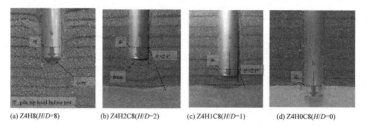

Figure 7. Observation of the soil near the pile tip after the centrifuge.

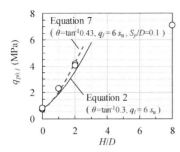

Figure 9. $q_{p0.1}$–H/D relationship.

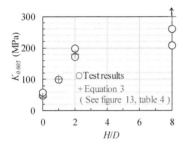

Figure 10. $K_{p0.005}$–H/D relationship.

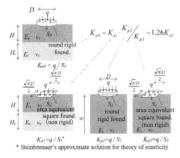

Figure 11. Method for predicting K_{p0}.

Table 4. Results of the prediction of K_{p0}.

H/D	σ_c' MPa	E_s MPa	E_c MPa	K_{p0} MPa	Ref. $S_s:S_c$
8	224	394	—	585	100:0
2	224	394	24.8	179	25:75
1	231	400	24.8	100	11:89
0	237	406	24.8	42.5	0:100

Note: $e_s = 0.64$, $v_s = 0.333$ $v_c = 0.429$.

$$K_{p0} = 1.26 K_{p1} \qquad (3)$$

where K_{p0} = initial stiffness from the $q_p - S_p/D$ relationship; K_{p1} = elastic stiffness derived according to an approximate settlement solution for theory of elasticity proposed by Steinbrenner (1936).

The deformation moduli of the bearing layer, E_s, and the lower layer, E_c, used in the present analysis are shown in Table 4 and are calculated as follows:

$$G_{s0} = 68600 \frac{(2.17 - e_s)^2}{1 + e_s} \left(\frac{\sigma_c'}{98.1} \right)^{0.5} \qquad (4)$$

$$E_{s0} = 2(1 + v_s) G_{s0} \qquad (5)$$

$$E_{c0} = 200 s_u \qquad (6)$$

where G_{s0} = the initial shear modulus of the bearing layer; e_s = the void ratio of the bearing layer (experimentally estimated to be 0.64); v_s = the Poisson's ratio of the bearing layer (assumed 0.333).

Equation 4 was previously proposed by Hardin & Richart (1963), and Equation 6 represents the average relationship between the value of s_u and E_{50} obtained through a series of triaxial compression tests of kaolin clay made by a similar method. Equation 3 corresponds to $K_{p0.005}$ except for in the case of $H/D=8$ when $K_{p0.005}$ was less than the corresponding analytical result, as shown figure 10. This is assumed to be because the stress on the soil surrounding the pile tip was large due to the high stiffness and the soil had considerably plasticity. However, Equation 3 can be used to predict the initial stiffness for cases of piles on a thin bearing layer in which $H/D \leq 2$.

4 PREDICTION OF LOAD–SETTLEMENT RELATIONSHIP OF PILE TIP

Given the deformation mechanism shown Figure 12, when the compression of soil is very small, q_p and S_p/D are linearly related to the shear stress, τ_c, on the slip surface of the lower layer and the shear strain, γ_c, respectively. It is well known that the relationship between τ_c and γ_c for the soil in a compression test, where the shear deformation occurs may be represented by a Kondner-type hyperbolic curve. Therefore, this is applied to predict the $q_p - S_p/D$ curve for piles on a thin bearing layer as follows (see Fig. 13):

$$q_p = \frac{S_p/D}{1/K_{p0} + 1/q_{p\max} \cdot S_p/D} \qquad (7)$$

where K_{p0} = the initial stiffness determined according to Equation 3; q_{pmax} = the maximum value of q_p as determined using Equation 2 with $q_l = 6 s_u$ and $\theta = \tan^{-1}(0.43)$.

Figure 14 shows a comparison between the prediction made using Equation 7 and experimental results for $1 \leq H/D \leq 2$ (Z4H2C8 and Z4H1C8). The prediction approximately agrees with the experimental results. Thus, the proposed method can be used to predict the load–settlement relationship for piles on thin bearing layers. Furthermore, $q_{p0.1}$ can be predicted using Equation 7 assuming $S_p/D = 0.1$. The predicted results of $q_{p0.1}$ corresponded to the experimental results, as shown in Figure 9 (dotted curve). Further, the predicted values are more accurate than those estimated via the modified projected-area method because they take into consideration both the strength and the deformation characteristics of the lower layer.

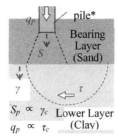

Figure 12. Assumed deformation.

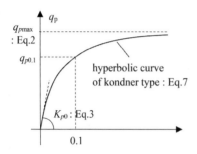

Figure 13. Method to predict $q_p - S_p/D$ relationship.

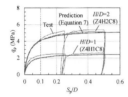

Figure 14. Prediction results.

5 CONCLUSIONS

The following conclusions were drawn:

1. For piles on a thin bearing layer, the load per unit area of a pile tip, q_p, increases with S_p/D (where S_p denotes the settlement of the pile tip and D denotes the pile diameter) and reaches its maximum value when $S_p/D > 0.2$; thereafter, it maintains approximately the same load until $S_p/D = 0.5$. This result indicates an ultimate state.

2. A punching failure with a truncated sand block under the pile (with a spread angle, θ, of approximately 23° in this experiment) occurs in the case of piles on thin bearing layers represented by $1 \leq H/D \leq 2$.

3. The maximum of q_p, q_{pmax}, increases with H/D. The projected-area method can be used to predict the increase of q_{pmax} for a pile on a thin bearing layer assuming a θ of 23° (that is, a 1:0.43 vertical:horizontal ratio) and a bearing capacity of six times the shear strength, s_u, for the lower layer.

4. In practical design applications, the ultimate bearing capacity of the pile tip, $q_{p0.1}$, increases with H/D when $S_p/D = 0.1$. The modified projected-area method (RTRI 1987) adequately predicts the increase of $q_{p0.1}$ assuming a θ of 17° (i.e. a 1:0.3 vertical:horizontal ratio) and a bearing capacity of six times s_u for the lower layer.

5. The modulus of the $q_p - S_p/D$ relationship in the initial stage, $K_{p0.005}$, which is taken as the secant modulus when $S_p/D = 0.005$, increases with H/D and reaches a maximum value with a sufficiently large H/D. In the case of piles on thin bearing layers, $K_{p0.005}$ can be predicted by Equations 3-6 using the approximate solution for theory of elasticity and the relationship for the deformation modulus of sand proposed by Hardin & Richart (1963) and for the clay layer in this experiment.

6. In the case of piles on thin bearing layers, a Kondner-type hyperbolic curve (Equation 7) given K_{p0} and q_{pmax} mentioned earlier correspond to experimentally determined $q_p - S_p/D$ relationship and the value of $q_{p0.1}$.

ACKNOWLEDGEMENTS

The authors wish to acknowledge Prof. Kabeyasawa, Earthquake Research Ins., The University of Tokyo, for his considerable technical direction in this paper.

REFERENCES

Hardin, B. O. & Richart, F. E. 1963. Elastic wave velocity in granular soils. *ASCE* 89: 33–65.

Hanna, A. M. & Meyerhof, G. G. 1980. Design charts for ultimate bearing capacity of foundations on sand overlying soft clay. *Can. Geotech. J.* 17: 300–303.

Horii, Y. Nagao, T. & Koyama, S. 2016. Centrifuge Modeling and Application of Piles on Thin Bearing Layers with Medium Depth. *Proc. 3rd GEOTEC HANOI 2016*: 75–82.

Lee, K. K. Randolph, M. F. & Cassidy, M. J. 2013. Bearing capacity on sand overlying clay soils: a simplified conceptual model. *Geotechnique* 63 (15): 1285–1297.

Matsui, T. & Oda, K. 1991. End bearing mechanism of bored pile on thin layer. *Proc. 9th Asian Region. Conf. on Soil Mechanics and Found. Eng. Bangkok, December 1991*.

Meyerhof, G. G. 1976. Bearing Capacity and Settlement of Pile Foundations. *ASCE* March: 197–228.

Nagura, K. Tanaka, M., Kawasaki, K. & Higuchi, Y. 1994. Development of an earthquake simulator for the TAISEI centrifuge. *Proc. Centrifuge 94*: 151–156.

Okamura, M. Takemura, J. & Kimura, T. 1998. Bearing Capacity Predictions of Sand Overlying Clay Based on Limit Equilibrium Methods. *Soils and Found.* 38 (1): 181–193.

Railway Technical Research Institute. 1987. *Design Standards for Railway Structures and Commentary*. (in Japanese) Tokyo: R.T.R.I.

Steinbrenner, W. 1936. Tafeln zur setzungsberechnung, bodenmechanik und neuzeitlicher strassenbau, schriftenreihe der strasse, Nr.3.

Teh, K. L. 2007. *Punchi-through of spudcan foundation in sand overlying clay*. PhD thesis. Singapore: National University of Singapore.

Yasufuku, N. Ochiai, H. Ohno, S. 2001. Pile End-Bearing Capacity of Sand Related to Soil Compressibility. *Soils and Found.* 41 (4): 59–71.

Yamaguchi, H. 1963. Practical formula of bearing value of two layered ground. *Proc. of 2nd Asian Regional Conf. on Soil Mechanics and Found. Eng.* 1:176–180.

Rigid pile improvement under rigid slab or footing under cyclic loading

O. Jenck, F. Emeriault, C. Dos Santos Mendes, O. Yaba, J.B. Toni & G. Vian
3SR Lab, Grenoble Alps University, Grenoble, France

M. Houda
College of Architecture and Design, Effat University, Jeddah, Saudi Arabia

ABSTRACT: The behaviour and the design of soil improvement by rigid piles is nowadays well understood under monotonic loading and for uniform surface loadings, while it appeals to additional research works for cyclic loadings and for non-uniform surface loading, such as those obtained under isolated footings. A study has been conducted on a 1g small-scale model, at a reduced length-scale equal to 1/10. Two types of surface loadings are studied: the case of a rigid slab covering the entire surface and the case of a square rigid footing covering the four central piles. Experiments have been conducted under monotonic and low frequency cyclic loading. The two cases highlight a different behaviour in terms of load redistribution, while a settlement accumulation is obtained during the cycles. One of the objectives of this study is the development of numerical models able to simulate the behaviour, applicable to real case conditions.

1 INTRODUCTION

Soft soil improvement using rigid piles is one of the techniques to improve compressible soil stratum in order to build transportation infrastructure, industrial platforms or buildings. It differs from the classical deep foundation scheme as the piles are not directly connected to the surface structure, but the surface loads go through a Load Transfer Platform (LTP), usually composed of a granular soil (Fig. 1). A large part of the load is transmitted to the pile head, in particular due to shear mechanisms occurring in the LTP. The LTP also permits to limit the surface differential settlement under the surface structure, thus to optimize its design.

The recommendations established by the French National Project on soil reinforcement by rigid inclusions (ASIRI project, IREX 2012) were focused on the case of monotonic loading. Yet, various cases of structures under cyclic loading are commonly encountered in practice which requires the understanding of this technique under cyclic and/or dynamic loading.

Experimental works on laboratory small-scale model of piled improved system have been previously performed under either normal gravity (Van Eekelen et al. 2012) or using centrifuge modelling (Okyay et al. 2013). These studies helped understanding the load transfer mechanisms and developing numerical and/or analytical models. However, none of these three-dimensional existing studies focused on the localized loading as given by a surface footing. In the literature, few studies concerning the behaviour under a footing exist (Combarieu 1990, who proposed an analytical design method) and the precise behaviour, in particular of the LTP, is not clearly known, all the more so as

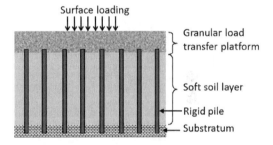

Figure 1. Schematic cross section of the system of pile improvement.

cyclic loading is applied. Experimental works combined with numerical modelling could help to address these issues.

In the present experimental study, a 1g small laboratory test set-up was used, with a scale factor in the length equal to 1/10th. The device has been first developed by Houda et al. (2016) to study the load transfer mechanism under uniform surface loading, either with a membrane acting directly on the LTP and thus providing a uniform pressure distribution, or on a rigid slab placed at the ground surface (uniform surface settlement condition). Monotonic loading as well as low frequency cyclic loading conditions were applied.

The aim of the present study was to assess the behaviour under localized surface loading, such as those encountered under isolated or strip footings, and to compare the behaviour of the structure with the case of the rigid slab covering the entire surface of the model.

2 1G SMALL SCALE MODEL

2.1 Test set-up

A 1-g three-dimensional physical model with a scale factor of 1/10th on length has been designed and developed (Fig. 2). Its objective was to study the behaviour of the soil reinforcement by rigid piles under quasi-static vertical cyclic loading. It consisted of a rigid square box made of aluminium and steel frames with internal dimensions of 1000 mm × 1000 mm. It contained 16 cylindrical aluminium piles 35 mm in diameter, arranged with a centre-to-centre spacing equal to 200 mm. The area ratio (proportion of the in-plane surface covered by the piles) was thus equal to 2.4 %. A 400 mm thick layer of soft material was placed on a fixed bottom steel plate, and a layer of gravel with a thickness equal to 100 mm was placed on top of the soft soil-pile composite layer (LTP layer). Two cases of surface loading conditions were considered in this study: i) vertical loading of a square footing of dimension 0.4 m × 0.4 m placed in the centre of the box (Fig. 3); ii) vertical loading of a rigid slab covering the entire surface of the model (Houda et al. 2016).

The system was instrumented with force sensors placed on top of 8 of the piles (F_1 to F_8) and with displacement sensors recording the settlements at the base of the LTP (D1 and D2). The force (F_J) and displacement of the jack acting on the footing or the pressure (Pm) in the membrane acting on the slab were controlled and/or recorded.

2.2 Materials

The compressible soil usually consists in clay and/or silty soils. In order to obtain a deformation in a very short period of time, it was here simulated by a mixture of Fontainebleau fine sand and expanded polystyrene (EPS) balls with an average unit weight equal to 4.5 kN/m³. Oedometer tests gave $C_c/(1+e_0) = 0.22$, where C_c is the compressibility index and e_0 the initial void ratio, considering EPS balls as void. Moreover, this kind of material, previously satisfactorily implemented in other research works related to soft

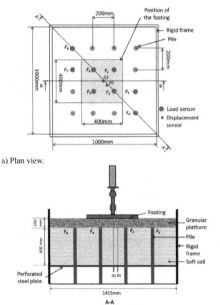

a) Plan view.

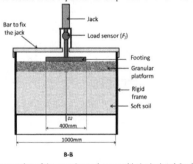

b) A-A cross-section of the experimental setup with the isolated footing.

c) B-B cross-section of the experimental setup with the isolated footing.

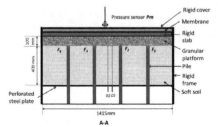

d) A-A cross-section of the experimental setup with the rigid slab

Figure 2. Schematic views of the experimental device.

Figure 3. Photograph of the experimental device with footing of dimension 0.4 m × 0.4 m.

Table 1. Experimental campaign.

Name of test	Type of test	Boundary condition
M_S	Monotonic	Slab
C1_S	Cyclic	Slab
C2_S	Cyclic with preloading	Slab
M_F	Monotonic	Footing
C1_F	Cyclic	Footing
C2_F	Cyclic with preloading	Footing

soil improvement with piles (Dinh 2010), permitted obtaining a material with homogeneous and repeatable compressibility characteristics from one test to the other.

The LTP was composed of a 2–5 mm in diameter gravel. The average unit weight after compaction was equal to 14.8 kN/m^3 corresponding to a density index $I_D = 0.75$. The friction angle and the cohesion have been determined from triaxial tests equal to $\varphi = 45°$ and $c = 0$ kPa.

2.3 Experimental campaign

Monotonic and quasi-static cyclic loadings have been considered for both loading surface conditions (footing and slab) and are summarized in Table 1. The test results obtained with the surface slab have been previously analysed and published by Houda et al. (2016). The experimental set-up and procedure were validated at that time. These test results are considered in the last section for comparison with the case of a localized surface loading as applied by the footing, presented in detail in the next section. Cyclic loading tests consisted in applying 50 cycles between $Pm = 10$ and 20 kPa for the tests with the slab and between $F_J = 1600$ and 3200 N in the jack for the tests with the footing. These values correspond to an equivalent average stress under the footing of square section 0.16 m^2 equal to respectively 10 and 20 kPa, allowing a comparison with the results obtained under the slab. Moreover, these values allowed staying within the bearing capacity domain of the LTP under the footing, identified by the monotonic tests as shown in the next section. For tests with preloading (C2_S and C2_F), a monotonic loading up to $Pm = 30$ kPa or force in the jack $F_J = 4800$ N was applied before applying the cyclic loading.

3 EXPERIMENTAL STUDY UNDER ISOLATED FOOTING

3.1 Monotonic loading

The first series of tests consisted in applying a monotonic loading to the footing, in order to assess the bearing capacity of the system, to verify the load distribution on the central pile heads and to verify the repeatability.

Figures 4 and 5 depict respectively the vertical force applied to the footing (F_J) according to its

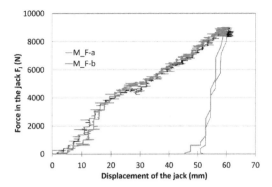

Figure 4. Force-displacement curve of the footing.

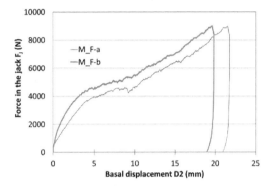

Figure 5. Force in the jack (F_J) according to the LTP basal displacement (D2).

vertical displacement and according to the LTP basal displacement (displacement D2). For each figure, both curves – corresponding to the same test configuration M_F – are well superimposed, showing the good repeatability of the tests. A change of behaviour is remarkable from a loading of the footing (F_J) equal to about 4500 N. This value seems to be the "bearing capacity" of the LTP.

Figure 6 represents the load recorded on the four central pile heads (F_1 to F_4), placed under the footing, for test M_F-b (the forces recorded on the lateral piles – F_5 to F_8 – were almost zero compared to the load values on the central piles). The four curves are smooth and superimposed up to a loading $F_J = 4500$ N, corresponding to the change of behaviour noticed on the previous figures. At this stage, there might have been a punching of the LTP by the piles, and beyond, a resumption of the load transfer from the footing to the piles, as the LTP was relatively thin (the precise mechanisms are detailed and illustrated using a Digital Image Correlation method in Houda et al. 2016, obtained under a rigid slab), explaining the consecutive increase of loading on the piles. This is accompanied with a scattering of the load distribution, attributed to the coarse nature of the LTP platform (compared to the pile size) undergoing shearing mechanisms and leading to a small dissymmetry of the system thus of the loading distribution (the forces on the pile heads

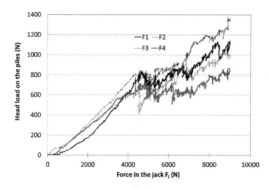

Figure 6. Head forces on the central piles (F_1 to F_4) according to the force in the jack (F_J).

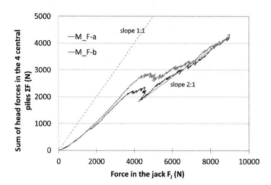

Figure 7. Sum of forces in the central piles (ΣF) according to the force in the jack (F_J).

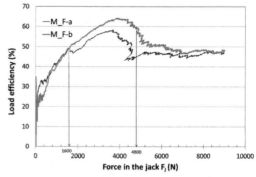

Figure 8. Load efficiency under monotonic loading.

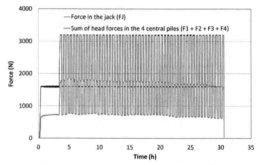

Figure 9. Evolution of loads during cyclic test C1_F.

display a variation up to + or − 20% compared to the average value, at the final stage of the loading).

Figure 7 enables to assess the repeatability and to perform a global analysis of the load transmitted to the piles under the footing. The sum of the head forces in the 4 central piles is $\Sigma F = F_1 + F_2 + F_3 + F_4$. Repeatability is satisfactory. After a disturbance of the system for a surface loading F_J approximately equal to 4500 N, the figure depicts a linear evolution of the forces on the piles with a slope equal to 0.5, showing that half of the additional load applied on the system by the footing was transmitted to the piles. The slope equal to 1:1 would correspond to the case where the entire force in the jack is transmitted to the pile heads.

The load efficiency is the proportion of the load applied on the footing transmitted to the piles (ratio of sum of loads on the pile heads to the force in the jack, *i.e.* $\Sigma F/F_J$). Figure 8 shows that the load efficiency increased up to about 60% and decreased down to 50% after reaching the bearing capacity of the LTP, remaining constant thereafter.

3.2 Cyclic loading

Cyclic tests were performed with a force applied on the footing (F_J) varying between 1600 and 3200 N (Fig. 9). These values stay within the bearing capacity domain of the LTP identified by the monotonic tests (below $F_J = 4500$ N) and are consistent with the equivalent tests performed under surface slab C1_S and C2_S (see next section). 50 cycles have been applied as for tests C1_S and C2_S. Repeatability results are not presented here, but it appeared very satisfactory both in terms of loads and displacements.

Figure 9 shows that the sum of loads of the pile heads (ΣF) then varied between 700 and 1800 N along the cyclic loading. The maximum and minimum values recorded a minor decrease throughout the cycles; the load efficiency for the maximum force in the jack was initially equal to 57% and decreased down to 49% after 50 cycles. The load transfer mechanisms through the LTP could nevertheless be qualified as persistent during the cycles.

Figure 10 highlights that the settlement at the base of the LTP, at mid-span between the piles (displacement D2, see Fig. 2), accumulated during the cycling loading. The settlement accumulation during the 50 cycles was equal to 2.5 mm.

For the tests with preloading before applying the 50 cycles, the forces on top of the piles slightly increased during the first 10 cycles, then minimum and maximum values for each cycle were stable (Fig. 11). The load efficiency was approximately equal to 50% at each cycle when the force in the jack $F_J = 3200$ N. The load transfer mechanisms were then persistent and load efficiency is similar to the one obtained for C1_F tests (without preloading).

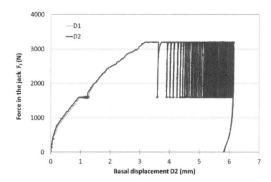

Figure 10. Force in the jack (F_J) according to the LTP basal displacement (D2) for cyclic test C1_F.

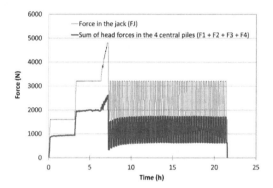

Figure 11. Evolution of loads during cyclic test C2_F.

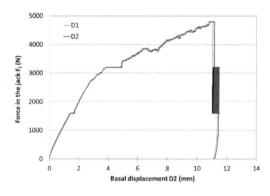

Figure 12. Force in the jack (F_J) according to the LTP basal displacement (D2) for cyclic test C2_F.

During the cycles, the basal settlement accumulation was very small, almost negligible – less than 0.2 mm (Fig. 12) compared to 2.5 mm in C1_F test, but the initial value before applying the cycles was much larger (11 mm instead of 3.5 mm).

4 COMPARISON WITH THE STUDY UNDER RIGID SLAB

Comprehensive experimental campaigns have been previously performed under uniform surface boundary conditions (Houda et al. 2016), either by covering the entire model surface with a rigid slab (uniform surface settlement) or by applying a uniform surface pressure (soft membrane under pressure Pm directly on top of the LTP). The results obtained for the tests with the rigid slab are compared with those obtained with the rigid footing, in order to assess the change of behaviour due to partial loading of the system.

For monotonic tests with slab (M_S), the load efficiency reached 95% for a pressure of $Pm = 10$ kPa in the loading membrane placed on top of the rigid slab and decreased down to 75% for $Pm = 30$ kPa. For the test with the footing, an average stress of 10 and 30 kPa is reached for a force in the jack F_J equal to 1600 and 4800 N respectively (footing of $0.16\,m^2$ surface). For these values of loading, test M_F gave load efficiency values of 50% and 60% respectively (see Fig.8). In fact, there was a possible load lateral diffusion in the LTP under the footing that cannot occur under the slab, so the loads were higher on the piles for the latter, leading to higher load efficiency values.

With the slab, cyclic tests have been performed with a membrane pressure Pm varying between 10 and 20 kPa, i.e corresponding to average stress values under the footing for a force in the jack F_J varying between 1600 and 3200 N. Table 2 reports the main results for the cyclic tests, for both slab and footing surface boundary condition. For C1_S as well as for C2_S tests, the load efficiency at each cycle peak (for $Pm = 20$ kPa) remained almost constant and equal to 82%. As for tests under footing, the basal settlement accumulation under slab was much less for the preloaded test (C2_S). However, the settlements were much larger under the slab than under the footing. In fact, for the case of the footing, the total surface load applied was much smaller than for the study with slab; and the deformation mechanisms in the LTP were different, due to the surface conditions in which the deformations were much more constrained under the slab than at the footing periphery. However, concerning the load efficiency evolution during the cycles, it appeared as stable, either under the rigid slab or under the footing.

Table 2. Comparison of results with slab and footing.

Test	Efficiency after 50 cycles	D2 accumulation
C1_S	83%	12 mm
C1_F	49%	2.5 mm
C2_S	83%	4 mm
C2_F	50%	0.2 mm

5 CONCLUSIONS

This paper analysed the behaviour of a system composed of a surface footing, loaded vertically, over a complex of soil improved by piles and a load transfer platform, under monotonic and quasi-static cyclic loadings. A 1g small scale model developed for the case of uniform surface loading conditions has been used after adaptation to this new configuration, and results have been compared to those obtained in the initial configuration, since they have been performed with the same stress level.

The repeatability of the tests is an essential indicator for the validation of the developed experimental set-up and procedure. It appeared satisfactory also for the case of the loading by a footing.

Both for the footing and surface slab cases, persistent load efficiency was obtained and a settlement accumulation has been recorded during the cycles. The settlement accumulation was strongly reduced when an initial preloading of the system was performed.

Nevertheless, the obtained results cannot be directly and quantitatively extrapolated to a real structure, as the similarity with a prototype is not strictly respected. One of the aims of performing 1g small scale physical modelling is to create an experimental base for numerical modelling. A numerical back analysis of the tests with homogeneous surface condition (slab or uniform pressure) has already been performed, providing encouraging results in the ability of the numerical model to take the cyclic loading behaviour into account (Houda et al. 2017). Numerical simulations are currently under progress to examine the possibilities of the proposed numerical model to assess the observed behaviour under an isolated footing. Moreover, the numerical modelling would permit to obtain additional results, as the horizontal forces on the piles and the precise load transfer mechanism from the footing to the piles through the LTP and displacement analysis. The next stage is then to perform equivalent numerical analysis on a real scale case, to be able to conclude on the quantitative behaviour of such system.

ACKNOWLEDGMENTS

The authors are grateful to the French Federation of Public Works (FNTP) for its support in developing the research study and the engineering school Grenoble INP-ENSE3 for giving the opportunity to students to perform the experimental campaign presented in this paper through a research project.

REFERENCES

Combarieu, O. 1990. Fondations superficielles sur sol amélioré par inclusions rigides verticals. Revue Française de Géotechnique 53, 33–44. In French.

Dinh, A.Q. 2010. Etude sur modèle physique des mécanismes de transfert de charge dans les sols renforcés par inclusions rigides. Application au dimensionnement (Study of the load transfer mechanisms inside the reinforced soils by rigid inclusions using a physical model). PhD thesis. Paris, France: Ecole des Ponts. In French.

Houda, M., Jenck, O., Emeriault, F. & Toni, J.B. 2016. Study of rigid pile reinforcement under vertical cyclic loading using a 1g laboratory small scale model. *Proc. Eur. Conf. on Physical Modelling in Geotechnics, Nantes, France, June 2016.*

Houda, M., Jenck, O. & Emeriault, F. Toni, J.B. 2017. Numerical back analysis of the behaviour of soft soil improved by rigid piles under cyclic loading. *Proc. Int. Conf. Soil. Mech. Geotech. Engg, Seoul, South Korea, Sept. 2017.*

IREX, 2012. Recommendations for the design, construction and control of rigid inclusion ground improvements: ASIRI National Project. Presses des Ponts, Paris, France.

Okyay, U.S., Dias, D, Thorel, L. & Rault, G. 2013. Centrifuge modeling of a pile-supported granular earth-platform. J. Geotech. Geoenv. Engg. 140(2).

Van Eekelen, S.J.M., Bezuijen, A. & Van Tol, A.F. 2012. Model experiments on piled embankments. Part I. Geotext. Geomembr. 32: 82–94.

Pull-out testing of steel reinforced earth systems: Modelling in view of soil dilation and boundary effects

M. Loli, I. Georgiou & A. Tsatsis
National Technical University of Athens, Greece

R. Kourkoulis & F. Gelagoti
Grid Engineers, Greece

ABSTRACT: Mechanically Stabilized Earth (MSE) retaining systems have become increasingly popular because they combine resilience with cost effectiveness and low environmental impact. Their capacity in lateral loading (pull-out resistance) is a critical design parameter, especially in cases where seismic loading has to be considered. The pull-out test is a very efficient method to identify this parameter and has become common-practice at least for the purpose of designing important infrastructure. The paper presents a modern pull-out testing device recently constructed at the laboratory of soil mechanics in the National Technical University of Athens. Results from a series of preliminary pull-out tests on steel meshes embedded in dense sand, as well as accompanying parametric numerical analyses, are shown to discuss the critical effects of: soil dilatancy, top boundary rigidity, and friction upon lateral boundaries. Discussion focuses on effective means to appropriately resolve these effects and increase the reliability of results.

1 INTRODUCTION

MSE walls have continually been a focus of study since their inception into standard engineering practice in the early 70's. Based on the simple concept of using metal strips, geotextiles, or geogrids to reinforce the soil mass, they provide a generally economical alternative to gravity retaining walls.

Constantly striving to remain competitive in the market, the industry of MSE walls relies on gaining a greater understanding of the interaction mechanisms that develop between the reinforcement and the backfill soil. Focus is placed on taking advantage of overstrength and resilience parameters that may allow moving towards a more cost efficient design in comparison to traditional theoretical solutions. Pull-out tests provide valuable information towards this objective.

A large-scale, pull-out testing device (Fig. 1) has been recently installed at the Laboratory of Soil Mechanics in the National Technical University of Athens. It permits testing inextensible (steel) strips and grids as well as geosynthetic (extensible) structures embedded into soil. The design has accounted for future preparation and testing of saturated soil specimens, yet, the herein presented results are limited to tests on steel reinforcement in dry sand.

Naturally, the design of this device has built upon existing knowledge regarding size optimization, boundaries, pull-out rate etc. provided by researchers that have conducted pull-out experiments in the past (Bergado et al. 1992, Farrag et al. 1993, Alfaro

Figure 1. View of the pull-out testing device used in this study with indicated dimensions (in millimetres).

et al. 1995, Ochiai et al. 1996, Sugimoto et al. 2001, Palmeira 2004, Texeira et al. 2007).

In addition to giving a detailed description of the newly developed pull-out testing facility, this paper elaborates on key modelling parameters that significantly affect the reliability of results. Recordings from a series of preliminary tests are presented to indicate the dramatic effect of restraining soil dilatancy. Furthermore, sensitivity to boundary effects related to rigidity and friction of internal walls are discussed.

2 EXPERIMENTAL METHOD

2.1 Set-up

Figure 2 displays a schematic cross-section of the pull-out device indicating its key components and instrumentation.

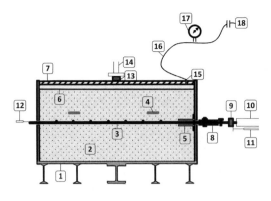

1. Box
2. Compacted soil
3. Reinforcement
4. Earth pressure cell
5. Horizontal sleeves
6. Air cushion
7. Rigid plate
8. Clamping system
9. Load cell
10. Pulling jack
11. Crack-meter
12. LVDTs
13. Load cell
14. Vertical Actuator
15. Air pressure in/outlet
16. Stiff tubing
17. Regulator
18. Air compressor

Figure 2. View of the pull-out testing device used in this study with indicated dimensions (in millimetres).

A rigid box (comprising of 20 mm thick steel plates) with internal dimensions 1500 mm × 950 mm × 750 mm contains the soil, within which the reinforcement is installed. A rigid steel lid is placed atop to carry the load applied by the vertical actuator. A horizontal slot of appropriate width has been machined on the front and rear walls to allow clamping, and controllable movement of the reinforcement. Horizontal and vertical reaction frames have been constructed, using steel beams, to safely transmit loads onto the ground floor.

2.2 Model preparation

Figure 3 shows photos from different stages of the model preparation process, which involves five stages: pluviation, compaction, installation of the reinforcement, placement of pressure cells, and application of vertical load (controlling overburden stress).

Pluviation is conducted manually, as shown in Figure 3a. The soil used for the herein presented set of preliminary tests is a dry, industrially produced, fine quartz sand with $D_{50} = 0.5$ mm and uniformity coefficient $C_u = 2.5$. Figure 4a shows its grain size distribution. The void ratios at the loosest and densest state have been measured as $e_{max} = 1.02$ and $e_{min} = 0.51$, and its strength properties (friction and dilation angles) have been characterized through direct shear test results (Fig. 4b).

Layers of sand (having thickness of approximately 200 mm) are poured into the container and compacted, using an electrical vibrator built in-house (Fig. 3b), until the desired density is achieved. A precedent calibration study is helpful in associating compaction duration with achieved density for the specific sand.

Figure 3. Photos from different stages of the model preparation: (a) pluviation; (b) compaction; (c) placement of the reinforcement; and (d) the pressure cells.

Moreover, density cups are installed at the bottom of the container to verify the soil density after every test.

Reinforcement is installed (Fig. 3c) after half of the soil specimen has been prepared. This stage is accompanied with connection of displacement transducers as detailed in the following. Attention is paid to positioning the reinforcement with precision in the centre of the box and within the clamp. The latter has been designed appropriately so as to ensure that the reinforcement is sufficiently fastened and that the pull-out force is uniformly distributed along its whole width (no sliding or torsional rotation is permitted).

2.3 Loading

When model preparation is complete, vertical loading is applied at the top of the lid by means of a manually controlled hydraulic actuator with a capacity of 300 kN. Force is applied using an oil piston controlled by a manually operated lever. Oil pressure may be increased through the lever and decreased by opening the relief valve.

When the desired level of overburden stress is reached, horizontal pull-out loading is applied at the reinforcement using an electronically controlled mechanical actuator (also with a capacity of 300 kN). Following recommendations from the literature (Farrag et al. 1993), displacement-controlled loading is applied at a low enough rate, as low as 1 mm/min, to minimize rate effects. Furthermore, the so called front wall effect (Lopes & Ladeira 1996; Raju et al., 1996; Bolt & Duszynska, 2000; Sugimoto et al., 2001) has been remediated thanks to the attachment of a 150 mm long sleeve on the front wall (see Fig. 2). This serves the purpose of transferring the point of pull-out load application further away from the rigid front wall boundary.

2.4 Instrumentation

Two earth pressure cells are installed about 200 mm above the reinforcement (Fig. 3d) in order to accurately

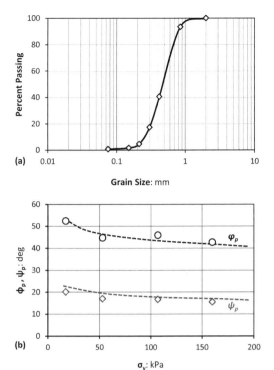

Figure 4. Properties of the backfill sand: (a) measured grain size distribution and (b) summary of direct shear-box test results indicating the variation of friction (φ) and dilation (ψ) angles with respect to overburden stress.

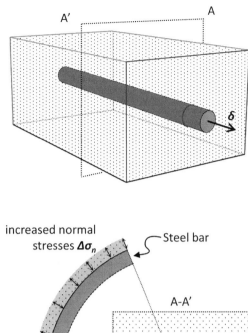

Figure 5. Mechanism of shear mobilization and formation of a dilative zone characterising pull-out tests in dense sand.

measure the actual soil pressure acting on the reinforcement level.

The pull-out force was measured by a ring-type load cell placed between the clamping system and the extraction jack. Similarly, a load cell sandwiched between the vertical actuator and the rigid box lid was used to monitor the vertical load. Reinforcement displacements were recorded by a crack-meter placed at the head and 3 to 5 (depending on the number of steel longitudinal bars) LVDTs attached at the rear of the steel grids. The crack-meter measures the longitudinal displacement applied by the horizontal actuator, while the LVDTs measure the axial displacement of the longitudinal bars. The difference between the displacement at the front and that at the back is attributable to elongation of the bars (negligible in the case of such inextensible systems).

3 IMPLICATIONS OF VERTICAL LOADING APPLICATION AND DILATANCY

Pulling a steel bar out of soil mobilizes shear strains within a zone in the vicinity of the soil-steel interface, according to the mechanism conceptually illustrated in Figure 5.

In the case of dense sand, the response of this zone is governed by dilatancy, naturally manifested by local enhancement of normal stresses and heave formation. In fact, for the relatively dilative sand considered herein, preliminary pull-out tests on steel bars subjected to low overburden stress (without vertical constrain by the top actuator) indicated significant heaving observable due to the upwards movement of the lid. What is more, heaving is considerably magnified when grids, rather than isolated bars, are pulled out. Owing to the soil in front of the transverse ribs being displaced over and under them, dilative response affects a significantly wider zone of soil.

However, the presence of the vertical (top) actuator inhibits any upward movement of the lid. As a result, dilation is at least partially restrained which leads to increased normal stresses. Remarkably large was this increase in a number of tests. For example, Figure 6a plots the evolution of stresses recorded by the two pressure cells (one in the front and one at the back) during pull-out testing of a 4-bay grid with 600 mm width.

It should be noted that such normal stress enhancement crucially jeopardises the results indicating unrealistic hardening (overstrength) in the force–displacement response. Furthermore, while the intention was to perform the test under constant and uniform

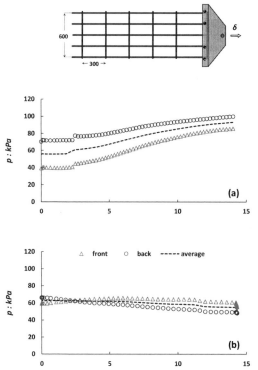
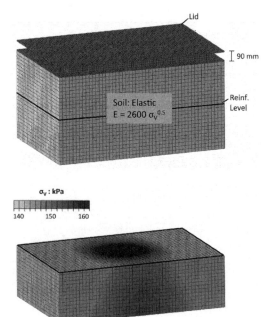

Figure 6. Variation of pressure cell readings during two different pull-out tests on a 600 mm wide steel grid with four bays: (a) without and (b) with air cushion.

Figure 7. Results from a numerical investigation of the lid – soil interaction effect on the distribution of vertical stresses acting on the reinforcement.

overburden stress (namely, 60 kPa in this case), there appears to be significant deviation between the two pressure cells measurements even prior to pulling (it should be noted that similar was more or less the case in a series of tests).

Non-uniform stress distribution is the result of interaction between the lid (essentially a quite stiff, yet not ideally rigid, footing) and the soil. 3-D Finite Element modelling using the ABAQUS code was useful in verifying this, and a parametric numerical study was carried out to investigate whether increasing the lid thickness (and hence rigidity) could improve the comparison. Figure 7 indicatively shows results in terms of vertical stress contour plots for the rather extreme case of using a 90 mm thick steel lid.

Considerable concentration of normal stresses in the middle of the reinforcement area lead to adopting an alternative solution. This solution involves using a tailor made inflatable cushion placed between the lid and the soil surface (Fig. 2). The cushion is filled with the air needed to acquire the desired stress level before testing, while an external regulator (manometer) is used to monitor its internal pressure. Not only does it produce a practically uniform initial stress field, but it also accommodates the heave occurring due to soil dilation.

Figure 6b demonstrates the effectiveness of this solution. Plotted are the recordings of the two pressure cells during pull-out testing of an identical (4-bay, 600 mm wide) grid where the cushion was used. Unlike what was the case in the first test (Figure 6a), remarkable stress regularity is achieved.

4 LATERAL BOUNDARY EFFECTS

Farrag et al. (1993), who performed pull-out tests on geosynthetic reinforcements, suggest that cushioning the geosynthetic amid layers of soil as thick as 300 mm is essential in limiting the effect of the top and bottom boundaries. The herein presented set-up has taken into account this recommendation (Fig. 2). On the other hand, dealing with lateral boundaries, is less straightforward due to the great variation in the width of the tested structures.

The significance of lateral boundary effects was investigated experimentally. Figure 8 compares the pull-out force versus displacement responses measured during testing of a 450 mm wide steel grid for two different boundary conditions. Rough boundaries refer to steel–sand interface conditions while smoothness is introduced through polishing of the internal walls. Evidently, friction increases the pull-out capacity by a significant amount. It should be noted that this effect would be even greater in the case of the previously considered, wider, grid. It should be noted that this result is in agreement with results published by other researchers (e.g. Palmeira and Milligan, 1989). Moreover, while reality in the field is expected to vary somewhere in between these two idealized (smooth vs. rough) boundary conditions, it is common notion

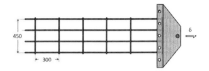
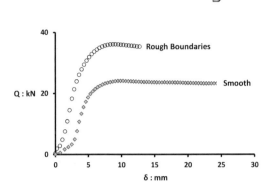

Figure 8. Effect of the frictional properties of lateral boundaries (internal walls) the pull-out force – displacement response of a 3 bay, 450 mm wide steel grid embedded in dense sand.

that design should be based on results for smooth boundaries for the sake of conservatism.

5 CONCLUSIONS

The paper has presented the set-up of a newly constructed pull-out testing device and the experimental method adopted for testing steel grids embedded in dense granular soils. The method has been based on a review of existing devices as well as a parametric study of key modelling parameters. This study involved a series of preliminary tests supplemented by 3-D numerical modelling using Finite Elements. Key findings may be summarised as follows:

(a) Restraining soil heaving due to dilative behaviour (for example, by inhibiting upward movement of the top lid as was the case herein) significantly increases the normal stress field that the reinforcement is subjected to. As a result, measured pull-out capacity is characterised by unrealistic hardening.
(b) Relative flexibility of the top lid produces stress localisations in the middle of the container potentially jeopardising the results.
(c) Both issues can be resolved by introducing an air cushion of controllable internal pressure between the lid and the soil surface.
(d) Lateral boundary effects are remediated using polished (smooth) internal wall surfaces for the sake of being reasonably conservative.

ACKNOWLEDGMENT

The work presented herein has been funded by the Greek State Scholarships Foundation (IKY), within the project "RESEARCH PROJECTS FOR EXCELLENCE IKY/SIEMENS".

REFERENCES

Alfaro, M.C., Miura N., & Bergado, D.T. 1995. Soil-geogrid reinforcement interaction by pullout and direct shear tests. *American Testing Journal*, ASTM ,18 (2), 157–167.

Bolt, A.F., Duszynska, A. 2000. Pull-out testing of geogrid reinforcement. *Proceedings of the Second European Geosynthetics Conference—EUROGEO 2000*, vol. 2, Bologna, Italy, pp. 939–943.

Bergado, D., Lu, K.H, Chai, J.C, Shivashankar, R., Alfaro, M., Loren, R. 1992. Pullout Tests Using Steel Grid Reinforcements with Low-Quality Backfill. *Journal of Geotechnical Engineering*, ASCE, 118(7), July, 1992.

Farrag, K., Acar, Y.B., Juran, I., 1993. Pull-out resistance of geogrid reinforcements. *Geotextiles and Geomembranes* 12, 133–159.

Lopes, M.L., Ladeira, M. 1996. Role of specimen geometry, soil height and sleeve length on the pullout behaviour of geogrids. *Geosynthetics International*, 3 (6), 701–719.

Ochiai, H., Otani, J., Hayashic, S. & Hirai T. 1996. The Pullout Resistance of Geogrids in Reinforced Soil. *Geotextiles and Geomembranes*, 14, 19–42.

Palmeira, E.M. 2009. Soil–geosynthetic interaction: Modelling and analysis. *Geotextiles and Geomembranes*, 27, 368–390.

Raju, D.M., Lo, S.C.R., Fannin, R.J., Gao, J. 1996. Design and interpretation of large laboratory pullout tests. *Proceedings of the Seventh Australia–New Zealand Conference on Geomechanics*, Adelaide, Australia, pp. 151–156.

Sugimoto, M., Alagiyawanna, A. M. N., & Kadoguchi, K. 2001. Influence of rigid and flexible face on geogrid pullout tests. *Geotextiles and Geomembranes*, 19(5), 257–328.

Teixeira, S.H.C., Bueno, B.S. & Zornberg, J.G. 2007. Pull-out Resistance of Individual Longitudinal and Transverse Geogrid Ribs. *Journal of Geotechnical and Geoenvironmental Engineering*, 133, (1), 37–50.

Pile jetting in plane strain: Small-scale modelling of monopiles

S. Norris & P. Shepley
Department of Civil & Structural Engineering, University of Sheffield, UK

ABSTRACT: Large monopiles are used extensively in offshore applications, however current installation methods have negative environmental consequences. Water jetting can be implemented to reduce the need for traditional installation techniques, saving time and reducing environmental concerns. Although water jetting has been used since the 1950s, its underlying mechanics are poorly understood. Thus, an innovative experimental methodology has been used to model a segment of a monopile foundation in plane strain, installing into medium dense sand. Through experimental work, both Stages 1 and 2 of a Shepley and Bolton (2014) jetting mechanism have been observed. PIV and instrumental data have been used to verify mechanisms, suggesting a localised fluidized region around piles during stage 1 and a global movement of soil medium during stage 2. Load tests of Jetted piles have shown a reduction in bearing capacity compared to non-jetted piles. Jetting properties at the base of the pile have been shown to have a significant impact upon the initial stiffness response and final performance of foundations.

1 INTRODUCTION

The use of water jetting during pile foundation installation is not a new concept, with methods documented in literature as far back as 1959 (Tsinker, 1998), yet there remains a considerable uncertainty regarding jetted installation best practice. It is clear that water jetting is beneficial in many foundation applications due to the technique reducing costs and increasing installation speeds. Offshore foundations, given their location in a large body of water, offer a unique opportunity to utilise straightforwardly water jetting during installation, delivering vast savings for installation costs in the process.

Over 75% of offshore structures in Europe are monopile foundations (Golightly, 2014). Monopiles differ from traditional piles, and can best be described as large diameter, thin walled steel caissons (Golightly, 2014). Offshore monopiles are installed at water depths of around 30 to 40 m (Gavin et al., 2011 & Li et al., 2016). Currently, most are installed by hammer driving. This technique has been subject to increasing criticism because of the negative environmental impacts caused by associated noise emissions (Saleem, 2011). Moreover, driven installations of larger monopile foundations (greater than 6 m in diameter) will unlikely conform to noise emission limits set out in countries such as Germany (Verbeek and Middendorp, 2012; Faijer, 2014). Therefore the use of water jetting either on its own or as a supplementary method could reduce the requirements for traditional driving, cutting costs and reducing noise (Gabr et al., 2014).

This paper presents an experimental investigation of the installation of jetted monopiles in sandy ground conditions through novel physical modelling. The visualization of water jetting in plane strain has been used to validate a governing mechanism. The paper also examines the effectiveness of using water jetting as a supplementary installation technique and the impact that jetting has on final pile performance.

2 BACKGROUND

Due to the relative novelty of water jetting, little direct literature relating to jetted monopiles is available. Jetting of traditional geometry piles has however been analysed by several authors. The state-of-the-art in water jetting was first presented by Tsinker (1988). He defined water jetting as the procedure of discharging a stream of water near or alongside the sides of a pile, loosening the surrounding soil and allowing the pile to self-install under its own weight.

Tsinker (1998) identified three distinct zones surrounding such a self-driven installation in sand. Directly under the pile base the natural structure of the sand is destroyed, with the soil completely liquefied and suspended in a zone of heavy mixing. Alongside the pile shaft, a vertical transport (return flow) zone of jetted water and sand rises to the surface. Outside of these zones an infiltration zone occurs, whereby an excess pore pressure gradient exists, created by water infiltration into the surrounding soil body.

Shepley and Bolton (2014) further examined the mechanics of jetting. In their experiments jetting was used as a supplementary method to jacked installation. The authors were able to identify several distinct installation stages. During Stage 1, jetting is able to fluidize the soil, creating a fully developed jetting mechanism with jetting zones as identified by Tsinker (1988). Throughout this time the pile requires zero jacking force to progress installation, effectively self-driving.

As the installation progresses, the resistance offered by the soil medium increases. This constricts the zones of disturbed jetted soil. Shepley and Bolton (2014) showed that to maintain Stage 1, the region of fluidized soil must extend to at least 2 pile radii from the pile centreline to continue the zero-load installation. Once the fluidized region was constricted below this volume, the load reappears onto the pile, albeit smaller than would be the case for a non-jetted installation. At this time the installation enters Stage 2 and further penetration requires supplementary driving methods to be implemented.

The maximum penetration depth for a Tsinker-style self-driven installation has been studied by Passini and Schnaid (2015). This also presents the maximum depth of Stage 1 as per Shepley and Bolton (2014) pile installations. Passini and Schnaid (2015) were able to show that the maximum depth is a function of fluid flow rate, soil particle size and relative density. They were also able to show that an increase in applied flow rate increases both installation depth and penetration velocity. Installation velocity has also been investigated by Gabr et al. (2014). They identified a need for increased fluid flow as soil shear resistance increases. Both works by Passini and Schnaid (2015) and Gabr et al. (2014) are able to successfully produce employable equations. However, each is reliant upon empirically fitted coefficients, preventing universal application.

An area of disagreement in the literature is the extent of the disturbed soil regions created by jetting. Passini and Schnaid (2015) were able to show that the disturbed jetting zone is constrained to 2 pile diameters (2D) from the edge of a pile, irrespective of sand density. A disturbed zone of 2D was also found in the work of Shepley (2013). Passini and Schnaid (2015) also found that a disturbed zone could extend to 6D if the installation was artificially paused. Meanwhile, Bhasi et al. (2010) studied the installation of jetted piles next to regular driven piles. They demonstrated that piles spaced greater than 8D apart felt no interaction due to the jetting of adjacent piles, giving an upper bound to the disruption from water jetting.

Shepley (2014) also studied the effect of altering the shape of jetted regions by altering nozzle arrangements. This work highlighted that a simple vertical jet placed centrally within a circular pile (as in other literary works, but difficult for the field application) performs more effectively in reducing installation loads compared to more complex angled and side exiting nozzles, despite their greater proximity to the edge of the disturbed region. When using non-vertical jets the disturbed soil around the shaft of the pile is also found to be significantly larger, leading to poorer ultimate structural performance.

Fluidization mechanics have been studied by several authors, including Kebede et al. (2014), who investigated impinging jets in sand. The authors identified that the excess pore pressure introduced by jetting was the limiting factor to the onset of fluidization. In their work they found that fluidization occurred when excess pressure introduced by jetting was equal to that of the total soil stress. Niven and Khalil (1998) also studied impinging jets into a granular medium and reported stable open symmetric jetting regions as found by Shepley and Bolton (2014), with further penetration leading to a period of asymmetric and unstable jetting regions before the region closes due to the greater total stresses surrounding the fluidized region.

3 EXPERIMENTS CONDUCTED

3.1 Materials

Leighton Buzzard Fraction C silica sand was used throughout the testing. Details of the sand are shown in Table 1.

3.2 Test pile

A 150 mm long, 135 mm wide and 15 mm thick aluminium plate was used to represent a small segment of a 6 m diameter monopile with a wall thickness of 100 mm. Figure 1(a) demonstrates the pile geometry and the four vertical jetting holes that were machined through the plate to provide water at the pile base. These holes were spaced at 34 mm centers, a similar distance to the 2D jetting extents discussed in previous literature. The flow nets and associated increased in pore water pressure combine at the midpoint between nozzles.

Table 1. Fraction C properties.

d10	637 μm
d60	937 μm
Model unit weight	15.3 ± 0.5 kN/m^3
Permeability	6.52×10^{-4} m/s

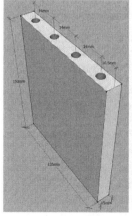

Figure 1. (a) Experimental pile and (b) jetting nozzle arrangement.

Interchangeable nozzles were fitted to the base of the pile, as shown in Figure 1(b). Nozzles of 2 mm and 3 mm diameter were used in order to vary the jetting properties, enabling an assessment of the sensitivity of observed mechanisms to flow variations at the same jetting pressure.

3.3 Experiment apparatus

A photograph of the experimental setup is included in Figure 2. The test chamber had an internal width of 400 mm, depth of 140 mm and height of 350 mm. The box was completely water tight with a glass and Perspex front viewing panel, designed to limit electrostatic attraction of sand on the Perspex window. A small 2.5 mm gap was present between the pile edge and the box at both sides of the pile. This boundary condition created a potential preferential flow route for water and allowed occasional sand particles to become interlocked in the void. This boundary had a minor negative effect on recorded results; however, the boundary has facilitated invaluable imagery and as such was a necessary compromise.

A syphon tube was placed on the inside of the left face of the box and allowed for the control of hydrostatic water level by removing extra water introduced by jetting, and providing some control over the water level.

A constant displacement motor was bolted to the top of the test box and attached to the centre of the pile. In each test the pile was driven at 0.11 mm/s whilst simultaneously jetting from the pile base. Jetting pressure was controlled in each test, with flow monitored using an inline rotameter. Installation was completed to a sand depth of 40 mm, where the pile was allowed to rest for 5 minutes (dissipating excess pore pressures) before an axial test was performed where the pile was further jacked at a speed of around 0.05 mm/s.

3.4 Experiment programme

The experimental programme used in this works can be seen in Table 2. Both nozzle pressure and nozzle diameter were varied. This has created a locus of data with unique pressure/flow jetting characteristics.

4 PLANE STRAIN MODELLING

The conducted experimental work utilized a novel method for monopile modelling whereby a small segment of the monopile circumference was modelled in isolation. The size of disturbed regions from previous work suggest a limit of 2D for solid tubular piles, and a similar (slightly larger) value can be assumed to govern wall structures. This suggests that monopile jetting will not cause a single fluidized region across the pile base area and instead a ring of fluidization will occur along the pile circumference.

Considering the above, it is reasonable to model the monopile behaviour through a plane strain segment approach. In this case, the segment is 0.7% of the pile circumference, ensuring it is reasonable to model the pile as a non-curved segment. Using a single element to represent a monopile prevents any potential plugging mechanism and representative lateral loading, which has limited the ability to model monopiles in plane strain previously, but given the focus on installation it is a fair simplification.

A series of point source nozzles were used to create an approximately plane strain pore pressure profile across the model breadth. The 34 mm nozzle spacing caused spherical flow nets to overlap at a distance of 1 pile width from the nozzle. These overlapping flownets create a more consistent pore pressure profile along the pile wall centreline, supporting the plane strain assumptions made during the modelling.

Figure 2. Experimental set up.

Table 2. Experiment programme.

Test	Jetting pressure (kPa)	Nozzle diameter (mm)
T1	75	2
T2	100	2
T3	75	2
T4	50	2
T5	100	2
T6	50	3
T7	75	3
T8	100	3
T9	110–175	3

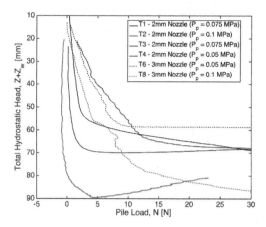

Figure 3. Pile installation results.

5 RESULTS AND DISCUSSION

5.1 *Installation behaviour*

Pile head loading was monitored throughout the installations. From this data there is a clear relationship between increased jetting pressure and a reduced installation load. Moreover, the results corroborate the staged installation first outlined by Shepley and Bolton (2014), where increased pile jetting intensity increases the depth at which a Stage 1 (zero head load) is observed. This behaviour is observed in Figure 3 where the pile load is plotted against the 'Total Hydrostatic Head' which was calculated by summing the total pile depth in the sand (Z) and the water depth above the surface (Z_w). This value demonstrates the sensitivity of the installation process to the hydrostatic pore water pressure in the sand body. Both tests T2 and T8 were subject to maximum jetting intensity, and showed the deepest transitions. (Test T2 appears to uninstall after transition because of a sharp decrease in water level occurring after this point. Tests T1 and T3 were completed under the same conditions, but the results differ due to sand becoming jammed between the pile and the window.)

When comparing like for like experiments little correlation can be seen between jacking requirements at an installed soil depth, often due to sand particles becoming lodged between the box walls and the pile. However, the effect of jetting can still be investigated through comparisons between jacking requirements and hydrostatic head at the pile base, as shown in Figure 3. Here, the transition between Stage 1 (where base resistance is completely eliminated by a fully developed fluidized region around the pile base) and Stage 2 (where jetting is insufficient to eliminate base resistance, but limited soil disruption continues) is seen for 2 mm nozzles using 0.075 MPa jetting pressure and found to occur at a similar depth (within 0.5D between the installations). Given the installation equipment, this observation is within experimental tolerances.

The installation results reinforce the findings of Kebede et al. (2014), where fluidization was reliant on hydrostatic head. The results also support the mechanisms derived by Shepley and Bolton (2014), where fluidized bulbs were presented in terms of hydrostatic head.

From the results, the transitional depth exhibits little sensitivity to changes in flow for a given pressure. This is best seen through tests T4 and T6, where very little difference in transition depth was observed. This conclusion must be treated with caution due to the low stress conditions and the low flow measured during the tests.

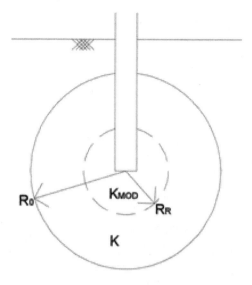

Figure 4. Cylindrical flow net around pile base.

5.2 *Mechanism observations*

Previous research has shown that an understanding of the disturbed soil region is critical in establishing jetted pile installation performance. From the work of Shepley and Bolton (2014), a plane strain jetting mechanism can be developed utilizing a cylindrical Laplacian flow net around the base of the pile, as seen in Figure 4. A modified soil permeability (K_{MOD}) is assumed for the fluidized bulb at the pile base, accounting for the increased permeability at very low effective stresses (Haigh et al., 2012).

Fluidized bulbs as per Figure 4 were observed throughout testing, and an example from test T8 is shown in Figure 5. The regions were shallower than proposed for spherical cases around axi-symmetric piles and non-circular (indicated by the black region whereby the sand has been expelled in the fluidized sand mixing). This disagrees with the suggested cylindrical mechanism schematic of Figure 4, but can be classed as sufficiently close to provide an analytical mechanism. The bulb is likely to have been affected by the up-flow mechanism, a limitation of the derived flow net mechanism, and localized loss of water at the boundary. The smaller disturbed region size is also likely to be caused by the point nozzle spacing, rather than a continuous jetting slit being implemented.

Figure 5. Test T8 fluidised bulb (circled).

Figure 7. Global soil movement during Stage 2 installation.

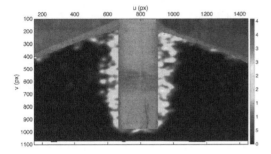

Figure 6. Test T9 vertical transport mechanism highlighted by plotting the soil strains.

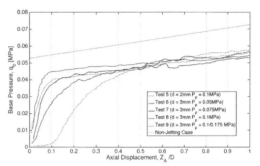

Figure 8. Load test performance of completed pile installations.

Vertical transport mechanisms, crucial to maintaining a stage 1 mechanism were also observed during the experiments. Figure 6 shows a fully developed Stage 1 mechanism from test T9. Utilizing GeoPIV (White et al., 2003) to analyze the images, an identifiable jetting region is recognized, with the regions subject to significant strains being part of the fluidized body. Shown around the sides of the pile is a surface return flow, in this case extending 0.6D from the edge of the pile. The figure indicates a likely return flow along the pile shaft rather than a cylindrical flow net controlling the jetting mechanism, hence the smaller fluidized region thickness around the pile. These initial results suggest that only a minor volume of soil around a monopile would be affected by water jetting, offering some promise for the methodology.

Beyond the clearly fluidized region, zero soil strain was observed (as shown in Figure 6). Following the collapse of the stage 1 jetting mechanism, this localization of strains ceased, and more typical global soil deformations were observed around the advancing pile as presented in Figure 7. This global movement requires the jacking force associated with Stage 2, and demonstrates that typical soil responses are governing installation during this stage, albeit with higher pore water pressures due to the jetting.

Coordination can be seen between the results of Section 5.1 and 5.2. Although a purely zero head load was rarely observed during installations, sharp increases in loading at transitional points were found immediately after the images showed the collapse of the vertical transport mechanism. This has been verified at the observation window and along the length of the pile through the vertical transport of sand. Observations of the total pile perimeter demonstrated that the vertical transport system was found to collapse at the boundaries before the main body of the pile and the onset of pile loading. This is simply a boundary effect, with the mechanism collapsing prematurely because of the nozzle distance from the boundary and the preferential flow path.

5.3 *Pile performance*

Axial load/displacement testing has been used to assess installed pile performance. This data provides an indication of the behavioral characteristics of jetted monopiles.

A clear relationship can be seen between final jetting performance and final installation jacking load. Piles with a higher final jacking load had a stiffer load response. This is best exemplified by test T11 in Figure 8, which was still in Stage 1 when the installation was completed. Test T11 then exhibited very low stiffness, taking 0.4 Z_A/D (where Z_A was the additional pile displacement during the load test) to achieve a similar load displacement curve to that of 0.2 Z_A/D for installations that terminated in Stage 2.

A comparison in Figure 8 is also made to a non-jetted installation. Compared to this case, all jetted piles feature a permanent reduction in bearing capacity, even at large displacements. This is most likely due to significant disturbance to soil structure along the pile shaft and the softening of soil below the pile base.

6 CONCLUSIONS

A successful study was conducted into the mechanisms governing water jetting, focusing on the potential utilization of jetting in offshore monopile installations and visualizing plane strain pile installations. The outcomes of this study can be summarized as follows:

- Through analysis of experimental imagery and sensor data, the proposed mechanism (based upon the works of Shepley and Bolton (2014)) has been validated through visual observations of Stage 1 and Stage 2.
- Analysis of experimental imagery has shown a localization of soil strain during a fully developed jetting mechanism. The collapse of this fluidized region and the commencement of pile load corroborated the suggested Stage 2 behavior.
- The region of disturbed soil around a jetted monopile extended to 0.6 times the wall thickness – many times smaller than the overall monopile diameter.

Despite testing at a wide variety of jetting pressures, the axial performance of jetted piles has been shown to be permanently reduced by the use of water jetting and the initial soil stiffness under pile load testing is shown to be proportional to the pile head load during installation. This suggests caution is required for determining the axial capacity of completed installations, but further investigation is required to determine the effects on lateral load capacity.

ACKNOWLEDGEMENTS

The authors wish to thank the technical staff of the Department of Civil and Structural Engineering at the University of Sheffield for their continuous support and resources provided throughout the project.

REFERENCES

Faijer, M. J. 2014. *Underwater noise caused by pile driving. Impacts on marine mammals, regulations and offshore wind developments.* An Hengelo.

Gabr, M. A., Borden, R. H., Denton, R. L. and Smith, A. W. 2014. An insertion rate model for pile installation in sand, *Geotechnical Testing Journal*, **37**(1), pp. 1–11.

Gavin, K., Igoe, D. and Doherty, P. 2011. Piles for offshore wind turbines: a state-of-the-art review, *Geotechnical Engineering*, 164(GE4).

Golightly, C. 2014. Tilting of monopiles: Long, heavy and stiff; pushed beyond their limits, *Ground Engineering*, January, pp. 20–23.

Haigh, S. K., Eadington, J. and Madabhushi, S. P. G. 2012. Permeability and stiffness of sands at very low effective stresses, *Géotechnique*, **62**(1), pp. 69–75.

Hameed, R. A., Gunaratne, M., Putcha, S., Kuo, C. and Johnson, S. 2000. Lateral load behavior of jetted piles, *Geotechnical Testing Journal*, **23**(3), pp. 358–368.

Kebede, Y. A., Gabr, M. A. and Kayser, M. F. 2014. Scour zone characterization by deep impinging jet, in *Proceedings of the ASME 2014 33rd International Conference on Ocean, Offshore and Arctic Engineering*. The American Society of Mechanical Engineers, pp. 1–7.

Lehane, B. M. and Gavin, K. G. 2001. Base Resistance of jacked Pipe Piles in Sand, *Journal of Geotechnical and Geoenvironmental Engineering*, **127**(6), pp. 473–480.

Li, L., Acero, W. G. and Gao, Z. 2016. Assessment of allowable sea states during installation of offshore wind turbine monopiles with shallow penetration in the seabed, *Journal of Offshore Mechanics and Arctic Engineering*, **138**(4), p. 041902:1-041902:17.

Middendorp, P. and Verbeen, G. E. H. 2012. At the cutting edge of pile driving and pile testing, in *The 9th Internation Conference on Testing and Design Methods for Deep Foundations*. Kanazawa, Japan.

Niven, R. K. and Khalil, N. 1998. In situ fluidisation by a single internal vertical jet, *Journal of Hydraulic Research*, **36**(2), pp. 199–228.

Passini, L. B. and Schnaid, F. 2015. Experimental investigation of pile installation by vertical jet fluidization in sand, *Journal of Offshore Mechanics and Arctic Engineering*, **137**(4), p. 042002:1-10.

Saleem, Z. 2011. Mitigation, Alternatives and modifications of Monopile foundation or its installation technique for noise. See www.vliz.be/imisdocs/publications/223688.pdf

Shepley, P. 2013. *Water injection to assist pile jacking*. PhD thesis, University of Cambridge.

Shepley, P. 2014. Optimisation of water injection nozzles to aid pile jacking, *ICE Journal of Geotechnical Engineering*, **168**(GE3), pp. 257–266.

Shepley, P. and Bolton, M. D. 2014. Using water injection to remove pile base resistance during installation, *Canadian Geotechnical Journal*, **51**(11), pp. 1273–1283.

Tsinker, G. P. 1988. Pile Jetting, *Journal of Geotechnical Engineering*, **114**(3), pp. 326–334.

White, D. J., Take, W. A. and Bolton, M. D. 2003. Soil deformation measurement using particle image velocimetry (PIV) and photogrammetry, *Géotechnique*, **53**(7), pp. 619–631.

Influence of geometry on the bearing capacity of sheet piled foundations

J.P. Panchal, A.M. McNamara & R.J. Goodey
Research Centre for Multi-Scale Geotechnical Engineering, City, University of London, UK

ABSTRACT: Bored concrete piles are commonly used to support moderate loads from buildings in urban areas. At the end of their 25-30 year lifespan these structures are decommissioned but their foundations are left in place. These cannot be inspected hence the bearing capacity cannot be accurately verified. A hybrid foundation comprising sheet piles and a pilecap to mobilise shaft friction and end bearing was demonstrated to be a feasible and sustainable alternative to cast in-situ concrete piles. This research investigated the influence of sheet pile geometry on ultimate bearing capacity. A centrifuge test at $50\,g$ was performed in over-consolidated clay where a square hybrid sheet pile group was axially loaded and vertical settlements recorded. Results indicated a square sheet pile group offers 70% greater capacity than a circular sheet pile group of similar surface area and 24% improved performance over the solid pile loaded in the same test. Analysis of results suggested that the ultimate bearing capacity of the square sheet pile group compared with a solid pile of equivalent base area were within 0.2%, emphasising the importance of shape on capacity and the feasibility of the hybrid system as a viable foundation solution.

1 INTRODUCTION

Bored piles are a common foundation type for moderately loaded structures and recent calls have been made to extend piling applications to low rise residential projects (Ground Engineering, 2018). Concrete piles comprise a significant volume of concrete and are rarely removed during demolition. These piles consequently cause obstructions to future developments and must be removed or avoided which can have large financial impact on a project and result in programme delays.

Significant efforts have been made to increase foundation solution sustainability, including the recent development and trialling of hollow piles (McNamara et al. 2014) which benefit from a significantly reduced volume of concrete.

CIRIA C653 (2007) recommends the reuse of existing piles, however it acknowledges that uncertainties remain in establishing pile integrity and reliability owing to difficulties in inspection.

Smaller diameter piles generate greater capacity from less concrete. However, the issue of breaking out, transporting and crushing concrete still remains before it can be recycled. A more sustainable solution may exist by use of sheet piles, which can be extracted following superstructure demolition. Their condition can be assessed on site and they can be reused immediately, saving time and eliminating muck away.

2 BACKGROUND

Previous centrifuge modelling studies were conducted (Panchal et al., 2016) with a sheet pile group arranged in circular formation. The diameter of the foundation at the neutral axis of the sheet pile group was 60 mm. The sheet pile shaft protruded above ground level and a resin pile cap was cast within the area enclosed by the sheet pile group. 5mm holes had been drilled along each sheet pile shaft at 30 mm centres with the aim of increasing frictional resistance.

Smooth and rough solid circular shafted model piles, 60 mm in diameter, were also tested to provide comparisons against the sheet pile foundations. The aim of this was to determine whether the hybrid pile system was a credible foundation solution offering comparable or improved bearing capacity over conventional solid piles.

Results, see Figure 1, indicated that the sheet pile group with cap arrangement offered a 22% increase in bearing capacity compared with the smooth shafted pile and only 12% lower capacity than a rough pile. This validated the idea that this hybrid pile arrangement was a reasonable alternative to conventional straight-shafted concrete piles with scope for further development.

One concern associated with this piling method however would be the contractor's ability to drive sheet piles in a circular arrangement on site, whilst maintaining verticality and interlock between sections. An alternative sheet pile arrangement was sought in order to evaluate the influence of the pile group shape on bearing capacity.

3 OBJECTIVES

This study aimed to determine the influence of sheet pile group shape on the ultimate bearing capacity.

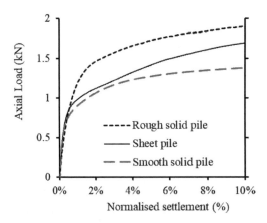

Figure 1. Centrifuge test results by Panchal et al. (2016).

An additional centrifuge test at 50 g was conducted to compliment earlier published literature (Panchal et al., 2016). This experiment focussed on modelling a comparable cross sectional area but varying the pile configuration in plan.

In designing a more buildable sheet pile formation the shaft area was inevitably altered. The purpose of this was to establish whether any change in ultimate bearing capacity was observed for an open ended capped sheet pile. In addition, an assessment was to be made to determine whether the end bearing or shaft friction was more critical in improving the capacity of a hybrid pile.

4 SOIL MODEL

Centrifuge experiments were conducted in a 300 mm deep 420 mm diameter steel cylindrical centrifuge tub. The final sample was required to be flush with the top of the tub which was achieved by bolting on a 300 mm deep cylindrical extension.

The walls of the tub were lubricated with water pump grease and a layer of porous plastic and filter paper were placed at the base. Herringbone channels cut into the base of the tub directed water towards two drainage taps.

Speswhite kaolin powder was mixed with distilled water to a water content of 120%, which is approximately twice its liquid limit. An industrial ribbon blade mixer was used to produce a uniform slurry. Slurry was carefully placed in the tub by means of a scoop whilst a palette knife was used to agitate the clay between each pour to prevent the entrapment of air. It was placed to a depth of approximately 550 mm before being sandwiched between another layer of porous plastic and filter paper.

The package was transferred to a hydraulic press where a loading platen, attached to a ram was lowered onto the sample. Pipes were connected to the drainage taps of the centrifuge tub and directed to a bucket. Holes drilled in the top of the platen also allowed water to seep up as the sample was loaded,

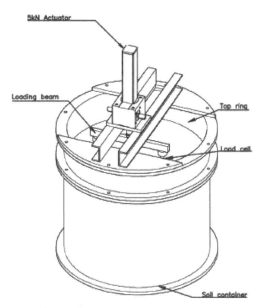

Figure 2. Centrifuge loading frame (Gorasia, 2013).

halving the drainage path length and accelerating the rate of consolidation. A reasonably stiff sample was achieved by gradually increasing the pressure on the sample from 20 kPa to 500 kPa over a period of a week before swelling it back to 250 kPa the day prior to testing. This produced a sample that protruded above the top of the centrifuge tub.

5 APPARATUS AND EQUIPMENT

The loading apparatus used in this experiment was designed and manufactured by Gorasia (2013) and is illustrated in Figure 2. It comprised a frame that bolted above the centrifuge tub and housed a lead screw actuator. A loading beam was connected to the actuator and could accommodate two load cells at each end and a number of LVDTs.

A rough solid circular shafted pile consisted of a 48 mmOD aluminium tube which was closed at the base and positioned in a 60 mm diameter open bore, the annulus of which was filled with resin during model making. Two 60 mm long rods had been drilled through the pile perpendicular to its length and served the purpose of centralising the pile in the bore before the resin was placed. A 10 mm thick 20 mm diameter Perspex spacer was glued to the base of the pile which also permitted resin to coat the base of the pile, see details in Figure 3.

The sheet pile foundation was fabricated from a single 0.5 mm thick stainless steel sheet that had been repeatedly pressed to form each rib. The sheet was then folded to form a square section with a perimeter of 246 mm. The sheet had been sandblasted to produce a suitably rough surface. Characteristics of all piles analysed in this paper are summarised in Table 1 at model scale.

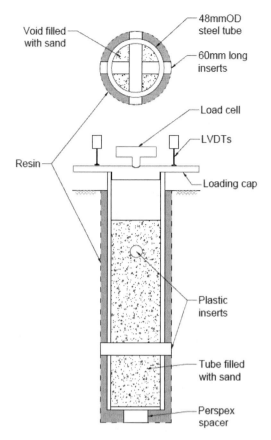

Figure 3. Rough solid circular pile arrangement (Panchal et al., 2016).

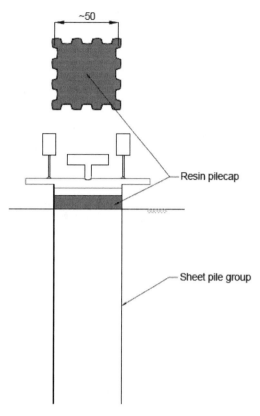

Figure 4. Square sheet pile with pilecap and loading cap arrangement.

Table 1. Model pile characteristics (all piles 180 mm long).

Pile type	Nominal diameter or length (mm)	Pile shaft perimeter (mm)	Pile base area (mm^2)
Rough solid	60	188	2827
Circular sheet	60	217	2827
Square sheet	50	246	3782
Comparisons of pile characteristics			
Square sheet vs rough solid pile		31%	34%
Circular sheet vs rough solid pile		15%	0%
Square sheet vs circular sheet pile		13%	34%

An aluminium square loading cap sat on the sheet pile and the underside had been machined to provide a lip in which the sheet pile could sit, see Figure 4. A loading cap rested on the pilecap on which the load cell and LVDTs were seated.

6 TESTING PROCEDURE

The centrifuge model was prepared at 1 g. Firstly, the tub was removed from the hydraulic press and the extension unbolted. The sample was trimmed using a wire cutter and palette knife until it was level with the top of the tub, giving a 300 mm deep sample. PlastiDip (an aerosol applied synthetic rubber membrane) was sprayed across the surface of the clay to prevent it from drying out during model making and when in-flight.

The loading frame was then placed on the tub and the beam lowered until the load cells indented the soil surface to mark out the centres of the piles. The square sheet pile was aligned such that the centre was approximately aligned with the indent and the hydraulic press embedded the sheet pile in a carefully controlled manner to a depth of 180 mm.

A pair of dividers was used to mark out the circumference of the pile before using a thin walled cutter and guide to form the 180 mm deep 60 mm diameter bore. The base of the bore was scraped and care was taken when placing the 48 mmOD hollow tube in the bore to ensure that the 60 mm plastic inserts did not scrape the edges of the bore. Sand was poured in the tube so that the final weight of the cast in-situ pile was equal to the weight of soil removed. This was necessary to prevent the pile from becoming buoyant during consolidation.

Two-part epoxy resin was thoroughly mixed before being carefully poured around the sides of the open bore to the top of the clay surface. Resin also formed the pilecap of the sheet pile and was contained within

Figure 5. Sealed sample with piles and resin cast in-place prior to securing the loading frame.

Figure 7. Recovered solid circular and sheet piles post-test.

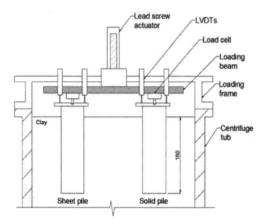

Figure 6. Cross section through centrifuge tub illustrating location of piles and instrumentation (Panchal et al., 2016).

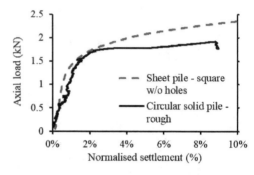

Figure 8. Results from current centrifuge test on a rough solid circular pile and a square sheet pile.

the sheet upstand, see Figure 5. Once the resin had cured the loading caps were placed on each of the piles before securing the loading frame to the centrifuge tub. The LVDTs were secured in position and rested on the loading cap plates. A bead of silicone grease was applied around the edge of the model to prevent it from drying out in-flight. The package was weighed and transferred to the centrifuge in preparation for testing.

A standpipe was connected to establish a water table 30 mm below the surface of the clay. It was intended that the sample would be left to consolidate at $50\,g$ overnight to allow excess pore pressures to dissipate, however problems with the centrifuge apparatus meant that the test was conducted immediately upon reaching $50\,g$ and a water table was not established.

Testing of the model involved loading the piles at a rate of 1 mm/minute to simulate an undrained loading event until settlement equivalent to 10% of the pile diameter was achieved. Figure 6 provides a schematic of the model and testing apparatus. Following the test, shear vane readings were taken and the piles were recovered (Figure 7).

7 TEST RESULTS

One centrifuge test was carried out as part of this study to compliment findings from Panchal et al. (2016). The centrifuge tub was sufficiently large to test two piles and avoid boundary effects.

A rough circular solid pile was tested alongside a square sheet pile with a resin pile cap. This permitted a means of comparing and analysing the bearing capacity of each pile in the same soil sample.

The results from this test have been plotted in Figure 8. A greater bearing capacity of 25% was achieved by the square sheet pile group in comparison to a rough circular solid pile, which was expected as the square sheet pile group was 31% larger in perimeter and 34% greater in base area.

However, results published by Panchal et al., (2016) showed that although the circular sheet pile group was 15% larger in perimeter than a solid circular pile it reduced the ultimate capacity by 12%. This suggests that a relatively smooth circular sheet pile group

Table 2. Summary of tests used in analysis.

Test	Pile type		$Q_{(ult)}$ (kN)
1 (Panchal et al., 2016)	Solid	Rough	1.90
	Sheet	Circular with holes	1.69
2	Sheet	Circular without holes	1.17
3	Solid	Rough	1.91
3	Sheet	Square without holes	2.37

does not offer significant benefits over a conventional circular bored concrete pile.

8 ANALYSIS OF RESULTS

Establishing whether any structural benefit existed in altering the shape of the sheet piled foundation relied on results from previous experiments. Published results (Panchal et al., 2016) of a rough solid pile and circular sheet pile group were compared against the measurements taken from this experiment. This provided a wide range of results, summarised in Table 2, from which observations were drawn. $Q_{(ult)}$ is defined as the ultimate pile bearing capacity at 10% settlement.

Results from all experiments were plotted in Figure 9. Although in the most recent test the model was not left to consolidate overnight, it was observed that the ultimate bearing capacities of the rough solid piles were comparable. Shear vane readings were also consistent between tests. This highlighted the reliability and consistency between tests and permits comparisons to be made between the circular and square sheet piled foundations.

Two variations of circular sheet piled foundations were previously investigated; with and without 5 mm holes drilled at 30 mm centres along the ribs. The results from both scenarios were also plotted on Figure 9 and the capacity of a sheet pile with holes was approximately 1.4-1.5 times greater throughout the duration of the test. This trend was applied to the square sheet pile group and the loads were multiplied by 1.45 to estimate the response of a square sheet pile with holes as illustrated by the dashed black line in Figure 9.

The ultimate bearing capacity ($Q_{(ult)}$) is a summation of the base capacity (Q_b) and shaft friction (Q_s) and is calculated using Equations (1) to (3).

$$Q_{(ult)} = Q_b + Q_s \quad (1)$$

$$Q_s = EA_s \, S_u \, \alpha \quad (2)$$

$$Q_b = EA_b \, (N_c \, S_u + \gamma \, H) \quad (3)$$

Where EA is the external area; S_u the undrained shear strength; α the adhesion factor; N_c the dimensionless factor governed by pile diameter and depth; L the pile length and γ the soil bulk unit weight.

Owing to the geometry of the piles in these tests, N_c equated to 9. The sheet piled foundations were

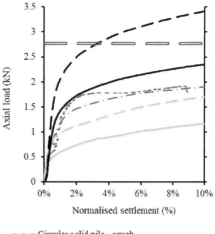

Figure 9. Results from current and previous centrifuge tests.

Table 3. Alpha values back analysed from centrifuge tests.

Test	Pile type		α values
1	Solid	Rough	0.522
	Sheet	Circular with holes	0.359
2	Sheet	Circular without holes	0.304
3	Solid	Rough	0.597
	Sheet	Square without holes	0.559

analysed as open ended tubular piles and following the ICP design methods (Jardine et al., 2005), Q_b was reduced by half to account for the plugging effect at the base of the pile.

The adhesion factors (α) were back analysed from the difference between the ultimate capacity measurement and base capacity calculation and are given in Table 3 and show a reasonable range of values (Bell & Robinson, 2012).

It is worth noting that the α value for the square sheet pile group was considerably higher than the values obtained for the circular piles. This was likely to have occurred owing to the square sheet pile group having a larger surface area than the solid shafted and circular sheet piles. In addition, the ribs on the square sheet pile group are less open than those of the circular sheet pile group. The ribs may have become plugged with soil hence loading the pile would mobilise a higher proportion of the soil strength as it sheared against

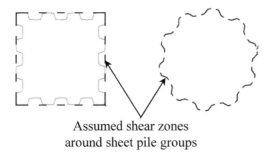

Assumed shear zones around sheet pile groups

Figure 10. Assumed shear zone around square sheet piles.

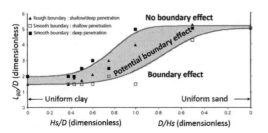

Figure 11. Boundary effects zones in centrifuge tests (Ullah et al., 2016).

the soil/steel and soil/soil interfaces, as illustrated in Figure 10.

To assess the performance of a square sheet pile group against a solid circular shafted rough pile of comparable size it was necessary to scale up the geometry of the rough pile used in this most recent test. The base area equalled that of the square sheet pile and α was taken as 0.597, as indicated in Table 3. $Q_{(ult)}$ was calculated as 2.77 kN which is 17% greater than the measured response of the square sheet pile. However, assuming that holes drilled through the sheet pile shaft increases the ultimate capacity to would offer an improved bearing capacity of 23% over the conventional concrete bored pile.

An assessment of the boundary effects was necessary to determine the validity of the results from these experiments. The stress bulb that forms below a shallow foundation is assumed to be equal to $2B$ (Boussinesq, 1885). In this model the piles were nominally 60 mm diameter (B) and 180 mm long in a 300 mm deep soil sample. This provided 120 mm clearance and satisfied the $2B$ end bearing boundary effects criteria.

Ullah et al. (2016) investigated the lateral boundary effects of modelling foundations in the centrifuge. Figure 11 maps the criteria for eliminating or reducing the boundary effects of a foundation in the model. The minimum recommended L/D dimension is 1.5 in a uniform clay sample and any value greater than 2 indicates that no boundary effects exist. In these experiments L/B was 1.75 and within the potential boundary effect zone. Although this may have had some influence on the results, consistency of boundary effects was achieved across tests owing to similarities in the experiment and apparatus set up.

9 CONCLUSION

One centrifuge test at $50\,g$ was conducted to measure the response of a rough solid circular pile and a sheet pile foundation in a square formation with a resin pilecap.

The results from this experiment were compared with those published in the literature (Panchal et al., 2016) to understand the influence of the foundation shape on its performance.

Results showed that the square sheet pile group achieved a 25% greater bearing capacity than the solid circular pile tested in this experiment. Analysis of these results showed that for a comparable pile base area similar capacities would be obtained.

The perimeter of the square sheet pile group was only 13% greater than the circular sheet pile group and this scenario offered a 40% increase in bearing capacity over an alternative sheet pile arrangement.

This investigation suggests that a square sheet pile group is a viable alternative to solid concrete piles and offers huge economic and sustainability benefits in pile construction, removal and reuse.

REFERENCES

Bell, A. and Robinson, C. 2012. Chapter 54 Single piles. ICE manual of geotechnical engineering: Volume II. January 2012, pp. 803–821.

Boussinesq, M. J. 1885. Application des potentils a l'etude de l'equilibre et du mouvement des solides elastiques, principalement au calcul des deformations et des pressions que produisent, dans ces solides, des efforts quelconques exerces sur une petite partie de leur surface ou de leur interieur: Memoire suivi de notes etendues sur divers points de physique mathematique et d'analyse. Gauthier Villars, Paris, pp. 722.

CIRIA 2007. C653: Reuse of Foundations. CIRIA, London.

Gorasia, R.J. 2013. Behaviour of ribbed piles in clay. PhD thesis, Research Centre in Multi-scale Geotechnical Engineering, City University London, UK.

Ground Engineering 2018. Call for end to strip foundations for housing, Ground Engineering Magazine, http://digitalissues.geplus.co.uk/GE/2018/JanFeb/index.html, pp. 7.

Jardine, R., Chow, F., Overy, R. and Standing, J. 2005. 4 Design Methods for Piles in Clay. ICP design methods for driven piles in sands and clays. January 2005, Thomas Telford, London, pp. 28–37.

McNamara, A.M., Suckling, T.P., McKinley, B. and Stallebrass, S. 2014. A field trial of a reusable, hollow, cast-in-situ pile. Proceedings of the Institution of Civil Engineers – Geotechnical Engineering 2014, Volume 167, Issue 4, pp. 390–401.

Panchal. J., McNamara. A. and Goodey, R. 2016. Bearing capacity of sheet piled foundations. Proceedings of the 2nd Asian Conference on Physical Modelling in Geotechnics. Asiafuge 2016 (ed. Ma, X.), Tongji University, Shanghai, China, pp. 1–6.

Ullah, S.N., Hu, X., Stanier, S. and White D. 2016. Lateral boundary effects in centrifuge foundation tests. International Journal of Physical Modelling in Geotechnics, Volume 17(3), pp. 144–160.

Kinematic interaction of piles under seismic loading

J. Pérez-Herreros & F. Cuira
TERRASOL (SETEC group), France

S. Escoffier
IFSTTAR, GERS, SV, Nantes, France

P. Kotronis
Ecole Centrale de Nantes, Université de Nantes, CNRS, GeM, Nantes, France

ABSTRACT: Under earthquake excitation, piles undergo loads arising both from the deformation of the surrounding soil (kinematic interaction) and the inertial forces due to the vibration of the superstructure (inertial interaction). Evaluating these interactions is an important issue for the seismic design of deep foundations and has been an active research topic since the early 1970s. A series of dynamic centrifuge tests has been conducted to examine the behaviour of free-head end-bearing single piles subjected to several sinusoidal and earthquake events. The soil profile consists of three horizontal layers alternating dense HN31 sand (bottom and top layers) and overconsolidated speswhite kaolinite. The reproducibility of the soil foundation is verified using bender element measurements. The seismic response of the piles is analysed in terms of maximum, minimum and residual bending moments.

1 INTRODUCTION

Few experimental studies exist on the behaviour of piles in clays under seismic loading (Meymand 1998, Boulanger et al. 1999, Banerjee 2009, Zhang et al. 2017).

As part of the development of a new macroelement approach for deep foundations under dynamic loads (Correia et al. 2012, Li et al. 2016), a series of experimental centrifuge tests is conducted to increase the existing database and to validate numerical models.

In order to understand the mechanisms of soil-pile kinematic interaction, the first phase of this study consisted of two tests with a free-head end-bearing single pile in a multilayer soil profile that was subjected to dynamic loads.

2 EXPERIMENTAL SETUP AND PROCEDURE

The two tests were performed in an ESB (Equivalent Shear Beam) container at 50 g using the IFSTTAR centrifuge (Corté & Garnier 1986). Figure 1 depicts the experimental layout of the tests performed.

The soil profile consists of three horizontal layers of saturated HN31 sand (bottom), overconsolidated speswhite kaolin clay (middle) and dry HN31 sand (top). The properties of clay and sand are summarized in Tables 1 and 2.

The sand layers were obtained by air pluviation with an automatic hopper system at 1 g gravity level (Ternet 1999). The relative density was controlled to be 81%. The sand layer at the bottom of the container is vacuum saturated with water to avoid desaturation at the bottom of the clay layer and liquefaction of the sand layer (due to scaling laws the simulated prototype has thus a higher permeability).

The overconsolidated speswhite kaolinite formation is prepared in three layers and is subjected to 160 kPa preloading pressure under 1 g gravity field. The OCR values in the middle of the layers 1, 2 and 3 (Fig. 1) are 1.55, 2.13 and 3.42 respectively.

Accelerometers, pore pressure transducers and bender elements are placed in the soil to measure the soil response at different depths (Fig. 1). Acceleration and pore pressure results are not presented in this article. Bender elements are used to compare the shear wave velocity at different levels between both containers. Their results are discussed in the following section.

The soil profile is subjected to four 1-50-1 g cycles followed by a stabilization phase at 50g prior to the application of dynamic loads (based on the work of Khemakhem, 2012).

An instrumented aluminium tubular pile was used in the tests. It is equipped with 14 equally spaced half-bridge strain gauges along the inner side of the pile shaft in order to measure bending moment. The uppermost strain gauges are located at the soil surface level. Axial forces in the pile are measured by means of two

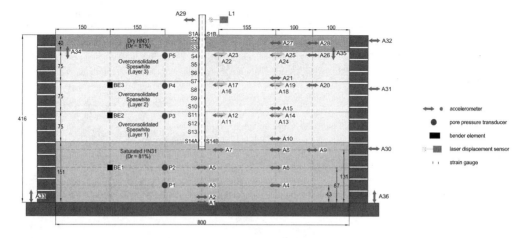

Figure 1. Experimental layout (dimensions are in mm, model scale).

Table 1. Properties of speswhite kaolin clay (Khemakhem 2012).

Symbol	Material properties	Value
γ_s	Unit weight of the grains	26.5 kN/m³
w_P	Plastic limit	30%
w_L	Liquid limit	55%
I_P	Plasticity index	25%
C_c	Compression index	0.33
C_s	Recompression index	0.06

Table 2. Properties of HN31 Hostun sand (Benahmed 2001).

Symbol	Material properties	Value
d_{50}	Average grain size	0.35 mm
C_u	Coefficient of uniformity	1.57
e_{min}	Minimum void ratio	0.656
e_{max}	Maximum void ratio	1
γ_s	Unit weight of the grains	26 kN/m³
$\gamma_{d,min}$	Minimum dry unit weight	13.047 kN/m³
$\gamma_{d,max}$	Maximum dry unit weight	15.696 kN/m³

Table 3. Properties of the instrumented pile.

Property	Hollow aluminium (model scale)	Reinforced concrete (prototype scale)
Outer diameter	18 mm	0.977 m
Length	335 mm	16.75 m
Young modulus	74 GPa	20 GPa
Flexural rigidity	1.43E-04 MNm²	8.95E02 MNm²
Yield moment	5.27E-05 MNm	6.59 MNm

Table 4. Characteristics of the selected inputs (prototype scale).

#	Input	PGA (g)	PGV (m/s)	Ia (m/s)	Duration (s)
1	Northridge	0.05	0.037	0.025	11.163
2	Landers	0.05	0.051	0.031	11.505
3	Northridge	0.3	0.220	0.897	11.163
4	Landers	0.3	0.306	1.109	11.505
5	Sine 1 Hz	0.1	0.157	1.417	16.348
6	Sine 3.2 Hz	0.1	0.049	0.443	5.112
7	Sine 1.8 Hz	0.1	0.087	0.787	9.089
8	Sine 2.4 Hz	0.1	0.065	0.588	6.808
9	Sine 1 Hz	0.3	0.470	12.760	16.348
10	Sine 3.2 Hz	0.3	0.146	3.986	5.112
11	Sine 1.8 Hz	0.3	0.261	7.087	9.089
12	Sine 2.4 Hz	0.3	0.195	5.295	6.808
13	Northridge	0.05	0.037	0.025	11.163
14	Landers	0.05	0.051	0.031	11.505

additional half-bridge strain gauges placed near the tip of the pile and at the level of soil surface. The properties of the pile are given in Table 3.

The pile was installed at 1 g gravity level using a hydraulic actuator. A low driving speed of 0.1 mm/s was used to allow dissipation of pore pressure in the surrounding soil. The pile is embedded one diameter in the saturated sand layer.

A total of two dynamic tests have been performed (identified C04 and C05), the second one being a reproducibility test. Each test involved 14 base shakings. Two types of input have been selected: sines with tapered parts and real broadband earthquakes Landers 1992 (Lucern Valley station) and Northridge 1994 (Tarzana station) records. In order to be in the capacity range of the shaker (Chazelas et al. 2008), earthquake signals have been filtered outside the frequency range of 0.4–6 Hz. The characteristics of these signals are given in Table 4 (PGA and PGV: peak ground acceleration and velocity, Ia: Arias Intensity). Figure 2 presents the frequency spectrum of the seismic inputs and the time representation of the 1 Hz signal.

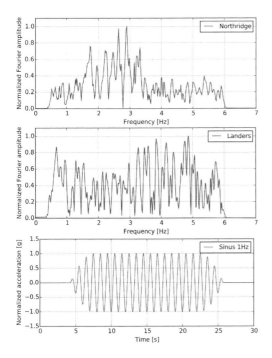

Figure 2. Frequency and time representation of seismic and 1Hz sine inputs (prototype scale).

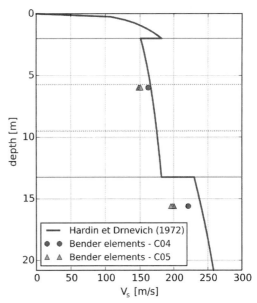

Figure 3. Small-strain shear wave velocity measures estimated from bender element tests compared to the Hardin & Drnevich (1972) empirical formulation.

It is found that there is not yet a consensus in the scientific literature as to the ideal parameter to estimate the intensity of an earthquake. Some authors consider that the most suitable is the Arias Intensity while others highlight the PGV or the PGD for strong earthquakes and the PGA for weak earthquakes (Auclair & Rey 2009).

The following criterion was applied in the choice of the order of application of the inputs: earthquakes are applied before the sines (the sinusoidal signals are generally more destructive) and in increasing order of the Arias Intensity (two weak and two strong earthquakes). Then the sines are applied in an order which is a function of the difference between the frequency of the signal and the response frequency of the soil column (estimated numerically from an elastic model at about 2.62 Hz). Two weak earthquakes are applied at the end to compare the response of the system before and after strong excitations.

3 ANALYSIS OF THE RESULTS

Unless otherwise indicated, the experimental results are given using prototype scale units.

3.1 Shear wave velocity profile

The shear wave velocity at 6 and 15.6 m depths is calculated from bender element measurements. The first arrival method based on the visual identification of the wave arrival to the receiver is used to identify the traveling time (Mitaritonna et al. 2010).

The shear wave velocities for the "virgin state" are obtained from measurement performed just before the first base shaking. The obtained values are compared to a theoretical profile in Figure 3.

The formula proposed by Hardin & Drnevich (1972) is used to estimate the maximum shear modulus profile:

$$G_{max} = 625 \frac{OCR^k}{0.3 + 0.7e^2} \sqrt{p_a \sigma'_m} \qquad (1)$$

where OCR = overconsolidation ratio; σ'_m = mean effective stress; and e = void ratio. The value of k depends on the plasticity index.

The shear wave velocity can be determined from the following equation:

$$V_s = \sqrt{G_{max}/\rho} \qquad (2)$$

An overall diminution of shear wave velocities is observed between C04 and C05 tests results. In the case of clay, C04 test measurements give an average velocity of 162.1 m/s where for C05 a value of 149.3 m/s is found. This represents a decrease of 7.9%. For HN31 sand, the trend is similar, with a decrease of 10.3% between C04 (221.2 m/s) and C05 (198.4 m/s). The small deviations observed suggest a good reproducibility of the shear wave velocity in the initial state.

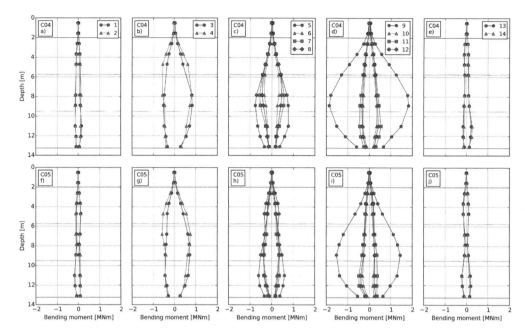

Figure 4. Measured maximum and minimum bending moments: (a–e) results from C04 test and (f–j) from C05. Curves are named following the numeration of base shakings given in Table 4.

The theoretical formula slightly overestimates the value of shear wave velocity for both clay and sand, giving shear wave velocity values of 152 m/s and 240 m/s respectively. It can be concluded that there is a good agreement between experimental results and the theoretical expression.

Several additional shear wave measurements were made through successive base shaking. Throughout both tests the shear wave velocity decreased on average by 28.6%. Much of this evolution is due to the 0.3 g sinusoidal base shakings (for instance 24.6% of decrease for C05 test). For the sand, bender element results show a slight increase in V_s values, being equal to +2.4% and +8.7% for C04 and C05 tests, respectively.

Change of V_S in sand may be explained by a densification of the granular structure. The evolution of clay, on the other hand, needs to be looked at more closely in order to identify the mechanism (or mechanisms) behind this important change of shear wave velocity.

3.2 Maximum bending moments in pile

Figure 4 shows the envelope curve of the maximum bending moments measured in the pile for every dynamic loading in C04 and C05 tests. Results are grouped according to PGA level and motion type.

With the exception of the 1Hz sine input, good reproducibility is observed between C04 and C05 tests results. The maximum bending moments are in the same order of magnitude and the envelope profiles have the same overall form.

For both tests the maximum bending moment has been recorded during the 1 Hz 0.3 g base shaking. The maximum value for the C04 and C05 tests were measured in the clay at a depth of 8.9 m and are respectively 1.87 MNm and 1.53 MNm. The difference represents a decrease of 18.2% between C04 and C05.

As previously mentioned the order of the sine base shaking was based on the PGA and by taking into account the numerically determined frequency response of the soil columns at low deformations. For the same PGA, the maximum moment decreases when the frequency of the sine input is increased (Figs 4c,d,i). These differences are due to resonance and amplitude effects. For 1Hz sine excitation, an important amplification of the response is observed between 0.1 g and 0.3 g shakings. On the contrary, results for 1.8 Hz, 2.4 Hz and 3.2 Hz are almost identical for 0.1 g and 0.3 g levels of PGA. It can be concluded that the effect of frequency in the maximum moment depends on the level of PGA and that this relation is highly non-linear, with almost no impact for frequencies 1.8 Hz to 3.2 Hz.

Near the pile tip, the bending moments were measured in the vicinity of the clay-sand interface. Except for the low amplitude earthquakes the bending moments found close to the pile tip are not negligible. The maximum values are found for 1 Hz sinusoidal base shaking (0.64 MNm for C04 and 0.61 MNm for C05). The near pile tip bending moment represents up to 70% of the maximum moment (in average) in the case of weak earthquakes applied at the end of both tests.

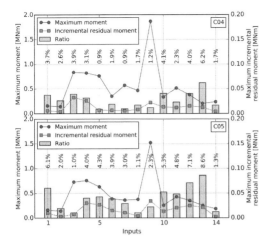

Figure 5. Maximum bending moment vs. maximum incremental residual bending moment for every input.

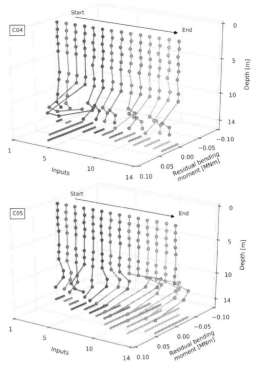

Figure 6. Evolution of cumulative residual bending moment during all 14 excitation in C04 and C05 tests.

3.3 Residual bending moments

The evolution of incremental residual bending moment, defined as the difference between the bending moment at the beginning and the end of base shaking (for each input) is presented in Figure 5. It can be observed that incremental residual bending moments remain low, with a maximum value of 0.035 MNm found for 1 Hz sine at 0.3 g for C05 test.

In order to estimate the importance of the residual moment induced by each dynamic loading, the ratio between maximum bending moment and incremental residual bending moment is considered. This ratio is given in Figure 5 and shows that the incremental residual moment remains limited in comparison to the maximum bending moment for all the cases, with maximum values of 6.2% and 8.6% found for Northridge 0.05 g input in C04 and C05 tests respectively.

Figure 6 presents the evolution of the cumulative residual bending moment, defined as the bending moment at the end of each shaking minus the initial electric offset at the beginning of first input, for C04 and C05 tests. It may be noticed that the evolution of cumulative residual bending moment takes place essentially between 9 m depth and the tip of the pile. This confirms the importance of the moment near the pile tip as mentioned before.

It is to be noted that the sign of the incremental residual moment has an impact on the evolution of cumulative residual bending moment as it is observed in Figure 6. In some cases the incremental residual moment would increase the accumulated residual moment from previous base shakings whereas in other cases it would decrease cumulated values (opposite signs).

For both tests, C04 and C05, the maximum ratios between the cumulative residual bending moment and the maximum moment are found for the last two base shakings (Northridge and Landers earthquakes at 0.05 g). In the case of C04 test, the ratios are 15.4% and 13.9% respectively. For C05 the corresponding values are 22.9% and 23.6%. This result shows that the cumulative residual bending moment is not negligible at the end of the base shakings series when weak signals are used after strong ones.

3.4 Comparison of bending moments at the beginning and at the end of the input series

The recorded maximum bending moment envelopes for the two weak earthquake inputs applied at the beginning and at the end of the input series are compared to study the influence of the input history in the response of the pile. This comparison is given in Figure 7.

It is found that the results remain in the same order of magnitude. The maximum and minimum bending moment profiles obtained for the end-of-series shots show larger peaks which surpass almost systematically the values recorded for the inputs applied at the beginning of each series.

In addition, it is observed that the moment near the tip of the pile is always greater at the end of the tests. This is accompanied by an increase of the ratio between near pile moment to maximum moment of about 18%. This means that the difference between the maximum moment near the pile tip and the maximum pile moment decreases as a result of the precedent large base shakings.

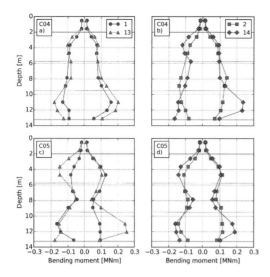

Figure 7. Comparison of recorded bending moments in pile at the beginning and at the end of the input series: (a-b) results from C04 test and (c-d) from C05. Curves are named following the numeration of base shakings given in Table 4.

The changes observed in the response of the pile are partly due to the evolution of the soil (as introduced in §3.1), which in the case of clay presents a smaller resistance at the end of the test and thus permits higher displacements of pile. The cumulative residual bending moment plays also an important role in these changes, imposing a preloaded initial state to the pile at the beginning of each shaking.

4 CONCLUSIONS

Dynamic centrifuge tests were conducted to study the behaviour of free-head end-bearing single piles in a layered soil. The following conclusions are drawn from the results of the study.

- Despite the complexity in the fabrication of the soil profile, the differences between the two tests in terms of shear wave velocity remain below 10.4%, which shows a satisfactory repeatability of the soil profile.
- The comparison in terms of maximum moment envelopes (overall shape of curves, evolution and maximum values), between C04 and C05 shows a good repeatability of the tests.
- The loading history influences the results of subsequent loadings, especially when low amplitude loads (e.g. 0.05 g earthquakes) are applied after strong inputs (e.g. 0.3 g sine).
- The maximum bending moment of the pile depends on the frequency content and the amplitude (PGA) of the input signal. The weight of each factor in the response is not linear and results show a high influence of PGA in 1 Hz sine results.
- Bending moments due to pile tip embedment in dense sand are not negligible, especially when strong motions are applied.

ACKNOWLEDGMENTS

The work presented in this paper has been conducted with the financial support of Terrasol company (SETEC group). Their support is greatly acknowledged.

REFERENCES

Auclair, S. & Rey, J. 2009. *Corrélation indicateur de mouvement du sol/intensité. Vers l'acquisition conjointe de données instrumentales et macrosismiques*. Report BRGM/RP-57785-FR.

Banerjee, S. 2009. Centrifuge and numerical modelling of soft clay-pile-raft foundations subjected to seismic loading. Ph.D. thesis, National University of Singapore.

Benahmed, N. 2001. *Comportement mécanique d'un sable sous cisaillement monotone et cyclique*. Ph.D. thesis, Ecole Nationale des Ponts et Chaussées.

Boulanger, R.W., Curras, C.J., Kutter, B.L., Wilson, D.W. & Abghari, A. 1999. Seismic soil-pile-structure interaction experiments and analyses. *Journal of Geotechnical and Geoenvironmental Engineering*, 125: 750–759.

Chazelas, J.L., Escoffier, S., Garnier, J., Thorel, L. & Rault, G. 2008. Original technologies for proven performances for the new LCPC earthquake simulator. *Bulletin of Earthquake Engineering*, 6: 723–728.

Correia, A.A., Pecker, A., Kramer, S.L. & Pinho, R. 2012. Nonlinear pile-head macro-element model: SSI effects on the seismic response of a monoshaft-supported bridge. 15th WCEE, Lisbon.

Corté, J.F. & Garnier, J. 1986. A centrifuge for research in geotechnics. *Bull Liaison Lab Ponts et Chaussées*, 145.

Hardin, B.O. & Drnevich, V.P. 1972. Shear modulus and damping in soils: design equations and curves. *Journal of the Soil Mechanics and Foundations Division*, 98: 667–692.

Khemakhem, M. 2012. *Etude expérimentale de la réponse aux charges latérales monotones et cycliques d'un pieu foré dans l'argile*. Ph.D. thesis, Ecole Centrale de Nantes.

Li, Z., Kotronis, P., Escoffier, S. & Tamagnini, C. 2016. A hypoplastic macroelement for single vertical piles in sand subjected to three-dimensional loading conditions. *Acta Geotechnica*, 11: 373–390.

Meymand, P.J. 1998. Shaking table scale model tests of non-linear soil-pile-superstructure interaction in soft clay. Ph.D. thesis, University of California, Berkeley.

Mitaritonna, G., Amorosi, A. & Cotecchia, F. 2010. Multidirectionnal bender element measurements in the triaxial cell: equipment set-up and signal interpretation. *RIG Rivista Ita-liana di Geotecnica*, 44: 50–69.

Ternet, O. 1999. Reconstitution and characterization of the sand samples. Application to the tests out of the centrifuge and calibration chamber. Ph.D. thesis, Université de Caen.

Zhang, L., Goh, S.H. & Yi, J. 2017. A centrifuge study of the seismic response of pile-raft systems embedded in soft clay. *Géotechnique*, 67: 479–49.

Behaviour of piled raft foundation systems in soft soil with consolidation process

E. Rodríguez
National University of Colombia, Colombia

R.P. Cunha
New Institute, Gouda, The Netherlands

B. Caicedo
Los Andes University, Colombia

ABSTRACT: This research aims to evaluate the behaviour of Piled Raft Foundation (PRF) systems, built on soft clay in consolidation process due to the structural loads and the reduction of the pore water pressure generated by pumping water from deep permeable layers. The experimental work was performed using reduced scale models in geotechnical centrifuge. It was observed that settlement controls are negligible beyond a number of piles. Furthermore, due to the reduction process of pore water pressure, the raft loses contact with the soil partially and, therefore, transfers a higher load to the piles into the PRF system. Thus, if the piles are designed for working very close to the ultimate load, the risk of a total system failure is greater.

1 INTRODUCTION

A significant number of buildings are being constructed over Piled Raft Foundation systems (PRF) (Poulos 2001; Garcia 2015). This system can reduce the full relaxation stresses during the excavation processes and control the differential and total settlement (El Mossallamy 2002). In this type of system, the load can be supported by both, the plate and the piles, with an independent safety factor for each element.

According to Banerjee (2009), several cities with soft soils (clayey) use this type of foundation: Shanghai, Bangkok, Mumbai, Kuala Lumpur, Jakarta, Singapore, Mexico and Bogotá. In these cities there are subsidence phenomena related to the reduction of pore pressures in the soil. As a consequence, when the total stress condition varies, the working conditions of the foundation system will also vary, causing damage, as reported by Bareño & Rodríguez (1999).

As a result of the subsidence due to the lowering of pore pressure, negative friction can be induced in the piles installed in these soils. This negative friction produces additional vertical loads to those on the project. These loads (*download*) are associated, in turn, to settlements (*downdrag*) which, in extreme conditions, can cause the element to fail (Leung et al. 2004).

Different studies have been carried out to understand the behaviour of individual piles constructed in soft soils with consolidation by load or pore pressure reduction. Most of these works are focused on piles working by tip (Rodriguez 2016).

A few experimental studies have been reported in the literature, using 1g scale model or geotechnical centrifuge models to analyse the behaviour of PRF systems or piles groups, constructed in soft soils (Thaher & Jessberger 1991; Horikoshi & Randolph 1996; Tran et al. 2012).

The results presented in this article are the product of the physical modelling of two PRF models in a geotechnical centrifuge. The two models used nine piles with different distributions.

The systems responses are analysed and discussed, based on the settlement data reports and the variation in the load distribution, for the different configurations.

2 EXPERIMENTAL PROGRAMME

2.1 Testing apparatus

The testing facility employed during this research is located at the Geotechnical Models Laboratory of Los Andes University. This facility consists of a beam geotechnical centrifuge with the characteristics shown in Table 1.

2.2 Geometry and scale of the models

The scale factor, the geometry and the elements of the models were selected based on the modelling box dimensions and the capacity of the geotechnical centrifuge. In this research, the models were evaluated at a scale of 1/200 (200 g).

Table 1. Geotechnical centrifuge characteristics.

Turning radius	1.90 m
Model boxes dimensions	40 × 50 × 50 cm
Gravitational field maximum	200 g
Maximum model weight	4.0 kN
Nominal power	3 HP
Channels for data acquisition	40

Table 2. Kaolin properties.

Parameter		Value
Specific gravity	Gs	2.68
Liquid limit	w_L (%)	54
Plasticity index	I_P (%)	33
Plastic limit	w_P (%)	21
Compression index	Cc	0.37
Swelling index	Cs	0.09
Vertical consolidation coefficient	*Cv (m²/s) × 10^{-6}	0.49–0.62

*Cv at 100 kPa vertical stress.

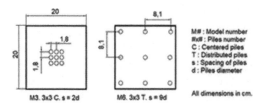

Figure 1. Geometrical configurations of the models.

The geometric configurations evaluated are shown in Figure 1. Aluminium of 1.3 cm of thickness and 20 × 20 cm of area was used for the plate. Aluminium tubes of 9 mm in diameter and 320 mm in length were used for the piles.

2.3 Soil profile

The profile proposed by Rincón & Rodríguez (2001) was used for the soil. The cohesion profile varies in depth between 10 kPa at the top and 40 kPa at the bottom. This profile represents a soft clay soil typical of Bogotá City.

2.4 Soil manufacturing

Commercial kaolin was used for the soil profile construction, with the characteristics shown in Table 2.

The procedure described by Rincón & Rodríguez (2001) was used for the soil fabrication. It consists of the following steps:

1. Prepare a kaolin slurry, with water content of 1.5 times the liquid limit;
2. Place this grout in a model box;
3. Establish a correlation curve between the applied load and non-drained resistance
4. Fabricate the layers of material in the modelling box to reproduce the proposed soil profile.

At the bottom of the modelling boxes, a layer of filtering sand was arranged to control the hydraulic load in the model. The hydraulic pressure was controlled by an external tank connected to the bottom of the box.

In order to carry out a faster consolidation process, intermediate sand filters were constructed between the clay layers. These filters have a thickness of 0,7 cm and are connected with lateral geotextiles arranged in the walls of the modelling boxes.

The loads over each layer were increased in stages, as in a normal consolidation test. Each load was maintained until reaching 90% of the total consolidation, controlling this process with the Taylor method.

During the entire process of soil manufacturing, the external hydraulic load was maintained by means of a constant water level in the external tank.

2.5 Testing stages

Based on the objective of this research, the phenomenon to be reproduced corresponds to the process of consolidation of a soil by external loads or by a reduction of pore pressure in depth. This reduction represents the subsidence phenomenon induced by the extraction of water from deep aquifers.

In order to generate the lowering of pore pressures to a specific depth, the modelling boxes presented a filtering lower layer (sand) under the layers of clay constructed with kaolin. Through this porous sand (permeable layer), the equilibrium condition of pore pressures can be guaranteed in the soil, when only the influence of the load is evaluated. It is also possible to represent the lowering of pore pressure by allowing water out of the modelling box and removing the connection with the external tank of water control.

The proposed testing stages are schematised in Figure 2. Stops were made during the testing due to the impossibility of making modifications to the model during the test.

2.6 Instrumentation, load and times

Based on the model scale and the parameters to be read, LVDTs gauges, load cells (on the plate and on the head of the piles) and piezometers in the soil were implemented. Figures 3 and 4, show the arrangement of the instruments for each model.

The load was applied in stages, until reaching the ultimate load, simulating a constructive process of at least one year.

Where **PP, Pz**: Piezometers; **C**: Load cell on the plate; **Er**: LVDT on the plate; **Es**: LVDT on the soil; **f**: Load cell on the pile head.

A loading pneumatic system was used to apply the load over the PRF system. The maximum admissible pressure to the plate, obtained by the Meyerhoff methodology, was 62.4 kPa. This pressure is equivalent to 2.46 kN on the model.

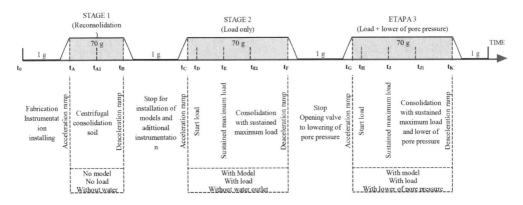

Figure 2. Model's time line.

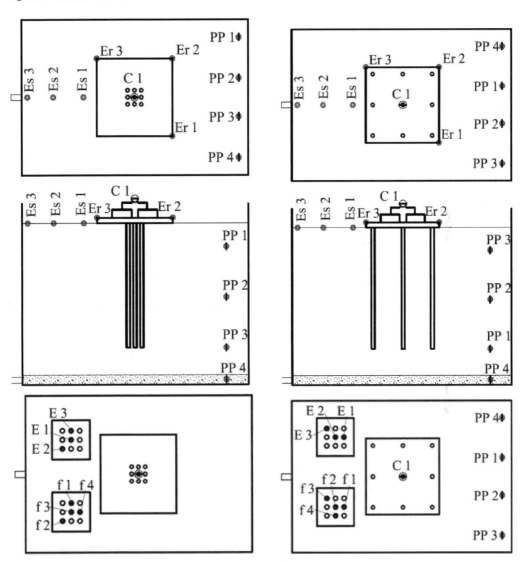

Figure 3. Instrumentation distribution and load assembly in M3 Model.

Figure 4. Instrumentation distribution and load assembly in M6 Model.

Table 3. Filling factors to the models (FF).

Parameter	M3	M6
d (cm)	63	63
N_p	9	9
s (cm)	126	567
A_R (m^2)	196	196
A_G (m^2)	63.5	12.86
FF	0,016	0,073

3 EXPERIMENTAL RESULTS AND DISCUSSION

3.1 Displacements

The measured displacements over both, the soil surface and the plate, were scaled to the dimensions of the equivalent prototype.

The Filling Factor (FF) in Equation 1, defined by Mandolini et al. (2013), was used to carry out an analysis in the final condition (long-term), which is the critical behaviour condition.

$$FF = \frac{A_G}{A_R}\frac{d}{s} \qquad (1)$$

where A_G: piles group area, defined by Sanctis et al. (2002) in Equation 2:

$$A_G = \left[\left(\sqrt{N_p}-1\right)s\right]^2 \qquad (2)$$

where A_R: plate area, s: piles spacing; d: piles diameter and, N_p: number of piles in the group.

The filling factors (FF) for the two PRF systems used in this paper are presented in Table 3.

The displacements profiles along the central axis in the longest direction of the model box are presented in Figure 5. In the two models, the displacement is controlled by the presence of the piles. For the model with distributed piles, the settlements are smaller in Stage 2. Both PRF models present similar values of final settlement. The largest settlements are presented in Stage 3, when the lowering of pore pressure occurs.

The influence of the piles presence in the displacements of the soil is evidenced due to the influence of the self weight of these elements. PRF systems with piles grouped in the centre of the plate present larger settlements, compared to those with piles distributed throughout the area. Similar observations presented by Cunha et al. (2000) indicate that settlement decreases with the inclusion of piles, especially when placed in the centre. However, Bisht & Singh (2012) pointed out that with smaller spacing than to 3d, the pressure bulbs (of stress) overlap and increase the settlement, as seemed to happen in this case.

3.2 Load distribution

The proportion of the final load supported by each pile was calculated with the data of the cell load placed in the piles head, it was possible to calculate the proportion of final load that is supported by each pile, as shown in Table 4. Piles in symmetrical positions are assumed to have similar loads to determine the total load on the piles (Lee 1993).

The proportion of load in the Stage 2, initially supported by the plate, is between 74 and 77%. This proportion is lower when the lowering of pore pressure occurs, reaching values between 46 and 54%. This variation indicates that there is a transfer of load from the plate to the piles.

Mandolini et al. (2013) presented a graph that relates the proportion of load supported by the plate with the filling factor of the PRF systems. The results

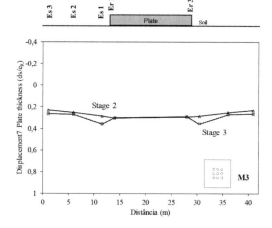

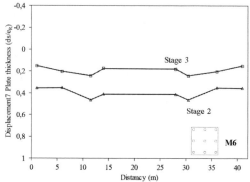

Figure 5. Displacement profiles in large axis model box.

Table 4. Load distribution.

Model	Stage 2		Stage 3	
	Load on pile group (kN)	Load on plate (kN)	Load on pile group (kN)	Load on plate (kN)
M3	1753.86 23%	5742.70 77%	3423.10 46%	3994.75 54%
M6	1916.69 26%	5442.14 74%	3962.80 54%	3356.67 46%

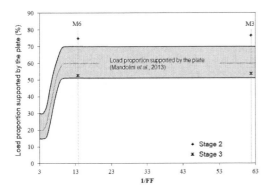

Figure 6. Load proportion supported by the plate.

of the load percentage assumed by the plate in this paper are drawn in the proposed graph of Figure 6.

Thus, the initial load values supported by the plate in the two PRF models, for the Stage 2, are less than 80% at the upper limit of the graph. When the lowering of pore pressure is applied, these values fall to the lower limit of the graph, with values close to 50%.

The increase of general load in the piles groups varies between 23 and 28%. The load transfer from the plate to the piles is associated with the settlements of the system and the possible partial loss of contact between the plate and the consolidated soil.

When comparing the maximum load supported by the piles with the theoretical load of project, it does not exceed 56%. However, the piles had a maximum increase of 16% in their loads between Stage 2 to Stage 3.

4 CONCLUSIONS

The inclusion of piles in Piled Raft Foundation systems allows the control of the settlements generated over the structure. It can be observed that in the process of subsidence or consolidation of a soil, represented in this paper by the lowering of pore pressures, a greater settlement of the soil mass is generated. This displacement is not accompanied by the structural system. As a consequence of the greater settlements of the soil, a separation of this and the plate can occur modifying the load conditions in the PRS.

A greater separation of piles does not necessarily control the settlements induced by the load or by the lowering of pore pressure and the piles are less efficient to control the settlements when consolidation processes are generated in the soil.

The proportion of load supported by the piles increases with the processes of lowering of pore pressure in the soil. As the spacing between the piles increases, the load assumed by the piles also increases. This phenomenon happens because the soil around the piles is mobilised to resist the applied load.

As the plate loses capacity to support load due to a possible separation effect between it and the soil, the piles assume a higher percentage of load. This can lead to the pile failure if the load design, as is common in many cases, is very close to the ultimate load or failure load pile.

ACKNOWLEDGEMENTS

The authors would like to acknowledge National University of Colombia, Los Andes University and COLCIENCIAS in Colombia and University of Brasilia and CNPq in Brazil, which collectively funded the doctoral research from which the information for the present article was extracted.

REFERENCES

Banerjee, S. 2009. Centrifuge and Numerical Modelling of Soft Clay-Pile-Raft Foundations Subject-ed to Seismic Shaking. PhD Thesis, University of Singapore, Singapore.

Bareño, E. & Rodríguez, E. 1999. Clays Shrinkage (In Spanish). *Undergraduate Thesis, National University of Colombia*, Bogotá D.C, Colombia.

Bisht, R. S. & Singh, B. 2012. Study on behaviour of piled raft foundation by numerical modelling. *Research Symposium on Engineering Advancements, SAITM–RSEA 2012*, Colombo, 1: 23–26.

EL-Mossallamy, Y. 2002. Innovative Application of Piled Raft Foundation in stiff and soft subsoil. *Deep Foundations 2002*, ASCE, Orlando, Florida: 426–440.

Garcia, J.R. 2015. Experimental and numerical analysis of piled raft executed in Campinas soils. (In Portuguese). *PhD Thesis, Campinas Statal University*, São Paulo, Brasil.

Horikoshi, K. & Randolph, M. 1996. Centrifuge modelling of piled raft foundations on clay. *Géotechnique* 46(4): 741–752.

Leung, C., Liao, B., Chow, Y., Shen, R. & Kog, Y. 2004. Behavior of pile subject to negative skin friction and axial load. *Soils and Foundations*. 44(6):17–26.

Mandolini, A., D.I Laora, R. & Mascarucci, Y. 2013. Rational Design of Piled Raft. *Procedia Engineering*, 57: 45–52.

Poulos, H.G. 2001. Piled raft foundations: design and applications. *Géotechnique*, 51(2): 95–113.

Rincón, C.L. & Rodríguez, E. (2001). Centrifuge Physical Modeling of a wall without anchors in a soft clay (In Spanish). *Master Tesis, Facultad de Ingeniería, Universidad de Los Andes*. Bogotá D. C. Colombia.

Rodriguez, E. 2016. Experimental analysis of the behavior of piled raft systems in soft soils with consolidation process. (In Portuguese). *PhD Thesis, University of Brasilia*, Brasília D.F., Brazil.

Thaher, M. & Jessberger, H.L., 1991. The behavior of pile-raft foundatios, investigated in centrifuge model tests. *Centrifuge 91, ISMFE*, Rotterdam, Germany: 225–234.

Tran, T. V., Teramoto, S., Kimura, M., Boonyatee, T. & Vinh, L. B. 2012. Effect of ground subsidence on load sharing and settlement of raft and piled raft foundations. *World Academy of Science, Engineering and Technology, International Journal of Civil, Envi-ronmental, Structural, Construction and Architectural Engineering*, 6(2): 120–127.

Displacement measurements of ground and piles in sand subjected to reverse faulting

C.F. Yao, S. Seki & J. Takemura
Department of Civil and Environmental Engineering, Tokyo Institute of Technology, Japan

ABSTRACT: Pile foundations are deemed to be vulnerable to large ground deformation caused by faulting, especially reverse faulting due to large discontinuous ground displacement under large earth pressure. As the relative positions of piles to the fault rupture is a key condition on pile damages, it is necessary to observe the behaviour of piles at different positions. In this paper, centrifuge model tests have been conducted to investigate the behaviour of single piles located to various points in sand subjected to reverse faulting. From the vertical and horizontal displacements, and rotation of pile caps measured by LDT scanning technique, and strains of piles, the internal forces and deflection of piles in the ground were measured. Furthermore vertical and horizontal displacements of ground surface and rupture propagation in sand were observed by the scanning and PIV techniques respectively. These displacement measurements of piles and ground could enable to observe the complicated soil–pile interaction. For example, when the fault rupture crosses the middle part of pile, it is likely be sheared-off near the cross point, while the pile behaves like a cantilever when the rupture crosses its upper part.

1 INTRODUCTION

Permanent ground deformations due to faulting can cause catastrophic damages to structures located within fault zone, which have been observed in Turkey in 1999 (e.g. Bray, J. D. 2001), Taiwan (Chi-Chi) also in 1999 (e.g. Chen et al. 2000) as well as New Zealand in 2010 (Van Dissen 2011). These severe damages of infrastructures motivated studies to explore mechanisms of structure-fault interaction, most of which focused on shallow foundations and pipelines (e.g. Bransby et al. 2008). However, pile foundations were observed more vulnerable to fault deformation than the shallow ones, especially in reverse fault that large discontinuous ground displacement induced under large earth pressure. Therefore, it is significant to investigate the behaviour of piles subjected to reverse faulting.

Centrifuge modelling is a good approach for deep insight into pile behaviour during faulting due to the lack of well-documented field data. During the experiments, the displacement measurements of ground and piles are the major concern, as precise and detailed data is important for the investigation of the pile-soil interaction mechanism subjected to fault deformation.

In this paper, centrifuge model tests have been conducted to investigate the behaviour of single piles located to various points in sand subjected to reverse faulting. Both LDT scanning technique and PIV techniques are applied to measure the displacement of ground and piles. The displacements and rotation of pile caps as well as the ground surface displacements are measured by LDT scanning technique, and rupture propagation and ground displacements are observed by Particle Image Velocimetry (PIV) techniques (White et al. 2003). Moreover, strain gauge measurement is employed to observe strains, internal forces and deflection of piles. On the basis of the above-mentioned measurements, fault rupture propagation, ground deformation and pile behaviour are explored.

2 DISPLACEMENT MEASUREMENTS METHOD

2.1 The fault simulator

A fault simulator was adopted to simulate the fault deformation, as illustrated in Figure 1. It has total dimensions of 800 mm long, 600 mm wide and 755 mm high. A container (item 2 in Fig. 1) for model ground is designed on the upper part of the fault simulator, with dimensions of 500 mm × (length) 300 mm (width) × 400 (height). The left sidewall and base (item 7 in Fig. 1) is moveable to motivate a reverse or normal fault deformation, with the maximum bedrock offset of 30 mm at the dip angle of 60°. A screw jack is installed to support the movable sidewall and base, and the speed of bedrock offset is controlled by an AC servo-motor (item 6 in Fig. 1). The front wall of the container is transparent for ground deformation observation. In front of the container, a digital camera (item 4 in Fig. 1) is mounted to record the images during faulting. Details of the simulator were reported by Takemura et al. (2010).

1. Observational window
2. Model ground container
3. Trapezoidal prim
4. Digital camera
5. Jigs supporting camera
6. Worm gear
7. Moveable side wall & base
8. Pentagon block
9. Stationary base
   & inclined guide
10. Brackets &
    outer side wall
11. Sandpaper

Figure 1. The fault simulator.

Table 1. Test cases.

Cases	Fault type	Dr. of sand
RP60	Reverse	60%
RP80	Reverse	80%

2.2 Test setup

All tests were performed under an acceleration of 50G. Hereafter, all data will be reported at prototype scale unless otherwise stated. There were two cases, namely RP60 and RP80, with relative density sand of 60% and 80% respectively (Table 1). Dry Toyoura sand ($G_s = 2.65$, $D_{50} = 0.19$ mm, $e_{max} = 0.973$, $e_{min} = 0.609$) was used to make the ground with the height of 11 m, as shown in Figure 2.

In each case, there were four piles, namely pile 1, 2, 3 and 4, located at various points with an interval of

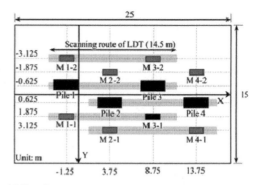

(a) Top view

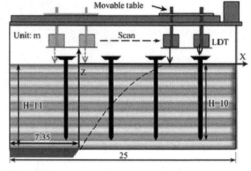

(b) Front view

Figure 2. Test setup.

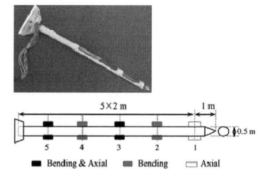

Figure 3. Model pile.

5 m. Two markers were embedded at both the front and back sides of each pile to obtain ground surface displacements by LDTs. The coordinate system is illustrated in Figure 2. The model pile made of a solid acrylic circular bar ($EI = 9.63 \times 10^6$ N.m^2) was 11 m long with an embedded length of 10 m (Fig. 3). Five pairs of strain gauges were pasted on the pile surface to measure the strains of piles during faulting.

2.3 Displacement measurements of ground and piles

LDT scanning technique was developed to access the multi-points displacement measurements of ground

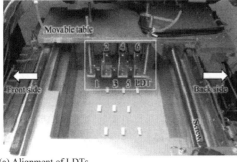

(a) Alignment of LDTs

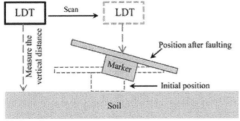

(b) Scanning process

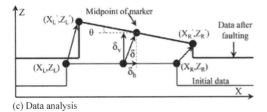

(c) Data analysis

Figure 4. LDT scanning technique.

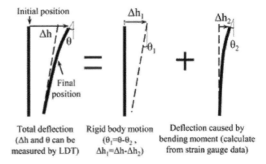

Figure 5. Calculation method of pile deflection.

surface and pile caps. Six LDTs are fixed on a moveable table (Fig. 4a) with a scanning distance of 14.5 m (Fig. 2a). A LVDT parallel to X-axis is installed to record the positions of LDTs at X coordinate during scanning, which was made every 100 mm (2 mm in model scale) offset.

When the LDTs scan from one side to the other side, it will record the distance from the LDTs to the ground surface and the corresponding X coordinate values (Fig. 4b), by which the profiles of ground surface can be captured (Fig. 4c). In data analysis, left and right edges of markers or pile caps are detected, from which the coordinates of the two edges, the horizontal and vertical displacements and the rotations of markers or pile caps can be obtained.

The calculation of pile deflection along the pile is based on data measured by LDTs and strain gauges. The total pile deflection includes two parts: rigid body motion and deflection resulted from bending moment, as illustrated in Figure 5. The total deflection (Δh) and rotation (θ) of pile top is measured by LDT scanning technique, which can provide a boundary condition for the calculation. The distribution of pile deflection caused by bending moment can be calculated directly by the measured strains. The pile top values measured by LDTs (Δh and θ) and strain gauges (Δh_2 and θ_2) provide basic data to calculate deflection ($\Delta h_1 = \Delta h - \Delta h_2$) and rotation ($\theta_1 = \theta - \theta_2$) at pile top caused by rigid body motion. The distribution of pile deflection resulted from rigid body motion can be obtained. Finally, the total pile deflection distribution equals the sum of deflection values caused by rigid body motion and bending moment.

In addition, PIV technique (White et al. (2003)) was applied to analyse the ground displacements, based on images captured by digital camera during faulting.

2.4 *Test procedures*

The test procedure was as follows: (1) set the piles by jigs, and pour Toyoura sand into the container by air pluviation method (inked sand was poured every m thickness in close proximity to observational window); (2) put the makers on the ground surface, and place the movable table holding six LDTs on the simulator; (3) mount the completed test setup on the swing platform, and run the centrifuge; (4) record initial data of LDTs, strain gauges and front view images when the acceleration of ground model reaches 50 G; (5) activate the fault simulator up to the maximum offset of 1.5 m and stop it at every fault offset increment of 0.05 m, then record all data at every fault offset increment of 0.05 m.

3 RESULTS AND DISCUSSIONS

3.1 *Accuracy of LDT scanning technique*

The accuracy of LDT scanning technique is one of the major concerns for discussion of pile behaviour. To make sure the accuracy of LDT scanning technique, comparisons of ground surface displacements measured by LDT scanning and PIV in Case RP60 are shown in Figure 6. Figure 6a shows good agreement, together with horizontal and vertical component of bedrock offset (δb_h, δb_v), for both horizontal and vertical displacements observed by LDT scanning and PIV. Although vertical displacements obtained by LDT scanning agree with that observed by PIV at X = 3.75 m (Fig. 6b), the horizontal displacements obtained by LDT scanning are a little larger than that of PIV. This difference is more obvious for both horizontal and vertical displacements in Figure 6c, especially

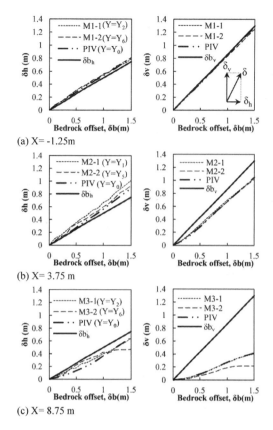

(a) X= −1.25 m

(b) X= 3.75 m

(c) X= 8.75 m

Figure 6. Comparison of ground surface displacements obtained from LDT scanning and PIV in Case RP60 ($Y_0 = 7.5$ m, $Y_1 = 3.125$ m, $Y_2 = 1.875$ m, $Y_5 = -1.875$ m, $Y_6 = -3.125$ m).

for Marker 3-2. It should be pointed out that the markers and observed points of PIV have different Y coordinate values. So the possible reason of the difference is that both the wall friction and piles could prevent the development of ground displacements. The differences of ground surface displacements at different Y coordinates become the most marked at X = 8.75 m, near the scarp the fault rupture. The difference at X = −1.25 m is very small because the influence of pile is very weak, as pile 1 is likely to move together with the hanging wall (the pile behaviour will be discussed later).

Although there is a difference for displacements observed by LDT scanning and PIV at X = 3.75 m and 8.75 m, the comparisons at X = −1.25 m have proved that LDT scanning has enough accuracy for multi-points displacement measurements.

3.2 Rupture propagation

Figure 7 shows the observed fault ruptures of Case RP60 and RP80 at $\delta b = 1.5$ m. The surface ruptures are fluctuated along the Y coordinate (Fig. 7a), which may be influenced by the piles and wall friction. For loose sand case, there is only one surface rupture, while in dense sand an additional half rupture induced.

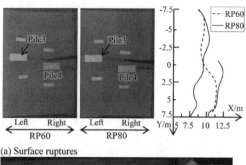

(a) Surface ruptures

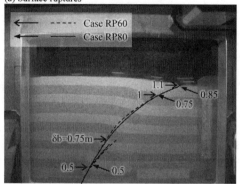

(b) Ruptures propagation

Figure 7. Fault rupture propagation in sand.

Rupture propagations in loose and dense sand (Fig. 7b) show small differences. The ruptures in dense sand propagated faster than those in loose sand, because larger stiffness of sand leads to earlier propagation (Lin et al. 2006). The offset ratio ($\delta b_v/H$, the vertical component of bedrock offset δb_v normalized by ground height H) for outcropping at ground surface was 10%, which was the same with that observed in a free-field test in loose sand (Dr. = 60%) conducted by Bransby et al. (2008). The influence width (horizontal distance from the fault tip to the rupture top W) was 0.984 H for dense sand, and was supported by results of Cole & Lade (1984) that the W/H value was 0.98 in dense sand (Dr. = 80%) for reverse rupture having dip angle of 60°.

3.3 Ground deformation

Figure 8a shows the ground surface profiles of each case at $\delta b = 0.5$ m, 1.0 m and 1.5 m observed by LDTs and PIV. The fault deformation caused large deformation on the hanging wall side, while the displacements were almost zero on the footwall side. A scarp was gradually induced with the increase of bedrock offset. Within the scarp area, the vertical displacement of dense sand was larger than that of loose sand due to earlier propagation.

Figure 8b shows the vertical displacements of sand at different depth in both cases at $\delta b = 1.0$ m, which are normalized by the vertical component of bedrock offset. Clear discontinuities could be observed at all depths, and they were steeper in dense sand than

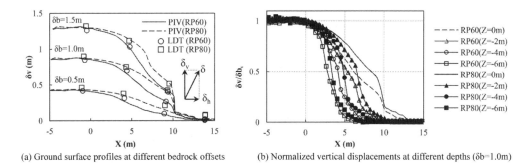

Figure 8. Observed ground displacements.

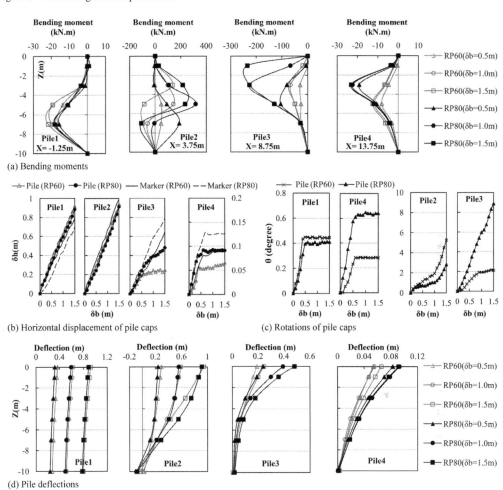

Figure 9. Behaviour of piles.

those in loose sand due to faster propagation speed in dense sand. Clear discontinuities were also observed in deeper locations in both cases.

3.4 Pile behaviour

With the above mentioned measurements, the behaviour of piles was depicted as shown in Figure 9. Bending moments are taken to be positive if tensile stress was induced on the footwall side.

When the piles were crossed at the middle part by the ruptures, i.e. Pile 2, the bending moments were mainly concentrated on the lower part of pile at $\delta b = 0.5$ m (Fig. 9a). However, as the bedrock offset increased, large positive and negative bending moment appeared at the lower side and the upper side of the rupture respectively. Although the horizontal pile cap displacement of loose sand was a little larger than that of dense sand (Fig. 9b), the bending moments in loose sand was smaller than that of dense sand due to smaller stiffness of the sand. The pile in loose sand had almost the same horizontal displacement of pile cap with that in dense sand (Fig. 9b), but the pile cap rotation (Fig. 9c). The piles were highly bent at the middle part under large bedrock offset (Fig. 9d), which indicates that piles are likely be sheared-off near the cross point of the rupture surface. It should be noted that the gauge spacing (40 mm in the model, 2 m in the prototype may not be enough to capture sharp bent caused by the rupture.

When piles were crossed at the upper part by the ruptures, i.e. Pile 3, the bending moment, horizontal displacement and rotation of pile cap as well as pile deflection were larger in dense sand than those in loose sand, , which can be attributed to the larger stiffness and displacement of dense sand than those of loose one. Bending moments were induced mainly at upper part of piles in each case (Fig. 9a), as the cross points of piles and rupture were near the pile top. The pile deflections both in loose and dense sand distributed like a cantilever. The lower part of the pile was strongly confined by soil and the upper part of pile was pushed by the upper sand in hanging wall.

As for piles located a little far from fault zone, Piles 1 and 4, they seemed just slightly affected by the fault deformation, as bending moments were very small compared to values of Piles 2 and 3 (Fig. 9a). Pile 1 moved together with the hanging wall, and Pile 4 slightly bent towards the footwall side (Fig. 9d).

4 CONCLUSIONS

Centrifuge tests were conducted to explore the behaviour of single piles in sand subjected to reverse faulting at various relative locations from the fault rupture. LDT scanning technique, PIV techniques as well as strain gauge measurement were employed to capture complicated pile behaviour and ground deformation. Following conclusions are drawn:

(1) The LDT scanning technique has enough accuracy for multi-points measurements of ground surface and pile caps.

(2) Piles located within the fault zone are strongly affected by the fault deformation, while very limited influences are experienced when they are a little far from the fault zone. When piles are crossed at the middle part by the ruptures, they are highly bent at the middle part, which are likely sheared-off near the cross point. When the rupture crosses the upper part of piles, the piles behave like a cantilever, which are strongly confined by the deeper surround soil and the upper part of piles bend towards the foot wall side.

(3) Piles located within fault zone experience larger bending moment in dense sand than those in loose sand due to large stiffness and dilation of dense sand.

REFERENCES

Bray, J. D. 2001. Developing mitigation measures for the hazards associated with earthquake surface fault rupture. In *Workshop on seismic fault-induced failures—possible remedies for damage to urban facilities. University of Tokyo Press* (pp. 55–79).

Bransby, M. F., Davies, M. C. R., El Nahas, A., & Nagaoka, S. 2008. Centrifuge modelling of reverse fault–foundation interaction. *Bulletin of Earthquake Engineering*, 6(4), 607–628.

Chen, C. C., Huang, C. T., Cherng, R. H., & Jeng, V. 2000. Preliminary investigation of damage to near fault buildings of the 1999 Chi-Chi earthquake. J. Earthquake Eng. Eng. Seism, 2, 79–92.

Cole Jr, D. A., & Lade, P. V. 1984. Influence zones in alluvium over dip-slip faults. *Journal of Geotechnical Engineering*, 110(5), 599–615.

Lin, M. L., Chung, C. F., & Jeng, F. S. 2006. Deformation of overburden soil induced by thrust fault slip. *Engineering Geology*, 88(1), 70–89.

Takemura, J., Ishii, Y., & Kusakabe, O. 2010. Development of a fault simulator in a centrifuge and preliminary study on a buried pipe subjected to a fault differential displacement. In *Proc. 3rd Asia Conference on Earthquake Engineering*, Bankoku, CD-ROM, P-080.

Van Dissen, R., Barrell, D., Litchfield, N., Villamor, P., Quigley, M., King, A., & Stahl, T. 2011. Surface rupture displacement on the Greendale Fault during the Mw 7.1 Darfield (Canterbury) earthquake, New Zealand, and its impact on man-made structures.

White, D. J., Take, W. A., & Bolton, M. D. 2003. Soil deformation measurement using particle image velocimetry (PIV) and photogrammetry. *Géotechnique*, 53(7), 619–631.

22. Walls and excavations

Centrifuge simulation of heave behaviour of deep basement slabs in overconsolidated clay

D.Y.K. Chan & S.P.G. Madabhushi
Department of Engineering, University of Cambridge, England

ABSTRACT: High demand on land in major cities is driving construction of basement structures to create additional space. Long-term heave of base slabs is a pertinent problem in deep basement construction in over-consolidated clay strata, such as the London clay. Sub-structures must be designed to withstand soil pressures and displacements that evolve gradually for many years after construction is complete. This paper discusses an ongoing research project using centrifuge modelling to quantify the development of long-term heave by shortening the time-scale through dimensional similarity. The excavation process is simulated by draining of a heavy fluid (sodium polytungstate) and a model basement structure is instrumented to record the evolution of heave movements with time. This paper presents the preliminary results of a centrifuge test, which captured the magnitude of short-term differential and total heave deformation, the changes in support loads in horizontal props, and the evolution of pore pressures around the basement structure. Challenges encountered in this experimental technique and plans for further experimental work are discussed.

1 INTRODUCTION

Urbanisation is driving the demand for deep basements to be created within a dense built-environment to accommodate new public infrastructure such as underground railway stations and shopping mall cellars. The construction of a deep basement inevitably causes upward ground movements due to the permanent removal of soil overburden and these movements need to be predicted and controlled. In most geological strata, these ground movements are small and they occur within the timescale of the excavation, so they are accommodated during construction. However, in over-consolidated clay strata such as London clay, the soil permeability is so low that these upward movements continue for many years after structural completion. This process is known as long-term heave and the basement structure must be designed to withstand it, leading to much conservatism in design.

Some previous work has been carried out to quantify the effect of long-term heave using site data, notably a site on Horseferry Road, London where a basement was built in London clay and its heave movement was monitored for 21 years. Figure 1 shows that the evolution of heave movement with time at this site agrees with one-dimensional consolidation theory and this finding has served as a guideline for the designs of many deep basements in London clay. Although a few sites have published their monitoring data, data availability remains scarce, and the complexity of live sites means that there is a need for experimental data to improve engineers' understanding of the

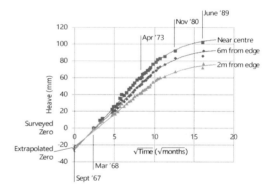

Figure 1. Monitoring data from Horseferry Road site. Continuous lines are 1D consolidation theory best-fit lines. (Courtesy of Sergio Solera of Mott MacDonald, Cambridge).

underlying mechanisms of deformation. Previous researchers have performed geotechnical centrifuge simulations of basement heave in over-consolidated clay because this method can replicate full-scale soil and structure stresses, but these studies have generally focused on the effect of specific methods of heave mitigation (Ohishi et al. (2000); McNamara & Taylor (2004)). There remains a need for an experimental study on the effect of the basement structure's stiffness on heave pressures and deformations, which is the focus of the ongoing study described in this paper.

2 MODEL DESIGN AND PREPARATION

The research discussed in this paper is part of an ongoing series of experimental investigations to model the post-construction basal heave behaviour of a deep basement. Each test would involve a basement structure, complete with a base slab, underlain by a stratum of saturated, over-consolidated clay. The rest of this paper will discuss the initial centrifuge testing attempts of this problem.

2.1 Design of basement model and excavation

The experiment used a rectangular basement model made of aluminium alloy, whose dimensions were specified with the configurations of typical permanent basement structures in mind, except there was no tension embedment extending beneath formation level. Tension embedment was avoided as the main purpose of the testing was to promote upward heave of the base slab. Each wall and slab plate is 5 mm thick; at $100\,g$ centrifugal acceleration this corresponds to a 1 m-thick prototype reinforced concrete element.

The general arrangement of the model is shown in Figures 2 & 3. The plan area of the model is $150\,\text{mm} \times 300\,\text{mm}$ (prototype $15\,\text{m} \times 30\,\text{m}$) and the buried depth is 150 mm (prototype 15 m). Two props, each made from a 6 mm aluminium rod and a load cell, crossed from one long edge of the basement box to the other, at a level 25 mm above the soil surface. A 5.4 kg brass block was placed upon the short edges of the basement model to simulate the weight of an associated superstructure. The presence of a surcharge would encourage differential heave in the base slab, rather than a simple uplift of the whole basement. The joints of the basement box were bolted together with joint stiffeners.

Excavation was simulated by extracting a heavy fluid (sodium polytungstate solution) of the same density as the soil surrounding the basement. The basement box was buried in the soil and filled with heavy fluid before spin-up, so the vertical stress under the basement matches in-situ conditions before excavation, albeit the use of a liquid imposes $K_0 = 1$ on the soil surrounding the basement, as discussed by Lam (2010). At $100\,g$, excavation was achieved by opening a set of valves to let the heavy fluid drain from the basement box to an external catch-tank under gravity (see Figures 2 & 5). The basement box was waterproofed with one layer of Plasti Dip rubber coating followed by two layers of Aquaseal tanking membrane. The use of a latex bag was not attempted because previous experiments had already shown that water-tightness between latex sheets and metal plumbing connections was unreliable under high fluid pressure.

2.2 Model soil

Long-term heave in over-consolidated clays is fundamentally caused by the swelling of the clay in response to the reduction in effective stress. This is characterised

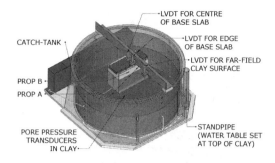

Figure 2. Isometric view of centrifuge model.

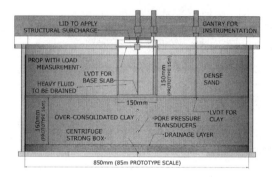

Figure 3. Cross-section of the centrifuge model, scale as shown.

Table 1. Comparison of swelling indices of clays.

Type of clay	Kaolin	London clay	K + B mix
κ (Schmidt method)	0.0485	0.0753	0.0707

by the swelling index κ. Although Speswhite kaolin is commonly used in physical models of soil-structure interaction in clay, preliminary investigations on the one-dimensional compression and swelling behaviour of clay samples showed that the swelling index of kaolin is much lower than that of London clay. Therefore, an artificial clay mix comprising 90% Speswhite kaolin and 10% calcium bentonite (K + B mix) was used to bring the swelling capacity of the clay in the centrifuge model to a level similar to that of London clay while preserving experimental repeatability. Table 1 presents the κ values for different clays.

The centrifuge model involved two layers of soil: 160 mm of K + B mix clay at the bottom, pre-consolidated to a vertical effective stress of 800 kPa; and 150 mm of dry, dense Hostun sand (density $1595\,\text{kg/m}^3$) on top. This gives the clay an over-consolidation ratio of 2.1–3.4 in centrifuge flight. The formation level of the basement model was set at 5 mm above the sand-clay boundary. This arrangement allows the water table of the centrifuge model to be drawn down to the formation level so that there

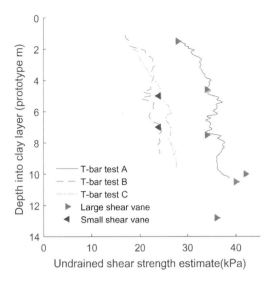

Figure 4. Undrained shear strength measurements in the K + B clay layer (measured at 1 g).

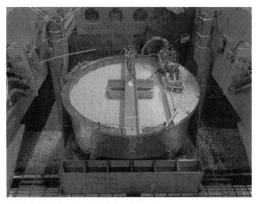

Figure 5. Photograph of centrifuge package assembled inside beam centrifuge.

would not be any significant flotation and to provide adequate drainage at the bottom of the slab, but it also ensures that the basement structure was predominantly underlain by saturated clay.

After the centrifuge test, T-bar tests and shear vane tests were undertaken to measure the undrained shear strength of the clay in the centrifuge model. The T-bar's diameter was 8 mm, its width was 40 mm, and it was driven into the soil at a rate of 13 mm/s (all model scale). All the strength measurements were taken from outside the footprint of the basement model. Figure 4 shows the results of these undrained shear strength tests, which reported values between 20–42 kPa. There is a general trend that undrained shear strength increased with depth, but there is also significant variation of strength for different samples at the same depth, which is presumed to be caused by uneven gain of moisture and loss of strength during the disassembly of the model from the centrifuge.

2.3 Model preparation and instrumentation

To prepare the clay layer, bentonite powder was added to water in a vacuum mixer, followed by kaolin powder, to achieve a homogeneous slurry at 125% water content. The slurry was poured into a 850 mm-diameter strong box, then put into an Enerpac consolidometer with double drainage, where the compressive load was doubled every 1–4 days until the vertical applied stress reached 400 kPa.

When the clay had consolidated sufficiently at 400 kPa such that there was no significant reduction in volume in 24 hours, the vertical load was removed to put the clay into suction, then pore pressure transducers (PPTs) were installed through the side ports of the strong box. The consolidometer load was put back to 400 kPa for a day, then increased to the desired pre-consolidation pressure of 800 kPa and kept at the same load until there was negligible pore pressure change in a day. The load was then decreased from 800 kPa to zero in 80 kPa steps, keeping adequate water supply throughout the process, so that the PPTs would not experience absolute tension during model preparation, mitigating the chances of delamination or instrument damage. The strong box, the consolidation procedure, and the PPTs installation procedure used in this centrifuge model were the same as that described in Faustin (2017) except that this model required a larger load decrement during PPT installation. Overall, the clay slurry compressed from an initial height of 400 mm to a final height of 185 mm upon removal from the consolidometer.

The top 25 mm of the K + B clay layer was scraped off to remove any surface contamination and reduce the thickness of the clay layer to the desired level. Then, a 5 mm layer of Hostun sand was laid upon the central part of the clay surface. The basement model was placed on top of this sand layer and connected via 1/4″ BSP ports on the side of the tub to the external heavy fluid receiver tank. The automatic sand-pourer described in Madabhushi et al. (2006) then poured the rest of the 150 mm sand layer into the strong box. Figure 5 shows the complete centrifuge package assembled inside the Cambridge geotechnical beam centrifuge just before testing, showing the catch-tank in front, the top layer of sand and the instrumentation gantry on top, and the basement model in the middle.

In addition to the aforementioned PPTs and load cells, the centrifuge model also included three linear variable differential transformers (LVDTs) to measure heave and settlement directly. The first LVDT measured displacement of the centre of the base slab; the second measured the side of the base slab; and the third measured the clay-sand boundary 150 mm (prototype 15 m) away from the side of the basement using an extension foot which was laid down before sand-pouring. Additional pressure transducers were used to monitor the depth of heavy fluid in the basement box and the depth of water in the standpipe which set the water table. Figures 2 & 3 show the arrangement of instrumentation.

3 PRELIMINARY RESULTS

3.1 Spin-up and re-consolidation

In the main centrifuge flight, the initial plan was to allow five hours of centrifuge reconsolidation at $100\,g$, but excavation was triggered prematurely at one hour after spin-up. This is because a cross-check of instrument readings had ascertained that there was a slow leak of heavy fluid from the basement box to the surrounding soil.

Figure 6 plots the variation of instrument readings throughout the centrifuge test, with axes in prototype scale using scale factors at $100\,g$. The response of instruments during centrifuge spin-up and consolidation is plotted towards the left of Figure 6. The negative pore pressure that the clay maintained on removal from the consolidometer turned into a high positive pore pressure as the weight of the soil generated compression during spin-up, while imperfections in the installation of the props caused one of them to gain compression and the other to gain tension during spin-up. Thereafter, the excess pore pressures dissipated slowly as the clay layer settled. Both props picked up compressive load, partly due to settlement of the basement box and partly due to the reduction in pressure inside the basement box from the slow leak.

3.2 Excavation phase

It is estimated that 15% of the heavy fluid had leaked from the basement box to the soil by the time excavation was triggered, bringing the fluid level in the basement box to 23 mm (prototype 2.3 m) below the sand surface level. The remaining fluid took about 30 s (3.5 days in prototype scale) to drain from the basement box to the catch-tank. This caused undrained heave of the basement and there was an immediate response from the instrumentation. Figure 7(a) shows that both the middle and the edge of the base slab recorded significant undrained heave, whereas the far-field clay surface showed no discernible instantaneous movement in response to excavation.

Figure 8 shows that the base slab underwent both total heave and differential heave. The centre of the base slab heaved up by 21 mm (0.21 mm model scale) during the excavation and continued to heave afterwards relative to the far-field clay surface. The edge of the base slab heaved up by 11 mm (0.11 mm model scale) immediately and eventually stabilised at 13 mm heave.

Gasparre (2005) tabulated the undrained vertical stiffness (E_u^v) for several London clay samples in triaxial extension. Her reported values clustered around $E_u^v = 200$ MPa. Taking this representative value of

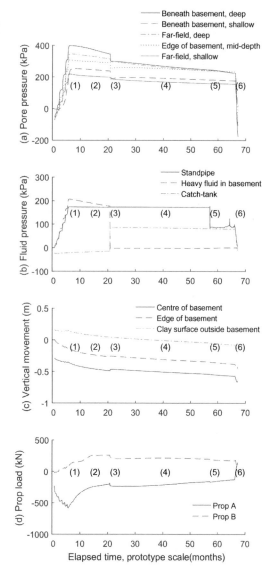

Figure 6. Graphs of instrument readings during centrifuge test, readings in prototype scale. (1) end of spin-up; (2) consolidation phase; (3) excavation; (4) post-excavation phase; (5) leak in water drainage; (6) spin-down.

undrained stiffness, the removal of 175 kPa of overburden from the excavation in this centrifuge test should lead to an undrained swelling strain of 0.09%. If this strain was uniform over the 16 m-deep (model scale 160 mm) clay stratum, the expected magnitude of heave would be 14 mm. Allowing for differential heave, the observed undrained heave of 21 mm at the centre 11 mm at the edge thus falls within the expected range. The magnitude of immediate heave also agrees with the observation that deep basement sites in London clay typically see a short-term heave displacement of 0.1%–0.2% of the excavation depth.

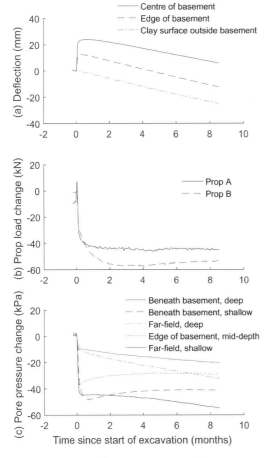

Figure 7. Change of instrument readings in response to excavation, relative to readings at the start of excavation; readings in prototype scale.

The compression in the prop loads decreased in the immediate response to excavation, which is perhaps counter-intuitive. This is because the slab-wall connections were deliberately made stiff, so the hogging deformation of the base slab was transferred into the walls as a prying movement. Figure 7(b) shows that the total drop in prop load was about 10 N in model scale, corresponding to a relief of 100 kN of prototype prop load between the two props.

The excavation caused pore pressures below the basement to drop sharply, as expected. PPTs beneath and near the footprint of the basement recorded short-term pressure drops of 38–49 kPa. The changes in pore pressures spread in all directions, with far-field, shallow-level PPTs also recording drops of about 10 kPa. Figure 7(c) shows that the pore pressures then recovered slightly as water was recharged into the clay layer, before continuing to decrease due to consolidation. Unfortunately, excavation was triggered before much consolidation had taken place due to the aforementioned technical difficulties, so the pore pressures remained well above hydrostatic throughout the centrifuge test.

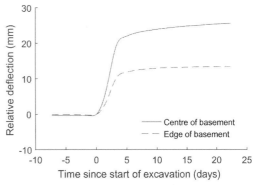

Figure 8. Short-term relative deflection of two positions of the base slab with respect to far-field clay surface, in prototype scale.

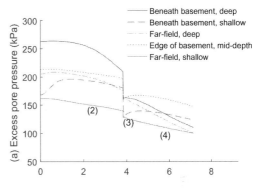

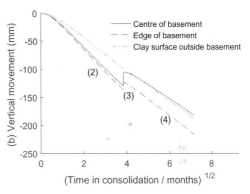

Figure 9. Graph of dissipation of excess pore pressures and changes in settlement against square-root of time in consolidation in prototype scale. (2) consolidation before excavation; (3) excavation; (4) post-excavation phase.

3.3 Post-excavation phase

The short term response lasted about 8 months (2000 s in model scale) from the start of excavation. By this time, the recovery of pore pressures was complete. The centrifuge test continued for another two hours, during which the bulk of the clay layer continued to compress and excess pore pressures from spin-up continued to dissipate. This is shown in Figure 9 which plots excess pore pressures relative to hydrostatic pressure and vertical movements of the three LVDT locations against $\sqrt{t}$, where time is measured from the end of spin-up, with units in prototype scale.

During the post-excavation phase, a slow downward trend of fluid pressure in the catch-tank raised concern. However, analysis of the data showed that this was caused by a drift of centrifuge speed. There was also some evaporation of water from the heavy fluid: the fluid recovered at the end of the centrifuge test was 6% denser than at the start. There was no sign of leakage of the heavy fluid out of the centrifuge package.

However, after 4.2 hours of centrifuge flight (corresponding to 4.8 years in prototype scale), a leak developed in the drainage fittings near the top of the clay, causing the water table to drop, and the test was stopped shortly afterwards.

4 FUTURE WORK

As discussed above, leakage was a significant challenge in this centrifuge model. The valves needed to hold shut at 5 bar pressure in the centrifuge and the first centrifuge flight was stopped soon after spin-up due to valve leakage; different valves were used in subsequent centrifuge flights. The basement box developed a slow-leak at $100\,g$ despite its waterproof coating. Towards the end of the main centrifuge flight, a water supply connection providing top drainage to the clay layer also began to leak, causing the water table to drop. Fortunately, there was no leakage from the catch-tank which had fully welded joints, so there was no significant loss of mass from the centrifuge package which could upset the balance of the centrifuge. In future centrifuge tests, structural connections between metal components will be welded where water-tightness is crucial.

The basement box used in this pilot test was specified with wall and slab stiffnesses commensurate to that of typical deep basement structures in London. Future tests will reduce the stiffness of the slab as the aim of this research project is to investigate the feasibility of using lighter designs of base slabs.

This pilot test only used two LVDTs to measure the heave movement of the base slab. The plan is to include further instrumentation such as strain gauges in future test to obtain more detailed measurements of the differential heave of the slab and the consequent bending in the base slabs and the basement walls.

5 CONCLUSIONS

- This paper presented early attempts in centrifuge modelling of heave deformations of deep basement structures in over-consolidated clay.
- Instrumentation in the model was able to capture differential heave movements, changes in pore pressure, and changes in prop loads caused by undrained heave in response to excavation.
- Further work will be undertaken to improve equipment reliability, so that long-term heave can be reproduced and quantified in the model.
- Future investigations will aim to vary the stiffness of the model structure and also include additional instrumentation to shed light on the differential heave behaviour of the base slab.

ACKNOWLEDGEMENTS

The authors would like to thank Dr. Yu Sheng Hsu, Hock Liong Liew and Sergio Solera of Mott MacDonald; and Dr. Wilson Kesse, Adam Locke and Corin Walford of Laing O'Rourke for their provision of past site data and information about current design practices.

This research project is supported by the EPSRC Centre for Doctoral Training in Future Infrastructure and Built Environment in the University of Cambridge.

REFERENCES

Faustin, N. E. (2017). *Performance of circular shafts and ground behaviour during construction.* PhD Thesis, Cambridge University.

Gasparre, A. (2005). *Advanced laboratory characterisation of London Clay.* PhD Thesis, Imperial College London.

Lam, S. Y. (2010). *Ground movements due to excavation in clay: physical and analytical models.* PhD Thesis, Cambridge University.

Madabhushi, S. P. G., N. E. Houghton, & S. K. Haigh (2006). A new automatic sand pourer for model preparation at University of Cambridge. In *Physical Modelling in Geotechnics.*

McNamara, A. M. & R. N. Taylor (2004). The influence of enhanced excavation base stiffness on prop loads and ground movements during basement construction. *Structural Engineer 82*(4), 30–36.

Ohishi, K., K. Azuma, M. Katagiri, & K. Saitoh (2000). Deformation behaviour and heaving analysis of deep excavation. In *Geotechnical Aspects of Underground Construction in Soft Ground,*, Tokyo.

Schmidt, B. (1966). Earth Pressures at Rest Related to Stress History. *Canadian Geotechnical Journal 3*(4), 239–242.

Soil movement mobilised with retaining wall rotation in loose sand

C. Deng & S.K. Haigh
Department of Engineering, University of Cambridge, UK

ABSTRACT: The soil displacements and stresses mobilised with retaining wall movement is significant for exploring better and more economic design methods in excavations. This paper describes a pair of centrifuge tests carried out to explore the relationship between sand displacements and retaining wall movement, modelling two types of wall rotation in loose sand. A new actuator system is designed to rotate the rigid retaining wall in the tests and soil displacements are observed and calculated through Particle Image Velocimetry (PIV). Sand displacements and strains in active and passive rotation are presented and comparison between the sand deformation in two types of wall rotation is discussed.

1 INTRODUCTION

With a huge increase in population and a dramatic shortage of land in urban areas, the development and utilization of underground space has become increasingly popular and consequently a growing number of underground constructions, such as subways, underground car parks and underground shopping centres, have been built in cities. Retaining walls are frequently used in excavations to resist earth pressure and water pressure behind the wall and to prevent soil collapse, which may cause inclination of adjacent buildings and movement of nearby tunnels. The design and construction of retaining walls are thus highly significant and hence substantial research has been carried out on retaining walls in excavations for many years.

The majority of research in this field focuses on the ground settlement induced by excavations as well as the displacement and bending moment of retaining walls. The main techniques used for retaining wall research are empirical and numerical methods.

Some empirical charts and forms, which are widely used in practical engineering due to their convenience and simplicity, are produced on the basis of field measurement and local experience (Terzaghi 1943, Peck 1969, Goldberg et al. 1976, Clough et al. 1989, Clough and ORourke 1990, Wong et al. 1997, Ng 1998, Moormann 2004). Unfortunately, those empirical methods are not universally applicable, because retaining wall displacements and bending moments as well as soil settlements are influenced largely by soil properties, which vary substantially in different places.

Numerical methods can work out the stress and strain of soil and the retaining wall at every point if constitutive models and boundary conditions are adequate, and consequently many scholars have researched the mechanism of excavations through numerical modelling (Simpson et al. 2008, Potts and Fourie 1984, Whittle 1993, Addenbrooke et al. 2000, Dinakar and Prasad 2014). However, numerous parameters are incorporated into numerical modelling including constitutive models and boundary conditions, for example the constitutive model for overconsolidated clays developed in Massachusetts Institute of Technology uses 15 parameters to characterise a given clay (Whittle and Kavvadas 1994), it is therefore extremely difficult and time-consuming for engineers to design a real project through such complicated numerical methods.

In conclusion, more fundamental questions should be explored by more advanced methods to establish a better understanding of the deformation mechanism in excavations, which can allow safer and/or more economic design methods. Therefore, the objective of this research is to establish the relationship between soil displacements and retaining wall rotation movement. This paper presents a pair of centrifuge tests simulating the rigid retaining wall rotation about the wall top and bottom points in loose sand and sand deformation in two tests is analysed and compared.

2 CENTRIFUGE TESTS

The rigid retaining wall rotation about the top and bottom points in loose sand is simulated in centrifuge tests, the details of which are described below.

2.1 *Test apparatus*

The rotation of a rigid retaining wall was conducted at 30 g on the Turner beam centrifuge as shown in Figure 1, at the Schofield Centre, Department of

Figure 1. Turner beam centrifuge at University of Cambridge.

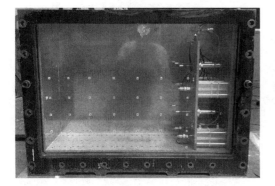

Figure 2. Empty container and actuator system.

Engineering, University of Cambridge, UK. The container used in the tests is 780 mm wide, 560 mm high and 200 mm thick and a transparent Perspex window is installed on the front side in order to record soil displacements and wall movement.

A new actuator system is designed to rotate the rigid retaining wall in the tests. Two actuators are installed 40 mm above the container bottom and another two are located 290 mm above the bottom as shown in Figures 2–4. The rotation about the top point of the retaining wall is conducted through closing the top two actuators and connecting the bottom two actuators with a piston, which can pull and push the actuator strokes to model the active and passive rotation respectively. The bottom two actuators are connected together in order to keep the oil pressures in two actuators the same, consequently keep the force from the two strokes the same, finally prevent the twist of the rigid retaining wall during the whole tests. Conversely, the wall rotation about the bottom point is realised by closing the bottom two actuators and connecting the top two with the piston. Specifically, the two working actuators are also connected with 2 bar air pressure in opposite channels to oil pressure in order to pull the strokes back more fluently during the active rotation.

2.2 Retaining wall and sand

The effective height of the rigid retaining wall is 250 mm (7.5 m at prototype scale) in Figure 4, while the wall is extended from the container bottom to the top edge in order to be rotated more conveniently. As shown in Figure 4, the top and bottom rotation points, which are the same levels with two pairs of actuators, are 40 mm and 290 mm above the container bottom respectively. The retaining wall, with a thickness

Figure 3. Package with retaining wall and sand in centrifuge tests.

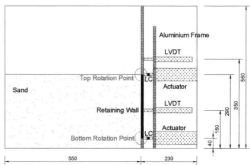

Figure 4. Schematic diagram of the package in centrifuge tests.

Table 1. Geotechnical properties of Hostun sand.

Item	Value	Unit
Maximum density	1620	kg/m^3
Minimum density	1318	kg/m^3
Maximum void ratio	1.01	–
Minimum void ratio	0.555	–
Friction angle	34	°
Specific gravity of solid	2.65	–
d_{50} size	480	Microns

of 10 mm (0.3 m at prototype scale) and width of 200 mm (6.0 m at prototype scale), is approximately rigid during the whole tests.

Dry Hostun sand, the geotechnical properties of which are shown in Table 1 (Flavigny et al. 1990), was used in the tests and pluviated into the container using an automatic sand-pouring machine in Schofield Centre, Department of Engineering, University of Cambridge (Madabhushi et al. 2006). The relative density of Hostun sand in the container is determined by the drop height and the flow rate, which is controlled by the diameter of the nozzle. Specifically, higher drop height causes higher relative density while larger nozzle diameter leads to lower relative density (Bouckovalas et al. 2015). A series of calibrations were conducted to finalize the two key parameters in order to obtain the loose Hostun sand with a relative density of 40%. Finally, the drop height was decided as 700 mm and the nozzle diameter was set as 9 mm during the sand pouring.

2.3 Instrumentation

Four actuators and four linear variable differential transformers (LVDT) are installed by an aluminium frame in the container in Figures 2–4. All the actuators are installed with load cells, which are hung on through steel sticks connected with actuator strokes, to measure the force applied on the rigid retaining wall. Two LVDTs are installed 150 mm above the container bottom and another two are set 350 mm above the bottom, so the wall rotation angle is obtained through the four LVDTs. One pore pressure transducer (PPT) is connected with the oil system including a piston and two working actuators to measure the oil pressure during the tests.

The front side of the container is transparent and two cameras were used to record the movement of sand during the tests. Particle Image Velocimetry (PIV), a velocity-measuring procedure initially developed in the field of fluid mechanics (Adrian 1991) and then adopted to geotechnical research (White and Take 2002), was used to analyse the sand deformation. The GeoPIV-RG (Stanier et al. 2015), a MatLab module implementing the PIV in a style applied to the analysis of geotechnical tests, was used to calculate and plot the displacement and strain of sand.

A Tekscan pressure mapping system (Palmer 1999) was used to measure the earth pressure behind the retaining wall during the tests. Tekscan sheet sensor transforms compressive pressure loads to a change in resistance, then scanning electronics collect sensor data and convert it to a digital signal (Brimacombe et al. 2009). The I-Scan software, which displays real-time pressure applied on the contact area (Bachus et al. 2006), was used to show and analyse the earth pressure mobilised with retaining wall rotation.

2.4 Test procedure

The test procedure of the wall rotation about the top point is divided into four stages as shown in Table 2. Initially, the centrifuge is swung up from 1 g to 30 g, consequently, the retaining wall is rotated anti-clockwise with the wall bottom moving 5 mm horizontally and then rotated clockwise with the wall bottom moving 40 mm in the opposite direction, which correspond to the active rotation and the passive rotation respectively. The retaining wall horizontal displacement is defined as negative during the active rotation in the tests. Finally, the centrifuge swings down and all the data is saved.

Table 2. Test procedure of rotation about top point.

Stage	Item	Wall bottom movement	G-level
1	Swing up	–	1 g to 30 g
2	Active rotation	0 to −5 mm	30 g
3	Passive rotation	−5 to 35 mm	30 g
4	Swing down	–	30 g to 1 g

Table 3. Test procedure of rotation about bottom point.

Stage	Item	Wall top movement	G-level
1	Swing up	–	1 g to 30 g
2	Active rotation	0 to −5 mm	30 g
3	Passive rotation	−5 to 35 mm	30 g
4	Swing down	–	30 g to 1 g

Similarly but conversely, the test procedure of the wall rotation about the bottom point is also divided into 4 stages and shown in Table 3. When the condition reaches 30 g, the retaining wall is rotated clockwise with the wall top moving 5 mm horizontally and then rotated anti-clockwise with the wall top moving 40 mm in the opposite direction. The retaining wall horizontal displacement is also defined as negative during the active rotation about the bottom point.

3 TEST RESULTS AND ANALYSIS

The test results and analysis described here will be concentrated on sand deformation with retaining wall rotation and the results from two types of wall rotation will be compared below. In this section, all quantities will be discussed at prototype scale and the depth in all figures is 7.5 m, which is the height of the retaining wall. The retaining wall is located at the right side in all figures and the sand with a width of 12 m is observed and analysed.

The sand resultant displacements with the active wall rotation about the top and bottom points are calculated through GeoPIV-RG and shown in Figures 5 and 6 respectively, which correspond to the horizontal displacements of the wall bottom and top as 0.15 m. Both sand resultant displacement surfaces are like triangles and the width of the influenced area at the sand surface is 8.2 m in the active rotation about the wall top, which is more than that in the rotation about the wall bottom with the value of 7.2 m. The maximum sand resultant displacement occurs at the top in Figure 6 because the maximum wall displacement is at the wall top and sand always follows the wall and then runs lower right in union, therefore, the sand resultant displacement surface is consisted of a series of isosceles right triangular wedges. However, the maximum sand resultant displacement is located at 5 m deep in Figure 5 because sand behind the wall flows down so rapidly that huge vertical displacement is mobilised during the active rotation about the wall top,

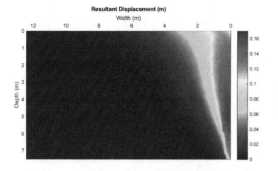

Figure 5. Sand resultant displacements with the active wall rotation about the top point.

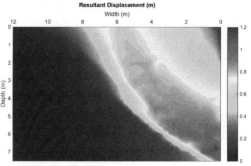

Figure 7. Sand resultant displacements with the passive wall rotation about the top point.

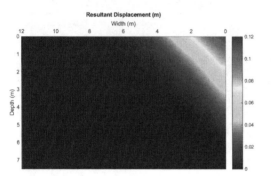

Figure 6. Sand resultant displacements with the active wall rotation about the bottom point.

Figure 8. Sand resultant displacements with the passive wall rotation about the bottom point.

consequently, the sand resultant displacement surface is comprised of several converted narrow triangular bands.

The sand resultant displacements with the passive wall rotation about the top and bottom points are calculated through GeoPIV-RG and shown in Figures 7 and 8 in accordance with the horizontal displacements of the wall bottom and top as 1 m respectively. Both sand resultant displacement surfaces are similar to wedges with the maximum displacements occurring at the wall bottom and wall top in Figures 7 and 8 separately, which correspond to a clear sliding from the wall bottom and a distinct upheaval near the wall top respectively. Specifically, the sliding surface originates from the wall toe and extends to the sand surface in Figure 7 and the sliding surface angle is 32°, which is extremely closed to the friction angle of Hostun sand. Interestingly, the maximum sand displacement at the soil surface is located at the place where is 6 m away from the wall top, while the sand near the wall top is not affected prominently. Conversely, a huge upheaval is mobilised near the wall top during the passive wall rotation about the bottom point and the angle of the resultant displacement surface wedge is also 32°. Deeper sand is not influenced obviously and the resultant displacement decreases with the distance to the wall top gradually.

The maximum shear strain, which is defined as Equation 1, is calculated by GeoPIV-RG in the two tests and shown in Figures 9–12. The gaps in Figures 9–12 are originated from sand subsets moving away from initial mesh areas during retaining wall rotation.

$$\gamma_{max} = \sqrt{(\epsilon_x - \epsilon_y)^2 + (\gamma_{xy})^2} \qquad (1)$$

where ϵ_x and ϵ_y are the normal strains in x and y directions respectively, and γ_{xy} is the shear strain (Beer et al. 2001).

The maximum shear strains with the active wall rotation about the top and bottom points are shown in Figures 9 and 10, which correspond to the sand resultant displacement surfaces in Figures 5 and 6 separately. A clear narrow band extends from the wall bottom to the middle level of the wall with the strain value dropping quickly in Figure 9, in accordance with the converted narrow triangular bands in Figure 5 due to the sand behind the wall flowing vertically during the active wall rotation about the top point. Differently, a prominent wedge situated in the top right corner in Figure 10, which coincides with the isosceles right triangular wedges in Figure 6, shows that the sand behind the retaining wall follows the wall moving. Although the maximum sand resultant displacements with the active wall rotation about the top and bottom points are similar, there is a huge difference between the peak values in Figures 9 and 10, which proves the sand behind the wall slides downwards dramatically during the rotation about the wall top. Conversely, the

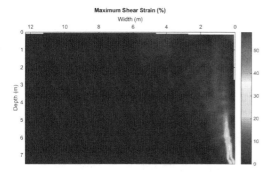

Figure 9. Maximum shear strain with the active wall rotation about the top point.

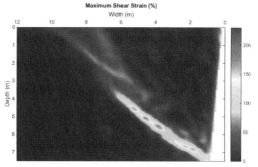

Figure 11. Maximum shear strain with the passive wall rotation about the top point.

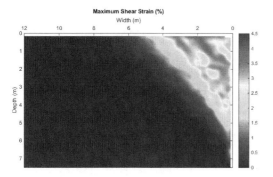

Figure 10. Maximum shear strain with the active wall rotation about the bottom point.

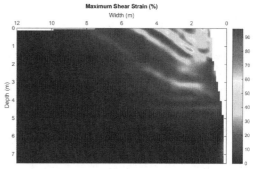

Figure 12. Maximum shear strain with the passive wall rotation about the bottom point.

maximum shear strain behind the wall decreases with depth gradually and the peak value is only 4.5% in Figure 10, which shows the sand behind the wall moves lower right in union with small relative displacements.

The maximum shear strains with the passive wall rotation about the top and bottom points are shown in Figures 11 and 12, which correspond to the sand resultant displacement surfaces in Figures 7 and 8 separately. Correspondingly, a clear sliding surface is also observed in Figure 11 and the angle of this area is calculated as 33°, which is nearly the same with that in Figure 7 because the sand relative displacement on the sliding surface is much larger than that in other places. The maximum shear strain of the sand above the sliding surface is minimal and consequently a conclusion can be made that the sand above the sliding surface moves upwards together like a rigid. Nevertheless, the band with huge maximum shear strains ends at the middle depth because the influence from the wall movement decreases with the distance to the wall toe. However, an obvious wedge including several parallel bands with large maximum shear strains is located in the top right corner in Figure 12, which shows that the sand in that area moves upper left with some relative displacements between sand layers instead of a rigid. Interestingly, the maximum shear strain values of these bands decrease with the distance to the wall top gradually, which is an evidence that the trend of relative movement between sand layers slows down.

4 CONCLUSIONS AND DISCUSSION

This paper described a new actuator system for rotating the rigid retaining wall in centrifuge tests and presented the results of two centrifuge tests performed in loose sand.

A new actuator system was designed to rotate the retaining wall about the wall top and bottom in the container to explore the sand movement mobilised with retaining wall rotation. PIV was used to calculate and analyse the sand resultant displacement and maximum shear strain during the active and passive wall rotation.

It has shown that the sand resultant displacement surface is consisted of several converted narrow triangular bands during the active wall rotation about the wall top, while a series of isosceles right triangular wedges are observed in the sand resultant displacement surface during the active wall rotation about the wall bottom. Correspondingly, a clear narrow band extends from the wall toe while an obvious triangular wedge occurs at the top right corner in the maximum shear strain surfaces with the active wall rotation about the top and bottom points respectively.

Additionally, a sliding surface extends from the wall toe to the sand surface and a clear band with huge maximum shear strains is recorded during the passive wall rotation about the top point, which shows the retaining wall rotation causes a sliding and the sand above the sliding surface moves upwards like a rigid. However, an upheaval is mobilised near the wall top during the passive wall rotation about the bottom point and there is some relative movement between sand layers.

Furthermore, the translation of the retaining wall will be simulated by the actuator system with connecting the four actuators together and the sand deformation with retaining wall translation will be explored. Additionally, the dense sand deformation mobilised with retaining wall rotation and translation will be also researched in the future.

ACKNOWLEDGEMENT

The authors would like to acknowledge the financial support from the China Scholarship Council.

REFERENCES

Addenbrooke, T., D. Potts, & B. Dabee (2000). Displacement flexibility number for multipropped retaining wall design. *Journal of geotechnical and geoenvironmental engineering 126*(8), 718–726.

Adrian, R. J. (1991). Particle-imaging techniques for experimental fluid mechanics. *Annual review of fluid mechanics 23*(1), 261–304.

Bachus, K. N., A. L. DeMarco, K. T. Judd, D. S. Horwitz, & D. S. Brodke (2006). Measuring contact area, force, and pressure for bioengineering applications: using fuji film and tekscan systems. *Medical engineering and physics 28*(5), 483–488.

Beer, F. P., E. R. Johnston Jr, & J. DeWolf (2001). Stress and strain–axial loading. *Plant, J.,(ed.), Mechanics of Materials, 3rd Ed., McGraw-Hill, New York*, 48–57.

Bouckovalas, G. D., D. K. Karamitros, G. S. Madabhushi, U. Cilingir, A. G. Papadimitriou, & S. K. Haigh (2015). Fliq: experimental verification of shallow foundation performance under earthquake-induced liquefaction. In *Experimental Research in Earthquake Engineering*, pp. 525–542. Springer.

Brimacombe, J. M., D. R. Wilson, A. J. Hodgson, K. C. Ho, & C. Anglin (2009). Effect of calibration method on tekscan sensor accuracy. *Journal of biomechanical engineering 131*(3), 034503.

Clough, G. & T. ORourke (1990). Constructed induced movements of in-situ walls. In *Proc. ASCE Specialty Conference, Cornell*.

Clough, G. W., E. M. Smith, & B. P. Sweeney (1989). Movement control of excavation support systems by iterative design. In *Foundation engineering: current principles and practices*, pp. 869–884. ASCE.

Dinakar, K. & S. Prasad (2014). Behaviour of tie back sheet pile wall for deep excavation using plaxis. *International Journal of Research in Engineering and Technology 3*(6), 97–103.

Flavigny, E., J. Desrues, & B. Palayer (1990). Note technique-le sable d'hostun" rf". *Revue française de géotechnique* (53).

Goldberg, D. T., W. E. Jaworski, & M. D. Gordon (1976). *Lateral Support Systems and Underpinning: Design and Construction*. Federal Highway Administration, Offices of Research & Development.

Madabhushi, S., N. Houghton, & S. Haigh (2006). A new automatic sand pourer for model preparation at university of cambridge. In *Proceedings of the 6th International Conference on Physical Modelling in Geotechnics*, pp. 217–222. Taylor & Francis Group, London, UK.

Moormann, C. (2004). Analysis of wall and ground movements due to deep excavations in soft soil based on a new worldwide database. *Soils and Foundations 44*(1), 87–98.

Ng, C. W. (1998). Observed performance of multipropped excavation in stiff clay. *Journal of geotechnical and geoenvironmental engineering 124*(9), 889–905.

Palmer, C. J. (1999). *Tactile pressure sensor technology applications to geotechnical engineering*. Ph. D. thesis, University of Massachusetts. Lowell.

Peck, R. B. (1969). Deep excavations and tunneling in soft ground. *Proc. 7th Int. Con. SMFE, State of the Art*, 225–290.

Potts, D. & A. Fourie (1984). The behaviour of a propped retaining wall: results of a numerical experiment. *Geotechnique 34*(3), 383–404.

Simpson, B., N. O'riordan, & D. Croft (2008). A computer model for the analysis of ground movements in london clay. In *The Essence of Geotechnical Engineering: 60 years of Géotechnique*, pp. 331–361. Thomas Telford Publishing.

Stanier, S. A., J. Blaber, W. A. Take, & D. White (2015). Improved image-based deformation measurement for geotechnical applications. *Canadian Geotechnical Journal 53*(5), 727–739.

Terzaghi, K. (1943). *Theoretical soil mechanics*, Volume 18. Wiley Online Library.

White, D. & W. Take (2002). Geopiv: Particle image velocimetry (piv) software for use in geotechnical testing.

Whittle, A. (1993). Evaluation of a constitutive model for overconsolidated clays. *Geotechnique 43*(2), 289–313.

Whittle, A. J. & M. J. Kavvadas (1994). Formulation of mit-e3 constitutive model for overconsolidated clays. *Journal of Geotechnical Engineering 120*(1), 173–198.

Wong, I. H., T. Y. Poh, & H. L. Chuah (1997). Performance of excavations for depressed expressway in singapore. *Journal of geotechnical and geoenvironmental engineering 123*(7), 617–625.

Lateral pressure of granular mass during translative motion of wall

P. Koudelka
Czech Academy of Sciences, Institute of Theoretical and Applied Mechanics

ABSTRACT: Novel equipment is introduced for researching lateral pressure (programmed and driven by two computers) that can move the front rigid wall arbitrarily slowly, applying one of three basic movements, measuring both components (normal, frictional) of contact pressures acting on the front and back walls and registering slip surfaces in the soil mass. Two complete histories of normal lateral pressure, both active and passive, are presented. The histories prove the behaviour of ideally noncohesive mass during wall translative motion by two doubles of the same experiments (passive pressure and active pressure) using the same material. The wall movement velocity was less than 0.005 mm/min.

1 INTRODUCTION

Fundamental works concerning lateral earth pressure have been carried out by Roscoe (1970), on the influence of strains in soil mechanics. Much of that work was about the lateral pressure of sand. Research was also carried out by James & Bransby (1970), on passive pressure. The work of Roscoe is linked to previous tests by Roscoe et al. (1963), James (1965), Bransby (1968), and Lord (1969). Furthermore, the works of Rowe & Brigs (1961) and Rowe & Peaker (1965) should be mentioned.

Physical research on lateral earth pressure in the 1970s (from the author's point of view) was largely aimed at measurements of real structure behaviour. An exception to this appears in research by Gudehuss (1980). In research similar to the work of Terzaghi (1943), based on laboratory investigations, Gudehuss drafted four theoretical dependence histories of the normal pressure coefficient K on wall movement for pressure acting on the wall rotated about the toe and translative moved. The dependencies are provided separately for active and passive pressure. The histories do not consider an upper limit (passive) of pressure at rest, K_{0p} (Pruška, 1973). All dependencies consider only a lower limit of pressure at rest, $K_0 = K_{0a}$ (Jáky 1943), including passive pressure histories.

The above-mentioned physical modelling was carried out under conditions of natural gravity (1G). However, significant changes in physical modelling technology took place in the 1980s. During the past several decades, physical modelling has been influenced mainly by centrifuge technology tests performed under multi-G gravity conditions.

During the past 20 years, one of the aims of the Czech Academy's research on soil mechanics has been similar to that of Roscoe's research in the 1960s (Roscoe, 1970). The principal objective is to develop an improved understanding of the stress–strain behaviour of a laterally retained soil mass, to obtain reliable predictions concerning the type and direction of retaining wall movement.

The research presented here deals with two problems of lateral pressure: passive pressure extremes and pressure at rest, proving the behaviour of two identical, ideally noncohesive masses for a given basic wall movement mode. Revised histories and theories on normal lateral pressure distributions and total forces are provided in detail according to the extent possible here.

2 EXPERIMENTAL EQUIPMENT

The experimental equipment was developed gradually from 1998 to 2010, during which the initial equipment was used to carry out certain initial experiments with active pressure (Koudelka, 2000). The first experiment with passive pressure, E3/0,2 (Koudelka & Koudelka, 2004), reconstruction, and novelisation followed in the period from 2003 to 2009 (Koudelka & Bryscejn, 2010; Koudelka et al., 2011). The presented experiments were performed using the last-stage equipment (Figure 1), which has been in operation since 8 April 2010. All experiments used the same equipment, except for the sets and places of bicomponent sensors that have been changed according to the wall movement modes.

The basic concept covered experimental equipment of medium-sized samples of $3.0-1.5 \times 1.2 \times 0.98$ m (length, height, and width, respectively), with the front wall arbitrarily movable in the required active direction within 300 mm, in the passive direction within 242 mm. (The inward wall movements generate passive lateral pressure; the outward movements generate active lateral pressure). The wall enables rotation about the toe or top, as well as a translative motion, with all three movements in both directions. A sample size of $3.0 \times 1.2 \times 0.98$ m was selected for the passive

Figure 1. Side view of experimental equipment part for sample: moved front wall on left, stable back wall on right.

Figure 2. Side view of tested sample: moved front wall and driving facilities on left, stable back wall on right.

pressure experiments, and $1.3 \times 1.2 \times 0.98$ m was used for the active pressure research.

The equipment enables movement of the front wall with an arbitrary slow velocity. A wall velocity of less than 0.005 mm/min was applied.

3 SOIL SAMPLE

The same 0.3-mm sharp quartz material was used repeatedly for all experiments, as well as for experiments E5/0,3, E6/0,3, E7/0,3 and E8/0,3. The masses were homogeneous and were compacted at the same medium level. A special exact instrument was used to provide the same energy in each layer. Despite this, smaller differences appear between the larger (E5/0,3 and E6/0,3) and smaller (E7/0,3 and E8/0,3) samples. The values for the basic material properties of the samples are summarised in Table 1. The friction values considered on the rear face of the wall are:

$$\delta_{ef} = \delta_r = 14.4°, \delta_0 = 0°, a_{ef} = a_r = 0 \text{ kPa}$$

where parameters δ_{ef}, δ_r, δ_0 are effective, residual or at rest wall–ground interface friction angle, a_{ef}, a_r, a_0 are effective, residual or at rest wall–ground interface adhesions.

Table 1. Material properties of samples of the experiments.

Mark	Unit	Passive motion		Active motion	
		E5/0.3	E6/0.3	E7/0.3	E8/0.3
δ	kg/m³	1556.9	1548.2	1474.2	1480.9
	kN/m³	15.273	15.187	14.462	14.527
ϕ_{ef}	°	43.1			
ϕ_r	°	38			
ϕ_0	°	31.3			
c_{ef}	kPa	4.5			
c_r	kPa	6.2			
c_0	kPa	2.9			
n	1	0.417	0.420	0.448	0.445
e	1	0.715	0.724	0.811	0.803
S_D	1	0.658	0.602	0.272	0.303
I_D	1	0.688	0.652	0.300	0.333

Note: Quantities are marked according to EC 7-1. Indexes – see text

All samples were provided by red strips (right site) and small black globules (left site) on sample contacts with the glass equipment sides (Figure 2) for visual monitoring processes into the samples.

4 EXPERIMENTAL PROGRAMME

The *passive* pressure experiments E5/0,3 and E6/0,3 were in progress in five motional phases applying a front wall translative motion. The first imperceptible motions of approximately $u_{0a} = 0.3$ mm out of the mass reached for a limit of active pressure at rest e_{0a}. Other motions into the mass returned the wall to its original position u_{or} (~0.3 mm), a further one u_{0p} at the approximate limit of passive pressure at rest e_{0p} (~0.7 mm), and then one at the supposed position (EN 1997-1) for half of the full passive pressure $e_{p/2}$ ($u_{p/2} = 54$ mm and 63 mm, respectively). The final maximal motion positions u_{max} were reached at 175.80 mm and 190.06 mm. The masses were observed also at rest (without front wall motions). Most of the rest phases persisted for approximately one week; however, the rest phases at the half passive pressure positions ($u_{p/2}$) remained for lengthy periods of 61 days and 60 days, respectively.
Where:

- Ex-or – original pressure by the experiment Ex – colour curve
- Ex-op – passive pressure at rest by experiment Ex colour curve
- Ex-px-yy – passive pressure of experiment Ex after motion of $u_{px} =$ yy – colour curves
- e_{0a} – active pressure at rest (Jáky) – black line
- e_{0a} – passive pressure at rest (Pruška) – black line
- e_a – active pressure (EC 7-1) – black line
- e_p - passive pressure (EC 7-1) – black line
- $e_{p/2}$ – half passive pressure (EC 7-1) – black line
- e_a – active pressure (EC 7-1) – black line

The values in frames in Figures are relevant to the wall motions u_x of the marked experiments.

The active pressure experiments E7/0,3 and E8/0,3 were in progress during three motional phases, applying an inverse front wall translative motion. The first imperceptible motions u_{0a} of about 0.3 mm again reached a limit of active pressure at rest.

Other motions moved the wall at the supposed position (EN 1997-1) for the full active pressure e_a (u_a − 1.36 mm and −1.31 mm) and at the final maximal positions of $u_{max} = -100.42/-57.46$ mm. Also, the masses were observed in the rest phases. Most rest phases persisted for approximately one week; however, the rest phases at the active pressure positions remained for 99 and 75 days, respectively.

5 RESULTS

The entire research (12 experiments) yielded a huge data quantity of 24 GB in a space-saving registering format *lfx*, except for the visual monitoring data. The pressure data of the four presented experiments with translative wall movement had a size of 7.33 GB.

5.1 Passive pressure — E5/0,3 and E6/0,3

Distributions of the normal components of lateral passive pressure, both in the original positions and Figure 4. Distributions of passive normal pressure at the supposed limit of passive pressure at rest, after front wall motions of $u_{0p} = 0.696$ mm and 0.822 mm following important wall movements, are illustrated in the following figures. The pressure distributions for the theoretically supposed motions mobilising passive pressure at rest, half of the full passive pressure, and full passive pressure are illustrated in Figures 3-6. Figure 4 shows differences between experimental curves E5-0p and E6-0p and a theoretical line of pressure at rest according to Pruška. The differences between the curves are probably caused by an imperceptible difference in the wall motions of approximately 0.13 mm. Figures 3 and 4 illustrate that soil masses at rest are very sensitive, even to very small wall motions.

Table C.2 of EN 1997-1 (Annex C) presents approximately relative (to wall height) intervals of motions of the wall to be mobilised by half (average of 0.75% and 1.2%) and full passive pressure (average of 4.5% to 8%). The derived motion values of medium compacted soils, such as of the used samples, are approximately 1.0% (for 10 mm) for mobilisation of the half passive pressure and 6.25% (for 62.5 mm) for full passive pressure mobilisation.

Both pressure distribution curves in Figure 5 practically reach supposed full pressure by ep-EC 7-1 (black dash-and-dot line) already following the wall movements of 55.377 mm and 51.701 mm. Then, the pressures increase more slowly during the further motions, probably to their upper limits around the wall motions of 180 mm and 190 mm (see Figure 6 and also the following important paragraph).

The pressure distribution according to the curve of E6-pxf-190-k (red dashed line) in Figure 6 proves the time instability of mobilised passive pressure

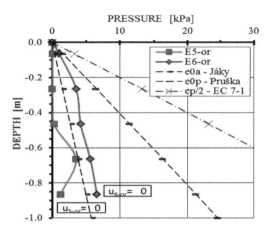

Figure 3. Distributions of pressure normal components before experiments E5/0,3 and E6/0,3, wall motions are of $u_{or} = 0$ mm.

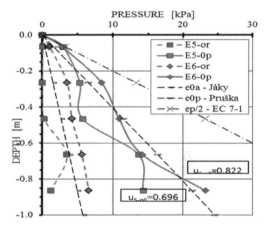

Figure 4. Distributions of passive pressure normal components at the supposed limit of passive pressure at rest. Front wall motions of $u = 0.696$ mm.

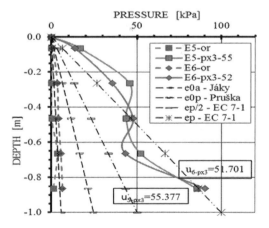

Figure 5. Distributions of passive pressure normal components at the supposed (EN 1997-1) position for a half of full passive pressure. Front wall motions of $u = 55.377$ mm.

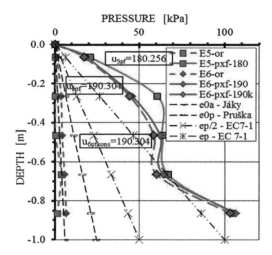

Figure 6. Distributions of passive pressure normal components at the last front wall positions. Front wall motions $u = 180.256$ mm.

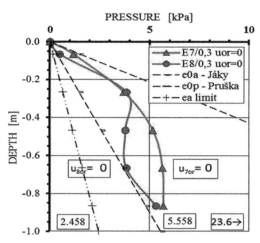

Figure 7. Distributions of pressure normal components before experiments E7/03 & E8/03. Wall motion $u = 0$ mm.

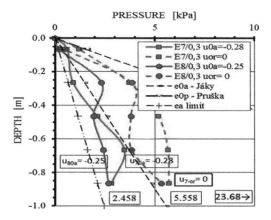

Figure 8. Distributions of minims of active pressure of the experiment E7/03&E8/03 after front wall motions $u = 0.283$ mm.

following the wall movement stop (without any motion). This distribution occurred after 93 days of the mass at rest (also Koudelka & Koudelka, 2003).

5.2 Active pressure — E7/0,3 and E8/0,3

The following figures, according to the results of the presented experiments, illustrate the distributions of normal components of lateral *active* pressure, both in the original positions and following important wall movements in sensitive scales.

Experiments E7/0,3 and E8/0,3 follow up the previous experiments with *passive* pressure (E5/0,3 and E6/0.3), in which the wall translative motion was applied *inward* the mass. The experiments E7/0,3 and E8/0,3 applied an opposite translative motion *out of* the mass. Of course, the direction of pressure acting on the wall was the same. The active pressure distributions presented in Figures 7 to 10 follow up the distribution histories of passive pressure in Figures 3 to 6. The scales of Figures 7-10 are more than 10 times as sensitive (see maximum theoretical values of active pressures e_a, e_{0a} and e_{0p} in frames below).

The pressures before experiments E7/0,3 and E8/0,3 in Figure 7 occurred in the area of pressure at rest (between both Jáky' and Pruška' limits), similarly to the pressure before experiment E6/0 in Figure 3. The pressure before experiment E5/0,3 (Figure 3) is below the limit of the considered Jáky *active* pressure at rest. The compactions of all samples were the same, and the reason for this pressure difference is unknown to date.

The sensitivity of a noncohesive mass at rest is significant. Imperceptible front wall motions of 0.283 mm and 0.247 mm were sufficient for pressure decreases under *active* pressure at rest. The pressure differences are obvious in Figure 8 when comparing the full curves (after the motions in the frames) to the dashed original pressure curves (before the experiments).

Comparing Figures 9 and 10 (where pressures are closer to full active pressure e_a), we find pressures/distributions not far from *active* pressure at rest e_{0a}; however, overall, the pressures do not reach the minimal values of the full active pressure e_a. Moreover, the pressures achieved wall motions of 1.357 mm and -1.313 mm, which are assumed according to EN 1997-1 for mobilisation of active pressure, increase during the following rest consolidation time of 99 days and 75 days, respectively. That is, there exists a *tendency of mobilised active* pressure *to the original values* (thick dashed curves) during an at-rest consolidation.

The limit/full active pressure (minimum) is mobilised as far as an approximate wall motion of 40 mm (4%), which is a significantly longer motion than the theoretically supposed motion with a value of approximately 0.14%, namely, 1.4 mm (compare Figures 9 and 10). However, it should also be taken

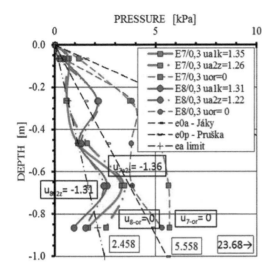

Figure 9. Distributions of active normal pressure of experiments E7/0,3 and E8/0,3 following front wall motions of $u = 1.357$ mm.

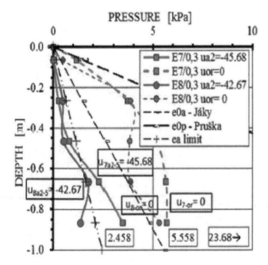

Figure 10. Distributions of the minima of active pressure of experiments E7/0,3 and E8/0,3 following front wall motions of $u = -45.677$ mm.

into account that pressures acting on the wall (particularly on its upper half) decrease because of a decrease in mass surface, owing to the long wall motion towards away from the mass. The mass behind the wall sank and uncovered the first upper pressure sensor (vanishing pressure). Pressures of the samples with plain surfaces, which covering all sensors, would be higher above, and the minimum value may also be higher.

Experiment E7/0,3 includes wall motions up to $u = -100.423$ mm. It is interesting but logical that, despite the decreased sample surface behind the wall, the pressure on the lower part of the wall remains between the theoretical values of Jáky' pressure at rest and active pressure (minimal values) as far as a wall

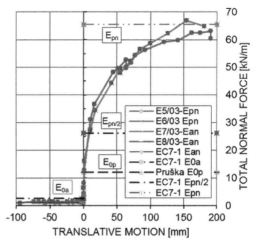

Figure 11. Complete graph of history of active and passive pressures according to experiments E5/0,3 and E7/0,3. The part around pressure at rest (of $u = 0$).

Table 2. Data of the total force histories of pressure at rest

Experiment	Active pressure			Passive pressure		
	E_{0a} kN/m'	k_{E0a} kN/mm	α_{E0a} °	E_{0p} kN/m'	k_{E0p} kN/mm	α_{E0p} °
E5/0.3	0.90	4.62	77.8	16.20	24.64	87.7
E6/0.3	2.07	7.59	82.5	12.81	5.22	79.2
E7/0.3	2.01	8.61	83.4	-	-	-
E8/0.3	2.02	6.56	81.3	-	-	-

Note: α_{E0} — angle of line inclination of total normal pressure force E_n in area of pressure at rest in k_{E0} — inclination $\tan(\alpha_{E0})$

motion of approximately $u = -70$ mm, namely 7%. A further point of view on lateral pressure behaviour is the possible analysis of total pressure forces in Section 6.

6 TOTAL PRESSURE HISTORIES

Five sensors placed in the front wall registered both pressure components (normal and friction) during the experiments and provided the pressure distributions. The distributions changed themselves in a complex and lively manner in the experimental processes, depending on the wall motion and creation of slip surface systems. As a result, an evaluation of the total effects, namely, the total pressure force and total pressure moment to the toe, is difficult to achieve according to distribution histories only. Therefore, particular pressures of the movable wall sensors were summarised to find total normal forces E_{pn} and E_{an} acting on the whole wall during the experiments. The behaviours of the total normal forces monitored during experiments E5/0,3 and E7/0,3 are provided below.

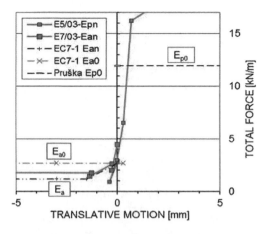

Figure 12. Detail of total force history of experiments E5/0,3 and E7/0,3 around the area of pressure at rest.

Figure 11 and Table 2 illustrate similar, almost identical, doubled histories of the normal lateral pressure during both experiments, which can be considered to prove the real behaviour of ideally noncohesive soil masses. A graph illustrates the most important interval of wall motions, determining that constructively applicable pressures are around the interval of pressure at rest. This interval appears to be (−2; +4) mm.

This fact directs attention to the problem of pressure at rest and its importance in geotechnical design. Values of the active and passive pressure differ in an orderly manner, and the active part is not synoptic. Figure 12 presents a detailed graph of the area around the zero wall motion, according to experiments E5/0,3 and E7/0,3. It can be observed that the results of both experiments concur strongly and (together with the results of experiments E6/0,3 and E8/0,3) prove the existence of the value interval of pressure at rest.

7 CONCLUSION

Brief information on lateral noncohesive soil pressure during the complete translative motion scale was presented. Most of the knowledge gained remains out of the provided framework (for example, deformations, slip surfaces, pressure time instability and details). Nevertheless, the following observations were made:

1. The theoretical upper passive limit of pressure at rest according to Pruška exists, and, because of this, pressure at rest is the almost singular pressure interval between Jáky's and Pruška's limits.
2. Lateral pressures following wall motion are not constant, but they are time instable and change themselves with a tendency to the original pressure.
3. Values of active/passive pressures (e_a/e_a) were attained as higher/lower than the theoretical values, respectively.

REFERENCES

Bransby, P.L. 1968. Deformation of soft clay. Ph.D. Thesis, University of Cambridge

Gudehus G.1980. Materialverhalten von Sand: Anwendung neuerer Erkentnise im Grundbau. Bauingenieur, 55(9): 351–359.

Jáky, J. 1944. A Nyugalmi nyomás tényeroje. *A Magyar's Mérnokés Épitész – Egylet Koylonye* 1944. 78(22): 355–358.

James, R.G. & Bransby, P.L. 1968. Experimental and theoretical investigations of a passive earth pressure problem. *Géotechnique* 20, No.1, 17–37.

James, R.G. 1965. Stress and strain fields in sand. Ph.D. thesis, University of Cambridge.

Koudelka, P. 2000. Non-linear bi-component lateral pressures and slip surfaces of granular mass. *Proc. IC GeoEng2000*, Melbourne, Technomic Publ. Co. Inc., Lanc./Basel, p.72 ps.8).

Koudelka, P. & Koudelka, T. 2003: Time Instability of Passive Pressure of Non-cohesive Materials. *Proc. 12th ARC SMGE* Singapore; Leung, Phoon, Chow, Yong & Teh, Singapore, ISBN 981-238-559-2 (Set), Vol.1, pp.801–804.

Koudelka, P. & Koudelka, T. 2004. History of Passive Noncohesive Mass and Its Consequences for Theory of Earth Pressure. *Proc. 5th IC Case Histories in Geotechnical Engineering, New York, University of Missouri-Rolla, Rolla (Missouri),* Shamsher Prakash, # 5.67.

Koudelka, P. & Bryscejn, J. 2010. Original Experimental equipment for Slow Processes of Lateral Pressure in Granular Masses. *48th ISC on Experimental Stress Analysis-2010. Proc.* ISBN 978-80-244-2533-7, P. Šmíd, P. Horváth, Habovský, 177-184. Web of Science, Thompson Reuters.

Lord, J.A. 1969. Stresses and strains in an earth pressure problem. Ph.D. thesis, University of Cambridge.

Pruška, L. 1973. Physical Matter of Earth Pressures and Its Application for Solution of Earth Pressures at Rest (in Czech). *Proc. IInd NS Advanced Foundation Method and Soil Mechanics,* Brno-Cz, Dúm techniky Brno, pp. 1–23.

Roscoe, K.H. 1970. The Influence of Strains in Soil Mechanics. *Géotechnique* 20(2), pp. 129–170.

Terzaghi, K. 1943. *Theoretical soil mechanics*. Willey 1943.

Deflection and failure of self-standing high stiffness steel pipe sheet pile walls embedded in soft rocks

V. Kunasegarm, S. Seki & J. Takemura
Tokyo Institute of Technology, Tokyo, Japan

ABSTRACT: In this study, a centrifuge modelling technique has been developed, in which the loading process can be simulated from design conditions to the ultimate failure conditions on an embedded wall in soft rock. A series of centrifuge tests has been carried out to investigate the influence of embedment depth on the stability of self-standing steel pipe sheet pile walls. Observed results reveal that with relatively small embedment, the wall can stand in the design condition with a reasonable safety margin against ultimate failure. The stiff sheet pile walls suffer by rigid body rotation about a pivot point and small increment in the embedment depth e.g., 0.5m, can significantly increase the wall stability.

1 INTRODUCTION

Cantilever sheet pile wall is a preferable retaining structures in vast range of civil engineering applications due to its simple retaining mechanism. For the application of self-standing retaining walls, relatively large stiffness and strength are required for the embedment ground, such as dense sand or soft rock to secure the passive resistance against the force from the retained soils. Furthermore, to control the wall deflection below the allowable limit, the walls with large flexural rigidities (EI) are required especially to sustain a large retained height of earth. Large diameter steel pipe sheet pile (SPSP) wall is a possible alternative of large stiffness wall.

Over the past decades, extensive investigations on the behavior of self-standing walls in clays and sandy soils have contributed to the development of design codes and the calibration of numerical models. However, the behavior of these types of wall in relatively hard mediums such as soft rocks is still not well understood. Poor understanding about the characteristics of soft rocks and traditional way of extrapolations from two distinct areas of geotechnics either soil mechanics or rock mechanics (Johnston 1991), often lead technically non-feasible and over conservative designs with non-economical solutions. Especially for the application of large diameter SPSP walls as a cantilevered structure, current design guidelines require relatively large embedment depths. Which hinders the application of self-standing large diameter SPSP walls due to the difficulty in installing such pile into hard mediums such as soft rocks.

Therefore, if the stability of retaining wall can be secured with relatively small embedment depth in hard mediums, its applicability can be increased, which contributing the reduction of construction time and cost. For an economical design method with the reduction of the embedment depth smaller than that of traditional design methods, both the serviceability and ultimate limits should be reasonably examined in the design process. In other words, the design conditions should secure the allowable displacement by a reasonable evaluation method and the reasonable margin of safety over the ultimate loading conditions must be incorporated to prevent a catastrophic failure. Therefore, the process from the design conditions to the failure should be well studied by physical means, such as a physical modelling. In this research, a centrifuge modelling technique on self-standing retaining wall has been developed, by which the behaviour of walls embedded in hard mediums from the design loading condition to the ultimate failure conditions can be simulated in a sequential manner under a constant centrifugal acceleration.

2 CENTRIFUGE MODELLING

2.1 Modelling of soft rocks for centrifuge study

Soft rocks can be divided in to sandstones and mudstones. Sandstone is a sedimentary rock formed by consolidation of sand cemented with clay and other minerals, while the original constituent of mudstone is clay or muds. Clay: Sand content, described in Table 1 indicates the compositions of model soft rocks which can be used to model soft sandstone or soft mud rock.

According to the engineering classification, the unconfined compressive strength (UCS) of 0.6 MPa −1.25 MPa defined as very soft rocks and 1.25 MPa −5.0 MPa given as soft rocks. Establishing the lower

Table 1. Physical and mechanical properties of ground.

Embedment medium	Sand rock	Mud rock
Water/Cement ratio (%)	395	510
Water content (%)	21.5	39
Clay: Sand (wt. %)	30:70	100:0
Bulk density (kg/m^3)	2060	1820
Dry density (kg/m^3)	1715	1320
UCS (MPa)	1.3	1.0
$E_{50\_\varepsilon ad}$ - $E_{50\_\varepsilon as}$ (MPa)	260-660	200-420

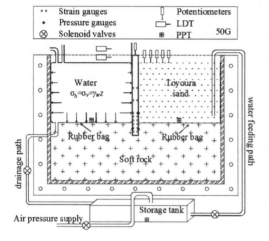

Figure 1. Experimental set up.

limit for the strength of soft rock is more complicated since some hard soils also exhibits higher UCS, however Terzaghi & Peck (1948) defined an SPT value above 50 and UCS greater than 0.4 MPa as the lower limit of soft rocks. Artificial soft rocks for centrifuge model, namely sand rock and mud rock were made by using sand-clay-cement mixtures at appropriate moisture contents and water–cement ratios as described in Table-1.

The mechanical properties of the model soft rocks were investigated through the unconfined compression (UC) tests which were conducted for the cylindrical moulded specimens. In this study, averaged external strain (ε_{ad}) and surface strain (ε_{as}) of the specimens were measured by using a dial gauge and a pair of strain gauges, respectively. Secant modulus (E_{50}) of the artificially made rock samples were estimated based on averaged external strain ($E_{50\_\varepsilon ad}$) and averaged strain gauge ($E_{50\_\varepsilon as}$) measurements.

2.2 Experimental setup and modelling procedure

Centrifuge model studies were conducted using TIT Mark III geotechnical centrifuge at 50 g centrifugal acceleration, the specifications of centrifuge facility are listed in Takemura et al. (1999). An illustration of a centrifuge model arrangement is given in Figure 1, which represents a self-standing sheet pile wall embedded in artificially made soft rock. In the centrifuge model, a model wall was placed throughout the breadth of the container to secure the plane strain deformations of the wall. Soft rock model was prepared in a container which had the internal dimensions of 700 mm in length, 150 mm in breadth and 500 mm in depth. The container was made up of removable steel frame on the rear face and a transparent acrylic panel stiffened by steel hollow frame in the front, both face panels were bolted with the main body to form a rigid container.

The model ground with the embedded sheet pile wall was prepared 14 days in advance to the test in order to achieve required strength and stiffness at 14 days curing time. Prior to the casting of soft rock sample, 0.5 mm thick Teflon sheets were pasted in the front and rear faces of the container and lubricated by silicon grease for easy detachment of wall from the hardened soft rock ground. Afterwards, the sheet pile wall was rigidly fixed upright at the centre of container with the help of a guide plate. Subsequently, the soft rock ground was constructed by compacting the sand-clay-cement mixture of each 30 mm thick layers on either sides of the sheet pile wall and the density of the ground was controlled by measuring the weight of each layer.

Upon completion of designed height, the model ground with the embedded wall was covered by wet towels and conserved under room temperature for 14 days curing. At the end of 14 days curing time the guide plate was unbolted from the container panels and both face panels of the container were detached. The Teflon sheet on the rear wall was then-replaced by a new one and lubricated with grease to minimise the friction. Meanwhile, the front transparent window, front and rear faces of model ground were coated with silicone grease and reformed the rigid container with lubricated interfaces. Also front and rear face edges of the wall were coated with sponge tapes, which prevents the intrusion of tiny particles in to the interface. Thereon, the rubber boxes to be filled with water and Toyoura sand were placed in the front and back sides of the wall respectively. The front one was a polypropylene made closed box and the back side one was an open box made of latex rubber. The friction between rubber boxes, the retaining wall and inner wall of the container was minimised by means of silicon-grease lubrication.

The rubber boxes were equipped with pair of pore pressure transduces (PPT) at their bottoms. After that, the retained dry Toyoura sand on the back side was made by using sand hopper at 80% relative density and the surface was levelled by means of vacuum. Detail mechanical properties of Toyoura sand are described in Tatsuoka et al. (1986). Water in the wall front rubber box was filled at the same level of the retained ground surface. A series of potentiometers on the levelled surface of the retained soil and a pair of laser displacement transducers (LDT) were utilised in the centrifuge model to monitor the surface settlements and to track the lateral movement of the wall top respectively.

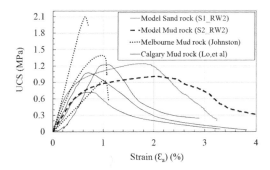

Figure 2. Stress-strain behaviour of centrifuge model soft rocks and natural soft rocks.

Table 2. Experiment conditions.

Series	Test code	H M	d_e M	EI_{wall} GNm^2/m	K_{0_50g}
1-Sand rock	S1_RW1	12	2.5	11.096	0.53
	S1_RW2	12	3.0	11.096	0.46
	S1_RW3	9	1.8	0.697	0.58
2-Mud rock	S2_RW1	12	2.5	11.096	0.55
	S2_RW2	12	3.0	11.096	0.54

H- Retained height, d_e Embedment depth
K_{0_50g} – initial earth pressure coefficients at rest @50g

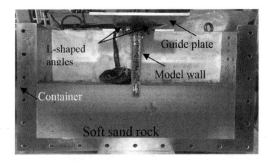

Figure 3. Typical model (S1_RW1) preparation for test.

Typical stress-strain behaviour of model soft rocks is illustrated in Figure 2 along with the natural soft rocks encountered in Melbourne (Johnston 1984) and Calgary (Lo, et al.2009). From the stress strain behaviour, up to 0.4% axial strain the model soft rocks exhibit similar stiffness with the natural soft rocks. However, the post peak behaviour of natural rocks is more brittle compare to the model rocks.

2.3 Test conditions and model sheet pile walls

This paper reports the results of five centrifuge model tests, which are divided into two test series, i.e., sand rock and mud rock series, as shown in Table 2. All dimensions and the properties of the model walls are given in prototype scale in Table 2. Model sheet pile walls were made up of aluminium (A5052-O) alloy plates having the yield strength of 95 MPa and the

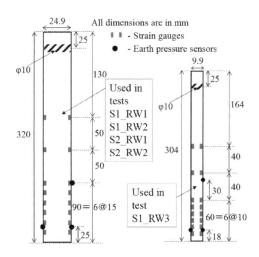

Figure 4. Model sheet pile wall sections.

Young's modulus of 69 GPa. Rigid and flexible sheet pile walls having the thickness of 24.9 mm and 9.9 mm in model scale are illustrated Figure 4. The transformed sections of rigid and flexible model sheet pile walls were designed to replicate the flexural rigidities of steel pipe sheet pile walls with the Diameter (Ø) of 2.5 m, thickness t = 25 mm and Ø = 1.0 m, t = 10 mm respectively.

2.4 Simulation of excavation and additional loading

A model ground after 14 days curing is shown in Figure 3. Upon completion of the model arrangements the model container was mounted on the centrifuge platform. An inflight drainage and water feeding system was then developed with the utilisation of solenoid valves and an external pressure board which was connected to the inflight storage tank as described in Figure 1. Subsequently, the model was gradually accelerated and the equilibrium conditions were achieved in flight at 50 g centrifugal acceleration at an effective radius of 2.09 m. While maintaining the equilibrium conditions over a certain period of spinning, unloading in the wall front was simulated by draining out the water from the rubber bag to the storage tank.

After the excavation process, the model was allowed to reach the equilibrium with a certain level of induced strains which was caused by the unloading. Thereon to study the behaviour of walls under extreme loading conditions, the additional load was applied by feeding the drained water from the storage tank to the backside rubber box which was done by raising the air pressure of the storage tank by manual operation. The excavation depth (Z_e) and the rise of water level (h_w) in the loading process as well as the change in tank pressure were monitored using PPTs at the bottoms of rubber boxes and storage tank as described in Figure 1. Meanwhile the wall top displacements and the surface settlements also were recorded using LDT and potentiometers respectively.

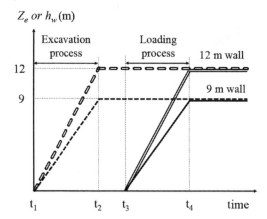

Figure 5. Sequence of excavation and loading.

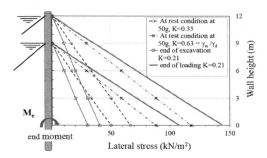

Figure 6. Expected lateral stress profiles for 9 m and 12 m walls at different stages.

Sequence of excavation and additional lateral loading as well as the possible lateral stress profiles, namely at rest pressure ($K = 0.35 = K_0$), and Rankine active pressure ($K = 0.21 = K_a$) are described in Figure 5 and Figure 6 respectively. The implemented loading mechanism in this study, can raise the thrust on the wall about three times from the dry condition to the submerged condition which enables to study the behaviour of walls embedded in relatively hard mediums under a large range of deformations.

Utilisation of a heavy liquid having the identical unit weight of back fill material could create the vertical stress similar to that of the sand. However, the unbalance of horizontal stresses at rest condition and the back ward wall movements caused by heavy liquid could be higher than those caused by water as a draining liquid. It is important to note that; the earth pressure coefficient (K) in the retained soil is expected to be higher than that of at rest coefficient ($K_0 = 1 - \sin\varphi' \sim 0.35$), which can be attributed to the backward movement of the wall due to relatively large water pressure than at rest earth pressure of the sand with $\gamma_d = 15.6\,\mathrm{kN/m^3}$.

End moments at the embedment level as shown in Figure 6, were calculated from measured strains at rest condition at 50 g prior to the excavation. Considering the moment equilibrium about the embedment level, estimated K-values at rest are presented in table 2.

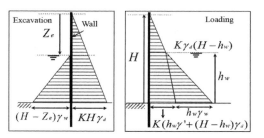

Figure 7. Earth pressure profiles during excavation and loading.

Which indicates the initial K-values in the range of 0.35–0.63. Figure 7 shows the assumed earth pressure profiles during the test. Corresponding applied moment loads at embedment levels during excavation and loading are described in equations 1, 2 respectively. Equation 3 indicates the moment load corresponding to a G-up stage to create the ultimate collapse of the wall especially for the case of S1_RW2. In this calculation only the stress change due to centrifugation was considered while the wall dimensions remain constant.

$$ML = \frac{(KH^3\gamma_d - (H - Z_e)^3\gamma_w)}{6} \quad (1)$$

$$ML = \frac{K\gamma_d(H - h_w)(H^2 + Hh_w + h_w^2) + h_w^3(K\gamma' + \gamma_w)}{6} \quad (2)$$

For equations 1&2 $H = 50*H_{model}$ where $N = 50\,g$
For equation 3 $H = N*H_{model}$ where $N > 50\,g$
H_{model} is the height of wall in model scale

$$ML = \frac{H^3(K\gamma' + \gamma_w)}{6} \quad (3)$$

3 RESULTS AND DISCUSSION

3.1 Observed behaviours

Calculated moment loads using the equations 1 & 2 for different K-values and the measured bending moments at the embedment levels are plotted against the excavation depth and loading height in Figure 8. From this observation a negative end moment prior to the excavation can be seen, which was caused by the water pressure in the wall front. Also it indicates the initial earth pressure coefficients are closer to 0.5 and the earth pressure mobilised in between K_a-K_0 at the end of excavation.

However, at the end of the loading stage the K-values are closer to K_a, relatively large lateral displacement of the wall and the induced strains in the embedded portion could be the cause for this behaviour.

Observed wall top displacements and rotations against the excavation depth and the loading heights

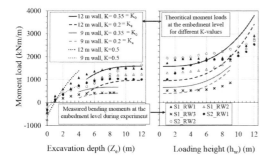
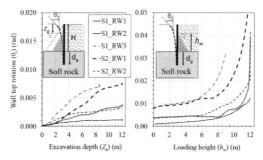

Figure 8. Theoretical and measured bending moments at the embedment level during excavation and loading.

Figure 10. Variation of measured wall top rotation with excavation depth and lading height.

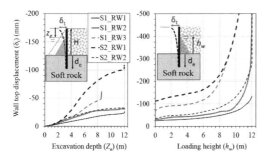
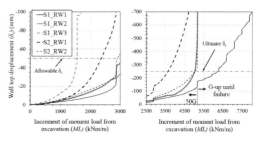

Figure 9. Variation of measured wall top displacement with excavation depth and loading height.

Figure 11. Wall top displacement with the increment of moment load at the embedment level.

are presented in Figures 9 & 10 respectively. Referring to the observed displacement and rotations a decreasing trend with excavation can be seen for the walls embedded in mud rocks and sand rocks. However, considering the similar embedment and loading conditions the walls embedded in mud rocks exhibits large displacements and rotations compare to the one in sand rocks, which can be attributed to the difference in the deformation characteristics and mechanical properties of sand and mud rocks.

3.2 Yielding and influence of embedment depth

Figure 11 describes the variation of wall top displacements against the applied moment load increment at the embedment level from the initiation of excavation. From Figure 11, the flexible wall used in S1_RW3 exhibits large plastic deformation under small increment of moment loads. Which clearly indicates the necessity of rigid walls to sustain against the large bending moments.

Considering the 2% and 10% of the pile diameter as allowable and ultimate limits of the wall top displacements it can be said that, rigid walls embedded in sand rock (S1_RW1 & S1_RW2) can secure the allowable displacements under design loads. However, the embedment in mud rock must be increased to sustain the design loads. Referring to the ultimate limit of wall top displacement and the behaviour of walls, it can be said that 0.5 m increment in the embedment depth significantly improved the stability of self-standing walls embedded in soft rocks. Considering the overall behaviour of S1_RW2 & S2_RW2 it can be concluded that, mud rock exhibits large plastic deformations soon after the initiation of yield deformation occurs, however the behaviour in sand rock is more ductile and no abrupt failure patterns can be observed as shown in Figure 14 even under the extreme loading conditions. Which clearly indicates that, 3 m embedment depth in sand rocks could provide additional margin of safety even after the application of ultimate loadings.

3.3 Bending moment-deflection and failure

Measured strains along the walls are illustrated in Figure 12 in terms of bending moments. Unlike the typical moment profiles in sand or clays; the walls embedded in soft rocks exhibits the peak bending exactly at the level of excavation, which can be attributed to the higher lateral confinement of relatively hard mediums even at smaller embedment depths. This phenomenon could result the yielding of low stiff sheet piles at the embedment level.

Referring to Figure 13 and observed pivot points in Table 3 it can be said that, unlike the walls in sand and clays in usual practice, the walls embedded in soft rocks rotates about the pivot point which is located far above from the bottom tip of the walls. Furthermore, from Figure 13, certain amount of deflection at wall top caused by the rotation of wall in the embedded portion especially for rigid walls.

Unlike rigid walls, a clear bending deformation of S1_RW3 can be conformed from the deflection profiles presented in Figure 13. However, the failure of flexible wall (S1_RW3) embedded in soft rock

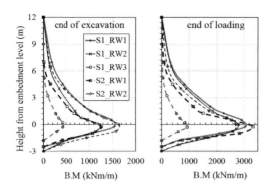

Figure 12. Bending moment profiles at the end of excavation and end of loading.

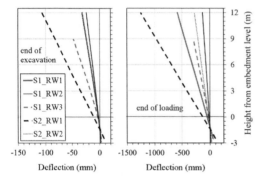

Figure 13. Deflection profiles at the end of excavation and the end of loading.

Table 3. Observed points of rotation at failure.

Test code	Pivot point (d_p) from embedment level m	d_p/d_e
S1_RW1	1.275	0.510
S1_RW2	1.562	0.521
S1_RW3	1.045	0.581
S2_RW1	1.450	0.580
S2_RW2	1.550	0.516

*d_e – Embedment depth.

is caused by the combination of bending and the large plastic deformations in the wall front embedded medium, which eventually lead towards the rotation of wall about the pivot point with an induced shear failure in the toe back during the wall collapse.

Figure 14 shows the observed deformation and failure patterns at the end of loading (top row) and up on removal of the wall (bottom row). A clear shear failure in the toe back can be seen for the cases of S1_RW1, S2_RW1 & S2_RW2 with the induced strains in the wall front of the embedded regime. Which clearly indicates the failure of walls by rigid body rotation about the pivot points as illustrated in Figure 13.

Considering the case of S1_RW2; from the observed wall deflections in Figure 13, an additional safety margin can be seen even after the experience of

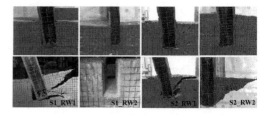

Figure 14. Observed deformation and failure patterns at the end of experiment (top) and up on removal of wall (bottom).

extreme loading conditions. Furthermore, from Figure 14, there is no clear plastic deformations can be seen in the embedded medium even after the application of large lateral load up to the centrifugation of 95 g. Which indicates that, small increment in the embedment depth e.g., 0.5 m, can significantly increase the stability of self-standing walls embedded in soft sand rock.

4 CONCLUSIONS

A centrifuge modelling system was successfully developed to study the behaviour of walls embedded in hard mediums from serviceability limit states to the ultimate failure.

Small increment (0.5 m) in the embedment depth significantly increased the stability and failure loads of self-standing walls embedded in soft rocks.

Stability of walls embedded in soft sand rocks can be secured with relatively small embedment depths than that of current design practices. However, the behaviour in mud rocks must be well studied with an increased embedment depth.

Artificial soft rocks with the similar stiffness of naturally existing mud rocks were modelled by using sand-clay-cement mixture at appropriate mixing conditions.

REFERENCES

Johnston, I.W. & Chiu, H.K. 1984. Strength of weathered Melbourne mudstone. *Journal of Geotechnical Engineering*, 110(7), pp. 875–898.

Johnston, I.W. 1991. Geomechanics and the emergence of soft rock technology. *Australian Geomechanics*, 21, pp. 3–26.

Lo, Kwan,Y., Micic, S., Lardner, T. & Janes, M. 2009. Geotechnical properties of a weak mudstone in downtown Calgary. *GeoHalifax '09*.

Takemura, J., Kondoh, M., Esaki, T., Kouda, M. & Kusakabe, O. 1999. Centrifuge model tests on double propped wall excavation in soft clay. *Soils and Foundations*, 39(3), 75–87.

Tatsuoka, F., Goto, S. & Sakamoto, M. 1986. Effects of some factors on strength and deformation characteristics of sand at low pressures. *Soils and foundations*, 26(1), pp. 105–114.

Terzaghi, K., Peck, R.B. & Mesri, G. 1948. Soil mechanics in engineering practice. John Wiley & Sons, New York, USA.

A new approach to modelling excavations in soft soils

J.P. Panchal, A.M. McNamara & S.E. Stallebrass
Research Centre for Multi-Scale Geotechnical Engineering, City, University of London, UK

ABSTRACT: Centrifuge modelling has been extensively used in the past to observe the ground response and mechanisms of movement around excavations. Owing to difficulties in performing in-flight excavations, some modellers use a range of cutters to form the main excavation trench and a void for the embedded length of the retaining wall before placing the model on the centrifuge swing. Preparation of an excavation using this technique in very soft clay, where the undrained shear strength is as low as 6 kPa at ground level is challenging as the surrounding soil can easily be disturbed. In addition, as the model reconsolidates in-flight the observed settlements can be considerable. This paper describes a novel process that was developed to prepare a soft soil sample and the apparatus used to model a deep excavation in soft soil. The technique ensured excellent consistency between tests, maintained stability of the excavation during model making and minimised disturbance to the surrounding soil.

1 INTRODUCTION

A rise in global development results in congested towns and cities as space above ground becomes a scarce resource. Underground spaces are therefore engineered to accommodate transport routes, basements and service trenches. In favourable ground conditions, deep excavations can be constructed with little impact on neighbouring structures if appropriate construction methods are adopted. However, where deposits of soft soil exist, ensuring the stability of an excavation is more complex and consequential ground movements can be excessive.

Engineers are often left to predict the ground response from the unloading of the overburden material based on empirical formulae from case studies. However, owing to the variables in each study it is challenging to accurately predict such ground movements arising at one particular site based on information from another. The excavation geometry, method of construction and soil stress history all have considerable impact on resultant ground movements, which increases the difficulty in predicting settlements.

2 BACKGROUND

There has been a steady rise in interest in the influence of deep excavations in very soft soils. Significant efforts have been made in modelling a wide variety of construction methods in such conditions and guaranteeing the repeatability of such tests is vital for providing a means of comparison between experiments.

Previous studies conducted by Lam (2010), Lam et al. (2012) and Yaodong (2004) used in-flight excavators which scraped away layers of soil before either actuating props or observing movements of a cantilevered excavation. Although these studies modelled the soil response to an excavation process, additional ground movements were permitted to occur that were a consequence of wall bending and rotation.

The research presented in this paper predominately focussed on modelling a variety of excavation techniques which had the potential to minimise ground movements in soft soils. The study required the segregation of lateral and basal heave movements whilst particular attention was paid to reducing the magnitude of heave.

Centrifuge modelling at 160g was used to simulate the vertical unloading of a formation level and soil movements were observed by on-board digital cameras and standard LVDTs.

3 OBJECTIVES

This paper describes the method that was successfully adopted for consistently producing a significant series of centrifuge experiments that modelled a deep excavation in very soft clay.

The complexity of preparing a soft soil sample means that it is often difficult to produce a clean and accurate model that is both representative of the prototype and repeatable. Model making techniques and bespoke apparatus were developed as part of this research project and will be presented in this paper.

4 SOIL MODEL

Plane strain centrifuge experiments were conducted in a 375 mm deep rectangular aluminium alloy strongbox with internal plan dimensions 550×200 mm. Speswhite kaolin clay was used as its properties are well established (Al-Tabbaa, 1987) and, owing to its relatively high permeability, results in a significantly reduced in-flight consolidation time compared to most natural clays.

The clay samples were made by initially creating a slurry consisting of Speswhite kaolin and distilled water, mixed to a water content of 120% in an industrial ribbon blade mixer. This is approximately twice the liquid limit of Speswhite kaolin.

Drainage channels cut in a herringbone formation in the base of the strongbox lead to drainage taps at either end of the strongbox, which allowed water to flow out of the sample during consolidation. Sheets of porous plastic and filter paper were positioned at the base of the box, over the drainage channels, before carefully placing the slurry with a scoop and palette knife. The slurry was agitated between each pour to minimise the risk of air entrapment.

To produce a sufficiently deep sample it was necessary to consolidate approximately 500mm of slurry. A 300mm deep extension was therefore bolted to the top of the strongbox to accommodate the total slurry depth. A greased O-ring between the extension and strongbox provided a watertight seal. The slurry was sandwiched between more sheets of filter paper and porous plastic before being transferred to a hydraulic press.

A platen attached to the piston ram fitted snugly inside the extension box. Holes in the top of the platen allowed water to drain from the top of the sample. Pipes connected to the base drainage taps were submerged in a bucket of water to prevent air from entering the sample. Allowing water to drain from the top and bottom of the sample halved the drainage path length and consequently reduced the overall consolidation time.

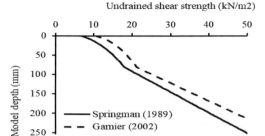

Figure 1. Estimation of undrained shear strength profile of centrifuge model.

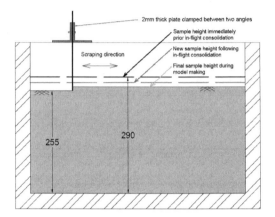

Figure 2. Scraper used to trim sample to correct depths.

profile of the model following 100 kPa consolidation at 1g is illustrated in Figure 1.

$$S_u = 0.19 \, \sigma'_v \, OCR^{0.59} \quad (1)$$

$$S_u = 0.22 \, \sigma'_v \, OCR^{0.706} \quad (2)$$

where σ'_v = vertical effective stress in the model and OCR = overconsolidation ratio, defined as $\sigma'_{v(max)}/\sigma'_{v(current)}$.

However, owing to the difficulties in preparing a consistent model in a sample following consolidation under a low effective stress, the sample was also consolidated in-flight at 160g before any attempt was made to construct the model.

In order to achieve this, the sample was removed from the hydraulic press and trimmed to a depth was 290 mm. Reasonable trimming of the sample could only be achieved by scraping the top of the soil surface in numerous passes thereby minimising soil disturbance. A tapered scraper was fabricated and this was dragged across the top of the strongbox walls to ensure that exactly the correct depth of soil remained. The scraper could be adjusted to provide a range of trimming heights, as illustrated in Figure 2.

To consolidate the sample in-flight it was necessary to provide a water feed to the model. A conventional overflow standpipe connected to the base drain was

5 SAMPLE CONSOLIDATION

In summary, the following steps were undertaken in preparing the sample and details of which will follow.

1. Consolidate slurry to 100 kPa
2. Trim sample to 290 mm and consolidate in-flight resulting in less than 20 mm surface settlement
3. Remove sample from centrifuge and trim remainder of sample to 255 mm (final model height)

The sample was initially subjected to a consolidation pressure of 25 kPa and this was gradually increased to 100 kPa over 2 days. The sample was left to consolidate under a maximum vertical effective stress of 100 kPa at 1g, typically for 10 days, to produce a very soft soil sample with an S_u lower than 40 kN/m² at depth at 160g.

An estimation of the soil undrained shear strength profile under centrifugal loading was determined using the equations given by Springman (1989) and Garnier (2002) as defined in Equations (1) and (2) respectively. The expected undrained shear strength

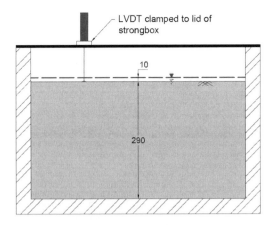

Figure 3. In-flight consolidation set up.

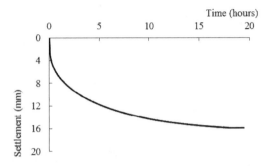

Figure 4. In-flight consolidation settement curve at 160g.

used. There was also a requirement to ensure the sample remained saturated whilst in-flight. The water table was therefore set at 300 mm; 10 mm above the top of the initial soil level. Evaporation of the free standing water was prevented by bolting a lid to the strongbox. This also served the purpose of housing an LVDT, the footing of which was positioned on the clay surface to monitor the settlement during in-flight consolidation, as shown in Figure 3. A typical settlement curve against time is shown in Figure 4.

6 TESTING APPARATUS

Modelling a deep excavation in soft soil ground conditions required careful consideration to permit repeatability of tests and accurate model making. Hence, a number of bespoke tools and pieces of apparatus were developed to aid in the preparation of the model.

Previous researchers (McNamara, 2001; Richards & Powrie, 1998) cut the excavation void and formed a narrow trench for the embedded length of the wall before inserting a rectangular cross sectioned aluminium wall. However, concerns regarding the ability of soft soil in supporting a shallow embedded wall led to the development of an alternative modelling solution and wall design.

To mitigate the risk of the wall collapsing before forming the excavation it was necessary to install the

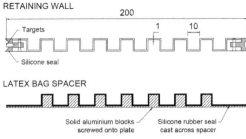

Figure 5. Plan of retaining wall and spacer used in tests.

wall prior to removing any soil. A crenulated wall following a similar concept to a steel sheet piled wall was therefore used in this series of tests so that it could be pushed into the soil. It was machined from a single 10 mm thick plate such that the wall was consistently 1 mm thick, as shown in Figure 5. Owing to the reduced thickness of the wall the stiffness was improved by machining the wall from a stainless steel plate compared with aluminium used previously (McNamara, 2001). At prototype scale this possessed equivalent stiffness to a reinforced concrete diaphragm wall that was approximately 2m thick. The installation of this wall effectively replicated that of a sheet pile wall and the total bulk unit weight of the wall and soil encased within the ribs was calculated as 27.5 kN/m^3.

To prevent the flow of water around the wall, silicone seals were cast onto the sides of the wall. These were housed within lightweight aluminium channels which were screwed into the stainless steel wall flanges, details of which are illustrated in Figure 5.

Guaranteeing that the wall was positioned vertically and in the same position in every test was vital to ensure repeatability and comparison of results. Consequently, a Perspex guide was fabricated which bolted onto the strongbox and the excavation cutting shelf. This guide supported the entire length of the wall and the ribs of the Perspex corresponded to the profile of the wall, ensuring that it was inserted vertically, see Figure 6.

An aluminium block (McNamara et al., 2009) was designed to support the exposed length of wall. This essentially acted as a very stiff prop in compression. A sliding capping beam attached to the stiffener and was seated over the wall upstand to provide tension propping in the event of passive failure.

Simulation of the excavation required the use of a pressurised latex airbag, which surcharged the formation level as the sample reconsolidated and was later deflated to model the vertical stress relief associated with excavations. However, owing to the shape of the retaining wall there was a risk that the bag would burst under contact with the wall at high pressure. Hence, a spacer was designed to slot into the ribs of the wall, providing a reaction for the bag and preventing water seepage into the excavation (Figure 5).

Figure 6. Perspex guide used for installation of wall.

Figure 7. Void formed to cater for wall silicone seals.

Figure 8. Plates used to remove layers of soil and form the excavation void.

7 MODEL MAKING PROCEDURE

Following in-flight consolidation the sample was removed from the centrifuge and all standing water mopped from the surface. The sample was trimmed to give a final height of 255 mm before a profiled template was placed over the excavation area. This template served the purpose of preventing the placement of PlastiDip (an aerosol applied synthetic rubber membrane) to the excavation void, as shown in Figure 6. This ensured that the PlastiDip was not pushed into the clay as the wall was installed.

The front face of the strongbox was removed and a thin layer of silicone oil was spread across the clay to prevent excessive drying out of the model. A bespoke cutting shelf was attached to the front of the model which supported the Perspex guide.

Careful consideration of the silicone seals along the edges of the wall was required, as the wall could not be directly pushed into the soil without forming a void for the seal channels. Thin walled circular and square brass cutters, 10 mm in diameter and width respectively, were used to clear an area to the correct depth for the seals, as shown in Figure 7. The wall seals were greased before installation and steel plates were subsequently used to remove layers of soil from the excavation area (Figure 8) to the formation level 75 mm (H) below ground level. The shelf ensured the correct dimension was achieved in every test, as illustrated in Figure 9.

Having formed the excavation void the cutting shelf was removed and steps were taken to install the remaining apparatus and instrumentation. A latex bag was positioned and connected to an air pressure transducer union through the strongbox wall. The bag spacer was pushed into the wall and the aluminium stiffener and capping beam were subsequently lowered into the excavation void. Care was taken to avoid disturbing the wall or formation level during this stage (Figure 10).

Figure 9. Clean and accurate excavation formed in each test.

Figure 10. Stiffener and capping beam fixed in place.

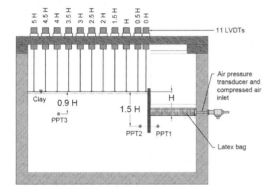

Figure 11. Location of PPT and LVDT instrumentation.

Figure 12. Instrumented model prior to centrifuge test.

Black glass ballotini beads, 1mm in diameter, were rolled into the front face of the model; this permitted observation of subsurface movements using onboard digital cameras. Analysis of ground movements was later achieved using particle image velocimetry (GeoPIV) developed by Stanier & White (2013).

Pore pressure transducers could not be installed prior to in-flight consolidation owing to the large settlements that occurred and the risk of both damaging the instrumentation and an inability to determine their exact position within the consolidated model. They were installed during the model making stage by coring through the backwall of the strongbox to the centreline of the model. PPTs had been saturated and de-aired in a vacuum chamber before being inserted in the model. They were subsequently backfilled with de-aired clay slurry mixed to a water content of 120%.

At least three PPTs were used across the tests; two were located at a depth of approximately $1.5H$ below ground level and $0.5H$ either side of the wall. The third PPT was located at a distance of $4H$ to measure far field pore pressure changes, as shown in Figure 11.

Surface settlements were measured along the centreline of the model using eleven LVDTs. They were positioned at $H/2$ intervals behind the retaining wall with an additional one being located directly behind the wall. These were clamped to a gantry which was bolted to the top of the strongbox, as shown in Figure 12.

Following the complete assembly of the model, the front silicone wall seal was greased to prevent water seepage. A thin layer of high viscosity silicone oil was applied to a Perspex window to reduce boundary effects before being bolted to the model prior testing. On average, the entire model making process typically took less than 4 hours to complete.

8 TESTING PROCEDURE

As the model was accelerated to $160g$ the pressure in the airbag was gradually increased to apply the correct overburden pressure at varying g levels. The model was left to reconsolidate whilst the pore pressure readings were monitored.

Data logging was carried out using LabView and at 90% consolidation the model was tested. This comprised decreasing the pressure in the airbag at a rate of 1 kPa/sec.

Consistency was achieved as the same operator controlled the air pressure valve in the centrifuge control room. Images were taken at one second intervals and were calibrated against the sample count of the data file. Following complete removal of the surcharge the model remained in-flight to monitor long-term settlements and pore pressure changes before removal from the centrifuge.

9 RELIABILITY OF MODELLING PROCEDURE

Immediately after the model was decelerated, shear vane readings were taken through the model at the following distances behind the wall; H, $2H$, $3H$ and $4H$. The undrained shear strength profiles from each test at distance $3H$ and the profile predicted using Springman (1989) have been plotted in Figure 13. The undrained shear strength at a distance from the excavation could

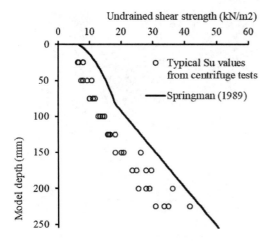

Figure 13. Undrained shear strength profile from centrifuge tests taken at a distance of $3H$ from the retaining wall.

reasonably be expected to remain uninfluenced, irrespective of the construction technique adopted in each test.

Figure 13 illustrates that the S_u profile generally follows the Springman trend and is fairly consistent with strengths measured in a number of tests. Variation in the measured and predicted S_u values is likely due to the development of cavitation when the model was decelerated and unloaded. This would result in a decreased S_u (Mair, 1979) as illustrated in Figure 13.

Water contents were taken at various stages of preparing the model and the soil bulk unit weight at the base of the excavation ranged between 15.87 kN/m^3 and 16.18 kN/m^3.

10 CONCLUSION

This research project focussed on modelling a variety of excavation techniques in very soft soil. A large series of centrifuge tests were conducted at 160g in clay where the undrained shear strength ranged from 6 kN/m^2 at ground level to 40 kN/m^2 at depth. The model making procedure adopted was developed to ensure the repeatability of tests and to permit comparison between them.

Consistency between tests was confirmed by the similarities in undrained shear strengths of soil at a distance of $3H$ from the excavation. The profile also followed the expected S_u profile derived from the Springman (1989) equation. Water contents taken from each test specimen before and after in-flight consolidation were within 1% of the average value, emphasising the consistency of this method of sample preparation for soft soil models.

REFERENCES

Al-Tabbaa, A. 1987. Permeability and stress-strain response of Speswhite kaolin, PhD Thesis. University of Cambridge, UK.

Garnier, J. 2002. Properties of soil samples used in centrifuge models. Physical Modelling in Geotechnics: ICPMG '02, Phillips, Guo & Popescu (eds.) 2002 Swets & Zeitlinges Lisse, ISBN 90 5809 389 1. pp 5–19.

Lam, S.S.Y. 2010. Ground movements due to excavation in clay: physical and analytical models. PhD Thesis, University of Cambridge.

Lam, S.Y., Elshafie, M.Z.E.B., Haigh, S.K. & Bolton, M.D. 2012. A new apparatus for modelling excavations. International Journal for Physical Modelling in Geotechnics, Vol. 12, No. 1, pp 24–38.

Mair, R.J. 1979. Centrifugal modelling of tunnelling construction in soft clay. PhD Thesis, University of Cambridge, UK.

McNamara, A.M. 2001. Influence of heave reducing piles on ground movements around excavations. PhD Thesis. City University London, UK.

McNamara, A.M., Goodey, R.J. & Taylor, R.N. 2009. Apparatus for centrifuge modelling of top down basement construction with heave reducing piles. International Journal of Physical Modelling in Geotechnics, Vol. 9, No. 1, pp 1–14.

Richards, D.J. & Powrie, W. 1998. Centrifuge model tests on doubly propped embedded retaining walls in overconsolidated kaolin clay, Geotechnique 48(6): 833–846.

Springman, S. M. 1989. Lateral loading on piles due to simulated embankment construction. PhD Thesis. University of Cambridge, UK.

Stanier, S.A & White D.J. 2013. Improved image-based deformation measurement for the centrifuge environment, Geotechnical Testing Journal, 36(6): 915–928.

Yaodong, Z. 2004. An embedded improved soil berm in excavation – Mechanisms and capacity. PhD Thesis. National University of Singapore, Singapore.

1g-modelling of limit load increase due to shear band enhancement

K.-F. Seitz & J. Grabe
*Insitute of Geotechnical Engineering and Construction Management,
Hamburg University of Technolgy, Germany*

ABSTRACT: Strengthening of just the shear bands is a patented idea to prevent the critical failure mechanism in order to increase the capacity of a geotechnical system with a minimum of material. 1g-model tests are used to examine the proposed method for limit load increase. Therefore the failure mechanism and bearing load of a geotechnical system without shear band enhancement are compared to those with shear band enhancement. Two basic geotechnical systems are examined using a narrow model container filled with dry sand: a shallow foundation with bearing failure and a retaining wall with horizontal wall movement resulting in an active earth wedge. The failure mechanisms are identified using PIV. The shear bands are enhanced by pressure controlled injection of a water-cement suspension. 1g-model tests show that shear band enhancement leads to a capacity increase for a simple geotechnical system.

1 INTRODUCTION

Guidelines for the design and the calculation of geotechnical systems e. g. for the determination of the bearing capacity of a shallow foundation or the active earth pressure on a retaining wall are based on localization of shear bands and failure mechanisms in the soil. However, shear bands have not been explicitly taken into consideration for the actual design. Therefore, this paper proposes a method to include the shear bands in the design. By strengthening just the shear bands the loading capacity of the system would be increased with a minimum of material.

A particular failure mechanism will evolve when a retaining wall is horizontally displaced (Coulomb 1776) or a shallow foundation is loaded until its limit load is reached (Prandtl 1920). When the failure mechanism is prohibited through shear band enhancement the earth pressure on the retaining wall or the bearing capacity of the shallow foundation are altered, see Figure 1. Based on the concept of one critical failure mechanism the limit load can be increased if this particular failure mechanism is disabled through targeted soil reinforcement.

The proposed method for limit load increase has been presented by Grabe & Seitz (2016) and Seitz & Grabe (2018). Grabe & Seitz (2016) show the results of 1g-model tests of a retaining wall, in which the shear band is enhanced. It is possible to reduce the loads resulting from active earth pressure by 29%. However, the applied method for shear band enhancement is not easily reproduced, therefore an automated procedure is developed for the examinations in this paper. Seitz & Grabe (2018) present an automated procedure for the numerical enhancement of shear bands using

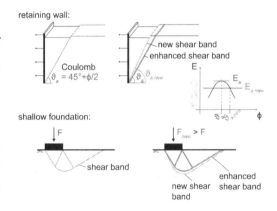

Figure 1. Concept for limit load increase due to shear band enhancement. Modified from Grabe & Seitz (2016).

OptumG2. Two geotechnical systems are analysed: a shallow foundation and a retaining wall. The numerically obtained results prove the capability of increasing limit loads.

In this paper 1g-model tests are used to analyse the proposed method for limit load increase. The failure mechanism and the bearing loads of two basic geotechnical systems with and without shear band enhancement are compared. Using a narrow model container filled with dry sand, a shallow foundation and a retaining wall with horizontal wall movement resulting in an active earth wedge are examined. The potential shear bands are enhanced by injecting a water-cement suspension using a newly developed method for automated enhancement.

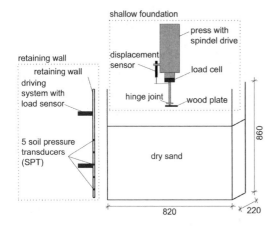

Figure 2. Testing stand for 1g-model test with equipment set-up for a retaining wall model and a shallow foundation model. Units in mm.

Table 1. Properties of model soil according to Kaya (2016).

grain density	ρ_d	[g/cm^3]	2.65
min. void ratio	e_{min}	[–]	0.526
max. void ratio	e_{max}	[–]	0.813
critical shear angle	φ_c	[°]	31

2 MODEL TEST SET-UP

2.1 Testing stand

The testing stand (see Figure 2) consists of a soil container with inner dimensions of 860, 220 and 820 mm in length, width and height. It has previously been used in Qiu (2012) and Grabe & Seitz (2016). For the presented analysis it has been modified in order to reduce the width of the container to 220 mm. The container's boundaries are made of 45 mm thick transparent acrylic plates, which are covered with glass to reduce the wall friction in planar direction. The testing stand can either be used for the model retaining wall or the model shallow foundation depending on the set-up.

The shallow foundation is a wooden plate measuring 220 × 50 × 10 mm. The load is exerted on the plate through a hinge joint which enables tilting of the wooden plate. The load application is eccentric to determine the direction of the failure mechanism. The eccentricity measures 25 mm.

The wall on the right of the testing set-up is a horizontally movable retaining wall made of steel. For the retaining wall tests the wall will be moved away from the sand within the container in order to induce the sliding of an active earth wedge towards the wall.

All model tests are set-up in order to achieve plane strain conditions. The shear bands may be observed from the front of the model container. The location of the shear bands is checked by a look from above on the soil surface, to verify that the shear bands are orthogonal to the view plane.

The tests are carried out under natural gravitational acceleration (1g-model tests). This leads to model errors, since e.g. the stresses within a small scale model are very low compared to a prototype scale. However, the presented 1g-model tests may serve as a proof of concept for the presented idea of shear band enhancement. They are designated for a qualitative comparison between the tests with shear band enhancement and those without shear band enhancement.

2.2 Model soil

The soil body consists of granular material, which is a round grained medium to coarse sand. Table 1 gives an overview of the soil's main properties according to Kaya (2016). The sand is filled into the soil container using the pluviation technique according to Gutberlet (2008). During pluviation the drop height between the outlet and the current surface is kept constant in order to obtain an equally dense soil sample. This method is chosen to ensure the reproducibility of the conducted model tests. The reproducibility is important to allow for a qualitative comparison of the model tests.

For the shallow foundation tests the sand is pluviated until it reaches a height of 300 mm. The retaining wall tests are carried of with a height of 400 mm. The relative soil density is calculated as $I_D = (e_{max} - e)/(e_{max} - e_{min})$ according to DIN 18126 (1996) with void ratio e. The shallow foundation tests are carried out at a relative density between 0.74 and 0.88 and for the retaining wall tests the relative density ranges between 0.45 and 0.63.

2.3 Instrumentation

The shallow foundation set-up uses a vertical press with a spindle drive in order to apply load on the wooden plate. The press is driven with a low constant vertical speed of 0.1 mm/s in order to ensure a static load-displacement behaviour. The load is measured with a load cell, which has a measuring range of 1 kN. The displacement of the press is measured with a displacement sensor.

The retaining wall has a driving system with three drives each including a force transducer. The drives are arranged in a triangular shape spanning over the former width of the container (920 mm). Only one of the drives is located directly behind the sand model and will therefore be the only one used for data analysis of the test results. Five strain-gauge-type soil pressure transducers are placed on the movable wall in the central zone of the model. They are positioned at 25, 75, 125, 225 and 275 mm beneath the surface of the model sand. The retaining wall is moved with a constant velocity of 50 μm/s.

2.4 PIV

The failure mechanism is detected using particle image velocimetry (PIV) (Take et al. 2003). The shallow foundation tests have been carried out using a Nikon D7000 camera and two light sources. The camera takes picture in 2" intervals. The pictures are analyzed using PIVLab (Thielicke & Stamhuis 2014).

For the retaining wall system the PIV set-up has been modified. A compact high speed monochrome camera (UX50) and two LEDs (ILA_5150 LPS System) are used. Pictures are taken every second and the LEDs are used as a continuous source of green light. The green light is chosen because the the camera has a high relative spectral response in this wavelength and it enables the reduction other light sources' influence due to a bandpass filter, which is placed between the lens and the sensor of the camera. GeoPIV-RG (Stanier et al. 2016) is used for the analysis.

3 SHEAR BAND ENHANCEMENT

Shear band enhancement refers to the enhancement of the soil sample at a specific loaction, namely were the potential shear bands arise. It does not refer to a certain technique of soil manipulation. In the presented study, the shear band enhancement is achieved by injecting a water-cement suspension into the soil. However, other injection techniques and materials are imaginable.

Reference tests without shear band enhancement are carried out in order to determine the location of the shear bands using PIV. In order to prove the reproducibility these reference tests have been conducted several times yielding similar results concerning the limit loads and the shear band location. Once the location has been determined the main tests with shear band enhancement can be carried out. For that the soil container is again pluviated with sand. Before the test is carried out by loading the foundation or moving the wall, the potential shear band is enhanced.

For the enhancement a water-cement mixture ($w/c = 0.54$) is injected into the soil. CEM I is used for the shallow foundation and CEM II for the retaining wall tests. However, CEM I and CEM II have similar characteristics within these model tests. A cartridge gun with a volume of 410 ml is filled with the water-cement mixture. The water-cement mixture is pressed through a system of tubes and a pipe into the soil. The pipe is pushed into the ground. When it has reached the final enhancement depth, the injection begins and the pipe is gradually torn out out the soil. Columns of mixed cement soil develop using this method. After 24 hours the columns have reached sufficient strength for testing.

Using the PIV technique the shear bands may be detected in the view plane. A look from above on the soil surface is carried out to verify that the shear band is orthogonal to the view plane. In order to maintain plane strain conditions, it is paid attention to enhance throughout the depth of the model. Therefore the enhancement columns are placed tangentially. They result in a enhanced plane. After carrying out the tests a look from above on the model surface confirms

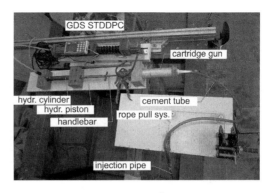

Figure 3. Water-cement mixture injection device consisting of a GDS pressure controller connected to a cartridge gun loaded with a water-cement mixture. Modified from Kistler (2017).

shear bands that are orthogonal to the view plane, when using enhancement.

The shallow foundation tests with shear band enhancement are prepared by manual injection of the water-cement mixture. The pipe is inserted into the soil in a defined grid. The material is injected by operation of the cartridge lever and simultaneous pulling of the pipe.

The retaining wall tests with shear band enhancement are prepared with an automatized injection of the water-cement mixture. Figure 3 shows the injection device, which has been developed for the present study. A GDS Standard Pressure Volume Controller (STDDPC) is used to inject the water-cement mixture with a constant flow rate. The pressures are logged during injection. The GDS device is connected to the cartridge with a hydraulic cylinder. The piston automatically moves the cartridge handlebar and the water-cement mixture is pressed via the tube into the injection pipe. As soon as the cement mixture reaches the bottom of the pipe, the pipe is pulled out by the rope pull out system. In order to do this, a crank needs to be operated at constant speed (system developed by Kistler (2017)). The mean pressure for injection is 653 kPa. With this method columns of about 1.5 cm diameter are produced. These columns are narrowly placed to obtain a planar enhancement at the location of the shear bands.

4 TEST RESULTS

4.1 Shallow foundation

For testing the shallow foundation three reference tests are carried out to determine the failure mechanism and the load displacement behaviour. As expected the failure mechanism develops to the right underneath the eccentrically loaded shallow foundation, compare Figure 4. The failure mechanism proves to be reproducible, except for slight differences in the measurements of the geometry. Therefore a determination of the area to be enhanced is possible.

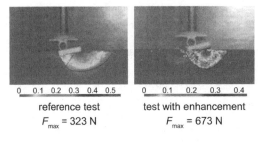

reference test
F_{max} = 323 N

test with enhancement
F_{max} = 673 N

Figure 4. Failure mechanism for the shallow foundation tests, analyzed with PIVlab (velocity magnitude in m/s). Left reference test 1 without shear band enhancement and right test with shear band enhancement.

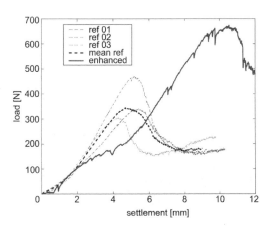

Figure 5. Load settlement curves of the reference tests without shear band enhancement and for the main test with shear band enhancement.

The enhancement within the full extent of the shear band is difficult. Especially the curved part may not be enhanced without disturbing the covering soil with the chosen method. In the case of the shallow foundation the shear bands emerge from the edges of the foundation and then develop to the surface, compare e.g. Murthy (2014). Therefore it is assumed, that the beginning of the shear band has the biggest impact on the failure mechanism and the corresponding bearing capacity. Consequently, only the straight parts of the shear bands underneath the shallow foundation are enhanced.

The failure mechanism of one reference test and the main test with shear band enhancement are shown in Figure 4 on the right. The results show that the enhancement of the beginning of the shear bands has an impact on the failure mechanism. The failure mechanism still develops on the right hand side. However, the extent of the failure mechanism has decreased. The inclination of the shallow foundation is higher at limit load than for the reference tests.

The load settlement-curve of the shallow foundation tests (Weirauch 2016) are depicted in Figure 5. It is striking, that the reached limit loads are higher, than theoretically expected. This may be explained by side wall friction within the model and the fact that the hinge joint is not modelled ideally. However, for a qualitative comparison of the model tests among each other the results are good enough, since the same deviations are expected in all the model tests, that are to be carried out. The results show that the bearing capacity of the shallow foundation could be increased by 331 N (197%), when comparing the averaged value of the reference tests and the test with enhancement. It can also be seen, that the deformation, which corresponds to the limit load, has increased significantly.

The tests show, that the enhancement of the beginning of the shear band successfully leads to a change in failure mechanism and the corresponding bearing capacity. However, it has to be stated, that the enhancement is quite large in comparison to the model dimensions. Therefore, the idea of the thin enhancement of only the shear band is not be fully implemented in the model tests, yet.

4.2 Retaining wall with horizontal wall movement

Four reference tests are carried out to determine the failure mechanism of the retaining wall induced through horizontal wall movement. The slip line inclination in the reference tests ranges between 64 and 70°, which is around 8° higher than theoretically expected. However, the results are in good agreement with the theoretical expectations considering the remaining side wall friction and the low confining pressures resulting from the small scale of the model having an influence on the dilational behaviour of the soil. The measured average inclination of 66° is used as a basis for the enhancement.

The enhancement is carried out with the above described method for automatized injection. Six tests with enhancement in the potential shear band are carried out. Figure 6 shows the test results analyzed with GeoPIV-RG. On the left hand side a reference test and on the right a main test with enhancement is shown. It can been observed that the earth wedge becomes smaller and the inclination of the slip line has been changed from 66 to 69° when the potential shear band is enhanced.

The load exerted on the retaining wall is measured throughout the wall displacement process. The corresponding load displacement curves are shown in Figure 7. Compared to analytically obtained values the loads obtained during the reference tests are lower. This can be explained by the side wall friction and the low confining pressures in the small scale model, compare above. The necessary relative wall displacement to induce active earth pressure is bigger than theoretically expected. This might be explained by the relatively fast wall movement. However, the overall shape of the load displacement curve is in good agreement with theory considering a dense sand.

The earth pressure at rest is increased by nearly 300 N through shear band enhancement, as it can be seen in Figure 7. During the enhancement process the soil is displaced when inserting the tube and injecting

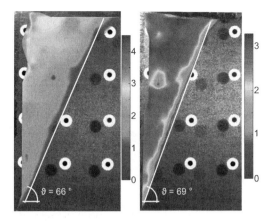

Figure 6. Failure mechanism of a reference test (left) and a main test with enhancement (right). Contour plot for the resultant displacement after horizontal wall movement [mm], analyzed with GeoPIV-RG.

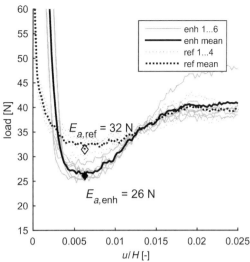

Figure 8. Load displacement curves of the retaining wall test. Focus on the load resulting from active earth pressure.

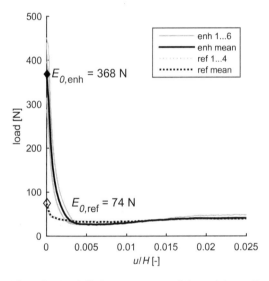

Figure 7. Load displacement curves of the retaining wall test. Focus on the load excerted on the retaining wall at rest.

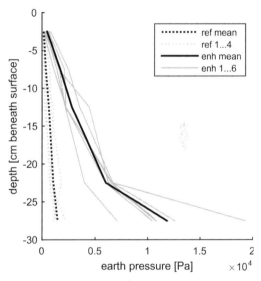

Figure 9. Earth pressures before horizontal wall movement for reference tests (dotted lines) and main tests (solid line) with shear band enhancement.

the cement mixture. This leads to higher pressures, which significantly increase the load exerted on the retaining wall.

When the retaining wall is displaced the loads decrease rapidly until the active earth pressure is reached. As it can be seen in Figure 8 the load resulting from active earth pressure is decreased through shear band enhancement from 32 to 26 N (18.8%).

Figure 9 shows the measured earth pressures at rest before the retaining wall is horizontally displaced. As expected it can be seen, that the earth pressures are significantly higher due to the increased pressure within the soil body arising from enhancement.

The earth pressures after the horizontal wall movement are shown in Figure 10. The reference tests and the main tests with enhancement have each been averaged. It can be seen, that the averaged earth pressures are reduced in the vicinity of the surface. However, deeper below the surface the enhancement is not able to reduce the earth pressures.

5 CONCLUSIONS

The model tests show that the critical failure mechanism can be changed by injecting cement into the soil. The limit load for the shallow foundation could be reduced significantly.

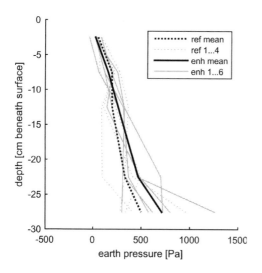

Figure 10. Earth pressures after horizontal wall movement for reference tests and main tests with shear band enhancement.

It is also possible to reduce the load arising from active earth pressure exerted on the retaining wall. However, the difference between the maximal forces of the reference models and the main models with shear band enhancement are very small. Further tests are to be carried out in the container with the original width. This should increase the influence of the testing stand because the loads exerted on the wall are expected to be much higher. A similar set-up has been used in Grabe & Seitz (2016) and showed good results. The earth pressure sensors do not provide a clear result. In future tests more earth sensors should be set-up in order to increase the quality of the monitoring.

Up to now, the enhancement results in relatively big columns compared to the model dimensions. Further test have to be carried out to reduce the size of the enhancement compared to the shear bands.

Even though, the results can only show the impact of shear band enhancement qualitatively, the present paper can serve as proof of concept. It proves, that shear band enhancement results in an increase of admissible limit load and should therefore be further investigated in future research.

Up to now, the results of the model tests are by far not scalable to prototype scale. The 1g model conditions result in low stress levels within the soil body, which has an impact on the dilation and the stiffness of the soil. In order to overcome these drawbacks of the current results, ng-model tests are planned to be carried out in the near future in cooperation with the Centre for Offshore Foundation Systems.

ACKNOWLEDGEMENT

The authors thank Selma Rachel Kistler and Manya Weirauch for assisting the experimental works and carrying out some of the model tests.

The work is part of the project "Increase of bearing capacity for geotechnical constructions through shear zone enhancement" (GR 1024/23-1), which is funded by the German Research Foundation (DFG).

REFERENCES

Coulomb, C. A. (1776). *Essai sur une application des règles de maximis & minimis à quelques problèmes de statique, relatifs à l'architecture*. Paris: De l'Imprimerie Royale.

DIN 18126 (1996). Deutsches Institut für Normung e.V. Bestimmung der Dichte nichtbindiger Böden bei lockerster und dichtester Lagerung.

Grabe, J. & K.-F. Seitz (2016). Optimization of geotechnical structures for states of serviceability and ultimate loads. In A. Zingoni (Ed.), *Insights and Innovations in Structural Engineering, Mechanics and Computation*, pp. 2048–2053. Boca Raton: CRC Press.

Gutberlet, C. (2008). *Erdwiderstand in homogenem und geschichtetem Baugrund - Experimente und Numerik*, Volume H. 78 of *Mitteilungen des Institutes und der Versuchsanstalt für Geotechnik der Technischen Universität Darmstadt*. Darmstadt: Inst. u. Versuchsanst. für Geotechnik.

Kaya, H. (2016). *Bodenverschleppung und Spaltbildung infolge der Einbringung von Profilen in Dichtungsschichten aus Ton*. Hamburg: Hamburg University of Technology.

Kistler, S. (2017). *Untersuchungen zur Verfestigung der Scherfuge des aktiven Erddruckgleitkeils*. Projektarbeit, Technische Universität Hamburg-Harburg, Hamburg.

Murthy, T. (2014). Evolution of deformation fields in 1-g model tests of footings on sand. In C. Gaudin and D. White (Eds.), *Physical modelling in geotechnics*, Leiden, The Netherlands, pp. 653–657. CRC Press/Balkema.

Prandtl, L. (1920). Über die Härte plastischer Körper. *Nachrichten von der königlichen Gesellschaft der Wissenschaften zu Göttingen, mathematisch-physikalische Klasse*.

Qiu, G. (2012). *Coupled Eulerian Lagrangian Simulations of selected soil-structure interaction problems*. Dissertation, Technische Universität Hamburg-Harburg, Hamburg.

Seitz, K.-F. & J. Grabe (2018). Numerical modelling of limit load increase due to shear band enhancement - to be published. In *Physical modelling in geotechnics: Proceedings of the 9th NUMGE Conference on Numerical Methods in Geotechnical Engineering in Porto, Portugal, 25-27 June 2018*.

Stanier, S. A., J. Blaber, W. A. Take, & D. J. White (2016). Improved image-based deformation measurement for geotechnical applications. *Canadian Geotechnical Journal* 53(5), 727–739.

Take, W. A., M. D. Bolton, & D. J. White (2003). Soil deformation measurement using particle image velocimetry (PIV) and photogrammetry. *Géotechnique* 53(7), 619–631.

Thielicke, W. & E. J. Stamhuis (2014). PIVlab – towards user-friendly, affordable and accurate digital particle image velocimetry in MATLAB. *Journal of Open Research Software 2*, 1202.

Weirauch, M. (2016). *Machbarkeitsstudie zur Scherfugenverfestigung bei Grundbruch im 1g-Modellversuch*. Bachelorarbeit, Technische Universität Hamburg-Harburg, Hamburg.

Concave segmental retaining walls

D. Stathas
EBP Schweiz AG, Zurich, Switzerland formerly Department of Civil & Environmental Engineering, Hong Kong University of Science and Technology, Kowloon, Hong Kong

L. Xu
Department of Civil Engineering & Engineering Mechanics, Columbia University, New York, USA

J.P. Wang
Department of Civil Engineering, National Central University, Zhongli, Taiwan

H.I. Ling & L. Li
Department of Civil Engineering & Engineering Mechanics, Columbia University, New York, USA

ABSTRACT: Segmental Walls (SWs) are earth retaining structures with an easy and fast construction method by using pre-fabricated modular blocks. However, such a wall face is usually planar. In this study, we designed curved SWs with a special block called Porcupine. A series of small-scale (1:10) model tests were conducted in the centrifuge facility at Columbia University, showing that a concave profile can significantly improve the overall performance of the gravity segmental walls (GSWs). When reinforcement is added, to increase structure's load capacity, the system's performance is dominated by the reinforcement stiffness and the wall facing geometry has no significant influence. Another highlight of this study is the use of low-cost 3D-printing in model preparation. To sum up, this study finds an easy, practical solution to improve the performance of GSWs by aligning the wall facing blocks in a curved geometry, also demonstrating how to utilize 3D printing in geotechnical model testing.

1 INTRODUCTION

1.1 Segmental walls

Segmental walls (SWs) are usually made of dry-cast concrete units stacked together to build an earth retaining structure. The absence of mortar between the blocks makes this type of structure easy to build with minimum construction time. Furthermore, SWs are environmental friendly structures and pleasant from aesthetic point of view (NCMA 2009). However, SWs face height limitation. Structures above 3.0 m height are not feasible with the usual block types or they would require heavier and bulkier blocks. Hence, soil reinforcement is usually employed to increase the retaining height. Reinforced soil segmental walls (RSWs) are complex systems with many interactions (soil-facing-reinforcement) and the optimization of their design is a popular research topic (Leshchinskly et al. 2014, Allen and Bathurst 2015).

1.2 Concave profile

To the best of our knowledge, the profile of SWs is always planar with conventional blocks assembled together. Analytical and partially numerical studies suggested that log-spiral or circular profiles could improve slope stability (Jeldes et al. 2014, Utili and Nova 2007, Vahedifard et al. 2016b). Furthermore, a circular geometry is expected to reduce reinforcement loads at a limit equilibrium state (Vahedifard et al. 2016a). The improvement in the performance, when a concave profile is employed, could be attributed to the mass reduction of the critical active wedge. Hence, a lower force is required to support the unstable soil mass behind the wall.

1.3 Porcupine block

A concave profile could be quite demanding from a construction point of view. Hence, it is necessary to find a realistic solution to build a concave wall. Porcupine is a modular block with a curved surface and multiple shear keys. The unique characteristics of this modular block allow the construction of complicated wall geometries. This type of block has been utilized in this study to build the model walls and achieve the desired wall geometry.

1.4 Scope of this study

This study aims to investigate the influence of a concave profile on the system's performance with and without reinforcement by means of physical modeling. Furthermore, the current research work suggests a practical and directly applicable solution to construct concave wall profiles. Finally, the preparation of the small-scale walls required the fabrication of numerous

Table 1. Scaling laws with scaling factor N.

Properties	Scaling factor
Mass in kg	$1/N^3$
Dimensions in m	$1/N$
Reinforcement stiffness in kN/m	$1/N$

block miniatures fulfilling similarity laws of physical modeling. This challenging task was tackled with the use of a 3D printer, which is also demonstrated as part of this study.

Figure 1. CAD drawing of miniature block with Autodesk 123D Design.

2 EXPERIMENTAL SETUP

2.1 3D-printed block miniatures

The porcupine block is the key-element of this study. The proper scaling of its unique geometry and mass according to scaling laws provided in Table 1 is essential for the current research. Hence, the block miniatures were manufactured by an UPBOX 3D-printer, which enables the accurate, fast and easy manufacturing of the objects.

UPBOX uses thermoplastic materials and extrusion technique to build 3D-printed objects. The thermoplastic used here is PLA. Autodesk 123D Design software was used to design the miniature with the desired appearance (Fig. 1). The miniature block was slightly modified from the original to provide a more symmetrical form, but the basic dimensions (Height/Length/Width = 15/30/20 mm) were similar to the prototype blocks and scaled down with a factor $N = 10$. In order to meet the mass requirements, the blocks were designed with two holes (7 mm diameter) placed symmetrically and filled with four lead balls (6 mm diameter), and gap filling material, resulting in an average density of the composite (i.e. PLA and Lead balls) miniature equal to that of dry cast concrete. The holes were covered with a cap that was also 3D-printed. The different stages described earlier are shown in Figure 2.

In total, four different types of porcupine blocks were printed (Fig. 3a). These are the regular block (Type A), a block with flat base (Type B) that was used as the bottom layer, blocks with reinforcement socket (Type C) - however, this feature was not utilized in the current study and it is related to future work - and a half width block (Type D) that were used every two rows at the edges (Fig. 3b). The miniature blocks can be stacked in three possible combinations (Position -1, 0, +1) as shown in Figure 4.

2.2 Model tests

The porcupine miniatures allow the construction of segmental walls with various batter angles. The usual retaining wall range is up to 20° batter angle (NCMA 2009). Therefore, two segmental walls with planar profile geometry and batter angles 9° and 18° were tested and compared.

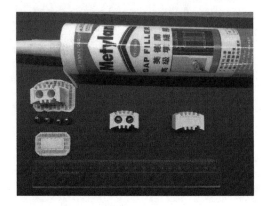

Figure 2. Manufacturing stages of block miniatures (empty block, lead balls, cap, gap-filling material, block filled with the lead balls, finished block filled with gap-filling material and covered with the cap).

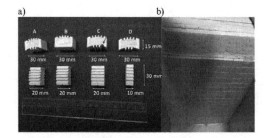

Figure 3. a) Different types of miniature blocks (normal, flat, with socket, half-width) b) Blocks positions in the model.

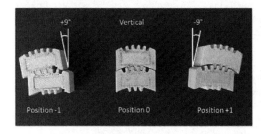

Figure 4. Possible stacking combinations.

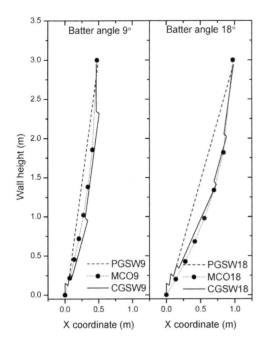

Figure 5. Wall geometry for 9° and 18° batter angle.

Table 2. Model tests.

Model name	Geometry (–)	Batter angle (°)	Reinforcement (–)
PGSW9	Planar	9	No
CGSW9	Concave	9	No
PGSW18	Planar	18	No
CGSW18	Concave	18	No
PRSW18	Planar	18	Yes
CRSW18	Concave	18	Yes

The next step is the design of concave profiles to compare with the benchmark planar models. Vahedifard et al. (2016a, b) proposed the mid-chord-offset (MCO) concept as a single variable design method for circular profiles in slopes and reinforced-soil retaining walls. This design method was followed in order to determine the ideal optimal circular profile (MCO9 and MCO18) for the two benchmark profiles (PGSW9 and PGSW18). Finally, the porcupine blocks were stacked with a pattern to follow the circular geometry resulting in two concave geometries (CGSW9 and CGSW18). Profile geometries of the models as well as the theoretical circular profile (MCO) are illustrated in Figure 5. The four model walls mentioned previously are gravity segmental walls (GSWs).

Two additional models were tested with the addition of soil-reinforcement materials. The batter angle for the reinforced-soil walls (RSWs) was 18° and identical to the geometry of the GSW models. Table 2 summarizes the model walls tested in this study.

The six model tests were prepared inside a rigid steel box with dimensions height/ width/ length = 50/

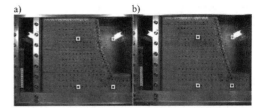

Figure 6. a) CRSW18 and b) PRSW18 model (red-dashed line: reinforcement) alternatively without reinforcement a) CGSW18 and b) PGSW18.

Table 3. Soil properties.

Nevada sand	Density ρ (gr/cm^3)	Friction angle φ (°)	Dilation angle ψ (°)
Backfill	1.636	33	3
Foundation	1.718	33	5

20/ 48 cm. One side of the box is made of transparent glass to allow the capture of images for later image analysis. The wall models were made 30 cm high (i.e. 3 m in prototype) on a 10 cm foundation layer. A 3 cm deep embedment was also placed in front of the wall models by backfilling with soil properties similar to the foundation. For RSW models the vertical spacing of reinforcement was 10 cm with a length of 21 cm (i.e. 70% wall height H). Figure 6 shows the finished model walls with 18 degrees batter angle ready for testing.

2.3 Model preparation

2.3.1 Soil properties
The model tests were conducted in the centrifuge facility of Columbia University. The soil used in the study was Nevada sand (average grain size $d_{50} = 0.15$ mm and specific gravity $G_s = 2.67$) with 5% water content. The soil was compacted in layers of 2 cm height to reach the desired density. The properties of soil used to build the foundation and backfill are given in Table 3. The soil angle of internal friction was determined from direct shear tests.

2.3.2 Reinforcement
The reinforcement used in the two RSWs models was a plastic grid with 2.2×2.2 mm aperture size. The initial tensile stiffness of the reinforcement was 210 kN/m, which would be 2100 kN/m in prototype (Table 1). That is a very stiff reinforcement, unrealistic for the wall height employed in this study. Hence, part of the reinforcement area was removed to reduce its stiffness to 69 kN/m (690 kN/m in prototype), which would be a more realistic value for our tests.

The direct placing of the reinforcement between the blocks among the keys, which is also the usual practice in the vast majority of reinforced-soil segmental walls, would result in loose contact between the blocks. Therefore, an alternative type of connection

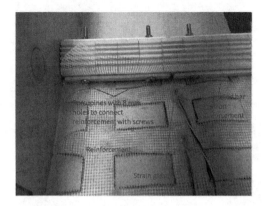

Figure 7. Reinforcement detail.

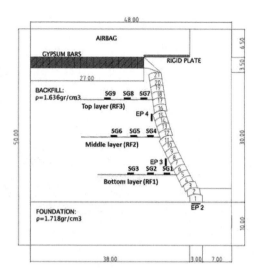

Figure 8. Model layout and instrumentation (dimensions in cm).

was employed in this study. 3D-printed bars with holes were glued on one edge of the model reinforcement. The bars were subsequently screwed with bolts in special block miniatures with a long hole in the middle. This type of connection was employed to resemble similar connections between modular blocks and reinforcement used in the field (Ling et al. 2000). Details of small-scale geogrids and connection used in these model tests are shown in Figure 7.

2.3.3 Model instrumentation

Soil movement was estimated with GeoPIV_RG (Stanier & White 2013, Stanier et al. 2015). Black pins with glued black markers on the pin-head were placed every 2 cm at the top of every compacted layer to increase the color contrast in soil and to improve the PIV analysis performance. Four static markers were also placed as reference points on the transparent glass of the rigid box, where the models were built. Additional thin layers of blue colored sand ware placed every 4 cm to help visualize the critical slip surface.

Apart from the image analysis, three small-size earth pressure transducers (i.e. EP2, EP3 and EP4 in Figure 8) were placed inside the soil to record the development of soil pressures. EP2 was placed under the toe to measure vertical pressures, whereas EP3 and EP4 were positioned at 0.3 and 0.7 H to record the lateral earth pressure.

Finally, the tensile forces in RSW models were measured with the help of strain gauges (SG). The apertures of the small-scale geogrid were filled with sealing material to create a solid base for the strain gauge. The gauges were glued on the sealed surfaces at 0.1, 0.3 and 0.5 L (where L is the length of reinforcement) distance from the back of the wall. In total nine strain gauges (SG1 to SG9 in Figure 8) were used in the model. Tensile tests were conducted to calibrate the reading from SG and to estimate the tensile force within reinforcement. Figure 9 shows the results from the reinforcement tensile tests with and without stain gauges up to 0.5% strain level. It can be observed that the stiffness of the reinforcement was slightly increased due to the strain gauges attachment.

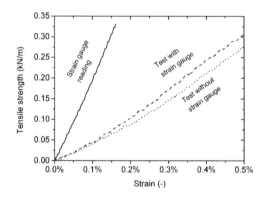

Figure 9. Tensile tests of reinforcement with and without strain gauge.

2.4 Testing program

The models presented in Section 2.2 were first spun up to 5 g and then up to 10 g. Afterwards an increasing surcharge load was applied on the top of the backfill until the collapse/failure of the system. Every loading stage was kept constant for almost two minutes. The surcharge load was applied with the help of an airbag fixed on the top of the rigid box. The airbag width was 16 cm and could not cover the model width (i.e. 20 cm). Thus, gypsum bars were placed between the airbag and soil-surface to transfer and distribute the load over the entire model width to ensure plane strain conditions.

3 EXPERIMENTAL RESULTS

3.1 CGSW9 and PGSW9

The planar profile (PGSW9) was initially designed at a limit state condition. Hence, when the centrifugal

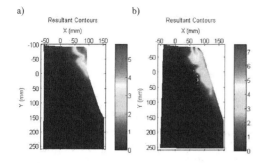

Figure 10. Soil deformation at failure point for a) CGSW18 and b) PGSW18.

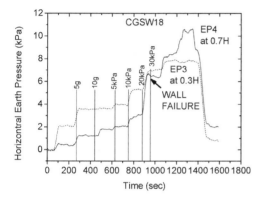

Figure 11. Horizontal earth pressure at different loading stages for CGSW18.

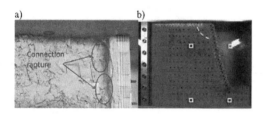

Figure 12. a) Connection rapture at RF1 of CSRW18 b) local soil failure above RF1 of PRSW18.

acceleration reached the level of 9.5 g the wall overturned. On the other hand, CGSW9 was able to reach the 10 g centrifugal acceleration without any problem. Furthermore, a surcharge load of 5 kPa was applied on the top of the backfill before showing signs of failure (i.e. overturning).

3.2 CGSW18 and PGSW18

Due to the increased batter angle, the two gravity walls with 18° batter angle did not collapse entirely like the 9° models. Hence, it was able to overload the structures beyond the failure point, which was determined from the large sudden soil settlements in the backfill soil. PGSW18 failed when the surcharge load reached the value of 20 kPa. The model with a concave profile

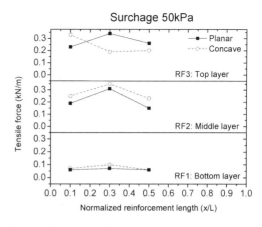

Figure 13. Strain distribution along reinforcement for planar and concave profile at 50 kPa.

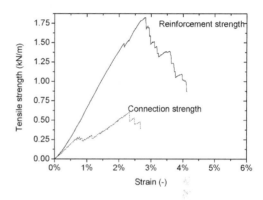

Figure 14. Strength-strain behavior of model geogrid and connection.

failed when air-pressure reached 30 kPa. Figure 10 shows the soil deformations (i.e. critical soil mass) estimated from GeoPIV_RG. The failure moment (i.e. surcharge load capacity) could be also identified from the measured value drop of earth pressure transducers (Fig. 11).

3.3 CRSW18 and PRSW18

The reinforcement strength governed the system's performance. Hence, the failure observed in CRSW18 was due to connection rupture (Fig. 12a) of RF3 at 75 kPa surcharge. The failure mode observed in PRSW18 was local soil failure above RF3 associated with toppling of the upper three blocks at 100 kPa (Fig. 12b).

The change in profile geometry has an influence on the tensile-force distribution along reinforcement layers. The location of maximum tensile force within RF3 for the concave profile was close to the reinforcement connection (Fig. 13), whereas for the planar profile was at 0.3 L distance from the wall facing. The former in combination with the reduced connection strength (Fig. 14) caused the connection rupture for CRSW18.

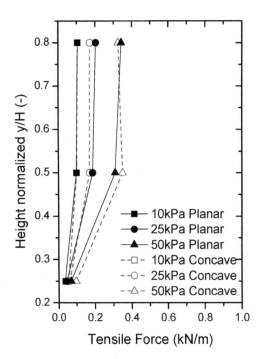

Figure 15. Strain distribution along reinforcement for planar and concave profile at 50 kPa.

A comparison of the performance of the two models at various loading stages (Fig. 15) shows that there is no significant difference at the total mobilized reinforcement force. This indicates that the performance improvement due to concave geometry is dominated from reinforcement stiffness and the loading conditions.

4 SUMMARY

The current study implemented the MCO concept to design and test SWs with and without reinforcement. Porcupine blocks were employed as a practical solution to build concave walls. Furthermore, porcupine block miniatures were manufactured with a 3D printer to enable the physical modeling. For GSWs the performance of the structure was improved significantly. The addition of reinforcement increased the wall's loading capacity, governed the system's performance and diminished the beneficial effect of the concave geometry.

5 CONCLUSIONS

The key findings of this study can be summarized in the following points:

- Concave profiles can improve the performance of gravity segmental walls in terms of absolute load capacity compared to planar geometries. Furthermore, they can reduce the construction cost due to less backfill volume (i.e. compaction and material cost)
- Reinforcement stiffness and loading conditions would govern the system's performance and dominate the positive effect of a concave geometry in RSWs
- Concave profile in RSWs affects the load distribution in the upper reinforcement layers. Block-reinforcement connection at the top layers of a concave profile is critical and should be designed carefully
- Block-reinforcement connection of the type used in this study should be avoided because it reduces the reinforcement strength and stiffness
- Commercial 3D-printers can assist the physical modelling in geotechnical engineering. The manufacturing of complicated miniatures could be achieved cheaply and fast.

REFERENCES

Allen, T.M. & Bathurst, R.J. 2015. Improved Simplified Method for Prediction of Loads in Reinforced Soil Walls. *Journal of Geotechnical and Geoenvironmental Engineering* 141(11): 04015049.

Jeldes, I.A., Vence, N.E., & Drumm, E.C. 2014. Approximate solution to the Sokolovskii concave slope at limiting equilibrium. *International Journal of Geomechanics* 15(2). doi:10.1061/(ASCE)GM.1943-5622.0000330.

Leshchinsky, D., Kang, B.J., Han, J. & Ling, H.I. 2014. Framework for limit state design of geosynthetic-reinforced walls and slopes. *Transportation Infrastructure Geotechnology* 1 (2): 129-164.

Ling, H.I., Cardany, C.P., Sun, L.X. & Hashimoto, H. 2000. Finite Element Study of a Geosynthetic-Reinforced Soil Retaining Wall with Concrete-Block Facing. *Geosynthetics International* 7(3): 163-188.

National Concrete Masonry Association (NCMA) 2009. *Design Manual for Segmental Retaining Walls (3rd ed.)*. Hendron, VA, USA: National Concrete Masonry Association.

Stanier, S.A. & White, D.J. 2013. Improved image-based deformation measurement for the centrifuge environment. *Geotechnical Testing Journal* 36(6): 915-927. doi: 10.1520/GTJ20130044.

Stanier, S., Blaber, J., Take, W., & White, D. 2015. Improved image-based deformation measurement for geotechnical applications. *Canadian Geotechnical Journal* 53(5): 727-739.

Utili, S., & Nova, R. 2007. On the optimal profile of a slope. *Soils and Foundations* 47(4): 717-729.

Vahedifard, F., Shahrokhabadi, S., & Leshchinsky, D. 2016a. Geosynthetic reinforced soil structures with concave facing profile. *Geotextiles and Geomembranes* 44(3): 358-365.

Vahedifard, F., Shahrokhabadi, S., & Leshchinsky, D. 2016b. Optimal profile for concave slopes under static and seismic conditions. *Canadian Geotechnical Journal* 53(9): 1522-1532.

A combined study of centrifuge and full scale models on detection of threat of failure in trench excavations

S. Tamate & T. Hori
National Institute of Occupational Safety and Health (JNIOSH), Tokyo, Japan

ABSTRACT: A combined study of both a centrifuge model test and a full scale model test was carried out to examine the method of detection for the potential threat of failure in trench excavations. A centrifuge model test was first conducted to ascertain the limit depth of trench. A full scale model test was next performed to understand the hazard in the usual trench excavation processes. Several sets of vertical excavations were carried out from the shoulder of a model slope to simulate the trench excavation. In addition, a new method to measure the shear strain, θ in the shallow subsurface was studied to detect an increase of the potential threat of the failure. A clear increases in θ appeared with the progress of excavations in the model. In particular, the relationship between θ and the displacement shows good agreement as a time dependent phenomena. Finally, this study discuses a safety measure of the simplified method by measuring, θ in the shallow section to provide warning to workers at the operations.

1 INTRODUCTION

Trench failures frequently cause construction accidents. Many workers are seriously injured by collapsed soil blocks. Retaining wall systems must be installed to support the trench walls for prevention of the failures. Therefore, a luck of installation is main cause of the accident. However, workers sometime cannot help to enter in excavated trenches prior to the installations. Consequently, the threat of labor accident exists at the process of trench excavations.

A delayed escape of workers is one of causes of such accidents. In addition, they could not notice the increased threat prior to the failure by just watching. It is also known that even a small collapse can cause serious injury to workers. Therefore, safety measures must be taken to save workers' lives. A combination study of both a centrifuge model test and a full scale model test was carried out to investigate the existing hazard in the trench excavations. In addition, adaptability of a developed simplified method of measurement around the shoulders was studied through the tests.

2 CENTRIFUGE MODEL TEST

2.1 Preparation of model

A centrifuge model test was carried out to simulate the trench failure in a slope model. 1/20th scale model slope composed of Kanto loam was prepared with the optimum moisture content to represent unsaturated condition of soil deposit in the shallow section.

Table 1 shows the physical properties of Kanto loam that is unsaturated volcanic cohesive soil in Japan. Kanto loam was uniformly deposited into an experimental container of 150 mm in width, 450 mm in length and 250 mm in depth. A 25 kPa was applied to create the shear strength in the shallow section of the slopes. A uniform pressure was applied on each layer through a loading plate.

Figure 1 shows a comparison between a prototype trench and an experimental model. A vertical wall of trench is modified by an excavated wall from the shoulder in a model slope. 120 mm of a width of the vertical wall D_1, 30 mm of a width of the remaining slope D_2, 135 mm of a height of the wall and 45 degree of an angle of slope are provide in the model for a centrifuge model test. Lubrication are conducted inside of the container to reduce the friction between soil and the walls. A Teflon board was placed in the back side of the wall. In the front side of a transparent wall, silicon grease was spread over on the face. In addition, plastic films with 10μm in thickness were placed over the grease to further reduce the friction. The films were separated vertically with stripe shapes so as to move the soil without any constraints.

Table 1. Soil properties of Kanto loam.

Density of soil particles ρ_s (g/cm^3)	2.759
Sand (0.075~2 mm) %	6.2
Silt (0.005~0.075 mm) %	45.3
Clay (Diameter<0.005 mm) %	48.5
Liquid limit ω_L (%)	158.3
Plastic limit ω_p (%)	97.7
Plasticity index I_p	60.6
Dry density ρ_{dmax} (g/cm^3)	0.665
Optimum water content ω_{opt} (%)	102.0

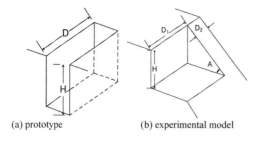

(a) prototype   (b) experimental model

Figure 1. Modelling of trench excavation in the slope.

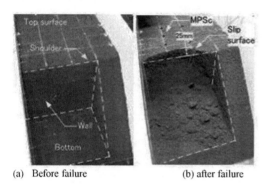

(a) Before failure   (b) after failure

Figure 2. Comparison of a part of the model trench wall.

2.2 Investigation of limit height of excavated wall

A geotechnical centrifuge of 2.38 m of effective radius, was used for the model tests. A centrifugal acceleration was gradually increased up to the number of 20.7 g where a trench wall failed. Figure 2 shows an outline of the collapsed wall. An edge of a slip surface on the top slightly curves around shoulder. A maximum distance from the shoulder to the edge was approximately 25 mm.

Several sets of small sensors, so called MPSc, were installed on the surface at a distance of 40 mm and 80 mm from the shoulder as shown in Figure 3. MPSc are composed of a steel plate with 0.5 mm thickness to measure the bending deformation that is reacted by shear deformation of the surrounding soil.

Figure 4 shows the relationship between an interpreted shear strain θ and a number of centrifugal acceleration n. In addition, an interpreted depth of trench H_p are expressed on the upper horizontal axis. A value of MPSc_1 are increased with a value of n. A negative value of θ means that a bending deformation of MPSc is appeared like a barrel shape as shown in Figure 3. The value of MPSc_1 is greater than that of MPSc_2.

This means that the shear strain developed largely near the wall. Moreover, the value of MPSc_1 is quickly built up where a number of n is greater than 20 g. A movement of the wall getting unstable was detected by measurements in the sub-surface around the shoulder. This method is applied in the following a full scale model tests. A limit depth of the trench was estimated as 2.8 m from a calculation.

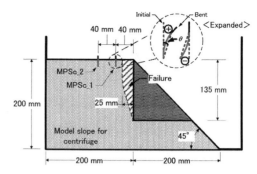

Figure 3. A profile of model slope for simulation of trench failure in centrifuge.

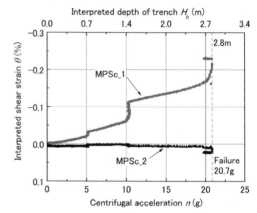

Figure 4. A relationship between an interpreted shear strain θ and a number of centrifugal acceleration n.

3 FULL SCALE MODEL TEST

3.1 Preparation of model and method of excavations

Figure 5 shows an outline of the preparation of the model slope. Kanto loam was spread over in 0.3 m of a layer thickness on a testing place in a laboratory. Static compressions were carried out by using an excavator. A radius R between the bucket and a fulcrum of rear wheels of crawlers was kept about 3.5 m so that a constant value of an acting load F was applied through an acting area the bucket A. 5 kN/m² of shear strength was mobilized in the model slope where 23 kN/m² of the static pressure p was applied by F as shown in Table 2. Therefore, uniform shear strength was reproduced in the both the model slopes of the centrifuge model and the full scale model.

A model slope of 3.0 m in height, 4.0 m in width, 2.8 m in length at the top and 45 degrees in inclination was made using the same soil material to represent the model of the centrifuge test as shown in Figure 6. Plastic sheets were placed to lubricate the friction between soil and concrete walls to have same boundary condition as of the centrifuge. Eight steps of vertical excavations from S1 to S8 were carried out by the excavator to make the slope unstable as shown in Figure 7. 3.3 m in width was excavated in the shoulder. 30 minutes of an interval time was provided to observe the movement after the excavations at each step. Catastrophic failure occurred after 23 minutes from the final excavation of S8b performed by 3.0 m in depth. This value was quite similar to the result that was 2.8 m of the limit depth in the centrifuge model test.

3.2 Development of sensor – Mini Pipe Shear strain meter (MPS) – for simplicity measurement

Mini Pipe Strain meter, MPS, was developed to measure shear strain in the shallow subsurface of slopes (Tamate 2010). MPS is composed of a compact rod that is 0.6 m in length, 10 mm in diameter, and 3.6 N in weight (see Figure 8). A screw point of 80 mm in length attached at the lower end enables the unit to penetrate into the ground without pre-boring. A taper end of 100 mm in length is used to provide lateral compression to the surrounding soil so that MPS reacts to the slope movement by its bending deformation. Quick installation is available in MPS by use of a hand operated drill. A battery cell operates alone around 20 days.

Figure 9 shows a schematic view of the distribution of strain in horizontal ε_x near the shoulder that is an example case of FEM analysis. A large increase in ε_x appeared near the slip surface, and its increment converged with increasing of the distance from the slip surface. However, it seems that small strain in the

Table 2. Conditions at static compression using construction machinery.

Weigh of an excavator W	116 kN
Radius R	3.5 m
Acting load F	33.8 kN
Acting area through bucket A	0.9 m × 1.6 m
Pressure at compression $p\,(=F/A)$	23 kN/m²

Figure 7. Excavation by an excavator.

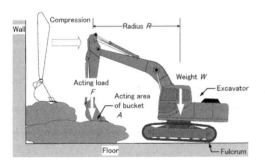

Figure 5. Static compression of soil material using an excavator.

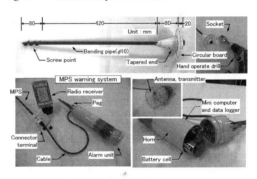

Figure 8. Mini Pipe Strain meter (MPS) and warning system.

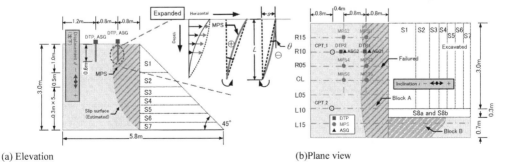

(a) Elevation

(b) Plane view

Figure 6. Positions of installed sensors and part of excavations.

1465

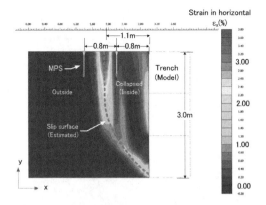
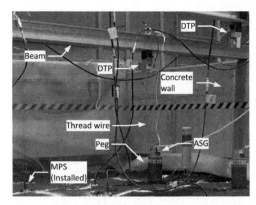

Figure 9. Distribution of axial strain of horizontal component.

Figure 10. Installation method of DTP to measure the settlement of the surface near the shoulder.

shallow subsurface was also mobilized, corresponding to a development of the slip surface.

This study attests an existence of shear strain in the shallow subsurface by MPS. In particular, an installed MPS bent a little to follow the curve of displacement as shown in Figure 6. Thus, MPS measured shear strain in the shallow subsurface areas that have previously been ignored. An interpreted shear strain θ (%) was defined as the ratio of the differential movement s to the effective length L of the MPS as shown in Equation 1. (Tamate et al. 2013). An outline of θ is also shown in Figure 6.

$$\theta(\%) = \frac{s}{L} \times 100 \qquad (1)$$

Extensometers (DTP), inclinometers (ASG), and MPSs were installed on a flat surface of the top prior to beginning of the excavations. Figure 6 shows the positions where the sensors were installed. Two sets of DTPs were installed at the right column (R10) at 0.8 m intervals so that increments of the displacement d could be measured. The sensor units of the DTPs were fixed on a beam bridging over both sides of concrete walls, while extended thread wires were connected to pegs on the top as shown in Figure 10. Two sets of ASGs were installed in the column of R10 at the same height as the DTPs so that increments of the inclination i were measured on the surface. In addition, six sets of MPSs were installed in the columns of CL, R05 and R15 in the same manner as ASGs so that increments of the interpreted shear strain θ were measured.

3.3 Experimental analyses of movement near the shoulder

Figure 11 shows a process of failure in the excavated wall. 5 sheets of picture were taken in about 5 seconds. Cracks on the top surface of the top were opened in Figure 11(a). A position of the failure block is also indicated in Figure 6(b). Figure 11(b) and (c) shows that a mass of the failure block was falling down. The mass was separated into small pieces in the (d). Collapsed soil spread over on the floor can be seen in the (e). As a remaining slope in the left side supported the trench wall, the shoulder did not move in parallel at a beginning of the failure. However, an entire of the trench wall finally failed.

It was impossible for people to detect an increase of potential threat prior to failure by just watching observations. In addition, it is dangerous as people, who work into trenches, are not aware of the increase of the potential threat by the enlargement of the cracks. Figure 12 shows a shape of the slope after the failure. Since three curves of R10, CL and L10 are almost coincided, the shape after the failure are reproduced likely as the plane strain condition. Therefore, final shapes of trench after the failure were slightly different between the centrifuge and the full scale. However, an inclination of the slip surface was similar to the result of centrifuge model so that vertical wall appears in an elevation between 1.6m and 3.0m. Collapsed soil traveled around 5m in distance.

Figure 13 shows reactions from three kinds of sensors, which are DTP, ASG and MPS, from the beginning of excavation to the failure. The elapsed time T is shown in the lower horizontal axis. Steps of excavation is also denoted as S1 and so on the upper horizontal axis. 15 minutes of an interval time was included to wait a convergence of increase between S8a and B8b. Two curves in each group of the sensors show data that was obtained in two different rows. Experimental results are calculated in an equivalent condition on distance from the shoulder. An excavated wall was failed at 4.6 hr, and 23 minutes had passed after S8b of the final excavation.

An increase in the reactions of sensors appears prior to the failure. A value of displacement d on DTP_1, which was 0.8 m far from the shoulder, shows a clear increase from S6. Moreover, the value is accelerated in its increase before a couple of minutes of the failure. Meanwhile, DTP_2 was positioned in an outside from an area of the failure so that a value of the DTP_2, which was 1.6 m far from the shoulder, does not show clear increase.

Values of the inclination i commonly decrease from the beginning of excavation S1. A decrease of ASG_2,

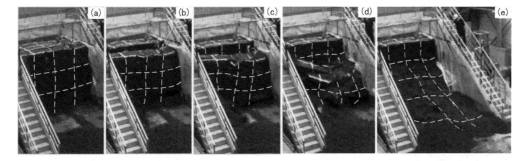

Figure 11. Process of failure of excavated wall in 5 seconds.

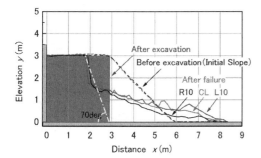

Figure 12. Shape of the slope after failure.

which was far from the shoulder, is larger reactions than that of ASG_1. These values of ASG are next turned into the increase from S6. In ASG_2, moreover, two times of a step increase are appeared prior to the failure.

A value of ASG_1 is quickly increased at the failure following the gradual increase. In ASG, a larger reaction was recorded in the far position from the shoulder. Therefore, DTP and ASG show an opposite results on the relationship between an amount of reaction and a distance from the shoulder. This reason would be because directions of reaction, which is either vertical or horizontal, are different by the sensors. Vertical displacement becomes larger near the shoulder though horizontal strain due to shear deformation is small. Horizontal strain develops along the slip surface, and an amount of the value decreases in depth as shown in Figure 9.

Two curves of MPS_3 and MPS_4 are commonly increased after S8a following a decrease in T between 0 and 4 hr. A coincidence is seen in a shape of the curves between ASGs and MPSs. In addition, an amount of the increase in MPS_3 is almost 0.1 % of θ as same as MPS_4. It seems that MPS takes advantages to obtain a clear reaction of ground movement at 1.6m far from the shoulder.

A decrease in value θ of MPS, whereas an increase in negative, means that a horizontal displacement distributes like a barrel shape with an increase of excavated depth. A positive increase means horizontal strain developed as a bowing shape in the distribution. After S8b, curves of DTP_1, ASG_1 and MPS_3 show a linear increase prior to the failure though an acting load in the model was already constant by a completion of the final excavation S8b. Accordingly, creep phenomena are detected in the measurements around shoulder of trench.

4 DISCUSSION

4.1 Detection of potential threat of failure by identification of creep phenomena

The left-side of Figure 14 shows the relation between the inversed number of the shear strain rate $1/v_\theta$ and t_e on a logarithmic scale on the vertical axis. In the figure, v_θ is defined as the per-minute value of the increment of θ. Therefore, it is considered as a type of velocity. When t_e was -10, the $1/v_\theta$ values of MPS_1 were distributed at higher values of 1000 min/%. This means that less 0.001%/min of v_θ appeared 10 min before the failure. However, $1/v_\theta$ shows a drastic decrease, whereas v_θ increased between -10 and 0 of t_e. The right-side of Figure 14 shows an expanded view by a linear scale on a vertical axis, showing $1/v_\theta$ values between 10 and 400 min/% for 5 min when t_e was between -5 and -1 min. The values of v_θ were interpreted as between 0.0025 and 0.1 %/min. Consequently, in the same manner as the third creep, the shear strain mostly accelerated in its increase. Accordingly, a clear increase in shear strain in the shallow subsurface was confirmed in the full-scale test model. In addition, it was proven that this phenomenon reflects an increase in potential threat of trench failure. Therefore, a couple of minutes could be provided for escape by identifying the creep phenomena.

4.2 Safety measures for labor safety in excavation working sites by simplicity monitoring

Excavating grounds are composed of either a naturally deposited soil or an artificial reclaimed soil. Shear strength of soil is not uniform even in small area of working sites. In addition, the shear strength cannot be identified by just watching. Consequently, threat of failure also exists in shallow trenches those are not supported by retaining walls. Many devices developed in the field of landslide prediction are generally used for a long term measurement in a wide area. This study proposes that monitoring is also required in small area

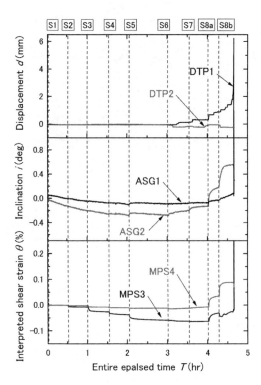

Figure 13. Reactions of sensors prior to failure.

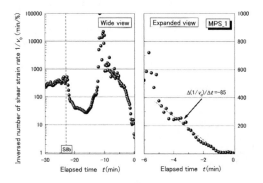

Figure 14. Inverse number of velocity of shear strain by MPS1.

of working sites of trenches and slopes for workers safety. As for DTP, fixed points in addition to moving points are required in measurement of precise displacement in sites. Additional preparation works are needed even though a duration of measurement is a short term in small construction sites. As movements of workers and operations of machineries mobilize vibration in the sites, it is difficult to measure the precise inclination on the surfaces. Therefore, conventional sensors cannot be adaptable in the working sites so that new devices of a simplicity method are required for safety. MPS has the advantage of simplicity and reactivity. Threat of trench failures are noticeable by a warning system comprising an MPS and an alarm unit was developed (Tamate and Hori 2017). The potential threat of accidents and workplace injuries can also be reduced by application of the monitoring.

5 CONCLUSIONS

A combined study of centrifuge and full scale models was carried out to attest the method of detection for the potential threat of failure in trench excavations. It was made sure that both conventional sensors and a developed MPS detected the threat of failure earlier than the watching. Small movement were measured by the sensors prior to failure. Accordingly, the monitoring enables the detection of residual threat—crucial to avoid overconfidence about safety. The method that uses an MPS has the advantage of simplicity and reactivity and can assist human observations on construction sites. Shear strain increases in the shallow subsurface as well as in the slip surface. It was determined that a couple of minutes could be provided for escape by identifying either the second or third creep. Accordingly, this study concludes the threat of people being injured by collapsing soil can be reduced by using the proposed method and sensor.

ACKNOWLEDGEMENT

This work was supported by JSPS KAKENHI Grant Numbers 16K01306.

REFERENCES

Tamate, S. 2010. *Penetration-type pipe strain gauge*, United States Patent, No.7,762,143 B2, Jul.27.

Tamate, S., and Hori, T. 2007 Study on Monitoring for Detection of Potential Risk of Slope Failure for Labor Safety. Geo-Institute of the American Society of Civil Engineers, Geo-risk, Vol. Geotechnical special publication No.284, 2017, pp. 267-279.

Tamate, S., Hori, T., Mikuni, C. and Suemasa, N. Experimental analyses on detection of potential risk of slope failure by monitoring of shear strain in the shallow section. 2013. *Proceedings of the 18th International Conference on Soil Mechanics and Geotechnical Engineering*, pp.1901-1904.

Dynamic behaviour on pile foundation combined with soil-cement mixing walls using permanent pile

K. Watanabe
Technical Research Institute, Obayashi Corporation, Tokyo, Japan

M. Arakawa & M. Mizumoto
Obayashi Corporation, Tokyo, Japan

ABSTRACT: Soil-cement mixing walls are often used for temporary structures as earth retaining walls when the ground is excavated. However, when soil-cement mixing walls are used as permanent piles, they are expected to support foundation structures. A centrifuge shaking table experiment was conducted on models of pile foundation structures to examine the effect of the presence of soil-cement mixing walls installed at the external peripherals on the responses of piles and structures, to understand the characteristics of foundation structures that use soil-cement mixing walls as permanent piles during earthquakes. The test results were reproduced via analysis to study a method of evaluating the horizontal resistance of a pile foundation structure using a soil cement column wall as its primary piles. This report describes the findings from the centrifuge shaking table experiment and analytical study.

1 INTRODUCTION

A soil-cement continuous column wall (hereafter, soil-cement mixing wall) is constructed by inserting core material, which acts as a stress transfer material, into soil-cement formed by injection of cement milk into the centre of the soil, and then churning and mixing the ground foundation. The soil-cement continuous column wall has previously been used as an earth retaining wall during excavations, and was treated as a temporary structure. At present, a method has not been established for evaluating its assumed seismic behaviour when bearing a building body load or something similar. In recent years, rationalization of the foundation structure, reduction of the environmental burden, and other needs have been rising, and as shown in Figure 1, studies have been proceeding into the use of soil-cement continuous column walls as permanent piles which are proposed by Watanabe et al. (2013), Watanabe et al. (2014) and Watanabe et al. (2015). In addition, with the previously temporary soil-cement mixing wall now being used as a permanent pile, it can be expected that eliminating the need for new work and reduce construction expenses.

There have been some studies on vertical bearing capacity where earth retaining walls are used as permanent piles, and these studies focused on full-scale load or construction tests. Kaneko (2004) is constructing high-strength soil-cement mixing walls, developing methods for their use as permanent piles, and evaluating their bearing capacities. Taya et al. (2009) are improving construction control methods and

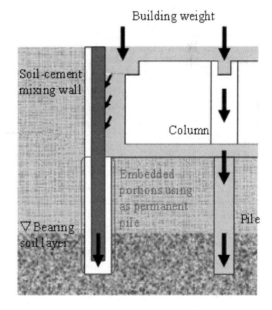

Figure 1. Concept of utilization for permanent pile of soil-cement mixing wall.

construction equipment, constructing high-strength soil-cement walls, and evaluating their bearing capacities. Both of these walls differ from the soil-cement walls in general use, in that they assume construction of high-strength soil-cement mixing walls.

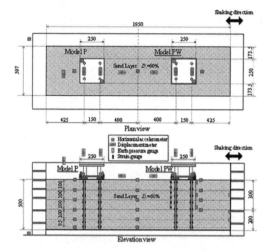

Figure 2. Schematic view of centrifuge shaking table test (Unit: mm).

Table 1. Model specifications.

	Model scale (1/50)	Prototype scale
Planner dimension of building	250 × 250 mm	12.5 × 12.5 m
Upper part mass	8.58 kg	1070t
Foundation	7.46 kg	933t
Natural frequency of building mass	556 Hz	11.1 Hz
Length of pile	500 mm	25.0 m
Length of soil-cement mixing wall	275 mm	13.75 m

Table 2. Input wave conditions.

Case	Frequency (Hz)	Maximum input acceleration (m/s^2)
S1	2.4	1.0
S2	2.4	3.1
S3	2.4	3.9
S4	2.4	7.0
R1	–	0.5
R2	–	2.4
R3	–	3.2
R4	–	6.0

Note: Case 'S' refers to Sinusoidal wave input whereas case 'R' refers to Rinkai wave input

Use of soil–cement mixing walls as permanent piles is promising from a foundation structure rationalization standpoint, but the horizontal seismic behaviour needs to be evaluated, as many aspects of the seismic behaviour of foundation structures that include the soil–cement mixing walls remain unknown. Accordingly, in order to determine the effect of presence/absence of soil–cement mixing walls on the response of foundation structures, centrifuge shaking table tests were conducted on the foundation structures, wherein the soil–cement mixing walls were used as permanent piles. This study presents the results of the test and the analytical study.

2 OUTLINE OF CENTRIFUGE SHAKING TABLE TEST

Figure 2 illustrates the layout of the models used for the centrifuge shaking table test. For this test, a shear box was used to produce model ground using the air-pluviation method so that the relative density of the ground D_r is 60%. The ground material used was No. 7 silica sand ($G_s = 2.645$) which has $\phi = 38.8$ deg. of internal friction angle and c = 2.9 kN/m^2 of cohesion. In the shear box, model P, which hypothesizes a normal pile foundation structure, and model PW, which arranges a soil cement column wall around the pile foundation, were installed.

Table 1 lists the specifications of the structure model. The piles were stainless steel pipes with 20 mm diameter and 0.5 mm skin thickness. The model piles have the smooth surface. The material of model pile is stainless steel and the tensile strength of that is 520 N/mm^2. Moreover, the soil cement column wall model which is the smooth surface was modelled using an aluminium plate with a thickness of 1.5 mm. The tensile strength of aluminium is 265 N/mm^2. The model scale for this test was 1/50 so that the centrifuge acceleration was set as 50G. Table 2 gives the input conditions for each shaking case. Based on the full-size conversion value, in cases S1 to S4 and in cases R1 to R5, 2.4 Hz sinusoidal waves and seaside waves, respectively, were the inputs, and in each case, the acceleration amplification was varied to conduct the shaking in steps. The acceleration with regard to the ground and the foundation footing, the displacement of the structure and the ground surface, the strain of the piles and the soil cement column wall, and the earth pressure acting on the soil cement column wall were measured. The strain and the soil pressure of the soil cement column wall were measured on the wall surface orthogonal to the shaking direction (below, out-plane wall). Further, all numerical values shown are values converted to full size.

When the stress produced in the piles is calculated from the test results, the pile's bending moment–curvature relationship (M–ϕ relationship) is necessary. Thus, a 4-point bending test of a member similar to the pile model used for the centrifugal model test was conducted separately. Figure 3 shows a layout of the bending test and Figure 4 shows the M–ϕ relationship obtained from the bending test results. Figure 4 shows that during the initial loading, the bending moment–curvature relationship was almost linear, but the increase in the curvature was accompanied by a decline of the gradient. The results of using a trilinear model to approximate the M–ϕ relationship obtained via the test results based on the above trends are also shown in Figure 4. The bending moment of the piles

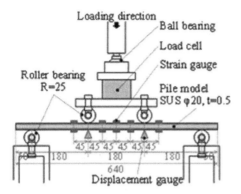

Figure 3. Schematic view of pile bending test (Unit: mm).

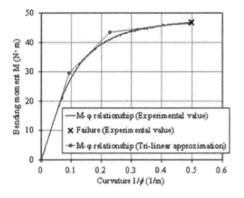

Figure 4. $M–\phi$ relationship based on pile bending test (Model scale).

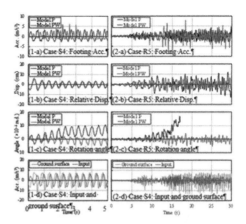

Figure 5. Time history curves at maximum shaking.

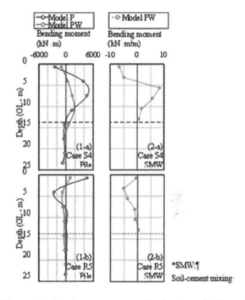

Figure 6. Maximum moment distribution of pile and soil-cement mixing wall.

3 RESULTS OF CENTRIFUGE SHAKING TABLE TEST

was calculated according to the results of the centrifuge shaking table test using the approximate $M–\phi$ relationship as shown in Figure 4.

Figure 5 shows the time history of the principle response during maximum loading of the sinusoidal waves and the seismic waves (Case S4 and Case R5). The foundation relative displacement was obtained by integrating the horizontal acceleration of the footing and the ground surface and subsequently differentiating both. The angle of rotation of the structure was obtained from the vertical displacement at both ends of the upper structure. With regard to the response acceleration of the foundation, the short period element is more dominant in the model PW as shown in Figures 5 (1-a) and (1-b). One reason for this is assumed to be the change in the response period properties of the structure–ground system occurring because of the presence of the soil cement column wall. In Figures 5 (2-a) and (2-b), the relative displacements of the foundation in model P and model PW are somewhat similar. Figures 5 (3-a) and (3-b) demonstrate that the residual rotational displacement occurs in model P (+) side and model PW (−) side, with this value tending to be higher in model P. Further, for Model P, in Figure 5 (3-b), a record just before the target of the laser displacement gauge slipped out of place is shown.

Figure 6 shows the bending moment distribution when the bending moment acting on the pile is the largest, and simultaneously, the bending moment distribution of the out-plane wall. The bending moment of the piles was calculated from the $M–\phi$ relationship and the value of the bending strain based on the trilinear model is shown in Figure 4. In the figure, among the four piles in each structure, the results for 1 pile are shown for the same shaking cases as in Figure 5. Figure 6 shows that in each shaking case, the maximum bending moment of the piles in model PW is smaller than that in model P. The bending moment distributions of the out-plane wall and the piles generally correspond, demonstrating that the soil cement

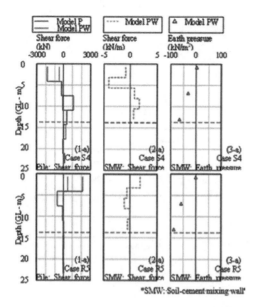

Figure 7. Maximum shear force and earth pressure distribution of pile and soil-cement mixing wall.

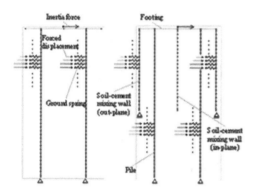

Figure 8. Summary of analysis model for response displacement method.

column wall contributes to the reduction of the bending moment of the piles. Figure 6 shows the shear force distribution when the shear force acting on the pile is at its maximum value, the shear force distribution of the out-plane wall simultaneously, and the dynamic element of the earth pressure acting on the out-plane wall. Figure 7 shows that in both model P and model PW, the shear force is maximum near the pile top, but this value in model PW is approximately 1/4 of its value in model P. The shear force distribution of the out-plane wall and piles roughly correspond, but the value of the shear force that the out-plane wall bears is much smaller than that borne by the piles. Therefore, it is assumed that the in-plane wall has a controlling impact on the manner in which the shear force is borne.

Moreover, the other shaking cases not shown in the figures confirmed a tendency for the pile stress of model P to be smaller than that of model PW as shown above.

Table 3. Analytical condition (Pile).

Poisson's ratio	0.30
Longitudinal elastic modulus (kN/m^2)	2.0 × 10^8 7.7 × 10^7
Outer diameter (m)	1.00
Inner diameter (m)	0.95

Table 4. Analytical condition (SMW).

Poisson's ratio	0.33
Longitudinal elastic modulus (kN/m^2)	7.0 × 10^8
Transverse elasticity modulus (kN/m^2)	2.6 × 10^7
Width (m)	12.50
Thickness (m)	0.075

4 OUTLINE OF ANALYTICAL STUDY BASED ON RESPONSE DISPLACEMENT METHOD

To study a method of examining the horizontal resistance of a soil cement column wall during an earthquake, an attempt was made to reproduce the test results using the response displacement method. Figure 8 is an outline of the analysis model. The piles, out-plane wall, in-plane wall, and the foundation footing of each structure model were modeled as beams, and each member was provided with the material properties listed in Table 3 and Table 4. The foundation footing was treated as a rigid body and provided with adequately large flexural rigidity.

The pile elements were given with regard to the M–ϕ relationship based on the trilinear model shown in Figure 4 as nonlinear properties. As stated in section 3, it is assumed that the impact of the in-plane wall controls the manner in which the shear force is borne by the soil cement column wall; thus, the out-plane wall and in-plane wall were modelled as follows. In brief, the out-plane wall and in-plane wall were both treated as elastic bodies with rectangular sections based on wall width and wall thickness. Therefore, the in-plane wall has significantly larger bending and shear rigidity than the out-plane wall. Moreover, it is assumed that the out-plane walls on both sides constrain the displacement of the in-plane wall so the loads are transmitted back and forth between the out-plane walls and the in-plane wall.

The building inertial force and forced displacement of the ground are applied to the analysis model as external forces. The building inertial force was obtained by calculating the product of the response acceleration of the foundation footing and the total mass of the superstructure and the foundation footing, and it was made to act on the centre of the foundation footing. The forced displacement of the ground was obtained by integrating the response acceleration at each depth of the ground model, and it was made to act on each node of the piles and the out-plane wall through a ground spring. The building inertial force and the forced displacement of the ground were, in each shaking case, the values at the time the bending moment of the piles

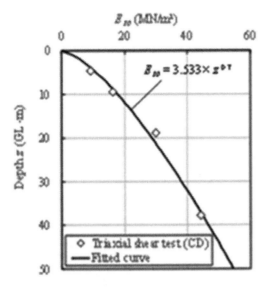

Figure 9. Deformation modulus distribution.

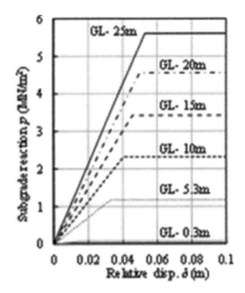

Figure 10. Soil spring considering confining pressure.

was at its maximum value. Moreover, the boundary condition at the tips of the piles and of the out-plane wall were horizontal rollers.

To evaluate the properties of the ground spring, a triaxial compression test (CD test) of the ground material used for the centrifuge shaking table test (silica sand No. 7, $D_r = 60\%$) was conducted. From triaxial compression tests, cohesion $c = 2.9$ (kN/m^2) and internal friction angle $\phi = 38.8$ (deg.) were obtained. Figure 9 shows the relationship of deformation modulus E_{50} obtained by the triaxial CD test with the corresponding depth z of the model ground (i.e. confining pressure dependency on deformation modulus). Equation (1) was obtained from the curve approximation of the relationship of deformation modulus E_{50} with z plotted in Figure 9. The deformation modulus of the ground at each depth was obtained by equation (1), and the modulus of subgrade reaction was evaluated based on equation (2) in the Recommendation for building foundation design (2001). The limit of the subgrade reaction was calculated using equation (3) proposed by Broms (1964). The coefficient of passive earth pressure K_p in equation (3) was set as $K_p = \tan^2(\pi/4 + \phi/2)$ using the angle of shear resistance ϕ.

$$E_{50} = 3533 \times z^{0.7} \quad (1)$$

$$k_{h0} = \alpha \cdot \xi \cdot E0 \cdot \overline{B}^{-3/4} \quad (2)$$

$$\frac{p_y}{\gamma B} = \kappa \cdot K_p \cdot \frac{z}{B} \quad (3)$$

where, E_{50}: Modulus of deformation (kN/m^2), z: Depth (m), k_{h0}: Standard horizontal modulus of subgrade reaction (kN/m^3), α: Constant according to the evaluation method (1/m), ξ: Coefficient considering impact of group piles, E_0: modulus of deformation (kN/m^2), $\overline{B}$: Dimensionless pile diameter, p_y: Plastic horizontal subgrade reaction (kN/m^2), γ: Unit weight of ground (kN/m^3), B: Pile diameter (m), κ: Coefficient considering impact of group piles, K_p: Coefficient of passive earth pressure.

Figure 10 shows an example of the ground spring properties depending on the confining pressure determined as above.

5 RESULTS OF ANALYSIS USING RESPONSE DISPLACEMENT METHOD

Figure 11 shows the analysis results of the pile stress distribution using the response displacement method.

The figure shows the forced displacement of the ground, the maximum bending moment distribution of the piles, and the shear force distribution simultaneously for each structural model in each shaking case. This analysis was conducted for the shaking cases (case S1 to S3, R1 to R3) before the bending moment of the piles approximated the ultimate value and reached the range where the bending strain increased abruptly.

Figure 11 shows that the analysis results for each shaking case generally reproduce the distribution of the test results and the value of the maximum stress. However, for (2-a) Case S3 Model P, (4-a) Case R3 Model P, and (4-b) Case R3 Model PW, the results of testing with regard to the pile stress vary between the two piles, and the analysis results tend to reproduce only one side. This is assumed to be caused by the residual rotational displacement of the structure discussed in section 3.

Moreover, Cases S2, S3, R2, and R3 also confirmed that the residual rotation displacement was produced in the structure as in Figure 4. Presumably, the axial burdens of the piles are, as a result of this, unequal, and the percentage of the bending moment and shear force borne vary.

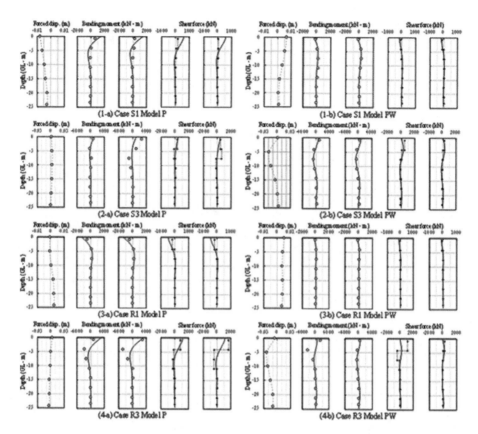

Figure 11. Analysis results of pile stress using response displacement method.

6 CONCLUSIONS

A centrifuge shaking table test of a foundation structure using a soil cement column wall as its primary piles was conducted to study a method to evaluate the pile stress during an earthquake. The following results were obtained from this study.

1. In both the sinusoidal wave and the seismic wave entry cases, the soil cement column wall lowered the pile stress.
2. It was confirmed that the proposed method applying the existing subgrade reaction evaluation formula to the simple response displacement method can approximately evaluate the maximum stress in the piles.
3. In a case where the rotational behaviour of a structure produces large axial force fluctuations in the piles, it is necessary to evaluate the pile stress via an analysis that considers the vertical elements.

REFERENCES

Architecture Institute of Japan 2001.*Reccomendation of Building Foundation Design.*

Broms, B.B. 1964. Lateral Resistance of Piles in Cohesionless Soils, *ASCE*, 90(SM3): 123–156.

Kaneko, O. 2004. Load-Displacement Relationships on Soil-cement H-shaped Pile, *Proceedings of 49th Geotechnical Symposium*: 127–132 (In Japanese).

Taya, Y. 2009. A Study on Soil-cement Mixing Wall Using Permanent Pile (Part1 and 2), *Proceedings of Japan National Conference on Architecture*, Structure 1: 667–670 (In Japanese).

Watanabe, K. 2013. In-situ Full Scale Load Test of Soil-cement Mixing Wall, *Proceedings of Japan National Conference on Architecture*, Structure 1: 487–488 (In Japanese).

Watanabe, K. 2014. A Study on Bonding Behavior on Soil-cement Mixing Wall Using Permanent Pile, *Proceedings of Japan National Conference on Architecture*, Structure 1: 437–438 (In Japanese).

Watanabe, K. 2015. In-situ Full Scale Load Test of Soil-cement Mixing Wall Using Permanent Pile, *Proceedings of Japan National Conference on Architecture*, Structure 1: 451–452 (In Japanese).

Centrifuge modelling of 200,000 tonnage sheet-pile bulkheads with relief platform

G.M. Xu, G.F. Ren, X.W. Gu & Z.Y. Cai
Geotechnical Engineering Department, Nanjing Hydraulic Research Institute, Nanjing, P.R. China

ABSTRACT: Two design schemes are proposed of 200,000 tonnage sheet-pile bulkheads with relief platform with forefront water depth of 20.5 m, the largest sheet-pile wharfs in China. The first scheme is of a rigid front wall of T-shaped reinforced concrete diaphragm with pile group of two rows of reinforced concrete cast-in-place piles. Second scheme is of flexible front wall of composite steel pipe pile with pile group of three rows of piles. Two series of geotechnical centrifuge model tests have been carried out to investigate their performance of sheet-pile bulkheads. It is shown that the bulkheads of two design schemes are verified to be feasible in that the displacements of front wall are within allowable limit. It is also found that characteristic bending moment values of front wall is greater than those of lateral pile group in first scheme, whereas the opposite is true in second scheme.

1 INTRODUCTION

Sheet-pile bulkheads have proven to be a viable and economical solution to waterfront construction (Tsinker 1997). The simplest sheet-pile bulkhead, which is traditionally named the sheet-pile bulkhead anchored by single layer of tie-rods, is composed of three parts: sheet-pile front wall, anchor wall and steel tie-rods. The front wall is for retaining soil and the anchor wall provides anchorage points for the tie-rods. It is known that the lateral load on the front wall will dramatically be increased with its forefront water depth. Since the lateral load carrying capacity is very limited for this traditional sheet pile bulkhead, the structure of sheet-pile bulkhead is mainly used in small and medium-sized terminal berths below 50,000 tonnage with forefront water depth less than 14 m (Li & Xu 2008). By introducing a lateral pile group of two or three rows of vertical piles in the soil behind the front wall to share one part of lateral load, the water depth in front of sheet-pile bulkhead can be enlarged by a large margin. This innovative sheet-pile bulkhead is named the sheet-pile bulkhead with relief platform (Xu et al. 2010).

The first sheet-pile bulkheads with relief platform are proposed for the construction of five new deep water berths of 100,000 tonnage with forefront water depth of 16.5 m in Jingtang Port area of Tangshan Port in China (Liu 2014). Combined with the optimization and verification study, a large number of geotechnical centrifugal model tests have been carried out. In the first place, the behaviours of two types of head fixity conditions, capped head and pinned head, are examined in two series of centrifuge model tests (Xu et al. 2016). The bending moments of piles and front wall, earth pressures on retaining side of front wall are measured together with tension forces of tie-rod and horizontal displacements of structural elements in the tests. It is found that the lateral pile group with capped head is more powerful than the pile group with pinned head in carrying earth pressure load, reducing tension forces of tie-rods and limiting horizontal displacements of front wall. However, it is also found that the location of maximum bending moments of piles with capped head occur close to the pile cap level, and the estimated values are equal to the value of pile's maximum allowable bending moment. Finally, the design scheme of the lateral pile group with pinned head is adopted which can eliminate excessive bending moment at pile head while tension force of tie-rod and inclination of front wall due to horizontal displacement can also be controlled within their allowable limits.

In the second place, the feasibility of the design scheme is verified by the parallel repetitive centrifugal model tests (Xu et al. 2008). The lateral bearing capacity of the lateral pile group is demonstrated by comparing with the working behaviour of traditional sheet pile bulkheads of same size anchored by single layer of tie-rods. The earth pressure acting on the front wall can be effectively reduced with the presence of the lateral pile group foundation, which significantly reduces the moment value of front wall, the lateral displacement of front wall at anchoring point and the internal force of tie-rods (Xu et al. 2010).

After the successful construction of the above five berths of 100,000 tonnage sheet-pile bulkheads with relief platform, a new berth of 200,000 tonnage with forefront water depth of 20.5 m, the largest sheet-pile bulkhead berth in China, is proposed to be built with the same structure of sheet-pile bulkhead with relief platform. And two design schemes are put forward to be verified by means of centrifuge modelling. In first scheme, a rigid front wall of T-shape reinforced concrete diaphragm is employed with lateral pile group of two rows of vertical reinforced concrete cast-in-place piles. In second scheme, a flexible front wall of composite steel pipe pile is employed with lateral pile group of three rows of vertical reinforced concrete cast-in-place piles. Flexural rigidity of section of front walls is 15.24 GN·m²/m and 1.37 GN·m²/m, respectively for the two schemes. The elevations of wharf surface, forefront mud surface and wall bottom end are +4.2 m, −20.5 m and −34 m, so the sheet-pile bulkhead of the front wall and the breast wall is 38.2 m in height with an embedded depth of 13.5 m. The relief platform of pile group is made of 1.0 m thick reinforced concrete plate but its breadths are 12.5 m and 14.05 m, respectively for the two schemes. The vertical reinforced concrete cast-in-place piles are 38 m in length but 1.2 m and 1.3 m in diameter, respectively for the two schemes. They are arrayed in a transverse spacing of 5 m and in a longitudinal spacing of 4 m. The anchor wall is made of 1.1 m thick reinforced concrete diaphragm which is 19.0 m in height. High strength steel tie-rods with a diameter of 70 mm are used to connect the breast wall and the anchor wall with 1.33 m in spacing.

In this paper, the working performance of 200,000 tonnage sheet-pile bulkheads with relief platform is examined by geotechnical centrifuge model tests, and the load sharing and shift mechanism between the front wall and the lateral pile group is also investigated.

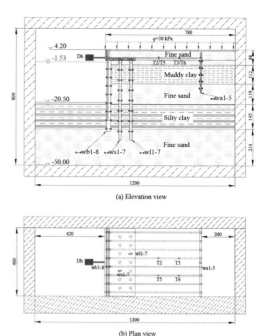

Figure 1. Model setup for first scheme (all dimensions in mm but elevations in m).

2 MODEL TESTS

2.1 Model setup

The centrifuge model tests are carried out at NHRI's large-scale geotechnical centrifuge, whose technical features are 200 g of maximum acceleration, 5.5 m of rotational radium, and 400 g-ton of capacity (Dou & Jin 1994). A plane-strain model box that is 1200 mm long, 400 mm wide and 800 mm deep, is used to accommodate the model simulating a length of sheet-pile bulkhead and its soil foundation, and a scaling factor of 80 is adopted for prototype to model, i.e. $N = 80$. The model setups are illustrated in Figures 1, 2, respectively for two schemes.

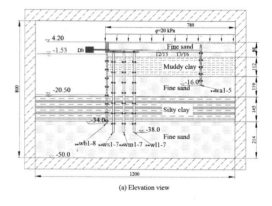

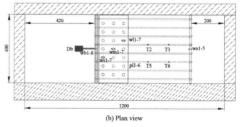

Figure 2. Model Setup for second scheme (all dimensions in mm but elevations in m).

2.2 Preparation of soil foundation

The model ground is made of disturbed in-situ soils from prototype. Although there are 13 layers of soils in prototype, a simplification is made into five main soil layers in model (Figures 1a, 2a). Thus, two kinds of soils are reconstituted to form soil ground. Three layers of fine sand are prepared into the model container in compaction method. The layers of muddy clay and silty clay are made from their slurries to cakes through consolidation under surcharge load in large-scale consolidometers. The undrained shear strength of clay cake, s_u, is measured by means of a miniature cone penetrometer. The main characteristic parameters of model ground are listed in Table 1.

Table 1. Properties of soil ground.

Soils	t mm	P kg/m³	w %	Su kPa
Fine sand	88	1970	23.5	
Muddy clay	112	1908	30.7	31~53
Fine sand	119	1970	23.5	
Silty clay	145	2000	22.9	112~121
Fine sand	214	1980	23.8	

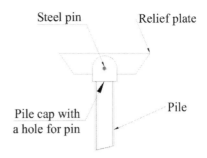

Figure 3. Sketch of connection between pile and relief plate.

2.3 Preparation of structural members

Model front wall, relief platform and cast-in-place piles and anchor wall are manufactured by using aluminium alloy, instead of by using reinforced concrete. Accordingly, modification has to be made to the dimensions of these structural members due to Young's modulus difference between aluminium and reinforced concrete (Xu et al. 2006). Considering that these structural components are flexural members, their thickness of model walls and plate is to be modified according to the following equation (1):

$$d_m^{al} = d_m^c \cdot \sqrt[3]{E_m^c/E_m^{al}} = d_p^c/N \cdot \sqrt[3]{E_m^c/E_m^{al}} \quad (1)$$

where, superscripts, al and c, denote aluminium and reinforced concrete, and subscripts, m and p, denote model and prototype, and d = thickness of flexural members; E = Young's modulus of flexural member. Regarding model cast-in-place piles to resist bending moment, they are prepared by hollow aluminium-alloy pipe, whose outer diameter is $1/N$ times prototype diameter and whose inner diameter is calculated according to the following equation (2):

$$b_m = \sqrt[4]{(a_m^4 - \frac{a_p^4}{n^4}\frac{E_p}{E_m})} = a_m \sqrt[4]{(1-\frac{E_p}{E_m})} \quad (2)$$

where, a and b, denote the outer and inner diameters of cast-in-place pile. Obviously, $b_p = 0$, and $a_m = a_p/80$.

A special connection mechanism is manufactured to simulate the prototype pinned head between piles and relief plate (Figure 3). A simplification is also made in the arrangement of steel tie-rods in that three tie-rods in prototype are represented by one tie rod in model with a cross section area three times the required one by the similitude law. This arrangement of tie-rods is shown in Figures 1b and 2b.

2.4 Instrumentation

To monitor the horizontal displacement of front wall, one laser sensor (Db in Figs. 1, 2), is mounted on the fixing frame from the model container cover, and its target of laser light is the anchoring point on the breast wall. There are 27 gauge points arranged to monitor the response of bending moment in first scheme, 8 locations of wb1-wb8 on front wall, 5 locations of wa1-wa5 on anchor wall, and 7 locations of ws1-ws7 on seaside pile and 7 locations of wl1-wl7 on landside pile (Figure 1). And there are 34 gauge points arranged in second scheme, 8 locations of wb1-wb8 on front wall, 5 locations of wa1-wa5 on anchor wall, and 7 locations of ws1-ws7 on seaside pile, 7 locations of wm1-wm7 on middle pile and 7 locations of wl1-wl7 on landside pile (Figure 2). Four gauge points are arranged along steel tie-rods, T2-T3 and T5-T6, to measure the internal tension force for each scheme (Figs. 1b, 2b).

2.5 Test procedure

The test procedure can be divided five stages. First, five layers of soil ground are prepared from the bottom up (Figs. 1a, 2a). Second, the front wall, anchor wall, and cast-in-place piles with relief platform are, in sequence, embedded in the soil ground. Then the steel tie-rods are installed and adjusted to maintain the front wall and anchor wall to the right degree of tension (Figs. 1a, 2a). Thirdly, the model is moved unto the centrifuge platform. The stress of model ground is restored to the original level of natural ground in prototype by putting it into flight of 80 g in order to bring soil and structural components into a good contact. Fourthly, as depicted in Figures 1a, 2a, the soil before the front wall above the dredge line is excavated at the floor of 1 g after the stoppage of centrifuge, and a basin of port is formed. Additionally, a layer of sand is placed evenly upon the surface of bulkhead so that it will produce a vertical surcharge of 20 kPa in flight of 80 g. Finally, upon the completion of excavation work in the lab floor, the model is brought into test by accelerating it to 80 g. The working behaviour of sheet pile bulkhead is observed by monitoring the response of all gauge points arranged in the front wall, anchor wall, cast-in-place piles and tie-rods.

3 TEST RESULTS AND ANALYSIS

The responses of bending moment of flexural members, internal tension force of tie-rods and horizontal displacement at anchoring point of breast wall are addressed in this paper. All these physical variables presented are in prototype scale by converting the measured values into the corresponding values in prototype scale in the light of scale ratios of similarity

Table 2. The main test results for two schemes.

Scheme	M_{fw} kN·m/m	M_{pg} kN·m/m	M_{aw} kN·m/m	T_{tr} kN	D_b mm
First	+3300/ −120	+860/ −910	−1000	570	74
Second	+800/ −230	+1250/ −2100	−1400	700	81

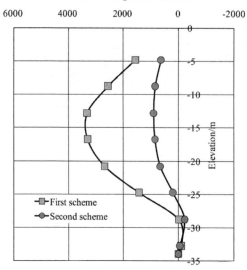

Figure 4. Unit width bending moment of front walls.

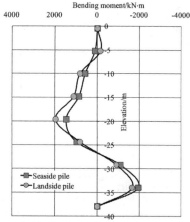

(a) Two rows of piles in first scheme

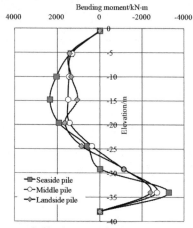

(b) Three rows of piles in second scheme

Figure 5. Bending moment of cast-in-place piles.

between model and prototype (Xu et al. 2006). A positive sign is assigned to the bending moment when the seaward side of its flexural member is under tension. The main test results for two schemes are listed in Table 2.

3.1 Bending moment of front wall

The distributions of bending moment of front walls are shown in Figure 4 for two schemes, and their characteristic bending moment values, i.e. maximum positive and negative bending moments, M_{fw}, are listed in Table 2. It is found that the bending moment of the upper large part of front wall is positive in sign, suggesting that its seaside is under tension, whereas its bending moment of the lower small part is negative in sign, indicating that its landside is under tension. And the maximum positive bending moment value is significantly greater than the absolute value of maximum negative moment for two front walls.

For the rigid front wall of T-shaped diaphragm in first scheme, its characteristic bending moment values are +3300 kN·m/m at El. −15 m and −120 kN·m/m at El. −32 m, respectively; for the flexible front wall of composite steel pipe pile in second scheme, its characteristic values are +800 kN·m/m at El. −13 m and −230 kN·m/m at El. −29 m, respectively. Obviously, the response intensity of bending moment of the flexible front wall is weaker than that of the rigid front wall. It is because the flexural rigidity of section of the flexible wall in second scheme is 1/11.12 times that of the rigid wall in first scheme.

3.2 Bending moment of cast-in-place piles

The distributions of bending moment of cast-in-place piles are shown in Figures 5a, b, respectively, for two schemes. Generally, the bending moment of the upper large part of pile is positive in sign, suggesting that its seaside is under tension, whereas the bending moment of the lower small part is negative in sign, indicating that its landside is under tension. Moreover, the distribution feature of bending moment of every row of piles in one scheme are very similar. In first scheme, the characteristic bending moment values of seaside pile are +1470 kN·m at El. −20 m, and −1960 kN·m at El. −34 m, respectively. In second scheme, the characteristic bending moment values of seaside pile are +2350 kN·m at El. −15 m, and −3280 kN·m at El. −34 m, respectively. Obviously, from the point of view of characteristic bending moment values, the response intensity of bending moment of lateral pile group in second schemes is much stronger than that of pile group in first scheme.

In order to evaluate the overall lateral bearing capacity of lateral pile group foundation, the characteristic bending moment values of each pile are converted into unit width values by dividing them by the longitudinal pile spacing, 4 m. Then the characteristic bending moment values of pile group, M_{pg}, are obtained by summing all unit width values of multiple piles in one scheme. It is found from Table 2 that the characteristic bending moment values of pile group are bigger than those of front wall in second scheme, indicating that the pile group in second scheme plays a greater role than the front wall, while the opposite is true in first scheme. There are two major reasons behind the difference in lateral bearing role of the pile group between two schemes. In the first place, there are three rows of cast-in-place piles connected with the relief plate in second scheme, one more row of piles than in first scheme, and the pile diameter in second scheme is 1.3 m, bigger than that of 1.2 m in first scheme. In other words, the pile group in second scheme is much stronger in overall flexural rigidity than that in first scheme. In the second place, the front wall in second scheme is much more flexible than the front wall in first scheme, therefore the bulkhead in second scheme will displace laterally toward seaside larger than that in first scheme. For this reason, the soil displacement around piles, especially the seaside piles in second scheme, will be larger than that in first scheme, which is suggestive of bigger relative displacement between pile and soil in second scheme.

3.3 Bending moment of anchor walls

The distributions of bending moment of anchor walls are shown in Figure 6 for two schemes, and their characteristic bending moments, M_{aw}, are listed in Table 2. It shows that the bending moment of the whole wall is negative in sign, indicating that its landside is under tension. And the maximum negative bending moment values are −1000 kN·m/m and −1400 kN·m/m at El. −13 m, respectively for two schemes.

3.4 Internal force of tie-rods

It shows in Table 2 that the average value of the internal tension force of each tie-rod, T_{tr}, are 570 kN and 700 kN, respectively for two schemes. They are less than the design value of the high strength steel rod with a diameter of 70 mm, which is 970 kN. It also shows that the tension force of tie-rods will become larger when the flexible front wall is used.

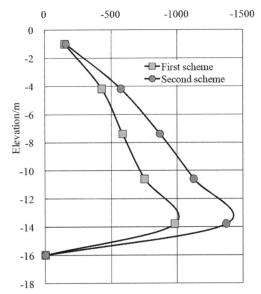

Figure 6. Unit width bending moment of anchor walls.

3.5 Displacement of front wall

It shows in Table 2 that the horizontal displacement of front wall at anchorage, D_b, are 74 mm and 81 mm, respectively for two schemes. The average values of front wall tilt are 0.22% and 0.24%, respectively, which are within the allowable limit of 0.5% (Tsinker 1997). It also shows that the horizontal displacement of front wall will become larger when it is flexible.

3.6 Comparison of two design schemes

As two main parts of the sheet-pile bulkhead with relief platform, the front wall and the lateral pile group will resist the soil movement toward seaside together, which is due to lateral unloading induced by dredging of port basin before the front wall and vertical loading on the wharf surface. It is known that the front wall in first scheme is more rigid than that in second scheme whereas the lateral pile group in first scheme is less rigid than that in second scheme. Therefore, it is seen above that the lateral displacement of front wall in first scheme is less than that in second scheme and the response intensity of bending moment of front wall in first scheme is stronger than that in second scheme, and the response intensity of bending moment of pile group is weaker than that in second scheme.

Such lateral load sharing and shift mechanism between the front wall and the lateral pile group works and depends on the comparison of their flexural rigidity. Based on the above performance of the front wall and the lateral pile group with their flexural rigidity, more options can be made in the design of the innovative sheet-pile bulkhead with relief platform. If the rigidity of front wall is very high, the overall flexural rigidity of pile group can be chosen to be low. On the contrary, if the rigidity of front wall is low, the overall flexural rigidity of lateral pile group must be chosen to be high as supplement.

4 CONCLUSIONS

Two design schemes of 200,000 tonnage sheet-pile bulkheads with relief platform are investigated by means of centrifuge modelling. First, the reasonability of two design schemes is validated in that the displacements of front wall at anchoring point under lateral unloading and vertical loading are within the allowable limit. Second, the lateral load sharing and shift mechanism is studied between the front wall and the lateral pile group.

1. With regard to first scheme in which the front wall is rigid T-typed reinforced concrete diaphragm, the front wall play a greater role than the lateral pile group with relief plate because the characteristic bending moment values of font wall are much bigger those of pile group and anchor wall along with small values of internal force of tie-rods and lateral displacement of front wall at anchoring point.
2. With regard to second scheme in which the front wall is flexible composite steel pipe pile, the lateral pile group of three rows of piles with relief plate play a greater role than the front wall because the characteristic bending moment values of pile group are bigger than those of front wall and anchor wall along with large values of internal force of tie-rods and lateral displacement of front wall at anchoring point.
3. As two main parts of sheet-pile bulkhead, the front wall and the lateral pile group resist together against soil movement due to lateral unloading by dredging before the front wall and vertical loading upon the wharf surface. The lateral load sharing and shift mechanism between two parts works and is dependent on the comparison of their flexural rigidity. If the rigidity of front wall is very high, the pile group with low overall flexural rigidity can be chosen. On the contrary, if the rigidity of front wall is low, the pile group with high overall flexural rigidity must be used.

REFERENCES

Dou, Y. & Jin, P. 1994. Development of NHRI-400g.t geotechnical centrifuge. In Leung, Lee & Tan (eds.), *Centrifuge 94*. Rotterdam: Balkema: 69–74.

Tsinker, G.P. 1997. Handbook of port and harbor engineering: geotechnical and structural aspects. Springer US: ITP International Thomson Publishing, Chapman and Hall Press.

Li, S.L. & Xu, G.M. 2008. Centrifuge modeling tests for sheet-pile bulkhead anchored by single layer of tie-rods. *Chinese Journal of Hydro-Science and Engineering* 2008(1): 67–72 (in Chinese).

Liu, Y.X. 2014. Design theories and methods for sheet-pile bulkhead. Beijing: China Communications Press Co. Ltd (in Chinese)

Xu, G.M., Cai, Z.Y., Zeng, Y.J., Li, J.L., Liu, Y.X. & Wu, L.D. 2006. Centrifuge modeling for a new type sheet pile bulkhead with barrier piles. In Ng, Zhang & Wang (eds.), *Physical modelling in Geotechnics – 6th ICPMG*. Rotterdam: Balkema:1125–1129.

Xu, G.M., Cai, Z.Y., Zeng, Y.J., Gu, X.W, Li, S.L, Liu, Y.X & Wu, L.D. 2010. Centrifuge modeling for an innovative sheet-pile bulkhead of diaphragm. *Chinese Journal of Rock and Soil Mechanics* 31(S1): 48–52 (in Chinese).

Xu, G. M. & Li, S. L. 2016. Experimental study of head fixity conditions of pile group in sheet-pile bulkhead. *Chinese Journal of Rock Mechanics and Engineering* 35(S1): 3365–3371 (in Chinese).

Author index

Abdoun, T. 293, 519, 847, 943, 949
Abrashitov, A.A. 203
Abuhajar, O.S. 221
Achmus, M. 1347
Adamidis, O. 113, 937
Adams, M.T. 1291
Aggarwal, P. 1113
Ahmed, U. 1025
Airey, D.W. 513
Ajagbe, W.O. 1353
Akoochakian, A. 1157
Akram, I. 1253
Al-Baghdadi, T. 695
Al-Defae, A.H. 241
Al-Fergani, M. 507
Alber, S. 113
Ali, U. 233
Almeida, M.C.F. 1031, 1093
Almeida, M.S.S. 1031, 1043, 1093
Alvarez Grima, M. 469
Anastasopoulos, I. 113, 1321
Apostolou, E. 1163
Arakawa, M. 1469
Archer, A. 489
Arnold, A. 119
Asaka, Y. 921
Ashtiani, M. 885
Askarinejad, A. 119, 461, 469, 987, 1119, 1149
Audrain, P. 465
Augarde, C. 695
Azúa-González, C. 469

Ball, J. 695
Barrett, J. 577
Bathurst, R.J. 1259
Bayton, S.M. 163, 533, 689, 719
Beber, R. 125
Becker, L.B. 527
Beckett, C.T.S. 823
Beddoe, R.A. 175
Beemer, R.D. 279
Beltrán-Rodriguez, L.N. 865
Bengough, A.G. 401, 425
Bezuijen, A. 285, 597, 815, 1037
Bhattacherjee, D. 337
Bienen, B. 33, 651, 669, 707, 719
Bilotta, E. 3, 955
Bisht, R.S. 1329
Bizzotto, T. 553

Black, J.A. 163, 507, 533, 545, 689, 719, 835
Blake, A. 695
Blanc, M. 465, 701, 719, 1031, 1043, 1341
Blom, C.B.M. 443
Boksmati, J. 343
Borghei, A. 247, 349
Boulanger, R.W. 21
Bowman, E.T. 377, 1075
Breen, J. 501
Brennan, A. 695
Brennan, A.J. 553, 1163, 1265
Bretschneider, A. 465
Broere, W. 317, 323, 443
Bronner, C.E. 21
Bronner, J.D. 21
Brown, M.J. 241, 553, 695
Burd, H.J. 725
Byrne, B.W. 725, 737

Cabrera, M. 1075
Cai, Z.Y. 583, 1143, 1187, 1475
Caicedo, B. 131, 155, 413, 1285, 1407
Carey, T. 293, 829
Cargill, A. 533
Caro, S. 155
Cassidy, M.J. 279, 615, 651, 731
Cassie, P. 725
Castillo, D. 155
Chalaturnyk, R.J. 1271
Chan, D.Y.K. 1421
Chandler, H. 553
Charles, J.A. 835
Chen, A.Z. 583
Chen, C.H. 873
Chen, C.H. 873
Chen, H. 909
Chen, S.S. 495
Chen, Y.M. 293, 407, 767, 929
Cheng, Y.B. 495
Chian, S.C. 355
Chow, S.H. 559
Chow, Y.K. 681
Coelho, P.A.L.F. 993
Coleman, J. 299
Colletti, J. 299
Combe, G. 841
Constantinou, M.C. 519
Coombs, W. 695
Cui, G. 137, 359

Cuira, F. 1401
Cumming-Potvin, D. 809
Cunha, R.P. 1407

da Silva, T.S. 1169
Dafni, J. 365
Darby, K.M. 21
Dasaka, S.M. 169
Dashti, S. 1005, 1199
Dave, T.N. 169
Davidson, C. 695
de Boorder, M. 1119
de Jager, R.R. 987
de Lange, D.A. 597
De, A. 779
DeJong, J.T. 21
DeJong, M.J. 437, 449
den Hamer, D.A. 285
Deng, C. 1427
Detert, O. 1175
Diakoumi, M. 743
Dijkstra, J. 317, 323
Dimitriadis, K. 903
Divall, S. 143, 539
Dobrisan, A. 125
Dobry, R. 519, 949
Doreau-Malioche, J. 841
Dos Santos Mendes, C. 1377

Ebeido, A. 1335
Ebizuka, H. 233
Egawa, T. 879
Ehrlich, M. 1223
Eichhorn, G.N. 785
El Haffar, I. 701, 1341
El Shafee, O. 943
El-Sekelly, W. 949
Elgamal, A. 1335
Elmrom, T. 163
Elshafie, M.Z.E.B. 791, 1169
Emeriault, F. 1377
Escobar, J. 131
Escoffier, S. 293, 1401
Exton, M.C. 371

Fagundes, D.F. 1031, 1043
Farrin, M. 797
Fasano, G. 955
Faustin, N.E. 791
Fioravante, V. 955
Fiumana, N. 603
Flora, A. 955
Forsyth, M. 559

Fourie, A.B. 823
Franza, A. 209
Frick, D. 1347
Frolovsky, Y.K. 1277
Fukutake, K. 215
Funahara, H. 961

Gaber, F. 377
Ganchits, V.V. 1277
Ganiyu, A.A. 1353
Garala, T.K. 1359
García-Torres, S. 1181
Garzón, L.X. 1285
Gaudin, C. 33, 279, 501, 603, 719
Gavras, A. 293
Geirnaert, K. 285
Gelagoti, F. 1383
Georgiou, I. 1383
Ghaaowd, I. 1365
Ghalandarzadeh, A. 565, 885
Ghayoomi, M. 247, 349
Giardina, G. 449
Giretti, D. 955
Girout, R. 1031, 1043
Gng, Z. 1149
Goodey, R.J. 43, 143, 191, 539, 853, 1395
Gorlov, A.V. 1277
Gourvenec, S. 51
Grabe, J. 669, 707, 1229, 1451
Gu, X.W. 495, 583, 1187, 1475
Gütz, P. 1347

Ha, J.G. 629, 897
Haigh, S.K. 125, 293, 639, 719, 785, 993, 1427
Hajialilue-Bonab, M. 797, 829
Halai, H. 191
Hallett, P.D. 401
Harry, S. 371
Hartmann, D.A. 1031
Hasebe, M. 921
Hasegawa, G. 1067
Hatanaka, Y. 891
Heins, E. 707
Hem, R. 431, 675
Heron, C.M. 137, 209, 359, 455
Hicks, M.A. 987
Higo, Y. 215
Hoffman, W. 299
Hong, J.Z. 495
Hori, T. 383, 1463
Horii, Y. 1371
Hossain, M.S. 329, 615
Hou, Y.J. 77, 1125
Houda, M. 1377
Houlsby, G.T. 737
Hu, L.M. 975
Hu, Y. 329, 615, 1309
Huang, B. 407
Huang, J.X. 975

Huang, M. 909, 1297
Huang, M.S. 761
Huang, Q. 519
Huang, X. 1365
Huang, Y.H. 1143
Hughes, F.E. 967
Hung, H.M. 1193
Hung, W.Y. 293, 975

Iai, S. 265, 1017
Ibsen, L.B. 623
Idinyang, S. 209
Imamura, S. 981
Indiketiya, S. 803
Iskander, M. 389, 859
Isobe, K. 879, 891
Iwamoto, T. 675
Iwasa, N. 1235

Jacobsz, S.W. 179, 185, 305, 311, 809, 1081
Jahnke, S.I. 311
Jegatheesan, P. 803
Jenck, O. 1377
Jeong, Y.H. 609
Jia, C.H. 1125
Jia, Y. 909
Jiang, Q. 395
Jo, S.B. 629, 1055
Jun, M.J. 615
Juneja, A. 1113, 1329

Kailey, P. 1075
Kamalzare, M. 1049
Kamchoom, V. 1131
Kanai, H. 1253
Kang, X. 1297
Kavand, A. 1157
Kearsley, E.P. 179, 185, 311, 809, 1081
Khaksar, R.Y. 565
Khan, I.U. 507
Kim, D.S. 293, 609, 629, 749, 897
Kim, J.H. 609, 749
Kim, N.R. 1055
Kim, Y.H. 615
Kimura, M. 663, 915, 1067
Kiriyama, T. 215
Kirkwood, P.B. 1199
Kishida, K. 915
Kita, K. 713
Kitazume, M. 1217
Klinkvort, R.T. 689, 719
Knappett, J.A. 87, 241, 401, 425, 475, 481, 695, 1011, 1265
Ko, K.W. 897
Kogure, T. 1253
Kokkali, P. 847
König, D. 1175
Koteras, A.K. 623
Kotronis, P. 1401

Koudelka, P. 1433
Kourkoulis, R. 1383
Kunasegarm, V. 1439
Kundu, S. 1205
Kutter, B. 293, 829
Kutter, B.L. 371
Kuwano, J. 1193, 1253
Kuwano, R. 233, 803, 1087, 1099
Kwa, K.A. 513

Lai, C.G. 955
Laporte, S. 175
Larrahondo, J.M. 865
Lawler, J. 943
Le Cossec, J. 903
Le, B.T. 853
Le, J. 559
Lee, F.H. 1329
Lee, M.G. 629
Lehane, B.M. 1309
Leung, A.K. 475, 481, 1131
Leung, C.F. 681, 1025
Li, J.C. 929
Li, L. 859, 1457
Li, L.F. 395
Li, Y.P. 635
Liang, F. 909
Liang, J.H. 1125
Liang, T. 401, 481
Liao, T.W. 975
Liel, A.B. 1005
Ling, D.S. 407
Ling, H.I. 1457
Linhares, R.M. 527
Liu, K. 293, 407
Liu, W. 495
Loades, K.W. 401, 425
Loli, M. 241, 1383
Louw, H. 1081
Lozada, C. 413
Lundberg, A.B. 317
Luo, F. 1137
Lutenegger, A.J. 1291

Ma, X.F. 419
Madabhushi, G.S.P. 437, 719, 937, 993, 1181, 1359
Madabhushi, S.P.G. 125, 293, 343, 967, 1421
Madabhushi, S.S.C. 125, 639
Maeda, K. 431
Maghsoudloo, A. 987
Mair, R.J. 449, 791
Mamaghanian, J. 1211
Manikumar, C.H.S.G. 1211
Manzari, M. 293
Marques, A.S.P.S. 993
Marques, F.L. 527
Marshall, A.M. 137, 209, 359, 455

Maruyama, K. 589
Mason, H.B. 371
Matsuda, S. 1217
Matsuda, T. 431
Mayall, R.O. 725
Mayanja, P. 1229
McAdam, R.A. 725
McCartney, J. 1365
McNamara, A.M. 191, 539, 1395, 1445
Meijer, G.J. 401, 425
Mikami, T. 1017
Mikasa, K. 645
Miles, S. 999
Mirmoradi, S.H. 1223
Mirshekari, M. 247
Miyamoto, J. 431, 571, 675
Miyata, Y. 1259
Miyazaki, Y. 915
Mizumoto, M. 1469
Mohan Gowda, K.T. 253
Molenkamp, F. 987
Momoi, M. 1217
Moormann, C. 197
Moradi, M. 565, 1157, 1247
Morikawa, Y. 589, 1259
Moug, D.M. 21
Movasat, M. 797
Mu, L. 1297
Muir Wood, D. 401

Nadimi, S. 853
Nagao, T. 1371
Nagula, S. 1229
Nakamoto, S. 1235
Nakase, H. 675
Néel, A. 465
Newson, T.A. 221
Ng, C.W.W. 489
Nicoll, B.C. 425
Niemann, C. 731
Nigorikawa, N. 921
Nishiura, D. 227
Nonoyama, H. 1259
Nordal, S. 1105
Norris, S. 1389

O'Loughlin, C.D. 33, 501, 559, 603, 651, 731
Ohara, Y. 1087
Okabayashi, K. 645
Okamura, M. 293
Olarte, J.C. 1005
Oliveira, F.S. 527
Oliveira, J.R.M.S. 1093
Omidvar, M. 859
Ong, D.E.L. 1025
Osman, M.H. 1353
Ota, M. 1099
Otsubo, M. 233
Ottolini, M. 323

Ovalle-Villamil, W. 259
Özcebe, A.G. 955

Panchal, J.P. 1395, 1445
Paramasivam, B. 1005
Parchment, J. 149
Park, H.J. 609, 629, 897
Park, S.G. 615
Pathmanathan, R. 803
Pearson, A. 1303
Pelekis, I. 437
Peng, R. 1125
Pérez-Herreros, J. 1401
Petryaev, A.V. 1277
Phillips, R. 577
Phoon, K.K. 1285
Piercey, G. 577
Pooresmaeili, A. 1247
Powell, T. 553
Prada-Sarmiento, L.F. 865
Pradhan, R.N. 1105
Pua, L.M. 155

Qi, S. 1011
Qin, C. 355

Ragni, R. 651
Rahadian, R. 443
Rammah, K.I. 1093
Ramos-Cañón, A.M. 865
Randolph, M.F. 707
Rashid, A.S.A. 1353
Raymond, A.J. 21
Razeghi, H.R. 1211
Ren, G.F. 495, 583, 1187, 1475
Reul, O. 731
Richards, D.J. 695, 743
Richards, I.A. 737
Ritchie, E.P. 143
Ritter, S. 449
Robinson, S. 1265
Rodríguez, E. 1407
Rosenbrand, E. 1037
Rotte, V.M. 1241
Ryan, C. 903

Sabermahani, M. 1247
Saboya, F. 465, 1365
Sakaguchi, H. 227
Sakellariadis, L. 1321
Salehi, A. 1309
Sánchez-Peralta, J.A. 865
Sánchez-Silva, M. 1285
Saran, R.K. 1061
Sasanakul, I. 259
Sassa, S. 589, 657
Sawada, K. 265
Sawamura, Y. 663, 915, 1067
Sayles, S. 743
Schanz, T. 1175
Schenkeveld, F. 461

Schiavon, J.A. 1315
Schindler, R. 1321
Schmoor, K.A. 1347
Seitz, K.-F. 1451
Seki, S. 1413, 1439
Sekita, K. 713
Seong, J.T. 749
Sera, R. 1087, 1099
Sett, K. 299
She, Y. 407
Sheil, B.B. 725
Shepley, P. 149, 1303, 1389
Shi, B.X. 583
Shi, J.Y. 635
Shibata, T. 663
Shields, L. 241
Shiraga, S. 1067
Siemens, G.A. 175
Silva, M. 841
Singla, R. 1113
Smit, M.S. 179, 185
Smith, C.C. 835
Soe, A.A. 1253
Song, G. 455
Song, L.B. 395
Sovso, S.L. 1309
Stallebrass, S.E. 143, 539, 1445
Stanier, S.A. 651
Stapelfeldt, M. 669
Stathas, D. 1457
Still, J. 999
Stone, K.J.L. 221, 743, 755, 903
Stone, N. 829
Stringer, M. 999
Sturm, A. 21
Sydrakov, A.A. 203

Taeseri, D. 1321
Takada, Y. 1017
Takahashi, H. 589
Takano, D. 1259
Take, W.A. 101
Takemura, J. 1235, 1413, 1439
Tamate, S. 383, 1463
Tang, H.W. 761
Tanghetti, G. 191
Tatari, A. 533
Taylor, R.N. 539, 853
Tehrani, F.S. 461
Tessari, A. 299, 847
Tessari, A.F. 545
Thakur, V. 1105
Thevanayagam, S. 519, 949
Thorel, L. 413, 465, 701, 719, 1031, 1043, 1315, 1341
Thusyanthan, N.I. 343
Tian, Y. 603, 731
Tibana, S. 1365
Tillman, A. 755
Tituaña-Puente, J.S. 865
Tobita, T. 271, 597

Tomita, N. 961
Toni, J.B. 841, 1377
Trejo, P.C. 1093
Trujillo-Vela, M.G. 865
Tsatsis, A. 1383
Tsuha, C.H.C. 1315
Tsurugasaki, K. 431, 571, 675
Tyagi, A. 1329

Ueda, K. 265, 293, 1017
Ueng, T.S. 873
Utsunomiya, T. 713

Valencia-Galindo, M.D. 865
van Beek, V.M. 1037
van der Woude, S. 443
van der Zon, J. 1119
van Zeben, J.C.B. 469
van 't Hof, C. 469
Vandenboer, K. 1037
Vaziri, M. 755
Vian, G. 1377
Viggiani, G. 841
Vincke, L. 285
Viswanadham, B.V.S. 253, 337, 1061, 1205, 1211, 1241
Vitali, D. 475

Wang, C. 1125
Wang, J.P. 1457
Wang, K. 1265
Wang, L. 695
Wang, L.J. 767, 929
Wang, N.X. 495
Wang, Y. 329, 1271
Wartman, J. 365
Watanabe, K. 1469
Wehr, J. 1163
Wen, K. 767
Wesseloo, J. 809
Whitehouse, R.J.S. 725
Wilschut, D. 443
Wilson, D.W. 21
Worbes, R. 197
Wu, W. 1075

Xia, P. 407
Xie, Y. 681
Xu, G.M. 495, 583, 1143, 1187, 1475
Xu, J.W. 419
Xu, L. 1457
Xu, T. 815

Yaba, O. 1377

Yamamoto, S. 227
Yamanashi, T. 879
Yao, C.F. 1413
Yeh, H. 371
Yifru, A.L. 1105

Zambrano-Narvaez, G. 1271
Zania, V. 719, 1309
Zayed, M. 1335
Zaytsev, A.A. 203, 1277
Zeghal, M. 293
Zhang, C. 1143
Zhang, C.R. 761
Zhang, G. 1137
Zhang, H. 909
Zhang, M. 395
Zhang, W. 119, 1119, 1149
Zhang, W.M. 495
Zhang, X. 481
Zhang, X.D. 1125
Zhang, Y. 1297
Zhang, Z. 355
Zhao, R. 475
Zhou, Y.G. 293, 407
Zhu, B. 767, 929
Zimmie, T.F. 779, 1049
Ziotopoulou, K. 21

Printed and bound by PG in the USA